ANALYSIS AND DESIGN OF HYBRID SYSTEMS 2006

A Proceedings Volume from the 2nd IFAC Conference
7–9 June 2006, Alghero, Italy

Edited by

CHRISTOS CASSANDRAS
Boston University
15 St Mary's St
Brookline, MA 02446
United States

ALESSANDRO GIUA
Universita di Cagliari
Dipartimento di Ingegneria, Elettrica ed Elettroni
Piazza D'Armi, Cagliari 09123
Italy

CARLA SEATZU
Universita di Cagliari
Dipartimento di Ingegneria, Elettrica ed Elettroni
Piazza D'Armi, Cagliari 09123
Italy

and

JANAN ZAYTOON
CReSTIC Research Centre,
University of Reims Champagne Ardenne,
Moulin de la Housse, Reims,
BP 1039, F-51687
France

Published for the
International Federation of Automatic Control
By
ELSEVIER LTD

Elsevier
The Boulevard, Langford Lane, Kidlington, Oxford OX5 1GB, UK
30 Corporate Drive, Suite 400, Burlington, MA 01803, USA

First edition 2006

British Library Cataloguing in Publication Data
A catalogue record for this book is available from the British Library

Library of Congress Cataloging-in-Publication Data
A catalog record for this book is available from the Library of Congress

ISBN–13: 978-0-08-044613-4
ISBN–10: 0-08-044613-2

Transferred to digital print 2008
Printed and bound by CPI Antony Rowe, Eastbourne

ADHS'06

2nd IFAC Conference on Analysis and Design of Hybrid Systems (Alghero, Italy) – June 7-9, 2006

Organized by

The Department of Electrical and Electronic Engineering of the University of Cagliari, Italy

IFAC Sponsors

- Technical Committee on Discrete Event and Hybrid Systems (TC 1.3) – *Main Sponsor*
- Technical Committee on Control Design (TC 2.1)
- Technical Committee on Manufacturing Plant Control (TC 5.1)
- Technical Committee on Fault Detection, Supervision & Safety of Technical Processes (TC 6.4)

National Organizing Committee

Chair: C. Seatzu (IT)
Members: M.P. Cabasino (IT), D. Corona (IT), N. Orani (IT), A. Pisano (IT), E. Usai (IT)

International Program Committee

Chairs: C. G. Cassandras (US), A. Giua (IT), J. Zaytoon (FR)
Industrial co-chair: A. Sangiovanni-Vincentelli (US)
Members:

H. Alla (FR)
P. Antsaklis (US)
A. Bemporad (IT)
A. Bicchi (IT)
F. Blanchini (IT)
R. Boel (BE)
J. Buisson (FR)
P. Caines (CA)
E.F. Camacho (ES)
P. Colaneri (IT)
J. Daafouz (FR)
J. Davoren (AU)
R. DeCarlo (US)
I. Demongodin (FR)
M.D. Di Benedetto (IT)
M. di Bernardo (IT)
M. Egerstedt (US)
S. Engell (DE)
A. Ferrara (IT)
G. Ferrari Trecate (FR)
L. Ferrarini (IT)
J.L. Ferrier (FR)
H. Guéguen (FR)
M. Heemels (NL)
T. Henzinger (US)
J. Imura (JP)
C. Iung (FR)
S. Kowalewski (DE)
V. Krebs (DE)
V. Lakshmikantham (US)
B. Lennartson (SE)
J.J. Lesage (FR)
J. Lunze (DE)
J. Lygeros (GR)
P. E. Miyagi (BR)
T. Moor (DE)
M. Morari (CH)
G. Morel (FR)
P.J. Mosterman (US)
G. Pappas (US)
J.C. Pascal (FR)
T. Parisini (IT)
M. Prandini (IT)
J. Raisch (DE)
N. Rakoto (FR)
A. Rantzer (SE)
C. Seatzu (IT)
M.P. Spathopoulos (UK)
O. Stursberg (DE)
Z. Sun (IE)
T. Suzuki (JP)
E. Usai (IT)
F. Vaandrager (NL)
C. Valentin (FR)
D.A. van Beek (NL)
A.J. van der Schaft (NL)
J.H. Van Schuppen (NL)
T. Villa (IT)
O. von Stryk (DE)
Y. Wardi (US)

Additional referees

A. Abate
A. Alessio
L. Benvenuti
J.-L. Boimond
P. Caravani
B. Castanier
E. De Santis
B. De Schutter
S. Di Cairano
S. Di Gennaro
A. Diamantopoulou
B. Djeridane
L. Doyen
S. Druhle
J. Duchoňová
A. D'Innocenzo
D. Fontanelli
S. Geist
N. Giorgetti
A. Girard
L. Greco
D. Gromov
A.A. Julius
A. Juloski
J. Kierzenka
J. Komenda
M. Lazar
S. Leirens
D. Li
S. Paoletti
P. Pepe
M. Petreczky
Ph. Planchon
V. Prabhu
M. Remelhe
A. Schild
K. Wulff
R. Zemouri
F. Zhang

Preface

This volume contains the proceedings of ADHS'06: the 2nd IFAC Conference on Analysis and Design of Hybrid Systems, organized in Alghero (Italy) on June 7-9, 2006.

ADHS is a series of triennial meetings that aims to bring together researchers and practitioners with background in control and computer science in order to provide a survey of the advances in the field of hybrid systems and of their ability to take up the challenge of analysis, design and verification of efficient and reliable control systems. ADHS'06 is the second Conference of this series after ADHS'03 in Saint Malo. The ADHS series follows the successful conference series on the Automation De Processus Mixed / Automation of Mixed Processes (ADPM'92 in Paris, ADPM'94 in Brussels, ADPM'98 in Reims, ADPM'2000 in Dortmund). The continuity between the ADHS conferences is guaranteed by the IFAC Technical Committee on Discrete Event and Hybrid Systems (TC 1.3) that is the main technical sponsor of the series.

We express our gratitude to the authors of the papers submitted to ADHS'06. Out of 96 submissions, 65 papers were selected after a careful reviewing process during which each paper received, on the average, three reviews. The high quality of the program is due to the authors of the submitted papers, to the members of the International Program Committee and to the referees who provided accurate and extensive reports.

Among the 65 papers accepted and published in this volume, 52 are coming from Europe, 7.5 from US/Canada, 2.5 from Latin America, 2 from Japan, 0.5 from Egypt, 0.5 from Australia.

Three distinguished speakers have been invited to give a one hour plenary talk during the conference. They are: Vadim I. Utkin, whose talk is titled *"Chattering Problem for Sliding Mode Control"*; Alberto Sangiovanni-Vincentelli, whose talk is titled *"System Theory Approaches for Embedded Systems"*; Claude Iung, whose talk is titled *"Optimal Control in Hybrid Systems"*. The invited speakers have contributed to the proceedings with an extended abstract of their talk.

Some of the papers presented at the conference have been submitted as part of invited sessions. All of them went through the normal review procedure. We are grateful to the organizers of these sessions that have timely put into focus interesting new research topics. They are: E. De Santis organizer of a session on *"Structural Analysis and Approximation of Hybrid Systems"*; S. Engell and O. Stursberg organizers of a session on *"Controller Design Based on Hybrid Models of Industrial Plants"*; A. Bemporad and F. Lamnabhi-Lagarrigue organizers of a session on *"Applications of Hybrid Control"*.

A special thank goes to C.G. Cassandras and P. Mosterman, organizers of an invited session on *"Hybrid Simulation Tools: Principles, Challenges and Applications"* that was also complemented by an interactive tool presentation. This session was composed by five presentations whose speakers have also contributed to the proceedings with an extended abstract of their talk.

We are grateful to the members of the Organizing Committee who took care of all important logistic and administrative details. In particular, we thank Maria Paola Cabasino for her help in preparing these proceedings. This conference benefits of support from the *Università di Cagliari*, from the *Comune di Alghero* and from *Akhela s.r.l.*

The Editors
C.G. Cassandras, A. Giua, C. Seatzu, J. Zaytoon

Table of Contents

ADHS'06

2nd IFAC Conference on Analysis and Design of Hybrid Systems
Alghero, Italy - June 7-9, 2006

Plenary Lectures

Chattering Problem for Sliding Mode Control Systems 1
V. Utkin, H. Lee

Challenges and opportunities for system theory in embedded controller desing ... 2
A. Sangiovanni Vincentelli

Optimal Control in Hybrid Systems 4
C. Iung, P. Riedinger

WA1 – Observers for Hybrid Systems

Convergent design of switched linear systems 6
R.A. van den Berg, A.Y. Pogromsky, J.E. Rooda

Observer design for a class of discrete time piecewise-linear systems 12
A. Birouche, J. Daafouz, C. Iung

Designing switched observers for switched systems using multiple Lyapunov functions and dwell-time switching 18
S. Pettersson

Critical states detection with bounded probability of false alarm and application to air traffic management 24
M.D. Di Benedetto, S. Di Gennaro, A. D'Innocenzo

WA2 - Continuous and Hybrid Petri Nets

Tracking control of join-free timed continuous Petri net systems 30
J. Xu, L. Recalde, M. Silva

On sampling continuous timed Petri nets: reachability "equivalence" under infinite servers semantics 37
C. Mahulea, A. Giua, L. Recalde, C. Seatzu, M. Silva

Modelling distributed manufacturing systems via first order hybrid Petri nets 44
M. Dotoli, M.P. Fanti, A.M. Mangini

Simulation of railway stations based on hybrid Petri nets 50
F. Kaakai, S. Hayat, A. El-Moudni

WB1 – Modeling and Simulation of Hybrid Systems

Modeling an impact control strategy using HyPA 56
P.J.L. Cuijpers, M.A. Reniers

Human skill modeling based on stochastic switched dynamics 64
T. Suzuki, S. Inagaki, N. Yamada

Building efficient simulations from hybrid bond graph models 71
C.D. Beers, E.-J. Manders, G. Biswas, P.-J. Mosterman

Robust control strategies for multi-inventory systems with average flow constraints 77
D. Bauso, F. Blanchini, R. Pesenti

WB2 - Control of Hybrid Systems 1

Hybrid constrained formation flying control of micro-satellites 83
F. Bacconi, A. Casavola, E. Mosca

A gradient-based approach to a class of hybrid optimal control problems 89
V. Azhmyakov, J. Raisch

Optimal mode-switching for hybrid systems with unknown initial state 95
H. Axelsson, M. Boccadoro, Y. Wardi, M. Egerstedt

Beyond the construction of optimal switching surfaces for autonomous hybrid systems 101
M. Boccadoro, M. Egerstedt, P. Valigi, Y. Wardi

WC1 - Structural Analysis and Approximation of Hybrid Systems (Invited)
Organizer: E. De Santis

Approximate simulation relations for hybrid systems 106
A. Girard, A.A. Julius, G.J. Pappas

Stabilizability based state space reductions for hybrid systems 112
E. De Santis, M.D. Di Benedetto, G. Pola

Reachability computation for uncertain planar affine systems using linear abstractions 118
O. Nasri, M.-A. Lefebvre, H. Guéguen

Exact differentiation via sliding mode observer for switched systems 124
H. Saadaoui, M. Djemai, N. Manamanni, T. Floquet, J.-P. Barbot

WC2 - Control of Hybrid Systems 2

Robust H-infinity control of uncertain discrete-time switching symmetric composite systems 130
L. Bakule

The elevator dispatching problem: hybrid system modeling and receding horizon control 136
K.S. Wesselowski, C.G. Cassandras

Robust piecewise linear sheet control in a printer paper path 142
B. Bukkems, J. de Best, R. van de Molengraft, M. Steinbuch

Stabilization of max-plus-linear systems using receding horizon control – The unconstrained case 148
I. Necoara, T.J.J. van den Boom, B. De Schutter, J. Hellendoorn

TA1 - Stochastic Hybrid Systems

Online classification of switching models based on subspace framework 154
K.M. Pekpe, S. Lecoeuche

Functional abstractions of stochastic hybrid systems 160
M.L. Bujorianu, H.A.P. Blom, H. Hermanns

Stochastic hybrid NETCAD systems for modeling call admission and routing control in networks 166
Z. Ma, P.E. Caines, R. Malhame

Parameter identification for piecewise deterministic Markov processes: a case study on a biochemical network 172
P. Kouretas, K. Koutoumpas, J. Lygeros

Using path integral short time propagators for numerical analysis of stochastic hybrid systems 179
G. Lichtenberg, P. Rostalski

TA2 - Controller Design Based on Hybrid Models of Industrial Plants (Invited)
Organizers: S. Engell, O. Stursberg

Challenges in start-up control of a heat exchange reactor with exothermic reactions; a hybrid approach 185
S. Haugwitz, P. Hagander

Feedback stabilization of the operation of an hybrid chemical plant 191
I. Simeonova, F. Warichet, G. Bastin, D. Dochain, Y. Pochet

A solar cooling plant: a benchmark for hybrid systems control 199
D. Zambrano, C. Bordons, W. García-Gabín, E. F. Camacho

Timed discrete event control of a parallel production line with continuous output 205
D. Gromov, S. Geist, J. Raisch

Dynamic optimization of an industrial evaporator using graph search with embedded nonlinear programming 211
C. Sonntag, O. Stursberg, S. Engell

TB1 - Diagnosis and Identification

Using neural networks for the identification of a class of hybrid dynamic systems 217
N. Messai, J. Zaytoon, B. Riera

Fault tolerant control design for switched systems 223
M. Rodrigues, D. Theilliol, D. Sauter

Discrete-event modelling and fault diagnosis of discretely controlled continuous systems 229
J. Lunze

Use of an object oriented dynamic hybrid simulator for the monitoring of industrial processes 235
N. Olivier, G. Hétreux, J.-M. Le Lann, M.-V. Le Lann

TB2 - Applications of Hybrid Control (Invited)
Organizers: A. Bemporad, F. Lamnabhi-Lagarrigue

Model predictive control of nonlinear mechatronic systems: an application to a magnetically actuated mass spring damper 241
S. Di Cairano, A. Bemporad, I. Kolmanovsky, D. Hrovat

Subtleties in the averaging of hybrid systems with applications to power electronics 247
L. Iannelli, K.H. Johansson, U. Jönsson, F. Vasca

Adaptive cruise controller design: a comparative assessment for PWA systems 253
D. Corona, B. De Schutter

Idle speed control - a benchmark for hybrid system research 259
A. Balluchi, L. Benvenuti, M.D. Di Benedetto, T. Villa, A.L. Sangiovanni-Vincentelli

TC1 - Hybrid Simulation Tools: Principles, Challenges and Applications (Invited)
Organizers: C.G. Cassandras, P. Mosterman

Simulation and verification of hybrid systems using Chi 265
D.A. van Beek, J.E. Rooda, R.R.H. Schiffelers

Hybrid system simulation with SIMEVENTS 267
C.G. Cassandras, M.I. Clune, P.J. Mosterman

HyVisual: a hybrid system modeling framework based on Ptolemy II 270
E.A. Lee, H. Zheng

TrueTime: simulation of networked computer control systems 272
D. Henriksson, A. Cervin, M. Andersson, K.-E. Arzen

CODIS - A framework for continuous/discrete systems co-simulation 274
G. Nicolescu, F. Bouchhima, L. Gheorghe

TC2 - Stability 1

On the finite-time stabilization of a nonlinear uncertain dynamics via switched control 276
G. Bartolini, A. Pisano, E. Usai

Search for period-2 cycles in a class of hybrid dynamical systems with autonomous switchings. Application to a thermal device 283
C. Quémard, J.-C. Jolly, J.-L. Ferrier

Stabilizability of bimodal piecewise linear systems with continuous vector field 290
K. Camlibel, M. Heemels, H. Schumacher

Global input-to-state stability and stabilization of discrete-time piece-wise affine systems 296
M. Lazar, W.P.M.H. Heemels

FA1 - Model Predictive Control

Feasible mode enumeration and cost comparison for explicit quadratic model predictive control of hybrid systems 302
A. Alessio, A. Bemporad

An efficient algorithm for predictive control of piecewise affine systems with mixed inputs 309
S. Leirens, J. Buisson

Explicit model predictive control of the boost DC-DC converter 315
A.G. Beccuti, G. Papafotiou, M. Morari

A new dual-mode hybrid MPC algorithm with a robust stability guarantee 321
M. Lazar, W.P.M.H. Heemels

Robust model predictive control for piecewise affine systems subject to bounded disturbances 329
J. Thomas, S. Olaru, J. Buisson, D. Dumur

FA2 - Stability 2

Stabilization of switched linear systems with unknown time varying delays 335
L. Hetel, J. Daafouz, C. Iung

Stabilizing dynamic controller of switched linear systems 341
S. Chaib, A. Benali, D. Boutat, J.-P. Barbot

Dynamic output feedback stabilization of continuous-time switched systems 347
J.C. Geromel, P. Colaneri

Practical stabilization of discrete-time linear LTI SISO systems under assigned input and output quantization 353
B. Picasso, A. Bicchi

Box invariance of hybrid and switched systems 359
A. Abate, A. Tiwari

FB1 - Verification and Safety

Performance verification of discrete event systems using hybrid model-checking 365
B. Denis, J.-J. Lesage, Z. Juárez-Orozco

Verification-integrated falsification of non-deterministic hybrid systems 371
S. Ratschan, J.-G. Smaus

An evaluation of two recent reachability analysis tools for hybrid systems 377
I. Ben Makhlouf, S. Kowalewski

Safety and reliability analysis of protection systems for power systems 383
L. Ferrarini, L. Ambrosi, E. Ciapessoni

A hybrid approach for safety analysis of aircraft systems 389
E. Villani, P.E. Miyagi

FB2 - Abstraction Based Approaches to Hybrid Control

Detecting and enforcing monotonicity for hybrid control systems synthesis 395
D. Gromov, J. Raisch

Hybrid system control using an on-line discrete event supervisory strategy 402
J. Millan, S. O'Young

Non-deterministic reactive systems, from hybrid systems and behavioural systems perspectives 409
J.M. Davoren, T. Moor

Control-invariance of sampled-data hybrid systems with periodically clocked events and jitter 417
Y. Tsuchie, T. Ushio

Author Index 423

ELSEVIER
IFAC
PUBLICATIONS

CHATTERING PROBLEM IN SLIDING MODE CONTROL SYSTEMS

Vadim Utkin * **Hoon Lee** *

* *Dept. of Electrical Engineering,*
The Ohio State University, USA
2015 Neil Avenue Columbus, OH 43210 USA

ABSTRACT

In practical applications of sliding mode control, engineers may experience undesirable phenomenon of oscillations having finite frequency and amplitude, which is known as 'chattering'. At the first stage of sliding mode control theory development the chattering was the main obstacle for its implementation. Chattering is a harmful phenomenon because it leads to low control accuracy, high wear of moving mechanical parts, and high heat losses in power circuits. There are two reasons which can lead to chattering.

- The chattering can be caused by fast dynamics which were neglected in the ideal model. These 'unmodeled' dynamics with small time constants are usually disregarded in models of servomechanisms, sensors and data processors.
- The second reason of chattering is utilization of digital controllers with finite sampling rate, which causes so called 'discretization chatter'. Theoretically the ideal sliding mode implies infinite switching frequency. Since the control is constant within a sampling interval, switching frequency can not exceed that of sampling, which lead to chattering as well.

Mechanism of chattering generating is demonstrated for control of inverted an pendulum with unmodeled actuator dynamics.

The efficient recipe for chattering suppression is use of asymptotic observers. The main idea of using an asymptotic observer to prevent chattering is to generate ideal sliding mode in the auxiliary loop including the observer. In the observer loop, sliding mode is generated in the control software; therefore, any unmodeled dynamics which cause chattering can be excluded.

As follows from analysis, based on the describing function method, the amplitude of chattering is proportional to magnitude of discontinuous control. Therefore the methods of chattering suppression can be developed such that the magnitude is decreased properly holding the establishment of sliding mode. First option is to decrease the magnitude along with the system states. The second one implies that the magnitude is the function of an equivalent control derived by a low-pass filter u_{eq}. The method can be applied for the plants subjected to unknown disturbances.

"Discretization chattering" in discrete-time systems is caused by discontinuities in control. Increasing a sampling frequency to decrease the chattering amplitude seems unjustified, since using a computer is adequate to control system dynamics if a sampling frequency corresponds to average, slow system motion rather than to a high frequency component.

First the definition of sliding mode is introduced embracing both discrete- and continuous-time systems. Then free of chattering design methodology is proposed for discrete-time systems. The fundamental difference is that the control should be a continuous function of the state.

Experimental results for sliding mode control of inductions motors with observers are discussed.

ELSEVIER
IFAC
PUBLICATIONS

CHALLENGES AND OPPORTUNITIES FOR SYSTEM THEORY IN EMBEDDED CONTROLLER DESIGN

Alberto Sangiovanni Vincentelli *

* *The Edgar L. and Harold H. Buttner Chair, Department of Electrical Engineering and Computer Science, University of California at Berkeley*

Abstract: Embedded controllers are essential in today electronic systems to assure that the behaviour of complex systems as cars, airplanes, trains, building security management systems, is compliant to strict safety constraints. I will review the evolution of embedded systems and the challenges that must be faced in their design. I will also present methodologies aimed at simplifying and speeding the design process. The role of hybrid systems in the development of embedded controllers will be outlined. Future applications such as wireless sensor networks in an industrial plant will also be presented. *Copyright © 2006 IFAC.*

Keywords: Embedded Systems, Systems Design, Systems Methodology, Control Applications, Distributed Control.

EXTENDED ABSTRACT

The ability of integrating an exponentially increasing number of transistors within a chip, the ever-expanding use of electronic embedded systems to control increasingly many aspects of the "real world", and the trend to interconnect more and more such systems (often from different manufacturers) into a global network, are creating a nightmarish scenario for embedded system designers. Complexity and scope are exploding into the three inter-related but independently growing directions mentioned above, while teams are even shrinking in size to further reduce costs. In this scenario the three challenges that are taking center stage are:

- *Heterogeneity and Complexity of the Hardware Platform.* The trends mentioned above result in exponential complexity growth of the features that can be implemented in hardware. The integration capabilities make it possible to build real complex system on a chip including analog and RF components. The decision of what to place on a chip is no longer dictated by the amount of circuitry that can be placed on the chip but by reliability, yield and ultimately cost (it is well known that analog and RF components force to use more conservative manufacturing lines with more processing steps than pure digital ICs). Even if manufacturing concerns suggest to implement hardware in separate chips, the resulting package may still be very small given the advances in packaging technology yielding the concept of System-in-Package (SiP). Pure digital chips are also featuring an increasing number of components. Design time, cost and manufacturing unpredictability for deep submicron technology make the use of custom hardware implementations appealing only for products that are addressing a very large market and for experienced and financially rich companies. Even for these companies, the present design methodologies are not yielding the necessary productivity forcing them to increase beyond reason the size of design and verification teams. These IC companies (for example Intel, AMD and TI) are looking increasingly to system design methods to allow them to assemble large chips out of pre-designed components and to reduce validation costs. In this context, the adoption of design models above RTL and of communication mechanism among components with guaranteed properties and standard interfaces is only a matter of time.

• *Embedded Software Complexity.* Given the cost and risks associated to developing hardware solutions, an increasing number of companies is selecting hardware platforms that can be customized by reconfiguration and/or by software programmability. In particular, software is taking the lion's share of the implementation budgets and cost. In cell phones, more than 1 Million lines of code is standard today, while in automobiles the estimated number of lines by 2010 is 100 Millions. The number of lines of source code of embedded software required for defense avionics systems is also growing exponentially. However, as this happens, the complexity explosion of the software component causes serious concerns for the final quality of the products and the productivity of the engineering forces. In transportation, the productivity of embedded software writers using the traditional methods of software development ranges in the few tens of lines per day. The reasons for such a low productivity are in the time needed for verification of the system and long redesign cycles that come from the need of developing full system prototypes for the lack of appropriate virtual engineering methods and tools for embedded software. Embedded software is substantially different from traditional software for commercial and corporate applications: by virtue of being embedded in a surrounding system, the software must be able to continuously react to stimuli in the desired way, i.e., within bounds on timing, power consumed and cost. Verifying the correctness of the system requires that the model of the software be transformed to include information that involve physical quantities to retain only what is relevant to the task at hand. In traditional software systems, the abstraction process leaves out all the physical aspects of the systems as only the functional aspects of the code matter.

• *Integration Complexity.* A standard technique to deal with complexity is decomposing "top-down" the system into subsystems. This approach, which has been customarily adopted by the semiconductor industry for years, has limitation as a designer or a group of designers has to fully comprehend the entire system and to partition appropriately its various parts, a difficult task given the enormous complexity of today's systems. Hence, the future is one of developing systems by *composing* pieces that all or in part have already been pre-designed or designed independently by other design groups or even companies. This has been done routinely in vertical design chains for example in the transportation vertical, albeit in a heuristic and ad hoc way. The resulting lack of an overall understanding of the interplay of the sub-systems and of the difficulties encountered in integrating very complex parts causes system integration to become a nightmare in the system industry. For example, Jurgen Hubbert, then in charge of the Mercedes-Benz passenger car division, publicly stated in 2003: "*The industry is fighting to solve problems that are coming from electronics and companies that introduce new technologies face additional risks. We have experienced blackouts on our cockpit management and navigation command system and there have been problems with telephone connections and seat heating.*"

I believe that in today's environment this state is the rule for the leading system OEMs let them operate in the transportation domain, in multimedia systems, in communication, rather than the exception. The source of these problems is clearly the increased complexity but also the difficulty of the OEMs in managing the integration and maintenance process with subsystems that come from different suppliers who use different design methods, different software architecture, different hardware platforms, different (and often proprietary) Real-Time Operating Systems. Therefore, there is a need for standards in the software and hardware domains that will allow plug-and-play of subsystems and their implementation while the competitive advantage of an OEM will increasingly reside on novel and compelling functionalities.

I will present a methodology to cope with some of these problems and that can use hybrid system modeling. I will review how this methodology can be applied to the design of embedded controllers for the automotive industry. Finally I will present the application of the methodology and of hybrid systems to the design of wireless sensor networks in an industrial environment.

ELSEVIER
IFAC
PUBLICATIONS

OPTIMAL CONTROL IN HYBRID SYSTEMS

Claude Iung, Pierre Riedinger

Institut National Polytechnique de Lorraine
CRAN UMR 7039 CNRS – UHP – INPL
ENSEM, 2, av. forêt de Haye,
54516, Vandoeuvre-Lés-Nancy, Cedex, France

Abstract: Necessary conditions for optimality have been established for hybrid systems. Unfortunately, these conditions lead to multi-point boundary value problems and they do not prevent from the combinatorial explosion when no constraints are given on the transitions. Different approaches have been proposed to approximate the solution, such as relaxed dynamic programming, non linear programming and sensitivity functions. *Copyright* © 2006 IFAC

Keywords: Hybrid systems, optimal control, sensitivity functions

EXTENDED ABSTRACT

In hybrid systems context, the necessary conditions for optimal control are now well known. These conditions mix discrete and continuous classical necessary conditions on the optimal control. The discrete dynamic involves dynamic programming methods whereas between the a priori unknown discrete values of time, optimization of the continuous dynamic is performed using the maximum principle (MP) or Hamilton Jacobi Bellmann equations(HJB). At the switching instants, a set of boundary tranversality necessary conditions ensure a global optimization of the hybrid system.

These theoretical conditions were applied to minimum time problem and to lincar quadratic optimization.

But it is practically very hard to perform such an optimization. The major raison is that discrete dynamic requires evaluating the optimal cost along all branches of the tree of all possible discrete trajectories.

Dynamic programming is then used, but the duration between two switchings and the continuous optimization procedure make the task really hard. This makes the complexity increasing and only problems with a poor coupling between continuous and discrete parts can be reasonably solved.

Nowadays, it seems obvious that only approximated solutions can be found. Various schemes have been imagined.

Recent works have proposed to solve optimal switching problems by using a fixed switching schedule. By switched systems we mean a class of hybrid dynamical systems consisting of a family of continuous (or discrete) time subsystems and a rule (to be determined) that governs the switching between them.

The optimization consists then in determining the optimal switching instants and the optimal continuous control assuming the number of switchings and the order of active subsystems already given.Then a nonlinear search method is used to determine the optimal solution.after the calculus of the derivatives of the value function with respect to the switching instants.

Relaxed Dynamic programming : a relaxed procedure based on upper and lower bounds of the optimal cost was recently introduced. It proved to give good results for piece-wise affine systems and to obtain a suboptimal state feedback solution in the case of a quadratic criteria

Algorithms based on the maximum principle for both multiple controlled and autonomous switchings with fixed schedule have been proposed. The algorithms use the transversality conditions at switching instants. Then, the authors develop a combinational search in order to determine the optimal switching schedule

Interesting results on state or output feedback have been given with the regions of the state space where an optimal mode switch should occur.

In complement of all the methods resulting from the resolution of the necessary conditions of optimality, we propose to use a multiple-phase multiple-shooting formulation which enables the use of standard constraint nonlinear programming methods. This formulation is applied to hybrid systems with autonomous and controlled switchings and seems to be of interest in practice due to the simplicity of implementation.

Sensitivity analysis is the key point of all the methods based on non linear programming. It can also be used to determine limit cycles and the optimal strategy to reach them.

All these items are discussed in the plenary session.

REFERENCES

Bemporad A., D.Corona, A.Giua and C.Seatzu (2003). Optimal State-Feedback Quadratic Regulation of Linear Hybrid Automata. In: ADHS03

Bemporad A., A.Giua and C.Seatzu (2002) An algorithm for the optimal control of continuous time switched linear systems. In: *IEEEComputer Society, Proceeding of the Sixth workshop on Discrete Event Systems*

Betts J.T. (2001). Practical methods for optimal control using nonlinear programming. In: Advanced design and control, Siam

Corona D., A. Giua and C. Seatzu (2003) Optimal Feedback Switching Laws for Homogeneous Hybrid Automata. In: proc. *IEEE Conference on Control and Decision.*

Daafouz J., P. Riedinger and C. Iung (2001) Static Output Feedback Control for Switched Systems.In: *proc. 40th IEEE Conference on Decision and Control.*

Egerstedt , Y. Wardi, and F. Delmotte (2003) Optimal Control of Switching Times. In *Switched Dynamical Systems. IEEE Conference on Decision and Control,.*

Egerstedt M., Y. Wardi and H. Axelsson (2006)Transition-time optimization for switched-mode dynamical systems. In: *IEEE Transactions in Automatic Control,* **vol 51**(1) 110-115

Flieller D.,P Riedinger and J.P.Louis (2006). Computation and stability of limit cycles in hybrid systems In: *Nonlinear Analysis*, **vol 64,** 352-367

Frank P.M (1978) Introduction to System sensitivity Theory Academic Press

Hedlund S and A Rantzer (1999) Optimal Control of Hybrid Systems. In : *Proceedings of 38th IEEE Conf. on Decision and Control.*

Hedlund S. and A. Rantzer (2002). Convex dynamic programming for hybrid systems. In: *IEEE Transactions in Automatic Control* **,vol 47**(9) 1536-1540.

Lincoln B. and A. Rantzer (2003). Relaxed Optimal Control of Piecewise Linear Systems. In: *ADSH 03.*

Lygeros.J., K. H. Johansson, S. N. Simic, J. Zhang and S.S. Sastry (2003). Dynamical Properties of Hybrid Automata. In: IEEE Transaction on Automatic Control, **vol 48**(1)

Rantzer A. and M. Johansson (2000). Piecewise Linear Quadratic Optimal Control In: *IEEE Transactions on Automatic Control,* **vol 45**(4) 629-637

Rantzer A. (2005). On Approximate Dynamic Programming in Switching Systems. In: proc.44th *IEEE Conference. on Decision and Control*

Riedinger P., F. Kratz, C. Iung and C. Zanne (1999). Linear Quadratic Optimization for Hybrid Systems. In: *proc. of the 38th IEEE Conference. on Decision and Control, 3059-3064*

Riedinger P., C. Iung,and F. Kratz (2003). An Optimal Control Approach for Hybrid Systems..In: *European Journal of Control,* **vol 9** (5), pp 449-458.

Riedinger P., J.Daafouz and.C. Iung (2003) Suboptimal switched controls in context of singular arcs. In: *42th IEEE Conference on Decision and Control,*

Riedinger,J.Daafouz and C. Iung (2005). About solving hybrid optimal control problems. In: *proc IMACS05*

Rosenwasser E.N (1967) General sensitivity equations of discontinuous systems. In: *Automation and Remote Control* **vol 28**, 400-404

Seatzu C., D.Corona, A.Guia and A.Bemporad (2006).Optimal Control of Continuous-Time Switched Affine Systems. In: *IEEE Transactions on Automatic Control* May 2006

Shaikh M.S.and P.E Caines (2003) On the optimal control of hybrid systems: Analysis and algorithms for trajectory and schedule optimization. In *proc. IEEE Conference on Control and Decision.*

Shaikh M.S.and P.E Caines (2003) On the optimal control of hybrid systems : Optimization of switching times and combinatoric location schedules. In *Proc. American Control Conference,* 2773-2778

Shaikh M.S.and P.E Caines (2003). On the optimal control of hybrid systems: Analysis and algorithms for trajectory and schedule optimization. In: *proc. IEEE Conference on Control and Decision.*

Sussmann H.J. (1999). A maximum principle for hybrid optimal control problems. In *proc. of the 38th IEEE Conf. on Decision and Control,* pp 425-430.

Wardi Y.,M. Egerstedt, M. Boccadoro and E. Verriest (2004). Optimal Control of Switching Surfaces In IEEE Conference on Decision and Control.

Xu X and P. J. Antsaklis (2001). An approach for solving Generalswitched Linear Quadratic Optimal Control Problems. In: *proc. 40th IEEE Conf. on Decision and Control.*

Xu. X.and P. J. Antsaklis (2002). Optimal Control of Switched Systems via Nonlinear Optimization Based on Direct Differentiations of Value Functions. In: *International Journal of Control, 75(16):1406-1426.*

ELSEVIER
IFAC
PUBLICATIONS

CONVERGENT DESIGN OF SWITCHED LINEAR SYSTEMS

R.A. van den Berg A.Y. Pogromsky J.E. Rooda

Eindhoven University of Technology
Department of Mechanical Engineering
P.O.Box 513, 5600 MB Eindhoven, The Netherlands.
[r.a.v.d.berg, a.pogromsky, j.e.rooda] @tue.nl

Abstract: This paper deals with the design of switching rules for switched linear systems with inputs, in such a way that the resulting closed-loop system is exponentially convergent. Two types of switching rules are addressed, that is state-based and observer-based rules. The developed theory is illustrated by two examples. *Copyright © 2006 IFAC*

Keywords: Convergent systems, switched systems, continuous time systems, exponential stability, performance evaluation, observer.

1. INTRODUCTION

A switched linear system is a hybrid/nonlinear system which consists of several linear subsystems and a switching rule that decides which of the subsystems is active at each moment in time. These systems have been a subject of growing interest in the last decades, see e.g. (Liberzon and Morse, 1999; DeCarlo *et al.*, 2000) and references therein. Because of the combination of multiple linear systems/controllers, a well-tuned switched linear system can achieve better performance then a single linear system, or can achieve certain control goals that cannot be realized by linear systems (Morse, 1996; Narendra and Balakrishnan, 1997; Feuer *et al.*, 1997).

Besides these extended possibilities that switched linear systems have with respect to linear systems, the design of such a switched system also brings along difficulties. For example, if all the linear subsystems of a switched system are stable, this does not automatically guarantee the stability of that switched system. A good example of this apparent contradiction is given in (Branicky, 1998). Another property that a linear time invariant (LTI) system with asymptotically stable homogeneous part has, but is not natural for a nonlinear/hybrid system, is that any solution of an LTI system with a bounded input converges to a unique solution that depends only on the input. Nonlinear/hybrid system that *do* possess this property are referred to as convergent. Solutions of convergent system "forget" their initial conditions and after some transient depend only on the system input, which can be a command or reference signal.

Convergency of nonlinear/hybrid systems is an interesting property, since it results in a limit solution that is independent of the initial conditions of the system. This is useful in for example synchronization problems (Pogromsky *et al.*, 2002). Another possible area of interest is the performance analysis of nonlinear systems. For general nonlinear systems simulation-based analysis is quite impossible, since all possible initial conditions need to be evaluated in order to obtain a reliable analysis. For a convergent system, however, this problem does not exist, since all initial conditions lead to the same limit solution. Therefore simulation can be used to analyse and optimize performance of convergent systems. This motivates studies related to the design of convergent systems.

The property that all solutions of a system "forget" their initial conditions and converge to some steady-state solution has been addressed in a number of publications, e.g. (Fromion *et al.*, 1996; Lohmiller and Slotine, 1998; Fromion *et al.*, 1999; Pavlov *et al.*, 2004; Angeli, 2002; Pavlov *et al.*, 2005*b*). In this paper, the focus lies on the convergent design of switched linear systems using only the switching rule as "design variable". Two different cases are considered. First, the case is considered in which the switching rule is based on static state feedback. Secondly, the case is considered in which full state information is not available. In this case a switching rule is discussed that is based on an observer.

The outline of this paper is as follows. In Section 2 a basic definition on stability is recalled that is required in the remainder of this article. Section 3 presents various definitions and properties of convergent systems. In Section 4 the design of a switching rule is discussed that makes the closed-loop switched linear system convergent. The main results of this section are presented in two theorems which give sufficient conditions under which such a switching rule can be found. Two examples are provided in Section 5 to illustrate these theorems. Section 6 concludes the paper.

2. PRELIMINARIES

In this article exponential stability will be considered. For the sake of completeness, this definition is given here.

Definition 1. A solution $\mathbf{x}(t, t_0, \bar{\mathbf{x}}_0)$ of a system $\dot{\mathbf{x}} = f(\mathbf{x}, t)$, defined for all $t \in (t_*, +\infty)$, is said to be *exponentially stable* if there exist positive δ, C, β such that $||\mathbf{x}_0 - \bar{\mathbf{x}}_0|| < \delta$ implies

$$||\mathbf{x}(t, t_0, \mathbf{x}_0) - \mathbf{x}(t, t_0, \bar{\mathbf{x}}_0)|| \leq Ce^{-\beta(t-t_0)}||\mathbf{x}_0 - \bar{\mathbf{x}}_0||$$

3. CONVERGENT SYSTEMS

In this section definitions and properties of convergent systems are presented. Those systems are very closely related to systems with globally exponentially stable solutions and the definitions presented here extend those given by Demidovich (Demidovich, 1967).

The following class of systems is considered

$$\dot{\mathbf{x}} = f(\mathbf{x}, \mathbf{w}(t)) \qquad (1)$$

with state $\mathbf{x} \in \mathbb{R}^n$ and input $\mathbf{w} \in \overline{\mathbb{PC}}_m$. Here, $\overline{\mathbb{PC}}_m$ is the class of bounded (for all $t \in \mathbb{R}$) piecewise continuous inputs $\mathbf{w}(t) : \mathbb{R} \rightarrow \mathbb{R}^m$. Assume that the function $f(\mathbf{x}, \mathbf{w})$ satisfies some regularity conditions to ensure the existence of a Filippov solution, see e.g. (Filippov, 1988), p.76.

Definition 2. System (1) is said to be *exponentially convergent* if there is a solution $\bar{\mathbf{x}}(t) = \mathbf{x}(t, t_0, \bar{\mathbf{x}}_0)$ satisfying the following conditions for every input $\mathbf{w}(t) \in \overline{\mathbb{PC}}_m$: (i) $\bar{\mathbf{x}}(t)$ is defined and bounded for all $t \in (-\infty, +\infty)$, (ii) $\bar{\mathbf{x}}(t)$ is globally exponentially stable for every input $\mathbf{w}(t) \in \overline{\mathbb{PC}}_m$.

The solution $\bar{\mathbf{x}}(t)$ is called a *limit solution.* As follows from the definition of convergency, any solution of a convergent system "forgets" its initial condition and converges to some limit solution which is independent of the initial conditions. For exponentially convergent systems this limit solution $\bar{\mathbf{x}}(t)$ is unique, i.e. it is the only solution defined and bounded for all $t \in (-\infty, +\infty)$ (Pavlov *et al.*, 2005*a*).

For system (1) consider a scalar continuously differentiable function $V(\mathbf{x})$. Define a time derivative of this function along solutions of system (1) as follows

$$\dot{V} = \frac{\partial V(\mathbf{x})}{\partial \mathbf{x}} \dot{\mathbf{x}}(t, t_0, \mathbf{x}_0) \quad \text{a.e.}$$

Definition 3. System (1) is called quadratically convergent if there exists a positively definite matrix $\mathbf{P} = \mathbf{P}^T > 0$ and a number $\alpha > 0$ such that for any input $\mathbf{w} \in \overline{\mathbb{PC}}_m$ for the function $V(\mathbf{x}_1, \mathbf{x}_2) = (\mathbf{x}_1 - \mathbf{x}_2)^T \mathbf{P}(\mathbf{x}_1 - \mathbf{x}_2)$ it holds that

$$\dot{V}(\mathbf{x}_1, \mathbf{x}_2, t) \leq -\alpha V(\mathbf{x}_1, \mathbf{x}_2). \qquad (2)$$

Lemma 4. (Pavlov *et al.*, 2005*a*) If system (1) is quadratically convergent, then it is exponentially convergent.

The proof of this lemma is based on the following result, which will be also used in the sequel.

Lemma 5. (Yakubovich, 1964) Consider system (1) with a given input $\mathbf{w}(t)$ defined for all $t \in \mathbb{R}$. Let $\mathcal{D} \subset \mathbb{R}^n$ be a compact set which is positively invariant with respect to dynamics (1).Then there is at least one solution $\bar{\mathbf{x}}(t)$, such that $\bar{\mathbf{x}}(t) \in \mathcal{D}$ for all $t \in (-\infty, +\infty)$.

Note that for convergent nonlinear systems performance can be evaluated in almost the same way as for linear systems. Due to the fact that the limit solution of a convergent system only depends on the input and is independent of the initial conditions, performance evaluation of one solution (i.e. one arbitrary initial state) for a certain input suffices, whereas for general nonlinear systems all initial states need to be evaluated to obtain a reliable analysis. This means that for convergent systems simulation becomes a reliable analysis tool and for example 'Bode-like' plots can be drawn to analyse the system performance. An example of simulation based performance analysis can be found in Section 5.1.

4. CONVERGENCY OF SWITCHED SYSTEMS

Consider the switched dynamical system

$$\begin{aligned} \dot{\mathbf{x}}(t) &= \mathbf{A}_i\mathbf{x}(t) + \mathbf{B}_i\mathbf{w}(t) \\ \mathbf{y}(t) &= \mathbf{C}_i\mathbf{x}(t) \end{aligned} \quad i = 1, \ldots, k \qquad (3)$$

where $\mathbf{x}(t) \in \mathbb{R}^n$ is the state, $\mathbf{w}(t) \in \overline{\mathbb{PC}}_m$ is the input, and $\mathbf{y}(t) \in \mathbb{R}^l$ is the output. These dynamics for example represent the system in Figure 1. Suppose the collection of matrices $\{\mathbf{A}_1, \ldots, \mathbf{A}_k\}$, $\{\mathbf{B}_1, \ldots, \mathbf{B}_k\}$, and $\{\mathbf{C}_1, \ldots, \mathbf{C}_k\}$ is given, and $\mathbf{A}_i$ is Hurwitz for all $i = 1, \ldots, k$. This implies for the system in Figure 1 that the plant and all linear controllers are already fixed. The general problem is to find a *switching rule* such that the closed-loop system is exponentially convergent. In this section, two kinds of switching rules are discussed. First, a switching rule is addressed that is based on static state feedback, i.e. $i = \sigma(\mathbf{x}, \mathbf{w})$. Secondly, the case is considered in which not the entire state can be measured, but just some output $\mathbf{y}$. For this case a switching rule is discussed that is based on an observer.

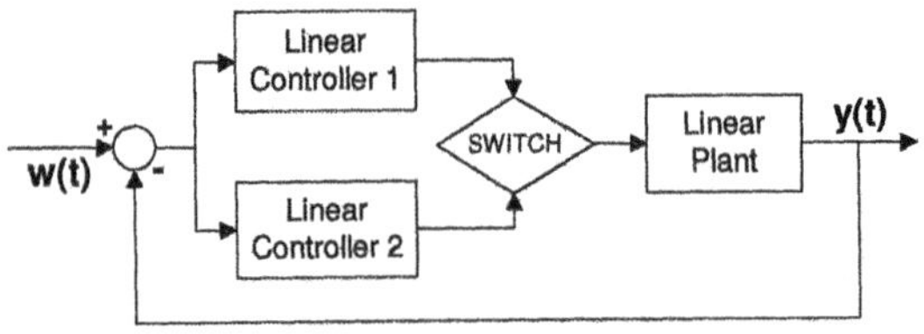

Fig. 1. Switched linear system.

4.1 Switching rule based on state feedback

Suppose a common Lyapunov matrix $\mathbf{P} = \mathbf{P}^T > 0$ exists that satisfies the following inequalities

$$\mathbf{A}_i^T\mathbf{P} + \mathbf{P}\mathbf{A}_i < 0, \quad i = 1, \ldots, k. \qquad (4)$$

Consider the following switching rule

$$\sigma(\mathbf{x}, \mathbf{w}) = \arg\ \min_i\{\mathbf{x}^T\mathbf{Z}_{ix}\mathbf{x} + \mathbf{x}^T\mathbf{Z}_{iw}\mathbf{w}\} \qquad (5)$$

in which $\mathbf{Z}_{iw} = 4\mathbf{P}\mathbf{B}_i$ and $\mathbf{Z}_{ix}$ are matrices to be defined.

Theorem 6. If there exist a solution $\mathbf{P} = \mathbf{P}^T > 0$ of (4) and $\mathbf{Z}_{1x}, \ldots, \mathbf{Z}_{kx}$ such that

$$\mathbf{Z}_{ix} \neq \mathbf{Z}_{jx} \text{ and/or } \mathbf{Z}_{iw} \neq \mathbf{Z}_{jw} \quad \forall i, j \leq k,\ i \neq j \qquad (6)$$

and for some $\varepsilon > 0$

$$\begin{bmatrix} \mathbf{P}\mathbf{A}_i + \mathbf{A}_i^T\mathbf{P} & -(\mathbf{A}_i^T\mathbf{P} + \mathbf{P}\mathbf{A}_j) \\ -(\mathbf{A}_j^T\mathbf{P} + \mathbf{P}\mathbf{A}_i) & \mathbf{P}\mathbf{A}_j + \mathbf{A}_j^T\mathbf{P} \end{bmatrix} + \begin{bmatrix} -(\mathbf{Z}_{ix} - \mathbf{Z}_{jx}) & 0 \\ 0 & \mathbf{Z}_{ix} - \mathbf{Z}_{jx} \end{bmatrix} \leq -\varepsilon \begin{bmatrix} \mathbf{I}_n & -\mathbf{I}_n \\ -\mathbf{I}_n & \mathbf{I}_n \end{bmatrix} \forall i, j \leq k,\ i \neq j \qquad (7)$$

then switching rule (5) with matrices $\mathbf{Z}_{1x}, \ldots, \mathbf{Z}_{kx}$ makes system (3) quadratically convergent.

Proof: Let $\mathbf{P}$ be a common Lyapunov matrix for the collection $\{\mathbf{A}_1, \ldots, \mathbf{A}_k\}$ and consider the Lyapunov function candidate

$$V(\mathbf{x}_1, \mathbf{x}_2) = (\mathbf{x}_1 - \mathbf{x}_2)^T\mathbf{P}(\mathbf{x}_1 - \mathbf{x}_2) \qquad (8)$$

If $\sigma(\mathbf{x}_1, \mathbf{w}) = \sigma(\mathbf{x}_2, \mathbf{w})$ the inequality

$$\dot{V} \leq -\alpha V, \quad \alpha > 0$$

is obviously satisfied. Let $\sigma(\mathbf{x}_1, \mathbf{w}) = p$ and $\sigma(\mathbf{x}_2, \mathbf{w}) = q$.

$$\begin{aligned} \dot{V} = {} & \mathbf{x}_1^T(\mathbf{A}_p^T\mathbf{P} + \mathbf{P}\mathbf{A}_p)\mathbf{x}_1 + \mathbf{x}_2^T(\mathbf{A}_q^T\mathbf{P} + \mathbf{P}\mathbf{A}_q)\mathbf{x}_2 \\ & - \mathbf{x}_1^T(\mathbf{A}_p^T\mathbf{P} + \mathbf{P}\mathbf{A}_q)\mathbf{x}_2 - \mathbf{x}_2^T(\mathbf{P}\mathbf{A}_p + \mathbf{A}_q^T\mathbf{P})\mathbf{x}_1 \\ & + 2\mathbf{x}_1^T\mathbf{P}(\mathbf{B}_p - \mathbf{B}_q)\mathbf{w} + 2\mathbf{x}_2^T\mathbf{P}(\mathbf{B}_q - \mathbf{B}_p)\mathbf{w} \end{aligned} \qquad (9)$$

The switching rule (5) implies the following constraint functions for mode p

$$S_1(\mathbf{x}, \mathbf{w}) = \mathbf{x}_1^T(\mathbf{Z}_{px} - \mathbf{Z}_{qx})\mathbf{x}_1 + \mathbf{x}_1^T(\mathbf{Z}_{pw} - \mathbf{Z}_{qw})\mathbf{w} \leq 0$$

and for mode q

$$S_2(\mathbf{x}, \mathbf{w}) = \mathbf{x}_2^T(\mathbf{Z}_{qx} - \mathbf{Z}_{px})\mathbf{x}_2 + \mathbf{x}_2^T(\mathbf{Z}_{qw} - \mathbf{Z}_{pw})\mathbf{w} \leq 0.$$

The system is quadratically convergent if for some $\varepsilon > 0$

$$\dot{V} \leq -\varepsilon \begin{bmatrix} \mathbf{x}_1 \\ \mathbf{x}_2 \end{bmatrix}^T \begin{bmatrix} \mathbf{I}_n & -\mathbf{I}_n \\ -\mathbf{I}_n & \mathbf{I}_n \end{bmatrix} \begin{bmatrix} \mathbf{x}_1 \\ \mathbf{x}_2 \end{bmatrix}$$

for all $(\mathbf{x}, \mathbf{w})$ that satisfy $S_1(\mathbf{x}, \mathbf{w}) \leq 0$ and $S_2(\mathbf{x}, \mathbf{w}) \leq 0$. Using the $\mathcal{S}$-procedure, the previous condition is satisfied if the following inequality holds

$$\dot{V} - S_1 - S_2 \leq -\varepsilon \begin{bmatrix} \mathbf{x}_1 \\ \mathbf{x}_2 \end{bmatrix}^T \begin{bmatrix} \mathbf{I}_n & -\mathbf{I}_n \\ -\mathbf{I}_n & \mathbf{I}_n \end{bmatrix} \begin{bmatrix} \mathbf{x}_1 \\ \mathbf{x}_2 \end{bmatrix}$$

This inequality is equivalent to (7). □

Remark 7. Note that (7) is an LMI with design variables $\mathbf{P}$ and $\mathbf{Z}_{1x}, \ldots, \mathbf{Z}_{kx}$, which can be solved efficiently using available LMI toolboxes.

4.2 Observer-based switching rule

Consider the observer for system (3)

$$\dot{\hat{\mathbf{x}}}(t) = \mathbf{A}_i\hat{\mathbf{x}}(t) + \mathbf{B}_i w(t) + \mathbf{L}_i\mathbf{C}_i(\mathbf{x} - \hat{\mathbf{x}}) \qquad (10)$$

with $i = 1, \ldots, k$, $\hat{\mathbf{x}}$ the estimate of state $\mathbf{x}$ and $\mathbf{L}_i \in \mathbb{R}^{n \times l}$ the observer gain matrix. Assume that a common Lyapunov matrix exists such that (4) is satisfied. Now consider the observer-based switching rule

$$\sigma(\hat{\mathbf{x}}, \mathbf{w}) = \arg \min_i \{\hat{\mathbf{x}}^T \mathbf{Z}_{ix} \hat{\mathbf{x}} + \hat{\mathbf{x}}^T \mathbf{Z}_{iw} \mathbf{w}\} \quad (11)$$

in which $\mathbf{Z}_{iw} = 4\mathbf{P}\mathbf{B}_i$ and $\mathbf{Z}_{ix}$ are matrices to be defined.

Theorem 8. If there exist a solution $\mathbf{P} = \mathbf{P}^T > 0$ of (4) and $\mathbf{Z}_{1x}, \ldots, \mathbf{Z}_{kx}$ such that conditions (6) and (7) are satisfied, *and* if there exist a $\mathbf{P}_2 = \mathbf{P}_2^T > 0$ and $\mathbf{L}_i$ for $i = 1, \ldots, k$, such that for all $i = 1, \ldots, k$

$$(\mathbf{A}_i - \mathbf{L}_i \mathbf{C}_i)^T \mathbf{P}_2 + \mathbf{P}_2 (\mathbf{A}_i - \mathbf{L}_i \mathbf{C}_i) < 0 \quad (12)$$

then switching rule (5) with matrices $\mathbf{Z}_{1x}, \ldots, \mathbf{Z}_{kx}$ makes system (3) exponentially convergent.

Proof: First it is proven that the state $\mathbf{x}(t)$ of system (3) either lies in a positive invariant compact set or converges exponentially in time to this set. Consider the Lyapunov function

$$V(\mathbf{x}) = \mathbf{x}^T \mathbf{P} \mathbf{x}$$

Since there exists a common $\mathbf{P}$ such that (4) is satisfied, it follows that

$$\dot{V}(\mathbf{x}) \leq -\alpha V + \beta^* |\mathbf{x}||\mathbf{w}| \leq -\alpha V + \beta \sqrt{V}$$

for some positive constants α, β^*, and β, and bounded input $\mathbf{w} \in \overline{\mathbb{PC}}_m$. Note that there exists a level set

$$\Omega = \left\{ \mathbf{x} \;\middle|\; V(\mathbf{x}) \leq \frac{\beta^2}{\alpha^2} \right\}$$

outside of which $\dot{V} < 0$. This implies that all initial $V(\mathbf{x}(0))$ within this level set remain within the set. All $V(\mathbf{x}(0))$ outside this set converge exponentially to this set as can be seen from

$$\dot{V} \leq -\alpha V + \beta \sqrt{V} \leq -\frac{1}{2}\alpha \left(V - \frac{\beta^2}{\alpha^2} \right)$$

Since V is a quadratic function of $\mathbf{x}(t)$, it can be concluded that $\mathbf{x}(t)$ also converges exponentially to the positively invariant compact set Ω.

Secondly it is proven that the estimation error $\mathbf{e}(t) = \mathbf{x}(t) - \hat{\mathbf{x}}(t)$ decreases exponentially towards zero as $t \to \infty$ if (12) holds for all $i = 1, \ldots, k$. Since both the observer (10) and system (3) use the same switching rule (11) the error dynamics become

$$\dot{\mathbf{e}} = \begin{cases} (\mathbf{A}_1 - \mathbf{L}_1 \mathbf{C}_1)\mathbf{e} & \text{for } \sigma(\hat{\mathbf{x}}, \mathbf{w}) = 1 \\ \vdots & \\ (\mathbf{A}_k - \mathbf{L}_k \mathbf{C}_k)\mathbf{e} & \text{for } \sigma(\hat{\mathbf{x}}, \mathbf{w}) = k \end{cases}$$

If there exists a common Lyapunov matrix $\mathbf{P}_2$ for all $(\mathbf{A}_i - \mathbf{L}_i \mathbf{C}_i)$, $i = 1, \ldots, k$, i.e. condition (12) is satisfied, then the equilibrium point $\mathbf{e} = 0$ is globally exponentially stable.

Finally consider the Lyapunov function and its derivative given in respectively (8) and (9). Let $\sigma(\hat{\mathbf{x}}_1, \mathbf{w}) = p$ and $\sigma(\hat{\mathbf{x}}_2, \mathbf{w}) = q$. The observer-based switching rule (11) implies the following constraint functions for mode p

$$S_1(\hat{\mathbf{x}}, \mathbf{w}) = \hat{\mathbf{x}}_1^T (\mathbf{Z}_{px} - \mathbf{Z}_{qx}) \hat{\mathbf{x}}_1 + \hat{\mathbf{x}}_1^T (\mathbf{Z}_{pw} - \mathbf{Z}_{qw}) \mathbf{w} \leq 0$$

and for mode q

$$S_2(\hat{\mathbf{x}}, \mathbf{w}) = \hat{\mathbf{x}}_2^T (\mathbf{Z}_{qx} - \mathbf{Z}_{px}) \hat{\mathbf{x}}_2 + \hat{\mathbf{x}}_2^T (\mathbf{Z}_{qw} - \mathbf{Z}_{pw}) \mathbf{w} \leq 0.$$

Substituting $\hat{\mathbf{x}}_i$ by $\mathbf{x}_i - \mathbf{e}_i$ gives

$$S_1(\hat{\mathbf{x}}_1, \mathbf{w}) = S_1(\mathbf{x}_1, \mathbf{w}) + S_1(\mathbf{e}_1, \mathbf{w}) - f(\mathbf{e}_1, \mathbf{x}_1),$$
$$S_2(\hat{\mathbf{x}}_2, \mathbf{w}) = S_2(\mathbf{x}_2, \mathbf{w}) + S_2(\mathbf{e}_1, \mathbf{w}) + f(\mathbf{e}_2, \mathbf{x}_2),$$

with

$$f(\mathbf{e}_i, \mathbf{x}_i) = \mathbf{x}_i^T (\mathbf{Z}_{qx} - \mathbf{Z}_{px}) \mathbf{e}_i + \mathbf{e}_i^T (\mathbf{Z}_{qx} - \mathbf{Z}_{px}) \mathbf{x}_i.$$

Subsequently, the $\mathcal{S}$-procedure is applied to obtain

$$\dot{V} - S_1(\hat{\mathbf{x}}_1, \mathbf{w}) - S_2(\hat{\mathbf{x}}_2, \mathbf{w}) \leq -\alpha_1 V + g(..)$$

with

$$g(..) = -S_1(\mathbf{e}_1, \mathbf{w}) - S_2(\mathbf{e}_1, \mathbf{w}) + f(\mathbf{e}_1, \mathbf{x}_1) - f(\mathbf{e}_2, \mathbf{x}_2)$$

Since $\mathbf{e}_i(t)$ tends exponentially towards zero as $t \to \infty$, $\mathbf{x}_i(t)$ lies in Ω or converges exponentially in time towards this set for $i = 1, 2$, and $\mathbf{w}(t)$ is bounded, function g tends exponentially towards zero as a function of time. Thus, using the switching rule (11) the following inequality is true

$$\dot{V} \leq -\alpha_1 V + \gamma e^{-\alpha_2 t}$$

where α_1, α_2, γ are some positive constants. This implies that $V(\mathbf{x}_1(t) - \mathbf{x}_2(t))$ reduces exponentially towards zero as $t \to \infty$ and therefore that system (3) is exponentially convergent. This completes the proof. □

Remark 9. Since there exists a common $\mathbf{P}$ for all $\mathbf{A}_i$, $i = 1, \ldots, k$, condition (12) can always be met (take $\mathbf{L}_i = 0$).

5. TWO EXAMPLES

The theory presented in the previous section is now illustrated my means of two examples. For both examples the system in Figure 1 is considered, of which the dynamics is given by

$$\begin{aligned} \dot{\mathbf{x}} &= \mathbf{A}_i \mathbf{x} + \mathbf{B}_i w(t), \quad i = 1, 2 \\ y &= \mathbf{C}\mathbf{x} \end{aligned} \quad (13)$$

with $\mathbf{x}(t) \in \mathbb{R}^3$ the state, $w(\cdot) \in \overline{\mathbb{PC}}_1$ the input, and

$$\mathbf{A}_1 = \begin{bmatrix} -5 & -8 & 3 \\ 10 & -2 & 0 \\ 9 & -1 & -6 \end{bmatrix}, \quad \mathbf{B}_1 = \begin{bmatrix} 14 \\ -6 \\ 7 \end{bmatrix},$$
$$\mathbf{A}_2 = \begin{bmatrix} -8 & -5 & -8 \\ 13 & -8 & 2 \\ -2 & 1 & -4 \end{bmatrix}, \quad \mathbf{B}_2 = \begin{bmatrix} 20 \\ -16 \\ 8 \end{bmatrix},$$
$$\mathbf{C} = \begin{bmatrix} 1 & 0 & 0 \end{bmatrix}.$$

In the first example this system is made quadratically convergent using a state dependent switching rule. For the obtained convergent system the performance is analyzed and compared to the performance of the corresponding linear systems. In the second example an observer-based switching rule is used to render a system exponentially convergent, when only the output y can be measured.

5.1 Performance of a convergent switched system

Consider system (13) with the given matrices. Using an LMI toolbox the following common Lyapunov matrix can be found

$$\mathbf{P} = \begin{bmatrix} 0.1973 & -0.0179 & 0.0073 \\ -0.0179 & 0.1653 & -0.0149 \\ 0.0073 & -0.0149 & 0.1932 \end{bmatrix} > 0$$

such that conditions (6) and (7) are satisfied, using $\mathbf{Z}_{ix} = \mathbf{A}_i^T\mathbf{P} + \mathbf{P}\mathbf{A}_i$, $i = 1, 2$. Switching rule (5) thus makes the system (13) quadratically convergent,

$$\begin{aligned}\dot{V} &\leq -6.6643(\mathbf{x}_1 - \mathbf{x}_2)^T\mathbf{P}(\mathbf{x}_1 - \mathbf{x}_2) \\ &\leq -6.6643V.\end{aligned}$$

Subsequently, the fact that

$$\begin{aligned}(\mathbf{x}_1 - \mathbf{x}_2)^T\mathbf{P}(\mathbf{x}_1 - \mathbf{x}_2) \geq \lambda_{\min}(\mathbf{P})(\mathbf{x}_1 - \mathbf{x}_2)^2 \\ (\mathbf{x}_1 - \mathbf{x}_2)^T\mathbf{P}(\mathbf{x}_1 - \mathbf{x}_2) \leq \lambda_{\max}(\mathbf{P})(\mathbf{x}_1 - \mathbf{x}_2)^2\end{aligned}$$

with $\lambda_{\min}(\mathbf{P})$ and $\lambda_{\max}(\mathbf{P})$ respectively the minimum and maximum eigenvalue of $\mathbf{P}$, leads to the following upper bound

$$\begin{aligned}|\mathbf{x}_1(t) - \mathbf{x}_2(t)| &\leq \sqrt{\frac{\lambda_{\max}}{\lambda_{\min}}}\,|\mathbf{x}_1(0) - \mathbf{x}_2(0)|\, e^{\frac{-6.6643}{2}t} \\ &\leq 1.3885\,|\mathbf{x}_1(0) - \mathbf{x}_2(0)|\, e^{-3.3321t}.\end{aligned} \tag{14}$$

In order to analyse the performance of this switched system, only one solution of the system needs to be evaluated, since the limit solution of this (convergent) system is independent of its initial conditions. In Figure 2 the performance of the switched system is compared with the performance of the two corresponding linear systems, i.e., $\dot{\mathbf{x}} = \mathbf{A}_1\mathbf{x} + \mathbf{B}_1 w(t)$ and $\dot{\mathbf{x}} = \mathbf{A}_2\mathbf{x} + \mathbf{B}_2 w(t)$. The performance measure applied here is the relative tracking error of the limit solution

$$\sqrt{\frac{\int_{t_l}^{t_l+T} (w(t) - y(t))^2\, dt}{\int_{t_l}^{t_l+T} w(t)^2 dt}}, \tag{15}$$

where T is a time period that is long enough to obtain a good average of the tracking error and t_l is a moment in time for which all considered solutions are close enough to the limit solution. The time t_l is in this example determined visually, but a bound can be calculated as well using (14).

The performance is evaluated for the following input signals

$$w(t) = \sin(bt), \quad b \in [10^{-2}, 10^3].$$

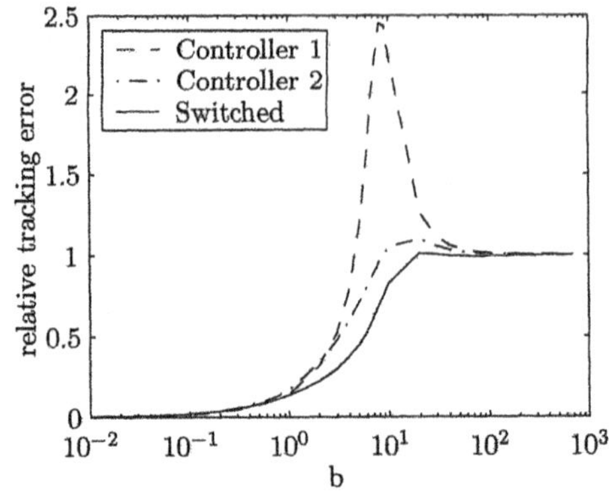

Fig. 2. Performance of switched system

From Figure 2 it can be concluded that for the considered performance measure (15) the switched system performs better than the linear systems for the input range $b \in [10^0, 10^2]$. This means that besides improvement of transient behaviour (see e.g. (Feuer *et al.*, 1997)), the use of switched control instead of linear control can sometimes provide better stationary behaviour.

5.2 Convergency using observer-based switching

In this example the effect of observer-based switching as opposed to state-based switching is shown. Consider again system (13) with the given matrices and consider (10) with gain matrices

$$\mathbf{L}_1 = \begin{bmatrix} 10 \\ 5 \\ 10 \end{bmatrix}, \quad \mathbf{L}_2 = \begin{bmatrix} 5 \\ 10 \\ -10 \end{bmatrix}$$

which are chosen in such a way that condition (12) is satisfied for some $\mathbf{P}_2$. Thereby all conditions for Theorem 8 are satisfied, which implies that that system (13) with observer-based switching is exponentially convergent for any initial condition $\mathbf{x}(0)$, any initial estimation error $\mathbf{e}(0) = \mathbf{x}(0) - \hat{\mathbf{x}}(0)$ and any input $w \in \mathbb{PC}_1$. In Figure 3 the convergency of the system output y is visualized for $\mathbf{x}(0) = [0; 0; 0]$, $w = \sin(5t)$, and several initial estimation errors $\mathbf{e}(0) = \{[10; 10; 10], [100; -100; 0], [-100, 0, 100]\}$ (respectively the dashed ,dash-dotted, and dotted line). Furthermore, the output of the system with state-based switching is plotted (solid line) to make a comparison with the observer-based switching. Note that only the transient solution is influenced by the choice of switching, the limit solution is identical for both types of switching. Therefore, if the performance analysis of Section 5.1 would

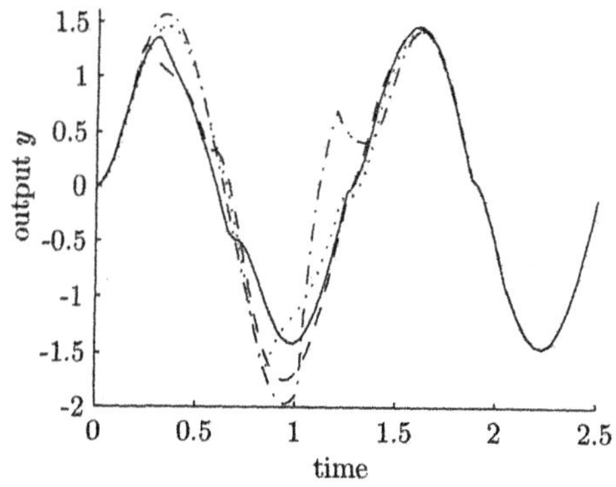

Fig. 3. Observer-based vs. state-based switching

be repeated for the same system but now with observer-based switching, then the results would be identical to those in Figure 2.

6. CONCLUSION

In this paper the following design problem for switched linear systems has been considered: under which conditions is it possible to design a switching rule that makes the resulting closed-loop system convergent? Two cases have been considered: state feedback and output feedback. Sufficient conditions have been found that guarantee the existence of a state-based switching rule which renders the closed-loop system quadratically convergent. For the case with output feedback, the switching rule is based on the state of an observer of the system, and sufficient conditions have been found for the existence of a switching rule that makes the closed-loop system exponentially convergent. By means of an example it has been illustrated that a simulation-based performance evaluation is feasible for a convergent switched system. In the example it has also been indicated that switched control can provide better limit behaviour than linear control.

ACKNOWLEDGEMENT

This work was partially supported by the Dutch-Russian mathematical program "Dynamics and Control of Hybrid Mechanical Systems" (NWO grant 047.017.018) and the HYCON Network of Excellence, contract number FP6-IST-511368.

REFERENCES

Angeli, D. (2002). A Lyapunov approach to incremental stability properties. *IEEE Trans. Automatic Control* **47**, 410–421.

Branicky, M.S. (1998). Multiple Lyapunov functions and other analysis tools for switched and hybrid systems. *IEEE Trans. Automat. Control* **43**(4), 475–482.

DeCarlo, R.A., M.S. Branicky, S. Pettersson and B. Lennartson (2000). Perspectives and results on the stability and stabilization of hybrid systems. *Proceedings of the IEEE* **88**(7), 1069–1082.

Demidovich, B.P. (1967). *Lectures on Stability Theory.* Nauka, Moscow.

Feuer, A., G.C. Goodwin and M. Salgado (1997). Potential benefits of hybrid control for linear time invariant plants. *Proceedings of ACC, New Mexico, USA* pp. 2790–2794.

Filippov, A.F. (1988). *Differential Equations with Discontinuous Right Hand Sides.* Kluwer Academic, Dordrecht.

Fromion, V., G. Scorletti and G. Ferreres (1999). Nonlinear performance of a PI controlled missile: an explanation. *Int. J. Robust Nonlinear Control* **9**, 485–518.

Fromion, V., S. Monaco and D. Normand-Cyrot (1996). Asymptotic properties of incrementally stable systems. *IEEE Trans. Automatic Control* **41**, 721–723.

Liberzon, D. and A.S. Morse (1999). Basic problems in stability and design of switched systems. *IEEE Control Systems Magazine* **19**(5), 59–70.

Lohmiller, W. and J.J.E Slotine (1998). On contraction analysis for nonlinear systems. *Automatica* **34**(6), 683–696.

Morse, A.S. (1996). Supervisory control of families of linear set-point controllers–part 1: Exact matching. *IEEE Trans. Automatic Control* **41**(10), 1413–1431.

Narendra, K.S. and J. Balakrishnan (1997). Adaptive control using multiple models. *IEEE Trans. Automatic Control* **42**(2), 171–187.

Pavlov, A., A. Pogromsky, N. van de Wouw and H. Nijmeijer (2005*a*). On convergence properties of piece-wise affine systems. *Proceedings of the 5th EUROMECH Nonlinear Oscillations Conference (ENOC).* Editor: D.H. van Campen, CD-ROM.

Pavlov, A., A. Pogromsky, N. van de Wouw, H. Nijmeijer and J.E. Rooda (2005*b*). Convergent piece-wise affine systems: Part II: Discontinuous case. *Proceedings of Conference on Decision and Control.* CD-ROM.

Pavlov, A., A. Pogromsky, N. v.d.Wouw and H. Nijmeijer (2004). Convergent dynamics, a tribute to Boris Pavlovich Demidovich. *Systems and Control Letters* **52**, 257–261.

Pogromsky, A., G. Santoboni and H. Nijmeijer (2002). Partial synchronization: from symmetry towards stability. *Physica D* **172**, 65–87.

Yakubovich, V.A. (1964). Matrix inequalities in stability theory for nonlinear control systems: I absolute stability of forced vibrations. *Automation and Remote Control* **7**, 905–917.

ELSEVIER
IFAC
PUBLICATIONS

OBSERVER DESIGN FOR A CLASS OF DISCRETE TIME PIECEWISE-LINEAR SYSTEMS

Abderazik Birouche, Jamal Daafouz, Claude Iung

Institut National Polytechnique de Lorraine,
CRAN UMR 7039 CNRS INPL UHP
ENSEM 2, Avenue de la Forêt de Haye 54516
Vandoeuvre-lès-Nancy Cedex, France.
Email:{Abdearzik.Birouche,Jamal.Daafouz,Claude.Iung}@ensem.inpl-nancy.fr

Abstract: In this paper we consider a class of discrete time piecewise-linear systems composed by linear discrete time LTI subsystems with autonomous switching. The aim is to propose a method for the synthesis of a hybrid observer. The approach proposed here consists on the combination of a discrete observer and a continuous observer. It is shown that under conditions related to minimum dwell time of each mode, one can express the switching law as a linear combination of the system input/output samples. The continuous observer is a piecewise-linear observer whose dynamics depend on the current active mode. Two estimation schemes are analyzed : on–line estimation and off–line estimation. *Copyright© 2006 IFAC.*

Keywords: Hybrid systems, Observer design, Piecewise-linear affine systems.

1. INTRODUCTION

Observer design for hybrid systems is an important and challenging problem. Applications for control or fault detection purposes is of first importance. Recently, many papers have considered such a problem. A discussion on observability conditions for switched linear systems is proposed in (Vidal *et al.*, 2004). Moving horizon estimation strategy is discussed in (Ferrari-Trecate *et al.*, 2002) for piecewise-linear systems. Piecewise-linear systems are an important class of hybrid dynamic systems which has attracted a growing interest (Bemporad *et al.*, 2000),(Johansson, 2003),(Liberzon, 2003). In (Balluchi *et al.*, 2002), a hybrid observer is proposed for systems with a hybrid automaton description. The scheme of the proposed observer consists of two blocs: a discrete observer, based on the discrete event dynamic framework, and a continuous observer, based on the classical state observation theory. The former identifies the current mode, while the latter produces an estimate of the evolution of the continuous state of the hybrid system. In this paper, we study the observation problem for a class of piecewise-affine hybrid systems where the switching law depends on the continuous state vector. The observation scheme proposed here consists on the combination of an active mode detection and continuous observer. This paper is organized as follows. First, we present a solution for detecting the switching time instant and the corresponding active mode. This method is the transposition in discrete time of the idea introduced in (Benali *et al.*, 2004) for the continuous time case. The adaptation to the discrete time case is not immediate and introduces some specificities. The continuous observer is based on the switching observer developed in (Daafouz *et al.*, 2002), whose dynamics depend on the active mode provided by the discrete observer. Two configurations are considered

[1] Work partially done in the framework of the HYCON Network of Excellence, contract number FP6-IST-511368

: on–line gain attribution, where the two parts of the observer work simultaneously, off–line gain attribution, where the hybrid observer works with a delay compared to the real system. We finish by an illustrative example and a conclusion.

2. PROBLEM STATEMENT

We consider the class of discrete time piecewise-linear systems given by :

$$\begin{cases} x_{k+1} = A_i x_k + B_i u_k & \text{if } Hx_k \in [a_i, a_{i+1}] \\ y_k = C_i x_k & \text{for } i = 1, 2, \cdots, s \end{cases} \quad (1)$$

where $x_k \in \mathbb{R}^n$ is the state vector, y_k is the output vector. k refers to the sample index. q_k is the active mode index, and the discrete state of the hybrid system, it takes its values in the finite set $Q = \{1, 2, \cdots, s\}$. Each triplet (A_i, B_i, C_i) characterizes the dynamics of the system in a region of the state space. s is the number of subsystems (also the number of regions). The switching strategy is specified by the linear form $Hx_k = \sum_{i=1}^{n} h_i x_k^i, H = [h_1, ..., h_n]$ which indicates the active subsystem and defines a partition of the state-space where each region is delimited by $Hx_k = a_i$, with $a_i \in \mathbb{R}$ and $a_1 < a_2 < ... < a_s$.

The hybrid observer has to provide an evaluation of the active mode (discrete state) $\widehat{q}_k \in Q$, and an estimate of the state vector $\widehat{x}_k$. The hybrid observer proposed here consists of two parts: *discrete observer* and *continuous observer* (see figure 1). Knowing the system input and output $(u_k, y_k, k = 1, ..., N)$ on a time horizon N, an evaluation $\widehat{q}$ of the active mode index $\hat{q}_k \in \{1, 2..., s\}$ is calculated. The result is used by the piecewise-linear observer to determine $\widehat{x}$ an estimation of the continuous state vector x.

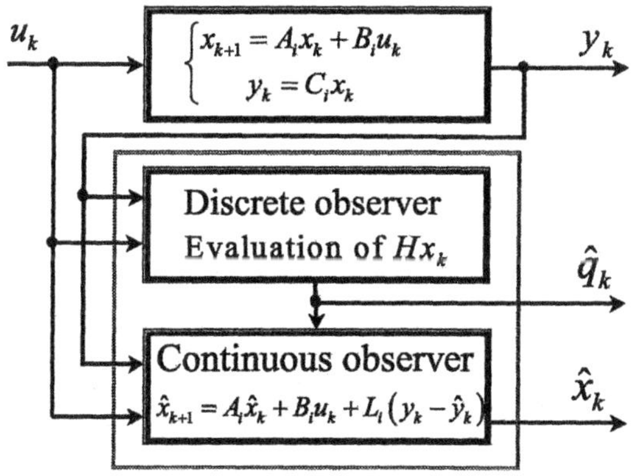

Fig. 1. Hybrid observer structure

The two parts of the observer are described in the following sections.

3. DISCRETE OBSERVER: ACTIVE MODE DETECTION

In this section we present a method for detecting the switching time instant and the corresponding active mode using the input/output data on a time horizon. First, the autonomous case is considered. A generalization is proposed in the next section for the non-autonomous case. We consider the following piecewise-linear system:

$$\begin{cases} x_{k+1} = A_i x_k & \text{if } Hx_k \in [a_i, a_{i+1}] \\ y_k = C_i x_k & \text{for } i = 1, 2, \cdots, s \end{cases} \quad (2)$$

The switching time instants are determined by the value of the quantity Hx_k. As we have only the collected output data y_k on a fixed horizon, the simplest idea is to express Hx_k function of these available outputs. This idea has been investigated in (Benali *et al.*, 2004) in the continuous time case and appears quite easy to exploit. It is possible to express $Hx(t)$ as a linear combination of the output $y(t)$ and its successive derivatives without any assumptions. We show that the transposition in the discrete time case cannot be used without additional assumptions.

As in the continuous cas, we try to express Hx_k as a linear combination of the outputs y_k taken on a time horizon of length N. The relation between the outputs and the switching function will be then written as :

$$Hx_k = [Hx_k]_\alpha = \sum_{j=0}^{N-1} \alpha_j y_{k+j} \quad (3)$$

where $\alpha = \begin{bmatrix} \alpha_0 & \alpha_1 & \cdots & \alpha_{N-1} \end{bmatrix}$ are output weighting coefficients to be computed.
Where in the continuous time case the successive output derivatives $y^{(n)}, n = 1, \cdots, n$ involve always the same active mode, in the discrete time case the samples $y_{k+j}, j = 1, \cdots, N$ do not necessarily correspond to a same active mode. In a consequence it is not possible to find α such that equation (3) is valid when a commutation occurs on $[0, ..., N-1]$.

The idea is, first to find α independent of the discrete state such that the equation (3) is valid whenever the discrete state keeps constant on the interval $[0, ..., N-1]$, second to detect the commutation when it occurs, at last find the new discrete state.

Before starting the theorem let us recall a few definitions associated with the problem of joint observability (Vidal *et al.*, 2004). We define G_k

$$G_k = \begin{pmatrix} C_1 & C_2 & \cdots & C_s \\ C_1 A_1 & C_2 A_2 & \cdots & C_s A_s \\ \cdots & \cdots & \cdots & \cdots \\ C_1 A_1^{k-1} & C_2 A_2^{k-1} & \cdots & C_s A_s^{k-1} \end{pmatrix}$$

Definition 1. (*The joint observability index*) The joint observability index is defined as

$$\mu = \max_k (rank(G))$$

Definition 2. (*The joint observability matrix*) G_μ is called joint observability matrix of the system $S_i = (A_i, C_i), i = 1, ..., s$

Lemma 1. There exist $\alpha^c = \begin{bmatrix} \alpha_0^c & \alpha_1^c & \cdots & \alpha_{\mu-1}^c \end{bmatrix}$ s.t : $Hx_k = [Hx_k]_{\alpha^c}$, where

$$[Hx_k]_{\alpha^c} = \sum_{j=0}^{\mu-1} \alpha_j^c y_{k+j} \tag{4}$$

whenever the discrete state $q \in Q$ keeps constant on $[k, ..., k+\mu-1]$ if and only if

$$rank\left(\begin{bmatrix} G_\mu^T & h^T \end{bmatrix}\right) = \mu \tag{5}$$

with $h = \underbrace{[H \; H \; ... \; H]}_{s \; times}$

If all $S_i = (A_i, C_i)$ is observable $\mu > n$.

Proof of lemma.1 :

If : if $rank\left(\begin{bmatrix} G_\mu^T & h^T \end{bmatrix}\right) = \mu$ there $h = \underbrace{[H \; H \; ... \; H]}_{s \; times}$ is a linear combination of the rows of the matrix G_μ, then exist $\begin{bmatrix} \alpha_0^c & \alpha_1^c & \cdots & \alpha_{\mu-1}^c \end{bmatrix}$ such that

$$H = \sum_{j=0}^{\mu-1} \alpha_j^c C_i A_i^j \quad \forall i \in Q \tag{6}$$

then $[Hx_k]_{\alpha^c} = \sum_{j=0}^{\mu-1} \alpha_j^c C_i A_i^j x_k = \sum_{j=0}^{\mu-1} \alpha_j^c y_{k+j}$ where y_{k+j} is the output of system S_i.

Only if : if there exist $\begin{bmatrix} \alpha_0^c & \alpha_1^c & \cdots & \alpha_{\mu-1}^c \end{bmatrix}$ s.t

$$Hx_k = \sum_{j=0}^{\mu-1} \alpha_j^c y_{k+j} \;, \forall x_k$$

while y_{k+1} is the output of any system S_i (q is constant on $[k, ..., k+\mu-1]$), then

$$Hx_k = \sum_{j=0}^{\mu-1} \alpha_j^c C_i A_i^j x_k, \forall x_k, \forall i \in Q$$

and $H = \sum_{j=0}^{\mu-1} \alpha_j^c C_i A_i^j$, $\forall i \in Q$ and h is linear combination of the rows G_μ. □

If $S_i = (A_i, C_i)$ is observable then x_k is linear combinaison of the outputs $y_k, y_{k+1}, ..., y_{k+\mu-1}$ and the Hx_k is also a linear combinaison of $y_k, y_{k+1}, ..., y_{k+\mu-1}$, so there exist $\alpha^i = \begin{bmatrix} \alpha_0^i & \alpha_1^i & \cdots & \alpha_{\mu-1}^i \end{bmatrix}$, for all $i \in Q$ s.t $[Hx_k]_{\alpha^i} = \sum_{j=0}^{\mu-1} \alpha_j^i y_{k+j} = \sum_{j=0}^{\mu-1} \alpha_j^i C_i A_i^j x_k$ then

$$H = \sum_{j=0}^{\mu-1} \alpha_j^i C_i A_i^j \tag{7}$$

Theorem 1. Suppose that the following assumptions are verified :

(1) $S_i, i = 1, ..., s$ observable.
(2) $rank\left(\begin{bmatrix} G_\mu^T & h^T \end{bmatrix}\right) = \mu$.
(3) the hybrid system exhibits transitions with time separation greater than or equal to some $\mu > 0$ (dwell time hypothesis).
(4) $C_i A_i \neq C_j A_j, \;\; \forall i, j \in Q$.
(5) $\alpha_{\mu-1}^c \neq \alpha_{\mu-1}^i$.

$$\text{Then :} \begin{cases} \text{If } [Hx_k]_{\alpha^c} = [Hx_k]_{\alpha^i}, \text{ then } q_k = i \\ \text{Else the switching instant is } t_c = k + \mu - 1 \end{cases}$$

Proof of theorem.1 :

First it is important to notice that we only prove the result on $[0, N]$, $N \geq \mu$. In fact when a commutation will be detected at time $k \in [0, N]$ we will reinitialize the detection procedure. $[Hx_0]_{\alpha_c}$ gives the discrete state $q_0 = i$ on $[0, \mu-1]$ thanks to lemma 1 and dwell-time hypothesis. Let us examine $[Hx_1]_{\alpha_c}$.

Case 1: $[Hx_1]_{\alpha_c} \neq [Hx_1]_{\alpha_{q0}}$ gives the information of a commutation at μ.

Case 2: $[Hx_1]_{\alpha_c} = [Hx_1]_{\alpha_{q0}}$. Assume that a commutation (from $q_0 = i$ to $q_1 = l \neq i$) occurs at μ then

$[Hx_1]_{\alpha^i} = \sum_{j=0}^{\mu-1} \alpha_j^i y_{j+1} = \sum_{j=0}^{\mu-1} \alpha_j^i C_i A_i^j x_1 + (\alpha_{\mu-1}^i C_l A_l A_i^{\mu-2} - \alpha_{\mu-1}^i C_i A_i^{\mu-1}) x_1$.

$[Hx_1]_{\alpha^c} = \sum_{j=0}^{\mu-1} \alpha_j^c y_{j+1} = \sum_{j=0}^{\mu-1} \alpha_j^c C_i A_i^j x_1 + (\alpha_{\mu-1}^c C_l A_l A_i^{\mu-2} - \alpha_{\mu-1}^c C_i A_i^{\mu-1}) x_1$.

we have $\sum_{j=0}^{\mu-1} \alpha_j^i C_i A_i^j x_k = \sum_{j=0}^{\mu-1} \alpha_j^c C_i A_i^j x_k$ then $[Hx_1]_{\alpha^c} - [Hx_1]_{\alpha^i} = (\alpha_{\mu-1}^c - \alpha_{\mu-1}^i)(C_l A_l - C_i A_i) A_i^{\mu-2} x_1 = 0$ only if $\alpha_{\mu-1}^c = \alpha_{\mu-1}^i$ or $(C_l A_l - C_i A_i) = 0$ which contradict the assumptions (4,5).

Remark : The previous theorem gives conditions under which the switching instant can be detected. The only situation where no conclusion can be made is $A_i^{\mu-2} x_1 \in Nul(C_l A_l - C_i A_i)$ which is far from being realistic.

3.1 Generalization to the non-autonomous case

According to the same procedure described for the autonomous piecewise-linear systems, we define Q_k expressed as a linear combination of the output samples taken on a time horizon μ:

$$Q_k = \sum_{j=0}^{\mu-1} \alpha_j y_{k+j} = \alpha Y_k \tag{8}$$

with $Y_k = \begin{bmatrix} y_k & y_{k+1} & \cdots & y_{k+\mu-1} \end{bmatrix}^T$. For each mode, we have :

$$Y_k = \mathcal{O}_i x_k + \Gamma_i U_k \tag{9}$$

where $\mathcal{O}_i$ the observability matrix and Γ_i is the Toeplitz matrix :

$$\mathcal{O}_i = \begin{bmatrix} C_i \\ C_i A_i \\ \vdots \\ C_i A_i^{\mu-1} \end{bmatrix}, \Gamma_i = \begin{bmatrix} 0 & \cdots & \cdots & 0 \\ C_i B_i & \ddots & \ddots & \vdots \\ \vdots & \ddots & \ddots & \vdots \\ C_i A_i^{\mu-2} B_i & \cdots & C_i B_i & 0 \end{bmatrix}$$

$U_k = \left[u_k \; u_{k+1} \; \cdots \; u_{k+\mu-1} \right]^T$. By substitution of equation (9) in (8), we obtain $Q_k = \alpha \mathcal{O}_i x_k + \alpha \Gamma_i U_k$ with $\alpha \mathcal{O}_i = \sum_{j=0}^{\mu-1} \alpha_j C_i A_i^j$. If $\alpha = \alpha^c$, then $Q_k = H x_k + \alpha^c \Gamma_i U_k$.

Using (8) with $\alpha = \alpha^c$, the switching function Hx_k is then given by

$$H x_k = \alpha^c Y_k + \beta_i^c U_k \tag{10}$$

where $\beta_i^c = -\alpha^c \Gamma_i$.

The term β_i^c in (10) states that s different evaluations of Hx_k are computed. As in the autonomous case, different coefficients α^i are used. For each subsystem, the quantity Hx_k is evaluated in two different ways, $[Hx_k]_{\alpha^c,\beta_i^c}$ and $[Hx_k]_{\alpha^i,\beta^i}$ given by

$$\begin{cases} [Hx_k]_{\alpha^c,\beta_i^c} = \alpha^c Y_k + \beta_i^c U_k \\ [Hx_k]_{\alpha^i,\beta^i} = \alpha^i Y_k + \beta^i U_k \end{cases}$$

where $\beta^i = -\alpha^i \Gamma_i$

The active subsystem $\widehat{q}_k$ and the switching instant t_c are given by

$$\begin{cases} \text{If } [Hx_k]_{\alpha^c,\beta_i^c} = [Hx_k]_{\alpha^i,\beta^i} \text{ then } \widehat{q}_k = i \\ \text{Else the switching instant is } t_c = k + \mu - 1 \end{cases}$$

4. HYBRID OBSERVER

In this section the continuous observer is analyzed considering the complete hybrid system obtained by composing the hybrid system and the hybrid observer. The continuous observer is a piecewise-linear observer of the form:

$$\begin{cases} \widehat{x}_{k+1} = A_i \widehat{x}_k + B_i u_k + L_i (y_k - \widehat{y}_k) \\ \widehat{y}_k = C_i \widehat{x}_k \qquad \text{if } \hat{q}_k = i \in Q \end{cases} \tag{11}$$

The hybrid observer has $(s \times s)$ situations of type $(q_i, \widehat{q}_i) \in Q \times Q$. To each $(q_k, \widehat{q}_k)$, the error dynamics is given by :

$$e_{k+1} = (A_i - L_i C_i) e_k \quad \text{if} \quad q_k = \hat{q}_k = i \tag{12a}$$

$$e_{k+1} = (A_i - L_i C_i) e_k + v_k \text{ if } q_k = j,\ \hat{q}_k = i \tag{12b}$$

with

$$v_k = ((A_j - A_i) - L_i (C_j - C_i)) x_k + (B_j - B_i) u_k \tag{13}$$

The equation (12a) describes the situation where the active mode is correctly identified, and the second equation (12b) describes a situation where the estimated mode is different from the actual active mode.

In this paper, we will study two possible estimation schemes :

(1) ***Off–line estimation***: The previously presented analyses a set $[y_1, ..., y_{t_f}]$ and gives the discrete state $\hat{q}_k = q_k \ \forall k \in [1, ..., t_f]$. The continuous observer uses the exact knowledge of q_k to construct $\hat{x}_k$.
(2) ***On–line estimation***: The hybrid observer and the hybrid system works simultaneously. In this case the input/output data provided by the hybrid system are analyzed by the hybrid observer in real time.

4.1 Off–line state estimation

To analyze the off–line estimation case, let us consider a time instant $t_f > \mu$. The input/output data information is available for $k \leq t_f$. Using a moving window of size μ, the discrete observer analyzes the data and provides $\widehat{q}_k$. The method proposed in the previous sections guarantees, under section 3, that $\widehat{q}_k = q_k$, this information is used by the continuous observer. The error dynamic is governed by the following equation :

$$e_{k+1} = (A_i - L_i C_i) e_k \tag{14}$$

The observer gains L_i are computed so that the estimated state $\widehat{x}_k$ converges towards the state of the system x_k for all the initial conditions i.e :

$$\forall e_0 \in \mathbb{R}^n \quad \lim_{k \to \infty} \|e_k\| = 0 \tag{15}$$

To obtain these observer gains, we use the LMI (Linear Matrix Inequalities (Boyd *et al.*, 1994) approach developed in (Daafouz *et al.*, 2002). An indicator vector $\xi_k = [\xi_k^1,\ \xi_k^2, \ldots, \xi_k^s]^T$ is defined as

$$\xi_k^i = \begin{cases} 1 \text{ if } \widehat{q}_k = i \\ 0 \quad \text{else} \end{cases} \tag{16}$$

The estimation error is :

$$e_{k+1} = \sum_{i=1}^{s} \xi_k^i (A_i - L_i C_i) e_k \tag{17}$$

Global convergence of (17) is ensured by selecting the gains $L_i; i = 1, \cdots, s$ such that the error stability condition established in (Daafouz *et al.*, 2002) are satisfied. It consists in finding positive definite matrices S_i and matrices F_i and G_i for $i = 1, \cdots, s$ solution of the following inequalities :

$$\begin{bmatrix} G_i + G_i^T - S_i & G_i^T A_i - F_i^T C_i \\ A_i^T G_i - C_i^T F_i & S_j \end{bmatrix} > 0 \quad (18)$$

with $j = 1, \cdots, s$. In this case, the matrices L_i are given by

$$L_i = G_i^{-T} F_i^T$$

The asymptotic stability is guaranteed by the Lyapunov function $V = \varepsilon^T \sum_{i=1}^{s} \xi_k^i P_i \varepsilon$ with $P_i = S_i^{-1}$.

4.2 On–line state estimation

We recall that to detect the active mode, we use the input/output samples available on a time horizon of size μ. Let us consider a switch occurring at time t_c. The discrete observer detects at $t_c + 1$ this commutation. However the new mode is only identified at the instant $t_c + \mu$. In the on–line state estimation case the continuous observer work simultaneously with the discrete observer, therefore on the interval $[t_c, t_c + \mu - 1]$, we must provide $\hat{q}_k$ for the continuous observer. Different schemes can be investigated during the interval time $[t_c, t_c + \mu - 1]$ as the discrete state is unknown and the continuous state cannot be correctly calculated. In this paper we propose to keep the discrete state with its last known value $t_c - 1$ and estimate the continuous state with the observer designed for value $q_{t_c - 1}$, with this option it can prove that the estimation error remains bounded.

The corresponding error dynamics is given by the equation (12).
A new indicator vector $\hat{\xi}_k = (\hat{\xi}_k^1, \hat{\xi}_k^2, \cdots, \hat{\xi}_k^s)^T$ is associated in order to establish a correspondence between discrete $\hat{q}_i$ and the triplet (A_i, B_i, C_i) :

$$\hat{\xi}_k^i = \begin{cases} 1 \text{ if } \hat{q}_k = i \\ 0 \quad \text{if } \hat{q}_k \neq i \\ \hat{\xi}_k^i = \hat{\xi}_{k-1}^i \text{ if } \hat{q}_k \text{ is not available.} \end{cases} \quad (19)$$

the estimation error is given by:

$$e_{k+1} = \sum_{i=1}^{s} \hat{\xi}_k^i (A_i - L_i C_i) e_k + \ v_k \quad (20)$$

This error cannot be asymptotically stable because of the term v_k. However, we can calculate the gain L_i, such that the error estimation is Input-to-State stable (Jiang *et al.*, 1999). For that, we use the result proposed in (Daafouz *et al.*, 2005). The gain L_i is calculated by solving the optimization problem :

$$\underset{\substack{P_i = P_i' \\ G_i = G_i' \\ F_i, \alpha}}{Min} \quad \eta$$

under

$$\begin{bmatrix} 1 - P_i & A_i' G_i - C_i' F_i' & A_i' G_i - C_i' F_i' \\ G_i A_i - F_i C_i & P_j - 2G_i & \mathbf{0} \\ G_i A_i - F_i C_i & \mathbf{0} & 2G_i - \eta \mathbf{1} \end{bmatrix} < \mathbf{0} \quad (21)$$

if the minimization problem has a solution $P_i^* \in \mathbb{R}^{n \times n}$, $G_i^* \in \mathbb{R}^{n \times n}$, $F_i^* \in \mathbb{R}^{n \times m}$ et $\eta^* \in]1, \infty[$, the gains L_i are given by :

$$L_i = G_i^{*-1} F_i^*$$

The Input-to-State stability is guaranteed by the Lyapunov function $V(e_k, \hat{\xi}_k^i) = e_k^T P_k e_k$ with $P_k = \sum_{i=1}^{s} \hat{\xi}_k^i P_i^*$. The estimation error verifies:

$$\|e_k\| \leq \sqrt{\eta^*} (1 - \frac{1}{\eta^*})^{k/2} \|e_0\| + \eta^* \|v\|_\infty \quad (22)$$

As the LMI resolution implies that η^* is necessarily greater than 1, when $k \longrightarrow \infty$ the estimation error is bounded $\|e_k\| \leq \eta^* \|v\|_\infty$ if v_k is bounded. exist $X > 0$ and $U > 0$, such that $\|x_k\|_\infty \leq X$, $\|u_k\|_\infty \leq U$ and

$\|v_k\|_\infty \leq V = \max_{i \neq j} \|((A_j - A_i) - L_i (C_j - C_i))\| X + \|(B_j - B_i)\| U$

5. ILLUSTRATIVE EXAMPLE

To illustrate the observation scheme proposed here, we consider a piecewise-linear system given by (1) with:

$$A_1 = \begin{bmatrix} 0.80 & 0.22 \\ -0.22 & 0.80 \end{bmatrix}, \quad A_2 = \begin{bmatrix} 0.79 & 0.29 \\ -0.29 & 0.50 \end{bmatrix}$$

$$B_1 = \begin{bmatrix} 0.20 \\ 0.20 \end{bmatrix}, \quad B_2 = \begin{bmatrix} 0.50 \\ -0.50 \end{bmatrix}$$

$$C_1 = C_2 = \begin{bmatrix} 1 & 1 \end{bmatrix}, \quad H = \begin{bmatrix} 0 & 1 \end{bmatrix}$$

The switching function Hx_k is defined by :

$$\begin{cases} \text{if} \quad (0.2 \leq Hx_k \leq 10) \quad \text{then} \quad i = 1 \\ \qquad \text{else} \quad i = 2 \end{cases}$$

For this system we find $\mu = 4$ and

$$\alpha^c = \begin{bmatrix} -11.87 & 39.65 & -45.18 & 18.65 \end{bmatrix}$$

In off–line state estimation, the gains of the observer are obtained solving the LMIs (18):

$$L_1 = \begin{bmatrix} -0.7660 \\ 2.5341 \end{bmatrix}, \quad L_2 = \begin{bmatrix} 0.5441 \\ 0.1030 \end{bmatrix}$$

Figure (2) shows the actual active mode (full line), the reconstructed active mode (stars $*$) and estimation error norm.

Notice that the modes are perfectly identified, and the error estimation norm decreases towards zero. A zoom shows that after the 15th sample the error norm is almost null.

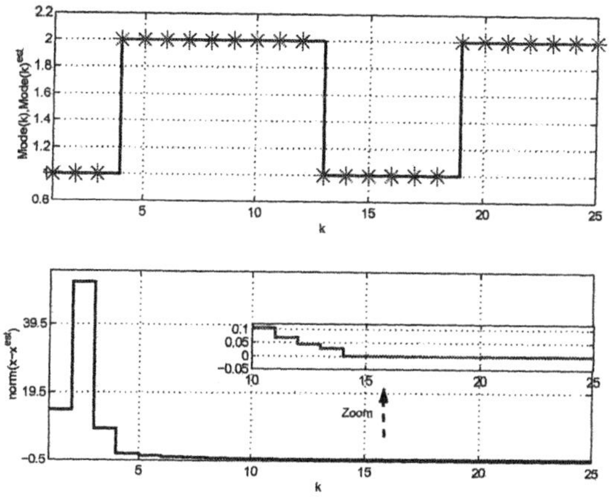

Fig. 2. off–line estimation case : mode evaluation (in the top) and the error estimation norm (in the bottom).

In on–line state estimation, the gains of the piecewise-linear observer are obtained by solving the optimization problem (21) :

$$\eta^* = 11.3122,, \; L_1 = \begin{bmatrix} -0.0621 \\ 1.3035 \end{bmatrix}, \; L_2 = \begin{bmatrix} 0.0524 \\ 0.8879 \end{bmatrix}$$

The actual mode (full line) and the evaluated mode (dotted line) and estimation error norm are given on the figure (3).

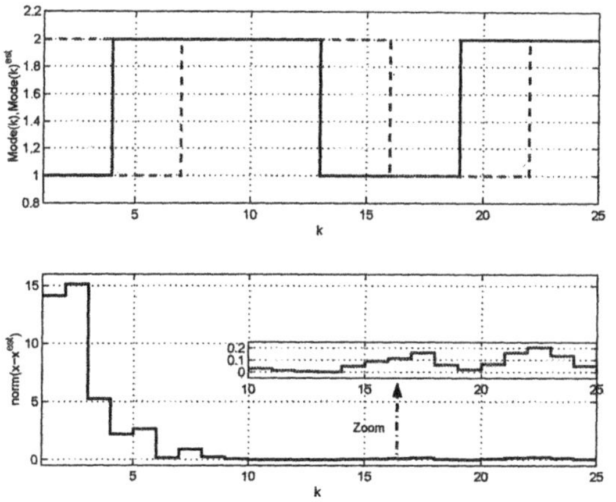

Fig. 3. on–line estimation case : mode evaluation (in the top) and the error estimation norm (in the bottom).

At the beginning of the estimate $k = 1$, we initialized the discrete state estimated at $\widehat{q}_k = 2, k = \{1, 2, 3\}$, whereas the true discrete state is $q_k = 1$. Since the acquisition of the 4th sample, the discrete observer provides $\widehat{q}_k = 1$, at the instant time $k = 5$ a switch occurs: $q_k = 2$. The discrete observer keeps its old value $\widehat{q}_k = 1$ for $k = \{5, 6, 7\}$. For the continuous component, we notes that starting from the sample $k = 10$ the norm of the estimation error is stabilized around zero. A zoom on the interval $[10, 25]$, shows that the error norm is not null but it is bounded.

6. CONCLUSION

In this paper, we propose a hybrid observer for a class of piecewise-linear systems. The association of a discrete state detection method and a piecewise-linear switched observer leads to a hybrid observer witch may operates off-line or on-line. In the off-line case, the observation error is guaranteed to converge toward zero whereas in the on-line case, it is guaranteed to be bounded.

REFERENCES

Balluchi, A., L. Benvenuti and A.L. Sangiovanni-Vincentelli (2002). Observers for hybrid systems with continuous state resets. *Proceedings of the 10th Mediterranean Conference on Control and Automation (MED2002), Lisbon, Portugal.*

Bemporad, A., G. Ferrari-Trecate and M. Morari (2000). Observability and controllability of piecewise affine and hybrid systems. *IEEE Trans. Autom. Contr.* **45**, 1864–1876.

Benali, A., D. Boutat and J.P. Barbot (2004). Une condition algébrique pour l'observabilité d'une classe de systèmes hybrides. *Proceedings of IEEE CIFA, Tunisie.*

Boyd, S., E. Feron L. El Ghaoui and V. Balakrishnan (1994). *Linear Matrix Inequalities in System and Control Theory.* Studies in Applied Mathematics.

Daafouz, J., G. Millerioux and C. Iung (2002). A poly-quadratique stability based approch for linear switched systems. *Special issue on switched, piecewise and polytopic linear systems. Int.J.Control.*

Daafouz, J., G. Millerioux and L. Rosier (2005). Observer design with guaranteed bound for lpv systems. *in Proceedings of IFAC World Congress on Automatic Control.*

Ferrari-Trecate, G., D. Mignone and M. Morari (2002). Moving horizon estimation for hybrid systems. *IEEE Trans. Autom Contr* **47**, 1663–1676.

Jiang, Z-P., E. Sontag and Y. Wang (1999). Input-to-state stability for discrete-time nonlinear systems. *Proc. 14th triennal World Congress.*

Johansson, M. (2003). *Piecewise linear control systems: Lecture notes in control and information sciences.* Springer. Germany.

Liberzon, D. (2003). *Switching in systems and control.* Birkhäuser Boston, Systems & control: foundations & applications,.

Vidal, R., R. Chiuso and S. Soatto (2004). Observability and identifiability of jump linear systems. *International Conference on Decision and Control.*

ELSEVIER
IFAC
PUBLICATIONS

DESIGNING SWITCHED OBSERVERS FOR SWITCHED SYSTEMS USING MULTIPLE LYAPUNOV FUNCTIONS AND DWELL-TIME SWITCHING

Stefan Pettersson *,[1]

* *Automatic Control, Chalmers University of Technology, S-412 96 Göteborg, Sweden*

Abstract: In this paper, observers are synthesized for switched linear systems, resulting in switched observers including state jumps. The synthesis problem involves multiple Lyapunov functions and is formulated as a linear matrix inequality problem. It is assumed that the current mode (active dynamics) of the switched linear is unknown, and it will be shown that the estimate of the continuous states will be bounded at worst, if the mode is wrongly estimated. If the active dynamics is estimated correctly within a certain time, and the dwell time of the switched linear system is lower bounded, it will be shown that the bound of the estimation error can be reduced significantly. *Copyright © 2006 IFAC*

Keywords: Switched systems; Hybrid systems; Observer design; Linear matrix inequalities (LMI); Lyapunov stability.

1. INTRODUCTION

A large class of systems is reasonably modelled by a family of continuous-time subsystems and logic rules that govern the switchings between them. In this paper, we are interested in the estimation problem of such *switched system*, and switched observers including state jumps are synthesized. The synthesis problem how to design the observer gains, or showing stability for existing observer gains, will be formulated as a linear matrix inequality problem.

Existing synthesis results can be divided into two categories depending on whether the discrete state (active dynamics or mode) of the switched system is known or not. The estimation problem simplifies significantly if the active dynamics is known (since the mode of the observer can change correspondingly), and synthesis results are proposed guaranteeing that the estimation error converges to zero if certain conditions are satisfied, see for instance (Alessandri and Coletta, 2001; Feron, 1996). However, if the active dynamics is unknown and needs to be estimated together with the continuous state, there are quite a few synthesis results. If the discrete mode is estimated from the continuous part of the switched system, there are not yet any synthesis results that are applicable to a large class of switched (linear) systems. Results so far only guarantee that the estimation error of the continuous state is bounded, but it cannot be shown that it goes to zero since there are no guarantees that the active dynamics is correctly estimated, cf. (Juloski *et al.*, 2002).

It is common to use a quadratic Lyapunov function when showing estimation error (stability) properties of the underlying switched observer, see for instance (Alessandri and Coletta, 2001; Juloski *et al.*, 2002). By using a common quadratic Lyapunov function, stability is guaranteed regardless of the mode switches in the system (and observer). However, the existing results are conservative since the estimation error might converge or be bounded without the existence of a common Lyapunov function. By introducing multiple quadratic Lyapunov functions, one for each observer mode, the conservatism can be relaxed, implying that a larger class of switched (linear) systems can be handled, see (Pettersson, 2005*a*; Pettersson, 2005*b*).

[1] Corresponding author S. Pettersson. Tel. +46317725146. Fax +46317721782. E-mail: stp@s2.chalmers.se

In this paper, we will study the estimation problem in the case when the active dynamics of the switched linear is unknown. A generic illustration of the observer is given in Figure 1, cf. (Balluchi *et al.*, 2002). The observer is divided into two parts:

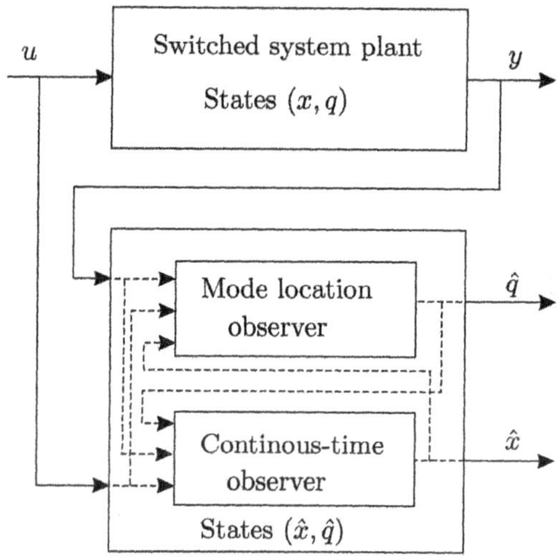

Fig. 1. Switched observer of a switched system.

the *Mode location observer* estimating the active dynamics and the *Continuous-time observer* estimating the continuous state of the switched system. We will focus on the continuous-time estimation problem in this paper, and not give any details of the *Mode location observer*. If the active dynamics needs to be estimated, we cannot show that the estimation error of the continuous-time state converges to zero since there is a possibility that a wrong location mode is estimated during a certain time which means that also the continuous state is estimated wrongly. However, it will be shown that the estimation error of the continuous state will be bounded, similar to the results in (Juloski *et al.*, 2002; Pettersson, 2005*b*). In this paper, we will improve the results regarding the precision of the bound of the estimation error, which is possible if it is assumed that the active dynamics is estimated correctly within a certain time, and the dwell time of the switched linear system is lower bounded. The first assumption is reasonable since otherwise the design of the *Mode location observer* is not very good. The second assumption is of no practical importance; see the comments in the next section.

The outline of this paper is: we start by defining the switched linear system model in the next section, followed by a detailed description of the switched observer with state jumps. In Section 4, the observer synthesis problem is formulated, followed by a section explaining how to solve the problem using linear matrix inequalities. Finally, the method is applied to an example.

2. SWITCHED LINEAR SYSTEM

The switched linear systems considered in this paper are described by the equations

$$\dot{x} = A_{q(t)}x + Bu, \quad y = Cx, \tag{1}$$

where $x \in \Re^n$ is the state vector, $u \in \Re^m$ is the input vector, $y \in \Re^p$ is the measurement (output) vector and $q(t)$ is an index function (discrete state) $q : [0\ \infty) \to I_N = \{1, \ldots, N\}$ deciding which one of the linear vector fields that is active at a certain time instant. Each of the indexes corresponds to a different model description and is referred to as a *mode* of the switched linear system. By *active dynamics* we mean the active subsystem, or model description, of (1).

The change of value of the index function occurs at certain times, which are defined by the set $\mathcal{T}$. One possibility is to define switch sets $S_{i,j} \subset \Re^n$, $(i,j) \in I_s$, where I_s is a set of tuples indicating which mode changes that might occur in the switched system. If $q(t) = i$ and the trajectory reach a state in $S_{i,j}$ at time t^+, then $q(t^+) = j$.

We will assume that there are only a finite number of mode changes in finite time. This does not exclude sliding motions, since if sliding motions occur in the switched system, new modes corresponding to the sliding modes are additionally introduced. The dynamics associated with the sliding mode is given by a (unique) vector field specified for instance by Fillipov's convex combination (Filippov, 1988). Then, a switched system with an equivalent dynamics is obtained, where there is a finite number of switches of the modes in finite time. The observer is designed for this equivalent switched system dynamics. To improve the results regarding the precision of the bound of the estimation error, we will later give results where the dwell time of the equivalent dynamics of a switched linear system is at least a time T.

3. SWITCHED OBSERVER WITH STATE JUMPS

The dynamics of the *Continuous-time observer* is defined as follows:

$$\dot{\hat{x}} = A_{\hat{q}(t)}\hat{x} + Bu + K_{\hat{q}(t)}(y - \hat{y}), \quad \hat{y} = C\hat{x}, \tag{2}$$

where $\hat{x} \in \Re^n$ is the estimate of the state vector x and $K_j \in \Re^{n \times p}$, $j \in I_N$, are the observer gains. The index function $\hat{q} : [0\ \infty) \to I_N = \{1, \ldots, N\}$ decides which one of the observer modes that is active at a certain time instant, and is the output of the *Mode location observer*, see Figure 1.

The purpose of the *Mode location observer* is to estimate the current mode q of the switched system, but we will not specify the details since we will focus on properties of the continuous state estimate in this paper. We merely indicate by $\hat{\mathcal{T}}$ the set of times when the *Mode location observer* switches mode, which are the times when $\hat{q}$ changes value. If the *Mode location observer* never estimates the correct discrete mode, the estimation error bound will be very conservative (related to the worst combination of system mode dynamics and observer mode dynamics), see (Pettersson, 2005*b*). However, if the *Mode location observer* estimates the correct discrete states part of the time, say within Δ time units, the bound of the estimation error can be reduced significantly, shown later on.

There is no guarantee that the estimation error will converge even if the active mode of the switched system is known and the observer gains K_i is designed such that the estimation error of each subsystem converges. What is further needed in the observer design is to properly update the estimated states of the observer, at the times in $\hat{\mathcal{T}}$ when the observer mode changes occur. If observer mode i is active and a mode change occur, the estimate $\hat{x}$ will abruptly be changed (jump) to $\hat{x}^+$, where $\hat{x}^+$ indicates the updated value of $\hat{x}$. More specifically, the estimated state jumps will be updated according to

$$\hat{x}^+ = T_1\hat{x} + T_2 y, \quad t \in \hat{\mathcal{T}},$$

which only depends on the observer states $\hat{x}$ and the measured value y. In the next section, we will show how to calculate T_1 and T_2, guaranteeing that the error between the estimated states and the states of the switched system is bounded.

4. OBSERVER SYNTHESIS

The estimation error dynamics obeys the equation

$$\dot{\tilde{x}} = \dot{x} - \dot{\hat{x}} = (A_{\hat{q}} - K_{\hat{q}}C_{\hat{q}})\tilde{x} + [A_q - A_{\hat{q}}]x.$$

Let us introduce multiple Lyapunov functions, one for each observer mode i,

$$V_i(\tilde{x}) = \tilde{x}^T P_i \tilde{x}, \quad i \in I_N,$$

where each $P_i \in \Re^{n\times n}$ is a symmetric matrix. The time derivative for the observer mode i, when the system state evolves according to mode j, becomes

$$\begin{aligned} \dot{V}_i(\tilde{x}) &= \tilde{x}^T([A_i - K_iC]^T P_i + P_i[A_i - K_iC])\tilde{x} \\ &\quad + \tilde{x}^T P_i(A_j - A_i)x + x^T(A_j - A_i)^T P_i\tilde{x}. \end{aligned} \tag{3}$$

We are now ready for the main theorem. If desirable, we can associate regions $x^TQ_ix \geq 0$ to the switched system (1) where mode i is possible, see (Pettersson and Lennartson, 2002). If not desirable, the $\mu_{i,j}$'s in the theorem is put to zero. The advantage of specifying regions where mode i is possible is to improve the bound given in the theorem. This is one form of relaxation which is similar to the one in (Juloski *et al.*, 2002).

Theorem 1. If there exist a solution to ($\epsilon \geq 0$, $\alpha > 0$, $\mu_{i,j} \geq 0$, $\gamma \geq 0$)

1. $\alpha I \leq P_i \leq \beta I, \qquad i \in I_N$
2. $\Gamma_{i,j} = \begin{bmatrix} \Gamma_{i,j}^{11} & \Gamma_{i,j}^{12} \\ (\Gamma_{i,j}^{12})^T & \Gamma_{i,j}^{22} \end{bmatrix} \leq 0, \qquad (i,j) \in I_s$
3. $P_j = P_i + d_{i,j}^T C + C^T d_{i,j}, \qquad (i,j) \in I_s$

where

$$\begin{aligned} \Gamma_{i,j}^{11} &= (A_i - K_iC)^T P_i + P_i(A_i - K_iC) + \gamma I \\ \Gamma_{i,j}^{12} &= P_i(A_j - A_i) \\ \Gamma_{i,j}^{22} &= \mu_{i,j}Q_j - \gamma\epsilon^2 I \end{aligned}$$

and the states of the hybrid observer is updated according to [2]

$$\begin{aligned} &\hat{x}^+ = (I - R_i^{-1}(CR_i^{-1})^\dagger C)\hat{x} + R_i^{-1}(CR_i^{-1})^\dagger y \\ &\forall t \in \hat{\mathcal{T}} \end{aligned} \tag{4}$$

then if for some $T_0 > 0$

$$\sup_{t>T_0} ||x(t)|| \leq x_{max}, \tag{5}$$

we have

$$\lim_{t\to\infty} \sup ||\tilde{x}(t)|| \leq \sqrt{\frac{\beta}{\alpha}}\epsilon x_{max}. \tag{6}$$

Furthermore, if the switched system (1) is in every mode at least a time T and it takes $\Delta \leq T$ to identify correct mode of the switched system, then we have

$$\lim_{t\to\infty} \sup ||\tilde{x}(t)|| \leq \left(\frac{e^{-\frac{\gamma}{\beta}(T-\Delta)} - e^{-\frac{\gamma}{\beta}T}}{1 - e^{-\frac{\gamma}{\beta}T}}\right)\sqrt{\frac{\beta}{\alpha}}\epsilon x_{max}. \tag{7}$$

Proof: We need to prove that the overall energy function $V(\tilde{x}(t))$ eventually is upper bounded by a constant. To do this, we will show that the energy decreases at the switching instants in $\hat{\mathcal{T}}$ when changing observer modes and that the energy in every observer mode eventually is upper bounded by a constant (regardless of the system mode). We begin by the first part and have to show that

$$(x - \hat{x}^+)^T P_j(x - \hat{x}^+) \leq (x - \hat{x})^T P_i(x - \hat{x}). \tag{8}$$

Let $\hat{x}^+$ be an arbitrary estimated state satisfying $y = C\hat{x}^+$. Since also $y = Cx$, we have

$$C(x - \hat{x}^+) = y - y = 0,$$

implying that $(x-\hat{x}^+)^T(d_{i,j}^TC + C^Td_{i,j})(x-\hat{x}^+) = 0$. Due to the relation in Condition 3, it means that (8) becomes

$$(x - \hat{x}^+)^T P_i(x - \hat{x}^+) \leq (x - \hat{x})^T P_i(x - \hat{x}), \tag{9}$$

and it remains to choose $\hat{x}^+$ satisfying $y = C\hat{x}^+$ such that this inequality is satisfied.

[2] $(*)^\dagger$ is the *pseudoinverse* of $(*)$, see (Strang, 1988).

Factorize P_i as $P_i = R_i^T R_i$, where $R_i \in \Re^{n \times n}$ is a symmetric positive definite matrix. This is always possible since P_i is a real symmetric positive definite (imposed by Condition 1) matrix, see (Strang, 1988). One choice is for instance,

$$R_i = V_i \sqrt{\Lambda_i} V_i^T,$$

where $V_i \in \Re^{n \times n}$ is the orthonormal eigenvectors of P_i and $\sqrt{\Lambda_i}$ is a diagonal matrix consisting of the square root of the (positive) eigenvalues of P_i. Now, Condition (9) is equivalent to show that

$$||R_i(x - \hat{x}^+)|| \leq ||R_i(x - \hat{x})||, \tag{10}$$

is fulfilled where $\hat{x}^+$ satisfies $y = C\hat{x}^+$.

We are now interested to find the updated value $\hat{x}^+$, lying on the hyper plane $y = C\hat{x}^+$, that minimizes the distance $||R_i(\hat{x}^+ - \hat{x})||$. This optimization problem can formally be defined as

$$\begin{aligned} &\min_{\hat{x}^+} ||R_i(\hat{x}^+ - \hat{x})|| \\ &\text{subject to: } C\hat{x}^+ = y \end{aligned} \tag{11}$$

which is geometrically illustrated in Figure 2.

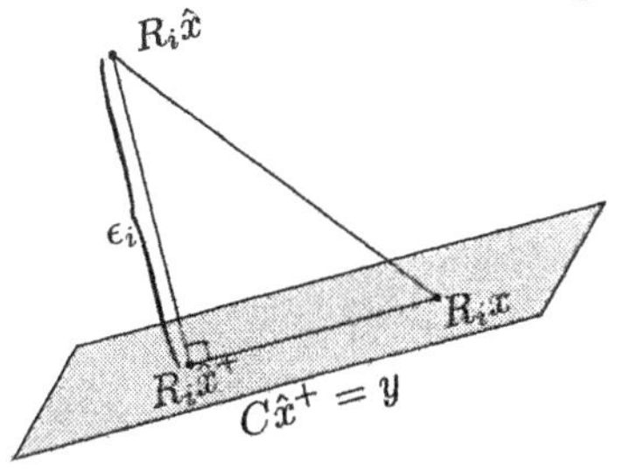

Fig. 2. The projection of $R_i\hat{x}$ onto the plane $C\hat{x}^+ = y$, resulting in the point $R_i\hat{x}^+$.

By introducing $\epsilon_i = R_i(\hat{x}^+ - \hat{x})$, we have $R_i\hat{x}^+ = \epsilon_i + R_i\hat{x}$, leading to the optimization problem

$$\begin{aligned} &\min ||\epsilon_i|| \\ &\text{subject to: } CR_i^{-1}\epsilon_i = y - C\hat{x} \end{aligned}$$

The solution to this problem, the minimum length least squares solution to $y - C\hat{x}$, is

$$\epsilon_i = (CR_i^{-1})^{\dagger}(y - C\hat{x}).$$

Hence, $R_i\hat{x}^+ = R_i\hat{x} + (CR_i^{-1})^{\dagger}(y - C\hat{x})$, which is equivalent to (4) after a multiplication of R_i^{-1} from the left.

It remains to show that the condition in (10) is satisfied for the state jump update (4). By construction, the vectors $R_i(\hat{x}^+ - \hat{x})$ and $R_i(x - \hat{x}^+)$ are orthogonal; otherwise ϵ_i would not be optimal. Hence, by Pythagoras' law

$$\begin{aligned} &||R_i(x - \hat{x})||^2 = ||R_i(x - \hat{x}^+) + R_i(\hat{x}^+ - \hat{x})||^2 = \\ &= ||R_i(x - \hat{x}^+)||^2 + \underbrace{2(x - \hat{x}^+)^T R_i^T R_i(\hat{x}^+ - \hat{x})}_{0} + \\ &+ ||R_i(\hat{x}^+ - \hat{x})||^2 \geq ||R_i(x - \hat{x}^+)||^2, \end{aligned}$$

where the inequality is true since $||R_i(\hat{x}^+ - \hat{x})|| \geq 0$. Hence, we have shown that (10) and consequently (8) is satisfied, ending the first part of the proof.

We now need to prove that the energy in every observer mode eventually is upper bounded by a constant (regardless of the system mode). By adding and subtracting $\gamma\tilde{x}^T\tilde{x}$, $-\gamma\epsilon^2 x^T x$, and $\mu_{i,j} x^T Q_i x$ (where $\mu_{i,j} \geq 0$), $\dot{V}_i$ in (3) becomes

$$\begin{aligned} \dot{V}_i(\tilde{x}) &= [\tilde{x}^T x^T]\Gamma_{i,j}[\tilde{x}^T x^T]^T - \mu_{i,j} x^T Q_i x \\ &\quad - \gamma\tilde{x}^T\tilde{x} + \gamma\epsilon^2 x^T x \\ &\leq -\gamma\tilde{x}^T\tilde{x} + \gamma\epsilon^2 x^T x \leq -\frac{\gamma}{\beta}V_i(\tilde{x}) + \gamma\epsilon^2 x_{max}^2, \end{aligned}$$

where the first and second inequality is due to Condition 2 and the fact that $-\mu_{i,j} x^T Q_i x \leq 0$ (since $\mu_{i,j} \geq 0$ and $x^T Q_i x \geq 0$ in regions where mode i of the switched system (1) is possible), and Condition 1 and (5) respectively. This differential inequality implies that

$$\begin{aligned} V_i(\tilde{x}(t)) &\leq e^{-\frac{\gamma}{\beta}(t-t_0)}V_i(\tilde{x}(t_0)) \\ &\quad + \beta\epsilon^2 x_{max}^2(1 - e^{-\frac{\gamma}{\beta}(t-t_0)}) \\ &\leq e^{-\frac{\gamma}{\beta}(t-t_0)}V_i(\tilde{x}(t_0)) \\ &\quad + \beta\epsilon^2 x_{max}^2(1 - e^{-\frac{\gamma}{\beta}(t-t_0)}), \end{aligned} \tag{12}$$

where $t_0 \geq T_0$. Consequently, the overall energy $V(\tilde{x}(t))$ decreases at the switching instants and is upper bounded by a constant. Due to Condition 1, we then have

$$\begin{aligned} ||\tilde{x}(t)|| \leq \Big(&e^{-\frac{\gamma}{\beta}(t-t_0)}V(\tilde{x}(t_0))/\alpha \\ &+ \frac{\beta}{\alpha}\epsilon^2 x_{max}^2(1 - e^{-\frac{\gamma}{\beta}(t-t_0)})\Big)^{\frac{1}{2}}. \end{aligned} \tag{13}$$

Hence, when $t \to \infty$ the exponential functions converge to zero implying that (6) is satisfied.

To prove that (7) is satisfied, we assume that the switched system (1) is in every mode at least a time T and it takes $\Delta \leq T$ to identify correct mode of the switched system. When the switched system changes mode at time, say, $t_0 \in \mathcal{T}$, to mode i, the observer is still in mode j during a time Δ. Hence, from (12) we have that

$$\begin{aligned} &V(\tilde{x}(t)) = V_j(\tilde{x}(t)) \leq e^{-\frac{\gamma}{\beta}(t-t_0)}V_j(\tilde{x}(t_0)) + \\ &\beta\epsilon^2 x_{max}^2(1 - e^{-\frac{\gamma}{\beta}(t-t_0)}),\ t_0 \leq t \leq t_0 + \Delta. \end{aligned}$$

At time $t_0 + \Delta$ ($\in \hat{\mathcal{T}}$), we identify correct mode of the system, and the observer changes correspondingly. The mode is correctly estimated at least in the time interval $t_0 + \Delta$ to $t_0 + T$ according to the assumptions. During this time interval,

$$\begin{aligned} &V(\tilde{x}(t)) = V_i(\tilde{x}(t)) \leq \\ &e^{-\frac{\gamma}{\beta}(t-t_0-\Delta)}V_i(\tilde{x}(t_0 + \Delta)),\ t_0 + \Delta \leq t \leq t_0 + T, \end{aligned}$$

since Condition 2 implies that $\Gamma_{i,i}^{11} < 0$ with $\epsilon = 0$. At the switch time $t_0 + \Delta$, the state $\hat{x}$ is updated according to (4), implying that $V_i(\tilde{x}(t_0 + \Delta)) \leq V_j(\tilde{x}(t_0 + \Delta))$. Combining this with the two inequalities, implies that we at time $t_0 + T$ have

$$
\begin{aligned}
V_i(\tilde{x}(t_0+T)) &\leq e^{-\frac{\gamma}{\beta}(T-\Delta)}V_i(\tilde{x}(t_0+\Delta)) \\
&\leq e^{-\frac{\gamma}{\beta}(T-\Delta)}V_j(\tilde{x}(t_0+\Delta)) \\
&\leq e^{-\frac{\gamma}{\beta}(T-\Delta)}\Big(e^{-\frac{\gamma}{\beta}\Delta}V_j(\tilde{x}(t_0)) \\
&\quad +\beta\epsilon^2 x_{max}^2(1-e^{-\frac{\gamma}{\beta}\Delta})\Big) \\
&\leq e^{-\frac{\gamma}{\beta}T}V_j(\tilde{x}(t_0)) \\
&\quad +\beta\epsilon^2 x_{max}^2(e^{-\frac{\gamma}{\beta}(T-\Delta)}-e^{-\frac{\gamma}{\beta}T})\Big).
\end{aligned}
$$

By repeatedly switchings, the energy converges (at worst) to the bound

$$
V_{max} = \beta\epsilon^2 x_{max}^2\left(\frac{e^{-\frac{\gamma}{\beta}(T-\Delta)}-e^{-\frac{\gamma}{\beta}T}}{1-e^{-\frac{\gamma}{\beta}T}}\right),
$$

which gives (7) due to Condition 1, ending the proof. ■

A sufficient condition for the existence of a solution to the inequalities in the theorem is that $\Gamma_{i,j}^{11} < 0$ in Condition 2. This is the formulation of the estimation problem assuming that the system mode is known. In this case, the estimation error obeys (cf. (13) in case when $\epsilon = 0$ and $t_0 = 0$)

$$
||\tilde{x}(t)|| \leq \sqrt{\frac{\beta}{\alpha}}e^{-\frac{\gamma}{2\beta}t}||\tilde{x}_0||,
$$

implying that $||\tilde{x}(t)||$ goes to zero as time goes to infinity regardless of the value of $x(t)$. When we do not know the mode, which is handled in the theorem, we cannot say that the estimation error goes to zero but is upper bounded according to (6), which depends on the largest value of $||x(t)||$. This bound is usually very conservative, indicated by the example later on, since it is obtained having the worst possible combination of observer mode and system mode. However, the result shows that if the active dynamics is estimated correctly within a certain time (Δ), and the dwell time of the switched linear system is lower bounded (by T), the bound of the estimation error can be reduced significantly according to (7). If $\Delta \to 0$ the bound becomes zero, but if we do not succeed to estimate the discrete state correctly until the next mode change of the switched system, implying that $\Delta \to T$, then we again get (6). Note that if the switched system stops to switch, and the correct mode is identified, the estimation error will converge to zero.

Except the properly updates according to (4), the theorem uses multiple Lyapunov functions, which increases the possibility to find the unknown variables satisfying the conditions in the theorem. Using a common Lyapunov function (corresponds to $d_{i,j} = 0$ in the theorem) to prove convergence, the energy decrease condition (9) is trivially satisfied by letting $\hat{x}^+ = \hat{x}$, i.e. no updates of the estimated states are necessary. However, also in this case, the updates of the estimated states according to (4) will improve the real convergence rate and should be used also in case when a common Lyapunov function is searched for.

5. SOLUTION USING LINEAR MATRIX INEQUALITIES

Theorem 1 has to be valid whether the observer gains K_i are decided *a priori* or not. The unknowns in Theorem 1 will be found by iteratively fixing ϵ to a value and search for the smallest β satisfying the conditions to find a low bound on the right-hand side of (6). If there is no solution for the fixed value of ϵ, the value is increased. Furthermore, without loss of generality, α is scaled to 1 to prevent the P_i's to be positive semi-definite.

Whether the observer gains K_i are decided *a priori* or not, Theorem 1 can be reformulated as a linear matrix inequality (LMI) problem in the unknown variables P_i, $d_{i,j}$, $\mu_{i,j}$, γ and possible K_i. In the case of unknown $K_i's$ they have to be constrained in some way to prevent them from being too large. One possibility is to introduce the condition

$$
\begin{bmatrix} \lambda_i^2 I_{p\times p} & W_i^T \\ W_i & I_{n\times n} \end{bmatrix} \geq 0, \quad i \in I_N
$$

implying that $K_i^T K_i \leq \lambda_i^2 I_{p\times p}$, cf. (Pettersson, 2005*a*; Pettersson, 2005*b*).

6. EXAMPLE

We now illustrate the observer synthesis in this paper in case of two modes of the (autonomous) switched linear system (1) given by

$$
A_1 = \begin{bmatrix} 1 & -5 \\ 0 & 1 \end{bmatrix}, A_2 = \begin{bmatrix} 1 & 0 \\ 5 & 1 \end{bmatrix}, B = \begin{bmatrix} 0 \\ 0 \end{bmatrix}, C = \begin{bmatrix} 1 \\ -2.4 \end{bmatrix}^T.
$$

The system and observer time switchings are indirectly given by specifying switch sets defined by linear hyper planes according to

$$
S_{i,j} = \{x \in \Re^n \mid s_{i,j}x = 0\},\ (i,j) \in \{(1,2),(2,1)\},
$$

$$
\hat{S}_{i,j} = \{\hat{x} \in \Re^n \mid \hat{s}_{i,j}\hat{x} = 0\},\ (i,j) \in \{(1,2),(2,1)\},
$$

where the switch planes of the observer are put equal to the switch planes of the system to mimic the switching behavior, according to

$$
s_{1,2} = \hat{s}_{1,2} = [1.56\ \ 1], \quad s_{2,1} = \hat{s}_{2,1} = [1\ \ -1.56].
$$

We assume that the design of the the observer gains is not known *a priori* but is a part of the synthesis problem. We will study the solution in case when $\lambda = \lambda_1 = \lambda_2 = 5$.

It should be noted that there does not exist a solution to Theorem 1 with a common P; hence, the suggested observer synthesis in this paper is less conservative than existing results using a common quadratic Lyapunov function. Solving the corresponding LMI problem of Theorem 1 with multiple Lyapunov functions, results in a solution

$$
K_1 = \left[1.79 - 1.02\right]^T,\ K_2 = \left[-3.44 - 3.54\right]^T,
$$

with $\alpha = 1$, $\beta = 5.70$, $\epsilon = 4.98$, $\gamma = 1.88$. According to (6) in the theorem, we therefore have the bound $||\tilde{x}(t)|| \leq 11.87x_{max}$. The switched system changes mode every 0.45 time units; hence $T = 0.45$. The bound can be improved according to (7) if we can identify the correct mode within a time $\Delta < T$.

Figure 3 shows a trajectory simulation x of the switched linear system, in the case when $x_{max} = 1$, together with the estimated states $\hat{x}$ updating the estimator states according to (4) at the switching instants. As can be seen from the figure, the estimated states converge to the switched linear states exactly. Consequently, the times $\Delta \to 0$.

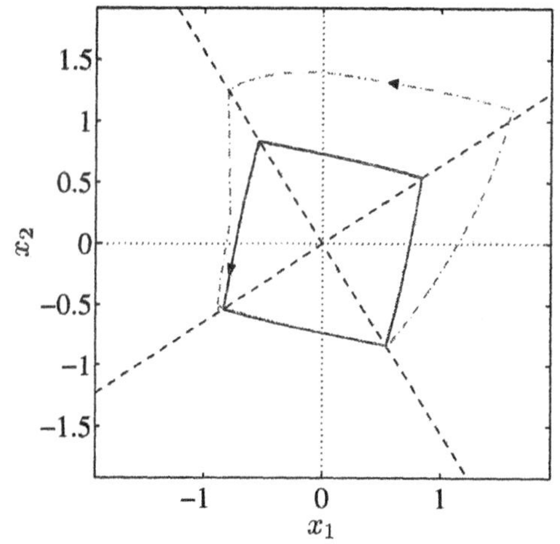

Fig. 3. The estimated states $\hat{x}$ (dash-dotted) converges to the switched linear system states x (solid line) using the projection.

To compare, a trajectory simulation of the estimated states $\hat{x}$ when not updating the estimator states according to (4) at the switching instants is shown in Figure 4. In this case, it can be seen that the estimated states converge to a limit cycle. Hence, it is advantageous to update the continuous estimator states at the switching instants.

7. CONCLUSIONS

In this paper, it has been shown how to estimate the continuous states of a switched linear systems by designing a switched observer including state jumps. By using multiple Lyapunov functions and properly update the continuous estimated states when the mode changes occur, an observer is synthesized by solving a linear matrix inequality problem. The bound of the estimation error is reduced compared to earlier result, if it can be shown that the active dynamics is estimated correctly within a certain time, and if the dwell time of the switched linear system is lower bounded. Future research will deal with the question how to guarantee that the estimation error actually goes to zero, as in the example.

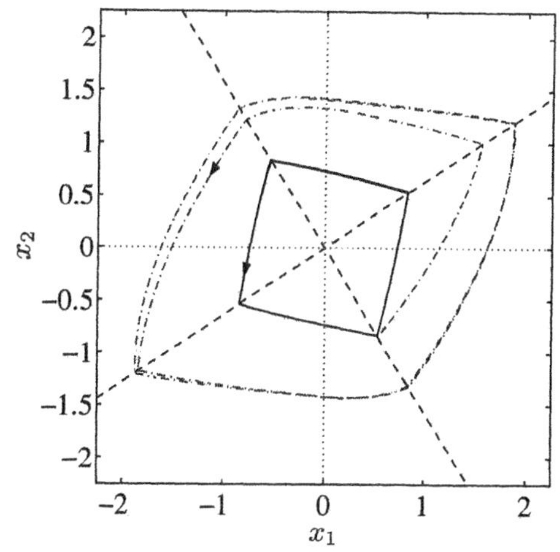

Fig. 4. The estimated states $\hat{x}$ (dash-dotted) converges not to the switched linear system states x (solid line) since projection is not used.

REFERENCES

Alessandri, A. and P. Coletta (2001). Switching observers for continuous-time and discrete-time linear systems. In: *Proc. of the American Control Conference*. Arlington, Virginia. pp. 2516–2521.

Balluchi, A., L. Benvenuti, M. D. Di Benedetto and A. L. Sangiovanni-Vincentelli (2002). Design of observers hybrid systems. In: *Lecture Notes in Computer Science 2289* (C. J. Tomlin and M. R. Greenstreet, Eds.). Springer. pp. 76–89.

Feron, E. (1996). Quadratic stabilizability of switched systems via state and output feedback. Technical Report CICP-P-468. MIT.

Filippov, A. F. (1988). *Differential Equations with Discontinuous Righthand Sides*. Kluwer Academic Publishers.

Juloski, A. Lj., W. P. M. H. Heemels and S. Weiland (2002). Observer design for a class of piecewise affine systems. In: *Proc. of the 41st IEEE Conference on Decision and Control*. Las Vegas, Nevada, USA. pp. 2606–2611.

Pettersson, S. (2005*a*). Observer design for switched systems using multiple quadratic lyapunov functions.. In: *13th Mediterranean Conference on Control and Automation*. Limassol, Cyprus. pp. 262–267.

Pettersson, S. (2005*b*). Switched state jump observers for switched systems. In: *Proc. of the 16th IFAC Word Congress*. Prague. pp. Tu-E12–TO/5.

Pettersson, S. and B. Lennartson (2002). Hybrid system stability and robustness verification using linear matrix inequalities. *Int. J. Control* **75**(16/17), 1335–1355.

Strang, G. (1988). *Linear Algebra and its Applications*. Harcourt Brace Jovanovich College Publishers.

ELSEVIER
IFAC
PUBLICATIONS

CRITICAL STATES DETECTION WITH BOUNDED PROBABILITY OF FALSE ALARM AND APPLICATION TO AIR TRAFFIC MANAGEMENT [1]

M.D. Di Benedetto * , **S. Di Gennaro** *
and A. D'Innocenzo *

* *Department of Electrical and Information Engineering and Center of Excellence DEWS, University of L'Aquila.*
E.mail: {dibenede, digennar, adinnoce}@ing.univaq.it

Abstract: The analysis of error propagation in an Air Traffic Management (*ATM*) environment is addressed. The theory of Hybrid Systems is used to model the error evolution, an observability problem for a Markov Chain with discrete output symbols associated to the transitions is stated, and a runtime observer is proposed for estimating the probability of a given discrete state to be active. Sufficient conditions are given for characterizing the decidability of the addressed observability problem. The results are related to previous works on location observability of deterministic hybrid systems, and are used to analyze an *ATM* case study, the "*clearance to change the flight plan*".

Keywords: Failure Detection, Markov Chains, Air Traffic Management

1. INTRODUCTION

Hybrid systems are a powerful tool for the analysis and control of Air Traffic Management (*ATM*) systems, as shown in the IST European Project HYBRIDGE (see http://www.nlr.nl/public/hosted–sites/hybridge). Each agent, in an *ATM* environment, executes a sequence of operations that may be characterized by different dynamics (Di Benedetto *et al.*, 2005): this is a typical hybrid context. Moreover, since we are dealing with human agents, the behavior is non–deterministic. The non–determinism of human agents is mainly due to *Situation Awareness*, which is defined in (Endsley, 1995), (Stroeve *et al.*, 2003) as "*the perception of elements in the environment, the comprehension of their meaning, and the projection of their status in the near future*". *Situation Awareness* may be wrong for wrong perception of relevant information, wrong interpretation of perceived information, or wrong prediction of a future state and propagation of error due to agents communication. Moreover, statistic data retrieved by the analysis of real cases of *ATM* procedures may be used to define specific error probability in *ATM* operations, thus it is reasonable to introduce a stochastic framework to analyze error propagations. In the context of error detection analysis, partially observable discrete event systems have been extensively studied in fault detection and supervisory control problems. (Yoo and Lafortune, 2001) analyze the diagnosability of partially observable discrete event systems, and propose a polynomial verification method. (Hadjicostis, 2002) discusses a probabilistic methodology for detecting functional changes in the state transition mechanism of a deterministic finite-state machine (FSM). Results are achieved by computing the deviation between the expected observations and the actual measurements, assuming to know an appropriate statistical characterization of the FSM input. In (Kennedy *et al.*, 1987) a decision feedback equalizer (DFE) operating on a noisy channel is considered, and it is shown how the results concerning a noiseless channel can be extended to yield tight bounds on the stationary error probability performance for the noisy case. Similar approaches were developed in (Aghasaryan *et al.*, 1997; Boubour *et al.*, 1997) for Petri nets. Observability of hybrid systems has been also analyzed in (Balluchi *et al.*, 2002), where a definition of observability of the current state (*current location observability*) has been provided and a procedure for the construction of an observer of the discrete and continuous states is proposed, and in (D'Innocenzo *et al.*, 2006), where *current location observability* of hybrid automata has been studied.

[1] This work was partially supported by European Commission under Project IST NoE HyCON contract n. 511368.

The above definitions of diagnosability and observability do not require a real time state estimation, while in safety-critical applications such as *ATM*, we need to determine the actual state of the system immediately, as a delay can lead to unsafe or even catastrophic behavior of the system. For this reason, we focus here on the concept of observability in prescribed time horizon, and we introduce a stochastic framework to model and test *Situation Awareness* error evolution in *ATM* operations. In Section 2, we define a class of stochastic Hybrid Systems. In Section 3, we propose a definition of observability with bounded probability of false alarm for this class of systems. We propose a design method for a runtime estimator of the discrete state of $\mathcal{S}$ on the basis of the measured outputs, and we give conditions for the system to be observable. In Section 4, we state sufficient conditions such that observability is decidable. In Section 5, we relate our results to previous works on location observability of deterministic hybrid systems (De Santis *et al.*, 2005). In Section 6 we present a case study, the *Clearance to change the flight plan*, where the developed methodologies are used to yield a conditioned probability distribution of the *SA* error evolution. Section 6 offers conclusions and tips for further work.

2. DEFINITIONS AND SETTING

We define a Markov Hybrid System as a tuple $\mathcal{S} = (Q \times X, Q_0 \times X_0, U, Y, S_q, \Sigma, E, \Psi, \eta, \Pi, \Pi_0)$ where:

- $Q = q_1, \cdots, q_N$ is the discrete state set;
- X is the continuous state space;
- Q_0 is the set of initial discrete states;
- X_0 is the set of initial continuous states;
- U is the continuous input space;
- Y is the continuous output space;
- S_q associates linear continuous dynamics A_q, B_q, C_q to each discrete state $q \in Q$;
- Σ is the finite set of input symbols;
- $E \subseteq Q \times \Sigma \times Q$ is a collection of edges;
- Ψ is the finite set of output symbols;
- η: $E \to \Psi$ is the output function;
- Π is a transition probability matrix with $\Pi_{ij} = \mathcal{P}[q(k+1) = q_j \mid q(k) = q_i]$ for each k;
- Π_0 is an initial probability distribution $(\mathcal{P}[q(0) = q_1] \cdots \mathcal{P}[q(0) = q_N])$, where $\Pi_{0i} = 0$ if $q_i \notin Q_0$.

A Markov Hybrid System is similar to a Hybrid Markov chain as proposed in (Shi *et al.*, 2004). However, in our model no guard functions are considered, and we do not assume that the embedded Markov Chain is irreducible and positive recurrent.

To define the executions of $\mathcal{S}$, we introduce a hybrid time basis $\tau = \{I_k\}_{k\geq 0} \in \mathcal{T}$ as a finite or infinite sequence of intervals $I_k = [t_k, t'_k]$ such that (Lygeros *et al.*, 1999)

(1) I_k is closed if τ is infinite; I_k might be right–open if it is the last interval of a finite sequence τ;
(2) $t_k \leq t'_k$ for all k and $t'_{k-1} \leq t_k$ for $k > 0$.

The cardinality $|\tau|$ of the hybrid time basis is the number of intervals I_k in τ.

An execution of $\mathcal{S}$ is a collection $\chi = (\tau, x, q)$, with x, q satisfying the continuous and discrete dynamics of $\mathcal{S}$. A string $\rho = q_0, \cdots, q_s$ is an execution of the discrete state q of $\mathcal{S}$ with a finite number of transitions $|\rho| - 1 = s$ if $q_0 \in Q_0$ and $\forall I_k \in \{I_1, \cdots, I_s\}, (q_{k-1}, q_k) \in E$. The discrete state execution is ruled by a discrete time Markov chain.

Let $\Upsilon(Q_0)$ be the set of all executions ρ of the discrete state of $\mathcal{S}$ with a finite number of transitions. Given $q \in Q$, let $\Upsilon_q(Q_0)$ be the set of all executions $\rho \in \Upsilon(Q_0)$ such that the last visited state is $q_s = q$. We associate to each execution ρ the observed output as the string $p = P(\rho) = \psi_1 \cdots \psi_s$ where $\psi_k = \eta(q(I_{k-1}), q(I_k))$ for $k = 1, \cdots, s$. We define $\mathcal{L}(\mathcal{S}) = \{P(\rho) \mid \rho \in \Upsilon(Q_0)\}$ the set of output strings that can be generated by all executions of the system $\mathcal{S}$. Given an output string $p = \psi_1 \cdots \psi_s$, we define

$$Reach_{\mathcal{S}}(Q_0, p) := \{q \in Q \mid \exists \rho \in \Upsilon_q(Q_0), P(\rho) = p\}$$

the set of all states that can be reached from an initial state in Q_0 and such that the observed output string is p.

Let $\mathcal{H} = (Q \times X, Q_0 \times X_0, U, Y, S_q, \Sigma, E, \Psi, \eta)$ be a Hybrid System defined by the same tuple of $\mathcal{S}$, except for the stochastic matrices Π and Π_0. The space of all executions of $\mathcal{S}$ and that of $\mathcal{H}$ coincide. However, the discrete execution is non deterministic on $\mathcal{H}$, while on $\mathcal{S}$ it is subtended by a probability space, denoted $(\Omega, \mathcal{F}, \mathcal{P})$, on which the stationary Markov chain $q(I_0), q(I_1), q(I_2), \cdots$ is defined. Ω is the space Υ of all executions ρ of the discrete space, and $\mathcal{F}$ the associated sigma–algebra. $\mathcal{P}$ is uniquely defined by the transition probability matrix Π and the initial probability distribution Π_0. Let $\pi_i(I_k) := \mathcal{P}[q(I_k) = q_i]$ and $\pi(I_{k+1}) = \Pi^T \pi(I_k)$ the corresponding dynamics. We now introduce a well known formalism that will be necessary in the following sections:

Let a Markov Hybrid System $\mathcal{S} = (Q \times X, Q_0 \times X_0, U, Y, S_q, \Sigma, E, \Psi, \eta, \Pi, \Pi_0)$ and the subsets $Q' \subset Q$, $E' \subset E \cap (Q' \times Q')$ be given; $\mathcal{S}' = (Q' \times X, Q'_0 \times X_0, U, Y, S_q, \Sigma', E', \Psi', \eta', \Pi', \Pi'_0)$ is the subsystem induced by (Q', E') on $\mathcal{S}$, where (Π', Π'_0) are normalized stochastic matrices.

3. P–OBSERVABILITY OF A DISCRETE STATE

In this section, we propose a definition of observability for a Markov Hybrid System, with respect to a given discrete state. We then propose a constructive procedure for an estimator of the discrete state and a verification procedure for observability. Finally, give conditions such that P–Observability is decidable.

Given a Markov Hybrid System $\mathcal{S}$, our goal is to use the discrete output string to compute the probability distribution of the current discrete state conditioned to a subset of trajectories, namely all the trajectories whose output is the measured output. Consider the probability space $(\Omega, \mathcal{F}, \mathcal{P})$. When an output string $p = \psi_1 \cdots \psi_s$ is generated up to time t_s, it is possible to define the set $\mathcal{G}(p) \subseteq \Omega$ of executions $\rho \in \Upsilon(Q_0)$ such that $P(\rho) = p$. $\mathcal{G}(p)$ is given by $\mathcal{G}_k(p)$ for $k = s$, where

$$\mathcal{G}_0(p) = \Omega$$

$$\mathcal{G}_k(p) = \mathcal{G}_{k-1}(p) \cap \left(\bigcup_{q \in Reach_{\mathcal{S}}(Q_0, p|_k)} \Upsilon_q(Q_0) \right)$$

where $p\,|_k = \psi_1 \cdots \psi_k$ is the truncation of p up to time k. Note that $\mathcal{G}(\epsilon) = \Omega$ and $\mathcal{G}(p\,|_k) = \mathcal{G}_k(p)$. Let us define

$$\mathcal{P}[q(I_{|p|}) = q_i \mid p\,]: \; = \mathcal{P}[\rho \in \Upsilon_{q_i}(Q_0) \mid P(\rho) = p\,] = \\ = \mathcal{P}[q(I_{|p|}) = q_i \mid \mathcal{G}(p)]$$

and let $q_c \in Q$ be a given critical state of $\mathcal{S}$, namely a discrete state associated to a behavior of the system which may lead to unsafe situations. We want to construct an observer of a critical state with the property that it always detects if the current state of $\mathcal{S}$ is q_c, and such that the probability that detection of q_c is a false alarm is bounded. Let $P \in [0,1]$ be the maximal probability of false alarm we accept to tolerate. We can formalize the property that such an observer exists by the following definition:

Definition 1. Given a Markov Hybrid System $\mathcal{S}$, a discrete state $q_c \in Q$ is P–Observable (observable with probability of false alarm P) if $\forall p \in \mathcal{L}(\mathcal{S})$

$$\mathcal{P}[q(I_{|p|}) = q_c \mid \mathcal{G}(p)] \in \{0\} \cup [1-P, 1].$$

This condition implies that, given the measured output of $\mathcal{S}$, either we are sure that we are not in a critical state (thus we don't have to worry) or the probability that the current state is critical is very high, and it is reasonable to give an alarm signal. Namely, if a discrete state q_c of $\mathcal{O}$ is P–Observable, we guarantee that it is always possible to detect if the current state is q_c, with a probability of generating a false alarm less than P. We obtain the limit case (0–Observability) when the information given by the output of the system is rich enough that $\mathcal{P}[q(I_{|p|}) = q_c \mid \mathcal{G}(p)]$ assumes only the values 1 or 0 for all $p \in \mathcal{L}(\mathcal{S})$, that is we know at each time instant with probability 1 if the current state is q_c or not.

We propose now a method for constructing a system whose input is the discrete output string $p \in \mathcal{L}(\mathcal{S})$, and whose output is the probability $\mathcal{P}[q(I_{|p|}) = q_i \mid p\,], \forall i = 1 \cdots N$. Note that such system uses the only discrete output of $\mathcal{S}$ to estimate the current discrete state. Consider a hybrid system $\mathcal{O} = (\hat{Q} \times \hat{X}, \hat{q}_0 \times \hat{x}_0, \hat{U}, \hat{Y}, \hat{S}_{\hat{q}}, \hat{\Sigma}, \hat{E}, \hat{R})$ such that:

- $\hat{Q} \subseteq 2^Q$ is the set of discrete states;
- $\hat{X} = [0,1]^N$ is the continuous state space, and $\hat{\pi} = [\hat{\pi}_1, \hat{\pi}_2, \cdots, \hat{\pi}_N]$ is the continuous state;
- $\hat{q}_0 = Q_0 \subseteq 2^Q$ is the initial discrete state of $\mathcal{O}$;
- $\hat{x}_0 = \Pi_0$ is the initial continuous state of $\mathcal{O}$, namely the initial probability distribution of each discrete state of $\mathcal{S}$;
- $\hat{U} = \varnothing$ is the continuous input space;
- $\hat{Y} = \hat{X}$ is the continuous output space;
- $\hat{S}_{\hat{q}}$ is such that $A_{\hat{q}} = B_{\hat{q}} = \mathbf{0}, C_{\hat{q}} = \mathbf{I}\ \forall \hat{q} \in \hat{Q}$;
- $\hat{\Sigma} = \Psi$ is the set of input symbols, namely the set of output symbols of $\mathcal{S}$;
- $\hat{E} = \hat{Q} \times \hat{\Sigma} \times \hat{Q}$ is the set of edges associated to an input symbol;
- $\hat{R} : \hat{E} \times \hat{X} \to \hat{X}$ is a deterministic non-linear reset function of the continuous state $\hat{\pi}$.

The discrete layer of $\mathcal{O}$ may be constructed as in (Di Benedetto *et al.*, 2005). Given an output string $p \in \mathcal{L}(\mathcal{S})$, the associated hybrid execution of $\mathcal{O}$ is unique ($\mathcal{O}$ is deterministic), and by construction of $\mathcal{O}$ $\hat{q}(I_{|p|}) = Reach_{\mathcal{S}}(Q_0, p) \subseteq 2^Q$ is the set of states of Q that may be active in the time interval $I_{|p|}$ accordingly to the observed output p. Note that given any execution of $\mathcal{S}$ and the associated execution of $\mathcal{O}$, the associated hybrid time bases $\tau_{\mathcal{H}}$ and $\tau_{\mathcal{O}}$ coincide.

Let us define the dynamics of the continuous state $\hat{\pi}_i(t)$ of $\mathcal{O}$. Note that, since $\hat{\pi}(t)$ is piecewise constant for each interval $I_k = [t_k, t_{k+1})$ and is only modified by reset functions, we refer to $\hat{\pi}(I_k)$ as the value on such intervals. The reset function $\hat{R}(\hat{e}, \hat{\pi})$ for each $(\hat{e}, \hat{\pi}) \in \hat{E} \times \hat{X}$ is defined in (Di Benedetto *et al.*, 2005), and the following is proved:

Proposition 1. (Di Benedetto *et al.*, 2005) Given a system S and the associated system O. Then, $\hat{\pi}_i(I_{|p|}) = \mathcal{P}[q(I_{|p|}) = q_i \mid \mathcal{G}(p)], \forall i = 1 \cdots N, \forall p \in \mathcal{L}(\mathcal{S})$.

Remark 1. Let $\hat{Q}_c = \{\hat{q} \in \hat{Q} \mid q_c \in \hat{q} \wedge |\hat{q}| > 1\}$: a discrete state q_i of a Markov Hybrid System $\mathcal{S}$ is P–Observable if the reach set of the hybrid state $(\hat{q}, \hat{\pi})$ of $\mathcal{O}$ has empty intersection with the set $\hat{Q}_c \times \{\hat{\pi}_i \in (0, 1-P)\}$.

For all proofs of this paper the reader is referred to (Di Benedetto *et al.*, 2006). We will now characterize decidability of the P–Observability problem for a Markov Hybrid System $\mathcal{S}$. We define the set $\mathcal{K}_{q_c}(\mathcal{S}) = \{p \in \mathcal{L}(\mathcal{S}) \mid q_c \in Reach_{\mathcal{S}}(Q_0, p) \wedge |Reach_{\mathcal{S}}(Q_0, p)| > 1\}$, namely the set of *bad output strings* that yield q_c indistinguishable from some other state in Q. Moreover, let $\mathcal{L}_{\hat{Q}_f}(\mathcal{O})$ be the language accepted by a non-deterministic finite automaton (*NFA*) with the same discrete topological structure as $\mathcal{O}$ and such that the set of final states is $\hat{Q}_f$.

Lemma 1. The following statements hold:

(i) $\mathcal{L}(\mathcal{S})$ is a regular language;
(ii) $\mathcal{K}_{q_c}(\mathcal{S}) = \mathcal{L}_{\hat{Q}_c}(\mathcal{O}) = \bigcup_{\hat{q} \in \hat{Q}_c} \mathcal{L}_{\hat{q}}(\mathcal{O})$
(iii) $\mathcal{K}_{q_c}(\mathcal{S})$ is a regular language, and $\mathcal{K}_{q_c}(\mathcal{S}) \subset \mathcal{L}(\mathcal{S})$.

Proposition 2. Given a Markov Hybrid System $\mathcal{S}$, a discrete state $q_c \in Q$ is P–Observable (observable with probability of false alarm P) if $\forall p \in \mathcal{K}_{q_c}(\mathcal{S})$, $\mathcal{P}[q(I_{|p|}) = q_c \mid p\,] \in \{0\} \cup [1-P, 1]$.

Let $\hat{q} = \{q_1, \cdots, q_m, q_{m+1}\} \in \hat{Q}_c$, where $q_c = q_{m+1}$ and $m \geq 1$ by definition of $\hat{Q}_c$. Given the output string p, let us define:

$$\theta(Q_0, q, p): \; = \sum_{\substack{\rho \in \Upsilon_q(Q_0) \\ P(\rho) = p}} \mathcal{P}[q(I_0) = \rho_0] \cdot \\ \cdot \left(\prod_{k=0}^{|p|} \mathcal{P}[q(I_{k+1}) = \rho_{k+1} \mid q(I_k) = \rho_k, \psi_k] \right) \quad (1)$$

where $\rho = \rho_0 \cdots \rho_{|p|+1}$. Note that 1 implies

$$\mathcal{P}[q(I_{|p|}) = q_c \mid p\,] = \frac{\theta(Q_0, q_c, p)}{\sum_{i=1}^{m+1} \theta(Q_0, q_i, p)} \quad (2)$$

Proposition 3. A state q_c is P–Observable for a system $\mathcal{S}$ if and only if $\forall \hat{q} \in \hat{Q}_c$ and $\forall p \in \mathcal{L}_{\hat{q}}(\mathcal{O})$ the following holds:

$$\theta_c(p) := \frac{\theta(Q_0, q_c, p)}{\sum_{i=1}^{m} \theta(Q_0, q_i, p)} \geq \frac{1-P}{P}. \quad (3)$$

If the cardinality of the language $\mathcal{L}_{q_c}(\mathcal{S})$ is finite, the P–Observability problem is decidable. If not, the following theorem gives sufficient conditions on the language $\mathcal{K}_{q_c}(\mathcal{S})$ to achieve decidability of the P–Observability problem.

Theorem 1. Given a Markov Hybrid System $\mathcal{S}$, let $A_{\hat{q}}$ be the regular expression that generates $\mathcal{L}_{\hat{q}}(\mathcal{O})$ for $\hat{q} \in \hat{Q}$. If $A_{\hat{q}}$ can be expressed in the form $A_1 + \cdots + A_M$ for each $\hat{q} \in \hat{Q}_c$, where $A_i = a_{i1}a_{i2}\cdots a_{in_i}$ and $a_{ij} \in \{\sigma, \sigma^*, \sigma + \sigma'\}$, $\sigma, \sigma' \in \Sigma$, then P–Observability of $\mathcal{S}$ is decidable.

4. P–OBSERVABILITY FOR $P = 0$

In this section, we introduce an equivalence relation between P–Observability of $\mathcal{S}$ and critical observability (De Santis *et al.*, 2005),(Di Benedetto *et al.*, 2005) of $\mathcal{H}$. More precisely, we prove that, given a system $\mathcal{S}$, the P–Observability conditions for $P = 0$ on the associated observer $\mathcal{O}_\mathcal{S}$ are equivalent to the critical observability conditions (De Santis *et al.*, 2005) on the associated observer $\mathcal{O}_\mathcal{H}$. Note that $\mathcal{O}_\mathcal{S}$ and $\mathcal{O}_\mathcal{H}$ have the same discrete dynamics, and therefore the same topological structure. We first recall the definition given in (De Santis *et al.*, 2005) for a non-deterministic hybrid system $\mathcal{H}$ w.r.t. a discrete state $q_c \in Q$:

Definition 2. (De Santis *et al.*, 2005) A hybrid system $\mathcal{H}$ is critically location observable w.r.t. a discrete state $q_c \in Q$ (q_c–critically location observable) if, for any initial state $q_0 \in Q_0$, the current state $q(k)$ can be detected from the output string p whenever $q(k) = q_c$.

Proposition 4. (De Santis *et al.*, 2005) A hybrid system $\mathcal{H}$ is q_c–critically location observable if and only if, for each discrete state $\hat{q}$ of the associated observer $\mathcal{O}$ such that $q_c \in \hat{q}$, then $|\hat{q}| = 1$.

We can now state the following:

Proposition 5. Given the systems $\mathcal{H}$ and $\mathcal{S}$, and the corresponding observers $O_\mathcal{H}$ and $O_\mathcal{S}$, the following are equivalent:

(1) q_c is P–Observable with $P = 0$ for $\mathcal{S}$;
(2) $\mathcal{H}$ is q_c–critically location observable.

5. CASE STUDY: CLEARANCE TO CHANGE THE FLIGHT PLAN

A Clearance to Change the Flight Plan involves a pilot of a flying aircraft and an air traffic controller. We assume that the procedure is started by a decision of the controller because of a conflict resolution. We describe now the agents involved and the specific behavior of each of them:

The **Flight Management System** (*FMS*) is a technical system that holds the flight plan, modeled as a list of positions s_i and an arrival times t_i. The *FMS* is configured by the *PF*, and controls the aircraft direction, speed and flight mode.The **Flight Data Processing System** (*FDPS*) is a system containing the flight plan, and is reconfigured by the controller. The **Aircraft** (*AC*) is totally controlled by the *FMS*. The **Pilot flying** (*PF*) interacts via VHF communication with the Controller, and can change the actual flight plan by re-configuring the *FMS* system. The **Air Traffic Controller** (*CO*) interacts via VHF communication with the *PF* and monitors the aircraft information (position, velocity, altitude, direction, aircraft code etc) on the *FDPS*.

A Clearance to Change the Flight Plan procedure starts when the Controller, to resolve a conflict, decides to ask the pilot to reconfigure the actual flight plan. The interaction between the *CO* and the *PF* may be assumed as a request by the *CO* to the *PF* to reconfigure the *FMS* with a new position and arrival time, and a confirm by the *PF*, who inserts the new data on the *FMS*. The Controller too configures the *FDPS* with the new coordinates. This simple operation may be affected by several errors, which can bring to an erroneous flight plan configuration and therefore to a risk situation. We suppose without loss of generality that the Controller decided for a secure flight plan, and that the *FMS* configuration is executed before the *FDPS* configuration. Furthermore it is assumed that the technical systems are operative, to set the focus on human *Situation Awareness* . The following errors are considered: Communication error, *FMS* configuration error and *FDPS* configuration error. An analysis of the propagation of *Situation Awareness* errors may be done by formalizing a stochastic system whose continuous dynamics are the aircraft dynamics given by the position and the velocity, and whose discrete states are all possible combinations of *Situation Awareness* values of the agents. More precisely, we define the *SA* of each agent involved in the procedure as its awareness of the flight plan. The information flow previously described can cause errors in the propagation of the *SA* among agents. The *Situation Awareness* of each agent may assume one of the following values:

(1) Former flight plan before the decision of the controller (**Old**)
(2) New flight plan decided by the controller (**New**)
(3) Erroneous flight plan due to communication error between ATC and *PF* ($\mathbf{E}_{COM}$)
(4) Erroneous flight plan due to erroneous programming of the *FMS* ($\mathbf{E}_{FMS}$)
(5) Erroneous flight plan due to erroneous programming of the *FDPS* ($\mathbf{E}_{FDPS}$)

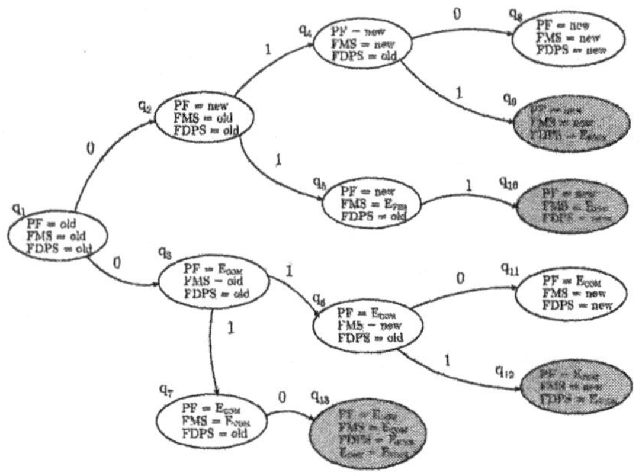

Fig. 1. System $\mathcal{S}$ modeling *Situation Awareness* error evolution

We suppose here, without loss of generality, that a communication error and a *FMS* programming error cannot happen simultaneously. This condition simplifies the number of states and transitions in the error evolution model. We consider the *SA* of the Pilot Flying (SA_{PF}), of the Flight Management System (SA_{FMS}), and of the Flight Data Processing System (SA_{FDPS}).

At the beginning of the Clearance to Change the Flight Plan, $SA_{PF} = SA_{PF} = SA_{PF} =$**Old**. This will be considered as the initial discrete state. Considering possible errors in the *SA* propagation, we can construct an automaton where each discrete state is a different value of the vector $(SA_{PF}, SA_{PF}, SA_{PF})$. The discrete states of the *SA* propagation model are all possible permutations of the considered agents' *SA*. We consider here only the most relevant states of this system, in order to avoid the generation of a too complex model. In such a system, the continuous aircraft dynamic associated with each location may be the same even in case of erroneous *FMS* configuration: e.g. if the correct altitude level given by the Air Traffic controller is 220 and the level understood by the pilot is 240, the rise dynamic of the aircraft may be identical. This means that the use of continuous dynamics to detect the current discrete state (Balluchi *et al.*, 2002) may not always solve the problem. Thus, in order to get extra discrete information from the system, we assume that it is possible to compare the flight plan configured on the *FMS* and the flight plan memorized in the *FDPS*: if they are equal, the system output is 0, otherwise it is 1.

A Clearance to Change the Flight-Plan procedure can be described by the following Markov Hybrid System $\mathcal{S}$, which models the *Situation Awareness* error evolution:

- $Q = \{q_1, \cdots, q_{13}\}$ is the set of discrete states (See Figure 1);
- $X = \mathbb{R}^3 \times \mathbb{R}^3$ is the continuous state space, where $x = (s, v)$ specifies the aircraft position s and the velocity v;
- $Q_0 \times X_0 = \{q_0\} \times \{x_0\}$, where q_1 is associated to the *Situation Awareness* vector $(SA_{PF}, SA_{PF}, SA_{PF}) =$ **(Old,Old,Old)** and x_0 are the aircraft continuous position and velocity when the Clearance to Change the Flight Plan procedure starts;
- U is the space of the continuous input control u on the velocity of the aircraft;
- Y is the space of the continuous output (the measure of the position of the aircraft);
- S_q is given by A_q, B_q, C_q:

$$A_q = \begin{bmatrix} 0 & I_3 \\ 0 & 0 \end{bmatrix}, B_q = \begin{bmatrix} 0 \\ I_3 \end{bmatrix}, C_q = [\, I_3 \; 0 \,]$$

  $\forall q \in Q$ are the continuous dynamics. The velocity vector v_{q_i} depends on the flight plan configured on the *FMS* and is controlled by u.
- $\Sigma = \{\sigma\}$ is a discrete disturbance event that triggers the actions of the agents.
- $\Psi = \{0, 1, \varepsilon\}$ where ε is the unobservable output, 0 indicates that the flight plan memorized in the *FMS* is equal to the flight plan memorized on the *FDPS* ($SA_{FMS} = SA_{FDPS}$), and 1 indicates that they are not equal ($SA_{FMS} \neq SA_{FDPS}$); *SA* stays for *Situation Awareness*.
- E, η are defined according to the automaton in Figure 1;
- Π is the transition probability matrix defined according to ATM statistics, which are usually estimated by airlines companies and ATM research centers. In this analysis, we do not assign numerical values to Π_{ij} since our aim here is to illustrate how the methodology proposed in the previous section can be applied to our case study.
- $\Pi_0 = [1 \; 0 \; \cdots \; 0]^T$.

The construction procedure previously described leads to a system $\mathcal{O}$, that shows that the discrete output obtained comparing the FMS and the FDPS data is not sufficient to achieve 0–Observability of the discrete state q_{13} of $\mathcal{S}$, since it is indistinguishable by q_8 and q_{11}. Therefore, additional discrete outputs must be introduced. Finding the set of extra discrete outputs necessary to obtain 0–Observability is a combinatorial problem on the set of edges E of the system $\mathcal{S}$, and may be trivially solved by adding all possible combinations of additional outputs to the set E, and verifying 0–Observability conditions on the system with the new outputs. To obtain P–Observability, a similar procedure can be followed. Since P–Observability conditions are weaker than deterministic critical observability conditions, the number of necessary number of additional outputs would be lower.

In the particular example considered, suppose $\rho = q_1, q_3, q_6, q_{11}$ be the execution of $\mathcal{S}$, and $p = P(\rho) = 010$ the associated output string. The discrete state of the system $\mathcal{O}$ is steered by p to $\hat{q}(I_3) = \{q_8, q_{11}, q_{13}\}$. Since the value Π_{13} (communication error probability, transition from q_1 to q_3) is certainly very low, P–Observability does not hold for a reasonable value of P. Note that the maximal probability of false alarm that can be accepted for fault detection in an ATM procedure is a design constraint. Thus, we have to add new output $\psi_{E_{COM}}$ to the transition (q_1, q_3). To generate the output $\psi_{E_{COM}}$, since the VHF speech communication cannot be measured, it is necessary to change the Clearance to Change the Flight Plan procedure, introducing a protocol for the flight plan data transmission, such that an error in the data transfer can be detected.

By adding further output symbols as done for $\psi_{E_{COM}}$, it is easy to see that 0–Observability of q_{13} is achievable by adding the discrete outputs $\psi_{E_{COM}} = \eta((q_1, q_3))$, $\psi_{E_{FMS}} = \eta((q_2, q_5))$ and

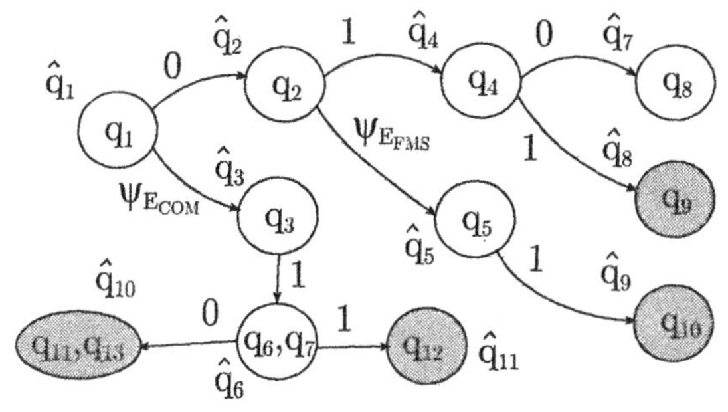

Fig. 2. System $\mathcal{O}'$ constructed from $\mathcal{H}$ with new output symbols

$\psi_{E_{FDPS}} = \eta((q_4, q_9)) = \eta((q_6, q_{12})) = \eta((q_7, q_{13}))$. However, note that the probability associated to the transition (q_3, q_6) is certainly very low, since it is very unlikely that a pilot wrongly understands a flight plan in the VHF communication and then configures the FDPS with the correct values. Hence, it is easy to see from Figure 2 that, by adding only the discrete outputs $\psi_{E_{COM}}$ and $\psi_{E_{FMS}}$, the critical discrete state q_{13} is P–Observable with a low value of P. This shows that the detection of only two errors out of three yields P–Observability with a low probability P of false alarm.

6. CONCLUSIONS

We showed that estimating and mitigating the probability of SA error in ATM can be supported by observability analysis. We proposed a definition of observability for a class of stochastic hybrid systems. For this class of systems, conditions for checking observability were given, and an algorithm to design an observer was illustrated. The equivalence between the observability notion presented here, with a zero probability of false alarm, and critical observability as defined in (De Santis *et al.*, 2005) was proven. This stochastic framework was then used to analyze error evolution in an ATM example. The framework proposed in this paper can be used for simulating ATM procedures and verifying "observability" - i.e. detectability - of dangerous operations. If the system is not observable with an acceptably low probability of generating a false alarm, the procedure must be changed with the introduction of new system outputs, and the verification procedure can be used on the resulting new system. Future research will focus on the minimization of the set of discrete outputs necessary to obtain P–Observability and on the extension of our results to continuous time Markov Chains.

7. ACKNOWLEDGEMENTS

The third author wishes to thank Rajeev Alur for interesting discussions on Regular Languages.

REFERENCES

Aghasaryan, A., E. Fabre, A. Benveniste, R. Boubour and C. Jard (1997). A petri net approach to fault detection and diagnosis in distributed systems. part ii : extending viterbi algorithm and hmm techniques to petri nets.. *Proceedings of the 36th Conference on Decision and Control, San Diego, California, USA* **2289**, 726–731.

Balluchi, A., L. Benvenuti, M.D. Di Benedetto and A.L. Sangiovanni-Vincentelli (2002). Design of observers for hybrid systems. *Hybrid Systems: Computation and Control 2002, Lecture Notes in Computer Science, C.J. Tomlin and M.R. Greensreet, Eds.* **2289**, 76–89.

Boubour, R., C. Jard, A. Aghasaryan, E. Fabre and A. Benveniste (1997). A petri net approach to fault detection and diagnosis in distributed systems. part i: application to telecommunication networks, motivations, and modelling.. *Proceedings of the 36th Conference on Decision and Control, San Diego, California, USA* pp. 720–725.

De Santis, E., M.D. Di Benedetto, S. Di Gennaro, A. D'Innocenzo and G. Pola (2005). Critical observability of a class of hybrid systems and application to air traffic management. *To Appear as a Book Chapter to Lecture Notes on Control and Information SciencesSpringer Verlag.*

Di Benedetto, M.D., S. Di Gennaro and A. D'Innocenzo (2005). Error detection within a specific time horizon and application to air traffic management. *Proceedings of the Joint 44th IEEE Conference on Decision and Control and European Control Conference (CDC-ECC'05), Seville, Spain* pp. 7472–7477.

Di Benedetto, M.D., S. Di Gennaro and A. D'Innocenzo (2006). Critical states detection with bounded probability of false alarm and application to air traffic management. Technical Report R.06-86. www.diel.univaq.it/tr/web/web_search_tr.php.

D'Innocenzo, A., M.D. Di Benedetto and S. Di Gennaro (2006). Observability of hybrid automata by abstraction. *Hybrid Systems: Computation and Control 2006, Lecture Notes in Computer Science* **3927**, 169–183.

Endsley, M.R. (1995). Towards a theory of situation awareness in dynamic system. *Human Factors* **37,1**, 32–64.

Hadjicostis, C.N. (2002). Probabilistic fault detection in finite-state machines based on state occupancy measurements. *Proceedings of the 41st IEEE Conference on Decision and Control, Las Vegas, Nevada, USA* pp. 3994–3999.

Kennedy, R.A., B.D.0. Anderson and R.R. Bitmead (1987). Tight bounds on the error probabilities of decision feedback equalizers. *IEEE Transactions on communications* **COM-35(10)**, October.

Lygeros, J., C. Tomlin and S. Sastry (1999). Controllers for reachability specications for hybrid systems. *Automatica, Special Issue on Hybrid Systems.*

Shi, L., A. Abate and S. Sastry (2004). Optimal control for a class of stochastic hybrid systems. *Proceedings of the 43rd IEEE Conference on Decision and Control, Atlantis, Paradise Island, Bahamas* pp. 1842–1847.

Stroeve, S., H.A.P. Blom and M. Van der Park (2003). Multi-agent situation awareness error evolution in accident risk modelling. *FAA-Eurocontrol, ATM2003.*

Yoo, T. and S. Lafortune (2001). On the computational complexity of some problems arising in partially-observed discrete-event systems. *Proceedings of the 2001 American Control Conference, Arlington , Virginia* pp. 25–27.

ELSEVIER
IFAC
PUBLICATIONS

TRACKING CONTROL OF JOIN-FREE TIMED CONTINUOUS PETRI NET SYSTEMS

Jing Xu, Laura Recalde, Manuel Silva

Instituto de Investigación de Ingeniería de Aragón (I3A)
Departamento de Informática e Ingeniería de Sistemas,
Centro Politécnico Superior, Universidad de Zaragoza,
Email:{jxu,lrecalde,silva}@unizar.es

Abstract: A new low-and-high gain algorithm is presented for tracking control of timed continuous Petri Net (contPN) systems working under infinite servers semantics. The inherent properties of timed contPN determine that the control signals must be non-negative and upper bounded by functions of system states. In the proposed control approach, LQ theory is first utilized to design a low-gain controller such that the control signals satisfy the input constraints. Based on the low-gain controller, a high-gain term is further added to improve the system transient performance. In order to guarantee global tracking convergence and smoothness of control signals, in our work, a new mixed tracking trajectory (state step and ramp) is utilized instead of a pure step reference signal.

1. INTRODUCTION

Petri Nets (PNs) are powerful mathematical tools with appealing graphical representations for the modeling, analysis and synthesis of Discrete Even Systems (DESs). However, like in all DESs, discrete PN systems suffer from the so called state explosion problem. One possible way to partially tackle this problem is to fluidify the discrete PN models. The resulting *continuous* Petri Net (contPN) systems have the potential for the applications of more analytical techniques which were originally developed for continuous and hybrid systems. Furthermore, analogous to discrete PN systems, time has been introduced to contPN systems, which leads to timed contPN systems. Steady state control is studied in (Mahulea *et al.*, 2005); assuming all transitions can be controlled, it can be solved by means of a LPP. Nevertheless, dynamic control of timed contPN systems needs further investigations and great improvements.

For control of timed contPN systems, we will begin from *Join-Free* (JF) timed contPN systems with step-tracking control target. For a JF timed contPN system, a linear differential equation describes its behavior, but it is subject to certain input constraints, i.e. the control signal must be non-negative and upper bounded by a function of system states. These constraints result in a *hybrid system*, more precisely a piecewise linear system. The main challenge in our work is to develop control laws under these special input constraints so that global tracking convergence can be ensured.

The input constraints can be treated as input saturations. As input saturation is one of the common phenomena encountered in control systems, hitherto lots of works have been done. In (Wredenhage and Bélanger, 1994), a kind of piecewise-linear control law was derived. However, such a design method will lead to a low-gain controller. To improve the control performance, a low-and-high gain approach was given in (Saberi *et al.*, 1996). Recently, several nonlinear control methods were further presented (Lin *et al.*, 1998), which mainly focus on the high-gain design in the low-and-high gain algorithms aiming to achieve better transient control performance. It should be pointed out that common assumptions in all these works are that the lower saturation bounds are negative constants and the upper saturation bounds are positive constants. However, in JF timed contPN systems, the lower saturation bound is zero and the upper saturation bound depends on system states. Therefore, the existing control strategies for systems with input saturations cannot be applied directly.

In this paper, a new low-and-high gain approach will be proposed for step-tracking control of JF timed contPN systems. The presented algorithm can ensure global asymptotical convergence in presence of the input constraints existing in JF timed contPN systems. According to initial marking, desired marking and net structure, a new reference trajectory is constructed. Based on the new tracking target, a low-gain controller is designed according to LQ theory so that the control signals are within the required regions. Since the upper bounds of the control inputs depend on the system states, the design method in (Wredenhage and Bélanger, 1994) fails to work. By combining the new tracking trajectory design with LQ method, a novel design scheme for the low-gain part is given in our work. Analogous to the works of (Saberi *et al.*, 1996), a high-gain part is further added to make better use of the available control authority, and consequently faster system response can be obtained. Note that in the high-gain term, the control vector of the low-gain part is maintained, which makes possible the analysis of system stability. Rigorous proof is provided to guarantee the global asymptotical convergence.

The paper is organized as follows. Section 2 introduces the required concepts of contPN systems and timed contPN systems. The control for JF timed contPN systems is formulated in Section 3. How to construct the new tracking trajectory is outlined in Section 4. Section 5 focuses on the development of control laws and the analysis of global asymptotical convergence property. An illustrative example is given in Section 6. Finally, Section 7 concludes the paper.

2. CONTINUOUS PETRI NET SYSTEMS

2.1 Untimed Continuous Petri Net Systems

A contPN system can be described as $\langle \mathcal{N}, \mathbf{m}_0 \rangle$, where $\mathcal{N} = \langle \mathbf{P}, \mathbf{T}, \mathbf{Pre}, \mathbf{Post} \rangle$ specifies the net structure ($\mathbf{P}$ and $\mathbf{T}$ are disjoint (finite) sets of places and transitions, and $\mathbf{Pre}$ and $\mathbf{Post}$ are incidence matrices), and $\mathbf{m}_0$ is the initial marking. $\mathcal{N}$ is assumed to be connected, while $\mathbf{P}$ and $\mathbf{T}$ have n and m elements, respectively. The marking $\mathbf{m}$ belongs to R^{+^n}, where R^+ is the set of non-negative real numbers, and both $\mathbf{Pre}$ and $\mathbf{Post}$ are of the size $n \times m$. For $w \in \mathbf{P} \cup \mathbf{T}$, the set of its input and output nodes are denoted as ${}^\bullet w$, and $w^\bullet$, respectively. $\mathcal{N}$ is JF (or rendez-vous free) iff $\forall t \in \mathbf{T}$, $|{}^\bullet t| = 1$. If $\mathbf{m}$ is reachable from $\mathbf{m}_0$ through a sequence $\boldsymbol{\sigma} \in \mathsf{R}^{+^m}$, the state equation is $\mathbf{m} = \mathbf{m}_0 + \mathbf{C} \cdot \boldsymbol{\sigma}$, where $\mathbf{C} = \mathbf{Post} - \mathbf{Pre}$ is the token flow matrix and $\boldsymbol{\sigma}$ is the firing count vector.

2.2 Timed Continuous Petri Net Systems

A timed contPN system can be represented as $\langle \mathcal{N}, \boldsymbol{\lambda}, \mathbf{m}_0 \rangle$, where $\boldsymbol{\lambda}[t] > 0$ is the internal firing rate of transition t. The state equation has an explicit dependence on time $\mathbf{m}(\tau) = \mathbf{m}_0 + \mathbf{C}\boldsymbol{\sigma}(\tau)$, where τ is time. Deriving with respect to it, $\dot{\mathbf{m}}(\tau) = \mathbf{C}\dot{\boldsymbol{\sigma}}(\tau)$ is obtained. Define $\mathbf{f}(\tau) = \dot{\boldsymbol{\sigma}}(\tau)$, which denote flows of transitions. The state equation is $\dot{\mathbf{m}}(\tau) = \mathbf{C}\mathbf{f}(\tau)$. For notation simplicity, τ will be omitted in the rest of the paper. For the definition of flow $\mathbf{f}$, different semantics have been introduced and the most important ones are *infinite servers* and *finite servers*. Infinite servers semantics will be considered in this work, which usually provides a much better approximation of discrete behaviors. Under infinite server semantics, $\mathbf{f}$ is the product of $\boldsymbol{\lambda}[t]$ and the instantaneous enabling of the transition, i.e. $\mathbf{f}[t] = \boldsymbol{\lambda}[t] \cdot enab(t, \mathbf{m}) = \boldsymbol{\lambda}[t] \cdot \min_{p \in {}^\bullet t}\{\mathbf{m}[p]/\mathbf{Pre}[p,t]\}$. As in JF nets any transition has only one input place, the flow can be expressed as $\mathbf{f} = \boldsymbol{\Phi} \cdot \mathbf{m}$, where $\boldsymbol{\Phi} \in \mathsf{R}^{+^{m \times n}}$ and $\boldsymbol{\Phi}[t,p] = \boldsymbol{\lambda}[t]/\mathbf{Pre}[p,t]$ if $p = {}^\bullet t$, $\boldsymbol{\Phi}[t,p] = 0$ otherwise. Moreover, each row of $\boldsymbol{\Phi}$ has only one non-zero element. $\forall j \in M \triangleq \{1, 2, \cdots, m\}$, the non-zero element for the j-th row of $\boldsymbol{\Phi}$ is denoted as $\phi_{j,i}$ where $i \in N \triangleq \{1, 2, \cdots, n\}$. Therefore, $\forall j \in M$, we have $f_j = \phi_{j,i} m_i$, where f_j is the j-th element of $\mathbf{f}$ and m_i is the i-th element of $\mathbf{m}$. The evolution of the marking can be written as follows:

$$\dot{\mathbf{m}} = \mathbf{C} \cdot \mathbf{f} = \mathbf{A} \cdot \mathbf{m}, \tag{1}$$

where $\mathbf{A} = \mathbf{C} \cdot \boldsymbol{\Phi}$. Finally the definition of conservativeness and some related properties for a JF timed contPN are listed as follows.

Definition 1. A PN is *conservative* iff $\exists \mathbf{y} > 0$, such that $\mathbf{y} \cdot \mathbf{C} = \mathbf{0}$. Any left non-negative annuller of matrix $\mathbf{C}$, i.e. $\mathbf{y}$, is called P-semiflow. A P-semiflow is *minimal* if 1 is the greatest common divisor of its elements, and its support does not strictly contain the support of other P-semiflow.

Property 1. (Teruel *et al.*, 1997) Let $\mathcal{N}$ be a JF net.

1.1 If $\mathcal{N}$ is strongly connected,

a) $\mathcal{N}$ has at most one minimal P-semiflow;

b) $\mathcal{N}$ is conservative iff it has one P-semiflow.

1.2 If $\mathcal{N}$ is conservative, it is consistent iff it is strongly connected.

Property 2. Let $\mathcal{N}$ be a conservative and strongly connected JF timed contPN. The matrix $\mathbf{A}$ for $\mathcal{N}$ defined in (1) has only one zero-eigenvalue. Moreover, the remaining eigenvalues of $\mathbf{A}$ are with negative real parts.

Proof: As $\mathcal{N}$ is conservative and strongly connected, from *Property 1.2*, $\mathcal{N}$ is consistent. From the *Proposition 1* in (Jiménez *et al.*, 2005), for the conservative and consistent JF PN $\mathcal{N}$, the matrix $\mathbf{A}$ has only one zero-valued eigenvalue. On the other hand, as $\mathbf{A}$ is a Metzler matrix and the eigenvalue of zero is the (Frobenius) dominant eigenvalue (Jiménez *et al.*, 2005), the remaining eigenvalues of $\mathbf{A}$ are with negative real parts. ∎

3. PROBLEM FORMULATION

We will restrict our research to conservative and strongly connected JF timed contPN systems and assume all the markings are observable. For concise expression, "timed contPN" will be written as "contPN". Like all the other systems, control actions can also be introduced to modify autonomous evolution of PN systems. The possible control action that can be applied to PN systems is to slow down their unforced firing flows. Hence, the forced flows of controlled transitions become $\mathbf{f}-\mathbf{u}$, where $\mathbf{u}$ is the control signal and must satisfy $\mathbf{0} \le \mathbf{u} \le \mathbf{f}$. Considering (1), a JF contPN system with a control action can be described as follows:

$$\dot{\mathbf{m}} = \mathbf{C}(\mathbf{\Phi m} - \mathbf{u}) \triangleq \mathbf{Am} - \mathbf{Bu}, \tag{2}$$

where $\mathbf{m} \in \mathsf{R}^{+n}$, $\mathbf{A} \in \mathsf{R}^{n \times n}$, $\mathbf{B} \triangleq \mathbf{C} \in \mathsf{R}^{n \times m}$ and $\mathbf{u} \in \mathsf{R}^{+m}$. It should be noted that the constraints on input $\mathbf{u}$ lead the closed system to be a piecewise linear system.

Our control objective is to construct control laws such that $\mathbf{m}$ and $\mathbf{u}$ respectively converge to desired values: $\mathbf{m}_d$ and $\mathbf{u}_d$. To satisfy reachability, $\mathbf{m}_d$ must fulfill that $\mathbf{y} \cdot \mathbf{m}_d = \mathbf{y} \cdot \mathbf{m}_0$ where $\mathbf{y} \in \mathsf{R}^n$ is the basis of P-semiflows. For $\mathbf{u}_d$, $\mathbf{0} \le \mathbf{u}_d \le \mathbf{\Phi m}_d$ must be satisfied, i.e. $\forall j \in M$, $0 \le u_{d,j} \le \phi_{j,i} m_{d,i}$ where $u_{d,j}$ and $m_{d,i}$ are the j-th and i-th elements of $\mathbf{u}_d$ and $\mathbf{m}_d$ respectively. Moreover, as $\mathbf{m}_d$ is constant, according to (2), the desired control input must be a solution of the following equation.

$$\mathbf{0} = \mathbf{Am}_d - \mathbf{Bu}_d. \tag{3}$$

In our work, the following assumption is made for $\mathbf{m}_0$ and $\mathbf{m}_d$.

Assumption 1. $\forall i \in N$, $m_{0,i} > 0$ and $m_{d,i} > 0$.

Remark 1. If we consider optimal steady states in practical manufacture systems, generally $\mathbf{m}_d > \mathbf{0}$ is valid. On the other hand, if some elements of $\mathbf{m}_0$ are zeros, as $\mathbf{m}_d > \mathbf{0}$, a firing sequence can always be found such that $\mathbf{m}_0[\boldsymbol{\sigma} > \mathbf{m}$ and $\mathbf{m} > \mathbf{0}$ (Recalde *et al.*, 1999). Then the control algorithm proposed in our work can be further applied.

4. DESIGN OF NEW TRACKING REFERENCE

To ensure global convergence and smoothness of the control signal, a step target $\mathbf{m}_d$ is replaced by the following reference trajectory $\mathbf{m}_r(\tau)$.

$$\mathbf{m}_r(\tau) = \begin{cases} \mathbf{m}_{r0} + \dfrac{\mathbf{m}_d - \mathbf{m}_{r0}}{h}\tau, & \tau \in [0, h] \\ \mathbf{m}_d, & \tau \in [h, \infty) \end{cases} \tag{4}$$

where $\mathbf{m}_r(\tau) \in \mathsf{R}^{+n}$, $\mathbf{m}_{r0} \triangleq \mathbf{m}_r(0)$ and $h > 0$ determines the time when $\mathbf{m}_r(\tau) = \mathbf{m}_d$. Here we choose $\mathbf{m}_{r0} = \mathbf{m}_0 + \delta(\mathbf{m}_d - \mathbf{m}_0)$, where $0 \le \delta < 1$ is a parameter to be designed. The parameter h is chosen such that, for given $\mathbf{m}_{r0}$ and $\mathbf{m}_d$, valid control actions $\mathbf{u}_{r0} \in \mathsf{R}^{+m}$ and $\mathbf{u}_{rh^-} \in \mathsf{R}^{+m}$ exist for the following equations.

$$\mathbf{Am}_{r0} - \mathbf{Bu}_{r0} = \frac{\mathbf{m}_d - \mathbf{m}_{r0}}{h} \tag{5}$$

$$\mathbf{Am}_d - \mathbf{Bu}_{rh^-} = \frac{\mathbf{m}_d - \mathbf{m}_{r0}}{h}, \tag{6}$$

where $0 \le \mathbf{u}_{r0} \le \mathbf{\Phi m}_{r0}$ and $0 \le \mathbf{u}_{rh^-} \le \mathbf{\Phi m}_d$.

Proposition 1. For given $\mathbf{m}_0$ and reachable $\mathbf{m}_d$, δ and h can always be found such that (5) and (6) are valid. Moreover, as δ increase, smaller h can be chosen.

Proof: Given $\mathbf{m}_0$ and reachable $\mathbf{m}_d$, $\boldsymbol{\sigma} \ge \mathbf{0}$ can always be found so that $\mathbf{m}_d = \mathbf{m}_0 + \mathbf{B}\boldsymbol{\sigma}$. Thus,

$$\mathbf{m}_d - \mathbf{m}_{r0} = \mathbf{B}(1-\delta)\boldsymbol{\sigma}. \tag{7}$$

Substituting (7) into (5) and (6) yields

$$\mathbf{B\Phi m}_{r0} - \mathbf{Bu}_{r0} = \mathbf{B}(1-\delta)\boldsymbol{\sigma}\frac{1}{h},$$

$$\mathbf{B\Phi m}_d - \mathbf{Bu}_{rh^-} = \mathbf{B}(1-\delta)\boldsymbol{\sigma}\frac{1}{h}.$$

Obviously, $\mathbf{u}_{r0} = \mathbf{\Phi m}_{r0} - (1-\delta)\boldsymbol{\sigma}\frac{1}{h}$ and $\mathbf{u}_{rh^-} = \mathbf{\Phi m}_d - (1-\delta)\boldsymbol{\sigma}\frac{1}{h}$ are solutions for the above equations. From $\mathbf{0} \le \mathbf{u}_{r0} \le \mathbf{\Phi m}_{r0}$, we have:

$$\mathbf{0} \le \boldsymbol{\sigma}\frac{1}{h} \le \frac{1}{1-\delta}\mathbf{\Phi m}_{r0} \tag{8}$$

$$\Rightarrow \mathbf{0} \le \boldsymbol{\sigma}\frac{1}{h} \le \mathbf{\Phi m}_0 + \frac{\delta}{1-\delta}\mathbf{m}_d. \tag{9}$$

From $\mathbf{0} \le \mathbf{u}_{rh^-} \le \mathbf{\Phi m}_d$, we can obtain that

$$\mathbf{0} \le \boldsymbol{\sigma}\frac{1}{h} \le \frac{1}{1-\delta}\mathbf{\Phi m}_d. \tag{10}$$

Therefore, the final constraints for h are:

$$\mathbf{0} \le \frac{\boldsymbol{\sigma}}{h} \le \min\{\mathbf{\Phi m}_0 + \frac{\delta \mathbf{m}_d}{1-\delta}, \frac{1}{1-\delta}\mathbf{\Phi m}_d\} \tag{11}$$

where, for $\mathbf{a}_1 \in \mathsf{R}^{+^n}$ and $\mathbf{a}_2 \in \mathsf{R}^{+^n}$, the i-th element of $\min\{\mathbf{a}_1, \mathbf{a}_2\}$ is defined as $\min\{a_{1,i}, a_{2,i}\}$. From *Assumption 1*, all the elements of $\mathbf{m}_0$ and $\mathbf{m}_d$ are strictly positive. Hence, $\min\{\mathbf{\Phi m}_0 + \frac{\delta}{1-\delta}\mathbf{m}_d, \frac{1}{1-\delta}\mathbf{\Phi m}_d\}$ is positive. Considering $\boldsymbol{\sigma} \ge \mathbf{0}$, for a given $0 \le \delta < 1$, a sufficiently large $h > 0$ can always be found such that (11) is valid. Moreover, for a chosen $\boldsymbol{\sigma}$, as δ increase, smaller h can be obtained. ∎

Proposition 1 clearly shows that the input constraints result in the constraints for h and, for a chosen $\boldsymbol{\sigma}$, smaller h can be realized by increasing δ. Hence, to obtain faster response, in (4), $\mathbf{m}_{r0}$ is designed instead of using $\mathbf{m}_0$ directly. However, larger δ will lead to larger initial error, which may destroy the tracking convergence. Hence, δ and h should be properly chosen such that both tracking convergence and possible fast response can be obtained.

To simplify the design procedure, in the proposed algorithm, δ will be designed first. Its design will be given in next Section. Now let us discuss how to calculate h for a given δ. According to (8) and (10), the constraints for h can be rewritten as $\mathbf{0} \le \frac{\boldsymbol{\sigma}}{h} \le \frac{\mathbf{\Phi}\min\{\mathbf{m}_{r0},\mathbf{m}_d\}}{1-\delta}$. From the definition of $\mathbf{m}_{r0}$, we have $\mathbf{0} < \frac{\mathbf{\Phi}\min\{\mathbf{m}_0,\mathbf{m}_d\}}{1-\delta} \le \frac{\mathbf{\Phi}\min\{\mathbf{m}_{r0},\mathbf{m}_d\}}{1-\delta}$. Hence, h can be designed to satisfy $\mathbf{0} \le \frac{\boldsymbol{\sigma}}{h} \le \frac{\mathbf{\Phi}\min\{\mathbf{m}_0,\mathbf{m}_d\}}{1-\delta}$. As $\min\{\mathbf{m}_0, \mathbf{m}_d\}$ is strictly positive, the solution of h always exists. Furthermore, h can be chosen as $h = \beta(1-\delta)$, where $\beta > 0$ and $\frac{\boldsymbol{\sigma}}{\beta} \le \mathbf{\Phi}\min\{\mathbf{m}_0, \mathbf{m}_d\}$. Consequently, $\mathbf{u}_{r0}$ and $\mathbf{u}_{rh^-}$ can be rewritten as follows:

$$\mathbf{u}_{r0} = \mathbf{\Phi m}_{r0} - \frac{\boldsymbol{\sigma}}{\beta} \tag{12}$$

$$\mathbf{u}_{rh^-} = \mathbf{\Phi m}_d - \frac{\boldsymbol{\sigma}}{\beta}. \tag{13}$$

Remark 2. For given $\mathbf{m}_0$ and $\mathbf{m}_d$, $\boldsymbol{\sigma}$ is designed first. Based on the chosen $\boldsymbol{\sigma}$, the minimum β such that $\frac{\boldsymbol{\sigma}}{\beta} \le \mathbf{\Phi}\min\{\mathbf{m}_0, \mathbf{m}_d\}$ is chosen.

5. TRACKING CONTROL OF JF CONTPN SYSTEMS

The control signal $\mathbf{u}$ is designed as follows:

$$\mathbf{u} = sat(\mathbf{u}_{lg} + \mathbf{u}_{hg}) + \mathbf{u}_r(\tau), \tag{14}$$

where $\mathbf{u}_{lg}$ is the low-gain part, $\mathbf{u}_{hg}$ is the high-gain term and $\mathbf{u}_r(\tau)$ is the reference control input defined as follows:

$$\mathbf{u}_r(\tau) = \begin{cases} \mathbf{u}_{r0} + \dfrac{\mathbf{u}_{rh^-} - \mathbf{u}_{r0}}{h}\tau, & \tau \in [0, h^-] \\ \mathbf{u}_d, & \tau \in [h^+, \infty) \end{cases},$$

where $\mathbf{u}_{r0}$ and $\mathbf{u}_{rh^-}$ are given in (12) and (13) respectively. Moreover, $\forall\ \mathbf{d} \in \mathsf{R}^m$, $sat(\mathbf{d}) \triangleq [sat(d_1), \cdots, sat(d_m)]^T$ and $\forall j \in M$, $sat(d_j)$ is

$$sat(d_j) = \begin{cases} \phi_{j,i}m_i - u_{r,j}, & \textbf{if } d_j \ge \phi_{j,i}m_i - u_{r,i} \\ d_j, & \textbf{if } -u_{r,j} < d_j < \phi_{j,i}m_i - u_{r,j} \\ -u_{r,j}, & \textbf{if } d_j \le -u_{r,j} \end{cases}.$$

Define $\mathbf{e} = \mathbf{m}_r(\tau) - \mathbf{m}$. From (2) and (4), we have

$$\dot{\mathbf{e}} = \begin{cases} \dfrac{\mathbf{m}_d - \mathbf{m}_{r0}}{h} - \mathbf{Am}_r(\tau) \\ \quad +\mathbf{Ae} + \mathbf{Bu} & \tau \in [0, h^-] \\ -\mathbf{Am}_d + \mathbf{Ae} + \mathbf{Bu} & \tau \in [h^+, \infty) \end{cases} \tag{15}$$

From the definition of $\mathbf{u}_r(\tau)$ and considering (5) and (6), for $\tau \in [0, h^-]$, it can be derived that

$$\begin{aligned} \mathbf{Bu}_r(\tau) &= \mathbf{Am}_{r0} - (\mathbf{m}_d - \mathbf{m}_{r0})\frac{1}{h} + [\mathbf{Am}_d - (\mathbf{m}_d \\ &\quad - \mathbf{m}_{r0})\frac{1}{h} - \mathbf{Am}_{r0} + (\mathbf{m}_d - \mathbf{m}_{r0})\frac{1}{h}]\frac{\tau}{h} \\ &= -(\mathbf{m}_d - \mathbf{m}_{r0})\frac{1}{h} + \mathbf{Am}_r(\tau). \end{aligned} \tag{16}$$

According to (3) and (16), the following is valid:

$$\mathbf{Bu}_r(\tau) = \begin{cases} -\dfrac{\mathbf{m}_d - \mathbf{m}_{r0}}{h} + \mathbf{Am}_r(\tau), & \tau \in [0, h^-] \\ \mathbf{Am}_d, & \tau \in [h^+, \infty) \end{cases}.$$

Substituting (14) into (15) and considering the above results, (15) can be rewritten as

$$\dot{\mathbf{e}} = \mathbf{Ae} + \mathbf{B}sat(\mathbf{u}_{lg} + \mathbf{u}_{hg}). \tag{17}$$

ContPNs with at least one P-semiflow are non-controllable, according to the classical linear control theory (Mahulea *et al.*, 2005). Hence, a transformation matrix $\mathbf{H} \in \mathsf{R}^{n \times n}$ is constructed to separate the system states into controllable and non-controllable parts. The first row of $\mathbf{H}$ is a basis of P-semiflow (here only one vector; *Property 1.1*) and the remaining rows are completed with elementary vectors such that $\mathbf{H}$ is full rank. Define $\bar{\mathbf{e}} = \mathbf{He}$. Then (17) becomes

$$\dot{\bar{\mathbf{e}}} = \bar{\mathbf{A}}\bar{\mathbf{e}} + \bar{\mathbf{B}}sat(\mathbf{u}_{lg} + \mathbf{u}_{hg}) \tag{18}$$

where $\bar{\mathbf{A}} = \mathbf{HAH}^{-1}$ and $\bar{\mathbf{B}} = \mathbf{HB}$. According to the definitions of $\mathbf{A}$, $\mathbf{B}$ and $\mathbf{H}$ and considering $\mathbf{y} \cdot \mathbf{C} = \mathbf{0}$, it can be derived that the first rows of both $\bar{\mathbf{A}}$ and $\bar{\mathbf{B}}$ are zeros, which leads to $\dot{\bar{e}}_1 = 0$.

From the definition of $\bar{\mathbf{e}}$ and $\mathbf{H}$, we have $\bar{e}_1(0) = \mathbf{y}(m_{d,1} - m_1(0))$. As $\mathbf{ym}_d = \mathbf{ym}(0)$, $\bar{e}_1(0) = 0$ can be obtained. Therefore, $\forall \tau \in [0, \infty)$, $\bar{e}_1(\tau) = 0$. Moreover, the controllable part of (18) is

$$\dot{\bar{\mathbf{e}}}_c = \bar{\mathbf{A}}_c \bar{\mathbf{e}}_c + \bar{\mathbf{B}}_c sat(\mathbf{u}_{lg} + \mathbf{u}_{hg}), \quad (19)$$

where $\bar{\mathbf{e}}_c \triangleq [\bar{e}_2, \cdots, \bar{e}_n]^T \in R^{n-1}$, $\bar{\mathbf{A}}_c \in R^{(n-1)\times(n-1)}$ and $\bar{\mathbf{B}}_c \in R^{(n-1)\times m}$. From the definition of $\bar{\mathbf{e}}$, we have $\mathbf{e} = \mathbf{H}^{-1}\bar{\mathbf{e}}$. As $\bar{e}_1 = 0$, $\mathbf{e}$ can be rewritten as $\mathbf{S}\bar{\mathbf{e}}_c$ where $\mathbf{S} \in R^{n\times(n-1)}$ is $\mathbf{H}^{-1}$ without the first column.

5.1 Design of $\mathbf{u}_{lg}$ and $\mathbf{u}_{hg}$

$\mathbf{u}_{lg}$ is designed to minimize the following quadratic performance criterion

$$J(\bar{\mathbf{e}}_c(0)) = \int_0^\infty (\bar{\mathbf{e}}_c^T \mathbf{Q} \bar{\mathbf{e}}_c + \gamma \mathbf{u}^T \mathbf{R} \mathbf{u}) d\tau, \quad (20)$$

where $\mathbf{Q} \in R^{(n-1)\times(n-1)}$ is a diagonal positive definite matrix, $\mathbf{R} = diag(r_1, \cdots, r_m)$ is positive definite and the $\gamma > 0$ is a parameter to be designed.

Define $\mathbf{c}_1 = \min\{\mathbf{u}_{r0}, \mathbf{u}_{rh^-}, \mathbf{u}_d\}$. Obviously, $\mathbf{c}_1 \geq \mathbf{0}$. The design of $\mathbf{u}_{lg}$ is classified into two cases.

Case 1. $\mathbf{c}_1 > \mathbf{0}$

The low gain controller is $\mathbf{u}_{lg} = -\mathbf{K}\bar{\mathbf{e}}_c$, where $\mathbf{K} = \frac{1}{\gamma}\mathbf{R}^{-1}\bar{\mathbf{B}}_c^T\mathbf{W}$ and $\mathbf{W}$ can be found from the following Riccati equation,

$$\mathbf{W}\bar{\mathbf{A}}_c + \bar{\mathbf{A}}_c^T\mathbf{W} - \frac{1}{\gamma}\mathbf{W}\bar{\mathbf{B}}_c\mathbf{R}^{-1}\bar{\mathbf{B}}_c^T\mathbf{W} + \mathbf{Q} = 0. \quad (21)$$

Case 2. $\mathbf{c}_1$ have zero-elements

To clearly explain the basic idea, assume only one element of $\mathbf{c}_1$, i.e. $\mathbf{c}_{1,z}$ $(z \in M)$, is zero. However, if $\mathbf{c}_1$ have several zero-elements, the design of $\mathbf{u}_{lg}$ can be derived analogously. In this Case, $\mathbf{W}$ is calculated from the following Riccati equation,

$$\mathbf{W}\bar{\mathbf{A}}_c + \bar{\mathbf{A}}_c^T\mathbf{W} - \frac{1}{\gamma}\mathbf{W}(\bar{\mathbf{B}}_c - \Delta\bar{\mathbf{B}}_c) \mathbf{R}^{-1}(\bar{\mathbf{B}}_c - \Delta\bar{\mathbf{B}}_c)^T\mathbf{W} + \mathbf{Q} = 0, \quad (22)$$

where $\Delta\bar{\mathbf{B}}_c \in R^{(n-1)\times m}$, the z-th column of $\Delta\bar{\mathbf{B}}_c$ is same as the z-th column of $\bar{\mathbf{B}}_c$ and all the remaining columns of $\Delta\bar{\mathbf{B}}_c$ are $\mathbf{0}$. As $\mathbf{A}_c$ is stable (from *Property 2*), the solution of $\mathbf{W}$ always exists. The low gain controller is $\mathbf{u}_{lg} = -\mathbf{K}\bar{\mathbf{e}}_c$, where $\mathbf{K} = \frac{1}{\gamma}\mathbf{R}^{-1}(\bar{\mathbf{B}}_c - \Delta\bar{\mathbf{B}}_c)^T\mathbf{W}$. From the definition of $\Delta\bar{\mathbf{B}}_c$, it can be derived that the z-th row of $\mathbf{K}$, i.e. $\mathbf{k}_z$, is $\mathbf{0}$. Hence, $\Delta\bar{\mathbf{B}}_c\mathbf{K} = \mathbf{0}$.

For the high-gain term, in both **Case 1** and **Case 2**, $\mathbf{u}_{hg} = -l\bar{\mathbf{B}}_c^T\mathbf{W}\bar{\mathbf{e}}_c$ where l is a positive constant.

5.2 Design of the parameters δ and γ

Define $\epsilon(\mathbf{W}, \rho) \triangleq \{\bar{\mathbf{e}}_c : \bar{\mathbf{e}}_c^T\mathbf{W}\bar{\mathbf{e}}_c \leq \rho\}$, where $\rho = \bar{\mathbf{e}}_c^T(0)\mathbf{W}\bar{\mathbf{e}}_c(0)$. δ and γ are designed off-line such that $\forall \bar{\mathbf{e}}_c \in \epsilon(\mathbf{W}, \rho)$ and $\forall j \in M$, $-\mathbf{k}_j\bar{\mathbf{e}}_c \geq -c_{1,j}$ and $-\mathbf{k}'_j\bar{\mathbf{e}}_c \leq c_{2,j}$, where $c_{1,j}$ is the j-th element of $\mathbf{c}_1$, $c_{2,j} \triangleq \min\{\frac{\sigma_j}{\beta}, \phi_{j,i}m_{d,j} - u_{d,j}\}$, $\mathbf{k}_j$ is the j-th row of $\mathbf{K}$, $\mathbf{k}'_j \triangleq \mathbf{k}_j - \phi_{j,i}\mathbf{s}_i$ and $\mathbf{s}_i$ is the i-th row of $\mathbf{S}$. Note that $c_{2,j} \geq 0$.

The following proposition implies the existence of the solutions of δ and γ.

Proposition 2. Let $\langle \mathcal{N}, \boldsymbol{\lambda}, \mathbf{m}_0 \rangle$ be a conservative and strongly connected JF contPN system. For given $\mathbf{Q}$ and $\mathbf{R}$, δ and γ can always be found such that $\forall \bar{\mathbf{e}}_c \in \epsilon(\mathbf{W}, \rho)$ and $\forall j \in M$, $-\mathbf{k}_j\bar{\mathbf{e}}_c \geq -c_{1,j}$ and $-\mathbf{k}'_j\bar{\mathbf{e}}_c \leq c_{2,j}$.

Proof: Same as the design of $\mathbf{u}_{lg}$, the proof also contains two cases.

Case 1. $\mathbf{c}_1 > \mathbf{0}$

$\forall j \in M$, the maximum values of $|-\mathbf{k}_j\bar{\mathbf{e}}_c|$ and $|-\mathbf{k}'_j\bar{\mathbf{e}}_c|$ subjected to $\bar{\mathbf{e}}_c^T\mathbf{W}\bar{\mathbf{e}}_c \leq \rho$ are as follows (Wredenhage and Bélanger, 1994):

$$\max_{\bar{\mathbf{e}}_c \in \epsilon(\mathbf{W},\rho)} |-\mathbf{k}_j\bar{\mathbf{e}}_c| = \sqrt{\rho}(\mathbf{k}_j\mathbf{W}^{-1}\mathbf{k}_j^T)^{1/2}, \quad (23)$$

$$\max_{\bar{\mathbf{e}}_c \in \epsilon(\mathbf{W},\rho)} |-\mathbf{k}'_j\bar{\mathbf{e}}_c| = \sqrt{\rho}(\mathbf{k}'_j\mathbf{W}^{-1}\mathbf{k}'^T_j)^{1/2}. \quad (24)$$

To satisfy the design requirements, we need to prove that, $\forall j \in M$,

$$\sqrt{\rho}(\mathbf{k}_j\mathbf{W}^{-1}\mathbf{k}_j^T)^{1/2} \leq c_{1,j}, \quad (25)$$

$$\sqrt{\rho}(\mathbf{k}'_j\mathbf{W}^{-1}\mathbf{k}'^T_j)^{1/2} \leq c_{2,j}. \quad (26)$$

Based on (25) and (26) and considering the definitions of ρ, $\mathbf{k}_j$ and $\mathbf{k}'_j$, δ_j and γ_j can be calculated for every $j \in M$. Therefore, $\delta = \min_{j\in M}\{\delta_j\}$ and $\gamma = \max_{j\in M}\{\gamma_j\}$. $\forall j \in M$, the existence of γ_j and δ_j can be disscused according to two cases:

A. $c_{2,j} = 0$

In this case, let $\delta_j = 0$ and γ_j can be any positive value.

B. $c_{2,j} > 0$

From the definition of $\mathbf{k}_j$ and $\mathbf{k}'_j$, it can be derived that any given γ, $\mathbf{k}_j$ and $\mathbf{k}'_j$ are finite constant matrices, i.e. irrelated with δ. Therefore, $\forall j \in M$, both $\mathbf{k}_j\mathbf{W}^{-1}\mathbf{k}_j^T$ and $\mathbf{k}'_j\mathbf{W}^{-1}\mathbf{k}'^T_j$ are constants. On the other hand, smaller δ_j will lead to smaller initial error which further results in smaller $\bar{\mathbf{e}}_c(0)$ and ρ. Therefore, as both $c_{1,j}$ and $c_{2,j}$ are strictly positive constants, a positive δ_j, which is small enough, can always be found such that (25) and (26) are valid.

Case 2. $\mathbf{c}_1$ have zero-elements

Assume $\mathbf{c}_{1,z} = 0$. δ and γ are designed off-line such that $\forall \bar{\mathbf{e}}_c \in \epsilon(\mathbf{W},\rho)$, $-\mathbf{k}_j\bar{\mathbf{e}}_c \geq -c_{1,j}$ ($\forall j \in \{1,\cdots,z-1,z+1,\cdots,m\}$) and $-\mathbf{k}'_j\bar{\mathbf{e}}_c \leq c_{2,j}$ ($\forall j \in M$). According to the result in **Case 1** and considering the strictly positiveness of $c_{1,j}$ ($\forall j \in \{1,\cdots,z-1,z+1,\cdots,m\}$), the existence of δ and γ can also be guaranteed. On the other hand, as $\mathbf{k}_z = \mathbf{0}$, $-\mathbf{k}_j\bar{\mathbf{e}}_c = 0$. Consequently, $-\mathbf{k}_z\bar{\mathbf{e}}_c \geq -c_{1,z}$ is always valid. Therefore, δ and γ can always be found such that $\forall \bar{\mathbf{e}}_c \in \epsilon(\mathbf{W},\rho)$ and $\forall j \in M$, $-\mathbf{k}_j\bar{\mathbf{e}}_c \geq -c_{1,j}$ and $-\mathbf{k}'_j\bar{\mathbf{e}}_c \leq c_{2,j}$. ■

5.3 Asymptotical convergence analysis

Theorem 1. Let $\langle \mathcal{N}, \boldsymbol{\lambda}, \mathbf{m}_0\rangle$ be a conservative and strongly connected JF contPN system. For any $\mathbf{m}_0 > \mathbf{0}$ and any reachable $\mathbf{m}_d > \mathbf{0}$ (*Assumption 1*) and $\mathbf{u}_d$, control law (14) with the parameters δ and γ designed in *Proposition 2* can ensure the global asymptotical convergence of both the system markings and the control signals.

Proof:

Case 1. $\mathbf{c}_1 > \mathbf{0}$

Assume $\bar{\mathbf{e}}_c \in \epsilon(\mathbf{W},\rho)$, where $\epsilon(\mathbf{W},\rho)$ is defined in *Proposition 2*. *Proposition 2* implies that δ and γ can always be found so that, $\forall \bar{\mathbf{e}}_c \in \epsilon(\mathbf{W},\rho)$ and $\forall j \in M$, $-\mathbf{k}_j\bar{\mathbf{e}}_c \geq -c_{1,j}$ and $-\mathbf{k}'_j\bar{\mathbf{e}}_c \leq c_{2,j}$. Define $V = \bar{\mathbf{e}}_c^T\mathbf{W}\bar{\mathbf{e}}_c$, where $\mathbf{W}$ is obtained from (21). Hence,

$$\dot{V} = \dot{\bar{\mathbf{e}}}_c^T\mathbf{W}\bar{\mathbf{e}}_c + \bar{\mathbf{e}}_c^T\mathbf{W}\dot{\bar{\mathbf{e}}}_c. \tag{27}$$

On the other hand, (19) can be rewritten as

$$\dot{\bar{\mathbf{e}}}_c = (\bar{\mathbf{A}}_c - \bar{\mathbf{B}}_c\mathbf{K})\bar{\mathbf{e}}_c + \bar{\mathbf{B}}_c\mathbf{v}, \tag{28}$$

where $\mathbf{v} = sat(\mathbf{u}_{lg}+\mathbf{u}_{hg})+\mathbf{K}\bar{\mathbf{e}}_c$. Substituting (28) into (27) and considering (21), we have

$$\begin{aligned}\dot{V} =& \bar{\mathbf{e}}_c^T(\mathbf{W}\bar{\mathbf{A}}_c + \bar{\mathbf{A}}_c^T\mathbf{W} - \frac{1}{\gamma}\mathbf{W}\bar{\mathbf{B}}_c\mathbf{R}^{-1}\bar{\mathbf{B}}_c^T\mathbf{W})\bar{\mathbf{e}}_c\\ &-\frac{1}{\gamma}\bar{\mathbf{e}}_c^T\mathbf{W}\bar{\mathbf{B}}_c\mathbf{R}^{-1}\bar{\mathbf{B}}_c^T\mathbf{W}\bar{\mathbf{e}}_c + 2\bar{\mathbf{e}}_c^T\mathbf{W}\bar{\mathbf{B}}_c\mathbf{v}\\ \leq& -\bar{\mathbf{e}}_c^T\mathbf{Q}\bar{\mathbf{e}}_c + 2\sum_{j=1}^{m}\bar{\mathbf{e}}_c^T\mathbf{W}\bar{\mathbf{b}}_{c,j}v_j\end{aligned} \tag{29}$$

where $v_j = sat(u_{lg,j}+u_{hg,j}) + \mathbf{k}_j\bar{\mathbf{e}}_c$ is the j-th element of $\mathbf{v}$. $\forall j \in M$, let us discuss $\bar{\mathbf{e}}_c^T\mathbf{W}\bar{\mathbf{b}}_{c,j}v_j$ in (29) according to the following three cases.

I. $-u_{r,i} < u_{lg,j} + u_{hg,j} < \phi_{j,i}m_i - u_{r,j}$

$u_{lg,j} + u_{hg,j}$ is not saturated. Hence,

$$\begin{aligned}\bar{\mathbf{e}}_c^T\mathbf{W}\bar{\mathbf{b}}_{c,j}v_j =& \bar{\mathbf{e}}_c^T\mathbf{W}\bar{\mathbf{b}}_{c,j}(-\mathbf{k}_j\bar{\mathbf{e}}_c - l\bar{\mathbf{b}}_{c,j}^T\mathbf{W}\bar{\mathbf{e}}_c + \mathbf{k}_j\bar{\mathbf{e}}_c)\\ =& -l({\bar{\mathbf{e}}_c}^T\mathbf{W}\bar{\mathbf{b}}_{c,j})^2 \leq 0.\end{aligned} \tag{30}$$

II. $u_{lg,j} + u_{hg,j} \leq -u_{r,j}$

$$\bar{\mathbf{e}}_c^T\mathbf{W}\bar{\mathbf{b}}_{c,j}v_j = \bar{\mathbf{e}}_c^T\mathbf{W}\bar{\mathbf{b}}_{c,j}(-u_{r,j} + \mathbf{k}_j\bar{\mathbf{e}}_c). \tag{31}$$

As $-\mathbf{k}_j\bar{\mathbf{e}}_c \geq -c_{1,j}$, considering the definitions of $c_{1,j}$ and $\mathbf{u}_r(\tau)$, we have

$$-\mathbf{k}_j\bar{\mathbf{e}}_c \geq -u_{r,j} \Rightarrow -u_{r,j} + \mathbf{k}_j\bar{\mathbf{e}}_c \leq 0. \tag{32}$$

On the other hand,

$$\begin{aligned}&u_{lg,j} + u_{hg,j} \leq -u_{r,j} \Rightarrow u_{hg,j} \leq -u_{r,j} - u_{lg,j}\\ &\Rightarrow -l\bar{\mathbf{b}}_{c,j}^T\mathbf{W}\bar{\mathbf{e}}_c \leq -u_{r,j} + \mathbf{k}_j\bar{\mathbf{e}}_c.\end{aligned} \tag{33}$$

From (32) and (33), it can be derived that $\bar{\mathbf{b}}_{c,j}^T\mathbf{W}\bar{\mathbf{e}}_c > 0$. Hence, from (31), $\bar{\mathbf{e}}_c^T\mathbf{W}\bar{\mathbf{b}}_{c,j}v_j < 0$.

III. $u_{lg,j} + u_{hg,j} \geq \phi_{j,i}m_i - u_{r,j}$

Similarly to the proof in *II*, $\bar{\mathbf{e}}_c^T\mathbf{W}\bar{\mathbf{b}}_{c,j}v_j < 0$ can be derived.

From *I*, *II* and *III*, $\dot{V} < -\bar{\mathbf{e}}_c^T\mathbf{Q}\bar{\mathbf{e}}_c$. Hence, $\epsilon(\mathbf{W},\rho)$ is a positively invariant region. As $\bar{\mathbf{e}}_c(0) \in \epsilon(\mathbf{W},\rho)$, $\bar{\mathbf{e}}_c(\tau) \in \epsilon(\mathbf{W},\rho)$ for all $\tau \geq 0$. Therefore, since $\dot{V} < -\bar{\mathbf{e}}_c^T\mathbf{Q}\bar{\mathbf{e}}_c$, $\bar{\mathbf{e}}_c$, $\bar{\mathbf{e}}$ and $\mathbf{e}$ asymptotically converge to zero. Furthermore, the convergence of $\bar{\mathbf{e}}_c$ leads to the convergence of $\mathbf{u}$ to $\mathbf{u}_r(\tau)$.

Case 2. $\mathbf{c}_1$ have zero-elements

The convergence analysis is quite similarly to that in **Case 1**. From (19), the error dynamics can be rewritten as follows:

$$\begin{aligned}\dot{\bar{\mathbf{e}}}_c =& [\bar{\mathbf{A}}_c - (\bar{\mathbf{B}}_c - \Delta\bar{\mathbf{B}}_c)\mathbf{K}]\bar{\mathbf{e}}_c + \bar{\mathbf{B}}_c[sat(\mathbf{u}_{lg} + \mathbf{u}_{hg})\\ &+\mathbf{K}\bar{\mathbf{e}}_c] - \Delta\bar{\mathbf{B}}_c\mathbf{K}\bar{\mathbf{e}}_c.\end{aligned} \tag{34}$$

As $\Delta\bar{\mathbf{B}}_c\mathbf{K} = \mathbf{0}$, (34) becomes

$$\dot{\bar{\mathbf{e}}}_c = [\bar{\mathbf{A}}_c - (\bar{\mathbf{B}}_c - \Delta\bar{\mathbf{B}}_c)\mathbf{K}]\bar{\mathbf{e}}_c + \bar{\mathbf{B}}_c\mathbf{v}. \tag{35}$$

Define the same Lyapunov function, considering the relationship (22), (29) can also be obtained here.

Since $\forall \bar{\mathbf{e}}_c \in \epsilon(\mathbf{W},\rho)$ and $\forall j \in M$, $-\mathbf{k}_j\bar{\mathbf{e}}_c \geq -c_{1,j}$ and $-\mathbf{k}'_j\bar{\mathbf{e}}_c \leq c_{2,j}$, analogously to **Case 1**, $\bar{\mathbf{e}}_c^T\mathbf{W}\bar{\mathbf{b}}_{c,j}v_j < 0$ ($\forall j \in M$) can be derived. Therefore, $\dot{V} < -\bar{\mathbf{e}}_c^T\mathbf{Q}\bar{\mathbf{e}}_c$. The global asymptotical convergence of $\mathbf{e}$ and $\mathbf{u}$ can be obtained. ■

6. ILLUSTRATIVE EXAMPLE

Consider the JF net in Figure 1 with $\boldsymbol{\lambda} = [1,1,2,1,2]^T$. Hence, $\boldsymbol{\Phi} = diag(1,1,1,1,1)$. The minimal P-semiflow is $\mathbf{y} = [1,1,1,2,2]^T$. Adding elementary vectors, $\mathbf{H}$ is chosen as $\mathbf{H} = \begin{bmatrix}1&1&1&2&2\\0&1&0&0&0\\0&0&1&0&0\\0&0&0&1&0\\0&0&0&0&1\end{bmatrix}$.

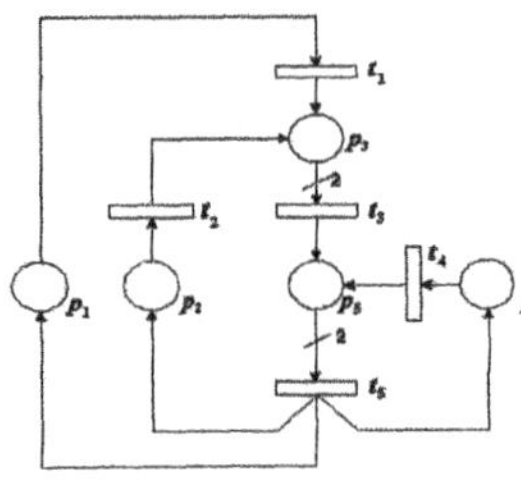

Fig. 1. Timed Join-Free Net System.

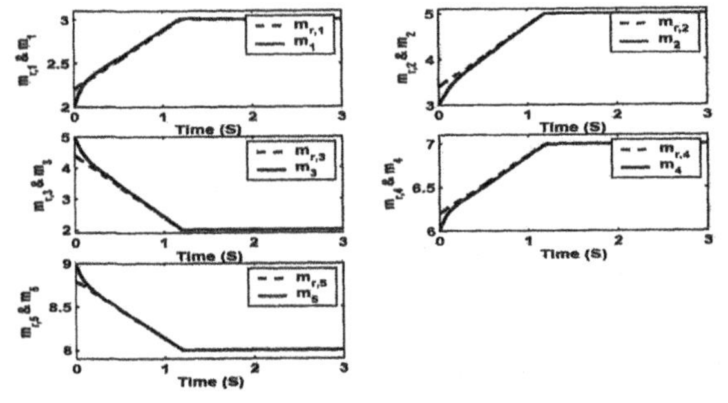

Fig. 2. Convergence of markings.

Assume an initial marking as $\mathbf{m}_0 = [2, 3, 5, 6, 9]^T$ and a desired marking $\mathbf{m}_d = [3, 5, 2, 7, 8]^T$. To maximize the flows of the steady state, the desired final control input $\mathbf{u}_d = [1, 3, 0, 5, 6]^T$.

Based on $\mathbf{m}_0$ and $\mathbf{m}_d$, $\boldsymbol{\sigma} = [1, 0, 2, 1, 2]^T + \alpha[1, 1, 1, 1, 1]^T$ where $\alpha \geq 0$. Here we randomly choose $\alpha = 1$, hence $\boldsymbol{\sigma} = [2, 1, 3, 2, 3]^T$. Considering $\boldsymbol{\sigma}$, $\mathbf{m}_0$ and $\mathbf{m}_d$, $\beta = 1.5$ is the minimum value such that $\frac{\boldsymbol{\sigma}}{\beta} \leq \boldsymbol{\Phi} \min\{\mathbf{m}_0, \mathbf{m}_d\}$. According to (12) and (13), it can be derived that $\mathbf{u}_{r0} = [0.6667, 2.3333, 3, 4.6667, 7]^T$ and $\mathbf{u}_{rh^-} = [1.6667, 4.3333, 0, 5.6667, 6]^T$. Hence, $\mathbf{c}_1 = [0.6667, 2.3333, 0, 4.6667, 6]^T$. As $\mathbf{c}_1$ has one zero-element, $\mathbf{u}_{lg}$ is calculated based on (22). Let $\delta = 0.2$ and $\gamma = 5$. It is easy to verify that $\forall j \in \{1, 2, 3, 4, 5\}$, both (25) and (26) are valid. Then, $h = \beta(1 - \delta) = 1.2$. For simplicity, choose $\mathbf{Q} = \mathbf{I}_{4\times4}$ and $\mathbf{R} = \mathbf{I}_{5\times5}$ for the low-gain design. For the high-gain term, any positive l can guarantee the tracking convergence. Generally, smaller l will lead to slower system response. However, due to the existence of input saturation, when l is sufficiently large, the system responses have little difference by further increasing l. In our simulation, we choose $l = 10$.

The simulation results are shown in Figures 2 and 3. Figure 2 illustrates the convergence of the markings under the designed control law. Figure 3 shows the control signals $\mathbf{u}$ and the state-related upper bound, i.e. $\phi_{j,i} m_i$. (Note the net structure determines that $\forall j \in \{1, 2, 3, 4, 5\}$, $i = j$.) It can be seen that $0 \leq u_j \leq \phi_{j,i} m_i$ and the final control signals converge to the desired ones.

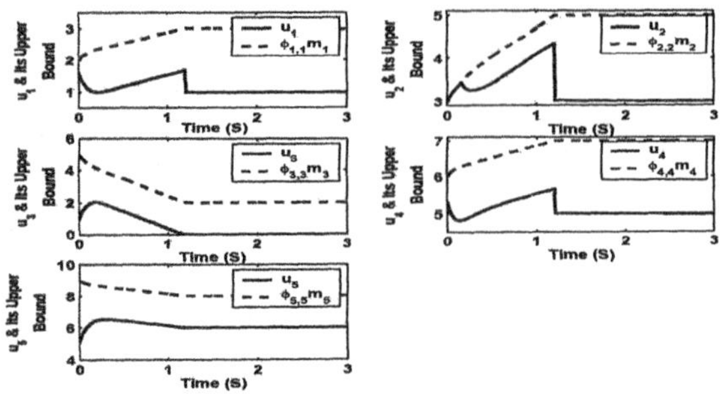

Fig. 3. Control signals.

7. CONCLUSION

The main concern of our work is to construct proper control laws for step-tracking of timed contPN systems in presence of the existing state-related input constraints. To guarantee global convergence and smoothness of states and control signals, the design method for a step-ramp tracking trajectory has been outlined. With the new tracking target, a novel low-and-high gain control method has been further proposed.

REFERENCES

Jiménez, E., J. Júlvez, L. Recalde and M. Silva (2005). On controllability of timed continuous Petri net systems: the join free case. In: *Proceedings of the 44th IEEE CDC and ECC 2005.* Seville, Spain.

Lin, Z., M. Pachter and S. Banda (1998). Toward improvement of tracking performance - nonlinear feedback for linear systems. *International Journal of Control* **70**(1), 1–11.

Mahulea, C., A. Ramírez, L. Recalde and M. Silva (2005). Steady state control, zero valued poles and token conservation laws in continuous net systems. In: *Proceedings of the International Workshop on Control of Hybrid and Discrete Event Systems.* Miami, USA.

Recalde, L., E. Teruel and M. Silva (1999). Autonomous continuous P/T systems. In: *Application and Theory of Petri Nets 1999* (S. Donatelli and J. Kleijn, Eds.). Vol. 1639 of *Lecture Notes in Computer Science.* Springer. pp. 107–126.

Saberi, A., Z. Lin and A. R. Teel (1996). Control of linear system with saturating actuators. *IEEE Transactions on Automatic Control* **41**(3), 368–378.

Teruel, E., J. M. Colom and M. Silva (1997). Choice-free Petri nets: A model for deterministic concurrent systems with bulk services and arrivals. *IEEE Transactions on Systems, Man, and Cybernetics* **27**(1), 73–83.

Wredenhage, G. F. and P. R. Bélanger (1994). Piecewise-linear LQ control for systems with input constraints. *Automatica* **30**(3), 403–416.

ELSEVIER
IFAC
PUBLICATIONS

ON SAMPLING CONTINUOUS TIMED PETRI NETS: REACHABILITY "EQUIVALENCE" UNDER INFINITE SERVERS SEMANTICS

Cristian Mahulea, * **Alessandro Giua,** **
Laura Recalde, * **Carla Seatzu,** ** **Manuel Silva.** *

* *Dep. de Informática e Ingeniería de Sistemas, Universidad de Zaragoza, Spain, {cmahulea,lrecalde,silva}@unizar.es*
** *Dip. di Ingegneria Elettrica ed Elettronica, Università di Cagliari, Italy, {giua,seatzu}@diee.unica.it*

Abstract: This paper addresses a sampling problem for timed continuous Petri nets under infinite servers semantics. Different representations of the continuous Petri net system are given, the first two in terms of piecewise linear system and the third one, for the controlled continuous Petri nets systems, in terms of a particular linear constrained system with null dynamic matrix. The last one is used to obtain the discrete-time representation. An upper bound on sample period is given in order to preserve important information of timed continuous nets, in particular the positiveness of the markings. The reachability space of the sampled system in relation to autonomous continuous Petri nets is also studied.

Keywords: Petri-nets, Continuous-time systems, Discrete-time systems.

1. INTRODUCTION

Discrete Petri nets (PNs) (Silva, 1993) are a mathematical formalism with an appealing graphical representation for the description of discrete-event systems, successfully used for modeling, analysis and synthesis of such systems. To study performance evaluation, timing should be introduced and timed PNs are obtained.

Discrete PNs may suffer the state explosion problem, when the number of tokens is large. As in the case of other formalisms (e.g. integer programming or queuing networks), continuous relaxation can provide a good approximation for discrete models under certain circumstances (Silva and Recalde, 2002).

Continuous Petri nets (contPNs) are a formalism in which the marking of each place is a non-negative real number (David and Alla, 2004) (Silva and Recalde, 2002). As in discrete case, timing can be associated to transitions resulting in timed contPNs. Controllers and observers can be designed for this class of systems but taking into account that probably they need to be implemented on some computer, the sampling of the continuous system is required.

For finite servers semantics, sampling is not a hard problem because the flow of the transitions is constant inside each invariant behavior (IB) state (David and Alla, 2004), and the times at which IB state changes occur can be computed. This

This work was partially supported by project CICYT and FEDER DPI2003-06376, in which Alessandro Giua and Carla Seatzu are also participating.

allows to tackle the problem as an event-driven control (Júlvez *et al.*, 2004). However, it seems that infinite servers semantics usually provides a much better approximation of the discrete system than finite servers semantics (Mahulea *et al.*, 2006). Under infinite servers semantics, there is not an easy way to compute the equivalent to these IB states, so sampling is an important issue.

In classical *Systems and Signal Theory*, it is well-known that the *Sampling theorem* (frequently known as the Nyquist-Shannon sampling theorem) provides an upper bound for the sampling period of limited bandwidth signals in order "not to loose information". Here, it is shown that sampling at "too low rate", spurious solutions can appear, in particular negative markings. In this paper, for timed contPNs, an upper bound on the sampling period is given in order to avoid spurious solutions. In other words, for the sampled timed contPNs, some "equivalence results" regarding the reachability space of sampled timed contPNs and (autonomous) contPNs are presented.

2. CONTINUOUS PETRI NETS

***Definition** 2.1.* A contPN system is a pair $\langle \mathcal{N}, \boldsymbol{m_0} \rangle$, where: (1) $\mathcal{N} = \langle P, T, \boldsymbol{Pre}, \boldsymbol{Post} \rangle$ is the net structure with set of places P, set of transitions T, pre and post incidence matrices $\boldsymbol{Pre}, \boldsymbol{Post} : P \times T \to \mathbb{N}$; and (2) $\boldsymbol{m_0} : P \to \mathbb{R}_{\geq 0}$ is the initial marking (or distributed state).

The number of places of a net is $n = |P|$ and the number of transitions is $m = |T|$. We also denote $\boldsymbol{m}(\tau)$ the marking at time τ and in discrete time we denote $\boldsymbol{m}(k)$ the marking at sampling instant k ($\tau = k \cdot \Theta$, where Θ is the sampling period). The token load contained in place p_i at marking $\boldsymbol{m}$ is denoted m_i. Finally, *preset* and *postset* of a node $X \in P \cup T$ are denoted ${}^\bullet X$ and $X^\bullet$, respectively.

A transition $t_j \in T$ is *enabled* at $\boldsymbol{m}$ iff $\forall p_i \in {}^\bullet t_j, m_i > 0$, and its *enabling degree* is

$$enab(t_j, \boldsymbol{m}) = \min_{p_i \in {}^\bullet t_j} \left\{ \frac{m_i}{Pre(p_i, t_j)} \right\}.$$

An enabled transition t can fire in any real amount $0 \leq \alpha \leq enab(t, \boldsymbol{m})$ leading to a new marking $\boldsymbol{m'} = \boldsymbol{m} + \alpha \boldsymbol{C}(\cdot, t)$, where $\boldsymbol{C} = \boldsymbol{Post} - \boldsymbol{Pre}$ is the *incidence matrix*; this firing is also denoted $\boldsymbol{m}[t(\alpha)\rangle \boldsymbol{m'}$.

In general, if $\boldsymbol{m}$ is reachable from $\boldsymbol{m_0}$ through a sequence $\sigma = t_{r_1}(\alpha_1) t_{r_2}(\alpha_2) \ldots t_{r_k}(\alpha_k)$, and we denote by $\boldsymbol{\sigma} : T \to \mathbb{R}_{\geq 0}$ the *firing count vector* whose component associated to a transition t_j is

$$\sigma_j = \sum_{h \in H(\sigma, t_j)} \alpha_h,$$

where $\quad H(\sigma, t_j) = \{h = 1, \ldots, k \mid t_{r_h} = t_j\}$,

then we can write: $\boldsymbol{m} = \boldsymbol{m_0} + \boldsymbol{C} \cdot \boldsymbol{\sigma}$, which is called the *fundamental equation*.

The basic difference between discrete and continuous PN is that the components of the markings and firing count vectors are not restricted to take value in the set of natural numbers but in the non-negative reals. The set of markings that are reachable with a finite firing sequence for a given system $\langle \mathcal{N}, \boldsymbol{m_0} \rangle$ is denoted as $RS^{un}(\mathcal{N}, \boldsymbol{m_0})$.

***Definition** 2.2.* Let $\langle \mathcal{N}, \boldsymbol{m_0} \rangle$ be a contPN system and $RS^{un}(\mathcal{N}, \boldsymbol{m_0})$ the set of *reachable markings*, i.e., the set of markings $\boldsymbol{m} \in \mathbb{R}^m_{\geq 0}$ such that a finite fireable sequence $\boldsymbol{\sigma} = t_{a_1}(\alpha_1) \cdots t_{a_k}(\alpha_k)$ exists, and $\boldsymbol{m_0} \xrightarrow{t_{a_1}(\alpha_1)} \boldsymbol{m_1} \xrightarrow{t_{a_2}(\alpha_2)} \boldsymbol{m_2} \cdots \xrightarrow{t_{a_k}(\alpha_k)} \boldsymbol{m_k} = \boldsymbol{m}$, where $t_{a_i} \in T$ and $\alpha_i \in \mathbb{R}^+$.

A relaxation of this space can be considered allowing an infinite firing sequence and lim-reachable space is obtained:

***Definition** 2.3.* Let $\langle \mathcal{N}, \boldsymbol{m_0} \rangle$ be a continuous system. A marking $\boldsymbol{m}$ is *lim-reachable* iff a sequence of reachable markings $\{\boldsymbol{m_i}\}_{i \geq 1}$ exists such that $\boldsymbol{m_0} \xrightarrow{\sigma_1} \boldsymbol{m_1} \xrightarrow{\sigma_2} \boldsymbol{m_2} \cdots \xrightarrow{\sigma_i} \boldsymbol{m_i} \cdots$ and $\lim_{i \to \infty} \boldsymbol{m_i} = \boldsymbol{m}$. The set of lim-reachable markings is denoted as $lim - RS^{un}(\mathcal{N}, \boldsymbol{m_0})$.

***Definition** 2.4.* A (deterministically) timed contPN system $\langle \mathcal{N}, \boldsymbol{\lambda}, \boldsymbol{m_0} \rangle$ is a contPN system $\langle \mathcal{N}, \boldsymbol{m_0} \rangle$ together with a vector $\boldsymbol{\lambda} : T \to \mathbb{R}_{>0}$, where λ_j is the firing rate of transition t_j.

Now, the fundamental equation depends on time: $\boldsymbol{m}(\tau) = \boldsymbol{m_0} + \boldsymbol{C} \cdot \boldsymbol{\sigma}(\tau)$, where $\boldsymbol{\sigma}(\tau)$ denotes the firing count vector in the interval $[0, \tau]$. Deriving it with respect to time the following is obtained: $\dot{\boldsymbol{m}}(\tau) = \boldsymbol{C} \cdot \dot{\boldsymbol{\sigma}}(\tau)$. The derivative of firing vector represents the *flow* of the timed model $\boldsymbol{f}(\tau) = \dot{\boldsymbol{\sigma}}(\tau)$. Depending on how the flow of the transition is defined many firing semantics are possible; the most used in literature are *finite servers semantics* (or *constant speed*) and *infinite server semantics* (or *variable speed*) (Recalde and Silva, 2001) (David and Alla, 2004).

This paper deals with *infinite server semantics* in which the flow of transition t_j is given by:

$$f_j = \lambda_j \min_{p_i \in {}^\bullet t_j} \left\{ \frac{m_i}{Pre(p_i, t_j)} \right\} \tag{1}$$

***Example** 2.5.* Let us consider the net system in Fig. 1. The flow of transitions are:

$$\begin{cases} f_1 = \lambda_1 \cdot \min \left\{ \dfrac{m_1}{2}, m_3 \right\} \\ f_2 = \lambda_2 \cdot \min \{m_1, m_2\} \end{cases}.$$

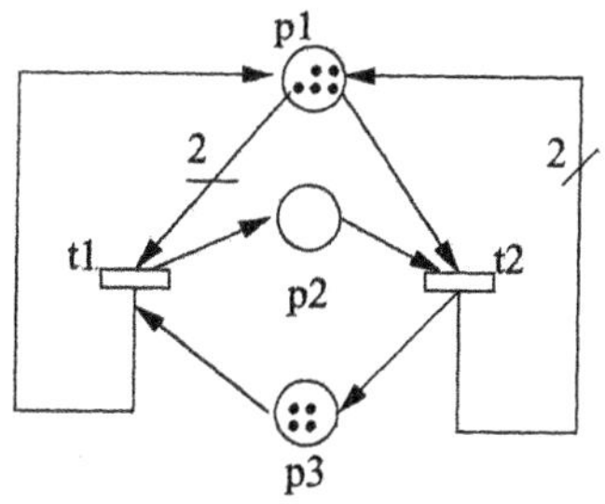

Fig. 1. Continuous PN system.

Thus, the state space representation of this unforced system ($\dot{m}(\tau) = C \cdot f(\tau)$) is:

$$\begin{cases} \dot{m}_1 = -\lambda_1 \cdot \min\left\{\frac{m_1}{2}, m_3\right\} + \lambda_2 \cdot \min\{m_1, m_2\} \\ \dot{m}_2 = \lambda_1 \cdot \min\left\{\frac{m_1}{2}, m_3\right\} - \lambda_2 \cdot \min\{m_1, m_2\} \\ \dot{m}_3 = -\lambda_1 \cdot \min\left\{\frac{m_1}{2}, m_3\right\} + \lambda_2 \cdot \min\{m_1, m_2\} \end{cases} \quad (2)$$

■

Because the flow of a transition depends on its enabling degree which is based on the minimum function, a timed contPN under infinite servers semantics is a piecewise linear system. In fact, if we define

$$s = \prod_{t \in T} |{}^\bullet t|,$$

the state space of a timed contPN can be *partitioned*[1] as follows: $R_1 \cup \cdots \cup R_s$, where each set R_k (for $k = 1, \ldots, s$) denotes a *region* (eventually empty) where the flow is limited by the same subset of places (one for each transition). For a given region R_k, we can define the *constraint matrix* $\mathbf{\Pi}_k : T \times P \to \mathbb{R}$ such that:

$$\mathbf{\Pi}_k(t_j, p_i) = \begin{cases} \dfrac{1}{Pre(p_i, t_j)}, \text{ if } (\forall \boldsymbol{m} \in R_k) \\ \dfrac{m_i}{Pre(p_i, t_j)} = \min\limits_{p_h \in {}^\bullet t_j} \left\{ \dfrac{m_h}{Pre(p_h, t_j)} \right\}; \\ 0, \text{ otherwise.} \end{cases} \quad (3)$$

Example 2.6. For the system sketched in Fig. 1, the flow of t_1 can be restricted by the marking of p_1 or p_3 and the flow of t_2 can be restricted by the marking of p_1 or p_2. The number of regions in this case is $s = 4$ and they are defined as follows:

- R_1: $\frac{m_1}{2} \leq m_3$ and $m_1 \leq m_2$ with $\mathbf{\Pi}_1 = \begin{bmatrix} \frac{1}{2} & 0 & 0 \\ 1 & 0 & 0 \end{bmatrix}$
- R_2: $\frac{m_1}{2} \leq m_3$ and $m_1 \geq m_2$ with $\mathbf{\Pi}_2 = \begin{bmatrix} \frac{1}{2} & 0 & 0 \\ 0 & 1 & 0 \end{bmatrix}$
- R_3: $\frac{m_1}{2} \geq m_3$ and $m_1 \leq m_2$ with $\mathbf{\Pi}_3 = \begin{bmatrix} 0 & 0 & 1 \\ 1 & 0 & 0 \end{bmatrix}$
- R_4: $\frac{m_1}{2} \geq m_3$ and $m_1 \geq m_2$ with $\mathbf{\Pi}_4 = \begin{bmatrix} 0 & 0 & 1 \\ 0 & 1 & 0 \end{bmatrix}$

■

[1] These partitions are disjoint except possibly on the borders.

If marking $\boldsymbol{m}$ belongs to R_k, we denote $\mathbf{\Pi}(\boldsymbol{m}) = \mathbf{\Pi}_k$ the corresponding constraint matrix. Furthermore, the firing rate of transitions can also be represented by a diagonal matrix $\mathbf{\Lambda} : T \times T \to \mathbb{R}_{>0}$, where

$$\Lambda(t_j, t_h) = \begin{cases} \lambda_j \text{ if } j = h \\ 0, \text{ otherwise} \end{cases}$$

Using this notation, the non-linear flow of the transitions at a given marking $\boldsymbol{m}$ (see eq. (1) for f_j) can be written as:

$$\boldsymbol{f} = \mathbf{\Lambda} \cdot \mathbf{\Pi}(\boldsymbol{m}) \cdot \boldsymbol{m} \quad (4)$$

We now consider net systems subject to external control actions, and assume that the only admissible control law consists in *slowing down* the firing speed of transitions (Silva and Recalde, 2004).

Definition 2.7. The flow of the forced (or controlled) timed contPN is denoted as $\boldsymbol{w}(\tau) = \boldsymbol{f}(\tau) - \boldsymbol{u}(\tau)$, with $0 \leq \boldsymbol{u}(\tau) \leq \boldsymbol{f}(\tau)$, $\boldsymbol{u}(\tau)$ represents the control input.

Therefore, the control input will be dynamically upper bounded by the flow of the corresponding unforced system. Under these conditions, the overall behavior of the system is ruled by the following system (Mahulea *et al.*, 2005):

$$\begin{cases} \dot{\boldsymbol{m}}(\tau) = \boldsymbol{C} \cdot [\mathbf{\Lambda} \cdot \mathbf{\Pi}(\boldsymbol{m}(\tau)) \cdot \boldsymbol{m}(\tau) - \boldsymbol{u}(\tau)] \\ 0 \leq \boldsymbol{u}(\tau) \leq \mathbf{\Lambda} \cdot \mathbf{\Pi}(\boldsymbol{m}(\tau)) \cdot \boldsymbol{m}(\tau) \end{cases} \quad (5)$$

This is a particular hybrid system: a piecewise linear system with autonomous switches and dynamic (or state-based) constraints in the input.

Example 2.8. Let us consider the net system in Fig. 1 with $\boldsymbol{\lambda} = [5, 1]^T$. It is ruled by the following set of systems of the form (5):

$$\boldsymbol{m} \in R_1 : \begin{cases} \dot{\boldsymbol{m}}(\tau) = \begin{bmatrix} -\frac{3}{2} & 0 & 0 \\ \frac{3}{2} & 0 & 0 \\ -\frac{3}{2} & 0 & 0 \end{bmatrix} \boldsymbol{m}(\tau) - \begin{bmatrix} -1 & 1 \\ 1 & -1 \\ -1 & 1 \end{bmatrix} \boldsymbol{u}(\tau) \\ 0 \leq \boldsymbol{u}(\tau) \leq \begin{bmatrix} \frac{5}{2} & 0 & 0 \\ 1 & 0 & 0 \end{bmatrix} \boldsymbol{m}(\tau) \end{cases}$$

$$\boldsymbol{m} \in R_2 : \begin{cases} \dot{\boldsymbol{m}}(\tau) = \begin{bmatrix} -\frac{5}{2} & 1 & 0 \\ \frac{5}{2} & -1 & 0 \\ -\frac{5}{2} & 1 & 0 \end{bmatrix} \boldsymbol{m}(\tau) - \begin{bmatrix} -1 & 1 \\ 1 & -1 \\ -1 & 1 \end{bmatrix} \boldsymbol{u}(\tau) \\ 0 \leq \boldsymbol{u}(\tau) \leq \begin{bmatrix} \frac{5}{2} & 0 & 0 \\ 0 & 1 & 0 \end{bmatrix} \boldsymbol{m}(\tau) \end{cases}$$

$$\boldsymbol{m} \in R_3 : \begin{cases} \dot{\boldsymbol{m}}(\tau) = \begin{bmatrix} 1 & 0 & -5 \\ -1 & 0 & 5 \\ 1 & 0 & -5 \end{bmatrix} \boldsymbol{m}(\tau) - \begin{bmatrix} -1 & 1 \\ 1 & -1 \\ -1 & 1 \end{bmatrix} \boldsymbol{u}(\tau) \\ 0 \leq \boldsymbol{u}(\tau) \leq \begin{bmatrix} 0 & 0 & 5 \\ 1 & 0 & 0 \end{bmatrix} \boldsymbol{m}(\tau) \end{cases}$$

$$
\boldsymbol{m} \in R_4 : \begin{cases} \dot{\boldsymbol{m}}(\tau) = \begin{bmatrix} 0 & 1 & -5 \\ 0 & -1 & 5 \\ 0 & 1 & -5 \end{bmatrix} \boldsymbol{m}(\tau) - \begin{bmatrix} -1 & 1 \\ 1 & -1 \\ -1 & 1 \end{bmatrix} \boldsymbol{u}(\tau) \\ 0 \leq \boldsymbol{u}(\tau) \leq \begin{bmatrix} 0 & 0 & 5 \\ 0 & 1 & 0 \end{bmatrix} \boldsymbol{m}(\tau) \end{cases}
$$

■

As a final remark, it should be noted that in this paper we assume that all transitions are controllable [2], i.e., may be slowed down. It may also be possible to extend the approach to deal with uncontrollability of certain transitions. If transition t_j cannot be controlled, then it is obvious that the control input must be $u_j = 0$ at every time instant.

3. A CONSTRAINED LINEAR REPRESENTATION OF CONTINUOUS PN

The system in the eq. (5) is a *piecewise linear* system with a dynamical constraint on the control input $\boldsymbol{u}$ that depends on the current value of the system state $\boldsymbol{m}$. In this section we provide an alternative expression that takes the form of a simple *linear* system with dynamical constraints on the control input.

Proposition *3.1.* Any piecewise linear constrained model of the form (5) can be rewritten, by suitably defining a matrix $\boldsymbol{G}$, as a linear constrained model of the form:

$$
\begin{cases} \dot{\boldsymbol{m}}(\tau) = \boldsymbol{C} \cdot \boldsymbol{w}(\tau) \\ \boldsymbol{G} \cdot \begin{bmatrix} \boldsymbol{w}(\tau) \\ \boldsymbol{m}(\tau) \end{bmatrix} \leq \boldsymbol{0} \\ \boldsymbol{w}(\tau) \geq \boldsymbol{0} \end{cases} \tag{6}
$$

that we call *continuous time controlled contPN* model, or *ct-contPN* model for short. The initial value of the state system is $\boldsymbol{m}(0) = \boldsymbol{m}_0 \geq \boldsymbol{0}$.

Proof: The equivalence of the dynamic equations immediately follows by replacing $\boldsymbol{w}(\tau) = \boldsymbol{f}(\tau) - \boldsymbol{u}(\tau)$ in (6) being $\boldsymbol{f}(\tau)$ defined as in (4).

Concerning the constraints on the input, we first observe that, by virtute of (4), constraints in (5) can be rewritten as $\boldsymbol{0} \leq \boldsymbol{w}(\tau) \leq \boldsymbol{f}(\tau)$, i.e., $\forall j = 1, \cdots, n$, and at any marking $\boldsymbol{m}$,

$$
0 \leq w_j \leq \lambda_j \min_{p_i \in {}^\bullet t_j} \left(\frac{m_i}{Pre(p_i, t_j)} \right)
$$

that is equivalent to the following set of equations

$$
0 \leq w_j \leq \lambda_j \frac{m_i}{Pre(p_i, t_j)} \qquad (\forall p_i \in {}^\bullet t_j).
$$

[2] We use "controllable" in the supervisory control sense. In (Mahulea *et al.*, 2005) the concept is referred as *control-feasible.*

All these equations can be combined as

$$
\boldsymbol{0} \leq \boldsymbol{Q} \cdot \boldsymbol{w} \leq \boldsymbol{R} \cdot \boldsymbol{m}
$$

where matrices $\boldsymbol{Q}$ $(q \times n)$ and $\boldsymbol{R}$ $(q \times m)$ have as many rows as there are "pre" arcs in the net, i.e., $q = \sum_{t \in T} |{}^\bullet t|$.

In particular, given a pre arc (p_i, t_j) the corresponding row of $\boldsymbol{Q}$ is the vector

$$
\Big[\underbrace{0 \; \cdots \; 0 \; 1}_{j} \; 0 \; \cdots \; 0 \Big],
$$

while corresponding row of $\boldsymbol{R}$ is the vector

$$
\Big[\underbrace{0 \; \cdots \; 0 \; \frac{\lambda_j}{Pre(p_i, t_j)}}_{i} \; 0 \; \cdots \; 0 \Big].
$$

If we let $\boldsymbol{G} = \begin{bmatrix} \boldsymbol{Q} & -\boldsymbol{R} \end{bmatrix}$ we obtain the constraints in the last two equations of (6). □

The system in eq. (6) is a linear system with a *dynamic-matrix* equal to $\boldsymbol{0}$ and an *input matrix* equal to the *token flow matrix* of the contPN. Note however, that there is still a dynamical constraint on the system inputs that depends on the value of the system state $\boldsymbol{m}$.

4. ON SAMPLED (OR DISCRETE-TIME) CONTINUOUS PETRI NETS MODELS

Let us obtain a discrete-time representation of continuous-time continuous Petri net under infinite servers semantics. Sampling should preserve the important information of the original model (for example the positiveness of the markings). This is studied in the next section through the equivalence of the reachability graph of the discrete-time model and the untimed model (not the reachability graph of discrete-time with continuous time). In this section the discretization is defined together with a bound for the sampling period.

The system given by the eq. (6) represents a continuous-time system and can be discretized. A first order discretization method is used here and we are proving that under some conditions, it ensures the reachability equivalence.

Definition *4.1.* Consider a ct-contPN as in eq. (6) and let Θ be a sampling period ($\tau = k \cdot \Theta$). The *discrete-time controlled contPN* or *dt-contPN* $\langle \mathcal{N}, \lambda, \boldsymbol{m}_0, \theta \rangle$ can be written as follows:

$$
\begin{cases} \boldsymbol{m}(k+1) = \boldsymbol{m}(k) + \Theta \cdot \boldsymbol{C} \cdot \boldsymbol{w}(k) \\ \boldsymbol{G} \cdot \begin{bmatrix} \boldsymbol{w}(k) \\ \boldsymbol{m}(k) \end{bmatrix} \leq \boldsymbol{0} \\ \boldsymbol{w}(k) \geq \boldsymbol{0} \end{cases} \tag{7}
$$

The initial value of the state of this system is $\boldsymbol{m}(0) = \boldsymbol{m}_0 \geq \boldsymbol{0}$.

The reachability space of dt-contPN can be defined as follows.

Definition 4.2. We denote $RS^{dt}(\mathcal{N}, \boldsymbol{m_0}, \Theta)$ the set of markings $\boldsymbol{m} \in \mathbb{R}_{\geq 0}$ such that there exists a finite input sequence $\boldsymbol{w} = \boldsymbol{w_1} \cdots \boldsymbol{w_k}$ and $\boldsymbol{m}(0) \xrightarrow{\boldsymbol{w_1}} \boldsymbol{m}(1) \xrightarrow{\boldsymbol{w_2}} \boldsymbol{m}(2) \cdots \xrightarrow{\boldsymbol{w_k}} \boldsymbol{m}(k) = \boldsymbol{m}$, where $\boldsymbol{0} \leq \boldsymbol{w}(k) \leq \boldsymbol{f}(k)$ $\forall k$, and $\boldsymbol{f}(k)$ is the flow of the unforced system at time $k \cdot \Theta$.

Example 4.3. Let us consider the net system in Fig. 1 with $\Theta = 1$, $\boldsymbol{\lambda} = [5, 1]^T$. Then the discrete-time representation is given by:

$$\begin{cases} \boldsymbol{m}(k+1) = \boldsymbol{m}(k) + \boldsymbol{C}\boldsymbol{w}(k) \\ w_1(k) - \frac{\lambda_1}{2} \cdot m_1(k) \leq 0 \\ w_1(k) - \lambda_1 \cdot m_3(k) \leq 0 \\ w_2(k) - \lambda_2 \cdot m_1(k) \leq 0 \\ w_2(k) - \lambda_2 \cdot m_2(k) \leq 0 \\ \boldsymbol{w}(k), \boldsymbol{m}(k+1) \geq \boldsymbol{0} \end{cases} \tag{8}$$

and

$$\boldsymbol{G} = \begin{bmatrix} 1 & 0 & -\frac{5}{2} & 0 & 0 \\ 1 & 0 & 0 & 0 & -5 \\ 0 & 1 & -1 & 0 & 0 \\ 0 & 1 & 0 & -1 & 0 \end{bmatrix} \tag{9}$$

It is important to stress that, although the evolution of a sampled contPN occurs in discrete steps, *discrete time evolutions* and *untimed evolutions* are not the same. As an example, while an untimed net can be seen evolving sequentially, executing a single transition firing at each step (because they are executed at the same time instant), a dt-contPN may evolve in concurrent steps where more than one transition fires. We denote such a concurrent step as follows:

$$\boldsymbol{m}[\{t_{i_1}(\alpha_1), t_{i_2}(\alpha_2), \ldots, t_{i_k}(\alpha_k)\}\rangle \boldsymbol{m}'.$$

In unforced ct-contPN under infinite servers semantics, the positiveness of the marking is ensured if the initial marking $\boldsymbol{m_0}$ is positive, because the flow of a transition goes to zero whenever one of the input places is empty (Silva and Recalde, 2004).

In a dt-contPN, this is not always true. Let us consider the net in Fig. 1, with $\boldsymbol{m_0} = [1.1, 3.9, 0.1]^T$, $\boldsymbol{\lambda} = [5, 1]^T$, $\Theta = 0.5$. Assume transition t_2 is stopped ($w_2(0) = 0$), then $m_3(1) = m_3(0) - \Theta \cdot w_1(0) = 0.1 - 0.5 \cdot w_1(0)$. But $w_1(0)$ is upper bounded by $\lambda_1 \cdot m_3(0) = 5 \cdot 0.1 = 0.5$. If the maximum value is chosen, then $m_3(1)$ will be negative!!!

This can be avoided if the sampling period is small enough. Let Θ be a sampling period such that for all $p \in P$ it holds that:

$$\sum_{t_j \in p^\bullet} \lambda_j \Theta < 1 \tag{10}$$

Proposition 4.4. Let $\langle \mathcal{N}, \boldsymbol{\lambda}, \boldsymbol{m_0}, \Theta \rangle$ be a dt-contPN system with $\boldsymbol{m_0} \geq \boldsymbol{0}$ and Θ verifying (10).

(1) Any marking reachable from $\boldsymbol{m_0}$ is non negative, i.e., $RS^{dt}(\mathcal{N}, \boldsymbol{m_0}, \Theta) \subseteq \mathbb{R}^m_{\geq 0}$.
(2) A place cannot be emptied with a finite sequence of firings, i.e., if $m_0(p) > 0$, then for all $\boldsymbol{m} \in RS^{dt}(\mathcal{N}, \boldsymbol{m_0}, \Theta)$ it also holds $m(p) > 0$.

Proof: Let us consider a place p_i with $p_i^\bullet = \{t_1, t_2, \cdots, t_j\}$ and $m_i(k) > 0$. Then: $m_i(k+1) = m_i(k) + \Theta C(i,:)\boldsymbol{w}(k) \geq m_i(k) - \Theta(\lambda_1 + \lambda_2 + \cdots + \lambda_j)m_i(k) = m_i(k)\left(1 - \sum_{t_j \in p^\bullet} \lambda_j \Theta\right) > 0$ □

In the rest of the paper we will assume that all nets are sampled with a sampling period Θ that satisfies (10).

Corollary 4.5. If a marking $\boldsymbol{m}$ is reachable in a dt-contPN system $\langle \mathcal{N}, \boldsymbol{\lambda}, \boldsymbol{m_0}, \Theta \rangle$ with Θ verifying (10) then is reachable in the underlying untimed contPN system $\langle \mathcal{N}, \boldsymbol{m_0} \rangle$ (i.e. $RS^{dt}(\mathcal{N}, \boldsymbol{m_0}, \Theta) \subseteq RS^{un}(\mathcal{N}, \boldsymbol{m_0})$).

In general the converse of Corollary 4.5 is not true: in fact, the second item of Proposition 4.4 shows that in a dt-contPN with Θ satisfying (10) it is never possible to empty a place (only at the limit, thus timed contPN can be deadlocked only at the limit), while this may be possible in an untimed net system. As an example, in the untimed net system in Fig. 1 from the marking shown it is possible to fire $t_1(2)t_1(0.5)$, thus emptying place p_1. This marking is clearly not reachable on the same net system if we associate to it a firing rate vector and choose a sampling period Θ satisfying (10).

In the next section, two relaxations can be done: (1) considering in the untimed case only those sequences that never empty a marked place or (2) allowing the lim-reachable markings of the discrete-timed model. These relaxations are the same as in continuous-time case (Mahulea *et al.*, 2005). So, in fact we will prove that under these relaxations and with the sampling period as in (10), the reachability space of the discrete-time model will be the same with reachability space of the continuous-time model.

5. REACHABILITY "EQUIVALENCE" BETWEEN SAMPLED AND CONTINUOUS MODELS

The condition (10) can be seen like a "kind of Sampling Theorem" for sampling linear-invariant systems: Θ should be small enough to maintain some properties as that in Proposition 4.4. But it does not mean that all information is preserved by sampling. The following result characterizes the reachability set of dt-contPN.

***Lemma** 5.1.* Let $\langle \mathcal{N}, \boldsymbol{\lambda}, \boldsymbol{m_0}, \Theta \rangle$ be a dt-contPN system and assume that in the underlying untimed net system it is possible from $\boldsymbol{m}$ to fire the sequence $\boldsymbol{m}[t_j(\alpha)\rangle \boldsymbol{m}'$ and that for a certain $a > 1$, for all $p \in {}^\bullet t_j$ it holds $m'(p) \geq m(p)/a$.

Then in $\langle \mathcal{N}, \boldsymbol{\lambda}, \boldsymbol{m_0}, \Theta \rangle$ marking $\boldsymbol{m}'$ is reachable from marking $\boldsymbol{m}$ with a finite sequence of length

$$r = \left\lceil \frac{a}{\Theta \lambda_j} \right\rceil .$$

Proof: Let us first prove by induction that the firing of a sequence $[t_j(\alpha\Theta\lambda_j/a)\rangle$ can at least be repeated $r-1$ times in the discrete time net.

(Basic step) It is immediate to observe that $t_j(\alpha\Theta\lambda_j/a)$ can be fired from $\boldsymbol{m}$, since $\Theta\lambda_j/a < 1$. The new marking is $\boldsymbol{m_1} = (\alpha\Theta\lambda_j/a) \cdot \boldsymbol{m}' + (1 - \alpha\Theta\lambda_j/a) \cdot \boldsymbol{m}$.

(Inductive step) Assume that at a given intermediate step $\boldsymbol{m_h} = \beta \boldsymbol{m}' + (1-\beta) \cdot \boldsymbol{m}$, with $0 < \beta < 1$. It can be observed that for all $p \in {}^\bullet t_j$, it holds $m_h(p) = \beta m'(p) + (1-\beta) m(p) \geq \beta \frac{m(p)}{a} + (1-\beta)\frac{m(p)}{a} = \frac{m(p)}{a}$, hence $t_j(\alpha\Theta\lambda_j/a)$ can be fired from $\boldsymbol{m_h}$, since $\Theta\lambda_j/a < 1$.

After $r-1$ firings $t_j(\alpha\Theta\lambda_j/a)$ can still be fired and it is sufficient to fire t_j for a quantity less or equal to that to reach $\boldsymbol{m}'$ in one step. □

According to the previous lemma, regardless of the initial token content in a place p, if an untimed sequence reduces the marking of p by at most a factor $1/a$, then an equivalent finite sequence exists in the dt-net system.

***Theorem** 5.2.* A marking $\boldsymbol{m}$ is reachable in a dt-contPN $\langle \mathcal{N}, \boldsymbol{\lambda}, \boldsymbol{m_0}, \Theta \rangle$ system (with Θ satisfying (10)) iff it is reachable in the underlying untimed contPN system $\langle \mathcal{N}, \boldsymbol{m_0} \rangle$ with a sequence that never empties an already marked place.

Proof: Mathematically, a sequence

$$\boldsymbol{m}[t_{i_1}(\alpha_1)\rangle \boldsymbol{m_1}[t_{i_2}(\alpha_2)\rangle \boldsymbol{m_2} \cdots [t_{i_k}(\alpha_k)\}\rangle \boldsymbol{m_k} = \boldsymbol{m}'$$

never empties a marked place if the following condition is verified

$$(\forall j = 1, \ldots, k), (\forall p \in {}^\bullet t_{i_j}) m_j(p) > 0 \qquad (11)$$

(If) Applying the previous Lemma for each $\boldsymbol{m_1}$, $\boldsymbol{m_2}$, $\cdots$, $\boldsymbol{m_k}$ implies that $\boldsymbol{m}'$ is reachable with a finite sequence.

(Only if) Assume there is a finite sequence that reaches $\boldsymbol{m}$ in the dt-contPN, then there exists an equivalent firing sequence for the untimed net system, according to Corollary 4.5. It is also immediate to observe that condition (11) holds because in the dt-contPN a place cannot be emptied with a finite sequence, according to Prop. 4.4 part 2. □

One may wonder what happens if a marking $\boldsymbol{m}$ is reachable in the untimed PN but there exists no sequence satisfying condition (11). In this case it can be easily proved that the marking is *lim-reachable* in the timed net, i.e., it is reachable with an unbounded sequence of steps. The result is formally proved in Theorem 5.3 by showing how such an infinite sequence may be determined.

***Theorem** 5.3.* If a marking $\boldsymbol{m}$ is reachable in the untimed contPN system $\langle \mathcal{N}, \boldsymbol{m_0} \rangle$, then it is lim-rechable in a dt-contPN system $\langle \mathcal{N}, \boldsymbol{\lambda}, \boldsymbol{m_0}, \Theta \rangle$ with Θ satisfying (10).

Proof: Assume that in the untimed net system $\boldsymbol{m_0}[t_{r_1}(\alpha_1)\rangle \boldsymbol{m_1}[t_{r_2}(\alpha_2)\rangle \boldsymbol{m_2} \cdots [t_{r_k}(\alpha_k)\rangle \boldsymbol{m_k} = \boldsymbol{m}$, and let us define $\sigma = t_{r_1}(\alpha_1) t_{r_2}(\alpha_2) \cdots t_{r_k}(\alpha_k)$.

We will prove that this sequence is equivalent to an infinite sequence $\sigma^1\sigma^2\cdots$ in which all the input places of the fired transitions are reduced by each firing by at most a factor $1/2$. Thus, applying Lemma 5.1, it can be fired in the discrete time net. This infinite sequence will fire each transition in σ, but in a smaller amount, and repeat the process. It will be seen that the amount of firing of each transition converges to the value in σ.

For each round, the sequence is defined as $\sigma^i = t_{r_1}(\beta_{i,1}\alpha_1) t_{r_2}(\beta_{i,2}\alpha_2) \cdots t_{r_k}(\beta_{i,k}\alpha_k)$ where

$$\beta_{i,1} = 1/2^i \qquad (i = 1, 2, \ldots),$$

$$\beta_{1,j} = 1/2^j, \qquad (j = 1, \ldots, k),$$

$$\beta_{i,j} = \frac{1}{2}\left(\sum_{l=1}^{i} \beta_{i,j-1} - \sum_{l=1}^{i-1} \beta_{i,j} \right) \qquad (i = 2, \ldots;\ j = 2, \ldots, k).$$

Intuitively, in the first round the proportion of firing is decreasing each time so that places are never emptied by more than one half. In the following rounds, it is taken into account how much the previous transitions in the sequence have been fired, and how much the actual transition has been fired until now, again to be sure that the reduction never exceeds one half.

Formally, consider an intermediate step in which $\sigma^1 \ldots \sigma^{i-1}$ and only part of σ^i, namely,

$t_{r_1}(\beta_{i,1}\alpha_1) t_{r_2}(\beta_{i,2}\alpha_2) \cdots t_{r_{j-1}}(\beta_{i,j-1}\alpha_{j-1})$, have been fired. If we denote $\boldsymbol{c_j} = \alpha_j \boldsymbol{C}(\cdot, t_{r_j})$ the actual marking can be described as

$$\boldsymbol{m_{i,j-1}} = \boldsymbol{m_0} + \left(\sum_{h=1}^{i} \beta_{h,1}\right) \boldsymbol{c_1} + \cdots + \left(\sum_{h=1}^{i} \beta_{h,j-1}\right) \boldsymbol{c_{j-1}} + \left(\sum_{h=1}^{i-1} \beta_{h,j}\right) \boldsymbol{c_j} + \cdots + \left(\sum_{h=1}^{i-1} \beta_{k,j}\right) \boldsymbol{c_k} = \left(1 - \sum_{h=1}^{i} \beta_{h,1}\right) \boldsymbol{m} + \left(\sum_{h=1}^{i} \beta_{h,1}\right) \boldsymbol{m_1} + \left(\sum_{h=1}^{i} \beta_{h,2}\right) \boldsymbol{c_2} + \cdots + \left(\sum_{h=1}^{i} \beta_{h,j-1}\right) \boldsymbol{c_{j-1}} + \left(\sum_{h=1}^{i-1} \beta_{h,j}\right) \boldsymbol{c_j} + \cdots + \left(\sum_{h=1}^{i-1} \beta_{k,j}\right) \boldsymbol{c_k} = \cdots =$$

$$\left(1-\sum_{h=1}^{i}\beta_{h,1}\right)\boldsymbol{m}+\left(\sum_{h=1}^{i}\beta_{h,1}-\sum_{h=1}^{i}\beta_{h,2}\right)\boldsymbol{m_1}+ \\ \cdots+\left(\sum_{h=1}^{i}\beta_{h,j-1}-\sum_{h=1}^{i-1}\beta_{h,j}\right)\boldsymbol{m_{j-1}}+ \\ \left(\sum_{h=1}^{i-1}\beta_{h,j}-\sum_{h=1}^{i-1}\beta_{h,j-1}\right)\boldsymbol{m_j}\cdots+ \\ \left(\sum_{h=1}^{i}\beta_{h,n-1}-\sum_{h=1}^{i-1}\beta_{h,k}\right)\boldsymbol{m_{k-1}}+ \\ \left(\sum_{h=1}^{i}\beta_{h,n}\right)\boldsymbol{m_k}$$

Hence, $\boldsymbol{m_{i,j-1}} \geq (\sum_{h=1}^{i}\beta_{h,j-1}-\sum_{h=1}^{i-1}\beta_{h,j})\boldsymbol{m_{j-1}}$ and so t_{r_j} can be fired half of this amount and no place looses more that one half of its token content.

With respect to the convergence to σ, it can be proved that $\beta_{i,j}=\frac{(i+j-2)!}{(j-1)!(i-1)!}\cdot\frac{1}{2^{i+j-1}}$, which is the probability mass distribution of the negative binomial of parameters j, $1/2$. Applying induction, the proof is based on the fact that the cumulative distribution function F_j can be immediately expressed as a regularized incomplete beta function, i.e., $F_j(h) = I_{1/2}(j, h+1)$, and that a regularized incomplete beta function enjoys the following property:

$$I_{1/2}(a,b)-I_{1/2}(a+1,b)=\frac{(a+b-1)!}{(a)!(b-1)!}\cdot\frac{1}{2^{a+b}}.$$

Observe that $\beta_{1,j}=\frac{1}{2^j}=\frac{(1+j-2)!}{(j-1)!(1-1)!}\cdot\frac{1}{2^{1+j-1}}$, and that $\beta_{i,1}=\frac{1}{2^i}=\frac{(i+1-2)!}{(1-1)!(i-1)!}\cdot\frac{1}{2^{i+1-1}}$.

Applying induction "following the rows", assume it holds for $\beta_{l,k}$, with $1 \leq l \leq i-1$ and $1 \leq k \leq n$, and for $\beta_{i,k}$, with $1 \leq k \leq j-1$. Let us prove it for $\beta_{i,j}$.

$$\beta_{i,j}=\frac{\sum_{l=1}^{i}\beta_{i,j-1}-\sum_{l=1}^{i-1}\beta_{i,j}}{2}= \\ \frac{\beta_{i,j-1}}{2}+\frac{\sum_{l=1}^{i-1}\beta_{i,j-1}-\sum_{l=1}^{i-1}\beta_{i,j}}{2}= \\ \frac{\beta_{i,j-1}}{2}+\frac{1}{2}\left(\sum_{l=1}^{i-1}\frac{\binom{l+j-3}{j-2}}{2^{l+j-2}}-\sum_{l=1}^{i-1}\frac{\binom{l+j-2}{j-1}}{2^{l+j-1}}\right) \\ =\frac{\beta_{i,j-1}}{2}+\frac{1}{2}\left(\sum_{l=0}^{i-2}\frac{\binom{l+j-2}{j-2}}{2^{l+j-1}}-\sum_{l=0}^{i-2}\frac{\binom{l+j-1}{j-1}}{2^{l+j}}\right)= \\ \frac{1}{2}\frac{\binom{i+j-3}{j-2}}{2^{i+j-2}}+\frac{1}{2}(I_{1/2}(j-1,i-1)-I_{1/2}(j,i-1))= \\ \frac{1}{2^{i+j-1}}\frac{(i+j-3)!}{(j-2)!(i-1)!}+\frac{1}{2^{i+j-1}}\frac{(i+j-3)!}{(j-1)!(i-2)!}= \\ \frac{1}{2^{i+j-1}}\frac{(i+j-2)!}{(j-1)!(i-1)!}$$

This means that the amount in which transition t_j is fired is α_j times a cumulative distribution function, and so in the limit it converges to α_j. □

6. CONCLUSIONS

In this paper we provide a study of contPNs under infinite servers semantics. First, different ways of describing the behavior of controlled contPNs are presented, starting with a min-based non-linear system (eq.(1) plus $\dot{\boldsymbol{m}} = \boldsymbol{C} \cdot \boldsymbol{f}$), continuing with a piecewise linear form (eq. (5)) and ending with a linear constrained form (eq. (6)).

The linear constrained system is then discretized and we provide a *Sampling theorem* giving an upper bound on sampling period. The purpose of the Sampling theorem presented here is to preserve reachability conditions (in particular non-negativity of markings), not to reconstruct the original signal from the sampled one.

The reachability space of the sampled system is studied in the last part of the paper and some relations between this space and the space of the underlying untimed contPN are provided. In practice, the sampling rate may be higher (like in Nyquist-Shannon sampling theorem) if signal reconstruction is required. But this is a topic to be considered in a future work. Anyhow the classical sampling theorem for linear systems should be respected for all embedded ones.

REFERENCES

David, R. and H. Alla (2004). *Discrete, Continuous and Hybrid Petri Nets.* Springer-Verlag.

Júlvez, J., A. Bemporad, L. Recalde and M. Silva (2004). Event-driven optimal control of continuous petri nets. In: 43rd *IEEE Conference on Decision and Control (CDC 2004).* Paradise Island, Bahamas.

Mahulea, C., A. Ramírez, L. Recalde and M. Silva (2005). Steady state control, zero valued poles and token conservation laws in continuous net systems. In: *Workshop on Control of Hybrid and Discrete Event Systems.* J.M. Colom, S. Sreenivas and T. Ushio, eds.. Miami, USA.

Mahulea, C., L. Recalde and M. Silva (2006). On performance monotonicity and basic servers semantics of continuous petri nets. In: *WODES'06:* 8th *International Workshop on Discrete Event Systems.* Michigan, USA.

Recalde, L. and M. Silva (2001). Petri Nets fluidification revisited: Semantics and steady state. *APII-JESA* **35**(4), 435–449.

Silva, M. (1993). Introducing Petri nets. In: *Practice of Petri Nets in Manufacturing.* Chapman & Hall.

Silva, M. and L. Recalde (2002). Petri nets and integrality relaxations: A view of continuous Petri nets. *IEEE Trans. on Systems, Man, and Cybernetics* **32**(4), 314–327.

Silva, M. and L. Recalde (2004). On fluidification of Petri net models: from discrete to hybrid and continuous models. *Annual Reviews in Control* **28**(2), 253–266.

ELSEVIER
IFAC
PUBLICATIONS

MODELLING DISTRIBUTED MANUFACTURING SYSTEMS VIA FIRST ORDER HYBRID PETRI NETS

Mariagrazia Dotoli, Maria Pia Fanti, Agostino Marcello Mangini

Dipartimento di Elettrotecnica ed Elettronica, Politecnico di Bari, Italy

Abstract: A Distributed Manufacturing System (DMS) is a collection of independent companies possessing complementary skills and integrated with transportation and storage systems. This paper proposes a new model for DMS employing first order hybrid Petri nets, i.e., Petri nets based on first order fluid approximation. More precisely, transporters and manufacturers are described by continuous transitions, buffers are continuous places and products are represented by continuous flows routing from manufacturers, buffers and transporters. Moreover, discrete events occurring stochastically in the system are considered to take into account the start of the retailer requests and the blocking of transports and raw material supply. With the aim of showing the model effectiveness, a DMS example is modelled and simulated under two different operative conditions. *Copyright © 2006 IFAC*

Keywords: manufacturing systems, Petri-nets, dynamic models, simulation, performance indices.

1. INTRODUCTION

A Distributed Manufacturing System (DMS) is a collection of independent companies possessing complementary skills and integrated with transportation and storage systems (Viswanadham and Raghavan 2000). Appropriate modelling and analysis of these highly complex systems are crucial for performance evaluation and comparison of competing DMS. However, few contributions face the problem of modelling the DMS in order to analyze the system performance measures and to optimize its functional objectives. Viswanadham and Raghavan (2000) model the system as a Discrete Event Dynamical System (DEDS), in which the evolution depends on the interaction of discrete events. Generalized Stochastic Petri Nets (GSPN) model a particular example of SC and determine the decoupling point location, i.e., the facility from which all finished goods are assembled after customer order confirmation. Moreover, in (Desrochers *et al.* 2005) a two product SC is modelled by complex-valued token Petri nets and the performance measures are determined by simulation. In addition, Dotoli and Fanti (2005) propose a GSPN model in order to describe in a modular way a generic DMS. However, the limit of this formalism is the modelling of products by means of discrete quantities (i.e., tokens). This assumption is not realistic in large systems with a huge amount of material flow. Since DMS are DEDS whose number of reachable states is very large, approximating fluid models can be used in this context as in manufacturing systems.

The aim of the paper is to propose a new model for DMS employing First Order Hybrid Petri Nets (FOHPN, Balduzzi *et al.* 2000), that include continuous places holding fluid and discrete places containing a non-negative integer number of tokens and transitions, where the latter are either discrete or continuous. Such a hybrid Petri net model is based on the framework proposed by Alla and David (1998) and presents the main key feature of allowing the Instantaneous Firing Speeds (IFS) of the continuous transitions to be chosen in a given range by a control agent. Moreover, the set of all admissible IFS vectors is explicitly characterized by the feasible solutions of a linear constraint set. Furthermore, an optimal IFS vector can be chosen according to a given objective function. Using such a modelling approach, this paper develops an FOHPN model of DMS by means of first-order fluid approximations. In particular, transporters and manufacturers are described by continuous transitions, buffers are continuous places and products are represented by continuous flows (fluids) routing from manufacturers, buffers and transporters. The model is built by using a modular approach based on the idea of the bottom-up methodology (Zhou and Venkatesh, 1998). A representative example, including the typical DMS elements, shows the effectiveness of the modelling technique that allows us to evaluate the system performance indices by the simulation.

The paper is structured as follows. Section 2 describes the structure and the dynamics of a generic DMS. Section 3 reports a brief overview of the FOHPN modelling formalism and Section 4 presents the modular DMS model. Section 5 describes and analyzes an example of DMS and a conclusion section closes the paper.

2. THE SYSTEM DESCRIPTION

A DMS may be described as a set of facilities with materials that flow from the sources of raw materials to subassembly producers and onwards to manufacturers and consumers of finished products. The DMS facilities are connected by transporters of materials, semi-finished goods and finished products. More precisely, the entities of a DMS can be summarized as follows.

1- *Suppliers:* a supplier is a facility that provides raw materials, components and semi-finished products to manufacturers that make use of them.
2- *Manufacturers and assemblers*: manufacturers and assemblers are facilities that transform input raw materials/components into desired output products.
3- *Logistics and transporters*: storage systems and transporters play a critical role in distributed manufacturing. The attributes of logistics facilities are storage and handling capacities, transportation times, operation and inventory costs.
4- *Retailers or customers*: retailers or customers are sink nodes of material flows.

Here, part of the logistics, such as storage buffers, is considered pertaining to manufacturers, suppliers and customers. Moreover, transporters connect the different stages of the production process.

The dynamics of the distributed production system is traced by the flow of products between facilities and transporters. Because of the large amount of material flowing in the system, we model a DMS as a hybrid system: the continuous dynamics models the flow of products in the DMS, the manufacturing and the assembling of different products and its storage in appropriate buffers. Hence, resources with limited capacities are represented by continuous states describing the amount of fluid material that the resource stores.

Moreover, we consider also discrete events occurring stochastically in the system, such as:

a) the blocking of the raw material supply, e.g. modelling the occurrence of labour strikes, accidents or stops due to the shifts;
b) the blocking of the transport operations due to the shifts or to unpredictable events such as jamming of transportation routes, accidents, strikes of transporters etc.;
c) the start of a request from the retailers.

3. FIRST-ORDER HYBRID PETRI NETS

3.1 The net structure and marking.

This section recalls the First Order Hybrid Petri Nets (FOHPN) formalism used in the following (Balduzzi *et al.* 2000).

A FOHPN is a bipartite digraph described by the six-tuple $PN=(P, T, Pre, Post, D, F)$. The set of places $P=P_d \cup P_c$ is partitioned into a set of discrete places P_d (represented by circles) and a set of continuous places (represented by double circles).

The set of transitions $T=T_d \cup T_c$ is partitioned into a set of discrete transitions T_d and a set of continuous transitions T_c (represented by double boxes). Moreover, the set of discrete transitions $T_d=T_I \cup T_E$ is further partitioned into a set of immediate transitions T_I (represented by bars) and a set of exponentially distributed transitions T_E (represented by boxes).

The matrices *Pre* and *Post* are the pre-incidence and the post-incidence matrices, respectively, of dimension $|P| \times |T|$. Note that the symbol $|A|$ denotes the cardinality of set A. Such matrices specify the net digraph arcs and are defined as follows:

$$Pre, Post : \begin{cases} P_c \times T \to \mathbb{R}^+ \\ P_d \times T \to \mathbb{N} \end{cases}.$$

We require that for all $t \in T_c$ and for all $p \in P_d$ it holds $Pre(p,t)=Post(p,t)$ (*well-formed nets*).

The function D: $T_t \to R^+$ specifies the timing associated to exponentially distributed timed transition $t_j \in T_E$. More precisely, we associate to each $t_j \in T_E$ the average firing delay $\delta_j = D(t_j)$. Moreover, the function $F : T_c \to R^+ \times R_\infty^+$ specifies the firing speeds associated to continuous transitions (we denote $R_\infty^+ = R^+ \cup \{\infty\}$). For any continuous transition $t_j \in T_c$ we let $F(t_j)=(V_{mj}, V_{Mj})$, with $V_{mj} \leq V_{Mj}$. Here, V_{mj} represents the minimum firing speed (mfs) and V_{Mj} the Maximum Firing Speed (MFS) of the generic continuous transition.

Given a FOHPN and a transition $t \in T$, the following sets of places may be defined: $\bullet t=\{p \in P: Pre(p,t)>0\}$, named pre-set of t; $t\bullet=\{p \in P: Post(p,t)>0\}$, named post-set of t. Moreover, the corresponding restrictions to continuous or discrete places are defined as ${}^{(d)}t=\bullet t \cap P_d$ or ${}^{(c)}t=\bullet t \cap P_c$. Similar notations may be used for pre-sets and post-sets of places. The incidence matrix of the net is defined as $\mathbf{C}(p,t)=Post(p,t)-Pre(p,t)$. The restriction of $\mathbf{C}$ to P_X and T_X (with X,Y$\in$ {c, d}) is denoted by $\mathbf{C}_{XY}$.

A marking

$$\mathbf{m} : \begin{cases} P_d \to \mathbb{N} \\ P_c \to \mathbb{R}^+ \end{cases}$$

is a function that assigns to each discrete place a non-negative number of tokens, represented by black dots, and to each continuous place a fluid volume; m_i denotes the marking of place p_i. The value of a marking at time τ is denoted by $\mathbf{m}(\tau)$. The restriction of $\mathbf{m}$ to P_d and P_c are denoted by $\mathbf{m}^d$ and $\mathbf{m}^c$, respectively. A FOHPN system $<PN,\mathbf{m}(\tau_0)>$ is a FOHPN with initial marking $\mathbf{m}(\tau_0)$.

The following statements rule the firing of the continuous and discrete transitions:

1- a discrete transition $t\in T_d$ is enabled at $\mathbf{m}$ if for all $p_i\in \bullet t$, $m_i\geq Pre(p_i,t)$;

2- a continuous transition $t\in T_c$ is enabled at $\mathbf{m}$ if for all $p_i\in {}^{(d)}t$, $m_i\geq Pre(p_i,t)$.

Moreover, we say that an enabled transition $t\in T_c$ is strongly enabled at $\mathbf{m}$ if for all places $p_i\in {}^{(c)}t$, $m_i>0$; we say that transition $t\in T_C$ is weakly enabled at $\mathbf{m}$ if for some $p_i\in {}^{(c)}t$, $m_i=0$.

In addition, if $<PN,\mathbf{m}>$ is an FOHPN system and $t_j\in T_c$ with instantaneous firing speed (IFS) v_j, it holds:

1- if t_j is not enabled then $v_j=0$;

2- if t_i is strongly enabled, then it may fire with any firing speed $v_j\in [V_{mj},V_{Mj}]$;

3- if t_j is weakly enabled, then it may fire with any firing speed $v_j\in [V_{mj},V_j]$, where $V_j\leq V_{Mj}$ depends on the amount of fluid entering the empty input continuous place of t_j.

We denote by $\mathbf{v}(\tau)\in [v_1(\tau)\ v_2(\tau)\ldots\ v_{|Tc|}(\tau)]^T$ the IFS vector at time τ. Any admissible IFS vector $\mathbf{v}$ at $\mathbf{m}$ is a feasible solution of the following linear set:

$$\begin{aligned} &V_{Mj}-v_j\geq 0 && \forall t_j\in T_e(\mathbf{m})\\ &v_j-V_{mj}\geq 0 && \forall t_j\in T_e(\mathbf{m})\\ &v_j=0 && \forall t_j\in T_\upsilon(\mathbf{m})\\ &\sum_{t_j\in T_e} C(p,t_j)v_j\geq 0 && \forall p\in P_e(\mathbf{m}), \end{aligned} \quad (1)$$

where $T_e(\mathbf{m})\subset T_c$ ($T_\upsilon(\mathbf{m})\subset T_c$) is the subset of continuous transitions that are enabled (not enabled) at $\mathbf{m}$ and $P_e(\mathbf{m})=\{p_i\in P_c \mid m_i=0\}$ is the subset of empty continuous places.

3.2 The net dynamics.

The hybrid dynamics of the net combines both time-driven and event-driven dynamics. We define *macro events* the events that occur when: i) a discrete transition fires or the enabling/disabling of a continuous transition takes place; ii) a continuous place becomes empty.

The equation that governs the time-driven evolution of the marking of a place $p_i\in P_C$ is:

$$\dot{m}_i(\tau)=\sum_{t_j\in T_c} C(p_i,t_j)v_j(\tau). \quad (2)$$

Now, if τ_k and τ_{k+1} are the occurrence times of two macro-events, we assume that within the time interval $[\tau_k,\tau_{k+1})$ (macro period) the IFS vector $v(\tau_k)$ is constant. Then the continuous behaviour of an FOHPN for $\tau\in[\tau_k,\tau_{k+1})$ is described by:

$$\begin{aligned} &\mathbf{m}^c(\tau)=\mathbf{m}^c(\tau_k)+\mathbf{C}_{cc}\mathbf{v}(\tau_k)(\tau-\tau_k)\\ &\mathbf{m}^d(\tau)=\mathbf{m}^d(\tau_k) \end{aligned} . \quad (3)$$

The evolution of the net at the occurrence of the macro-events is described by:

$$\begin{aligned} &\mathbf{m}^c(\tau_k)=\mathbf{m}^c(\tau_k^-)+\mathbf{C}_{cd}\sigma(\tau_k)\\ &\mathbf{m}^d(\tau_k)=\mathbf{m}^d(\tau_k^-)+\mathbf{C}_{dd}\sigma(\tau_k) \end{aligned}, \quad (4)$$

where $\sigma(\tau)$ is the firing count vector associated to the firing of the discrete transition t_j.

4. THE MODULAR DMS MODEL

Petri net modelling and synthesis is a very important research area that attracted much attention in the past (Zhou and Venkatesh, 1998). Based on the idea of the bottom-up approach, this section proposes a modular FOHPN model to describe a DMS. Such a method can be summarized in two steps: decomposition and composition. Decomposition involves dividing a system into several subsystems. In DMS this division can be performed based on the determination of distributed system facilities (i.e., suppliers, manufacturers, transporters and customers). All these subsystems are modelled by FOHPN. On the other hand, composition involves the interacting of these sub-models into a complete model, representing the whole DMS. The following FOHPN modules model each individual subsystem composing the DMS.

1- *The supplier module.* The supplier is modelled as a continuous transition and two continuous places (see Fig 1). The continuous place p_S represents the row material buffer of finite capacity C_S and p'_S represents the corresponding available capacity such that $m_S+m'_S=C_S$. Moreover, the continuous transition t_S models the arrival of raw material in the system at a bounded rate v_S with $F(t_S)=(V_{Smin},V_{Smax})$. In addition, we consider the possibility that the providing of raw material is blocked. This situation is represented by a discrete event modelled by two exponentially distributed transitions and two discrete places. In particular, place $p_k\in P_d$ models the operative state of the supplier and $p'_k\in P_d$ is the non-operative state (see Fig 1). The blocking and the restoration of the raw material supply corresponds to the firing of transitions t_k and t'_k, respectively. For the sake of clarity, Fig. 1 depicts the transition t_T modelling the transport operation.

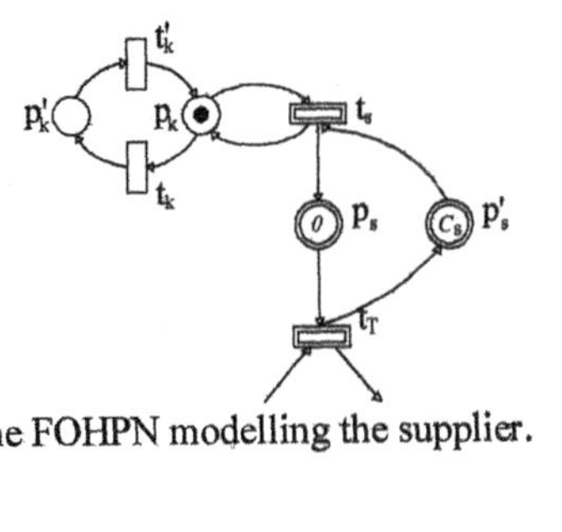

Fig. 1. The FOHPN modelling the supplier.

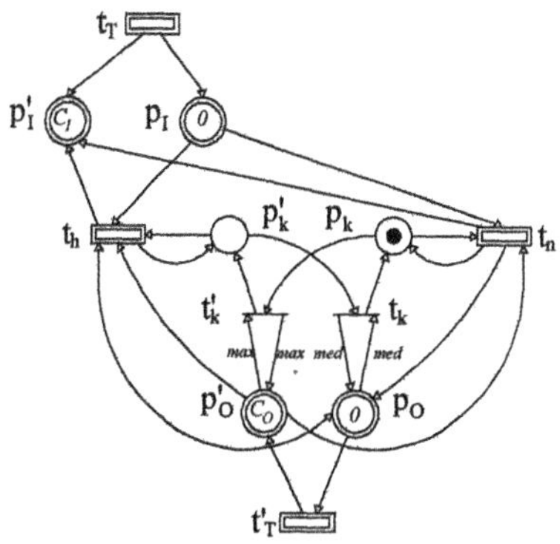

Fig. 2. The FOHPN modelling the manufacturer and the assembler.

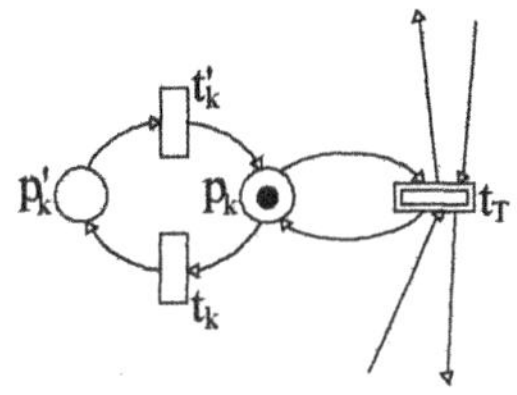

Fig. 3. The FOHPN modelling logistics.

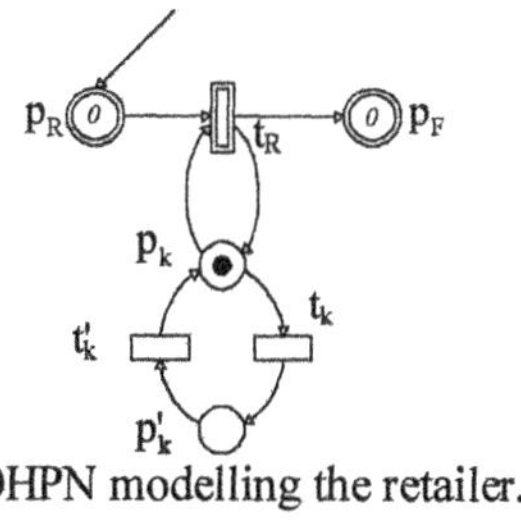

Fig. 4. The FOHPN modelling the retailer.

2- *The manufacturer and assembler module.* The manufacturers and assemblers are modelled by the FOHPN shown by Fig. 2. More precisely, place p_I is the input buffer of finite capacity C_I storing the input goods of a particular type. The corresponding place p'_I models the available buffer space so that $m_I+m'_I=C_I$. Analogously, the output buffer of capacity C_O storing the output product of a particular type is modelled by the place p_O representing the occupied buffer level and by place p'_O modelling the corresponding available capacity, with $m_O+m'_O=C_O$. To manage the production in function of the output inventory of the product, we assume that the production rate of the facility changes in function of the available space in the output buffer. Indeed, two different rates are considered for the production: a nominal rate $F(t_n)=(0,V_{mn})$ associated with the continuous transition t_n and a high rate $F(t_h)=(V_{mn},V_{Mh})$ associated with the continuous transition t_h. If it holds $m'_O(\tau)<max$ (i.e., the inventory is high enough and the buffer free space is not too high) then transition t'_k is disabled and the manufacturer works at the nominal rate with $m_k=1$ and $m_k=0$ with $p_k,p_k \in P_d$ as depicted in Fig. 2. On the contrary, when the buffer level of output parts decreases and hence the available buffer space reaches the maximum value $m'_O(\tau)=max$, the immediate transition t'_k is enabled and can fire. After the firing of t'_k, it holds $m_k=0$ and $m_k=1$, so that the facility works at the higher rate associated with the continuous transition t_h and the output buffer p_O is replenished more rapidly. Analogously, if the level of this buffer reaches the medium value $m_O(\tau)=med$, then the transition t_k is enabled and can fire, leading the facility rate at the nominal value. Note that transitions t_T and t'_T of Fig. 2 depicts the transports facilities.

3- *The logistics module.* The transporters connecting the different facilities are modelled by a continuous transition t_T that describes the flow of material from a facility to a subsequent one at a bounded rate $0 \le v_T \le V_{MT}$. Moreover, the random stop of the material transport is represented by two places $p_k,p'_k \in P_d$ and two exponentially distributed transitions $t_k,t'_k \in T_d$. If place $p_k \in P_d$ is marked, then the transport is operative. On the contrary, if transition t_k fires, then the transporters are not operative and the place $p'_k \in P_d$ becomes marked. When transition t'_k fires, the transporters are established again.

4- *The retailer module.* Finally, we consider the model of the retailers, represented by a continuous buffer place p_R of infinite capacity associated with each final product type. Moreover, a continuous transition t_R models the acquisition of final products by the retailer. However, we consider that the requests of the products are managed by discrete random events expressed by two discrete timed transitions t_k and t'_k and two places p_k and p'_k. If $m_k=1$ then the flow of material is enabled. On the contrary, if transition t_k fires, then a token in p'_k means that the retailer does not require any product and t_R is inhibited. The continuous place p_F models the system output and collects all the products requested by the retailer. Figure 4 depicts the retailer module.

5. AN EXAMPLE OF DMS

We consider a system producing two product types E and F by the four stages depicted in Fig. 5. The first stage includes two component suppliers S1 and S2, the second stage is composed by two subassembly manufacturers M1 and M2, the third stage is composed by two assemblers A1 and A2 and the last stage is constituted by the retailers R1 and R2. Moreover, six logistics service providers T1 to T6 suitably connect the DMS facilities that are located in different geographical sites. The buyers order two brands of products (E and F). Such products are obtained by two assemblers that assemble two types of products (C and D) obtained from two manufacturers. The subassemblies C and D are in turn produced by the manufacturers of the second stage, which receives the components of type A and B by the first supplier stage.

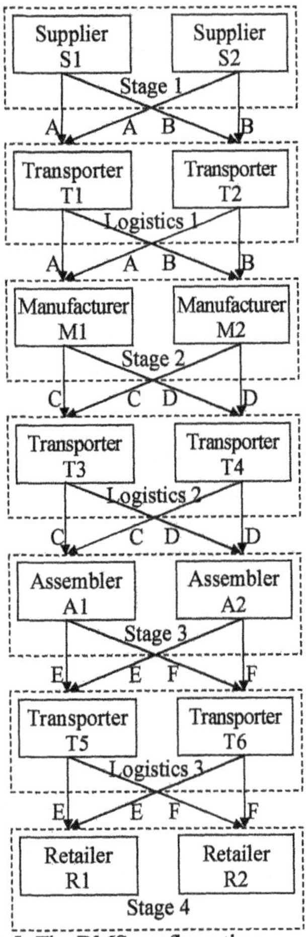

Fig. 5: The DMS configuration.

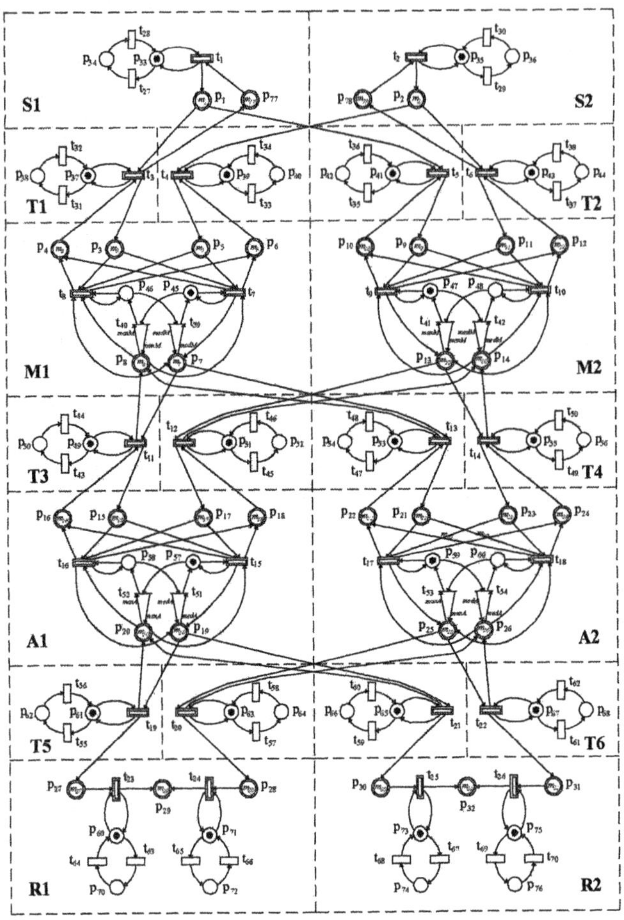

Fig. 6: The DMS model.

Table 1: Firing speeds and average firing delay of continuous and discrete transitions for Cases 1 and 2.

Continuous transitions	$[V_{min}, V_{max}]$ (Case 1)	$[V_{min}, V_{max}]$ (Case 2)	Discrete transitions	Average firing delay (hours)
t_1	[0, 8]	[0, 4]	t_{27}	14
t_2	[0, 10]	[0, 5]	t_{29}	14
t_3 t_6 t_{14} t_{22}	[0, 10]	[0, 5]	t_{28}	10
t_4 t_7 t_{13} t_{17} t_{19}	[0,8]	[0, 3]	t_{30}	10
t_5 t_{12} t_{20}	[0,9]	[0, 4]	t_{31} t_{37}	9
t_8	[8,20]	[0, 6]	t_{47} t_{49}	9
t_9	[0, 12]	[0, 2]	t_{55} t_{57}	9
t_{10}	[12,30]	[0, 6]	t_{32} t_{38}	15
t_{11} t_{21}	[0, 11]	[0, 3]	t_{48} t_{50}	15
t_{15}	[0, 6]	[4, 15]	t_{56} t_{58}	15
t_{16}	[6, 15]	[0, 4]	t_{33} t_{35}	10
t_{17}	[0, 8]	[4, 12]	t_{43} t_{45}	10
t_{18}	[8, 20]	[0, 3]	t_{59} t_{61}	10
t_{23}	[0, 3]	[0, 4]	t_{34} t_{36}	14
t_{24} t_{25}	[0, 4]	[4, 10]	t_{44} t_{46}	14
t_{26}	[0, 5]	[0, 6]	t_{60} t_{62}	14

We model the DMS in Fig. 5 by using in a modular way the elementary modules described in Section 3. Figure 6 shows the merged FOHPN model and depicts each facility module in dashed line squares.

To analyze the DMS behaviour, we simulate the FOHPN in two different cases that correspond to two different operative conditions. The data relative to Case 1 and 2 are shown in Table 1. In particular, Case 1 corresponds to a system with high production rate of each manufacturer and assembler. On the other hand, in Case 2 manufacturers and assemblers exhibit lower production rates than in Case 1 but the same transportation speed and buffer capacities. Moreover, Table 1 reports for each $t_j \in T_c$ the minimum and maximum firing speeds and for each $t_j \in T_E$ the average firing delay. In addition, Table 2 shows for Case 1 and Case 2 further data necessary to fully describe and simulate the system: the buffer capacities for the inventories of each stage, the initial markings of continuous places and the values of the edge weights.

To analyze the system dynamics, we define some performance indices assumed as relevant measures for the DMS analysis: i) the throughput T_i with i=1,2 of retailer Ri with i=1,2 respectively, i.e., the average number of products obtained by each retailer in a time unit; ii) the system throughput $T=T_1+T_2$; iii) the average input stocks in manufacturers Mi with i=1,2 (I_{Mi} i=1,2) and in assemblers Ai with i=1,2 (I_{Ai} i=1,2) during the run time TP; iv) the average output inventory in Mi with i=1,2 (O_{Mi} i=1,2) and in Ai with

$i=1,2$ (O_{Ai} $i=1,2$) during the run time TP; v) the average system inventory SI, i.e., the amount of product storage in each buffer during the run time TP; vi) the average lead time LT that is as follows:

$$LT=SI/T \quad (5)$$

The results are obtained by simulating the FOHPN in the Matlab environment for a simulation run of TP=480 hours. In particular, Fig. 7 shows the average inventories in manufacturers and in the assemblers in Cases 1 and 2. The figure shows that, also thanks to the dependence of manufacturers and assemblers production rates with their output inventories, the DMS is able to keep stocks at a satisfactorily high level, so that the demand is efficiently satisfied and inventory is not excessive. Moreover, as expected the DMS input stocks are always higher than the corresponding output inventories. However, the inventories of Case 1 are usually higher than the inventories in Case 2 because of the higher production rates.

In addition, Table 3 reports the average lead times and the system inventories for Cases 1 and 2. As expected, these two performance indices are higher in Case 1, exhibiting a higher level of congestion caused by the faster system production. Moreover, Fig. 8 reports the throughput of retailers R1 and R2 and the system throughput T. We remark that the throughput values in Case 1 are higher than the corresponding values in Case 2: this is expected, because manufacturers work in Case 1 with higher production rates.

CONCLUSIONS

The paper considers First-Order Hybrid Petri Nets (FOHPN) to model Distributed Manufacturing Systems (DMS), which are new emerging company networks, very complex to describe and manage. Combining continuous and discrete dynamics, FOHPN appears a promising formalism, able to capture the different properties of such discrete event systems, characterized by a large number of states.

Future research will apply the FOHPN conflict resolution policy in order to optimize a given DMS objective function under different management policies.

REFERENCES

Alla, H., and R. David, (1998). Continuous and Hybrid Petri Nets. J. *of Circuits, Systems and Computers*, **8** (1), 159-188.

Balduzzi, F, A. Giua, and G. Menga, (2000). Modelling and Control with First order Hybrid Petri Nets. *IEEE Trans. on Robotics and Automation*, **4** (16), 382 – 399.

Desrochers, A., T.J. Deal and M.P. Fanti, (2005) Complex-Valued Token Petri nets. *IEEE Transactions on Automation Science and Engineering*, **2**(4), 309-318.

Dotoli, M. and M.P. Fanti, (2005). A Generalized Stochastic Petri Net Model for Distributed Manufacturing System Management. *44th IEEE Conf. on Decision and Control and European Control Conference*, Seville, (Spain), 12-15.

Viswanadham, N. and S. Raghavan, (2000). Performance analysis and design of supply chain: a Petri net approach. *Journal of the Operational Research Society*, **51** (10), 1158-1169.

Zhou, M.C. and K. Venkatesh (1998). *Modeling, Simulation and Control of Flexible Manufacturing Systems. A Petri Net Approach.* World Scientific, Singapore.

Table 2: Initial marking of continuous places, capacities and edge weights for Cases 1 and 2.

Initial marking	Product units	Capacities
m_1 m_2	20	$C_1, C_2=200$
m_3 m_5 m_9 m_{11}	20	$C_4, C_6, C_{10}, C_{12}=100$
m_7 m_{13}	20	$C_8, C_{14}=80$
m_{15} m_{21} m_{17} m_{23}	20	$C_{16}, C_{22}, C_{18}, C_{24}=100$
m_{19} m_{25}	20	$C_{20}, C_{26}=80$
m_{27} m_{28} m_{29}	0	
m_{30} m_{31} m_{32}	0	
	Edge weights	
$max_M=190$		$max_A=190$
$med_M=50$		$med_A=50$

Table 3: Overall lead time and system inventory for Cases 1 and 2.

	Case 1	Case 2
LT (hours)	190.60	167.87
SI (product units)	445.80	218.38

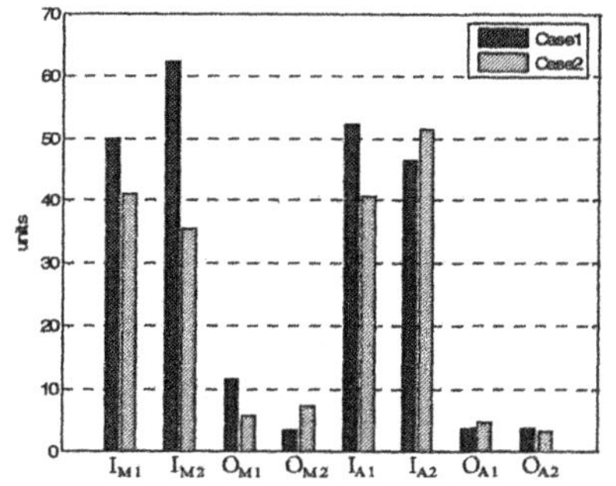

Fig. 7. The average inventory in manufacturers and in assemblers in Cases 1 and 2.

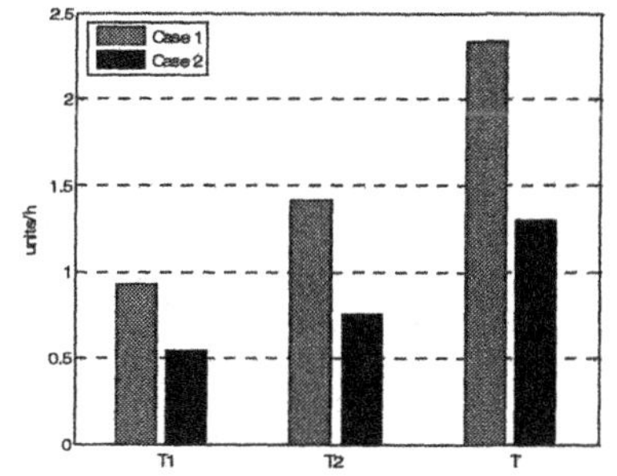

Fig. 8. Throughputs (units/hours) in Cases 1 and 2.

ELSEVIER
IFAC
PUBLICATIONS

SIMULATION OF RAILWAY STATIONS BASED ON HYBRID PETRI NETS

Fateh KAAKAI[(1)(2)], Said HAYAT [(1)], Abdellah El MOUDNI [(2)]

(1) Institut National de Recherche sur les Transports et leur Sécurité (INRETS)
20, rue Elisée Reclus, BP : 317, F-59666 VILLENEUVE d'ASCQ Cedex FRANCE.
Phone : (+33) (0) 3 20 43 83 11 - Fax : (+33) (0) 3 20 43 83 59 - Email : said.hayat@inrets.fr

(2) Université de Technologie de Belfort-Montbéliard (UTBM)
Laboratoire Systèmes et Transports (S.e.T) Site de Belfort, 90010 BELFORT Cedex France
Phone : (+33) 3 84 58 33 42, Fax : (+33)3 84 58 33 42
Email : fateh.kaakai@utbm.fr, abdellah.elmoudni@utbm.fr

Abstract: The purpose of this paper is to present a simulation model for Railway Stations (RS). Firstly, through a systemic description, we have shown that a RS can be considered as a Hybrid Dynamical System (HDS). Then, in order to represent the main intrinsic characteristics of RS, we have built a hybrid Petri net-based model. This model has been analyzed for i) checking purposes, and ii) evaluating some qualitative properties of the system. Finally, for illustrating the suggested methodology of study, a typical subway station has been simulated through a rush period scenario. Owing to a quantitative analysis of the marking evolution of the hybrid model, it is possible to make a performance evaluation of the station facilities and to underline some structural and functional limits of the studied station. Copyright© 2006 IFAC.

Keywords: Transport, stations, hybrid, Petri-nets, model, performance evaluation, design.

1. INTRODUCTION

The standard EN 13816 published in 2002 by the European Committee for Standardization (ECS) and the Transit Capacity and Quality of service Manual [2] are some of the works which deal with the *service quality in public transportation networks.* According to these, the maximal density of travellers into public transports or on waiting/queuing areas of station platforms must not be greater than 4 persons per m² for comfort and security purposes.

Nevertheless, we can observe that some Railway Stations (RS), especially into busy multimodal hubs, don't respect this limit during rush hours. When the travelers' concentration becomes too much high – for instance, 670 000 travelers/day transit into of the biggest multimodal hubs of Paris ("Chatelet-Les-Halles") (Luquet, 1998) - the most fragile persons (children, old people, etc.) can be seriously affected by the movements of the crowd when the subway or the train arrives at the station: *jostles, discomforts, falls, trampling,* etc. Anyway, these dangerous situations deteriorate the service quality of public transports and make them less attractive.

In order to avoid serious accidents and respect the service quality standard, it is necessary to have an efficient tool able to evaluate the performances of existing stations and to design more precisely new secure facilities able to absorb these peaks of travelers during rush periods. Therefore, the first step is to well understand, thanks to an appropriate model, the *internal operation* of RS and the *behavior of this system* when peaks of travelers are applied as input data. Then, this model must be analyzed in order to establish formal methods able to evaluate structural and behavioral properties of the station facilities.

2. SYSTEMIC DESCRIPTION OF RAILWAY STATIONS

Railway stations are stop points of a railway transportation line where a whole of facilities is available in order to allow the parking of the public transport and the landing and the boarding of travelers.

2.1. Structural description

A "railway station" consists of (i) a platform with its input and output gates, and (ii) a public transport (train, subway, tram, etc.) when it is well parked at the platform level. The *platform* is characterized by its surface capacity C_{SMax}, the number of input (resp. output) gates N_{IG} (resp. N_{OG}) and the capacity of an input gate C_{IG} (resp. output gate C_{OG}) which corresponds to the maximal number of travelers who can simultaneously go through this gate.

The *public transport* is characterized by the number of its vehicles N_V, the maximal capacity of a vehicle C_{VMax} expressed in number of travellers, the number of landing-boarding gates per vehicle $N_{LBG/V}$, and the capacity of a landing-boarding gate C_{LBG}. The maximal capacity of the public transport, also expressed in number of travellers, is given by $C_{PTMax} = N_V \cdot C_{VMax}$.

2.2. Functional description

Firstly, we can observe that a railway station has a periodic operation as shown in Fig. 1. The operating period H_k of a station corresponds to the public transport headway, i.e. the time interval between two successive arrivals (at t_k and t_{k+1}) of the public transport at the considered station.

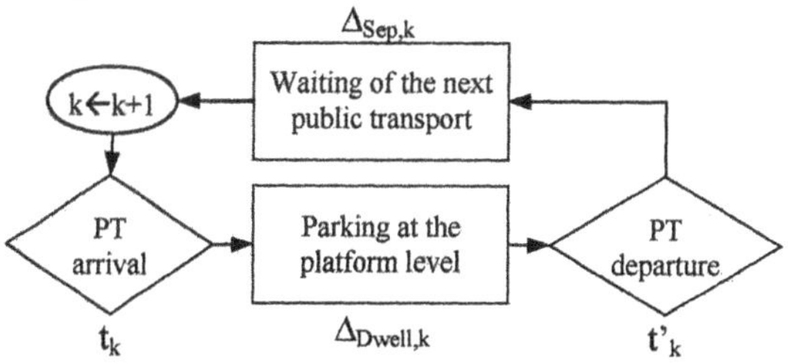

Fig. 1. Periodical operation of a railway station.

The duration of an operating period is given by (1):

$$\tau_k = t_{k+1} - t_k = \tau_{Dwell,k} + \tau_{Sep,k} \qquad (1)$$

where $\tau_{Dwell,k}$ and $\tau_{Sep,k}$ are the duration of the subphases $\Delta_{Dwell,k}$ and $\Delta_{Sep,k}$ respectively (Fig. 1).

Consequently, it is sufficient to study this system on only one operating period $H_k = [t_k \; t_{k+1}[$ which will be identified in the rest of the paper by the index k.

Now, let us describe in details the successive steps (represented in Fig. 1 and Fig. 2) of an operating period H_k:

At $t = t_k$, the public transport arrives and parks into the station. It is characterized by an initial occupancy rate $\eta_{Busy,k} = C_{Busy,k}/C_{PTMax,k}$ and a landing rate $\eta_{Land,k} = C_{Land,k}/C_{Busy,k}$. The number of waiting travelers into the public transport who don't land (because they have another destination) is equal to (2):

$$C_{WTV,k} = [(1 - \eta_{Land,k}) \cdot \eta_{Busy,k} \cdot C_{PTMax,k}] \qquad (2)$$

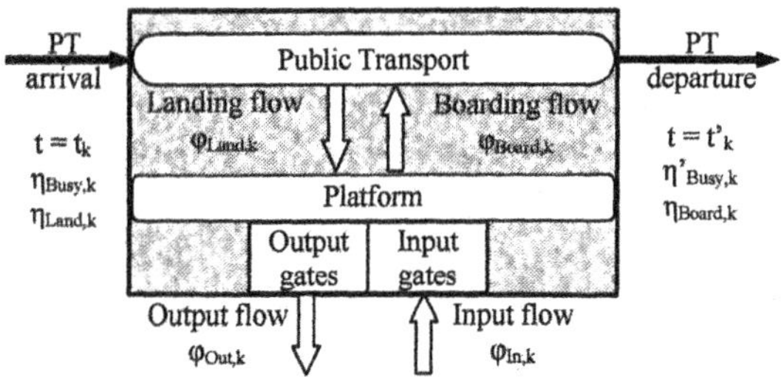

Fig. 2. Functional description of a railway station during an operating period H_k.

In the same way, $C_{WTP,k}$ denotes the initial number of travelers on the platform who are waiting in order to board into the public transport.

During the dwell time $\Delta_{Dwell,k} =]t_k , t'_k[$, an exchange of travelers takes place: $C_{Land,k} = (\eta_{Land,k} \cdot \eta_{Busy,k} \cdot C_{PTMax,k})$ travelers are going to land on the platform at a certain flow denoted $\varphi_{Land,k}$ [travelers / time unit] and then leave the station through the output gates at a certain flow $\varphi_{Out,k}$. At the same time, travelers enter into the station at a certain flow denoted $\varphi_{In,k}$ in order to board into the public transport at a certain flow $\varphi_{Board,k}$. Since the public transport has a limited capacity, the number of boarded travelers must respect the *maximal boarding constraint* (3) obtained by considering the free capacity of the public transport when the passengers' landing is completed:

$$C_{Board,k} \le C_{BMax,k} \quad \text{with} \quad C_{BMax,k} = [1 - \eta_{Busy,k} \cdot (1 - \eta_{Land,k})] \cdot C_{PTMax,k} \qquad (3)$$

At $t = t'_k$, the public transport leaves the station with a final occupancy rate $\eta'_{Busy,k} = C'_{Busy,k}/C_{Max,k}$ and a boarding rate $\eta_{Board,k} = C_{Board,k}/C'_{Busy,k}$.

Lastly, during the separation phase $\Delta_{Sep,k} =]t'_k , t_{k+1}[$ until the next public transport arrival at $t = t_{k+1}$, the station is only composed of the platform with its input and output gates. Thus, only two dynamics animate the system: the inflow of travelers ($\varphi_{In,k}$) and the outflow of travelers ($\varphi_{Out,k}$).

Table 1: Distinct groups of travelers into the station

Station	Groups of travelers	Abbrev.	Symbol
Public Transport (PT)	G_{T1}: travelers who are going to land on the platform	Landing travelers	Lng
	G_{T2}: travelers who have boarded into the PT	Boarded travelers	Bed
	G_{T3}: travelers who don't land because they have another destination	Waiting travelers	Wng
Platform	G_{P1}: travelers who have landed on the platform	Landed travelers	Led
	G_{P2}: travelers who are going to board into the PT	Boarding travelers	Bng

Four intrinsic dynamics which completely characterize the station operation have been identified: (i) *inflow* of travelers into the station, (ii) *boarding* of travelers into the public transport, (iii) *landing* of travelers on the platform, and (iv) *outflow* of travelers from the station. These different dynamics contribute to generate distinct groups of travelers into the public transport (3 groups) and on the platform (2 groups), as shown in Table 1.

2.3. Synthesis of the systemic description

Through this systemic description, we have observed that the operation of railway stations is governed by two kinds of phenomenon: (i) *external discrete events* corresponding to the successive arrivals and departures of the public transport (timetable), and (ii) *continuous dynamics* of the travelers into the station. Moreover, we can remark that some interactions exist between the discrete events and the continuous dynamics. For example, the public transport arrivals and departures control the start and the end of the landing and boarding flows of travelers. Thus, by definition, a railway station is a Hybrid Dynamical System (HDS). Furthermore, several intrinsic characteristics of this hybrid dynamical system have been described (see Fig. 2), notably the functional aspects of *synchronization* (of the public transport on the timetable), *parallelism* (of the flows of travellers), *resource sharing* and *limited capacities* of the public transport and the platform.

Among the several models dedicated to HDS (Antsaklis, *et al.*, 1996; Zaytoon, 2001; Engell, *et al.*, 2002), Hybrid Petri Nets constitute a suitable model for our study because they allow: (i) a graphical and modular representation of the different parts of the system, (ii) an easy and efficient modeling of the intrinsic characteristics quoted above, (iii) a qualitative study of the structural and behavioral properties of the model, and (iii) a quantitative performance evaluation which does not require an exhaustive enumeration of the state space.

3. MODELING

Firstly, we are going to precise the modeling framework of this paper. Then, we will present the suggested hybrid Petri net-based model. The basic concepts of hybrid Petri nets are assumed to be known to the reader. Otherwise, the reader may refer to (David, 1997; Alla and David, 1998; David and Alla, 2001).

3.1 Modeling framework

In this paper, we do not consider the disturbances which can affect the transportation network. Thus, the public transports are supposed to *respect* their timetables. We also consider that the landing and the boarding of travelers take place *simultaneously* (parallel approach) during the dwell phase ($\Delta_{Dwell,k}$) of the public transport at the platform level.

3.2 Hybrid Petri net of railway stations

The suggested hybrid Petri net-based model of railway stations is shown in Fig. 3. The Table 2 gives the initial markings of the discrete and continuous places and the expressions of the weights ($\neq$1) associated to some arcs of the net. Below, a description of the nodes of the hybrid model is given.

A. Public transport arrivals and departures

T_{Arr} (resp. T_{Dep}) is a timed discrete transition which represents the successive public transport arrivals at the station (resp. the successive public transport departures from the station). The time $\tau_{Sep,k}$ associated with T_{Arr} corresponds to the separation time introduced in section 2.2 (see Fig. 1). The time $\tau_{Dwell,k}$ associated with T_{Dep} represents the duration of the public transport dwell phase (see Fig. 1). These times $\tau_{Sep,k}$ and $\tau_{Dwell,k}$ can be considered as constant or variable during the period of study of the station (hour, day, week, etc.).

P_{Sep} and P_{Dwell} are two discrete places which models the periodicity of the station operation. A token in P_{Sep} corresponds to the phase $\Delta_{Sep,k}$ of Fig. 1 whereas a token in P_{Dwell} corresponds to the phase $\Delta_{Dwell,k}$ of Fig. 1.

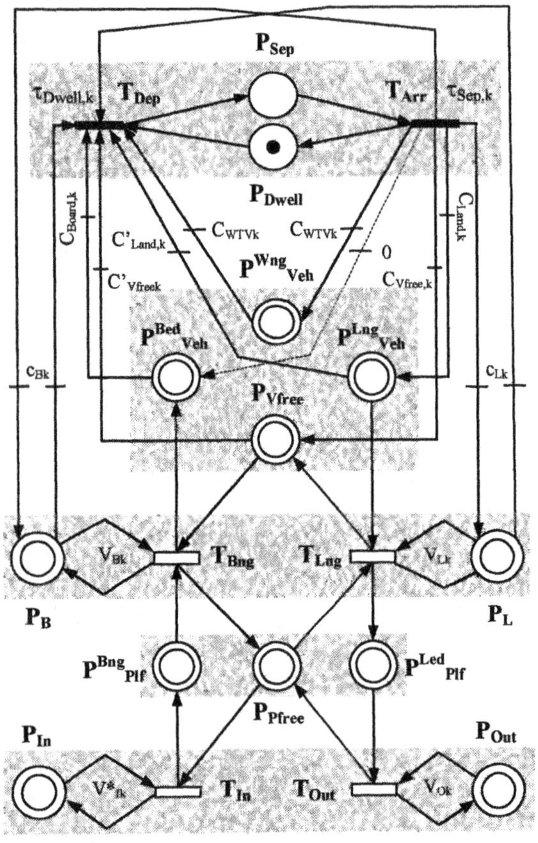

Fig. 3. Suggested hybrid Petri net of railway stations. The dotted arc between T_{Arr} and $P^{Bed}{}_{Veh}$ has a weight equal to zero. It has been introduced in the net just for underlining the symmetry of the model.

Table 2 : Initial marking and weights at $t = t_k$

	Public Transport	Platform
Initial Marking	$m^{Lng}{}_{Veh} = \eta_{Land,k} \cdot C_{PTMax,k}$	$m^{Led}{}_{Plf} = 0$
	$m^{Bed}{}_{Veh} = 0$	$m^{Bng}{}_{Plf} = C_{WTPk}$
	$m^{Wng}{}_{Veh} = C_{WTVk} = C_{Busy,k} - C_{Land,k}$	$m_{Pfree} = C_{SMax} - C_{WTPk}$
	$m_{Vfree} = C_{PTMax,k} - C_{Busy,k}$	$m_{In} = N_{IG} \cdot N_{IT/G}$
	$m_L = m_B = ½ \cdot N_{Vk} \cdot N_{LBG/V} \cdot C_{LBG}$	$m_{Out} = N_{OG} \cdot N_{OT/G}$
	$m_{Sep} = 0$, $m_{Dwell} = 1$	
Weights	$C_{Land,k} = \eta_{Land,k} \cdot C_{PTMax,k}$	
	$C_{WTVk} = C_{Busy,k} - C_{Land,k}$	
	$C_{Board,k}$ = value of $m^B{}_{Veh}$ at $t = t'_k$	
	$C_{Vfree,k} = C_{PTMax,k} - C_{Busy,k}$	
	$C'_{Vfree,k}$ = value of m_{Vfree} at $t = t'_k$	
	$C'_{Land,k}$ = value of $m^L{}_{Veh}$ at $t = t'_k$	
	$C_{Lk} = C_{Bk} = ½ \cdot N_{Vk} \cdot N_{LBG/V} \cdot C_{LBG}$	

B. Capacity of the public transport

The global capacity of the public transport (continuous place P_{Veh} which doesn't appear in the net) can be divided into four subcapacities (four continuous subplaces) according to the distinct groups of travelers shown in Table 1 and such that:

$$\begin{cases} P_{Veh} = P^{Lng}_{Veh} \cup P^{Bed}_{Veh} \cup P^{Wng}_{Veh} \cup P_{Vfree} \\ P^{Lng}_{Veh} \cap P^{Bed}_{Veh} \cap P^{Wng}_{Veh} \cap P_{Vfree} = \{\emptyset\} \end{cases} \tag{4}$$

where

- P^{Lng}_{Veh} corresponds to the public transport capacity taken up by the landing travelers.
- P^{Bed}_{Veh} corresponds to the public transport capacity taken up by the boarded travelers.
- P^{Wng}_{Veh} corresponds to the public transport capacity taken up by the waiting travelers.
- P_{Vfree} is a continuous place which represents the free welcome capacity ($C_{Vfree,k}$) of the public transport expressed in number of travelers.

C. Gates of the public transport

T_{Ldg} (resp. T_{Bdg}) is a continuous transition which models the travelers' landing (resp. the travelers' boarding) through the public transport gates. The maximal firing speed of T_{Ldg} (resp. T_{Bdg}) is denoted V_{Lk} (resp. V_{Bk}).

P_L (resp. P_B) is a continuous place which represents the capacity of the whole landing (resp. boarding) gates of the public transport $C_{Lk} = ½ \cdot N_{Vk} \cdot N_{LBG/V} \cdot C_{LBG}$ (resp. $C_{Bk} = ½ \cdot N_{Vk} \cdot N_{LBG/V} \cdot C_{LBG}$).

D. Platform capacity

P_{Plf} is a continuous place which represents the station platform capacity; this global continuous place (which doesn't appear in the model) can be divided into three continuous subplaces according to Table 1 and such that:

$$\begin{cases} P_{Plf} = P^{Led}_{Plf} \cup P^{Bng}_{Plf} \cup P_{Pfree} \\ P^{Led}_{Plf} \cap P^{Bng}_{Plf} \cap P_{Pfree} = \{\emptyset\} \end{cases} \tag{5}$$

where

- P^{Led}_{Plf} corresponds to the platform surface taken up by the *landed travelers*.
- P^{Bng}_{Plf} corresponds to the platform surface taken up by the *boarding travelers*.
- P_{Pfree} is a continuous place corresponding to the free welcome surface (C_{Sfree}) of the platform.

E. Gates of the platform

T_{In} (resp. T_{Out}) is a continuous transition which models the mean inflow (resp. the outflow) of travelers into (resp. from) the station. The maximal firing speed of T_{In} (resp. T_{Out}) is denoted V_{Ik} (resp. V_{Ok}).

P_{In} (resp. P_{Out}) is a continuous place which represents the maximal capacity of the whole input gates: $C_I = N_{IG} \cdot C_{IG}$ (resp. output gates: $C_O = N_{OG} \cdot C_{OG}$, see section 2).

3.3 Firing speeds of the continuous transitions

Several algorithms have been proposed in the literature for computing the instantaneous speeds of continuous transitions (David and Alla, 2001). In this study, we will use the CCPN-based algorithm for which the firing speed vector of the HPN remains constant during each invariant behavior phase (i.e. time interval between two events which change the behavior of the model). For more details, see (Alla and David, 1998).

Concerning the maximal firing speeds, they often only depend on structural and/or operating parameters of the *studied system*. In our case, we have identified two kinds of parameters which limit the amplitude of the travelers' flows into the station: the capacities of the gates and the crossing time (τ_{Ck}) which is the time required by a traveler for going through a gate. The crossing time will only be studied according to the period nature: it will be *minimum* ($\tau_{CMin,k}$) during off-peak periods and *maximal* ($\tau_{CMax,k}$) for rush periods. Average values of $\tau_{CMin,k}$ and $\tau_{CMax,k}$ must be measured on the field.

Moreover, in order to consider the travelers' inflow variations during the day, especially the difference between rush hours and off-peak periods, a real random variable denoted $A^* \in [0\ 1]$ has been introduced at the level of the input transition T_{In}. This random coefficient will take, according to the period nature, the following suggested values:

Off-peak period $$A^* = r(1)/K_1 \tag{6}$$

Rush period $$A^* = K_2 + r(1 - K_2) \tag{7}$$

where

- r is a generator function of uniformly distributed random numbers such that $0 \le r(x) \le x$.
- K_1 and K_2 are two real numbers. A typical value for K_1 is $\sqrt{2}$ whereas K_2 must be comprised between 0 and 1.

The maximal firing speeds of the hybrid Petri net of Fig. 3 are given by (8), (9) and (10):

$$V_{Lk} = V_{Bk} = ½ \cdot N_{Vk} \cdot N_{LBG/V} \cdot C_{LBG} / \tau_{Ck} \tag{8}$$

$$V^*_{Ik} = A^* \cdot N_{IG} \cdot C_{IG} / \tau_{Ck} \tag{9}$$

$$V_{Ok} = N_{OG} \cdot C_{OG} / \tau_{Ck} \tag{10}$$

Now, we are going to analyze the hybrid Petri net for, on the one hand, checking the validity of the suggested model and its modeling framework (coherence of the assumptions on which the model is based) and on the other hand, finding some important qualitative properties of the modeled system.

4. ANALYSIS OF THE HYBRID PETRI NET

In order to prove that the assumptions on which is based the suggested hybrid model are coherent with the real operation of a station, we can at first verify the global conservation of the number of travelers (marks) into the model. In other terms, we have to prove the following implications:

$$\forall t \in \Delta_{Dwell,k} = [t_k \quad t'_k[, \quad (4) \Rightarrow m_{Veh}^{Lng} + m_{Veh}^{Bed} + m_{Veh}^{Wng} + m_{Vfree} = C_{PTMax,k} \tag{11}$$

$$\forall t \in H_k = [t_k \quad t_{k+1}[, \quad (5) \Rightarrow m_{Plf}^{Led} + m_{Plf}^{Bng} + m_{Pfree} = C_{SMax} \tag{12}$$

The equations (11) and (12) correspond to marking invariants which can be found owing to algebraic methods. Indeed, the marking invariants of a Petri net are the solutions of (13):

$$^{T}F \cdot W_k = 0 \tag{13}$$

with $^{T}F = (q_1, \cdots, q_7)$ a vector of positive integers and W_k the incidence matrix of the hybrid Petri net. The details of the place invariant computation are not presented in this article since they are easy to develop. The linear combination of (11) and (12) allows concluding that the hybrid Petri net is *overall conservative* and thus *structurally bounded.* Another important issue for transportation systems consists in the research of deadlocks and the identification of states which induce deadlocks. Deadlocks can be identified when no movement of travelers (marks) is possible from a specific situation (given marking). For our model, we can identify a deadlock state which is reached when $m_{Vfree} = m_{Pfree} = 0$ at $t = t_k$. Indeed, if we consider a rush period scenario for which both the public transport and the station platform are initially saturated, then all continuous transitions are dead, and thus, the station is completely deadlocked. Fortunately, this specific situation is very rare in reality.

5. SIMULATION

We begin this section with a short recall of the simulation methodology of hybrid Petri nets. Then, the real data required by the simulation will be presented. However, most of the time, several real parameters are not available (even for public transport managers). Consequently, an artificial generation of data will be suggested and illustrated through a simulation scenario. Lastly, the simulation results will be analyzed for performance evaluation purposes.

5.1. Simulation methodology

The simulation of a hybrid Petri net is based on the determination of its Invariant Behavior States (IBS) owing to the fundamental equation (14). An IB-state is such that the marking of the discrete places and the instantaneous speed vector of the continuous transitions remain constant as long as the system is in the same IB-state.

$$\forall t \in IBS_j = [t_{ij} \quad t_{fj}[\tag{14}$$

$$m(t - t_{ij}) = m(0) + W_k \cdot \left(\sigma(t - t_{ij}) + \int_{t_{ij}}^{t_{fj}} v(u - t_{ij}) \cdot du \right)$$

In (14), t_{ij} and t_{fj} are the initial time and the final time of an IBS j. The vector $\sigma(t-t_{ij})$ represents the firing number of each discrete transition between t_{ij} and $t \leq t_{fj}$ (the components of σ associated with continuous transitions are null). The vector $v(u-t_{ij})$ symbolizes the instantaneous firing speeds associated with continuous transitions (the components of v associated with discrete transitions are null). For more details, see (David, 1997; Alla and David, 1998; David and Alla, 2001).

5.2 Required real data for the simulation

Table 3 and
Table 4 recapitulate the required data for the simulation. The following structural and functional parameters have been introduced during the systemic description and in the modeling section.

Table 3 : Characteristics of a subway station
(Mainly taken from http://www.metro-pole.net/)

	CHARACTERISTICS	EXAMPLES OF VALUES
PUBLIC TRANSPORT	Inter-arrival time (Headway H_k)	5 min.
	Dwell time ($\tau_{Dwell,k}$)	4 min.
	Numb. of vehicles of the PT (N_{Vk})	5
	Max. density of travelers (d_{VMax})	5 trav. / m²
	Max. capacity of a vehicle (C_{VMax})	200 trav.
	Number of gates per vehicle ($N_{LBG/V}$)	3
	Capacity of a gate (C_{LBG})	3 trav.
	Max. crossing time (τ_{CMax})	10 sec.
	Min. crossing time (τ_{CMin})	3 sec.
	Occupancy rate ($\eta_{Busy,k}$)	80 %
	Landing rate ($\eta_{Land,k}$)	65 %
PLATFORM	Surface (S_{Plf})	300 m²
	Max. density of travelers (d_{PMax})	5 trav. / m²
	Number of input gates (N_{IG})	2
	Capacity of an input gate (C_{IG})	4 trav.
	Number of output gates (N_{OG})	3
	Capacity of an output gate (C_{OG})	4 trav.
	Initial number of travelers (C_{WTPk})	200 trav.
	Max. crossing time (τ_{CMax})	10 sec.
	Min. crossing time (τ_{CMin})	3 sec.

Table 4 : Input data of the hybrid Petri net

DATA	MIN	MAX	UNITS
Landing and Boarding flows $V_{Lk} = V_{Bk}$	270	900	trav./min
Input flow V_{Ik}	48	160	trav./min
Output flow V_{Ok}	72	240	trav./min
Number of landing travelers ($C_{Land,k}$)	520		trav.
Maximal boarding constraint (C_{BMax})	720		trav.

5.3 Generation of artificial data

When some real data are not easy to find, a well known solution is to generate artificial data in order to fill in this lack of information. In our case, the occupancy rates ($\eta_{Busy,k}$) and the landing rates

($\eta_{Land,k}$) are two sets of statistical data which are not always available. That's why we suggest generating them using the following relations:

Off-peak period
$$\begin{cases} \eta^*_{Busy,k} = r(1)/B_1 \\ \eta^*_{Land,k} = r(1)/L_1 \end{cases} \quad (15)$$

Rush period
$$\begin{cases} \eta^*_{Busy,k} = B_2 + r(1-B_2) \\ \eta^*_{Land,k} = L_2 + r(1-L_2) \end{cases} \quad (16)$$

where

- r is a generator function of uniformly distributed random numbers such that $0 \le r(x) \le x$.
- B_1, B_2, L_1 and L_2 are four real values. A typical value for B_1 and B_2 is $\sqrt{2}$ whereas L_1 and L_2 must be comprised between 0 and 1.

5.4 Application

In order to illustrate the methodology of study presented in this paper, we have simulated (with MALAB) a typical station of the subway network of Paris during one hour and through a rush period scenario. We have used the simulation data presented in Table 3 and Table 4. Moreover, the parameters used for the generation of artificial occupancy and landing rates are: $B_1 = \sqrt{2}$, $B_2 = 0.9$, $L_1 = 1$ and $L_2 = 0.8$. We can observe on Fig. 4 for this rush period scenario that the public transport could become *saturated* before the end of the dwell time because of *an insufficient welcome capacity* (number of vehicles). In the same way, because of *successive accumulations* of travelers on the platform, which underline *an insufficient evacuation capacity* of the station (insufficient number of output gates for instance), the platform could reach its capacity and could remain saturated during the 8th operating period (H_8 = [35 40] (min.)). These observations allow concluding that this station is not adapted to rush periods since its facilities are not able to absorb peaks of travelers.

6. CONCLUSION

The contributions of this work are: (i) the physical description of railway stations as being a hybrid dynamical system, (ii) the new modeling approach based on hybrid Petri nets, and (iii) the performance evaluation of the station facilities owing to a quantitative analysis of the marking evolution of the model. The authors also believe that this simulation model can be successfully used for designing the station facilities by controlling intrinsic parameters of the model.

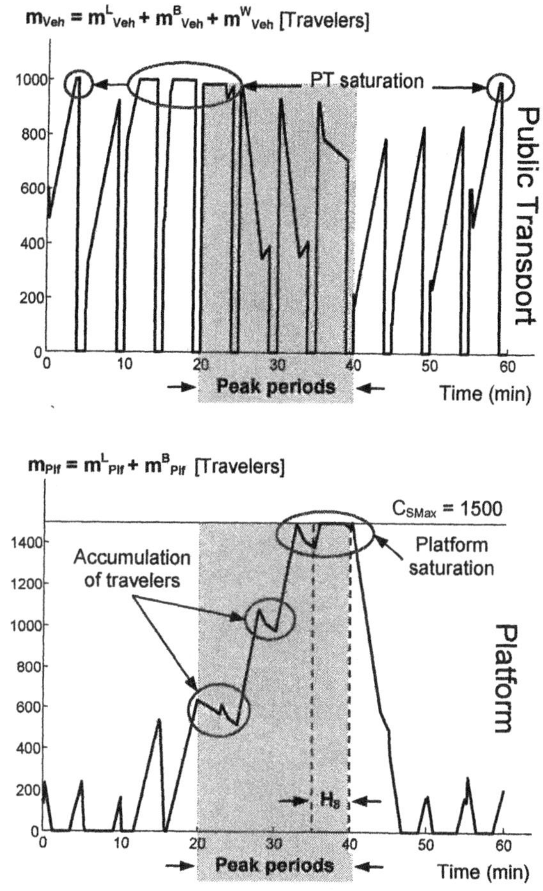

Fig. 4. Evolution of the total number of travelers into the public transport and on the platform for a rush period scenario.

REFERENCES

Alla, H. and R. David (1998). Modelling and analysis tool for discrete events systems: continuous Petri net. *Performance Evaluation*, v 33, n 3, pp. 175-199.

Antsaklis, P., X. Koutsoukos and J. Zaytoon (1996). On hybrid control of hybrid systems: a survey. *JESA*, vol. 32, no 9-10, pp. 1023-1045.

David, R. (1997). Modeling of hybrid systems using continuous and hybrid Petri nets. *International Workshop on Petri Nets and Performance Models*, pp. 47-58.

David, R. and H. Alla (2001). On Hybrid Petri Nets. *Journal of Discrete Event Dynamic Systems: Theory and Applications*, **Vol. 11**, pp. 9-40.

ECS, http://www.cenorm.be/cenorm/index.htm

Engell, S., G. Frehse and E. Schnieder (2002) *Modelling, Analysis, and Design of Hybrid Systems*. LNCIS, Ed. Springer-Verlag, Germany.

Kaakai, F., S. Hayat and A. El Moudni (2005). Formal design of public transport stations owing to Hybrid Event Graphs", in Proc. of IEEE Intelligent Transportation Systems Conf., Vienne.

Kittelson & Associates *et al.*, Transit *Capacity and Quality of service Manual* [Online]. Available: http://www4.trb.org/trb/crp.nsf/All+Projects/TCRP+A-15, 1nd edition, 2003.

Luquet, F. (1998). Multimodality: urban hub management. *Revue RATP Savoir-Faire*, N° 26, pp. 31-35.

Murata, T. (1989). Petri Nets: Properties, analysis and applications. *Proc. IEEE*, vol. 77, pp. 541–580.

Zaytoon J. (2001). Systèmes dynamiques hybrides. Traité IC2. (Hermès (Ed)), Paris.

ELSEVIER
IFAC
PUBLICATIONS

MODELING AN IMPACT CONTROL STRATEGY USING HYPA

Pieter J.L. Cuijpers *,[1] **Michel A. Reniers** *

* *Technische Universiteit Eindhoven (TU/e), Eindhoven*

Abstract: We analyze a control strategy for the pick-and-place module of a component mounting device. We use a combination of techniques from bondgraph-theory, systems theory, process algebra and differential algebra to achieve this, and we show how the hybrid process algebra HyPA aides us in combining these techniques and in using them on a common hybrid model of the device.

Keywords: hybrid systems, process algebra, bondgraph theory, impact control

1. INTRODUCTION

In this paper, we describe and analyze a control strategy for the pick-and-place module of a component mounting device, see figure 1, using the hybrid process algebra HyPA (Cuijpers and Reniers, 2005). The objective of this control strategy, designed by Philips CFT and Assembleon, is to bring a component to a PCB (Printed Circuit Board) as quickly as possible, and press it onto the PCB with sufficient force to make it stick. This all should be done without damaging the component. The focus of our analysis, is to show under which conditions safety of the controller can be guaranteed. In other words, we aim to find conditions under which it is certain that the component is not damaged.

The modeling and analysis of the pick-and-place module is carried out in the hybrid process algebra HyPA. This process algebra focuses on the description of so-called hybrid systems, i.e. on models of systems in which both continuous and discrete behavior occur. In the case of the pick-and-place module, hybrid modeling turns out to be useful, because it allows us to abstract from the precise mechanics of impact and other relatively fast behavior. This makes the dynamic behavior of the system much simpler which, rather unexpectedly, allows us to find analytic bounds for safety for this particular case study.

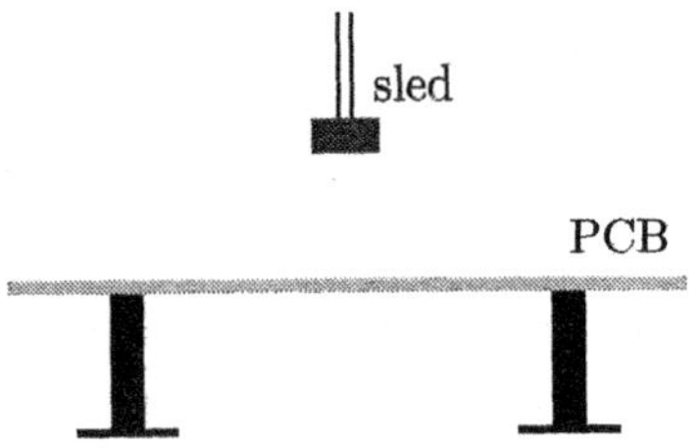

Fig. 1. A pick-and-place module

Apart from providing the user with a framework to specify hybrid systems, HyPA also allows algebraic reasoning about the compositional structure of such systems, i.e. on the way in which a hybrid system is composed from subsystems. One of the strengths of HyPA, is in finding different abstract representations of a system. In this paper, we first specify the system as a parallel composition of more-or-less standard components from bondgraph theory (Karnopp *et al.*, 1990; Mosterman *et al.*, 1998; Cuijpers *et al.*, 2004), and then rewrite it into a semi-linear representation (Usenko, 2002; van de Brand *et al.*, 2006) that

[1] We would like to express our gratitude to Progress/STW (Grant EES5173), Philips CFT and Assembleon, for their financial and material support of this case study, and of our project on the development of hybrid systems theory.

gives more insight in the functional behavior of the system. The exact functional behavior is extracted from this last representation using analysis of the differential equations and difference equations that occur in it.

In a sense, the focus of the process algebraic approach on compositional structure is supplemental to the kind of analysis of hybrid behavior as described in, for example, (Mosterman *et al.*, 1998; Febbraro *et al.*, 2001; Alur and Dill, 1994; Heemels *et al.*, 2001; Henzinger, 1996; van der Schaft and Schumacher, 2000). These latter approaches often consider a hybrid model that is already written in a specific form, and are able to give a very detailed analysis of such a model. The process algebraic approach provides a way of switching between different compositional representations of the model, i.e. between the different forms necessary for certain kinds of analysis. Recently, a number of other process algebras for hybrid systems have been introduced as well (Rounds and Song, 2003; Bergstra and Middelburg, 2005; van Beek *et al.*, 2006), which differ on a number of subtle details that are outside the scope of this paper. We suppose that these process algebras are equally suitable for the purpose of modeling and analyzing a system as presented in this paper, but it may be interesting to verify this in the future.

The structure of this paper is as follows. In section 2, we briefly introduce the hybrid process algebra HyPA. In section 3, we discuss the central model for this case study, namely that of the pick-and-place module. In section 4, we extend it with a model of a measuring and control strategy proposed by Philips CFT (Mateboer, 1999). In section 5, we formalize the analysis goal of this paper, i.e. we formalize what we mean by safety of a system. Finally, in section 6, we discuss the results of the actual analysis of the controlled pick-and-place module, and in section 7, we give some conclusive remarks and recommendations for future work.

2. HYBRID PROCESS ALGEBRA

In this section, the part of the syntax of HyPA that is used in this paper is discussed. The discussion presented here is adapted from (van de Brand *et al.*, 2006). For a more detailed work on this algebra see (Cuijpers and Reniers, 2005; Cuijpers, 2004).

The syntax of HyPA is an extension of the process algebra ACP (Baeten and Weijland, 1990), with the disrupt operator from LOTOS (Brinksma, 1985) and with variants of the flow clauses and re-initialization clauses from the event-flow formalism introduced in (van der Schaft and Schumacher, 2000). The signature of HyPA contains the following constant and function symbols:

(1) discrete actions $a \in \mathcal{A}$,
(2) flow clauses $c \in C$,
(3) a family of process re-initialization operators $d \gg \_$ where $d \in D$,
(4) alternative composition $\_ \oplus \_$,
(5) sequential composition $\_ \odot \_$,
(6) disrupt $\_ \blacktriangleright \_$ and left-disrupt $\_ \rhd \_$,
(7) parallel composition $\_ \parallel \_$,
(8) a family of encapsulation operators $\partial_H(\_)$, where $H \subseteq \mathcal{A}$, and predicate encapsulation operators $\partial_{P_m}(\_)$, where $P_m \in \mathcal{P}_m$.

The binding order of these operators is as follows: $\odot$, $\blacktriangleright$, $\rhd$, $d \gg$, $\parallel$, $\oplus$, where sequential composition binds strongest and alternative composition binds weakest. These constants and operators are described informally below.

Flow clauses are used to model continuous, never terminating, physical behavior by describing how the model variables $\mathcal{V}_m$ are allowed to change through time. A flow clause is a pair $(V \mid P_f)$ of a set of model variables $V \subseteq \mathcal{V}_m$ and a flow predicate $P_f \in \mathcal{P}_f$. This flow predicate may, for example, contain differential equations and algebraic equations. The set V models which variables are *not* allowed to jump at the beginning of a flow.

A process re-initialization $d \gg p$ models the behavior of p where the model variables are submitted to a discontinuous change as specified by the re-initialization clause d. A re-initialization clause describes a set of *re-initializations*, where a re-initialization is a pair of valuations representing the values of the model variables prior to and immediately after the re-initialization. The set of all re-initializations $Val \times Val$ is denoted $\mathcal{R}$.

A re-initialization clause is a pair $[V \mid P_r]$ of a set of model variables $V \subseteq \mathcal{V}_m$ and a re-initialization predicate $P_r \in \mathcal{P}_r$. The set V models which variables are allowed to change. Note that this is precisely opposite to flow clauses, where V denotes those variables that do *not* change (initially). Predicate P_r models the discontinuous changes. In a predicate, x^- denotes the value of a variable x before re-initialization, x^+ denotes the value of a variable x after re-initialization, and x' denotes the difference between x^+ and x^-, i.e., $x' = x^+ - x^-$. Also, P_r^- denotes the predicate P_r with all occurrences of x replaced by x^-.

The alternative composition $p \oplus q$ models a (non-deterministic) choice between the processes p and q. The sequential composition $p \odot q$ models a sequential execution of processes p and q. The process q is executed after (successful) termination of the process p.

The disrupt $p \blacktriangleright q$ models a kind of sequential composition where the process q may take over execution from process p at any moment, without waiting for its termination. This composition is essential for modeling two flow clauses executing one after the other, since the behavior of flow clauses never terminates. The left-disrupt $p \rhd q$ first executes a part of the process p and then behaves as a normal disrupt.

The parallel composition $p \parallel q$ models concurrent execution of p and q. The intuition behind this concurrent execution is that discrete actions are executed in an interleaving manner, with the possibility of synchronization of actions (as in ACP, where synchronization is called communication), while flow clauses are forced to synchronize, and can only synchronize if they accept the same solutions. The synchronization of actions takes place using a (partial, commutative and associative) communication function $| \in \mathcal{A} \times \mathcal{A} \mapsto \mathcal{A}$. For example, if the actions a and a' synchronize, then the resulting action is $a'' = a \,|\, a'$. Actions cannot synchronize with flow clauses, and in a parallel composition between those, the action executes first. Re-initializations synchronize only if the processes on which they act synchronize.

Encapsulation $\partial_H(p)$ models that the discrete actions from the set $H \subseteq \mathcal{A}$ are blocked during the execution of process p. This operator is often used in combination with the parallel composition to model that synchronization between discrete actions is enforced. Predicate encapsulation $\partial_{P_m}(p)$ models that all actions and flows that (at some point in time) satisfy the predicate on model variables P_m, are blocked. In (Cuijpers, 2004) and the present paper, it is used to formalize and analyze safety requirements on processes.

Terms can be constructed using variables from a given set of process variables $\mathcal{V}_p$ (with $\mathcal{V}_p \cap \mathcal{V}_m = \emptyset$), as usual. Finally, all processes should be interpreted in the light of a set E of recursive definitions of the form $X : p$, where X is a process variable and p is a term.

For some examples of modeling using HyPA we refer to (Cuijpers *et al.*, 2004; Cuijpers, 2004; Man *et al.*, 2005).

3. THE PICK-AND-PLACE MODULE

In this section, we give a model for the pick-and-place module based on bondgraph modeling.

In (Mateboer, 1999), a model of the pick-and-place module is used that contains differential and algebraic equations for the collision mechanics. Simulations are performed, to see how changes in the characteristics of the PCB influence the

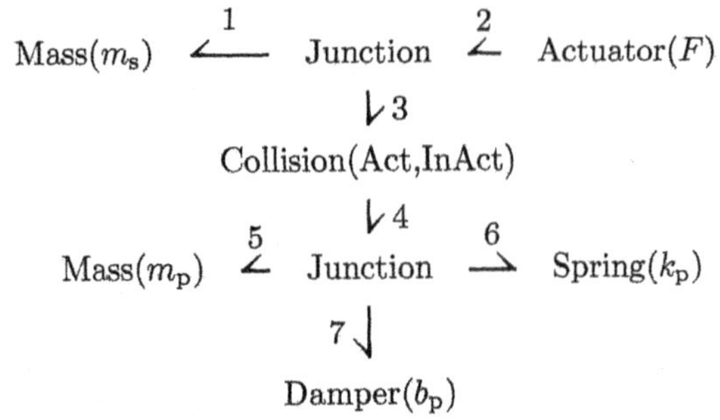

Fig. 2. Bondgraph of the pick-and-place module

performance of the controller. One of the conclusions of that report is that, if the characteristics of the collision mechanics are dominant over the characteristics of the PCB (an assumption that is reasonable in practice), then the impact behavior has ended before it is detected.

This relatively fast impact behavior, is the first reason why we abstract from the precise impact mechanics in our model, replacing it by discontinuous behavior. The second reason is that the parameters of the collision may vary wildly as different components and PCBs are used. Abstraction from the precise mechanics, means that we are robust against those variations.

In figure 2, a bondgraph model (Karnopp *et al.*, 1990) of the impact module is depicted. A bondgraph gives a graphical representation of the energy exchange in a physical system. In our example, the components of which the bondgraph consists are a mass (m_s) that models the sled, a force (F) that models the actuator through which the sled is controlled, a mass-spring-damper system (m_p, k_p, b_p) modeling the flexible PCB, and junctions (including a special collision junction) that model the exchange of energy between the other components.

Classically, the semantics of a bondgraph is given by a set of differential and algebraic (so called *constitutive*) equations for each of the bondgraph components (Karnopp *et al.*, 1990). However, inconsistencies tend to arise in these equations whenever connections between components are made or broken, which is what happens during collisions. In (Mosterman *et al.*, 1998), a method was proposed to deal with these inconsistencies while performing simulations of a system, and switching junctions such as our collision component were introduced to model the making and breaking of connections. In (Cuijpers *et al.*, 2004; Cuijpers, 2004) an algebraic semantics for hybrid bondgraphs was given using so called *hybrid constitutive processes* in the process algebra HyPA. These constitutive processes contain the usual constitutive differential and algebraic equations for continuous behavior, and in addi-

Table 1. Physical meaning of the variables (and constants).

Variable	Physical meaning	Unit
a	acceleration	$\frac{m}{s^2}$
b	damping constant	$\frac{Ns}{m} = \frac{kg}{s}$
E	energy	$J = \frac{kgm^2}{s^2}$
F	force	$N = \frac{kgm}{s^2}$
k	elasticity / spring constant	$\frac{N}{m} = \frac{kg}{s^2}$
m	mass	kg
p	momentum	$\frac{kgm}{s}$
v	velocity	$\frac{m}{s}$
x	position	m

tion contain constitutive difference equations that describe the discontinuities in behavior due to the aforementioned inconsistencies. Thus, the semantics of a hybrid bondgraph is formed by a parallel composition of constitutive hybrid processes for each of the bondgraph components. The constitutive hybrid processes associated with our impact system are explained below, and summarized in table 2.

The classical constitutive equations of bondgraphs are based on the insight that in many physical systems energy exchange (i.e. *power*) between two components is the product of two variables called *generalized effort* and *generalized flow*. In turn, generalized effort and generalized flow are the time-derivatives of the two state-variables of a physical component, which are *generalized momentum* and *generalized displacement* respectively. In mechanical systems, such as the one under study, *energy* (E) is a combination of kinetic and potential energy, generalized effort is better known as *force* (F), generalized flow is called *velocity* (v), generalized momentum is called *momentum* (p) and generalized displacement is called *position* (x). The behavior of each mechanical component can be expressed using these five variables, and their fundamental relations ($\dot{E} = F \cdot v$, $\dot{x} = v$ and $\dot{p} = F$) are captured in the Bond process in table 2. Note, that we have adapted the notation in this paper to that of the mechanical domain (see table 1).

In figure 2, each of the bonds (i.e. the half-arrows) is identified by a number. With each bond i variables E_i, p_i, and x_i, F_i and v_i, representing energy, momentum, position, force and velocity, of that bond, are associated. The direction of the bonds defines the positive direction of the flow of energy. The positive direction for the physical variables (e.g. which direction of movement is considered a 'positive' velocity) should be chosen in accordance to this. In relation to figure 1, all 'upward' displacements, velocities and forces are considered positive, contrary to what one might intuitively expect.

The continuous behavior of the mechanical components is described using the usual constitutive equations for conservation of energy and momentum. The components are additionally described by the equation $F = m \cdot a$ for a mass, $F = b \cdot v$ for a damper and $F = k \cdot x$ for a spring. The Actuator(F) process acts as a source of a constant force F. The discontinuous behavior of these mechanical components can be derived using partial integration from these laws. Partial integration of energy E as a function of momentum p, for example, gives us that the change in energy E' depends on the momentum p^- prior to the change and the momentum p^+ after the change, satisfying $E' = \frac{(p^+)^2 - (p^-)^2}{2 \cdot m}$. See (Cuijpers *et al.*, 2004) for a more complete treatment of the theory behind these derivations.

The Junction process serves to describe preservation of momentum and energy. Note our slight abuse of notation regarding the direction in which variables are to be interpreted. Firstly, for a set C of bonds, V_C denotes the set $\{E_c, p_c, x_c \mid c \in C\}$ of state-variables. Secondly, whenever we have an inward (outward) half-arrow in the bondgraph, we write a corresponding $+$ $(-)$ sign in the set C to signify the direction of the bond, and we write a $+$ $(-)$ in the corresponding constitutive equations whenever a $\pm$ sign occurs in these equations (typically, it will occur in front of a F_c or v_c).

The Collision process describes how junctions are made and/or broken as a result of collisions between masses (in (Cuijpers *et al.*, 2004), the Collision process is called a $(1/E)$-junction). The predicates Act and InAct describe the situation that the two masses are connected and act as one mass and the situation where the two masses are not connected, respectively (see table 2). Indeed, these two predicates are not local, in the sense that they use variables of components that are not directly connected by bonds. From a bondgraph perspective, this is still correct, since there is no exchange of energy defined through Act and InAct. However, the occurrence of non-local variables in the definition of Collision is, admittedly, confusing. A more elaborate variant of the bondgraph formalism exists that takes such exchange of information into account as well. It is for example used in (Mosterman *et al.*, 1998), but it is beyond the scope of this paper to treat such variants here.

4. CONTROL STRATEGY

In this section, we discuss the control strategy that is applied to the pick-and-place module. We base our discussion on the control strategy as it was first suggested by Philips CFT (Mateboer, 1999).

The control strategy consists of two phases. In the first phase, the sled is brought down using

Table 2. Definitions for the constitutive hybrid processes.

$$\text{Module} : \text{Mass}_1(m_s) \parallel \text{Bond}_1 \parallel \text{Actuator}_2(F) \parallel \text{Bond}_2 \parallel \text{Bond}_3 \parallel \text{Bond}_4 \parallel \text{Mass}_5(m_p) \parallel \text{Bond}_5 \parallel \text{Spring}_6(k_p) \parallel \text{Bond}_6 \parallel \text{Damper}_7(b_p) \parallel \text{Bond}_7 \parallel \text{Junction}_{\{-1,2,-3\}} \parallel \text{Collision}_{\{3,-4\}} \parallel \text{Junction}_{\{4,-5,-6,-7\}}$$

$$\text{Mass}_i(m) : \left(\left[E_i, p_i, x_i \;\middle|\; E_i' = \frac{(p_i^+)^2-(p_i^-)^2}{2\cdot m} \wedge x_i' = 0 \right] \gg \left(E_i, p_i, x_i \;\middle|\; p_i = m\cdot v_i \right)\right) \blacktriangleright \text{Mass}_i(m)$$

$$\text{Spring}_i(k) : \left(\left[E_i, p_i, x_i \;\middle|\; E_i' = k\cdot \frac{(x_i^+)^2-(x_i^-)^2}{2} \wedge p_i' = 0 \right] \gg \left(E_i, p_i, x_i \;\middle|\; x_i = \tfrac{1}{k}\cdot F_i \right)\right) \blacktriangleright \text{Spring}_i(k)$$

$$\text{Damper}_i(b) : \left(\left[E_i, p_i, x_i \;\middle|\; E_i' = 0 \wedge p_i' = 0 \wedge x_i' = 0 \right] \gg \left(E_i, p_i, x_i \;\middle|\; F_i = b\cdot v_i \right)\right) \blacktriangleright \text{Damper}_i(b)$$

$$\text{Actuator}_i(F) : \left(\left[E_i, p_i, x_i \;\middle|\; E_i' = 0 \wedge p_i' = 0 \right] \gg \left(E_i, p_i, x_i \;\middle|\; F_i = F \right)\right) \blacktriangleright \text{Actuator}_i(F)$$

$$\text{Bond}_i : \left(\left[E_i, p_i, x_i \;\middle|\; \mathit{true} \right] \gg \left(E_i, p_i, x_i \;\middle|\; \dot{E}_i = F_i\cdot v_i \wedge \dot{p}_i = F_i \wedge \dot{x}_i = v_i \right)\right) \blacktriangleright \text{Bond}_i$$

$$\text{Junction}_C : \left(\left[V_C \;\middle|\; \begin{array}{l} \sum_{c\in C} \pm E_c' = 0 \wedge \sum_{c\in C} \pm p_c' = 0 \\ \forall_{c,c'\in C}\; x_c' = x_{c'}' \end{array} \right] \gg \left(V_C \;\middle|\; \begin{array}{l} \sum_{c\in C} \pm F_c = 0 \\ \forall_{c,c'\in C}\; v_c = v_{c'} \end{array} \right)\right) \blacktriangleright \text{Junction}_C$$

$$\begin{aligned} \text{Collision}_C : \Bigg(& \left[V_C \;\middle|\; \begin{array}{l} \text{Act}^- \wedge \sum_{c\in C} \pm E_c' = 0 \\ \sum_{c\in C} \pm p_c' = 0 \wedge \forall_{c,c'\in C}\; x_c' = x_{c'}' \end{array} \right] \gg \left(V_C \;\middle|\; \begin{array}{l} \text{Act} \wedge \sum_{c\in C} \pm F_c = 0 \\ \forall_{c,c'\in C}\; v_c = v_{c'} \end{array} \right) \\ \oplus & \left[V_C \;\middle|\; \text{InAct}^- \wedge \forall_{c\in C}\; E_c' = 0 \wedge x_c' = 0 \right] \gg \left(V_C \;\middle|\; \text{InAct} \wedge \forall_{c\in C}\; F_c = 0 \right) \\ \oplus & \left[V_C \;\middle|\; \begin{array}{l} \text{InAct}^- \wedge \sum_{c\in C} \pm E_c' \geq 0 \\ \sum_{c\in C} \pm p_c' = 0 \wedge \forall_{c,c'\in C}\; x_c' = x_{c'}' \end{array} \right] \gg \left(V_C \;\middle|\; \begin{array}{l} \text{Act} \wedge \sum_{c\in C} \pm F_c = 0 \\ \forall_{c,c'\in C}\; v_c = v_{c'} \end{array} \right) \\ \oplus & \left[V_C \;\middle|\; \begin{array}{l} \text{Act}^- \wedge \sum_{c\in C} \pm E_c' \geq 0 \\ \sum_{c\in C} \pm p_c' = 0 \wedge \forall_{c,c'\in C}\; x_c' = x_{c'}' \end{array} \right] \gg \left(V_C \;\middle|\; \text{InAct} \wedge \forall_{c\in C}\; F_c = 0 \right) \\ \oplus & \left[V_C \;\middle|\; \text{InAct}^- \wedge \forall_{c\in C} \pm E_c' \geq 0 \wedge x_c' = 0 \right] \gg \left(V_C \;\middle|\; \text{Act} \wedge \sum_{c\in C} \pm F_c = 0 \wedge \forall_{c,c'\in C}\; v_c = v_{c'} \right) \\ \oplus & \left[V_C \;\middle|\; \text{Act}^- \wedge \forall_{c\in C} \pm E_c' \geq 0 \wedge x_c' = 0 \right] \gg \left(V_C \;\middle|\; \text{InAct} \wedge \forall_{c\in C}\; F_c = 0 \right)\Bigg) \blacktriangleright \text{Collision}_C \end{aligned}$$

$$\text{Act} \equiv (x_1 = x_5 \wedge v_1 > v_5) \vee (x_1 = x_5 \wedge v_1 = v_5 \wedge \tfrac{1}{m_s}\cdot(F_1 + F_3) \geq \tfrac{1}{m_p}\cdot(F_2 - F_4))$$

$$\text{InAct} \equiv x_1 \leq x_5$$

Table 3. Definitions for the controller, the sensor and the system.

$$\text{System} : \left[x_5^- = v_5^- = 0 \wedge \mathit{margin} \wedge \mathit{consistent} \right] \gg (\text{Module} \parallel \text{Controller} \parallel \text{Sensor})$$

$$\text{Controller} : \left(\begin{array}{l} -F_{\text{sat}} \leq F = -K\cdot(v_1 + v_{\text{seek}}) \leq F_{\text{sat}} \\ \vee \\ -F_{\text{sat}} \geq -K\cdot(v_1 + v_{\text{seek}}) \wedge F = -F_{\text{sat}} \\ \vee \\ -K\cdot(v_1 + v_{\text{seek}}) \geq F_{\text{sat}} \wedge F = 0 \end{array} \right) \blacktriangleright \text{impact?} \odot \left(F = -F_{\text{sat}} \right)$$

$$\text{Sensor} : \left(\mathit{margin} \right) \blacktriangleright \left[\begin{array}{c} v_1 \\ \mathit{clck} \end{array} \;\middle|\; \begin{array}{l} \mathit{margin}^- \\ \neg\mathit{margin}^+ \\ \mathit{clck}^+ = 0 \end{array} \right] \gg \left(\begin{array}{c} v_1 \\ \mathit{clck} \end{array} \;\middle|\; \begin{array}{l} \dot{\mathit{clck}} = 1 \\ \mathit{clck} \leq t_{\text{detect}} \end{array} \right) \blacktriangleright \left[0 < \mathit{clck}^- \leq t_{\text{detect}} \right] \gg \text{impact!}$$

$$\mathit{margin} \equiv -v_{\text{seek}} - v_{\text{detect}} \leq v_1 \leq -v_{\text{seek}} + v_{\text{detect}}$$

$$\mathit{consistent} \equiv \left(\begin{array}{l} p_1 = m_s\cdot v_1 \wedge p_5 = m_p\cdot v_5 \wedge x_6 = \frac{1}{k}\cdot F_6 \wedge F_2 = F \wedge F_7 = b\cdot v_7 \\ F_2 - F_1 = F_3 = F_4 = F_5 + F_6 + F_7 \wedge v_1 = v_2 = v_3 \wedge v_4 = v_5 = v_6 = v_7 \end{array} \right)$$

a proportional feedback law or P-controller (Dorf and Bishop, 1995). This means, that the controller measures the velocity v_1 of the sled and attempts to keep it at a constant value v_{seek}. It does this by changing the applied force F proportionally, by a constant factor K, to the difference between v_1 and v_{seek}. Note, that the controller synchronizes the values of v_1 and F with the sled and actuator components, respectively. In the second phase, the sled is pushed onto the PCB with a constant force, by bringing the force-actuator into saturation (F_{sat}). Naturally, in the actual system, the sled is brought up after placing the component on the PCB, but this is outside the scope of our studies at the moment. Here, we focus on one placement only. Our two phase control strategy is captured in the process Controller in table 3.

The controller switches from the first phase to the second when an impact is detected (the action impact?). A logical first step to detect this impact is by measuring the distance between sled and PCB. However, this approach needs an additional sensor, and turned out to be to expensive to be used in the final product. Instead, impact detection (the action impact!) is carried out by detecting an abrupt change in the velocity of the sled. The drawback of using a velocity sensor for impact detection, is that the detection is never immediate. We need to allow some time t_{detect} between impact and detection, and use an internal clock *clck* in our model, to measure this time. In order to make the model of the sensor more realistic, we also include a detection margin (with $v_{\text{detect}} > 0$) on the velocity (see the process Sensor in table 3). The sensor and the controller communicate over a channel according to impact! | impact? = impact.

For the impact detection to work correctly, it is important that the PCB is initially at rest $x_5 = v_5 = 0$. Furthermore, in order for an impact to be detectable, v_1 should be higher than a certain threshold. Lastly, for technical reasons, the initial state of the system should be physically consistent. The initial value of the physical variables should be chosen such that there is a possible solution to the constitutive equations. This is captured in the predicate *consistent*. In the remainder of this paper, we call the system as a whole System (see table 3).

5. SAFETY REQUIREMENTS

In (Mateboer, 1999) it is assumed that the placed component will be damaged when there is too great a force (F_{max}) acting on it. Using that assumption, a maximum impact velocity $v_{\text{max}} = \frac{F_{\text{max}}}{\sqrt{k_i \cdot m_s}}$ is calculated (where k_i is the internal elasticity of the component), at which the component can be brought down on the PCB. In our model, we have abstracted from the internal forces on the component, but we can still think about safety of the collision by assuming that there is a maximum impact velocity. This is reflected in the following predicate, that describes the condition under which a component remains undamaged:

$$S \equiv (x_1 = x_5 \Rightarrow v_1 - v_5 \geq -v_{\text{max}}).$$

Following the outline for safety analysis introduced in (Cuijpers, 2004), the analysis in the remainder of this paper is aimed at finding parameter values for K, F_{sat}, t_{detect} and v_{detect} such that the following condition holds:

$$\partial_{\neg S}\,(\text{System}) \leftrightarrows \text{System}.$$

Intuitively, one may say that a system is safe if it never goes into a state in which S is false. Since a change in the value of variables is always visible on the transitions, we may also say that no transitions may occur on which S is false. If (and only if) no transitions can occur on which S is false, then the process System is equivalent [2] to the $\neg S$-encapsulation of System, i.e. the process System with all $\neg S$-transitions removed.

In this paper, we only study damage that results directly from impact. Damage can also occur, for example, after an inelastic impact when $F_{\text{sat}} > F_{\text{max}}$. The study of this and other kinds of damage is left as a topic for future research.

6. ANALYSIS

Due to space considerations, we can only present the outline of our analysis of the pick-and-place module here. This outline is intended to give an overview of the different techniques that were used. For the details of the analysis we refer to (Cuijpers, 2004).

Recall that we set out to investigate under which constraints the condition $\partial_{\neg S}$ (System) $\leftrightarrows$ System holds. In this condition, System is formed by a parallel composition of various components. Since the encapsulation operator, in general, does not distribute over parallel compositions, a logical first step in our analysis was to obtain an equivalent representation of System without parallel compositions. This was done using a technique similar to *process algebraic linearization* (Usenko, 2002; van de Brand *et al.*, 2006) [3]. From this, we obtained an equivalent process description of the form

$$\text{System} \leftrightarrows \begin{bmatrix} x_5^- = v_5^- = 0 \\ \textit{margin}^- \end{bmatrix} \gg (\text{Seek} \blacktriangleright \text{Detect} \rhd \textit{impact} \odot \text{Bounce}),$$

in which Seek, Detect and Bounce are linear subprocesses that each describe the complete systems behavior in one of the three stages of control. The Seek process describes the systems behavior as the sled is brought down, the Detect process describes the systems behavior between collision and the detection of collision. The *impact* action describes the detection of the collision, and the Bounce process describes the systems behavior after detection, when a constant force is applied. Indeed, as the name suggests, the system sometimes performs bouncing behavior in this phase.

As parallel compositions are removed from the system description, elaborate flow clauses arise which describe the physical behavior of the system

---

[2] As $\leftrightarrows$ here means *initially stateless bisimulation* in the sense of (Mousavi *et al.*, 2005). This equivalence is a straightforward variant of bisimulation for transition systems in which data plays a role, but in (Cuijpers and Reniers, 2005) it was shown to have congruence problems with respect to parallel composition.

[3] As $\leftrightarrows$ is not a congruence for parallel composition, we used the stronger notion of *robust bisimulation* (Cuijpers and Reniers, 2005) (also called *stateless bisimulation* in (Mousavi *et al.*, 2005)) for this linearization. The use of two different equivalences indicates that the analysis takes place on two different levels of abstraction. Robust bisimulation is used to reason on an architectural level about the composition of components, while initially stateless bisimulation is used to reason about the behavioral aspects of the system as a whole.

using a set of differential and algebraic equations. These equations were solved using Mathematica, in order to find conditions under which each of the subprocesses, Seek, Detect and Bounce, are safe.

For the subprocesses Seek it was easy to show that the initial conditions $v_{\mathrm{max}} - v_{\mathrm{seek}} \geq v_{\mathrm{detect}}$ and $v_{\mathrm{detect}} \leq \frac{m_{\mathrm{p}}}{m_{\mathrm{s}}} v_{\mathrm{seek}} \leq v_{\mathrm{seek}}$ are sufficient to guarantee safety up to the point of collision.

Showing safety of Detect and Bounce turned out to be feasible only when we assume that collisions between sled and PCB are completely inelastic. Worse, if completely elastic collisions are considered possible, realistic parameter values can be found for which damage might occur at the second impact, i.e. after the first bounce! This is why we aimed at finding conditions under which no second impact could occur.

After changing the Collision process slightly, removing the options for partly elastic collision as described by the fourth and sixth alternative in the definition of Collision in table 2 , we repeated the linearization and found, again, a process of similar form.

But even the assumption of inelastic collisions turned out to be insufficient to guarantee that the sled and PCB stick together after the first impact. The Bounce process still admits bouncing behavior, because after the collision both masses move on together and a counter force builds up that eventually launches the sled of the PCB again. As it turned out, the force with which the sled is pressed onto the PCB should also be large enough to prevent such disconnection. We found a lower bound for this force

$$F_{\mathrm{sat}} \leq \frac{k_{\mathrm{p}} m_{\mathrm{s}}^2 \, v_{\mathrm{max}}}{(m_{\mathrm{s}}+m_{\mathrm{p}})\left(\theta m_{\mathrm{s}}+\omega\sqrt{m_{\mathrm{p}}(m_{\mathrm{p}}+2m_{\mathrm{s}})}\right)},$$

where $\omega = \frac{\sqrt{4k_{\mathrm{p}}(m_{\mathrm{s}}+m_{\mathrm{p}})-b_{\mathrm{p}}^2}}{2(m_{\mathrm{s}}+m_{\mathrm{p}})}$ and $\theta = \frac{b_{\mathrm{p}}}{2(m_{\mathrm{s}}+m_{\mathrm{p}})}$, which may be used as a requirement on the Controller process to guarantee that the sled is pressed to the PCB with sufficient strength, i.e. to guarantee that the Bounce subprocess is safe. Simulations have pointed out that this lower bound can be strengthened further, in some cases up to 60%, by taking the influence of damping into account. However, without insight in how sensitive the lower bound is for changes in the parameters, we must for the time being use the formula given above. Paradoxically, this means that the force that is suggested by Philips CFT in (Mateboer, 1999) should be *increased* by a factor 2 to ensure that the component mounting device is safe. As we have no insight in the maximum static load on the components, we can not determine whether this recommendation is reasonable or not.

Finally, the detection of the impact should be fast enough to guarantee that the pressing phase of the controller is activated before the disconnection of sled and PCB can no longer be prevented. We derived that we need

$$t_{\mathrm{detect}} \leq \frac{\frac{1}{2}\pi - \arctan\left(\frac{k_{\mathrm{p}}-b_{\mathrm{p}}\theta}{b_{\mathrm{p}}\omega}\right)}{\omega},$$

as a requirement on the Controller to guarantee that the Detect process is safe.

As a last step in our analysis, we need to show that the combination of safe initial conditions of the Seek, Detect and Bounce subprocesses forms a safe initial condition for the System process. For this, we used the special distribution laws for encapsulation and reinitialization given in (Cuijpers, 2004), that allow reasoning over $\leftrightarrows$ with respect to initial conditions.

7. CONCLUDING REMARKS

In this paper, we have build a hybrid model of a pick-and-place module, using bondgraph components described in the hybrid process algebra HyPA. Furthermore, we have analyzed a control strategy aimed at bringing a component to a PCB as quickly as possible, and press it onto the PCB with sufficient force to make it stick, but without breaking it.

We have shown how HyPA allows us to combine process algebraic analysis and system theoretic analysis, in the sense that these two kinds of analysis have been executed as separate operations on a common hybrid model. Regarding the use of process algebraic methods, we can conclude that the division of the system as a whole into manageable subprocesses certainly helped in the analysis of the pick-and-place module. Another strong point, is that we were able to describe impact detection as a separate sequential process. This allowed a separate formal treatment of the different phases of the process, which was useful because timing issues that played a role in the Detect process would otherwise have interfered with the search for a constraining force in the Bounce process. Also the collision dynamics were modeled as a separate component, which allowed us to easily change the model when it became apparent that we needed to assume inelastic collisions. This would not have been possible if we had not given semantics to the bondgraph paradigm using hybrid constitutive processes.

A weak point in the analysis, is that we needed to solve many of the differential equations that occur in the subprocesses. This made it hard to generalize our results to an analysis method, because many differential equations that occur in practice cannot be solved analytically. On the other hand, it was exactly the use of hybrid constitutive processes that allowed us to describe the physical behavior of the module using only simple differential equations in the first place. If we had not used the hybrid constitutive processes to model the module at exactly the right level of abstraction, the differential equations would certainly have been too difficult to solve. In particular, the hybrid approach allowed us to use basic conservation laws only to describe the impact behavior, instead of resorting to complex differential equations.

In the analysis discussed in this paper, HyPA was used to transfer analysis results from system theory to process algebra and back. The core of the safety-analysis was a proof that the system satisfies a certain process algebraic condition. Proving that this condition holds was done along process algebraic lines of reasoning, but each iteration required a system theoretic proof about the differential equations involved. In (Cuijpers, 2004), this method was already outlined, but as a topic for future research we propose to search for more and better combined proof methods of this kind, with a focus on hybrid topics like safety, stability, liveness and freedom of deadlocks.

ACKNOWLEDGEMENTS

Many thanks go to Paul van den Bosch, Jan Friso Groote and Ka Lok Man, for proof reading a preliminary version of this paper.

REFERENCES

Alur, R. and D.L. Dill (1994). A theory of timed automata. *Theoretical Computer Science* **126**(2), 183–235.

Baeten, J.C.M. and W.P. Weijland (1990). *Process Algebra.* Vol. 18 of *Cambridge Trancts in Theoretical Computer Science.* Cambridge University Press. Cambridge.

Bergstra, J.A. and C.A. Middelburg (2005). Process algebra for hybrid systems. *Theoretical Computer Science* **335**(2-3), 215–280.

Brinksma, E. (1985). A tutorial on LOTOS. In: *Proc. Protocol Specification, Testing and Verification V* (Michel Diaz, Ed.). Amsterdam, Netherlands. pp. 171–194.

Cuijpers, P.J.L. (2004). Hybrid Process Algebra. PhD thesis. Technische Universiteit Eindhoven (TU/e). Eindhoven, Netherlands.

Cuijpers, P.J.L. and M.A. Reniers (2005). Hybrid process algebra. *Journal of Logic and Algebraic Programming* **62**(2), 191–245.

Cuijpers, P.J.L., J.F. Broenink and P.J. Mosterman (2004). Constitutive hybrid processes. In: *Conference on Conceptual Modeling and Simulation.* Genua, Italy.

Dorf, R.C. and R.H. Bishop (1995). *Modern Control Systems.* Series in Electrical and Computer Engineering: Control Engineering. Addison-Wesley.

Febbraro, A. Di, Giua, A. and Menga, G., Eds. (2001). *Special Issue on Hybrid Petri Nets.* Vol. 11 of *Discrete Event Dynamic Systems.*

Heemels, W.P.M.H., B. De Schutter and A. Bemporad (2001). On the equivalence of classes of hybrid dynamical models. In: *Proc. Conference on Decision and Control.* Orlando, Florida. pp. 364–369.

Henzinger, T.A. (1996). The theory of hybrid automata. In: *Proceedings of the 11th Annual IEEE Symposium on Logic in Computer Science (LICS 1996).* IEEE Computer Society Press. pp. 278–292.

Karnopp, D.C., D.L. Margolis and R.C. Rosenberg (1990). *System Dynamics: A Unified Approach.* John Wiley and Sons, Inc.

Man, K., M.A. Reniers and P.J.L. Cuijpers (2005). Case studies in the hybrid process algebra hypa. *International Journal of Software Engineering and Knowledge Engineering* **15**(2), 299–305.

Mateboer, A.J. (1999). Eindverslag phi-z. Technical Report CTB595-99-3044. Philips CFT. Eindhoven, Netherlands.

Mosterman, P., G. Biswas and O. Otter (1998). Simulation of discontinuities in physical system models based on conservation principles. In: *Proceedings of the 1998 Summer Computer Simulation Conference.* Reno, Nevada. pp. 320–325.

Mousavi, M.R., M.A. Reniers and J.F. Groote (2005). Notions of bisimulation and congruence formats for SOS with data. *Information and Computation (I&C)* **200**(1), 104–147.

Rounds, W.C. and H. Song (2003). The ϕ-calculus: A language for distributed control of reconfigurable embedded systems. In: *Hybrid Systems: Computation and Control, 6th International Workshop, HSCC 2003* (F. Wiedijk, O. Maler and A. Pnueli, Eds.). Vol. 2623 of *Lecture Notes in Computer Science.* Springer-Verlag. pp. 435–449.

Usenko, Y.S. (2002). Linearization in μCRL. PhD thesis. Technische Universiteit Eindhoven (TU/e).

van Beek, D.A., K.L. Man, M.A. Reniers, J.E. Rooda and R.R.H. Schiffelers (2006). Syntax and consistent equation semantics of hybrid chi. *Journal of Logic and Algebraic Programming* **68**(1-2), 129–210.

van de Brand, P., M.A. Reniers and P.J.L. Cuijpers (2006). Linearization of hybrid processes. *Journal of Logic and Algebraic Programming* **68**(1-2), 54–104.

van der Schaft, A.J. and J.M. Schumacher (2000). *An Introduction to Hybrid Dynamical Systems.* Vol. 251 of *Lecture Notes in Control and Information Sciences.* Springer-Verlag. London.

ELSEVIER
IFAC
PUBLICATIONS

HUMAN SKILL MODELING BASED ON STOCHASTIC SWITCHED DYNAMICS

Tatsuya Suzuki * Shinkichi Inagaki * Naoyuki Yamada *

* *Nagoya University,*
Dept. of Mechanical Science and Engineering,
Furo-cho, Chikusa-ku, Nagoya

Abstract: This paper presents a modeling and analysis strategy of human skill based on stochastic switched dynamics. As a fundamental mathematical model, the Stochastic Switched ARX model (SS-ARX model) is introduced. Then the modeling and analysis strategy of human skill is proposed based on the stochastic switched impedance model which can be regarded as one of the SS-ARX model. Finally, the developed strategy is applied to peg-in-hole task which involves interesting dexterous human skill, and the effectiveness of the proposed strategy is discussed. *Copyright © 2006 IFAC*

Keywords: Human skill, Switched impedance model, Parameter estimation

1. INTRODUCTION

A system, in which an operator and artificial machine play in an interactive manner, is called a man-machine cooperative system. There are so many practical examples of this kind of systems such as, a power extender, automobile and so on. The goals of the man-machine cooperative systems are to enable the operator and artificial machine to work harmoniously. In order to realize this requirement, development of the 'intelligent assist' which does not conflict with the operator's intention must be addressed. Although this problem may include several subproblems, the most important one is the understanding of the operator's behavior by the artificial machine [Nechyba,1997] [Hannaford,1991] [Hirana,2004].

In order to model the operator's behavior, the conventional techniques such as the nonlinear regression models, the neural network and fuzzy systems have been used [Sjoberg,1995][Narendra,1990]. These techniques, however, have some problems as follows: (1) the obtained model often results in too complicated model, (2) this makes it impossible to understand the physical meaning of the operator's behavior. When we look at the human behavior, it is often found that the operator appropriately switches some simple primitive skills. The switching of primitive skills may be caused by operator's decision making. This consideration strongly motivates us to model the human behavior as a Hybrid Dynamical Systems (HDS). By regarding the operator's primitive skill and switching scenario as the continuous and discrete part of HDS, the understanding of the human behavior can be recasted as problem of the parameter estimation in HDS framework. Although many literatures have dealt with the expression, stability analysis, control, verification and identification [Ferrari,2003] [Bemporad,2004] of the HDS in the control and computer science communities, the application of the HDS model to the analysis of the human behavior has not been fully discussed yet.

Roughly speaking, HDS can be classified into two classes. The first one is the HDS where the transition between discrete states (modes) is specified by means of deterministic logics, and the second one is the HDS where the transition is specified by transition probabilities. In [Kim,2005], we have applied the HDS with deterministic mode change to the modeling of the driving behavior. Although this work can capture the motion and decision making aspects in the human behavior, it can not be suitable for the complex behavior analysis due to its high computational cost. This drawback is more emphasized when we consider the real-time application.

In this paper, first of all, a Stochastic Switched AutoRegressive eXogenous (SS-ARX) model is introduced. This model can be regarded as a natural extension of the standard Hidden Markov Model (HMM) [Rabiner,1989][Hannaford,1991] where different ARX model is allocated to each discrete state of the HMM. The significant advantages of using SS-ARX model as the behavior model is described as follows: (1) it can calculate the likelihood of the behavior with reasonable computational burden, (2) it can take into consideration the input and output signals of the human behavior, and (3) it can reflect the stochastic variance in the human behavior. Then, we develop the modeling and analysis strategy of human skill based on the stochastic switched impedance model, and finally, apply it to a peg-in-hole task which involves interesting dexterous human skill.

2. BRIEF REVIEW OF ARX MODEL

As a preliminary for SS-ARX model, the conventional ARX model is briefly reviewed.

The standard ARX model is described by the following difference equation:

$$\begin{aligned} y_t = c_1 y_{t-1} + c_1 y_{t-2} + \cdots + c_n y_{t-n} \\ + d_0 u_t + d_1 u_{t-1} + \cdots + d_m u_{t-m} + e_t \quad (1) \end{aligned}$$

where y_t and u_t are an output and an input of the system at t. They are supposed to be scalar-valued signals. Also, n and m are order of the ARX model, and $c_1, c_2, \cdots, c_n, d_0, d_1, \cdots, d_m$ are parameters. e_t is called an equation error, and is supposed to have a Gaussian distribution with variance σ.

By using the following vector form:

$$\boldsymbol{\theta} = (c_1, c_2, \cdots, c_n, d_0, d_1, \cdots, d_m)^T \quad (2)$$

$$\boldsymbol{\psi}_t = (y_{t-1}, y_{t-2}, \cdots, y_{t-n}, u_t, u_{t-1}, \cdots, u_{t-m})^T \quad (3)$$

equation (1) is rewritten as follows:

$$y_t = \boldsymbol{\psi}_t^T \boldsymbol{\theta} + e_t. \quad (4)$$

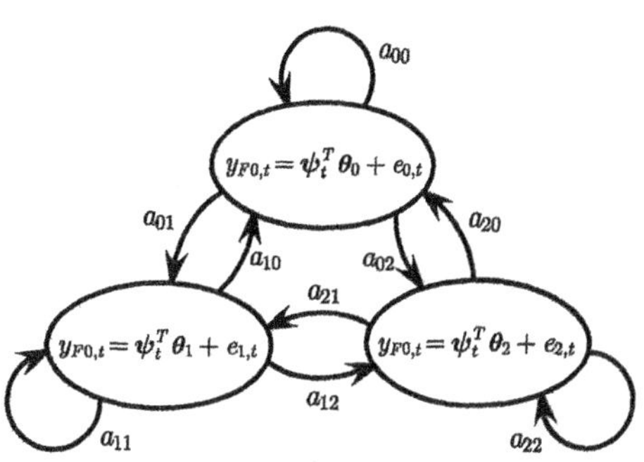

Fig. 1. SS-ARX model (three states)

3. PARAMETER ESTIMATION FOR SS-ARX MODEL

SS-ARX is defined as the system in which an ARX model is switched to other one according to the state transition probability as shown in Fig.1. This model can be regarded as the model in which the ARX model and HMM are combined.

3.1 Parameters in SS-ARX model

The parameters in SS-ARX model are specified as follows:

- Set of discrete states $S[= S_i, (i=0,1,\cdots,N)]$
- a_{ij}: State transition probability $(i=0,1,\cdots,N;$ $j=0,1,\cdots,N)$
- π_i: Initial state probability $(i=0,1,\cdots,N)$
- $\boldsymbol{\theta}_i$: Parameters in ARX model assigned to S_i $(i=0,1,2,\cdots,N)$
- σ_i: Variances of equation error in ARX model assigned to S_i $(i=0,1,2,\cdots,N)$

$N+1$ denotes the number of discrete states. In the following, we denote the set of parameters in the SS-ARX model by $\lambda=(\pi_i, a_{ij}, \boldsymbol{\theta}_i, \sigma_i)$.

3.2 Three fundamental problems

To address several problems listed below, the measured signal and its occurrence probability are defined for SS-ARX model as follows: First of all, a measured signal o_t at time t is defined as the combination of the output y_t and the regressor $\boldsymbol{\psi}_t$, that is, $o_t = (y_t, \boldsymbol{\psi}_t)$. Then, its occurrence probability $b_i(o_t)$ is defined by the assumption of the Gaussian distribution of the equation error, and is given by

$$b_i(o_t) = \frac{1}{\sqrt{2\pi}\sigma_i} \exp\left\{ -\frac{(\boldsymbol{\psi}_t^T \boldsymbol{\theta}_i - y_t)^2}{2\sigma_i^2} \right\}. \quad (5)$$

Based on these definitions, the following three fundamental problems can be addressed for SS-ARX model.

(1) Evaluation problem
In evaluation problem, the probability that the measured signal sequence $O=(o_0, o_1, \cdots, o_t, \cdots, o_T)$ occurs from the model $\lambda = (\pi_i, a_{ij}, \boldsymbol{\theta}_i, \sigma_i)$ is calculated. This problem can be solved by applying Forward algorithm [Rabiner,1989].
(2) Decoding problem
In decoding problem, the most likely underlying state sequence $\boldsymbol{s}=(s_0, s_1, \cdots, s_t, \cdots, s_T)$, which yields the measured signal sequence $O = (o_0, o_1, \cdots, o_t, \cdots, o_T)$, is found for the model $\lambda = (\pi_i, a_{ij}, \boldsymbol{\theta}_i, \sigma_i)$. This state estimation can be realized by applying Viterbi algorithm [Rabiner,1989].
(3) Estimation problem
In estimation problem, the model parameter $\lambda = (\pi_i, a_{ij}, \boldsymbol{\theta}_i, \sigma_i)$, which gives the highest occurrence probability for the measured signal sequence $O = (o_0, o_1, \cdots, o_t, \cdots, o_T)$, is estimated.

The solution for problems (1) and (2) are same as ones for the standard HMM. However, the parameter estimation algorithm for the SS-ARX model requires some extension to the one for the standard HMM. In the following subsection, the concrete parameter estimation algorithm for the SS-ARX model is derived.

3.3 Parameter estimation

Here, we assume that L measured signal sequences are collected for the parameter estimation of SS-ARX model.

3.3.1. EM algorithm First of all, we consider an unmeasurable state sequence

$$\boldsymbol{s} = (s_0, s_1, \cdots, s_t, \cdots, s_T) \tag{6}$$

and measurable signal sequences

$$\boldsymbol{O}_l = (o_{l,0}, o_{l,1}, \cdots, o_{l,t}, \cdots, o_{l,T}) \tag{7}$$

(where l represents the index of the measured signal sequence). The maximization of the likelihood value of the $\boldsymbol{s}$ and $\boldsymbol{O}_l$, $\sum_{l=1}^{L} L(\boldsymbol{s}, \boldsymbol{O}_l; \lambda) = \sum_{l=1}^{L} P(\boldsymbol{s}, \boldsymbol{O}_l|\lambda)$ is achieved by introducing the EM (Expectation and Maximization) algorithm.

Generally, the EM algorithm tries to find the parameter λ' which maximizes the following Q function:

$$Q(\lambda, \lambda') = \sum_{l=1}^{L} E[\log\{P(\boldsymbol{s}, \boldsymbol{O}_l|\lambda')\}\,|\boldsymbol{O}_l, \lambda] \tag{8}$$

$$= \sum_{l=1}^{L} \sum_{\boldsymbol{s}} P(\boldsymbol{s}|\boldsymbol{O}_l, \lambda) \log\{P(\boldsymbol{s}, \boldsymbol{O}_l|\lambda')\} \tag{9}$$

by executing following procedures iteratively.

(1) Specify an initial parameter λ.
(2) Find the λ' which maximizes the $Q(\lambda, \lambda')$.
(3) Substitute λ' for λ, and iterate (2) until $\lambda'=\lambda$ holds.

3.3.2. Parameter estimation algorithm The parameters of SS-ARX model before and after the update are supposed to be given by $\lambda = (\pi_i, a_{ij}, \boldsymbol{\theta}_i, \sigma_i)$, and $\lambda'=(\pi'_i, a'_{ij}, \boldsymbol{\theta'}_i, \sigma'_i)$. From (9),

$$Q(\lambda, \lambda') = \sum_{l=1}^{L} \left\{ \sum_{\boldsymbol{s}} \frac{1}{P(\boldsymbol{O}_l|\lambda)} P(\boldsymbol{s}, \boldsymbol{O}_l|\lambda) \times \log\{P(\boldsymbol{s}, \boldsymbol{O}_l|\lambda')\} \right\}. \tag{10}$$

Now, we replace $\frac{1}{P(\boldsymbol{O}_0|\lambda)}, \frac{1}{P(\boldsymbol{O}_1|\lambda)}, \cdots, \frac{1}{P(\boldsymbol{O}_L|\lambda)}$ in (10) by $k_0, k_1, \cdots, k_L$.

Since $k_0, k_1, \cdots, k_L$ are constant, the maximization of $Q(\lambda, \lambda')$ implies the maximization of $\tilde{Q}(\lambda, \lambda')$ given by

$$\tilde{Q}(\lambda, \lambda') = \sum_{\boldsymbol{s}} P(\boldsymbol{s}, \boldsymbol{O}_l|\lambda) \log\{P(\boldsymbol{s}, \boldsymbol{O}_l|\lambda')\}. \tag{11}$$

By using the definition, $\tilde{Q}(\lambda, \lambda')$ can be decomposed as follows:

$$\tilde{Q}(\lambda, \lambda') = \tilde{Q}_1(\lambda, \pi'_i) + \tilde{Q}_2(\lambda, a'_{ij}) + \tilde{Q}_3(\lambda, \boldsymbol{\theta'}_i, \sigma'_i) \tag{12}$$

where

$$\tilde{Q}_1 = \sum_{l=1}^{L} \sum_{i=0}^{N} k_l \pi_i b_i(o_{l,0}) \log\{\pi'_i\} \beta(l, i, 0) \tag{13}$$

$$\tilde{Q}_2 = \sum_{l=1}^{L} \sum_{t=1}^{T} \sum_{i=0}^{N} \sum_{j=0}^{N} k_l \log\{a'_{ij}\} \times \alpha(l, i, t-1) a_{ij} b_j(o_{l,t}) \beta(l, j, t) \tag{14}$$

$$\tilde{Q}_3 = \sum_{l=1}^{L} \sum_{t=0}^{T} \sum_{i=0}^{N} k_l \log\{b'_i(o_{l,t})\} \alpha(l, i, t) \beta(l, i, t). \tag{15}$$

The forward probability $\alpha(l, i, t)$ and backward probability $\beta(l, i, t)$ in (13), (14) and (15) are defined as follows:

$$\alpha(l, i, t) = \sum_{s_0=0}^{N} \sum_{s_1=0}^{N} \cdots \sum_{s_{t-1}=0}^{N} \pi_{s_0} b_{s_0}(o_{l,0}) \times a_{s_0 s_1} b_{s_1}(o_{l,1}) \times \cdots \times a_{s_{t-1} s_t} b_{s_t}(o_{l,t}) \tag{16}$$

$$\beta(l, i, t) = \sum_{s_{t+1}=0}^{N} \sum_{s_{t+2}=0}^{N} \cdots \sum_{s_T=0}^{N} a_{s_t s_{t+1}} b_{t+1}(o_{l,t+1})$$

$$\times a_{s_{t+1}s_{t+2}} b_{s_{t+2}}(o_{l,t+2}) \times \cdots \times a_{s_{T-1}s_T} b_{s_T}(o_{l,T}). \tag{17}$$

The meaning of $\alpha(l,i,t)$ is the probability for SS-ARX model λ to generate the lth measured signal subsequence $\boldsymbol{O}_l = (o_{l,0}, o_{l,1}, \cdots, o_{l,t})$ until t and reach the state S_i at t (i.e. $s_t = S_i$). Also, the meaning of $\beta(l,i,t)$ is the probability for SS-ARX model λ to generate the lth measured signal subsequence $\boldsymbol{O}_l = (o_{l,t+1}, o_{l,t+2}, \cdots, o_{l,T})$ starting from S_i at t (i.e. $s_t = S_i$) and reach the final state at T.

Then, by maximizing $\tilde{Q}_1(\lambda, \pi'_i)$, $\tilde{Q}_2(\lambda, a'_{ij})$ and $\tilde{Q}_3(\lambda, \boldsymbol{\theta}'_i, \sigma'_i)$, $\tilde{Q}(\lambda, \lambda')$ can be maximized. λ' which maximizes the $\tilde{Q}(\lambda, \lambda')$ can be obtained as follows:

$$\pi'_i = \frac{\sum_{l=1}^{L} k_l \pi_i b_i(o_{l,0}) \beta(l,i,0)}{\sum_{i=0}^{N} \sum_{l=1}^{L} k_l \pi_i b_i(o_{l,0}) \beta(l,i,0)} \tag{18}$$

$$a'_{ij} = \frac{\sum_{t=1}^{T} \sum_{l=1}^{L} k_l \alpha(l,i,t-1) a_{ij} b_j(o_{l,t}) \beta(l,j,t)}{\sum_{j=0}^{N} \sum_{t=1}^{T} \sum_{l=1}^{L} k_l \alpha(l,i,t-1) a_{ij} b_j(o_{l,t}) \beta(l,j,t)} \tag{19}$$

$$\boldsymbol{\theta}'_i = \left\{ \sum_{t=0}^{T} \sum_{l=1}^{L} k_l \boldsymbol{\psi}_{l,t} \boldsymbol{\psi}_{l,t}^T \alpha(l,i,t) \beta(l,i,t) \right\}^{-1} \times \left\{ \sum_{t=0}^{T} \sum_{l=1}^{L} k_l \boldsymbol{\psi}_{l,t} y_{l,t} \alpha(l,i,t) \beta(l,i,t) \right\} \tag{20}$$

$$\sigma'^2_i = \frac{\sum_{t=0}^{T} \sum_{l=1}^{L} k_l |\boldsymbol{\theta}'^T_i \boldsymbol{\psi}_{l,t} - y_{l,t}|^2 \alpha(l,i,t) \beta(l,i,t)}{\sum_{t=0}^{T} \sum_{l=1}^{L} k_l \alpha(l,i,t) \beta(l,i,t)}. \tag{21}$$

By iterating three steps in the EM algorithm together with (18), (19), (20) and (21), the parameter λ is locally maximized.

Note that eq. (20) can be regarded as the weighted least mean square solution in which the weight parameters are specified by $\alpha(l,i,t)\beta(l,i,t)$, i.e. the probability that the $o_{l,t}$ is generated from the state S_i.

Also, the parameter estimation algorithm described in this section can be easily extended to the multiple output case.

4. APPLICATION TO ANALYSIS OF HUMAN SKILL

4.1 Task environment

The task environment addressed in this paper is depicted in Fig.2. The peg is supposed to move only in $x-z$ plane.

Peg and hole are made of Aluminum and rubber, respectively. The clearance of the hole is 0.1[mm]. The operator executes the task based on the scenario depicted in Fig.3.

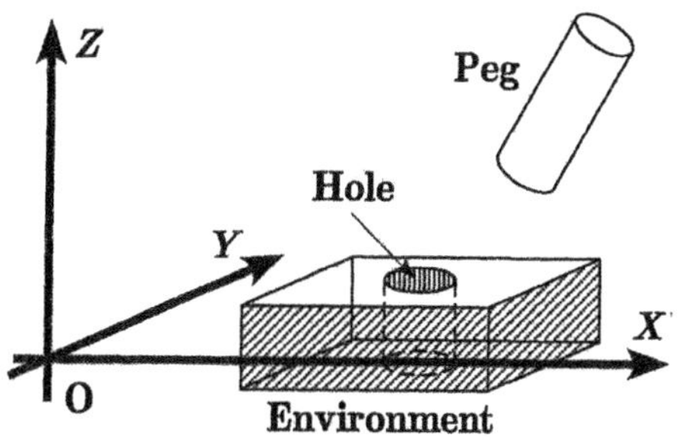

Fig. 2. Peg-in-hole task

In this scenario, since the switchings from (1) to (2) and (4) to (5) are considered to be deterministic, the transition from (2) to (4) were analyzed based on stochastic switched dynamics.

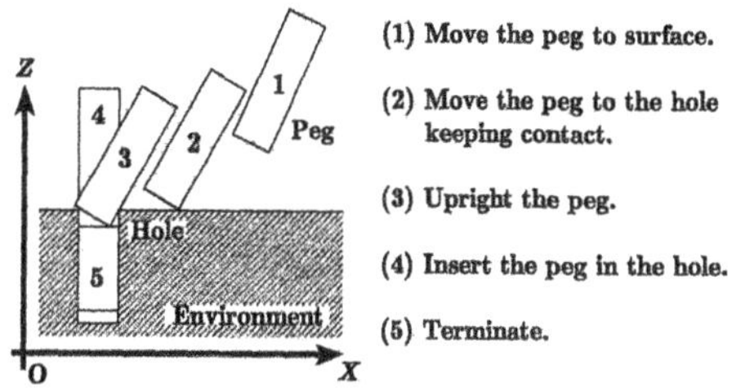

Fig. 3. Motion of peg

4.2 Impedance model representing human skill

4.2.1. Impedance model In the field of robotics, the impedance model has been considered as typical dynamical model to represent the human skill. In this work, therefore, we define the ARX model at each discrete state by impedance model to represent the dexterous human skill. The impedance model considered for the peg-in-hole task is shown in Fig.4, and can be described by (22), (23) and (24).

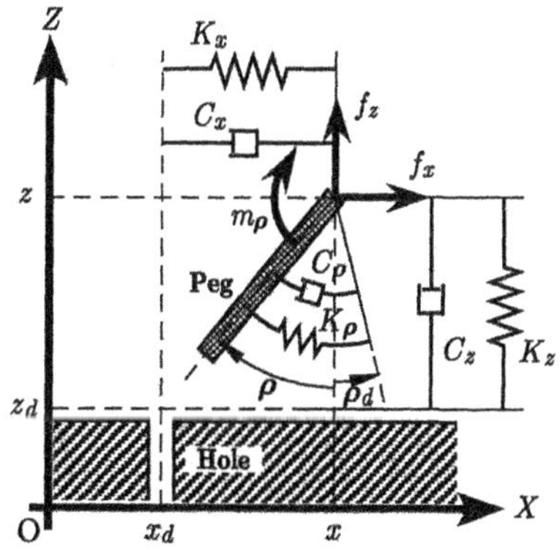

Fig. 4. Impedance model

$$M_{xi}\ddot{x} + C_{xi}\dot{x} + K_{xi}x + D_{xi} = f_x \tag{22}$$

$$M_{zi}\ddot{z} + C_{zi}\dot{z} + K_{zi}z + D_{zi} = f_z \tag{23}$$

$$M_{\rho i}\ddot{\rho} + C_{\rho i}\dot{\rho} + K_{\rho i}\rho + D_{\rho i} = m_\rho. \tag{24}$$

In Fig. 4, eqs. (22), (23) and (24), x, z, ρ denote position of peg, f_x, f_z, m_ρ denote interactive force, and M_{xi}, M_{zi}, $M_{\rho i}$, C_{xi}, C_{zi}, $C_{\rho i}$, K_{xi}, K_{zi}, $K_{\rho i}$ denote impedance parameters at ith discrete state. Also, $D_{xi} = -K_{xi}x_d + f_{xdi}$, $D_{zi} = -K_{zi}z_d + f_{zdi}$, $D_{\rho i} = -K_\rho \rho_d + m_{\rho di}$. x_d, z_d, ρ_d denote reference position, and f_{xdi}, f_{zdi}, $m_{\rho di}$ denote virtual reference force.

Note that virtual reference force varies from discrete state to state, and D_{xi}, D_{zi} and $D_{\rho i}$ are also unknown.

4.2.2. Transformation to discrete-time model Since equations (22), (23) and (24) are model in continuous time domain, their discrete-time model are derived with preserving parameters in continuous time model (See [Wada,1993] for detail). The transformation from continuous-time model to its discrete-time model is derived as follows: In the following, p represents one of the coordinate, i.e. p is a substitute for one of x, z and ρ. Note that D_{pi} is replaced by $D_{pi}d$ by introducing a parameter $d = 1$ in the following.

$$F_{pF}(z) = M_{pi}P_{F0}(z) + C_{pi}P_{F1}(z) + K_{pi}P_{F2}(z) + D_{pi}D_F(z) \quad (25)$$

where

$$F_{xF}(z) = \frac{(\frac{\Delta}{2})^2(1+z^{-1})^2}{\{(1+\nu\frac{\Delta}{2})-(1-\nu\frac{\Delta}{2})z^{-1}\}^2}F_x(z) \quad (26)$$

$$D_F(z) = \frac{(\frac{\Delta}{2})^2(1+z^{-1})^2}{\{(1+\nu\frac{\Delta}{2})-(1-\nu\frac{\Delta}{2})z^{-1}\}^2}D(z) \quad (27)$$

$$P_{Fk}(z) = \frac{(\frac{\Delta}{2})^k(1+z^{-1})^k(1-z^{-1})^{2-k}}{\{(1+\nu\frac{\Delta}{2})-(1-\nu\frac{\Delta}{2})z^{-1}\}^2}P(z) \quad (28)$$

($k=0,1,2$). $F_p(z)$, $D(z)$ and $P(z)$ are z-transform of f_p, d and p. ν is a time constant of the filter, and was set to be $\nu = 20$ in this work. Finally, by applying inverse z-transform to (25), and adding equation error $e_{pi,t}$, we obtain

$$f_{pF,t} = M_{pi}p_{F0,t} + C_{pi}p_{F1,t} + K_{pi}p_{F2,t} + D_{pi}d_{F,t} + e_{pi,t} \quad (29)$$

$$= \psi_{p,t}^T \theta_{pi} + e_{pi,t}. \quad (30)$$

Thus, the discrete-time impedance model can be obtained with preserving the parameters in continuous time domain.

Now , the parameter vectors θ_{xi}, θ_{zi} and $\theta_{\rho i}$ and regressor vectors $\psi_{x,t}$, $\psi_{z,t}$ and $\psi_{\rho,t}$ are defined for each coordinate as follows:

$$\theta_{xi} = (M_{xi}, C_{xi}, K_{xi}, D_{xi})^T \quad (31)$$

$$\theta_{zi} = (M_{zi}, C_{zi}, K_{zi}, D_{zi})^T \quad (32)$$

$$\theta_{\rho i} = (M_{\rho i}, C_{\rho i}, K_{\rho i}, D_{\rho i})^T \quad (33)$$

$$\psi_{x,t} = (x_{F0,t}, x_{F1,t}, x_{F2,t}, d_{F,t})^T \quad (34)$$

$$\psi_{z,t} = (z_{F0,t}, z_{F1,t}, z_{F2,t}, d_{F,t})^T \quad (35)$$

$$\psi_{\rho,t} = (\rho_{F0,t}, \rho_{F1,t}, \rho_{F2,t}, d_{F,t})^T. \quad (36)$$

4.2.3. Stochastic switched impedance model We consider the left-to-right model with five discrete states as the model for peg-in-hole task (Fig. 5).

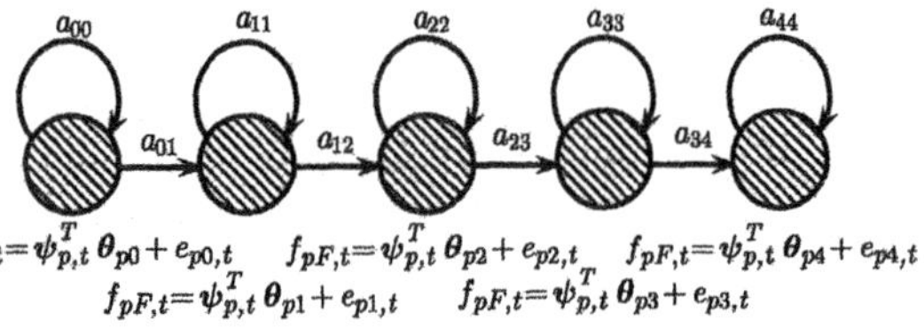

Fig. 5. Stochastic switched impedance model (left to right)

4.3 Acquisition of data and result of parameter estimation

4.3.1. Acquisition of data The operator manipulated the data acquisition tool shown in Fig. 6. The position (x[mm], z[mm], ρ[rad]) and interactive force (f_x[N], f_z[N], m_y[Nm]) are measured by three potentiometers and force sensor. Sampling interval was set to be 2[msec], it took about 5[sec] to complete the task.

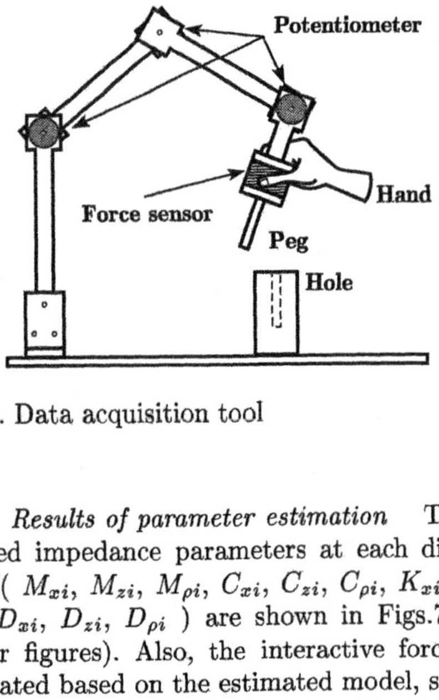

Fig. 6. Data acquisition tool

4.3.2. Results of parameter estimation The estimated impedance parameters at each discrete state (M_{xi}, M_{zi}, $M_{\rho i}$, C_{xi}, C_{zi}, $C_{\rho i}$, K_{xi}, K_{zi}, $K_{\rho i}$, D_{xi}, D_{zi}, $D_{\rho i}$) are shown in Figs.7 to 9 (upper figures). Also, the interactive force was calculated based on the estimated model, switching points (obtained by applying Viterbi algorithm) and the filtered measured position data in (29) in order to verify the parameter estimation. The calculated force profiles are depicted in lower figures in Figs.7 to 9 together with the estimated switching points represented by vertical dash lines. The profiles of the real force data also represent the data obtained by applying the low-pass filter $1/(1+\nu^2)$ to the original measured force data. In these figures, the real and calculated force data agree well with each other. This implies that the proposed estimation technique works well.

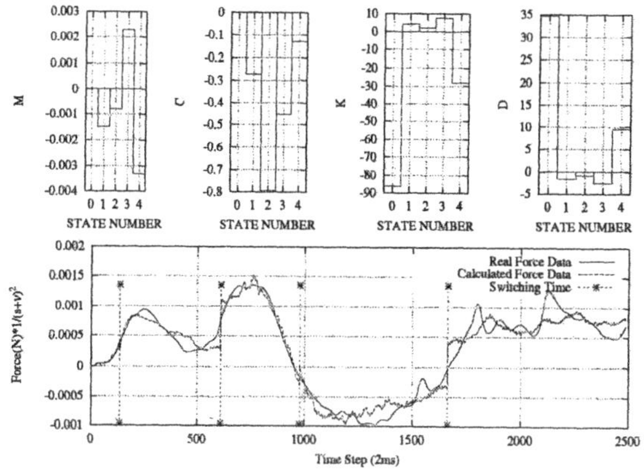

Fig. 7. Estimated parameter and comparison of real force and calculated force (X-axis)

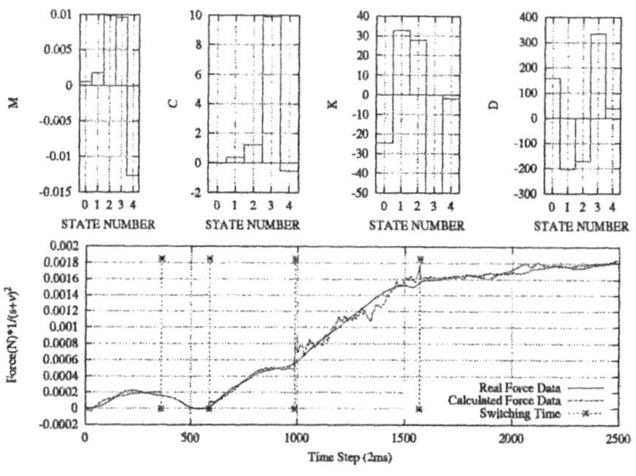

Fig. 8. Estimated parameter and comparison of real force and calculated force (Z-axis)

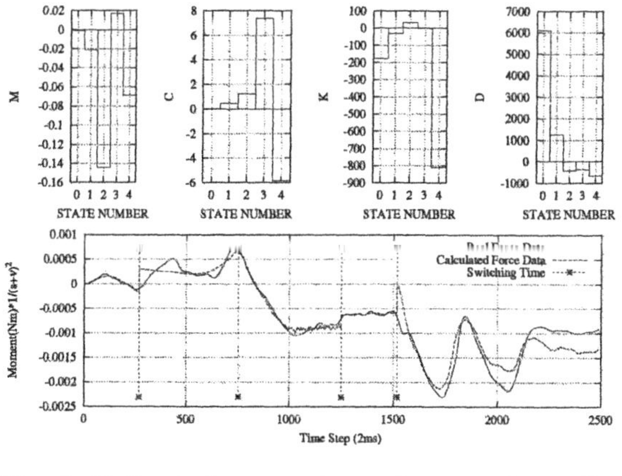

Fig. 9. Estimated parameter and comparison of real force and calculated force (around Y-axis)

4.4 Discussion and application

We can see the several interesting properties among obtained parameters. First of all, estimated switching points seem to coincide with the instance of change of geometrical constraints for the peg. As for the impedance parameters, although some parameter show negative value, the task was performed safely. This can be explained as follows: The stability of the task is specified by overall closed-loop dynamics given by Fig.10. In

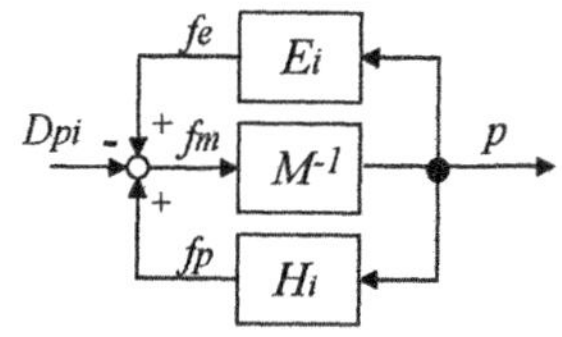

Fig. 10. Block diagram of whole system

Fig.10, each block is given by

$$M = M_m s^2 \tag{37}$$

$$H_i = M_{pi} s^2 + C_{pi} s + K_{pi} \tag{38}$$

$$E_i = C_{ei} s + K_{ei} \tag{39}$$

where H_i indicates ith dynamics of human skill (estimated in the previous subsection) , M represents the mass of the handling tool, i.e. the peg and force sensor, and E_i represents the dynamics of the environment at ith discrete state which consists of the dynamics of hole and force sensor. p represents the coordinate (x, z and ρ). This consideration allows us to have negative impedance parameter in realization of stable peg insertion.

The direct application of the obtained results is an implementation of the target impedance models for the industrial robot together with the switching scenario of them. This kind of 'Skill transfer' is one of the most promising idea to realize the dexterous manipulation. The other interesting application is a quantitative evaluation (recognition) of the skill based on the similarity measure for the stochastic processes. This kind of measure can be a criterion for the design of artificial skill [Nechyba,1997][Hirana,2004].

Also, it will be quite interesting to investigate the obtained impedance parameters for understanding of physical meaning of human skill. Our developed strategy can be a good first step to tackle this problem.

5. CONCLUSIONS

In this paper, first of all, a Stochastic Switched AutoRegressive eXogenous (SS-ARX) model was introduced and its parameter estimation algorithm has been derived.. Then, we have developed the modeling and analysis strategy of human skill based on the stochastic switched impedance model which can be regarded as one of the SS-ARX model. Finally, the developed strategy was applied to peg-in-hole task which involves interesting dexterous human skill, and the effectiveness of the proposed strategy was discussed.

REFERENCES

[Nechyba,1997] *M.C. Nechyba and Y. Xu : Human Control Strategy: Abstraction, Verification and Replication, IEEE Control Systems Magazine, vol. 17, no. 5, pp. 48-61, 1997*

[Hannaford,1991] *B.Hannaford and P.Lee: Hidden Markov Model Analysis of Force/Torque Information in Telemanipulation, The Int. J. of Robotics Research, 10-5, 528/539, 1991*

[Hirana,2004] *K.Hirana, T.Suzuki, et.al: Quantitative Evaluation for Skill Controller Based on Comparison with Human Demonstration, IEEE Trans. on Cont. Syst. Tech., Vol.12, No.4, pp.609-619, 2004*

[Sjoberg,1995] *J.Sjoberg, Q.Zhang, et.al : Nonlinear Black-Box Modeling in System Identification: a Unified Overview, Automatica, vol.31, no.12, pp.1691-1724, 1995*

[Narendra,1990] *K.S.Narendra and K.Pathasarathy: Identification and Control of Dynamical Systems Using Neural Networks, IEEE Trans. on Neural Networks, vol.1, no.1, pp.4-27, 1990*

[Ferrari,2003] *G.Ferrari-Trecate, M.Morari et.al; A clustering technique for the identification of piecewise affine systems, Automatica, 39, pp.205-217, 2003*

[Bemporad,2004] *J.Roll, A.Bemporad, et.al; Identification of piecewise affine systems via mixed-integer programming, Automatica, 40, pp.37-50, 2004*

[Kim,2005] *J.H.Kim, S.Hayakawa, T.Suzuki, et.al; Modeling of Driver's Collision Avoidance Maneuver based on Controller Switching Model IEEE Trans. of Syst. Man and Cyber., Vol.35, No.6, pp.1131-1143, 2005*

[Suzuki,2005] *T.Suzuki, S.Yamada, et.al: Modeling of Drivers Collision Avoidance Behavior based on Hybrid System Model -An Approach with Data Clustering- Proc. of IEEE Int. Conf. on Syst. Man and Cyber. pp.3817-3822, 2005*

[Rabiner,1989] *L.R.Rabiner; A Tutorial on Hidden Markov Models and Selected Applications in Speech Recognition, IEEE, Vol.77, 1989*

[Wada,1993] *S.Sagara, Z.J.Yang, K.Wada; Identification of Continuous Systems Using Digital Low-pass Filters, Int.J.Syst.Sci.,22-7,1159-1176, 1991*

[Kim,2004] *J.H.Kim, S.Hayakawa, T.Suzuki et.al; Modeling of Human Driving Behavior based on Expression as Hybrid Dynamical System, Trans.of SICE, Vol.40, No.2, pp.180-188, 2004*

[Balakrishnan,2004] *H.Balakrishnan et.al; Inference Methods for Autonomous Stochastic Linear Hybrid Systems" LNCS2993, HSCC 2004, pp.64-79, 2004*

Copyright © Analysis and Design of Hybrid Systems
Alghero, Italy 2006

ELSEVIER
IFAC
PUBLICATIONS

BUILDING EFFICIENT SIMULATIONS FROM HYBRID BOND GRAPH MODELS

Christopher D. Beers * Eric-Jan Manders * Gautam Biswas *
Pieter J. Mosterman **

** Dept. of EECS and ISIS, Vanderbilt University, Nashville, TN*
*** The MathWorks, Natick, MA*

Abstract: Embedded systems and their corresponding hybrid models are pervasive in engineering applications, therefore, systematic mathematical analysis using these models has become an important research area. Our approach to hybrid modeling with Hybrid Bond Graphs allows for seamless integration of physical system principles with discrete computational structures, but simulating the hybrid behaviors can be difficult and computationally expensive. In this paper, we develop a methodology that transforms Hybrid Bond Graphs into computational block diagrams and incrementally modifies the block diagram when mode changes occur. This forms the basis for a computationally efficient hybrid simulation algorithm. *Copyright © 2006 IFAC*

Keywords: Hybrid Bond Graphs, Causality, Hybrid Systems, Simulation

1. INTRODUCTION

Modeling and simulation play a central role in the design and operation of modern engineering systems that are made up of a large number of interacting components. Many of these systems are *hybrid*, i.e., they combine continuous and discrete behaviors. Hybrid simulation schemes must correctly handle discrete mode transitions that involve model reconfiguration and discontinuous updates to the system state variables. Recent research has begun to address the mathematical complexity of hybrid system simulation schemes [1].

A particularly intuitive physics-based modeling paradigm are Bond Graphs (BGs) [9]. They provide a uniform lumped parameter, energy-based topological framework across multiple physical domains (e.g., electrical, fluid, mechanical, and thermal). Hybrid Bond graphs (HBGs) extend the BGs by incorporating local switching functions that enable the reconfiguration of energy flow paths in the model ([15]). This allows for seamless integration of energetic modeling and model reconfiguration to handle hybrid behaviors.

The inherent causal structure in BG models provides the basis for efficient conversion of BGs to computation models (e.g., [3, 9]). For HBGs, the computation model is more complex because junction switching during behavior generation results in dynamic updating of the causal assignments and the computational structure during execution. This paper presents an efficient method for constructing a simulation model for HBGs. Specifically, we create block diagram models, where run time changes in model configuration are handled by reconfiguring the data flow through the blocks of the model. We demonstrate the technique by creating a computational model in a commercially available simulation environment.

The choice of block diagram models is motivated by our work on simulation testbeds for fault detection and isolation and integrated systems health monitoring. For these applications, the simulation environment has to be designed in a way that component parameters in the model can be accessed and changed at run time to emulate degradations and faults in the system. The change profile for the parameters can take on a number of time-varying forms. In our work, the component parameters are in 1-1 correspondence with BG parameters. and they map directly to individual blocks in the BG model.

2. HYBRID BOND GRAPHS (HBG) OVERVIEW

BGs are topological models that represent energy exchange pathways in physical processes [9]. The generic elements in BGs are energy storage, dissipative, transformation, and input-output elements. The dynamics of physical system behavior is captured by the transfer of energy between the components via the connected bonds and two idealized connections, called the 0- (common effort) and the 1- (common flow) junctions.

Introducing discrete behavior into continuous BGs has been investigated by several researchers [4, 11, 19]. HBGs introduce discrete changes in system configuration as idealized *controlled junctions* with two states *on* and *off* [15]. A Finite State Machine (FSM) implements the junction *control specification*. When the controlled junction is on, it behaves like a conventional junction. In the off state, all bonds incident on the junction are de-activated by enforcing a 0 effort or flow at the junction. Fig. 1 illustrates the discrete behavior of a controlled 1-junction. The system mode at any time is determined by a parallel composition of modes of the individual switched junctions.

To illustrate the concepts developed in this paper, we will use the bond graph model illustrated in Fig. 2. The system has a single source of effort, $v(t)$, that can be switched on or off, two capacities, C1 and C2, two inertias, L1 and L2, and two dissipators, R1 and R2. The switching junctions in the HBG (1_a and 1_b) are indexed with a subscript, and have a FSM (with the same subscript) attached with a dotted line. Two-port TF- and GY-elements are not included in our model, but they do not explicitly influence the causality in the system.

Computation and analysis with BG models is facilitated by the determination of causality, i.e., the input/output relationship between the effort and flow variables corresponding to BG elements. A standard algorithm for assigning causality to bonds is the Sequential Causality Assignment Procedure (SCAP) [9]. Modified causality assignment algorithms have been developed in other work [19].

Fig. 2 shows the causality information for configuration with both junctions on. All of the components

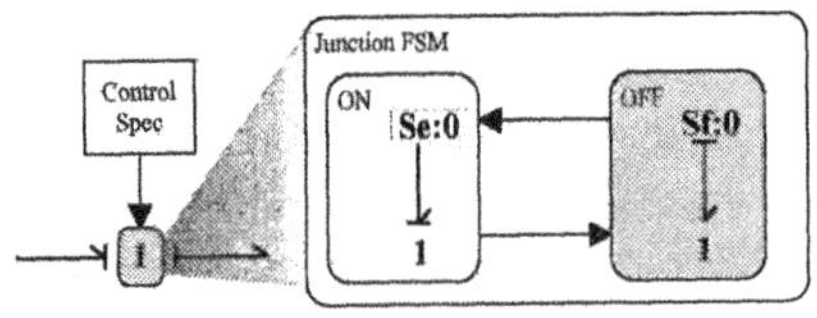

Fig. 1. Controlled junction as a Finite State Machine.

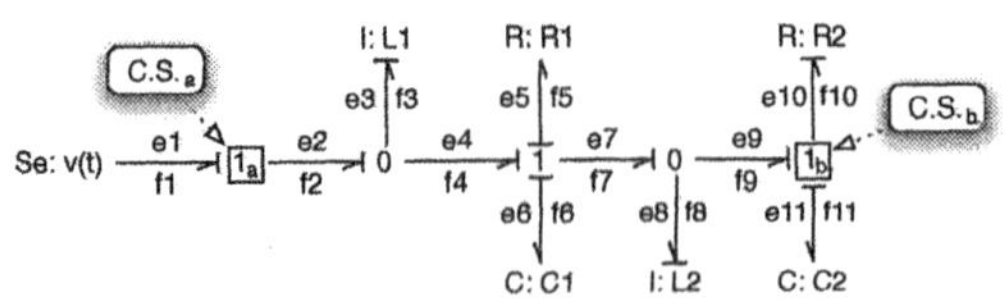

Fig. 2. Example system Hybrid Bond Graph.

in this model are in *integral causality*, which means that the constituent relations of the energy storage elements is formulated in their integral form (as opposed to the derivative form). In this paper, we make the assumption that all components will be in integral causality, and issues in dealing with derivative causality are discussed in section 5.

3. HBG COMPUTATIONAL MODEL

Computation models for continuous dynamic systems are represented by ordinary differential equations (ODEs), differential algebraic equations (DAEs), block diagrams (transfer functions), and signal flow graphs. In this work for the reasons described earlier, we adopt the block diagram representation.

3.1 Creating a causal model: Block Diagrams

We introduce the notion of a *determining bond* associated with every active (i.e., on) HBG junction. By definition, every active 0- (1-) junction in a valid bond graph will have one bond that determines the value of the effort (flow) for that junction. We label that bond as the determining bond for the particular junction. All other bond effort (flow) values are dependent and set equal to this effort (flow) value. Similarly, the flow (effort) value on a determining effort (flow) bond is an algebraic sum of the flow (effort) values of the other bonds that are connected to this 0- (1-) junction.

The determining bond plays a crucial role in mapping a HBG to a block diagram structure. Fig. 3 illustrates the mapping of a BG junction to a computational structure. The determining bond on the 1-junction, labeled with a 1, sets the flow at the junction (i.e., $f2 = f3 = f1$), and computes the dependent effort from the independent ones (i.e., $e1 = - e2 - e3$). The inputs to the adder blocks must carry appropriate signs depending on the orientation of the bond. In general, the algebraic constraints at a junction are of two forms, an equality constraint, represented by a dot, and a summation constraint, represented by an adder block. These relations are illustrated for both 0- and 1-junctions in Fig. 3.

Converting a BG model to a block diagram is a straightforward procedure when there are no algebraic loops and no elements are in derivative causality. First, each bond is replaced by two links, i.e., the effort and flow variables for the bond. Next, each junction is

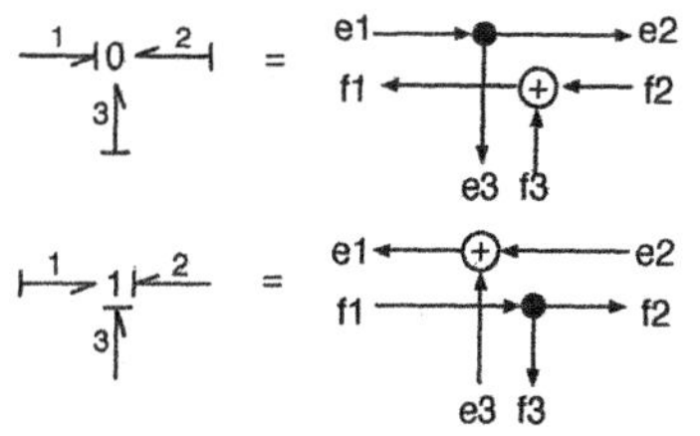

Fig. 3. Computational structures for bond graph junctions.

replaced by the algebraic constraints they impose (see Fig. 3). The individual blocks for the other elements are now connected using the algebraic constraints imposed by the junctions. The choice of block depends on the assigned causality. The 1-1 mapping from bond graph elements to corresponding block diagram fragment can be found in most bond graph texts (e.g., see [9]).

The determining bonds establish the independent effort (flow) variables and the form of the algebraic equation for the corresponding flow (effort) variables at 0- (1-junctions). Therefore, the determining bond sets the block diagram structure, i.e., the direction of flows and efforts through the blocks. In our example, the determining bond for the leftmost 1-junction is bond 2, therefore, f2 is the driving flow, and it establishes f1 = f2. In addition, this also sets e2 = e1. The rest of the link and junction connections can be derived in a similar manner. Fig. 4 shows the resulting block diagram for our example system.

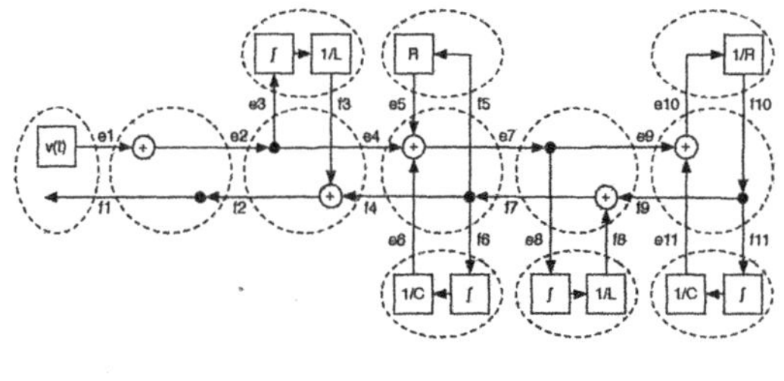

Fig. 4. Example HBG to block diagram conversion with both junctions active. The dotted circles encompass the original bond graph junctions and elements.

3.2 *Efficient model transformation*

For HBGs, the block diagram structure must handle junction switching. This is realized functionality as a control flow graph that dynamically reconfigures the computational block diagram when mode switches occur. If derivative causality was allowed, additional computational structures would be needed to update the system state at mode transitions.

Given a HBG with n switching junctions, there are potentially 2^n unique possible switching junction configurations. Pre-enumeration of all block diagram configurations offline and then selecting the appropriate one at run-time when junction configurations change is exponential in the number of switching junctions, and clearly a waste of space. On-line construction of the complete block diagram after each junction switch is space-efficient but wasteful in terms of computation time. Our solution to this problem is to construct a structurally adaptable block diagram model, and update the data flow paths through this model to match the causal structure when a junction switch occurs. Since we make the assumption that the system remains in integral causality, we exploit the locality principle for the propagation of causality changes through the model. This scheme may be combined with a caching mechanism that avoids having to recalculate causal assignment updates for discrete modes that have occurred previously.

3.3 *Run-time operation*

When junction switches occur in a HBG model the following changes are made to the existing block diagram to generate the block diagram for the new mode.

(1) Update the active HBG structure based on the junctions that change state. This procedure is described in Section 2.
(2) Evaluate the changes in the determining bonds for the junctions in the HBG structure, and propagate these changes to derive the block diagram structure for the new mode.

We illustrate the structural changes that result when the determining bond changes at a junction using our example. Consider the 1_a junction switching from on to off (Fig. 5). This event requires a new determining bond for the neighboring 0-junction. Since we assume preferred causality assignments, the I-element on this junction cannot switch its causal stroke, therefore, bond 4 must become the determining bond. This propagates to the adjacent 1-junction, and further until the 1-junction labeled 1_b, where the R-element absorbs the change by switching its causality assignment. Fig. 6 shows the resulting block diagram after the mode switch.

If the determining bond at a switching junction does not change, the effects of the mode switch do not propagate to adjacent junctions. For example, when the 1-junction labeled 1_b is switched off the determining bond for the adjacent 0-junction does not change. Consequently, there are no additional changes to the block diagram.

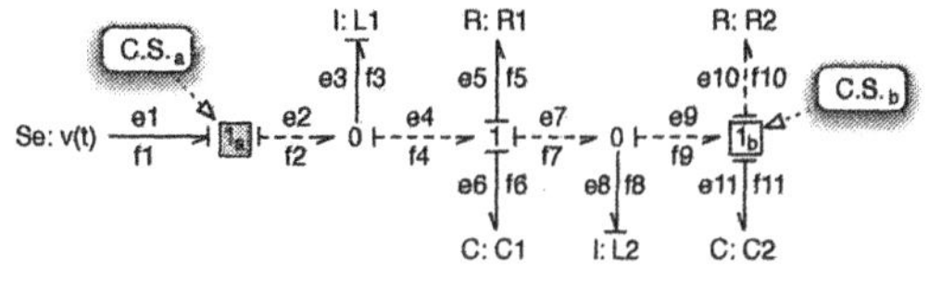

Fig. 5. HBG for the example system with junction 1_a off.

These examples illustrating the different effects of mode changes on determining bonds, indicate that an efficient scheme for incremental block diagram updating can be developed. The following lemmas make this explicit.

Lemma 1. Consider a junction switching from on to off or off to on. If the switch does not cause any change in the determining bond at its adjacent junctions, the only change in the junction structure of the block diagram is within the switching junction itself.

Lemma 2. Consider a junction switching from on to off or off to on. If the switch changes the determining bond at an adjacent junction, the change must be propagated, and the block diagram must be updated correspondingly.

Lemma 3. Consider two or more junctions switching simultaneously. Valid causal assignments are obtained regardless of the order in which the switches are evaluated.

The proofs for these lemmas are easily derived. Exploiting the fact that causality assignments determine the signal flow for the block diagram model, and that changes in causality have corresponding changes in the block diagram, we can minimize the computations required to generate an updated block diagram after a mode switch by identifying the points at which these will cause changes in the block diagram configuration, and then propagating these changes locally. This contrasts with the approach where one would rebuild the block diagram structure from scratch when mode changes occur. A general algorithm for our incremental updating procedure is described next.

A mode change is described by one or more junctions switching their state. When a 0- (1−) junction is deactivated, it provides a zero effort (flow) source with the causal strokes on each of the bonds away from (toward) the junction. When a junction is reactivated, it functions as a normal junction and its determining bond is updated. When a junction switches on or off, there are two possibilities: (i) the junction switching its state does not change the determining bond at any of its adjacent junctions, (ii) the switch changes the determining bond at an adjacent junction (Algorithm 1). For case (ii), if the adjacent junction

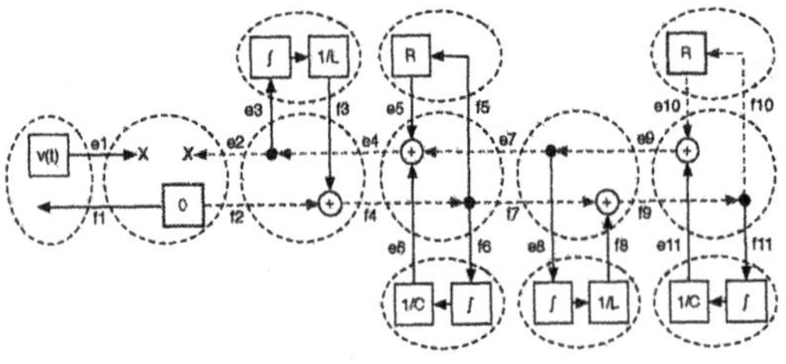

Fig. 6. Block diagram for the example system with junction 1_a off.

contains a R-element and it's causality may be flipped to remove an extra determining bond or create a new determining bond, the propagation stops. Otherwise, the bond connected to another adjacent junction has to be flipped, and the propagation continues. Algorithm 2 describes the iterative causality propagation scheme that calls the *JunctionUpdate* algorithm. An initial causal assignment is generated before entering the mode switch event loop.

Algorithm 1 JunctionUpdate

if *InvalidCausality* **then**
  if *NumDeterminingBonds*=2 **then**
    if *DeterminingBond*=R-element **then**
      Change the R-element's causality
    else
      Change causality of an unchanged adjacent junction
      Call *JunctionUpdate* on adjacent junction
    end if
  else
    if R-element at junction **then**
      Change the R-element's causality
    else
      Change the causality of a previously unchanged adjacent junction
    end if
  end if
end if

Algorithm 2 Causality Propagation

Initial Causality Assignment
Find determining bonds
loop
  if *ModeSwitch* **then**
    Update causality of bonds at the switching junction
    for all causality changes at adjacent junctions **do**
      Call *JunctionUpdate* on adjacent junction(s)
    end for
  end if
end loop

4. IMPLEMENTATION

For this work we implement the block diagram simulation models using Simulink®[13]. The Simulink environment provides all the primitives to implement the block diagram structure for a bond graph, and the bond graph elements. For hybrid junctions, we must implement the control structure as well as the dataflow structure. Rather than implementing a switching junction using discrete Simulink blocks, or using Stateflow extensions to Simulink, we implement the switching junction as custom written S-functions in C/C++. The S-function implements the dataflow machinery for the junction, as well as the evaluation of the control specification for the junction. For each bond connected to the junction, the S-function adds an input/output signal pair. The mapping of these signals to the effort/flow variables is determined dynamically by Algorithms 1 and 2. Note that we rely directly on the capabilities of the Simulink environment to detect the zero crossings, which define the mode changes.

An S-function obtains its current determining bond from a global data structure, which is updated at each junction switch. This data structure is updated by the control structure, represented explicitly in the Simulink model as the *CausalityUpdate* block. The *CausalityUpdate* block, when triggered by a junction switch, executes the *CausalityPropagation* algorithm that may reconfigure the junction input-output relations by updating the global data structure. Thus the physical structure of the Simulink model remains static, and all dynamic updates to the data flow through the Simulink model are handled by the C/C++ code that implements the controlled junctions.

The simulation model is created programmatically using the process described above through the Simulink Application Programming Interface. In other work, we have developed a physical system modeling environment that supports the building of HBG models [12]. An automatic model translation procedure creates the builder network that is then passed to the Simulink code generator. The Simulink model for the example (Fig. 2) is shown in Fig. 7.

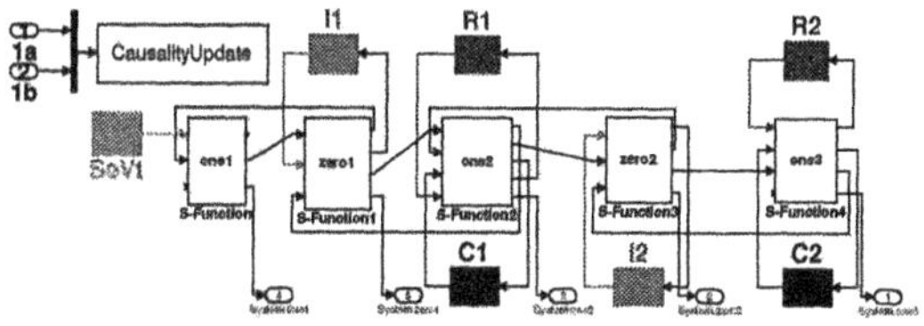

Fig. 7. Example system generated Simulink model.

The simulation results for the example system are shown in Fig. 8. All four configurations of the bond graph are visited, and the efforts and flows are graphed, along with the switching junction state.

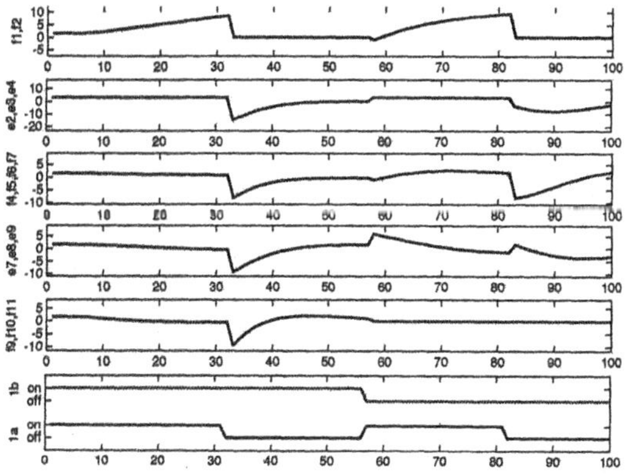

Fig. 8. Simulation outputs for a simple control sequence through each discrete state.

5. RELATED WORK

The concept of a determining bond introduced in this paper builds up on the methods presented in [9] on the efficient derivation of analytic and simulation models from BGs by exploiting the inherent causality relations in the underlying BG model. Extensions to hybrid systems have been studied by a number of researchers (e.g., [5, 2, 17, 10]), but few have discussed efficient methods for reassigning causality and regenerating the computational model at runtime when junction switches occur. As we have discussed earlier, our focus is on block-diagram based simulation models to facilitate component-oriented diagnostic analysis of systems [14]. [18] have looked at the notion of switching power junctions, and the explicit switching of effort-deciding or flow-deciding bonds when model configurations change. They define the notion of causality invariance and apply their method to special classes of electrical systems. The approach presented in this paper is more general, and the algorithms we have developed apply to any HBG model where the system remains in integral causality for all modes of operation. Similar work on switching power systems is reported in [7, 11], but both papers focus on equation generation based on Hamiltonian methods, which is somewhat tangential to our approach on block diagram-based simulation models. Buisson, et al. [4] use the notion of switched bond graphs, a very similar approach to our HBG models. But unlike the approach described in this paper, they do not use an incremental approach to causality re-assignment when junction switches occur. Their focus too is equation generation when mode changes occur using implicit state equation methods. [6] deal with switching junctions and causality re-propagation in hybrid systems, but their focus has mainly been on situations where energy-storage elements switch into derivative causality. The methods presented in this paper do not deal with situations where the system model goes into derivative causality.

HyBrSim is an experimental application for HBG modeling and simulation ([16]). The simulation algorithm includes mechanisms for performing event detection and location based on a bisectional search, and the algorithm can handle runtime causality changes (including derivative causality) when junctions switch on and off. HyBrSim runs in interpreted mode and numerical simulation of continuous-time behavior uses a forward Euler integration algorithm. HyBrSim can also generate C-code from the designed HBG for compiled simulation.

A very different simulation implementation can be taken as well, where the causal changes are handled based on the use of an implicit formulation of the junction equations ([16]). This eliminates the difficulty of explicitly handling causal changes during runtime. Instead the complexity is handled by the implicit equation solver, which typically relies on an algebraic equations solver. However, it is not clear how this method scales up for larger models, and when multiple junctions switch simultaneously.

6. CONCLUSIONS

The work presented in this paper uses physical system modeling semantics as defined by BGs and HBGs to impose semantic structure on hybrid computational models in Simulink. Other elegant computational approaches, such as Ptolemy and HyVisual [8] possess these semantics in a mathematical framework, but do not link these semantics to physical system principles. Therefore, we believe that our approach for building computational models from HBGs provides a comprehensive framework for starting from component-oriented physical system models and deriving efficient computational models for hybrid systems. In the future, we will extend our modeling approach and computational model generation schemes to handle situations of derivative causality, and propose a more formal model of computation that is linked to physical principles.

ACKNOWLEDGMENTS

This work is supported in part by grants from NSF ITR grant number: CCR-0225610, NSF SGER-052067, and NASA ALS contract number: NCC 9-159.

REFERENCES

[1] P. Antsaklis. A brief introduction to the theory and applications of hybrid systems. *Proc IEEE*, 88(7):879–887, 2000.

[2] W. Borutzky, J.F. Broenink, and K.C.J. Wijbrans. Graphical description of physical system models containing discontinuities. In A. Pave, editor, *Modelling and Simulation, Proc. of the European Simulation Multiconference*, pages 208–214, Lyon, France, June 1993.

[3] Jan F. Broenink. 20-sim software for hierarchical bond-graph block-diagram models. *Simulation Practice and Theory 7*, 7(5–6):481–492, 1999.

[4] J Buisson, H Cormerais, and P-Y Richard. Analysis of the bond graph model of hybrid physical systems with ideal switches. *Proc Instn Mech Engrs Vol 216 Part I: J Systems and Control Engineering*, pages 47–63, 2002.

[5] R. Cacho, J. Felez, and C. Vera. Deriving simulation models from bond graphs with algebraic loops. the extension to multibond graph systems. *J Franklin Institute*, 337:579–600, 2000.

[6] K Edström and J Strömberg. Aspects on simulation of switched bond graphs. *Proc. of the 35th Conf. on Decision and Control*, 1996.

[7] G. Escobar, A. van der Schaft, and R. Ortega. A hamiltonian viewpoint in modeling of switching power converters. *Automatica*, 1999.

[8] Christopher Hylands, Edward A. Lee, Jiu Liu, Xiaojun Liu, Stephen Neuendorffer, and Haiyang Zheng. Hyvisual: A hybrid system visual modeler. Technical Report Technical Memorandum UCB/ERL M03/1, University of California, Berkeley, CA, January 2003.

[9] D. C. Karnopp, D. L. Margolis, and R. C. Rosenberg. *Systems Dynamics: Modeling and Simulation of Mechatronic Systems*. John Wiley & Sons, Inc., New York, third edition, 2000.

[10] F. Lorenz and H. Haffaf. Combinations of discontinuities in bond graphs. In *Proc. Intl. Conf Bond Graph Modeling Simulation*, pages 56–64, Las Vegas, NV, January 1995.

[11] M. Magos, C. Valentin, and B. Maschke. Physical switching systems: From a network graph to a hybrid port hamiltonian formulation. In *Proc IFAC conf Analysis and Design of Hybrid Systems*, Saint Malo, France, June 2003.

[12] E.-J. Manders, G. Biswas, N. Mahadevan, and G. Karsai. Component-oriented modeling of hybrid dynamic systems using the Generic Modeling Environment. In *Proc of the 4th Workshop on Model-Based Development of Computer Based Systems*, Potsdam, Germany, March 2006. IEEE CS Press.

[13] MathWorks. MATLAB®/Simulink® The MathWorks, Inc, Natick, MA. (http://www.mathworks.com/products/simulink/).

[14] P. J. Mosterman and G. Biswas. Diagnosis of continuous valued systems in transient operating regions. *IEEE Trans. Syst., Man Cybern. A*, 29 (6):554–565, 1999.

[15] P. J. Mosterman and G. Biswas. A theory of discontinuities in physical system models. *J Franklin Institute*, 335B(3):401–439, 1998.

[16] P.J Mosterman. Hybrsim - a modeling and simulation environment for hybrid bond graphs. *Journal of Systems and Control Engineering - Part I*, 216, 1:35–46, 2002.

[17] U. Söderman, J. Top, and J. Stromberg. The conceptual side of mode switching. In *Proc. System, Man and Cybernetics*, 1993.

[18] Amod C. Umarikar and L. Umanand. Modelling of switching systems in bond graphs using the concept of switched power junctions. *J Franklin Institute*, 342:131–147, 2005.

[19] J. van Dijk. *On the role of bond graph causality in modelling mechatronic systems*. PhD dissertation, University of Twente, The Netherlands, 1994.

ELSEVIER
IFAC
PUBLICATIONS

ROBUST CONTROL STRATEGIES FOR MULTI–INVENTORY SYSTEMS WITH AVERAGE FLOW CONSTRAINTS

Dario Bauso * **Franco Blanchini** **,[1] **Raffaele Pesenti** *

* *Dipartimento di Ingegneria Informatica, Università di Palermo, Viale delle Scienze, I-90128 Palermo, ITALY, blanchini@uniud.it*
** *Dipartimento di Matematica ed Informatica, Università degli Studi di Udine, Via delle Scienze 206, 33100 Udine, ITALY, dario.bauso(pesenti)@unipa.it*

Abstract: In this paper we consider multi–inventory systems in presence of uncertain demand. We assume that i) demand is unknown but bounded in an assigned compact set and ii) the control inputs (controlled flows) are subject to assigned constraints. Given a long–term average demand, we select a nominal flow that feeds such a demand. In this context, we are interested in a control strategy that meets at each time all possible current demands and achieves the nominal flow in the average. We provide necessary and sufficient conditions for such a strategy to exist and we characterize the set of achievable flows. Such conditions are based on linear programming and thus they are constructive. In the special case of a static flow (i.e. a system with 0–capacity buffers) we show that the strategy must be affine. The dynamic problem can be solved by a linear-saturated control strategy (inspired by the previous one). We provide numerical analysis and illustrating examples. *Copyright © 2006 IFAC*

Keywords: Inventory control, Robust control, Bounded disturbances, Manufacturing systems, Linear programming.

1. INTRODUCTION

Multi–inventory systems (Hadley and Whitin, 1963) are met in several different contexts, such as manufacturing (Boukas *et al.*, 1995; Chase and Ramadge, 1992), network routing (Iftar and Davison, 1990), communications (Ephremides and Verdú, 1989), water distribution (Larson and Keckler, 1969), logistics and traffic control (Moreno and Papageorgiou, 1995). Hence, their control is of relevant economic interest. The control concerns storage and processing operations and aims at meeting the external demand of finished products (Forrester, 1961).

In this work we simultaneously consider the two following aspects.

- *Instantaneous fluctuations* — These are assumed unknown due to the large number of unpredictable factors that influence the demand. The control must face all possible variations, within prescribed limits, in order to meet the demand.
- *Long term information* — The long–term average demand, henceforth also called nominal

[1] This paper is the conference version of (Bauso *et al.*, 2006). Corresponding author F. Blanchini. Tel. +39 0432 558466. Fax +39 0432 558499.

demand, should be faced, in the average, by the nominal flow, whenever possible.

Therefore we are seeking for a stabilizing strategy capable of balancing the flow in the long run. The main results of the paper are reported next.

- We first consider static strategies (i.e. we assume 0–capacity buffers). We provide necessary and sufficient conditions for the existence of a strategy which is able to meet all the possible demands and assures the desired flow average, whenever the demand meets its nominal average. Such a static strategy is affine
- We characterize the set of all flows corresponding to the nominal demand which can be achieved in the average.
- We show that the very conditions, valid in the static case, are sufficient for the existence of a dynamic strategy, based on the feedback of the buffer levels. The proposed feedback strategy is a linear-saturated dynamic control.

2. PROBLEM FORMULATION

Consider the following continuous time system

$$\dot{x}(t) = Bu(t) - w(t), \tag{1}$$

where $x(t) \in \mathbb{R}^n$ is a vector whose components are the buffer levels, $u(t) \in \mathbb{R}^m$ is the controlled flow vector, B is the controlled process matrix and $w(t) \in \mathbb{R}^n$ is an exogenous (uncontrolled) input, typically modeling demand, whose value is externally determined. To model backlog $x(t)$ may be less than zero.

We assume that u and w are subject to the next constraints

$$u(t) \in \mathcal{U} = \{u : u^- \leq u \leq u^+\}, \tag{2}$$

where u^- and u^+ are assigned vectors and the expression is to be intended component-wise. We assume that w is constrained as follows

$$w(t) \in \mathcal{W}, \tag{3}$$

where $\mathcal{W}$ is a polytope. We also introduce the following assumptions.

Assumption 1. Matrix B has full row rank.

Given a vector function of time $f : \mathbb{R}^+ \to \mathbb{R}^n$ we introduce the following notation

$$Av[f] = \lim_{T\to\infty} \frac{1}{T} \int_0^T f(t)\, dt. \tag{4}$$

Function $Av[f]$ will be referred to as the deterministic average of f, henceforth the *average*, and we will always assume that such a value exists whenever considered.

Assumption 2. The set $\mathcal{W}$ includes $\bar{w} = Av[w]$ in its relative interior [2].

We will consider static and dynamic stabilizing policies for the system according to the following definitions.

Definition 3. The function $\Phi : \mathbb{R}^n \to \mathbb{R}^m$ is a *static balancing strategy* if for $u(t) = \Phi(w(t))$,

$$Bu(t) = w(t),$$

and $u(t) \in \mathcal{U}$, for all $w(t) \in \mathcal{W}$, for all $t \geq 0$.

Definition 4. Given $\epsilon > 0$ and a reference value $\bar{x}$, an *ϵ-stabilizing strategy* is a feedback control for which there exists a continuous positive function $\phi(t)$, monotonically decreasing and converging to 0 as $t \to \infty$ such that for all $w(t) \in \mathcal{W}$ and for all $x(0)$, the conditions $u(t) \in \mathcal{U}$ and

$$\|x(t) - \bar{x}\| \leq \max\{\|x(0)\|\phi(t), \epsilon\}$$

hold true.

As a preliminary result, we introduce the following basic conditions (Blanchini *et al.*, 2000).

Theorem 5. For the considered system

i there exists a static balancing strategy as in Definition 3 if and only if

$$\mathcal{W} \subseteq B\mathcal{U}; \tag{5}$$

ii there exists a feedback stabilizing strategy as in Definition 4 if and only if

$$\mathcal{W} \subseteq int\{B\mathcal{U}\}. \tag{6}$$

Henceforth, we assume that the appropriate necessary and sufficient condition is met (depending on which kind of strategy we are considering). Assume to apply either a balancing or an ϵ-stabilizing strategy. As a consequence, $x(t)$ remains constant or bounded. Then, by integrating (1) we have that, necessarily,

$$\lim_{T\to\infty} \frac{1}{T} \int_0^T [Bu(t) - w(t)]\, dt = \lim_{T\to\infty} \frac{1}{T} [x(t) - x(0)] = 0,$$

which implies that the average value of w is equal to the average value of Bu

$$B\ Av[u(t)] = Av[w(t)]. \tag{7}$$

[2] we mean that $\bar{w}$ is an interior point of $\mathcal{W}$ with respect to the smallest linear subspace including it, for instance given a vector $v \neq 0$, 0 is in the relative interior of a segment joining v and $-v$

Formally, the problem is the following.

Problem 6. Assume that the average $\bar{w} \in \mathcal{W}$ is given. Consider the feasible flow $\bar{u} \in \mathcal{U}$ such that

$$B\bar{u} = \bar{w}.$$

Provide a yes–no answer to the question: does there exist a static balancing (or dynamic ϵ–stabilizing) strategy such that whenever $Av[w] = \bar{w}$ then $Av[u] = \bar{u}$? In the case of a positive answer we will say that $\bar{u}$ is achievable.

In the following sections we will solve constructively the problem for both static and dynamic strategies.

3. STATIC STRATEGIES

In this section we consider the case in which the controlled flow is a function of the demand w so that $Bu(t) = w(t)$.

For the simple notations we work under the following assumption.

Assumption 7. The nominal average "demand" is zero, i.e. $\bar{w} = Av[w] = 0 \in \mathcal{W}$.

Then we can translate the problem by writing the new model

$$\dot{x}(t) = B(u(t) - u_0) - [w(t) - \bar{w}] = B\delta u(t) - \delta w(t)$$

and by translating the constraints as

$$u^- - u_0 \leq \delta u(t) \leq u^+ - u_0, \qquad \delta w(t) \in \mathcal{W} - \bar{w}.$$

where $Av[\delta w] = 0$.

Theorem 8. Under Assumption 1 and 2 let condition (5) be satisfied. Then there exists a static balancing strategy that achieves the average $Av[u] = 0$ whenever $Av[w] = 0$ if and only if there exists a "tall" matrix D $m \times n$ such that

$$BD = I \tag{8}$$
$$u^- \leq Dw^{(i)} \leq u^+, \quad i = 1, \ldots, s. \tag{9}$$

where $w^{(i)}$ are the vertices of $\mathcal{W}$. Moreover, if such necessary and sufficient conditions are satisfied, then the static strategy is linear

$$u(t) = Dw(t). \tag{10}$$

PROOF. See the proof in (Bauso *et al.*, 2006). □

The previous theorem allows us to check a single candidate $\bar{u}$ we fixed to zero. We can now characterize the *set of achievable average flows*, namely the set of all vectors such that $Av[w] = 0$ implies $Av[u] = \bar{u} \in \mathcal{U}$.

Corollary 9. The set of all achievable average flows, provided that a suitable static balancing strategy is applied, is made up by all the vectors $\bar{u} \in ker[B]$ such that there exists a matrix D, $m \times n$, with

$$BD = I \tag{11}$$
$$u^- \leq Dw^{(i)} + \bar{u} \leq u^+, \quad i = 1, \ldots, s. \tag{12}$$

In this case the static strategy is affine

$$u = Dw + \bar{u}.$$

PROOF. It follows immediately from the theorem by applying the translation $u - \bar{u}$. □

We have seen that as long as a strategy achieving the average exists, this has to be linear (or affine taking into account possible translations on w).

4. DYNAMIC STRATEGIES

Here we show how to achieve an average flow by a dynamic stabilizing strategy. show, in the next subsection, that conditions (8) and (9) are sufficient for the existence of a dynamic ϵ–stabilizing strategy of the form

$$\begin{aligned} \dot{y}(t) &= f(y(t), x(t), w(t)) \\ u(t) &= g(y(t), x(t), w(t)). \end{aligned} \tag{13}$$

To provide results about necessity of (8) and (9) we need to better characterize the class of dynamic strategies by additional assumptions (see, e.g., (Bauso *et al.*, 2006)).

4.1 Sufficiency of the conditions

Let assumptions (8) and (9) be satisfied and consider the corresponding matrix D. Equation (8) means that D is a right inverse of B and it is a standard property of linear algebra that this is equivalent to the existence of two matrices C and F which "square" B and D producing two matrices inverse to each other, namely such that

$$\begin{bmatrix} B \\ C \end{bmatrix} \begin{bmatrix} D & F \end{bmatrix} = I. \tag{14}$$

Consider the following augmented system

$$\begin{aligned} \dot{x}(t) &= Bu(t) - w(t) \\ \dot{y}(t) &= Cu(t). \end{aligned} \tag{15}$$

The additional dynamic variable $\dot{y}(t) = Cu(t)$ has the goal of keeping trace of the load unbalancing with respect to the desired average 0.

The first step is to show that under (8) and (9), the extended system (15) satisfies the stabilizability conditions (6) as well (in the extended state–space), precisely for all $w \in \mathcal{W}$ there exists $u \in \mathcal{U}$ such that

$$\begin{bmatrix} w \\ 0 \end{bmatrix} = \begin{bmatrix} B \\ C \end{bmatrix} u,$$

or equivalently that, for all $w \in \mathcal{W}$, there exists $u \in \mathcal{U}$ such that

$$u = \begin{bmatrix} D & F \end{bmatrix} \begin{bmatrix} w \\ 0 \end{bmatrix} = Dw.$$

The existence of such u is an immediate consequence of (9). Indeed, it is easy to verify that, if $\mathcal{W} \in int\{B\mathcal{U}\}$, then the u which corresponds to w is in the interior of the extended set. Then the problem can be solved as follows.

- Determine D such that (8) and (9) are satisfied.
- Determine C and F such that (14) is satisfied.
- Design a control which stabilizes (15).

Observe that Theorem 5 applied to the extended system (15) guarantees the existence of such a stabilizing control.

Here we propose a new strategy based on a variable transformation. In the following we exploit (for the first time) the structure of the set $\mathcal{U}$. Consider the new variable $z(t)$ defined as

$$z(t) = \begin{bmatrix} D & F \end{bmatrix} \begin{bmatrix} x(t) \\ y(t) \end{bmatrix}, \quad \begin{bmatrix} x(t) \\ y(t) \end{bmatrix} = \begin{bmatrix} B \\ C \end{bmatrix} z(t)$$

This variable satisfies the equation

$$\dot{z}(t) = u(t) - Dw(t). \tag{16}$$

The new system (16) is decoupled in its state variable, precisely it is equivalent to

$$\dot{z}_i(t) = u_i(t) - D_i w(t), \tag{17}$$

where we have denoted by D_i the ith row of D and where $u_i^- \leq u_i \leq u_i^+$. Denote by

$$\rho_i^- = \min_{w \in \mathcal{W}} D_i w,$$
$$\rho_i^+ = \max_{w \in \mathcal{W}} D_i w,$$

The stabilizability conditions are equivalent to the fact that for all $w \in \mathcal{W}$

$$u_i^- < \rho_i^- < \rho_i^+ < u_i^+.$$

Henceforth, without restriction, we consider the single–buffer case, namely the scalar system

$$\dot{z}(t) = u(t) - r(t),$$

with

$$\rho^- \leq r(t) \leq \rho^+, \quad u^- \leq u(t) \leq u^+.$$

Define the saturated control (see Fig. 1)

$$u(t) = sat_{[u^-, u^+]}(-\kappa z(t)) \tag{18}$$

with $\kappa > 0$ and where

$$sat_{[\alpha,\beta]}(\zeta) = \begin{cases} \beta, & \text{if } \zeta > \beta, \\ \zeta, & \text{if } \alpha \leq \zeta \leq \beta, \\ \alpha, & \text{if } \zeta < \alpha. \end{cases}$$

We will use the same notation (18) for the multi-input control derived applying the formula component–by–component. Note that this control

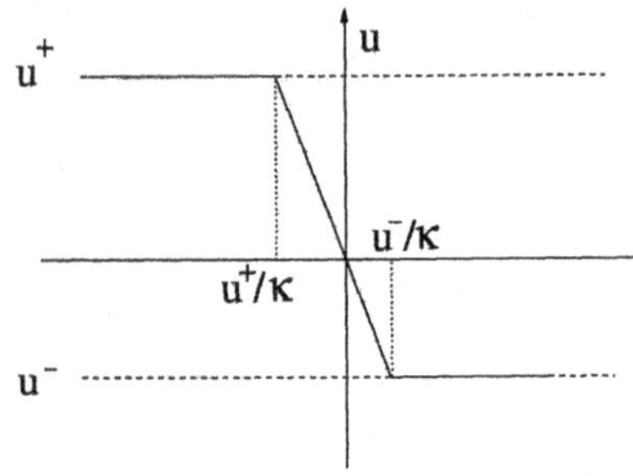

Fig. 1. The function (18)

function is Lipschitz continuous. For $\kappa \to \infty$, the control (18) converges to the bang bang control

$$bb_{[u^-,u^+]}(\zeta) = \begin{cases} u^+, & \text{if } \zeta > 0, \\ 0, & \text{if } \zeta = 0, \\ u^-, & \text{if } \zeta < 0, \end{cases}$$

which is of the type considered in (Blanchini *et al.*, 2000).

Theorem 10. The variable $z(t)$ with the control (18) converges to the interval $[-u^+/\kappa, -u^-/\kappa]$ (which includes 0 as an interior point). Therefore the global system converges to the corresponding hyper–box (i.e. that delimited by $-u_i^+/\kappa \leq z_i \leq -u_i^-/\kappa$, $i = 1, 2, \ldots, m$).

PROOF. The proof derives from the fact that, for $z \geq -u^-/\kappa$, we have that the control is saturated to its lower level $u = u^-$, then

$$\dot{z} = u^- - r \leq u^- - \rho^- < 0. \tag{19}$$

Conversely for $z \leq -u^+/\kappa$ we have that $u = u^+$, then

$$\dot{z} = u^+ - r \geq u^+ - \rho^+ > 0. \tag{20}$$

Therefore $z(t)$ reaches the interval in finite time and is ultimately confined in it. □

As a consequence of the previous theorem we have that, choosing κ large enough, we can bound z in an arbitrarily small interval. Therefore we achieve ϵ–stability. We have now to show that the controller so obtained satisfies the average requirement. Indeed variable $z(t)$ remains bounded so

arcs	1	2	3	4	5	6	7	8	9
upper bounds	3	2	3	3	3	3	3	5	5

Table 1. Controlled flows constraints

nodes	1	2	3	4	5
upper bounds	0	2	3	2	2
averages	0	1	2	1	1

Table 2. Demand bounds

$\|z(t) - z(0)\| \leq \xi$. By integrating (16) we have that

$$\frac{1}{T}\int_0^T u(t)dt - \frac{1}{T}\int_0^T Dw(t)dt = \frac{z(T) - z(0)}{T} \to 0$$

as $T \to \infty$. This yields

$$Av[u] = Av[Dw],$$

that is all we need to claim that sufficiency of (8) and (9) is proved.

Example 11. Let us solve Problem 6 for the system depicted in Fig. 2 (B is then the incidence matrix of the network). Table 1 summarizes the

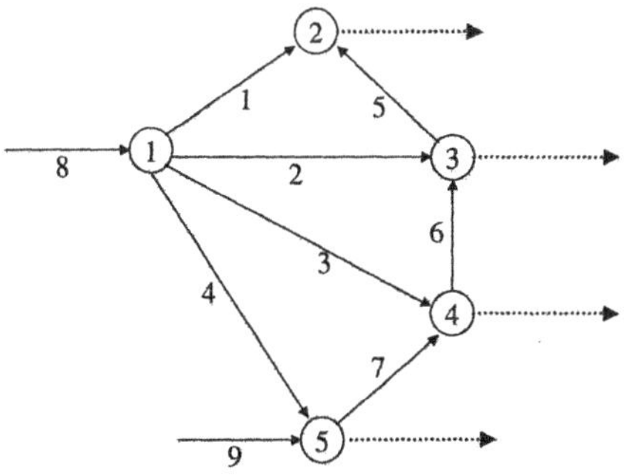

Fig. 2. Example of a system with 5 nodes and 9 arcs.

controlled upper flows constraints (the lower constraints are all set to 0) whereas Table 2 the demand bounds and the long–term average demands. Now, given the nominal demand $\bar{w} = [0\ 1\ 2\ 1\ 1]$ and the nominal balancing flow $\bar{u} = [1\ 1\ 1\ 0\ 0\ 1\ 1\ 3\ 2]' \in \mathcal{U}$ (which is $\bar{w} = B\bar{u}$) we have to determine whether $\bar{u}$ is an achievable average flow, namely, it is such that if $Av[w] = \bar{w}$ then $Av[u] = \bar{u}$. A possible matrix $D = \hat{D}$ satisfying (8)(9) is

$$\hat{D} = \begin{bmatrix} 0 & 1 & 0 & 0 & 0 \\ 0 & 0 & 0.5 & 0 & 0 \\ -0.1 & 0 & 0.5 & 0 & 0 \\ -0.2 & 0 & 0 & 0 & 0 \\ 0 & 0 & 0 & 0 & 0 \\ 0 & 0 & 0.5 & 0 & 0 \\ 0.1 & 0 & 0 & 1 & 0 \\ 0.6 & 1 & 1 & 0 & 0 \\ 0.4 & 0 & 0 & 1 & 1 \end{bmatrix}. \quad (21)$$

We simulate the system with dynamic strategy (18) (we initialize $x(0) = 0$, $y(0) = 0$, and set $\kappa = 4$). Fig. 3 displays the average flow $Av[\delta u(t)]$ and the variable $z(t)$. In agreement with the expected results, the simulated average flow $Av[u]$ tends to the prescribed average flow $\bar{u}$ and the variable z converges to the interval $[-\delta u^+/\kappa, -\delta u^-/\kappa]$ Fig. 4 shows that the fluctuations of the buffer lengths are confined within a pre-specified neighborhoods of 0. (we remind that $x = 0$ is the desired buffer level and thus negative values do not necessarily imply backlog).

REFERENCES

Bauso, D., F. Blanchini and R. Pesenti (2006). Robust control policies for multi-inventory systems with average flow constraints. *Automatica, Special Issue on Optimal Control Applications to Management Sciences. To appear.*

Blanchini, F., S. Miani and W. Ukovich (2000). Control of production-distribution systems with unknown inputs and system failures. *IEEE Transactions on Automatic Control* **45**(6), 1072–1081.

Boukas, E.K., H. Yang and Q. Zhang (1995). Minimax production planning in failure–prone manufacturing systems. *Journal of Optimization Theory and Applications* **82**(2), 269–286.

Chase, C. and P.J. Ramadge (1992). On real–time policies for flexible manufacturing systems. *IEEE Transactions on Automatic Control* **37**(4), 491–496.

Ephremides, A. and S. Verdú (1989). Control and optimization methods in communication networks. *IEEE Transactions on Automatic Control* **34**, 930–942.

Forrester, J. W. (1961). *Industrial dynamics.* The M.I.T. Press; John Wiley & Sons.

Hadley, G. and T.M. Whitin (1963). *Analysis of Inventory Systems.* Prentice-Hall.

Iftar, A. and E.J. Davison (1990). Decentralized robust control for dynamic routing of large scale networks. In: *Proceedings of the American Control Conference.* San Diego. pp. 441–446.

Larson, R. E. and W. G. Keckler (1969). Applications of dynamic programming to the control of water resource systems. *Automatica* **5**, 15–26.

Moreno, J. C. and M. Papageorgiou (1995). A linear programming approach to large-scale linear optimal control problems. *IEEE Transactions on Automatic Control* **40**, 971–977.

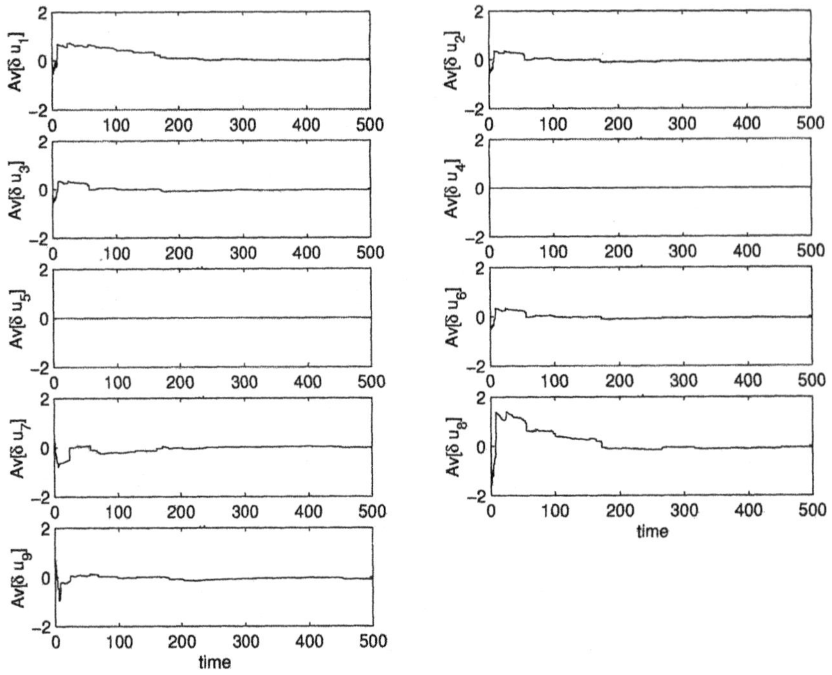

Fig. 3. The average flow $Av[\delta u]$ with dynamic strategy (18) and $\kappa = 4$.

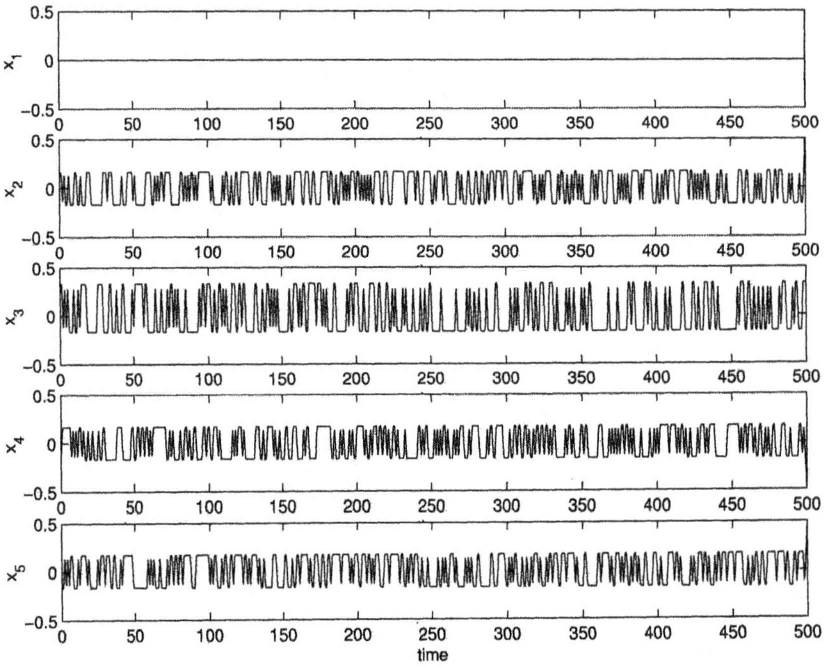

Fig. 4. The buffer length x with dynamic strategy (18) and $\kappa = 4$.

ELSEVIER
IFAC
PUBLICATIONS

HYBRID CONSTRAINED FORMATION FLYING CONTROL OF MICRO-SATELLITES

F. Bacconi [*,1] A. Casavola [**,2] E. Mosca [*,1]

[*] *Dipartimento di Sistemi e Informatica*
Università degli Studi di Firenze
[**] *Dipartimento di Elettronica, Informatica e Sistemistica*
Università della Calabria

Abstract: This paper proposes a centralized solution to the problem of formation reconfiguration and keeping for fleets of satellites in the presence of persistent disturbances and under input-saturation and formation accuracy constraints. Relative position and attitude dynamics are considered. For suitable initial conditions the proposed control scheme produces system evolutions that fulfill the coordination constraints at any time and satisfy desirable control performance. This is accomplished by using a bank of Command Governor units in a hybrid framework. An example is provided in order to exhibit the effectiveness of the technique. *Copyright © 2006 IFAC*

Keywords: formation flying control, leader following, saturation constraints, accuracy constraints, hybrid command governor.

1. INTRODUCTION

In recent years an increasing number of space missions have used small satellites. As a consequence of the significant reduction of the costs related to launches, maneuvers and maintenance, it is expected that a great deal of the future space systems will be based in fleets of micro-satellites (Esper *et al.*, 2003). Indeed, it can increase the overall efficiency, performance and survivability, in comparison to single, large vehicles, even in the presence of large instrumentation and payloads.

The aim of the present work is to propose a control technique for relative positions and attitudes in a fleet of micro-satellites with respect to a formation center, using a leader-following approach. The geometry of a formation can be subject to changes, and the relative states of the satellites can be required to fulfill stringent accuracy constraints. Thus, constrained control techniques might be necessary (Yeh *et al.*, 2000). Moreover, actuator saturation is inevitably present and has to be taken into account with small space vehicles, since small size thrusters cannot supply large torques (Hu and Lin, 2001).

Several control techniques based on convex optimization have been proposed to face this kind of problems, see e.g. (Tillerson *et al.*, 2002). However, a control strategy based on conceptual tools of model-based predictive control (MPC) (Manikonda *et al.*, 1991; Mayne *et al.*, 2000) appears particularly suitable. In this work we consider one of the simplest MPC techniques referred as the command governor approach (Bemporad *et al.*, 1997; Casavola *et al.*, 2000; Casavola *et al.*, 2006). It consists of adding to a primal compensated system a nonlinear device, called Command Governor (CG), whose action is based on the current state, set-point, and the prescribed

[1] Via S. Marta, 3 - 50139 Firenze, Italy - {bacconi, mosca}@dsi.unifi.it
[2] Via P. Bucci, 41C - 87037 Arcavacata di Rende (CS) Italy - casavola@deis.unical.it

constraints. The CG selects at each time instant a virtual sequence among a family of linearly parameterized command sequences, by solving a convex constrained quadratic optimization problem, and feeds the primal system according to a receding horizon control philosophy.

In (Bacconi *et al.*, 2004) we illustrated the advantages related to this approach with respect to others MPC control techniques. There, we assumed small angular displacements between the satellites and the formation center. However, this is acceptable only in the presence of small angle maneuvers. Here, we remove this assumption and describe the attitude of each member of the fleet by a nonlinear model. Since single CG based control laws cannot handle the complete range of possible angles, we consider a bank of controllers, each one designed with respect to a pre-established orbital reference frame. A centralized hybrid control scheme is then adopted.

Thus, the whole scheme retains the properties of a supervisory switching control and the presence of the CGs aims at enlarging the dynamic range where each compensated system can operate linearly. Moreover, the CGs do not modify the primal control system dynamics, since they operate only on the input signal, whenever necessary.

This produces a particularly simple structure, especially suitable for small formations. On the other hand, with large signals, there is a performance degradation. However, in the presence of limited computing power and energy, when the direct use of a bank of predictive controllers is not allowed, since it requires a quite massive amount of flops per sampling time, the use of the present CG hybrid schemes is widely justified.

The paper is organized as follows: in section 2 the mathematical model of the satellites is described. In section 3 the control problem is stated. The Command Governor approach and some variations suitable for the problem at hand are described in section 4. Moreover, a hybrid CG technique is stated in section 5 and a supervisory switching logic is introduced as well. Finally, in section 6 an example is presented where a coordinated large angle maneuver is requested to a couple of micro-satellites in an Earth Observation mission.

2. MATHEMATICAL MODEL

We consider a formation of micro-satellites in LEO orbit, using a description based on the leader following approach. The mathematical model of each member of the formation must consequently describe the attitude of a reference frame $\{B\}$ fixed on the vehicle's body, with respect to an appropriately defined orbiting reference frame $\{O\}$ and the position of $\{B\}$ into another, possibly coinciding with $\{O\}$, orbiting reference frame $\{C\}$ (Fig. 1). The latter is assumed to be centered on

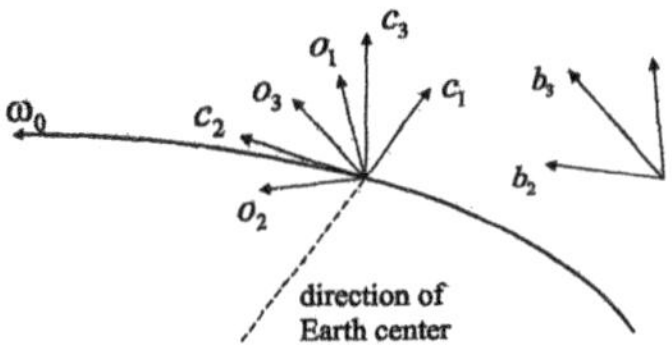

Fig. 1. Reference frames.

an orbiting point, intended as the formation center (virtual satellite). In this paper, the formation center reference frame $\{C\}$ is selected with the first axis in the opposite direction of Earth center. The second axis is assumed aligned with the orbit and the third axis completes an orthonormal right oriented versors set. Moreover, it is assumed no rotational motion of $\{O\}$ in $\{C\}$.

We introduced a linear model for the attitude of each member of the fleet in (Bacconi *et al.*, 2004) by means of the Euler's angles θ_i describing the orientation of $\{O\}$ in the body frame $\{B\}$ (Sidi, 1997), under the hypothesis of small angle maneuvers. Following the same procedure, the absolute angular velocity of $\{B\}$ can be rewritten as the sum of the velocity of $\{B\}$ with respect to $\{O\}$ and the velocity of $\{O\}$ with respect to ECI, defined as $\omega^o = [0, 0, \omega_0]'$ (all the variables expressed in $\{B\}$ coordinates):

$$\omega^b = \omega^{bo} + \omega^o = \omega^{bo''} + \omega^{o''o'} + \omega^{o'o} + \omega^o$$

with O' and O'' the reference frames obtained from O after a sequence of rotations of angles θ_1 and θ_2, respectively. By means of the rotation matrix $\mathcal{R}^{bo}$, that leads $\{O\}$ coinciding with $\{B\}$, we obtain (Bacconi *et al.*, 2004)

$$\omega^{bo} = \left[s_3\dot{\theta}_2 + c_2c_3\dot{\theta}_1, c_3\dot{\theta}_2 - c_2s_3\dot{\theta}_1, \dot{\theta}_3 + s_2\dot{\theta}_1\right] \tag{1}$$

For brevity we have replaced $\sin(\theta_i) = s_i$ and $\cos(\theta_i) = c_i$. Furthermore, also ω^o needs to be expressed in body coordinates. On this subject, notice that $\omega^o = \omega^c$. Hence,

$$\omega^o = \mathcal{R}^{bo}(\mathcal{R}^{oc} \cdot [0,\ 0,\ \omega_0]')$$

with $\mathcal{R}^{bo}$ and $\mathcal{R}^{oc}$ the two rotation matrices that lead $\{C\}$ coinciding with $\{O\}$ and then $\{O\}$ coinciding with $\{B\}$. Now, since both the velocity ω_0 of $\{C\}$ in ECI and the attitude of $\{O\}$ in $\{C\}$ are assumed constant and known, we can define the angular velocity of $\{C\}$ in $\{O\}$ coordinates as

$$\omega_R = \mathcal{R}^{oc} \cdot [0,\ 0,\ \omega_0]'$$

Consequently

$$\omega^o = \mathcal{R}^{bo}\omega_R = \mathcal{R}^{bo} \cdot [\omega_{R1},\ \omega_{R2},\ \omega_{R3}]'$$

It follows that ω^o corresponds to:

$$\begin{bmatrix} c_2c_3\omega_{R1} + (c_1s_3 + s_1s_2c_3)\omega_{R2} + (s_1s_3 - c_1s_2c_3)\omega_{R3} \\ -c_2s_3\omega_{R1} + (c_1c_3 - s_1s_2s_3)\omega_{R1} + (s_1c_3 + c_1s_2s_3)\omega_{R3} \\ s_2\omega_{R1} - s_1c_2\omega_{R2} + c_1c_2\omega_{R3} \end{bmatrix} \quad (2)$$

Finally, adding (1) and (2), taking the first derivative of ω^b and substituting into the well known Euler's equations, yields to a nonlinear model $\ddot{\theta} = f(\theta, \dot{\theta}, \omega_R, \tau)$. We do not assume small angular deviation between the formation center $\{C\}$ and $\{B\}$. On the other hand we assume small angles between $\{O\}$ and $\{B\}$ aiming at describing all the possible attitudes of a member of the fleet with a suitable set of frames $\{O_i\}$. As will be described in Sect. 5, it allows at addressing the whole problem in a hybrid framework. Thus, considering $\cos(\theta_i) \simeq 1$, $\sin(\theta_i) \simeq \theta_i$, and disregarding nonlinear terms, the Euler's equations become

$$\begin{aligned} \ddot{\theta}_1 =& (\omega_{R3}^2 - \omega_{R2}^2)J_1\theta_1 + \omega_{R3}(1+J_1)\dot{\theta}_2 + \omega_{R1}\omega_{R2}J_1\theta_2 \\ &+ \omega_{R2}(J_1 - 1)\dot{\theta}_3 - \omega_{R1}\omega_{R3}J_1\theta_3 + \tau_1 + M_1 \\ \ddot{\theta}_2 =& \omega_{R3}(J_2 - 1)\dot{\theta}_1 - \omega_{R1}\omega_{R2}J_2\theta_1 + (\omega_{R1}^2 - \omega_{R3}^2)J_2\theta_2 \\ &+ \omega_{R1}(1+J_2)\dot{\theta}_3 + \omega_{R2}\omega_{R3}J_2\theta_3 + \tau_2 + M_2 \\ \ddot{\theta}_3 =& \omega_{R2}(1+J_3)\dot{\theta}_1 + \omega_{R1}\omega_{R3}J_3\theta_1 + \omega_{R1}(J_3 - 1)\dot{\theta}_2 \\ &- \omega_{R2}\omega_{R3}J_3\theta_2 + (\omega_{R2}^2 - \omega_{R1}^2)J_3\theta_3 + \tau_3 + M_3 \end{aligned} \quad (3)$$

where $J_i = (I_j - I_k)/I_i$ $(i, j, k \in \underline{3},\ i \neq j \neq k$ and I_i are the principal moments of inertia of the spacecraft).

It can be rewritten in a more compact way, that will be useful in the following, as

$$\dot{\theta} = \Phi_\theta \theta + G_\theta(\tau + M) \quad (4)$$

introducing $\theta = [\dot{\theta}_1\ \theta_1\ \dot{\theta}_2\ \theta_2\ \dot{\theta}_3\ \theta_3]'$ as the state space vector corresponding to the Euler's angles, $\tau = [\tau_1\ \tau_2\ \tau_3]'$ as the control torques vector and $M = [M_1\ M_2\ M_3]'$ as the disturbance torques vector.

Next, we consider the position model of each member of the formation. We take into account fleets of micro-satellites in low orbits and close proximity. Thus, the motion of each spacecraft with respect to the formation center, can be described by the Hill's equations (Sidi, 1997; Tillerson *et al.*, 2002). They consist in the following linear model

$$\begin{aligned} m\ddot{p}_1 &= 3\omega_0^2 m p_1 + 2\omega_0 m \dot{p}_2 + f_1 + n_1 \\ m\ddot{p}_2 &= -2\omega_0 m \dot{p}_1 + f_2 + n_2 \\ m\ddot{p}_3 &= -\omega_0^2 m p_3 + f_3 + n_3 \end{aligned} \quad (5)$$

where m is the mass of the satellite. For simplicity, acting as above, we summarize (5) in the state space equation

$$\dot{p} = \Phi_p p + G_p(f + n) \quad (6)$$

where $p = [\dot{p}_1\ p_1\ \dot{p}_2\ p_2\ \dot{p}_3\ p_3]'$ is the state space vector corresponding to the relative-coordinates of the satellite, $f = [f_1\ f_2\ f_3]'$ are the actuator forces acting along the positive axes directions and $n = [n_1\ n_2\ n_3]'$ are the components of disturbance forces.

It is worth pointing out that here, with the assumption of small displacements between $\{B\}$ and $\{O\}$, there is no difference in representing inputs and disturbances either in body or orbital coordinates. Moreover, we neglect the effects of components $p_1 \neq 0$ on the angular velocity. Hence, ω_0 is assumed constant.

3. PROBLEM FORMULATION

Combining equations (4) and (6) gives a linear time-continuous system:

$$\dot{s} = A_c s + B_c(u + \xi) \quad (7)$$

where $s = [p'\ \theta']' \in \mathbb{R}^{12}$, $u = [f'\ \tau']' \in \mathbb{R}^6$, $\xi = [n'\ M']' \in \mathbb{R}^6$ and matrices A_c and B_c direct consequence. Hence, with sampling period T, the ZOH sampled dynamical model of each satellite takes the form

$$s(t+1) = As(t) + B(u(t) + d(t)) \quad (8)$$

where $d(t)$ represents relative disturbance forces and torques accumulated during a sampling period.

The objective of the control problem is to drive each component of the fleet to a desired position and attitude defined in $\{C\}$ along a pre-specified path. Notice that it encompasses collision avoidance constraints. Further, maneuvers are made thanks to a combination of small jet actuators subject to input saturation constraints of the form

$$|u_i(t)| \leq u_{i_{max}}, \quad i = 1, \cdots, 6 \quad (9)$$

Furthermore, we want to handle formation accuracy constraints, i.e. state-related constraints

$$|y_i(t) - r_i(t)| < \varepsilon, \quad i = 1, \cdots, 6 \quad (10)$$

$\forall\ t \in \mathbb{Z}_+$, with r the reference signal and y some suitably selected output. Consequently, we propose a control strategy based on a Command Governor approach. It consists in designing a primal control law that does not take the constraints into account, and an external unit capable of taking care of constraints fulfillment by modifying, whenever necessary, the reference. Since the primal controller is designed for the linear model (8), any simple control strategy can be selected. In this paper we solve the unconstrained control problem using a simple linear quadratic LQ regulator (Mosca, 1995). Therefore, our attention can be focused on the closed-loop unconstrained system

$$\begin{cases} x(t+1) = \Phi x(t) + Gr(t) + G_d d(t) \\ y(t) = H_y x(t) \end{cases} \quad (11)$$

4. COMMAND GOVERNOR APPROACH

The basis theory related to Command Governor is extensively described in (Bemporad *et al.*, 1997;

Casavola *et al.*, 2000; Albertoni *et al.*, 2003) and (Bacconi *et al.*, 2004). Briefly, the closed-loop state-space description of a plant regulated by a primal controller and CG unit is

$$\begin{cases} x(t+1) = \Phi x(t) + Gg(t) + G_d d(t) \\ y(t) = H_y x(t) \\ c(t) = H_c x(t) + Lg(t) + L_d d(t) \end{cases} \tag{12}$$

In particular, $x(t) \in \mathbb{R}^n$ is the state which includes plant and compensator states (if any), $g(t) \in \mathbb{R}^m$, which would be typically $g(t) = r(t)$ if no constraints were present (no CG present), is the CG output, viz. a suitably modified version of the reference signal $r(t) \in \mathbb{R}^m$. Moreover, $d(t) \in \mathbb{R}^{n_d}$ is an exogenous disturbance satisfying $d(t) \in \mathcal{D}$, $\forall\, t \in \mathbb{Z}_+$, with $\mathcal{D}$ a specified convex and compact set such that $0_{n_d} \in \mathcal{D}$, $y(t) \in \mathbb{R}^m$ is the output, viz. a performance related signal which is required to track $r(t)$ and $c(t) \in \mathbb{R}^{n_c}$ is the vector to be constrained.

The main idea of the CG technique is to choose at each time instant a constant *virtual command* $v(\cdot) \equiv w$, with minimal distance from the reference of value $r(t)$, such that the corresponding virtual evolution fulfill the constraints with a certain margin δ over a semi-infinite horizon. It can be summarized in the following problem: solve

$$g(t) = \arg \min_{w \in \mathcal{V}(x(t))} \|w - r(t)\|^2_\Psi \tag{13}$$

at any time instant. Here, $\Psi = \Psi' > 0_p$, $\|w\|^2_\Psi := x'\Psi x$ and $\mathcal{V}(x(t))$ is the set of signals w such that the virtual evolution of the system satisfy the constraints for any time instant. Such a command is applied, a new state is measured and the procedure is repeated.

It has been shown in (Bemporad *et al.*, 1997) that the problem is convex, the minimizer in (13) uniquely exists at each $t \in \mathbb{Z}_+$ and the overall system is asymptotically stable.

Moreover, in order to accommodate constraints in the form of (10) that transform the third equation in (12) in

$$c(t) = H_c x(t) + L_g g(t) + L_d d(t) + L_r r(t) \tag{14}$$

we introduced in (Bacconi *et al.*, 2004) a suitable parameterization of the reference trajectory

$$\alpha(t) = \alpha(t-1) + \Delta(t) \qquad \Delta(t) \in [0,1] \tag{15}$$

with $r(\alpha(t))$ any point $\in \mathbb{R}^m$ between $r(t-1)$ and $r(t)$ along the reference trajectory. In particular, $g(t) = r(\alpha(t))$ corresponding to the nominal point $r(t)$ in the reference trajectory, when $\Delta(t) = 1$, i.e. no constraints present.

The new system, arising from (12) after the introduction of (15) is

$$\begin{cases} x(t+1) = \Phi x(t) + Gr(\alpha(t)) + G_d d(t) \\ y(t) = H_y x(t) \\ c(t) = H_c x(t) + Lr(\alpha(t)) + L_d d(t) \\ \alpha(t) = \alpha(t-1) + \Delta(t) \end{cases} \tag{16}$$

where $L = L_g + L_r$. Therefore, the CG problem is

$$\begin{aligned} \Delta(t) &:= \arg \max_{r(\alpha(t-1)+\Delta) \in \mathcal{V}(x(t))} \Delta \\ w(t) &:= r(\alpha(t-1) + \Delta(t)) \end{aligned} \tag{17}$$

and all the properties pertaining the CG approach described above are restored.

Finally, notice that for the problem at hand, attitude paths are defined in combination with position paths. Thus, the applied control law has to select the most restrictive Δ between the one resulting from the attitude maneuver and the one resulting from the position maneuver.

5. HYBRID COMMAND GOVERNOR

The assumption of small angles, that leads to system (3), might not match with some demanding applications. Experimental results show that the linearized models have significant discrepancy with respect to non linear models for deviations of Euler's angles larger than 10 degrees.

Hence, when the small-angles assumption does not hold, we propose the use of a bank of linearized models in the form of (3), each one representing the attitude relative to a specific reference frame $\{O_i\}$, described in $\{C\}$ by a rotation matrix $\mathcal{R}^{o_i c}$. The position components are not affected by this problem and just one single model can be used along with each member of the given set of attitude models.

Consequently, referring to each vehicle in the fleet, a single CG unit can be designed for each linearized model and a suitably designed supervisory unit can take care of orchestrating the switching among the CG candidates during the on-line operations. The overall technique il termed *hybrid CG control scheme* (HCG). A similar approach have been previously used in (Albertoni *et al.*, 2003). In order to provide a control law for every possible situation, it requires a correct definition of the linearized systems.

Consider the following set of reference set-points r which are desired to be tracked without offset

$$r \in \Xi \subset \mathbb{R}^m$$

Assume that $\Xi \not\subset \mathcal{W}^\delta$, where $\mathcal{W}^\delta$ is the set of signals w such that the steady state virtual evolution of the system satisfy the constraints with a margin δ (Bemporad *et al.*, 1997). Thus, the requirement that all set-points in Ξ will be tracked without error cannot be satisfied. A way to overcome this limitation is that of covering the set Ξ with a collection of $\mathcal{W}^\delta_i$, $i = 1, \cdots, l$ with overlapping interior corresponding to l different CGs such that

$$\Xi \subset \bigcup_{i=1}^{l} \mathcal{W}^\delta_i \tag{18}$$

and Interior$\{\mathcal{W}_i^\delta \cap \mathcal{W}_j^\delta\} \neq 0$, for at least a pair $(i,j) \in \{1,\dots l\}$. Clearly, CG_i operates properly when initial and final set-points belong to $\mathcal{W}_i^\delta$. If the final set-point belongs to a different set $\mathcal{W}_j^\delta$, a procedure for switching between CG_i and CG_j has to be defined. To this end, let us consider the output admissible set $\mathcal{Z}_i^\delta \subset \mathbb{R}^m \times \mathbb{R}^n$ for CG_i. It consists of the set of all pairs $[r,x]'$ whose evolutions satisfy the constraints for all $t \in \mathbb{Z}_+$. Hence, we can define the set of all states which can be steered to feasible equilibrium points without constraints violation

$$\mathcal{X}_j^\delta := \{x \in \mathbb{R}^n : \begin{bmatrix} w \\ x \end{bmatrix} \in \mathcal{Z}_i^\delta \text{ for at least one } w\}$$

Now, if (i,j) is such that Interior$\{\mathcal{W}_i^\delta \cap \mathcal{W}_j^\delta\} \neq 0$ then also Interior$\{\mathcal{X}_i^\delta \cap \mathcal{X}_j^\delta\} \neq 0$. Thus, one can a-priori define a convenient *transition reference* $r_{ij} \in$ Interior$\{\mathcal{W}_i^\delta \cap \mathcal{W}_j^\delta\}$ such that $\overline{x}_{ij} \in$ Interior$\{\mathcal{X}_i^\delta \cap \mathcal{X}_j^\delta\}$, where $\overline{x}_{ij}$ is the equilibrium disturbance-free steady-state corresponding to r_{ij} (using a worst case approach). Finally, $[r_{ij}, \overline{x}_{ij}]' \in \{\mathcal{Z}_i^\delta \cap \mathcal{Z}_j^\delta\}$ and the transfer strategy is simply defined. Assume to be at instant $\bar{t}$, be using CG_i and let $r(\bar{t}) \in \mathcal{W}_i^\delta$, $r(\bar{t}+1) \in \mathcal{W}_j^\delta$ with $\{\mathcal{W}_i^\delta \cap \mathcal{W}_j^\delta\} \neq 0$. Hence, a possible switching logic is as follows:

1. Solve and apply

$$g(\bar{t}+k) = \arg \min_{w \in \mathcal{V}_i(x(\bar{t}+k))} \|w - r(\bar{t})\|_\Psi^2, \; k = 1,\dots,\bar{k}$$

2. At $t = \bar{t} + \bar{k}$, as soon as

$$x(t) \in \text{Interior}\{\mathcal{X}_i^\delta \cap \mathcal{X}_j^\delta\} \tag{19}$$

switch to CG_j and solve

$$g(t) = \arg \min_{w \in \mathcal{V}_j(x(t))} \|w - r(\bar{t}+1)\|_\Psi^2, \; t \geq \bar{t} + 1 + \bar{k}$$

The illustrated scheme, inspired by (Gilbert and Kolmanovsky, 1999), is motivated by the fact that for any $x \in \mathbb{R}^n$ the state evolution will enter in Interior$\{\mathcal{X}_i^\delta \cap \mathcal{X}_j^\delta\}$ within a finite number of time instants.

Finally, we propose a criterion to select the CG_j at time instant $t = \bar{t} + \bar{k}$, based on the Euclidean norm between the state and the linearization point. The supervisor switches to model K_j centered in $\{O_j\}$ where CG_j corresponds to

$$j = \min_j \|\theta(\{O_j\}) - \theta(t)\| \tag{20}$$

An analysis of the properties of the proposed switching criterion is under development. Of course, other possibilities for the switching logic exist, which could be more effective for some applications.

6. SIMULATIONS

We refer to a reconfiguration maneuver as depicted in Fig. 2 regarding three vehicles orbiting around Earth at a distance of about 600 Km (LEO) at the velocity $\omega_0 = 0.0011\ rad/s$. They could form, as an example, an Earth Observing System with two slaves satellites pointing to a master, e.g. number III. For each micro-satellite

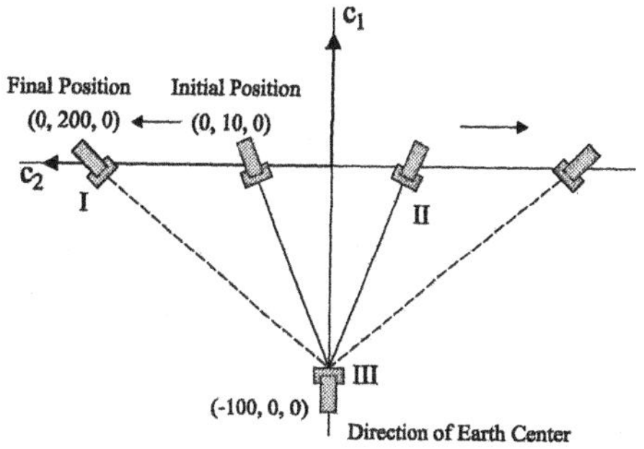

Fig. 2. Reconfiguration maneuver.

of the formation, we assume $m = 150\ Kg$ and $I = diag(35,16,25)\ Kg \cdot m^2$. Further the following saturation constraints are supposed: $|f|_{max} = 5 \cdot 10^{-2} N$, $|\tau|_{max} = 2 \cdot 10^{-3} Nm$. The maximum amplitudes of disturbances are assumed to be $|n|_{max} = 6 \cdot 10^{-3} N$ and $|M|_{max} = 2 \cdot 10^{-4} Nm$ and a value of $\delta = 10^{-4}$ is selected.

Besides saturation constraints we want to consider accuracy constraints $|p_i(t) - r_i(\alpha(t))| < 0.5\ m$ for each of the position components and $|\theta_i(t) - r_i(\alpha(t))| < 0.1\ deg$ for each of the attitude components.

The results of the application of the hybrid CG controller to the first satellite are illustrated in Fig. 3 and 4 (constraint boundaries in horizontal dash lines and instants of switching in vertical dash lines). Figures related to the second satellite

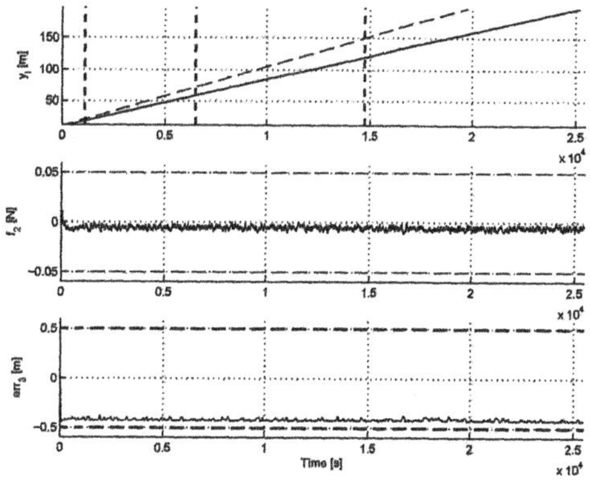

Fig. 3. Relative position, control force and error in coordination accuracy for $p_2(t)$ under HCG.

are mirror images. The constraints that influence the values of $\Delta(t)$ (17) are the ones related to the position accuracy. This is evident in Fig. 3 and Fig. 4 (bottom). Input forces and torques, on the contrary, take values close to zero during the entire reconfiguration. In the present simulation, a different linear model has been associated to the system, for Euler's angles of $(0,0,0)\ rad$,

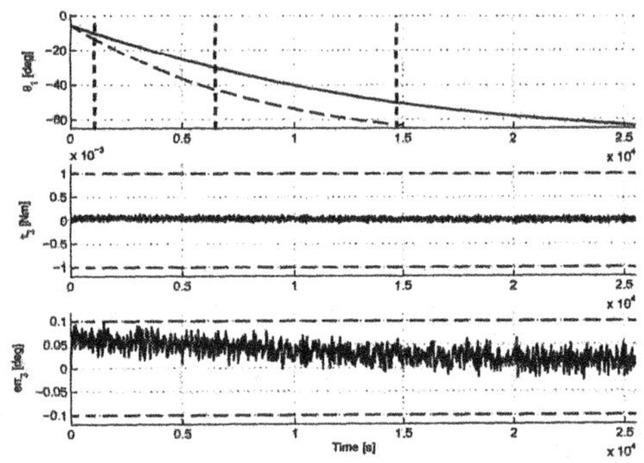

Fig. 4. Euler's angle, control torque and error in coordination accuracy for $\theta_3(t)$ under HCG.

$(0, 0, 20)$ rad, $(0, 0, 40)$ rad and so on.

Notice that, because of the use of the CG, the reference trajectory is tracked with a speed typically lower than the planned velocity. This is evident in Figs. 3 and 4.

If the hybrid CG scheme is not used, the formation accuracy constraints related to the position are not fulfilled.

7. CONCLUSIONS

The problem of reconfiguration for fleets of satellites subject to persistent disturbances and under input saturation and coordination accuracy constraints has been addressed.

The proposed control scheme is based on a bank of controllers composed by primal LQ control laws and nonlinear Command Governor units. The LQ controllers compensate the system without taking into account the presence of constraints whereas the CGs are used to generate a suitably modified version of the reference signal, capable of producing evolutions that fulfill the constraints at each time instant. A supervisory unit has been presented, capable of switching between the CG units of the bank, according to the attitude of the satellites.

Finally, the algorithm has been applied to reconfigure an Earth Observing System, forcing two satellites to cover a pre-specified trajectory defined with respect to a third one, fulfilling on it some accuracy constraints and preventing input saturation from occurring.

Acknowledgments: this work was partially supported by MIUR (Italian Ministry for Education, University and Research) under the Project *"Fault detection, diagnosis and control reconfiguration: methods and operational tools for supervisory industrial automation"*.

REFERENCES

Albertoni, A., A. Ballucchi, A. Casavola, C. Gambelli, E. Mosca and A.L. Sangiovanni Vincentelli (2003). Hybrid command governor approach for idle speed control in gasoline direct injection engines. In: *Proc. of 2003 IEEE American Control Conference.* pp. 773–778.

Bacconi, F., A. Casavola and E. Mosca (2004). A command governor approach to formation flying control problems. In: *Proc. of 16th IFAC Symposium on Automatic Control in Aerospace.*

Bemporad, A., A. Casavola and E. Mosca (1997). Nonlinear control of constrained linear systems via predictive reference menagement. *IEEE Transactions on Automatic Control* **42**, 340–349.

Casavola, A., E. Mosca and D. Angeli (2000). Robust command governors for constrained linear systems. *IEEE Transactions on Automatic Control* **45**, 2071–2077.

Casavola, A., E. Mosca and M. Papini (2006). Supervision of networked dynamical systems under coordination constraints. *IEEE Transaction on Automatic Control* **51**, 421–437.

Esper, J., S. Neeck, J. A. Slavin, J. Leitner, W. Wiskombe and F. H. Bauer (2003). Nano/micro satellite constellations for earth and space science. *Acta Astronautica* **52**, 785–791.

Gilbert, E.G. and I.V. Kolmanovsky (1999). Setpoint control of nonlinear systems with state and control constraints: A lyapunov-function, reference governor approach. In: *Proc. of 37th IEEE Coference on Decision and Control.* pp. 2507–2512.

Hu, T. and Z. Lin (2001). *Control Systems with Actuator Saturations.* Bikhäuser. Boston.

Manikonda, V., P.O. Arambel, M. Gopinathan, R.K. Mehra and F.Y. Hadaegh (1991). A model predictive control-based approach for spacecraft formation keeping and attitude control. In: *Proc. of the American Control Conference.* pp. 4528–4262.

Mayne, D. Q., J. B. Rawlings, C. V. Rao and P. O. M. Scokaert (2000). Constrained model predictive control. *Automatica* **36**, 789–814.

Mosca, E. (1995). *Optimal, Predictive, and Adaptive Control.* Prentice Hall. Englewood Cliffs, NJ.

Sidi, M. J. (1997). *Spacecraft Dynamics and Control: A Practical Engineering Approach.* Cambridge University Press.

Tillerson, M., G. Inalhan and J.P. How (2002). Co-ordination and control of distributed spacecraft systems using convex optimization techniques. *International Journal of Robust and Nonlinear Control* **12**, 207–242.

Yeh, H. H., E. Nelson and A. Sparks (2000). Nonlinear tracking control for satellite formations. In: *Proc. of the 39th IEEE Conference on Decision and Control.* pp. 328–333.

ELSEVIER
IFAC
PUBLICATIONS

A GRADIENT-BASED APPROACH TO A CLASS OF HYBRID OPTIMAL CONTROL PROBLEMS

Vadim Azhmyakov * **Jörg Raisch** *,**

* *Fachgebiet Regelungssysteme, Institut für Energie- und Automatisierungstechnik, Technische Universität Berlin, Einsteinufer 17, D-10587 Berlin, Germany.*
e-mail: azhmyakov@control.tu-berlin.de
** *Systems and Control Theory Group, Max Planck Institute for Dynamics of Complex Technical Systems, Sandtorstr. 1, D-39106 Magdeburg, Germany.*
email: raisch@control.tu-berlin.de and raisch@mpi-magdeburg.mpg.de

Abstract: We investigate optimal control problems for a class of non-stationary hybrid systems with autonomous location transitions. Using the Lagrange approach and the technique of the reduced gradient, we derive necessary optimality conditions for the considered class of problems. These optimality conditions are closely related to a variant of the Hybrid Maximum Principle and can be used for constructive optimization algorithms. *Copyright © 2006 IFAC*

Keywords: hybrid systems, optimal control, computational algorithms.

1. INTRODUCTION

The optimal control of hybrid systems has become a new focus of nonlinear control theory (see e.g., [4,5,6,12,13,17,18]). Hybrid control systems are mathematical modes of heterogeneous systems consisting of a continuous part, a finite number of continuous controllers and a discrete supervisor. For a hybrid optimal control problem, the main tool toward the construction of optimal trajectories is the Hybrid Maximum Principle [6,12,13,17,18]. This result generalizes the classical Pontryagin Maximum Principle [3,10,15]. It is well-known that the standard proof of the Pontryagin Maximum Principle is based on the techniques of "needle variations" (see e.g., [10,15]). The character of a general hybrid optimal control problem changes the possibility of using the standard needle variations [13]. Therefore, a variant of the Hybrid Maximum Principle for a hybrid optimal control problem can be proved only under some restrictive assumptions (see e.g., [12,13,17,18]). In effect, these assumptions guarantee that the classical needle variations are still admissible variations. Furthermore, in the context of a practical implementation of the Hybrid Maximum Principle we need to construct a simultaneous solution of a large-dimensional boundary-value problem and of a family of sophisticated auxiliary minimization problems. This is a complicated problem, especially in the case of many-dimensional systems with a lot of switchings.

In this paper, we consider a class of non-stationary hybrid control systems with autonomous (uncontrolled) location transitions. For general theory of hybrid systems and basic definitions we refer to, e.g., [2,11,16]. Using an approach based on Lagrange-type techniques and on reduced gradients, we obtain a set of first-order necessary optimality conditions for the above class of nonlinear hybrid optimal control problems. The explicit computation of the corresponding reduced gradients provides also a basis for applications of some effective gradient-based optimization algorithms to the given hybrid optimal control problems.

The remainder of the paper is organized as follows. Section 2 contains the initial hybrid optimal control problem and some basic facts. Section 3 is devoted to the concepts of reduced gradients for optimal control problems in abstract and specific hybrid settings. In Section 4 we present the necessary optimality conditions and propose a gradient-based computational approach to the initial problem. Section 5 summarizes the article.

2. PROBLEM FORMULATION

We start by introducing a variant of the standard definition of hybrid systems [12,13,17,18].

Definition 1. A hybrid system is a 7-tuple

$$\{Q, M, U, F, \mathcal{U}, I, \mathcal{S}\},$$

where

- Q is a finite set of discrete states (called *locations*);
- $M = \{M_q\}_{q\in Q}$ is a family of smooth manifolds, indexed by Q;
- $U \subseteq \mathbb{R}^m$ is a set of admissible control input values (called *control set*);
- $F = \{f_q\}$, $q \in Q$ is a family of maps
$$f_q : [0,1] \times M_q \times U \to TM_q,$$
where TM_q is the tangent bundle of M_q (see e.g., [7,10]);
- $\mathcal{U}$ is the set of all admissible control functions;
- $I = \{I_q\}$ is a family of subintervals of $[0,1]$ such that the length of each I_q is less than 1;
- $\mathcal{S}$ is a subset of Ξ, where
$$\Xi := \{(q, x, q', x') \ : \ q, q' \in Q, x \in M_q, x' \in M_{q'}\}$$

A hybrid system from Definition 1 is defined on the time-interval $[0,1]$. Note that in contrast to the general definition of a hybrid system [12,13,17,18], the control set U from Definition 1 is the same for all locations. Moreover, in the sense of this definition the set $\mathcal{U}$ is also independent of a location. Let us assume that U is a compact set and

$$\mathcal{U} := \{u(\cdot) \in \mathbb{L}^2_m([0,1]) \ : \ u(t) \in U \text{ a.e. on } [0,1]\},$$

where $\mathbb{L}^2_m([0,1])$ is the standard Lebesgue space of all square-integrable functions $u : [0,1] \to \mathbb{R}^m$. We now introduce some additional hypothesis for the vector fields f_q, $q \in Q$:

- all functions $f_q(t,\cdot,\cdot)$ from F are differentiable,
- f_q, $\partial f_q/\partial x$, $\partial f_q/\partial u$ are continuous and there exist constants $C_q < \infty$ such that
$$||\frac{\partial}{\partial x} f_q(t,x,u)|| \leq C_q,$$
$$q \in Q, \ (t,x,u) \in [0,1] \times M_q \times U.$$

For $q, q' \in Q$ one can also define the *switching set*

$$S_{q,q'} := \{(x, x') \in M_q \times M_{q'} \ : \ (q, x, q'x') \in \mathcal{S}\}.$$

from location q to location q'. The intervals I_q, $q \in Q$ indicate the lengths of time intervals on which the system can stay in location q. We say that a location switching from q to q' occurs at a *switching time* $t^{switch} \in [0,1]$. We now consider a hybrid system with $r \in \mathbb{N}$ switching times $\{t_i\}$, $i = 1, ..., r$, where

$$0 = t_0 < t_1 < ... < t_r < t_{r+1} = 1.$$

Note that the sequence of switching times $\{t_i\}$ is not defined a priory. A hybrid control system remains in location $q_i \in Q$ for all $t \in [t_{i-1}, t_i[$, $i = 1, ..., r+1$. Let $t_i - t_{i-1} \in I_{q_i}$ for all $i = 1, ..., r+1$. A hybrid system (in the sense of Definition 1) that satisfies the above assumptions is denoted by $\mathcal{HS}$.

Definition 2. Let $u(\cdot) \in \mathcal{U}$ be an admissible control for a hybrid control system $\mathcal{HS}$. Then a *continuous trajectory* of $\mathcal{HS}$ is an absolutely continuous function

$$x : [0,1] \to \bigcup_{q\in Q} M_q$$

such that $x(0) = x_0 \in M_{q_1}$ and

- $\dot{x}(t) = f_{q_i}(t, x(t), u(t))$ for almost all $t \in [t_{i-1}, t_i]$ and all $i = 1, ..., r+1$;
- the switching condition $(x(t_i), x(t_{i+1})) \in S_{q_i, q_{i+1}}$ holds if $i = 1, ..., r$.

The vector $\mathcal{R} := (q_1, ... q_{r+1})^T$ is a *discrete trajectory* of the hybrid control system $\mathcal{HS}$.

Definition 2 describe the dynamic of a hybrid control system $\mathcal{HS}$. Since $x(\cdot)$ is an absolutely continuous function, Definition 2 describe a class of hybrid systems without impulse components of the continuous trajectories. Therefore, the corresponding switching sets $S_{q,q'}$ (and $S_{q_i,q_{i+1}}$) are defined for $x = x'$ (for $x(t_i) = x(t_{i+1})$).

Under the above assumptions for the given family of vector fields F, for each admissible control $u(\cdot) \in \mathcal{U}$ and for every interval $[t_{i-1}, t_i]$ (for every location $q_i \in \mathcal{R}$) there exists a unique absolutely continuous solution of the corresponding differential equation. This means that for each $u(\cdot) \in \mathcal{U}$ we have a unique absolute continuous trajectory of $\mathcal{HS}$. Moreover, the switching times $\{t_i\}$ and the discrete trajectory $\mathcal{R}$ for a hybrid control system $\mathcal{HS}$ are also uniquely defined. Therefore, it is reasonable to introduce the following concept.

Definition 3. Let $\mathcal{HS}$ be a hybrid control system as defined above. For an admissible control $u(\cdot) \in \mathcal{U}$, the triplet $\mathcal{X}^u := (\tau, x(\cdot), \mathcal{R})$, where τ is the set of the corresponding switching times $\{t_i\}$, $x(\cdot)$ and $\mathcal{R}$ are the corresponding continuous and discrete trajectories, is called *hybrid trajectory* of $\mathcal{HS}$.

Let $\phi : \mathbb{R}^n \to \mathbb{R}$ be a continuously differentiable function. Given a hybrid control system $\mathcal{HS}$ we denote the following Mayer-type hybrid optimal control problem by OCP:

$$\begin{aligned} &\text{minimize } \phi(x(1)) \\ &\text{subject to } \dot{x}(t) = f_{q_i}(t, x(t), u(t)) \text{ a.e. on}[t_{i-1}, t_i] \quad (1) \\ &i = 1, ..., r+1, \; x(0) = x_0 \in M_{q_1}, \; u(\cdot) \in \mathcal{U}. \end{aligned}$$

Evidently, (1) is the problem of minimizing the Mayer cost functional $J(X) := \phi(x(1))$ over all hybrid trajectories X of $\mathcal{HS}$. Note that we study OCP (1) in the absence of target (endpoint) and state constraints. For necessary optimality conditions for (1) in the form of a Hybrid Maximum Principle we refer to [5,17].

3. REPRESENTATION OF FUNCTIONAL DERIVATIVES

3.1 The General Gradient Formula

Let us first examine an abstract optimal control problem (the generalization of (1)) which involves a control variable v along with a state variable ξ

$$\begin{aligned} &\text{minimize } T(\xi, v) \\ &\text{subject to } P(\xi, v) = 0, \qquad (2) \\ &(\xi, v) \in \Omega, \end{aligned}$$

where $T : X \times Y \to R$ is a cost functional, X, Y are real Banach and Hilbert spaces and

$$P : X \times Y \to X$$

is a given mapping. By Ω we denote here a nonempty subset of $X \times Y$.

Definition 4. We say that an admissible pair

$$(\hat{\xi}, \hat{v}) \in \tilde{\Omega} := \{(\xi, v) \in Q \mid P(\xi, v) = 0\}$$

is a local solution of (2) if

$$T(\hat{\xi}, \hat{v}) \leq T(\xi, v) \;\; \forall (\xi, v) \in W_{(\hat{\xi}, \hat{v})} \subset \Omega,$$

where $W_{(\hat{\xi}, \hat{v})} \subset X \times Y$ is a neighborhood of $(\hat{\xi}, \hat{v})$.

All derivatives considered in this papers are Fréchet derivatives. We assume that the mappings T and P are continuously differentiable and that the state equation $P(\xi, v) = 0$ can be solved with respect to ξ, i.e.

$$\xi = \omega(v),$$

where $\omega : Y \to X$ is a differentiable function. In this case the functional $T(\xi, v)$ can be represented as a functional depending only on v, namely,

$$T(\xi, v) = T(\omega(v), v) = \tilde{T}(v).$$

Note that the introduced abstract optimal control problem (2) is of primary importance in many applications. Not only the hybrid optimal control problem (1) but also an ODE- and a PDE-optimal control problem or an optimal control problem with integral equations can also be formulated (in various ways) as an abstract problem (2). Moreover, a usual finite-dimensional approximation of an infinite-dimensional OCP has the form of the minimization problem (2). In the above cases the condition $P(\xi, v) = 0$ represents the corresponding "state equation" of a specific optimal control problem. Assume that the abstract problem (2) is *regular* (see [10]). Define the Lagrangian of problem (2)

$$\mathcal{L}(\xi, v, p) := T(\xi, v) + \langle p, P(\xi, v)\rangle_X,$$

where $p \in X^*$ and $\langle p, \cdot \rangle_X : X \to \mathbb{R}$. Here X^* is the (topological) dual space to X. For the generalized Lagrange Multiplier Rule see e.g., [10,1]. We use the standard notation

$$T_\xi, \; T_v, \; P_\xi, P_v, \; \mathcal{L}_\xi, \; \mathcal{L}_p, \; \mathcal{L}_u$$

for the partial derivatives of the functions T, P and $\mathcal{L}$. Moreover, we introduce the adjoint operators

$$T^*_\xi, \; T^*_v, \; P^*_\xi, P^*_v, \; \mathcal{L}^*_\xi, \; \mathcal{L}^*_p, \; \mathcal{L}^*_u$$

to the corresponding derivatives (linear operators) and also consider the adjoint operator $\nabla\tilde{T}^*(v)$ to $\nabla\tilde{T}(v)$. In the context of (2) we now formulate an immediate consequence of the above solvability assumption for the state equation $P(\xi, v) = 0$. Note that a usual solvability criterion for this follows from an appropriate variant of the Implicit Function Theorem [7,1].

Theorem 5. Let T and P be continuously Fréchet differentiable and let the state equation in (2) be solvable. Assume that there exists the inverse operator

$$(P^*_\xi)^{-1} \in L((X^* \times Y^*), X^*)$$

to P_ξ. Then the gradient $\nabla\tilde{T}^*(v)$ can be found by solving the following equations

$$\begin{aligned} &P(\xi, v) = \mathcal{L}^*_p(\xi, v, p) = 0, \\ &T^*_\xi(\xi, v) + P^*_\xi(\xi, v)p = \mathcal{L}^*_\xi(\xi, v, p) = 0, \qquad (3) \\ &\nabla\tilde{T}^*(v) = T^*_v(\xi, v) + P^*_v(\xi, v)p = \mathcal{L}^*_v(\xi, v, p). \end{aligned}$$

Let us sketch the proof of Theorem 5. Differentiating the functional $\tilde{T}$ and state equation in (2) we obtain

$$\begin{aligned} &P_\xi(\xi, v)\nabla\omega(v) + P_v(\xi, v) = 0, \\ &\nabla\tilde{T}(v) = T_v(\xi, v) + T_\xi(\xi, v)\nabla\omega(v). \end{aligned}$$

The existence of $(P^*_\xi)^{-1}$ implies the formula

$$\nabla\omega(v) = -(P_\xi)^{-1}(\xi, v)P_v(\xi, v).$$

Hence

$$\nabla\tilde{T}(v) = T_v(\xi, v) - T_\xi(\xi, v)(P_\xi)^{-1}(\xi, v)P_v(\xi, v),$$

and

$$\nabla\tilde{T}^*(v) = T^*_v(\xi, v) - P^*_v(\xi, v)(P^*_\xi)^{-1}(\xi, v)T^*_\xi(\xi, v). \quad (4)$$

On the other hand, we can calculate p from the second (adjoint) equation in (3) and substitute it to the third (gradient) equation. In this manner we also obtain the given relation (4). Note that a related result was also obtained in [19] for classical optimal control problems.

3.2 Reduced Gradients for Hybrid Optimal Control Problems

Consider a hybrid control system $\mathcal{HS}$. For an admissible control function $u(\cdot) \in \mathcal{U}$ we obtain the corresponding hybrid trajectory $\mathcal{X}^u$. For every interval $[t_{i-1}, t_i]$ from τ we can define the characteristic function of $[t_{i-1}, t_i]$

$$\beta_{[t_{i-1},t_i)}(t) = \begin{cases} 1 & \text{if } t \in [t_{i-1}, t_i) \\ 0 & \text{otherwise.} \end{cases}$$

Using the introduced characteristic functions, we rewrite the state differential equations from Definition 2 for the continuous trajectory $x(\cdot)$ in the following form

$$\dot{x}(t) = \sum_{i=1}^{r+1} \beta_{[t_{i-1},t_i)}(t) f_{q_i}(t, x(t), u(t)), \tag{5}$$

where $x(0) = x_0$. Under the above assumptions for the family of vector fields F, the right-hand side of the obtained differential equation (5) satisfies the conditions of the extended Caratheodory Theorem (see e.g., [9]). Therefore, there exists a unique (absolutely continuous) solution of (5). We now apply the abstract Theorem 5 to the hybrid OCP (1). In the case of the hybrid control system $\mathcal{HS}$ we have

$$X = \mathbb{W}_n^{1,\infty}([0,1]),\ Y = \mathbb{L}_m^2([0,1]).$$

By $\mathbb{W}_n^{1,\infty}([0,1])$ we denote here the Sobolev space of all absolutely continuous functions with essentially bounded derivatives. Let us introduce the operator

$$P : \mathbb{W}_n^{1,\infty}([0,1]) \times \mathbb{L}_m^2([0,1]) \to \mathbb{W}_n^{1,\infty}([0,1]) \times \mathbb{R}^n,$$

where

$$P(x(\cdot), u(\cdot))\Big|_t := \begin{pmatrix} \dot{x}(t) - \sum_{i=1}^{r+1} \beta_{[t_{i-1},t_i)}(t) f_{q_i}(t, x(t), u(t)) \\ x(0) - x_0 \end{pmatrix}.$$

Evidently, the operator equation $P(x(\cdot), u(\cdot)) = 0$ is consistent with the state equation from the abstract optimization problem (2). Consider a regular OCP (1) and introduce the *Hamiltonian*

$$H(t, x, u, p) = \langle p, \sum_{i=1}^{r+1} \beta_{[t_{i-1},t_i)}(t) f_{q_i}(t, x, u) \rangle.$$

where $p \in \mathbb{R}^n$. Since every admissible control $u(\cdot)$ determines a unique hybrid trajectory $\mathcal{X}^u$, the following cost functional $\tilde{J} : \mathcal{U} \to \mathbb{R}$ such that $\tilde{J}(u(\cdot)) := J(\mathcal{X}^u)$ is well-defined. The differentiability of the given function ϕ implies the differentiability of $\tilde{J}$. The corresponding derivative is denoted by $\nabla\tilde{J}$. In the particular case of OCP (1) the evaluation of the adjoint operator $\nabla\tilde{J}^*$ to $\nabla\tilde{J}$ is relatively easy. We now present our main result that follows from Theorem 5.

Theorem 6. Consider a regular OCP (1). The gradient $\nabla\tilde{J}^*(u(\cdot))$ can be found by solving the equations

$$\begin{aligned} &\dot{x}(t) = H_p(t, x(t), u(t), p(t)),\ x(0) = x_0, \\ &\dot{p}(t) = -H_x(t, x(t), u(t), p(t)),\ p(1) = -\phi_x(x(1)), \\ &\nabla\tilde{J}^*(u(\cdot))(t) = -H_u(t, x(t), u(t), p(t)), \end{aligned} \tag{6}$$

where $p(\cdot)$ is an absolutely continuous function (an "adjoint variable").

Proof. The Lagrangian of the regular problem (1) can be written as

$$\mathcal{L}(x(\cdot), u(\cdot), \hat{p}, p(\cdot)) = \phi(x(1)) + \langle \hat{p}, x(0) - x_0 \rangle +$$
$$+ \langle p(t), \dot{x}(t) - \sum_{i=1}^{r+1} \beta_{[t_{i-1},t_i)}(t) f_{q_i}(t, x(t), u(t)) \rangle dt,$$

where the adjoint variable here contains two components $\hat{p} \in \mathbb{R}^n$ and $p(\cdot)$. If we differentiate the Lagrange function with respect to the adjoint variable, then we obtain the first equation from (6)

$$\begin{aligned} \dot{x}(t) &= \sum_{i=1}^{r+1} \beta_{[t_{i-1},t_i)}(t) f_{q_i}(t, x(t), u(t)) = \\ &= H_p(t, x(t), u(t), p(t)), \end{aligned}$$

with $x(0) = x_0$. Consider the term

$$\int_0^1 \langle p(t), \dot{x}(t) \rangle dt.$$

From the integration by part we have

$$\int_0^1 \langle p(t), \dot{x}(t) \rangle dt = \langle p(1), x(1) \rangle - \langle p(0), x(0) \rangle -$$
$$- \int_0^1 \langle \dot{p}(t), x(t) \rangle.$$

Hence

$$\begin{aligned} &\mathcal{L}(x(\cdot), u(\cdot), \hat{p}, p(\cdot)) = \phi(x(1)) + \langle p(1), x(1) \rangle + \\ &+ \langle \hat{p} - p(0), x(0) \rangle - \langle \hat{p}, x_0 \rangle - \int_0^1 \langle \dot{p}(t), x(t) \rangle dt + \\ &+ \int_0^1 \langle p(t), \sum_{i=1}^{r+1} \beta_{[t_{i-1},t_i)}(t) f_{q_i}(t, x(t), u(t)) \rangle dt. \end{aligned} \tag{7}$$

If we differentiate $\mathcal{L}$ in (7) with respect to $x(\cdot)$, we can use Theorem 5 and compute $\mathcal{L}_x$, $\mathcal{L}_x^*$. Thus we obtain the second relation in (6). Using (7), we also write

$$\mathcal{L}_u(x(\cdot), u(\cdot), \hat{p}, p(\cdot))v(\cdot) = -\int_0^1 H_u(t, x(t), u(t))v(t)dt$$

for every $v(\cdot) \in \mathbb{L}_m^2([0,1])$. By Theorem 5, we obtain the last relation in (6)

$$\nabla\tilde{J}^*(u(\cdot))(t) = \mathcal{L}_u^*(x(\cdot), u(\cdot), \hat{p}, p(\cdot)) = -H_u(t, x(t), u(t)).$$

The proof is finished. □

The formulated result allows the explicit computation of the gradient $\nabla\tilde{J}$ (or $\nabla\tilde{J}^*$) in a sophisticated minimization problem (1).

4. NECESSARY OPTIMALITY CONDITIONS AND THE COMPUTATIONAL ASPECT

In Section 3 we have developed explicit formulae for the reduced gradient of the cost functional in (1). To make a step forward in the study of the given OCP we will discuss the necessary optimality conditions for (1) and some related numerical aspects. Let us formulate an easy consequence of Theorem 6.

Theorem 7. Assume that OCP (1) has an optimal solution $(u^{opt}(\cdot), x^{opt}(\cdot))$ such that $u^{opt}(t) \in \text{int}\{U\}$, where $\text{int}\{U\}$ is the interior of the set U. Then $(u^{opt}(\cdot), x^{opt}(\cdot))$ can be found by solving the following equations

$$\begin{aligned}
&\dot{x}^{opt}(t) = H_p(t, x^{opt}(t), u^{opt}(t), p(t)),\\
&x(0) = x_0,\\
&\dot{p}(t) = -H_x(t, x^{opt}(t), u^{opt}(t), p(t)), \qquad (8)\\
&p(1) = -\phi_x(x^{opt}(1)),\\
&H_u(t, x^{opt}(t), u^{opt}(t), p(t)) = 0.
\end{aligned}$$

Clearly, the conditions (8) from Theorem 7 present a necessary optimality conditions for a special case of problem (1). Note that the last equation in (8) is consistent with the usual optimality condition $\nabla \tilde{J}^*(u(\cdot))(t) = 0,\ t \in [0, 1]$ if the optimal control takes values in an open bounded control set.

Alternatively, Theorem 6 and Theorem 7 provide a basis for a wide class of the gradient-based optimization algorithms for (1). We now assume that the control set U has a so-called box-form, namely,

$$U := \{u \in \mathbb{R}^m \ : \ b_-^j \leq u_j \leq b_+^j,\ j = 1, ..., m\},$$

where $b_-^j, b_+^j, j = 1, ..., m$ are constants. Let us consider, for example, the standard gradient algorithm in $\mathbb{L}_m^2([0, 1])$ (see e.g., [8,14])

$$\begin{aligned}
&u^{k+1}(t) = u^k(t) - \gamma_k \nabla \tilde{J}^*(u^k(\cdot))(t),\ t \in [0, 1]\\
&b_-^j \leq u_j^{k+1}(t) \leq b_+^j,\ j = 1, ..., m,\ k = 0, 1, ... \qquad (9)\\
&u^0(\cdot) \in \mathcal{U},
\end{aligned}$$

where γ_k is a step-size of the gradient algorithm and $\{u^k(\cdot)\} \subset \mathbb{L}_m^2([0, 1])$ is the sequence of iterations. Note that in general cases an admissible iterative control $u^{k+1}(\cdot)$ can also be obtained by a projection

$$u^{k+1}(t) = P_U(u^k(t) - \gamma_k \nabla \tilde{J}^*(u^k(\cdot))(t)).$$

Here P_U is a projection operator on the control set U. For the Projected Gradient Algorithm and for convergence properties of (9) and of some related gradient-type optimization procedures see e.g., [14,8].

Let us now present an implementable computational scheme that follows from our consideration presented above.

Algorithm 1. 1) Choose an admissible initial control $u^0(\cdot) \in \mathcal{U}$ and the corresponding continuous trajectory $x^0(\cdot)$. Set $k = 0$.

2) Given a $x_{q_i}^k(\cdot)$ define q_{i+1}^k, $i = 1, ..., r + 1$ and

$$t_{i+1}^k := \min\{t \in [0, 1] \ : \ x_{q_i}^k(t) \bigcap S_{q_i, q_{i+1}} \neq \emptyset\}.$$

3) For the determined hybrid trajectory

$$\mathcal{X}^{u^k} = (\tau^k, x^k(\cdot), \bar{q}^k)$$

solve the above equations (6) and define the gradient $\nabla \tilde{J}^*(u^k(\cdot))(t)$ $\forall t \in [0, 1]$ of the cost functional.

4) Using $\nabla \tilde{J}^*(u^k(\cdot))$, compute the iteration $u^{k+1}(t)$ by using a gradient-type method. Increase k by one and go to Step (2).

Note that the switching times, number of switches and switching sets in the given OCP (1) are assumed to be unknown. Using the iterative structure of the proposed Algorithm 1, one can compute the corresponding approximations of the optimal trajectory, optimal switching times and optimal switching sets. Let us now present the following convergence result for Algorithm 1.

Theorem 8. Assume that the data for the regular OCP (1) satisfies all hypotheses of Section 2 and that (1) has an optimal solution $(u^{opt}(\cdot), x^{opt}(\cdot))$. Let $\{\mathcal{X}^{u^k}\}$ be a sequence of hybrid trajectories generated by Algorithm 1. Then $\{\mathcal{X}^{u^k}\}$ is a minimizing sequence for (1), i.e.,

$$\lim_{k \to \infty} J(\mathcal{X}^{u^k}) = \phi(x^{opt}(1)).$$

Finally, note that Theorem 8 can be proved with the help of the dominated convergence theorem and the standard properties of a gradient minimization algorithm in Hilbert spaces.

5. CONCLUDING REMARKS

In this paper, we have developed a new approach to a class of hybrid optimal control problems of the Mayer type. This approach is based on explicit formulae for the reduced gradient of the cost functional of the given hybrid optimal control problem. The corresponding relations make it possible to formulate first-order necessary optimality conditions for the considered hybrid optimal control problems and provide a basis for effective computational algorithms. The idea of reduced gradients can also be used for some linearization procedures of the initial optimal control problem. Note that linearization techniques have been recognized for a long time as a powerful tool for solving optimization problems. The approach proposed in this paper can be extended to some other classes of hybrid optimal control problems. Finally, note that it seems to be possible to derive necessary ϵ-optimality conditions (the ϵ-Hybrid Maximum Principle) by means of the presented techniques from Theorem 6.

REFERENCES

[1] Azhmyakov, V. (2005). *Stable Operators in Analysis and Optimization*, Peter Lang, Berlin.

[2] Benedetto, M.D. and A. Sangiovanni-Vincentelli (2001). *Hybrid Systems Computation and Control.* Springer, Heidelberg.

[3] Berkovitz, L.D. (1974). *Optimal Control Theory.* Springer, New York.

[4] Branicky, M.S., V.S. Borkar and S.K. Mitter (1998). A unifed framework for hybrid control: model and optimal control theory. *IEEE Trans. Automat. Contr.* **43**, pp. 31-45.

[5] Caines, P. and M.S. Shaikh (2005). Optimality zone algorithms for hybrid systems computation and control: From exponential to linear complexity. In: *Proceedings of the 13th Mediterranean Conference on Control and Automation*. pp. 1292-1297, Limassol.

[6] Cassandras, C., D.L. Pepyne and Y. Wardi (2001). Optimal control of a class of hybrid systems. *IEEE Trans. Automat. Contr.* **46**, pp. 398-415.

[7] Clarke, F.H., Yu.S. Ledyaev, R.J.Stern and P.R. Wolenski (1998). *Nonsmooth Analysis and Control Theory*, Springer, New York.

[8] Fiacco, A.V. and G. McCormick, (1968). *Nonlinear Programming: Sequential Unconstrained Minimization Techniques*, Wiley, New York.

[9] Filippov, A.F. (1988). *Differential Equations with Discontinuous Right-Hand Sides*. Kluwer Academic Publishers, Dordrecht.

[10] Ioffe, A.D. and V.M. Tichomirov (1979). *Theory of Extremal Problems*. North Holland, Amsterdam.

[11] Morse, A.S., C.C. Pantelides and S. Sastry eds. (1999). Special issue on hybrid systems. *Automatica* **35**.

[12] Piccoli, B. (1998). Hybrid systems and optimal control. In: *Proceedings of the 37th IEEE Conference on Decision and Control*. pp. 13-18. Tampa.

[13] Piccoli, B. (1999). Necessary conditions for hybrid optimization. In: *Proceedings of the 38th IEEE Conference on Decision and Control*. pp. 410-415. Phoenix.

[14] Polak, E. (1997). *Optimization*, Springer, New York.

[15] Pontryagin, L.S., V.G. Boltyanski, R.V. Gamkrelidze and E.F. Mischenko (1962). *The mathematical Theory of Optimal Processes*. Wiley, New York.

[16] Raisch, J. (1999). *Hybride Steuerungssysteme*. Shaker, Aachen.

[17] Shaikh, M.S. and P.E. Caines (2004). On the hybrid optimal control problem: The hybrig maximum principle and dynamic programming theorem. *IEEE Trans. Automat. Contr.* submitted.

[18] Sussmann, H. (1999). A maximum principle for hybrid optimal control problems. In: *Proceedings of the 38th IEEE Conference on Decision and Control*. pp. 425-430. Phoenix.

[19] Teo, K.L., C.J. Goh and K.H. Wong (1991). *A Unifed Computational Approach to Optimal Control Problems*, Wiley, New York.

ELSEVIER
IFAC
PUBLICATIONS

OPTIMAL MODE-SWITCHING FOR HYBRID SYSTEMS WITH UNKNOWN INITIAL STATE

Henrik Axelsson * **Mauro Boccadoro** ** **Yorai Wardi** *
Magnus Egerstedt *

* *{henrik,ywardi,magnus}@ece.gatech.edu*
Electrical and Computer Engineering
Georgia Institute of Technology
Atlanta, GA 30332
** *boccadoro@diei.unipg.it*
Dipartimento di Ingegneria Elettronica e dell'Informazione
Universita di Perugia
06125, Perugia – Italy

Abstract: This paper concerns an optimal control problem defined on a class of switched-mode hybrid dynamical systems. Such systems change modes whenever the state intersects certain surfaces that are defined in the state space. These switching surfaces are parameterized by a finite dimensional vector called the *switching parameter*. The optimization problem we consider is to minimize a given cost-functional with respect to the switching parameter under the assumption that the initial state of the system is not completely known. Instead, we assume that the initial state can be anywhere in a given set. We will approach this problem by minimizing the worst possible cost over the given set of initial states using results from minimax optimization. The results are then applied in order to solve a navigation problem in mobile robotics. *Copyright © 2006 IFAC*

Keywords: Hybrid Systems, Switching Modes, Optimal Control, Gradient Descent, Numerical Algorithms, Minimax Techniques, Mobile Robotics

1. INTRODUCTION

Over the last couple of decades, a lot of effort has been directed towards optimal control of hybrid systems (Branicky *et al.*, 1998; Bemporad *et al.*, 2002; Guia *et al.*, 1999; Hedlund and Rantzer, 1999; Hristu-Varsakelis, 2001; Xu and Antsaklis, 2002; Caines and Shaikh, 2005; Attia *et al.*, 2005). Hybrid systems are complex systems that are characterized by discrete logical decision making at the highest level and continuous variable dynamics at the lowest level. Examples when these systems arise include situations where a control module has to switch its attention among a number of subsystems (Lincoln and Rantzer, 2001; Rehbinder and Sanfridson, 2000; Walsh *et al.*, 1999) or collect data sequentially from a number of sensory sources (Brockett, 1995; Egerstedt and Wardi, 2002; Hristu-Varsakelis, 2001).

The type of hybrid system under consideration in this paper can be described by the following equation

$$\dot{x}(t) \in \{f_\alpha(x(t), u(t))\}_{\alpha \in A}, \qquad (1)$$

where $x(t) \in \mathbb{R}^n$, $u(t) \in \mathbb{R}^k$, and $\{f_\alpha : \mathbb{R}^{n+k} \to \mathbb{R}^n\}_{\alpha \in A}$ is a collection of continuously differentiable functions, parameterized by α belonging to some given set A. The time t is confined to a given finite-length interval $[0, T]$. A supervisory controller is normally engaged for dictating the switching law, i.e. the rule for switching among the functions f_α in the right-hand side of (1).

This paper addresses a particular class of hybrid systems, called switched autonomous systems, where the continuous-time control variable is absent and the continuous-time dynamics change at discrete times (*switching-times*). For these systems, the authors have derived gradient expressions for the cost functional with the respect to the switching times when the initial state $x_0 \in \mathbb{R}^n$ is fixed. In particular, (Egerstedt *et al.*, 2006) presented a gradient and an algorithm that finds optimal switching-times, for when to switch between a given set of modes, for the case when the switching-times are controlled directly. Furthermore, (Boccadoro *et al.*, 2005*a*) considered the case when a switch between two different modes occurs when the state trajectory intersects a switching surface, defined by $g(x(t),a) = 0$, and parameterized by the parameter a. Reference (Boccadoro *et al.*, 2005*a*) can be thought of as the starting point of this paper, as we consider a similar problem but instead of optimizing with respect to a given fixed initial condition $x_0 \in \mathbb{R}^n$, we will assume that the initial state can be anywhere within a given set $S \subset \mathbb{R}^n$. In order to find a good value of the switching parameter a, independent of where in S we start, we will use the gradient formula presented in (Boccadoro *et al.*, 2005*a*) and find the optimal a such that we will minimize the worst possible cost for all trajectories starting in S. Hence, we have a minimax problem.

At this point, it should be noted that although we will focus on the case where switches occur when the state trajectory intersects a switching surface, the algorithm that will be presented in order to solve the minimax problem would also solve the free switching-time problem with only minor modifications. Hence, this paper presents a way to get rid of the dependence of the initial condition under the assumption that the initial state belongs to a given set.

Once the theoretical underpinnings have been presented, the results will be applied to a navigation problem in mobile robotics.

The robotics problem considered in this paper was also investigated in (Boccadoro *et al.*, 2005*a*), for a fixed initial state. However, for many applications the initial state is not known. An example of this is robotic systems that get their position from a Global Positioning System (GPS). Typically there is a nontrivial error associated with these systems. Hence, if the GPS indicates that the robot is at a point (x,y) the robot can be anywhere within the interval $(x-\Delta, x+\Delta) \times (y-\Delta, y+\Delta)$, for some positive constant Δ. As a result, solving the parameter optimization problem for a given fixed initial state might not give a good solution if the robot's position is given by a GPS.

The outline of this paper is as follows: In Section 2, the problem at hand is introduced together with some previous results relating to the gradient formula. Section 3 presents our solution using a minimax strategy. Simulation results for the robotics application are presented in Section 4 and conclusions are given in Section 5.

2. PROBLEM FORMULATION & PREVIOUS RESULTS

The state trajectory of the underlying system is given by the following equation

$$\dot{x}(t) = f_i(x(t)), \ t \in [\tau_{i-1}, \tau_i)), \ i \in \{1, \ldots, N+1\}, \quad (2)$$

where we assume that the system switches N times. The modal functions are chosen from a given set $\{f_\alpha\}_{\alpha \in A}$. However, we assume that the switching times are not controlled directly. Instead, a switch occurs whenever the state trajectory intersects a *switching* surface. This problem was initially considered in (Boccadoro *et al.*, 2005*b*) for a fixed initial state. We will follow the presentation of (Boccadoro *et al.*, 2005*b*) in order to set the stage for our minimax problem when the initial state in not completely known.

We assume that the switching times and the modal functions are determined recursively in the following way. Given f_i and $\tau_{i-1} > 0$ for some $i = 1, 2, \ldots$, let $A(i) \subset A$ be a given finite set of modes, labelled *the set of modes enabled by* f_i. Hence, there might be a restriction on the mode sequence. For every $\alpha \in A(i)$, we let $S_\alpha \subset \mathbb{R}^n$ be the $n-1$ dimensional surface enabling the switch to mode α. Then, the next switch is defined by

$$\tau_i = \min\{t > \tau_{i-1} \ : \ x(t) \in \cup_{\alpha \in A(i)} S_\alpha\} \quad (3)$$

and we note that it is possible to have $\tau_i = \infty$. If $\tau_i < \infty$ then we pick $\tilde{\alpha} \in A(i)$ such that $x(\tau_i) \in S_{\tilde{\alpha}}$, and we set $f_{i+1} = f_{\tilde{\alpha}}$. The system is initialized by setting $\tau_0 = 0$ and choosing what mode the system should start with.

The time when the state trajectory intersects a surface defines τ_i, and the index of the surface $S_{\tilde{\alpha}}$ defines f_{i+1}. In this paper, the surfaces $S_{\tilde{\alpha}}$ are defined by the solution points of parameterized equations from $\mathbb{R}^n$ to $\mathbb{R}$. We denote the parameter by a and suppose that $a \in \mathbb{R}^k$ for some integer $k \geq 1$. For every $\alpha \in A$, we let $g_\alpha : \mathbb{R}^n \times \mathbb{R}^k \to \mathbb{R}$ be a continuously differentiable function. For a given fixed value of $a \in \mathbb{R}^k$, denoted here by a_α, the switching curve S_α is defined by the solution points x of the equation $g_\alpha(x, a_\alpha) = 0$. Note that under mild assumption, S_α is a smooth $(n-1)$ dimensional manifold in $\mathbb{R}^n$, and a_α can be viewed as a control parameter of the surface. Using the terminology defined earlier, we will replace the index α by i; thus, S_i is the solution set of the equation

$$g_i(x, a_i) = 0, \quad (4)$$

which is parameterized by the control variable $a_i \in \mathbb{R}^k$. To summarize, the system changes dynamics whenever the state trajectory intersect a switching curve $g(x,a) = 0$ parameterized by a control variable a, as illustrated in Figure 1.

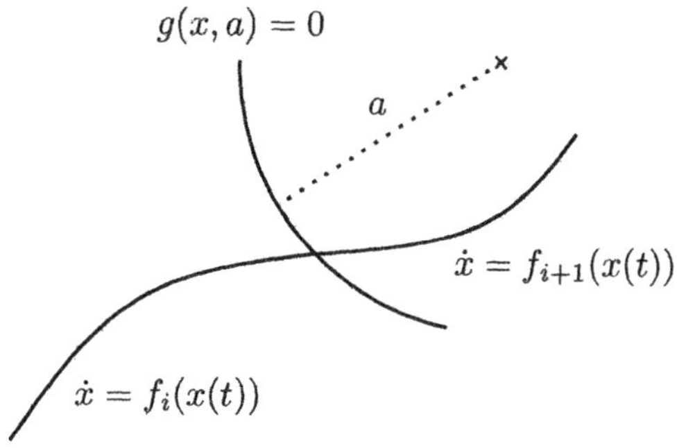

Fig. 1. Mode switching occur when the state trajectory intersect a switching surface. In this case, the switching surface is a circle parameterized by the radius a.

In order to minimize a cost criterion of the form

$$J=\int_0^T L(x(t))dt, \tag{5}$$

where $L:\mathbb{R}^n\to\mathbb{R}$, we need to determine the optimal switching surface parameters a since the state trajectory depends on a. To this end, (Boccadoro *et al.*, 2005*b*) presented an expression of the gradient of the cost functional with respect to switching surface parameter a. This gradient was presented under the assumption that the functions f_i, g_i, $i=1,\dots,N+1$, and L where continuously differentiable with respect to all its variables. Furthermore, it was assumed that f_i $i=1,\dots,N+1$, was uniformly Lipschitz.

We define $x_i=x(\tau_i)$, and the terms R_i and L_i by

$$R_i=f_i(x_i)-f_{i+1}(x_i), \tag{6}$$

and

$$L_i=\frac{\partial g_i}{\partial x}(x_i,a_i)f_i(x_i), \tag{7}$$

where we recognize L_i as the Lie derivative of g_i in the direction of f_i.

Now, in order to ensure that the gradient exists, the following assumption is presented;

Assumption 1. For all $i=1,\dots,N$, $L_i\neq 0$.

Given Assumption 1, reference (Boccadoro *et al.*, 2005*b*) derived the following expression for the derivative $\frac{dJ}{da_i}$.

Proposition 2.1. The following equation is in force,

$$\frac{dJ}{da_i}=-\frac{1}{L_i}p(\tau_i^+)R_i\frac{\partial g_i}{\partial a_i}(x_i,a_i). \tag{8}$$

where the costate equation is given by

$$\dot{p}(t)=-\left(\frac{\partial f_{i+1}}{\partial x}(x(t),t)\right)^T p(t)-\left(\frac{\partial L}{\partial x}(x(t))\right)^T; \quad t\in[\tau_i,\tau_{i+1}),\ i=1,\dots,N, \tag{9}$$

with terminal condition $p^T(t_N)=0$ when the final time is fixed, and reset conditions

$$p(\tau_i^-)=(I-\frac{1}{L_i}R_i\frac{\partial g_i}{\partial x}(x_i,a_i))^T p(\tau_i^+),\ i=1,\dots,N. \tag{10}$$

Proof: See (Boccadoro *et al.*, 2005*b*).

Having presented the expression for the gradient, as derived in (Boccadoro *et al.*, 2005*b*), we can now proceed to present the minimax solution to our switching surface parametrization problem.

3. MINIMAX OPTIMIZATION

Given a set of possible initial points $S\subset\mathbb{R}^n$, a set of switching surfaces parameterized by some vector a, and an instantaneous cost L, the total cost, starting at $x_0\in S$, is given by

$$J_{x_0}(a)=\int_0^T L(x(t))dt, \tag{11}$$

where T is a fixed final time and subscript x_0 indicates the initial condition. Our problem, denoted by P_S, can be stated as

P_S: Given a set of initial states S and a set of switching surfaces parameterized by a, find the surface parameter a such that

$$\max\{J_x(a)\mid x\in S\} \tag{12}$$

is minimized.

As mentioned earlier, the theory of minimax optimization and consistent approximations (Polak, 1997) will be utilized in order to implement and solve this problem.

Given a set of possible initial states $S\subset\mathbb{R}^n$, we will choose a sequence of sets of initial points, $\{\mathbb{X}_i\}_{i=0}^{\infty}$. This sequence will satisfy the following three conditions: Firstly, $\mathbb{X}_i\subset S$ $i=1,2\dots$; secondly, the number of elements in $\mathbb{X}_i$ is bigger than the number of elements in $\mathbb{X}_{i-1}$; thirdly, every point in S will be arbitrarily close to a point in $\mathbb{X}_i$, as i goes to infinity. Choosing $\{\mathbb{X}_i\}_{i=0}^{\infty}$ in this way enables us to find the solution to (12) by solving a sequence of optimization problems, each one with a different set of initial states.

For each $\mathbb{X}_i$ we will find the optimal switching parameter a_i^o that minimizes $\max\{J_x(a_i)\mid x\in\mathbb{X}_i\}$ through a gradient descent algorithm, as described below. After we have found the optimal a_i^o, we will solve $\max\{J_x(a_{i+1})\mid x\in\mathbb{X}_{i+1}\}$ by initializing a_{i+1} to a_i^o. This gives a good starting point for the gradient descent algorithm.

For each $\mathbb{X}_i$ we will find the optimal a_i^o by executing the following gradient descent algorithm with Armijo step size (Armijo, 1966). We assume that $\mathbb{X}_i$ have $N(i)$ elements, i.e. $\mathbb{X}_i=\{x_1,\dots,x_{N(i)}\}$ for some $x_1,\dots,x_{N(i)}$ in $S\subset\mathbb{R}^n$.

Algorithm 3.1 Gradient Projection Algorithm with Armijo Stepsize

Given: The Armijo constants α, β in $(0,1)$. Two constants $\delta > 0$, and $\varepsilon > 0$ and the set of initial points $\mathbb{X} = \{x_1, \ldots, x_N\} \subset S$.
Initialize: Choose a feasible initial guess on the switching surface parameter a.
Step I: Calculate the maximum cost for the given set of initial states, denoted

$$F(\mathbb{X}, a) = \max_x \{J_x(a) | x \in \mathbb{X}\}, \quad (13)$$

where J_x is given by (11). Let $I(\mathbb{X}, a)$ denote the index set of *active constraints*, i.e.

$$I(\mathbb{X}, a) = \{j \in \{1, \ldots, N\} \mid F(\mathbb{X}, a) - J_j(a) < \varepsilon\}. \quad (14)$$

Calculate the generalized gradient

$$\partial F(\mathbb{X}, a) = conv\{\nabla J_j(a) \mid j \in I(\mathbb{X}, a))\}, \quad (15)$$

where *conv* denotes the *convex hull.* Find the point in $\partial F(\mathbb{X}, a)$ closest to the origin and denote it by h. If $||h|| < \delta$ then STOP. Else, goto Step II.
Step II: Calculate the step-length λ according to Armijo's rule i.e.

$$\lambda = \max\{z = \beta^k; \ k \geq 0 \mid F(\mathbb{X}, a - zh) - F(\mathbb{X}, a) \leq -\alpha z ||h||^2\}.$$

Update a according to $a = a - \lambda h$, goto Step I. ■

A few remarks concerning Algorithm 3.1 are due.

Remark 3.1. The index set of active constraints, $I(\mathbb{X}, a)$, is introduced in order to determine what initial states in $\mathbb{X}$ we should take into consideration for a given a. If the index of an initial state is in the index set, then the gradient of the cost associated with that initial state is current in the calculation of the generalized gradient, $\partial F(\mathbb{X}, a)$. If $\varepsilon = 0$ in (14), i.e., we only optimize with respect to the initial state corresponding to the maximal cost, it is conceivable that we can only take a very small descent step since the index set changes when a changes.

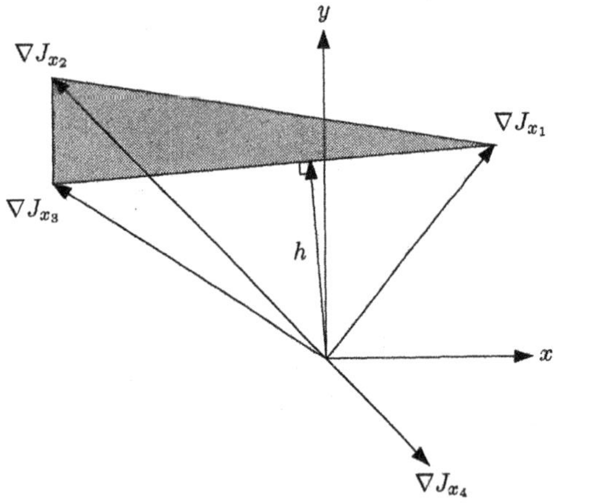

Fig. 2. Calculation of h given four initial states and their respective gradients. x_1 through x_3 are active initial states.

Remark 3.2. In order to find the optimal a for a given set of initial states, we would have to set the constants δ and ε to 0. However, doing this when we solve for a sequence of initial states, $\{\mathbb{X}_i\}_{i=0}^{\infty}$, would not give any additional benefit, instead we only require that for each consecutive problem we will solve, δ and ε will decrease, and in the limit when $i \to \infty$, they will be zero.

Remark 3.3. Solving for h is a standard quadratic optimization problem over a convex set, and can be solved using a variety of optimization algorithms.

Remark 3.4. In the robotics example presented in Section 4, a simple constraint is introduced on a. Hence we need to initialize a to be in the set of feasible points.

In order to illustrate the calculation of h, a simple example is presented. Assume that we have four different initial states, x_1 through x_4 in $\mathbb{R}^2$. In Figure 2, their respective gradients are plotted and it is assumed that x_1 through x_3 are active initial states for the given switching surface parameter a. The shaded region in Figure 2 corresponds to the convex hull of the gradients of the active initial states, and h is the closest vector in this set from the origin.

Having presented Algorithm 3.1 and the remarks that follow it, we are now in the position to present Algorithm 3.2 that will solve problem P_S.

Algorithm 3.2 Minimax optimization for unknown initial state:
Given: A sequence of initial sets $\{\mathbb{X}_i\}_{i=0}^{\infty} \in S \subset \mathbb{R}^n$, where $\mathbb{X}_i = \{x_1, \ldots, x_{N(i)}\}$ and $N(i) > N(i-1)$. Two positive sequences $\{\varepsilon_i\}_{i=0}^{\infty}$ and $\{\delta_i\}_{i=0}^{\infty}$ such that in the limit when $i \to \infty$, both are 0.
Init: Set $i = 0$, pick a feasible initial guess on a_0.
Step I: Use Algorithm 3.1 to optimize over a with $\mathbb{X} = \mathbb{X}_i$, $\delta = \delta_i$, $\varepsilon = \varepsilon_i$. Initialize a with a_{i-1} if $i \neq 0$, and with a_0 if $i = 0$.
Step II: Set a_i to a given from Algorithm 3.1. Increase i by one, goto Step I.

4. NUMERICAL EXAMPLE

In order to show the usefulness of Algorithm 3.2, we consider a mobile robot navigation problem. The task of the robot is to get to a goal point $x_g \in \mathbb{R}^2$ while avoiding an obstacle located at $x_{ob} \in \mathbb{R}^2$. It has to do this by switching between two different behaviors, one *go-to-goal* and one *obstacle-avoidance* behavior. These different behaviors are denoted by f_g and f_o respectively. We model the robot having unicycle dynamics

$$\begin{cases} \dot{x}_1 = v\cos(\phi), \\ \dot{x}_2 = v\sin(\phi), \\ \dot{\phi} = f_q(x_1, x_2, \phi), \end{cases} \quad (16)$$

where (x_1,x_2) is the position of the robot, ϕ is its heading, and $q \in \{g,o\}$ is the current behavior the robot evolves according to. We assume that the translational velocity v is constant. Our control variable is then given by the switching surface parameters of the goal and avoid obstacle guards that dictate what behavior the robot should evolve according to. A standard pair of "approach-goal" and "avoid-obstacle" behaviors are given by

$$f_g(x_1,x_2,\phi) = c_g(\phi_g - \phi), \tag{17}$$

$$f_o(x_1,x_2,\phi) = c_o(\pi + \phi_{ob} - \phi). \tag{18}$$

Here, c_g and c_o are the gains associated with each behavior, and ϕ_g and ϕ_{ob} are the angles to the goal and nearest obstacle respectively. Both of these angles are measured with respect to the x-axis and can be expressed as

$$\phi_g = \arctan(\frac{x_{g_2} - x_2}{x_{g_1} - x_1}), \tag{19}$$

$$\phi_{ob} = \arctan(\frac{x_{ob_2} - x_2}{x_{ob_1} - x_1}), \tag{20}$$

where (x_{g_1},x_{g_2}) and (x_{ob_1},x_{ob_2}) are the Cartesian coordinates of the goal and the nearest obstacle respectively.

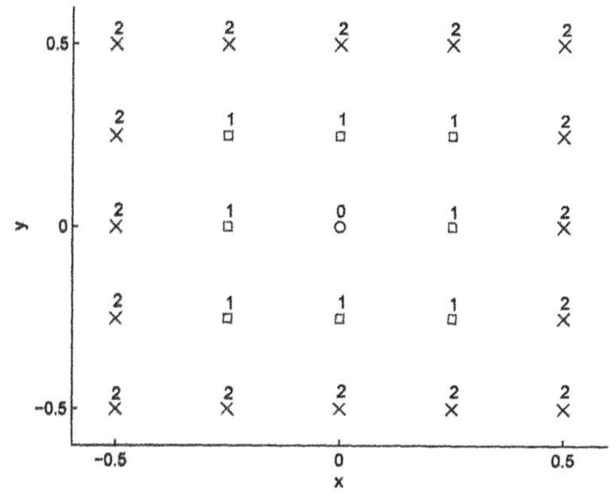

Fig. 3. Initial states used: $\mathbb{X}_i$ contains the points with index $i, i-1, \ldots, 0$.

The instantaneous cost L is given by

$$L(x(t)) = \rho||x_g - x(t)||^2 + \alpha e^{-\frac{||x_{ob}-x(t)||^2}{\beta}}, \tag{21}$$

where ρ is the gain of the goal attraction term, α is the gain of the obstacle avoidance term, and β is a shaping parameter that affects the range of the obstacle avoidance term.

For a given initial position $x_0 \in \mathbb{R}^3$ the total cost is given by (11). However, many mobile robots get their position from GPS readings which has an error associated with them. In our example, we assume that the robot get the initial position $x_0 = (0,0,\cdot)^T$ from the GPS and that the error associated with the GPS is 0.5 meters (note that GPS do not give the direction of a stationary robot). In order to simplify our exposition, we assume that the robot is always directed towards the goal, hence we will only show the (x_1,x_2) components in $\mathbb{X}_i$, $i = 0,1,2$. This is a reasonable assumption if the robot can see the goal, which we assume.

Due to the error in the GPS reading, the robot can be anywhere in the interval $[-0.5,-0.5] \times [0.5,0.5]$. Therefore we initialize Algorithm 3.2 with only one initial state, $\mathbb{X}_0 = (0,0)^T$, and we then extend the set of initial states, in a somewhat arbitrary fashion, as shown in Figure 3. In this example, we stop the algorithm after its third iteration, i.e. when $||h|| < \delta_2$, therefore we do not define $\mathbb{X}_i$ for $i = 3,4,\ldots$.

The switching surfaces for when to switch from f_g to f_o, and when to switch from f_o to f_g, are given by two circles with radius a_1 and a_2 respectively, where we require $a_1 \leq a_2$. Both circles are centered at the obstacle $x_o = (2,1.25)^T$. At this point it should be noted that having circular guards might not correspond to an optimal guard shape.

We initialize a to be $(1,1.5)^T$ and for the constants in L, we set $\rho = 0.01$, $\alpha = 10$ and $\beta = 0.1$ and we use $c_g = c_o = 1$ for the feedback gains in (17) and (18). The velocity of the robot is set to $v = 0.5$ and the goal is located at $x_g = (4,4)^T$. For the constants in the Armijo procedure, we use $\alpha = \beta = 0.5$. The sequences of ε_j and δ_j used is given by $\delta_j = \frac{\delta_{j-1}}{2.5}$ with $\delta_0 = 0.25$, and $\varepsilon_j = \frac{\varepsilon_{j-1}}{2.5}$ with $\varepsilon_0 = 0.1$

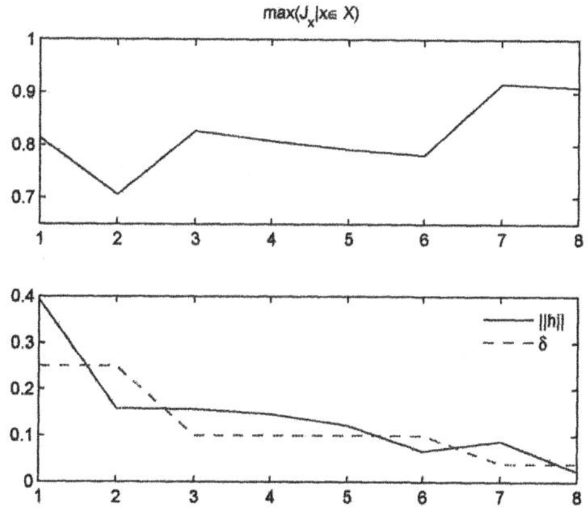

Fig. 4. (a) Change in maximum cost. (b) $||h||$ and δ as a function of the number of gradient descent iterations in Algorithm 3.2.

A plot of how the cost changes together with the norm of h and δ is shown in Figure 4. As can be seen in the figure, Algorithm 3.2 effectively reduces the maximum cost for a given set of initial states. Once the norm of h falls below δ, we update δ, ε and the set of initial states, $\mathbb{X}$.

Once we have updated $\mathbb{X}_0$ to $\mathbb{X}_1$ after iteration three, we see that the maximum of the cost increases, just as should be expected since $\mathbb{X}_1$ has more initial states that $\mathbb{X}_0$. Figure 5(a) shows how the switching surface parameters change. At the optimum, $a_1 = a_2$, i.e. both radii are the same.

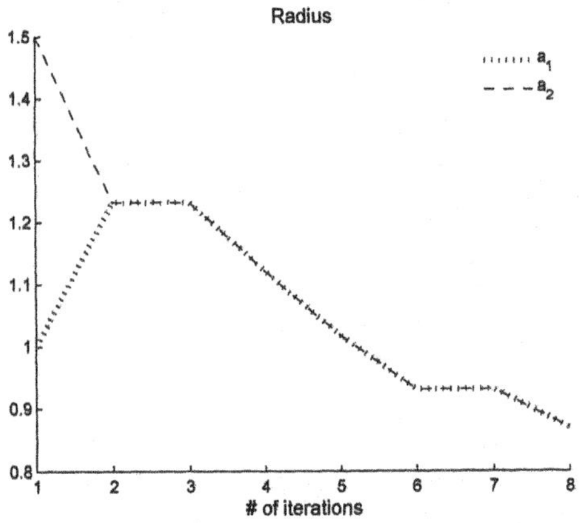

Fig. 5. Change in $a = (a_1, a_2)^T$ as a function of the number of gradient descent iterations in Algorithm 3.2. a_1 is the radius of the *obstacle-avoidance* switching surface, a_2 is the radius of the *go-to-goal* switching surface.

5. CONCLUSIONS

This paper presented a way of getting rid of the dependence on the initial condition when optimizing over when to switch between different modes in a switched-mode system. The dependence on the initial condition was dealt with by minimizing the switching parameter over the maximum cost for a given set of initial states. The only assumption made was that the initial state was confined to a given region in the state space.

REFERENCES

Armijo, L. (1966). Minimization of functions having lipschitz continuous first-partial derivatives. *Pacific Journal of Mathematics* **16**, 1–3.

Attia, S.A., M. Alamir and C. Canudas de Wit (2005). Sub optimal control of switched nonlinear systems under location and switching constraints. In: *IFAC World Congress.*

Bemporad, A., F. Borrelli and M. Morari (2002). Piecewise Linear Optimal Controllers for Hybrid Systems. In: *Hybrid Systems: Computation and Control*, M. Greenstreet and C. Tomlin, Editors, Springer-Verlag Lecture Notes in Computer Science, Number 2289. pp. 105–117.

Boccadoro, M., M. Egerstedt and Y. Wardi (2005*a*). Obstacle avoidance for mobile robots using switching surface optimization. *IFAC World Congress*, Prague, The Czech Republic.

Boccadoro, M., Y. Wardi, M. Egerstedt and E. Verriest (2005*b*). Optimal control of switching surfaces in hybrid dynamical systems. *Journal of Discrete Event Dynamic Systems* **15**, 433–448.

Branicky, M.S., V.S. Borkar and S.K. Mitter (1998). A unified framework for hybrid control: Model and optimal control theory. *IEEE Transactions on Automatic Control* **43**, 31–45.

Brockett, R. (1995). Stabilization of motor networks. In: 35th *IEEE Conference on Decision and Control*. pp. 1484–1488.

Caines, P. E. and M. S. Shaikh (2005). Optimality zone algorithms for hybrid systems computation and control: From exponential to linear complexity. In: *Proceedings of the 2005 International Symposium on Intelligent Control/ 13th Mediterranean Conference on Control and Automation*, Cyprus. pp. 1292–1297.

Egerstedt, M. and Y. Wardi (2002). Multi-process control using queuing theory. In: 41th *IEEE Conference on Decision and Control*, Las Vegas, NV.

Egerstedt, M., Y. Wardi and H. Axelsson (2006). Transition-time optimization for switched-mode dynamical systems. *IEEE Transactions on Automatic Control* **51**, to appear.

Guia, A., C. Seatzu and C. Van der Mee (1999). Optimal control of switched autonomous linear systems. In: 38th *IEEE Conference on Decision and Control*, Phoenix, AR. pp. 1816–1821.

Hedlund, S. and A. Rantzer (1999). Optimal control of hybrid systems. In: 38th *IEEE Conference on Decision and Control*, Phoenix, AR. pp. 1972–3977.

Hristu-Varsakelis, D. (2001). Feedback control systems as users of shared network: Communication sequences that guarantee stability . In: 40th *IEEE Conference on Decision and Control*, Orlando, FL. pp. 3631–3631.

Lincoln, B. and A. Rantzer (2001). Optimizing linear systems switching. In: 40th *IEEE Conference on Decision and Control*, Orlando, FL. pp. 2063–2068.

Polak, E. (1997). *Optimization Algorithms and Consistent Approximations.* Springer-Verlag. New York, New York.

Rehbinder, H. and M. Sanfridson (2000). Scheduling of a limited communication channel for optimal control. In: 38th *IEEE Conference on Decision and Control, Sidney, Australia.*

Walsh, G., H. Ye and L. Bushnell (1999). Stability analysis of networked control systems. In: *American Control Conference*. pp. 2876–2880.

Xu, X. and P. Antsaklis (2002). Optimal control of switched autonomous systems. In: 41th *IEEE Conference on Decision and Control*, Las Vegas, NV.

ELSEVIER
IFAC
PUBLICATIONS

BEYOND THE CONSTRUCTION OF OPTIMAL SWITCHING SURFACES FOR AUTONOMOUS HYBRID SYSTEMS

Mauro Boccadoro * Magnus Egerstedt **
Paolo Valigi * Yorai Wardi **

* boccadoro@diei.unipg.it
Dipartimento di Ingegneria Elettronica e dell'Informazione
Università di Perugia
06125, Perugia - Italy
** {magnus,ywardi}@ece.gatech.edu
Electrical and Computer Engineering
Georgia Institute of Technology
Atlanta, GA 30332

Abstract: In this paper we report of a technique to design optimal feedback control laws for hybrid systems with autonomous (continuous) modes. Existing techniques design the optimal switching surfaces based on a singular sample evolution of the system; hence providing a solution dependent on the initial conditions. On the other hand, the optimal switching times can be found, providing an an open loop control to the system, but those also are dependent on the initial conditions. The technique presented relies on a variational approach, giving the derivative of the switching times with respect to the initial conditions, thus providing a tool to design programs/algorithms generating switching surfaces which are optimal for any possible execution of the system. *Copyright © 2006 IFAC*

Keywords: Hybrid Systems, Switching Surfaces, Optimal Control, Variational Methods

1. INTRODUCTION

Consider a switched system with autonomous continuous dynamics,

$$\dot{x}(t) = f_{q(t)}(x(t)), \qquad (1)$$
$$q^+(t) = s(x(t), q(t)). \qquad (2)$$

where (1) describes the continuous dynamics of the state variable $x \in \mathcal{X} \subseteq \mathbb{R}^n$ and (2) describes the discrete event dynamics of the system. Given an initial condition $x_0 := x(t_0)$, the switching law (2) determines the switching instants t_i, $i = 1, 2, \ldots$, and thus the intervals where a certain modal function is active, as well as the initial condition for the o.d.e. which defines the evolution under the next mode. The discrete variable q is piecewise constant in time and belongs to a finite or countable set Q, hence, it can be expressed in terms of the index i as $q(i)$. In terms of such index the dynamics of a switched system is:

$$\dot{x}(t) = f_i(x(t)), \quad t \in (t_{i-1}, t_i] \qquad (3)$$
$$i^+ = s(x(t), i, t). \qquad (4)$$

with the understanding that $f_i := f_{q(i)}$, for a given map $q(i)$, i.e., in this case (4) only expresses the occurrence of the ith switch, the specification of the next active mode being given by the map $q(i)$.

[1] This work has been partially supported by MIUR under grant PRIN 2005092439

Since the continuous modes are autonomous, the evolution of the system is determined by the active modes, according to (4). When the function s does not depend by the (continuous) state variable x, the switching instants are determined as exogenous inputs, and the system is controlled in open loop (timing control); when s is dependent only on the state variables, the switching law is given in a feedback form, and it may be defined by switching surfaces in the state space.

To formulate the problem we are interested with, consider a simple execution of (3,4) with only one switch, starting at $x(t_0) = x_0$ with mode 1, switching to mode 2 at time t_1, an exogenous switch, and terminating either at a *fixed final time* t_2 or in correspondence of a *terminal manifold* defined by a function $g(x)$, so that t_2 satisfies $g(x(t_2)) = 0$. For ease of reference, denote such two sets of possible executions by χ_t and χ_g, respectively.

To fix notation, let the explicit representation of the evolution determined by mode i be given by $x(t) = \varphi_i(t, s, x(s))$, hence,

$$x(t) = \begin{cases} \varphi_1(t, t_0, x_0) & t \in [t_0, t_1] \\ \varphi_2(t, t_1, x(t_1)) & t \in (t_1, t_2] \end{cases} \tag{5}$$

Also, let $x_i := x(t_i)$, and $R := f_1(x_1) - f_2(x_1)$. In this paper the following conventions will be used: 1) vectors are columns; 2) the derivative of a scalar, e.g. L, w.r.t. a vector x is a row vector:

$$L_x := \frac{dL}{dx} = \left[\frac{\partial L}{\partial x_1}, \ldots, \frac{\partial L}{\partial x_n}\right]. \tag{6}$$

(hence L_x^T is a column vector). The Hessian matrix is denoted by L_{xx}. If f is a (column) vector, function of the vector x i.e.,

$$f = [f^{(1)}(x), \ldots, f^{(n)}(x)]^T$$

then

$$f_x := \frac{\partial f}{\partial x} = \begin{bmatrix} \frac{\partial f^{(1)}}{\partial x_1} & \cdots & \frac{\partial f^{(1)}}{\partial x_n} \\ \vdots & \ddots & \vdots \\ \frac{\partial f^{(n)}}{\partial x_1} & \cdots & \frac{\partial f^{(n)}}{\partial x_n} \end{bmatrix}$$

According to this convention, if c, t are scalar quantities, x, y, z, are vectors, and M is a square matrix, the usual chain rule applies to $c(x(t))$ and $c(x(y))$, i.e. $\frac{dc}{dt} = c_x \dot{x}$, $c_y = c_x x_y$ ($\dot{v}$ stays for $\frac{dv}{dt}$); also:

$$\frac{d}{dz}[cy] = y^T c_z + c y_z, \tag{7}$$

$$\frac{d}{dz}[x^T y] = y^T x_z + x^T y_z, \tag{8}$$

$$\frac{d}{dz}[M(t(z))y(t(z))] = (M_t y + M y_t) t_z. \tag{9}$$

2. PROBLEM FORMULATION

For those systems described above, when the optimal control problem to minimize a cost function

$$J = \int_{t_0}^{t_2} L(x(t))dt \tag{10}$$

is formulated, for some continuously differentiable function L, and such that L_{xx} is symmetric, then it is known that when $t_1 = t_1^*$, a (locally) optimal switching time, it satisfies the following condition, see e.g. (Egerstedt *et al.*, 2003):

$$c(t_1^*) := p^T(t_1^*)R(x_1^*) = 0 \tag{11}$$

where $p^T(t)$, for $t \in [t_1^*, t_2]$ is given by:

$$p^T(t) = \int_t^{t_2} L_x(x(s))\Phi_2(s,t)ds + p^T(t_2)\Phi_2(t_2,t) \tag{12}$$

with Φ_i the transition matrix of the linearized time-varying system $\dot{z}(t) = \frac{\partial f_i(x(t))}{\partial x} z(t)$, and $p^T(t_2) = 0$ for fixed final time and

$$p^T(t_2) = -\frac{L(x_2)g_x(x_2)}{\mathcal{L}_2}, \tag{13}$$

for an evolution ending at a terminal manifold, where $\mathcal{L}_2 := g_x(x_2)f(x_2)$, the Lie derivative of g along f_2 evaluated at x_2.

Assuming to start from a perturbed initial condition $\tilde{x}_0 = x_0 + \delta x_0$; it is possible to use the information of optimality of t_1^*, as a switching time, to determine $\tilde{t}_1^*$; in other words: what is the dependence of the optimal switching time on the initial conditions?

This problem is motivated by the determination of optimal switching surfaces, which tend to solve optimal control problems for autonomous system via the synthesis of feedback laws, which may be pursued for specifications of stability or optimal control. Relevant application of such technique may arise in many areas such as behavior based robotics (Arkin, 1998), or manufacturing systems (Khmelnitsky and Caramanis, 1998) to cite a few.

Computational methods exist and are based on the optimization of parametrized switching surfaces (Boccadoro *et al.*, 2005). However, the choice of the optimal values for such parameters depend on the particular trajectory chosen to run an optimization program, and thus, fundamentally, on the initial conditions (remind that systems with no continuous inputs are being considered).

An interesting reference for this type of approach is (Giua *et al.*, 2001), which addressed a timing optimization problem, and discovered the special structure of the solution for linear quadratic problems. Indeed, in that case it is possible to identify homogeneous regions in the continuous state space, whose boundaries, when reached, determine the optimal switches, thus providing a feedback solution to a problem which is formulated in terms of an open loop strategy.

Here we explicitly investigate the relation existing between optimal switching times and initial conditions, studying how the condition of optimality (11) that switching times must satisfy, vary in dependence of the initial conditions.

3. OPTIMAL SWITCHING TIMES V/S INITIAL CONDITIONS

It is well known that, under mild assumptions, executions of switched systems are continuous w.r.t. the initial conditions (Broucke and Arapostathis, 2002). If we assume that also the dependence of c on t_1^* as well as t_1^* on x_0 is such, we may characterize function t_1^* by deriving (11) w.r.t. x_0 and setting this derivative to zero. In fact, if starting from $\tilde{x}_0 = x_0+\delta x_0$, it results $\tilde{t}_1^* = t_1^*+\delta t_1^*$; then, by continuity, $0 = c(\tilde{t}_1^*) = c(t_1^*) + \frac{dc}{dx_0}\delta x_0 + o(\delta x_0)$. Hence, settting $\frac{dc}{dx_0} = 0$, to satisfy optimality condition for $\tilde{t}_1^*$, yields a formula for the variational dependence of t_1^* on x_0. To go further, the superscript * will be dropped (hence assuming that t_1, x_1 etc. are relative to optimal executions) in order to reduce the notational burden.

By (8) we have that

$$\frac{dc}{dx_0} = R^T\frac{dp(t_1)}{dx_0} + p^T(t_1)\frac{dR}{dx_0} \tag{14}$$

To calculate $\frac{dp(t_1)}{dx_0}$, account for the following result, which is readily verified:

$$\frac{d}{dx}\int_{a(x)}^{b(x)} f(s,x)ds = \int_a^b \frac{df}{dx}(s,x)ds + f(b,x)b_x - f(a,x)a_x \tag{15}$$

Then, considering the simpler case of fixed final time (so that t_2 is not a function of x_0), by (12, 8, 15)

$$\frac{dp(t_1)}{dx_0} = \int_{t_1}^{t_2} [\Phi_2^T(s,t_1)L_{xx}(x(s))\frac{dx(s)}{dx_0} + \frac{d\Phi_2^T(s,t_1)}{dt_1}L_x^T(x_s)\frac{dt_1}{dx_0}]ds \\ -\Phi_2^T(t_1,t_1)L_x^T(x_1)\frac{dt_1}{dx_0} \tag{16}$$

To compute $\frac{dx(s)}{dx_0}$ notice that $x(t_1) = \varphi_1(t_1,t_0,x_0)$, hence $x(s) = \varphi_2(s,t_1,\varphi_1(t_1,t_0,x_0))$ for $s \in [t_1,t_2]$, thus,

$$\frac{dx(s)}{dx_0} = \frac{\partial x(s)}{\partial t_1}\frac{dt_1}{dx_0} + \frac{\partial x(s)}{\partial x_1}\frac{\partial x_1}{\partial t_1}\frac{dt_1}{dx_0} + \frac{\partial x(s)}{\partial x_1}\frac{\partial x_1}{\partial x_0} \tag{17}$$

Now, $\partial x(s)/\partial t_1 = -f_2(x(s))^2$, $\partial x(s)/\partial x_1 = \Phi_2(s,t_1)$, $\partial x_1/\partial x_0 = \Phi_1(t_1,t_0)$, $\partial x_1/\partial t_1 = f_1(x_1)$, $\Phi_2(t_1,t_1) = I$,

$$\frac{d}{dt_1}\Phi_2(s,t_1) = -\Phi_2(s,t_1)\frac{\partial f_2(x_1)}{\partial x} \tag{18}$$

(to be transposed). It results:

$$\frac{d}{dx_0}p(t_1) = (I_1 - I_2 - I_3 - K)\frac{dt_1}{dx_0} + I_4 \tag{19}$$

where

$$I_1 = \int_{t_1}^{t_2} \Phi_2^T(s,t_1)L_{xx}(x(s))\Phi_2(s,t_1)f_1(x_1)ds$$

$$I_2 = \int_{t_1}^{t_2} \Phi_2^T(s,t_1)L_{xx}(x(s))f_2(x(s))ds$$

$$I_3 = \int_{t_1}^{t_2} f_{2x}^T(x_1)\Phi_2^T(s,t_1)L_x^T(x(s))ds$$

$$I_4 = \int_{t_1}^{t_2} \Phi_2^T(s,t_1)L_{xx}(x(s))\Phi_2(s,t_1)\Phi_1(t_1,t_0)ds$$

$$K = L_x^T(x_1)\frac{dt_1}{dx_0} \tag{20}$$

To handle these, integrate by parts I_2 (letting $\frac{dt_1}{dx_0}$), taking into account that

$$\int L_{xx}(x(s))f_2(x(s))ds = L_x^T(x(s))$$

we have

$$\int_{t_1}^{t_2} \Phi_2^T(s,t_1)L_{xx}(x(s))f_2(x(s))ds = \\ -I_3 + \Phi_2^T(s,t_1)L_x^T(x(s))\Big|_{t_1}^{t_2} = \\ -I_3 + \Phi_2^T(t_2,t_1)L_x^T(x_2) - K \tag{21}$$

This leads to the cancellation of I_3 and K in (19).

To complete, let's compute $dR(x_1)/dx_0$. Again, notice that $x_1 = x(t_1) = x[t_1(x_0),x_0]$, hence,

$$\frac{d}{dx_0}R(x_1) = \frac{\partial R}{\partial x}(x_1)\left[\frac{\partial x_1}{\partial t_1}\frac{dt_1}{dx_0} + \frac{\partial x_1}{\partial x_0}\right] = \\ \frac{\partial R}{\partial x}(x_1)\left[f_1(x_1)\frac{dt_1}{dx_0} + \Phi_1(t_1,t_0)\right] \tag{22}$$

Multiplying this by $p^T(t_1)$, (19) by R^T from the left and summing up we finally obtain:

$$\frac{dc(t_1)}{dx_0} = \\ [R^T(Qf_1 - \Phi_2^T(t_2,t_1)L_x^T(x_2)) + p^T(t_1)R_xf_1]\frac{dt_1}{dx_0} \\ + [R^TQ + p^T(t_1)R_x]\Phi_1(t_1,t_0) \tag{23}$$

where $f_1 := f_1(x_1)$, and

$$Q := \int_{t_1}^{t_2} \Phi_2^T(s,t_1)L_{xx}(x(s))\Phi_2(s,t_1)ds \tag{24}$$

which is a kind of quadratic form co-costate. Notice that the term multiplying $\frac{dt_1}{dx_0}$ above, is a

[2] For time invariant dynamics, $[\varphi(s,t+h,x) - \varphi(s,t,x)]/h = [\varphi(s-h,t,x) - \varphi(s,t,x)]/h = -f(x(s)) + o(h)$.

scalar. So, if we know that t_1^* is a local optimum for an evolution starting from x_0, then, assuming to start from $\tilde{x}_0 = x_0 + \delta x_0$, we simply must switch at $t_1^* + \delta t_1^* + o(\delta x_0)$. According to (23),

$$\delta t_1^* = \frac{-[R^T Q + p^T(t_1)R_x]\Phi_1(t_1, t_0)\, \delta x_0}{R^T(Qf_1 - \Phi_2^T(t_2, t_1)L_x^T(x_2)) + p^T(t_1)R_x f_1} \tag{25}$$

4. ENABLING CRITERIA FOR THE DESIGN OF THE OPTIMAL SWITCHING SURFACES

To put in use Eq. (25) assume that *one* optimal switching time has been derived for a certain "sample" evolution of the system, e.g. one starting in $\hat{x}_0$. Then the optimal switching surfaces are defined by the optimal switching *states* yielded by the variation on the optimal switching times when initial conditions different than $\hat{x}_0$ are considered. However, it must be paid attention to the fact that the formula derived above works for a fixed final time: indeed for the case of evolution ending at a terminal manifold the following result holds,

Theorem 1. Consider a nominal and a perturbed execution of the set χ_g, $x(\cdot)$ and $y(\cdot)$, respectively, the first starting at x_0 and the latter starting from a point y_0 which lies on the nominal trajectory; i.e., assume that it exists a duration δt_0 such that $y_0 = \varphi_1(t_0 + \delta t_0, t_0, x_0)$. Then, the following relation holds:

$$t_1^*(y_0) = t_1^*(x_0) - \delta t_0 \tag{26}$$

for all $\delta t_0 < t_1^* - t_0$

Proof The optimal evolution may be split into the trajectory from x_0 to y_0 and from y_0 onwards. Hence, by the principle of optimality, this second branch of the evolution must be itself optimal, so that the optimal switching state is the same. The result follows by time-invariance of the system. □

Remark 1. Theorem 1 easily extends to negative δt_0, i.e., if y_0 is chosen such that the evolution starting from y_0 will reach x_0 we must *add* the time needed to reach x_0 from y_0 to the optimal (nominal) switching time. □

In case of fixed terminal time the optimal switching state may vary because the perturbed trajectory described in Theorem 1 above, switching at $t_1^* - \delta t_0$, reaches the point $x(t_2)$ (of the nominal trajectory) at time instant $t_2 - \delta t_0$, thence "visits" additional states from $t_2 - \delta t_0$ to t_2 (in other words $\tilde{x}(\cdot)_{(t_2-\delta t_0, t_2]}$ is a set of states not visited by $x(\cdot)$). Such remnants of the perturbed trajectory add further costs, so that two different trajectories, even if the starting point of one of them lies in the trajectory of the other, cannot really be properly compared, in terms of optimal switching states.

This point is evident also from (25): take an i.c. $y_0 = \varphi_1(t_0 + \delta t_0, t_0, x_0)$ very close to x_0, so that $\delta x_0 = f_1(x_0)\delta t_0 + o(\delta t_0)$. Substituting such δx_0 in (25), we have that its numerator (plus higher order terms) is:

$$\begin{aligned} -[R^T Q + p^T(t_1)R_x]\Phi_1(t_1, t_0)\delta x_0 = \\ -[R^T Q f_1 - p^T(t_1)R_x f_1]\delta t_0 \end{aligned} \tag{27}$$

where $\Phi_1(t_1, t_0)f_1(x_0) = f_1(x_1)$ is due to the fact that vector fields obey their variational dynamics[3]. Hence,

$$\delta t_1^* = \frac{-[R^T Q f_1 + p^T(t_1)R_x f_1]\, \delta t_0}{R^T(Qf_1 - \Phi_2^T(t_2, t_1)L_x^T(x_2)) + p^T(t_1)R_x f_1} \tag{28}$$

In this case, condition (26) is equivalent to $\delta t_1^* = -\delta t_0$, and for this to be verified, denominator and numerator should have had the same terms, opposed in sign. Here, the only term making the difference, preventing (26) to hold (as expected) is

$$-R^T \Phi_2^T(t_2, t_1)L_x^T(x_2). \tag{29}$$

Remark 2. Notice, however, that for a case similar to those considered in (Giua *et al.*, 2001), where the the final dynamic mode is linear, stable and the terminal time tends to infinity, we have that the additional term (29) vanishes. Accordingly, optimal switching surfaces are well defined also for such situations, and could be possibly characterized using (25).

In summary, in force of Theorem 1 and Remark 2, the objective to characterize optimal switching surfaces independent of the initial conditions should be pursued considering evolutions ending at terminal manifolds or those evolutions of the family χ_t with the restrictions illustrated above, since in such cases variations in the switching times define soundly optimal switching *states* as well.

Theorem 1 and the discussion that follows, also give an hint about the set of initial conditions that should be considered to set such procedure. Indeed, it seems reasonable account only for that set of initial conditions which are transversal to the flow defined by the vector field of the initial dynamics (here f_1) which contains $\hat{x}_0$. Such set of initial condition is a surface itself and can be described by $s(x) = 0$ where s is a $\mathbb{R}$-valued function such that $s(\hat{x}_0) = 0$ and such that $s_x(x)$ is collinear with $f_1(x)$, so that s would be a kind

[3] Indeed, the variational system $\dot{z}(t) = \frac{\partial f(x(t))}{\partial x} z(t)$ has the solution $z(t) = f(x(t))$, which can be seen from the chain rule $\dot{f} = f_x f$

of *potential* of the vector field f_1. This choice is justified by the fact that the components of the variation δx_0 on some x_0 which are tangent to the flow yield no difference on the optimal switching *state*, hence giving no relevant information to the construction of a switching surface which is optimal for the executions determined by any possible initial condition (i.e., the *optimal switching surface*).

5. CONCLUSION AND FUTURE WORKS

This paper presents the first steps to design a new method to determine optimal switching surfaces for hybrid systems with autonomous modes. The idea is to characterize the variations in the optimal switching times corresponding to variations in the initial conditions, and to apply this formula for transverse shifts in the initial conditions, according to the considerations following the result stated in Theorem 1.

At the time of the first submission the formula relative to evolutions ending at a terminal manifold was not given, but successive studies led to its derivation. This new result, together with the analysis carried out in this paper, which identified those situations where an optimal switching surface independent of initial condition is well defined, allows to pursue the program, outlined above, based on the investigation of the effect of transverse variations in the initial condition on the switching states.

Future work will be devoted to further characterize the analytical properties of optimal switching surfaces, and develop efficient numerical procedures to generate the optimal switching surfaces, in force of the results given here.

REFERENCES

Arkin, R.C. (1998). *Behavior Based Robotics*. The MIT Press. Cambridge, MA.

Boccadoro, M., Y. Wardi, M. Egerstedt and E. Verriest (2005). Optimal control of switching surfaces in hybrid dynamical systems. *JDEDS* **15**(4), 433 - 448.

Broucke, M. and A. Arapostathis (2002). Continuous selections of trajectories of hybrid systems. *Systems and Control Letters* **47**, 149–157.

Egerstedt, M., Y. Wardi and F. Delmotte (2003). Optimal control of switching times in switched dynamical systems. In: *42nd IEEE Conference on Decision and Control (CDC '03)*. Maui, Hawaii, USA.

Giua, A., C. Seatzu and C. Van Der Mee (2001). Optimal control of switched autonomous linear systems. In: 40th *IEEE Conf. on Decision and Control (CDC 2001)*. Orlando, FL, USA. pp. 2472–2477.

Khmelnitsky, E. and M. Caramanis (1998). One-machine n-part-type optimal setup scheduling: analytical characterization of switching surfaces. *IEEE Trans. on Automatic Control* **43**(11), 1584–1588.

ELSEVIER
IFAC
PUBLICATIONS

APPROXIMATE SIMULATION RELATIONS FOR HYBRID SYSTEMS [1]

Antoine Girard * A. Agung Julius * George J. Pappas *

* *Department of Electrical and Systems Engineering*
University of Pennsylvania
Philadelphia, PA 19104
{agirard,agung,pappasg}@seas.upenn.edu

Abstract: Approximate simulation relations have recently been introduced as a powerful tool for the approximation of discrete and continuous systems. In this paper, we extend this notion to hybrid systems. Using the so-called simulation functions, we develop a computationally effective characterization of approximate simulation relations which can be used for hybrid systems approximation. An example of application in the context of safety verification is shown.

Keywords: Approximation of hybrid systems, Approximate simulation relation.

1. INTRODUCTION

Approximation of purely discrete systems has traditionally been based on language inclusion and equivalence with notions such as simulation or bisimulation relations (Clarke *et al.*, 2000). These concepts have been useful for simplifying problems such as safety verification or controller synthesis. More recently, they have been extended to the framework of continuous and hybrid systems (Pappas, 2003; Haghverdi *et al.*, 2005) allowing the approximation of systems in a unified (discrete/continuous) manner.

When dealing with continuous and hybrid systems, typically observed over the real numbers with possibly noisy observations, the usual notions based on *exact* language inclusion is quite restrictive and not robust. The notion of language approximation is much more adequate in this context. In (Girard and Pappas, 2005*c*), we proposed a framework for system approximation based on approximate versions of simulation relations. Instead of requiring that the observations of a system and its approximation are and remain equal, we require that they are and remain arbitrarily close. This approach not only defines more robust relations between systems but also allows more significant complexity reduction in the approximation process. In (Girard and Pappas, 2005*a*; Girard and Pappas, 2005*b*), this framework has been applied to constrained linear systems and nonlinear autonomous systems. Computational methods have been developed to quantify the distance between two systems. In (Julius *et al.*, 2006; Julius, 2006), the theoretical and computational frameworks have been extended to handle stochastic dynamical and hybrid systems (with purely stochastic jumps). Related work on approximate versions of simulation and bisimulation relations has been done for quantitative transition systems (de Alfaro *et al.*, 2004) or labeled Markov processes (Desharnais *et al.*, 2004).

In this paper, we apply our approximation framework to hybrid systems. Using the so-called simulation functions (Girard and Pappas, 2005*c*), we

[1] This research is partially supported by the Région Rhône-Alpes (Projet CalCel) and the NSF Presidential Early CAREER (PECASE) Grant 0132716.

develop a computationally effective characterization of approximate simulation relations which can be used for hybrid systems approximation. An example of application in the context of safety verification is shown.

2. APPROXIMATION OF TRANSITION SYSTEMS

In this section, we summarize the notion of approximate simulation relations for labeled transition systems as developed in (Girard and Pappas, 2005*c*). Labeled transition systems allow to model in a unified framework, discrete, continuous and hybrid systems. They can be seen as graphs, possibly with an infinite number of states or transitions.

Definition 2.1. A labeled transition system with observations is a tuple $T = (Q, \Sigma, \rightarrow, Q^0, \Pi, \langle\langle . \rangle\rangle)$ that consists of:

- a set Q of states,
- a set Σ of labels,
- a transition relation $\rightarrow \subseteq Q \times \Sigma \times Q$,
- a set $Q^0 \subseteq Q$ of initial states,
- a set Π of observations, and
- an observation map $\langle\langle . \rangle\rangle : Q \rightarrow \Pi$.

A state trajectory of T is a sequence of transitions,

$$q^0 \xrightarrow{\sigma^0} q^1 \xrightarrow{\sigma^1} q^2 \xrightarrow{\sigma^2} \dots, \text{ where } q^0 \in Q^0.$$

For a given initial state and sequence of labels, there may exist several state trajectories of T. Thus, the systems we consider are possibly non-deterministic. The associated external trajectory

$$\pi^0 \xrightarrow{\sigma^0} \pi^1 \xrightarrow{\sigma^1} \pi^2 \xrightarrow{\sigma^2} \dots, \text{ where } \pi^i = \langle\langle q^i \rangle\rangle$$

describes the evolution of the observations under the dynamics of the labeled transition system. The set of external trajectories of the labeled transition system T is called the language of T. The subset of Π reachable by the external trajectories of T is noted Reach(T). An important problem for transition systems is the safety verification problem which consists in checking whether the reachable set Reach(T) intersects a set of observations Π_U associated with unsafe states.

Exact simulation relations between two labeled transition systems require that their observations are (and remain) identical (Clarke *et al.*, 2000). Approximate simulation relations are less rigid since they only require that the observations of both systems are (and remain) arbitrarily close. Let $T_1 = (Q_1, \Sigma_1, \rightarrow_1, Q_1^0, \Pi_1, \langle\langle . \rangle\rangle_1)$ and $T_2 = (Q_2, \Sigma_2, \rightarrow_2, Q_2^0, \Pi_2, \langle\langle . \rangle\rangle_2)$ be two labeled transition systems with the same set of labels ($\Sigma_1 = \Sigma_2 = \Sigma$) and the same set of observations ($\Pi_1 = \Pi_2 = \Pi$). Let us assume that the set of observations Π is a metric space; d_Π denotes the metric on Π.

Definition 2.2. A relation $\mathcal{S}_\delta \subseteq Q_1 \times Q_2$ is a δ-approximate simulation relation of T_1 by T_2 if for all $(q_1, q_2) \in \mathcal{S}_\delta$:

(1) $d_\Pi(\langle\langle q_1 \rangle\rangle_1, \langle\langle q_2 \rangle\rangle_2) \leq \delta$,
(2) $\forall q_1 \xrightarrow{\sigma}_1 q_1'$, $\exists q_2 \xrightarrow{\sigma}_2 q_2'$ such that $(q_1', q_2') \in \mathcal{S}_\delta$.

Note that for $\delta = 0$, we have the usual notion of *exact* simulation relation (Clarke *et al.*, 2000).

Definition 2.3. T_2 approximately simulates T_1 with the precision δ (noted $T_1 \preceq_\delta T_2$), if there exists $\mathcal{S}_\delta$, a δ-approximate simulation relation of T_1 by T_2 such that for all $q_1 \in Q_1^0$, there exists $q_2 \in Q_2^0$ such that $(q_1, q_2) \in \mathcal{S}_\delta$.

If T_2 approximately simulates T_1 with the precision δ then the language of T_1 is approximated with precision δ by the language of T_2.

Theorem 2.4. (Girard and Pappas, 2005*c*) If $T_1 \preceq_\delta T_2$, then for all external trajectories of T_1,

$$\pi_1^0 \xrightarrow{\sigma^0} \pi_1^1 \xrightarrow{\sigma^1} \pi_1^2 \xrightarrow{\sigma^2} \dots,$$

there exists an external trajectory of T_2 with the same sequence of labels

$$\pi_2^0 \xrightarrow{\sigma^0} \pi_2^1 \xrightarrow{\sigma^1} \pi_2^2 \xrightarrow{\sigma^2} \dots$$

such that for all $i \in \mathbb{N}$, $d_\Pi(\pi_1^i, \pi_2^i) \leq \delta$.

Approximation of transition systems based on approximate simulation relations is useful for solving the safety verification problem. Indeed, from Theorem 2.4, it is straightforward that if T_2 approximately simulates T_1 with the precision δ and Reach$(T_2) \cap \mathcal{N}_\Pi(\Pi_U, \delta) = \emptyset$ (where $\mathcal{N}_\Pi(., \delta)$ denotes the δ-neighborhood for the metric d_Π), then Reach$(T_1) \cap \Pi_U = \emptyset$. Therefore, the safety of T_1 can be verified using the approximate system T_2.

3. HYBRID SYSTEMS AS TRANSITION SYSTEMS

In this section, we show that hybrid systems can be formulated as transition systems. A hybrid system is defined as a tuple $H = (L, n, p, E, F, Inv, G, R, Q^0)$ where

- L is a finite set of locations or discrete states. $|L|$ denotes the number of elements of L. Without loss of generality, $L = \{1, \dots, |L|\}$.
- $n : L \rightarrow \mathbb{N}$, where for every $l \in L$, n_l is the dimension of the continuous state space in

the location l. The set of states of the hybrid system is

$$Q = \bigcup_{l \in L} \{l\} \times \mathbb{R}^{n_l}.$$

- $p : L \to \mathbb{N}$, where for every $l \in L$, p_l is the dimension of the continuous observation of the hybrid system in the location l. The set of observations of the hybrid system is

$$\Pi = \bigcup_{l \in L} \{l\} \times \mathbb{R}^{p_l}.$$

- $E \subseteq L \times L$ is the set of events or discrete transitions.
- $F = \{F_l | \ l \in L\}$ defines the continuous dynamics in each location. For each $l \in L$, F_l is a triple (f_l, g_l, U_l) where $f_l : \mathbb{R}^{n_l} \times U_l \to \mathbb{R}^{n_l}$, $g_l : \mathbb{R}^{n_l} \to \mathbb{R}^{p_l}$ and $U_l \subseteq \mathbb{R}^{m_l}$ is a compact set of internal inputs accounting for disturbances and modelling uncertainties. While the discrete part of the state is l, the continuous part evolves according to

$$\begin{cases} \dot{x}(t) = f_l(x(t), u(t)), \ u(t) \in U_l \\ y(t) = g_l(x(t)). \end{cases}$$

- $Inv = \{Inv_l | \ l \in L\}$ defines an invariant set in each location. For each $l \in L$, $Inv_l \subseteq \mathbb{R}^{n_l}$ constrains the value of the continuous part of the state while the discrete part is l.
- $G = \{G_e | \ e \in E\}$ defines the guard for each discrete transition. For each $e = (l, l') \in E$, $G_e \subseteq Inv_l$. The discrete transition e is enabled when the continuous part of the state is in G_e.
- $R = \{R_e | \ e \in E\}$ defines the reset map for each discrete transition. For each $e = (l, l') \in E$, $R_e : G_e \to 2^{Inv_{l'}}$. When the event e occurs, the continuous part of the state is reset using R_e.
- $Q^0 \subseteq Q$ is the set of initial states:

$$Q^0 = \bigcup_{l \in L} \{l\} \times I_l^0, \text{ with } I_l^0 \subseteq Inv_l.$$

The semantics of a hybrid system is well established (see for instance (Alur *et al.*, 2000)) and is not defined here. In the spirit of (Alur *et al.*, 1995), we can derive from H a nondeterministic transition system $T = (Q, \Sigma, \to, Q^0, \Pi, \langle\langle . \rangle\rangle)$ where the set of states Q, the set of observations Π, and the set initial states Q^0 are the same than in H. The set of labels is $\Sigma = \mathbb{R}^+ \cup \{\tau\}$. The observation map is given by

$$\langle\langle (l, x) \rangle\rangle = (l, g_l(x)).$$

The transition relation $\to$ is given by:

(1) *continuous transitions* :
For $t \in \mathbb{R}^+$, $(l, x) \xrightarrow{t} (l, x')$ iff there exists a locally measurable function $u(.)$ and an absolutely continuous function $z(.)$ such that $z(0) = x$, $z(t) = x'$ and for all $s \in [0, t]$,

$$\dot{z}(s) = f_l(z(s), u(s))$$

with $u(s) \in U_l$ and $z(s) \in Inv_l$.

(2) *discrete transitions* :
$(l, x) \xrightarrow{\tau} (l', x')$ iff $(l, l') = e \in E$, $x \in G_e$ and $x' \in R_e(x)$.

The set of observation Π of a hybrid system is equipped with the following metric d_Π:

$$d_\Pi\left((l_1, y_1), (l_2, y_2)\right) = \begin{cases} \|y_1 - y_2\|, & \text{if } l_1 = l_2 \\ +\infty, & \text{if } l_1 \neq l_2 \end{cases}$$

In the following, we show that our approximation framework based on approximate simulation relations can be applied to hybrid systems.

4. APPROXIMATE SIMULATION RELATIONS FOR HYBRID SYSTEMS

Let $H_i = (L_i, n_i, p_i, E_i, F_i, Inv_i, G_i, R_i, Q_i^0)$, $(i = 1, 2)$ be two hybrid systems and $T_i = (Q_i, \Sigma_i, \to_i, Q_i^0, \Pi_i, \langle\langle . \rangle\rangle_i)$, $(i = 1, 2)$ be the associated transition systems. We assume that T_1 and T_2 have the same set of observations $\Pi_1 = \Pi_2 = \Pi$. Particularly, this implies that the set of locations and the dimensions of the continuous observations are the same for both systems (*i.e.* $L_1 = L_2 = L$, $p_1 = p_2 = p$). [2] We will further assume that the discrete dynamics of both systems are the same (*i.e.* $E_1 = E_2 = E$). The goal of the approximation process presented here is then essentially to simplify the continuous dynamics of the hybrid system H_1. In this section, we establish sufficient conditions so that H_2 approximately simulates H_1 and provide a method to evaluate the precision of the approximate simulation relation.

4.1 Simulation functions

Let $l \in L$, let $n_{1,l}$, $n_{2,l}$ be the dimensions of the continuous part of the state of H_1 and H_2 in the location l. Let $F_{1,l} = (f_{1,l}, g_{1,l}, U_{1,l})$ and $F_{2,l} = (f_{2,l}, g_{2,l}, U_{2,l})$ be the continuous dynamics of H_1 and H_2 associated to the location l. We define the following notations:

$$x = \begin{bmatrix} x_1 \\ x_2 \end{bmatrix}, \ f_l(x, u_1, u_2) = \begin{bmatrix} f_{1,l}(x_1, u_1) \\ f_{2,l}(x_2, u_2) \end{bmatrix},$$
$$g_l(x) = g_{1,l}(x_1) - g_{2,l}(x_2).$$

In (Girard and Pappas, 2005*c*), we showed that approximate simulation relations could be characterized efficiently using the notion of simulation

[2] The approximation of a hybrid systems by another hybrid systems with a smaller number of locations has been considered for systems with purely stochastic jumps (Julius, 2006). We will also consider this type of approximation for hybrid systems with non stochastic jumps in the future.

function. In our context, this can be instantiated as follows.

Definition 4.1. $V_l : \mathbb{R}^{n_{1,l}} \times \mathbb{R}^{n_{2,l}} \to \mathbb{R}^+$ is a simulation function of $F_{1,l}$ by $F_{2,l}$ if for all $x \in \mathbb{R}^{n_{1,l}} \times \mathbb{R}^{n_{2,l}}$,

$$V_l(x) \geq g_l(x)^T g_l(x), \quad (1)$$

$$\max_{u_1 \in U_{1,l}} \min_{u_2 \in U_{2,l}} \nabla V_l(x)^T f_l(x, u_1, u_2) \leq 0. \quad (2)$$

Remark 4.2. The concept of simulation function is related to robust control Lyapunov functions (Freeman and Kokotovic, 1996; Liberzon *et al.*, 2002), though they slightly differ in spirit. Indeed, considering the input u_1 as a disturbance and the input u_2 as a control variable, the interpretation of equation (2) is that for all disturbances there exists a control such that the simulation function decreases during the evolution of the system. In this context, u_2 may have full knowledge (and be a function) of u_1. In comparison, a robust control Lyapunov function would require that there exists a control u_2 such that for all possible (and unknown) disturbances u_1 the function decreases during the evolution of the system. Therefore, robust control Lyapunov functions require stronger conditions than simulation functions.

Methods for the computation of simulation functions have been proposed for the class of constrained linear systems (Girard and Pappas, 2005*a*) and autonomous nonlinear systems (Girard and Pappas, 2005*b*). These methods are based on linear matrix inequalities, sum of squares programs and static games and are thus computationally effective. The computation of simulation functions for constrained linear systems has been implemented in the Matlab toolbox MATISSE[3].

Simulation functions satisfy the following property which will be useful in characterizing approximate simulation relations for hybrid systems.

Proposition 4.3. For all $(x_1, x_2) \in \mathbb{R}^{n_{1,l}} \times \mathbb{R}^{n_{2,l}}$, for all inputs $u_1(.)$, there exists an input $u_2(.)$ such that

$$\forall t \in \mathbb{R}^+, V_l(z_1(t), z_2(t)) \leq V_l(x_1, x_2) \quad (3)$$

where

$$\dot{z}_i(t) = f_{i,l}(z_i(t), u_i(t)),\ z_i(0) = x_i,\ i = 1, 2.$$

Proof: Let us remark that

$$\dot{V}_l(z(t)) = \nabla V_l(z(t))^T f_l(z(t), u_1(t), u_2(t))$$

where $z(t) = [z_1(t)\ z_2(t)]^T$. Then, from equation (2), it is clear that for all inputs $u_1(.)$, there exists an input $u_2(.)$ such that $\dot{V}_l(z(t)) \leq 0$. ■

4.2 Approximate simulation relations

In this section, we give a characterization of approximate simulation relations for hybrid systems. Let us assume that for each location $l \in L$, there exists a simulation function V_l of the continuous dynamics $F_{1,l}$ by $F_{2,l}$. We define the following sets, for all $x_1 \in \mathbb{R}^{n_{1,l}}$, $\beta \geq 0$,

$$\mathcal{N}_l(x_1, \beta) = \{x_2 \in \mathbb{R}^{n_{2,l}} |\ V_l(x_1, x_2) \leq \beta\}.$$

Theorem 4.4. Let $\beta_1, \ldots, \beta_{|L|}$ be positive numbers such that

(a) for all $l \in L$, $\mathcal{N}_l(Inv_{1,l}, \beta_l) \subseteq Inv_{2,l}$,
(b) for all $e = (l, l') \in E$, $\mathcal{N}_l(G_{1,e}, \beta_l) \subseteq G_{2,e}$,
(c) for all $l \in L$,

$$\beta_l \geq \max_{x_1 \in I^0_{1,l}} \min_{x_2 \in I^0_{2,l}} V_l(x_1, x_2),$$

(d) for all $e = (l, l') \in E$,

$$\beta_{l'} \geq \max_{\substack{x_1 \in G_{1,e} \\ V_l(x_1, x_2) \leq \beta_l}} \left(\max_{x_1' \in R_{1,e}(x_1)} \min_{x_2' \in R_{2,e}(x_2)} V_{l'}(x_1', x_2') \right).$$

Let $\delta = \max(\sqrt{\beta_1}, \ldots, \sqrt{\beta_{|L|}})$. Then, the relation $\mathcal{S}_\delta \subseteq Q_1 \times Q_2$ defined by

$$\mathcal{S}_\delta = \{(l_1, x_1, l_2, x_2) |\ l_1 = l_2 = l,\ V_l(x_1, x_2) \leq \beta_l\}$$

is a δ-approximate simulation relation of T_1 by T_2 and $T_1 \preceq_\delta T_2$.

Proof: Let $(l_1, x_1, l_2, x_2) \in \mathcal{S}_\delta$, then $l_1 = l_2 = l$ and $V_l(x_1, x_2) \leq \beta_l$. From equation (1), we have that $\|g_{l,1}(x_1) - g_{l,2}(x_2)\| \leq \sqrt{\beta_l} \leq \delta$. Hence, the first property of Definition 2.2 holds.
Let $(l_1, x_1) \xrightarrow{t} (l_1, x_1')$, then there exists an input $u_1(.)$ and a function $z_1(.)$ such that $z_1(0) = x_1$, $z_1(t) = x'$ and for all $s \in [0, t]$, $u_1(s) \in U_{1,l}$, $z_1(s) \in Inv_{1,l}$ and

$$\dot{z}_1(s) = f_{l,1}(z_1(s), u_1(s)).$$

From Proposition 4.3, we know that there exists an input $u_2(.)$ and a function $z_2(.)$ such that $z_2(0) = x_2$, and for all $s \in [0, t]$, $u_2(s) \in U_{2,l}$,

$$\dot{z}_2(s) - f_{l,2}(z_2(s), u_2(s))$$

and $V(z_1(s), z_2(s)) \leq V(x_1, x_2) \leq \beta_l$. Then, assumption (a) of Theorem 4.4 insures that for all $s \in [0, t]$, $z_2(s) \in Inv_{l,2}$. Let $x_2' = z_2(t)$, we have $(l_2, x_2) \xrightarrow{t} (l_2, x_2')$ and since $V_l(x_1', x_2') \leq \beta_l$, $(l_1, x_1', l_2, x_2') \in \mathcal{S}_\delta$.
Let $(l_1, x_1) \xrightarrow{\mathcal{T}} (l_1', x_1')$, then there exists $e = (l_1, l_1')$ such that $x_1 \in G_{1,e}$ and $x_1' \in R_{1,e}(x_1)$. Assumption (b) of Theorem 4.4 ensures that $x_2 \in G_{2,e}$. From assumption (d) of Theorem 4.4, we

[3] MATISSE: Metrics for Approximate Transltion Systems Simulation and Equivalence, Available from `http://www.seas.upenn.edu/~agirard/Software/MATISSE`

have that there exists $x'_2 \in R_{2,e}(x_2)$, such that $V_{l'}(x'_1, x'_2) \le \beta_{l'}$ where $l' = l'_1$. Then, $(l_2, x_2) \xrightarrow{\tau} (l'_2, x'_2)$ with $l'_2 = l'$ and $(l'_1, x'_1, l'_2, x'_2) \in \mathcal{S}_\delta$. Therefore, $\mathcal{S}_\delta$ is a δ-approximate simulation relation of T_1 by T_2.
Finally, let $(l_1, x_1) \in Q_1^0$, then $x_1 \in I_{1,l}^0$ where $l = l_1$. From assumption (c) of Theorem 4.4, there exists $x_2 \in I_{2,l}^0$, such that $V_l(x_1, x_2) \le \beta_l$. Then, $(l_2, x_2) \in Q_2^0$ with $l_2 = l$ and $(l_1, x_1, l_2, x_2) \in \mathcal{S}_\delta$. Then $T_1 \preceq_\delta T_2$. ∎

Assumption (d) can be interpreted as a condition of non-propagation of the approximation error through the reset maps. It is clear that the scalars $\beta_1, \dots, \beta_{|L|}$ cannot be chosen independently. Thus, it is not necessarily the case that such numbers exist. There are two cases where we can guarantee easily the existence of these numbers. First, if we consider memoryless resets (*i.e.* $R_{1,e}(x_1) = R_{1,e}$ and $R_{2,e}(x_2) = R_{2,e}$ for all $e \in E$), then we can see that $\beta_1, \dots, \beta_l$ can be chosen independently. Second, if the graph (L, E) does not contain any cycle, then there is no circular dependency between $\beta_1, \dots, \beta_{|L|}$ and thus it is easy to compute numbers such that the fourth assumption holds.

4.3 Approximation of hybrid systems

Based on Theorem 4.4, we can define a procedure to approximate a hybrid systems H_1 by another hybrid system H_2 with simpler continuous dynamics and to compute the precision of the approximate simulation relation of T_1 by T_2.

First, in each location $l \in L$, we approximate the continuous dynamics $F_{1,l}$ by a *simpler* continuous dynamics $F_{2,l}$. If $F_{1,l}$ is a large linear system, then $F_{2,l}$ may be chosen as a smaller linear system (Girard and Pappas, 2005*a*). If $F_{1,l}$ is a nonlinear system, then $F_{2,l}$ may be chosen as a linear system (Girard and Pappas, 2005*b*). The goal of this approximation is to reduce the complexity of analysis tasks such as reachability computation. Then, we compute a simulation function V_l of the continuous dynamics $F_{1,l}$ by $F_{2,l}$. Note that such a function always exists if $F_{1,l}$ and $F_{2,l}$ are asymptotically stable. In the case of nonstable systems, a simulation function exists if the unstable subsystem of $F_{2,l}$ exactly simulates the unstable subsystem of $F_{1,l}$ (Girard and Pappas, 2005*a*).

The second part of the procedure consists in choosing the initial sets $I_{2,l}^0$ and the reset maps $R_{2,e}$ and computing scalars $\beta_1, \dots, \beta_{|L|}$ satisfying the assumptions (c) and (d) of Theorem 4.4. Then, we set the invariants $Inv_{2,l} = \mathcal{N}_l(Inv_{1,l}, \beta_l)$ and the guards $G_{2,e} = \mathcal{N}_l(G_{1,e}, \beta_l)$ where $e = (l, l')$. From Theorem 4.4, we know that $T_1 \preceq_\delta T_2$ with $\delta = \max(\sqrt{\beta_1}, \dots, \sqrt{\beta_{|L|}})$.

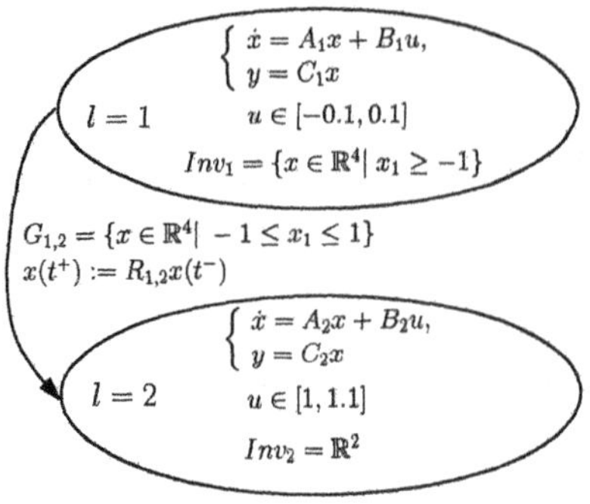

Fig. 1. Example of hybrid system

5. EXAMPLE

In this section, we illustrate our approximation framework in the context of a safety verification problem. Let us consider the hybrid system described in Figure 1. In each location, the continuous linear dynamics are given by the following matrices:

$$A_1 = \begin{bmatrix} -0.5 & 3 & 0 & 0 \\ -3 & -0.5 & 1 & 0 \\ 0 & 0 & -0.7 & 8 \\ 0 & 0 & -8 & -0.7 \end{bmatrix}, B_1 = \begin{bmatrix} 0 \\ 1 \\ 0 \\ 0 \end{bmatrix}, C_1^T = \begin{bmatrix} 1 & 0 \\ 0 & 1 \\ 0 & 0 \\ 0 & 0 \end{bmatrix}$$

$$A_2 = \begin{bmatrix} -0.5 & 1 \\ 0 & -1 \end{bmatrix}, B_2 = \begin{bmatrix} -10 \\ 5 \end{bmatrix}, C_2^T = \begin{bmatrix} 1 & 0 \\ 0 & 1 \end{bmatrix}.$$

The linear reset map is given by $R_{1,2} = \begin{bmatrix} 0 & 1 & 0 & 0 \\ 1 & 0 & 0 & 0 \end{bmatrix}$. The set of initial states is

$$Q^0 = \{1\} \times ([4, 5] \times [4, 5] \times [0.9, 1.1] \times \{0\}).$$

Let us consider the safety verification problem where the set of unsafe sets is

$$\Pi_U = \{2\} \times \{y \in \mathbb{R}^2 \mid (y_1 + 10)^2 + (y_2 + 1)^2 \le 1\}.$$

In order to solve the safety verification problem, we will use a two dimensional approximation of the continuous dynamics in location 1, given by the following matrices:

$$A'_1 = \begin{bmatrix} -0.5 & 3 \\ -3 & -0.5 \end{bmatrix}, B'_1 = \begin{bmatrix} 0 \\ 1 \end{bmatrix}, C'_1 = \begin{bmatrix} 1 & 0 \\ 0 & 1 \end{bmatrix}.$$

The two dimensional dynamics in location 2 will be kept unchanged. The reset map of the approximate hybrid system is given by the matrix $R'_{1,2} = \begin{bmatrix} 0 & 1 \\ 1 & 0 \end{bmatrix}$. The initial set is

$$Q'^0 = \{1\} \times ([4, 5] \times [4, 5]).$$

Simulation functions between the continuous dynamics are computed using the toolbox MATISSE. This essentially consists in solving a set of linear matrix inequalities and quadratic programs. Then, we compute β_1 and β_2 such that assumptions (c) and (d) of Theorem 4.4 hold. The invariant of location 1 and the guard of the transition $(1, 2)$ are bloated according to these numbers. The precision of the approximate simulation relation between the original hybrid system and its approximation is $\delta = 0.3877$. We computed the reachable sets of both systems using zonotope based reachability algorithms (Girard, 2005) implemented in MATISSE. We can see on Figure 2

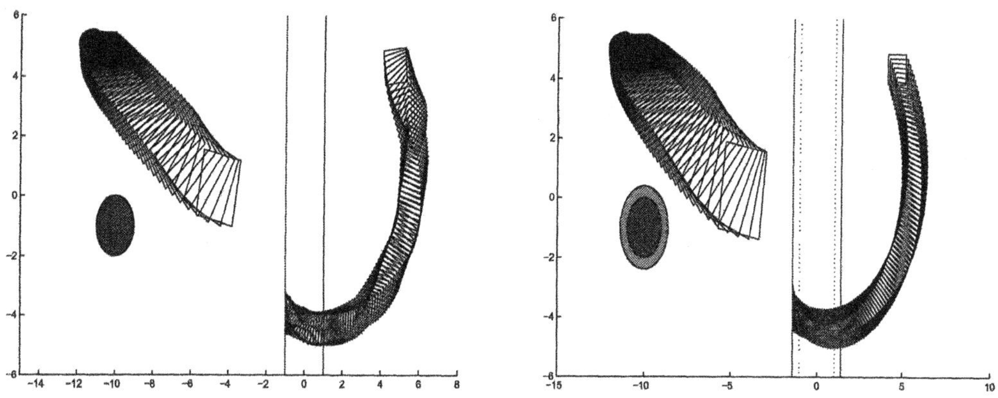

Fig. 2. Reachable sets of the original hybrid system (left) and of its approximation (right). We can see that the approximation allows to conclude that the original system is safe.

that the reachable set of the approximate hybrid system does not intersect the bloated unsafe set. Hence, this allows to conclude that the original hybrid system is safe.

6. CONCLUSION

In this paper, we extended the notion of approximate simulation relations to hybrid systems. We developed an effective characterization of approximate simulation relations based on simulation functions. We showed how our framework could be used to approximate hybrid systems and an example in the context of safety verification was shown. Future work includes developing more systematic methods to compute approximate simulation relations for hybrid systems as well as implementing these methods in MATISSE.

REFERENCES

Alur, R., T.A. Henzinger, G. Lafferriere and G.J. Pappas (2000). Discrete abstractions of hybrid systems. *Proceedings of the IEEE* **88**, 971–984.

Alur, Rajeev, Costas Courcoubetis, Nicolas Halbwachs, Thomas A. Henzinger, Pei-Hsin Ho, Xavier Nicollin, Alfredo Olivero, Joseph Sifakis and Sergio Yovine (1995). The algorithmic analysis of hybrid systems.. *Theor. Comput. Sci.* **138**(1), 3–34.

Clarke, E. M., O. Grumberg and D. A. Peled (2000). *Model Checking*. MIT Press.

de Alfaro, L., M. Faella and M. Stoelinga (2004). Linear and branching metrics for quantitative transition systems. In: *ICALP'04*. Vol. 3142 of *LNCS*. Springer. pp. 1150–1162.

Desharnais, J., V. Gupta, R. Jagadeesan and P. Panangaden (2004). Metrics for labelled markov processes. *Theor. Comput. Sc.* **318**(3), 323–354.

Freeman, R. A. and P. V. Kokotovic (1996). Inverse optimality in robust stabilization. *SIAM J. Control and Optimization* **34**(4), 1365–1391.

Girard, A. (2005). Reachability of uncertain linear systems using zonotopes. In: *Hybrid Systems: Computation and Control*. Vol. 3414 of *LNCS*. Springer. pp. 291–305.

Girard, A. and G. J. Pappas (2005*a*). Approximate bisimulations for constrained linear systems. In: *Proc. IEEE Conference on Decision and Control and European Control Conference*.

Girard, A. and G. J. Pappas (2005*b*). Approximate bisimulations for nonlinear dynamical systems. In: *Proc. IEEE Conference on Decision and Control and European Control Conference*.

Girard, A. and G. J. Pappas (2005*c*). Approximation metrics for discrete and continuous systems. Technical Report MS-CIS-05-10, Dept. of CIS, University of Pennsylvania.

Haghverdi, E., P. Tabuada and G. J. Pappas (2005). Bisimulation relations for dynamical, control, and hybrid systems. *Theor. Comput. Sc.* **342**(2-3), 229–262.

Julius, A.A. (2006). Approximate abstraction of stochastic hybrid automata. In: *Hybrid Systems: Computation and Control*. Vol. 3927 of *LNCS*. Springer. pp. 318–332.

Julius, A.A., A. Girard and G.J. Pappas (2006). Approximate bisimulation for a class of stochastic hybrid systems. In: *Proc. American Control Conference*.

Liberzon, D., E. D. Sontag and Y. Wang (2002). Universal construction of feedback laws achieving ISS and integral-ISS disturbance attenuation. *Systems and Control Letters* **46**, 111–127.

Pappas, G. J. (2003). Bisimilar linear systems. *Automatica* **39**(12), 2035–2047.

ELSEVIER
IFAC
PUBLICATIONS

STABILIZABILITY BASED STATE SPACE REDUCTIONS FOR HYBRID SYSTEMS [1]

Elena De Santis, Maria Domenica Di Benedetto, Giordano Pola

Center of Excellence DEWS
Dept. of Electrical Engineering and Computer Science
University of L'Aquila
Monteluco di Roio, 67040 L'Aquila, Italy
{desantis,dibenede,pola}@ing.univaq.it

Abstract: In this paper, we focus on a particular subclass of hybrid systems, the class of linear switching systems. We propose hybrid state space decompositions, based on hybrid invariant subspaces, which reduce the computational effort required for checking the structural property of asymptotic stabilizability.

Keywords: Switching systems, asymptotic stabilizability, state space reduction, Kalman decomposition.

1. INTRODUCTION

In this paper, we focus on a subclass of hybrid systems, the class of linear switching systems [3], where the continuous dynamics and the reset functions are linear and the transitions depend only on an event that acts as a discrete disturbance. The continuous dynamics are given by a linear dynamical control system (whose dynamical matrices depend on the current discrete state) and therefore an input function can be designed for controlling purposes.

Stability issues of hybrid systems have been extensively investigated in the last years (see e.g. in [2], [14], [9], [13] and references therein). However checking stabilizability of switching systems is not an easy task in general (see e.g. [9]) and a complete characterization of stabilizability properties of switching systems is still missing. This is the reason why in this paper we focus on some structural reductions of the hybrid state space, which allow the original problem to be split into simpler subproblems. Moreover connections to the well-known Kalman decomposition of linear dynamical control systems are also established. Dual results on detectability based state space reductions have been recently established in a companion paper [5].

The organization of the paper is as follows. We first recall some definitions of switching systems and stabilizability in Section 2. Then we define in Section 3 invariant hybrid subspaces, thereby extending to the hybrid framework the notions given in [1] for the linear case, and we propose an algorithm for the computation of the minimal invariant hybrid subspace containing a given hybrid subspace. In Section 4, by means of this minimal hybrid subspace, we define a state space transformation of the system, which allows stating conditions for stabilizability. Based on this result and on [7], the given system is decoupled into controlled and autonomous linear switching subsystems. The asymptotic stabilizability of the first ones and the asymptotic stability of the latter ones

[1] This work has been partially supported by the HYCON Network of Excellence, contract number FP6-IST-511368 and by Ministero dell'Istruzione dell'Universita' e della Ricerca under Projects MACSI and SCEF (PRIN05).

imply the asymptotic stabilizability of the given system. Some concluding remarks are offered in Section 6. The results are given without proof for lack of space. A full version of this paper can be found in [8].

2. SWITCHING SYSTEMS

In this section, we formally introduce the class of *linear switching systems* and the structural property of asymptotic stabilizability.

The hybrid state ξ of a linear switching system is composed of two components: the discrete state q_i, belonging to a finite set Q and the continuous state x, belonging to a linear space $\mathbb{R}^{n_i}$, whose dimension n_i depends on q_i. The evolution of the discrete state is governed by a Finite State Machine (FSM); a transition $e = (q_i, \sigma, q_h)$ may occur at time t from the discrete state q_i to the discrete state q_h, if the discrete disturbance σ occurs at time t. The evolution of the continuous state is described by a set of linear dynamical systems, whose matrices depend on the current discrete state q_i. Whenever a transition e occurs, the continuous state x is instantly reset to a new value $R(e)x$, where $R(e)$ is a matrix depending on the transition e. More formally,

Definition 1. A linear switching system S is a tuple

$$(\Xi, \Theta, S, E, R),$$

where:

- $\Xi = \bigcup_{q_i \in Q} \{q_i\} \times \mathbb{R}^{n_i}$ is the hybrid state space, where:
  - $Q = \{q_i, i \in J\}$ is the set of discrete states, $J = \{1, 2, \ldots, N\}$;
  - $\mathbb{R}^{n_i}$ is the continuous state space associated with the discrete state $q_i \in Q$;
- $\Theta = \Sigma \times U$ is the hybrid input space, where:
  - $\Sigma = \{\sigma_h, h \in J_1\}$ is the set of discrete disturbances, $J_1 = \{1, 2, \ldots, N_1\}$;
  - $U = \mathbb{R}^m$ is the continuous input space;
- S is a map associating to any discrete state $q_i \in Q$ the following linear dynamical control system:

$$\dot{x}(t) = A_i x(t) + B_i u(t),$$

  where $x(t) \in \mathbb{R}^{n_i}$ is the continuous state, and u is the continuous input function;
- $E \subset Q \times \Sigma \times Q$ is a collection of transitions;
- R is a function that associates to any $e = (q_i, \sigma, q_h) \in E$ the reset matrix $R(e) \in \mathbb{R}^{n_h \times n_i}$.

A linear switching system S is said to be *autonomous* if $U = \{0\}$.

We now formally define the semantics of linear switching systems. First of all we assume throughout the paper that *the discrete disturbance is not available for measurements*, thus yielding a non-deterministic system, and that the class of admissible continuous inputs is the set $\mathcal{U}$ of piecewise continuous control functions $u : \mathbb{R} \to U$. As defined in [10], a *hybrid time basis* τ is an infinite or finite sequence of sets $I_j = \{t \in \mathbb{R} : t_j \leq t \leq t'_j\}$, with $t'_j = t_{j+1}$; set $card(\tau) = L + 1$. If $L < \infty$, then t'_L can be finite or infinite. A hybrid time basis τ is said to be *finite*, if $L < \infty$ and $t'_L < \infty$ and *infinite*, otherwise. Given a hybrid time basis τ, any time instant t'_j is called *switching time.* Since linear switching systems are time invariant, we assume without loss of generality that $t_0 = 0$ in any hybrid time basis. Throughout the paper, we assume that there is a minimum time separation between two consecutive switching times:

Assumption 1 (*Minimum dwell time*) There exists a real $\delta_m > 0$, called minimum dwell time [11], such that for any hybrid time basis τ, $t'_j - t_j \geq \delta_m$.

The existence of a minimum dwell time is a widely used assumption in the analysis of switching systems (e.g. [11], [9], [6] and the references therein), and models the inertia of the system to react to an external (discrete) input. Denote by $\mathcal{T}$ the set of all hybrid time bases satisfying Assumption 1. The temporal evolution of a linear switching system can be now defined as follows.

Definition 2. (Switching system execution) An execution χ of a linear switching system S is a collection $(\xi_0, \tau, \sigma, u, \xi)$ with $\xi_0 \in \Xi$, $\tau \in \mathcal{T}$, $\sigma : \mathbb{N} \to \Sigma$, $u \in \mathcal{U}$, $\xi : \mathbb{R} \times \mathbb{N} \to \Xi$. The hybrid state evolution ξ is defined as follows:

$$\begin{aligned}
&\xi(0,0) = \xi_0, \\
&\xi(t,j) = (q(j), x(t,j)), t \in I_j, j = 0, 1, ..., L, \\
&\xi(t_{j+1}, j+1) = (q(j+1), R(e_j)x(t'_j, j)), \\
&j = 0, 1, ..., L,
\end{aligned}$$

where $q : \mathbb{N} \to Q$, $e_j = (q(j), \sigma(j), q(j+1)) \in E$ and $x(t,j)$ is the solution at time t of the dynamical system $S(q(j))$, with initial time t_j, initial condition $x(t_j, j)$ and continuous input u.

Remark 1. The class of linear switching systems is related to the class of *linear switched systems*, which has been extensively studied in the literature (see e.g. [13] and the references therein). While in a switching system transitions are caused by discrete disturbances, in a switched system they are caused by discrete inputs (i.e. discrete controls). A formal definition of switched systems can be obtained from Definition 1 by assuming that Σ is the set of discrete inputs. The semantics of switched systems is formally specified by Definition 2, where $\sigma : \mathbb{N} \to \Sigma$ is a discrete input function. The notion of switched systems obtained by Definition 1 generalizes the models of [13], where transitions are defined between every pair of discrete states and the reset matrix is the identity.

Given $\mathcal{S}$ and an execution χ, set $\eta(t) = \xi(t,j)$, $t \in [t_j, t'_j)$, $j = 0, 1, ..., L$. We assume that *the hybrid state evolution is available for control synthesis*: the set

$$\mathcal{Y} = \{\eta|_{[0,t]}, \eta : \mathbb{R} \to \Xi, t \geq 0\}$$

embeds all the information on the hybrid state evolution available for control purposes. A *control strategy* φ is a function $\varphi : \mathcal{Y} \to U$ such that the function defined by $u(t) = \varphi(\eta|_{[0,t]})$, $t \geq 0$ belongs to $\mathcal{U}$. A switching system $\mathcal{S}$ together with a control strategy φ is called *controlled switching system* and its executions with $u(t) = \varphi(\eta|_{[0,t]})$, $t \geq 0$ are called *controlled executions.*

We can now formally introduce our definition of asymptotic stabilizability. Let be

$$\mathcal{B} = \bigcup_{q_i \in Q} \{q_i\} \times \mathcal{B}_i,$$

where $\mathcal{B}_i = \{x \in \mathbb{R}^{n_i} : \|x\|_{n_i} \leq 1\}$ for any $i \in J$ and set $\varepsilon\mathcal{B} := \bigcup_{q_i \in Q} \{q_i\} \times \varepsilon\mathcal{B}_i$ for any $\varepsilon \geq 0$.

Definition 3. (*Asymptotic Stabilizability*) A linear switching system $\mathcal{S}$ is asymptotically stabilizable if there exists a control strategy φ such that $\forall \varepsilon > 0$ and for all controlled executions of $\mathcal{S}$ with initial hybrid state in $\mathcal{B}$, there exists $\hat{t} > 0$ such that:

$$\xi(t,j) \in \varepsilon\mathcal{B}, \quad \forall t \in I_j \cap [\hat{t}, \infty), \, \forall j = \hat{\jmath}, ...L,$$

where $\hat{\jmath} = \min\{j : \hat{t} \in I_j\}$. The control strategy φ is called stabilizing. If the condition above holds with $\varepsilon = 0$, then $\mathcal{S}$ is called controllable.

Remark 2. From the definition above, it is easy to see that a linear switching system $\mathcal{S}$ with minimum dwell time $\delta_m > 0$ is controllable if and only if any linear system $S(q)$, $q \in Q$ is controllable.

An asymptotically stabilizable autonomous linear switching system is said to be *asymptotically stable.*

Since our purpose is to reduce the state space while preserving stabilizability (hence an asymptotic property), we consider only executions of infinite duration.

3. INVARIANT HYBRID SUBSPACES

Aim of this section is to introduce an invariant linear hybrid subspace that will be the basis upon which stabilizability analysis for linear switching systems can be performed.

The notion of invariant linear subspace for switching systems can be defined as follows.

Definition 4. A set

$$\Omega = \bigcup_{i \in J'} \{q_i\} \times \Omega_i \subset \Xi$$

is a hybrid linear subspace of Ξ, if $J' = J$ and Ω_i is a linear subspace of $\mathbb{R}^{n_i}$, for any $i \in J$.

For shortness, a hybrid linear subspace will be simply called subspace.

Definition 5. Given a switching system $\mathcal{S}$, a set

$$\Omega = \bigcup_{i \in J} \{q_i\} \times \Omega_i \subset \Xi,$$

is $\mathcal{S}-$invariant if, for any initial hybrid state $\xi_0 \in \Omega$ and for any execution $\chi = (\xi_0, \tau, \sigma, u, \xi)$ with $u(t) = 0, \forall t \geq 0$,

$$\xi(t,j) \in \Omega, \, \forall t \in I_j, \, \forall j = 0, 1, ..., L.$$

The following result gives a necessary and sufficient condition for a subspace to be $\mathcal{S}-$invariant.

Proposition 1. Given a switching system $\mathcal{S}$, a subspace $\bigcup_{i \in J} \{q_i\} \times \Omega_i$ is $\mathcal{S}-$invariant if and only if for any $i \in J$ the following conditions hold:

- $A_i \Omega_i \subset \Omega_i$;
- $R(e)\Omega_i \subset \Omega_h$, for any $e = (q_i, \sigma, q_h) \in E$.

Since the intersection of any two $\mathcal{S}-$invariant subspaces is an $\mathcal{S}-$invariant subspace, the minimal $\mathcal{S}-$invariant subspace containing a given subspace is well defined.

Let

$$\mathcal{G} = \bigcup_{i \in J} \{q_i\} \times \mathcal{G}_i \quad (1)$$

be the minimal $\mathcal{S}-$invariant subspace containing

$$\mathcal{H} = \bigcup_{i \in J} \{q_i\} \times \text{Im}(B_i).$$

For any $i \in J$, let

$$\mathcal{C}_i = (\, B_i \; A_i B_i \; \ldots \; A_i^{n_i - 1} B_i \,)$$

be the controllability matrix associated with the linear system $S(q_i)$ and set

$$\mathcal{R} = \bigcup_{i \in J} \{q_i\} \times \mathcal{R}_i,$$

where $\mathcal{R}_i = \text{Im}(\mathcal{C}_i)$. The following result holds.

Lemma 2. The set $\mathcal{G}$ is the minimal $\mathcal{S}-$invariant subspace that contains the hybrid subspace $\mathcal{R}$.

The following result illustrates a procedure for computing $\mathcal{G}$ in a finite number of steps.

Theorem 3. Given $\mathcal{S}$, define the sequence of subspaces $\Omega_i^k \subset \mathbb{R}^{n_i}$, $k = 0, 1, 2, \ldots$, $i \in J$, as

$$\Omega_i^0 = \mathcal{R}_i,$$
$$\Omega_i^k = \sum_{h=0}^{n_i-1} (A_i)^h \, \Phi_i^k$$
$$\Phi_i^k = \sum_{j \in J_i} R((q_j, \sigma, q_i))\Omega_j^{k-1} + \Omega_i^{k-1}$$

where $J_i = \{j \in J : (q_j, \sigma, q_i) \in E\}$. The sequence $\{\Omega_i^k, i \in J\}_{k=0,1,2,\ldots}$ converges in $k^* \leq \sum_{i=1}^N n_i$ steps and

$$\mathcal{G} = \bigcup_{i \in J} \{q_i\} \times \Omega_i^{k^*}.$$

By definition, the discrete evolution of the switching system $\mathcal{S} = (\Xi, \Theta, S, E, R)$ is described by the FSM (Q, Σ, E). We recall that the FSM is said to be *strongly connected* if there exists a path between any pair of discrete states in Q. We conclude this section by giving the following result.

Proposition 4. If $n_i = n$ for any $i \in J$, if $R(e) = I$ for any $e \in E$, and if (Q, Σ, E) is strongly connected, then

$$\mathcal{G} = Q \times \widehat{\mathcal{G}},$$

where $\widehat{\mathcal{G}} \subset \mathbb{R}^n$ is the minimal linear subspace of $\mathbb{R}^n$ satisfying for any $i \in J$ the following conditions:

$$A_i\widehat{\mathcal{G}} \subset \widehat{\mathcal{G}}; \quad \mathrm{Im}\,(B_i) \subset \widehat{\mathcal{G}}.$$

Remark 3. The subspace $\widehat{\mathcal{G}}$ coincides with the 'multiple controllable subspace', as defined in [13] in the framework of switched linear systems (see also Remark 1).

4. STATE SPACE REDUCTIONS BASED ON STABILIZABILITY

It is well-known that a linear system S is asymptotically stabilizable if and only if a suitable subsystem extracted from S is asymptotically stable. In the context of general switching systems, stabilizability conditions become a bit more involved.

In this section, we show how to extract from a given linear switching system $\mathcal{S}$, a number of subsystems so that the stabilizability of some of them and the asymptotic stability of the remaining ones imply the stabilizability of $\mathcal{S}$. This reduces the computational effort required for checking the property under consideration.

Our procedure is based on the reduction of the state space of the linear switching system $\mathcal{S}$ by means of the invariant hybrid subspace $\mathcal{G}$, as defined in the previous section.

Given the hybrid invariant subspace $\mathcal{G}$ as in (1), let $\mu_i \leq n_i$ be the dimension of $\mathcal{G}_i$ and define a hybrid state space transformation for $\mathcal{S}$, as follows. For each $i \in J$, consider the matrix:

$$T_i = \left(b_1^i \ldots b_{\mu_i}^i \; v_1^i \ldots v_{n_i-\mu_i}^i\right) \in \mathbb{R}^{n_i \times n_i},$$

where the vectors $b_1^i, \ldots, b_{\mu_i}^i$ are a basis for $\mathcal{G}_i$ and the vectors $v_1^i, \ldots, v_{n_i-\mu_i}^i$ are such that T_i is full rank. Then the matrices:

$$\widehat{A}_i = T_i^{-1} A_i T_i,$$
$$\widehat{B}_i = T_i^{-1} B_i, i \in J$$
$$\widehat{R}(e) = T_h^{-1} R(e) T_i, e = (q_i, \sigma, q_h),$$

take the form:

$$\widehat{A}_i = \begin{pmatrix} A_i^{(11)} & A_i^{(12)} \\ 0 & A_i^{(22)} \end{pmatrix}, \quad \widehat{B}_i = \begin{pmatrix} B_i^{(1)} \\ 0 \end{pmatrix},$$
$$\widehat{R}(e) = \begin{pmatrix} R_e^{(11)} & R_e^{(12)} \\ 0 & R_e^{(22)} \end{pmatrix},$$

where $A_i^{(11)} \in \mathbb{R}^{\mu_i \times \mu_i}$. The switching system obtained after the hybrid state space transformation is algebraically equivalent [12] to the switching system $\mathcal{S}$. Note that, in general, the pair $(A_i^{(11)}, B_i^{(1)})$ is not controllable.

We introduce the following technical assumption that will be removed at the end of this section.

Assumption 2 For any $i \in J$, $0 \leq \mu_i < n_i$.

Under Assumption 2, we can define the following autonomous linear switching system (uncontrollable subsystem of $\mathcal{S}$):

$$\mathcal{S}_{un} = (\Xi_{un}, \Theta, S_{un}, E, R_{un}),$$

where:

- $\Xi_{un} = \bigcup_{i \in J} \{q_i\} \times \mathbb{R}^{n_i - \mu_i}$;
- for any $q_i \in Q$, $S_{un}(q_i)$ is described by the equation:
$$\dot{z}(t) = A_i^{(22)} z(t);$$
- for any $e \in E$, $R_{un}(e) = R_e^{(22)}$.

The following result gives a relationship between stabilizability properties of $\mathcal{S}$ and stability properties of $\mathcal{S}_{un}$.

Theorem 5. If Assumption 2 holds, then $\mathcal{S}$ is asymptotically stabilizable only if $\mathcal{S}_{un}$ is asymptotically stable.

A stronger result can be assessed under the following additional assumption that will be removed at the end of this section:

Assumption 3 For any $i \in J$, $0 < \mu_i \leq n_i$.

Note that if $\mu_i = 0$, then $B_i = 0$ and any continuous state in $\mathcal{G}_h$ is reset to the origin after any transition of the form $(q_h, \sigma, q_i) \in E$.

Under Assumption 3, we can define the linear switching system (controlled subsystem of $\mathcal{S}$):

$$\mathcal{S}_c = (\Xi_c, \Theta, S_c, E, R_c),$$

where:

- $\Xi_c = \bigcup_{i \in J} \{q_i\} \times \mathbb{R}^{\mu_i}$;
- for any $q_i \in Q$, $S_c(q_i)$ is described by the equation:

$$\dot{z}(t) = A_i^{(11)} z(t) + B_i^{(1)} u(t);$$

- for any $e \in E$, $R_c(e) = R_e^{(11)}$.

On the basis of the above decomposition, we now show that the asymptotic stabilizability of $\mathcal{S}$ can be reduced to the asymptotic stabilizability of $\mathcal{S}_c$ and the asymptotic stability of $\mathcal{S}_{un}$.

Theorem 6. If Assumptions 2 and 3 hold, then $\mathcal{S}$ is asymptotically stabilizable if and only if $\mathcal{S}_c$ is asymptotically stabilizable and $\mathcal{S}_{un}$ is asymptotically stable.

The result above clearly links to the classical *Kalman decomposition* of linear systems. We now show that the Kalman decomposition–based stabilizability characterization of linear systems can be extended to switching systems. We first need to introduce a particular class of controls. A control strategy φ is said to be a *static hybrid linear state feedback*, if for any discrete state $q_i \in Q$, there exists a matrix $K_i \in \mathbb{R}^{m \times n_i}$ such that:

$$\varphi(\eta|_{[0,t]}) = K_i x(t, j),$$
$$\eta(t) = (q_i, x(t, j)).$$

A switching system $\mathcal{S}$ is said to be *asymptotically stabilizable via static hybrid linear state feedback* if it is asymptotically stabilizable and the stabilizing control strategy is a static hybrid linear state feedback.

The following result shows that, under appropriate assumptions, the switching system $\mathcal{S}$ is asymptotically stabilizable if and only if $\mathcal{S}_{un}$ is asymptotically stable.

Proposition 7. If Assumptions 2 and 3 hold and if

$$\mathcal{G} = \mathcal{R}, \tag{2}$$

then $\mathcal{S}$ is asymptotically stabilizable if and only if $\mathcal{S}_{un}$ is asymptotically stable. Moreover, in this case, $\mathcal{S}$ is asymptotically stabilizable via static hybrid linear state feedback.

Even if condition (2) is not satisfied, some conditions on the switching system are given in [4], under which asymptotic stabilizability of $\mathcal{S}$ is implied by asymptotic stability of $\mathcal{S}_{un}$.

We conclude this section by removing Assumptions 2 and 3. We illustrate our result by means of a procedure that reduces step by step the computational effort required for checking stabilizability of linear switching systems.

In the following, 'controllable location' means a discrete state $q_i \in Q$ whose associated linear system $S(q_i)$ is controllable. A strongly connected component of the linear switching system $\mathcal{S}$ is a linear switching subsystem, whose FSM is a strongly connected component of the FSM associated with $\mathcal{S}$; such a system will be called maximal when its discrete state space is the maximal subset of Q having the property above.

Given a linear switching system $\mathcal{S} = (\Xi, \Theta, S, E, R)$, define the restriction of $\mathcal{S}$ to a subset Q' of Q as a linear switching system:

$$\mathcal{S}' = (\Xi', \Theta, S', E', R'),$$

where:

$$\Xi' = \bigcup_{q_i \in Q'} \{q_i\} \times \mathbb{R}^{n_i};$$
$$S'(q) = S(q), \ \forall q \in Q';$$
$$E' = \{(q_i, \sigma, q_h) \in E : q_i, q_h \in Q'\};$$
$$R'(e) = R(e), \ \forall e \in E'.$$

Removing locations in $Q'' \subset Q$ from $\mathcal{S}$ means defining the restriction of $\mathcal{S}$ to $Q' = Q \backslash Q''$.

Procedure (Stabilizability-based Reduction)

(1) Given a linear switching system $\mathcal{S}$, let Q_1 be the set of discrete states $q \in Q$ such that $S(q)$ is not controllable.
(2) If $Q_1 = \varnothing$ then **STOP**: $\mathcal{S}$ **is controllable.** Otherwise let $\mathcal{S}_1$ be the restriction of $\mathcal{S}$ to Q_1.
(3) Compute the maximal strongly connected components $\mathcal{F}_i$, $i \in J^1$, of $\mathcal{S}_1$ ($\mathcal{S}_1$ is asymptotically stabilizable if and only if each $\mathcal{F}_i$ is asymptotically stabilizable [7]); let $J_{\mathcal{F}_i}$ be the index set associated with the discrete states of $\mathcal{F}_i$, for any $i \in J^1$.
(4) Compute the invariant subspace $\mathcal{G}^{(i)} = \bigcup_{h \in J_{\mathcal{F}_i}} \{q_h\} \times \mathcal{G}_h^{(i)}$, for each strongly connected component $\mathcal{F}_i$. Let $\mathcal{S}_c^{(i)}$ be the controlled subsystem of $\mathcal{F}_i'$, $i \in J^1$, where $\mathcal{F}_i'$ is obtained by removing the locations q_h with $\mathcal{G}_h^{(i)} = \{0\}$ from $\mathcal{F}_i$.
(5) If $\mathcal{S}_c^{(i)}$ is not asymptotically stabilizable for some $i \in J^1$, then **STOP**: $\mathcal{S}$ **is not asymptotically stabilizable.**
(6) Remove the locations q_h, $h \in J_{\mathcal{F}_i}$, for which $\mathcal{G}_h^{(i)} = \mathbb{R}^{n_h}$ (for any execution with initial discrete state q_h the hybrid state remains in $\mathcal{G}^{(i)}$, for any control action. Since $\mathcal{S}_c^{(i)}$ is asymptotically stabilizable, then q_h can be removed [7]). Let Q_2 be the reduced discrete state space.
(7) If $Q_2 = \varnothing$ then **STOP**: $\mathcal{S}$ **is asymptotically stabilizable.** Otherwise let $\mathcal{S}_2$ be the restriction of $\mathcal{S}_1$ to Q_2.
(8) Compute the maximal strongly connected components $\widetilde{\mathcal{F}}_i$, $i \in J^2$, of $\mathcal{S}_2$.

(9) Compute the invariant subspace $\widetilde{\mathcal{G}}^{(i)}$, for each $\widetilde{\mathcal{F}}_i$. Let $\widetilde{\mathcal{S}}_{un}^{(i)}$, $i \in J^2$ be the uncontrolled subsystems of $\widetilde{\mathcal{F}}_i$.

(10) **STOP**: Return $\left\{\widetilde{\mathcal{S}}_{un}^{(i)}, i \in J^2\right\}$.

On the basis of the procedure above we can give the last result that generalizes Theorem 6 to the case where Assumptions 2 and 3 are not satisfied. Since controllability implies stabilizability, and controllability is easy to check (cf. Remark 2), in the following theorem we assume that $\mathcal{S}$ is not controllable.

Theorem 8. A noncontrollable linear switching system $\mathcal{S}$ is asymptotically stabilizable if and only if the linear switching system $\mathcal{S}_c^{(i)}$ is asymptotically stabilizable $\forall i \in J^1$ and the linear switching system $\widetilde{\mathcal{S}}_{un}^{(i)}$ is asymptotically stable $\forall i \in J^2$.

This last theorem decomposes the problem of checking stabilizability of a given linear switching system into simpler subproblems. In particular, the given system is decoupled into controlled and autonomous linear switching subsystems. The asymptotic stabilizability of the first ones and the asymptotic stability of the latter ones imply the asymptotic stabilizability of the given system.

Remark 4. The decoupling of Theorem 8 into controlled and autonomous subsystems is not possible, in general, in the case of switched systems, since the transitions are controlled. In fact, for the special class of switched systems where the continuous dynamical systems share the same matrix A, i.e. $A(q) = A, \forall q \in Q$, it was shown in [13] that the asymptotic stability of $\mathcal{S}_{un}$ (that in fact reduces to an autonomous linear system) implies the stabilizability of the given switched system.

5. CONCLUSIONS

In this paper, we considered linear switching systems and proposed some state space decompositions, based on hybrid invariant subspaces, which yield a complexity reduction in checking stabilizability. The given system is decoupled into controlled and autonomous linear switching subsystems. The asymptotic stabilizability of the first ones and the asymptotic stability of the latter ones imply and is implied by the asymptotic stabilizability of the given system.

REFERENCES

[1] Basile G., Marro G., Controlled and conditioned invariant subspaces in linear system theory, *J. Optim. Theory Appl.* 3(5), 1969.

[2] Branicky M.S., Multiple Lyapunov function and other analysis tools for switched and hybrid systems. *IEEE Trans. on Automatic Control*, 43:475–482, 1998.

[3] De Santis E., Di Benedetto M.D., Berardi L., Computation of maximal safe sets for switching linear systems, *IEEE Trans. on Automatic Control*, 49(2):184–195, 2004.

[4] De Santis E., Di Benedetto M.D., Pola G., Can linear stabilizability analysis be generalized to switching systems?, Proc. of Mathematical Theory of Networks and Systems (MTNS 04), Leuven (Belgium), July 5–9, 2004. (also available from www.diel.univaq.it/tr/web/web_search_tr.php).

[5] De Santis E., Di Benedetto M.D., Pola G., Detectability based state space reductions for hybrid systems, 17–th International Symposium on Mathematical Theory of Network and Systems, Kyoto, Japan, July 24–28, 2006.

[6] De Santis E., Di Benedetto M.D., Pola G., Digital Idle Speed Control of Automotive Engine: A Safety Problem for Hybrid Systems, Nonlinear Analysis, Special Issue Hybrid Systems and Applications, 2006. To Appear.

[7] De Santis E., Di Benedetto M.D., Theory and computation of discrete state space decompositions for a class of hybrid systems, (submitted) (also available from www.diel.univaq.it/tr/web/web_search_tr.php), 2006.

[8] De Santis E., Di Benedetto M.D., Pola G., Stabilizability based state space reduction for hybrid systems, (also available from www.diel.univaq.it/tr/web/web_search_tr.php), 2006.

[9] Liberzon D., Switching in Systems and Control, Birkhauser, Boston, MA, Volume in series Systems and Control: Foundations and Applications. ISBN 0-8176-4297-8, June 2003.

[10] Lygeros J., Tomlin C., Sastry S., Controllers for reachability specifications for hybrid systems, *Automatica*, Special Issue on Hybrid Systems, 35:349–370, 1999.

[11] Morse A.S., Supervisory control of families of linear set–point controllers– part 1: exact matching. *IEEE Trans. on Automatic Control*, 41(10):1413–1431, 1996.

[12] Pola G., van der Schaft A.J., Di Benedetto M.D., Equivalence of Switching Linear Systems by Bisimulation, *International Journal of Control*, 79(1):74–92, January 2006.

[13] Sun Z., Ge S.S., Switched Linear Systems – Control and Design, Communication and Control Engineering Series, Springer, 2005.

[14] Ye H., Michel A.N., Hou L., Stability theory for hybrid dynamical systems. *IEEE Trans. on Automatic Control*, 43:461–474, 1998.

Copyright © Analysis and Design of Hybrid Systems
Alghero, Italy 2006

ELSEVIER
IFAC
PUBLICATIONS

REACHABILITY COMPUTATION FOR UNCERTAIN PLANAR AFFINE SYSTEMS USING LINEAR ABSTRACTIONS

Othman Nasri
Marie-Anne Lefebvre
Hervé Guéguen

Supélec - IETR
BP 81127 Cesson-Sévigné Cedex, France

Abstract: Reachability computation is the central problem arising in the verification of hybrid or continuous systems. One approach, among others, to compute an over approximation of the reachable space is to split the continuous state space and to abstract the continuous dynamics in each resulting cell by a linear differential inclusion for which the reachable space may be computed with polyhedra. A previous work proposed to use characteristics of the affine continuous dynamics to guide the polyhedral partition. This paper presents an extension of this approach to uncertain planar systems where one parameter of the model may take its value in a polytope. It is shown that the result for all values of the parameter may be deduced from the computation for a finite number of values. An algorithm that performs the reachability computation and determines the minimum number of values of the parameter required at each step is proposed and exemplified. *Copyright © 2006 IFAC*

Keywords: Hybrid systems, reachability, abstractions

1. INTRODUCTION

Reachability computation is the central problem in the verification of hybrid or continuous systems (Guéguen and Zaytoon, 2004) and has become a major research issue in hybrid systems. Most approaches to solve this problem are based on a combination of numerical integration and geometrical algorithms (Girard, *et al.*, 2006; Henzinger, *et al.*, 2000). However it is also possible to use hybridization methods to perform this computation. The basic idea, introduced by (Henzinger, *et al.*, 1998), consists in splitting the continuous state space into cells and abstracting the continuous dynamics in each cell, by a linear differential inclusion for which the reachable space may be computed with polyhedra (Frehse, 2005). One key point is then to find a trade off between the number of cells that are introduced and the accuracy of the over-approximation. The choice of the hyperplanes that define the cells is also important and it is possible to use structural properties of the continuous dynamics to guide this choice (Lefebvre and Guéguen, 2006).

For affine systems, defined by equation (1), it is then possible (Lefebvre and Guéguen, 2006) to use left eigenvectors of matrix $\mathbf{A}$ to define the hyperplanes that split the continuous regions. This approach leads to interesting results but is limited to systems where the model is exactly known.

$$\dot{\mathbf{x}} = \mathbf{A}\mathbf{x} + \mathbf{b} \tag{1}$$

This paper is a first step towards the extension of the proposed approach to models where the parameter $\mathbf{b}$ of equation (1) is unknown and takes its values in a polytope. As a first attempt, it only considers planar systems where matrix $\mathbf{A}$ is non singular. In the next section, the approach for certain planar systems is briefly presented in order to illustrate the basic ideas. The principles and important properties for uncertain systems are presented in section 3. The proposed algorithm to compute an over-approximation of the reachable space is then presented in section 4. Finally the application of the algorithm to an example is shown in section 5.

2. ABSTRACTION BASED REACHABILITY FOR PLANAR REGULAR CERTAIN SYSTEMS

The dynamics of the systems considered in this section is specified by equation (1) where the dimension of the state space is two, matrix $\mathbf{A}$ is not singular and vector $\mathbf{b}$ is perfectly known. It is then possible to define the equilibrium point of the system by equation (2).

$$\mathbf{x}_e = \mathbf{A}^{-1}\mathbf{b} \tag{2}$$

This dynamics is associated with a polytopic region of the state space, denoted *Inv*, and the aim of the reachability calculation is to compute the set of points of this region that may be reached by the dynamics, from a given region denoted *Init*, namely the set:

$$\{\mathbf{x} / \exists \mathbf{x}_0 \in Init, \exists t \geq 0 \; s.a. \; \mathbf{x} = \Phi(t) \; with \; \Phi(0) = \mathbf{x}_0 \; and \; \forall \tau \leq t \; \dot{\Phi}(\tau) = \mathbf{A}\Phi(\tau) + \mathbf{b} \; and \; \Phi(\tau) \in Inv\} \tag{3}$$

The initial region *Init* will be considered as a polytope and it is then possible to compute the reachable space as the convex hull of the reachable space from each of its vertices. As it is difficult to have an explicit representation of the set defined by (3), hybridization methods aim at abstracting the continuous dynamics by polytopic differential inclusions on cells defined by splitting the region *Inv* with hyperplanes.

When applied to the category of systems considered in this paper, the method described in (Lefebvre and Guéguen, 2006) leads to consider families of hyperplanes, orthogonal to some vectors $\{\mathbf{q}_l\}$ generated by linear combinations of two left eigenvectors of matrix $\mathbf{A}$ if they are real, or two given vectors otherwise. Vectors $\{\mathbf{q}_l\}$ are chosen and ordered in such a way that each cell is defined by the intersection of the region *Inv* with the sector defined by (4).

$$\mathbf{q}_i^T(\mathbf{x}-\mathbf{x}_e) \geq 0 \wedge \mathbf{q}_{i+1}^T(\mathbf{x}-\mathbf{x}_e) \leq 0 \tag{4}$$

It is then possible to characterize the vector field in each point of this cell by (5), which can be used to abstract the dynamics. In (5), vectors γ are defined with respect to the vector field on the boundaries of the cell and are characterized by (6).

$$\gamma_i^T \dot{\mathbf{x}} \geq 0 \wedge \gamma_{i+1}^T \dot{\mathbf{x}} \leq 0 \tag{5}$$

$$\gamma_l = \left(\mathbf{A}^T\right)^{-1} \mathbf{q}_l \tag{6}$$

The computation of the reachable space from a point $\mathbf{x}_0$ within the cell is then straightforward and its result is given by the conjunction of (4) and (7).

$$\gamma_i^T(\mathbf{x}-\mathbf{x}_0) \geq 0 \wedge \gamma_{i+1}^T(\mathbf{x}-\mathbf{x}_0) \leq 0 \tag{7}$$

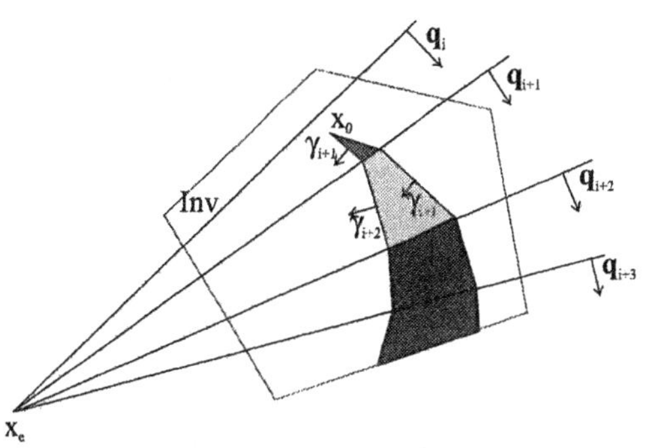

Fig 1. First steps of the reachable space computation with abstraction

Another simplification is induced by the choice of the separating hyperplanes: if the vector $\mathbf{q}_i$ is not a left eigenvector, the boundary it defines is crossed in only one direction, and if $\mathbf{q}_i$ is a real left eigenvector, then the corresponding hyperplane cannot be crossed. It is then possible, to choose the vectors $\{\mathbf{q}_l\}$ so that $\mathbf{q}_i$ characterizes the boundary through which the state trajectories enter the cell and $\mathbf{q}_{i+1}$ the one through which they leave it. So, state trajectories evolve from the cell specified by $(\mathbf{q}_i, \mathbf{q}_{i+1})$ to the one specified by $(\mathbf{q}_{i+1}, \mathbf{q}_{i+2})$. The first steps of the reachability computation are illustrated in figure 1. The intersection of the polyhedron defined by (7) with the outgoing boundary (specified by $\mathbf{q}_{i+1}$) has two vertices that are used as starting points for the computation within the next cell.

3. BASIC PRINCIPLES FOR UNCERTAIN SYSTEMS

It is now assumed that the dynamics of the system is still characterized by equation (1) but that vector $\mathbf{b}$ is unknown and may be characterized by equation (8), where $\mathbf{b}_0$ and $\mathbf{b}_1$ are 2 known vectors and α is the unknown parameter. It is also supposed that the value of this parameter is fixed. The problem is then to find a method to compute an over-approximation of the reachable space from an initial point whatever the value of α is.

$$\mathbf{b}_\alpha = (1-\alpha)\mathbf{b}_0 + \alpha\mathbf{b}_1 \; with \; \alpha \in [0,1] \tag{8}$$

When the system is characterized by equations (1) and (8), it is possible to associate an equilibrium point to each value of α. This point may be computed from the equilibrium points associated to $\mathbf{b}_0$ and $\mathbf{b}_1$ according to equation (9).

$$\mathbf{x}_{e_\alpha} = (1-\alpha)\mathbf{x}_{e_0} + \alpha\mathbf{x}_{e_1} \tag{9}$$

It is then possible to associate to each value of α the set of cells defined by the set of vectors $\{\mathbf{q}_l\}$, that are the intersection of the region *Inv* with the region characterized by (10). From now on, each of these cells that depends on the value of α and on the pair $(\mathbf{q}_i, \mathbf{q}_{i+1})$ will be denoted by $S_{i,\alpha}$ and its boundaries by $I_{i,\alpha}$ and $I_{i+1,\alpha}$ (11).

$$\mathbf{q}_i^T(\mathbf{x}-\mathbf{x}_{e_\alpha}) \geq 0 \wedge \mathbf{q}_{i+1}^T(\mathbf{x}-\mathbf{x}_{e_\alpha}) \leq 0 \qquad (10)$$

$$I_{i,\alpha} = \{\mathbf{x} / \mathbf{q}_i^T(\mathbf{x}-\mathbf{x}_{e_\alpha}) = 0\} \qquad (11)$$

For each point of the cell $S_{i,\alpha}$, the vector field may be abstracted by a differential inclusion in the region, characterized by equation (5), that does not depend on the value of α but only on the pair $(\mathbf{q}_i, \mathbf{q}_{i+1})$. The basic principle of the algorithm proposed below is based on the property that the differential inclusion associated to a cell $S_{i,\alpha}$ does not depend on the value of α but only on the index i.

It is then possible to express three properties that are useful to design an algorithm to compute an over-approximation of the reachable space.

Property 1. If α_0 and α_1 are such that, for all values of α between α_0 and α_1, an incoming point $\mathbf{x}_\alpha$ within the region $S_{i,\alpha}$ is aligned with the incoming points $\mathbf{x}_0$ and $\mathbf{x}_1$ within the regions S_{i,α_0} and S_{i,α_1}, the relative outgoing points of the reachable space within $S_{i,\alpha}$ are also aligned with the relative outgoing points of S_{i,α_0} and S_{i,α_1}.

This property can be illustrated by figure 2 and proved by considering that it is possible to find $\beta \in [0,1]$ such that $\alpha = \beta\alpha_0 + (1-\beta)\alpha_1$. As $\mathbf{x}_0 \in I_{i,\alpha_0}$, $\mathbf{x}_\alpha \in I_{i,\alpha}$ and $\mathbf{x}_1 \in I_{i,\alpha_1}$, it is possible to deduce from (11) and (9) that if $\mathbf{x}_0$, $\mathbf{x}_\alpha$ and $\mathbf{x}_1$ are aligned then

$$\mathbf{x}_\alpha = \beta\mathbf{x}_0 + (1-\beta)\mathbf{x}_1$$

Then it is possible to compute the outgoing point by $\mathbf{q}_{i+1}^T(\mathbf{y}_\alpha - \mathbf{x}_{e_\alpha}) = 0 \wedge \gamma_i^T(\mathbf{y}_\alpha - \mathbf{x}_\alpha) = 0$ and check that the solution is given by $\mathbf{y}_\alpha = \beta\mathbf{y}_0 + (1-\beta)\mathbf{y}_1$ where $\mathbf{y}_0$ and $\mathbf{y}_1$ are the relative outgoing points of sectors S_{i,α_0} and S_{i,α_1}.

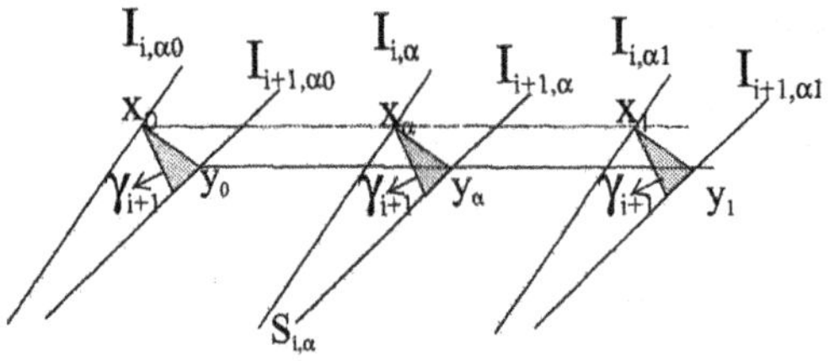

Fig 2. Characteristics of outgoing point y_α with respect to incoming points and value of α.

Property 2. If α_0 and α_1 and $(\mathbf{q}_i, \mathbf{q}_{i+1})$ are such that the assumption of property 1 is verified, then the over approximation of the reachable space from the set of entering points $\mathbf{x}_\alpha$ is

$$\bigcup_{\alpha \in [\alpha_0, \alpha_1]} A_{i,\alpha}(\mathbf{x}_\alpha) = convexhull\left(A_{i,\alpha_0}(\mathbf{x}_0), A_{i,\alpha_1}(\mathbf{x}_1)\right) \qquad (12)$$

where $A_{i,\alpha}(\mathbf{x}_\alpha)$ is characterized by the conjunction of (10) and (13).

$$\gamma_i^T(\mathbf{x}-\mathbf{x}_\alpha) \geq 0 \wedge \gamma_{i+1}^T(\mathbf{x}-\mathbf{x}_\alpha) \leq 0 \qquad (13)$$

This property is directly deduced from property 1. It is very important because it allows to compute the union of reachable region for a continuous variation of the parameter α from the computation for two values.

Property 3. If α_0 and α_1 are such that $\mathbf{x}_{init} \in S_{i,\alpha_0} \cap S_{i,\alpha_1}$ then $\forall \alpha \in [\alpha_0, \alpha_1]$ $\mathbf{x}_{init} \in S_{i,\alpha}$ and the vertices of the intersection of $A_{i,\alpha}(\mathbf{x}_{init})$ with $I_{i+1,\alpha}$ are aligned for all values of α between α_0 and α_1.

The first part of this property is proved by considering that if $\alpha = \beta\alpha_0 + (1-\beta)\alpha_1$ then $\mathbf{q}_i^T(\mathbf{x}_{init} - \mathbf{x}_{e\alpha}) = \beta\mathbf{q}_i^T(\mathbf{x}_{init} - \mathbf{x}_{e\alpha_0}) + (1-\beta)\mathbf{q}_i^T(\mathbf{x}_{init} - \mathbf{x}_{e\alpha 1})$ is positive and that $\mathbf{q}_{i+1}^T(\mathbf{x}_{init} - \mathbf{x}_{e\alpha})$ is negative with equivalent considerations.

The second part of the property is illustrated in figure 3 and is proved by considering that these vertices are the intersection of $\gamma_i^T(\mathbf{x}-\mathbf{x}_{init}) = 0$ (or $\gamma_{i+1}^T(\mathbf{x}-\mathbf{x}_{init}) = 0$), that does not depend on α, with $I_{i+1,\alpha}$.

So if α_0 and α_1 are such that the assumption of property 3 is verified, then the incoming points in the next region $S_{i+1,\alpha}$ are all aligned and property 2 can be used to compute the reachable region within this set of cells. This can be iterated as long as the considered regions do not intersect the boundary of *Inv* and leads to the algorithm presented in the next section.

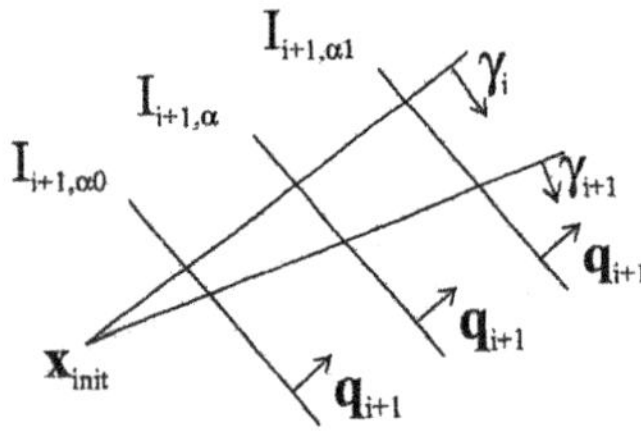

Fig 3. Initialisation

4. REACHABILITY ALGORITHM

This algorithm aims at computing an over approximation of the reachable space from an initial point that is valid for all values of the unknown parameter. It is based on properties 2 and 3 of the previous section that allow to compute the global set with a continuous variation of the parameter from the computation for a finite number of values of this parameter. It is summarized in algorithm 1.

Algorithm 1: Global reachability calculus

Input: ($\mathbf{A}$, $\mathbf{b}_0$, $\mathbf{b}_1$) continuous dynamics with the domain of the unknown parameter $\mathbf{b}$, *Inv* the considered region of the state space, $\mathbf{x}_{init}$ the initial point,

Result: $R(\mathbf{x}_{init})$ the over approximation of the reachable space from $\mathbf{x}_{init}$, within the region *Inv*, with respect to the dynamics constraint and valid for all the values of the parameter.

Step 1: decomposition of the region of interest.

Step 2: computation of the initial value of the index *i* of the cells $S_{i,\alpha}$ and of the initial set of values for the parameter α and initialisation of incoming points.

Repeat

Step 3: computation of the reachable region within the set of cells $S_{i,\alpha}$ for the index *i*: $\bigcup_k R_{i,k}$.

Step 4: computation for the next iteration of the index of cells $S_{i,\alpha}$, of the set of pertinent values of the parameter α, and of the incoming points.

Until *end*

4.1 Initialization phase

The initialization of the calculus consists in the two first steps of algorithm 1. The first step may be summarized by the interface of algorithm 2. The details for this step may be found in (Lefebvre and Guéguen, 2006). The important point is that the vectors are chosen and the set $\{\mathbf{q}_l\}$ is ordered so that cells are characterized by (10) and $\mathbf{q}_i$ characterizes the border of cell $S_{i,\alpha}$ through which trajectories enter the cell.

Algorithm 2: decomposition

Input: ($\mathbf{A}$, $\mathbf{b}_0$, $\mathbf{b}_1$), *Inv*,

Result: $\{\mathbf{q}_l\}$ the set of vectors defining the cells, $\{\gamma_l\}$ the set of vectors defining the differential inclusions and $(\mathbf{x}_{e0}, \mathbf{x}_{e1})$ the equilibrium points associated to $\mathbf{b}_0$ and $\mathbf{b}_1$.

The step 2 of the algorithm is based on property 3 of the previous section. It is summarized by algorithm 3.

Algorithm 3: initialisation of the loop

Input: $\{\mathbf{q}_l\}$, $(\mathbf{x}_{e0}, \mathbf{x}_{e1})$, $\mathbf{x}_{init}$,

Result: V_alp_init the ordered set of values of the parameter α that will have to be considered, the initial index *i*, *V_alp* the initial set of values of the parameter to be considered in the next step, for each value of the parameter the associated incoming points.

Step 1: computation of *V_alp_init*.

Step 2: for each value of *V_alp_init*, computation of the index *j* such that $\mathbf{x}_{init} \in S_{j\alpha}$ and initialisation of the index *i* to the minimum of these indexes.

Step 3: choice of values of *V_alp_init* such that $\mathbf{x}_{init} \in S_{i\alpha}$ to initialise *V_alp* and for each value of *V_alp* initialisation of the incoming points $\mathbf{p}_{1,\alpha}$ and $\mathbf{p}_{2,\alpha}$ to $\mathbf{x}_{init}$.

Step one consists in computing for each vector $\mathbf{q}_l$ the value of $\alpha \in [0,1]$, if it exists, such that $x_{init} \in I_{l\alpha}$ and to add this value to *V_alp_init* that is initialized to $\{0,1\}$. At the end of this step, two consecutive values in *V_alp_init* verify assumption of property 3.

4.2 Iterative phase

The first step of the iterative phase (step 3) is the computation of the reachable region within the cells. Firstly, for each value of *V_alp*, the region $A_{i,\alpha}$ characterized by (14), is computed. Then, property 2 is used to compute for each pair (α_k, α_{k+1}) of consecutive values in *V_alp*, $R_{i,k}$ (15) the reachable region within $S_{i,\alpha}$ for $\alpha \in [\alpha_k, \alpha_{k+1}]$. Finally the value of $R(\mathbf{x}_{init})$ after this iteration is computed (16).

$$A_{i,\alpha} = convexhull(A_{i,\alpha}(\mathbf{p}_{1,\alpha}), A_{i\alpha}(\mathbf{p}_{2,\alpha})) \quad (14)$$

$$R_{i,k} = convex\_hull(A_{i,\alpha_k}, A_{i,\alpha_{k+1}}) \cap Inv \quad (15)$$

$$R(\mathbf{x}_{init}) = R(\mathbf{x}_{init}) \cup \left(\bigcup_k R_{i,k} \right) \quad (16)$$

Step 4 is the preparation for the next iteration and may be summarized by algorithm 4.

Algorithm 4: preparation for the next iteration

Input: *V_alp*, $\{A_{i,\alpha}\}$ the set of reachable regions for each value in *V_alp*, $\mathbf{q}_{i+1}$, *Inv*, *V_alp_init*.

Result: *i* the index of the cells, *V_alp* the set of values of the parameter, and for each value the associated incoming points, $\mathbf{p}_{1,\alpha}$ and $\mathbf{p}_{2,\alpha}$.

Step 1: for each element of *V_alp*, computation of $O_\alpha = A_\alpha \cap I_{i+1,\alpha}$ the set of possible outgoing points.

Step 2: for all pairs (α_k, α_{k+1}) of consecutive elements of *V_alp* computation of Out_k the convex hull of O_{α_k} and $O_{\alpha_{k+1}}$.

Step 3: for all regions Out_k computation of Int_k the intersection with the region *Inv*.

If Int_k**=**Out_k nothing

If $Int_k=\emptyset$ the relative values of α are deleted from *V_alp*

else computation of the values of α such that the vertices of Int_k belongs to $I_{i+1,\alpha}$ and insertion of these values in *V_alp*.

Step 4: if $\mathbf{q}_{i+1}$ is a left eigenvector, setting of *V_alp* to empty. Increment of index *i*.

Step 5: computation for all elements of *V_alp* of the incoming points $(p_{1,\alpha}, p_{2,\alpha})$ in the cell $S_{i,\alpha}$.

Step 6: if $\mathbf{x}_{init} \in S_{i\alpha}$ for some value of *V_alp_init*, insertion of the relevant value to *V_alp* and setting of the associated incoming points to $\mathbf{x}_{init}$.

Step 7: for each three-tuple $(\alpha_j, \alpha_{j+1}, \alpha_{j+2})$ of consecutive values in *V_alp*, if the 3 points $(p_{1,\alpha_j}, p_{1,\alpha_{j+1}}, p_{1,\alpha_{j+2}})$ are aligned, and so are the points $(p_{2,\alpha_j}, p_{2,\alpha_{j+1}}, p_{2,\alpha_{j+2}})$, then deletion of the value α_{j+1} from *V_alp*.

Step 3 of this algorithm 4 is central in the approach because it ensures that at each step all the pairs of consecutive elements of the set *V_alp* verify the assumptions of property 2. The calculus for continuous variation of α from the computation for a finite number of values of this parameter, performed at step 4 of the global algorithm, is then valid. The case when the intersection Int_k is neither empty nor equal to Out_k is illustrated in figure 4. As the outgoing domain Out_k intersects the boundary of the region *Inv*, in the next step the incoming points for all values of α between α_1 and α_2 will not be aligned. A new value α* is then introduced, such that for all α between α_1 and α* on the one hand, and between α* and α_2 on the other hand, the incoming points are aligned and property 2 can be used in the next step. Step 7 of algorithm 4 ensures that the set of values *V_alp* is minimal as it deletes intermediate values that are not mandatory to guaranty this line condition.

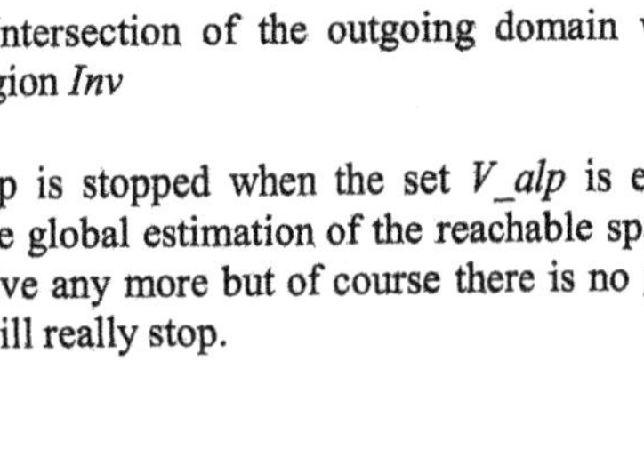

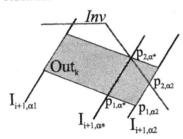

Fig 4. Intersection of the outgoing domain with the region *Inv*

The loop is stopped when the set *V_alp* is empty or when the global estimation of the reachable space does not evolve any more but of course there is no guaranty that it will really stop.

5. EXAMPLE

In order to illustrate this algorithm, its application to the computation of the reachable space from the initial point $x_{init}^T = [4 \quad 5]$ for the system specified by the following values is considered. For this system, matrix **A** has two real left eigenvectors that are used to generate the decomposition and 15 vectors $\mathbf{q}_i$ are considered.

$$\mathbf{A} = \begin{bmatrix} 0 & 1 \\ -4 & -5 \end{bmatrix}, \; \mathbf{b}_0 = \begin{bmatrix} 0 \\ 4 \end{bmatrix}, \; \mathbf{b}_1 = \begin{bmatrix} -1 \\ 19 \end{bmatrix}$$

$$Inv: \begin{bmatrix} 1 & 0 \\ -1 & 0 \\ 0 & 1 \\ 0 & -1 \\ 3 & -2 \end{bmatrix} \mathbf{x} \le \begin{bmatrix} 10 \\ -10 \\ 10 \\ -1 \\ 15 \end{bmatrix}$$

Step 2 of algorithm 1 then computes that the algorithm is initialized with:

$$V\_alp\_init = \{0, 0.25, 0.9077, 1\}$$
$$i = 4$$
$$V\_alp = \{0.9077, 1\}$$

The first two iterations of the reachability calculus are shown in figure 5. The results for the first iteration is shown in figure 5.a. For this step, the considered cells are $S_{4,\alpha}$ for the 2 values of the parameter α. As for $\alpha = 0.9077$, $\mathbf{x}_{init}$ belongs to the outgoing border of the cell, the reachable space within this cell is only this point and the global reachable space is computed for $\alpha = 1$. For the second iteration, the considered cells are $S_{5,\alpha}$ and there are 3 values for α. For $\alpha = 0.25$ the reachable space is one point, for $\alpha = 0.9077$ it is computed from $\mathbf{x}_{init}$, and for $\alpha = 1$ it is computed from the vertices stemming from the previous iteration. The polytopes A_α for this iteration are represented in figure 5.b and the reachable region for all values of α between 0.25 and 1 (R_1) is drawn in figure 5.c.

The result of the global computation is displayed in figure 6. In this case, the global computation stops and the reachable region is bounded by a line defined by the equilibrium points of the system for all possible values of α.

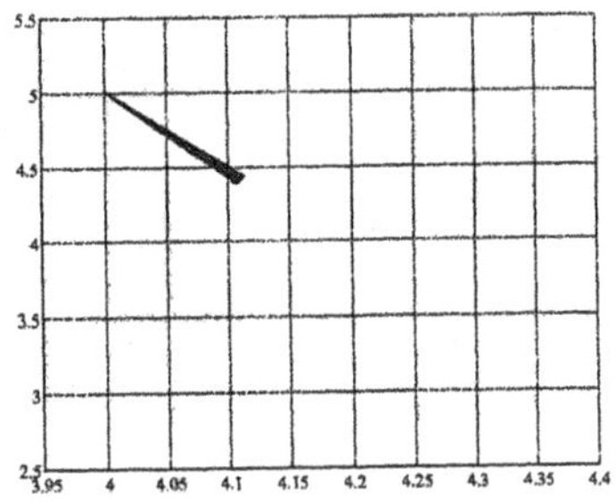

a)

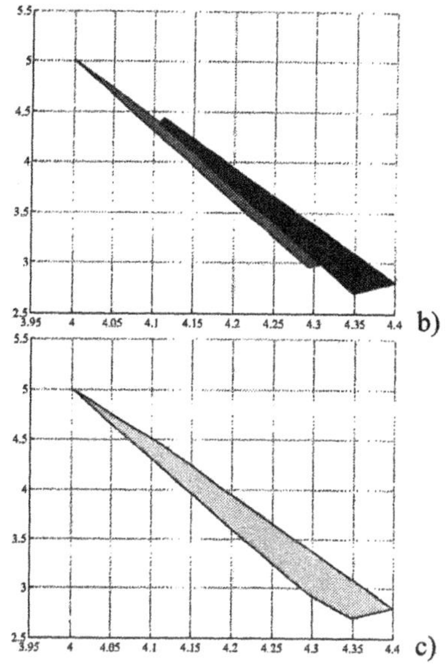

Fig. 5. First iterations of the reachability Computation

The sets of values of α that are considered at each iteration of the calculus are summarized in table 1. Three phases may be seen. In a first phase (iterations 1, 2, 3) at each iteration a new value from the initial set computed at step 2 of the global algorithm is added. During the second phase (iterations 4, 5, 6), the set does not change. Then, at the end of iteration 6, the boundary of the region *Inv* is crossed for some values of α as shown in figure 7. The values of α associated to the vertices of Int_1 and Int_2 are then introduced by step 3 of algorithm 4. From this iteration to the end of the computation, the algorithm adapts the values of α according to this step 3 and to step 7 of algorithm 4.

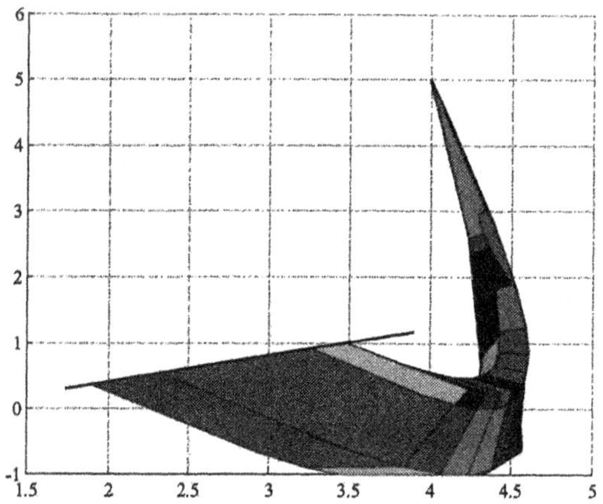

Fig. 6. Global reachable region

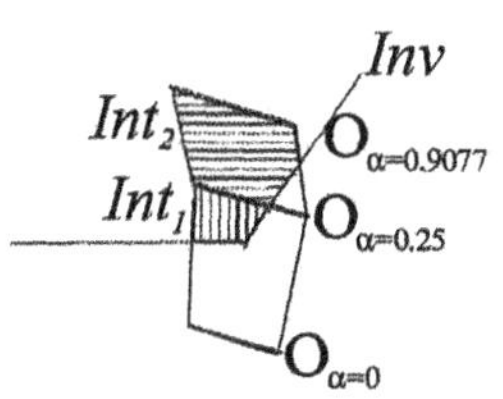

Fig. 7 Details of the first intersection with the boundary of *Inv*.

Table 1 Values of parameters α with respect to the calculus index

1	0.9077	1				
2	0.25	0.9077	1			
3	0	0.25	0.9077	1		
4	0	0.25	0.9077	1		
5	0	0.25	0.9077	1		
6	0	0.25	0.9077	1		
7	0.1275	0.1667	0.25	0.3653	0.9077	1
8	0.2971	0.3725	04987	0.9077	1	
9	0.3611	0.4848	0.5231	0.9077	1	
10	0.3611	0.5329	0.9077	1		
11	0.3611	0.5329	09077	1		
12	0.3611	0.5329	0.9077	1		
13	0.3611	0.5329	1			

CONCLUSION

In this paper, a method for reachability analysis of uncertain affine systems has been presented. It extends a previous work which allows the reachable space to be easily computed when the dynamics of the system under study is completely known. This approach also proved to be suited to uncertain systems as it allows to compute the over-approximation of the reachable space for uncertain systems where a parameter takes its values in a continuous space using a finite number of particular values. The set of these values is adapted at each iteration in order to limit the over-approximations of the reachable space to those linked, on the one hand, to uncertainty and, on the other hand, to the abstraction technique.

In this work, it has been assumed that the unknown parameter was fixed. Future work will consider reachability computation when this parameter may evolve with time, for example to take into account systems with bounded inputs (Girard, *et al.*, 2006).

REFERENCES

Frehse, G. (2005). PHAver: Algorithmic verification of hybrid systems past HyTech. in *HSCC2005: Hybrid Systems Computation and Control* (M. Morari and L. Thiele (Eds.))

Girard, A., C. Le Guernic and O. Maler (2006). Efficient computation of reachable sets of linear time-invariant systems with imputs. in *HSCC 2006:Hybrid Systems Computation and Control*

Guéguen, H. and J. Zaytoon (2004). On the formal verification of hybrid systems. *Control Engineering Practice,* **12,** 1253-1267

Henzinger, T., P. Ho and H. Wong-Toi (1998). Algorithmic analysis of non-linear hybrid systems. *IEEE Trans on Automatic Control,* **43,** 540-554

Henzinger, T., B. Horowitz, M. R. and W.-T. H. (2000). Beyond Hytech: Hybrid systems analysis using interval numerical methods. in *HSCC2000: Hybrid Systems Computation and Control* (N. Lynch and B. Krogh (Eds.))

Lefebvre, M.-A. and H. Guéguen (2006). Hybrid Abstraction of Affine Systems. *Nonlinear Analysis: Hybrid Systems and Applications, Elsevier,* doi:10.1016/j.na.2005.12.016,

ELSEVIER
IFAC
PUBLICATIONS

EXACT DIFFERENTIATION VIA SLIDING MODE OBSERVER FOR SWITCHED SYSTEMS

H. Saadaoui*, M. Djemaï*, N. Manamanni, T. Floquet***, J-P. Barbot***

* *Equipe Commande des Systèmes (ECS), ENSEA, 6 Av. du Ponceau, 95014 Cergy-Pontoise Cedex, France.*
{djemai,saadaoui,barbot}@ensea.fr
** *CReSTIC, University of Reims, Moulin de la Housse BP 1039, 51687 Reims cedex 2 - France,*
noureddine.manamanni@univ-reims.fr
*** *LAGIS, UMR CNRS 8146, Ecole Centrale de Lille, Cité Scientifique, BP 48, 59651 Villeneuve d'Ascq Cedex, France,* floquet@ec-lille.fr

Abstract: The main topic of this paper is the problem of observer synthesis for switched systems, which includes, as a specific case, the design of observers based on high order sliding mode technique. High order sliding mode is used to overcome the occurring chattering phenomena which induces some irrelevant decision of switching between the subsystems when the trajectory is in the neighborhood of the switching manifold. Moreover, in this paper, after presenting the general structure of the step by step differentiator, well show the step by step finite time convergence of the estimation error and the discrete state estimation. Two simulation examples illustrate the efficiency of the proposed approach.

Keywords: Hybrid system, Switched systems, Non linear observer, Sliding mode, Exact differentiator, Finite time convergence.

1. INTRODUCTION

Switched systems are a class of Hybrid Systems (HS) which consist of several subsystems that switch according to a given switching law (Antsaklis, 2000).

A rich and thorough bibliography deals with stability problems of switched systems, see (Branicky, 1998), (Liberzon, 2000), (Michel and Sun, 2003) and references therein. More recently, various researchers have studied observability and observer design for such systems. Some sufficient geometrical conditions to analyze the observability of hybrid dynamical systems were given in (Boutat et *al.*, 2004). These conditions are refined for the particular class of piecewise linear and nonlinear systems. The so-called extended joint observability matrix was proposed in (Vidal et *al.*, 2003), to analyze the observability of jump linear systems. In (Sontag, 1979), Sontag introduced a set of observability related definitions and examined the implications among the various concepts of observability. In the same way, other works deal with the hybrid observer design.

Indeed, in (Balluchi et *al.*, 2002) a methodology was presented for the design of dynamical observers of hybrid systems that reconstruct the discrete state and the continuous state from the knowledge of the continuous and discrete outputs. The design of linear observers for a class of linear hybrid systems was addressed in (De laSen and

Luo, 2000). Two observer prototypes based on the prediction errors were proposed.

Despite an abundant literature on the design of linear observers for hybrid systems, only few works are concerned with the design of nonlinear hybrid observers for hybrid systems (see for example (Lin et *al.*, 2002)and (Pettersson, 2005)).

Within this context, the problem of designing sliding mode observers for non linear hybrid systems without jump was discussed in (Djemaï et *al.*, 2005) and (Saadaoui et *al.*, 2005). Nevertheless, when the trajectory is in the neighborhood of the switching manifold, a chattering phenomena occurs and induces some irrelevant decision of switching between the subsystems. To overcome this problem, it is proposed to design high order sliding modes (Fridman and Levant, 1996).

Hence, the main purpose of this paper lies in nonlinear observer design for a class of HS. The considered class is assumed to be bounded state in finite time without, jumps and without Zeno phenomenon. The observer design is discussed by using a triangular input observer form introduced in (Barbot et *al.*, 1996) and (Drakunov and Utkin, 1995). The idea consists in using the step by step procedure. Another contribution of the paper lies in the convergence analysis of the estimation error in the general case.

The paper is organized as follows: Section 2 recalls some observability notions for hybrid systems. The high order sliding mode observer design and the convergence analysis are detailed in section 3. In section 4, two illustrative examples are discussed to show the performances of the developed algorithm.

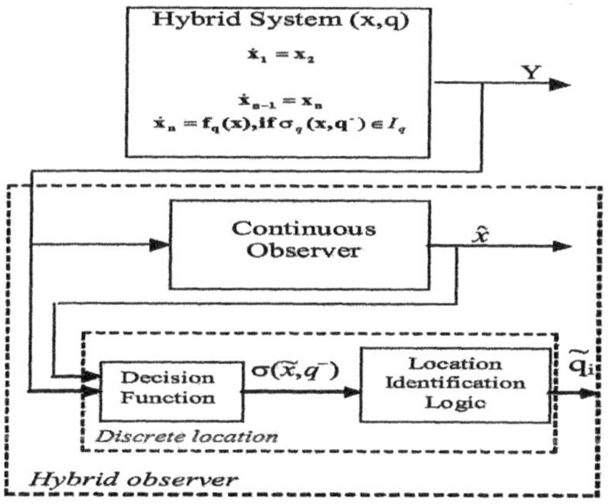

Fig. 1 : Hybrid observer structure

2. RECALLS ON OBSERVABILITY STUDY

Recently, many researchers have approached the study of observability of hybrid systems in general and switched systems in particular. In (Vidal et *al.*, 2003), was considered the case of autonomous switched systems. A definition of observability based on the concept of indistinguishability of continuous initial states and discrete states evolutions from the output in free evolution was given. In (Boutat et *al.*, 2004), both cases of linear and nonlinear switched systems were considered and some algebraic and geometrical conditions of observability for such class were stated.

In what follows is recalled the main result of (Boutat et *al.*, 2004), on the observability of the class of hybrid system considered in this paper. The proof of the theorem can be found in the cited reference. Let us consider the dynamical systems formed with two dynamics interconnected by a switching function:

$$\begin{cases} \dot{x} = F_1(x) \text{ and } y = h_1(x) \text{ if } \sigma(x) \le 0 \\ \dot{x} = F_2(x) \text{ and } y = h_2(x) \text{ if } \sigma(x) > 0 \end{cases} \tag{1}$$

where $F_i(x)$ are smooth vector fields, $h_i(x)$ are smooth outputs and $\sigma(x)$ is a smooth switching function.

Assumption 1. We assume throughout this paper that:

a) All the evolution duration of each subsystem of (1) are measurable.
b) For $i = 1 : 2$ the codistribution:

$$\left\{dh_i, dL_{F_i}h_i, ..., dL_{F_i}^{(n-1)}h_i\right\}$$

has rank n, this implies that is locally weakly observable.

Assumption 1 a) means that systems with Zeno phenomenon are not considered.

Under assumption 1, if we know which of the subsystem evolves, one can conclude on the observability of the global system (1). Hence, when considering the observability coordinates (z^j, $j = 1 : 2$) defined by:

$$z_{i+1}^j = L_{F_j}^{(i)} h_j \qquad \text{for } 0 \le i \le n-1$$

where $L_{F_j}^{(i)} h_j$ is the ith Lie derivative of h_j in the direction of F_j, and using the Fliess's observability canonical form, each subsystem of (1) can be written as:

$$\begin{cases} \dot{z}_i^1 = z_{i+1}^1 & \text{for } i = 1 : n-1 \\ \dot{z}_n^1 = g_1(z_1^1, z_2^1, ..., z_n^1) \end{cases} \tag{2}$$

if $\sigma_1 := \phi^{-1}(z_1^1, z_2^1, ..., z_n^1) \le 0$, and

$$\begin{cases} \dot{z}_i^2 = z_{i+1}^2 & \text{for } i = 1 : n-1 \\ \dot{z}_n^2 = g_2(z_1^2, z_2^2, ..., z_n^2) \end{cases} \tag{3}$$

if $\sigma_2 := \phi^{-1}(z_1^2, z_2^2, ..., z_n^2) > 0$.

The approach to analyze the observability of (1), presented in (Boutat et *al.*, 2004), is based on the comparison of g_1 and g_2 on the one hand and σ_1 and σ_2 on the other hand. For this, we need to evaluate such functions in terms of the same variables. These variables are given naturally by the output y and its successive time derivatives $y^{(i)} = \frac{d^i y}{dt^i}$ for $i = 1 : n-1$.
Let us consider the two submanifolds:

$$\mathcal{M} = \{v \in \mathbb{R}^n \ / \ g_1(v) = g_2(v)\}$$
$$\mathcal{S} = \{v \in \mathbb{R}^n \ / \ \sigma_1(v) = \sigma_2(v)\}$$

and finally, the submanifold of common singularities of subsystems of system (1):

$$\mathcal{L} = \{x \in \mathbb{R}^n \ / \ F_1(x) = F_2(x) = 0\}$$

The main result is recalled in the following theorem.

Theorem 1. i) If $\mathcal{M}$ is a discrete set then system (1) is observable for any switch σ for which we have $\sigma(\mathcal{L}) \leq 0$ or else $\sigma(\mathcal{L}) > 0$.
ii) If dynamics (2) and (3) are transverse to $\mathcal{M}$ except on a discrete subset then the system is observable for any switch σ for which we have $\sigma(\mathcal{L}) \leq 0$ or else $\sigma(\mathcal{L}) > 0$.
iii) If $\mathcal{S} = \mathbb{R}^n$ then system (1) is observable.

The reader can refer to (Boutat et *al.*, 2004) for proof and more details. Some algebraic sufficient conditions on the observability of piecewise linear systems can also be found.

3. HYBRID OBSERVER

In this paper will be designed a step by step sliding mode observer. The idea consists in using the concept of equivalent vector (see (Drakunov, 1992) and (Drakunov and Utkin, 1995)) in an iterative way. Both systems (2) and (3) are in the so-called canonical observer form of a nonlinear autonomous system:

$$\begin{cases} \dot{x}_1 = x_2 \\ \dots \\ \dot{x}_{n-1} = x_n \\ \dot{x}_n = f_q(x) \quad if \quad \sigma_q(x) \in I_q \end{cases} \tag{4}$$

where $y = x_1$, $\sigma_q(x)$ for $q = 1, .., p$ is a switching function and I_q is a domain of validation of a subsystem q. The considered class of systems is assumed to be bounded state in finite time without jumps and does not concern Zeno phenomena.

The consideration of such class of systems (4) is not restrictive. In fact, most of lagrangian systems, for example, are written in the considered form.

Remark 1. The assumption of a canonical form (4) without jump is quite restrictive, because the diffeomorphism linked to the sub-system $q = 1$ is not generally the same than the diffeomorphism linked to the sub-system $q = 2$. Then at each switch from $q = 1$ to 2, the state in the observability canonical form (4) ((Fliess, 1990)) jump even if the state in the original (1) coordinates does not jump. This jump may be easily taken into account by a new structure of sliding mode which will be given in a forthcoming paper.

In (Djemaï et *al.*, 2005), a step by step first order sliding mode observer was mainly employed for the following reasons: the finite time convergence and the ability to take naturally into account the variable structure of the HS. Nevertheless, some difficulties occur due to the chattering phenomena. It induces some irrelevant decision of switching between the subsystems when the trajectory is in the neighborhood of the switching manifold. This problem was bypassed by using a low pass filter during the computation of the equivalent vector; unfortunately, this solution introduces a delay. In this work, a relevant solution for the case of switched systems is given. It consists in using exact and robust second order sliding mode differentiators (Super Twisting Algorithm, see (Fridman and Levant, 1996)).

The "Super Twisting Algorithm" (Figure 2) is given by the following structure:

$$\sum\nolimits_{obs} = \begin{cases} u(e_1) = u_1 + \lambda_1 \left|e_1\right|^{\frac{1}{2}} sign(e_1) \\ \dot{u}_1 = \alpha_1 sign(e_1) \end{cases} \tag{5}$$
$$\lambda_1, \alpha_1 > 0$$

where $e_1 = x_1 - \hat{x}_1$ and λ_1, α_1 are positive

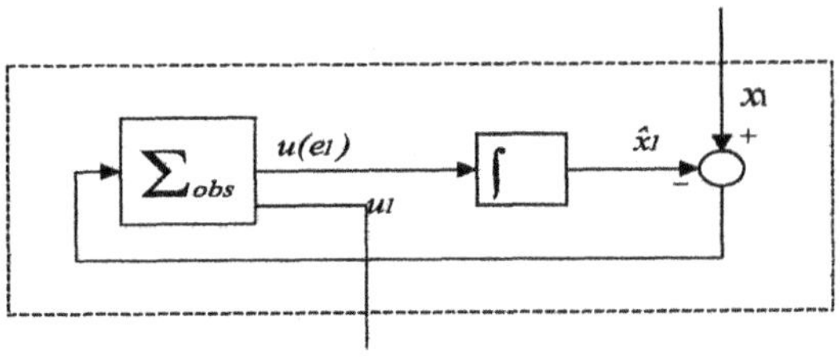

Fig. 2 : Super Twisting Algorithm Structure

parameters, and u_1 is the differentiator output where:

$$sign(e_1) = \begin{cases} +1 & \text{if} \quad e_1 > 0 \\ -1 & \text{if} \quad e_1 < 0 \\ \in [-1, 1] & \text{if} \quad e_1 = 0 \end{cases}$$

The step by step exact differentiator applied to (4), leads to the following form:

$$
\begin{cases}
\dot{\hat{x}}_1 = \tilde{x}_2 + \lambda_1 |\tilde{e}_1|^{1/2} sign(\tilde{e}_1) \\
\dot{\tilde{x}}_2 = \alpha_1 sign(e_1) \\
\dot{\hat{x}}_2 = E_1 \left[\tilde{x}_3 + \lambda_2 |\tilde{e}_2|^{1/2} sign(\tilde{e}_2)\right] \\
\dot{\tilde{x}}_3 = E_1 \alpha_2 sign(\tilde{e}_2) \\
\quad \vdots \\
\dot{\hat{x}}_{n-1} = E_{n-2} \left[\tilde{x}_n + \lambda_{n-1} |\tilde{e}_{n-1}|^{1/2} sign(\tilde{e}_{n-1})\right] \\
\dot{\tilde{x}}_n = E_{n-2} \alpha_{n-1} sign(\tilde{e}_{n-1}) \\
\dot{\hat{x}}_n = E_{n-1} \left[\tilde{\theta} + \lambda_n |\tilde{e}_n|^{1/2} sign(\tilde{e}_n)\right] \\
\dot{\tilde{\theta}} = E_{n-1} \alpha_n sign(\tilde{e}_n)
\end{cases} \tag{6}
$$

where $\tilde{e}_i = \tilde{x}_i - \hat{x}_i$, with $\tilde{x}_1 = x_1$ for $i = 1, .., n$, and the E_i for $i = 1, ...n-1$ are defined as

$$E_i = 0 \text{ if } \tilde{e}_i = \tilde{x}_i - \hat{x}_i \neq 0, \quad \text{else} \quad E_i = 1 \tag{7}$$

In practice

$$E_i = 0 \text{ if } \tilde{e}_i = \tilde{x}_i - \hat{x}_i > \epsilon, \quad \text{else} \quad E_i = 1$$

The structure of the step by step differentiator for a system of order n in canonical form is given in figure (3), where each bloc B_i for $i = 1, ..., n-1$, is only valid when $E_i = 1$.

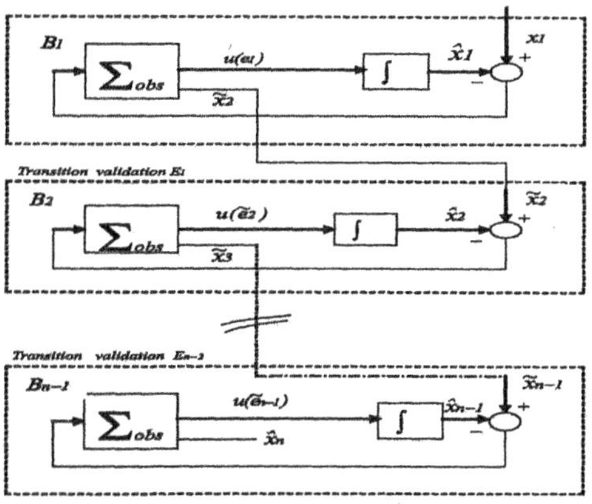

Fig. 3 : Structure of step by step differentiator

3.1 Convergence analysis

The convergence of the observation error is obtained in $(n-1)$ step in finite time. The idea consists in using the step by step observer such as described here after: The $(n-1)$ first steps consist in reconstructing the state vector and after that, under conditions of theorem 1, recovering in which state (location) q, the system evolves. Another feature of the differentiator (5) is the fact that the output u_1 does not depend directly on discontinuous functions but on an integrator output. So high frequency chattering, which can be very harmful for the system (see for example chaotic system known for its extreme sensitivity to noise (Djemaï et *al.*, 2005)), can be avoided. These properties are importants since the switching function can be obtained in a continuous way and without delays and chattering.

Theorem 2. Consider the system (4), assumed to be bounded state in finite time $t < \infty$, and the observer (6) based on the Super Twisting Algorithm (5). For any bounded initial conditions $x(0)$, $\hat{x}(0)$, there exists a choice of λ_i and α_i such that the state observer $\hat{x}$ converges in finite time $T_{fs} \ll \tau_q$ (dwell time(De Santis et *al.*, 2005)) to x and $\tilde{\theta}$ converge also in finite time to $f_q(x)$.

Proof. The proof is given in (Saadaoui et *al.*, 2005) for the case of $n = 2$ (see also (Levant , 1998) and (Davila et *al.*, 2005)). Figure 4 illustrates the finite time convergence behavior of the proposed observer. The demonstration is based on the error trajectory for each quadrant in the worst cases. In the case of $n > 2$, the convergence is ensured step by step following this order : $(\dot{e}_1 = e_2, e_1) \to (0,0)$ in finite time T_1 in the first step. $(\dot{e}_2 = e_3, e_2) \to (0,0)$ in finite time T_2 in the second step. And $(\dot{e}_i = e_i + 1, e_i) \to (0,0)$ in finite time T_i in the step i. Finally, $(\dot{e}_{n-1} = e_n, e_{n-1}) \to (0,0)$ in finite time T_{n-1} in the step $(n-1)$. The finite time convergence of the full state x is:

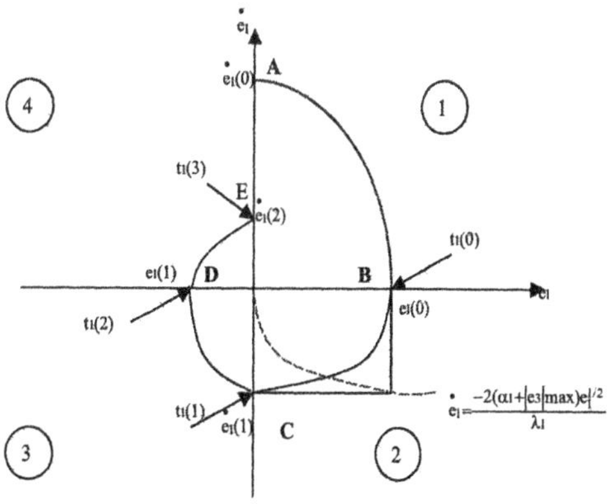

Fig. 4 : Majoring curve for finite time convergence observer

$$T_{fs} = \sum_{j=1}^{n-1} T_j \tag{8}$$

■

- Discrete time observer

This section, is concerned with the discrete part of the designed observer. Let us consider the system (4), the task of the discrete time observer is to locate which dynamic of the system is in evolution? In some cases, the knowledge of the system's output is sufficient to estimate the current location (i.e., if $\sigma(x) = \sigma(y)$). If this is not the case, some additional information obtained by using the continuous part of the observer, may be useful or are necessary to estimate the current location.

In our case, the discrete-observer receives as input: the observed state $\hat{x}$, the output y and the the information E_{n-1}. Its task is to provide an estimation $\hat{q}$ of the discrete location q of the hybrid plant at the current time. Contrarily to the general case; here, the continuous observer doesn't need to know the discrete location q. This is the main property of the canonical form (4). Indeed, the second order sliding mode observer (6) has to know only the output $y = x_1$ and also the $f_q(x)$ upper bound, noted g^+.

Thus, one can announce the following corollary:

Corollary 1. If the observer is sufficiently fast (i.e., $T_{fs} \ll \tau_q$), then for $t \geq T_{fs}$, one has $x = \hat{x}$ and $\tilde{\theta} = f_q(x)$. Then under conditions of theorem 1, the discrete state q is known.

Proof is a direct consequence of theorem 1 and 2.

4. SIMULATIONS AND COMMENTS

In (Djemaï et *al.*, 2005), when a low pass filter was used for $\tilde{x}_2$ and $\tilde{x}_3$ during the computation of the switching condition $\sigma(\tilde{x})$. The results showed a delay occurring for the switching decision and a chattering phenomenon. Indeed, the first order differentiator generates high levels of chattering even if its output is filtered through a low pass filter. A delay appears between the switching indicators S calculated on the basis of $\sigma(x)$ and S_o calculated on the basis of $\sigma(\tilde{x})$. When using sliding mode differentiators, the observer performances are presented, but the delay is completely removed and no chattering phenomenon occurs.

Example 1. Let us consider the following system put in the triangular input observer form:

$$\dot{x}_1 = x_2, \; \dot{x}_2 = x_3$$
$$\dot{x}_3 = \begin{cases} -\cos(30x_2) + 0.4 & \text{if} \quad x_2 < 0 \\ -40\cos(300x_3 + \pi/2) - 0.5 & \text{if} \quad x_2 \geq 0 \end{cases} \tag{9}$$

with $y = x_1$.

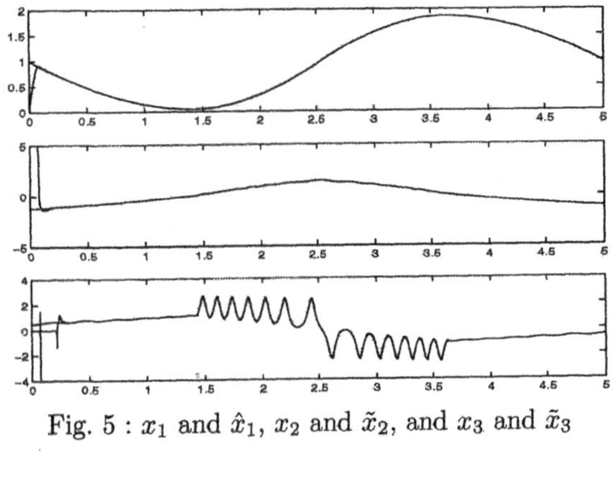

Fig. 5 : x_1 and $\hat{x}_1$, x_2 and $\tilde{x}_2$, and x_3 and $\tilde{x}_3$

The performances of the proposed hybrid observer are shown in figures 5 (estimated state, dashed line; system state, solid line).

Example 2. Let us consider the following switched systems:

$$\dot{x}_1 = x_2$$
$$\dot{x}_2 = x_3$$
$$\dot{x}_3 = \begin{cases} P_1(x) & if \quad \sigma(x) \geq 1 \\ P_2(x) & if \quad |\sigma(x)| < 1 \\ P_3(x) & if \quad \sigma(x) \leq -1 \end{cases}$$

with: $P_1(x) = -\frac{1800}{49}x_1 - \frac{55}{7}x_2 - \frac{25}{7}x_3 + \frac{2700}{49}$; $P_2(x) = \frac{100}{49}x_1 - \frac{36}{7}x_2 - \frac{6}{7}x_3$, and $P_3(x) = -\frac{1800}{49}x_1 - \frac{55}{7}x_2 - \frac{25}{7}x_3 + \frac{2700}{49}$. The switching condition: $\sigma(x) = -\frac{1}{100}(x_3 + x_2) - x_1$

The figure 6 highlights the efficiency of the proposed observer and shows the finite time step by step convergence.

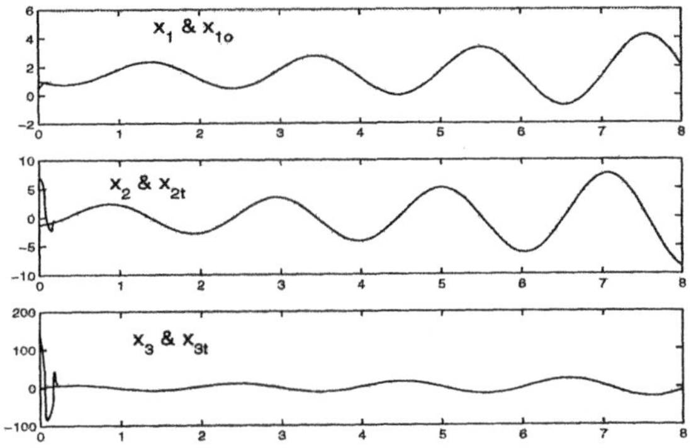

Fig. 6 : $(x_1, \hat{x}_1)$, $(x_2, \tilde{x}_2)$, and $(x_3, \tilde{x}_3)$

Figure 7 shows respectively the switching function $\sigma(x)$ and $\sigma(\tilde{x})$ and the corresponding switching indicator S, and S_o. There is no problem of delay between S (switching indicator calculated on the basis of the real states x_1, x_2, and x_3) and S_o (switching indicator calculated on the basis of the observed states x_1, $\tilde{x}_2$, and $\tilde{x}_3$). Also, it can be noted that when $\sigma(\tilde{x})$ is near -1 around $t = 5s$, there is no undesirable chattering phenomenon and no irrelevant decision of switching between the subsystems when the trajectory is in the neighborhood of the switching manifold as it was the case when classical sliding mode was used with a filter (see figure 8 and paper (Djemaï et *al.*, 2005)).

5. CONCLUSION

This paper has dealt with the design of non linear observer for hybrid systems. The considered systems concern switched non linear systems without jump and without zeno phenomenon. It was

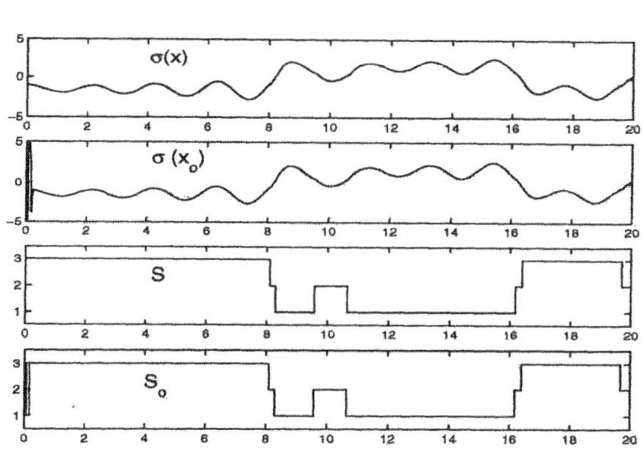

Fig. 7 : Switching surfaces $\sigma(x)$ and $\sigma(\tilde{x})$, and switching indicators S and So

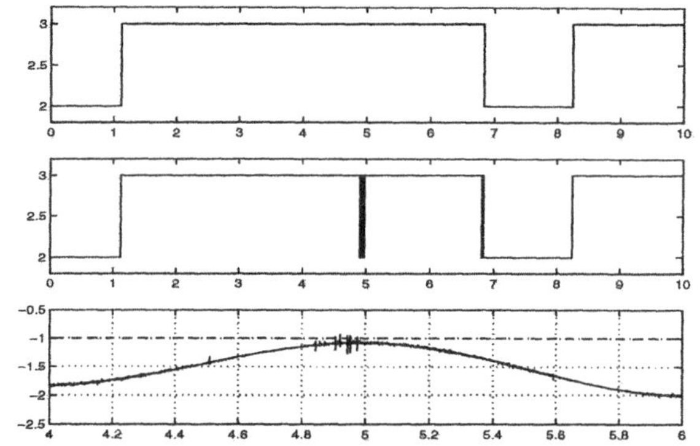

Fig. 8 : Switching indicators S and So obtained with classical sliding modes and zoom on $\sigma(\tilde{x})$

shown that the use of high order sliding mode technique; in our case exact differentiator of order 2 (Super twisting algorithm), leads to enhance the observer performances. Indeed, the robustness is guaranteed without introduction of any delays or chattering phenomena. Another contribution of the paper deals with the performed convergence analysis in the general case. We have showed that the finite time convergence of the full state is ensured and that the estimation of the discrete state is guaranteed.

REFERENCES

Antsaklis, P.J., editors (2000). *Proc. of the IEEE, Special Issue on Hybrid Systems: Theory and Applications, 43(2).*

Barbot, J.P., T. Boukhobza and M. Djemai. (1996). Sliding mode observer for triangular input form. *IEEE-CDC.*

Balluchi, A., L. Benvenuti, M.D. Di Benedetto and A.L. Sangiovanni-Vincentelly. (2002). Design of Observer for Hybrid Systems. *In LNCS 2289, pp. 77-89, Springer Verlag.*

Branicky, M.S., (1998). Multiple Lyapunov Fonctionsand Other Analysis Tools for Switched and Hybrid Systems. *IEEE Trans Auto. Cont., 43(2), pp. 461-474.*

Boutat, D., A. Benali and J.P. Barbot. (2004). About the observability of piecewise dynamical systems. *NOLCOS-2004. (CD-ROM).*

Davila, J., L. Fridman and A. Levant (2005). Second-Order Sliding-Mode Observer for Mechanical Systems.*IEEE Trans Auto. Cont., 50, pp. 41785-1790.*

Djemai, M., N. Manamanni and J. P. Barbot. (2005). Sliding Mode Observer For Triangular Input Hybrid System. *IFAC World congres Prague, 2005.*

De la Sen, M., and N. Luo. (2000). Design of linear observers for a class of linear hybrid systems. *Int. Journal of Systems Science, 31 No 9, pp. 1077-1090.*

De Santis, E., M.D. Di Benedetto, and G. Girasole. (2005). Digital Idle Speed Control of Automative Engines using Hybrid Models. *IFAC World Congress, Prague, July, 2005.*

Drakunov, S. (1992). Sliding mode observer based on equivalent control method. *IEEE-CDC, Tucson, USA, pp. 2368–2369.*

Drakunov, S., and V. Utkin. (1995). Sliding mode observer. tutorial. *IEEE-CDC, pp. 3376–3379.*

Fliess M. (1990). Generalized Controller Canonical Forms for Linear and Nonlinear Dynamics. *IEEE Trans on Automatic Control, 35, No 9, pp. 994–1000.*

Fridman, L., A. Levant. (1996), Sliding modes of higher order as a natural phenomenon in control theory. In Robust control via variable structure & Lyapunov techniques. *LNCIS 217. F. Garofalo, L. Glielmo Ed. Springer Verlag London, pp 107, 133. 1996.*

Liberzon, D.,(2003). On stabilization of linear systems with limited information. *IEEE Trans on Automatic Control, 48(2), Feb. 2003, pp. 304-307.*

Lin, L., Linawati, Lie Josa and E. Ambikarajah. (2002). A hybrid state estimation scheme for power systems. *Asian Pacific Conf. 1, pp. 555-558.*

Levant A., Robust exact differentiation via sliding mode technique, Automatica, vol. 34, pp. 379-384, 1998.

Michel, A.N., and Y. Sun. (2003). Stability analysis of discontinuous dynamical systems determined by semigroups. *IEEE-CDC, Hawai, USA, pp. 1669-1674.*

Pettersson, S. (2005). Observer Design for Switched Systems using Multiple Quadratic Lyapunov Functions. *IEEE Int. Symp. Mediterrean Conf. on Cont. and Automation, pp. 262 - 267.*

Saadaoui, H., N. Manamani, M. Djemai, J. P. Barbot and T. Floquet. (2005). Exact differentiation and Sliding mode observer for switched Lagrangian systems . *To appear in Nonlinear Analysis: Theory, Methods & Applications, special issue: Hybrid Systems and Applications, pp. 1-20, 2005.*

Sontag, E.D. (1979). On the observability of polynomial systems : finite-time problems. *SIAM J. Cont. and Optim., 17(1), pp. 139-151.*

Vidal, R., A. Chiuso, S. Soatto and S.S. Sastry. (2003). Observability of linear hybrid systems. *In Hybrid Systems: Computation and Control, LNCS in Springer Verlag, Vol. 2623, pp. 526-539.*

Copyright © Analysis and Design of Hybrid Systems
Alghero, Italy 2006

ELSEVIER
IFAC
PUBLICATIONS

ROBUST H_∞ CONTROL OF UNCERTAIN DISCRETE-TIME SWITCHING SYMMETRIC COMPOSITE SYSTEMS

Lubomír Bakule

Institute of Information Theory and Automation
Academy of Sciences of the Czech Republic
182 08 Prague 8, Czech Republic
bakule@utia.cas.cz

Abstract: Reduced–order design of decentralized H_∞ static output switching controllers is presented for uncertain discrete–time switching symmetric composite systems with state-dependent switching rule. First, a reduced-order control design model is constructed with the dimension equal to a subsystem's dimension of the original system. Then, quadratically stabilizing switching controller with a given H_∞ disturbance attenuation level is designed using linear matrix inequalities (LMI) for the design model. It is proved that when this controller together with the corresponding switching rule are implemented into each subsystem of the original system then such decentralized controller quadratically stabilizes the overall closed–loop system with H_∞ norm bound γ. The switching is decentralized into independent switching rules operating only on local subsystems states. *Copyright© 2006 IFAC*

Keywords: Large–scale systems, switching, decentralized control, discrete–time systems, uncertainty, H-infinity, reduced–order models.

1. INTRODUCTION

Multi–controller switched schemes provide an effective and powerful mechanism to cope with highly complex systems and/or systems with large uncertainties. There are real world systems which are not stabilizable by means of any individual continuous state feedback controller. For such systems, multi–controller switching among smooth controllers provides a good conceptual framework to solve the problem. As a special but very important class of switched systems, switched linear systems provide an attractive framework which bridges the gap between linear systems and the highly complex and/or uncertain systems. Switched linear systems are relatively easy to handle as many powerful tools from linear analysis are applicable to cope with these systems. Such systems are enough accurate to represent many practical engineering systems with complex dynamics. The study of switched linear systems provides additional insights to some long–standing problems, such as robust, adaptive, and intelligent control, gain scheduling, or multi–rate digital control. The recent results in switched systems has benefited many real world systems such as power systems, automotive control, air traffic control, network and congestion control.

The importance of multi–controller switched schemes is underlined in large scale complex systems when implementing low–cost low–order local controllers. It motivates the development of new control design methods which include the solution using multi–controller switched schemes mainly for large scale complex systems. Symmetric composite systems represent an important class of these systems.

1.1 *Prior work*

There are available many various results on switched systems (Leonessa *et al.* 2000), (Liberzon 2003),

[1] This work was supported in part by the Academy of Sciences of the Czech Republic under Grant A2075304 and in part by the Czech Ministry of Education, Youth, and Sport under Grant LA 282.

(Sun and Ge 2005), (Cheng 2004), (Pettersson 2003), (Varaiya 1993), (Wong and Brockett 1982), but there are only few results on uncertain switched systems. (Ji and Wang 2005) and (Savkin and Evans 2002) present results on norm bounded uncertainties. One of important problems in uncertain switched systems is the design of switching rules which guarantee quadratic stability and performance. Such switching rules must be independent of uncertainties. A state-dependent switching rule satisfying this requirement which is called the min-projection strategy presents (Ji and Wang 2005) and (Pettersson 2003). (Ji and Wang 2005) deals with robust H_∞ control for switched state feedback and static output switched feedback. All these references deal with a centralized switching rule.

Motivation for studying symmetric composite systems arises in very different application areas. Real world system examples can be found in parallel systems such as flow splitting parallel reactors with combined precooling (Hovd and Skogestad 1994), electric power systems operating in parallel (Bakule and Lunze 1988), (Lunze 1992), industrial manipulators with several degrees of freedom (Vukobratovic and Stokic 1992), flexible structures (Trächtler 1991), space crystal furnace (Ebert 1999), homogeneous interconnected systems such as seismic cables (El-Sayed and Krishnaprasad 1981). More exhaustive survey of other applications present (Hovd and Skogestad 1994), (Yang and Zhang 1996), and (Bakule 2005). Low–order control design for delay–less uncertain parameter symmetric composite system state space models consider (Hovd and Skogestad 1994), (Bakule and Rodellar 1996), and (Wang and Zhang 2000).

One of new open research directions is the inclusion of switching issues into the formulation and solution of the output switching controllers for a class of uncertain switching symmetric composite systems. The paper extends the results on the low–order control design for symmetric composite systems in (Bakule and Rodellar 1996), (Yang and Zhang 1995), (Bakule 2003). While (Bakule and Rodellar 1996), (Yang and Zhang 1995), and (Bakule 2003) include models with rank-one uncertainties, (Bakule 2005) considers norm bounded uncertainties. It essentially simplifies the solution.

To the author's best knowledge, the problem of low–order switching static output controller design with H_∞ norm bound γ including decentralized state–dependent switching rule for uncertain switching symmetric composite systems has not been solved up to now.

1.2 *Outline of the paper*

This paper presents complexity-reduced procedure for decentralized switching output controller design guaranteeing the level of disturbance attenuation γ together with decentralized state–dependent switching rule for uncertain switching discrete–time symmetric composite systems. The original system is reduced to a low–order control design problem preserving structural properties of the overall problem. Its order equals to a subsystem's order of the given global system. The switching static output controller is designed for this reduced system using the well known LMI for the selected switching rule. It is proved that when such switching controller is implemented into each subsystem together with the local switching rule, then the resulting decentralized switching controller quadratically stabilizes the overall system with a given bound of disturbance attenuation.

2. PROBLEM FORMULATION

Notation. In this paper $\|w(t)\|_2 = \sqrt{\int_0^\infty w^T(t)w(t)\,dt}$ denotes the 2-norm for $w : [0,\infty] \mapsto \mathbb{R}^p$ belonging to the space $L_2^p[0,\infty]$, provided that $\|w(t)\|_2 < \infty$.

Consider an uncertain switching linear symmetric composite system consisting of N subsystems, where the ith structural subsystem is described as follows

$$\begin{aligned} x_i(t+1) &= (A_r + \Delta A_i(t))x_i(t) + B_r u_i(t) + B_w w_i(t) \\ &\quad + s_{zi}(t) \\ v_i(t) &= C_v x_i(t) + D_v u_i(t) \\ y_i(t) &= C x_i(t) \qquad\qquad i = 1, \ldots, N \end{aligned} \tag{1}$$

where x_i, u_i, w_i, s_{zi}, v_i, and y_i are n-, m-, m_w-, p_s-, p_v-, and p_y-dimensional vectors of the subsystem states, control inputs, exogenous inputs, interconnection inputs, controlled and measured outputs, respectively. The continuous function $r = r(x_i,t) : \Re^n \times \Re^+ \to \{1,...,\kappa\}$ is the switching rule for the ith subsystem to be designed for all i. $r(x_i,t) = k$ means that the kth switching subsystem is activated for the ith structural subsystem. It is evident that there are N identical switching rules where each subsystem has assigned one local state–dependent switching rule operating independently from other rules. Interconnections are described in the form

$$s_{zi} = \sum_{j=1}^{N} L_{ij} y_{zj} \tag{2}$$

where y_{zj} is the p_z-dimensional vector of the interconnection output from the subsystem j which is related to the state vector in the form

$$y_{zj} = C_z x_j \tag{3}$$

The interconnection matrices L_{ij} have the following structure

$$L_{ii} = 0 \qquad L_{ij} = L_q + \Delta L_{qij}(t) \qquad (i \neq j) \tag{4}$$

$A_r, B_r, B_w, C, C_v, D_v$, and L_q, are constant nominal matrices. $\Delta A_i(t)$ and $\Delta L_{qij}(t)$ are norm bounded uncertainties which admit the following structure

$$\begin{aligned} \Delta A_i(t) &= D_A F_{Ai}(t) E_A \\ \Delta L_{qij}(t) &= D_L F_{Lij}(t) E_L \end{aligned} \tag{5}$$

$D_A, \ldots, E_L$ are constant matrices. Uncertainties are lumped in unknown Lebesgue measurable functions $F_{(*)}$ satisfying the bounds $F_{(*)}^T F_{(*)} \leq I$ for all $t \geq 0$. I denotes a unit matrix of appropriate dimensions.

Introduce the following inequality which is related to the system (1)–(5)

$$\|v(t)\|_2 \leq \gamma \|w(t)\|_2 \qquad (6)$$

where $v(t) = (v_1^T(t), \ldots, v_N^T(t))^T$ and analogously $w(t) = (w_1^T(t), \ldots, w_N^T(t))^T$.

Denote for (1)–(5) $x(t) = (x_1^T(t), \ldots, x_N^T(t))^T$, $u(t) = (u_1^T(t), \ldots, u_N^T(t))^T$, and $\bar{r}(x,t) = (r(x_1,t), \ldots, r(x_N,t))$. It means that the switching in the global system is realized at N different and independent places.

Definition 1. Consider the system (1)–(5). This system is *quadratically stabilizable via switched output feedback* if there exist a switching rule $\bar{r}(x,t)$ and an associated static output feedback $u_i = K_{r(x_i,t)} y_i$ with $K_k, k \in \Lambda, \Lambda = \{1, \ldots, \kappa\}$ for all i independently on all admissible uncertainties and such that the resulting closed–loop system with this feedback and $w(t) = 0$ is quadratically stable.

Definition 2. Consider the system (1)–(5). This system is *quadratically stabilizable with H_∞ disturbance attenuation via switched output feedback* if it is quadratically stabilizable via switched output feedback and under zero initial conditions the relation (6) holds for any non-zero $w(t)$ and for all admissible uncertainties.

The goal is to find a global decentralized static output feedback switching controller and a decentralized switching rule quadratically stabilizing the system (1)–(5) for any admissible uncertainties as follows

$$\begin{aligned} u_i &= K_{r(x_i,t)} y_i \\ r(x_i,t) &= \arg\min_{k \in \Lambda} (x_i^T \Omega_k x_i) \qquad i = 1, \ldots, N \end{aligned} \qquad (7)$$

where x_i is the n-dimensional controller state of the subsystem i. K_k is the constant matrix of the feedback controller acting at the kth switching mode with $k \in \Lambda$. The switching rule is selected as a quadratic form, where Ω_k is a constant matrix corresponding to the kth switching mode to be determined for each k. The set of these matrices is supposed independent of i. Notice that the set Λ of gain matrices to be determined is identical for all subsystems, thus taking advantage of the symmetric structure of the large scale composite system to reduce the control design complexity.

2.1 *The Problem*

Given a system (1)–(5) and a positive number γ. The goal is to derive a complexity-reduced procedure for designing a static output feedback decentralized switching memory–less controller and a decentralized switching rule (7) for the system (1)–(5) such that the closed–loop system (1)–(5), (7) is quadratically stable with H_∞ disturbance attenuation bound γ satisfying (6) for all admissible uncertainties.

3. MAIN RESULTS

This section presents a constructive procedure for the design of matrices K_k and Ω_k for all $k \in \Lambda$.

First, a low–order control design problem is constructed.

$$\Delta A_a(t) = D_a F_a(t) E_a \qquad (8)$$

where $D_a, \ldots, E_a$ are constant block matrices given by decomposing the matrix $\frac{N}{2} L_q C_z$ into the form

$$\frac{N}{2} L_q C_z = D_a E_a \qquad (9)$$

$F_a(t)$ in (8) is unknown Lebesgue measurable function satisfying $F_a(t)^T F_a(t) \leq I$ for all $t \geq 0$.

Define the n-dimensional system using the uncertainties (8) as follows

$$\begin{aligned} x_m(t+1) &= (A_{mr} + \Delta A_m(t)) x_m + B u_m(t) + B_w w_m(t) \\ v_m(t) &= C_v x_m(t) + D_v u_m(t) \\ y_m &= C x_m \end{aligned} \qquad (10)$$

where the nominal matrices are defined by the expressions

$$A_{mr} = A_r + (\frac{N}{2} - 1) L_q C_z \qquad (11)$$

with the uncertainties given as follows

$$\begin{aligned} \Delta A_m(t) &= D_A F_A(t) E_A + (\frac{N}{2} - 1) D_L F_L(t) E_L \\ &+ D_a F_a(t) E_a = D_{Am} F_{Am}(t) E_{Am} \end{aligned} \qquad (12)$$

$F_{(.)}(t)$ in (12) are unknown Lebesgue measurable functions satisfying standard norm bounded conditions.

Introduce the following inequality which is related to the system (10)–(12) for a given $\gamma > 0$

$$\|v_m(t)\|_2 \leq \gamma \|w_m(t)\|_2. \qquad (13)$$

Consider a static output switching stabilizing memory–less controller with the switching rule for the control problem (10)–(12) in the form

$$\begin{aligned} u_m &= K_{r(x_m,t)} y_m \\ r(x_m,t) &= \arg\min_{k \in \Lambda} (x_m^T \Omega_k x_m) \end{aligned} \qquad (14)$$

The matrices K_k for all $k \in \Lambda$ can be determined by the procedure as follows. Suppose that the matrix C in (10) has a full rank.

Theorem 3. Consider the system (10)–(12) and a positive constant γ as given in (13). The switched static controller (14) quadratically stabilizes the closed–loop system (10)–(12), (14) with H_∞ disturbance attenuation γ for all admissible uncertainties if there exist a matrix $Q > 0$, matrices N_k, V_k for all $k \in \Lambda$, and a scalar $\eta > 0$ such that for some scalars $\alpha_1, \ldots, \alpha_\kappa$ the following LMI

$$R(A_{mr}) = \begin{pmatrix} W & \Gamma & 0 & \Upsilon & \Xi \\ \bullet & \Phi & \Psi & 0 & 0 \\ \bullet & \bullet & \Theta & 0 & 0 \\ \bullet & \bullet & \bullet & -I & 0 \\ \bullet & \bullet & \bullet & \bullet & \eta I \end{pmatrix} < 0 \qquad (15)$$

and the relation

$$CQ = V_k C \qquad k \in \Lambda \qquad (16)$$

are satisfied. The blocks in (15) mean $W = \sum_{k=1}^{\kappa} \alpha_k Q$, $\Gamma = (\sqrt{\alpha_1} Q A_{m1}^T + C^T N_1^T B_1, \ldots, \sqrt{\alpha_\kappa} Q A_{m\kappa}^T + C^T N_\kappa^T B_\kappa)$, $\Upsilon = (\sqrt{\alpha_1} Q C_v^T + C^T N_1^T D_v, \ldots, \sqrt{\alpha_\kappa} Q C_v^T + C^T N_\kappa^T D_v)$, $\Xi = (\sqrt{\alpha_1} Q E_{Am}^T, \ldots, \sqrt{\alpha_\kappa} Q E_{Am}^T)$, $\Phi = \mathrm{diag}(-Q + \eta D_{Am} D_{Am}^T, \ldots, -Q + \eta D_{Am} D_{Am}^T)$, $\Psi = \mathrm{diag}(B_w, \ldots \ldots, B_w)$, $\Theta = \mathrm{diag}(-\gamma^2 I, \ldots, -\gamma^2 I)$.

The output feedback gain matrices are given by

$$K_k = \frac{1}{\sqrt{\alpha_k} N_k V_k^{-1}} \qquad (17)$$

and the switching rule in (14) has the form

$$\begin{aligned} \Omega_k = {} & (A_{mk} + B_k K_k C)^T (Q - \gamma^{-2} B_w B_w^T - \eta D_{Am} D_{Am}^T)^{-1} \\ & (A_{mk} + B_k K_k C) + \eta^{-1} E_{Am}^T E_{Am} - Q^{-1} + (C_v \\ & + D_v K_k C)^T (C_v + D_v K_k C) \end{aligned} \qquad (18)$$

Remark. Theorem 3 is based on Lemma 4 and Theorem 3 in (Ji and Wang 2005). This result is convenient for direct computations of the switching static output controller with the given switching rule. The following theorem states the main result.

Theorem 4. Given the switching symmetric composite system (1)–(5) and a positive constant γ. Construct the reduced control design system (10)–(12). Select the controller matrices K_k and Ω_k in (14) satisfying (15) and (15) for the system (10)–(12). Then implementing the matrices K_k, Ω_k into (7), the global closed loop overall system (1)-(5), (7) is quadratically stable with H_∞ disturbance attenuation γ by (6).

The proof is given in the Appendix.

Remark. A common Lyapunov function is used in Theorem 3. However, the methodology presented by Theorem 4 can be directly extended on the case of multiple Lyapunov functions to reduce the conservatism caused by a single Lyapunov function, see e.g. (Pettersson 2003).

4. CONCLUSION

The paper contributes by a new complexity–reduced control design method for low–order switching H_∞ static output feedback controllers guaranteeing the level of disturbance attenuation for a class of switching uncertain discrete–time symmetric composite systems. The structural properties of this class of large scale systems are used for the construction of low–order switching design system. The switching H_∞ output control together with the selected switching rule designed for this switching design model is consequently implemented as local identical switching controllers including local switching rules into the given original system. The procedure ensures the quadratic stability of the global switched closed–loop system with given bound of disturbance attenuation.

5. REFERENCES

Bakule, L. (2003). Decentralized control of time-delayed uncertain symmetric composite systems. In: *Proceedings of the 4th International Conference on Control and Automation.* Montreal, Canada. pp. 399–403.

Bakule, L. (2005). Complexity-reduced guaranteed cost control design for delayed uncertain symmetrically connected systems. In: *Proceedings of the American Control Conference.* Portland, Oregon, USA. pp. 2590–2595.

Bakule, L. and J. Lunze (1988). Decentralized design of feedback control for large–scale systems. *Kybernetika* **24**(3–6), 1–100.

Bakule, L. and J. Rodellar (1996). Decentralized control design of uncertain nominally linear symmetric composite systems. *IEE Proceedings-Control Theory and Applications* **143**, 630–536.

Cheng, D. (2004). Stabilization of planar switched systems. *Systems & Control Letters* **51**, 79–88.

Ebert, W. (1999). Towards delta domain in predictive control-an application to the space crystal furnace titus. In: *Proceedings of the 1999 International Conference on Control Applications.* Banff, Canada. pp. 391–396.

El-Sayed, M. and P.S. Krishnaprasad (1981). Homogeneous interconnected systems: An example. *IEEE Transactions on Automatic Control* **26**, 894–901.

Hovd, M. and S. Skogestad (1994). Control of symmetrically interconnected plants. *Automatica* **30**, 957–973.

Ji, Z. and L. Wang (2005). Robust h∞ control and quadratic stabilization of uncertain discrete-time switched linear systems. In: *Proceedings of the American Control Conference.* Portland, Oregon, USA. pp. 24–29.

Leonessa, A., W.M. Haddad and V.-S. Chellaboina (2000). *Hierarchical Nonlinear Switching Control Design with Applications to Propulsion Systems.* Springer-Verlag. London.

Liberzon, D. (2003). *Switching in Systems and Control*. Birkhäser. Boston.

Lunze, J. (1992). *Feedback Control of Large-Scale Systems*. Prentice Hall. London.

Pettersson, S. (2003). Synthesis of switched linear systems. In: *Proceedings of the IEEE Conference on Decision and Control*. Maui, Hawaii, USA. pp. 5283–5288.

Savkin, A.V. and R.J. Evans (2002). *Hybrid Dynamical Systems*. Birkhäser. Boston.

Sun, Z. and S.S. Ge (2005). *Switched Linear Systems*. Springer-Verlag. Berlin.

Trächtler, A. (1991). Entwurf strukturbeschränkter rückführungen an symmetrischen systemen. *Automatisierungstechnik* **39**, 239–244.

Varaiya, P. (1993). Smart cars on smart roads: Problems of control. *IEEE Transactions on Automatic Control* **AC-48**(1), 195–207.

Vukobratovic, M. and D.M. Stokic (1992). *Control of Manipulator Robots: Theory and Applications*. Springer-Verlag. Berlin.

Wang, Y.H. and S.Y. Zhang (2000). Robust control for nonlinear similar composite systems with uncertain parameters. *IEE Proceedings-Control Theory Applications* **147**, 80–84.

Wong, W.S. and R.W. Brockett (1982). Systems with finite communication bandwidth constraints-part 1: State estimation problem. *IEEE Transactions on Automatic Control* **AC-267**(5), 1071–1085.

Yang, G.H. and S.Y. Zhang (1995). Stabilizing controllers for uncertain symmetric composite systems. *Automatica* **31**, 337–340.

Yang, G.H. and S.Y. Zhang (1996). Robust stability and stabilization of uncertain composite systems with circulant structures. In: *Proceedings of the 13th Triennial World Congress*. San Francisco, USA. pp. 5a–02–6.

APPENDIX

The Appendix presents the proof of Theorem 4. It requires to introduce selected preliminaries. Denote the global system description of the system (1)–(5) as follows

$$\begin{aligned} x(t+1) &= (\overline{A}_r + \Delta\overline{A}(t))x(t) + \overline{B}u(t) + \overline{B}_w w(t) \\ v(t) &= \overline{C}_v x(t) + \overline{D}_v u(t) \\ y(t) &= \overline{C}x(t) \end{aligned} \tag{19}$$

where x, u, w, v, and y are $n \times N$-, $m \times N$-, $m_w \times N$-, $p_v \times N$, and $p_y \times N$-dimensional vectors of the subsystem states, control inputs, exogenous inputs, controlled and measured outputs, respectively. The continuous function $r = r(x,t) : \Re^{n \times N} \times \Re^+ \to \{1,...,\overline{\kappa}\}$ is the switching rule for the global system. The nominal matrices are defined as follows

$$\begin{aligned} &\overline{A}_r = (\overline{A}_{rij}) \quad \overline{A}_{rii} = A_r \quad \overline{A}_{ij} = L_{ij}C_z \\ &\overline{B}_r = \mathrm{diag}(B_r,...,B_r) \quad \overline{B}_w = \mathrm{diag}(B_w,...,B_w) \\ &\overline{C}_v = \mathrm{diag}(C_v,...,C_v) \quad \overline{D}_v = \mathrm{diag}(D_v,...,D_v) \\ &\overline{C} = \mathrm{diag}(C,...,C) \end{aligned} \tag{20}$$

The uncertainty terms have the form

$$\Delta\overline{A}(t) = \overline{D}_A \overline{F}_A(t) \overline{E}_A \tag{21}$$

The constant matrices are defined as follows

$$\begin{aligned} \overline{D}_A &= \mathrm{diag}(\overline{D}_1,...,\overline{D}_N) \\ \overline{D}_i &= (D_L...D_L \; D_A \; C_L...D_L) \\ \overline{E}_A &= \mathrm{diag}(\overline{E}_1 ...,\overline{E}_N) \\ \overline{E}_i &= (E_L...E_L \; E_A \; E_L...E_L) \end{aligned} \tag{22}$$

D_A is located at the ith position in $\overline{D}_i$. The uncertainty structure is lumped in uncertainty functions in the form

$$\begin{aligned} \overline{F}_A(t) &= \mathrm{diag}(F_{A1},...,F_{AN}) \\ F_{Ai} &= \mathrm{diag}(F_{Li1},... \\ &\quad ...,F_{Lii-1},F_{Ai},F_{Lii+1},...,F_{LiN}) \end{aligned} \tag{23}$$

Uncertainties $\overline{F}_A(t)$ are unknown Lebesgue measurable functions satisfying standard norm bounded conditions.

Consider a static output switching stabilizing full order controller with the switching rule for the system (19)–(23) in the form

$$\begin{aligned} u &= \overline{K}_{\overline{r}(x,t)} y \\ \overline{r}(x,t) &= \arg\min_{l \in \overline{\Lambda}} (x^T \overline{\Omega} x) \end{aligned} \tag{24}$$

where $\overline{\Lambda} = \{1,..,\kappa^N\}$. $\overline{\Omega} = \mathrm{diag}(\Omega,...,\Omega)$ with $\Omega \in \{\Omega_1,...,\Omega_\kappa\}$ which are selected according to the switching rule. To simplify the notation, the symbol Ω is used for the set of available matrices in the switching rule for any structural subsystem. It is identical for all these subsystems.

The matrices $\overline{K}_l = \mathrm{diag}(K_{r(x_1,t)},...,K_{r(x_N,t)})$ for all $l \in \overline{\Lambda}$ can be determined by the procedure as follows.

Definition 5. Consider the system (19)–(23). This system is *d-quadratically stabilizable with H_∞ disturbance attenuation via switched output feedback* if it is quadratically stabilizable with H_∞ disturbance attenuation via switched output feedback with a block diagonal gain matrix and decentralized switching rule. Each local gain has its own switching rule.

Theorem 6. Consider the system (19)–(23) and a positive constant γ as given in (6). The switched static controller (24) d-quadratically stabilizes the closed–loop system (19)–(24) with H_∞ disturbance attenuation γ for all admissible uncertainties if there exist a block diagonal matrix $\overline{Q} > 0$, block diagonal matrices

$\overline{N}_l, \overline{V}_l$ for all $l \in \overline{\Lambda}$ with $n \times n$ diagonal blocks, and a scalar $\eta > 0$ such that for some scalars $\alpha_1, ..., \alpha_{\overline{\kappa}}$ the following LMI

$$\overline{R}(\overline{A}_r) = \begin{pmatrix} \overline{W} & \overline{\Gamma} & 0 & \overline{\Upsilon} & \overline{\Xi} \\ \bullet & \overline{\Phi} & \overline{\Psi} & 0 & 0 \\ \bullet & \bullet & \overline{\Theta} & 0 & 0 \\ \bullet & \bullet & \bullet & -I & 0 \\ \bullet & \bullet & \bullet & \bullet & \eta I \end{pmatrix} < 0 \qquad (25)$$

and the relation

$$\overline{CQ} = \overline{V}_l \overline{C} \qquad l \in \overline{\Lambda} \qquad (26)$$

are satisfied. The blocks in (25) mean $\overline{W} = \sum_{l=1}^{\overline{\kappa}} \alpha_l \overline{Q}$, $\overline{\Gamma} = (\sqrt{\alpha_1} \overline{QA}_1^T + \overline{C}^T \overline{N}_1^T \overline{B}_1, ..., \sqrt{\alpha_{\overline{\kappa}}} \overline{QA}_{\overline{\kappa}}^T + \overline{C}^T \overline{N}_{\overline{\kappa}}^T \overline{B}_{\overline{\kappa}})$, $\overline{\Upsilon} = (\sqrt{\alpha_1} \overline{QC}_v^T + \overline{C}^T \overline{N}_1^T \overline{D}_v, ..., \sqrt{\alpha_{\overline{\kappa}}} \overline{QC}_v^T + \overline{C}^T \overline{N}_{\overline{\kappa}}^T \overline{D}_v)$, $\overline{\Xi} = (\sqrt{\alpha_1} \overline{QE}_A^T, ..., \sqrt{\alpha_{\overline{\kappa}}} \overline{QE}_A^T)$, $\overline{\Phi} = \mathrm{diag}(-\overline{Q} + \eta \overline{D}_A \overline{D}_A^T, ..., -\overline{Q} + \eta \overline{D}_A \overline{D}_A^T)$, $\overline{\Psi} = \mathrm{diag}(\overline{B}_w, ..., \overline{B}_w)$, $\overline{\Theta} = \mathrm{diag}(-\gamma^2 I, ..., -\gamma^2 I)$.

The output feedback gain matrices are given by

$$\overline{K}_l = \frac{1}{\sqrt{\alpha_l} \overline{N}_l \overline{V}_l^{-1}} \qquad (27)$$

and the switching rule in (24) has the form

$$\begin{aligned} \overline{\Omega}_l = {} & (\overline{A}_l + \overline{B}_1 \overline{K}_l \overline{C})^T (\overline{Q} - \gamma^{-2} \overline{B}_w \overline{B}_w^T - \eta \overline{D}_A \overline{D}_A^T)^{-1} \\ & \overline{A}_l + \overline{B}_l \overline{K}_l \overline{C}) + \eta^{-1} \overline{E}_A^T \overline{E}_A - \overline{Q}^{-1} + (\overline{C}_v \\ & + \overline{D}_v \overline{K}_l \overline{C})^T (\overline{C}_v + \overline{D}_v \overline{K}_l \overline{C}) \end{aligned} \qquad (28)$$

Consider a real $sn \times sn$ matrix $T(n,s)$ in the form

$$T(n,1) = I,$$

$$T(n,s) = \begin{pmatrix} I & 0 & \dots & 0 & I \\ 0 & I & \dots & 0 & I \\ \vdots & \vdots & \ddots & \vdots & \vdots \\ 0 & 0 & \dots & I & I \\ -I & -I & \dots & -I & I \end{pmatrix}, \qquad s > 1, \qquad (29)$$

where I denotes here $n \times n$ identical matrix. Denote

$$\begin{aligned} \overline{T}(i) &= \mathrm{diag}[T(n, N-i)I, ..., I], \quad i = 0, ..., N-1, \\ G &= \overline{T}(0)\overline{T}(1) \quad ... \quad \overline{T}(N-1). \end{aligned} \qquad (30)$$

The following theorem presents the way how to use the structural properties of symmetric composite systems to construct reduced-order systems with equivalent dynamic properties (Yang and Zhang 1995).

Theorem 7. Consider the matrix $\tilde{A}$ in the system (20)–(23) and any given $J = \mathrm{diag}[J_o, ..., J_o]$, where J, J_o are $Nn \times Nn$, $n \times n$ matrices, respectively. Then it holds

$$\begin{aligned} G^{-1} \tilde{A} G &= \mathrm{diag}(A_s, ..., A_s, A_c), \\ G^T \tilde{A} G &= \mathrm{diag}(2A_s, ..., 6A_s ... N(N-1)A_s, NA_c), \\ G^{-1} J (G^{-1})^T &= \mathrm{diag}(\frac{1}{2} J_o, \frac{1}{6} J_o, ..., \frac{1}{N(N-1)} J_o), \\ G^T J G &= \mathrm{diag}(2J_o, ..., N(N-1)J_o, NJ_o) \end{aligned} \qquad (31)$$

Proof of Theorem 4. Consider the system (19)–(23) with $A_s = A - L_q C_z$, $A_c = A_s + NL_q C_x$ and two particular cases of the uncertainties (8) such as $\Delta A_a(t) = \frac{N}{2} L_q C_z$ and $\Delta A_a(t) = -\frac{N}{2} L_q C_z$. It leads using (31) to the relations

$$A_{mr} + \Delta A_m(t) = A_s + \Delta A_a(t) + \Delta A(t) = A_c - \Delta A(t) \qquad (32)$$

Suppose that the gain matrices K_k in (17) satisfy the conditions (15)–(16) in Theorem 4 for a given matrix $Q > 0$ and a constant γ. Then by Theorem 4 it is sufficient to show that these gain matrices when implemented into the decentralized controller (7) together with the switching rule lead to quadratic stability with H_∞ disturbance attenuation γ satisfying (6) for the closed–loop system (1)–(5), 7). The notion of the quadratic stability with H_∞ disturbance attenuation γ of the system (1)–(6) is equivalent to the notion of feasible solution by Theorem 4.

Denote $\bar{G} = \mathrm{diag}(G, G, G, G, G)$, where G defined by (31). Now, we get by using Theorem 7 when applying only standard operations on all terms of the inequality (25) in Theorem 6

$$\begin{aligned} & \bar{G}^T \bar{R}(\bar{A}_r) \bar{G} \\ & = \mathrm{diag}[2R(A_s), ..., N(N-1)R(A_s), NR(A_c)] \end{aligned} \qquad (33)$$

If $R(A_{mr}) < 0$ holds by Theorem 3, then also $R(A_s) < 0$ and $R(A_c) < 0$ hold because the uncertainties in (10), (12) include both systems with the matrices A_s, A_c as special cases. The matrix $\bar{G}_s$ is nonsingular. Therefore, the relation $\bar{G}^T \bar{R}(\bar{A}_r) \bar{G} < 0$ holds.

The last item concerns the relation between the switching rules (28) and (18). The way of reasoning is analogous to the transformation of LMIs in (33). It results in the relation

$$\begin{aligned} & \bar{G}^T \bar{\Omega}(\bar{A}_r) \bar{G} \\ & = \mathrm{diag}[2\Omega(A_s), ..., N(N-1)\Omega(A_s), N\Omega(A_c)] \end{aligned} \qquad (34)$$

which are included in $\Omega(A_{mr})$ as indicated in (18). It leads to N independent switching rules where one switching rule is assigned to one structural subsystems. The necessity to select scalars α_l in Theorem 6 reduces accordingly to κ identical scalars for each structural subsystem.

Thereby, the closed loop system (1)–(5), (7) is d-quadratically stabilized with H_∞ disturbance attenuation γ satisfying (6). Q.E.D.

ELSEVIER
IFAC
PUBLICATIONS

THE ELEVATOR DISPATCHING PROBLEM: HYBRID SYSTEM MODELING AND RECEDING HORIZON CONTROL

K. S. Wesselowski, C. G. Cassandras [1]

Center for Information and Systems Engineering
Boston University

Abstract: Elevator dispatching is a combinatorially hard problem with many control constraints, time-varying traffic patterns, partial state information, and random effects. We develop a hybrid system model of a building with multiple elevator cars and apply a receding horizon control approach, bypassing some of the problem's complexities. Thus, we obtain a "universal" dispatcher which is robust with respect to changing traffic patterns and avoids the problem of having to switch among different controllers when these patterns change over the course of a day (as currently done). Moreover, simulation results show that the performance of this approach improves upon that of state-of-the-art dispatchers.

Keywords: Hybrid Systems, Receding Horizon Control, Elevator Dispatching

1. INTRODUCTION

In an elevator system, the "dispatcher" is a centralized, upper-level controller which determines where each car is to stop, either to load passengers (in response to "hall calls" at various floors) or to unload them (in response to "car calls"). Effective elevator dispatching must meet certain quality-of-service criteria and gives rise to a problem of great practical importance which remains open because of several difficulties.

A multi-car elevator system is a hybrid system that combines the time-driven motion dynamics of the cars with the event-driven dynamics imposed by the dispatching controller. The state space of such a system is enormous; as an example, a building with 18 floors and 6 cars has a state space size of the order of 10^{44}, which is approximately the same size as that of the game of chess. The dispatching problem is further complicated by numerous elevator operating constraints and by partial state information, e.g., the controller knows about the existence of a hall call at some floor, but not the number of passengers there.

Because of these difficulties, the problem is often decomposed into a set of archetypal passenger arrival patterns, with control algorithms customized for a specific pattern. For example, the *uppeak traffic pattern* occurs during the morning rush hour at a typical office building, when hall calls occur at the lobby only and result in car calls to all floors. Given a cost function based on specified quality-of-service criteria, the dispatching problem can, in principle, be solved through dynamic programming. In practice, however, this is only possible for very limited cases under specific modeling assumptions, such as the uppeak scenario (Pepyne and Cassandras, 1997) and the parking problem (Brand and Nikovski, 2004). Other dynamic programming approaches, e.g. (Nikovski and Brand, 2003), may not handle all traffic scenarios well and/or impose a high computational burden. To avoid the intractability of dynamic

[1] The authors' work is supported in part by NSF under grant DGE-0221680 and by AFOSR under grant FA9550-04-1-0208.

programming approaches, many heuristic algorithms have been developed, most based on the idea of dynamically assigning a car to serve a certain sector of the building, e.g., the Dynamic Load Balancing (DLB) algorithm in (Cassandras *et al.*, 1990) and other sectoring schemes, e.g., (Chan *et al.*, 1997). Efforts to use fuzzy logic to control switching between algorithms depending on the traffic patterns (Powell and Sirag, 1993), or attempts to implement various reinforcement learning techniques using neural networks (Siikonen, 1997), (Crites and Barto, 1996), have yielded limited success.

The goal of this work is to explore a new approach to elevator dispatching based on a formal hybrid system model in conjunction with a receding horizon controller. Our motivation is twofold. First, there is a continuing need to demonstrate that substantially better performance is achievable by new controllers relative to state-of-the-art heuristics. Second, with existing algorithms customized to specific traffic patterns, dispatchers are forced to switch among algorithms by detecting when it is optimal to do so. It is, therefore, desirable to have a "universal" dispatching controller which is equally effective in all situations and involves few, if any, tunable parameters. The approach we propose is characterized by this feature, while also showing (based on simulation experiments) substantial performance improvements compared to state-of-the-art dispatching algorithms. By using a hybrid model of the elevator system, our approach allows us to conveniently calculate the travel times of cars to various hall or car calls. Moreover, this model facilitates the use of the cooperative receding horizon control developed in (Li and Cassandras, 2006).

2. A HYBRID SYSTEM MODEL

Consider a multi-car elevator system with C cars, each with capacity P passengers, operating in an N-floor building, where floor 1 is the ground floor (or lobby) and floor N is the top floor. The system can be represented by a graph we will refer to as the *spine model* shown in Figure 1, where each floor is associated with three nodes: two spike nodes and one spine node. Letting $M = 2N$, the spike nodes are indexed by $i = 1, \dots, M$ in a clockwise direction, where nodes $i = 1, \dots, N$ on the left-hand side are associated with upward-bound passengers, and nodes $i = N+1, \dots, M$ on the right-hand side are associated with downward-bound passengers. Nodes $i = 1, \dots, N-1$ can be origins for upward-bound passengers, and nodes $i = 2, \dots, N$ can be destinations for upward-bound passengers. Similarly, nodes $i = N + 1, \dots, M-1$ and $i = N+2, \dots, M$ can be origins and destinations, respectively, of downward-bound passengers. The spine nodes, indexed by $i = M+1, \dots, M+N$ (in the center of the model in Figure 1) represent *decision points* for stopping at floors $1, \dots, N$ respectively. That is, a car reaching node $M+i$ at top speed may choose to decelerate and stop at node i (if moving upwards) or node $M - i + 1$ (if moving downward); otherwise it continues moving at top speed and skips floor i.

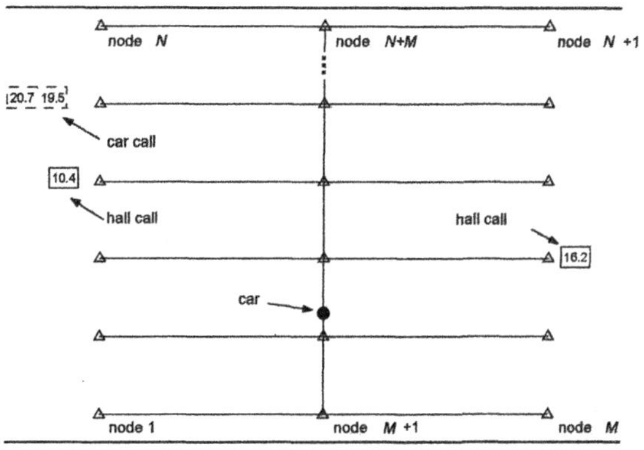

Fig. 1. The "spine model" of an elevator system.

Cars move on the arcs connecting the nodes: on vertical arcs, cars move between floors at top speed with a corresponding arc travel time D; on horizontal arcs (or "spikes"), cars decelerate as they approach a floor or accelerate away from some floor with a corresponding arc travel time d. A car located at a spike node is stopped at the corresponding floor either serving upward-bound passengers (for $i = 1, \dots, N$) or downward-bound passengers (for $i = N+1, \dots, M$). We represent hall calls and car calls by numbers in boxes next to the appropriate spike node. A hall call is generated by the first passenger at a node who presses an up or down button. Its box is bordered by a solid line and lists all passenger waiting times (note that additional passengers may arrive at the same node after the initial hall call is initiated). A car call is generated when a car answers a hall call and a passenger entering this car selects a destination floor. A car call is represented by a box bordered in a dashed line shown next to the destination node and lists all times since corresponding passengers in that car issued the call. For this model, we assume that all acceleration/deceleration times (required for a car to go between a complete stop and full speed) are fixed and equal and all top speeds are fixed and equal to a value v_{max}. For simplicity, we also assume that the vertical distance travelled by a car during acceleration is less than half the vertical distance between floors.

The state of this hybrid system consists of a continuous (real-valued) component for the C cars and a discrete (non-negative integer-valued) component for all hall and car calls present. The state of car $j = 1, \ldots, C$ is a vector $[e_j, v_j]$ where e_j is the elevation of car j and $v_j = \dot{e}_j$ is its velocity. The state of hall calls in the system is described by the hall call queue lengths $h(i)$, $i = 1, \ldots, M$ and by $a^h(i,l)$, $l = 1, \ldots, h(i)$, the arrival time of each passenger waiting at the hall call at node i. The state of car calls is described by the car call queue lengths $\psi_j(i)$, $i = 1, \ldots, M$, $j = 1, \ldots, C$, and by $a_j^c(i,l)$, $l = 1, \ldots, \psi_j(i)$, the arrival times of each passenger in car j with destination node i.

The dynamics of the system depend on an *event set* E and a *set of control actions* U imposed by the dispatcher. Events in E correspond to the arrival of cars at various decision points on the spine model, along with an exogenous event for passenger arrivals at (hall-call) nodes. The type of event that occurs dictates the feasible transitions, or decisions, for each car. There are $(6C+4)(N-1)$ events in E.

Next, we define a small set of control actions, $U = \{STOP, GOUP, GODN, LOAD\}$, which may be taken upon the occurrence of certain events. For example, a control action must be taken when some upward-moving car j arrives at a spine node $i \in \{M+2, \ldots, M+N\}$ defining a decision point. The dispatcher must select: (*i*) $STOP$ if the car is to be routed to the corresponding spike node $i \in \{2, \ldots, M\}$, or (*ii*) $GOUP$ if the car is to be routed to spine node $i+1$.

A different set of commands may be feasible when a car is at a spike node. For example, when a car finishes a loading operation at a spike node i, the dispatcher must select: (*i*) $GOUP$ (or $GODN$) if the car is to immediately depart node $i \in \{1, \ldots, N-1\}$ (or $i \in \{N+1, \ldots, M-1\}$), or (*ii*) $STOP$ if the car is to remain at node i with the doors open to possibly accept more passengers. This construction allows for the car to idle and a "timeout" to be implemented to prevent a loaded passenger from waiting more than a set amount of time before departing.

The dispatcher may issue a command to a parked (idling) car as a result of some event that is exogenous to that car. This event may be a passenger arrival event or an event associated with some car other than the parked car. In either case, the dispatcher may issue a command to the parked car to either (*i*) $LOAD$ if the event occurs at the floor where the car is located, or (*ii*) $GOUP$ (or $GODN$) if the car is to immediately depart.

Clearly, the choice of control action at a decision point depends on the full state of the system, denoted by $\mathbf{x}$, at that time. For example, if a car arrives at a spike node i with $\psi_j(i) > 0$, then j is obligated to serve the associated car calls with destination i, therefore $STOP$ must be

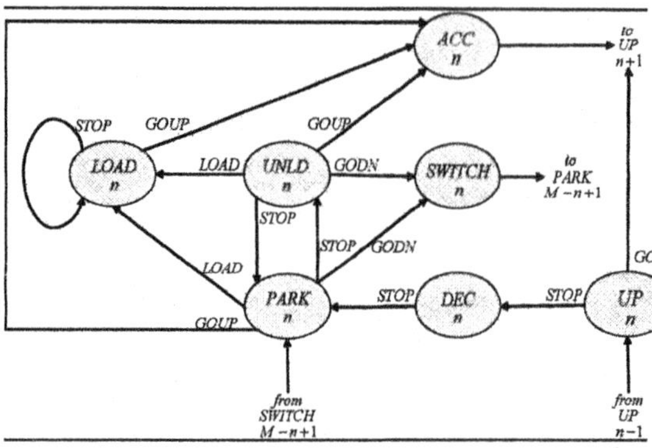

Fig. 2. State transition diagram for a car at or approaching floor $n = 2, \ldots, N-1$.

selected from the feasible control action set $U = \{STOP, GOUP, GODN, LOAD\}$. An $UNLOAD$ command is not necessary, because passengers must be allowed to depart at their destination.

The complete state transition diagram for the hybrid automaton model of this system is extremely complex and depends on the specific dispatching control scheme used. However, we can obtain a state transition diagram for each individual car based on the observation that dispatching decisions depend only on certain discrete states defining a discrete event system abstracted from the hybrid automaton. For example, the state transition diagram shown in Figure 2 corresponds to a car moving upward at some floor $n < N$. For simplicity, the control actions are shown, but not events that may precipitate transitions.

Note that all hall call and car call states are also abstracted out of this subsystem, thus allowing for all possible control actions as defined above. A state transition diagram for the entire system would include the union of C state transition diagrams, each with N subsystems as shown in Figure 2, along with all car call and hall call states. Which of these transitions and states are feasible depends on the dispatching controller defined (see next section). As an example, consider Figure 2 for some upward-moving car j with the added information that $\psi_j(n) > 0$ and $\sum_{i=n+1}^{N} \psi_j(n) > 0$, i.e., there is a car call with destination n and there are additional car calls for higher floors. In this case, for example, transitions to the $SWITCH$ state are not feasible.

3. THE RECEDING HORIZON DISPATCHING CONTROLLER

The basic idea of the Receding Horizon Controller (RHC) used for dispatching cars in an elevator system is similar in spirit to that developed in (Li and Cassandras, 2006) for the control of coordinated teams of fully-actuated autonomous vehicles moving freely in a 2-dimensional space.

Receding horizon schemes are also associated with model-predictive control (e.g., (Mayne and Michalska, 1990)). In the case of dispatching, the optimal control problem is combinatorially intractable and further complicated by random events due to uncertain passenger arrivals and stochastic loading/unloading times. Typically, dispatching algorithms assign cars to floors as new hall calls and car calls are generated. In contrast, the RHC we propose issues commands from the set $U = \{STOP, GOUP, GODN, LOAD\}$ that are applied to a car whenever it comes to a decision point as a result of the events defined in the previous section. There is no attempt to commit a car to a floor except at the last possible point possible (i.e., when the car arrives at the spine node for that floor). This allows for taking into account unexpected events that may alter a prior car-to-floor assignment. Moreover, the control action set U is much smaller than the set of all possible car-to-floor assignments for C cars and N floors.

The RHC is invoked whenever any new event occurs in the system at some time t, at which point it solves an optimization problem using all available current state information $\mathbf{x}_t$ and a given objective function defined over the time interval $[t, t + H(\mathbf{x}_t)]$, where $H(\mathbf{x}_t)$ is called the *planning horizon*. However, the control actions determined by the RHC at t are executed only until the next system event is observed (typically much sooner than $t + H(\mathbf{x}_t)$), at which point a new optimization problem is solved with all updated state information. For vehicles moving freely in a simple two-dimensional space, as in (Li and Cassandras, 2006), the notion of Euclidean distance between two points is well-defined and simple to use. In contrast, each car of an elevator system is constrained to move in the graph topology of the spine model of Figure 1. Thus, we need to adapt the RHC from a Euclidean space to a graph.

The first step is to develop an appropriate (non-Euclidean) distance metric for a graph. Let $d(r, s)$ be the *distance* between points r, s both belonging to the same arc on the graph as the time required for a car to travel from r to s. Depending on the arc in the spine model, this is simply the time to travel from r to s at speed $v_{\max}$ (spine arcs) or subject to the acceleration/deceleration parameters given (spike arcs). If $r = s$ is a node of the graph, then $d(r, r)$ is the time spent by a car at a node, e.g., while loading/unloading. This time generally depends on both the car and the node states. This motivates a car-dependent definition of the metric for nodes, $d_j(r, r, \boldsymbol{x}_t)$, which also incorporates dependence on the system state $\mathbf{x}_t$.

The *neighborhood* of a point r on the graph, $\mathcal{N}(r)$, is the set of all nodes with arcs connecting them to r. For any two points r, s, a *path* $\mathbf{p}(r, s)$ from r to s is a sequence of nodes $\mathbf{p}(r, s) = \{n_1, \dots, n_L\}$ such that $n_1 \in \mathcal{N}(r)$, $n_L \in \mathcal{N}(s)$, and $n_{l+1} \in \mathcal{N}(n_l)$, $l = 1, \dots, L-1$. We can then define the distance (or length) of a path as

$$d_j(r, s, \mathbf{x}) = d_j(r, n_1, \mathbf{x}) + \tag{1}$$
$$\sum_{l=1}^{L-1} [d_j(n_l, n_l, \mathbf{x}) + d(n_l, n_{l+1})] + d_j(n_L, s, \mathbf{x})$$

Of particular interest are paths from a car j's location at time t, denoted by c_j^t, to nodes where there are hall calls or car calls to serve. We shall refer to these as "target nodes" for car j and denote the corresponding set at time t by R_j^t. For any $r \in R_j^t$, a car-to-target node path is

$$\mathbf{P}_j^t(r) \equiv \mathbf{p}(c_j^t, r), \quad r \in R_j^t \tag{2}$$

Given the path length in (1), we define the *target time* $\tau_j^t(c_j^t, r)$ as the time when car j reaches r along that path, i.e., $\tau_j(c_j^t, r) = t + d_j(c_j^t, r, \mathbf{x})$.

Finally, each path has a corresponding *path input sequence* $\boldsymbol{\mathcal{U}}_j^t(r)$, where the ith element, $u_j(r, i)$, is the ith feasible control action involved along the path $\mathbf{P}_j^t(r)$, as a result of events that would occur as car j travels on the path to target r.

Car trajectory construction under RHC. In the simple 2-dimensional space used in (Li and Cassandras, 2006), the RHC's function is to solve an optimization problem whose solution gives optimal vehicle headings at time t. Given a vehicle speed v and a planning horizon H, the set of points on a circle of radius H/v around the vehicle's current position defines its feasible *horizon points*. The RHC determines the optimal heading that leads to some corresponding optimal horizon point, $x_H^*(t)$. The optimal RHC trajectory is a straight line to $x_H^*(t)$, and the RHC trajectory's target time for a target at point y is simply $t + H + \|y - x_H^*(t)\| / v$, where $\|\cdot\|$ is the Euclidean metric.

In our spine model setting, the set of feasible horizon points has to be defined accordingly. Given a path $\mathbf{P}_j^t(r)$ as in (2), a horizon point is a point that corresponds to the car's future position on this path at time $t + H(\mathbf{x}_t)$. Formally, we define a horizon point $\omega_j(r)$ for some target node $r \in R_j^t$ as a point uniquely given by

$$d_j(c_j^t, \omega_j(r), \mathbf{x}_t) = H(\mathbf{x}_t)$$
$$d_j(c_j^t, n_k, \mathbf{x}_t) \leq H(\mathbf{x}_t) \leq d_j(c_j^t, n_{k+1}, \mathbf{x}_t)$$

for some $n_k \in \mathbf{P}_j^t(r)$. The horizon point $\omega_j(r)$ belongs to some arc (n_k, n_{k+1}) on the path to target node r such that the distance from c_j^t to $\omega_j(r)$ is the horizon $H(\mathbf{x}_t)$. Note that it is possible for $\omega_j(r)$ to coincide with a node in $\mathbf{P}_j^t(r)$ since cars can spend a positive amount of time at a node in the spine model. The set of all horizon points for a car j is denoted by $\Omega_j^t = \cup_{r \in R_j^t} \{\omega_j(r)\}$. We can then define a path $\mathbf{p}(\omega, r)$ from any horizon point

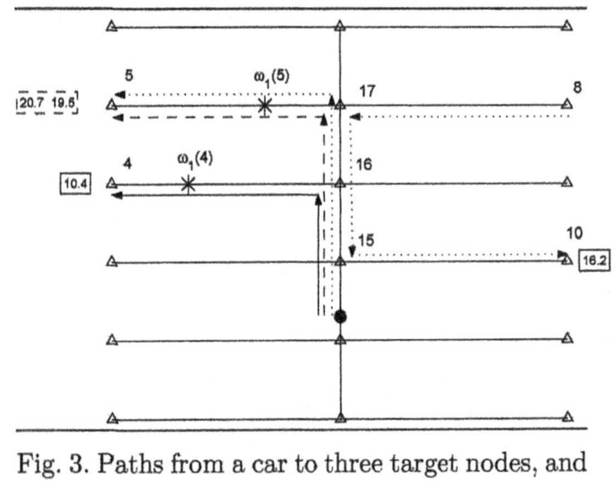

Fig. 3. Paths from a car to three target nodes, and resulting horizon points.

$\omega \in \Omega_j^t$ to any target $r \in R_j^t$ and an associated target time

$$\tau_j(\omega, r) = t + H(\mathbf{x}_t) + d_j(\omega, r, \mathbf{x}_{t+H(\mathbf{x}_t)})$$

As an example of paths from cars to target nodes, consider Figure 3, where a car is carrying passengers upward in a building with $N = 6$ floors. Since the car cannot switch directions until the car call at node 5 is served, the path to the downward-bound car call at $r = 10$ is $\mathbf{P}_1^t(10) = \{15, 16, 17, 5, 8, 17, 16, 15, 10\}$, as shown by the dotted line. Figure 3 also shows how a given planning horizon $H(\mathbf{x}_t)$ locates the horizon points $\omega_1(4)$ and $\omega_1(5) = \omega_1(10)$ in the set Ω_1^t.

The final issue we address before discussing the optimization problem that the RHC must solve is that of selecting an appropriate planning horizon $H(\mathbf{x}_t)$. Motivated by the analysis in (Li and Cassandras, 2006), we shall also choose $H(\mathbf{x}_t)$ to be the shortest distance (in time units) from any car to any target node, i.e.,

$$H(\mathbf{x}_t) = \min_{r \in R_j^t, j=1,\ldots,C} d_j(c_j^t, r, \mathbf{x}_t)$$

Intuitively, $H(\mathbf{x}_t)$ provides the earliest time at which the system workload could be reduced by any car.

RHC objective function. To construct the objective function for the RHC, we will concentrate here on minimizing the waiting times of passengers. Recall that Ω_j^t is the set of all feasible horizon points for car j at time t, and $\tau_j(\omega, r)$ is the earliest possible time that a passenger in the hall call queue of some node r can be served if car j chooses horizon point $\omega \in \Omega_j^t$. Then, the *hall-call waiting time function*, denoted by $W_j(\omega, r)$, is

$$W_j(\omega, r) = \sum_{l=1}^{h(r)} \left[\tau_j(\omega, r) - a^h(r, l)\right]$$

where r is a spike node with $h(r) > 0$ and $a^h(r, l) \leq t$ is the arrival time of the lth passenger in the hall call r queue. Similarly, the *car-call system time function*, denoted by $Y_j(\omega, r)$, is

$$Y_j(\omega, r) = \sum_{l=1}^{\psi_j(r)} \left[\tau_j(\omega, r) - a_j^c(r, l)\right]$$

where r is a spike node with $\psi_j(r) > 0$ and $a^c(r, l) \leq t$ is the arrival time of the lth passenger in the car call queue for destination r originating at some node previously visited by car j.

The *relative proximity function* $q_j(\boldsymbol{w}, r) \in [0, 1]$, where r is a hall call target, and $\boldsymbol{w} = [\omega_1, \ldots, \omega_C]$, $\omega_j \in \Omega_j^t$, $j = 1, \ldots, C$ is intended to engender cooperation between the cars. It was introduced by (Li and Cassandras, 2006), but here we must adapt it to our particular topology. The function $q_j(\boldsymbol{w}, r)$ is monotonically non-increasing in the *relative distance function*, denoted by $\delta_j(\boldsymbol{w}, r)$. It depends on how close ω_j is to target r, compared to all other horizon points. We order the elements of $\boldsymbol{w}$ so that $d_{j_1}(\omega_{j_1}, r, \boldsymbol{x}_{t+H(\boldsymbol{x}_t)}) \leq \ldots \leq d_{j_C}(\omega_{j_C}, r, \boldsymbol{x}_{t+H(\boldsymbol{x}_t)})$ and using the cars with shortest distances, j_1 and j_2, we define

$$\delta_{j_i}(\boldsymbol{w}, r) = \frac{d_{j_i}(\omega_{j_i}, r, \boldsymbol{x}_{t+H(\boldsymbol{x}_t)})}{\sum_{k=j_1,j_2} d_k(\omega_k, r, \boldsymbol{x}_{t+H(\boldsymbol{x}_t)})}$$

and $\delta_{j_i}(\boldsymbol{w}, r) = 0$ for $i = 3, \ldots, C$ if applicable. A simple and convenient choice of relative proximity function is

$$q_j(\boldsymbol{w}, r) = \begin{cases} 1 & \text{if } \delta_j \leq \Delta \\ \dfrac{(1-\Delta) - \delta_j}{1 - 2\Delta} & \text{if } \Delta < \delta_j < 1 - \Delta \\ 0 & \text{if } \delta_j \geq 1 - \Delta \end{cases}$$

where for brevity we let $\delta_j \equiv \delta_j(\boldsymbol{w}, r)$, and $\Delta \in [0, 0.5)$ is a constant adjustable to quantify the extent of car cooperation.

The RHC's function is to select $\boldsymbol{w}$, i.e., a horizon point $\omega_j \in \Omega_j^t$ for each car j, so as to minimize

$$J(\boldsymbol{w}) = \sum_j \sum_{r \in R_j^t} [W_j(\omega_j, r)]^2 q_j(\boldsymbol{w}, r) + Y_j(\omega_j, r)$$

The minimizing vector of horizon points is $\boldsymbol{w}^* = \arg\min J(\boldsymbol{w})$. Each element ω_j^* of $\boldsymbol{w}^*$ is associated with a path input sequence $\mathcal{U}_j^t(\omega_j^*)$, and the control input applied to car $j = 1, \ldots, C$ at time t is the first element $u_j(\omega_j^*, 1)$ of $\mathcal{U}_j^t(\omega_j^*)$. In the example of Figure 3, the optimal first control action is the one associated with leading the car to either $\omega_1(4)$ or $\omega_1(5)$ from its current position, whichever minimizes $J(\boldsymbol{w})$; here it happens that $u_1(\omega_1(4), 1) = u_1(\omega_1(5), 1) = GOUP$. A simple calculation shows that the size of the decision space for the RHC optimization problem is $O\big((2 + 2\lceil d/D \rceil)^C\big)$ and independent of N.

Implementation. In the actual system, the complete state is unknown. The RHC is required to estimate the number of passengers associated with each call based on the known elapsed time since the call was issued and a known (or estimated) passenger arrival rate at each node (determining these rates may itself be a difficult problem.) In addition, if $h(i) = 0$ for some i, an anticipatory element can be added to the RHC by estimating the length of a currently empty hall call queue in the future (specifically, at time $\tau_j(\omega, i)$ for some $\omega \in \Omega_j^t$). Another complication is that cars whose passenger capacity P is reached must be prevented from responding to hall calls.

4. NUMERICAL RESULTS

The RHC and several competing controllers were implemented in a simulation environment developed using MATLAB. Here, we limit the results shown to two different traffic patterns: a two-hour lunchtime traffic scenario (about 1,500 passengers delivered), with a combination of uppeak, downpeak, and interfloor traffic, and a two-hour pure uppeak case. For both tests, each car has a capacity of $P = 16$ passengers, top speed $v_{max} = 1$ m/sec, and acceleration $a = 0.4$ m/sec^2.

For the lunchtime scenario, we compare the performance of the RHC to that of the DLB algorithm of (Cassandras *et al.*, 1990) in a simulated building with $N = 10$ and $C = 4$. Table 1 shows averages over 10 simulation runs where the RHC compares favorably to the DLB algorithm. The passenger arrivals at each floor were simulated according to a time-varying Poisson process.

Table 1. Lunchtime scenario.

Controller	Avg. Waiting Time, sec	Avg. Sys. Time, sec	Avg. Sq. Sys. Time, sec^2
RHC	17.47	44.96	2,670
DLB	20.11	52.52	3,686

A comparison of the performance of the RHC to that of an Uppeak Threshold Controller (UTC) is shown in Table 2 for a simulated building with $N = 20$ and $C = 4$, using a constant Poisson arrival rate of 5 passengers per minute at the lobby. It was shown in (Pepyne and Cassandras, 1997) that the optimal dispatching policy in the pure uppeak case (with no interfloor or downpeak component) is a threshold-based policy, assuming Markovian arrival and service processes. The UTC based on this idea uses a vector $\boldsymbol{T} = [T_1, \ldots, T_C]$, where each element T_i is the threshold (number of passengers) for dispatching a car loading at the lobby when there are i cars available at the lobby. The RHC has the advantage that it uses the location information for other cars, and is able to dispatch a car from the lobby when a returning car gets close enough to place a horizon point at node 1. When applying the RHC, an expected arrival rate $\hat{\lambda}_1$ at node 1 is used to estimate future arrivals and effectively achieve a thresholding behavior. As seen in Table 2, the UTC under $T = [1, 1, 1, 1]$ and the RHC under $\hat{\lambda}_1 = 0$ psgr/min have the same behavior since they both ignore future arrivals (of course, determining a $\hat{\lambda}_1$ that achieves the best possible performance, in this case $\hat{\lambda}_1 = 0.37$ psgr/min, remains a side estimation issue). What is important is that the RHC achieves the same uppeak performance as the UTC *without switching to a specialized operating mode*, substantiating the claim that the RHC is a robust "universal" controller.

Table 2. Uppeak scenario.

Controller	Avg. Sys. Time, sec	Avg. Sq. Sys. Time, sec^2
RHC, $\hat{\lambda}_1 = 0.00$ psgr/min	62.57	5,079
RHC, $\hat{\lambda}_1 = 0.37$	58.49	4,205
UTC, $T = [1, 1, 1, 1]$	62.57	5,079
UTC, $T = [2, 1, 1, 1]$	61.26	4,707

REFERENCES

Brand, M. and D. Nikovski (2004). Optimal parking in group elevator control. In: *Intl. Conf. on Robotics and Automation.* Vol. 1. pp. 1002–1008.

Cassandras, C. G., T. E. Djaferis, J. Lewis and D. P. Looze (1990). Dispatching through dynamic load balancing (dlb): the "noontime" scenario. Tech. Report, Dept. of ECE, U. of Massachusetts.

Chan, W. L., A. T. P. So and K. C. Lam (1997). Dynamic zoning in elevator traffic control. *Elevator World* pp. 136–139.

Crites, R. H. and A. G. Barto (1996). Improving elevator performance using reinforcement learning. In: *Advances in Neural Information Processing Systems 8* (D. S. Touretzky, M. C. Mozer and M. E. Hasselmo, Eds.). MIT Press. pp. 1017–1023.

Li, W. and C. G. Cassandras (2006). A cooperative receding horizon controller for multi-vehicle uncertain environments. *IEEE Trans. on Auto. Control* **51**(2), 242–257.

Mayne, D. Q. and L. Michalska (1990). Receding Horizon Control of Nonlinear Systems. *IEEE Trans. on Auto. Control* **AC-35**(7), 814–824.

Nikovski, D. and M. Brand (2003). Marginalizing out future passengers in group elevator control. In: *Proc. of the 19th Conf. on Uncertainty in Artificial Intelligence.* pp. 443–450.

Pepyne, D. and C. Cassandras (1997). Optimal dispatching control for elevator systems during uppeak traffic. In: *IEEE. Trans. on Control Systems Technology.* Vol. 5. pp. 629–642.

Powell, B. A. and D. J. Sirag (1993). A new way of thinking about the complexities of dispatching elevators. *Elevator World* pp. 78–84.

Siikonen, M.-L. (1997). Elevator group control with artificial intelligence. Report A67, Systems Analysis Lab., Helsinki U. of Tech.

ELSEVIER
IFAC
PUBLICATIONS

ROBUST PIECEWISE LINEAR SHEET CONTROL IN A PRINTER PAPER PATH [1]

Björn Bukkems, Jeroen de Best,
René van de Molengraft, Maarten Steinbuch

Dynamics and Control Technology Group,
Department of Mechanical Engineering
Technische Universiteit Eindhoven,
P.O. Box 513, 5600 MB Eindhoven, The Netherlands
Phone: +31 40 247 2841, Fax: +31 40 246 1418
B.H.M.Bukkems@tue.nl, J.J.T.H.d.Best@student.tue.nl,
M.J.G.v.d.Molengraft@tue.nl, M.Steinbuch@tue.nl

Abstract: This paper presents a control design approach for robust sheet control in a printer paper path. The overall design question is formulated in terms of a hierarchical control set-up with a low level motor control part and a high level sheet control part. The sheet dynamics, subject to disturbances and parameter uncertainties, are captured in the piecewise linear modeling formalism. Based on this model, the control design yields robust sheet feedback controllers and a closed-loop system capable of rejecting disturbances up to a prescribed level. The effectiveness of the control design is shown on an experimental paper path setup.

Keywords: Piecewise Linear Systems, Hierarchical Control, Robust Control, Linear Matrix Inequalities.

1. INTRODUCTION

The design of a reliable sheet handling mechanism is a central issue in the development of today's cut sheet printer paper paths. An example of such a paper path is shown in Figure 1. Sheets enter this paper path at the Paper Input Module (PIM) and are transported to the Image Transfer Station (ITS) where the image is printed onto the sheet at high pressure and high temperature. After the print has been made, sheets can either re-enter the first part of the paper path for back side printing or they can go to the finisher (FIN). The transportation of sheets is done via pinches. A pinch is a set of rollers consisting of two parts: one part that is actuated by a motor and one part that is used to apply sufficient normal force to prevent the sheet from slipping. As can be seen from Figure 1, pinches can be driven either individually or grouped together in sections.

One of the objectives of the printer's sheet handling mechanism is to accurately deliver sheets to the ITS. Each sheet must synchronize with its corresponding image with respect to both the ITS entry time and the constant printing velocity to achieve a high printing quality. One way to realize the desired printing quality is using a high precision mechanical design. An alternative approach is to exploit the power of closed-loop sheet control. In this approach, the tolerances on the mechanical parts of the paper path are allowed

[1] This work has been carried out as part of the Boderc project under the responsibility of the Embedded Systems Institute. This project is partially supported by the Netherlands Ministry of Economic Affairs under the Senter TS program.

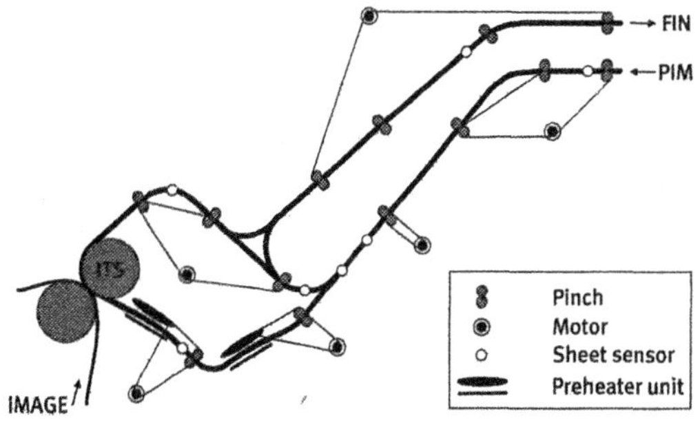

Fig. 1. Schematic representation of a paper path.

to be larger and less effort and money have to be put in constructing a very stiff frame and drive train, since robustness against disturbances and uncertainties in the mechanical design is achieved by sheet feedback control. To realize a sheet feedback control system, the sheet position has to be known. This can, for example, be realized by adding position sensors, possibly in combination with model-based observer techniques.

Known results on sheet feedback control can be found in (Kruciński, 2000; Cloet, 2001; Rai and Jackson, 1998), where robustness against perturbations and disturbances is not taken into account in the control design. In this paper we present a model-based sheet feedback control design procedure that takes into account both system uncertainties and disturbances. Based on the results presented in (Feng, 2002; Chen *et al.*, 2004) we guarantee both stability and performance of the closed-loop system. To synthesize controllers for the sheet tracking problem, we formulate the system in terms of its error dynamics. Experiments will confirm the robustness and disturbance attenuation of the closed-loop system.

The remainder of this paper is organized as follows: in Section 2 the system under consideration will be discussed in more detail and the problem statement will be given. In Section 3 we will discuss the controller design method for the uncertain paper path system that is subject to disturbances. In Section 4 we will present the experimental setup that has been used to validate the proposed control design approach in practice. The validation experiments will be presented in Section 5, and conclusions and recommendations will come at the end.

2. SHEET FEEDBACK CONTROL PROBLEM

In this paper, the focus will be on sheet feedback control design in a basic paper path, shown in Fig 2. By considering this basic version, the essence of the control problem becomes clear. As a result, the switching nature of the system, caused by the consecutive changing of the driving pinch,

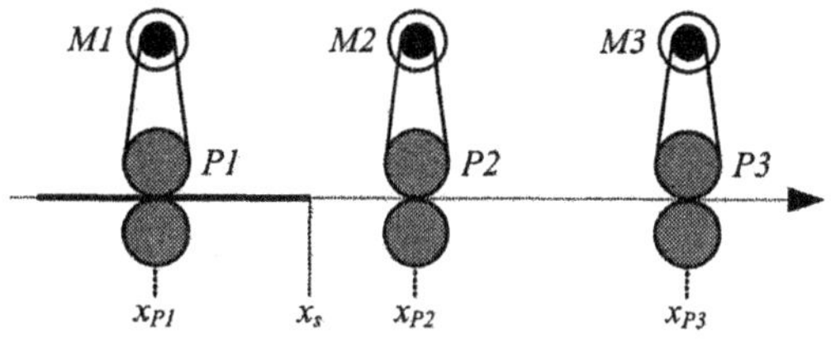

Fig. 2. Schematic representation of the printer paper path.

naturally arises in the control design and a structured design approach can be carried out. Since we consider the motion of sheets only when they are in the paper path, the PIM and FIN are not taken into account. The considered paper path consists of three pinches (*P1*, *P2*, and *P3*) only, each of which is driven by a separate motor (*M1*, *M2*, and *M3*, respectively). The locations of the three pinches in the paper path are represented by x_{P1}, x_{P2}, and x_{P3}, respectively. These locations are chosen such that the distance between two pinches is equal to the sheet length L_s, so the sheet can only be in one pinch at the same time. No slip is assumed between the sheet and the pinches and the coupling between the pinches and motors is assumed to be infinitely stiff. The mass of the sheet is assumed to be zero, which simplifies modeling of the sheet dynamics. The sheet position, defined as x_s, is assumed to be measured.

We adopt a hierarchical, cascaded control structure for the sheet feedback control design. This control layout consists of low level motor control loops and a high level sheet control loop for tackling disturbances and uncertainties at the motor level and at the sheet level, respectively. The control goal we adopt for the basic paper path case study is the design of stable and robust high level feedback controllers (HLCs) that track the desired reference trajectory. Regarding this reference motion task, possible choices are absolute reference tracking control and inter-sheet spacing control (Kruciński, 2000; Cloet, 2001). In this paper the first one is chosen and it is required that sheets are able to track a second-order sheet reference trajectory $x_{s,r}$.

The closed-loop linear motor dynamics in the Laplace domain can be represented by

$$\Omega_{Mi}(s) = T_i(s)\Omega_{Mi,r}(s), i \in \mathcal{I}, \qquad (1)$$

with $T_i(s)$ the complementary sensitivity function of controlled motor i, which maps the input of the low level closed-loop system (the motor reference velocity $\omega_{Mi,r}(t)$, with Laplace transform $\Omega_{Mi,r}(s)$), to its output (the actual motor velocity $\omega_{Mi}(t)$). Furthermore, $\mathcal{I} = \{1, 2, 3\}$ represents the index set of sheet regions. Since the bandwidth of the low level control loops is required to be significantly higher than the bandwidth of the high level control loop (Stephanopoulos, 1984), we

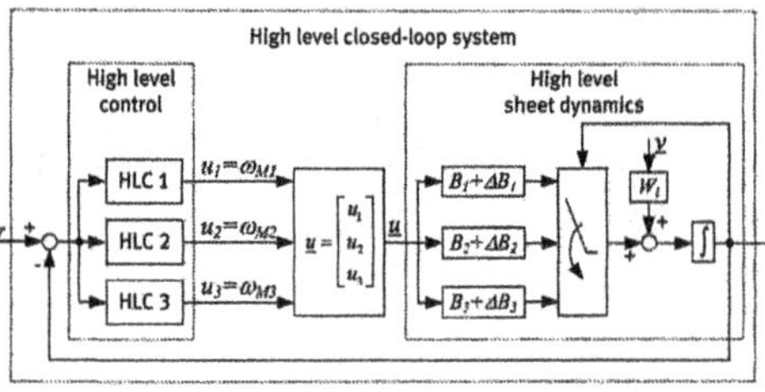

Fig. 3. Block diagram of the total control system.

assume perfect tracking behavior of the controlled motors, i.e., $T_i(s) = 1, \forall i \in \mathcal{I}$.

Under the assumption of ideal behavior in the low level control loop, the inputs u_i of the high level sheet dynamics will be directly generated by the HLCs. This is shown in Fig. 3, which represents the block diagram of the control system at hand. Since at each time instant the sheet is only driven by one pinch, the input of the sheet dynamics will change when the sheet arrives at the next pinch. This switching behavior can be easily captured in the PWL modeling formalism. The sheet velocity is derived from the motor velocities via straightforward holonomic kinematic constraint relations that describe the relation between motor velocity and pinch velocity, and pinch velocity and sheet velocity, respectively. The nominal high level sheet model, i.e. the sheet model without parameter uncertainties and disturbances, is:

$$\dot{x}_s = B_i \underline{u} \text{ for } x_s \in \mathcal{X}_i, i \in \mathcal{I}\,, \tag{2}$$

with the input matrices B_i defined as $B_1 = \begin{bmatrix} n_1 r_{P1} & 0 & 0 \end{bmatrix}$, $B_2 = \begin{bmatrix} 0 & n_2 r_{P2} & 0 \end{bmatrix}$, and $B_3 = \begin{bmatrix} 0 & 0 & n_3 r_{P3} \end{bmatrix}$, respectively. In these definitions, n_i represents the transmission ratio between motor i and pinch i and r_{Pi} represents the radius of the driven roller of pinch i. Furthermore, $\underline{u}$ is the column with inputs of the high level sheet dynamics: $\underline{u} = \begin{bmatrix} \omega_{M1} & \omega_{M2} & \omega_{M3} \end{bmatrix}^T$. The partitioning of the state space into the three regions is represented by $\{\mathcal{X}_i\}_{i \in \mathcal{I}} \subseteq \mathbb{R}$. Here, $\mathcal{X}_1 = \{x_s | x_s \in [x_{P1}, x_{P2})\}$, $\mathcal{X}_2 = \{x_s | x_s \in [x_{P2}, x_{P3})\}$, and $\mathcal{X}_3 = \{x_s | x_s \in [x_{P3}, x_{P3} + L_s)\}$.

In case parameter uncertainties and external disturbances are present, the high level PWL sheet model becomes:

$$\dot{x}_s = (B_i + \Delta B_i)\,\underline{u} + W_i \underline{v} \text{ for } x_s \in \mathcal{X}_i, i \in \mathcal{I}, \tag{3}$$

where ΔB_i is the constant uncertainty term of the i-th subsystem. In this model, this term can represent, for example, an uncertainty in the transmission ratio between motor i and pinch i or an uncertainty in the radius of the driven roller of pinch i. Column $\underline{v}$ acts as disturbances on the sheet velocity. These disturbances are scaled by the matrices W_i, defined as $W_1 = \begin{bmatrix} w_1 & 0 & 0 \end{bmatrix}$, $W_2 = \begin{bmatrix} 0 & w_2 & 0 \end{bmatrix}$, and $W_3 = \begin{bmatrix} 0 & 0 & w_3 \end{bmatrix}$, respectively.

3. ROBUST SHEET CONTROLLER DESIGN

In this section, we present the H_∞ controller synthesis method for the PWL sheet model (3), subject to uncertainties and disturbances. Since we are dealing with a tracking problem, the system is formulated in terms of its tracking error dynamics (Franklin *et al.*, 2002):

$$\begin{aligned} \dot{\underline{q}} &= F\underline{q} + (G_i + \Delta G_i)\,\underline{\mu} + V_i \underline{\nu} \\ &\quad \text{for } (x_{s,r} - \begin{bmatrix} 1 & 0 & 0 \end{bmatrix} \underline{q}) \in \mathcal{X}_i, i \in \mathcal{I} \\ z &= H\underline{q}. \end{aligned} \tag{4}$$

In this notation, the state vector $\underline{q}$ is defined as $\underline{q} = \begin{bmatrix} e_s & \dot{e}_s & \ddot{e}_s \end{bmatrix}^T$, with $e_s = x_{s,r} - x_s$ the tracking error. The control input $\underline{\mu}$ and the disturbance $\underline{\nu}$ are defined as $\underline{\mu} = \ddot{\underline{u}}$ and $\underline{\nu} = \ddot{\underline{v}}$, respectively. The system matrix is defined as $F = \begin{bmatrix} 0 & 1 & 0 \\ 0 & 0 & 1 \\ 0 & 0 & 0 \end{bmatrix}$, whereas the input matrix and the uncertainty term of the input matrix are defined as $G_i = \begin{bmatrix} 0_{3\times1} & 0_{3\times1} & -B_i^T \end{bmatrix}^T$ and $\Delta G_i = \begin{bmatrix} 0_{3\times1} & 0_{3\times1} & -(\Delta B_i)^T \end{bmatrix}^T$, respectively. Furthermore, the disturbance gain matrix and the output matrix in error space are defined as $V_i = \begin{bmatrix} 0_{3\times1} & 0_{3\times1} & -W_i^T \end{bmatrix}^T$ and $H = \begin{bmatrix} 1 & 0 & 0 \end{bmatrix}$, respectively. Regarding the uncertainty term ΔG_i, we make the assumption that its upper bound is known *a priori*:

$$[\Delta G_i]\,[\Delta G_i]^T \leq E_{Gi} E_{Gi}^T, i \in \mathcal{I}, \tag{5}$$

where E_{Gi} is a constant matrix satisfying (5) with the same dimensions as ΔG_i.

Given this notation in error space, the controller synthesis can be carried out. For this purpose, the approach presented in (Feng, 2002; Chen *et al.*, 2004) is slightly adjusted. Switching of the error dynamics from the one regime to the other does not depend on e, $\dot{e}$, or $\ddot{e}$, but on the sheet position x_s. As a result, using piecewise quadratic Lyapunov functions is not trivial for our tracking control case and, hence, we will use a common quadratic Lyapunov function in our analysis and synthesis.

The goal of the control design procedure is to find a feedback control law that stabilizes the uncertain PWL system (4). Furthermore, the controller should result in a guaranteed performance in the H_∞ sense. This means that, given a prescribed level of disturbance attenuation $\gamma > 0$, the induced L_2-norm of the operator from $\underline{\nu}$ to the controlled output z should be smaller than γ under zero initial conditions for all nonzero $\underline{\nu} \in L_2$ (Feng *et al.*, 2002; Feng, 2002; Chen *et al.*, 2004):

$$\|z\|_2 < \gamma \|\underline{\nu}\|_2. \tag{6}$$

In this case, the closed-loop error dynamics are said to be globally stable with disturbance attenuation γ. The control law we propose to realize the goal is based on static state feedback:

$$\underline{\mu} = -K\underline{q}. \tag{7}$$

Substitution of (7) into (4) yields the closed-loop error dynamics:

$$\begin{aligned} \dot{\underline{q}} &= A_{Ci}\underline{q} + V_i\underline{\nu} \\ &\quad \text{for } \left(x_{s,r} - \begin{bmatrix} 1 & 0 & 0 \end{bmatrix} \underline{q}\right) \in \mathcal{X}_i, i \in \mathcal{I}, \\ z &= H\underline{q}, \end{aligned} \tag{8}$$

with $A_{Ci} = F - (G_i + \Delta G_i) K$.

Given this closed-loop system, we can now present the following theorem for the controller design, which is based on Theorem 2 in (Chen *et al.*, 2004):

Theorem 3.1. Given a constant $\gamma > 0$, the PWL system (8) is globally stable with disturbance attenuation γ if the following matrix inequalities are satisfied:

$$0 < P = P^T \tag{9}$$

$$0 > \begin{bmatrix} \Omega_i & PH^T & Q^T \\ HP & -I & 0 \\ Q & 0 & -\epsilon I \end{bmatrix}, i \in \mathcal{I}, \tag{10}$$

with

$$\begin{aligned} \Omega_i = &PF^T + FP - Q^T G_i^T - G_i Q + \\ &+ \gamma^{-2} V_i V_i^T + \epsilon E_{Gi} E_{Gi}^T. \end{aligned} \tag{11}$$

Moreover, the controller gain for each subsystem is given by:

$$K = QP^{-1}. \tag{12}$$

For the proof of this theorem the reader is referred to (Chen *et al.*, 2004). As can be seen from Theorem 3.1, the conditions (9) and (10) are Linear Matrix Inequalities (LMIs) in the variables P, Q, ϵ, and γ^{-2}. This is in contrast with (Chen *et al.*, 2004), where products of ϵ and γ^{-2} exist. This results from the fact that we assume no uncertainty in the scaling matrix V_i of the disturbances.

4. EXPERIMENTAL SETUP

To experimentally validate the proposed control design approach, we use the paper path setup depicted in Figure 4. As can be seen in the figure, the setup consists of a PIM and a paper path with five pinches. In our experiments, only the second, third, and fourth pinch will be used. For the sake of notation, in the remainder of this paper we will refer to these pinches as pinch 1, pinch 2 and pinch 3, respectively. Each pinch is connected to a motor via a gear belt. The nominal transmission ratios between the motors and pinches are $n_1 = 0.49$, $n_2 = 0.47$, and $n_3 = 0.5$, respectively, and the pinch radii are $14 \cdot 10^{-3}$ m. The motors are 10 W DC motors, driven by power amplifiers with built-in current controllers. The angular positions of the motor shafts are measured using optical incremental encoders with a resolution of 2000 increments per revolution. Both the amplifiers and the encoders are connected to a PC-based control system. This system consists of a Pentium 4 host computer running RTAI/Fusion Linux and Matlab/Simulink and three TUeDACS USB I/O devices (van de Molengraft *et al.*, 2005). The sheets are guided through the paper path via thin steel wires and their position is measured using optical mouse sensors, which are directly connected to the host computer via USB.

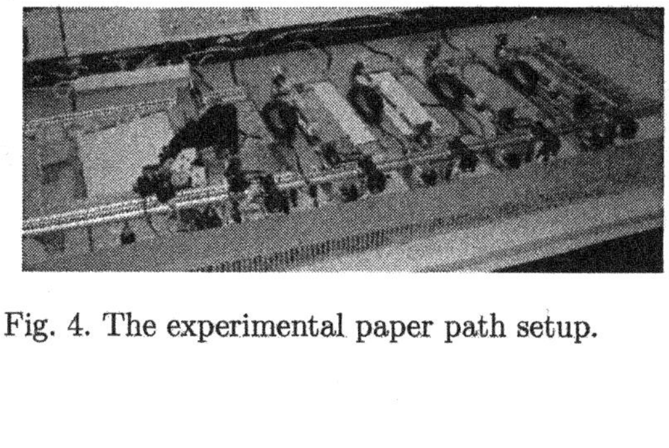

Fig. 4. The experimental paper path setup.

5. EXPERIMENTAL RESULTS

5.1 Control Design Results

To show the robustness against uncertain system parameters, the transmission ratios between the motors and pinches can be varied in the experimental setup. Besides the nominal ratios, a ratio of 0.53 can be implemented in the setup by changing the gear wheels. This results in maximum variations of the transmission ratios of $\Delta n_1 = 0.05$, $\Delta n_2 = 0.07$, and $\Delta n_3 = 0.03$, respectively. These variations will take the role of uncertain deviations from the nominal ratios. Given these deviations, the uncertainty terms ΔB_i in (3) become $\Delta B_1 = \begin{bmatrix} \Delta n_1 r_{P1} & 0 & 0 \end{bmatrix}$, $\Delta B_2 = \begin{bmatrix} 0 & \Delta n_2 r_{P2} & 0 \end{bmatrix}$, and $\Delta B_3 = \begin{bmatrix} 0 & 0 & \Delta n_3 r_{P3} \end{bmatrix}$, respectively. From ΔB_i, also the uncertainty terms on the input matrices of the open-loop error dynamics (4) can be calculated, which in turn are used to calculate the upper bounds on these terms according to (5), yielding $E_{Gi}E_{Gi}^T(k,l) = (\Delta n_i)^2 r_{Pi}^2$ for $k = l = 3$ and $E_{Gi}E_{Gi}^T(k,l) = 0$ for $k = l \neq 3$ and $k \neq l$, with k and l the row and column index, respectively. Given these matrices and by choosing $\gamma = 0.048$ the controller gain can be calculated according to Theorem 3.1:

$$K = 1 \cdot 10^5 \begin{bmatrix} -1.63 & -0.51 & -0.08 \\ -1.71 & -0.53 & -0.08 \\ -1.58 & -0.49 & -0.08 \end{bmatrix}. \tag{13}$$

The value of γ has been chosen such that the bandwidth of the high level control loop is significantly lower than the bandwidth of the low level control loops, as will be discussed in Section 5.2.

5.2 Low Level Motor Control

In this subsection, the low level control of motor 1 and its influence on the high level sheet dynamics is discussed. Although not shown, similar results were obtained for motors 2 and 3.

In the design procedure of the sheet feedback controllers we assumed perfect tracking behavior of the controlled motors, i.e. we assumed an infinite bandwidth of the motor control loops. Furthermore, we assumed an infinitely stiff coupling between the pinches and the motors. In a practical environment, however, these assumptions do not hold. Moreover, a digital implementation will cause a delay in the loop which will limit the attainable bandwidth. Based on identified motor dynamics, PID feedback controllers have been designed using loopshaping techniques (Franklin *et al.*, 2002). The controller parameters are tuned such that a bandwidth of 50 Hz has been realized. This can be seen in Figure 6, which depicts the Frequency Response Function (FRF) of the loopgain. Here, the bandwidth is defined as the frequency at which the 0 dB line of the open-loop FRF is crossed.

The rubber belt that connects the motor with the driven roller of the pinch has a limited stiffness, as can be observed from Figure 5, which shows the FRF of the transmission between motor 1 and pinch 1. It can be seen that the assumption on the infinite stiff coupling between motor and pinch only holds for frequencies up to approximately 100 Hz. In this frequency range, the measured transmission ratio coincides with the nominal transmission ratio of 0.49 (≈ -6.3) dB. For higher frequencies, the flexibility becomes dominant.

Given the high level sheet model and the HLCs, together with the controlled motor-pinch dynamics, the loopgain of the first subsystem can be derived. This loopgain is the transfer function

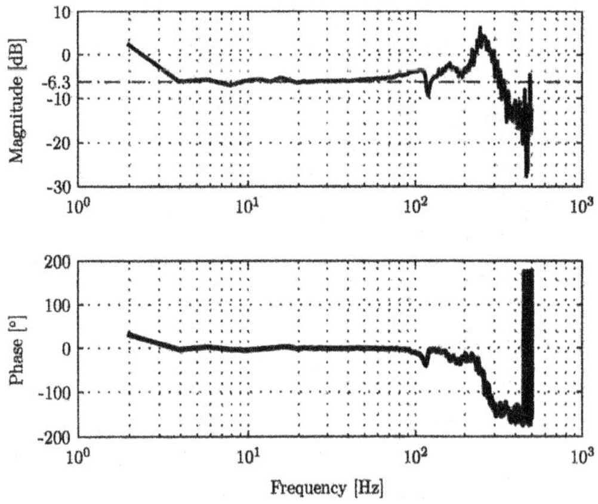

Fig. 5. FRF of the transmission between motor 1 and pinch 1 (solid), and the nominal transmission ratio (dashed)

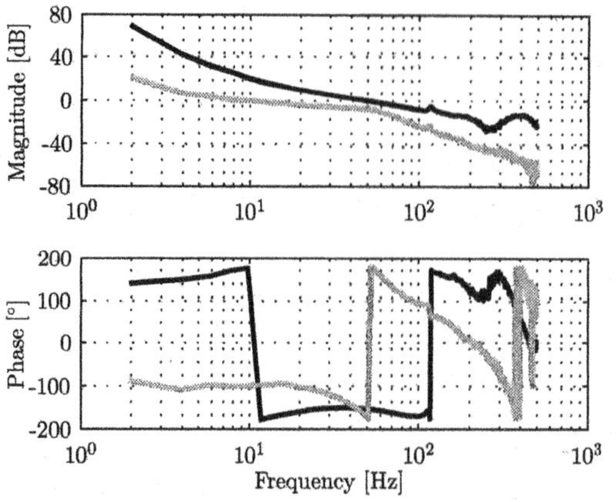

Fig. 6. FRF of the loopgain of the first motor control loop (black) and the FRF of loopgain of the first subsystem, including low level control (gray).

from the sheet tracking error to the actual sheet position. The FRF of this loopgain is also shown in Figure 6. It can be seen that a bandwidth of approximately 10 Hz has been realized. This is a factor 5 lower than the bandwidth of the motor control system, as required in a cascade control structure (Stephanopoulos, 1984). Furthermore, the phase lag at 10 Hz is approximately 90°. From this we can conclude that the first subsystem is stable.

In the control design procedure for the regulation of the PWL error dynamics, stability was proven for the case of perfect low level motor behavior. However, we want to apply the calculated controller (13) also in practical cases where we have to deal with non-ideal low level behavior, and still guarantee that the overall switched system is stable. In general, sheets have to be transported from the PIM to the FIN via the ITS (see Figure 1 and assume no backside printing) and the sheet reference velocity will be positive (Cloet *et al.*, 2001). To realize sequential switching in practice, we have to avoid negative actual sheet velocities to prevent the sheet from moving back to the previous pinch. Since the sheet control loops have a relatively low bandwidth and the sheet controllers can immediately respond to sheet tracking errors, negative sheet velocities are not likely to occur. Consequently, stability of the high level PWL system at hand is plausible in case of stable individual subsystems. So far, this reasoning has been approved by experimental results. However, a mathematically founded stability proof is subject of future research.

5.3 Validation Results

In the experimental validation of the control design, the focus is on the robustness of the system against parameter uncertainties. Therefore, the implemented transmission ratios are $n_1 = 0.49$, $n_2 = 0.53$, and $n_3 = 0.49$, i.e. the ratios of the second and third subsystem deviate from the nominal values. For the sheet motion task, a constant velocity of 0.27 ms^{-1} is chosen that has to be tracked throughout the entire paper path. The corresponding sheet reference motion $x_{s,r}$ is therefore a ramp function. Since no feedforward control input has been used, all three pinches are standing still until a sheet enters the first pinch. Due to the difference between the initial reference velocity and the actual initial velocity, the sheet error starts increasing when the sheet enters the first pinch, as can be seen in Figure 7. However, this error is decreased by the sheet controller in the first regime. It can be seen that some overshoot is present. Furthermore, it can be seen that the error increases when the sheet enters pinches two and three. This is due to the deviation of the transmission ratios with respect to the nominal values. Also these increases are controlled towards zero. From Figure 7, we can conclude that the closed loop system is stable and robust for parameter uncertainties within the specified bounds.

The difference between the experimentally obtained tracking error and the one obtained from simulation is also depicted in Fig. 7. It can be seen that there is a close match between both responses. This close match justifies the assumption on ideal low-level motor dynamics in the controller synthesis approach.

6. CONCLUSIONS AND FUTURE WORK

In this paper, a control design approach for robust sheet control in a printer paper path has been presented. The use of cheap optical mouse sensors as sheet position sensors has enabled the practical validation of the control design. Experiments show that a stable and robust closed-loop system has been obtained. This gives an opportunity in industrial applications to use less expensive mechanics with larger tolerances, and still to achieve the desired printing quality. Future research will focus on control design for cases in which pinches are coupled into sections, driven by one motor, and cases in which more than one pinch can influence the sheet motion. Furthermore, a mathematically founded stability proof of the total system in case of non-ideal motor control will be carried out.

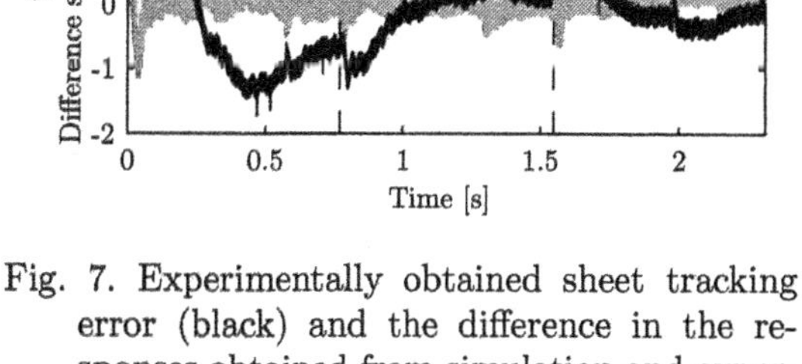

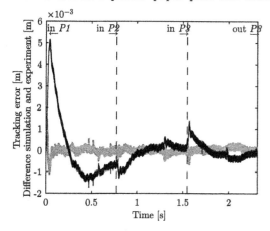

Fig. 7. Experimentally obtained sheet tracking error (black) and the difference in the responses obtained from simulation and experiment (gray).

REFERENCES

Chen, M., C.R. Zhu and G. Feng (2004). Linear-matrix-inequality-based approach to H_∞ controller synthesis of uncertain continuous-time piecewise linear systems. *IEE Proc.-Control Theory Appl.* **151(3)**, 295–300.

Cloet, Carlo (2001). *A Mechatronics Approach to Copier Paperpath Design.* Ph.D. thesis, University of California Berkeley, CA, USA.

Cloet, Carlo, Masayoshi Tomizuka and Roberto Horowitz (2001). Design requirements and reference trajectory generation for a copier paperpath. In: *Proceedings of the 2001 IEEE/ASME International Conference on Advanced Intelligent Mechatronics.* pp. 911–916.

Feng, Gang (2002). Controller design and analysis of uncertain piecewise-linear systems. *IEEE Trans. Circuits Syst. I* **49(2)**, 224–232.

Feng, Gang, G.P. Lu and S.S. Zhou (2002). An approach to H_∞ controller synthesis of piecewise linear systems. *Communications in Information and Systems* **2(3)**, 245–254.

Franklin, Gene F., J. David Powell and Abbas Emami-Naeini (2002). *Feedback control of dynamic systems.* Prentice Hall. Upper Saddle River, New Jersey, USA.

Kruciński, Martin (2000). *Feedback Control of Photocopying Machinery.* Ph.D. thesis, University of California Berkeley, CA, USA.

Rai, Sudhendu and Warren B. Jackson (1998). A hybrid hierarchical control architecture for paper transport systems. In: *Proc. of the 37th IEEE Conference on Decision and Control.* Tampa, Florida, USA. pp. 4249–4250.

Stephanopoulos, George (1984). *Chemical Process Control.* Prentice Hall. Englewood Cliffs, New Jersey, USA.

van de Molengraft, René, Bram de Kraker and Maarten Steinbuch (2005). Integrating experimentation into control courses. *IEEE Control Syst. Mag.* **25(1)**, 40–44.

ELSEVIER
IFAC
PUBLICATIONS

STABILIZATION OF MAX-PLUS-LINEAR SYSTEMS USING RECEDING HORIZON CONTROL — THE UNCONSTRAINED CASE

I. Necoara * T.J.J. van den Boom * B. De Schutter * J. Hellendoorn *

* *Delft Center for Systems and Control, Delft University of Technology, Mekelweg 2, 2628 CD Delft, The Netherlands, email: {i.necoara,b.deschutter,t.j.j.vandenboom}@tudelft.nl*

Abstract: Max-plus-linear (MPL) systems are a class of discrete-event systems that can be described by models that are linear in the max-plus algebra. MPL systems arise in the context of e.g. manufacturing systems, telecommunication networks, railway networks, and parallel computing. We derive a receding horizon control scheme for MPL systems that guarantees a priori stability (in the sense of boundedness of the normalized state) of the closed-loop system in the unconstrained case. We also discuss the main properties of the resulting receding horizon controllers. *Copyright © 2006 IFAC.*

Keywords: Stabilizing controllers, discrete-event systems, receding horizon control.

1. INTRODUCTION

Discrete-event systems (DES) are event-driven dynamical systems (i.e. the state transitions are initiated by events, rather than a clock). In the last couple of decades there has been an increase in the research on DES that can be modeled as max-plus-linear (MPL) systems (Baccelli *et al.*, 1992; Heidergott *et al.*, 2005). There are two main directions in MPL DES control: one direction uses optimal control based on residuation theory (Cottenceau *et al.*, 2001; Maia *et al.*, 2003; Menguy *et al.*, 1998; Menguy *et al.*, 2000), and the other a receding horizon control (RHC) based approach (De Schutter and van den Boom, 2001). Although there are several papers on optimal and RHC control for MPL DES, the literature on the stabilizing controller for this class of systems is relatively sparse. In fact, to the authors' best knowledge, the only papers explicitly dealing with stabilizing control of MPL DES are (Maia *et al.*, 2003; van den Boom *et al.*, 2005).

Receding horizon control (RHC), also known as model predictive control, is an attractive feedback strategy for linear or nonlinear processes subject to input and state constraints (Mayne *et al.*, 2000). The essence of RHC is to determine a control pro le that optimizes a cost criterion over a prediction window and then to apply this control pro le until new process measurements become available. Feedback is incorporated by using these measurements to update the optimization problem for the next step.

This paper considers the problem of designing a stabilizing receding horizon controller for the class of MPL DES. We consider a trade-off between tracking a reference state trajectory and just-in-time production for the so-called unconstrained case, in which only the constraint that the input (i.e., the sequence of feeding times) should be nondecreasing is taken into account. In this particular case we derive a stable RHC scheme for which the analytic solution exists. The main advantage of this paper compared to most of the results on RHC of MPL DES is the fact that we guarantee *a priori* stability of the closed-loop system. Moreover, the conditions that we will derive in this paper are less strict than those of (van den Boom *et al.*, 2005) (where output tracking is considered). We also prove several properties of the RHC controllers, and we characterize a whole class of stabilizing controllers for MPL DES.

2. MAX-PLUS ALGEBRA AND MPL DES

2.1 Max-plus algebra

De ne $\varepsilon := -\infty$ and $\mathbb{R}_\varepsilon := \mathbb{R} \cup \{\varepsilon\}$. The max-plus-algebraic (MPA) addition ($\oplus$) and multiplication ($\otimes$) are de ned as (Baccelli *et al.*, 1992): $x \oplus y := \max\{x, y\}$, $x \otimes y := x + y$, for $x, y \in \mathbb{R}_\varepsilon$. For matrices $A, B \in \mathbb{R}_\varepsilon^{m\times n}$ and $C \in \mathbb{R}_\varepsilon^{n\times p}$ we have $(A \oplus B)_{ij} := A_{ij} \oplus B_{ij}, (A \otimes C)_{ij} := \bigoplus_{k=1}^{n} A_{ik} \otimes C_{kj}$, for all i, j. De ne $\mathcal{E}_{m\times n}$ as the $m \times n$ MPA zero matrix: $(\mathcal{E}_{m\times n})_{ij} := \varepsilon$, for all i, j; E_n is the $n \times n$ MPA identity matrix: $(E_n)_{ii} := 0$, for all i and $(E_n)_{ij} := \varepsilon$, for all i, j with $i \neq j$. For $A \in \mathbb{R}_\varepsilon^{n\times n}$ we de ne $A^* := E_n \oplus A \oplus \cdots \oplus A^{\otimes^n} \oplus A^{\otimes^{n+1}} \oplus \cdots$ For a positive integer n, we denote $\underline{n} := \{1, 2, \cdots, n\}$. Given $x \in \mathbb{R}_\varepsilon^n$ we de ne $\|x\|_\oplus := \max_{i\in\underline{n}} x_i$ and $\|x\|_\infty := \max_{i\in\underline{n}} |x_i|$. A matrix $\Gamma \in \mathbb{R}_\varepsilon^{n\times m}$ is *row-finite* if for any row $i \in \underline{n}$, $\max_{j\in\underline{m}} \Gamma_{ij} \neq \varepsilon$.

We denote with $x \oplus' y := \min\{x, y\}$ and $x \otimes' y := x + y$ (the operations $\otimes$ and $\otimes'$ differ only in that $(-\infty) \otimes (+\infty) := -\infty$, while $(-\infty) \otimes' (+\infty) := +\infty$). The matrix multiplication and addition for $(\oplus', \otimes')$ are de ned similarly as for $(\oplus, \otimes)$. It can be shown that the following relations hold for any matrices A, B and vectors x, y of appropriate dimensions over $\mathbb{R}_\varepsilon$:

$$A \otimes' (B \otimes x) \geq (A \otimes' B) \otimes x, \tag{1a}$$

$$((-A^T) \otimes' A) \otimes x \geq x \tag{1b}$$

$$x \leq y \Rightarrow\ A \otimes x \leq A \otimes y \text{ and } A \otimes' x \leq A \otimes' y. \tag{1c}$$

Lemma 2.1. (Baccelli *et al.*, 1992) (i) The inequality $A \otimes x \leq b$ has a unique largest solution given by $x_{\text{opt}} = (-A^T) \otimes' b = -(A^T \otimes (-b))$ (by the largest solution we mean that $\forall x : A \otimes x \leq b \Rightarrow x \leq x_{\text{opt}}$).

(ii) $x = A \otimes x \oplus b$ has a solution $x = A^* \otimes b$. If $A_{ij} < 0$ for all i, j, then the solution is unique.

2.2 MPL systems

DES with only synchronization and no concurrency can be modeled by an MPA model of the form (Baccelli *et al.*, 1992) [1]

$$x_{\text{sys}}(k) = A_{\text{sys}} \otimes x_{\text{sys}}(k-1) \oplus B_{\text{sys}} \otimes u_{\text{sys}}(k), \tag{2}$$

where $x_{\text{sys}}(k) \in \mathbb{R}_\varepsilon^n$ represents the state, $u_{\text{sys}}(k) \in \mathbb{R}_\varepsilon^m$ is the input and where $A_{\text{sys}} \in \mathbb{R}_\varepsilon^{n\times n}$, $B_{\text{sys}} \in \mathbb{R}_\varepsilon^{n\times m}$ are the system matrices. In the context of DES k is an event counter while $u_{\text{sys}}, x_{\text{sys}}$ are dates (feeding times and processing times, respectively). A typical constraint that appears in the context of DES where the input represents times, is that the signal u_{sys} should be increasing: $u_{\text{sys}}(k+1) - u_{\text{sys}}(k) \geq 0$.

Let λ^* be the largest MPA eigenvalue of A_{sys} ($\lambda \in \mathbb{R}_\varepsilon$ is an MPA eigenvalue of A_{sys} if there exists an MPA eigenvector $v \in \mathbb{R}_\varepsilon^n$, $v \neq \mathcal{E}_{n\times 1}$ such that $A_{\text{sys}} \otimes v = \lambda \otimes v$). In the next section we will consider a reference signal that the state should track of the form

$$r_{\text{sys}}(k) = x_{\text{sys,t}} + k\rho. \tag{3}$$

Since through the term $B_{\text{sys}} \otimes u_{\text{sys}}$ it is only possible to create delays in the starting times of activities, we should choose $\rho \geq \lambda^*$. If $\lambda^* > \varepsilon$ (in practical applications we even have $\lambda^* \geq 0$), then there exists an MPA invertible matrix $P \in \mathbb{R}_\varepsilon^{n\times n}$ such that [2] the matrix $\bar{A} = P^{\otimes^{-1}} \otimes A_{\text{sys}} \otimes P$ satis es $\bar{A}_{ij} \leq \lambda^*$ for all $i, j \in \underline{n}$ (De Schutter, 1996). We make the following change of coordinates $\bar{x}(k) = P^{\otimes^{-1}} \otimes x_{\text{sys}}(k)$. We denote with $\bar{B} = P^{\otimes^{-1}} \otimes B_{\text{sys}}$ and $\bar{u}(k) = u_{\text{sys}}(k)$. In the new coordinates the system (2) becomes:

$$\bar{x}(k) = \bar{A} \otimes \bar{x}(k-1) \oplus \bar{B} \otimes \bar{u}(k).$$

We now consider the normalized system with $x(k) = \bar{x}(k) - \rho k$, $u(k) = \bar{u}(k) - \rho k$, $A = \bar{A} - \rho$ (i.e. by subtracting in the conventional algebra all entries of $\bar{x}, \bar{u}$ and of $\bar{A}$ by ρk and ρ, respectively) and $B = \bar{B}$. The normalized system can be written as:

$$x(k) = A \otimes x(k-1) \oplus B \otimes u(k). \tag{4}$$

The MPL system (4) is *controllable* if and only if (iff) each component of the state can be made arbitrarily large by applying an appropriate controller to the system initially at rest. It can be checked that the system is controllable iff the matrix $\Gamma_n := [B \;\; A \otimes B \cdots A^{\otimes^{n-1}} \otimes B]$ is row- nite [3].

The following key assumption will be used throughout the paper:

Assumption A: We assume that $\rho > \lambda^* \geq 0$ and that the system is controllable.

The conditions of this assumption are quite weak and are usually met in applications. Note that from Assumption **A** it follows that $A_{ij} < 0$, for all $i, j \in \underline{n}$. In the new coordinates the state should be regulated to the desired target $x_t := P^{\otimes^{-1}} \otimes x_{\text{sys,t}}$.

Since $A_{ij} < 0$ for all i, j, we have $A^* = E_n \oplus A \oplus \cdots \oplus A^{\otimes^{n-1}}$ (see (Baccelli *et al.*, 1992, Theorem 3.20)). Note that for any nite, constant input u there exists a state equilibrium x (i.e. $x = A \otimes x \oplus B \otimes u$), viz. $x = A^* \otimes B \otimes u$. Note that x is unique (see Lemma 2.1 (ii)) and nite (since Γ_n is row- nite). We associate to x_t the largest equilibrium pair $(x_{\text{el}}, u_{\text{el}})$ satisfying $x_{\text{el}} \leq x_t$. From the previous discussion it follows that $(x_{\text{el}}, u_{\text{el}})$ is unique, nite and given by:

$$u_{\text{el}} = (-(A^* \otimes B))^T \otimes' x_t,\ x_{\text{el}} = A^* \otimes B \otimes u_{\text{el}}. \tag{5}$$

---

[1] In general there is also an output equation of the form $y(k) = C \otimes x(k)$, but in this paper we assume that all the states can be measured (i.e. $C = E_n$). Note however that the results of this paper can also be extended to take the output into account.

[2] $P^{\otimes^{-1}}$ denotes the MPA inverse of P: $P \otimes P^{\otimes^{-1}} = E_n$.

[3] This definition is equivalent to the one used in (Baccelli *et al.*, 1992), where the system is called controllable if all states are connected to some input.

3. STABILIZING MPL DES CONTROLLERS FOR THE UNCONSTRAINED CASE

3.1 Stabilizing control for MPL DES

In this section we consider the normalized system (4), where A satis es $A_{ij} < 0$ for all i, j (according to Assumption **A**) and with the constraint that the original input signal (u_{sys}) should be nondecreasing:

$$u(k+1) - u(k) \geq -\rho, \ \ \forall k \geq 0. \tag{6}$$

Given a desired target $x_t \in \mathbb{R}^n$, let (x_{el}, u_{el}) be the largest equilibrium pair satisfying $x_{el} \leq x_t$ (cf. (5)). We de ne also an upper bound on x_t: $x_{ub} = A^* \otimes x_t \geq x_t$, $u_{ub} = (-B)^T \otimes' (A^* \otimes x_t)$. These pairs are uniquely determined and nite. Note that $u_{el} \leq u_{ub}$ and whenever x_t is an equilibrium state (i.e. there exists a nite u_t such that $x_t = A \otimes x_t \oplus B \otimes u_t$) then $x_{el} = x_{ub} = x_t$ and consequently $u_{el} = u_{ub} = u_t$.

Definition 3.1. Given a state feedback controller $\mu : \mathbb{R}^n_\varepsilon \to \mathbb{R}^m_\varepsilon$, then the closed-loop system $x(k) = A \otimes x(k-1) \oplus B \otimes \mu(x(k-1))$ is *stable* iff the state remains bounded, i.e. for every $\delta > 0$ there exists a real-valued function $\theta(\delta) > 0$ such that $\|x(0) - x_{el}\|_\infty \leq \delta$ implies $\|x(k) - x_{el}\|_\infty \leq \theta(\delta)$ for all $k \geq 0$.

Now we formulate the control problem that we will solve in the sequel:

Problem 1: Design a state feedback controller $\mu : \mathbb{R}^n_\varepsilon \to \mathbb{R}^m_\varepsilon$ for the MPL system (4) such that the closed-loop system is stable.

3.2 Stabilizing state feedback controller

Assume we are at event step k. Given the previous [4] state $x(k-1)$ and input $u(k-1)$, we de ne two controllers: a feedback controller

$$u^{f}(k) := (-B^T) \otimes' (A \otimes x(k-1) \oplus B \otimes (u(k-1) - \rho) \oplus x_t) \tag{7}$$

and a constant controller:

$$u^{c}(k) := u_{el} \oplus (u(k-1) - \rho). \tag{8}$$

Later on, we will show that under some conditions the RHC controller lies between these two controllers. Let us now study the (stabilizing) properties of these two controllers. Note that $u^f(k)$ satis es the constraint (6). Indeed, from (1c) it follows that $u^f(k) \geq (-B^T) \otimes' (B \otimes (u(k-1) - \rho))$ and from (1a) and (1b) we conclude that $u^f(k) \geq u(k-1) - \rho$. Using similar arguments we can prove that $u^f(k) \geq u_{el}$, for all $k \geq 1$. Similarly, $u^c(k)$ satis es the constraint (6) and $u^c(k) \geq u_{el}$, for all $k \geq 1$.

With the controller (7), the closed-loop normalized system (4) becomes

$$x^{f}(k) = A \otimes x^{f}(k-1) \oplus B \otimes u^{f}(k), \tag{9}$$

where the initial conditions $x^f(0) = x(0)$ and $u^f(0) = u(0)$ are given. Note that $u^c(k) = u_{el} \oplus (u(0) - \rho k)$ and the corresponding closed-loop system, for $x^c(0) = x(0)$ and $u^c(0) = u(0)$ is given by:

$$x^{c}(k) = A \otimes x^{c}(k-1) \oplus B \otimes u^{c}(k). \tag{10}$$

First let us note that:

$$\begin{cases} x^f(k) \leq A \otimes x^f(k-1) \oplus B \otimes (u^f(k-1) - \rho) \oplus x_t \\ x^f(k) \geq A \otimes x^f(k-1) \oplus B \otimes (u^f(k-1) - \rho) \oplus B \otimes u_{el} \end{cases} \tag{11}$$

Indeed, from Lemma 2.1 we have $B \otimes u^f(k) \leq A \otimes x^f(k-1) \oplus B \otimes (u^f(k-1) - \rho) \oplus x_t$ and thus $x^f(k) \leq A \otimes x^f(k-1) \oplus B \otimes (u^f(k-1) - \rho) \oplus x_t$. The second inequality is straightforward (recall that $u^f(k) \geq u^f(k-1) - \rho$ and $u^f(k) \geq u_{el}$ and using the monotonicity property (1c) it follows that $x^f(k) \geq A \otimes x^f(k-1) \oplus B \otimes (u^f(k-1) - \rho) \oplus B \otimes u_{el}$. The following inequality is also useful: since $x^f(k-1) \geq B \otimes u^f(k-1)$ it follows that

$$B \otimes (u^f(k-1) - \rho) = (B \otimes u^f(k-1)) - \rho \leq x^f(k-1) - \rho \tag{12}$$

We have (see (Necoara *et al.*, 2006; Necoara, 2006) for the proof):

Lemma 3.2. $u^f(k) \geq u^c(k)$, $x^f(k) \geq x^c(k)$, $\forall k \geq 0$.

The stabilizing properties of the two state feedback controllers are summarized in the next theorem:

Theorem 3.3. The following statements hold:

(i) For any initial condition $x^f(0) = x(0)$ and $u^f(0) = u(0)$ there exists a nite K^f such that $x^f(k) \leq x_{ub}$ and $u_{el} \leq u^f(k+1) \leq u_{ub}$, for all $k \geq K^f$.

(ii) For any initial condition $x^c(0) = x(0)$ and $u^c(0) = u(0)$ there exists a nite K^c such that $x^c(k) = x_{el}$ and $u^c(k) = u_{el}$, for all $k \geq K^c$.

(iii) The closed-loop systems (9) and (10) are stable.

PROOF. (i) From (11) and (12) it follows that: $x^f(k) \leq A \otimes x^f(k-1) \oplus B \otimes (u^f(k-1) - \rho) \oplus x_t \leq A \otimes x^f(k-1) \oplus (x^f(k-1) - \rho) \oplus x_{ub}$. By induction it is straightforward to prove that:

$$x^f(k) \leq \bigoplus_{t=0}^{k} (A^{\otimes^{k-t}} \otimes (x^f(0) - t\rho)) \oplus x_{ub}. \tag{13}$$

Recall that $A_{ij} < 0$ for all $i, j \in \underline{n}$. Then, it is well-known that (Baccelli *et al.*, 1992):

$$A^{\otimes^k} \otimes x^f(0) \to \mathcal{E}_{n\times 1} \ \text{as} \ k \to \infty. \tag{14}$$

We denote with $z_0 = x^f(0)$ and iteratively $z_k = \bigoplus_{t=0}^{k} (A^{\otimes^{k-t}} \otimes x^f(0) - t\rho) = \max\{A^{\otimes^k} \otimes x^f(0), z_{k-1} - \rho\}$. From (14) and $\rho > 0$ it follows that

$$z_k \to \mathcal{E}_{n\times 1} \ \text{as} \ k \to \infty. \tag{15}$$

Therefore, there exists a nite integer K^f such that $\bigoplus_{t=0}^{k} (A^{\otimes^{k-t}} \otimes (x^f(0) - t\rho)) \leq x_{ub}$ for any $k \geq K^f$. In conclusion, $x^f(k) \leq x_{ub}$ for any $k \geq K^f$.

[4] Timing aspects and the interplay between event steps and time steps are discussed in (van den Boom and De Schutter, 2002).

Now consider k satisfying $k \geq K^f$. Therefore, $x^f(k) \leq x_{ub}$. We obtain $A \otimes x^f(k) \leq A \otimes x_{ub} \leq x_{ub}$. Similarly, from (12) we have $B \otimes (u^f(k) - \rho) \leq x^f(k) - \rho \leq x_{ub}$. Using now (1c) we obtain:

$$u^f(k+1) \leq (-B^T) \otimes' x_{ub} = u_{ub}.$$

By induction, using the same procedure it follows that $u^f(K^f + l) \leq u_{ub}$, for all $l \geq 1$. On the other hand $u^f(k) \geq u_{el}$ for all $k \geq 1$. We conclude that $u_{el} \leq u^f(K^f + l) \leq u_{ub}$, for all $l \geq 1$.

(ii) Since $\rho > 0$, $u^c(k) = u_{el}$ for k large enough. Also,

$$x^c(k) = A^{\otimes^k} \otimes x^c(0) \oplus (\bigoplus_{t=1}^{k} A^{\otimes^{k-t}} \otimes B \otimes (u^c(0) - t\rho)) \oplus (\bigoplus_{t=1}^{k} A^{\otimes^{k-t}} \otimes B \otimes u_{el}).$$

From (14) we have $A^{\otimes^k} \otimes x^c(0) \to \mathcal{E}_{n\times 1}$ as $k \to \infty$. So, $\bigoplus_{t=1}^{k} A^{\otimes^{k-t}} \otimes B \otimes (u^c(0) - t\rho) \to \mathcal{E}_{n\times 1}$ as $k \to \infty$ (this can be proved in a similar way as (15)). Since $x_{el} = \bigoplus_{t=1}^{n} A^{\otimes^{n-t}} \otimes B \otimes u_{el}$, it follows that there exists a $K^c \geq n$ such that $x^c(k) = x_{el}$ and $u^c(k) = u_{el}$, for all $k \geq K^c$.

(iii) Let us now prove stability of the closed-loop systems (9) and (10). Let $\delta > 0$ and consider $\|x(0) - x_{el}\|_\infty \leq \delta$. From $u^c(k) \geq u_{el}$ it follows that $x^c(k) \geq x_{el}$ for all $k \geq n$. Since the system is controllable (by Assumption **A**), for any $1 \leq k \leq n-1$ and any $i \in \underline{n}$, one of the two following conditions are satis ed:

$$x_i^c(k) \geq B_{ij} + (u_{el})_j, \text{ with } B_{ij} \neq \varepsilon \quad (16)$$

$$\exists\, p \in \underline{n} \text{ s. t. } x_i^c(k) \geq (A^{\otimes^p})_{lj} + x_j^c(k-p), \text{ with} (A^{\otimes^p})_{lj} \neq \varepsilon. \quad (17)$$

Note that $x_j^c(k-p)$ is either equal to $x_j^c(0)$ or satis es (16).

Hence, for *any* $k \geq 0$ and for *any* index $i \in \underline{n}$ we have $(x_{el} - x^c(k))_i \leq \theta_1(\delta) := \max\{0, (x_{el})_{i_1} - B_{i_1 j} - (u_{el})_j, (x_{el})_{i_2} - (A^{\otimes^p})_{li_1} - x_{i_1}(0), (x_{el})_{i_3} - (A^{\otimes^p})_{li_1} - B_{i_1 j} - (u_{el})_j\}$ for some indices i_1, i_2, i_3, j.

So from $x^c(k) \leq x^f(k) \leq z_k \oplus x_{ub}$ it follows that:

$$\begin{aligned}
\|x^f(k) - x_{el}\|_\infty &= \max_{i\in\underline{n}}\{(x^f(k) - x_{el})_i, (x_{el} - x^f(k))_i\}\\
&\leq \max_{i\in\underline{n}}\{((z_k \oplus x_{ub}) - x_{el})_i, (x_{el} - x^c(k))_i\}\\
&\leq \max_{i\in\underline{n}}\{(z_k - x_{el})_i, (x_{ub} - x_{el})_i, \theta_1(\delta)\}\\
&\leq \max_{i\in\underline{n}}\{(z_k - x_{el})_i, \theta_2(\delta)\}\\
&\leq \max_{i,j}\{(A^{\otimes^j} \otimes x(0) - (k-j)\rho - x_{el})_i, \theta_2(\delta)\}\\
&\leq \max_{i,j}\{(A^{\otimes^j} \otimes x(0) - (k-j)\rho - A^{\otimes^j} \otimes x_{el})_i, \theta_2(\delta)\}\\
&\leq \max_{i,j}\{(A^{\otimes^j} \otimes x(0) - A^{\otimes^j} \otimes x_{el})_i, \theta_2(\delta)\}\\
&\leq \max_{i}\{(x(0) - x_{el})_i, \theta_2(\delta)\} \leq \theta(\delta)
\end{aligned}$$

with $\theta_2(\delta) = \max\{\max_{i\in\underline{n}}(x_{ub} - x_{el})_i, \theta_2(\delta)\}$ and $\theta(\delta) = \max\{\delta, \theta_1(\delta)\}$, and where for the last transition we have used that fact that from standard properties of the max operator (recall that by de nition $\varepsilon - \varepsilon = \varepsilon$) it follows that: $a^T \otimes x - a^T \otimes y \leq \|x - y\|_\oplus$, for any $a \in \mathbb{R}_\varepsilon^n$ and $x, y \in \mathbb{R}^n$.

An immediate consequence of Theorem 3.3 is:

Proposition 3.4. For any input signal $u(\cdot)$ ful lling the constraint (6) and $u^c(k) \leq u(k) \leq u^f(k)$, the corresponding trajectory satis es $x^c(k) \leq x(k) \leq x^f(k)$, for all k and consequently $u(\cdot)$ is stabilizing. Moreover, there exists a nite K such that $x_{el} \leq x(k) \leq x_{ub}$, for all $k \geq K$.

3.3 Stabilizing receding horizon controller

Given the state and the input at event step $k-1$, the following cost function is introduced:

$$J(x(k-1), \tilde{u}(k)) = \sum_{j=0}^{N-1}\sum_{i=1}^{n} \max\{x_i(k+j|k-1) - (x_t)_i, 0\} - \beta \sum_{j=0}^{N-1}\sum_{i=1}^{m} u_i(k+j|k-1).$$

where N is the prediction horizon, $x(k+j|k-1)$ is the system state at event step $k+j$ as predicted at event step $k-1$, based on the MPL difference equation (4), the state $x(k-1)$ and the future input sequence

$$\tilde{u}(k) = [u^T(k|k-1) \cdots u^T(k+N-1|k-1)]^T.$$

In the context of DES the rst term of J expresses the tardiness (i.e. the delay with respect to the desired due date target x_t), while the second term maximizes the feeding times. We de ne the following receding horizon control (RHC) based optimization problem:

$$J^*(x(k-1)) = \min_{\tilde{u}(k)} J(x(k-1), \tilde{u}(k)) \quad (18)$$

$$\text{s.t.} \begin{cases} x(k+j|k-1) = A \otimes x(k+j-1|k-1) \oplus \\ B \otimes u(k+j|k-1) \\ u(k+j|k-1) - u(k+j-1|k-1) \geq -\rho \\ \forall j \in \{0, \cdots, N-1\}. \end{cases} \quad (19)$$

where $x(k-1|k-1) = x(k-1), u(k-1|k-1) = u(k-1)$. Let $\tilde{u}^\natural(k)$ be the optimal solution of the optimization problem (18) (19). Using the receding horizon principle at event counter k we apply the input $u^{\mathrm{RHC,N}}(k) = u^\natural(k|k-1)$. The evolution of the closed-loop system obtained from applying the receding horizon controller is denoted with

$$x^{\mathrm{RHC,N}}(k) = A \otimes x^{\mathrm{RHC,N}}(k-1) \oplus B \otimes u^{\mathrm{RHC,N}}(k),$$

with given initial conditions $x^{\mathrm{RHC,N}}(0) = x(0)$, $u^{\mathrm{RHC,N}}(k) = u(0)$.

Let us de ne the matrices

$$\tilde{D} = \begin{bmatrix} B & \mathcal{E} & \cdots & \mathcal{E} \\ A \otimes B & B & \cdots & \mathcal{E} \\ \vdots & \vdots & \ddots & \vdots \\ A^{\otimes^{N-1}} \otimes B & A^{\otimes^{N-2}} \otimes B & \cdots & B \end{bmatrix}, \tilde{C} = \begin{bmatrix} A \\ A^{\otimes^2} \\ \vdots \\ A^{\otimes^N} \end{bmatrix}$$

and the vectors $\tilde{u}(k) = [u^T(k-1) - \rho \cdots u^T(k-1) - N\rho]^T$, $\tilde{u}_{\rm el} = [u_{\rm el}^T \cdots u_{\rm el}^T]^T$, $\tilde{x}_{\rm t} = [x_{\rm t}^T \cdots x_{\rm t}^T]^T$ and $\tilde{x}(k) = [\tilde{x}^T(k|k-1) \cdots \tilde{x}^T(k+N-1|k-1)]^T = \tilde{C} \otimes x(k-1) \oplus \tilde{D} \otimes \tilde{u}(k) \oplus \tilde{x}_{\rm t}$.

Now we give some properties of the receding horizon controller (see (Necoara *et al.*, 2006; Necoara, 2006) for the proofs):

Lemma 3.5. $u^{\rm c}(k) \le u^{\rm RHC,N}(k), x^{\rm c}(k) \le x^{\rm RHC,N}(k)$, $\forall k \ge 0$.

Proposition 3.6. Assume $\beta < \frac{1}{mN}$ and consider the maximization problem

$$\tilde{u}^\sharp(k) = \arg\max_{\tilde{u}(k)} \sum_{j=0}^{N-1} \sum_{i=1}^{m} u_i(k+j|k-1) \tag{20}$$

$$\text{s.t.} \begin{cases} \tilde{D} \otimes \tilde{u}(k) \le \tilde{x}(k) \\ u(k+j|k-1) - u(k+j-1|k-1) \ge -\rho, \\ \forall j \in \{1, \cdots, N-1\}. \end{cases} \tag{21}$$

Then $\tilde{u}^\sharp(k)$ is also the optimal solution of the optimization problem (18) (19).

De ne $\tilde{u}^{*N}(k) := (-\tilde{D}^T) \otimes' \tilde{x}(k)$. The following proposition provides an analytic solution to the optimization problem (20) (21).

Proposition 3.7. The optimization problem (20) (21) has an unique solution given by:

$$\begin{cases} u^\sharp(k+N-1|k-1) = u^{*N}(k+N-1|k-1) \\ u^\sharp(k+j|k-1) = \min\{u^{*N}(k+j|k-1), \\ u^\sharp(k+j+1|k-1) + \rho\}, \end{cases} \tag{22}$$

for $j = N-2, \cdots, 0$.

Lemma 3.8. Any feasible solution $\tilde{u}_{\rm feas}(k)$ of (20) (21) satis es $\tilde{u}_{\rm feas}(k) \le \tilde{u}^\sharp(k)$.

The next theorem characterizes the stabilizing properties of the receding horizon controller:

Theorem 3.9. Given a prediction horizon N such that $\beta < \frac{1}{mN}$, the following inequalities hold

$$\begin{cases} u^{\rm c}(k) \le u^{\rm RHC,N}(k) \le u^{\rm f}(k) \\ x^{\rm c}(k) \le x^{\rm RHC,N}(k) \le x^{\rm f}(k) \end{cases} \tag{23}$$

and thus the receding horizon controller stabilizes the system (4).

PROOF. The left-hand side of inequalities (23) follows from Lemma 3.5.

The right-hand side of inequalities (23) is proved using induction. For $k = 0$ we have $u^{\rm RHC,N}(0) = u^{\rm f}(0) = u(0)$ and $x^{\rm RHC,N}(0) = x^{\rm f}(0) = x(0)$. Let us assume that $u^{\rm RHC,N}(k-1) \le u^{\rm f}(k-1)$ and $x^{\rm RHC,N}(k-1) \le x^{\rm f}(k-1)$ are valid and we prove that they also hold for k. Since $x(k|k-1) = A \otimes x(k-1) \oplus B \otimes (u(k-1) - \rho) \oplus x_{\rm t}$ and $B \otimes u^\sharp(k|k-1) \le \tilde{x}(k|k-1)$, we have

$$u^{\rm RHC,N}(k) \le \tag{24}$$
$$(-B^T) \otimes' (A \otimes x(k-1) \oplus B \otimes (u(k-1) - \rho) \oplus x_{\rm t})$$

From (24) and our induction hypothesis we have:

$$B \otimes u^{\rm RHC,N}(k) \le$$
$$A \otimes x^{\rm RHC,N}(k-1) \oplus B \otimes (u^{\rm RHC,N}(k-1) - \rho) \oplus x_{\rm t} \le$$
$$A \otimes x^{\rm f}(k-1) \oplus B \otimes (u^{\rm f}(k-1) - \rho) \oplus x_{\rm t}$$

On the other hand, $u^{\rm f}(k)$ is the largest solution of

$$B \otimes u^{\rm f}(k) \le A \otimes x^{\rm f}(k-1) \oplus B \otimes (u^{\rm f}(k-1) - \rho) \oplus x_{\rm t}$$

From Lemma 2.1 it follows that $u^{\rm RHC,N}(k) \le u^{\rm f}(k)$. Then, $x^{\rm RHC,N}(k) = A \otimes x^{\rm RHC,N}(k-1) \oplus B \otimes u^{\rm RHC,N}(k) \le A \otimes x^{\rm f}(k-1) \oplus B \otimes u^{\rm f}(k) = x^{\rm f}(k+1)$.

The stabilizing property of the receding horizon controller follows from Proposition 3.4.

4. EXAMPLE

We consider the following system:

$$x_{\rm sys}(k) = \begin{bmatrix} \varepsilon & 1 & \varepsilon & \varepsilon \\ \varepsilon & \varepsilon & 2 & \varepsilon \\ \varepsilon & \varepsilon & \varepsilon & 3 \\ 4 & \varepsilon & \varepsilon & \varepsilon \end{bmatrix} \otimes x_{\rm sys}(k-1) \oplus \begin{bmatrix} 2 \\ \varepsilon \\ \varepsilon \\ \varepsilon \end{bmatrix} \otimes u_{\rm sys}(k)$$

For this example we have a (unique) MPA eigenvalue $\lambda^* = 2.5$, and a corresponding MPA eigenvector $v^* = [0\ 1.5\ 2\ 1.5]^T$. We consider the due date signal $r_{\rm sys}(k) = [17\ 15\ 1\ 10]^T + 4.5k$ (so $\rho = 4.5$), and the initial condition $x_{\rm sys}(0) = [20\ \ 31.5\ \ 42\ \ 51.5]^T$ and $u_{\rm sys}(0) = 20$. The system and reference signal de ned above satisfy Assumption **A**.

Now we design stabilizing state feedback and receding horizon controllers for this system. For the RHC controllers we consider the prediction horizons $N = 2$ and $N = 5$, and a weight $\beta = 0.1$ that satis es the conditions of Proposition 3.6 and Theorem 3.9. In Figures 1 and 2 we have plotted respectively the control signals and the state trajectories for the closed-loop controlled normalized system. Clearly, all controllers are stabilizing. Moreover, the constant controller and the RHC controller with $N = 5$ also make all states less than the target states. This is not always the case for the state feedback controller and for the RHC controller with $N = 2$ (so in the latter case the prediction horizon is clearly selected too small). Also note that $u^{\rm c}(k) \le u^{\rm RHC,N}(k) \le u^{\rm f}(k)$ for all k and for $N = 2, 5$. Furthermore, $u^{\rm RHC,5}(k) \le u^{\rm RHC,2}(k)$ and $x^{\rm RHC,5}(k) \le x^{\rm RHC,2}(k)$ for all k. In fact, these two inequalities hold in general as it can be shown that if $\beta < \frac{1}{mN}$ then for two prediction horizons $N_1 < N_2$ we have $u^{\rm RHC,N_2}(k) \le u^{\rm RHC,N_1}(k)$, $x^{\rm RHC,N_2}(k) \le x^{\rm RHC,N_1}(k)$ for all k (see (Necoara *et al.*, 2006)).

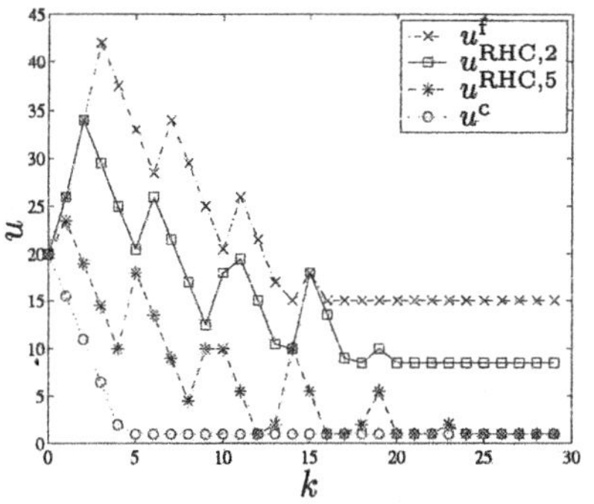

Fig. 1. The state feedback, constant and RHC control signals for the normalized system.

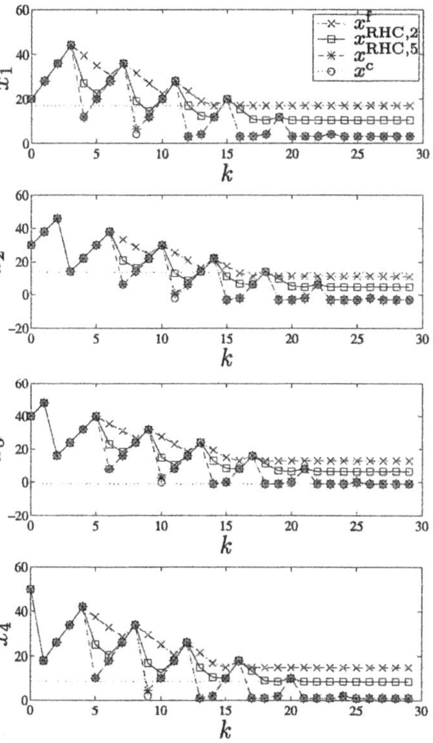

Fig. 2. The evolution of the states for the state feedback, constant and RHC controllers for the normalized system. The dotted lines correspond to the target state x_{t}.

5. CONCLUSIONS AND FUTURE RESEARCH

We have discussed the problem of stabilization of max-plus-linear (MPL) discrete-event systems. We have de ned a stabilizing constant controller and a stabilizing state feedback controller that could be considered as a lower and upper bound respectively for the receding horizon control (RHC) controllers. For the RHC controllers we have considered a trade-off between minimizing the tardiness and maximizing the input times. Using only the constraint that expresses that the input signal should be nondecreasing and provided the trade-off weight is small enough, we have derived an analytic expression for the RHC controller and proved that stability can be achieved in nite time.

We have also discussed also the main properties of the state feedback, constant , and RHC controllers.

ACKNOWLEDGEMENTS

Research supported by STW projects "MPC for hybrid systems" (DMR.5675) and "Multi-agent control of large-scale hybrid systems" (DWV.6188), European NoE HYCON (FP6-IST-511368), and European IST project SICONOS (IST-2001-37172).

REFERENCES

Baccelli, F., G. Cohen, G.J. Olsder and J.P. Quadrat (1992). *Synchronization and Linearity*. John Wiley & Sons. New York.

Cottenceau, B., L. Hardouin, J. Boimond and J. Ferrier (2001). Model reference control for timed event graphs in dioid. *Automatica* **37**(8), 1451 1458.

De Schutter, B. (1996). Max-Algebraic System Theory for Discrete Event Systems. PhD thesis. K.U.Leuven, Belgium.

De Schutter, B. and T. van den Boom (2001). Model predictive control for max-plus-linear discrete event systems. *Automatica* **37**(7), 1049 1056.

Heidergott, B., G.J. Olsder and J. Woude (2005). *Max Plus at Work*. Princeton Univ. Press. Princeton.

Maia, C. A., L. Hardouin, R. Santos-Mendes and B. Cottenceau (2003). Optimal closed-loop control of timed event graphs in dioids. *IEEE Trans. on Automatic Control* **48**(12), 2284 2287.

Mayne, D.Q., J.B. Rawlings, C.V. Rao and P.O.M. Scokaert (2000). Constrained model predictive control: Stability and optimality. *Automatica* **36**(7), 789 814.

Menguy, E., J.L. Boimond and L. Hardouin (1998). Optimal control of discrete event systems in case of updated reference input. In: *Proc. IFAC Conf. on System Structure and Control (SSC'98)*. Nantes, France. pp. 601 607.

Menguy, E., J.L. Boimond, L. Hardouin and J.L. Ferrier (2000). A rst step towards adaptive control for linear systems in max algebra. *Discrete Event Dynamic Systems: Theory and Applications* **10**(4), 347 367.

Necoara, I. (2006). Model Predictive Control for Max-Plus-Linear and Piecewise Af ne Systems. PhD thesis. Delft Univ. of Technology, The Netherlands.

Necoara, I., T.J.J. van den Boom, B. De Schutter and J. Hellendoorn (2006). Stabilization of max-plus-linear systems using receding horizon control: The unconstrained case Extended report. Tech. Rep. 06-007a. Delft Center for Systems and Control, Delft Univ. of Techn., The Netherlands.

van den Boom, T. and B. De Schutter (2002). Properties of MPC for max-plus-linear systems. *European Journal of Control* **8**(5), 453 462.

van den Boom, T.J.J., B. De Schutter and I. Necoara (2005). On MPC for max-plus-linear systems: Analytic solution and stability. In: *Proc. 44th IEEE Conf. on Decision and Control, and the European Control Conf. 2005 (CDC-ECC'04)*. Seville, Spain. pp. 7816 7821.

ELSEVIER
IFAC
PUBLICATIONS

ONLINE CLASSIFICATION OF SWITCHING MODELS BASED ON SUBSPACE FRAMEWORK

K. M. Pekpe * S. Lecœuche **

** LAGIS - UMR CNRS 8146*
Polytech'Lille - IMA 2 Bd Langevin, 59 655 Villeneuve d'Ascq
Phone: +33(0)320434565, Fax : +33(0)320337189
Email : Midzodzi.Pekpe@polytech-lille.fr
*** LAGIS - UMR CNRS 8146*
Département G.I.P., Mines de Douai, 59508 Douai Cedex
Phone: +33(0)327712445, Fax : +33(0)327712980
Email: lecoeuche@ensm-douai.fr

Abstract: The paper deals with the modelling of switching systems and focuses on the characterization of the local functioning modes using online clustering approach. The considered system is represented as a weighted sum of local linear models where each model could have its own structure. That implies that the parameters and the order of the switching system could change when the system switches. The presented method consists in two steps. First, an online estimation method of the Markov parameters matrix of the local linear models is established. Secondly, the labelling of theses parameters is done using a dynamical decision space worked out with learning techniques, each local model being represented by a cluster. The paper ends with an example, in view to illustrate the method performances.

1. INTRODUCTION

An online classification method for switching systems is proposed in this paper. Switching systems are a particular class of hybrid systems which can be considered as the weighted sum of the linear local models (or local models) with normalized weights. In this modelling, only one model is active at each time (binary weights). On the one hand, several methods are proposed for offline clustering of the hybrid systems (Breiman, 1993), (Bemporad *et al.*, 2001), (Ferrari-Trecate *et al.*, 2003), and (Pekpe *et al.*, 2004). A clustering method based on hyperplane determination is proposed in (Breiman, 1993), but this method consists of the estimation of two hyperplanes continuously joined together. The method proposed in (Bemporad *et al.*, 2001) uses a mixed integer programming which is NP-hard in the worst case, then this is practically applicable only when the number of the data is very small (see (Ferrari-Trecate *et al.*, 2003)). Moreover, the order of the system cannot change in the two above methods. In 2004, (Vidal, 2004) proposes an identification of pwarx hybrid models with unknown and different orders. Also, a recent offline method is proposed in (Pekpe *et al.*, 2003) for data classification and parameter identification of switching state-space models with variable structure. This method uses a change detection technique to estimate the switches and a subspace method to estimate the Markov parameters (state representation) of the local models.

On the other hand, online data classification does not seem to have a great attention in hybrid systems community. Recently, (Lecœuche *et al.*, 2006) presents an approach based on the combination of a recursive identification technique and a classifier set for non stationary environment. But, this approach is limited to non-stationnary systems with a fixed structure.

The method proposed in this paper supposes that the number and the order of the local linear models are unknown/ Moreover, the order could change for each local model. In fact, no *a priori* knowledge, even the number of models, is required, the knowledge on the switching model is gained using online estimation and continuous learning. The approach developped here is based on a similar way than (Lecœuche *et al.*, 2006) : one stage for the estimation and one stage for the classification.

The paper structure is as follow. After presenting the overall formulation of the problem in the second section, the third and fourth sections are respectively dedicated to the presentation of the online estimation method and to the online clustering technique. The last section presents first results based on simulated switching system.

2. PROBLEM FORMULATION

The output of the switching system is represented as a weighted sum of the outputs of h local models:

$$y_k = \sum_{s=1}^{h} p_{s,k} . y_{s,k} \quad (1)$$

where the scalar k represents the time index, the vector $y_k \in R^\ell$ the output of the system, the vector $y_{s,k} \in R^\ell$ the output of local model s and the scalar $p_{s,k}$ the weight associated to $y_{s,k}$. For each time k, the weights verify the following condition :

$$p_{s,k} \in \{0,1\} \text{ and } \sum_{s=1}^{h} p_{s,k} = 1, \ \forall k \quad (2)$$

Each local model is supposed to be linear and described by the state space model of order n_s :

$$\begin{aligned} x_{s,k+1} &= \mathbf{A_s} x_{s,k} + \mathbf{B_s} u_k + v_{s,k} \\ y_{s,k} &= \mathbf{C_s} x_{s,k} + \mathbf{D_s} u_k + w_{s,k} \end{aligned} \quad (3)$$

where the process noises $v_{s,k} \in R^{n_s}$ and the measurement noises $w_{s,k} \in R^\ell$ of the sth local model are zero mean white noises which are uncorrelated with the inputs $u_k \in R^m$. The vector $x_{s,k} \in R^{n_s}$ represents the state vector of model s.

The h local models are assumed to be stable and being active during a minimal time τ (Hespanha and Morse, 1999), this parameter is called the dwell time.

From available measurements of inputs u_k and outputs y_k, our goal is to characterize the current functioning mode using online classification of the regressors vectors and, in other words, to estimate the weight $p_{s,k}$.

This task is done in two stages described hereafter :

- *Online local model Markov parameters estimation* First, the online estimation of the Markov parameters matrix of the local models is fulfilled. This black-box estimation uses the state space representation, which is particularly adapted to MIMO processes and does not require any canonical parameterization. Moreover, the use of the Markov parameters suits for the estimation of the system parameters with variable structure or order. The presented method is based on the FIR modelling of the local models. It is shown that the regressors of a FIR model lie in the hyperplane which orthogonal matrix is the Markov parameters matrix. Markov parameters of each local model are estimated by least squares method from data derived from a sliding window.

- *Determination of the current functioning mode* This classification is based on a dynamical decision space obtained by online learning techniques. The novelty of the proposed approach consists in exploiting a specific clustering technique making possible the continuous modelling of the functioning modes.These ones are modelled by an online neural network technique (Lecœuche and Lurette, 2003). When a new observation is presented at this algorithm (new estimation of Markov parameters matrix), the decision space is updated according to the information brought by this observation and the current functioning mode is determined (the closest linear local model). Then, the characterization of the current functioning mode is given in term of membership degree of the identified parameters vector to updated classes representing the actual local linear modes.

These two stages will be more precisely presented in the following two sections.

3. LOCAL MODELS PARAMETERS ESTIMATION

The online estimation of the local models Markov parameters matrix is discussed in this section. First, it establishes that the regressors of one local model belong to the same hyperplane and an orthogonal matrix of this hyperplane is the Markov parameters matrix. Then, the online estimation of the Markov parameters matrix by least squares method is proposed.

3.1 *Local model hyperplane equation*

The aim of this paragraph is to perform the online estimation of the Markov parameters of the local models. To reach this goal, it is established first that all the regressors built from the data resulting from the same local model belong to the same hyperplane. This could be proved from the expression of the local model output according to the state and the inputs multiplied by the Markov parameters. This expression being developed, the state influence is deleted by weighting it by a high power of the local state matrix ($\mathbf{A}_s$) which is supposed to be steady. This implies that the local model output is equal to the inputs multiplied by the Markov parameters, similar to the approximation by the FIR model. This equality can be rewritten as the orthogonality of the regressors (which contains the local model output and the inputs) and the Markov parameters matrix.

In the following, it is supposed that the system stays in each mode during a minimal time (or dwell time) τ.

The sth local model being active, its output can be expressed (see relation (3)) as:

$$\begin{aligned} y_{s,k} = \mathbf{C}_s \mathbf{A}_s^{i-2} \mathbf{B}_s u_{k-i+1} + \ldots + \mathbf{C}_s \mathbf{B}_s u_{k-1} \\ + \mathbf{D}_s u_k + \mathbf{C}_s \mathbf{A}_s^{i-2} v_{s,k-i+1} + \ldots + \mathbf{C}_s v_{s,k-1} + w_{s,k} \\ + \mathbf{C}_s \mathbf{A}_s^{i-1} x_{s,k-i+1} \end{aligned} \quad (4)$$

As the local model is assumed to be stable, the term $\mathbf{C}_s\mathbf{A}_s^{i-1}x_{s,k-i}$ could be neglected for high values of "i". This term is considered negligible if:

$$\left\|\mathbf{C}_s\mathbf{A}_s^{i-1}x_{s,k-i+1}\right\| \leq \sqrt{variance(w_{s,k})} \quad (5)$$

Thus, (4) becomes:

$$y_{s,k} = \mathbf{C}_s\mathbf{A}_s^{i-2}\mathbf{B}_s u_{k-i+1} + \ldots + \mathbf{C}_s\mathbf{B}_s u_{k-1} + \mathbf{D}_s u_k + \beta_{s,k} \quad (6)$$

with:

$$\beta_{s,k} = \mathbf{C}_s\mathbf{A}_s^{i-1}x_{s,k-i+1} + \mathbf{H}_{s,i}^{v}\bar{v}_{s,k-1} + w_{s,k} \quad (7)$$

where

$$\mathbf{H}_{s,i}^{v} = \left(\mathbf{C}_s\mathbf{A}_s^{i-2} \ \ \mathbf{C}_s\mathbf{A}_s^{i-3} \ \ \ldots \ \ \mathbf{C}_s\mathbf{A} \ \ \mathbf{C}_s \right) \in R^{\ell\times m(i-1)}$$
$$\bar{v}_{s,k-1} = \left(v_{s,k-i+1}^T \ \ \cdots \ \ v_{s,k-2}^T \ \ v_{s,k-1}^T \right)^T \in R^{\ell(i-1)\times n_s} \quad (8)$$

the vector $\beta_{s,k}$ is the perturbation due to the noises and the approximation of the local model by a FIR model. The previous equation can be written as:

$$y_{s,k} = \mathbf{H}_{s,i}^{v}\left(u_{k-i+1}^T \ \ \cdots \ \ u_{k-1}^T \ \ u_k^T \right)^T + \beta_k \quad (9)$$

Indeed, equation (9) gives the FIR approximation of the local model, it can be rewritten as :

$$\mathbf{M}_s z_{s,k} = \beta_k \quad (10)$$

where $z_{s,k}$ is the regressors vector of the local model:

$$z_{s,k} = \left(u_{k-i+1}^T \ \ \cdots \ \ u_{k-1}^T \ \ u_k^T \ \ y_{s,k}^T \right)^T \in R^{(mi+\ell)} \quad (11)$$

and $\mathbf{M}_s$ is a orthogonal matrix of the regressors, this matrix is called also as the augmented Markov parameters matrix:

$$\mathbf{M}_s = \left(\mathbf{H}_{s,i} \ \ -\mathbf{I}_\ell \right) \in R^{n_s\times(mi+\ell)} \quad (12)$$

where $H_{s,i} \in R^{\ell\times mi}$ is the Markov parameter matrix and defined as:

$$\mathbf{H_{s,i}} = \left(\mathbf{C}_s\mathbf{A}_s^{i-2}B_s \ \ \mathbf{C}_s\mathbf{A}_s^{i-3}\mathbf{B}_s \ \ \ldots \ \ \mathbf{C}_s\mathbf{B}_s \ \ \mathbf{D}_s \right) \quad (13)$$

The sub-script s denotes the index of the active local model, this sub-script disappears if the local model index is not indicated. In the deterministic case, all the regressors $z_{s,k}$ lie in the hyperplane:

$$\mathbf{M}_s z_{s,k} = 0 \quad (14)$$

But this relation changes in the presence of the noises, that makes the modelling problem more difficult.

3.2 *Online estimation of the local models Markov parameters*

The local model hyperplane equation has been established in the previous paragraph, least squares estimation of a orthogonal matrix of the hyperplane is proposed now. From the hyperplane equation (10), the orthogonal matrix ($\mathbf{M}_s$) is estimated. This estimation can be done if there are mi independent regressors $z_{s,k}$. These regressors are independent if the inputs are persistently excited of order mi. Consider the regressors matrices $\underline{z}_{s,k}$:

$$\underline{z}_{s,k} = \left(z_{s,k-\rho+1} \ \ \cdots \ \ z_{s,k+1} \ \ z_{s,k} \right) \quad (15)$$

$$\underline{z}_{s,k} = \begin{pmatrix} \mathbf{U}_k \\ \underline{y}_{s,k} \end{pmatrix} \quad (16)$$

with

$$\underline{y}_{s,k} = \left(y_{s,k-\rho+1} \ \ \cdots \ \ y_{s,k-1} \ \ y_{s,k} \right) \in \mathbb{R}^{\ell\times\rho} \quad (17)$$

$$\mathbf{U}_k = \begin{pmatrix} u_{k-\rho-i+2} & \cdots & u_{k-i} & u_{k-i+1} \\ u_{k-\rho-i+3} & \cdots & u_{k-i+1} & u_{k-i+2} \\ \vdots & \cdots & \vdots & \vdots \\ u_{k-\rho+1} & \cdots & u_{k-1} & u_k \end{pmatrix} \in R^{mi\times\rho} \quad (18)$$

and ρ is an integer equals or greater than the number of the row (mi) of the matrix $\mathbf{U}_k$. The unique matrix which is orthogonal (and defined by equation (12)) to this regressors matrix is determined. To estimate the matrix $\mathbf{M}_s$, it is enough to determine the Markov parameters matrix $\mathbf{H}_{s,i}$ (12). The following theorem gives the least squares estimation of this matrix.

Theorem 3.1. Under the following conditions :

- the matrix $\mathbf{U}_k \in R^{mi\times\rho}$ is full row rank,
- the local model s is stable,

the Markov parameters matrix is given by :

$$\underline{y}_{s,k}\mathbf{U}_k^T(\mathbf{U}_k\mathbf{U}_k^T)^{-1} = \mathbf{H}_{s,i} + \underline{\mathbf{B}}_{s,k}\mathbf{U}_k^T(\mathbf{U}_k\mathbf{U}_k^T)^{-1} \quad (19)$$

where the perturbation matrix $\underline{\mathbf{B}}_{s,k}$ is defined as:

$$\underline{\mathbf{B}}_{s,k} = \left(\beta_{s,k-\rho+1} \ \ \cdots \ \ \beta_{s,k-1} \ \ \beta_{s,k} \right) \in \mathbb{R}^{\ell\times\rho} \quad (20)$$

The mathematic expectation of the matrix $\underline{y}_{s,k}\mathbf{U}_k^T(\mathbf{U}_k\mathbf{U}_k^T)^{-1}$ is:

$$\begin{aligned} \mathbf{E}[\underline{y}_{s,k}\mathbf{U}_k^T(\mathbf{U}_k\mathbf{U}_k^T)^{-1}] &= \mathbf{E}[\mathbf{H}_{s,i}] + \mathbf{E}[\underline{\mathbf{B}}_{s,k}\mathbf{U}_k^T(\mathbf{U}_k\mathbf{U}_k^T)^{-1}] \\ &= \mathbf{H}_{s,i} + \mathbf{E}[\underline{\mathbf{B}}_{s,k}]\mathbf{U}_k^T(\mathbf{U}_k\mathbf{U}_k^T)^{-1} \end{aligned}$$

whereas (see equation (7)):

$$\mathbf{E}[\underline{\mathbf{B}}_{s,k}] = \mathbf{C}_s\mathbf{A}_s^{i-1}\left(x_{s,k-\rho-i+2} \ \ \cdots \ \ x_{s,k-\rho-i+2} \ \ x_{s,k-\rho-i+1} \right)$$

which can be neglected if integer i is great enough, or be considered as a deterministic perturbation if the term $\left\|\mathbf{C}_s\mathbf{A}_s^{i-1}x_{s,k-i+1}\right\|$ is not negligible.

If the consecutive regressors

$$z_k = \left(u_{k-i+1}^T \ \ \cdots \ \ u_{k-1}^T \ \ u_k^T \ \ y_k^T \right)^T, \ k = 1, \ldots\rho, \ \rho \geqslant mi \quad (21)$$

are generated by the same local model "s" (i.e. $y_k = y_{s,k}$) then the matrices $\underline{y}_{s,k}$ (resp. $\underline{z}_{s,k}$) and $\underline{y}_k$ (resp. $\underline{z}_k$) are equal. That suppose the dwell time τ should be greater than $i+\rho$ (see figure (1)). In this case the local output matrix (for local model s) $\underline{y}_{s,k}$ can be replace in the theorem by the global output matrix $\underline{y}_{s,k}$

$$\underline{y}_k = \left(y_{k-\rho+1} \ \ \cdots \ \ y_{k-1} \ \ y_k \right) \in \mathbb{R}^{\ell\times\rho} \quad (22)$$

If the matrix $\underline{y}_k$ is built with a data from two local models then the system is in transient mode. In the next section, the clustering method used to determine the class of the estimated matrix is given.

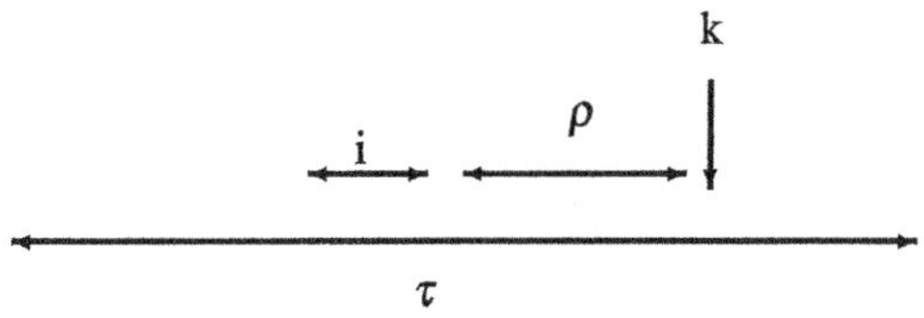

Fig. 1. Illustration of the dwell time

4. CLUSTERING METHOD

The modes modelling tool treats the data extracted from the identified Markov parameters matrices as they arrive. The new information is incorporated continuously in order to redefine the structure of the functioning modes (also named linear local models) and thus to model continuously the decision space. The taking into account of the various situations related to non stationarity environment requires the setting of specific adaptation rules through a continuous learning process. In the area of machine learning, some techniques exist with architectures exploiting incremental learning (Deng and Kasabov, 2003), (Eltoft and deFigueiredo, 1998), (Mouchaweh, 2004). Most of these algorithms present some disadvantages related to a coarse classes modelling and/or limited adaptation capacities in non stationary environment. In order to fill these gaps, two neural algorithms for the dynamic classification of the evolutionary data have been previously developed (Lecœuche and Lurette, 2003; Amadou-Boubacar and Lecœuche, 2005). These algorithms use a multi-prototype approach making possible to accurately model the structure of complex classes. In this paper, the AUDyC network based on a Gaussian modelling is used. Each functioning mode corresponds to a label of a complex class Ω which could be defined by an assembly of Gaussian prototypes Φ. The activation function of each hidden neuron determines the membership degree μ_j^t of the observation X_t to the prototype Φ_j. In order to obtain a fine classes representation, this one is based on the Mahanalobis distance

$$\mu_j^t = \exp\left(\frac{1}{2}\cdot(X_t - \overline{X}_j)^T * \Sigma_j^{-1} * (X_t - \overline{X}_j)\right) \quad (23)$$

where $\overline{X}_j$ and Σ_j are respectively the center and the covariance matrix of the prototype. The use of the membership function allows the implementation of the learning rules. With the first acquisition X_1, the network is initialized: creation of the first prototype Φ_1 constituting the first class Ω_1 (first functioning mode). The prototype is parameterized by its center $\overline{X}_1 = X_1$ and an initial covariance matrix Σ_{ini} beforehand selected. Then, according to new acquisitions, various situations can arise by comparison of the membership degree with two fixed thresholds μ_{min} and μ_{max} (resp. limit of prototype and class membership). Each case leads to a specific procedure (see Table 1).

Then, the AUDyC learning process is established in three principal phases:

Table 1. Classifier Adaptation rules.

	If	Then
1	$\mu_j^t < \mu_{min} \forall j \in \{1..J\}$	Creation $\Phi_{new} \in \Omega_{new}$
2	$\exists \Phi_j \in \Omega_i, \mu_{min} < \mu_j^t < \mu_{max}$	Creation $\Phi_{new} \in \Omega_i$
3	$\exists \Phi_j, \mu_{max} < \mu_j^t$	Adaptation $\widehat{\Phi}_j = \Phi_j(X_t)$
4	$\exists \Phi_l \in \Omega_p \cup$ $\exists \Phi_m \in \Omega_q, \mu_{min} < \mu_{l,m}^T < \mu_{max}$	Ambiguity $X_t \in \chi_{amb}$

4.1 *First phase: classification*

The classification stage corresponds to the creation and adaptation of prototypes and classes. In cases 1 and 2 of table 1, the observation X_t is not close to any existing prototype. These cases are similar to a distance rejection which could be used to detect the novelty in the multiclass environment. If the observation is not sufficiently close to any class (case 1), it leads to the creation of a new prototype and a new class corresponding to a new system mode. In the case 2, a new prototype is created and affected to the nearest class in order to contribute to a better definition of the mode model. In situation 3, the observation is rather close to a prototype to take part in its definition. The functioning mode adaptation is then carried out by using the following recursive equations

$$\overline{X}_j^t = \overline{X}_j^{old} + \frac{1}{N_p}(X_t - X_{t-N_p+1}) \quad (24)$$

$$\Sigma_j^t = \Sigma_j^{old} + \Delta X \begin{pmatrix} \frac{1}{N_p} & \frac{1}{N_p(N_p-1)} \\ \frac{1}{N_p(N_p-1)} & \frac{-(N_p+1)}{N_p(N_p-1)} \end{pmatrix} \Delta X^T \quad (25)$$

with $\Delta X = \left[X_t - \overline{X}_j^{old} \quad X_{i-N+1} - \overline{X}_j^{old}\right]$, N_p : prototype size.

4.2 *Second phase: fusion*

The case 4 of table 1 depicts the case of the rejection in ambiguity when an observation is sufficiently close to two or several prototypes (e.g. l, m) to contribute to their structure. The fusion procedure consists in evaluating the similarity of two densities by using an acceptance criterion based on the Kullback-Leibler distance (Zhou and Chellappa, 2004). When this criterion is higher than a threshold, the different classes (e.g. p, q) merge onto a unique new functioning mode.

4.3 *Third phase: evaluation*

The evaluation phase is significant to eliminate the parasite prototypes and classes possibly created by the noise influence. To detect not-representative modes, this phase is based on the cardinality of the models.

For more details on the AUDyC network, the reader can consult (Lecœuche and Lurette, 2003), (Lurette, 2003) and (Amadou-Boubacar *et al.*, 2005) which give theoretical and practical analysis of the AUDyC.

5. ILLUSTRATION EXAMPLE

Consider a switching system which is the sum of three models. The state-space model is used and each matrix ($\mathbf{A}_s, \mathbf{B}_s, \mathbf{C}_s$ and $\mathbf{D}_s$) is specific to each local model. The matrices of the local models are defined below :

$$\mathbf{A}_1 = \begin{pmatrix} 0{,}4 & 0{,}1 & 0 \\ 0{,}8 & 0{,}4 & 0 \\ 0 & 0 & 0{,}8 \end{pmatrix}, \mathbf{B}_1 = \begin{pmatrix} 1{,}5 & 0{,}9 \\ 1 & -1 \\ -1{,}5 & 2{,}3 \end{pmatrix}$$

$$\mathbf{C}_1 = \begin{pmatrix} 0{,}8 & 1{,}1 & 2 \\ -1{,}3 & 0{,}7 & 1{,}7 \\ 1{,}5 & 0{,}7 & -0{,}9 \end{pmatrix}, \mathbf{D}_1 = \begin{pmatrix} 0{,}4 & 0{,}8 \\ -0{,}6 & 1{,}4 \\ 1{,}3 & -0{,}75 \end{pmatrix}$$

$$\mathbf{A}_2 = \begin{pmatrix} 0{,}4 & 0{,}6 \\ 0{,}5 & 0{,}1 \end{pmatrix}, \mathbf{B}_2 = \begin{pmatrix} 1{,}5 & 0{,}9 \\ 1 & -1 \end{pmatrix}$$

$$\mathbf{C}_2 = \begin{pmatrix} 0{,}8 & 1{,}1 \\ -1{,}3 & 0{,}7 \\ 1{,}5 & 0{,}7 \end{pmatrix}, \mathbf{D}_2 = \begin{pmatrix} 0{,}4 & 0{,}8 \\ -0{,}6 & 1{,}4 \\ 1{,}3 & -0{,}75 \end{pmatrix}$$

$$\mathbf{A}_3 = \begin{pmatrix} 0{,}3 & 0{,}2 & 0 \\ 0{,}8 & 0{,}2 & 0 \\ 0 & 0 & -0{,}75 \end{pmatrix}, \mathbf{B}_3 = \mathbf{B}_1$$

$$\mathbf{C}_3 = \mathbf{C}_1, \mathbf{D}_3 = \mathbf{D}_1$$

The changes of the system dynamics are summarized as follows: $s = 1$ over the intervals $[1,099]$, $[300,399]$ and $[600,699]$, $s = 2$ over the intervals $[100,199]$, $[400,499]$ and $[700,800]$ and $s = 3$ over the intervals $[200,299]$ and $[500,599]$. The figure 2 gives the index of the active local model according to the time.

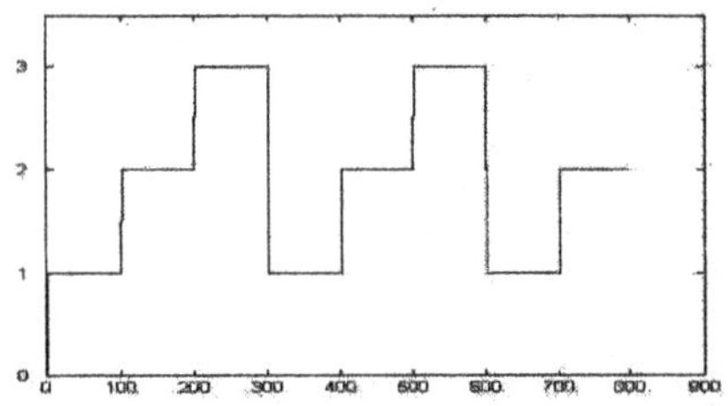

Fig. 2. The index of the active local model

The inputs u_k are Pseudo-Random Binary Sequences (PRBS) with variable amplitudes. The signal to noise ratio of the outputs with respect to the measurement noises is 25db. The process noises v_k are a Gaussian white noises and have a covariance $var(v_k) \simeq 7 \times 10^{-4} I_3$; the covariance matrix of the inputs is $var(u_k) \simeq 7 \times 10^{-2} I_2$. During the simulation, 900 input-output data have been processed on the system.

From, these IO data, the estimation of the Markov parameters matrix on sliding window is achieved by theorem *3.1*. The size of the sliding window p is chosen equal 35 and the integer i is fixed equal 13.

These identified parameters are sent to the clustering tool. The values of the identified matrix $\hat{\mathbf{H}}_{\mathbf{s},\mathbf{i}}$ constitute the X_t observation vector. In fact, this vector consists of all the monitored parameters. According to the complexity of the application, its dimension could be reduced (Markov parameters selection or reduction) or increased by adding, for example, complementary physical information. In this paper, in order to present tight and comprehensible results, we have applied a space reduction in order to obtain a 3D representation.

From the X_t information, the dynamical classifier updates the decision space (functioning modes models) and determines the current functioning mode. For the whole of this study, the AUDyC parameters are fixed as follows: $\Sigma_{ini} = 2.5, \mu_{min} = 0.1, \mu_{max} = 0.3, NP_{min} = 40, N_P = 500$ and $N_{amb} = 5$. For more information about the choice of the parameters, the reader can refer to (Amadou-Boubacar *et al.*, 2005). Figure 3.a illustrates the final representation space of the X_t raw data (the circle are non-classified data) and the figure 3.b gives the final classes locations.

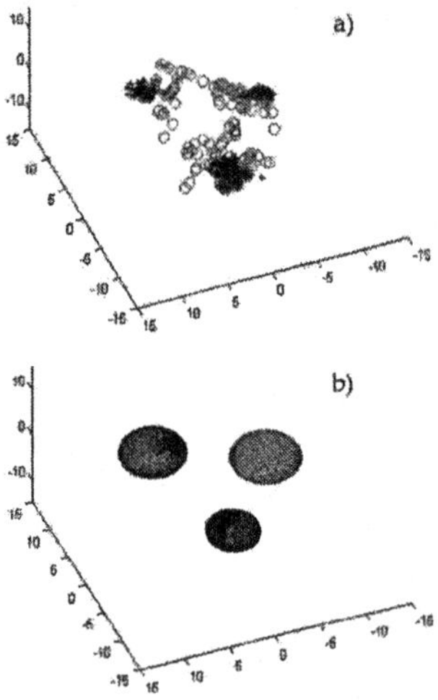

Fig. 3. Classified data and Final decision space

In fact, the data are online classified and the decision space is built and updated in a recursive way. Any *a priori* knowledge is required. From the first similar data, the classifier creates the first class corresponding to the first local model. When a switch occurs, according to the choice of the membership threshold, the data generated after the switch are not recognized as a known class. After a delay corresponding to a stabilization of these data (when the number of stable observations exceeds the NP_{min} threshold) in a particular area of the decision, a new class that characterizes the second functioning mode is created. All the new classes are created in this way (cf. Table 1).

At each time, the current functioning mode is determined by using the membership degree of the observation (membership ratio rule according to a threshold θ). On figure 4 where θ equals 0.25, it can be noticed that the observations located between modes are non classified and the class creation (3 first situations) is effective after an extra delay corresponding to NP_{min}. When the mode is already known, the decision is done quicker (e.q way through mode 3 to mode 1). In this case, the delay of recognition is mainly due to the convergence of the Markov parameters matrices estimation and to the noise influence.

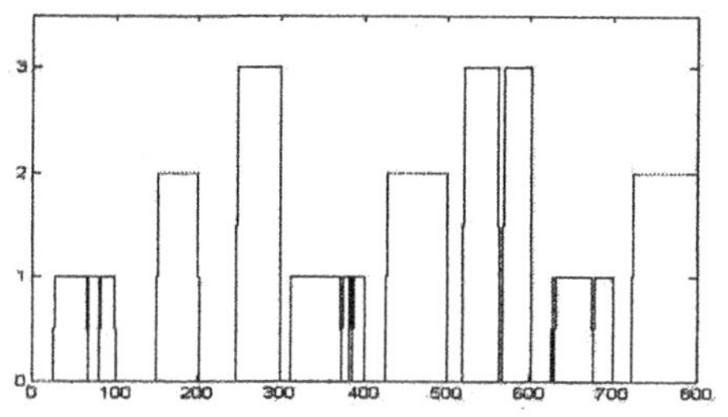

Fig. 4. Classification of the Markov Parameters Matrices; Membership threshold = 0.25

The figure 5 illustrates the behaviour of this method for a small value of θ. In this case, the membership is forced to known classes.

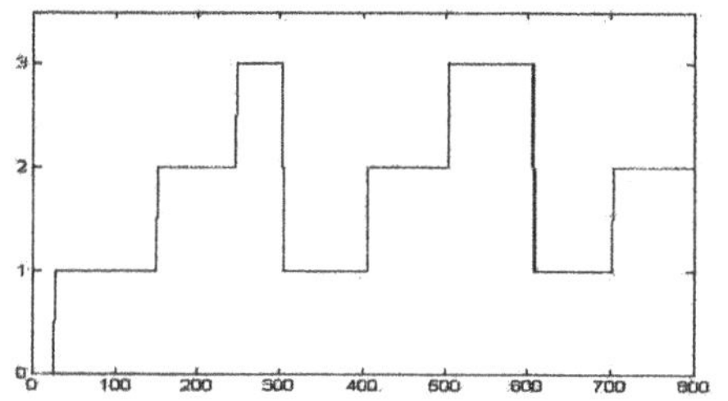

Fig. 5. Classification of the Markov Parameters Matrices; Membership threshold = ε

One can notice, even if this is not presented on the figures, that the classes parameters are continuously adapted. This allows a better accurate definition of the decision space and make possible the modelling of non-stationary local linear models (Lecœuche *et al.*, 2006). Of course, from this point, expert knowledge should be introduced in order to define if this evolution is "normal" (running in periods, chemical transition...) or if a progressive failure appears to the process.

6. CONCLUSION

In this paper, a new approach has been presented for the online determination of the functioning mode of a switching system. This approach is based on a recursive estimation tool coupled with a dynamic classification algorithm. The method is dedicated to switching systems where each linear local model could have its own structure (parameters and order). The interest of this approach is also to take into account commutations and evolutions of modes without *a priori* knowledge thanks to the use of a dynamical classifier.

7. REFERENCES

Amadou-Boubacar, H. and S. Lecœuche (2005). A new kernel-based algorithm for online clustering. In: *7th International Conference on Artificial Neural Networks*. Warsaw, Poland. pp. 350–358.

Amadou-Boubacar, H., S. Lecœuche and S. Maouche (2005). Audyc neural network using a new gaussian densities merge mechanism. In: *7th International Conference on Adaptive and Natural Computing Algorithms*. Coimbra, Portugal. pp. 155–158.

Bemporad, A., J. Roll and L. Ljung (2001). Identification of hybrid systems via mixed-integer programming. In: *13th International Conference on Artificial Neural Networks*. Orlando, FL. pp. 786–792.

Breiman, L. (1993). Hinging hyperplanes for regression, classification, and function approximation. *IEEE Transaction on Information Theory* **39**(3), 999–1013.

Deng, D. and N. Kasabov (2003). On-line pattern analysis by evolving self-organizing maps. *Neurocomputing* **51**, 87–103.

Eltoft, T. and R. deFigueiredo (1998). A new neural network for cluster-detection-and-labeling. *IEEE Transactions on Neural Networks* **9**, 1021–1035.

Ferrari-Trecate, G., M. Muselli, D. Liberati and M. Morari (2003). A clustering technique for the identification of piecewise affine systems. *Automatica* **39**(2), 205–217.

Hespanha, J. P. and A. S. Morse (1999). Stability of switched systems with average dwell-time. In: *38th CDC*. Phoenix, USA. pp. 2655–2660.

Lecœuche, S. and C. Lurette (2003). Auto-adaptive and dynamical clustering neural network. In: *13th International Conference on Artificial Neural Networks*. Istanbul, Turkey. pp. 350–358.

Lecœuche, S., G. Merc'ere and H. Amadou-Boubacar (2006). Modelling of non stattionary systems based on a dynamical decision space. In: *14th IFAC symposium on System Identification*. Newcastle, Australia.

Lurette, Christophe (2003). Dévelopement d'une technique neuronale auto-adaptative pour la classification de donnèes non-stationnaires.. PhD thesis. Lille University.

Mouchaweh, M. (2004). Diagnostic in real time for evolutionary processes in using pattern recognition and possibility theory. *International Journal of Computational Cognition* **2**(1), 79–112.

Pekpe, K. M., K. Gasso, G. Mourot and J. Ragot (2003). Subspace identification of switching model. 13th IFAC Symposium on System Identification. Rotterdam, The Netherlands.

Pekpe, K. M., K. Gasso, G. Mourot and J. Ragot (2004). Identification of switching systems using change detection technique in the subspace framework. 43rd IEEE Conference on Decision and Control. Atlantis, Bahamas.

Vidal, René (2004). Identification of pwarx hybrid models with unknown and possibly different orders. In: *IEEE American Conference on Control*. Boston, MA, USA.

Zhou, S. K. and R. Chellappa (2004). Kullback-leibler distance between two gaussian densities in reproducing kernel hilbert space. In: *International Symposium on Information Theory*. Chicago, USA.

Copyright © Analysis and Design of Hybrid Systems
Alghero, Italy 2006

ELSEVIER
IFAC
PUBLICATIONS

FUNCTIONAL ABSTRACTIONS OF STOCHASTIC HYBRID SYSTEMS

Manuela L. Bujorianu * **Henk A.P. Blom** ** **Holger Hermanns** ***

* *Faculty of Computer Science, Unversity of Twente, The Netherlands, email: L.M.Bujorianu@cs.utwente.nl*
** *National Aerospace Laboratory NLR, The Netherlands*
*** *Department of Computer Science, Universitat des Saarlandes, Germany*

Abstract: The verification problem for stochastic hybrid systems is quite difficult. One method to verify these systems is stochastic reachability analysis. Concepts of abstractions for stochastic hybrid systems are needed to ease the stochastic reachability analysis. In this paper, we set up different ways to define abstractions for stochastic hybrid systems, which preserve the parameters of stochastic reachability. A new concept of stochastic bisimulation is introduced and its connection with equivalence of stochastic processes is established.

Keywords: Markov models, probability function, verification, hybrid systems

1. INTRODUCTION

The investigation of stochastic hybrid systems (SHS) has recently received significant attention (Bujorianu, 2004), (Blom, 2003), (Pola *et al.*, 2003). The need for probabilistic modelling is motivated mainly by the partial knowledge about very complex systems, as well as, by the possibility of simplifying deterministic models introducing probabilistic reasoning. There are several dimensions of probabilistic reasoning over hybrid systems: (1) probabilistic quantification of discrete transitions; (2) stochastic reasoning about continuous evolution; (3) probabilistic aspects of the interaction between continuous and discrete dynamics. Because of these multiple dimensions, there is need to develop approaches towards the verification of SHS. Until now, concrete steps to solve the verification problems were made only for ad-hoc or particular classes of SHS models.

The novelty of the approach presented in this paper is to study the verification problem of SHS as bisimulation for a stochastic realization problem and to define abstractions of SHS that 'preserve' the probabilities used to define the stochastic reachability problem. Our abstractions focus on achieving numerical evaluation of the probability bounds even with the price of sacrificing the model expressivity. Ideally, abstractions for stochastic reachability analysis would provide state spaces in the real line, making available efficient algorithms from numerical analysis.

An important problem in the development of SHS is the preservation of stochastic properties. An abstraction can affect the probabilities and implicitly, the reachability analysis. Therefore, it is crucial to develop a theory of 'bisimulation' of SHS that preserves the relevant reach set probabilities of SHS considered. The aim of this paper is to initiate the development of such bisimulations/abstractions for SHS in a general setting.

The paper is structured as follows. In the next section, we give the motivation of this work and the formulation of the problem. Then, we present a short background on SHS, their semantics as stochastic processes and different operator methods which can be used to characterize stochastic processes. The main body of the paper is constituted by Sect.4. The main goal of this section is to define concepts of stochastic bisimulation and abstractions for SHS, which do not conduct to the equivalence of the SHS realizations. The attempt to define bisimulation that preserves different stochastic parametrizations of the SHS realizations is a wrong track, since it is nothing else, but another way, to present equivalence of stochastic processes. This section is divided in five parts. In the first subsection we define a space reduction technique based on quadratic forms associated to stochastic processes. The use of this technique is motivated by the formulation of the stochastic reachability for SHS. The goal of stochastic reachability analysis is to measure the set of trajectories, which reach a given target set until a given time horizon. Often, in practice, the target set is described as a level set for a 'nice' real-valued function defined on the whole state space. This function might be a norm, an observably function, or a weight function, etc. Applying this function to the paths of stochastic process that constitutes the SHS realization gives rise to a new stochastic process with the state space in the real line. Even the given process is Markovian and the above function has some measurability properties the new process might not be Markovian (Rogers and Pitman, 1981). To 'hide' this drawback we use the induced quadratic form, which intuitively is the composition between the quadratic form of the initial process and the above function. Usually, the new quadratic form has good properties such that there exists a Markov process associated to it. This new process is the best candidate to represent an abstraction of the initial process. We call it *functional abstraction.* In this way, the SHS realizations are 'approximated' by stochastic processes with a much smaller state space. The stochastic parameters of the induced process can be easily derived. Using this method, we formally define, in the following subsections, new concepts of stochastic bisimulation and functional abstractions of SHS. We have to underlie that the functional abstractions 'preserve' the continuous and jumping parts of an SHS. The paper ends with some conclusions.

2. PROBLEM FORMULATION

In this section we briefly present the stochastic reachability problem for stochastic hybrid systems and starting from it we derive the main ideas for defining abstractions for SHS.

Stochastic Reachability. Consider $M = (x_t, P_x)$ being a strong Markov process, the realization of a stochastic hybrid system (see definitions below). For this strong Markov process we address a verification problem consisting of the following stochastic reachability problem. Given a set $A \in \mathcal{B}(X)$ and a time horizon $T > 0$, let us to define (Bujorianu and Lygeros, 2003), (Bujorianu, 2004):

$$Reach_T(A) = \{\omega \in \Omega \mid \exists t \in [0,T] : x_t(\omega) \in A\}$$
$$Reach_\infty(A) = \{\omega \in \Omega \mid \exists t \geq 0 : x_t(\omega) \in A\}. \quad (1)$$

These two sets are the sets of trajectories of M, which reach the set A (the flow that enters A) in the interval of time $[0,T]$ or $[0,\infty)$. The reachability problem consists of determining the probabilities of such sets. The reachability problem is well-defined, i.e. $Reach_T(A)$, $Reach_\infty(A)$ are indeed measurable sets. Then the probabilities of reach events are: $P(T_A < T)$ or $P(T_A < \infty)$, where $T_A = \inf\{t > 0 | x_t \in A\}$ and P is a probability on the measurable space $(\Omega, \mathcal{F})$ of the elementary events associated to M. P can be chosen to be P_x (if we want to consider the trajectories, which start in x) or P_μ (if we want to consider the trajectories, which start in x_0 given by the distribution μ). Recall that $P_\mu(A) = \int P_x(A)d\mu$, $A \in \mathcal{F}$.

Usually a target set A is a level set for a given function $F : X \to \mathbb{R}$, i.e. $A = \{x \in X | F(x) > l\}$. The probability of the set of trajectories, which hit A until time horizon $T > 0$ can be expressed as $P[\sup\{F(x_t) > l \mid t \in [0,T]\}]$.

Problem Formulation. Define a new stochastic process $M^\#$ such that the reach set probabilities are preserved.

Idealy, the above argument shows that $F(x_t)$ would represent the best candidate for defining a possible abstraction for M, which preserves the reach set probabilities. The main difficulty is that $F(x_t)$ is a Markov process only for special choices of F (Rogers and Pitman, 1981). The problem is how to choose F well.

3. PRELIMINARIES

In this section we give the necessary background for stochastic hybrid systems, their semantics, some stochastic analysis tools and Dirichlet forms.

3.1 Stochastic Hybrid Systems

Many practical systems such as automobiles, chemical processes, and autonomous vehicles are best described by dynamics that comprise continuous state evolution within a mode of operation and discrete transitions from one mode to another, either controlled or autonomous. Such systems often interact with their environment in the presence of uncertainty and variability. SHS can model complex dynamics, uncertainty, multiple modes of operations and support high-level control specifications that are required for design of (semi-)autonomous applications. Several modelling paradigms for SHS have been already proposed in literature. A stochastic hybrid scheme that allows the continuous flows at each discrete location to be characterized by stochastic differential equations is described in (Hu *et al.*, 2000). An extension of this model that satisfies the strong Markov property is presented in (Blom, 2003). Methods to study the reachability problem for SHS have been addressed in (Bujorianu and Lygeros, 2003). Dynamically Coloured Petri Nets and Communicating Piecewise Deterministic Markov Processes as compositional specifications for SHS in (Everdij and Blom, 2005) with emphasis on modelling concurrency. Applications of SHS to large distributed systems have been studied for air traffic management systems (Pola *et al.*, 2003) and for communication networks (Hespanha, 2004).

3.2 Semantics of Stochastic Hybrid Systems

The executions of a stochastic hybrid system H form a stochastic process. Let us consider $M = (\Omega, \mathcal{F}, \mathcal{F}_t, x_t, P_x)$, the realization (or semantics) of H. Under mild assumptions on the parameters of H, M can be viewed as a family of Markov processes with the state space $(X, \mathcal{B})$, where X is the union of modes and $\mathcal{B}$ is its Borel σ-algebra. Let $\mathcal{B}^b(X)$ be the lattice of bounded positive measurable functions on X. The meaning of the elements of M can be found in any source treating continuous-parameter Markov processes (Davis, 1993). Suppose we have given a σ-finite measure μ on $(X, \mathcal{B})$.

In the following, some operator characterizations (used, in this paper, to define abstractions for SHS) of stochastic processes are given.

Operator Semigroup / Resolvent. Let $p_t(x, A) = P_x(x_t \in A)$, $A \in \mathcal{B}$ be the transition probability function. The meaning of this is the probability that, if $x_0 = x$, x_t will lie in the set A. The operator semigroup $\mathcal{P}$ is defined by $P_t f(x) = \int f(y) p_t(x, dy) = E_x f(x_t), \forall x \in X$, where E_x is the expectation w.r.t. P_x. The operator semigroup $(P_t)_{t>0}$ is, in fact, the collection of all first order moments, which can be associated with the family of random variables $\{x_t | t > 0\}$. The operator resolvent $\mathcal{V} = (V_\alpha)_{\alpha \geq 0}$ associated with $\mathcal{P}$ is $V_\alpha f(x) = \int_0^\infty e^{-\alpha t} P_t f(x) dt$, $x \in X$. Let denote by V the initial operator V_0 of $\mathcal{V}$, which is known as the *kernel operator* of the Markov process M. The operator resolvent $(V_\alpha)_{\alpha \geq 0}$ is the Laplace transform of the semigroup. The strong *generator* $\mathcal{L}$ is the derivative of P_t at $t = 0$. Let $D(\mathcal{L}) \subset \mathcal{B}_b(X)$ be the set of functions f for which the following limit exists (denoted by $\mathcal{L}f$): $\lim_{t \searrow 0} \frac{1}{t}(P_t f - f)$. A *quadratic form* $\mathcal{E}$ can be associated to the generator of a Markov process in a natural way. Let $L^2(X, \mu)$ be the space of square integrable μ-measurable extended real valued functions on X, w.r.t. the natural inner product $< f, g >_\mu = \int f(x) g(x) d\mu(x)$. A quadratic form $\mathcal{E}$ is defined as a closed form: $\mathcal{E}(f, g) = - < \mathcal{L}f, g >_\mu, f \in D(\mathcal{L}), g \in L^2(X, \mu)$. This leads to another way of parameterizing Markov processes. Instead of writing down a generator one starts with a quadratic form. As in the case of a generator it is typically not easy to fully characterize the domain of the quadratic form. For this reason one starts by defining a quadratic form on a smaller space and showing that it can be extended to a closed form in subset of $L^2(\mu)$. When the Markov process can be initialized to be stationary, the measure μ is typically this stationary distribution (see (Davis, 1993), p.111). More generally, μ does not have to be a finite measure. If M is a right Markov process then $\mathcal{E}$ is a regular Dirichlet form (Ma and Rockner, 1990), (Fukushima, 1980).

Dirichlet Forms. A *coercive closed form* (Albeverio *et al.*, 1993) is a quadratic form $(\mathcal{E}, D(\mathcal{E}))$ with $D(\mathcal{E})$ dense in $L^2(X, \mu)$, which satisfies the: (i) closeness axiom, i.e. its symmetric part is positive definite and closed in $L^2(X, \mu)$, (ii) continuity axiom (*Sector condition*). $\mathcal{E}$ is called *Dirichlet form* if, in addition, it satisfies the third axiom: (iii) contraction condition (*Dirichlet property*), i.e. $\forall u \in D(\mathcal{E})$, $u^* = u^+ \wedge 1 \in D(\mathcal{E})$ and $\mathcal{E}(u \pm u^*, u \mp u^*) \geq 0$. See (Ma and Rockner, 1990), (Fukushima, 1980), (Albeverio *et al.*, 1993).

Let $(\mathcal{L}, D(\mathcal{L}))$ be the generator of a coercive form $(\mathcal{E}, D(\mathcal{E}))$ on $L^2(X, \mu)$, i.e. the unique closed linear operator on $L^2(X, \mu)$ such that $1 - \mathcal{L}$ is onto, $D(\mathcal{L}) \subset D(\mathcal{E})$ and $\mathcal{E}(u, v) = < -\mathcal{L}u, v >$ for all $u \in D(\mathcal{L})$ and $v \in D(\mathcal{E})$. Let $(T_t)_{t>0}$ be the strongly continuous contraction semigroup on $L^2(X, \mu)$ generated by $\mathcal{L}$ and $(G_\alpha)_{\alpha>0}$ the corresponding strongly continuous contraction semigroup. A right process M with the state space X is *associated* with a Dirichlet form $(\mathcal{E}, D(\mathcal{E}))$ on $L^2(X, \mu)$ if the semigroup (P_t) of the process M is a μ-version of the form semigroup (T_t). Only those Dirichlet forms (called *quasi-regular Dirichlet forms*), which satisfy some regularity conditions can be associated with some right Markov processes and viceversa (Th.1.9 (Albeverio *et al.*, 1993)). Prop. 4.2 from (Albeverio *et al.*, 1993) states that two right Markov processes M and M' with state space X associated with a common quasi-regular Dirichlet form $(\mathcal{E}, D(\mathcal{E}))$ are stochastically equivalent (Ma and Rockner, 1990), (Fukushima, 1980). That means a quasi-regular Dirichlet form characterizes a class of stochastically equivalent right Markov processes.

4. ABSTRACTIONS OF STOCHASTIC HYBRID SYSTEMS

The idea is to apply a "state space reduction" technique based on the general 'induced Dirichlet forms' method to achieve abstractions for SHS. With this technique, the realizations of SHS are 'approximated' by a one-dimensional stochastic process with a much smaller state space.

4.1 Induced Dirichlet Forms

First, we define the concept of *induced Dirichlet form* (introduced in (Iscoe and McDonald, 1990) only for symmetric Dirichlet forms) and prove some properties (relations between generators, operator semigroups, kernel operators) of this concept, which will be used further in defining a new concept of stochastic bisimulation between Markov processes.

Let $M = (\Omega, \mathcal{F}, \mathcal{F}_t, x_t, P_x)$ be a right Markov process with the state space X. Now assume that X is a Lusin space (i.e. it is homeomorphic to a Borel subset of a compact metric space) and $\mathcal{B}(X)$ or $\mathcal{B}$ is its Borel σ-algebra. Note that for the majority of the stochastic hybrid system models the state space is a Lusin space (Pola *et al.*, 2003). Assume also that μ is a σ-finite measure on $(X, \mathcal{B})$ and μ is a stationary measure of the process M. Let $X^{\#}$ another Lusin space (with $\mathcal{B}^{\#}$ its Borel σ-algebra) and $F : X \to X^{\#}$ be a measurable function. Let $\sigma(F)$ be the sub-σ-algebra of $\mathcal{B}$ generated by F. If μ is a probability measure then the projection operator between $L^2(X, \mathcal{B}, \mu)$ and $L^2(X, \sigma(F), \mu)$ is the conditional expectation $E_\mu[\cdot|F]$. Recall that E_μ is the expectation defined w.r.t. P_μ. We denote by $\mu^{\#}$ the image of μ under F, i.e. $\mu^{\#}(A^{\#}) = \mu(F^{-1}(A^{\#}))$, for all $A^{\#} \in \mathcal{B}^{\#}$. In general, anything associated with $X^{\#}$ will carry an #-superscript in this section.

Let $\mathcal{E}$ be the Dirichlet form on $L^2(X, \mu)$ associated to M. F induces a form $\mathcal{E}^{\#}$ on $L^2(X^{\#}, \mu^{\#})$ by

$$\mathcal{E}^{\#}(u^{\#}, v^{\#}) = \mathcal{E}(u^{\#} \circ F, v^{\#} \circ F);$$

for $u^{\#}, v^{\#} \in D[\mathcal{E}^{\#}]$, where

$$D[\mathcal{E}^{\#}] = \{u^{\#} \in L^2(X^{\#}, \mu^{\#}) | u^{\#} \circ F \in D[\mathcal{E}]\}.$$

It can be shown (Prop.1.4 (Iscoe and McDonald, 1990)), under a mild condition on the conditional expectation operator $E_\mu[\cdot|F]$ that $\mathcal{E}^{\#}$ is a Dirichlet form. If, in addition, $\mathcal{E}^{\#}$ is quasi-regular then we can associate with it a right Markov process $M^{\#} = (\Omega, \mathcal{F}, \mathcal{F}_t, x_t^{\#}, P_x^{\#})$ be a right Markov process with the state space $X^{\#}$. The process $M^{\#}$ is called the *induced Markov process* w.r.t. to the proper map F. If the image of M under F is a right Markov process then $x_t^{\#} = F(x_t)$. The process $M^{\#}$ might have some different interpretations like a refinement of discrete transitions structure, or an approximation of continuous dynamics or an abstraction of the entire process. It is difficult to find a practical condition to impose on F, which would guarantee that $\mathcal{E}^{\#}$ is also quasi-regular. To circumvent this problem, it is possible to restrict the original domain $D[\mathcal{E}^{\#}]$ and impose some regularity conditions on F (Iscoe and McDonald, 1990).

Assumption 1. Suppose that $\mathcal{E}^{\#}$ is a quasi-regular Dirichlet form.

Let $(\mathcal{L}, D(\mathcal{L}))$ and $(\mathcal{L}^{\#}, D(\mathcal{L}^{\#}))$ be the generators of $\mathcal{E}$ and $\mathcal{E}^{\#}$, respectively.

Proposition 1. Under assumption 1, the generators $\mathcal{L}$ and $\mathcal{L}^{\#}$ are related as follows

$$\mathcal{L}(u^{\#} \circ F) = \mathcal{L}^{\#} u^{\#} \circ F, \forall u^{\#} \in D(\mathcal{L}^{\#}).$$

Theorem 2. Under assumption 1, for all $A^{\#} \in \mathcal{B}^{\#}(X^{\#})$ and for all $t > 0$ we have $p_t^{\#}(Fx, A^{\#}) = p_t(x, F^{-1}(A^{\#}))$, where $(p_t^{\#})$ and (p_t) are the transition functions of $M^{\#}$ and M, respectively.

Corollary 3. Under assumption 1, the semigroups $(P_t^{\#})$ and (P_t) of $M^{\#}$ and M are related by $P_t^{\#} u^{\#} \circ F = P_t(u^{\#} \circ F), \forall u^{\#} \in \mathcal{B}^b(X^{\#})$.

Remark 1. In the terminology of (Bujorianu *et al.*, 2005), we can say that $M^{\#}$ simulates M.

4.2 Stochastic Bisimulation

In this subsection we define a new concept of stochastic bisimulation for SHA. This concept is defined as measurable relation (Strubbe and Schaft, 2005), which induces equivalent Dirichlet forms on the quotient spaces. In comparison with (Strubbe and Schaft, 2005), in defining stochastic bisimulation, we do not impose the equivalence of the quotient processes, which might not have Markovian properties (Rogers and Pitman, 1981), but we impose the equivalence of the induced Markov processes (that can differ from the quotient processes) associated with the induced Dirichlet forms.

Let $(X, \mathcal{B}(X))$ and $(Y, \mathcal{B}(Y))$ be Lusin spaces and let $\mathcal{R} \subset X \times Y$ be a relation such that $\Pi^1(\mathcal{R}) = X$ and $\Pi^2(\mathcal{R}) = Y$. We define the equivalence relation on X that is induced by the relation $\mathcal{R} \subset X \times Y$, as the transitive closure of $\{(x, x') | \exists y$ s.t. $(x, y) \in \mathcal{R}$ and $(x', y) \in \mathcal{R}\}$. Analogously, the induced (by $\mathcal{R}$) equivalence relation on Y can be defined. We write $X/_{\mathcal{R}}$ and $Y/_{\mathcal{R}}$ for the sets of equivalence classes of X and Y induced by $\mathcal{R}$. We denote the equivalence class of $x \in X$ by $[x]$. Let $\mathcal{B}^{\#}(X) = \mathcal{B}(X) \cap \{A \subset X |$ if $x \in A$ and $[x] = [x']$ then $x' \in A\}$ be the collection of all Borel sets, in which any equivalence class of X is either totally contained or totally not contained. It can be checked that $\mathcal{B}^{\#}(X)$ is a σ-algebra. Let $\pi_X : X \to X/_{\mathcal{R}}$ be the mapping that maps each $x \in X$ to its equivalence class and let $\mathcal{B}(X/_{\mathcal{R}}) = \{A \subset X/_{\mathcal{R}} | \pi_X^{-1}(A) \in \mathcal{B}^{\#}(X)\}$. Then $(X/_{\mathcal{R}}, \mathcal{B}(X/_{\mathcal{R}}))$, which is a measurable space, is called the quotient space of X w.r.t. $\mathcal{R}$. The quotient space of Y w.r.t. $\mathcal{R}$ is defined in a similar way. We define a bijective mapping $\psi : X/_{\mathcal{R}} \to Y/_{\mathcal{R}}$ as $\psi([x]) = [y]$ if $(x, y) \in \mathcal{R}$ for some $x \in [x]$ and some $y \in [y]$. We say that the relation $\mathcal{R}$ is *measurable* if X and Y if for all $A \in \mathcal{B}(X/_{\mathcal{R}})$ we have $\psi(A) \in \mathcal{B}(Y/_{\mathcal{R}})$ and vice versa, i.e. ψ is a homeomorphism (Strubbe and Schaft, 2005). Then the real measurable functions defined on $X/_{\mathcal{R}}$ can be identified with those defined on $Y/_{\mathcal{R}}$ through the homeomorphism ψ. We can write $\mathcal{B}^b(X/_{\mathcal{R}}) \overset{\psi}{\cong} \mathcal{B}^b(Y/_{\mathcal{R}})$. These functions can be thought of as real functions defined on X or Y measurable w.r.t. $\mathcal{B}^{\#}(X)$ or $\mathcal{B}^{\#}(Y)$.

Assumption 2. Suppose that $X/_{\mathcal{R}}$ and $Y/_{\mathcal{R}}$ with the topologies induced by projection mappings are Lusin spaces.

Let M and W be two right Markov processes with the state spaces X and Y. Assume that μ (resp. ν) is a stationary measure of the process M (resp. W). Let $\mu/_{\mathcal{R}}$ (resp. $\nu/_{\mathcal{R}}$) the image of μ (resp. ν) under π_X (resp. π_Y). Let $\mathcal{E}$ (resp. $\mathcal{F}$) the quasi-regular Dirichlet form corresponding to M (resp. W). The equivalence of the induced processes can be used to define a new bisimulation between Markov processes, as follows.

Under assumptions 1 and 2, a measurable relation $\mathcal{R} \subset X \times Y$ is a *bisimulation between M and W* if the mappings π_X and π_Y define the same induced Dirichlet form on $L^2(X/_{\mathcal{R}}, \mu/_{\mathcal{R}})$ and $L^2(Y/_{\mathcal{R}}, \nu/_{\mathcal{R}})$, respectively. This bisimulation definition states that M and W are bisimilar if $\mathcal{E}/_{\mathcal{R}} = \mathcal{F}/_{\mathcal{R}}$. Here, $\mathcal{E}/_{\mathcal{R}}$ (resp. $\mathcal{F}/_{\mathcal{R}}$) is the induced Dirichlet form of $\mathcal{E}$ (resp. $\mathcal{F}$) under the mapping π_X (resp. π_Y). Clearly, this can be possible iff $\mu/_{\mathcal{R}} = \nu/_{\mathcal{R}}$.

Assumption 3. Suppose that $\mathcal{E}/_{\mathcal{R}}$ and $\mathcal{F}/_{\mathcal{R}}$ are quasi-regular Dirichlet form.

Denote the Markov process associated to $\mathcal{E}/_{\mathcal{R}}$ (resp. $\mathcal{F}/_{\mathcal{R}}$) by $M/_{\mathcal{R}}$ (resp. $W/_{\mathcal{R}}$).

Proposition 4. Under assumptions 1,2, 3, M and W are stochastic bisimilar w.r.t. $\mathcal{R}$ iff the processes $M/_{\mathcal{R}}$ and $W/_{\mathcal{R}}$ are $\mu/_{\mathcal{R}}$-equivalent.

Let H and H' be two stochastic hybrid systems, with the realizations M and W, strong Markov processes defined on the state spaces $(X, \mathcal{B}(X))$ and $(Y, \mathcal{B}(Y))$, respectively. H and H' are *bisimilar* if there exists a bisimulation relation under which their realizations M and W are bisimilar.

4.3 Weak Stochastic Bisimulation

The way to define the concept of stochastic bisimulation, in the previous subsection, presents two main difficulties: assumptions 2 and 3. It seems difficult to find a practical condition to impose on $\mathcal{R}$, which would guarantee that these two assumptions are fulfilled. There exist conditions that ensure that the quotient space of an analytic space is again analytic, but it might be difficult to find out necessary conditions on $\mathcal{R}$, which ensure that the quotient space of a Lusin space is again Lusin. When we pass from the quasi-regular Dirichlet forms of the initial processes it might be hard to deal with conditions on the projection mappings π_X and π_Y, which assure the regularity of the induced Dirichlet forms.

With these arguments in mind, in this subsection we will introduce a weaker version of stochastic bisimulation and the concept of *functional abstraction* of a stochastic hybrid system.

Let M and W be two right Markov processes with the state spaces X and Y, as in the previous subsections. Suppose we have given two weight (measurable) functions $F : X \to \mathbb{R}$ and $G : Y \to \mathbb{R}$ (F or G can be the function used to define the target sets in context of the stochastic reachability problem. Let $\mathcal{E}^{\#}$ (resp. $\mathcal{F}^{\#}$) be the induced Dirichlet form of $\mathcal{E}$ (resp. $\mathcal{F}$) through the mapping F (resp. G).

Assumption 4. Suppose that $\mathcal{E}^{\#}$ and $\mathcal{F}^{\#}$ are quasi-regular Dirichlet forms.

M and W are *(weak) stochastic bisimilar* if the induced Dirichlet forms are equal, i.e. $\mathcal{E}^{\#} = \mathcal{F}^{\#}$.

The advantage of this new definition is that we do not need anymore the property to be Lusin of the quotient spaces. On the other hand the induced processes are one-dimensional stochastic processes whose state spaces are much smaller ones.

4.4 Functional Abstractions

The stochastic reachability definition gives the idea to introduce the following concept of functional abstraction for SHS.

Given a right Markov process M defined on the Lusin state space $(X, \mathcal{B})$, and $F : X \to \mathbb{R}$ a measurable weight function, suppose that Ass.1 is fulfilled. The process $M^{\#}$ associated to the induced Dirichlet form $\mathcal{E}^{\#}$ under function F is called a *functional abstraction of* M.

Let H a stochastic hybrid system and M its realization. Suppose that M is a right Markov process defined on the Lusin state space $(X, \mathcal{B})$.

Any stochastic hybrid system $H^{\#}$ whose realization is a functional abstraction of M is called a *functional abstraction of* H.

Let M be a right Markov process, thought as the realization of an SHA, H.

Proposition 5. If M is a diffusion (resp. jump process) then any functional abstraction $M^{\#}$ of M is a diffusion (resp. jump process).

Since the realization of an SHA is a stochastic process, which can be viewed an interleaving between some diffusion processes and a jump process (Bujorianu and Lygeros, 2004) we can write the following corollary of Prop.5.

Proposition 6. Any functional abstraction of an SHA is again an SHA.

5. CONCLUSIONS

In this paper, motivated by problem of stochastic reachability for SHS we have introduced

- a new concept of stochastic bisimulation for SHS from a functional viewpoint, i.e. this bisimulation is defined using the parameters, which appear in the reachability problem formulation and preserves the bounds of reach set probabilities;

- functional abstractions for SHS, defined again from a functional perspective.

The main tool used in defining of these new concepts is constituted by the quadratic forms associated to the realizations of the SHS. The quadratic form technique makes possible to obtain abstractions of SHS realizations, which are one-dimensional stochastic processes with a much smaller state space. The meaning of the induced stochastic process might be different depending on the context: refinement of discrete transitions structure, or approximation of continuous dynamics or, finally, abstraction of the entire process.

REFERENCES

Albeverio, S., Z.M. Ma and M. Rockner (1993). Quasi-regular dirichlet forms and markov processes. *J.Funct.An.* **111**, 118–154.

Blom, H.A.P. (2003). Stochastic hybrid processes with hybrid jumps. In: *ADHS*.

Bujorianu, M.L. (2004). Extended stochastic hybrid systems. In: *HSCC* (R. Alur and G. Pappas, Eds.). pp. 234–249. Number 2993 In: *LNCS*. Springer.

Bujorianu, M.L. and J. Lygeros (2003). Reachability questions in piecewise deterministic markov processes. In: *HSCC* (O. Maler and A. Pnueli, Eds.). pp. 126–140. Number 2623 In: *LNCS*. Springer.

Bujorianu, M.L. and J. Lygeros (2004). General stochastic hybrid systems: Modelling and optimal control. In: *43th CDC*.

Bujorianu, M.L., J. Lygeros and M.C. Bujorianu (2005). Bisimulation for general stochastic hybrid systems. In: *HSCC* (M. Morari and L. Thiele, Eds.). pp. 198–216. Number 3414 In: *LNCS*. Springer.

Davis, M.H.A. (1993). *Markov Models and Optimization*. Chapman & Hall. London.

Everdij, M.H.C. and H.A.P Blom (2005). Pdmp represented by dynamically coloured petri nets. *Stochastics* **77**, 1–29.

Fukushima, M. (1980). *Dirichlet Forms and Markov Processes*. N. Holland.

Hespanha, J.P. (2004). Stochastic hybrid systems: Application to communication network. In: *HSCC* (R. Alur and G. Pappas, Eds.). pp. 387–401. Number 2993 In: *LNCS*. Springer.

Hu, J., J. Lygeros and S. Sastry (2000). Towards a theory of stochastic hybrid systems. In: *HSCC* (N. Lynch and B. Krogh, Eds.). pp. 160–173. Number 1790 In: *LNCS*. Springer.

Iscoe, I. and D. McDonald (1990). Induced dirichlet forms and capacitary inequalities. *Ann. Prob.* **18**(3), 1195–1221.

Ma, M. and M. Rockner (1990). *The Theory of Non-Symmetric Dirichlet Forms and Markov Processes*. Springer. New York.

Pola, G., M.L. Bujorianu and J. Lygeros (2003). Stochastic hybrid models: An overview with applications to air traffic management. In: *ADHS*.

Rogers, L.C.G. and J.W. Pitman (1981). Markov functions. *Ann. Prob.* **9**(4), 573–582.

Strubbe, S.N. and A.J. van der Schaft (2005). Bisimulation for communicating pdps. In: *HSCC* (M. Morari and L. Thiele, Eds.). pp. 623–640. Number 3414 In: *LNCS*. Springer.

ELSEVIER
IFAC
PUBLICATIONS

STOCHASTIC HYBRID NETCAD SYSTEMS FOR MODELING CALL ADMISSION AND ROUTING CONTROL IN NETWORKS

Zhongjing Ma * **Peter E. Caines** ** **Roland Malhame** ***

* *mazhj@cim.mcgill.ca*
** *peterc@cim.mcgill.ca*
*** *roland.malhame@polymtl.ca*

Abstract: In this paper a stochastic hybrid systems framework is established for the formulation of call admission control (CAC) and routing control (RC) problems in networks. The hybrid state process of the underlying system is a piecewise deterministic Markovian process (PDMP) evolving deterministically between random event instants at which times the state jumps to another state value. The random events in the system correspond to the arrival of call requests or the departure of connections. The resulting NETCAD stochastic state space systems framework permits the formulation and analysis of centralized optimal stochastic control with respect to specified utility functions [3,8,10,11].

Keywords: Call Admission Control, Routing Control, Stochastic Hybrid Systems, Piecewise Deterministic Markovian Processes, Point Processes

1. INTRODUCTION

Call admission control (CAC) and routing control (RC) in telecommunication networks have been topics of active research for decades (see e.g. [1,5,6,12]). In the 1960s, Benes pioneered routing control in telephone networks, providing a general mathematical formulation and the extensive associated analysis for telephone systems presented in [1].

In this paper, CAC and RC in networks are modeled as stochastic control problems for the so-called NETwork Connection Assignment and Departure (NETCAD) systems. The state process of the underlying NETCAD systems has some particular characteristics: (1) the state process has two parts: the first is a piecewise constant integer-valued point process [2], while the second is a variable dimension real-valued piecewise deterministic process; (2) it is a piecewise deterministic Markov process [4] with respect to a Markovian control law, where the state value evolves deterministically between the random event instants and jumps to some other state value at random event instants subject to controlled transition probabilities. NETCAD systems may be viewed as stochastic hybrid systems generalizing the class of deterministic hybrid systems which was defined in [13] and the references therein; as indicated above, the state process is composed of two components: a discrete component, denoting the connections along the set of (origin-destination) routes in the NETCAD network, and a continuous component constituting the vector of ages of the call requests and active connections in the NETCAD system.

The distinction between the work in this paper and that found in standard telecommunication texts and papers (see e.g. [1,5,6,12]) is that here a network system is represented within a formal stochastic systems framework with a specified class of input stochastic processes and a stochastic hybrid state space process with a controlled evolution equation while such a fine low level analysis is not formulated in [1,5,6,12]. This permits the formulation of an optimal stochastic control theory for NETCAD systems in the current work [8].

The paper is organized as follows. In Section 2, we present the formal definition of the networks upon which NETCAD systems are based; in Section 3, we formulate the network connection assignment and departure (NETCAD) systems and the Markov property of the state process is proved. Section 4 contains the conclusions and outlines future work.

2. THE NETWORK OF A NETCAD SYSTEM

2.1 NETCAD Networks

A NETCAD network is a capacitated network $Net(\mathbb{V}, \mathbb{L}, \mathbb{C})$ as defined below. Based upon this notion, a NETCAD system is defined in Definition 3.7 at the end of Section 3.

Definition 2.1. A *network*, or *graph*, $Net(\mathbb{V}, \mathbb{L})$ consists of a set of vertices $\mathbb{V} = \{v_1, \cdots, v_V\}$, $V \in \mathbb{Z}_1$, and a set of lines $\mathbb{L} = \{l_1, \cdots, l_L\}, L \in \mathbb{Z}_1$, where each line $l \in \mathbb{L}$ is an ordered pair $(v', v'') \in \mathbb{V} \times \mathbb{V}$ of distinct vertices.
A *network* $Net(\mathbb{V}, \mathbb{L})$ *with (line) capacities* $\mathbb{C} = \{c_s \equiv c(l_s) : 1 \leq s \leq L, c_s \in \mathbb{Z}_1\}$, shall be denoted by $Net(\mathbb{V}, \mathbb{L}, \mathbb{C})$. □

Definition 2.2. A *route*, r in the network $Net(\mathbb{V}, \mathbb{L})$, connecting a vertex $o \in \mathbb{V}$ to a vertex $d \in \mathbb{V}$, $d \neq o$, is a finite sequence of vertices $r = (v'_1, \cdots, v'_k)$, such that

$$v'_1 = o,\ v'_k = d,$$
$$v'_i \neq v'_j,\ \text{for } i \neq j,$$
$$(v'_i, v'_{i+1}) \in \mathbb{L},\ \text{for } i = 1, \cdots, k-1.$$

The *set of routes* in the network $Net(\mathbb{V}, \mathbb{L})$ is denoted by $\mathcal{R}$, and we denote R as the cardinality of $\mathcal{R}$, i.e. $R = |\mathcal{R}|$. □

Fig.1 is an illustration of 3 distinct routes between node v_1 and v_8, which are $(v_1, v_2, v_5, v_4, v_8)$, (v_1, v_4, v_8) and (v_1, v_3, v_7, v_8) respectively in a network.

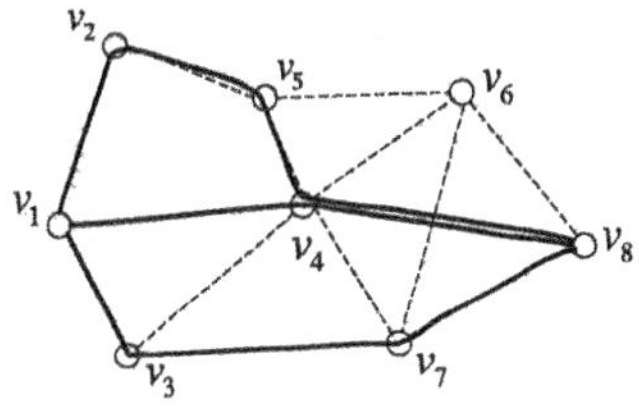

Fig. 1. Distinct routes in a network

Definition 2.3. The *feasible set* of *origin destination vertex pairs*, denoted by $\mathbb{V}^{\triangle}$, is defined as

$$\mathbb{V}^{\triangle} = \Big\{ \langle o, d \rangle \in \mathbb{V} \times \mathbb{V};\ \exists\, r \in \mathcal{R},$$
$$\text{s.t. } r = (v'_1, \cdots, v'_j), v'_1 = o, v'_j = d,\ o \neq d \Big\} \quad (1)$$

□

Remark: Call requests from any node to itself are excluded in this paper.

Definition 2.4. The *admissible set* of *connections*, denoted by $\mathcal{N}$, in $\mathcal{R}$ in the network with capacities $Net(\mathbb{V}, \mathbb{L}, \mathbb{C})$, is defined as

$$\mathcal{N} = \Big\{ \mathbf{n} = (\mathbf{n}_r) \in \mathbb{Z}_+^R :$$
$$\sum_{r \in \mathcal{R};\ l_s \in r} \mathbf{n}_r \leq c_s,\ \forall s, 1 \leq s \leq L \Big\} \quad (2)$$

□

We observe that in the definition of $\mathcal{N}$, for each fixed l_s, the set of $r \in \mathcal{R}$ appearing in the sum is the set of routes each of which contains l_s as a line.
Since the routes in $\mathcal{R}$ are in one-to-one correspondence with the index of the components of a vector in $\mathbb{Z}_+^R \subset \mathbb{R}^R$, we shall by abuse of notation let $r \in \mathcal{R}$ also denote the integer indexing the corresponding vector component in $\mathbb{R}^R$.

2.2 A Simple Example of a Capacitated Network

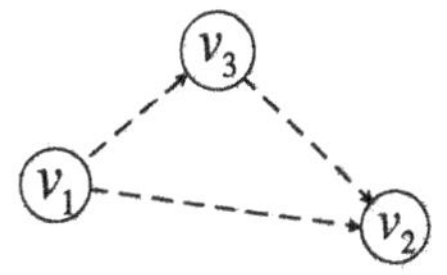

Fig. 2. A three node capacitated network

We consider a simple network $Net(\mathbb{V}, \mathbb{L}, \mathbb{C})$, see Fig.2, with

$$\mathbb{V} = \{v_1, v_2, v_3\}$$
$$\mathbb{L} = \{l_1 = (v_1, v_2), l_2 = (v_1, v_3), l_3 = (v_3, v_2)\}$$
$$\mathbb{C} = \{c_l = 2;\ l \in \mathbb{L}\}$$

Hence the set of routes, $\mathcal{R}$, is defined as

$$\mathcal{R} = \{r_1 = (v_1, v_3, v_2), r_2 = (v_1, v_2),$$
$$r_3 = (v_1, v_3), r_4 = (v_3, v_2)\},$$

and the admissible connections set, $\mathcal{N}$, is defined as

$$\mathcal{N} = \{\mathbf{n} = (\mathbf{n}_{r_1}, \mathbf{n}_{r_2}, \mathbf{n}_{r_3}, \mathbf{n}_{r_4}) \in \mathbb{Z}_+^4;$$
$$\sum_{r_i \in \mathcal{R};\ l \in r_i} \mathbf{n}_{r_i} \leq 2,\ \forall l \in \mathbb{L}\}$$

3. THE NETWORK CONNECTION ASSIGNMENT AND DEPARTURE (NETCAD) SYSTEMS

3.1 The NETCAD System Framework

We consider the following:

(1) The probability space $(\Omega, \mathcal{F}, P)$ carries the family of independent $\mathbb{R}_+$ valued random variables

$$\{\tau_k^{\langle o,d\rangle}, \tau_j^c;\ k, j \in \mathbb{Z}_1, \langle o,d\rangle \in \mathbb{V}^\triangle\}, \quad (3)$$

where $\tau_k^{\langle o,d\rangle}$ denotes the length of the interval between the $(k-1)$th and the kth $\langle o,d\rangle$ call request; τ_j^c denotes the lifetime of the jth allocated connection in the network.

(2) For each $\langle o,d\rangle \in \mathbb{V}^\triangle$, the random variables $\{\tau_k^{\langle o,d\rangle}, k \in \mathbb{Z}_1\}$ are assumed to have a common arbitrary distribution $A^{\langle o,d\rangle}(t)$ with density function $a^{\langle o,d\rangle}(t)$, i.e.

$$\mathbb{P}(\tau_k^{\langle o,d\rangle} \le t) = A^{\langle o,d\rangle}(t) = \int_{-\infty}^{t} a^{\langle o,d\rangle}(s)ds. \quad (4)$$

(3) The random variables $\{\tau^{c_j}, j \in \mathbb{Z}_1\}$ are assumed to have a common arbitrary distribution $B(t)$ with density function $b(t)$, i.e.

$$\mathbb{P}(\tau^{c_j} \le t) = B(t) = \int_{-\infty}^{t} b(s)ds. \quad (5)$$

Remarks: For each $\langle o,d\rangle \in \mathbb{V}^\triangle$, we shall denote the call (connection) request process of $\langle o,d\rangle$ by $N^+_{\langle o,d\rangle}$. Then from the above assumption, we observe that, for each $\langle o,d\rangle \in \mathbb{V}^\triangle$, the $N^+_{\langle o,d\rangle}$ process is an autonomous, i.e. control independent, point process in $\mathbb{R}_+$ with independent inter-event times $\{\tau_k^{\langle o,d\rangle}, k \in \mathbb{Z}_1\}$. In other words, for each $\langle o,d\rangle \in \mathbb{V}^\triangle$, the call request process $N^+_{\langle o,d\rangle}$ is a point process with independent interarrival times distributed $A^{\langle o,d\rangle}(.)$.

Definition 3.1. The *sub-state space*, $Z_{\mathbf{n}}$, with respect to $\mathbf{n} = (\mathbf{n}_{r_1}, \cdots, \mathbf{n}_{r_R}) \in \mathcal{N}$, is defined as the following collection of index and age pairs:

$$Z_{\mathbf{n}} = \{z \equiv (\mathbf{n}; \zeta);\ \zeta \in \mathbb{R}_+^{m(\mathbf{n})}\}, \quad \text{where} \quad (6)$$

$$m(\mathbf{n}) \triangleq |\mathbb{V}^\triangle| + \sum_{r \in R} \mathbf{n}_r,$$

$$\zeta \triangleq \Big(\{\overbrace{\zeta^{\langle o,d\rangle_1}, \zeta^{\langle o,d\rangle_2}, \cdots, \zeta^{\langle o,d\rangle_{|\mathbb{V}^\triangle|}}}^{|\mathbb{V}^\triangle|}\}_{od},$$

$$\{\overbrace{\zeta^{c_{r_1,1}}, \cdots, \zeta^{c_{r_1,\mathbf{n}_{r_1}}}}^{\mathbf{n}_{r_1}}\}_{r_1}, \cdots,$$

$$\{\overbrace{\zeta^{c_{r_R,1}}, \cdots, \zeta^{c_{r_R,\mathbf{n}_{r_R}}}}^{\mathbf{n}_{r_R}}\}_{r_R}\Big),$$

where the following constraints necessarily hold:

$$\zeta^{c_{r_i,1}} > \cdots > \zeta^{c_{r_i,\mathbf{n}_{r_i}}} \ge 0,\ \forall i \in \{1, \cdots, R\},$$

and where $\zeta^{\langle o,d\rangle_i}$ denotes the elapsed time since an $\langle o,d\rangle_i$ call request and $\zeta^{c_{r_i,j}}$ denotes the age of connection $c_{r_i,j}$.

The *state space*, denoted by Z, is defined as

$$Z \triangleq \dot{\bigcup}_{\mathbf{n} \in \mathcal{N}} Z_{\mathbf{n}}. \quad (7)$$

□

Remarks:

(1) *We set, for each $\langle o,d\rangle \in \mathbb{V}^\triangle$, a unique index number i, $i \in \{1, 2, \cdots, |\mathbb{V}^\triangle|\}$, and denote this $\langle o,d\rangle$ pair by $\langle o,d\rangle_i$, i.e. for each i, $i \in \{1, 2, \cdots, |\mathbb{V}^\triangle|\}$, there is a unique $\langle o,d\rangle \in \mathbb{V}^\triangle$ and $\langle o,d\rangle_i \equiv \langle o,d\rangle$.*

(2) *Since there are $\mathbf{n}_{r_i}$ connections along route r_i, then each of these connections can be uniquely denoted by $c_{r_i,j}$, $j(i) \in \{1, \cdots, \mathbf{n}_{r_i}\}$ and its age is denoted by $\zeta^{c_{r_i,j}}$.*

(3) *Specifically, the sequence of $\mathbf{n}_{r_i}$ connections along route r_i can be indexed by their age, or time since birth time, such that $\zeta^{c_{r_i,1}} > \zeta^{c_{r_i,2}} > \cdots > \zeta^{c_{r_i,\mathbf{n}_{r_i}}}$, corresponding to the fact that the earlier a connection was established, the smaller is its index number.*

Here we give an example to illustrate the definition of the state z.

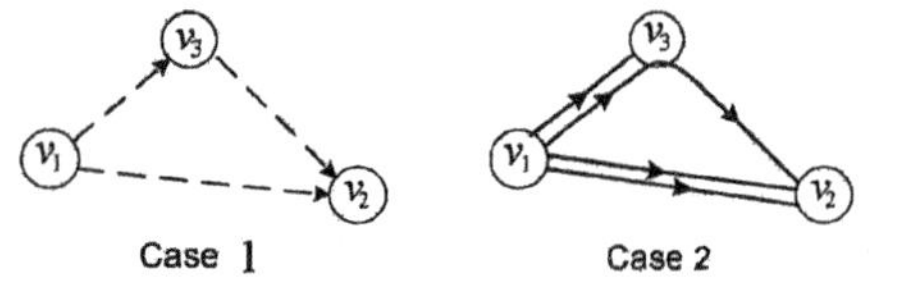

Fig. 3. Network state values with respect to different connection allocations

In Case 1 in Fig.3 there is no connection in the network, and the corresponding state value z takes the following form:

$$z = \left(\begin{bmatrix} 0 \\ 0 \\ 0 \\ 0 \end{bmatrix}, \begin{bmatrix} \zeta^{\langle v_1,v_2\rangle} \\ \zeta^{\langle v_1,v_3\rangle} \\ \zeta^{\langle v_2,v_3\rangle} \end{bmatrix} \right)$$

In Case 2 in Fig.3 there are respectively one, two and one connections along the route $r_1 \equiv (v_1, v_3, v_2)$, $r_2 \equiv (v_1, v_2)$ and $r_3 \equiv (v_1, v_3)$. The corresponding state value z takes the following form:

$$z = \left(\begin{bmatrix} 1 \\ 2 \\ 1 \\ 0 \end{bmatrix}, \begin{bmatrix} \zeta^{\langle v_1,v_2\rangle} \\ \zeta^{\langle v_1,v_3\rangle} \\ \zeta^{\langle v_2,v_3\rangle} \end{bmatrix}, \begin{bmatrix} \zeta^{c_{r_1,1}} \\ \zeta^{c_{r_2,1}} \\ \zeta^{c_{r_2,2}} \\ \zeta^{c_{r_3,1}} \end{bmatrix} \right)$$

Definition 3.2. The *(call request and connection departure) event set*, $E_{\mathbf{n}}$, with respect to a connection vector value $\mathbf{n} \in \mathcal{N}$, s.t. $\mathbf{n} = (\mathbf{n}_{r_1}, \cdots, \mathbf{n}_{r_R})$, is defined as:

$$E_{\mathbf{n}} = e_{\mathbf{n}}^0 \dot{\bigcup} E_{\mathbf{n}}^+ \dot{\bigcup} E_{\mathbf{n}}^-, \quad \text{where} \quad (8)$$

$$E_{\mathbf{n}}^{+} = \dot{\bigcup}_{\langle o,d\rangle_q \in \mathbb{V}^{\triangle}} e^{+}_{\langle o,d\rangle_q}$$

$$E_{\mathbf{n}}^{-} = \dot{\bigcup}_{r_i \in \mathcal{R}} \Big\{ \dot{\bigcup}_{j\in\{1,\cdots,\mathbf{n}_{r_i}\}} e^{-}_{c_{r_i,j}} \Big\},$$

where $\dot{\bigcup}$ denotes the disjoint union of the indicated entities and

(1) $e^0_{\mathbf{n}} \equiv \mathbf{0} = \begin{bmatrix} 0 \\ \cdots \\ 0 \end{bmatrix} \in \mathbb{R}^M$ denotes absence of call request or connection departure event and $\mathbf{0}$ is the zero vector in $\mathbb{R}^M$, and

$$M \equiv M(\mathbf{n}) = |\mathbb{V}^{\triangle}| + \sum_{r\in R} \mathbf{n}_r;$$

(2) $e^{+}_{\langle o,d\rangle_q} \equiv \mathbf{1}_m = \begin{bmatrix} 0 \\ \cdots \\ 0 \\ 1 \\ 0 \\ \cdots \end{bmatrix} \in \mathbb{R}^M$ denotes the call request (event) from vertex o to vertex d, with $\langle o,d\rangle = \langle o,d\rangle_q$ and $\mathbf{1}_m$ is the m-th unit vector in $\mathbb{R}^M$, where $m = q$;

(3) $e^{-}_{c_{r_i,j}} \equiv \mathbf{1}_m = \begin{bmatrix} 0 \\ \cdots \\ 0 \\ 1 \\ 0 \\ \cdots \end{bmatrix} \in \mathbb{R}^M$ denotes the connection $c_{r_i,j}$ departure (event) and $\mathbf{1}_m$ is the m-th unit vector in $\mathbb{R}^M$, and

$$m = |\mathbb{V}^{\triangle}| + \sum_{l=1}^{i-1} \mathbf{n}_{r_l} + j.$$

The *(total) event set*, E, is defined as

$$E \triangleq \dot{\bigcup}_{\mathbf{n}\in\mathcal{N}} E_{\mathbf{n}}. \tag{9}$$

□

Definition 3.3. The *(extended) sub-state space*, X_n, with respect to a *connection vector*, $\mathbf{n} = (\mathbf{n}_{r_1}, \cdots, \mathbf{n}_{r_R}) \in \mathcal{N}$, is defined as the following: $X_{\mathbf{n}} \triangleq Z_{\mathbf{n}} \times E_{\mathbf{n}}$.

The *(total extended) state space*, X, is defined as: $X \triangleq \dot{\bigcup}_{\mathbf{n}\in\mathcal{N}} X_{\mathbf{n}}$. □

Definition 3.4. The *feasible control (value) set*, $U(x)$, with respect to an extended state value $x = (z, e) \equiv (\mathbf{n}, \zeta, e) \in X$, s.t. $\mathbf{n} = (\mathbf{n}_{r_1}, \cdots, \mathbf{n}_{r_R})$ and an event $e \in E^{+}_{\mathbf{n}}$, is defined as:

$$U(x) = \mathbf{0}^{\langle o,d\rangle}(x) \,\dot{\bigcup}\, \mathbf{1}_r^{\langle o,d\rangle}(x) \,\dot{\bigcup}\, -\mathbf{1}_c(x), \tag{10}$$

where

$$\mathbf{0}^{\langle o,d\rangle}(x) = \dot{\bigcup}_{\langle o,d\rangle_q \in \mathbb{V}^{\triangle}} 0^{\langle o,d\rangle_q}(x),$$
$$\text{with } 0^{\langle o,d\rangle_q}(x) \equiv 0^{\langle o,d\rangle_q(x)}(x),$$

$$\mathbf{1}_r^{\langle o,d\rangle}(x) = \dot{\bigcup}_{\langle o,d\rangle_q \in \mathbb{V}^{\triangle}} \Big\{ \dot{\bigcup}_{\substack{r\in\mathcal{R}_{\langle o,d\rangle_q} \\ \mathbf{n}+\mathbf{1}_r\in\mathcal{N}}} 1_r^{\langle o,d\rangle_q}(x) \Big\},$$
$$\text{with } 1_r^{\langle o,d\rangle_q}(x) \equiv 1_r^{\langle o,d\rangle_q(x)}(x),$$

$$-\mathbf{1}_c(x) = \dot{\bigcup}_{r_i\in\mathcal{R}} \Big\{ \dot{\bigcup}_{j\in\{1,\cdots,\mathbf{n}_{r_i}\}} -1_{c_{r_i,j}}(x) \Big\},$$
$$\text{with } -1_{c_{r_i,j}}(x) \equiv -1_{c_{r_i,j}(x)}(x),$$

where

(1) $0^{\langle o,d\rangle_q}(x) \equiv \mathbf{0} \in \mathbb{R}^R$ denotes the fact that the call request $e^{+}_{\langle o,d\rangle_q}$ is rejected and $\mathbf{0}$ is the zero vector in $\mathbb{R}^R$;

(2) $1_r^{\langle o,d\rangle_q}(x) \equiv \mathbf{1}_i = \begin{bmatrix} \cdots \\ 0 \\ 1 \\ 0 \\ \cdots \end{bmatrix} \in \mathbb{R}^R$ denotes that the call request $e^{+}_{\langle o,d\rangle_q}$ is accepted and is allocated on the route $r_i = r$, s.t. $r = \{v_{q_1}, \cdots, v_{q_m}\} \in \mathcal{R}$, $\langle v_{q_1}, v_{q_m}\rangle = \langle o,d\rangle_q$ and $\mathbf{n} + \mathbf{1}_r \in \mathcal{N}$;

(3) $-1_{c_{r_i,j}}(x) \equiv -\mathbf{1}_i = -\begin{bmatrix} \cdots \\ 0 \\ 1 \\ 0 \\ \cdots \end{bmatrix} \in \mathbb{R}^R$ denotes that a connection in r_i departs.

The *control (value) set* U is defined as

$$U = \dot{\bigcup}_{x\in X} U(x). \tag{11}$$

Here we give an example to display a feasible control with respect to an index $n \in \mathcal{N}$. See Fig.4.

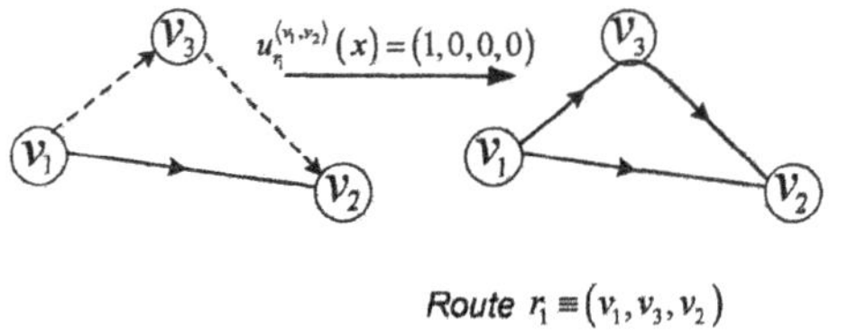

Fig. 4. A feasible control with respect to a state value $((0,1,0,0),\zeta)$

Suppose that $x = ((0,1,0,0),\zeta) \in X$ and $e = (1,0,0,0,0)$, i.e. a call request $\langle v_1, v_2\rangle$ occurs, a feasible control can be $u_{r_1}^{\langle v_1,v_2\rangle}(x) = (1,0,0,0)$, i.e. the call from v_1 to v_2 is allocated to the route $r_1 \equiv (v_1, v_3, v_2)$.

Definition 3.5. When z and e depend in a progressively measurable way on $(\Omega, \mathcal{F}, \mathbb{P})$, we refer to $z = \{z(t,\omega);\ t \in [0,T], \omega \in \Omega\}$ and $e = \{e(t,\omega);\ t \in [0,T], \omega \in \Omega\}$ as *state* and *event processes*; and x as the *(extended) state process*

$$x = \Big\{ x(t,\omega) \triangleq (z(t^-,\omega), e(t,\omega)), t \in [0,T],$$
$$\omega \in \Omega \Big\}, \text{ with } z(t^-,\omega) \equiv \lim_{s\uparrow t} z(s,\omega). \tag{12}$$

We define $\mathcal{F}_t \triangleq \bigvee_{r\in[0,t]} \sigma(x_r) \in \mathcal{F}$, i.e. $\mathcal{F}_.$ is the *natural filtration* extended by the process x, where $\sigma(x_r)$ denotes the σ field generated by the random variable x_r. □

Definition 3.6. The set of *state dependent*, or *Markovian (measurable), control laws* is denoted by $\mathcal{U}[0,T]$, $0 \le T < \infty$, and is given by,

$$\mathcal{U}[0,T] = \{u : [0,T] \times \Omega \to U;$$
$$\text{s.t. } u_t \text{ is } \sigma(x_t) \text{ measurable, } t \in [0,T]\} \tag{13}$$
$$\mathcal{U}[0,\infty) = \cup_{T\ge 0} \mathcal{U}[0,T] \tag{14}$$

□

Definition 3.7. A family of state processes $\{x_t \triangleq (z_{t-}, e_t)\}$ taking values in X in a capacitated network $Net(\mathbb{V}, \mathbb{L}, \mathbb{C})$, together with a family of feasible controls u, is called a *network connection assignment and departure (stochastic) system*, or a *NETCAD system*, for short. □

Definition 3.8. We term a *sequence* of *event instants* $\{t_j(\omega)\}$ in $\mathbb{R}_+$

$$0 \le t_1(\omega) < \cdots < t_j(\omega) < t_{j+1}(\omega) < \cdots, \quad (\Omega, \mathcal{F}, \mathbb{P}), \ \omega \in \Omega, \tag{15}$$

at which random call request or connection departure event occurs as a sequence of *(random) event instants* $t_\cdot : \mathbb{Z}_+ \times \Omega \to \mathbb{R}_+$. The sequence $\tau_\cdot : \mathbb{Z}_1 \times \Omega \to \mathbb{R}_+$, with $\tau_{k+1} \triangleq t_{k+1}(\omega) - t_k(\omega)$, where $t_0(\omega) \equiv 0$ is defined as the sequence of *event intervals* (associated to $t_\cdot(\omega)$). □

Definition 3.9. *State response* or *transition equation*, with a measurable Markov control law $u \in \mathcal{U}[0, \infty)$, i.e. u_t is $\sigma(x_t^u)$ measurable, for the evolution of the state process

$$x^u : [0, \infty) \times \Omega \to X, \tag{16}$$

with initial state value $x_{t_0}^u \equiv (z_0^u, e_0^u) = ((\mathbf{n}, \zeta), \mathbf{0})$, is given by

$$x_t^u = (z_t^u, e_t^u), \quad t_{i-1} \le t < t_i, \ i \in \mathbb{Z}_1 \tag{17}$$

$$\begin{aligned} z_t^u &= z_{t_{i-1}}^u + \int_{t_{i-1}}^t (\mathbf{0_n}, \mathbf{I}_\zeta) dr \\ &= \left(\mathbf{n}_{t_{i-1}}, \zeta_{t_{i-1}} + [t - t_{i-1}]\, \mathbf{I}_\zeta\right), \end{aligned} \tag{18}$$

where $t_{i-1} \le t < t_i$, $i \in \mathbb{Z}_1$.

$$e_t^u = \begin{cases} e \in E_{\mathbf{n}}, & t = t_i, \ i \in \mathbb{Z}_1, \quad \mathbf{n} \equiv \mathbf{n}(x_{t_{i-1}}^u) \\ \mathbf{0}, & \text{otherwise} \end{cases} \tag{19}$$

$$\begin{aligned} x_{t_i^+}^u &\equiv \lim_{s \downarrow t_i} x_s = \lim_{s \downarrow t_i} (z_{s-}^u, e_s^u) = \left(\lim_{s \downarrow t_i} z_{s-}^u, \lim_{s \downarrow t_i} e_s^u\right) \\ &= \left(\lim_{s \downarrow t_i} \lim_{r \uparrow s} z_r^u, \mathbf{0}\right) = \left(z_{t_i^+}^u, \mathbf{0}\right) = \left(z_{t_i}^u, \mathbf{0}\right), \end{aligned} \tag{20}$$

where $\mathbf{0}$ denotes the zero vector with some proper dimension, $\mathbf{I}_\zeta$ denotes $(1, \cdots, 1)$ with a dimension depending on the value of ζ.

The state transition equation at the event instant t_i is the following:

$$z_{t_i}^u = \left(\mathbf{n}_{t_i^-} + u_{t_i}(x_{t_i}^u),\ \mathbf{A}[\mathbf{I}_{M \times M} - e_{t_i} e_{t_i}^T]\, \zeta_{t_i^-}\right), \tag{21}$$

where M is the dimension of the vector $\zeta_{t_i^-}$ and

$$\begin{aligned} \mathbf{A} &\equiv \mathbf{A}(x_{t_i}^u, u_{t_i}(x_{t_i}^u)) \\ &= \begin{cases} \mathbf{A}^+_{(M+1) \times M}, & \text{if } u_{t_i}(x_{t_i}^u) > 0 \\ \mathbf{A}^-_{(M-1) \times M}, & \text{if } u_{t_i}(x_{t_i}^u) < 0 \\ \mathbf{I}_{M \times M}, & \text{otherwise} \end{cases} \end{aligned}$$

$$\mathbf{A}^+_{(M+1) \times M} = \begin{bmatrix} \mathbf{I}_{m \times m} & \mathbf{0}_{(m+1) \times (M-m)} \\ \mathbf{0}_{(M-m+1) \times m} & \mathbf{I}_{(M-m) \times (M-m)} \end{bmatrix},$$

$$\text{where } m = |\mathbb{V}^\Delta| + \sum_{j=1}^{l} \mathbf{n}_{r_j}(t_i^-), u_{t_i}(x_{t_i}^u) = 1^{r_l}_{\langle o,d \rangle_q}$$

$$\begin{aligned} &\mathbf{A}^-_{(M-1) \times M} \\ &= \begin{bmatrix} \mathbf{I}_{m \times m} & \mathbf{0}_{m \times (M-m)} \\ \mathbf{0}_{(M-m-1) \times (m+1)} & \mathbf{I}_{(M-m-1) \times (M-m-1)} \end{bmatrix}, \end{aligned}$$

$$\text{where } m = |\mathbb{V}^\Delta| + \sum_{j=1}^{l-1} \mathbf{n}_{r_j}(t_i^-) + [k-1]$$

$$\text{and } u_{t_i}(x_{t_i}^u) = -1_{c_{r_l,k}}$$

Remarks: $\mathbf{I}_{j \times j}$, $j \in \mathbb{Z}_1$, *denotes the* j*-dimension identity matrix.* □

3.2 An Example of a NETCAD System

Considering the capacitated network defined in Section 2.2, we specify a realization of the controlled state process z^u during $[0, t_2)$ to help the audiences to understand the state transition procedure.

Suppose that $z_0^u = (\mathbf{n}_0^u, \zeta_0^u) = \left(\begin{bmatrix} 0 \\ 0 \\ 0 \\ 0 \end{bmatrix}, \begin{bmatrix} a \\ b \\ c \end{bmatrix}\right)$, $a, b, c \in \mathbb{R}_+$, then for $0 < t \le t_1$,

$$z_{t^-}^u = \left(\begin{bmatrix} 0 \\ 0 \\ 0 \\ 0 \end{bmatrix}, \begin{bmatrix} t+a \\ t+b \\ t+c \end{bmatrix}\right), \quad z_{t_1^-}^u = \left(\begin{bmatrix} 0 \\ 0 \\ 0 \\ 0 \end{bmatrix}, \begin{bmatrix} t_1+a \\ t_1+b \\ t_1+c \end{bmatrix}\right),$$

Remarks: during $[0, t_1)$, *the dimension of the vector* ζ_t *is* 3, *since there is no connection in the network during this interval.*

Suppose at t_1, $e_{t_1} = \begin{bmatrix} 1 \\ 0 \\ 0 \end{bmatrix}$ and $u_{t_1}(x_{t_1}^u) = \begin{bmatrix} 0 \\ 1 \\ 0 \\ 0 \end{bmatrix}$, i.e. an $\langle o, d \rangle_1 \equiv \langle v_1, v_2 \rangle$ call request occurs at t_1 and this call request is allocated to the route r_2, then

$$\begin{aligned} z_{t_1}^u &= \left(\mathbf{n}_{t_1^-} + u_{t_1}(x_{t_1}^u),\ \mathbf{A}[\mathbf{I}_{3 \times 3} - e_{t_1} e_{t_1}^T] \zeta_{t_1^-}\right) \\ &= \left(\begin{bmatrix} 0 \\ 0 \\ 0 \\ 0 \end{bmatrix} + \begin{bmatrix} 0 \\ 1 \\ 0 \\ 0 \end{bmatrix}, \mathbf{AB} \begin{bmatrix} t_1+a \\ t_1+b \\ t_1+c \end{bmatrix}\right), \quad \text{where} \\ &\qquad \mathbf{A} = \mathbf{A}^+_{4 \times 3} = \begin{bmatrix} 1 & 0 & 0 \\ 0 & 1 & 0 \\ 0 & 0 & 1 \\ 0 & 0 & 0 \end{bmatrix} \\ &\qquad \mathbf{B} = \begin{bmatrix} 1 & 0 & 0 \\ 0 & 1 & 0 \\ 0 & 0 & 1 \end{bmatrix} - \begin{bmatrix} 1 \\ 0 \\ 0 \end{bmatrix} \begin{bmatrix} 1 & 0 & 0 \end{bmatrix} \\ &= \left(\begin{bmatrix} 0 \\ 1 \\ 0 \\ 0 \end{bmatrix}, \begin{bmatrix} 0 \\ t_1+b \\ t_1+c \\ \hline 0 \end{bmatrix}\right). \end{aligned}$$

Then we obtain that, for any $t_1 < t \le t_2$,

$$z_{t^-}^u = z_{t_1}^u + \int_0^{t-t_1} (\mathbf{0_n}, \mathbf{1}_\zeta) dr = \left(\begin{bmatrix} 0 \\ 1 \\ 0 \\ 0 \end{bmatrix}, \begin{bmatrix} t-t_1 \\ t+b \\ t+c \\ \hline t-t_1 \end{bmatrix}\right),$$

$$\text{and } z_{t_2^-}^u = \left(\begin{bmatrix} 0 \\ 1 \\ 0 \\ 0 \end{bmatrix}, \begin{bmatrix} t_2-t_1 \\ t_2+b \\ t_2+c \\ \hline t_2-t_1 \end{bmatrix}\right).$$

Remarks: A connection was established in the network at t_1. *So* $Dim(\zeta_{t_1}) = Dim(\zeta_{t_1^-}) + 1 = 3 + 1 = 4$, *i.e during* $[t_1, t_2)$ *the dimension of the vector* ζ_t *is* 4. □

3.3 Hybrid Nature of the NETCAD System

The state process z^u of the NETCAD system is composed of the two parts, such that

$$z_t^u = \begin{pmatrix} \mathbf{n}_t \\ \zeta_t \end{pmatrix}, \quad \text{at any instant } t,\ t \in [0, T], \quad (22)$$

where $\mathbf{n}$, a discrete process, keeps unchanged between the random event instants and is transferred to a state value $\mathbf{n}' \in \mathcal{N} \subset \mathbb{Z}_+^R$, with some controlled transition probability; while ζ, a continuous process, evolves deterministically between the random event instants and is transferred to a state value $\zeta' \in \mathbb{R}^{M(\mathbf{n}')}$, where the dimension of ζ' is dependent on the value of $\mathbf{n}'$.

3.4 Markov Property of the State Process

Lemma 3.1. [9] For each event instant, t_j, $j \in \mathbb{Z}_+$ and $t \in \mathbb{R}_+$, $\{t_j \leq t\} \in \mathcal{F}_t$, i.e. t_j is a stopping time of the filtration $\mathcal{F}_\cdot$. □

Theorem 3.1. [9] For all $t, s \geq 0$ and any $\Gamma \in \sigma(X)$, where $\sigma(X)$ denotes the σ-field generated by X,

$$\mathbb{P}(x_{t+s}^u \in \Gamma \,|\mathcal{F}_t) = \mathbb{P}(x_{t+s}^u \in \Gamma \,|\sigma(x_t^u)), \quad (23)$$

i.e. with extended state feedback the overall closed loop NETCAD system generates a Markov state process x^u. □

4. CONCLUSION

The stochastic state space dynamical systems framework for call request and routing in what are termed NETCAD networks has been introduced in this paper. A feature of the resulting stochastic NETCAD systems is that they are hybrid stochastic systems with variable dimension state processes; for these processes certain properties, such as the Markovian property and piecewise continuity, have been established.

The NETCAD framework permits the formulation and analysis of centralized optimal stochastic control with respect to specified utility functions; in particular, this entails the derivation of the Hamilton-Jacobi-Bellman equation for optimally controlled NETCAD systems [3,8,10,11]. In addition this framework provides the foundation for the current work [7] on decentralized suboptimal control based upon state aggregation and estimation.

5. ACKNOWLEDGEMENTS

The authors gratefully acknowledge many discussions of this work with Lorne G. Mason.

REFERENCES

[1] V.E. Benes, "Mathematical theory of connecting networks and telephone traffic", Academic Press, New York, 1965.

[2] R. Boel and P. Varaiya, "Optimal Control of Jump Processes", *SIAM J. Control and Optimization*, vol. 15, No.1, January 1977.

[3] P.E. Caines, Z. Ma and L.G. Mason, "Stochastic Control of Communication Networks". *Mediterranean Conference on Complex Systems*, Aiya Napa, Cyprus, July, 2005.

[4] M.H.A. Davis, "Markov Models and Optimization", Chapman & Hall, London, 1993.

[5] Z. Dziong and L.G. Mason, "Call Admission and Routing in Multi-Service Loss Networks", *IEEE Trans. Commun.*, vol. 42, pp. 2011-2022, Feb./Mar./Apr. 1994.

[6] F.P. Kelly and R.J. Williams, "Dynamic routing in stochastic networks", *IMA Volumes in Mathematics and its Applications*, 71. Springer-Verlag, New York, 1995. 169-186.

[7] Z. Ma, "Call Admission Control and Routing Control in Integrated Communication Networks via Dynamic Programming", Masters Thesis, McGill U., Feb. 2005.

[8] Z. Ma, P.E. Caines and R.P. Malhamé, "Optimal Stochastic Control of Networks: Call Admission and Routing for Simple Networks" . Submitted to *International Conference on Intelligent Systems And Computing: Theory And Applications (ISYC'06).*

[9] Z. Ma, P.E. Caines and R.P. Malhamé, "Stochastic Control of Network Systems I: NETCAD State Space Structure & Dynamics", Submitted to *45th IEEE Int. Conf. Decision and Control.*

[10] Z. Ma, P.E. Caines and R.P. Malhamé, "Stochastic Control of Network Systems II: NETCAD Optimal Control & the HJB Equation", Submitted to *45th IEEE Int. Conf. Decision and Control.*

[11] Z. Ma, P.E. Caines and R.P. Malhamé, "Viscosity Solutions of HJB Equations for CAC and RC Problems in NETCAD Systems", Research Report Dept. ECE, McGill U., Feb, 2006.

[12] P. Marbach, O. Mihatsch, and J.N. Tsitsiklis, "Call Admission Control and Routing in Integrated Services Networks Using Neuro-Dynamic Programming", *IEEE Trans. Commun.*, vol. 18, no.2, pp. 197-207, Feb. 2000.

[13] M. S. Shaikh and P. E. Caines, "On the optimal control of hybrid systems: Analysis and algorithms for trajectory and schedule optimization". In Proc. *42nd IEEE Int. Conf. Decision and Control*, pp 2144-2149, Maui, Hawaii, 2003.

ELSEVIER
IFAC
PUBLICATIONS

PARAMETER IDENTIFICATION FOR PIECEWISE DETERMINISTIC MARKOV PROCESSES: A CASE STUDY ON A BIOCHEMICAL NETWORK

Panagiotis Kouretas[†], **Konstantinos Koutoumpas**[†], **and John Lygeros**[†]

[†]*Department of Electrical and Computer Engineering, University of Patras, Rio, Patras, GR 26500, Greece*

Abstract: The first part of the paper focuses on the use of Piecewise Deterministic Markov Processes (PDMP) for the modeling of biochemical networks. As a case study, the production of the peptide antibiotic *B.Subtilis* is considered. The second part of the paper is devoted to the problem of identifying unkown parameters of the model equations, based on experimental data. For the identification task we propose the use of Genetic Algorithms. An example demonstrates the feasibility of our aproach. *Copyright © 2006 IFAC*

Keywords: Stochastic Hybrid Models, Piecewise Deterministic Markov Processes, Subtilin Production, Genetic Algorithms, Biochemical Networks

1. INTRODUCTION

As far as biology is concerned, the last 20 years marked an astonishing advancement due to pioneering, high-throughput techniques (e.g. micro-arrays) that allowed biologists to shed light to numerous aspects of their field. However researchers realized that the perspective of information gathering was insufficient to provide explanations to difficult questions and lead to the next big step. That possibly explains the recent tendency to express biological processes in terms of a complete system. While in the past, biologists would only observe a target organism, nowadays the first priority is to infer the underlying mechanism that produces the observation.

Such attempts gave birth to a new field called systems biology that applies the aforementioned way of thinking: first a large number of information is collected, then a model for the underlying model is proposed and finally its fitness is validated by a new series of experiments. The general course comprises the thorough observation of the biological process so as to understand all of its characteristics. Then the appropriate model family has to be selected, and finally, each parameter must be refined.

In what follows, we focus our attention on the already studied organism *Bacillus subtilis*. In this paper we emphasize on the conversion of the model that describes the biological process to a valid PDMP model. Piecewise Deterministic Markov Processes are a sub category of stochastic hybrid models (SHM). SHMs were initially developed to capture all the characteristics of systems containing a combination of digital and analogue components and found a wealth of applications in cases such as automated highway systems, air-traffic management systems, manufacturing systems, robotics and real-time communication networks.

Apparently, PDMP seem to be applicable to system biology, apart from being friendly to study and implement. Reasons for choosing this class of SHM will be analyzed later on. Immediately after the expression of the bio-process as a PDMP model, we encounter the difficulty of defining each parameter of the model. The complexity of testing all possible parameter configurations is exponential and would require tremendous computational time and resources. Our work takes into consideration what is biologically feasible and observable in order to identify all the parameter values. Contrary to what is currently present in the literature, our ultimate goal is to infer the parameters and give a valid PDMP model for the target biological process with respect to the quantitative aspect.

2. STOCHASTIC HYBRID MODELS

The great interest of research community for the field of stochastic hybrid systems in recent years led to the introduction of different types of stochastic hybrid models. The main difference between these classes of stochastic hybrid models lies in the way the stochasticity appears [1]. In some models continuous evolution may be governed by stochastic differential equations, while in others not. Likewise, some models include forced transitions, which take place whenever the continuous state tries to leave a given set, others only allow transitions to take place at random times (spontaneously at a given possibly state-dependent rate), while others allow both. Finally the destination of discrete transitions may be given by a probability kernel. In this report the model that will be analyzed is the Piecewise Deterministic Markov Processes [2], which is the class of models that will be used for B-subtilis. Apart from the PDMP there are also the Switched Diffusion Processes (SDP) [3] and the Stochastic Hybrid Systems (SHS) [4]. An overview of them can be found in [1].

2.1 Piecewise Deterministic Markov Processes (PDMP)

PDMPs are a class of non-linear continuous-time stochastic hybrid processes which covers a wide range of non-diffusion phenomena. PDMP involve a hybrid state space, with both continuous and discrete states. The particularity of this model is that randomness appears only in the discrete transitions; between two consecutive transitions the continuous state evolves according to a non-linear ordinary differential equation. Transitions occur either when the state hits the state space boundary, or in the interior of the state space, according to a generalized Poisson process. Whenever a transition occurs, the hybrid state is reset instantaneously according to a probability distribution which depends on the hybrid state before the transition. We introduce formally PDPM following the notation of ([5], [6]). Let Q be a countable set of discrete states, and let $d : Q \to \mathbb{N}$ and $X : Q \to \mathbb{R}^{d(\cdot)}$ be two maps assigning to each discrete state $i \in Q$ an open subset of $\mathbb{R}^{d(i)}$. We call the set

$$\mathcal{D}(Q,d,X) = \bigcup_{i\in Q} \{i\}\times X(i) = \{(i,x) : i \in Q, x \in X(i)\}$$

the hybrid state space of the PDMP and $\alpha = (i,x) \in \mathcal{D}(Q,d,X)$ the hybrid state. We define the boundary of the hybrid state space as

$$\partial\mathcal{D}(Q,d,X) = \bigcup_{i\in Q} \{i\} \times \partial X(i).$$

where as usual $\partial X(i)$ denotes the boundary of the open set $X(i)$.

A vector field f on the hybrid state space $\mathcal{D}(Q,d,X)$ is a function $f : \mathcal{D}(Q,d,X) \to \mathbb{R}^{d(\cdot)}$ assigning to each hybrid state $(i,x) \in \mathcal{D}$ a direction $f(i,x) \in \mathbb{R}^{d(i)}$. The flow of f is a function $\Phi : \mathcal{D}(Q,d,X)\times\mathbb{R} \to \mathcal{D}(Q,d,X)$ with

$$\Phi(i,x,t) = \begin{bmatrix} \Phi_Q(i,x,t) \\ \Phi_X(i,x,t) \end{bmatrix},$$

$\Phi_Q(i,x,t) \in Q$ and $\Phi_X(i,x,t) \in X(i)$, such that (i,x), $\Phi(i,x,0) = i,x$ and for all $t \in \mathbb{R}$, $\Phi_Q(i,x,t) = i$ and

$$\frac{d}{dt}\Phi_X(i,x,t) = f(\Phi(i,x,t)) \qquad (2.1)$$

Let

$$\Gamma((Q,d,X),f) =$$

$$\{\alpha \in \partial\mathcal{D}(Q,d,X) \mid \exists(\alpha',t) \in \mathcal{D}(Q,d,X) \times \mathbb{R}^+, \alpha = \Phi(\alpha',t)\}$$

Denotes the part of the boundary of $\mathcal{D}$ that can be reached from $\mathcal{D}$ under f and let

$$\bar{\mathcal{D}}(Q,d,X) = \mathcal{D}(Q,d,X) \cup \Gamma((Q,d,X),f)$$

Let $\bar{\mathcal{D}} = Q \times \mathbb{R}^\infty$ and

$$\mathcal{B}(\bar{\mathcal{D}}) = \sigma\left(\bigcup_{i\in Q} \{i\} \times \mathcal{B}(i)\right)$$

be the smallest σ - algebra on $(\bar{\mathcal{D}}$ containing all sets of the form $i \times A$ with $A \in B(i)$ a Borel subset of $X(i)$. It can be shown that the space $(\bar{\mathcal{D}}, \mathcal{B}(\bar{\mathcal{D}}))$ is a Borel Space and $\mathcal{B}(\bar{\mathcal{D}})$ is a sub-σ-algebra of its Borel σ-algebra.

We can now introduce the following definition.

Definition 1. (**Piecewise Deterministic Markov Process**). A Piecewise Deterministic Markov Process (PDMP) is a collection $H = ((Q,d,X), f, Init, \lambda, R)$ where

- Q is a countable set of discrete variables;
- $d : Q \to \mathbb{N}$ is a map giving the dimensions of the continuous state spaces;
- $X : Q \to \mathbb{R}^{d(\cdot)}$ maps each $i \in Q$ into an open subset $X(i)$ of $\mathbb{R}^{d(i)}$;
- $f : \mathcal{D}(Q,d,X) \to \mathbb{R}^{d(\cdot)}$ is a vector field;
- $Init : \mathcal{B}(\bar{\mathcal{D}}) \to [0,1]$ is an initial probability measure on $(\bar{\mathcal{D}}, \mathcal{B}(\bar{\mathcal{D}}))$, with $Init(\mathcal{D}^c) = 0$;
- $\lambda : \bar{\mathcal{D}}(Q,d,X) \to \mathbb{R}^+$ is a transition rate function;
- $R : \mathcal{B}(\bar{\mathcal{D}}) \times \bar{\mathcal{D}}(Q,d,X) \to [0,1]$ is a transition measure, with $R(\mathcal{D}^c,(i,x)) = 0$ for all $(i,x) \in (\bar{\mathcal{D}}(Q,d,x)$.

To define the PDMP executions we introduce the notion of exit time $t^* : \mathcal{D} \to \mathbb{R}^+ \cup \{\infty\}$,

$$t^*(i,x) = \inf\{t > 0 : \Phi(i,x,t) \notin \mathcal{D}\}$$

and of survival function $F : \mathcal{D} \times \mathbb{R}^+ \rightarrow [0, 1]$,

$$F(i,x,t) = \begin{cases} \exp\left(-\int_0^t \lambda(\Phi(i,x,\tau))d\tau\right) & \text{if } t < t^*(i,x) \\ 0 & \text{if } t \geq t^*(i,x). \end{cases}$$

The executions of the PDMP can be thought of as being generated by the following algorithm.

Algorithm 1. **(Generation of PDMP Executions)**
set $T = 0$
select $\mathcal{D}$-valued random variable $(\hat{i}, \hat{x})$ according to $Init$
repeat
select $\mathbb{R}^+$-valued random variable $\hat{T}$ such that $P(\hat{T} > t) = F(\hat{i}, \hat{x}, t)$
set $(i_t, x_t) = \Phi(\hat{i}, \hat{x}, t - T)$ for all $t \in [T, T + \hat{T})$
select $\mathcal{D}$-valued random variable $(\hat{i}, \hat{x})$ according to $R(., \Phi(\hat{i}, \hat{x}, \hat{T}))$
set $T = T + \hat{T}$
until true

To ensure the process is well-defined, the following assumption is introduced in [2].

Assumption 1. The sets $X(i)$ are open. For all $i \in Q$, $f(i, .)$ is globally Lipschitz continuous. $\lambda : \bar{\mathcal{D}}(Q, d, X) \rightarrow \mathbb{R}^+$ is measurable. For all $i, x \in \mathcal{D}$ there exists $\varepsilon > 0$ such that the function $t \rightarrow \lambda(\Phi(i, x, t))$ is integrable for all $t \in [0, \varepsilon)$. For all $A \in \mathcal{B}(\bar{\mathcal{D}})$, $R(A, \cdot)$ is measurable.

All random extractions in Algorithm 1 are assumed to be independent. To ensure that (i_t, x_t) is defined on the entire $\mathbb{R}^+$ it is necessary to exclude Zeno executions [6]. The following assumption is introduced in [2] to accomplish this.

Assumption 2. Let $N_t = \sum_i I_{(t \geq T_i)}$ be the number of jumps in $[0, t]$. Then $\mathbb{E}[N_t] < \infty$ for all t.

Under Assumptions 1, 2 it can be shown that the Algorithm 1 defines a strong Markov process [2], continuous from the right with left limits.

3. MODEL OF SUBTILIN PRODUCTION

In order to display the descriptive power of PDMP, we develop a model for the system that governs Subtilin production by *B. subtilis* bacterium in terms of Piecewise Deterministic Markov Processes(PDMP). This form of stochastic hybrid model was selected because it coincides the inherent characteristics of the model. In the following sections, we will focus on special characteristics, and we will provide a consummate PDMP description.

3.1 Subtilin production

Subtilin is an antibiotic released by *B. subtilis* as a way to confront difficult environmental conditions. The factors that govern subtilin production can be divided into internal (the physiological states of the cell) and external (local population density,nutrient levels, aeration, environmental signals in general). Roughly speaking, a high concentration of nutrients in the environment results in an increase in *B. subtilis* population without a remarkable change in subtilin concentration. Subtilin production starts when the amount of nutrient falls under a threshold because of excessive population growth [8]. *B. subtilis* produces subtilin and uses it as a weapon to increase its food supply, by eliminating competing species; in addition to reducing the demand for nutrients, the decomposition of the organisms killed by subtilin releases additional nutrients in the environment.

According to the simplified model for the subtilin production process, developed in [7], subtilin derives from the peptide SpaS. Responsible for the production of SpaS is the activated protein SpaRK, which is composed in turn by the binding of the SigH protein to upstream genes of SpaS protein. Finally, the composition of SigH is turned on whenever the nutrient concentration falls below a certain threshold.

3.2 Model equations

A stochastic hybrid model for this process was proposed in [7]. The equations of the model were developed based on the qualitative description of Section 3.1. The model comprises 5 continuous states: the population of *B. subtilis*, x_1, the concentration of nutrients in the environment, x_2, and the concentrations of the SigH, SpaRK and SpaS molecules (x_3, x_4 and x_5 respectively).

The model also comprises $2^3 = 8$ discrete states, generated by three binary switches, which we denote by S_3, S_4 and S_5. Switch S_3 is deterministic: it goes ON when the concentration of nutrients, x_2, falls below a certain threshold (denoted by η), and OFF when it rises over this threshold. The other two switches are stochastic. In [7] this stochastic behavior is approximated by a discrete time Markov chain, with constant sampling interval Δ. Given that the switch S_4 is OFF at time $k\Delta$, the probability that it will be ON at time $(k+1)\Delta$ depends on the concentration of SpaS at the time $k\Delta$, $x_3(k\Delta)$. More specifically, this probability is

$$a_0(x_3) = \frac{cx_3}{1 + cx_3},$$

where c is a model constant. Notice that the probability of switching ON increases to 1 as x_3 gets higher. Conversely, given that the switch S_4 is ON at time $k\Delta$, the probability that it will be OFF at time $(k + 1)\Delta$ is

$$a_1(x_3) = \frac{1}{1 + cx_3}.$$

Notice that this probability increases to 1 as x_3 gets smaller. The dynamics of switch S_5 are similar, with the concentration of SpaRK, x_4, replacing x_3.

The continuous dynamics for the *B. subtilis* population x_1 are given by

$$\dot{x}_1 = rx_1(1 - \frac{x_1}{D_\infty(x_2)}).$$

Under this equation, x_1 will tend to converge to D_∞, the steady state population for a given nutrient amount. D_∞ depends on x_2 and is given by

$$D_\infty(x_2) = min\{\frac{x_2}{X_0}, D_{max}\}.$$

X_0 and D_{max} are constants of the model; the latter represents constraints on the population because of space limitations and competition within the population.

The continuous dynamics for x_2 are governed by:

$$\dot{x}_2 = -k_1x_1 + k_2x_3$$

where k_1 denotes the rate of nutrient consumption per unit of population and k_2 the rate of nutrient production due to the action of subtilin. In reality, the second term is proportional to the average concentration of SpaS, but for simplicity we follow [7] and assume that the average concentration is proportional to the concentration of SpaS for a single cell.

The continuous dynamics for the remaining three states depend on the discrete state, i.e. the state of the three switches. In all three cases the equations take the form:

$$\dot{x}_i = \begin{cases} -l_ix_i & \text{if } S_i \text{ is OFF} \\ k_i - l_ix_i & \text{if } S_i \text{ is ON.} \end{cases}$$

It is easy to see that the concentration x_i decreases exponentially toward zero whenever the switch S_i is OFF and tends exponentially toward k_i/l_i whenever S_i is ON. Note that the model is Piecewise Affine (PWA) with the exeption of nonlinear x_1 and the stochastic terms used to describe switch behavior.

3.3 PDMP formalism

We now try to express the model for subtilin production using the PDMP formalism. As presented earlier, a Piecewise Deterministic Markov Process is a collection $H = ((Q, d, X), f, Init, \lambda, R)$. We saw that the subtilin production has 8 discrete states. Therefore, the set Q is a countable set, the cardinality of which is 8. Let

$$Q = q_0, ...q_7,$$

so that the index (in binary) of each discrete state reflects the state of the three switches. For example, state q_0 corresponds to binary 000, i.e. all three swithes being OFF. Likewise, state q_5 corresponds to binary 101, i.e. switches S_3 and S_5 being ON and switch S_4 being OFF. In the following discussion, the state names $q_0, ...q_7$ and the binary equivalents of their indices will be used interchangeably.

A wildcard, *, will be used when in a statement the position of some switch is immaterial; e.g. 1** denotes that something holds when S_3 is ON, whatever the values of S_4 and S_5 may be.

The discussion in the previous section suggests that there are 5 continuous states and all of them are active in all discrete states. Therefore, the dimension of the continuous state space is constant $d(q) = 5$, for all $q \in Q$. The definition of the survival function suggests that the open sets $X(q) \subseteq \mathbb{R}^5$ are used to force discrete transitions to take place at certain values of state. In the subtilin production model outlined above the only forced transitions are those induced by the deterministic switch S_3; S_3 has to go ON whenever x_2 falls under the threshold η and has to go OFF whenever it rises over this threshold. These transitions can be forced by defining

$$X(0 * *) = \mathbb{R} \times (\eta, \infty) \times \mathbb{R}^3$$

and

$$X(1 * *) = \mathbb{R} \times (-\infty, \eta) \times \mathbb{R}^3$$

The above elements completely determine the hybrid state space, $\mathcal{D}(Q, d, X)$, of the PDMP. As far as the vector field is examined, we have to specify the direction $f(a)$ that is assigned to each state $a = (q, x)$. The vector field is not dependent on the value of X but depends on the discrete states. Therefore, we have:

$$f(q_0, x) = \begin{bmatrix} rx_1(1 - \frac{x_1}{D_\infty}) \\ -k_1x_1 + k_2x_3 \\ -l_3x_3 \\ -l_4x_4 \\ -l_5x_5 \end{bmatrix} \quad f(q_1, x) = \begin{bmatrix} rx_1(1 - \frac{x_1}{D_\infty}) \\ -k_1x_1 + k_2x_3 \\ -l_3x_3 \\ -l_4x_4 \\ k_5 - l_5x_5 \end{bmatrix}$$
$$f(q_2, x) = \begin{bmatrix} rx_1(1 - \frac{x_1}{D_\infty}) \\ -k_1x_1 + k_2x_3 \\ -l_3x_3 \\ k_4 - l_4x_4 \\ -l_5x_5 \end{bmatrix} \quad f(q_3, x) = \begin{bmatrix} rx_1(1 - \frac{x_1}{D_\infty}) \\ -k_1x_1 + k_2x_3 \\ -l_3x_3 \\ k_4 - l_4x_4 \\ k_5 - l_5x_5 \end{bmatrix}$$
$$f(q_4, x) = \begin{bmatrix} rx_1(1 - \frac{x_1}{D_\infty}) \\ -k_1x_1 + k_2x_3 \\ k_3 - l_3x_3 \\ -l_4x_4 \\ -l_5x_5 \end{bmatrix} \quad f(q_5, x) = \begin{bmatrix} rx_1(1 - \frac{x_1}{D_\infty}) \\ -k_1x_1 + k_2x_3 \\ k_3 - l_3x_3 \\ -l_4x_4 \\ k_5 - l_5x_5 \end{bmatrix}$$
$$f(q_6, x) = \begin{bmatrix} rx_1(1 - \frac{x_1}{D_\infty}) \\ -k_1x_1 + k_2x_3 \\ k_3 - l_3x_3 \\ k_4 - l_4x_4 \\ -l_5x_5 \end{bmatrix} \quad f(q_7, x) = \begin{bmatrix} rx_1(1 - \frac{x_1}{D_\infty}) \\ -k_1x_1 + k_2x_3 \\ k_3 - l_3x_3 \\ k_4 - l_4x_4 \\ k_5 - l_5x_i \end{bmatrix} \tag{3.1}$$

Regarding the initial state of the model, for simplicity reasons, as well as for biological common sense, we assume that executions start always from the q_0 discrete state. Also, we require that the probability distribution Init satisfies

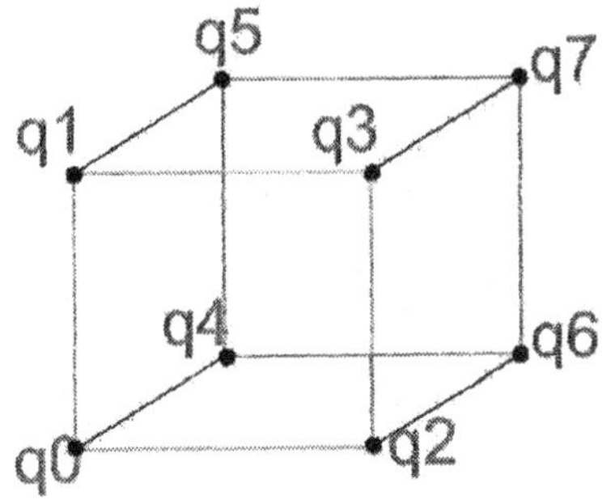

Fig. 1. discrete state space

$Init(0 ** \times x \in \mathbb{R}^5 | x_2 \leq \eta) = 0,$
$Init(1 ** \times x \in \mathbb{R}^5 | x_2 \geq \eta) = 0.$

The initial state should reflect any other constraints imposed by biological intuition. For example, since x_1 reflects the *B.Subtilis* population, it is reasonable to assume that they $x_1 \geq 0$. Another reasonable constraint is that initially $x_1 \leq D_\infty(x_2)$. Finally, since continuous states $x_2, ..., x_5$ reflect concentrations, it is reasonable to assume that they also start with non-negative values. These constraints can be imposed if we require that for all $q \in Q$

$$Init(\{q\} \times \{x \in \mathbb{R}^5 | \\ x_1 \in (0, D_\infty(x_2)) \text{ and } min(x_2, x_3, x_4, x_5) > 0\}) = 1 \tag{3.2}$$

All probability distributions that respect the above constraints are considered acceptable for our model.

The main problem we confront when trying to express the subtilin production model as a PDMP is the need to define the λ function. Intuitively, this function indicates the "tendency" of the system to jump and switch its discrete state. The rate function λ will govern the spontaneous transitions of the switches S_4 and S_5 (switch S_3 is governed by a forced transition). To present the design of an appropriate λ function we focus on discrete state q_6. Figure 2 summarizes the discrete transitions out of state q_6. Simultaneous switching of more than one of the switches S_3, S_4, S_5 is not allowed. This is a reasonable assumption, since simultaneous switching of two or more switches is a null event in the unerlying probability space. q_6 corresponds to binary 110, i.e. switches S_3 and S_4 being ON and S_5 being OFF. Of the three transitions out of q_6, the one to q_2 ($S_3 \rightarrow$ OFF) is forced and does not feature in the construction of the rate function. For the remaining two transitions, we define two separate rate functions, $\lambda_{S_4 \rightarrow OFF}(x)$ and $\lambda_{S_5 \rightarrow ON}(x)$. These functions need to be linked somehow to the transition probabilities of the discrete time Markov chain with sampling period Δ used to model the probabilistic switching in [7]. The survival function of states that the probability that the switch S_4 remains ON throughout the interval $[(k-1)\Delta, k\Delta]$ is equal to

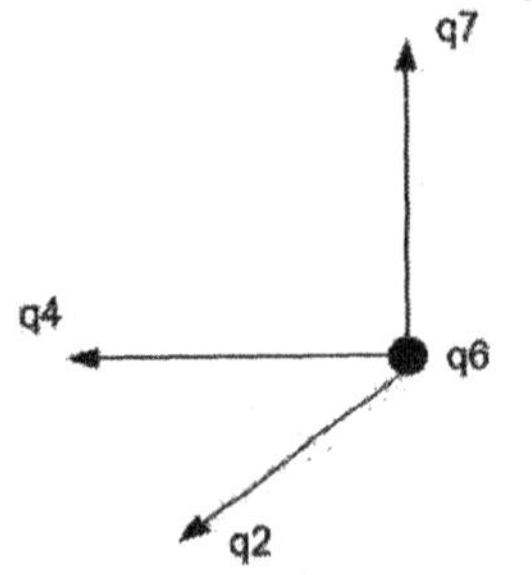

Fig. 2. possible transitions

$$exp\left(\int_{(k-1)\Delta}^{k\Delta} \lambda_{S_4 \rightarrow OFF}(x(\tau))d\tau\right)$$

This propability should be equal to $1 - a_0(x_3((k-1)\Delta))$. Assuming that Δ is small enough, we have that $1 - a_0(x_3((k-1)\Delta)) \approx exp(-\Delta\lambda_{S_4 \rightarrow OFF}(x(k\Delta)))$. Selecting $\lambda_{S_4 \rightarrow OFF}(x) = \frac{ln(1+cx_3)}{\Delta}$ achieves the desired effect. Likewise, we define $\lambda_{S_5 \rightarrow ON}(x) = \frac{ln(1+cx_4) - ln(cx_4)}{\Delta}$ and set the transition rate for discrete state q_6 to $\lambda(q6, x) = \lambda_{S_4 \rightarrow OFF}(x) + \lambda_{S_5 \rightarrow ON}(x)$

The functions $\lambda_{S_4 \rightarrow OFF}(x)$ and $\lambda_{S_5 \rightarrow ON}(x)$ take non-negative values and are therefore good candidates for rate functions. In a similar way, we define rate functions $\lambda_{S_5 \rightarrow OFF}(x)$ (replacing x_3 by x_4) and $\lambda_{S_4 \rightarrow ON}(x)$ (replacing x_4 by x_3) and use them to define the transition rates for the remaining discrete states. In order to complete the PDMP model we need to define the probability distribution for the state after a discrete transition. The only difficulty here is removing any ambiguities that may be caused by simultaneous switches. We do this by introducing the a priority scheme: when the forced transition has to take place, it does, else either of the spontaneous transitions can take place. For state q_6 this leads to

$$R(q_6, x) = \delta_{(q_2,x)}(q, x) \text{ if } (q_6, x) \in \mathcal{D} \text{ else}$$
$$R(q_6, x) = \frac{\lambda_{S_4 \rightarrow OFF}(x)}{\lambda(q6, x)}\delta_{(q_4,x)}(q, x) + \frac{\lambda_{S_5 \rightarrow ON}(x)}{\lambda(q6, x)}\delta_{(q_7,x)}(q, x) \tag{3.3}$$

Here $\delta_{(\hat{q},\hat{x})}(q, x)$ denotes the Dirac measure concentrated at $(\hat{q}, \hat{x})$. If desired, the two components of the measure R can be written together using the indicator function, $I_\mathcal{D}(q, x)$, of the set $\mathcal{D}$. It is easy to see that this probability measure satisfies Assumpion 1.

The above discussion shows that the PDMP model also satisfies most of the conditions of Assumption 1. The only problem may be the non-Zeno condition.

4. PARAMETER IDENTIFICATION

4.1 Problem formulation

In this section we will concentrate on formulating mathematically the problem, in order to make it solvable

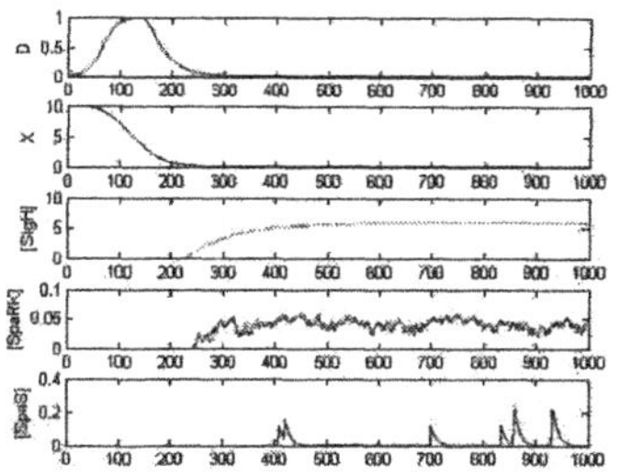

Fig. 3. Execution for randomly selected values

by the use of Genetic Algorithms. First of all, we must clarify that our intention is to estimate only the parameters involved in the differential equations governing the function $f(x)$ of the PDMP formalism. We have to emphasize at this point, that only the first two curves of the Fig 3 (showing the evolution of x_1 and x_2) are to be used, because they comprise the only observable data of the system. The other three curves indicate the evolution of intracellular concentrations, and are not readily available for measurement. Therefore, the problem can be expressed as: Is it possible to exploit the curves expressing the food and the population evolution in order to reveal the values of the parameters of the underlying model? In what follows, we present our effort to estimate the values by measuring the distance of a generated curve (based on random selection of parameters) from its target-curve (the original one). That is, we treat these curves as bearing all the information hidden in the system in the form of parameter values.

Figure 3 shows the execution of the five continuous states of the aforementioned system in case of random selection of the target values, so as to become evident that the a priori knowledge of the structure of the system is not sufficient to guarantee reasonable results.

The exact set of parameters we want to identify are the five values of synthesis rates k_1 to k_5, three degradation rates l_3 to l_5, the constant r, the time interval δ and the threshold η.

4.2 Proposed solution based on GA

The genetic algorithm is a method for solving optimization problems. They are based on natural selection, and are inspired by the Darwinian optimization process that governs evolution in real life. The genetic algorithm first creates and then modifies a set of individual solutions. At each step, the genetic algorithm must select a subset of individuals from the given population for mating reasons. The selected individuals produce the population of the next generation. Over successive generations, we expect the population to evolve toward a better solution, according to a fitness function. Researchers have proposed numerous slight modifications (concerning a GA's evolution). However, the three main types of rules of a GA are the Selection rules, the Crossover rules and the Mutation rules. The next paragraph describes implementation topics of our genetic algorithm application.

The simulation was held on the MATLAB environment for ease of use [9]. We exploited the conveniences provided by the version 1.0.1 of the Genetic Algorithm toolbox to obtain our results. In order to run the algorithm, we must firstly designate all the parameters of the algorithm informing properly all the fields of the structure *gaoptimset*.

CreationFcn: our selection is the function *gacreationuniform* for we want to initialize a uniform population

CrossoverFraction: denotes the fraction of the population to be created from one generation to the next by the crossover. It is set to the default value 0.8.

CrossoverFcn: informs the GA in what way to create crossover children. We selected *crossoverscattered*

PopInitRange: Vector needed to inform of the boundary values. In our case it coincides with the default value [0;1] because the majority of the parameters must operate inside this area.

PopulationType: The data type of the population could be 'bitstring','doubleVector', or customized. The first two do not serve our purposes, consequently we had to use the 'custom' option because our values must be positive double values. Besides that, we had to render the selected creation and mutation functions operational with our data type, including changes in the source code of the respective functions

MutationFcn: Governs the way that mutations are generated. *Gaussian* mutation was selected in order to produce generally small perturbations.

PopulationSize: This number shows the number of the population. The greater the number, the more precise our search of the state space (but also the more slow the search becomes).

SelectionFcn: The function that selects individuals for mating purposes is set to the commonly used *selectionroulette*. This is a selection operator in which the chance of an individual getting selected is proportional to its fitness. This is where the concept of survival of the fittest comes into play.

The **fitness function** is perhaps the most important parameter that we can define. It was noted earlier that the trajectories of the food as well as the population evolution have to be considered as evaluation means. If we simply calculate the mean square distance between the original trajectories and those generated by a different set of parameters, we do not take into consideration the stochastic nature of the model. The results include different sets of parameters that reproduce satisfactorily the observed trajectories.

The following formula describes our fitness function. That is, for every set of parameters we calculate the evolution of x_1 and x_2. Then these curves are sampled with sampling interval δ. Small letters are used to denote the generated curves, while capital letters denote the target curves.

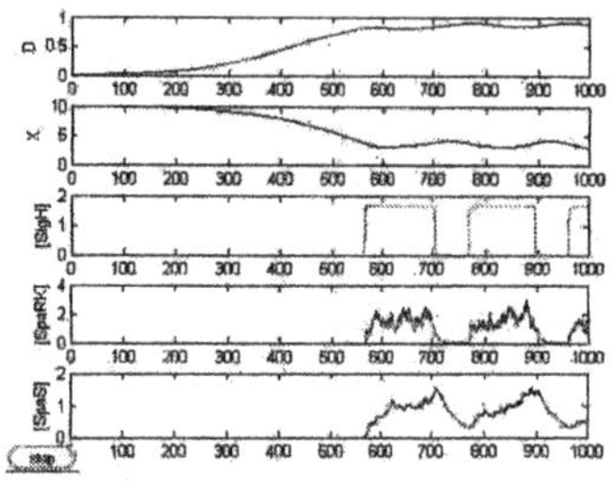

Fig. 4. Results generated by simple fitness function

$$\sum_k[(x_1(k\delta) - X_1(k\delta))^2 + (x_2(k\delta) - X_2(k\delta))^2]$$

Consequently, the previous formula measures the square distance,at specific time points of the evolution, between the target curves (which have to be observed experimentally) and the original ones. Results are shown in Figure 4.

4.3 Results for B.Subtilis model

The PDMP model for Subtilin production comprises randomness that clearly leads each execution to differ from every other (same set of parameters and initial conditions). Thus, we believe that a sole execution is not adequate to fully characterize a given set of values.
Our intention is then to run multiple experiments in order to calculate multiple measurements of mean square distances between the original and the under-examination trajectories. The mean arithmetic value serves then as a fitness indicator.

$$\sum_{i=1}^{2} \frac{\sum_k[(x_1^i(k\delta) - X_1(k\delta))^2 + (x_2^i(k\delta) - X_2(k\delta))^2]}{2}$$

We now extend the fitness function shown before, in order to capture more than one experiments. More specifically, we simulate the evolution that is forced by a set of parameters and then calculate the simple square distance. Then we re-simulate the evolution of the system, and calculate its distance once more. Due to stochastic nature of the system, its evolution (and therefore its distance from the target curves) will differ from execution to execution. In our example $i = 2$, showing that two executions of the system are taken into consideration. Generated results are shown in Figure 5

4.4 Conclusion

PDMPs are a class of models that capture inherent randomness. Their stochastic nature makes it difficult to apply parameter identification techniques to find the parameter values of the $f(x)$ function. However, genetic algorithms were able, only by measuring the mean square distance of the population and food trajectories, to estimate a fair set of parameters, that produce executions, matching the original curves.

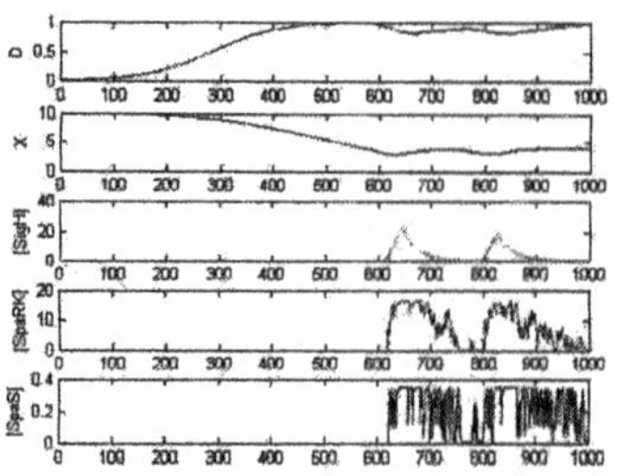

Fig. 5. Results generated by "multiple" fitness function

The fact that an increasing number of researchers try to describe biological processes in terms of stochastic hybrid models, along with the friendly nature of the PDMP formalism, makes us believe that the latter will be further used in the near future in the field of systems biology. We expect genetic algorithms to continue to serve as a means of parameter identification technique for such cases.

5. ACKNOWLEDGMENT

Research supported by the European Commission under the project HYGEIA, NEST-4995. The authors wish to thank the anonymous referees for their careful reading of the manuscript and their fruitful comments.

REFERENCES

1. G. Pola, M. Bujorianu, J. Lygeros, and M. Benedetto, Stochastic hybrid models: An overview, Science,vol. 2,.
2. M. Davis, Markov Processes and Optimization. London: Chapman and Hall, (1993)
3. M. Ghosh, A. Arapostathis, and S. Marcus, Ergodic control of switching diffusions, SIAM Journal on Control Optimization, vol. 35, pp. 19521988.
4. J. Hu, J. Lygeros, and S. Sastry, Towards a theory of stochastic hybrid systems, no. 1790, pp. 160173, 2000.
5. M. Bujorianu and J. Lygeros, Reachability questions in piecewise deterministic markov processes, in Submitted to HSCC.
6. J. Lygeros, K. H. Johansson, S. N. Simic, J. Zhang, and S. Sastry, Dynamical properties of hybrid automata, IEEE Transactions on Automatic Control, 2003.
7. J. Hu, W. Wu, and S. Sastry, Modeling subtilin production in bacillus subtilis using stochastic hybrid systems, The Seventh International Workshop on Hybrid Systems: Computation and Control, 2004.
8. T. Stein and S. Borchert and P. Kiesau and S. Heinzmann and S. Kloss and M. Helfrich and K.D. Entian, *Dual Control of Subtilin Biosynthesis and Immunity in Bacillus Subtilis*, Molecular Microbiology. Vol. 44, 2002, 403-416.
9. www.mathworks.com/access/helpdesk/help/techdoc/matlab.shtml

ELSEVIER
IFAC
PUBLICATIONS

USING PATH INTEGRAL SHORT TIME PROPAGATORS FOR NUMERICAL ANALYSIS OF STOCHASTIC HYBRID SYSTEMS

Gerwald Lichtenberg *[1] Philipp Rostalski *[2]

* *Institute of Control Systems, TU Hamburg-Harburg*

Abstract: Algorithms to approximate the evolution of probability density functions for stochastic hybrid systems rely on the knowledge of appropriate short time propagators. It is shown that a path integral propagator known for continuous stochastic systems can be adapted to the hybrid case. With this propagator, the *HybPathTree* algorithm performs well concerning precision and computational effort, e.g. in reachability analysis. *Copyright © 2006 IFAC*

Keywords: Stochastic Hybrid Systems, Reachability Analysis, Path Integrals

1. INTRODUCTION

The prediction of probability density functions for stochastic hybrid systems is relevant to a variety of technical disciplines. State of the art methods and applications can be found e.g. in (Blom and Lygeros 2005). Even for simple systems, only numerical methods can be applied. For pure continuous stochastic systems many approaches are known which introduce a Markov chain that approximates the probability density at grid points via finite difference schemes (Kushner and Dupuis 2001). Path integrals methods can be applied to compute Markov chain approximations of continuous stochastic systems, which give better results than these classical methods (Wehner and Wolfer 1983).

The novelty of this paper can be seen from two perspectives: On the one hand, the path integral based PATHTREE algorithm (Ingber *et al.* 2001) is extended to a class of stochastic hybrid systems. On the other hand, existing numerical methods for analysis of stochastic hybrid systems are improved by generating the transition probabilities of the approximating Markov chain by a short time propagator resulting from path integrals.

After introducing the model class and the notion of path integrals in Sections 2 and 3, the stochastic reachability problem is given in Section 4. For its solution, extensions of path integrals for hybrid systems are derived in Section 5, which are the basis of Markov chain approximations given in Section 6 that are applied to two tank systems in Section 7.

2. MODELING

In this paper, we are using a slight adaption of the description of a generalized stochastic hybrid system given in (Bujorianu and Lygeros 2003). Let Q denote the set of discrete states and $X^q \subset \mathbb{R}^{d(q)}$ an open $d(q)$-dimension subspace assign to each of those discrete states $q \in Q$. We will refer to those as continuous state spaces. The closure of X^q is defined as $\bar{X}^q = X^q \cap \partial X^q$ with ∂X^q denoting the boundary. The hybrid state space can now be defined as $\mathcal{H} = \bigcup_{q\in Q}\{q\} \times X^q$ with $\bar{\mathcal{H}} = \bigcup_{q\in Q}\{q\} \times \bar{X}^q$ and $\partial\mathcal{H} = \bigcup_{q\in Q}\{q\} \times \partial X^q$ being its closure and boundary respectively.

[1] Current Affiliation: Automatic Control Laboratory, Swiss Federal Institute of Technology, Zürich.

The definition of a generalized stochastic hybrid system can be restated as

Definition 1. A Generalized Stochastic Hybrid System (GSHS) is a collection

$$\mathcal{M} = ((Q, d, \mathcal{X}), a, \sigma, Init, \lambda, R) ,$$

where

- $Q = \{1, 2, \ldots N_Q\}$ is a countable set of discrete states,
- $d : Q \rightarrow \mathbb{N}$ is a map giving the dimension of the continuous state spaces, $\mathcal{X} : Q \rightarrow \mathbb{R}^{d(\cdot)}$ maps each $q \in Q$ into an open subset $X^q \subset \mathbb{R}^{d(q)}$ (continuous state space assigned to the discrete state $q \in Q$),
- $a : Q \times X^q \rightarrow \mathbb{R}^{d(\cdot)}$ is a vector field (describing the system dynamics in each discrete state),
- $\sigma : Q \times X^q \rightarrow \mathbb{R}^{d(\cdot) \times m}$ is a X^d-valued matrix, $m \in \mathbb{N}$ (describing the variance of the noise in each discrete state),
- $Init : \mathcal{B}(\mathcal{H}) \rightarrow [0, 1]$ is a probability measure on $\mathcal{H}$ (distribution of the initial state), where $\mathcal{B}(\mathcal{H})$ is the space of σ-algebras generated by $\mathcal{H}$ (Borel σ algebra).
- $\lambda : \bar{\mathcal{H}} \rightarrow \mathbb{R}^+$ is a transition rate function giving probabilistic changes of discrete state,
- $R : \bar{\mathcal{H}} \times \mathcal{B}(\bar{\mathcal{H}}) \rightarrow [0, 1]$ is a transition measure describing the distribution of the continuous state after a jump.

In the following, we will assume that there are no spontaneous jumps, i.e. a jump-rate $\lambda = 0$ and we will call models of this class stochastic hybrid systems (SHS). Thus, for the transition measure $R : \partial\mathcal{H} \times \mathcal{B}(\bar{\mathcal{H}}) \rightarrow [0, 1]$ holds.

The execution of such a stochastic hybrid system can be defined as follows, compare (Bujorianu and Lygeros 2003):

Definition 2. A stochastic process $h(t) = (q(t), \boldsymbol{x}(t))$ is called a SHS execution if there exists a sequence of stopping times $T_0 = 0, \leq T_1 \leq T_2 \leq \ldots$ such that for each $k \in \mathbb{N}$:

- $h_0 = (q_0, x_0^{q_0})$ is a $Q \times X$-valued random variable extracted according to the probability measure *Init*;
- For $t \in [T_k, T_{k+1})$, $q_t = q_{T_k}$ is constant and $\boldsymbol{x}(t)$ is a (continuous) solution of the SDE
  $dx^{(i)} = a^{(i)}(q_{T_k}, x(t))dt + \sigma^{(i,j)}(q_{T_k, x(t)})d\Gamma^{(j)}(t)$
  where $\Gamma^{(j)}(t)$ is the m-dimensional standard Wiener process
- $T_{k+1} = T_k + S^{i_k}$ where S^{i_k} is the stopping time of the process, i.e. the time where $x(t)$ first hits the boundary of ∂X^q,
- The probabilistic distribution of $x(T_{k+1})$ is governed by the law $R((q_{T_k}, \boldsymbol{x}(T_{k+1}^-)), \cdot)$.

This modelling approach can be seen as a collection of several continuous time, stochastic differential equations and its domains together with jumps. These are executed whenever the continuous state $\boldsymbol{x}$ reaches certain areas of the continuous state-space (the guards).

3. PATH INTEGRALS

The concept of path integrals will be first introduced for a continuous dynamical system, given by the following stochastic differential equation

$$dx^{(i)} = a^{(i)}(\boldsymbol{x})dt + \sigma^{(i,j)}(\boldsymbol{x})d\Gamma^{(j)}(t) \qquad (1)$$

with $i = 1 \ldots n, j = 1 \ldots m$, also known as *Langevin Equation*. The stochastic term is modelled with a m-dimensional standard Wiener noise process $\Gamma(t)$.

In the following, eqn. (1) should be interpreted in the sense of Stratonovich. For many applications, the probability $p(\boldsymbol{x}, t)$ of the state vector $\boldsymbol{x}$ at time t is of interest.

With the knowledge of some initial probability distribution $p(\boldsymbol{x}_0, t_0)$ at a certain time t_0, the conditional probability distribution $p(\boldsymbol{x}, t)$ at a later time t is given by its generator which in this simple case comes as the *Fokker-Planck Equation*

$$\frac{\partial p(\boldsymbol{x}, t \,|\, \boldsymbol{x}_0, t_0)}{\partial t} = L_{\mathrm{FP}}(\boldsymbol{x}, t) \cdot p(\boldsymbol{x}, t \,|\, \boldsymbol{x}_0, t_0) \, . \quad (2)$$

with the *Fokker-Planck Operator*

$$L_{FP}(\boldsymbol{x}, t) = -\sum_{i=1}^{n} \frac{\partial}{\partial x^{(i)}} D_1^{(i)}(\boldsymbol{x}, t) + \\ + \frac{1}{2} \sum_{i=1}^{n} \sum_{j=1}^{m} \frac{\partial^2}{\partial x^{(i)} \partial x^{(j)}} D_2^{(i,j)}(\boldsymbol{x}, t) \, . \quad (3)$$

The term $D_1^{(i)}(\boldsymbol{x}, t)$ is the i'th component of the *drift* that holds information about the underlying deterministic movement of the system, while $D_2(\boldsymbol{x}, t)$ gives the *diffusion*, i.e. the additional noise. For a multi-dimensional stochastic process, the drift operator

$$D_1^{(i)}(\boldsymbol{x}, t) = \\ a^{(i)}(\boldsymbol{x}, t) + \frac{1}{2} \sum_{k=1}^{n} \sum_{j=1}^{m} \sigma^{(k,j)}(\boldsymbol{x}, t) \frac{\partial}{\partial x^{(k)}} \sigma^{(i,j)}(\boldsymbol{x}, t) \, ,$$

and the diffusion operator

$$D_2^{(i,j)}(\boldsymbol{x}, t) = \sum_{k=1}^{n} \sigma^{(i,k)}(\boldsymbol{x}, t) \sigma^{(j,k)}(\boldsymbol{x}, t) \, .$$

can be derived from the Langevin equation (1).

The concept of path integrals uses the fact, that the conditional probability on the left hand of (2) can be recursively applied, if the so called *short time propagator*

$$p(\boldsymbol{x}, t + \Delta t \,|\, \tilde{\boldsymbol{x}}, t) \;, \tag{4}$$

which gives the conditional transition probabilities for short times Δt is known.

The conditional probability at a constant final time t_e is given by the so called *path integral*

$$p(\boldsymbol{x}_e, t_e \,|\, \boldsymbol{x}_0, t_0) = \lim_{N\to\infty} \underbrace{\int \!\! \cdots \!\! \int}_{N \text{ times}}$$

$$\prod_{i=0}^{N-1} d\boldsymbol{x}(t_i) \prod_{i=0}^{N-1} p\left(\boldsymbol{x}(t_{i+1}), t_{i+1} \,|\, \boldsymbol{x}(t_i), t_i\right) \tag{5}$$

by taking the limit to an infinite number N of (infinitely small) time steps Δt constrained by the fact that the final time t_e has to be constant.

Usually, the propagator

$$p(\tilde{\boldsymbol{x}}, t + \Delta t \,|\, \boldsymbol{x}, t) = (2\,\pi)^{-\frac{n}{2}} [\det(\Delta t D_2(\boldsymbol{x}, t))]^{-\frac{1}{2}} \cdot$$
$$\exp\Big(-\frac{1}{2} \sum_{l,k}^{n} \Big[(\tilde{x}^{(l)} - x^{(l)} - D_1^{(l)}(\boldsymbol{x}, t)\Delta t) \frac{1}{\Delta t}$$
$$(D_2^{-1}(\boldsymbol{x}, t))^{(l,k)} (\tilde{x}^{(k)} - x^{(k)} - D_1^{(k)}(\boldsymbol{x}, t)\Delta t) \Big] \Big) \tag{6}$$

is used, but this propagator is not unique (Wehner and Wolfer 1983). With any propagator (4) satisfying eqn. (2) up to order Δt^2 eqn. (5) is a solution to the Fokker-Planck equation (Risken 1989).

For some processes, the limit of (6) can be computed analytically in closed-form if the short time propagators are appropriate, e.g. in the case of an Ornstein Uhlenbeck Process. In general, it is only possible to find numerical approximations.

4. STOCHASTIC REACHABILITY PROBLEM

Reachability analysis of hybrid systems is a topic with a lot of recent research effort (Bujorianu and Lygeros 2003, Bujorianu 2004). Usually there are two sets of states defined called target states and unsafe. The probability that a target state is reached while all unsafe states are avoided has to be computed to assess the reachability of a hybrid system (Koutsoukos and Riley 2006). In contrast to this formulation, here the only interest lies in the computation of the probability that some target states are reached at a certain time.

We assume in the sequel, that the execution of a GSHS admits a smooth probability density. This is the case e.g. for systems with bijective and deterministic reset maps $R(\cdot,\cdot)$ and eqi-dimensional continuous state-spaces X^q, i.e. $d(q) = d \,\forall\, q \in Q$ (Bect *et al.* 2006).

$$p(\boldsymbol{h}(t), t) = \text{Prob}\left(\{\tilde{\boldsymbol{x}}(t) \in [\hat{\boldsymbol{x}}(t), \hat{\boldsymbol{x}}(t) + \boldsymbol{dx}] \right. \tag{7}$$
$$\left. \wedge\, \tilde{q}(t) = q(t)\}\right) \;, \tag{8}$$

with

$$\oint_{\mathcal{H}} p(\boldsymbol{h}(t), t)\boldsymbol{dh} = 1 \,, \tag{9}$$

where we have introduced the symbol

$$\oint_{\mathcal{H}} \boldsymbol{dh} = \sum_{q(t)\in Q} \int_{X^1} \cdots \int_{X^{N_Q}} d\boldsymbol{x} \cdots d\boldsymbol{x} \,.$$

As in the case of continuous systems, the evolution of probability densities can be described by means of a partial differential operator, the so called infinitesimal generator of the stochastic process,

$$L_{\text{SHS}}(\boldsymbol{h}, t) = -\sum_{i=1}^{n} \frac{\partial}{\partial x^{(i)}} D_1^{(i)}(\boldsymbol{h}, t) + \tag{10}$$
$$+\frac{1}{2} \sum_{i=1}^{n} \sum_{j=1}^{n} \frac{\partial^2}{\partial x^{(i)} \partial x^{(j)}} D_2^{(i,j)}(\boldsymbol{h}, t) \,, \tag{11}$$

with boundary condition

$$p(\boldsymbol{h}, t) = \oint_{\mathcal{H}} p(\tilde{\boldsymbol{h}}, t) R(\boldsymbol{h}, \boldsymbol{d}\tilde{\boldsymbol{h}}) \boldsymbol{d}\tilde{\boldsymbol{h}} \,, \tag{12}$$

for all $\boldsymbol{h} \in \partial\mathcal{H}$.

The generator describes the behavior of the process in the interior of the state space. The interconnections of the different discrete states $q \in Q$ is given by the boundary conditions on each continuous domain ∂X^q. A mathematically rigorous introduction of infinitesimal generators can be found in (Bujorianu and Lygeros 2004) or much earlier (Feller 1952).

The conditional probability

$$p(\tilde{\boldsymbol{h}}, t \mid \boldsymbol{h}_0, t_0) =$$
$$\text{Prob}\left(\boldsymbol{h}(t) \in [\tilde{\boldsymbol{h}}, \tilde{\boldsymbol{h}} + \boldsymbol{dh}] \;\middle|\; \boldsymbol{h}(t_0) = \boldsymbol{h}_0 \right) \tag{13}$$

gives the probability that a stochastic hybrid system comes from the hybrid state $\boldsymbol{h}_0 = (q_0, \boldsymbol{x}_0)^{\mathrm{T}}$ at time t_0 to a state $\boldsymbol{h}(t) \in [\boldsymbol{h}, \boldsymbol{h} + \boldsymbol{dh}]$ with $\boldsymbol{dh} = (0, \boldsymbol{dx})^{\mathrm{T}}$ at time $t > 0$.

The probability density of the hybrid state at time $\tilde{t}$ depends on the conditional probability and the initial density $p(\boldsymbol{h}(t), t)$ of the hybrid state $\boldsymbol{h} \in \mathcal{H}$ at time t. This is given by the equation

$$p(\boldsymbol{h}(\tilde{t}), \tilde{t}) = \oint_{\mathcal{H}} p(\boldsymbol{h}(\tilde{t}), \tilde{t} \,|\, \boldsymbol{h}(t), t)\, p(\boldsymbol{h}(t), t)\, \boldsymbol{dh(t)}$$

The stochastic reachability problem can now be posed as the determination of the probability

$$\text{Prob}\left(\boldsymbol{h}(T) \in \mathcal{H}_e \mid p(\boldsymbol{h}_0, t_0)\right) \tag{14}$$

that the hybrid system reaches an area $\mathcal{H}_e$ of the hybrid state space at time t under the knowledge of the initial hybrid state distribution $p(\boldsymbol{h}_0, t_0)$.

This probability (14) can be derived formally by calculating

$$p(\boldsymbol{h}(T),T) = \oint_{\mathcal{H}_e}\oint_{\bar{\mathcal{H}}} p(\boldsymbol{h}(T),T \mid \boldsymbol{h}_0,t_0)\, p(\boldsymbol{h}_0,t_0)\, \boldsymbol{dh}_0 \boldsymbol{dh} \,.$$

5. HYBRID SYSTEMS PATH INTEGRALS

Using basic properties, an equation

$$p(\boldsymbol{h}(t_3),t_3) = \oint_{\bar{\mathcal{H}}} p(\boldsymbol{h}(t_3),t_3|\boldsymbol{h}(t_2),t_2)p(\boldsymbol{h}(t_2),t_2|\boldsymbol{h}(t_1),t_1)\, \boldsymbol{dh}(t_2) \tag{15}$$

that is similar to the Chapman-Kolmogorov equation for continuous systems can be derived.

The existence of path integrals for stochastic hybrid systems was claimed first in (Prasanth 2003). A generalization of the idea of Fokker-Planck equations for stochastic hybrid systems is described in (Bect *et al.* 2006).

If one iterates eqn. (15) while the time steps get smaller, one gets in the limit for the long time conditional probability

$$p(\boldsymbol{h}(t),t \mid \boldsymbol{h}(t_0),t_0) = \lim_{N\to\infty} \oint_{\bar{\mathcal{H}}} \cdots \oint_{\bar{\mathcal{H}}} \prod_{i=0}^{N-1} p(\boldsymbol{h}(t_{i+1}),t_{i+1} \mid \boldsymbol{h}(t_i),t_i)\, p(\boldsymbol{h}(t_0),t_0)\, \boldsymbol{dh}(t_i) \tag{16}$$

for stochastic hybrid systems, assumed that $t_{i+1} = t_i + \triangle t$ and $t = \lim_{N\to\infty} t_N$ holds.

This is similar to eqn. (5) but it will be very difficult to find propagators $p(\boldsymbol{h}(t_{i+1}),t_{i+1} \mid \boldsymbol{h}(t_i),t_i)$. As the limit of the integrals (5) only is computable in rare and very simple cases even for continuous systems, numerical methods have to be found to approximate the hybrid path integral (16). Known numerical methods for continuous path integrals can be extended which is discussed in the next section.

6. MARKOV CHAIN APPROXIMATIONS

By introducing finitely many mesh points $\bar{\boldsymbol{h}}_i = (\bar{q}_i, \bar{\boldsymbol{x}}_i) \in \mathcal{H}$, each representing a finite generalized volume $\triangle H_i = (\bar{q}_i, \triangle V_i)$ that partitions the hybrid state space $\mathcal{H}$ (in fact the continuous state spaces $X^{\bar{q}_i}$), we can approximate the continuous probability density $p(\boldsymbol{h}(t),t)$ by a discrete probability vector $\boldsymbol{p}(t) = (p_1(t), \dots p_{N_d}(t))^T$. The probabilities $p_i(t)$ can be calculated as

$$p_i(t) = \oint_{\triangle H_i} p(\boldsymbol{h}(t),t)\boldsymbol{dh} = \int_{\triangle V_i} p\left((\bar{q}_i, \boldsymbol{x}(t))^T, t\right) \boldsymbol{dx} \,.$$

The evolution of these approximated probability densities is given as the iteration of a Markov chain

$$\boldsymbol{p}(t+\triangle t) = \boldsymbol{T}^{\mathcal{H}} \left(\boldsymbol{p}(t) + \boldsymbol{T}^{\partial\mathcal{H}} \boldsymbol{p}(t)\right) \,. \tag{17}$$

The elements of the transition matrix $\boldsymbol{T}^{\mathcal{H}}$ are given by

$$T^{\mathcal{H}}_{i,j} = \frac{1}{\triangle V_j} \oint_{\triangle H_j} \oint_{\triangle H_i} p(\tilde{\boldsymbol{h}}(t+\triangle t), t+\triangle t | \boldsymbol{h}(t), t) \boldsymbol{dh}\, \boldsymbol{d\tilde{h}} \,, \tag{18}$$

i.e. by evaluation of the integral of the short time propagator (6) over the volume $\triangle H_j$ of all possible start points next to the mesh point j and the volume $\triangle H_i$ of end points respectively.

The term $\boldsymbol{T}^{\partial\mathcal{H}}$ represents the transitions from the boundaries due to the reset probability $R(\cdot,\cdot)$ given as

$$T^{\partial\mathcal{H}}_{i,j} = \frac{1}{\triangle V_j} \oint_{\triangle H_j} \oint_{\triangle H_i} R(\boldsymbol{h}(t), \boldsymbol{d\tilde{h}}(t)) \boldsymbol{dh}(t) \,. \tag{19}$$

This is an extension of an algorithm for continuous systems called PATHINT, for which local consistency and convergence have to be proven still (Wehner and Wolfer 1983).

The computational problem of this HYBPATHINT algorithm lies in the nonsparseness of the matrix $\boldsymbol{T} = \boldsymbol{T}^{\mathcal{H}} + \boldsymbol{T}^{\mathcal{H}}\boldsymbol{T}^{\partial\mathcal{H}}$, since the number of volumes grows exponentially with the order of the system.

This problem is overcome with the algorithm HYBPATHTREE, which extends the PATHTREE algorithm for continuous systems (Ingber *et al.* 2001). This algorithm only uses transitions from $\bar{\boldsymbol{h}}_i$ to the nearest neighbors, i.e. a predefined subset of mesh points $\bar{\boldsymbol{h}}_j$ with the property that $\bar{q}_i = \bar{q}_j$, and whose volumes $\triangle V_j$ all lie next to the volume $\triangle V_i$.

This leads to a sparse transition matrix $\boldsymbol{T}$ with good approximation properties. It should be mentioned that these algorithms are similar to the Markov Chain approximation algorithms in (Kushner and Dupuis 2001), although they are mainly dealing with continuous systems and use different short-time-propagators.

7. EXAMPLES

Figure 1 shows the setup for the first example which is an extended version of a classical example in hybrid systems analysis, (Chase *et al.* 1993).

The switching rules for the outflow are

- Switch to the next tank if the current tank gets empty.
- If a tank gets full, immediately switch to that tank.

- Keep lowering the current tank if none of the above conditions are fulfilled

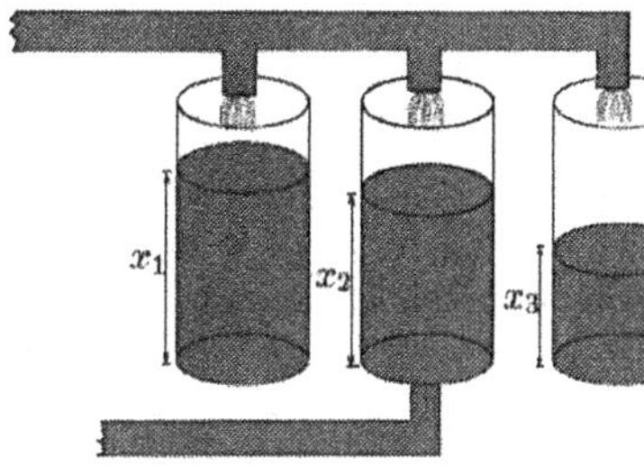

Fig. 1. Tanks System – Switched Outflow

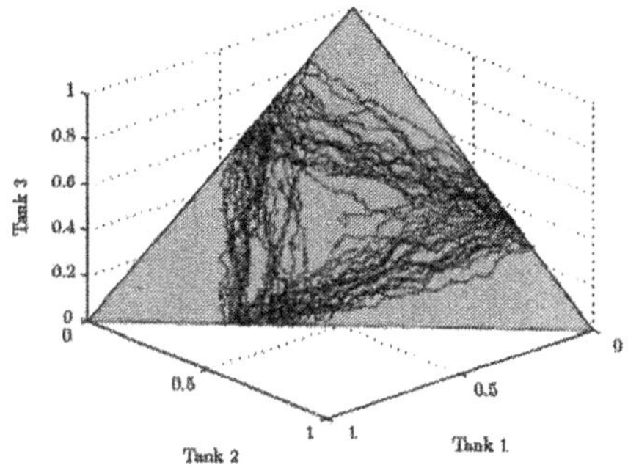

Fig. 2. Switched Outflow – Random Trajectory

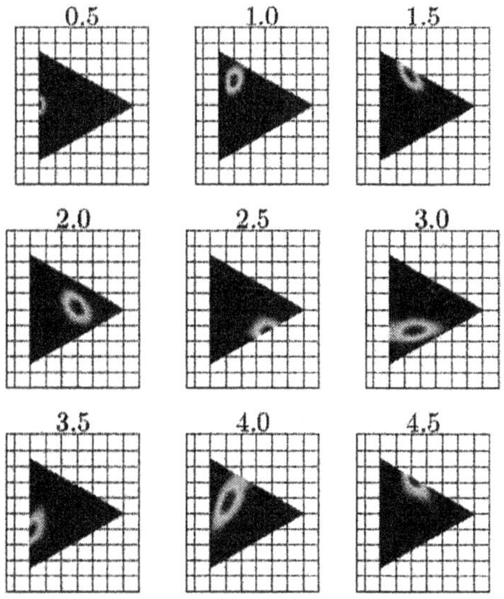

Fig. 3. Switched Outflow – HybPathTree

Changing the systems configuration as shown in Figure 4 changes the dynamics of the system completely. For the deterministic case the switched outflow system exhibits periodic behavior whereas the behavior of the switched inflow system is chaotic.

The switching conditions for the inflow are

- Switch to the next tank if the current tank gets full.
- If a tank gets empty, immediately switch to that tank.
- Keep filling the current tank if none of the above conditions are fulfilled

In order to avoid Zeno-behaviour in this case, we add the condition that if two tanks are empty at the same time, the tank with the smaller number will be filled first and the switching is locked for a small but non-zero time T_{fill}.

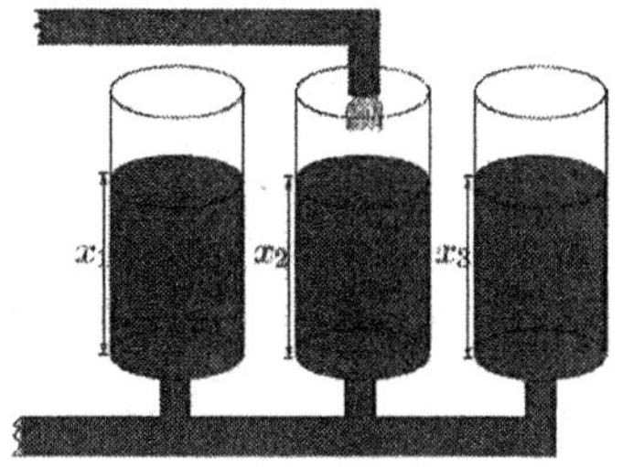

Fig. 4. Tanks System – Switched inflow

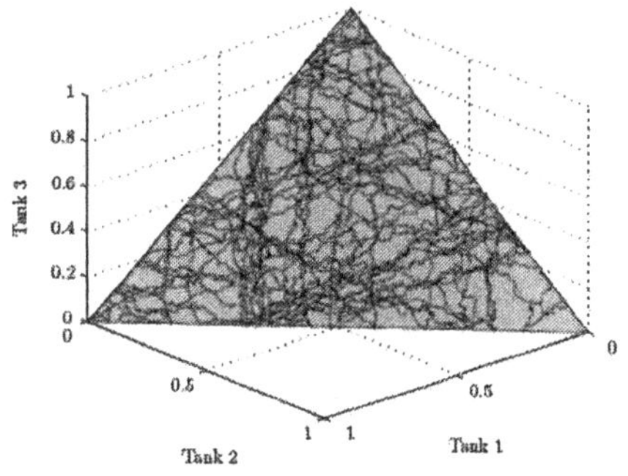

Fig. 5. Switched Inflow – Random Trajectory

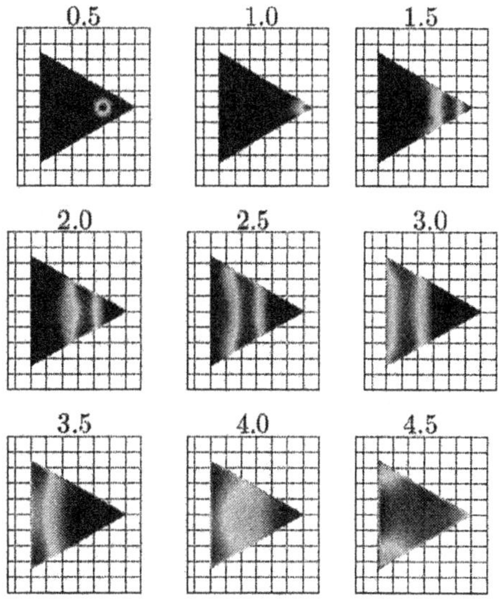

Fig. 6. Switched Inflow – HybPathTree

For both systems, the continuous state space is restricted by an invariant condition for the overall volume that can be expressed by $\sum_{i=1}^{3} x_i = 1$. The accessible state space is only two dimensional and has a triangular shape (see Figure 2). Additional Gaussian noise that doesn't contradict this conservation law is assumed. This reflects pressure fluctuations between concurring outflows or inflows respectively. Sample random trajectories are displayed in Figures 2 and 5. It is assumed further that all tanks have the same level at time 0.

The results of the HYBPATHTREE algorithm are given in Figure 3 and Figure 6. The brightness indicates the value of the sum of probability distribution $\sum_q^3 p(\{q\} \times X^q)$ shown on the continuous state space $X = X^q \forall q \in Q$. Dark areas indicate either high or low values of the probability density whereas light areas give the areas of middle probability. The densities are given for the times $0.5, 1.0, 1.5, \ldots, 4.5$ displayed at the top of each graph. Axis labeling is omitted for the sake of simplicity - the triangles show all of the allowed 2-D state space. Computations times are some minutes on a standard PC. With the help of Monte-Carlo methods pictures can be produced with a similar resolution but with an order of magnitude higher computation times.

Figure 3 shows that the behavior of the stochastic switching outflow system is approximately periodic, similar to the deterministic case. At any time, it is very unlikely that the system is in a state where two tanks are empty, i.e. in one of the corners of the triangles. In contrast to that, Figure 6 shows that the stochastic switched inflow system after completely filling one tank (movement to the right corner until time 1) has a large ambiguity of the current state. This leads to the movement of the red area from time 1.5 to 3.5 to the left border of the triangle. The distribution shows that at time 4.5, the state can nearly be anywhere with the same probability.

8. CONCLUSIONS

It has been shown how path integral methods can be adapted to numerical prediction of probability densities for stochastic hybrid systems. The benefit for reachability analysis is ease of computation due to a sparse Markov chain structure and a good approximation quality because of precise short time propagators.

ACKNOWLEGDEMENT

The authors would like to thank Prof. Lester Ingber for his support and discussions.

REFERENCES

Bect, J., H. Baili and G. Fleury (2006). Generalized fokker-planck equation for piecewise-diffusion processes with boundary hitting resets. In: *submitted to: 17th International Symposium on Mathematical Theory of Networks and Systems.*

Blom, H. and J. Lygeros (2005). Hybridge final project report. Technical report. European Commission.

Bujorianu, M. (2004). Extended stochastic hybrid systems and their reachability problem. In: *Hybrid Systems: Computation and Control: 7th International Workshop*. Pennsylvania. pp. 234–249.

Bujorianu, M. and J. Lygeros (2003). Reachability questions in piecewise markov processes. In: *Hybrid Systems: Computation and Control: 6th International Workshop*. Prague. pp. 126–140.

Bujorianu, M. and J. Lygeros (2004). General stochastic hybrid systems. In: *IEEE Mediterranean Conference on Control and Automation MED 04*. Turkey.

Chase, C., J. Serrano and P. Ramadge (1993). Periodicity and chaos from swiched flow systems; contrasting examples of discretely controlled continuous systems. *IEEE Transaction on Automatic Control.*

Feller, W. (1952). The parabolic differential equations and the associated semi-groups of transformations. *The Annals of Mathematics, 2nd Ser.* **55**(3), 468–519.

Hu, J., J. Lygeros and S. Sastry (2000). Toward stochastic hybrid systems. In: *Hybrid Systems: Computation and Control* (B.H. Krogh N. Lynch, Ed.). Springer. pp. 160–173.

Ingber, L., C. Chen, R. Mondescu, D. Muzzall and M. Renedo (2001). Probability tree algorithm for general diffusion processes. *Phys. Rev. E 64*.

Koutsoukos, X. and D. Riley (2006). Computational methods for reachability analysis of stochastic hybrid systems. In: *HSCC 2006* (A. Tiwari J.P. Hespanha, Ed.). Springer. pp. 377–391.

Kushner, H. and P. Dupuis (2001). *Numerical Methods for Stochastic Control Problems in Continuous Time.* 2 ed.. Springer-Verlag.

Prasanth, R. (2003). Analysis of stochastic hybrid systems using path integrals. In: *Signal Processing, Sensor Fusion and Target Recognition XII* (Ivan Kadar, Ed.). SPIE, the international Society of Optical Engineering. pp. 324–333.

Risken, H. (1989). *The Fokker-Planck Equation.* Springer-Verlag, Heidelberg.

Wehner, M. and W. Wolfer (1983). Numerical evaluation of path integral solution to the Fokker-Planck equations I. *Phys. Rev. A* **27**, 2663–2670.

ELSEVIER
IFAC
PUBLICATIONS

CHALLENGES IN START-UP CONTROL OF A HEAT EXCHANGE REACTOR WITH EXOTHERMIC REACTIONS; A HYBRID APPROACH

Staffan Haugwitz and Per Hagander

Department of Automatic Control, Faculty of Engineering
Lund University
Box 118, SE-22100 Lund, Sweden
Email: {staffan.haugwitz | per.hagander}@control.lth.se

Abstract: In this paper, the control of a continuous heat exchange reactor is investigated from a hybrid perspective with focus on the start-up phase and the transition to the optimal operating point. The temperature sensitive exothermic reaction leads to the possibility of multiple steady states and in combination with safety constraints forms an interesting challenge for a safe and efficient start-up. A series of MPC controllers are developed with a switching logic that transfers the process from initial rest to continuous optimal operation mode. The control procedure is verified in simulations with a full nonlinear model of the Open Plate Reactor, an improved heat exchange reactor being developed by Alfa Laval AB. The case study can be seen as a benchmark problem for start-up control of exothermic reactions. *Copyright @ 2006 IFAC*

Keywords: reactor start-up, hybrid control, heat exchange reactor, process control, model predictive control, exothermic reaction

1. INTRODUCTION

For industrial production of temperature sensitive exothermic reactions, safe and efficient start-up control is important.

Normally the reactor operates in a continuous mode around an optimal operating point. However, there are several different control modes associated with the production. In this paper, we will focus on the start-up mode, the continuous operation mode and the transition in between. We will show that it may be necessary to switch between several controllers with different optimization criteria to allow a safe and efficient start-up and thereafter optimal production.

The chemical reactor and the exothermic reaction form a highly nonlinear process and for some operating conditions also an unstable process. It is therefore essential that the reactor is started in a safe and accurate fashion. As shown later in Section 4, there can possibly be multiple steady states for a given process input, that is, applying the same process input may lead to different steady states depending on the current state of the process. The start-up phase is therefore non-trivial and much effort has to be used to ensure a safe operation.

In this paper, a hybrid control approach is used, where different controllers are used to transfer the process from initial conditions to a target region at the optimal steady state operating point.

The reactor studied in this paper is the Open Plate Reactor (OPR), currently being developed by Alfa Laval AB. It is a continuous heat exchange reactor, where the key concept is to combine efficient micromixing with improved heat transfer into one operation and it is further described in Section 2. The modeling and control of the OPR in the continuous operation mode is discussed in (Haugwitz and Hagander, 2006).

Many processes studied from a hybrid control point of view has a mixture of both continuous and discrete inputs, see e.g. (Stursberg, 2004), whereas the OPR only has continuous physical inputs. Instead, hybrid control may be required due to the multiple steady states of the OPR with both stable and unstable modes and the distributed actuation of the multiple inlet ports. This is further described in Sections 4 and 5.

2. THE OPEN PLATE REACTOR

The OPR consists of a number of reactor plates, in which the reactants mix and react. On each side of a reactor plate there is a cooling plate, through which cold water is circulated. In this paper a simple first order exothermic reaction is considered, see Eq. 1.

$$A + B \rightarrow C + D + \text{heat} \tag{1}$$

In Figure 1, a schematic figure of the first rows of a reactor plate is shown. The reactant A flows into the reactor from the upper left inlet. Between the inlet and the outlet, the reactants are forced by inserts to flow in horizontal channels in alternating directions. The inserts are specifically designed to enhance the mixing and at the same time the heat transfer capacity. The concept relies on an open and flexible reactor configuration. The type of inserts and the number of rows in the reactor plate, which determines the residence time, can be adjusted, based on the type and rate of the chosen reaction.

The reactant B can be added through multiple inlet ports, typically in the beginning and in the middle of the reactor. Temperature sensors can be mounted arbitrarily inside the reactor, specifically after each inlet port. To acquire accurate measurements of the temperature profile along the flow direction of the reactor, as many as 10 temperature sensors can be used. There can also be other sensors, such as pressure or conductivity sensors. The signals from the internal sensors are then used in the control system for emergency supervision and process control.

2.1 *Modelling*

A model of the OPR can be derived from first principles, with partial differential equations (PDE) for heat transfer, reaction kinetics, mass, energy and chemical balances, see for example (Thomas, 1999) or (Fogler, 1992).

The multiple consecutive horizontal channels inside the OPR in Figure 1, can be approximated as a continuous tubular reactor with axial dispersion with multiple inlet ports of reactant B along the reactor.

To simplify analysis of the PDE, which is an infinite dimensional system, the spatial derivative is approximated, using a first order backward difference method, as a finite system of ordinary differential equations (ODE). The model is discretized using n elements of equal size, where n can be a design parameter and may

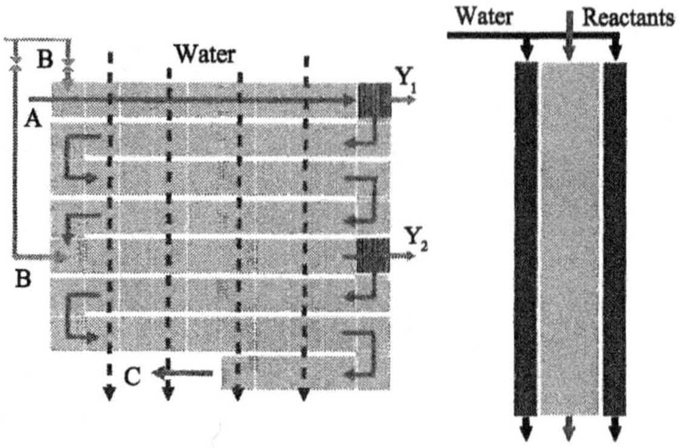

Fig. 1 Left: A schematic of a few rows of a reactor plate. Reactant A enters at top left and reactant B is added through several inlet ports along the reactor. Y_1 and Y_2 are internal temperature sensors used for process control and supervision. The cooling water flows from top to bottom in separate cooling plates. Right: The plate reactor seen from the side, with the reactor part in the middle and cooling plates on each side.

then be chosen such that the numerical dispersion approximates the actual dispersion of the reactor. The nonlinear model of the OPR is described in more detail in (Haugwitz and Hagander, 2006).

3. PROCESS OPERATION

There are four main control signals of the OPR; the feed flow rates of reactant B added at the two inlet ports, u_{B1} and u_{B2}, the inlet temperature of the cooling water T_{cool} and finally the inlet temperature T_{feed} of the reactant A, which constitutes the main part of the total reactor flow.

Of these four control variables, the two flow rates, u_{B1} and u_{B2}, are the most important, since they have the largest control gain from control signal to reactor temperature and they have the fastest actuator dynamics. Therefore, from a safety point of view, it is desirable to include the feed flow rates as control variables. However, changes in the flow rates may lead to stoichiometric imbalance so they should be used very carefully. T_{feed} and T_{cool} have lower control gains and slower actuator dynamics compared to the flow dynamics of u_{B1} and u_{B2}, but do not effect the chemical balance.

In some cases the total feed flow rate of reactant B is fixed to guarantee stoichiometric relations with reactant A, which flow is fixed. Then the control variable "feed flow distribution" u_B is used, that is, how large fraction of the total amount of reactant B that is fed through the first inlet port. The remainder $1 - u_B$ is then fed through the second inlet port.

4. START-UP DYNAMICS

As described in Section 2.1, the reactor is a distributed parameter system and is normally described with non-linear partial differential equations (PDE). Analysis of these infinite dimensional systems is quite difficult. Some results can e.g. be found in (Laabissi *et al.*, 2002) and references therein.

From (Laabissi *et al.*, 2002), it is known that there may be multiple steady states for a tubular reactor with axial dispersion. Similar effects may also be seen when there is significant heat storage and heat conduction in the axial direction inside the reactor. More references can be found in e.g. (Gray and Scott, 1990). The spatial discretization of the PDE into a system of ODEs used in Section 2.1 can also be viewed as the well known "tanks-in-series"-approximation, see e.g. (Fogler, 1992). To visualize the start-up dynamics, a simplified analysis is made based on this approximation. The numerical values used in this paper for the hypothetical reaction are found in Table 1.

In Figure 2, the process is simulated in open loop, for two different cases. The upper plot shows the reactor temperature around the first inlet port. The middle point shows the feed temperature of the reactant A and the lower plot shows the conversion of the reaction at the reactor outlet. Reactant B is added at time $t = 0$ through only one inlet port at the beginning of the reactor. The input conditions are identical for the two simulation cases, except for the feed temperature of reactant A. In the first case (dashed), the feed temperature remains constant at $T_{feed} = 20°\text{C}$. Almost no reaction occurs and the conversion is only 14 %. In the second case (solid) the reactant A is pre-heated during 30 seconds, but this is enough to temporarily increase the reaction rate. The exothermic reaction then releases heat itself, so when the pre-heating stops the reaction continues.

Another interesting part of Figure 2 is the sudden temperature increase around the inlet port at time $t = 46$ s as the pre-heating increases the reaction rate. The temperature increases 100°C within 1 second, which reveals the potential dangers during start-up. In fact, the heat release is so large that the safety limit at 150°, beneath which it is safe to operate, is violated quite rapidly. Without pre-heating, the conversion stays around 14 %, whereas with pre-heating it increases very quickly to almost 100% . It is then clear that with pre-heating the process is permanently moved from one stable equilibrium point to another, even though the input signals return to the same values. "Ignition" is said to have occurred when the process moves from an equilibrium at lower temperature to an equilibrium at a higher temperature. Note that there is a flow delay from the inlet port at the reactor inlet to the reactor outlet, which explains the delay from temperature increase at the inlet to the increase of the conversion at the reactor outlet.

Figure 2 shows that pre-heating of the reactants may

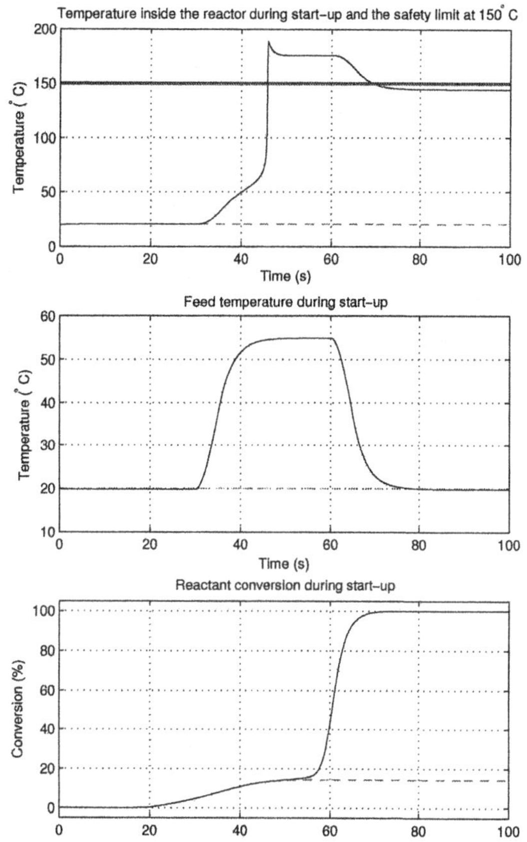

Fig. 2 The plot shows possible multiple steady states for the open loop system. If the steady state values of the inputs are applied from start, almost no reaction occurs inside the reactor (dashed). When a short time of pre-heating is done (solid), another steady state is reached.

Table 1 The data values used in simulations

Variable/Parameter	Value
Activation energy, E_a	77000 J/mol
Pre-exponential factor, k_0	2e7 $m^3/(mol\ s)$
Heat of reaction, ΔH	1.17e6 $J/(mol\ m^3)$
Feed inlet temperature, T_{feed}	20° C
Cooling temperature, T_{cool}	20°C

be necessary to start the reaction. However, it may be dangerous to start feeding B before the pre-heating gives favorable conditions for the reaction rate. Even with closed-loop feedback control using the feed temperature as manipulated variable, there may still be safety issues due to slow actuator dynamics and model uncertainties. Therefore not only feedback control, but also the sequence of the start-up actions is critical as will be explained in the next section.

Ignition and safety aspects require pre-heating of the

reactant, but on the other hand constant excessive pre-heating may not be desirable. When there are hard constraints in the reactor temperature, pre-heating may decrease the production capacity of the process. This means that we can feed less reactant B in the first inlet port if the temperature there is already high due to pre-heating. In addition, less pre-heating means less energy input being required.

To summarize, pre-heating of reactant A may be needed to start the reaction, the order of the start-up actions is important and the use of pre-heating should in steady state be as low as possible to allow high production rate of the reactor.

5. START-UP CONTROL OF THE OPR

The main rule is that no feeding of B should be made before the reactor temperature is such that the reaction starts immediately when adding B. If B is being fed and the reactor temperature is too low to allow sufficient reaction rate, there will be large quantities of unreacted chemicals inside the reactor. Then the risk of run-away reaction increases as seen in Figure 2, where feeding was started before the pre-heating was initiated. In this section one possible start-up control sequence will be presented.

During start-up the following control variables are available:

- u_{B1}, feed flow of B into the first inlet port
- u_{B2}, feed flow of B into the second inlet port
- T_{cool}, inlet temperature of the cooling water
- T_{feed}, inlet temperature of reactant A

The feed flows are defined as ratios of the total nominal flow of reactant B, that is, ranges from 0 to 1. In continuous operation, this flow is fixed to keep stoichiometric relations with reactant A, which has a fixed flow. However, during start-up this constraint is relaxed to improve safety and flexibility.

The following variables are to be controlled according to given reference values:

- T_1, the temperature in the reactor around the first inlet port
- T_2, the temperature in the reactor around the second inlet port

Since there are four control variables available to control only two temperatures, our control strategy uses two additional reference signals $u_{B1,ref}$ and $u_{B2,ref}$ for the two feed flow variables u_{B1} and u_{B2}. This penalizes deviations in these control signals from desired values during start-up. However, the controller can still use them to avoid violating constraints.

The start-up can be divided into the following steps, which are also graphically sketched in Figure 3:

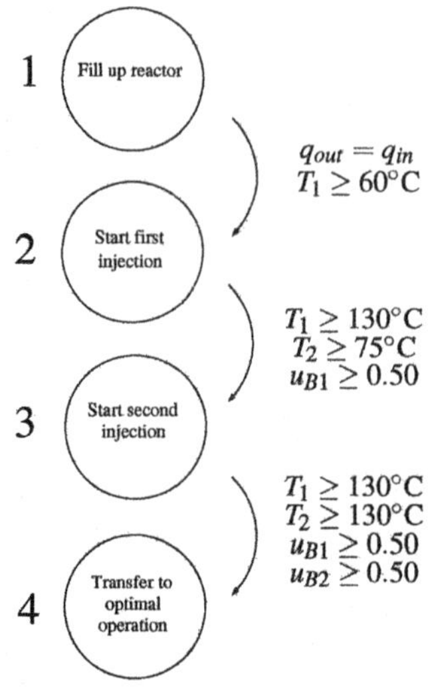

Fig. 3 State machine to illustrate the different steps during start-up and the guards corresponding to each transition. Note that the transitions are one-directional.

1. Fill the reactor with reactant A, which is pre-heated to 60°C. Wait until the outflow q_{out} equals the inflow q_{in}.

2. Start feeding reactant B in the first inlet port. The following reference values are used: $T_{1,ref} = 135°C$ and $u_{B1,ref}$ increases along a ramp from 0 to 0.5, which corresponds to 50% of the reactant B being added there. These references aim at obtaining a safe and reliable ignition of the reaction. The reference $T_{2,ref} = 80°C$ aims at giving good temperature conditions for the second inlet port, similar to the pre-heating before the first inlet port. And finally we are not allowed to feed anything into the second point until there has been a safe ignition at the first inlet port and there are favorable temperature conditions for the second inlet port, therefore $u_{B2,ref} = 0$.

3. Start feeding also in the second inlet port. As before $T_{1,ref} = 135°C$ and $u_{B1,ref} = 0.5$. $T_{2,ref}$ is set to 135°C and $u_{B2,ref}$ increases along a ramp from 0 to 0.5.

4. When the reactor temperatures have converged and feed flows reached their recommended values, the transition phase to the continuous operation mode begins. The controlled variables are now outlet concentrations of A and B, which should be controlled to zero. To fulfill stoichiometric relations, the feed flows of B are no longer controlled individually, but instead the distribution, u_B, between them is used, where $u_{B1} = u_B$ and $u_{B2} = 1 - u_B$. In addition, a reference value for the feed temperature $T_{feed,ref}$, is used to emphasize a low amount of pre-heating.

Whenever the guard conditions in Figure 3 have been fulfilled, a switch is initiated to the next step. During the entire start-up procedure, there is a critical safety constraint $T_{max} = 150°C$, so that all reactor temperatures should stay below that. This safety limit can be derived from by-product formation, cooling capacities or further exothermic side reactions.

5.1 *Implementing the start-up control*

It is possible to derive an open-loop start-up control procedure. However, it is non-trivial to find the best stationary operating point, while respecting the temperature constraints. In addition, it is difficult to find suitable open loop control trajectories to take the process from initial rest - through the ignition phase - to the optimal operating point. Model errors and disturbances may for some trajectories lead to hazardous operating conditions.

Therefore, the start-up procedure described above should be implemented with feedback control. The actual procedure is generic and is not restricted to any specific controllers. Alternatives can for example be PI-controllers with selectors or multivariable Model Predictive Control. Regardless of the chosen controller type, there will be a set of different controllers, one for each step of the start-up, see Figure 3.

In this paper, we have chosen to use MPC, due to its multivariable nature and its capacity to handle state constraints.

5.2 *Model Predictive Control*

A standard linear MPC controller was designed for each of the steps 2,3 and 4 in Figure 3, based on the notations of (Maciejowski, 2002) and algorithms from (Åkesson, 2003). The nonlinear process model was linearized at the switching point for each controller. Output feedback is implemented with an extended Kalman Filter (EKF), since only the temperatures are measurable. The concentrations inside the reactor and the feed concentrations are estimated by the EKF to improve robustness towards disturbances in the feed conditions. Due to lack of space, the details of the linear control design with MPC and EKF for the OPR can be found in (Haugwitz and Hagander, 2006).

6. SIMULATIONS

In Figures 4 and 5, the closed-loop start-up control procedure is simulated with the nonlinear process model. It is assumed that reactant A flows through the reactor and is being pre-heated to 60°C at time $t = 0$. This means that the switch between step 1 and 2, sketched in Figure 3, occurs at $t = 0$.

The MPC controller begins adding reactant B into the first inlet port at $t = 0$ s. As the feed flow increases the

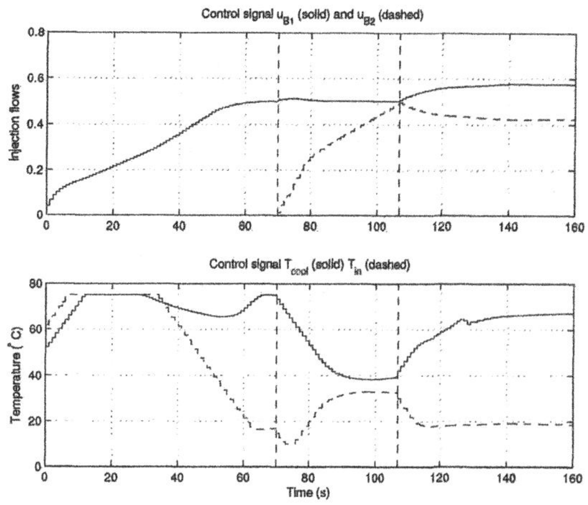

Fig. 4 Control signals during start-up procedure and transition to continuous operation mode. The vertical dashed lines indicate the switching times 70 s and 107 s.

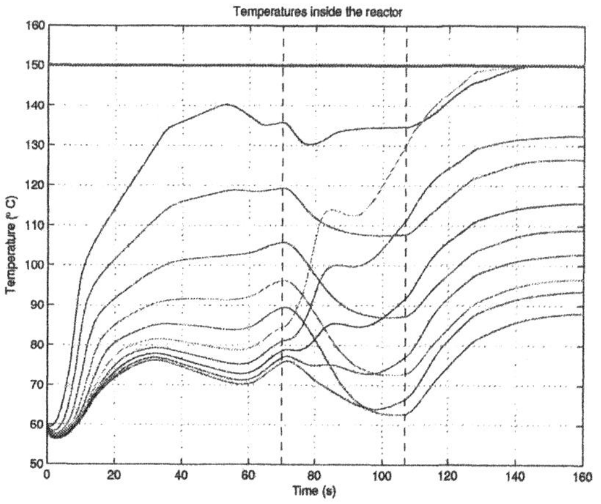

Fig. 5 Temperatures inside the reactor with the dashed line being the temperature at the first inlet port and the dash-dot line the temperature at the second inlet port. The vertical dashed lines indicate the switching times 70 s and 107 s.

controller has to decrease the pre-heating (with T_{feed}) to follow the temperature reference at $T = 135°C$. After 40 seconds, the temperature at the first inlet port has reached 135°C and the reaction has safely been ignited.

At $t = 70$ s the controller switches from step 2 to step 3 as the guard conditions from Figure 3 have been fulfilled. A MPC controller with different tuning parameters is used, but the main difference is that the feed flow rate at the second inlet port is now also included as a control variable. Again the aim is to track temperature references, now at the first and second inlet ports, and if possible follow control signal references for the two feed flows. As the second feed increases, the cooling is intensified and more preheating is again required.

At $t = 107$ s the guard conditions for transition between step 3 and 4 are fulfilled. The process has reached the pre-defined target region for the start-up procedure. Thus the controller switches mode from start-up mode to continuous operation mode. Another MPC controller is used, now with the aim of optimizing the conversion in the reactor. This is carried out by increasing the reactor temperature to the highest temperature allowed at which it is safe to operate. A secondary objective is to decrease the use of pre-heating. Therefore some reactant B is redistributed from the second to the first inlet port, which in turn decreases the need of cooling, thus saving energy.

At the end of the start-up, step 3, the conversion had reached 96.8% and after step 4, the transition to continuous operation mode, the controller has carefully adjusted the control signals, so that the conversion reaches 98.6 %. This is mainly due to operation closer to the reactor temperature limitations.

7. SUMMARY & CONCLUSIONS

The start-up control of a heat exchange reactor has been studied. The multiple steady states and multiple inlet ports along the reactor form a process with challenging dynamics. As seen in open loop simulations, the order of the start-up procedures is of great importance as well as the need for closed loop control.

The main rule is that the feed of reactant B should only start when there are favorable conditions for the reaction to ignite, especially in terms of reactor temperature. In this case, this means that the temperature of the reactants at each inlet port should be carefully controlled. To ensure safe start-up, a hybrid controller is presented, which step by step transfers the process from initial conditions to an operating point where the chemical conversion is maximized. In each step, a separate optimization criteria is used to ensure a safe transition using Model Predictive Control. MPC is also essential to handle safety constraints on the reactor temperatures. An extended Kalman Filter is designed to allow output feedback as concentration measurements are not generally available and to improve robustness for variations in feed conditions.

This case study can be seen as a benchmark problem for start-up control of exothermic reactions.

8. FUTURE WORK

To reduce the sensitivity to model errors from linearization, the start-up control procedure is currently being extended to nonlinear Model Predictive Control. It is then possible to take advantage of the available nonlinear process model. It will also be an interesting benchmark for NMPC, to see how it handle large transitions and the complicated ignition dynamics.

It would also be beneficial to use better and faster numerical algorithms to solve the optimization problems. This will allow faster sampling rate, which is desirable with the very fast ignition dynamics of the process.

Another approach in a different direction is to implement the same start-up control sequence using simpler controllers, such as Proportional-Integral controllers with selectors to allow constraint handling. This may be an interesting alternative as it is easier to implement in a industrial environment and may for several cases perform almost as good as the presented MPC solution.

Finally, it would be desirable to validate the process model with the real process with focus on parameters important for the start-up dynamics, such as thermal inertias, mass and heat dispersion coefficients and heat conduction in the axial direction.

9. ACKNOWLEDGMENTS

The authors gratefully acknowledge funding within the HYCON-project of the European Union, http://www.ist-hycon.org/.

10. REFERENCES

Åkesson, J. (2003): "Operator interaction and optimization in control systems." Licentiate thesis ISRN LUTFD2/TFRT--3234--SE. Department of Automatic Control, Lund Institute of Technology, Sweden. MPCTools available at http://www.control.lth.se/user/johan.akesson/mpctools/

Fogler, S. (1992): *Elements of chemical reaction engineering*. Prentice Hall.

Gray, P. and S. Scott (1990): *Chemical oscillations and instabilities*. Oxford University Press.

Haugwitz, S. and P. Hagander (2006): "Modeling and control of a novel heat exchange reactor, the open plate reactor." *To appear in Control Engineering Practice*.

Laabissi, M., M. Achhab, J. Winkin, and D. Dochain (2002): "Equilibrium profiles of tubular reactor nonlinear models." *In Proceedings of 15th Int. Symposium on Mathematical Theory of Networks and Systems*.

Maciejowski, J. M. (2002): *Predictive Control with Constraints*. Pearson Education Limited.

Stursberg, O. (2004): "Dynamic optimization of processing systems with mixed degrees of freedom." *In Proceedings of 7th Int. Symposium on Dynamics and Control of Process Systems*, p. ID 164.

Thomas, P. (1999): *Simulation of Industrial Processes*. Butterworth-Heinemann.

Copyright © Analysis and Design of Hybrid Systems
Alghero, Italy 2006

ELSEVIER
IFAC
PUBLICATIONS

FEEDBACK STABILIZATION OF THE OPERATION OF AN HYBRID CHEMICAL PLANT

I. Simeonova * F. Warichet ** G. Bastin *
D. Dochain * Y. Pochet ***

* CESAME - UCL, Bâtiment Euler, Av. G. Lemaitre, 4
1348 Louvain la Neuve, Belgium
(simeonova, bastin, dochain) @ inma. ucl.ac. be
** CORE - UCL, Voie du Roman Pays, 34
1348 Louvain - la - Neuve, Belgium
warichet @ core. ucl.ac. be
*** CORE - UCL and IAG - UCL, Voie du Roman Pays,
34 1348 Louvain - la - Neuve, Belgium
pochet @ core. ucl.ac. be

Abstract: This paper deals with the feedback stabilization of the operation of a simple hybrid chemical plant at an optimally scheduled operation. The optimally scheduled plant operation is obtained as a solution of a cyclic discrete time scheduling optimization problem and it is open loop unstable. The goal is to illustrate how the stability problem of an hybrid chemical process can be solved by using simple P and PI - like control laws. The sensitivity of the closed loop plant operation to the choice of the controllers parameters is also presented.

Keywords: chemical process, hybrid system, hybrid automaton, open loop scheduling, feedback control, stabilization.

1. INTRODUCTION

This communication deals with the feedback stabilization of an hybrid chemical plant at an optimally scheduled operation which is open-loop unstable. The goal is to assess the efficiency of simple P and PI-like controllers for the stabilization of an hybrid system, through a simple but realistic case-study. The considered plant consists of two tanks (a batch chemical reactor and a buffer tank). The overall system is hybrid in the sense that it is made up of both continuous and discrete event processes. It is modelled by means of the hybrid automaton formalism (Willems, 2003), (Lygeros *et al.*, 2003), (Bemporad and Heemels, 2005).
For this chemical plant, we first address the determination of a cyclic optimal schedule that maximizes the plant productivity. This optimization problem is solved by using a discrete time periodic scheduling method.
However, the state trajectory produced by the optimal schedule is unstable. This fact is easily emphasized by observing that under an arbitrarily small constant disturbance, the actual plant trajectory steadily diverges from the optimal one. As a result, the plant operation becomes not only sub-optimal but even infeasible. Our concern in this communication is to illustrate how this stability problem can be solved by using simple P and PI control laws. The performance of the control is illustrated through various simulation experiments carried out in the Matlab/Simulink/Stateflow environment. With the

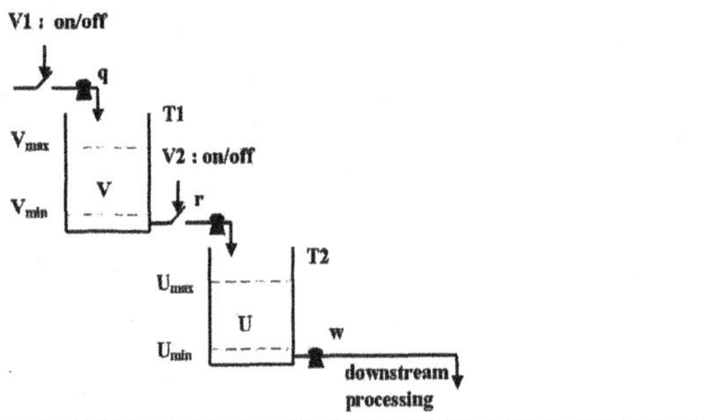

Fig. 1. Hybrid chemical plant

considered control laws, it is shown that the operation of the plant is stabilized in the sense that the plant trajectory converges towards a limit cycle which is close to the optimal cyclic schedule. The dependence of the closed loop performance on the choice of the controller parameters is graphically illustrated. The organization of the paper is as follows. The hybrid chemical plant and its hybrid automaton model are presented in Section 2. Section 3 is concerned with the statement and the solution of the optimal cyclic scheduling problem. In Section 4, we present the feedback P and PI controllers. The closed loop performance and the control tuning are then assessed with a benchmark simulator of the plant. The results are given in Section 5. Some final remarks and directions for future work are presented in Section 6.

2. DESCRIPTION OF THE HYBRID CHEMICAL PLANT AND ITS HYBRID AUTOMATON MODEL

The considered hybrid chemical plant is depicted in figure (1). It consists of two tanks: (**T1** and **T2**) and two valves: (**V1** and **V2**). The first tank (**T1**) is a batch chemical reactor which is automatically operated with four successive operation modes: filling with raw material, production by chemical reaction, discharging (i.e. harvesting of final product), cleaning and waiting for the next operation. However, although it is produced in a discontinuous way, the final product has to be continuously delivered to the downstream processing stage. Therefore, there is an intermediate buffer tank **T2** between the batch reactor and the downstream processing plant which is discontinuously fed from the reactor, but continuously withdrawn. The plant is actually an hybrid system that combines time driven and event driven dynamics. The plant is modeled using the hybrid automaton formalism (Willems, 2003), (Lygeros *et al.*, 2003),(Bemporad and Heemels, 2005) which is presented hereafter. It should be remarked that the considered plant is a simplified version of the benchmark chemical plant that we have described in a former publication (Simeonova *et al.*, 2005*b*).

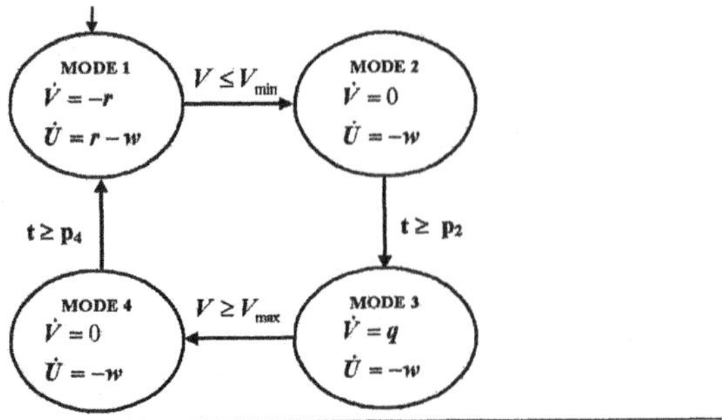

Fig. 2. Hybrid automaton diagram of the plant

2.1 Hybrid automaton model of the hybrid plant

The process carried out in the plant follows a sequence of four successive modes (or phases), namely: Mode 1, Mode 2, Mode 3 and Mode 4. This set of four modes represents the discrete state variables of the considered hybrid process. The hybrid automaton diagram of the plant is depicted in figure (2). The vertices of this graph represent the modes of the automaton while the edges represent the time and/or continuous state event driven transitions between the modes. The process behavior during each mode is characterized by a set of continuous differential equations (mass balances) with continuous state variables: V $[m^3]$ - volume of **T1** and U $[m^3]$ - volume of **T2**. A short description of the process dynamics in each mode of the plant operation is as follows.

Initially the plant is in Mode 1, that means: **V1** is closed (off) and **V2** is open (on). The initial volume of **T1**, V_0 is equal to the maximal **T1** volume, $V_{\max}$ $[m^3]$ and the initial volume of **T2**, U_0 is equal to the minimal **T2** volume, $U_{\min}$ $[m^3]$.

- **Mode 1**: During this mode the material in **T1** is discharging with an output flow rate r $[m^3/h]$. Simultaneously **T2** is filling with the same input flow rate r $[m^3/h]$ and in the same time is discharging with an output flow rate w $[m^3/h]$. This is modeled as follows:

$$\dot{V} = -r \quad \dot{U} = r - w \tag{1}$$

  This mode lasts until the volume of the tank **T1** reaches a minimal value $V_{\min}$, then the plant process moves to Mode 2 (figure (2)).
- **Mode 2**: During this mode **V1** and **V2** are closed (off) and therefore **T1** is in stand by, while the material of **T2** is discharging with the output flow rate w $[m^3/h]$, and therefore:

$$\dot{V} = 0 \quad \dot{U} = -w \tag{2}$$

  Mode 2 lasts for a given time duration p_2 $[h]$, assigned by the operator, then the plant goes to Mode 3 (figure (2)).
- **Mode 3**: During this mode **V1** is open (on) and **V2** is closed (off). **T1** is filling with raw material with an inflow rate q $[m^3/h]$ while

T2 is discharging with the output flow rate w $[m^3/h]$. This is modeled as follows:

$$\dot{V} = q \quad \dot{U} = -w \tag{3}$$

The process lasts until the volume of **T1** reaches a maximal value $V_{\max}$, then the plant enters in Mode 4 (figure (2)).

- **Mode 4**: During this mode **V1** and **V2** are closed (off) and the chemical reaction proceeds in **T1** while the material of **T2** is discharging with an output flow rate w $[m^3/h]$. Therefore:

$$\dot{V} = 0 \quad \dot{U} = -w \tag{4}$$

After a certain time duration p_4 $[h]$, the reaction is completed and the plant goes to Mode 1 (figure (2)), and the operation cycle starts again.

We first present the open loop discrete time scheduling optimization of the hybrid chemical plant.

3. DISCRETE TIME SCHEDULING OPTIMIZATION IN OPEN LOOP

We consider the open loop cyclic scheduling problem consisting of maximizing the flow rate (the productivity) w $[m^3/h]$ of material leaving tank **T2**. Open loop means here that there is no feedback controller. Cyclic means that the overall process operation must take the form of a periodic repetition of a basic operation pattern.

First, we define the constant parameters and the variables of the problem. The parameters are: the minimum and maximum volume of the tanks **T1** and **T2** denoted by $V_{\min}$ $[m^3]$, $V_{\max}$ $[m^3]$, $U_{\min}$ $[m^3]$ and $U_{\max}$ $[m^3]$; the minimum $w_{\min}$ $[m^3/h]$ and maximum rate $w_{\max}$ $[m^3/h]$ of material leaving **T2**; the rate of material entering **T1**, q $[m^3/h]$ and the rate of material leaving **T1** and entering **T2**, r $[m^3/h]$; the processing time for each mode p_i for $i \in [1,4]$ where $p_1 = \frac{V_{\max}-V_{\min}}{r}$ $[h]$ and $p_3 = \frac{V_{\max}-V_{\min}}{q}$ $[h]$, while p_2 and p_4 are assigned by the operator.

The variables of the problem are the rate of material leaving tank **T2** denoted by w_i for each mode $i \in [1,4]$ and the value of the stock in **T2** denoted by U_i at the beginning of each mode $i \in [1,4]$.

The constraints of the problem are the following :

$$\begin{aligned} U_1 &= U_{\min} \\ U_1 &= U_4 - w_4 p_4 \\ U_2 &= U_1 + r p_1 - w_1 p_1 \\ U_3 &= U_2 - w_2 p_2 \\ U_4 &= U_3 - w_3 p_3 \\ U_{\min} &\leq U_i \leq U_{\max} \ \forall i \end{aligned} \tag{5}$$

These constraints guarantee that the schedule is cyclic. In order to have a relatively smooth material transfer to the downstream processing, we impose also that the w_i for $i \in [1,4]$ do not vary too much. This is guarantied by the following constraints:

$$|w_i - w_j| \leq \epsilon \qquad \forall i,j \in [1,4], i \neq j \tag{6}$$

where ϵ is a small value. It is also imposed that:

$$w_{\min} \leq w_i \leq w_{\max} \ \forall i \tag{7}$$

The objective function is:

$$\max \sum_{i=1}^{4} w_i p_i \tag{8}$$

It is maximized with respect to the values w_i $[m^3/h]$ and U_i $[m^3]$ under the inequality constraints (5– 7).

The scheduling problem is solved under the following conditions for its constant parameters: $V_{\min} = 10$ $[m^3]$, $V_{\max} = 40$ $[m^3]$, $U_{\min} = 70$ $[m^3]$, $U_{\max} = 125$ $[m^3]$, $w_{\min} = 1$ $[m^3/h]$, $w_{\max} = 30$ $[m^3/h]$, $q = 30$ $[m^3/h]$, $r = 30$ $[m^3/h]$, $p_1 = 1$ $[h]$, $p_2 = 4$ $[h]$, $p_3 = 1$ $[h]$, $p_4 = 6$ $[h]$, $\epsilon = 0.2$ $[m^3/h]$.

A cyclic schedule solution of the resulting linear program is represented in figures (3) and (4).

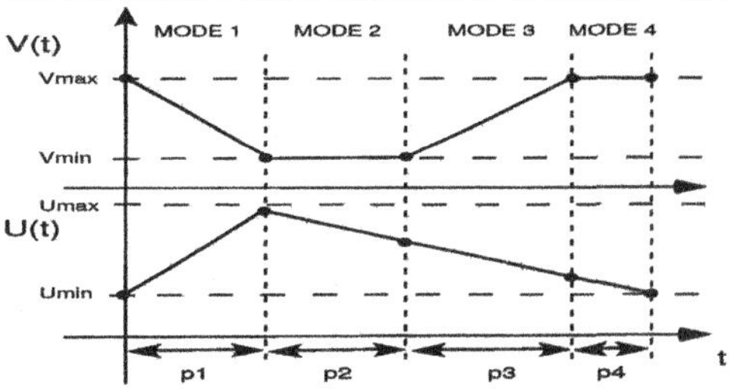

Fig. 3. Evolution of the material in the two tanks

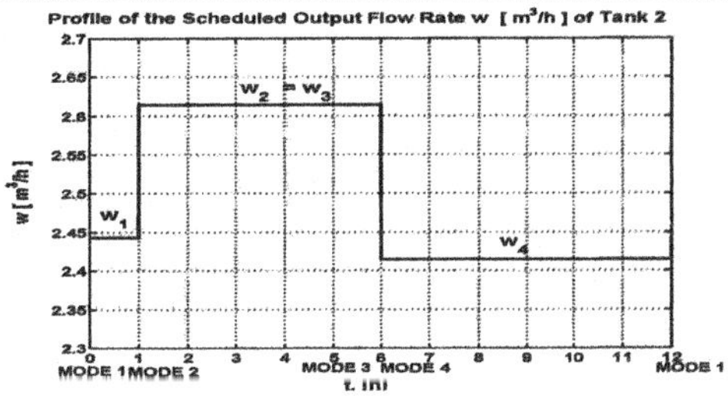

Fig. 4. Scheduled output flow rate of **T2**

This optimal solution of the scheduling problem is however not robust in cases of disturbances. The actual plant operation diverges from the optimal schedule in presence of disturbances. As a result the real plant performance becomes sub - optimal and the schedule may even become infeasible. This means that the plant operation is not stable when driven by the optimal schedule.

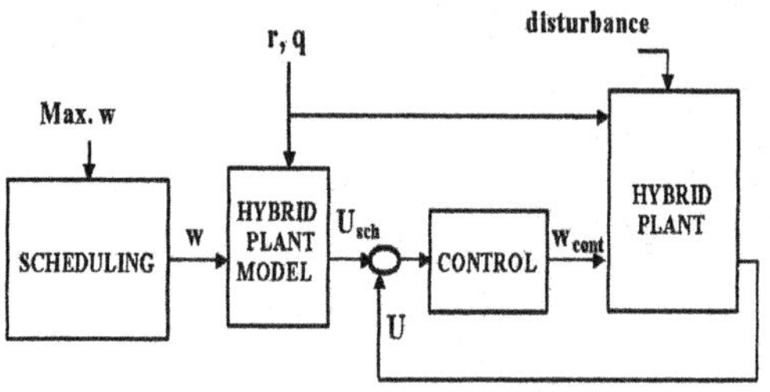

Fig. 5. Feedback control strategy

4. FEEDBACK STABILIZATION OF THE HYBRID CHEMICAL PLANT OPERATION

Let us recall that the objective of this paper is to illustrate how the stability problem of the plant operation can be solved by using simple P and PI control laws. The sensitivity of the closed loop plant operation to the choice of the controller parameters is also presented. The general structure of the feedback control strategy is presented in figure (5).

4.1 Statement of the closed loop control problem

In order to stabilize the operation of the considered hybrid chemical plant in the presence of disturbances, we consider the problem of controlling the volume of **T2**, U $[m^3]$ at the scheduled volume profile, U_{sch} $[m^3]$ (figure (5)) by acting on the output flow rate of material leaving **T2**, w $[m^3/h]$. It is assumed that the volume of **T2**, U $[m^3]$ is a measured output variable for the plant. In order to solve the control problem we first use a simple continuous feedback P control law which is written as follows:

$$w_{cont} = w - K_p.e_U \qquad (9)$$

where $e_U = [U_{sch} - U]$ $[m^3]$ is the error between the scheduled and the real volume profile of **T2**, w $[m^3/h]$ is the scheduled output flow rate of material leaving **T2** (figure (4)), K_p is the coefficient of the P controller. As a second step a simple continuous feedback PI control law is used. It has the form:

$$w_{cont} = w - K_p.\left[e_U + \frac{1}{\tau_I}.\int e_U dt\right] \qquad (10)$$

here K_p and τ_I $[h]$ are the coefficients of the PI controller. With these control laws, as shown in the next Section, the plant operation is stabilized in the sense that the plant trajectory converges towards a limit cycle which is close to the optimal cyclic schedule. Similarly to the constraint (6) in the scheduling problem here we impose that w_{cont} $[m^3/h]$ is in some interval $[w_{cont}^{min}\ w_{cont}^{max}]$. It is assumed that w_{cont}^{min} and w_{cont}^{max} are two times smaller and bigger, respectively to the minimal and the maximal values of w $[m^3/h]$. In order to illustrate the efficiency of both feedback control strategies, a Simulator of the hybrid plant has been developed in a Matlab / Simulink / Stateflow environment (Simeonova *et al.*, 2005*a*).

5. SIMULATION RESULTS

The hybrid plant is simulated under the following conditions for the plant constant parameters: $V_{\min} = 10$ $[m^3]$, $V_{\max} = 40$ $[m^3]$, $V_0 = 40$ $[m^3]$, $U_0 = 70$ $[m^3]$, $p_2 = 4$ $[h]$, $p_4 = 6$ $[h]$, $q = 30$ $[m^3/h]$, $r = 30$ $[m^3/h]$.
The overall operation process of the hybrid chemical plant is presented hereafter.

5.1 Hybrid chemical plant operation

Four case studies of the plant operation are successively considered: scheduled plant operation in the absence of disturbances, plant operation in the presence of process disturbance, plant operation in the presence of process disturbance under P and PI control, respectively.

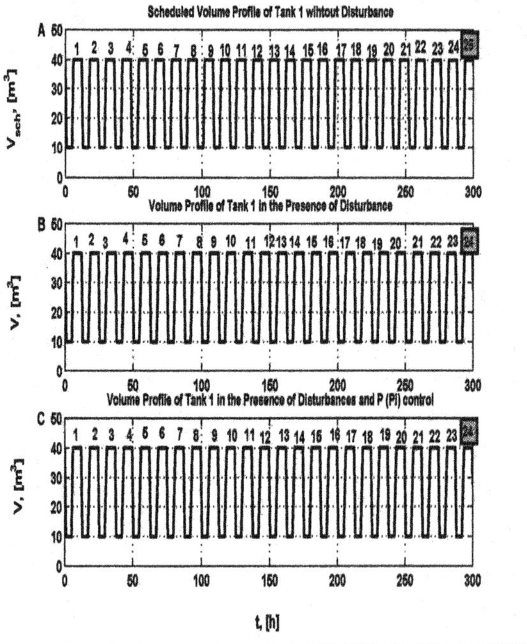

Fig. 6. V $[m^3]$: Volume profile of **T1**

- Scheduled Plant Operation in the Absence of Disturbance

The optimal scheduled plant operation is obtained after the application of the optimal scheduled flow rate w $[m^3/h]$ (figure (4)) to the plant Simulator

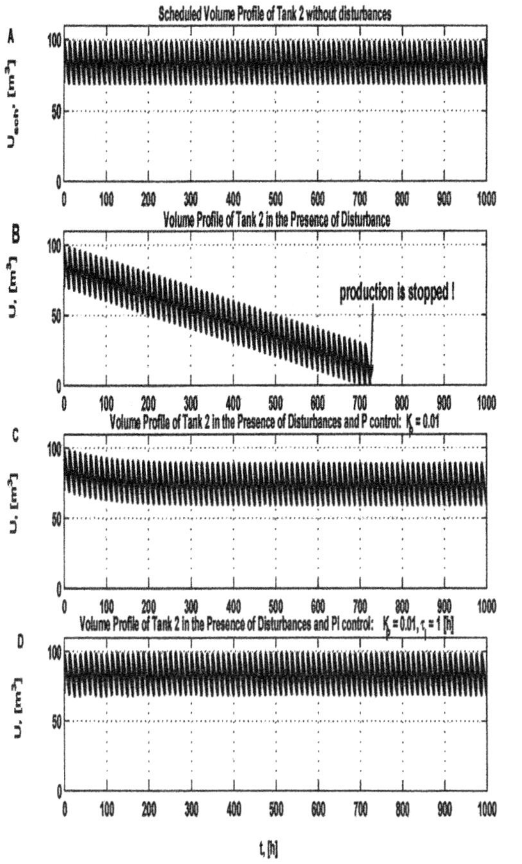

Fig. 7. U $[m^3]$: Volume profile of **T2**

in the absence of disturbances. In figure (6A) and (7A) are given respectively the scheduled volume profiles of **T1**, V_{sch} $[m^3]$ and **T2**, U_{sch} $[m^3]$. As expected the values of V_{sch} are in the interval $[V_{min}\ V_{max}]$ $[m^3]$. In figure (6A) it is also observed that there are 25 batches produced during the time period of approximately 300 $[h]$. In figure (7A) it is seen that the values of U_{sch} are in the approximate interval [70 100] $[m^3]$. Figure (8A) gives a phase plan representation of the optimal cycle. This cycle is actually unstable in open loop. Let us now consider the case when there is a small constant disturbance in the plant process.

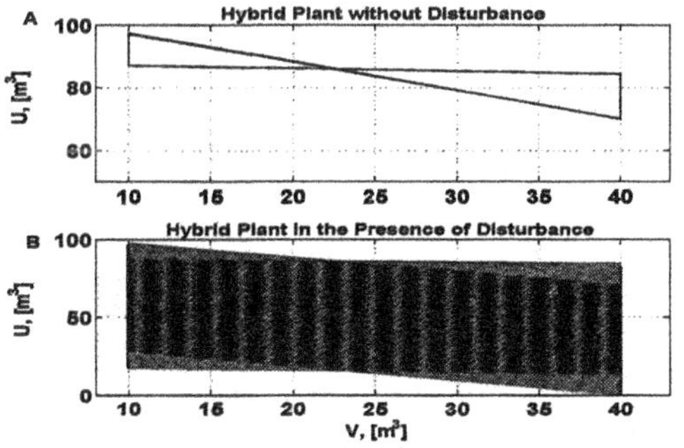

Fig. 8. Phase plane representation of the hybrid process in open loop

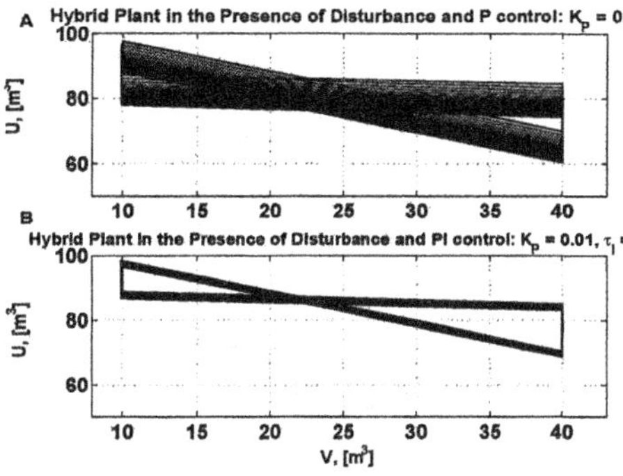

Fig. 9. Phase plane representation of the hybrid process in closed loop

- Plant Operation in the Presence of a Process Disturbance

Let at $t = 17$, $[h]$ the input flow rate $q = 30$ $[m^3/h]$ of material entering **T1** is incidentally decreased. Its new value is: $\widetilde{q} = 20$ $[m^3/h]$. The values of the other input variables are unchanged. As a consequence the real plant operation diverges from the optimal scheduled cycle. In figure (6B) it is observed that the values of the volume of **T1**, V stay in the interval $[V_{min}\ V_{max}]$ $[m^3]$. This is natural because the values V_{min} and V_{max} are used as a switching criteria for the hybrid process carried out in the plant. However due to the disturbance, there is a time delay in the evolution of the volume of **T1** and as a result the number of batches produced at $t \approx 300$ $[h]$ is 24 instead of 25 as scheduled. In figure (7B), we can clearly see the overall plant instability: the actual volume of **T2**, U $[m^3]$ steadily deviates from the scheduled volume (figure 7A). At time $t_{stop} \approx 724$ $[h]$, **T2** is totally empty and the production is stopped. In figure (8B) the plant instability is also observed: the plant trajectory progressively diverges from the optimal cycle (figure 8A). The periodically increasing - oscillatory behavior of the volume error in **T2**, e_U $[m^3]$ is shown in figure (10A). It has a basic period of 12 $[h]$, within an envelope of approximately 300 $[h]$.
In order to stabilize the plant operation and to avoid production stopping, the continuous time feedback P and PI control laws are implemented.

- Stabilization of the Plant Operation Under Process Disturbance by P Control

The P control law is tuned with the following controller parameter: $K_p = 0.01$. For both controllers $[w_{cont}^{min}\ w_{cont}^{max}] = [1.2\ \ 5.2]$ $[m^3/h]$.
In figure (7C) it is observed that, due to the use of the P control law, the volume profile of **T2**, U $[m^3/]$ is stabilized in the approximate interval [60 90] $[m^3]$ and the process in no longer interrupted at t_{stop} $[h]$. The settling time of the controller is $t \approx 200$ $[h]$. In figure (9A) it is seen that the plant trajectory converges towards a

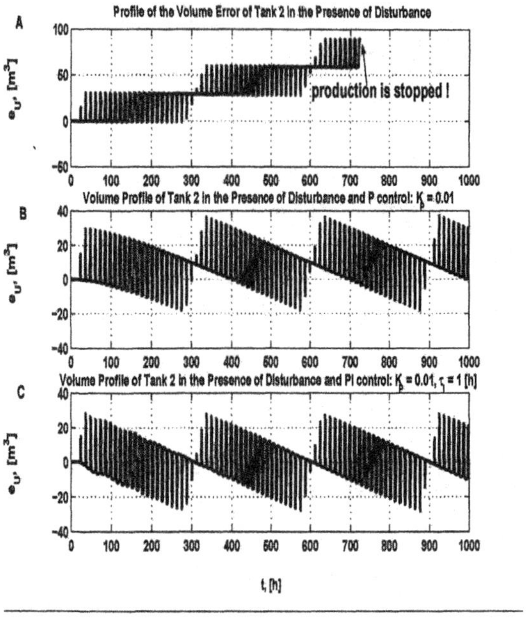

Fig. 10. e_U $[m^3]$: Error profile of **T2**

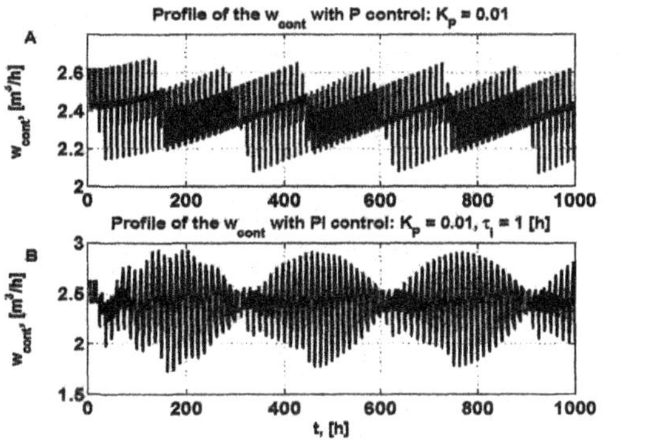

Fig. 11. Controlled output flow rate of **T2**

limit cycle which is not very far from the optimal cyclic schedule. The availability of a static error compared with the optimal cycle (8A) is observed. Figure 11A shows the corresponding profile of the controlled output flow rate of **T2**, w_{cont} $[m^3/h]$. The range of its values is not very different from the scheduled one (figure (4)).

Let us now consider the effect of the change of K_p parameter of the P control law, on the closed loop plant stability. In figure (12A,B,C) it is seen that if we decrease the values of the coefficient K_p, $K_p = 0.001$ the plant process becomes unstable. The volume error of **T2**, e_U $[m^3]$ has a periodically increasing time evolution (figure (12A)). If the value of K_p is increased, $K_p = 0.08$ the plant behavior becomes slightly oscillatory. This is observed from the volume profile of **T2**(figure (12F)). Moreover the values of the output flow rate w_{cont} $[m^3/h]$ (figure (12E)) increase a lot compared to the scheduled ones (figure (4)). Naturally both effects are undesirable.

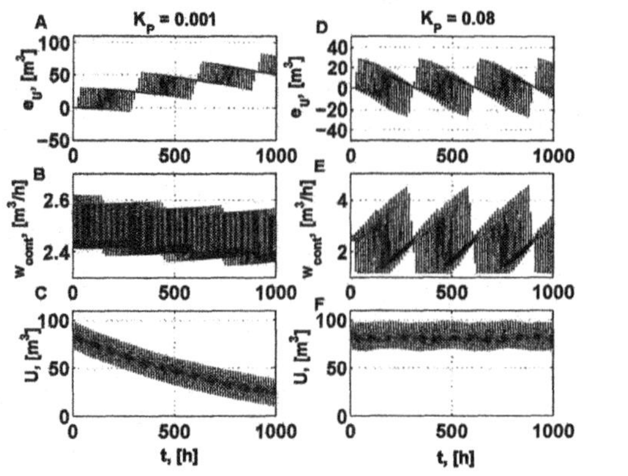

Fig. 12. Influence of K_p on the plant stability with P controller

- Stabilization of the Plant Operation Under Process Disturbance by PI Control

The PI control law is tunned with the following controller parameters: $K_p = 0.01$, $\tau_I = 1$ $[h]$.
The time evolution of the volume of **T1**, V $[m^3]$ after the application of PI control law is the same as in both previous cases (figure 6C). As observed in figure (7D), similarly to the case when the P control law is used, by means of the PI control law, the volume profile of **T2**, U $[m^3]$ is stabilized and the process in no longer interrupted at t_{stop} $[h]$. Moreover in this figure it is also seen that there is no static error. In figure (9B) it is seen that the plant operation is stabilized in the sense that the plant trajectory converges towards a limit cycle which is closer to the optimal cycle (figure 8A) compared to the case when the P control law is used (figure (9A)). In figure (10C) is observed that the volume error of **T2**, e_U $[m^3]$ stays in smaller interval compared to the case when P control is applied. As seen in figure (11B) the range of the controlled output flow rate of **T2**, w_{cont} $[m^3/h]$ is slightly increased with respect to the P controlled one (figure (11A)).

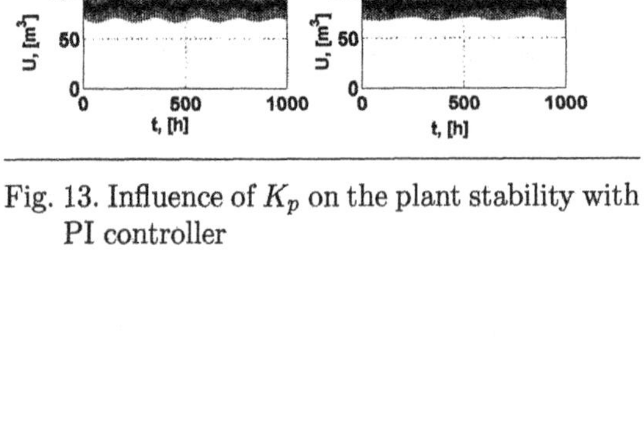

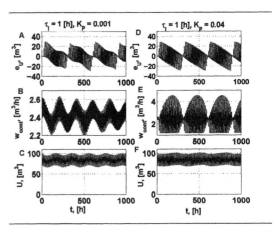

Fig. 13. Influence of K_p on the plant stability with PI controller

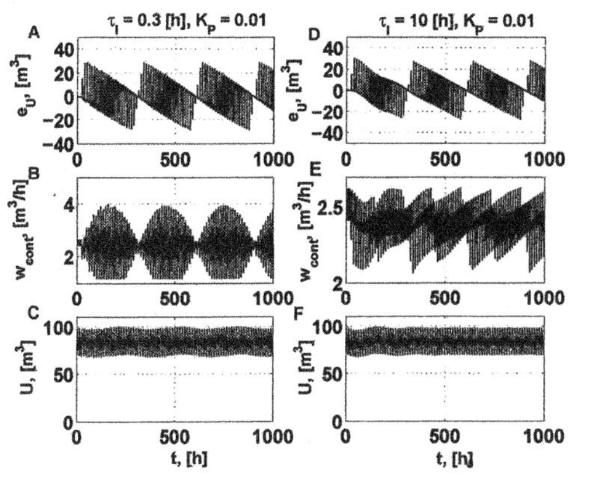

Fig. 14. Influence of τ_I on the plant stability with PI controller

Let us now consider the effect of the change of K_p parameter of the PI control law. In figure (13A, B, C) it is seen that if we decrease the values of the coefficient K_p, $K_p = 0.001$ the plant behavior becomes strongly oscillatory which is definitely undesirable. If the value of the coefficient K_p of the PI controller is increased, $K_p = 0.04$ the plant volume (13F) has a slightly oscillatory behavior compared to the case when $K_p = 0.01$. Moreover the range of the output flow rate is increased a lot compared to the scheduled once (figure (4)). Actually this behavior is similar to the case when we increase the value K_p of the P control law. Let us now observe the effect of the change of τ_I $[h]$. If the value of τ_I $[h]$ is decreased, $\tau_I = 0.3$ $[h]$ the plant behavior is slightly oscillatory and the range of w_{cont} $[m^3/h]$ is increased similarly to the case when we increase the coefficient K_p of the P and PI controllers. If the value of τ_I $[h]$ is increased, $\tau_I = 10$ $[h]$ instead of $\tau_I = 1$ $[h]$ the settling time of the controller is increased but the controlled output flow rate of **T2**, w_{cont} $[m^3/h]$ (figure 14E) is closer to the scheduled once compared to the case when $K_p = 0.01$.

Let us now see what is the quantity of material produced from the hybrid plant.

- Quantity of Produced Material

In figure (15A,B) the dotted line represents the optimal scheduled plant production W_{opt30} $[m^3]$ in the absence of disturbances during $t = 1000$ $[h]$ of operation. The solid line represents the closed loop quantity of material produced by P and PI control, respectively. As seen due to the availability of unknown disturbance the difference between the scheduled and actual quantity of produced material is progressively increasing. In figure (15C,D) the dotted line is the optimal plant production W_{opt20} $[m^3]$ supposing that at $t = 17$ $[h]$ (the time when the disturbance appears) a rescheduling is done knowing that the new value of q is $\tilde{q} = 20$ $[m^3/h]$. As observed the actual plant production converges to the optimally rescheduled one.

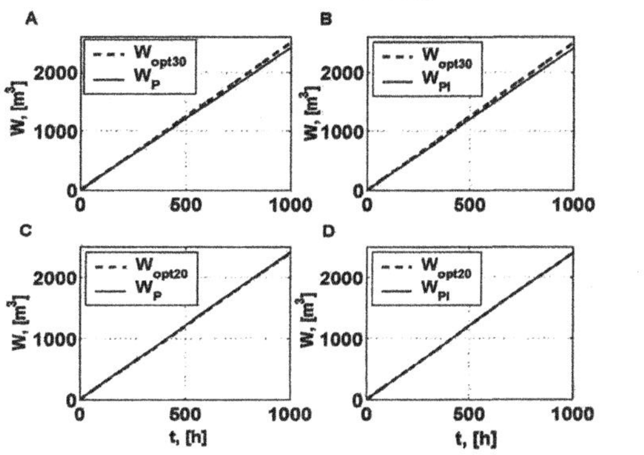

Fig. 15. $W[m^3]$: Quantity of produced material

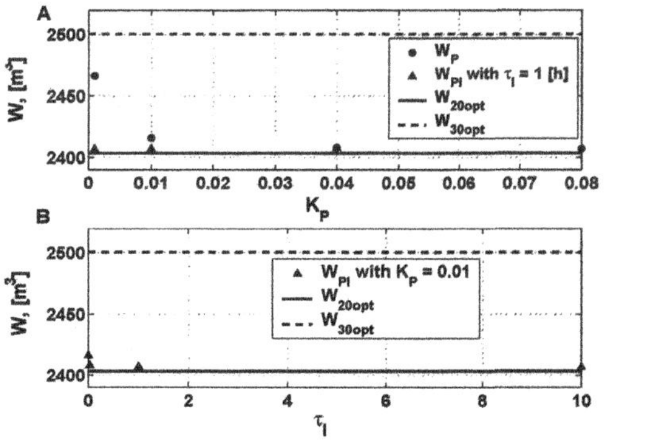

Fig. 16. Influence of the controller parameters on the closed loop plant production

Let us now see what is the influence of the controller parameters on the quantity of produced material. In figure (16A,B) the dotted line is the initial optimal scheduled quantity of produced material, W_{opt30} $[m^3]$ in the absence of disturbances after $t = 1000$ $[h]$ of operation. In the same figure the solid line is the optimal quantity of produced material, W_{opt20} $[m^3]$ assuming that at $t = 17$ $[h]$ (the time when the disturbance appears) the rescheduling is done knowing that the new value of q is $\tilde{q} = 20$ $[m^3/h]$. In figure (16A) the dotes illustrate the dependence of the production W_p $[m^3]$, after the application of P control, on the coefficient $K_p = \{0.001\ 0.01\ 0.4\ 0.8\}$. As seen the bigger is the coefficient K_p the closer is the value of the quantity W_p to the rescheduled quantity value, W_{opt20} $[m^3]$. In contrast the smaller is the value of K_p the closer is W_p $[m^3]$ to the initial optimal value W_{opt30} $[m^3]$. In the same figure the triangles represent the dependence of the production, obtained after the application of PI control, on the coefficient $K_p = \{0.001\ 0.01\ 0.4\}$. As seen this produced quantity is closer the the optimal rescheduled value W_{opt20} $[m^3]$ compared to the P control only. It is also seen that it does not depend a lot on the choice of K_p. In figure (16B) it is observed that the bigger is the parameter τ_I $[h]$

the closer is the productivity W_p $[m^3]$ to W_{opt20} $[m^3]$. The quantity W_{PI} in both cases are far from the initial optimal quantity W_{op30} $[m^3]$.

6. CONCLUSIONS AND FUTURE WORK

In this communication, we have dealt with the feedback stabilization of an hybrid chemical plant at an optimally scheduled operation which is open-loop unstable. The goal has been to illustrate how this stabilization problem can be solved by using simple P and PI control laws. The performance of the control has been illustrated through various simulation experiments carried out in the Matlab/Simulink/Stateflow environment. With the considered control laws, it is shown that the operation of the plant is stabilized in the sense that the plant trajectory converges towards a limit cycle which is close to the optimal cyclic schedule. An interesting challenge that is under study is to analyze the stability of these limit cycles through the fixed point stability of the corresponding Poincarre maps (Hiskens, 2001) (Girard, June 2003).

Acknowledgments: Work partially done in the frameworks of the "Fonds de Recherche SOLVAY", "HYCON Network of Excellence,contract number FP6-IST-511368" and "Belgian Programme on Interuniversity Attraction Poles, initiated by the Belgian Federal Science Policy Office"

REFERENCES

Bemporad, A. and M. Heemels (2005). *Lecture Notes from the First HYCON Ph.D. Summer School on Hybrid Systems.* http://www.dii.unisi.it/hybrid/school/.

Girard, A. (June 2003). Computation and stability analysis of limit cycles in piecewise linear hybrid systems. *Proceedings of the IFAC Conf. ADHS 2003, Saint Malo - Brittany - France* pp. 217–222.

Hiskens, I.A. (2001). Stability of limit cycles in hybrid systems. *Proceedings 34th IEEE International Conference in Systems Science, Hawaii.*

Lygeros, J., K. Johansson, S. Simic, J. Zhang and S. Sastry (2003). Dynamical properties of hybrid automata. *IEEE Transactions on automatic control* **48**, 2–17.

Simeonova, I., F. Warichet, G. Bastin, D. Dochain and Y. Pochet (2005*a*). Feedback control of the operation of an hybrid chemical plant. *Extended version of this paper, in preparation.*

Simeonova, I., F. Warichet, G. Bastin, D. Dochain and Y. Pochet (2005*b*). On line scheduling of chemical plants with parallel production lines and shared resources : a feedback implementation. *CD-Rom Proceedings IMACS World Congress, Paris.*

Willems, J.R. (2003). *Lecture Notes in Hybrid Systems.* UCL - Belgium, Graduate School in Systems and Control.

ELSEVIER
IFAC
PUBLICATIONS

A SOLAR COOLING PLANT: A BENCHMARK FOR HYBRID SYSTEMS CONTROL

Darine Zambrano * Carlos Bordons **
Winston García-Gabín * Eduardo F. Camacho **

* *Escuela de Ingeniería Eléctrica, Universidad de Los Andes, Venezuela*
** *Departamento de Ingenieria de Sistemas y Automatica, Escuela Superior de Ingenieros, Universidad de Sevilla, España*

Abstract: This paper describes the hybrid model of a solar cooling plant. This model considers all possible operating modes of the process, which are modelled as a finite state machine whose transition conditions are given by the discrete variables. The discrete variables are the electrovalves and pumps. The model has been written as a mixed logical dynamical system and is simulated using Stateflow/Simulink Matlab. The model has been validated using real data from the plant. This plant is being used as a benchmark for hybrid control experiences by many European researchers in the framework of the HYCON Network of Excellence.

Keywords: Hybrid system, dynamic models, solar energy.

1. INTRODUCTION

The use of solar energy in air conditioning systems is one of the most evident and not sufficiently exploited applications of this source of renewable energy. The use of solar radiation for cooling allows a time synchronization between solar offer and cooling demand since cold air is in general more necessary when the solar radiation is high, reducing this way the need of storage systems, which is one of the drawbacks of the use of solar energy for other application such as heating.

There exist different procedures for cold production which have been tested using solar radiation as primary source (see (Sayigh and (editors), 1992)). One of the most successful methods is by means of an absorption machine, which produces chilled water when hot water is injected into its generator. As the water is heated by the sun, this type of air conditioning systems reduces conventional energy consumption and therefore contributes to the maintenance of a non-polluted air. The plant includes buffer storage and an auxiliary energy system to be used in the case that the solar energy supply is not enough.

However, the operation of this type of system presents some particular issues that must be addressed by the control strategy. Firstly, the primary energy (sun) cannot be manipulated, as is the case of any conventional thermal process. Secondly, there exist big disturbances on the process, mainly due to the change in environmental conditions. There are also dead-times associated to fluid transportation, which are variable depending on the operating conditions. Finally, the cooling demand is variable, since it depends on the occupancy and use of the space (laboratories in this case).

In addition to this, its dynamics changes depending of the source of energy (solar, auxiliary heater, storage tank or a combination of them) that is used to feed the absorption machine. Several control strategies have been tested on different solar power plants (see, for instance, (Camacho *et al.*, 1997; Lemos *et al.*, 2002; Silva *et al.*, 2002)), but none of them takes the hybrid nature of the process into account.

This process presents many continuous and discrete variables and presents multiple operating modes. The operating modes are defined by the components that provide thermal energy to the absorption machine. For these characteristics, this process has to be modelled as a hybrid system. The hybrid model represents the continuous and discrete components, and allows configuration of the operating modes. The models have been obtained using energy balances, physical laws, and models based in practical data, and are validated with real data.

This paper shows the development of a hybrid model of the whole plant, which is validated through experiments performed on the real plant. A simulation model is obtained as a useful tool for analysis and controller design.

This plant is used in the framework of the Network of Excellence HYCON, whose objective is to establish a durable community of leading researchers and practitioners working in hybrid systems control. This solar plant is a performance evaluation platform for testing control technologies of hybrid systems. Another benchmark problem is included in HYCON: an idle speed motor control system (Balluchi *et al.*, 2006).

The paper is organized as follows: Section 2 presents a description of the solar cooling plant, the principal systems and the operating modes are described. The hybrid models of the components are derived in section 3. Section 4 is devoted to the implementation and experimental validation of the model and finally the conclusions of the work are presented in section 5.

2. PLANT DESCRIPTION

The solar air conditioning plant is located at Seville (Spain) and is used to cool the Laboratories of the System Engineering and Automation Department of the University of Seville. It consists of a solar field that produces hot water which feeds an absorption machine which generates chilled water and injects it into the air conditioning system, achieving a cooling power of 35 kW.

A general scheme of the plant is shown in Fig. 1, showing its main components: the solar system, composed of a set of flat solar collectors, the accumulation system, composed of two tanks storing hot water, and the cooling machine. There also exist an auxiliary gas-fired heater that can supply energy in those situations where solar radiation is not enough, and a load simulator (a heat pump) that allows to perform tests for different load profiles.

The control objective is to supply chilled water to the air distribution system at the demanded tempera-

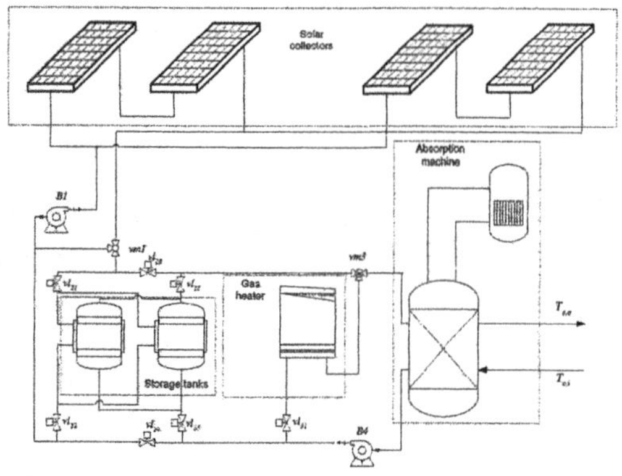

Fig. 1. Plant scheme

ture minimizing auxiliary energy (gas) consumption and fulfilling operational constraints in the absorption machine. Also, the storage energy in the tanks is considered at the end of the day, since it can be used in the following day when the solar radiation is low. The primary energy source (solar radiation) cannot be manipulated and has to be treated as a measurable disturbance. This implies that the control system must keep the cooling machine working at the desired operating point and this is achieved by keeping the machine inlet water temperature at the given set-point. The inlet water is the mix of the water coming from the solar system, the storage tanks and the water coming from the gas-fired heater in case additional energy is needed. Additionally, the temperature of the water in the solar system can be controlled adjusting the water flow inside the solar field.

The hybrid nature of the plant comes from the use of two different energy sources (solar and gas), which can be combined or used independently. Besides, thermal energy coming from a storage tank can be added to the system. The plant can be reconfigured on-line manipulating electrovalves and pumps (on/off) to allow selecting the components for energy supply.

2.1 *Systems description*

This section shows the main components of the plant.

Solar system: The primary source of energy is solar radiation, which is used by the solar collectors to increase the temperature of the circulating water. The solar field is composed of 151 m^2 of flat collectors which work in the range of $60°C$ to $100°C$ and supply a nominal power of 50 kW.

Accumulation system: The accumulation system is composed of two tanks of 2500 L each one, that work in parallel. This system acts as a buffer, storing hot water to be used in transient situations where solar radiation does not allow to obtain the desired temperature at the end of the hot water circuit.

Auxiliary energy system: As a complement to the solar energy supplied by the field, an auxiliary energy system consisting of a gas-fired heater of 60 kW can be used in case solar radiation is not enough. This heater transfers additional thermal energy to the water coming from the collectors. The existing heater has a built-in ON/OFF controller, which makes its outlet temperature rather oscillatory.

Cooling system: The cooling system is an absorption machine that works with water as cooling fluid and a water solution of lithium bromide ($H_2O-LiBr$). The correct operating of the machine requires that its inlet temperature is inside the range of $75°C-100°C$ for chilled water production. The machine has four different circuits: evaporator, generator, condenser and absorber, where the energy exchanges for chilled water production take place. The experimental results are similar to (Syed *et al.*, 2005). The critical variables that have influence on operation are:

- Condenser temperature, which establishes condenser and generator pressures.
- Evaporator temperature, which establishes evaporator and absorber pressures.
- Generator temperature, that together with condenser pressure fixes the solution concentration that leaves the generator.
- Absorber temperature, that together with the evaporator pressure fixes the solution concentration that enters the generator.

Continuous manipulated variables: The continuous variables are the pump velocity (v_{B1}) and the mix valve position (vp_{vm3}).

Discrete manipulated variables: The logic manipulated variables are the electrovalves and pumps,

- vl_{21}/vl_{22} allows connect the solar collectors with the tanks
- vl_{23}/vl_{24} allows connect the solar collectors with the absorption machine/gas heater
- vl_{25}/vl_{26} allows connect the tanks with the absorption machine/gas heater
- vl_{31} allows circulate water through the gas heater
- $B4$ allows pump water from absorption machine (constant velocity)
- $B1$ allows pump water through solar collectors (variable velocity)
- $vm1$ allows flow the water to solar collectors or to the other components

2.2 *Operating modes of the process*

The plant evolves among several operating modes during its daily operation. A logic variable "l_i" is associated to each operating mode. The operating modes are selected using pumps and ON/OFF valves and are the following:

(1) Recirculation $\Rightarrow l_1 = 1$: All water flow through the solar collectors. In this mode, the temperature of the water is elevated.
(2) Loading the tanks with hot water $\Rightarrow l_2 = 1$: The water in the solar collectors flows through the tanks. When a little solar radiation is appeared, then the stored hot water is used.
(3) Using the solar collectors $\Rightarrow l_3 = 1$: The water is heated in the solar collectors. The water flows to absorption machine.
(4) Using the solar collectors and gas heater $\Rightarrow l_4 = 1$: The water is heated in the solar collectors and the gas heater. The gas heater is used when the absorption machine input temperature is inadequate. The gas heater on must be avoided.
(5) Using the gas heater $\Rightarrow l_5 = 1$: The water is heated in the gas heater, and then it flows through the absorption machine.
(6) Using the tanks and gas heater $\Rightarrow l_6 = 1$: The water of absorption machine is given by the tanks. The gas heater is used when the absorption machine input temperature is not enough. The gas heater on must be avoided too.
(7) Using tanks $\Rightarrow l_7 = 1$: When little solar radiation is appeared, then the heat stored in the tanks is used to operate the absorption machine.
(8) Loading the tanks and using the gas heater $\Rightarrow l_8 = 1$: The tanks are loaded with heated water by the solar collectors. The water of absorption machine is given by the gas heater.
(9) Recirculation and using the gas heater $\Rightarrow l_9 = 1$: The water is recirculated through the solar collectors. The water to the absorption machine from the gas heater only.
(10) Using the solar collectors and loading tank $\Rightarrow l_{10} = 1$: The water from the solar collectors is divided into the tanks and the absorption machine.
(11) Using the solar collectors and gas heater, and loading tanks $\Rightarrow l_{11} = 1$: The water from the solar collectors is divided into the tanks and the absorption machine. The inlet water to the absorption machine is from the solar collectors and gas heater.

3. HYBRID MODEL

A hybrid model has been developed using a Finite State Machine (FSM). The transition conditions are given by the discrete manipulated variables (1),

$$
\begin{aligned}
c1 &= \sim vl_{21} \wedge \sim vl_{23} \wedge \sim vl_{25} \wedge \sim vl_{31} \wedge \sim B4 \wedge B1 \wedge \sim vm1 \\
c2 &= vl_{21} \wedge \sim vl_{23} \wedge \sim vl_{25} \wedge \sim vl_{31} \wedge \sim B4 \wedge B1 \wedge vm1 \\
c3 &= \sim vl_{21} \wedge vl_{23} \wedge \sim vl_{25} \wedge \sim vl_{31} \wedge B4 \wedge B1 \wedge vm1 \\
c4 &= \sim vl_{21} \wedge vl_{23} \wedge \sim vl_{25} \wedge vl_{31} \wedge B4 \wedge B1 \wedge vm1 \\
c5 &= \sim vl_{21} \wedge \sim vl_{23} \wedge \sim vl_{25} \wedge vl_{31} \wedge B4 \wedge \sim B1
\end{aligned} \tag{1}
$$

$c6 = \sim vl_{21} \wedge \sim vl_{23} \wedge vl_{25} \wedge vl_{31} \wedge B4 \wedge \sim B1$

$c7 = \sim vl_{21} \wedge \sim vl_{23} \wedge vl_{25} \wedge \sim vl_{31} \wedge B4 \wedge \sim B1$

$c8 = vl_{21} \wedge \sim vl_{23} \wedge \sim vl_{25} \wedge vl_{31} \wedge B4 \wedge B1 \wedge vm1$

$c9 = \sim vl_{21} \wedge \sim vl_{23} \wedge \sim vl_{25} \wedge vl_{31} \wedge B4 \wedge B1 \wedge \sim vm1$

$c10 = vl_{21} \wedge vl_{23} \wedge \sim vl_{25} \wedge \sim vl_{31} \wedge B4 \wedge B1 \wedge vm1$

$c11 = vl_{21} \wedge vl_{23} \wedge \sim vl_{25} \wedge vl_{31} \wedge B4 \wedge B1 \wedge vm1$

In the figure 2 is shown the FSM of the solar plant implemented in Stateflow. The state output is the operating mode of the process.

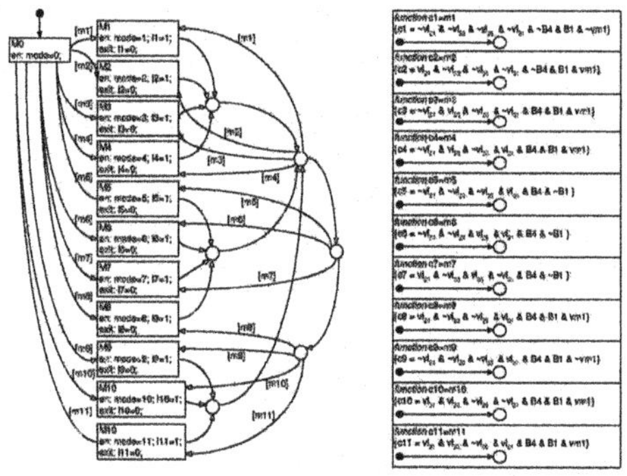

Fig. 2. Finite state machine of the solar plant

The states are the operating modes of the process (see 2.2). The state outputs are the logic variables $l_1, l_2, \ldots, l_{11}$. This logic variables are included in the model equations of each subsystem. The model equations have been written as a Mixed Logic Dinamycal (MLD) system (Bemporad and Morari, 1999).

The logic variables l_i are included in the energy balance equations. The variables determine when a system is on. When a system is not working, the environmental losses are modelled. The input temperatures and the flows of each system are select by the logic variables.

3.1 *Solar collectors model*

A solar collector can be modeled as a heat exchanger that absorbs part of the solar radiation, warms up the circulating fluid and has some thermal losses. The equation that represents its behaviour can be obtained by means of energy balance and is given by:

Solar collector output temperature

$$[C_A L_1 + C_E(1 - L_1)]\frac{dT_{sc,m}}{dt} = [F(I_G + U(T_{amb} - T_{sc,m}))$$
$$+\frac{2\dot{m}_{sc}C_p}{A_C}(T_{sc,i} - T_{sc,m})]L_1 + [U_E(T_{amb} - T_{sc,m})](1 - L_1)$$
$$T_{sc,o} = 2T_{sc,m} - T_{sc,i} \qquad (2)$$

Solar collectors logic condition

$$L_1 = l_1 + l_2 + l_3 + l_4 + l_8 + l_9 + l_{10} + l_{11} \qquad (3)$$

Solar collector input temperature

$$T_{sc,i}(t) = T_{sc,o}(t - t_m)(l_1 + l_9) + T_{tc,o}(t - t_m)(l_2 + l_8)$$
$$+T_{g,o}(t - t_m)(l_3 + l_4) + \frac{(\dot{m}_{sc} - \dot{m}_g)T_{sc,o} + \dot{m}_g T_{g,o}}{\dot{m}_{sc}} l_{10}$$
$$+\frac{(\dot{m}_{sc} - \dot{m}_g + \dot{m}_{gh})T_{sc,o} + (\dot{m}_g - \dot{m}_{gh})T_{g,o}}{\dot{m}_{sc}} l_{11} \qquad (4)$$

where $T_{sc,m}$ is the average temperature of the water inside the collector ($^\circ C$), $T_{sc,i}$ is the inlet water temperature ($^\circ C$), T_{amb} is the ambient temperature ($^\circ C$), C_A is the heat capacity per surface unit of the collector ($\frac{J}{^\circ C m^2}$), C_p is the heat capacity of water ($\frac{J}{k_g {}^\circ C}$), U is the coefficient of losses to the atmosphere ($\frac{W}{^\circ C m^2}$), I_G is the irradiation on the collector surface ($\frac{W}{m^2}$), A_C the collector area (m^2), U_E is the coefficient of losses to the ambient and C_E is the time constant when B1 pump is off, F is the efficiency factor, $\dot{m}_{gh}$ is the water mass through the gas heater ($\frac{kg}{s}$), $\dot{m}_g$ is the water mass in the generator ($\frac{kg}{s}$) and $\dot{m}_{sc}$ is the water mass flow through the collector ($\frac{kg}{s}$). The output temperature of the collector, $T_{sc,o}$, is obtained assuming that the temperature distribution is linear.

3.2 *Accumulation model*

The equations that model the process are based on the energy balance. The tanks inner temperature has been considered. The inlets tanks are not open simultaneously. The transport delay of the output temperatures has been estimated in function of the flows intensity. The equations are:

Tanks temperature

$$C_p V_t \rho \cdot \frac{dT_t}{dt} = \dot{m}_{tsc} C_p (T_{tsc,i} - T_t)$$
$$+\dot{m}_{tam} C_p (T_{tam,i} - T_t) + U_t A_t (T_{amb} - T_t) \qquad (5)$$

Tanks output temperature

$$T_{tsc,o}(t) = T_t(t - t_{0sc})(l_2 + l_8 + l_{10} + l_{11}) \qquad (6)$$
$$T_{tam,o}(t) = T_t(t - t_{0am})(l_6 + l_7)$$

Tanks input temperature

$$T_{tsc,i} = T_{sc,o}(l_2 + l_8 + l_{10} + l_{11}) \qquad (7)$$
$$T_{tam,i} = T_{g,o}(l_6 + l_7)$$

where U_t is the tank loss coefficient per surface unit ($\frac{W}{^\circ C \cdot m^2}$), A_t are the surfaces of the control volume in contact with the environment (m^2), V_t are the slices volumes (m^3), ρ is water density ($\frac{kg}{m^3}$), T_t is the tanks intern temperature ($^\circ C$), $\dot{m}_{tsc}$ is the flow from solar fields ($\frac{kg}{s}$), $\dot{m}_{tam}$ is the flow from absorption machine, $T_{tsc,i}$ is the input temperature from solar collectors ($^\circ C$), $T_{tam,i}$ is the input temperature from absorption machine ($^\circ C$), $T_{tsc,o}$ is the output temperature to solar collectors ($^\circ C$), $T_{tam,i}$ is the output temperature to absorption machine ($^\circ C$), $T_{g,o}$ is generator outlet temperature ($^\circ C$), t_{0sc} is the transport delay to output solar collectors (s), t_{0am} is the transport delay to absorption machine (s).

3.3 *Gas heater model*

A complete model can see in (Claquin *et al.*, 1994). Heater equations model two temperatures evolution: gas temperature (T_{ch}), and outlet water temperature ($T_{gw,o}$):

Water output temperature

$$[C_a L_2 + C_E(1-L_2)]\frac{dT_{gw,o}}{dt} = [U_{aw}(T_{amb} - T_{gw,o}) + U_{gw}(T_{ch} - T_{gw,o}) + C_p \dot{m}_{gh}(T_{gw,i} - T_{gw,o})]L_2 l_{12} + U_E(T_{amb} - T_{gw,o})(1-L_2) \quad (8)$$

Gas heater logic condition

$$L_2 = l_4 + l_5 + l_6 + l_8 + l_9 + l_{11} \quad (9)$$

Gas output temperature

$$[C_g L_2 + C_E(1-L_2)]\frac{dT_{ch}}{dt} = [U_{ag}(T_{amb} - T_{ch}) + U_{gw}(T_{gw,o} - T_{ch}) + P]L_2 l_{12} + U_E(T_{amb} - T_{ch})(1-L_2) \quad (10)$$

$$P = \dot{Q}_{on} l_{13} + \dot{Q}_{off}(1 - l_{13}) \quad (11)$$

$$l_{12} = \begin{cases} 1 & \dot{m}_{gh} \geq 1500 l/h \\ 0 & \dot{m}_{gh} < 1500 l/h \end{cases}; l_{13} = \{ 1 \quad 86^\circ C \leq T_{gw,o} \leq 94^\circ C$$

Gas input temperature

$$T_{gw,i}(t) = T_{g,o}(t - t_m)L_2 \quad (12)$$

where $T_{gw,i}$ is the heater inlet water temperature, C_a is water heat capacity ($\frac{W}{^\circ C}$), C_g is gas heat capacity ($\frac{W}{^\circ C}$), U_{aw} heat transfer coefficient between ambient and water ($\frac{W}{^\circ C}$), U_{gw} heat transfer coefficient between gas and water ($\frac{W}{^\circ C}$), U_{ag} heat transfer coefficient between ambient and gas ($\frac{W}{^\circ C}$), U_E is the coefficient of losses to the ambient and C_E is the time constant when the gas heater is off, and $\dot{Q}$ is the burner power (W).

3.4 *Absorption machine model*

The equations that describes the behavior of water temperature are given by:

Generator output temperature

$$[C_{gm} L_3 + C_E(1-L_3)]\frac{dT_{g,o}}{dt} = [U_{agm}(T_{amb} - T_{gm})] + [\dot{m}_g(T_{g,i} - T_{g,o}) + Q_G]L_3 l_{14} \quad (13)$$

$$Q_G = (Q_T - Q_M)/(1 + COP_n) \quad (14)$$

Evaporator output temperature

$$[C_{em} L_3 + C_E(1-L_3)]\frac{dT_{e,o}}{dt} = [Uae(T_{amb} - T_{e,o})] + [\dot{m}_e(T_{e,i} - T_{e,o}) + Q_E]L_3 l_{14} \quad (15)$$

$$Q_E = C_p \dot{m}_g (T_{g,i} - T_{g,o}) COP_n \quad (16)$$

Absorption machine logic condition

$$L_3 = l_3 + l_4 + l_5 + l_6 + l_7 + l_8 + l_9 + l_{10} + l_{11} \quad (17)$$

$$l_{14} = \begin{cases} 1 & 75^\circ C \leq T_{g,i} \leq 100^\circ C \\ 0 & T_{g,i} < 75^\circ, T_{g,i} > 100^\circ C \end{cases} \quad (18)$$

Generator input temperature

$$T_{g,i} = T_{sc,o}(l_3 + l_{10}) + T_{gh,o}(l_5 + l_8 + l_9) + \frac{\dot{m}_{sc} T_{sc,o} + \dot{m}_{gh} T_{gh,o}}{\dot{m}_{sc} + \dot{m}_{gh}} l_4 + T_{tam,o} l_7 + \frac{(\dot{m}_g - \dot{m}_{gh})(T_{tam,o} l_6 + T_{sc,o} l_{11}) + \dot{m}_{gh} T_{gh,o}(l_6 + l_{11})}{\dot{m}_g} \quad (19)$$

where $\dot{m}_e$ is the water mass in the evaporator ($\frac{kg}{s}$), $T_{g,i}$ is inlet generator temperature ($^\circ C$), $T_{e,o}$ is outlet evaporator temperature ($^\circ C$), $T_{e,i}$ is inlet evaporator temperature ($^\circ C$), C_{gm} and C_{em} are the heat capacities in the generator and evaporator ($\frac{W}{^\circ C}$), U_{agm} and U_{ae} are the thermal losses coefficients, Q_G and Q_E is the exchanged power in the generator and evaporator respectively, COP_n is the coefficient of performance.

3.5 *Flows models*

Finally, the solar collectors $\dot{m}_{sc}$, gas heater $\dot{m}_{gh}$ and absorption machine $\dot{m}_g$ flows have been identified. The flows depends of the pump velocity v_{B1} and the valve position vp_{vm3}. The identification is based in the real data collected for each operating mode.

$$\dot{m}_{sc} = f_{s1}(v_{B1})(l_1 + l_9) + f_{s2}(v_{B1})(l_2 + l_8) + f_{s3}(v_{B1}, vp_{vm3})l_3 + f_{s4}(v_{B1}, vp_{vm3})l_4 + f_{s10}(v_{B1}, vp_{vm3})l_{10} + f_{s11}(v_{B1}, vp_{vm3})l_{11} \quad (20)$$

$$\dot{m}_{gh} = f_{g4}(v_{B1}, vp_{vm3})l_4 + f_{g5}(vp_{vm3})(l_5 + l_8 + l_9) + f_{g6}(vp_{vm3})l_6 + f_{g10}(v_{B1}, vp_{vm3})l_{10} + f_{g11}(v_{B1}, vp_{vm3})l_{11} \quad (21)$$

$$\dot{m}_g = f_{mg3}(v_{B1}, vp_{vm3})l_3 + f_{mg4}(v_{B1}, vp_{vm3})l_4 + f_{mg5}(vp_{vm3})(l_5 + l_8 + l_9) + f_{mg6}(vp_{vm3})l_6 + f_{mg7}(vp_{vm3})l_7 + f_{mg10}(v_{B1}, vp_{vm3})l_{10} + f_{mg11}(v_{B1}, vp_{vm3})l_{11} \quad (22)$$

$$\dot{m}_{tsc} = \dot{m}_{sc}(l_1 + l_8) + (\dot{m}_{sc} - \dot{m}_g)l_{10} + (\dot{m}_{sc} - \dot{m}_g + \dot{m}_{gh})l_{11} \quad (23)$$

$$\dot{m}_{tam} = (\dot{m}_g - \dot{m}_{gh})l_2 + \dot{m}_g l_7 \quad (24)$$

4. MODEL VALIDATION

In order to validate the hybrid model, different experiments have been realized in the real plant. Firstly, the validation of the subsystems was done individually. Therefore, the complete model was validated.

The figure 3 shows the validation in the recirculation mode ($l_1 = 1$), the real data have been collected on 12/27/05. The real solar collectors temperature (solid line) and the simulated solar collectors temperature (dash line) are shown, the ambient temperature (dash-dot line) too. In this mode, the collectors temperature behavior depend on the solar radiation. The solar radiation shows a cloudy day, the simulation is between 10:45 and 14:00 hours. The model is able to follow the solar radiation variation. The solar radiation is a important variable, it is the principal energy source and a disturbance. The ambient temperature has little variations. The flow in the solar collectors are showed, the solid line is the simulated flow and the dash line is the real flow. The real flow has fluctuations in spite of the pump velocity is fixed. In this case, the solar energy is not enough for the absorption machine operation.

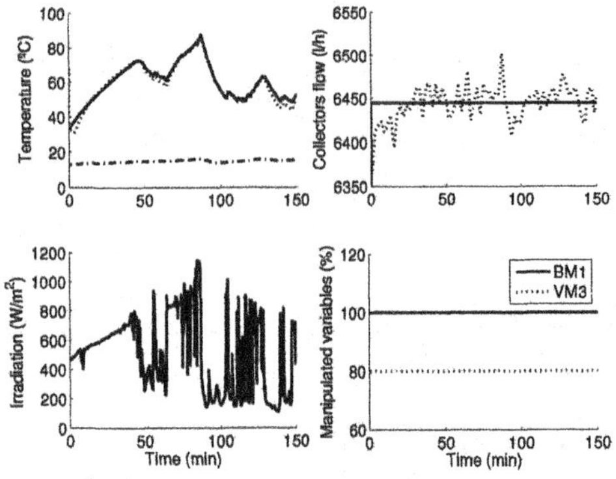

Fig. 3. Recirculation mode

Figure 4 shows the validation using the gas heater ($l_5 = 1$). The experiment was done the 12/01/05, at 15:45 hours. The real (dash line) and simulated (solid line) generator input temperature are shown. The temperatures have an adequate value for the absorption machine operation ($T_{g,i} > 75°C$) in at 11'15" and at 12'00" minutes respectively. When the generator output temperature is higher than $75°C$, the refrigeration process began. The evaporator input temperature (dash-dot line) is given by a load simulator, the initial value is $30°C$. The real (dash line) and simulated (solid line) evaporator output temperature have a similar behavior. In each time instant, the chilled water decreases $5°C$ approx . The evaporator flow is shown, the mixed valve position vp_{vm3} is constant after of 3 minutes, the real flow (dash line) has little variations.

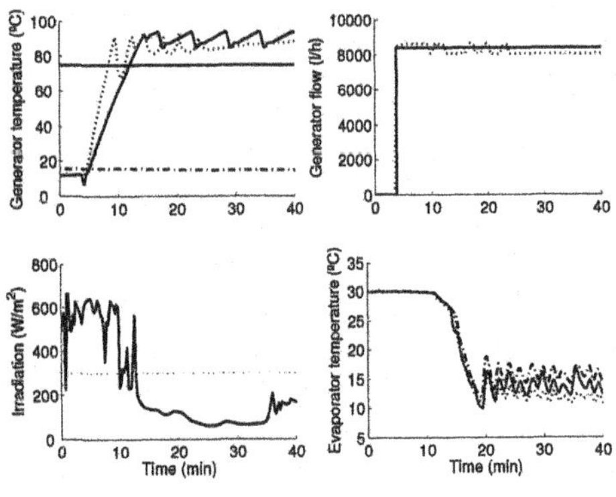

Fig. 4. Using the gas heater

The simulation compares the real data with the simulated data of the principal subsystems of the solar plant.

5. CONCLUSIONS

An hybrid model of the solar plant has been realized. The principal subsystems of the plant are modelled using energy balance equations, and the flow models are realized through identification using real data. The complete hybrid model is written using the logic variables corresponding to each operating mode. The hybrid model simulates the real behaviour of the solar plant according to the discrete and continuous manipulated variables. The hybrid model has been validated for different operating modes with satisfactory results. Future research will be devoted to develop the hybrid controller for this process.

ACKNOWLEDGMENTS

Supported by the Programme Alβan, the European Union Programme of High Level Scholarships for Latin America scholarship N° E05D053808VE and European Union project HYCON)(FP6-511368).

6. REFERENCES

Balluchi, Andrea, Luca Benvenuti, Maria D. Di Benedetto, Claudio Lemma, Pierpaolo Murrieri, Tiziano Villa and Alberto L. Sangiovanni-Vincentelli (2006). Idle speed control: a benchmark for hybrid system design. *Submitted to 2nd IFAC Conference on Analysis and Design of Hybrid Systems.*

Bemporad, Alberto and Manfred Morari (1999). Control of systems integrating logic, dynamics and constraints. *Automatica* **35**, 407–427.

Camacho, E.F., M. Berenguel and F.R. Rubio (1997). *Advanced Control of Solar Plants.* Springer. London.

Claquin, Sandrine, Alain Carriere and François Rocaires (1994). Modelling and application of adaptive control to a gas heater. *Proceedings of the Third IEEE Conference on Control Applications.*

Lemos, J. M., E. Mosca, R. N. Silva and P. O. Shirley (2002). Industrial applications of predictive adaptive control based on multiple identifiers. *15th IFAC World Congress.*

Sayigh, A. M. and J.C. McVeigh (editors) (1992). *Solar Air Conditioning and Refrigeration.* Pergamon Press. USA.

Silva, R. N., L. M. Rato and J.M. Lemos (2002). Observer based non uniform sampling predictive controller for a solar plant. *15th IFAC World Congress.*

Syed, A., M. Izquierdo, P. Rodriguez, G. Maidment, J. Missenden, A. Lecuona and R. Tozer (2005). A novel esperimental investigation of a solar cooling system in Madrid. *International Journal of Refrigeration.*

ELSEVIER
IFAC
PUBLICATIONS

TIMED DISCRETE EVENT CONTROL OF A PARALLEL PRODUCTION LINE WITH CONTINUOUS OUTPUT [1]

Dmitry Gromov * Stephanie Geist * Jörg Raisch*, **

* *Fachgebiet Regelungssysteme*
Technische Universität Berlin, Germany
Email: *gromov@control.tu-berlin.de*
** *Systems and Control Theory Group*
Max-Planck-Institut für Dynamik komplexer technischer Systeme, Magdeburg, Germany

Abstract: In this paper we present an approach to formulate and solve certain scheduling tasks using timed discrete event control methods. To demonstrate our approach, we consider a special class of systems: a cyclically operated chemical plant with parallel reactors using common resources and a continuous output. This problem was motivated by a benchmark proposed within the EU Network of Excellence HYCON. For this class of systems, we show how to pose the control problem within a discrete event framework by modelling system components as multirate timed automata. Safety and nonblocking are investigated. These properties have to be achieved in the presence of a class of bounded errors/disturbances. *Copyright © 2006 IFAC*

Keywords: Parallel Production Line, Scheduling, Timed Automata

1. INTRODUCTION

In this contribution, we investigate a *"parallelised" production line with resource constraints and continuous output.* Such a plant has been proposed as a case study by the Université Catholique de Louvain for the EU Network of Excellence HYCON (Simeonova *et al.*, 2005). In this example two parallel reactors sharing resources as, e.g., reactants, hot steam, cool water, are considered. The reactors are discharged into a shared storage tank that has a continuous outflow. The plant is cyclically operated. For this hybrid system, the goal is to assure non-conflicting work of these reactors and to prevent over- and underfilling of the tank.

We present an approach to the scheduling problem guaranteeing non-blocking and safety despite disturbances, using a timed automata formulation of the problem. Motivated by the HYCON case study we consider a generalised problem consisting of an arbitrary number of parallel reactors, an arbitrary number of shared resources and one storage tank.

Describing scheduling problems in a discrete event framework allows a very intuitive way of problem formulation. All system components including the resources can be considered as subsystems which can be easily described by timed automata and subsequently composed to form the overall problem.

[1] Work partially done in the framework of the HYCON Network of Excellence, contract number FP6-IST-511368

The advantage of using a formal approach, as opposed to heuristic strategies, is that desired properties can be guaranteed.

It is well known that only certain classes of timed automata systems are computationally tractable. For these classes, however, there are various numerical and symbolical methods for analysis and computation. Numerical methods are described in (Pettersson, 1999; Bengtsson and Yi, 2004; Bozga *et al.*, 1998) and symbolical methods are presented, for instance, in (Asarin *et al.*, 1995).

This paper is arranged as follows: in Section 2, we give a formal description of the overall problem. A short introduction to multirate timed automata is given in Section 3. In Section 4, the modelling of the system components is described. In Section 5, we address control issues and give some remarks on implementation. Finally, in Section 6, we apply our approach to the HYCON benchmark example described in (Simeonova *et al.*, 2005).

2. PROBLEM STATEMENT

Figure 1 presents a schematic view of the chemical plant considered in the sequel. The system consists of n parallel reactors with k common resources, e.g., reactants, cold/hot water supplies and pumps. The reactors are discharged into one tank that acts as an output buffer and has the continuous output flow $F_{out,t}$. The volumetric flow $F_{out,r}$ during the discharging of a reactor is fixed, the output flow of the tank $F_{out,t}$ can be adjusted within a given range. In each reactor the same process is performed. The goal is to assure non-conflicting use of resources and to keep the level of the tank volume between given values v_{max} and v_{min}. Furthermore, due to safety reasons, the tank outflow may not be interrupted.

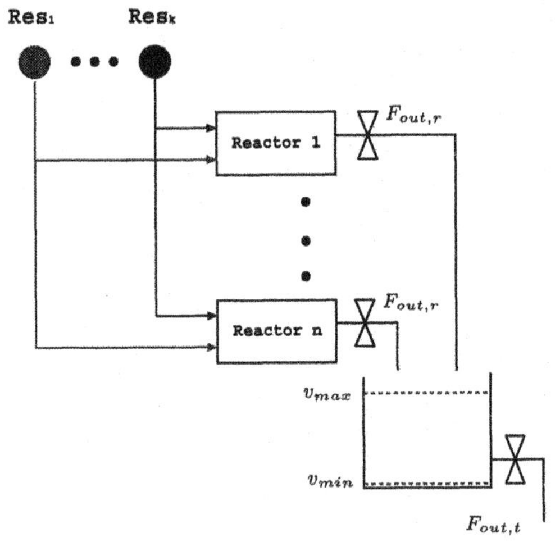

Fig. 1. A parallelised production line with resource constraints

A production cycle in the j-th reactor consists of a set of operations: $O_j = \{o_{ij}\}$, $i = 1 \ldots m$, e.g., heating, cooling, reaction, discharging and so one, which, in turn, are characterised by their processing times d_{ij}. The temporal ordering of these operations is given. We will also consider a case when the processing times of operations are not fixed, but an upper and lower bound is known: $d_{ij} \in [d_i^* - \underline{d}_i; d_i^* + \overline{d}_i]$. These deviations in processing times may be caused by disturbances.

There are a set of resources R and sets of "resource-sensitive" operations $O'_j \subset O_j$, $j = 1, \ldots, n$. A map $r_j : O'_j \to R$ associates a resource to each operation $o_{ij} \in O'_j$. Here we assume that these maps are injective, i.e. resources are used only once within the reaction cycle of a reactor but multiple use by different reactors is allowed.

Due to technological or safety requirements some operations must be processed without delay. These operations are grouped into tasks $K_j^l = \{o_{ij}^l\}, K_j^{l_1} \cap K_j^{l_2} = \emptyset$. Each task must contain at least one resource-sensitive operation, i.e. $K_j^l \cap O'_j \neq \emptyset$ where, within a task, delays are not permitted. We assume that an isolated operation also forms a task if it is resource-sensitive. Otherwise, it can be joined with the neighbour task. Hence, each operation belongs to some task, $\bigcup_{j,l} K_j^l = O = \bigcup_j O_j$. We denote the set of all tasks by $K = \{K_j^l\}, j = 1, \ldots, n, l = 1, \ldots, r$.

3. MULTIRATE TIMED AUTOMATA

Timed automata (TA) (Alur and Dill, 1994) are finite automata augmented with continuous clocks whose values grow uniformly at each discrete state. The set of clock variables is denoted by X. A clock valuation ν for the set X assigns a real value to each clock. Clocks can be reset to zero at certain transitions. There are also *clock constraints* $\Phi(X)$ defined over X in the following way:

$$\phi := x \leq c \mid x < c \mid x \geq c \mid x > c \mid \phi_1 \wedge \phi_2.$$

That means that each clock constraint can be represented as an union of inequalities. A transition may be equipped with a clock constraint which is interpreted as an enabling or guard condition. Clock constraints attached to locations can be interpreted as invariants. In this paper, a transition condition with upper time limit ∞ occurs only for the transitions whose switching can be controlled. Hence, we can assume that these transitions switch at the first possible time instant.

We consider an extended class of timed automata, namely *multirate timed automata* (Alur

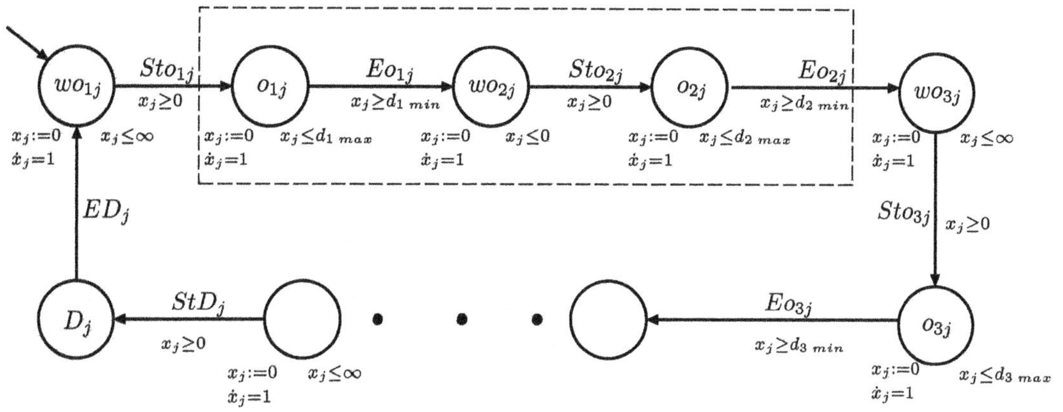

Fig. 2. A timed automaton model of reactor j

et al., 2000). A multirate TA is a tuple $\mathcal{T} = (L, L_0, \Sigma, X, I, E, c, \lambda)$, where

- L is a finite set of locations, and $L_0 \subseteq L$ is a set of initial locations,
- Σ is a finite set of labels (events),
- X is a finite set of clock variables,
- I is a map that associates some clock constraints in $\Phi(X)$ to each location, i.e. $I : L \to \Phi(X)$,
- $E \subseteq L \times \Sigma \times \Phi(X) \times L$ is a set of transitions,
- $c : L \times X \to \mathbb{R}$ is a function that defines the rate of change of each clock in a certain location. Thus, the clock dynamics in location l can be described by a simple differential equation $\dot{x}_i = c_{l,i} = const.$ If $c_{l,i}$ is equal to 1 for all indices l, i, we deal with the standard timed automaton.
- $\lambda : E \to 2^X$ associates to each transition a set of clocks to be reset.

Furthermore, we will use the standard definition of the product for timed automata (Alur and Dill, 1994). Let $A_1 = (L_1, L_{01}, \Sigma_1, X_1, I_1, E_1, f_1, \lambda_1)$ and $A_2 = (L_2, L_{02}, \Sigma_2, X_2, I_2, E_2, f_2, \lambda_2)$ be two timed automata. Assume that the clock sets X_1 and X_2 are disjoint. Then, the product, denoted $A_1||A_2$, is the timed automaton $(L, L_0, \Sigma, X, I, E, f, \lambda)$, where $L = L_1 \times L_2$, $L_0 = L_{01} \times L_{02}$, $\Sigma = \Sigma_1 \cup \Sigma_2$, $I = I_1 \cap I_2$ and $X = X_1 \cup X_2$. The transition structure E and the reset function λ are defined by the following rules:

(1) $\sigma \in \Sigma_1 \cap \Sigma_2$: for every $e_1 = (l_1, \sigma, \phi_1, l_1') \in E_1$ and $e_2 = (l_2, \sigma, \phi_2, l_2') \in E_2$, $e = ((l_1, l_2), \sigma, \phi_1 \wedge \phi_2, (l_1', l_2')) \in E$ and $\lambda(e) = \lambda_1(e_1) \cup \lambda_2(e_2)$.

(2) $\sigma \in \Sigma_1 \setminus \Sigma_2$: for every $e_1 = (l_1, \sigma, \phi_1, l_1') \in E_1$ and $l_2 \in L_2$, $e = ((l_1, l_2), \sigma, \phi_1, (l_1', l_2)) \in E$ and $\lambda(e) = \lambda_1(e_1)$.

(3) $\sigma \in \Sigma_2 \setminus \Sigma_1$: for every $e_2 = (l_2, \sigma, \phi_2, l_2') \in E_2$ and $l_1 \in L_1$, $e = ((l_1, l_2), \sigma, \phi_2, (l_1, l_2')) \in E$ and $\lambda(e) = \lambda_2(e_2)$.

4. TIMED AUTOMATON MODEL OF THE PLANT

4.1 Reactors

The first step is the modelling of the reactors using timed automata. Since the operation sequence is identical in each reactor, they can be described in a uniform way, as shown in Fig. 2.

There are two types of locations: wo_{ij} and o_{ij} which mean "wait before i-th operation starts in reactor j" and "i-th operation is active in reactor j". The events Sto_{ij} and Eo_{ij} denote start and end of the i-th operation in the j-th reactor, respectively. Fig.2 is to be interpreted as follows: in location o_{ij}, the progress of time is measured by a clock modelled by $\dot{x}_j = 1$, and the clock is reset to zero when the location is entered, i.e. when event Sto_{ij} occurs. We are only allowed to stay in the location if $x_j \leq d_{i\,max}$ holds (invariant). The event Eo_{ij} may only occur if $x_j \geq d_{i\,min}$ holds (guard). Hence, the transition between location o_{ij} to location $wo_{(i+1)j}$ has to happen when $d_{i\,min} \leq x_j \leq d_{i\,max}$. In location wo_{ij}, there are two possibilities: either invariant and guard of the outgoing transition enforce an immediate switch to the next location (an example for this case is location wo_{2j} in Fig.2), or invariant and guard allow an arbitrary stay within the location (an example for this case is location wo_{3j} in Fig.2). If a wo_{ij}-location is of the former type, $o_{(i-1)j}$ and o_{ij} belong to the same task (see the dashed box in Fig.2).

The last operation, denoted by D_j, is the discharging of reactor j. Note that the operation "discharging" always represents a task since the output tank can be interpreted as an external resource.

4.2 Resources

The next step is to model the restrictions on resource availability. The simplest way is to build a finite automaton for each resource R_i and the corresponding operations $o_{ij} = r_j^{-1}(R_i) \in O_j'$, $j = 1, \ldots, n$ as shown in Fig.3. The depicted timed automaton represents a simple rule: a resource sensitive operation can be simultaneously carried out in one reactor only.

To enforce uniqueness of the solution, a sequence of reactors is predetermined. As all reactors are equal, this does not restrict generality.

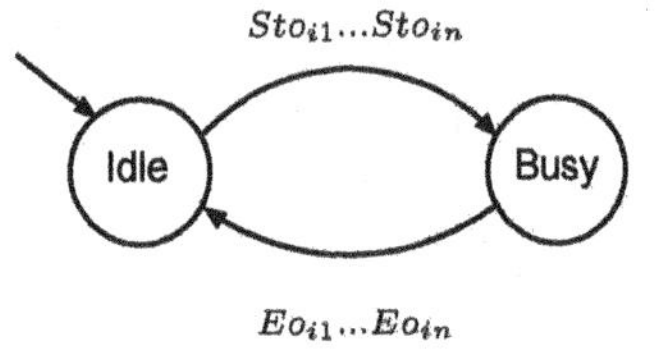

Fig. 3. A finite automaton modelling the availability of resource i

4.3 Output tank

The last element of the plant is the output tank. Its timed automaton model model is presented in Fig. 4. The transitions StD and ED denote "start discharging" and "discharging is finished". The clock variable v models the amount of liquid in the tank. Here, $a = F_{out,r} - F_{out,t}$ and $b = -F_{out,t}$, where $F_{out,r}$ is the volumetric rate of the flow from any reactor j to the tank, whereas $F_{out,t}$ is the volumetric rate of the output flow from the tank. If the value of v becomes too small or too big, the automaton goes to one of the locations modelling a forbidden situation.

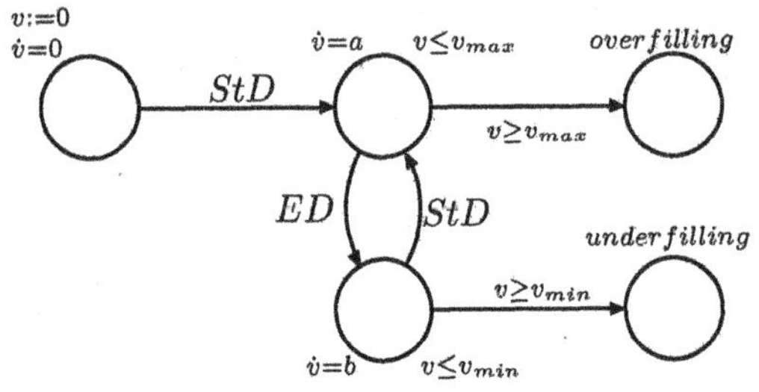

Fig. 4. A timed automaton modelling the output tank

5. CONTROL ISSUES

Two specifications have to be enforced: 1. the closed-loop system must be non-blocking; 2. the closed-loop system must be safe, i.e. the locations *underfilling* and *overfilling* must be rendered unreachable..

In the following we present several possible control strategies to enforce the specifications. Note that we can provide a formal guarantee for the specifications to hold, even if the actual design process contains some heuristics.

5.1 Non-blocking

It is obvious that the synchronous product of the n reactor models (Fig.2) and the k resource availability models (Fig.3) may give rise to blocking. This may happen if a resource, say R_i, is being allocated by an operation o_{ij} in reactor j, an operation $o_{(i-1)k}$ is finished in reactor k and operation o_{ik} belonging to the same task as $o_{i-1,k}$ attempts to allocate R_i. In this situation, o_{ik} must start at the same time as $o_{(i-1)k}$ finishes, which is clearly impossible as the corresponding resource is being used by another reactor. To avoid this situation, the start of tasks has to be delayed appropriately. This is being done by assigning one timed automaton to each task $K_i^l = \{o_{\mu j}^l\}$, $\mu = 1, 2, \ldots$. We assume that the index μ describes the temporal ordering of operations within the task (Fig.5). These automata are similar to the resource models, but with additionally introduced time constraints. The first location is added because in the beginning of the process the operation can start immediately.

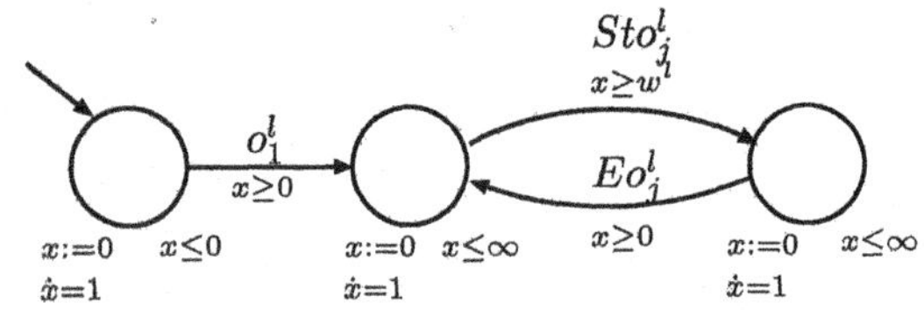

Fig. 5. A timed automaton modelling resource availability

The parameter w^l can be found as a solution of a simple scheduling problem. There is a fixed relation between the temporal position of an operation within the task and within the overall sequence of operations in the reactor. The latter is denoted by i and given by $i = \sum_{q=1}^{l-1} |K_j^q| + \mu$. The start and end times of $o_{\mu j}^l$ are $s_{\mu j}^l$ and $f_{\mu j}^l$ and its durations is $d_{\mu j}^l$. The goal is to ensure that

$$\min_{\mu : o_{\mu j}^l \in O_j'} (s_{\mu,j+1}^l - f_{\mu j}^l) = 0, \qquad (1)$$

where $s_{\mu 1}^l = s_{11}^l + \sum_{k=1}^{\mu-1} d_{k1}^l$, $f_{\mu 1}^l = s_{\mu 1}^l + d_{\mu j}^l$,

$s_{\mu,j}^l = f_{1(j-1)}^l + w^l + \sum_{k=2}^{\mu-1} d_{kj}^l$, $d_{\mu j}^l = d_{\mu}^l$, $j > 1$.

This can be done easily. The situation becomes more complicated if we suppose that the processing durations are only known unprecisely, i.e.,

$d^l_{\mu j} \in [d^*_i - \underline{d}_i; d^*_i + \overline{d}_i]$. Condition (1), then takes the form

$$\min_{\mu : o^l_{\mu j} \in O'_j} \quad \min_{d^l_{\mu j}, d^l_{\mu(j+1)} \in [d^*_i - \underline{d}_i; d^*_i + \overline{d}_i]} (s^l_{\mu(j+1)} - f^l_{\mu j}) = 0. \qquad (2)$$

The main difference, compared to the previous case is that the processing time of the first operation is a priori unknown. Thus, w^l is calculated in a worst-case fashion and is therefore, conservative.

5.2 Safety

In a first step, we form the synchronous product of the reactor models (Fig.2), the extended resource availability models (Fig.5) and the tank model (Fig.4). As we have previously determined suitable waiting times w^l, in this step only nonblocking schemes are considered. If all processing times are known, using standard verification procedures, we can check whether the locations "Overfilling" and "Underfilling" can be reached for a given fixed outflow rate. If yes, it is an easy exercise to suitably adjust the outflow rate.

The analysis shows that for fixed and physically reasonable processing times a fixed outflow can be determined such that safety is guaranteed.

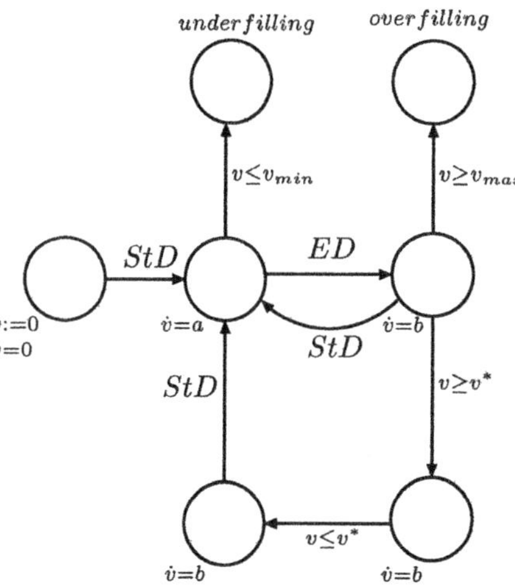

Fig. 6. A modified timed automaton of the output tank

If processing times are known imprecisely, one needs an additional degree of freedom. One possibility to achieve an additional degree of freedom is to introduce additional waiting times for the discharging of the reactors. This is illustrated in Fig.6. The figure shows that the start of a discharging process is only allowed if the liquid volume is below a certain threshold. In this way, overfilling can be avoided. To avoid underfilling, we need the maximal cycle duration $t_{c,max}$, i.e. the maximal time between two discharging operations in the same reactor provided the processing times of all operations take maximal values. $t_{c,max}$ can be obtained by applying verification procedures to the product of reactor models and resource availability models for uncertain processing times. Then, the outflow of the output tank can be set to

$$F_{out,t} = \frac{nF_{out,r}d_d}{t_{c,max}},$$

where d_d is the duration of the discharging operation.

An alternative is to switch the output rate online between several values $F_{out,ti}, i = 1, q$. In this case, the tank model has to be modified according to Fig.7.

To implement the necessary verification procedures we can use slightly modified versions of standard algorithms (see, e.g. (Pettersson, 1999; Bengtsson and Yi, 2004; Bozga *et al.*, 1998) and references therein). This procedure can be considered as the computation of the set of all reachable states of the timed automaton under consideration. Obviously, this set will consist of a set of locations and sets of clock valuations associated with these locations. Fortunately, these sets of clock valuations can be represented through the union of parallelograms on the space X, which makes the procedure computationally tractable.

6. EXAMPLE

We now consider the specific example described in detail in (Simeonova *et al.*, 2005). The plant consists of two reactors. In each reactor the following sequence of operations is performed: filling (d^*_1 = 0.17h), heating (d^*_2 = 0.45h), temperature regulation (d^*_3 = 3.44h), cooling (d^*_4 = 0.92h) and discharging (d^*_5 = 0.17h). The operations filling, heating and cooling are resource-sensitive. The set of operations has been partitioned into three tasks: K^1_j={filling}, K^2_j={heating, temperature regulation, cooling} and K^3_j={discharging}, $j = 1, 2$. The time for heating is only known imprecisely: $d_2 \in [d^*_2 - \underline{d}_2, d^*_2 + \overline{d}_2]$ where $\underline{d}_2 = \overline{d}_2 = 0.13$h. The minimal and maximal volume of liquid in the tank is $V_{min} = 0$ and $V_{max} = 50\text{m}^3$.

We applied the method presented in the previous section to obtain a solution which guarantees safety and nonblocking for all possible variations of parameters. For example, the particular schedule for the worst case $d_2 = d^*_2 + \overline{d}_2$ is shown in Fig.8.

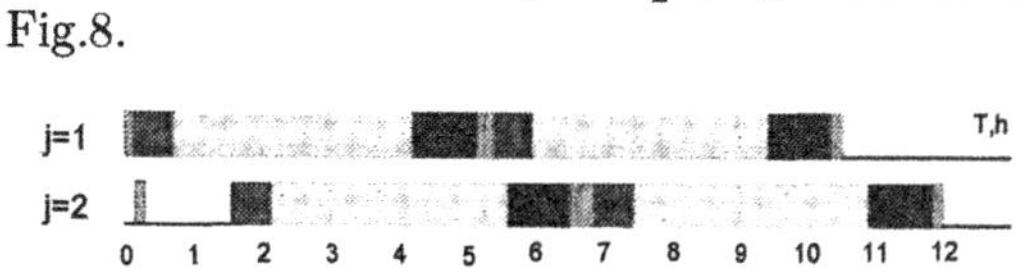

Fig. 8. Resulting schedule for $d_{2j} = d^*_2 + \overline{d}_2$, $j = 1, 2$

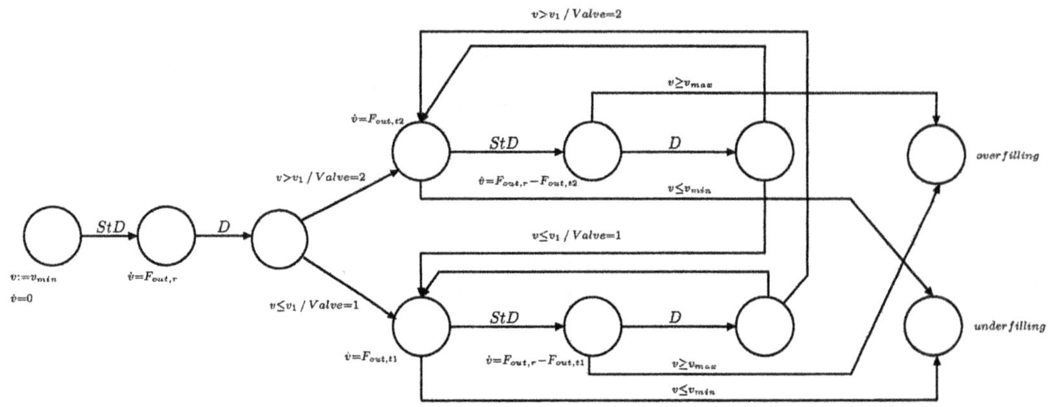

Fig. 7. A timed automaton modelling the output tank

The maximal admissible constant tank outflow is $F_{out,t} = 10.23\text{m}^3/\text{h}$. This results in the change of the liquid volume in the tank as shown in Fig.9.

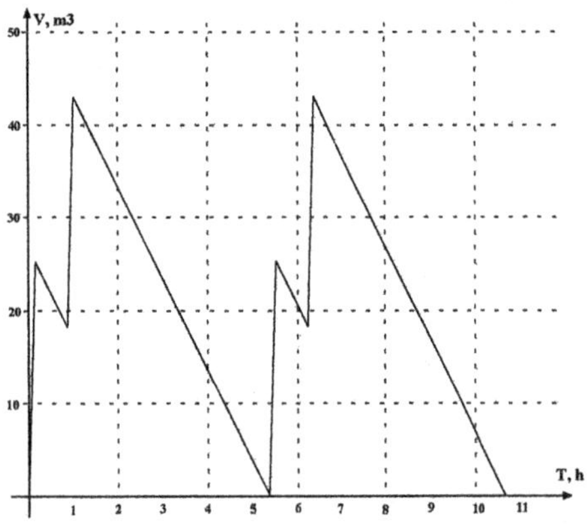

Fig. 9. Liquid volume in the tank

7. CONCLUSIONS AND FUTURE WORK

In this contribution, we investigated the use of timed automata for the scheduling of a class of parallel production lines. We have addressed the case when uncertainties regarding certain operating times are present. Although the prescribed approach contains heuristic elements in the design procedure, we can guarantee that non-blocking and safety are obtained. We have applied this procedure to a specific process which has been suggested as a benchmark problem within the EU Network of excellence HYCON.

REFERENCES

Alur, R. and D.L. Dill (1994). A theory of timed automata. *Theoretical Computer Science* **126**, 183–235.

Alur, R., T.A. Henzinger, G. Lafferriere and G.J. Pappas (2000). Discrete abstractions of hybrid systems. *Proceedings of the IEEE* **88**(7), 971–984.

Asarin, E., O. Maler and A. Pnueli (1995). Symbolic controller synthesis for discrete and timed systems. In: *Hybrid Systems II*. pp. 1–20. LNCS 999. Springer.

Bengtsson, J. and W. Yi (2004). Timed automata: Semantics, algorithms and tools. In: *Lecture Notes on Concurrency and Petri Nets* (W. Reisig and G. Rozenberg, Eds.). LNCS 3098. Springer–Verlag.

Bozga, M., C. Daws, O. Maler, A. Olivero, S. Tripakis and S. Yovine (1998). Kronos: A model-checking tool for real-time systems.. In: *Computer Aided Verification, CAV '98, Vancouver, Canada* (A.J. Hu and M.Y. Vardi, Eds.). LNCS 1427. Springer. pp. 546–550.

Pettersson, P. (1999). Modelling and Verification of Real-Time Systems Using Timed Automata: Theory and Practice. PhD thesis. Uppsala University.

Simeonova, I., F. Warichet, G. Bastin, D. Dochain and Y. Pochet (2005). On line scheduling of chemical plants with parallel production lines and shared resources : a feedback implementation. In: *Proceedings IMACS World Congress, Paris*.

ELSEVIER
IFAC
PUBLICATIONS

DYNAMIC OPTIMIZATION OF AN INDUSTRIAL EVAPORATOR USING GRAPH SEARCH WITH EMBEDDED NONLINEAR PROGRAMMING

Christian Sonntag [*,1] **Olaf Stursberg** [**]
Sebastian Engell [*]

[*] *Process Control Laboratory (BCI-AST)*
University of Dortmund, 44221 Dortmund, Germany.
[**] *Industrial Automation Systems (EI-LSR)*
Technical University of Munich, 80290 Munich, Germany.

Abstract: In this paper, we consider the task to start the operation of an industrial evaporation system. Rigorous modelling gives rise to a hybrid automaton with large nonlinear DAE-models that describe the continuous evolution in the discrete locations. The optimization problem is solved by a hierarchical procedure that consists of a branch-and-bound algorithm with embedded nonlinear dynamic optimization over a finite look-ahead horizon. Important elements of the algorithm are the introduction of a dynamic choice of the time intervals over which the controls are constant and of tailored penalty functions in order to obtain solutions which are close to infeasible trajectories. *Copyright © 2006 IFAC*

Keywords: Automaton, hybrid systems, large-scale models, nonlinear programming, optimal control.

1. INTRODUCTION

Chemical processing systems usually exhibit nonlinear and, in case of changes of the physical state of the substances (e.g. from liquid to vapour), switched dynamics. In addition, the states and the inputs are constrained, and a considerable number of actuated inputs usually is of discrete nature, e.g. valves which can be either open or closed. Besides continuous feedback controllers, there are logic (switching) controls that establish sequential procedures and initiate exception routines if malfunctions of sensors or actuators or serious disturbances occur. An important function of logic controls is for example the safe shut-down of the plant in case of dangerous situations. In order to describe such automated processing plants, hybrid models with nonlinear dynamics are appropriate. However, the computation of optimal controls for such models is very challenging due to the usually large number of discrete degrees of freedom in connection with the nonlinear dynamics. As an example, the hybrid model of a simplified industrial evaporation system is described in this paper, and the task of optimal start-up of this plant is formulated. The evaporator is a simplified version of a benchmark example that was developed within the EU Network of Excellence HyCon in cooperation with an industrial partner.

Due to the complexity of the hybrid model, the application of techniques which evaluate the optimality conditions, see e.g. (Sussmann, 1999; Bran-

---

[1] Corresponding author: c.sonntag@bci.uni-dortmund.de.
This work is supported by the EU-funded NoE HyCon and the Graduate School of *Production Engineering and Logistics* at University of Dortmund.

icky *et al.*, 1998; Shaikh and Caines, 2003), seems not very promising. It is preferable instead to either employ techniques based on simpler approximating models , e.g. (Bemporad *et al.*, 2000; Lincoln and Rantzer, 2001; Stein *et al.*, 2004), or to use numeric methods for optimizing the original dynamics such that fast convergence to a good solution is obtained, e.g. (Buss *et al.*, 2000; Barton and Lee, 2002). The approach for solving the optimal control problem for the evaporator presented in this paper is based on the technique introduced in (Stursberg, 2004*a*; Stursberg, 2004*b*). It consists of a graph search algorithm that fixes discrete degrees of freedom, and embedded nonlinear programming (NLP) is used to select locally optimal continuous inputs. The hybrid dynamics is evaluated within the NLP by hybrid simulation. For the evaporator, however, this algorithm does not produce satisfying result without further tuning – hence, two modifications are introduced here: the first determines a time discretization for selecting the inputs based on the progress in the state space, and the second introduces specific penalty functions that allow for trajectories of low costs which are close to the boundary of the feasible state set.

2. A HYBRID MODEL OF THE EVAPORATOR

In the processing industries, liquids are often concentrated by evaporating volatile solvents such that non-volatile components (often the products) are enriched in the liquid phase. Fig. 1 shows a scheme of the industrial-scale evaporation system considered in this paper (as a particular instance of the multi-stage system in (Sonntag and Stursberg, 2005)). During start-up, the initially empty evaporation vessel (A) is filled with cold liquid feed consisting of a non-volatile product, water, and alcohol. The heat supply to the vessel is realized by condensing hot steam in a heat exchanger (B). When the liquid starts to boil, vapor can be drained from the top of the vessel. If the product concentration meets desired purity requirements, the product can be continuously drained from the bottom of the vessel. The operation is controlled using four valves and two pumps (C). While the vapor drain and the liquid feed can be adjusted continuously by V_{V1} and V_1, the valves controlling the inflow of hot steam (V_{V2}) and the product drain (V_2) are switched between two discrete settings.

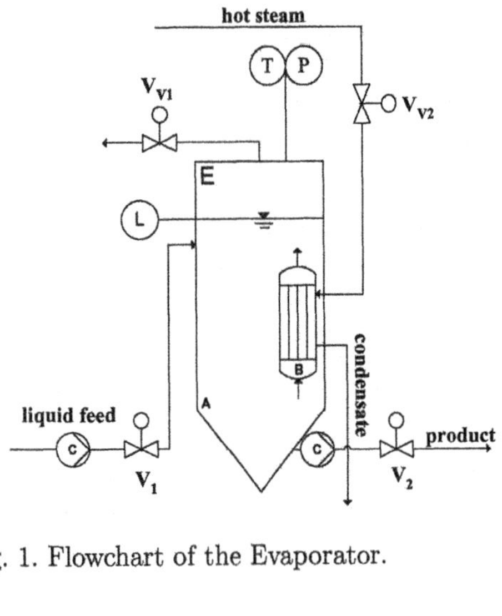

Fig. 1. Flowchart of the Evaporator.

2.1 Hybrid Automaton with DAE Dynamics

To model the evaporator, we extend the hybrid automaton specified in (Stursberg, 2004*a*) by an input availability mapping and continuous dynamics given as DAEs:

Definition 1. A hybrid automaton with DAE dynamics $HA = (Z, z_0, X, x_0, Y, U, V, \psi, inv, \Theta, g, j, f, l)$ consists of:

- the finite set of *locations* $Z = \{z_1, \ldots, z_{n_z}\}$;
- an *initial location* $z_0 \in Z$;
- the *continuous state space* $X \subseteq \mathbb{R}^n$ on which the vector of (differential and algebraic) *continuous variables* $x(t) \in X$ is defined (for simplicity of notation, we do not distinguish between variables and their evaluations here);
- an initialization of the continuous variables $x_0 \subseteq X$;
- the *continuous output space* $Y \subseteq \mathbb{R}^m$ on which the vector of *continuous output variables* $y(t) \in Y$ is defined;
- the *space of continuous inputs* $U = [u_1^-, u_1^+] \times \ldots \times [u_{n_u}^-, u_{n_u}^+]$ with $u_j^-, u_j^+ \in \mathbb{R}$, and the continuous inputs are denoted by $u(t) \in U$;
- a finite set of discrete inputs $V = \{v_1, \ldots, v_{n_d}\}$ with discrete inputs $v(t) \in V$ for which $v_j \in \mathbb{R}^{n_v}$;
- the *input availability mapping* $\psi : Z \to \mathcal{U} \times \mathcal{V}$, $\mathcal{U} \subseteq U$, $\mathcal{V} \subseteq V$ that assigns sets of available continuous and discrete inputs to each location $z \in Z$;
- the invariant function $inv : Z \to 2^X$ which assigns an *invariant set* $inv(z) \subseteq X$ to each $z \in Z$; we require that $x_0 \in inv(z_0)$;
- the set of *discrete transitions* $\Theta \subseteq Z \times Z$, and each transition is a pair $\theta = (z_i, z_j) \in \Theta$;
- the function $g : \Theta \to 2^X$ that associates a *guard set* $g(\theta) \subseteq X$ with each transition $\theta \in \Theta$; for each $z \in Z$, the guard sets of all transitions $\theta = (z, \bullet)$ are required to be pairwise disjoint;
- the *jump function* $j : \Theta \times X \to X$ which assigns an update $j(\theta, x) \in X$ of the continuous variables to each $\theta \in \Theta$ and $x \in g(\theta)$;
- the *flow function* $f : Z \times X \times U \times V \to \mathbb{R}^n$ that determines the continuous evolution in each

location $z \in Z$. For given input trajectories $u(t)$ and $v(t)$ and a location $z \in Z$, we assume that a unique solution of $M\dot{x}(t) = f(z, x(t), u(t), v(t))$ exists for an initialization $x(0) \in inv(z)$, $t \in \mathbb{R}^{\geq 0}$, and $M \in \mathbb{R}^{n \times n}$, $M_{ij} = 0 \; \forall \; i \neq j$; $i, j \in \{1, \ldots, n\}$; note that f constitutes a system of semi-explicit differential-algebraic equations (DAEs) if M is not regular;

- and the *output function* $l : X \rightarrow Y$ which uniquely maps the evaluation of the continuous variables to output variables $y(t) \in Y$. ◇

Let $\Sigma = \bigcup_{z \in Z} \bigcup_{x \in inv(z)} (z, x)$ denote the set of valid *hybrid states* $\sigma = (z, x)$ of HA with $z \in Z$, $x \in inv(z)$. $T = \{t_0, t_1, t_2, \ldots\}$ is the ordered set of time points $t_k \in \mathbb{R}^{\geq 0}$ which contains the initial time $t_0 = 0$ and all points of time at which an input changes or a transition occurs. For $t_k \in T$, the hybrid states, inputs, and outputs are written as: $\sigma_k := (z_k, x_k)$, $u_k := u(t_k)$, $v_k := v(t_k)$ with $(u_k, v_k) \in \psi(z_k)$, and $y_k := y(t_k)$.

A continuous input trajectory defined on T is a sequence $\phi_u = (u_0, u_1, u_2, \ldots)$, and a discrete input trajectory is given by $\phi_v = (v_0, v_1, v_2, \ldots)$, meaning that u_k and v_k are piecewise constant on $[t_k, t_{k+1}[$. For given ϕ_u and ϕ_v, a sequence $\phi_\sigma = (\sigma_0, \sigma_1, \sigma_2, \ldots)$ of hybrid states $\sigma_k = \sigma(t_k) = (z_k, x_k)$ is a deterministic *feasible run* of HA iff:

- $\sigma_0 = (z_0, x_0)$, with $x_0 \in inv(z_0)$, and
- $\sigma_{i+1} = (z_{i+1}, x_{i+1}) \in \Sigma$ follows from $\sigma_i = (z_i, x_i) \in \Sigma$ according to:
  - $\overline{x}_{i+1} = x(t_{i+1})$ is the continuous state obtained from solving $M\dot{x}(t) = f(z_i, x(t), u(t), v(t))$ for $t \in [t_i, t_{i+1}]$ starting from $x_i = x(t_i)$, and $x(t) \in inv(z_i)$ must apply for all $t \in [t_i, t_{i+1}]$, and $x(t) \notin g(\theta)$ for $t \in [t_i, t_{i+1}[$ and for all $\theta = (z_i, \bullet)$.
  - if $\overline{x}_{i+1} \in g(\theta)$, $\theta = (z_i, z')$: $z_{i+1} := z'$, $x_{i+1} := j(\theta, \overline{x}_{i+1}) \in inv(z_{i+1})$; else $x_{i+1} := \overline{x}_{i+1}$, $z_{i+1} := z_i$. ◇

2.2 HA Model for the Evaporator

Since product can be drained from the vessel only if the target region (see Sec. 3) is reached, it is sufficient to introduce four locations, i.e. $Z = \{z_{NE}, z_{NET}, z_E, z_{ET}\}$, where 'E' stands for evaporating, 'NE' for not evaporating, and 'T' for target. The continuous dynamics of all locations is modeled by a DAE system with four differential equations (modeling the the total masses of the three components m_1 (product), m_2 (water), and m_3 (alcohol), and the total inner energy U), and 13 algebraic equations that were derived using the assumption that the system is in thermodynamical equilibrium during evaporation. The algebraic variables are the mass fractions of the three components in the liquid phase (w_1, w_2, w_3) and in the vapor phase (ξ_1, ξ_2, ξ_3), the temperature in the evaporator T, the total mass of the liquid m_{liq}, the volume of the vapor phase V_{vap}, the energy transfer between the heat exchanger and the evaporator $\dot{Q}$, the mean temperature T_m and the pressure P_{HE} in the heat exchanger, and the flow rate of hot steam into the heat exchanger $\dot{F}_{HE}$. If the system is in location z_{NE}, the variables ξ_1, ξ_2, ξ_3, and V_{vap} are set to zero since a vapor phase does not exist. As an example for the high degree of complexity and nonlinearity of the DAE-system, the equation describing the equilibrium conditions for the mass of the alcoholic component looks like:

$$m_3 = m_{liq} \cdot w_3 + \frac{w_3 \cdot P_3^0(T) \cdot V_{vap}}{R \cdot T \cdot \left[\frac{w_1}{M_1} + \frac{w_2}{M_2} + \frac{w_3}{M_3}\right]} \quad (1)$$

with the universal gas constant R, the molecular weights M_1, M_2, and M_3, and $P_3^0(T)$ is a third-order polynomial in T.

Using the justifiable assumption that the liquid mass m_{liq} and the mass fractions w_1, w_2, and w_3 remain constant for any transition θ, all hybrid states $\sigma = (z, x)$ of the evaporator are uniquely determined by a reduced vector:

$$x_{red} = [w_1, w_2, T, L, P]. \quad (2)$$

The output variables L and P, which can be uniquely determined from the state variables, represent the liquid level and the pressure in the evaporator [2]. While L follows from the mass and density of the liquid, P is given by:

$$P = \frac{\frac{w_1}{M_1} \cdot P_1^0(T) + \frac{w_2}{M_2} \cdot P_2^0(T) + \frac{w_3}{M_3} \cdot P_3^0(T)}{\frac{w_1}{M_1} + \frac{w_2}{M_2} + \frac{w_3}{M_3}}, \quad (3)$$

where $P_1^0(T)$, $P_2^0(T)$, and $P_3^0(T)$ are third-order polynomials in T.

A bounded and reduced continuous state space can be defined as:

$$X_{red} = [0, 0.98] \times [0, 1] \times [300, 440] \times [0, 100] \times [0, 5], \quad (4)$$

and $x_{red} \in X_{red}$. (Units are omitted for abbreviation.) During start-up, the upper bounds given in Eq. 4 must not be exceeded for safety reasons.

The continuous inputs u_1 and u_2 represent the settings of the valves V_{V1} and V_1, and are defined on the range [0 %, 100 %], where 0 % means completely closed. For all $z \in Z$, both continuous inputs are available. A discrete input is defined as the vector $v_j = (V_2, V_{V2})$ and the discrete input space by $V = \{(0\,\%, 80\,\%), (0\,\%, 100\,\%), (11.5\,\%, 80\,\%), (11.5\,\%, 100\,\%)\}$. If $z \in \{z_{NE}, z_E\}$, the product drain valve must remain closed, such

[2] The dynamics of the process is nevertheless determined by the higher-order DAE-system since the latter cannot be solved explicitly for x_{red}, i.e. the use of x_{red} does not reduce the model complexity.

that the available discrete inputs are reduced to $\mathcal{V}_{z_{NE},z_E} = \{(0\,\%, 80\,\%), (0\,\%, 100\,\%)\}$. Otherwise, all discrete inputs are available.

Between all pairs of locations, transitions in both directions can occur, i.e. $|\Theta| = 12$. The transitions can be divided into three classes: (a) transitions occurring if the product concentration reaches or leaves the target region (see Sec. 3), (b) transitions occurring when the liquid begins or stops to evaporate, and (c) transitions for which both is true. While the guard condition for (a) only depends on the current value of w_1, the guard set for (b) is assumed to be given by $P = 0.4$, with P defined according to Eq. 3. The guard for (c) is the conjunction of the two previous ones. The invariants of the locations are bounded by the set of states defined by the guard conditions and the boundaries in Eq. 4. Although the state variables ξ_1, ξ_2, ξ_3, and V_{vap} are reset with discrete transitions of the hybrid model, the variables in x_{red} are not affected due to the modeling assumptions described above.

3. THE OPTIMAL CONTROL PROBLEM

The following type of optimal control problem is considered in this paper: Given are an initial state $\sigma_0 \in \Sigma$ and a target set $\Sigma_t \subset \Sigma$ with $\Sigma_t = \{(z_t, x) | \exists_1 z_t \in Z : x \in X_t \subset inv(z_t)\}$. It is assumed that the ordered set of time points $T = \{t_0, t_1, t_2, \ldots, t_f\}$ is finite, and that the continuous and discrete inputs can only be changed at $t_k \in T_s \subset T$, while ϕ_σ remains defined on T. The set $\Phi_{u,s}$ contains all possible continuous input trajectories $\phi_u = (u_0, u_1, u_2, \ldots)$ defined on T_s, and $\Phi_{v,s}$ contains all possible $\phi_v = (v_0, v_1, v_2, \ldots)$. The control task is then to determine input trajectories ϕ_u^* and ϕ_v^* that lead to a feasible run ϕ_σ^* of HA from σ_0 into Σ_t such that a cost function Ω is minimized:

$$\min_{\phi_u \in \Phi_{u,s}, \phi_v \in \Phi_{v,s}} \Omega\,(t_f, \phi_\sigma) \tag{5}$$
$$s.t.\ \phi_\sigma = (\sigma_0, \ldots, \sigma_f) \text{ with } \sigma_0 = (z_0, x_0),$$
$$\sigma_f := (z(t_f), x(t_f)) \in \Sigma_t.$$

The initial state of the evaporator is given by $x_{0,red} = [0.12, 0.85, 327, 1, 0.282]$, denoting a state with very low product concentration, low level, and no evaporation ($z_0 := z_{NE}$). The control task is to drive the system into the location $z_t = z_{ET}$ and the target region $X_{t,red} = [0.8, 0.84] \times [0.16, 0.2] \times [370, 420] \times [60, 64] \times [0.5, 4]$ in a time-optimal fashion.

Using the cost function $\Omega = t_f$ in (5) and optimizing over the complete time horizon $[t_0, t_f]$ is often computationally intractable since $|T_s|$ choices for u and v lead to an exponential growth of the solution space with increasing $|T_s|$. In this case, a substitute for Ω may be chosen which allows for an appropriate cost evaluation of trajectories also over shorter time horizons. A possible choice is a cost function which combines t_f with a notion of distance of any intermediate state σ_k to the target region Σ_t. It was found for the evaporator, however, that this choice for Ω does provide only solutions for which t_f is somewhat worse than the optimal value. Reasoning about the expected behavior of the system led to the following heuristically-chosen cost function:

$$\Omega(x_{s,red}) = \begin{cases} \alpha \cdot |w_{s,1} - w_{s,1,t}| + \beta \cdot |L_s - L_{s,max}| \\ \quad \text{if } |w_{s,1} - w_{s,1,t}| > 0.09, \\ \beta \cdot |w_{s,1} - w_{s,1,t}| + \alpha \cdot |L_s - L_{s,t}| \\ \quad \text{if } |w_{s,1} - w_{s,1,t}| \leq 0.09. \end{cases} \tag{6}$$

Note that before evaluating $\Omega(x_{s,red})$, all variables of x_{red} are scaled to the range $[0,1]$ according to $x_{s,red} = D^{-1} \cdot x_{red} - c$, with $D = \text{diag}(0.98, 1, 140, 100, 5)$ and $c = [0, 0, 2.1429, 0, 0]^T$. In Eq. 6, $w_{s,1}$ is the scaled product concentration, $w_{s,1,t} = 0.804$ the scaled target concentration, L_s the scaled liquid level, $L_{s,max} = 1$ the upper bound for L_s, $L_{s,t} = 0.62$ is the scaled target level, and α and β denote weighting constants. The motivation for this choice of $\Omega(x_{s,red})$ is that the fastest way to increase $w_{s,1}$ should be to maximize the energy transfer $\dot{Q}$ between the heat exchanger and the evaporation vessel (and thus, the evaporation rate of the volatile components). The energy transfer increases with the difference between the liquid temperature T and the temperature of the hot steam. If the liquid level L is held close to its maximum value, a minimal temperature of the liquid is achieved. After the product concentration has reached a threshold which is close to its target value, the level has to be driven into the target region. In order to push the increase of $w_{s,1}$ which exhibits slow dynamics, the constant weights were chosen according to $\frac{\alpha}{\beta} = 4$ if $|w_{s,1} - w_{s,1,t}| > 0.09$. For $|w_{s,1} - w_{s,1,t}| \leq 0.09$, the weights were chosen according to $\frac{\beta}{\alpha} = 4$ to put the focus on driving L_s into the target.

4. THE OPTIMIZATION APPROACH

The approach used here for optimizing the start-up of the evaporator is based on the graph search algorithm presented in (Stursberg, 2004*a*; Stursberg, 2004*b*). This section first reviews the basic principle, and then describes necessary modifications to solve the case study.

4.1 Graph Search with Embedded NLP

The main idea of the approach is to separate the optimization of the discrete and continuous degrees of freedom by encoding the discrete choices in an acyclic graph. For each node of the graph, optimal values for the continuous degrees of freedom are determined using nonlinear programming in which numerical simulation of the hybrid model over a constant time horizon is employed to evaluate the cost for the corresponding evolution of the system. Each node of the graph is characterized by a structure $n = (\phi_\sigma, \phi_u, \phi_v, c_a, c_p)$ which contains the state (ϕ_σ) and input (ϕ_u, ϕ_v) trajectories by which the hybrid state $\sigma_k = (z_k, x_k)$ assigned to the node was reached, the cost c_a that was accumulated on the path into σ_k, and, to determine the most promising nodes for further investigation, a prediction c_p of the cost for the remaining path into the target region. Two different techniques are used to prune the search graph: upper bounds of the accumulated cost are iteratively determined to remove branches that will lead to provably inferior solutions. Furthermore, if the hybrid states of two nodes are in close neighborhood, a cost comparison criterion is applied to prune the locally inferior node. The neighborhood of the hybrid state (z_k, x_k) of a node is modeled using an ellipsoidal set defined by

$$(x - x_k)^T \cdot P_v \cdot (x - x_k) \leq \epsilon, \qquad (7)$$

with a small ϵ.

4.2 Progress-Dependent Simulation Times

Due to the large variations of the gradients over X, and in particular due to the fact that $|\dot{L}| \gg |\dot{w}_1|$, the use of a constant simulation time is a bad choice for the evaporator. For example, large time periods drive L outside the permitted range for most inputs, while such steps are necessary to obtain the required change in w_1. Rather than using a fixed time to advance between nodes, we employ a criterion for terminating the simulation that is bound to the progress in X. If $\sigma_k = (z_k, x_k)$ is the initial hybrid state, the simulation is terminated if the continuous state trajectory reaches the boundary of a hyper-ellipsoid[3] defined by $(x - x_k)^T \cdot P_p \cdot (x - x_k) = r^2$.

4.3 Exclusion of Infeasible State Trajectories

As discussed in Sec. 3, $L_s(t)$ should be held close to 100 % for parts of the start-up procedure. In order to permit this behavior while preventing the generation of infeasible state trajectories ($L_s(t) >$ 1 would terminate the simulation), the NLP step is modified as sketched in Fig. 2: It shows two state trajectories leading into the neighborhood of an invariant boundary. If $x_{red,k}$ is reached

[3] In the general case, this criterion is used only within $inv(z_k)$; for the evaporator, however, it can be applied even for transitions occurring within $[t_k, t_{k+1}]$ for $t_k, t_{k+1} \in T_s$.

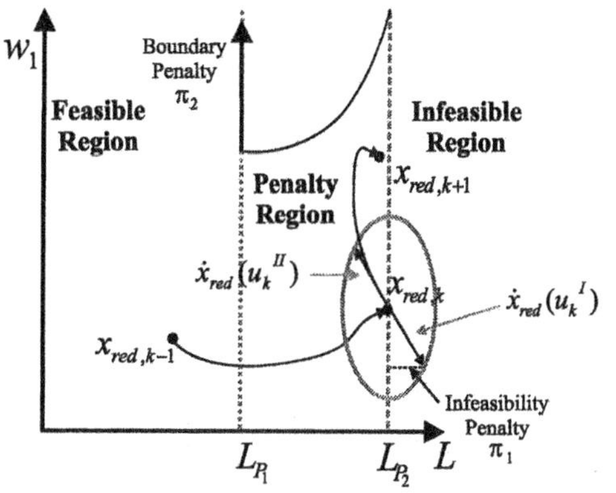

Fig. 2. Penalty computation for hybrid states near to invariant boundaries.

from $x_{red,k-1}$ during simulation, the latter is stopped and a new node is created. When the NLP step is then carried out for $x_{red,k}$, it is likely that the NLP solver uses a guess for u_k which leads into the infeasible region (marked as u_k^I in Fig. 2). In this case, the simulation would immediately stop again, and the solver would lack the information about the performance of this guess. To avoid this problem, the point $x_{red,k}$ is first projected onto an ellipsoid surrounding $x_{red,k}$ by using a numerical approximation of the gradient $\dot{x}_{red}(z, x_{red,k}, u_k^I, v_k)$. The distance of the resulting point to the boundary of the invariant then constitutes an infeasibility penalty π_1. As a result, the NLP solver would prefer a choice for u_k, for which $\dot{x}_{red}$ points in direction of the feasible region; then $\pi_1 := 0$. In Fig. 2, the choice u_k^{II} leads to $\dot{x}_{red}(z, x_{red,k}, u_k^{II}, v_k)$ and a feasible next state $x_{red,k+1}$. As an additional penalty, states close to the boundary of $inv(z_k)$ are penalized by a function $\pi_2 = a \cdot (\frac{L(t) - L_{P_1}}{L_{P_2} - L_{P_1}})^b$ with positive constants a and b (see Fig. 2). The two penalties π_1 and π_2 are only used within the NLP step, but not to assess the nodes within the search tree.

5. OPTIMIZATION RESULTS

Fig. 3 shows a projection of the optimization result for the start-up of the evaporator into the (w_1, T, L)-space. Nodes that have been explored are marked by an 'x', and pruned nodes are shown as a '+' (neighborhood) or an 'o' (cost). The solid line represents the state trajectory of the best solution found. The following parameterization of the search algorithm was used: The neighborhood ellipsoids (according to Eq. 7) were parameterized as $P_v = \text{diag}(10^6, 10^6, 1, 100, 4 \cdot 10^4)$, $\epsilon = 1$, and the ellipsoid for the progress criterion were chosen small in the directions of w_1 and w_2, i.e. $P_p = \text{diag}(1600, 1600, 0.01, 0.0025, 100)$, $r = 2$. Fig. 4 shows the continuous input trajectory corresponding to the best found solutions: The inputs are determined by the algorithm such that the vapor drain through V_{V1} is maximized while L is held at 100 % using appropriate settings for V_1. Towards the end, V_1 is closed to drive the liquid level into the target region. V_2 is always closed since w_1 is the last variable to reach the target region, and V_{V2} remains open. During the optimization, 600 nodes were investigated, and 96 solutions were found. The complete optimization run took around 6 hours on a PC with Pentium-IV 2.8 GHz. The best solution was determined after 746 seconds, and using the input trajectories defined by this solution, the target region is reached after 13937 seconds.

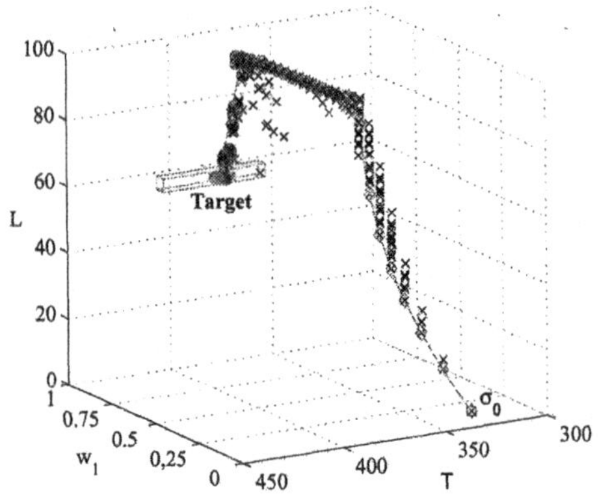

Fig. 3. Explored nodes and the state trajectory of the best solution.

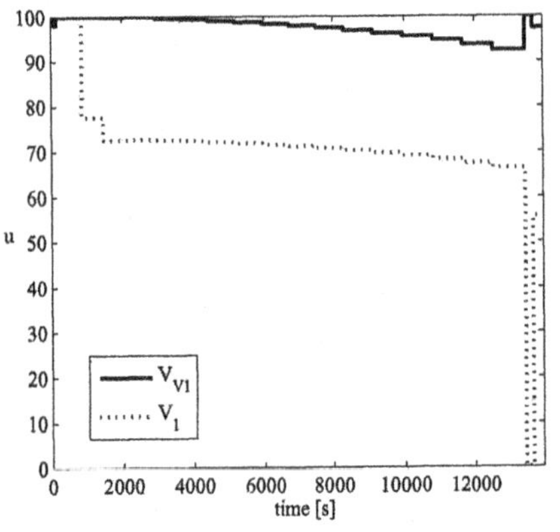

Fig. 4. Continuous input trajectories for the best solution.

6. CONCLUSIONS

The problem of optimizing the start-up of a hybrid model of an industrial-scale evaporator with complex nonlinear continuous dynamics was tackled using a graph search algorithm with embedded nonlinear programming. To obtain feasible and good solutions with the algorithm, two modifications had to be introduced: (a) a state-dependent termination criterion for the embedded hybrid simulation, and (b) a modified cost function for the NLP step to rule out infeasible evolutions while allowing optimization near or on the boundary of invariants.

The next steps of this research are as follows: We try to develop schemes that are able to find feasible solutions for similarly challenging problems without the requirement of first determining specifically tuned cost functions (as described in Sec. 3). In addition, we aim at extending the solution concept presented here to the multi-stage evaporator as described in (Sonntag and Stursberg, 2005).

REFERENCES

Barton, P.I. and C.K. Lee (2002). Modeling, simulation, sensitivity analysis and optimization of hybrid systems. *ACM Trans. Modeling and Comp. Simulation* **12**(4), 256–289.

Bemporad, A., N. A. Bozinis, V. Dua, M. Morari and E. N. Pistikopoulos (2000). Model predictive control: A multiparametric programming approach. In: 10th *ESCAPE Symposium*. pp. 301–306.

Branicky, M. S., V. S. Borkar and S. K. Mitter (1998). A unified framework for hybrid control: Model and optimal control theory. *IEEE Trans. Automatic Control* **43**(1), 31–45.

Buss, M., O. von Stryk, R. Bulirsch and G. Schmidt (2000). Towards hybrid optimal control. *Automatisierungstechnik* **9**, 448–459.

Lincoln, B. and A. Rantzer (2001). Optimizing linear system switching. In: 40th *IEEE CDC*. pp. 2063–2068.

Shaikh, M.S. and P.E. Caines (2003). On the optimal control of hybrid systems. In: *Hybrid Systems: Comp. and Control*. Vol. 2623 of *LNCS*. Springer. pp. 466–481.

Sonntag, C. and O. Stursberg (2005). Optimally controlled start-up of a multi-stage evaporation system. Technical report for the network of excellence HyCon. University of Dortmund.

Stein, O., J. Oldenburg and W. Marquardt (2004). Continuous reformulations of discrete - continuous optimization problems. *Comp. and Chemical Eng.* **28**(10), 1951–1966.

Stursberg, O. (2004*a*). Dynamic optimization of processing systems with mixed degrees of freedom. In: *Proc.* 7th *Int. IFAC Symp. Dyn. and Control of Process Systems*. number 164.

Stursberg, O. (2004*b*). A graph-search algorithm for optimal control of hybrid systems. In: 43rd *IEEE CDC*. pp. 1412–1417.

Sussmann, H.J. (1999). A maximum principle for hybrid optimal control problems. In: 38th *IEEE CDC*. pp. 425–430.

ELSEVIER
IFAC
PUBLICATIONS

USING NEURAL NETWORKS FOR THE IDENTIFICATION OF A CLASS OF HYBRID DYNAMIC SYSTEMS

N. Messai, J. Zaytoon and B. Riera

CReSTIC, Université de Reims Champagne-Ardenne,
Moulin de la Housse BP 1039, 51687 REIMS cedex 2 France
e -mail : {nadhir.messai; janan.zaytoon, bernard.riera}@univ-reims.fr

Abstract: This paper addresses the problem of the identification of Hybrid Dynamic System (HDS) by focusing the attention on the identification of a global model that predicts the continuous outputs of the HDS. The proposed approach considers the identification of HDS in terms of the architectures and the learning algorithms developed for Feed-Forward neural networks. *Copyright © 2006 IFAC*

Keywords: Neural Network, Hybrid Systems, Identification.

1. INTRODUCTION

Over the past years, the study of Hybrid Dynamic systems (HDS), which combines continuous and discrete dynamics, has attracted increased attention. Most works that deal with the control, the analysis and the diagnosis of hybrid systems are based on the assumption that the model of the hybrid system is available (Branicky, 1996, Coquempot *et al.,* 2004). Hence, it seems that little attention has been paid to the problem of obtaining a model from a given input-output data generated by a hybrid system (Bemporad *et al.*, 2001; Bemporad *et al.,* 2004, Ferrari-Trecate *et al.*, 2003 Hoffmann and Engell, 1998; Juloski *et al.*, 2004; Simon and Engell, 2001; Vidal *et al.*, 2004).

Early work on the identification of hybrid systems employ statistical methods to detect the discrete change of the continuous dynamics (Hoffmann and Engell, 1998). Based on this work, Simon and Engell (2001) decompose the I/O data obtained by hybrid linear systems on wavelets and characterize the discrete events (i.e. the switching) as maxima in the coefficients of the transform. Once, the switching points are detected, the set of data is partitioned into sections in which a constant set of model parameters is estimated to describe the continuous dynamics.

Most of the proposed approaches for the identification of HDS concern the classes of switched linear and PieceWise Affine system. They can be classified into the Mixed_Integer Programming approach (Bemporad *et al.*, 2001), the clustering-based approach (Ferrari-Trecate *et al.,* 2003), the bounded-error approach (Bemporad *et al.*, 2004), the Bayesian approach (Juloski *et al.*, 2004) and the algebraic approach (Vidal *et al.*, 2004).

In (Bemporad *et al.*, 2001), it is shown that the identification problem can be reformulated for two subclasses of PieceWise Affine systems. These reformulations lead to the proposition of algorithms based on Mixed-Integer Linear or Quadratic Programming, which are guaranteed to converge to a global optimum. However, this approach is computationally affordable only for a few measured data because the complexity of the used algorithms is NP-hard. The four other approaches deal particularly with the class of PieceWise Affine AutoRegressive eXogenous (PWARX) system, i.e., models in which the regressor space is partitioned into polyhedra with affine ARX sub-models for each polyhedron. The basic steps that these approaches perform are: the estimation of the parameters, the classification of the data attributed to each mode and the estimation of the polyhedral regions. However, the clustering-based approach (Ferrari-Trecate *et al.*, 2003), the bounded-error approach (Bemporad *et al.*, 2004) and the Bayesian approach (Juloski *et al.*, 2004) require that the ARX submodels orders are fixed. Furthermore, both of the clustering-based approach and the

Bayesian approach require a priori knowledge of the number of the modes.

Unfortunately, the existing identification algorithms deal only with special linear classes of hybrid models. This is a significant limitation, because, to the best of our knowledge, there is no work addressing the case in which the dynamics can be nonlinear and the number of modes, the model parameters and the switching sequence are unknown.

The aim of this paper is to investigate an alternative method that seems to be promising for handling this more challenging case. However, we will restrict this first study to a class of HDS, which is characterised by continuous inputs, continuous outputs and binary discrete inputs. In the proposed approach, we consider the plant as a nonlinear black-box model and try to capture its behaviour globally. In this context, we will consider Feed-Forward neural networks (NNs) as global parametric models in order to show that the behaviours of the considered class of HDS can be predicted with NNs. This permits to obtain global parametric models of HDS without needing to cluster the data or to know the current mode. To the best of our knowledge, and apart from the few remarks addressed by Ferrai-Trecate and Muselli (2002), no works have addressed the issue of using NNs in the context of HDS identification.

The proposed approach has the advantage of considering the identification problem of HDS in terms of the architectures and the learning algorithms developed for NNs. It can therefore deal with system nonlinearities and can be used to track the behaviours of the HDS without a priori knowledge about the current mode. However, this approach will result in a black-box model with a large number of parameters Furthermore, although the obtained NNs represent average models that can fairly approximate a given HDS, they are not able to predict the behaviours of this system with a similar precision in all the modes. Finally, the obtained NNs are not adapted for some control and analysis problems, but they can be very useful to deal with the model-based diagnosis of HDS (Messai *et al.*, 2006).

In this paper, the Feed-Forward-neural-networks based approach for the identification of HDS is presented in Section 2 and, then, this approach is illustrated with the help of a benchmark example in Section 3.

2. IDENTIFICATION OF THE PARAMETERS OF THE NEURAL NETWORKS

Before the presentation of the identification procedure, let us, firstly, attempt to explain the mechanism by which NNs can learn the behaviours of HDS. In fact, the output of the neural network depends on the number of the hidden neurons and the activation of these hidden neurons. Hence, if the set of the hidden neurons is divided into several groups and if each of these groups of neurons is active in a distinct mode of the HDS, then the neural network will fairly approximate the I/O data of the HDS. Of course, the real mechanism is more complex since several subgroups of hidden neurons can be combined to reproduce other modes of the HDS.

As a simple example that illustrates this idea, consider the non linear hybrid system represented by Figure 1. This system, which switch for the function f_1 to the function f_2 when the input U is equal to 20, could be approximated by a NN involving two groups of hidden neurons (figure 2 and 3). The first group is composed of three hidden neuron, which are active for all the values of the inputs $U \leq 20$, (fig. 2). Therefore, the sum of their outputs will reproduce the function f_1 when $U \leq 20$ and this sum will be equal to 1 when $U > 20$. On the other hand, the three hidden neurons of the second group are saturated when $U \leq 20$ and the sum of their three outputs will reproduce the function f_2 when $U > 20$ (fig. 3). Consequently, the output of a NN with an output neuron using a linear activation function will exactly reproduce the HS if $w_{i,i\in\{1,2,..,6\}} = 1$ and $b = -1$, where $w_{i,i\in\{1,2,..,6\}}$ are the weights between the hidden neurons and the output neuron and b is the bias of the output neuron.

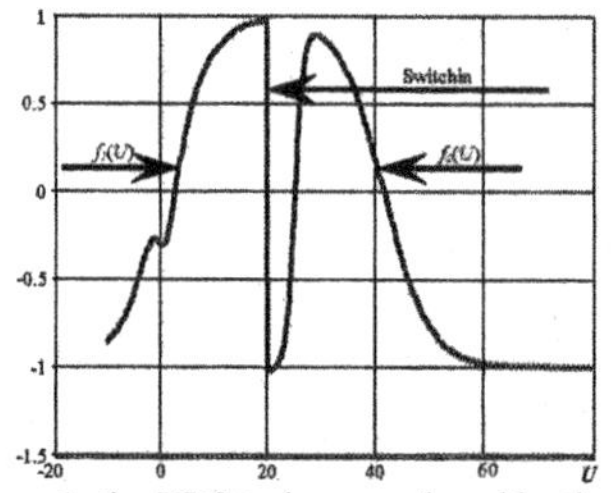

Figure 1: the HDS to be reproduced by the NN

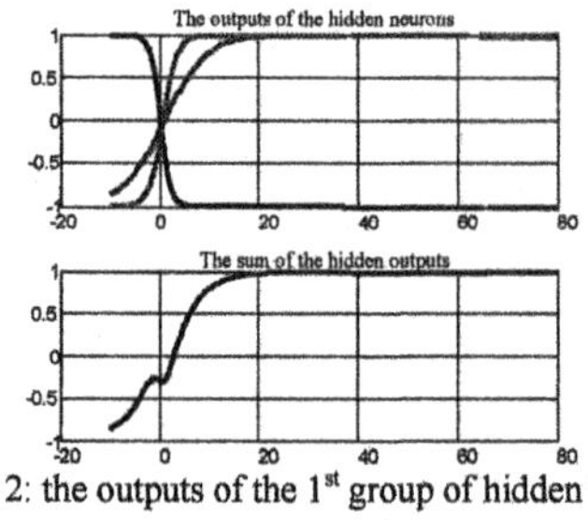

Figure 2: the outputs of the 1st group of hidden neurons

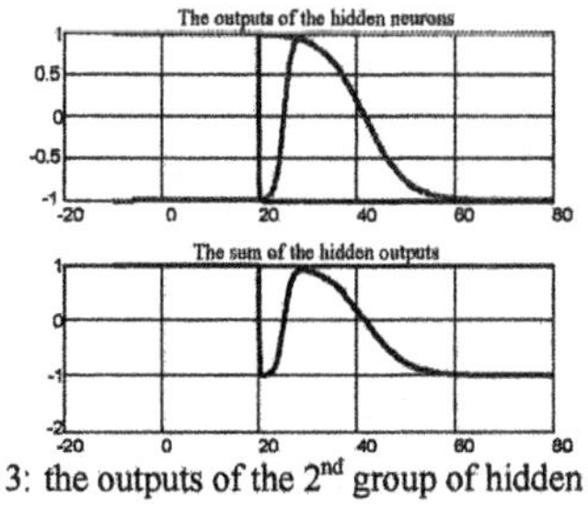

Figure 3: the outputs of the 2nd group of hidden neurons

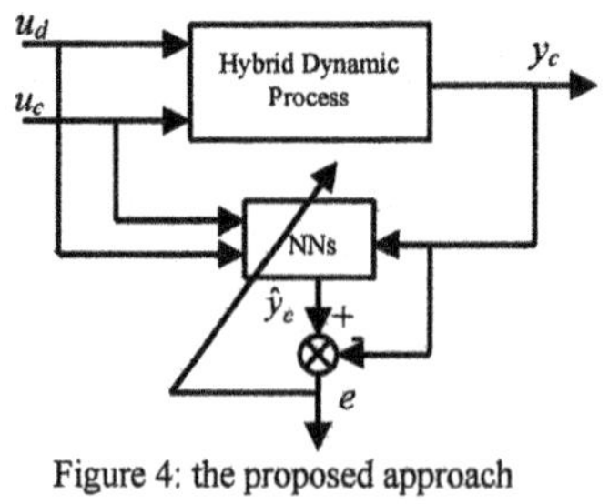

Figure 4: the proposed approach

The proposed identification approach is depicted in figure 4 where the hybrid process is characterised by: the binary discrete inputs (u_d), the continuous inputs (u_c) and the continuous outputs (y_c). The NNs predict the continuous outputs in terms of the past measured input/output variables and without using the state variables. Furthermore, neither the information about the current mode nor the number of the modes are needed to work out the NNs. This configuration is of practical interest, science accurate models of complex hybrid system are often difficult to obtain and only data measured by both continuous and binary sensors are available.

Feed-forward neural networks, with a hidden layer using sigmoidal activation function and an output layer with a linear one, can be used to extract powerful models from experimental data (Carotenuto et al., 1998, Cybenko, 1989). Therefore, the NNs that will be used to predict the continuous outputs y_c are formed by:

- An output layer, containing linear neurons, that provide the continuous outputs,
- A hidden layer containing neurones with sigmoidal activation function,
- An input layer that receives the data from the sensors of the real world hybrid systems.

To begin with the identification procedure, suppose that the input/output data is observed with a sampling period T and that the state changes of the binary inputs are only taken into account at the sampling instants. Then, consider:

$u_c^k = [u_c(1), u_c(2), u_c(3) \dots u_c(k)]$,

$u_d^k = [u_d(1), u_d(2), u_d(3) \dots u_d(k)]$,

$y_c^k = [y_c(1), y_c(2), y_c(3) \dots y_c(k)]$,

where u_c^k, u_d^k and y_c^k represent, respectively, the continuous inputs, the discrete inputs and the continuous outputs at instants kT; $k \in \{1,2,\dots\}$. The dimensions of the vectors $u_c(k)$, $u_d(k)$ and $y_c(k)$ are respectively given by the number of the continuous inputs, the number of the discrete inputs and the number of the continuous outputs.

In order to model a HDS by NNs, which predict the continuous outputs, we propose to write the relation between $(u_c^{k-1}, u_d^{k-1}, y_c^{k-1})$ and $y_c(k)$ in the form:

$$y_c(k) = g(u_c^{k-1}, u_d^{k-1}, y_c^{k-1}) + e(k) \qquad (1)$$

where g is an unknown function and e is an additive term, indicating that $y(k)$ can not be exactly determined from the previous observations.

Although, the equation (1) can be used to model the HDS, it remains very general to be exploitable and the function $g(u_c^{k-1}, u_d^{k-1}, y_c^{k-1})$ will therefore be decomposed into two functions ϕ and h:

$$\varphi^k = \phi(u_c^{k-1}, u_d^{k-1}, y_c^{k-1}) \qquad (2)$$

$$y_c(k) = h(\varphi^k, \theta) + e(k) \qquad (3)$$

where φ^k is the regressors vector, θ is the parameters vector to be identified and h is a function that expresses the relation between the regressors and the outputs of the NNs. Note that in the case of feed Feed-Forward neural networks, the function h is decomposed into a set of sigmoidal function f representing the neurone of the hidden layer.

At this stage, the identification problem of the HDS, will be a problem of choosing the optimal structure of the NNs. This problem involves the choice of the inputs (i.e., the regression vector), the number of the hidden neurones and the number of the outputs neurones.

Concerning the choice of the number of the output neurons, two alternatives are possible. The first one is to build a single NN with a number of output neurons equal to the number of the continuous outputs of the HDS to be identified. Each output neuron of this NN corresponds to a continuous output of the HDS. The second alternative is to associate a NN with each of the continuous outputs of the system. In this case the number of NNs will be equal to the number of the continuous outputs of the HDS to be identified and each NN will predict one of the continuous outputs.

To choose the regressors, φ^k is decomposed into two parts: the regressors of the continuous measured and/or estimated variables, φ_c^k, and the regressors of the binary discrete inputs, φ_d^k.

To select φ_d^k, we propose to associate a large number of input neurones with each discrete input and to use a pruning algorithm, such as the optimal Brain Surgeon (OBS) algorithm (Reed, 1993), to remove the parameters, and subsequently the input neurons, that are not needed. Hence, if we have N_d discrete inputs, each of which is associated with n_{ed} neurones, the initial regressors φ_d^k will be composed of $n_{ed}.N_d$ elements. These elements represent the values of all the discrete inputs between the instant $(k-1)T$ and the instant $(k-n_{ed})T$. Then, the initial regressors φ_d^k will be optimised at the end of the training by the pruning algorithm.

On the other hand, the determination of φ_c^k is derived from the general linear model given by (Chen *et al.*, 1990, Ljung, 1987):

$$A(q^{-1})y_c(k) = q^{-n_k}\frac{B(q^{-1})}{F(q^{-1})}u_c(k) + \frac{C(q^{-1})}{D(q^{-1})}e(k) \qquad (4)$$

where n_k is the delay and the polynomials $A(q^{-1})$, $B(q^{-1}), C(q^{-1}), D(q^{-1})$ and $F(q^{-1})$ are respectively characterised by the orders n_a, n_b, n_c, n_d and n_f. The identification procedure starts with a regressor vector composed of a large identical number of delayed inputs and delayed outputs. The delay n_k is then estimated by modelling the system for various values of n_k and by choosing the value which will correspond to the model providing the smallest residual criterion.

Finally, the regression vector is obtained by: i) modelling the system with ascending values of the orders of the polynomials used in the equation 4, and ii) choosing the model providing the smallest value of the final prediction error as suggested by Akaike (Ljung, 1987):

$$V_{FPE} = \frac{N+d}{N-d}\sum_{k=1}^{N}(e(k))^2 \qquad (5)$$

where, N is the number of data in the training data set and d is the number of weights and the bias (i.e., the parameters) of the NN (Fig. 5).

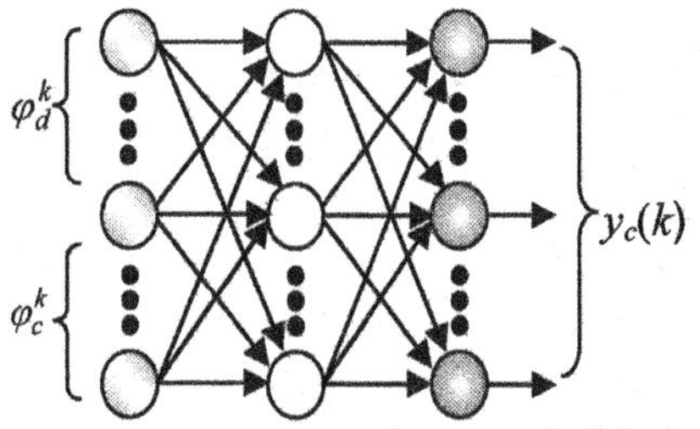

Figure 5: structure of the NN model

3. APPLICATION EXAMPLE

In order to illustrate the proposed approach let us consider the two tanks flow system depicted in the figure 6. This system is the benchmark of the Specific Action on the diagnosis of hybrid systems (AS193) of CNRS[1] and GDR MACS[2]. Although, this system can be perfectly described by a set of mathematical equations, we will consider it as a black-box system equipped with some sensors that provide the I/O data. Consequently, the behaviour of this system will be firstly simulated using a mathematical model of the system. Then, the data obtained in simulation are considered as a set of I/O data measured by fictive sensors and provided to the NNs that predict the next continuous outputs.

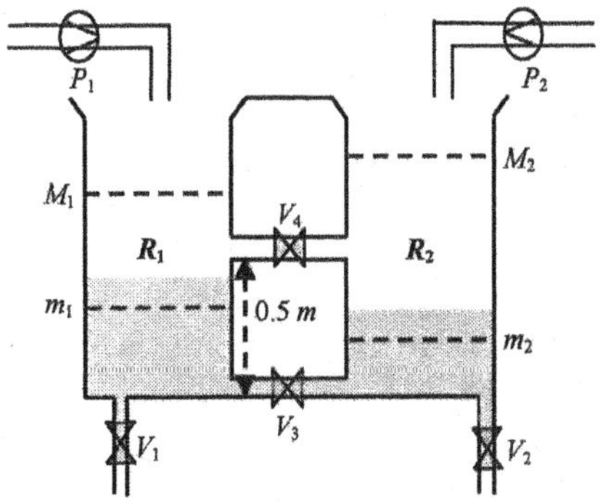

Figure 6: the two tanks system

3.1 Description of the system

The system consists of two cylindrical tanks, R_1 and R_2. Each tank has an input pipe and an output pipe. The input pipes use built-in identical on-off pumps (P_1 and P_2) and the output pipes are controlled by two on-off electro valves, V_1 and V_2. The tanks are connected to each other by means of two pipes, located at the bottom of the tanks and at 0.5 *m* height. These pipes are also controlled by on-off electro valves, V_3 and V_4. Finally, each tank is equipped with an analogue sensor that measures the level of the fluid.

The system can be represented as a HDS with two continuous variables, h_1 and h_2, representing the height of the fluid in the tanks and six binary inputs: P_1, P_2, V_1, V_2, V_3 and V_4.

For the purpose of simulation let us consider that the dynamics of the pumps are very fast. Hence, we can suppose that the input flows are constants when the pumps are on and are null when the pumps are off.

According to the Torricelli model, the dynamics of the system can be described by the following equations:

$$Q_{P1} = D.P_1, \qquad P_1 \in \{0,1\} \qquad (6)$$

$$Q_{P2} = D.P_2, \qquad P_2 \in \{0,1\} \qquad (7)$$

$$Q_1 = A\sqrt{2.g.h_1}V_1, \qquad V_1 \in \{0,1\} \qquad (8)$$

$$Q_2 = A.\sqrt{2.g.h_2}V_2, \qquad V_2 \in \{0,1\} \qquad (9)$$

$$Q_3 = .\alpha.\sqrt{|h_1 - h_2|}V_3, \qquad V_3 \in \{0,1\} \qquad (10)$$

$$Q_4 = \alpha.\sqrt{|\sup(h_1,0.5) - \sup(h_2,0.5|}V_4, \qquad V_4 \in \{0,1\} \qquad (11)$$

$$S\dot{h}_1 = Q_1 - Q_3 - Q_4 \qquad (12)$$

$$S\dot{h}_2 = -Q_2 + Q_3 + Q_4 \qquad (13)$$

where $\alpha = A.\sqrt{2.g}sign(h_1 - h_2)$, D is the constant input flow of both pumps (D=$10^{-4}m^3/Sec$), $Q_{Pi,i\in\{1,2\}}$ is the input flow of tank i, $Q_{i,i\in\{1,2\}}$ is the output flow of tank i, Q_3 is the flow in the pipe C_1 and Q_4 is the flow in the pipe C_2.

In order to avoid either the draining or the overflow of the tanks, the electro valves V_1, V_3 and V_4 as well as the pump P_1 are driven by an algorithm which guarantees the following levels of the fluids in the tanks R_1 and R_2:

$$m_1 \le h_1 \le M_1 \qquad (14)$$

$$m_2 \le h_2 \le M_2 \qquad (15)$$

with M_1=0.6*m*, M_2=0.75*m*, m_1=0.4*m* and m_2=0.2*m*.

Finally, to further complicate the modelling tasks, the pump P_2 and the electro valve V_2 are considered as perturbations which cannot be controlled. Hence, P_2 and V_2 were opened and closed according to two Nearly Random Binary Sequences of length: $l_{p2} \in [10,30]$ and $l_{v2} \in [30,50]$.

Although the number of modes of this benchmark example is known (64 modes), the identification approach is designed to deal with unknown number of modes. Hence, the knowledge about the number of modes and the current mode will not be used to build the NNs.

3.2 The NN identification results

The above hybrid system is modelled by two feed-forward neural networks according to the approach presented in section 2. Each NN predicts the level of the liquid in the corresponding tank.

[1] Centre National de la Recherche Scientifique
[2] Groupement de Recherche "Modélisation, Analyse et Conduite des Systèmes dynamiques" of CNRS

During the simulations phase, two data sets, each containing $N = 5000$ data, were generated. The first data set, called the identification data set, is used to find the optimal structure of the NNs (i.e., the number of neurons in each layer and the parameters of each NN) and the second data set, called the validation data set, is used to verify the accuracy of the obtained structures when other data are used.

These data sets that have been used to build the NNs were obtained by:

- Fixing the sampling period to T=1 *Sec*,
- Simulating the behaviours of the system according to the equations 6 to 13. The initial conditions were: $h_0 = h_1 = 0\,m$ for the identification data set and $h_0 = h_1 = 1\,m$ for the validation data sets,
- Adding Gaussian noises with zero mean and a standard deviation $\sigma = 0.01$ to the data sets. These noises were added to evaluate the accuracy of the obtained model in the presence of noise.

In order to obtain the structure of the NNs, 5 input neurons were associated with each binary input at the beginning of the identification procedure. Consequently the initial vector φ_d^k was composed by 30 elements that represent the 5 last states of the valves and the last 5 states of the pumps. Note that this initial φ_d^k is considered sufficient because we have observed that the OBS algorithm eliminates some inputs at the end of each training.

The identification procedure is initialised with a regressor vector, φ_c^k, composed of a sufficiently high number of delayed outputs (n_a=5) in order to determine the delay n_k. Once this delay is obtained, the regressors are selected by modelling the system with ascending values of the orders and by choosing the orders of the model providing the smallest value of the final prediction error.

During the identification phase, the parameters of the networks were estimated by means of Levenberg-Marquardt algorithm (Declerq and Dekeyser, 1995) with 20 different initializations and a maximum of *5000* iterations; these several trials were performed in order to select the model with the best performances. Then, starting from a structure with 12 hidden neurons, the Optimal Brain Surgeon (OBS) algorithm (Reed, 1993) was used to remove the parts that are not needed. Hence, after several simulations, according to the approach described in section two, we have opted for two NNs with:

- 21 input neurons that receive the last three outputs of the considered sensor (continuous variables), the last three binary states of the valves and the last three binary state of the pumps,
- 8 hidden neurons,
- 1 output neuron using a linear activation function. The outputs of this neuron correspond to the predicted liquid level in the considered tank.

Figures 7a, 7b, 7c, 8a, 8b and 8c present, respectively, the identification and the validation results of the selected models. The upper part of figures 7a and 8a depicts the predicted and the measured liquid levels. The analysis of these figures shows that it is difficult to distinguish between these levels because the liquid levels predicted by both the NNs closely match the measured data. The lower parts of figures 7a and 7b show that the residuals, which represent the differences between the measured and the predicted levels, are very small and indicate that the relative errors do not exceed 5% of the measured levels. The same remarks apply to figures 8a and 8b. These results validate the identified models and indicate that the NNs are able to estimate the level of the liquid for all the modes of the considered systems.

Other results, not presented here, indicated that the values of the cross correlation functions between the inputs and the outputs are lower than the practical threshold given by Landau (2001). These results pointed out the independence between the residuals and the inputs and show that the NNs are able to reproduce all the dynamics of the system. Furthermore, the analysis of the autocorrelation functions of the residuals has shown that the values of these functions belong to the 99% confidence intervals. These results confirm the independence between the residues and indicate that the residuals can be considered as a white noise. Finally, figures 7c and 8c confirm the analysis of the autocorrelation plots and indicate that the distributions of the residuals are similar to the distributions of the noise that was added to the data sets.

4. CONCLUSION

A methodology to build black-box models of HDS has been proposed. According to this methodology, the behaviours of HDS can be predicted by feed-forward neural networks that track all the modes of the system and the determination of the structure of these neural networks can be viewed as a system identification problem. This approach was illustrated with a simulation example and the obtained results provide strong evidence of the good performances of the obtained model. Several problems remain open, such as the proposition of residual criteria and validation method, which guarantee the same validity of the NNs for all the modes of the HDS.

REFERENCES

Bemporad A., Garulli A., Paoletti S. and Vicino, A. (2004) Data classification and parametr estimation for the identification of piecewise affine models. 43rd IEEE Conference on Decision and Control, Paradise Island, Bahamas pp. 20-25

Bemporad A., Roll J. and Ljung L. (2001) Identification of Hybrid Systems via Mixed-Integer Programming. *40th IEEE Conference on Decision and Control*, Orlando, Florida USA, pp. 786-792

Branicky M. (1996) General hybrid dynamic system modelling: analysis and control. *Lecture note. Hybrid System III*, **vol.4066**, pp. 187-200

Carotenuto R., Fedele L., Tornic M., (1998) An innovative neural networks model for bench testing hybrid means of transport, *Proc. of NEURAP'98*, Marseille, France, pp. 141-148

Chen S., Billings S., Grant P., (1990) Non-linear systems identification using neural networks, *International Journal of Control*, **vol.51**, pp.1191-1214

Cocquempot V., El Mezyani T., and Staroswiecki M (2004) Fault Detection and Isolation for Hybrid Systems using Structured Parity Residuals. *5th IEEE Asian Control Conference*, Melbourne, Australie, pp. 1204-1212

Cybenko G. (1989): Approximation by superposition of sigmoidale function. *Mathematics of Control Signal and Systems* vol. 2, pp. 524-532

Declercq F., Dekeyser R., (1995) Using Levenberg-Marquardt minimization in neural model based predicive control, *International Workshop on Artificial Intelligence in Real-time Control*, Bled, Slovenia, pp.275-279

Ferrari-Trecate G., Muselli M., Liberati D. and Morari M. (2003) Aclustering technique for the identification of piecewise affine and hybrid systems. *Automatica*, **vol. 39**, pp. 205-217

Ferrari-Trecate G., Muselli M (2002): A new learning method for piecewise linear regression. *LNCS*, **vol. 2415**, pp. 444-449

Hoffmann I. and Engell S. (1998) Identification of hybrid systems. *IEEE American Control Conference*, Philadelphia, USA, pp.711-712

Juloski A., Weiland s. and Heemels W. (2004) A Bayesian approach to identification of hybrid systems. *43rd IEEE Conference on Decision and Control*, Paradise Island, Bahamas pp. 13-19

Landau I.D., (2001) *Identification des systèmes*, Traité IC2 Section Systèmes Automatisés, Editors: Landau I.D. and Voda A.B., Hermes, Paris

Ljung L. (1987) *System identification: Theory for the user*. Prentice-Hall, Englewood Cliffs, N.J

Messai N., Thomas P., Lefebvre D. and Riera B (2006) Fault detection for HDS by means of neural networks: application to two tanks hydraulic system. *To appear in Proc. of SAFPROCESS'06*

Reed R., (1993) Pruning algorithms a survey, *IEEE Transaction on Neural Network*, **vol.4**, pp.740-747

Simon S. and Engell S. (2001) Using wavelets for the identification of hybrid systems. *APII-JESA*, **vol. 35**, pp. 535-552

Vidal R., Soatto S., Ma Y., Sastry S. (2004) Identification of PWARX hybrid models with unknown and possibly different orders. *IEEE American Control Conference*, pp. 547-552

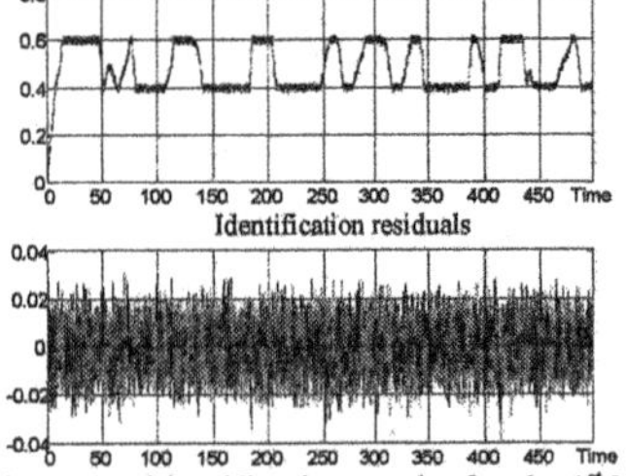

Figure 7a: identification results for the 1st NN

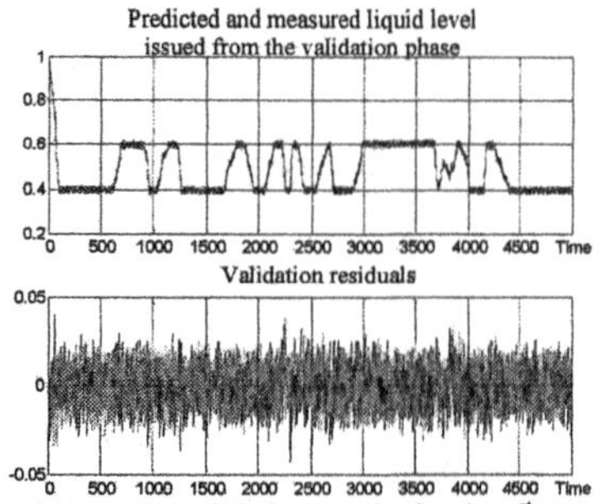

Figure 7b: validation results for the 1st NN

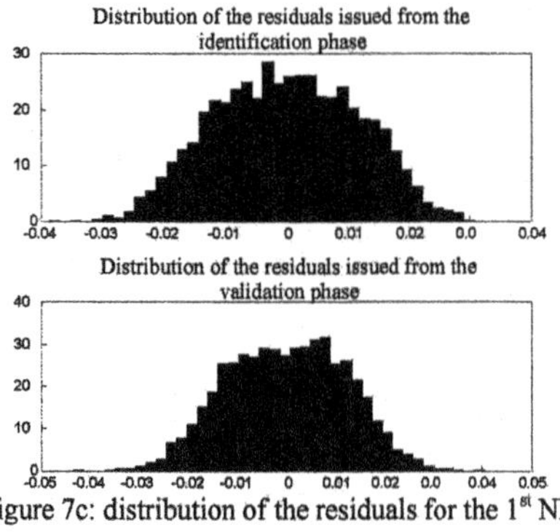

Figure 7c: distribution of the residuals for the 1st NN

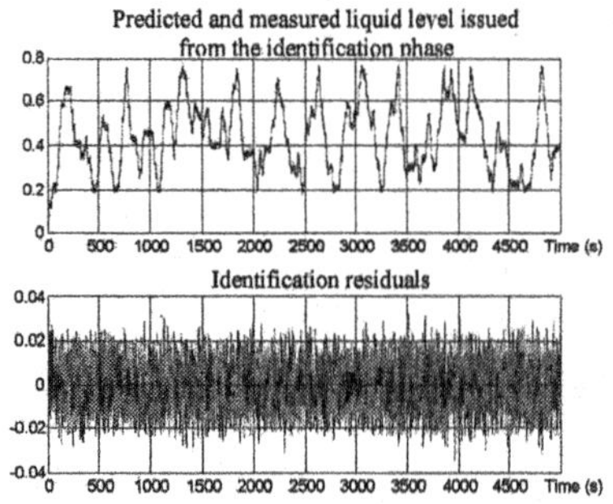

Figure 8a: identification results for the 2nd NN

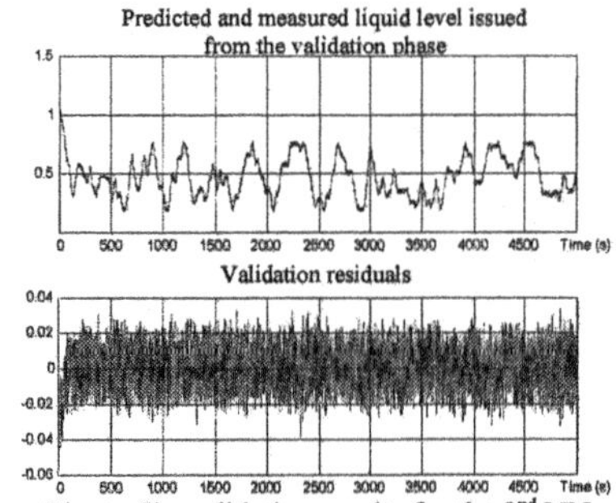

Figure 8b: validation results for the 2nd NN

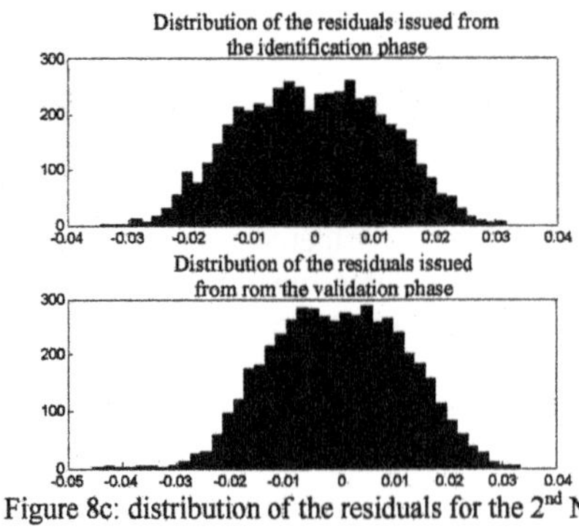

Figure 8c: distribution of the residuals for the 2nd NN

ELSEVIER
IFAC
PUBLICATIONS

FAULT TOLERANT CONTROL DESIGN FOR SWITCHED SYSTEMS

Mickaël Rodrigues, Didier Theilliol and Dominique Sauter

Centre de Recherche en Automatique de Nancy,
CRAN-UMR-CNRS 7039 Université Henri Poincaré
Nancy I, B.P. 239, 54506 Vandoeuvre-lès-Nancy Cedex,
France. Phone: +33 383 684 480 - Fax: +33 383 684 462
E-mail: mickael.rodrigues@cran.uhp-nancy.fr

Abstract: In this paper, we develop a Fault Tolerant Control (FTC) strategy designed to preserve closed-loop stability in spite of actuator failures. The main contribution is the design of a linear output feedback control function for a class of discrete-time switched linear systems with explicit modelling of multiple actuator failures. Assuming an on-line and real-time Fault Detection and Isolation (FDI) scheme such that, each time there is a change in status of one or more of the p many actuators, the FDI scheme will update a p-length vector of real parameters in the interval [0, 1] recording the degree of loss of effectiveness (with 0 for normal operation and 1 for total failure), we can perform controller redesign on-line using LMI techniques. We conclude the paper with an example illustrating the performance of our proposed FTC strategy. *Copyright © 2006 IFAC*

Keywords: Fault Tolerant Control, Switched Systems, Actuator Failure, LMI, Output Feedback, Stability

1. INTRODUCTION

In recent years, the study of switched systems has received a growing attention in control theory and practice. By switched systems we mean a class of hybrid dynamical systems consisting of a family of continuous (or discrete) time subsystems and a rule that governs the switching between them. A survey of basic problems in stability and design of switched systems is given in (Liberzon and Morse, 1999) where some contributions are summarized. Most of these contributions deal with stability analysis or design of state feedback based control laws in fault-free case (Mignone *et al.*, 2000), (Daafouz and Bernussou, 2002), (Chadli *et al.*, 2002), (Daafouz *et al.*, 2003) but rarely in faulty case. This paper aims to consider multiple actuator faults on switched systems and to ensure a pole placement of the closed-loop system eigenvalues.

The objective of Fault Tolerant Control system (FTC) is to maintain current performances closed to desirable performances and preserve stability conditions in the presence of component and/or instrument faults. Accommodation capability of a control system depends on many factors such as the severity of the failure, the robustness of the nominal system, and the actuators redundancy. Various approaches for FTC have been suggested in the literature (Noura *et al.*, 2000), (Blanke *et al.*, 2003) and (Zhang and Jiang, 2003) but usually deal with linear systems.

A FTC system is characterized by an on-line and a real-time FDI process and control reconfiguration mechanism. There are two types of control reconfiguration strategies, one is real-time control redesign (Wu *et al.*, 2000), (Theilliol *et al.*, 2002), (Rodrigues, 2005), and the other is to switch among pre-designed multiple controllers (Zhang and Jiang, 2001), (Zhang *et al.*, 2005). However, controllers switching underline the fact that many faulty system representations had to be identify so as to synthesize off-line pre-computed and stabilized controllers. These identifications are sometimes difficult to obtain. Moreover, the most of research works in switched systems do not deal with actuator faults, so we develop an output feedback synthesis which can ensure closed-loop system stability despite of faults. This paper extends some previous results in polytopic framework with only state feedback synthesis (Rodrigues *et al.*, 2005) or in LPV case (Rodrigues *et al.*, 2006) under multiple actuator faults.

The paper is organized as follows. The section II gives the problem statement of Fault Tolerant Control in switched systems. In section III, we design a controller synthesis for each actuator and we generate an output feedback control law for switched systems both in fault-free and faulty cases. An illustrative example is given in section IV to underline the FTC method. Finally, concluding remarks are given in the last section. In this paper, the following notations are used: I_n denotes an identity matrix of dimension $n \times n$. The dimensions are sometimes omitted in which cases they can be implied from the context.

2. PROBLEM STATEMENT

2.1 Switched systems representation in faulty-case

Let consider the following switched system with multiple actuator faults (Rodrigues, 2005):

$$x_{k+1} = \sum_{j=1}^{N} \alpha_k^j [A_j x_k + B_j (I_p - \gamma) u_k]$$
$$y_k = \sum_{j=1}^{N} \alpha_k^j [C_j x_k] \quad (1)$$

where $x \in \mathbb{R}^n$ represents the state vector, $u \in \mathbb{R}^p$ is the input vector, $y \in \mathbb{R}^m$ is the output vector. $\gamma \triangleq diag[\gamma_1, \gamma_2, \dots, \gamma_p]$, $\gamma_i \in \mathbb{R}$, such that $\gamma_i = 1$ represents a total lost, a failure of i-th actuator,$i \in [1, \dots, p]$ and $\gamma_i = 0$ implies that i-th actuator operates normally. $A_j \in \mathbb{R}^{n \times n}$, $B_j \in \mathbb{R}^{n \times p}$, $C_j \in \mathbb{R}^{m \times n}$ are invariant matrices defined for the jth mode with $j \in [1, \dots, N]$.

The switched system is scheduled through switching functions designed as follows: α_k^j $\forall j \in [1, \dots, N]$ lie in a convex set $\Omega = \{\alpha_k^j \in \mathbb{R}^N, \alpha_k = [\alpha_k^1 \ \dots \alpha_k^N]^T,\ \alpha_k^j \geq 0 \ \forall j$ and $\sum_{j=1}^{N} \alpha_k^j = 1\}$. These switching functions are supposed to be available in real time depending on parameters measurement and not corrupted by faults. The switching functions α_k^j are defined as follows:

$$\alpha_k^j = \begin{cases} 1 & \text{when the switched system is described} \\ & \text{by the j-th mode such that:} \\ & A(\alpha) = \sum_{j=1}^{N} \alpha_k^j A_j = A_j \\ 0 & \text{otherwise} \end{cases} \quad (2)$$

So, if $\alpha_k^j = 1$ the system is described by the jth mode with $[A_j, B_j, C_j]$.

We propose to develop an Output Feedback for switched systems with actuator failures. Consider the matrix b_j^i representing total faults in all actuators except the i-th:

$$B_j^i = [0, \dots, 0, b_j^i, 0, \dots, 0] \quad (3)$$

and $B_j = [b_j^1, b_j^2, \dots, b_j^p,]$ with $b_j^i \in \mathbb{R}^{n \times 1}$. It is assumed that each column of B_j is full column rank whatever the model j. As (Maki *et al.*, 2001), the following assumption is considered.
Assumption 1: The pairs $(A_j, b_j^i), \forall i = 1, \dots, p$ are assumed to be controllable $\forall j = 1, \dots, N$. □

2.2 Principles of Fault Tolerant Control Strategy

Without loss of generality, it is assumed that when actuator fault occurs on the system, the matrix γ can be decomposed as:

$$\gamma = \begin{pmatrix} \gamma_{p-h} & 0 \\ 0 & I_h \end{pmatrix} \quad (4)$$

with γ_{p-h} a diagonal matrix where its elements $\gamma_{p-h}^i, i \in [1, \dots, p]$ are different from 1 which represent the number of actuators not out of order ($\gamma^i \neq 1$), and I_h represents the number h of actuators totally failed. Let define $\Gamma \triangleq (I_p - \gamma)(I_p - \gamma)^+$ which represents only totally failed actuators and the corresponding matrix partitions of B such as:

$$B = [B_{p-h} \quad B_h] \quad (5)$$

$B_{p-h} \in \mathbb{R}^{n x (p-h)}$ and $B_h \in \mathbb{R}^{n x h}$ and Γ:

$$\Gamma = \begin{pmatrix} I_{p-h} & 0 \\ 0 & 0_h \end{pmatrix} \quad (6)$$

We will present a control law which is able to vanish actuator faults into the state space representation (1) and to ensure closed-loop stability despite of multiple actuator failures. Based on a multiplicative fault representation defined in (1), the new control law u_{FTC} must vanish all actuator faults on the system such that:

$$u_{FTC} = [I - \gamma]^+ u_{nom} \quad (7)$$

Let introduce the set of indexes of all actuators that are not out of order, i.e.

$$\Phi \triangleq \{i : i \in (1, \dots, p), \gamma^i \neq 1\} \quad (8)$$

and note that $u_{FTC} = [I-\gamma]^{+}u_{nom} = -[I-\gamma]^{+}Fy_k = -F_{FTC}y_k$ where F is a nominal controller gain and F_{FTC} the new controller gain. So, this specific control law in the state space representation (1) leads to:

$$\begin{aligned} B_j(I-\gamma)u_{FTC} &= B_j(I-\gamma)(I-\gamma)^{+}u_{nom} \\ &= B_j\Gamma u_{nom} = \sum_{i\in\Phi} B_j^i u_{nom}^i \end{aligned} \quad (9)$$

which avoids actuator fault effect and where $\sum_{i\in\Phi} B_j^i$ represents the actuators not out of order, i.e. $\sum_{i\in\Phi} B_j^i = B_{p-h}$ and u_{nom}^i the i-th element of u_{nom}. Due to the fact that each pair $(A_j, b_j^i), \forall i = 1,\ldots,p$ are assumed to be controllable for all $\forall j = 1,\ldots,N$, the system still remains controllable despite of actuator failures.

3. FAULT TOLERANT CONTROL FOR SWITCHED SYSTEMS

3.1 Control law synthesis in fault-free case

Let us recall the multiplicative actuator fault representation on a polytopic system as follows:

$$\begin{aligned} x_{k+1} &= \sum_{j=1}^{N} \alpha_k^j [A_j x_k + \sum_{i=1}^{p} B_j^i (I_p-\gamma)u_k] \\ y_k &= Cx_k \end{aligned} \quad (10)$$

where α_k^j represents the switching functions.

Assumption 2: The matrix $C = C_j, \forall j \in [1\ldots N]$ is full row rank. □

Assumption 3: Any actuator can fail, but at least one actuator is not lost, which means that the situation $\gamma^1 = \cdots = \gamma^p = 1$ is excluded. □

In the nominal case, the linear output feedback can be expressed such as:

$$u_k = -Fy_k \quad (11)$$

with $y_k = Cx_k$ and $F \in \mathbb{R}^{p\times m}$ is the output feedback controller. In the nominal case ($\gamma = 0$), the representation (10) with a controller $u_k = -Fy_k$ is rewritten as:

$$\begin{aligned} x_{k+1} &= \sum_{j=1}^{N} \alpha_k^j [A_j x_k + B_j(I-\gamma)(-Fy_k)] \\ &= \sum_{j=1}^{N} \alpha_k^j (A_j - B_jFC)x_k \end{aligned} \quad (12)$$

with the output feedback controller F to determine so as to vanish actuator faults on the system. We want to establish the stability of the closed-loop system with a $\mathcal{LMI}$ pole placement. In order to achieve some desired transient performance, a pole placement should be considered. For many problems, exact pole assignment may not be necessary, it suffices to locate the pole of the closed loop system in a sub-region of the complex left half plane (Chilali and Gahinet, 1996) and (Rodrigues *et al.*, 2005).

So, let define a disk region $\mathcal{LMI}$ $\mathcal{D}$ included in the unit circle. The pole placement of the closed-loop system (12) for all the models $j \in [1\ldots N]$ in a $\mathcal{LMI}$ region, can be expressed as the following (Chilali and Gahinet, 1996):

$$\begin{pmatrix} -rX & qX + (A_jX - B_jFCX)^T \\ qX + (A_jX - B_jFCX) & -rX \end{pmatrix} < 0 \quad (13)$$

However these inequalities are no longer linear with regard to the unknown matrices $X = X^T > 0$ and $F, \forall j \in [1\ldots N]$. So, the solution is not guaranteed to belong to a convex domain and the classical tools for solving sets of matrix inequalities cannot be used. It constitutes the major difficulty of output feedback design.

We propose to transform $\mathcal{BMI}$ conditions (13) in X and $F, \forall j \in [1\ldots N]$, in $\mathcal{LMI}$ conditions which will be used to synthesize directly a stabilized output feedback. Controllers gains F_i are synthesized for each actuator in order to define an output feedback even if there are failures in the system.

Theorem 1. Consider the system (10) in fault-free case ($\gamma_k = 0$), defined $\forall j \in [1\ldots N]$. Let assume that for each j each pairs (A_j, b_j^i) are controllable and suppose it is possible to find matrices $X_i = X_i^T > 0$, M and V_i $\forall i = [1,\ldots,p]$ such that $\forall i = [1,\ldots,p], \forall j = [1,\ldots,N]$:

$$\begin{pmatrix} -rX_i & qX_i + (A_jX_i - B_j^iV_iC)^T \\ qX_i + A_jX_i - B_j^iV_iC & -rX_i \end{pmatrix} < 0 \quad (14)$$

with

$$CX_i = M_iC, \quad \forall i = [1,\ldots,p], \forall j = [1,\ldots,N] \quad (15)$$

The control law with output feedback $u_k = -Fy_k$ allows to place the eigenvalues of the closed-loop system (10) in a predetermined $\mathcal{LMI}$-region with $FM = V$, $F = \sum_{i=1}^{p} G_iV_i(CC^T(C\sum_{i=1}^{p} X_iC^T)^{-1})$ or $F = VCC^T(CXC^T)^{-1}$, with $G_i = B_j^{i+}B_j^i, \forall i = [1,\ldots,p], \forall j = [1,\ldots,N]$. ■

Proof. This proof is similar to (Rodrigues *et al.*, 2005). Summation of (14) for the actuators set $i \in [1,\ldots,p]$ of the system (10) $i = 1,\ldots,p$ gives for one model $j, \forall j = [1,\ldots,N]$:

$$\sum_{i=1}^{p} \begin{pmatrix} -rX_i & qX_i + (A_jX_i - B_j^iV_iC)^T \\ qX_i + A_jX_i - B_j^iV_iC & -rX_i \end{pmatrix} < 0 \quad (16)$$

Let denote $X = \sum_{i=1}^{p} X_i$ (with $X = X^T > 0$) to obtain

$$\begin{pmatrix} -rX & qX + (A_jX - \sum_{i=1}^{p} B_j^iV_iC)^T \\ qX + (A_jX - \sum_{i=1}^{p} B_j^iV_iC) & -rX \end{pmatrix} < 0 \quad (17)$$

$\forall i = [1, \ldots, p], \forall j = [1, \ldots, N]$. Now, denote the l-th row of the matrix V_i as V_i^l, $i = 1, \ldots, p$ and $l = 1, \ldots, p$, and $G_i = B_j^{i+} B_j^i$ is a matrix equals to zero except in entry (i, i) where there is a one:

$$V_i^l = G_l V_i \tag{18}$$

Therefore,

$$\sum_{i=1}^{p} B_j^i V_i C = \sum_{i=1}^{p} [0, \ldots, 0, b_j^i, 0, \ldots, 0] V_i^i C$$
$$= B_j \sum_{i=1}^{p} V_i^i C \tag{19}$$

leading to

$$\sum_{i=1}^{p} B_j^i V_i C = B_j \sum_{i=1}^{p} V_i^i C = B_j (\sum_{i=1}^{p} G_i V_i C) \tag{20}$$

By taking $V = \sum_{i=1}^{p} G_i V_i$, equation (20) becomes

$$\sum_{i=1}^{p} B_j^i V_i C = \sum_{i=1}^{p} B_j^i V_i C = B_j V C \tag{21}$$

we get $\forall i = [1, \ldots, p], \forall j = [1, \ldots, N]$

$$\begin{pmatrix} -rX & qX + (A_j X - B_j V C)^T \\ qX + (A_j X - B_j V C) & -rX \end{pmatrix} < 0 \tag{22}$$

With the changes of variables $V = FM$ and $CX = MC$ which substituted in $\mathcal{LMI}$ (22), leads to

$$\begin{pmatrix} -rX & qX + (A_j X - B_j F C X)^T \\ qX + (A_j X - B_j F C X) & -rX \end{pmatrix} < 0 \tag{23}$$

$\forall i = [1, \ldots, p], \forall j = [1, \ldots, N]$. These inequalities (23) are $\mathcal{BMI}$s (13) which could not be solve with classical tools. By multiplying each $\mathcal{LMI}$ (22) by α_k^j and summing all of them, we obtain

$$\begin{pmatrix} -rX & qX + \sum_{j=1}^{N} \alpha_k^j (A_j X - B_j V C)^T \\ qX + \sum_{j=1}^{N} \alpha_k^j (A_j X - B_j V C) & -rX \end{pmatrix} < 0 \tag{24}$$

it is equivalent to

$$\begin{pmatrix} -rX & qX + (A(\alpha) X - B(\alpha) V C)^T \\ qX + (A(\alpha) X - B(\alpha) V C) & -rX \end{pmatrix} < 0 \tag{25}$$

with $A(\alpha) = \sum_{j=1}^{N} \alpha_k^j A_j$ and $B(\alpha) = \sum_{j=1}^{N} \alpha_k^j B_j$. Due to the fact that matrix C is supposed to be full row rank, we deduce from (15) there exists a non-singular matrix $M = CXC^T(CC^T)^{-1}$ and then after variables changes $F = VM^{-1} = \sum_{i=1}^{p} G_i V_i (CC^T (C \sum_{i=1}^{p} X_i C^T)^{-1})$. So, quadratic $\mathcal{D}$-stability is ensured by solving (24) with a linear output feedback $u_k = -F y_k$. □

3.2 Control law synthesis in faulty case

By considering the system (10) and based on the previous synthesis control law, the FTC method can be developed in this section under assumption that actuator fault estimation $\hat{\gamma}$ is suitable and known without uncertainty, i.e $\hat{\gamma} = \gamma$.

Theorem 2. Consider the system (10) in actuator faulty case ($\gamma^i \neq 0$) under assumption 3, defined for all modes $j, j = 1, \ldots, N$. Let introduce the set of indexes of all actuators that are not completely lost, i.e.

$$\Phi \triangleq \{i : i \in (1, \ldots, p), \gamma^i \neq 1\}$$

The Fault Tolerant Control law with a linear output feedback is equivalent to

$$u_{FTC} = -(I - \gamma)^+ (\sum_{i \in \Phi} G_i V_i (CC^T (C \sum_{i \in \Phi} X_i C^T)^{-1})) y_k$$
$$= -(I - \gamma)^+ F_{rec} y_k = -F_{FTC} y_k \tag{26}$$

with $G_i = B_j^{i+} B_j^i$, applied to the faulty system (10) allows to constrain pole in prescribed $\mathcal{LMI}$ region. The output feedback control law $u_k = -F_{FTC} y_k$ allows to place the eigenvalues of the closed-loop system in predetermined $\mathcal{LMI}$ region with $F_{rec} M = V$, $F_{rec} = \sum_{i \in \Phi} G_i V_i (CC^T (C \sum_{i \in \Phi} X_i C^T)^{-1}$ or $F_{rec} = VCC^T(CXC^T)^{-1}$, with $G_i = B_j^{i+} B_j^i, \forall i = [1, \ldots, p], \forall j = [1, \ldots, N]$. ■

Proof.
Applying the new control law (26) to the faulty system (10), leads to the following equation

$$B_j (I - \gamma) u_{FTC} = -B_j \Gamma (\sum_{i \in \Phi} G_i V_i (CC^T (C \sum_{i \in \Phi} X_i C^T)^{-1})) y_k \tag{27}$$

with $\Gamma = (I - \gamma)(I - \gamma)^+$ defined as

$$\Gamma = \begin{pmatrix} I_{p-h} & 0 \\ 0 & O_h \end{pmatrix} \tag{28}$$

Γ is a diagonal matrix which contains only entries zero (representing total faults) and one (no fault), see section 2.2. Since $B_j \Gamma = \sum_{i \in \Phi} B_i^j$ models only the actuators that are not completely lost, then performing the summations in the proof of Theorem (1) over the elements of Φ shows that $\sum_{i \in \Phi} G_i V_i (CC^T (C \sum_{i \in \Phi} X_i C^T)^{-1})$ is the output feedback gain matrix for the faulty system $(A_j, \sum_{i \in \Phi} B_i^j, C)$. □

The control law u_{FTC} of the system (26) is realized as:

$$u_{FTC} = -K_{FTC} y_k \tag{29}$$

with

$$K_{FTC} = (I - \gamma)^+ \sum_{i \in \Phi} G_i V_i (CC^T (C \sum_{i \in \Phi} X_i C^T)^{-1}) \tag{30}$$

4. ILLUSTRATIVE EXAMPLE

Let consider the following switched system which consider actuator faults:

$$x_{k+1} = \sum_{j=1}^{N} \alpha_k^j A_j x_k + \sum_{j=1}^{N} \alpha_k^j B_j (I - \gamma_k) u_k$$
$$y_k = C x_k \tag{31}$$

with the set of matrices described as follows:

$$A_1 = \begin{bmatrix} 0.75 & 0 & 0 & 0 \\ 0 & 0.85 & 0 & 0 \\ 0 & 0 & 0.75 & 0 \\ 0 & 0 & 0 & 0.9 \end{bmatrix}, B_1 = \begin{bmatrix} 0.95 & 0.95 \\ 1 & 1 \\ 0.9 & 0.9 \\ 1 & 1 \end{bmatrix}$$

$$A_2 = \begin{bmatrix} 0.75 & 0 & 0 & 0 \\ 0 & 0.85 & 0 & 0 \\ 0 & 0 & 0.8 & 0 \\ 0 & 0 & 0 & 0.9 \end{bmatrix}, B_2 = \begin{bmatrix} 1.05 & 1.05 \\ 1 & 1 \\ 0.9 & 0.9 \\ 1 & 1 \end{bmatrix}$$

$$A_3 \begin{bmatrix} 0.75 & 0 & 0 & 0 \\ 0 & 0.85 & 0 & 0 \\ 0 & 0 & 0.70 & 0 \\ 0 & 0 & 0 & 1.1 \end{bmatrix}, A_4 = \begin{bmatrix} 0.75 & 0 & 0 & 0 \\ 0 & 0.85 & 0 & 0 \\ 0 & 0 & 0.8 & 0 \\ 0 & 0 & 0 & 1.1 \end{bmatrix}$$

$$B_3 = \begin{bmatrix} 0.95 & 0.95 \\ 1 & 1 \\ 1.1 & 1.1 \\ 1 & 1 \end{bmatrix}, B_4 = \begin{bmatrix} 1.05 & 1.05 \\ 1 & 1 \\ 1.1 & 1.1 \\ 1 & 1 \end{bmatrix}, C = \begin{bmatrix} 0 & 1 & 0 & 0 \\ 0 & 0 & 1 & 0 \\ 0 & 0 & 0 & 1 \end{bmatrix}$$

The switched system (31) is described by 4 modes defined previously. The modes of such switched system are represented through matrices A_j, B_j, C with parameters $\alpha_k^j = 1, \alpha_k^i = 0, i \neq j$. The following matrices are directly linked with Theorems (1) and (2), with parameters $q = -0.05$ and $r = 0.93$:

$$V_1 = \begin{bmatrix} -0.0863 & -0.0889 & -0.1051 \\ 0 & 0 & 0 \end{bmatrix}$$

$$M_1 = M_2 \begin{bmatrix} 0.9951 & -0.0004 & 0.0629 \\ -0.0004 & 1.0000 & 0.0054 \\ 0.0629 & 0.0054 & 0.1895 \end{bmatrix}$$

and

$$X_1 = X_2 \begin{bmatrix} 1 & 0 & 0 & 0 \\ 0 & 0.9951 & -0.0004 & 0.0629 \\ 0 & -0.0004 & 1 & 0.0054 \\ 0 & 0.0629 & 0.0054 & 0.1895 \end{bmatrix}$$

$$V_2 = \begin{bmatrix} 0 & 0 & 0 \\ -0.0863 & -0.0889 & -0.1051 \end{bmatrix}$$

with $F = VM^{-1} = \sum_{i=1}^{p} G_i V_i (CC^T (C \sum_{i=1}^{p} X_i C^T)^{-1})$:

$$F = \begin{bmatrix} 0.0265 & 0.0430 & 0.2674 \\ 0.0265 & 0.0430 & 0.2674 \end{bmatrix}, G_1 = \begin{bmatrix} 1 & 0 \\ 0 & 0 \end{bmatrix}, G_2 = \begin{bmatrix} 0 & 0 \\ 0 & 1 \end{bmatrix}$$

At the same sample $k = 2$, the first actuator is out of order and a fault of 90% loss of effectiveness appears on the second actuator. The matrix γ is equal to

$$\gamma = \begin{bmatrix} 1 & 0 \\ 0 & 0.9 \end{bmatrix}, k \geq 2$$

The figure (1) represents respectively in fault-free case: the output vector evolution in (a), state vector evolution in (b), evolution of the second actuator in (c) and the first actuator in (d), and finally parameters evolution α_k^j in (e). The closed-loop system is stable without any fault. The figure (2) represents the same characteristics evolution as in figure (1) with an actuator failure on the 1st actuator and a fault of 90% in the 2nd actuator at sample $k = 2$. This figure illustrates the instability of the closed-loop system in faulty-case.

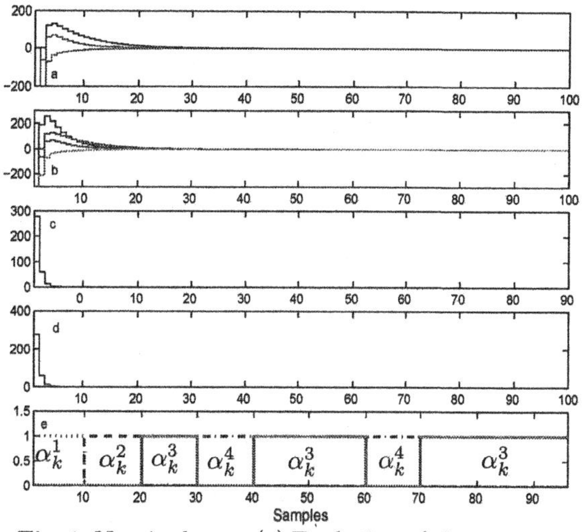

Fig. 1. Nominal case: (a) Evolution of the outputs, (b) Evolution of the states, (c) Evolution of the 2nd actuator, (d) Evolution of the 1st actuator, (e) Evolution of parameters α_k^j

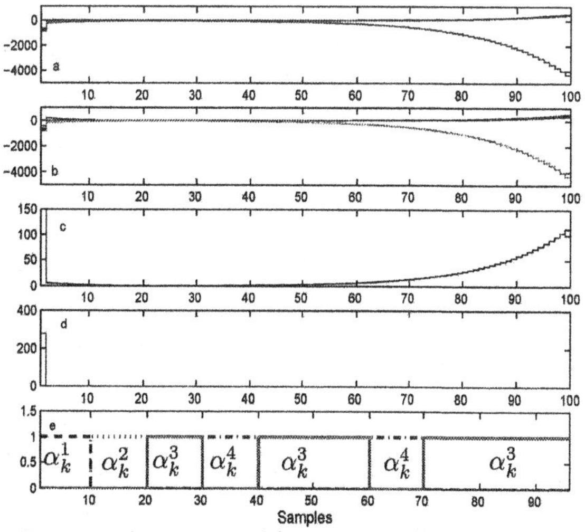

Fig. 2. In faulty case: (a) Evolution of the outputs, (b) Evolution of the states, (c) Evolution of the 2nd actuator, (d) Evolution of the 1st actuator, (e) Evolution of parameters α_k^j

The figure (3) underlines the contribution of Theorems (1) and (2) applied to switched systems. The Fault Tolerant Control law remains the system stable despite of multiple actuator faults. In order to simulate a time delay provided from FDI block, the new control law is only applied at sample $k = 15$. (Shin, 2003) discusses issues with a time delay in FTC reconfiguration. The reader could refer to this NASA/NIA Report for more information on time delay in reconfiguration.

5. CONCLUSION

The method developed in this paper emphasises the importance of the Fault Tolerant Control for switched systems. A controller is designed for

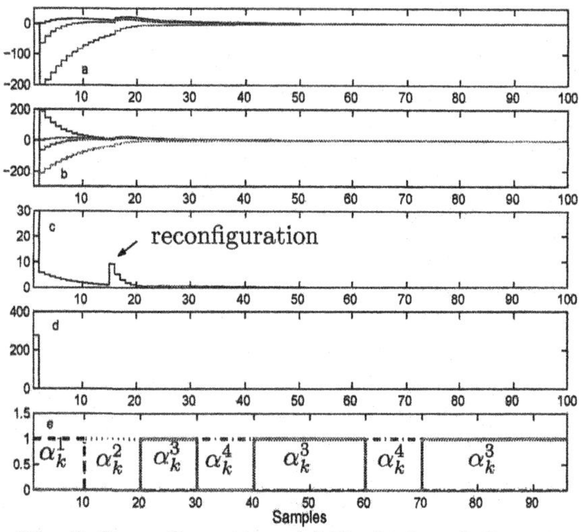

Fig. 3. Reconfiguration: (a) Evolution of the outputs, (b) Evolution of the states, (c) Evolution of the 2nd actuator, (d) Evolution of the 1st actuator, (e) Evolution of parameters α_k^j

each separate actuator through an $\mathcal{LMI}$ pole placement in fault-free case and faulty case. It allows the system to continue operating safely, to avoid stopping it immediately and to ensure stability despite the presence of actuator failures. The synthesis of a linear output feedback takes into account the information provided by a FDI scheme. An illustrative example with switched system has been presented to demonstrate the effectiveness of the scheme.

REFERENCES

Blanke, M., M. Kinnaert, J. Lunze and M. Staroswiecki (2003). *Diagnosis and Fault Tolerant Control.* Edts Springer-Verlag.

Chadli, M., D. Maquin and J. Ragot (2002). An lmi formulation for output feedback stabilization in multiple model approach. In: *Proc. of the 41'st IEEE Conf. on Decision and Control, Las Vegas, USA.* pp. 311–316.

Chilali, M. and P. Gahinet (1996). H_∞ design with pole placement constraints: an LMI approach. *IEEE Trans. on Automatic Control* **41**(3), 358–367.

Daafouz, J. and J. Bernussou (2002). Robust dynamic output feedback control for switched systems. In: *Proc. of the 41'st IEEE Conf. on Decision and Control, Las Vegas, USA.* pp. 4389–4394.

Daafouz, J., P. Riedinger and C.Iung (2003). Stabilizing switched control design with performances. In: *Proc. IFAC Conference on Analysis and Design of Hybrid Systems, ADHS'03, Saint-Malo, France.*

Liberzon, D. and A. Stephen Morse (1999). Basic problems in stability and design of switched systems. *IEEE Control Systems* **19**(5), 59–70.

Maki, M., J. Jiang and K. Hagino (2001). A stability guaranteed active fault-tolerant control system against actuator failures. In: *Proc. of the 40th IEEE Conference on Decision and Control, Orlando, Florida.*

Mignone, D., G. Ferrari-Trecate and M. Morari (2000). Stability and stabilization of piecewise affine and hybrid systems: An lmi approach. In: *Proc. of the 39'th IEEE Conf. on Decision and Control,Sydney, Australia.*

Noura, H., D. Sauter, F. Hamelin and D. Theilliol (2000). Fault-tolerant control in dynamic systems: Application to a winding machine. *IEEE Control Systems Magazine* pp. 33–49.

Rodrigues, M. (2005). Diagnostic et commande active tolérante aux défauts appliqués aux systèmes décrits par des multi-modèles linéaires. Phd thesis. Centre de Recherche en Automatique de Nancy, UHP. Nancy, France.

Rodrigues, M., D. Theilliol and D. Sauter (2005). Design of an Active Fault Tolerant Control and Polytopic Unknown Input Observer for Systems described by a Multi-Model Representation. In: *Proc. 44th IEEE Conference on Decision and Control and European Control Conference ECC, Sevilla, Spain.*

Rodrigues, M., D. Theilliol and D. Sauter (2006). Active Actuator Fault Tolerant Control Design for Polytopic LPV Systems. In: *Proc. Proc. 6th IFAC Symposium Safeprocess, Beijing, China.*

Shin, J-Y. (2003). Parameter transient behavior analysis on fault tolerant control system. Technical Report NASA-CR-2003-212682-NIA Report No. 2003-05. National Institute of Aerospace. Hampton, Virginia, USA.

Theilliol, D., H. Noura and J.C. Ponsart (2002). Fault diagnosis and accommodation of three-tank system bsaed on analytical redundancy. *ISA Transactions* **41**, 365–382.

Wu, N. E., Y. Zhang and K. Zhou. (2000). Detection, estimation and accommodation of loss of control effectiveness. *Int. J. of Adaptive Control and Signal Processing* **14**(7), 775–795.

Zhang, Y. and J. Jiang (2001). Integrated active Fault-Tolerant Control using IMM approach. *IEEE Transactions on Aerospace and Electronics Systems* **37**(4), 1221–1235.

Zhang, Y. and J. Jiang (2003). Bibliographical review on reconfigurable Fault-Tolerant Control systems. In: *Proc. IFAC Symposium Safeprocess, Washington. D.C, USA, CD-Rom.*

Zhang, Y., J. Jiang, Z. Yang and A. Hussain (2005). Managing performance degradation in Fault Tolerant Control Systems. In: *Proc. 16th IFAC World Congress, Prague, Czech Republic.*

ELSEVIER
IFAC
PUBLICATIONS

DISCRETE-EVENT MODELLING AND FAULT DIAGNOSIS OF DISCRETELY CONTROLLED CONTINUOUS SYSTEMS

Jan Lunze *

* *Ruhr-Universität Bochum*
Institute of Automation and Computer Control
44780 Bochum, Germany

Abstract: The papers presents five diagnostic strategies for discretely controlled continuous systems, which differ with respect to the abstraction of the model and measurement information used. From the original hybrid model, four more abstract representations are derived, which have the form of embedded maps, semi-Markov processes, timed automata or nondeterministic automata, respectively. The validity of the diagnostic result is ensured by the claim that the models should be complete and, hence, in the appropriate fault case, consistent with the input-output sequence of the discretely controlled system. In this way a hierarchy of models and of diagnostic results is obtained. *Copyright © 2006 IFAC*

Keywords: Hybrid system, discretely controlled system, fault diagnosis, model hierarchy, embedded map, semi-Markov process, timed automaton.

1. INTRODUCTION

Discretely controlled continuous systems comprise an important class of hybrid systems, where the continuous system represents a technological process whose operation mode is switched by a feedback controller (Fig. 1). The dynamics is characterised by discrete mode changes $q(t)$ and a continuous state movement $\mathbf{x}(t)$.

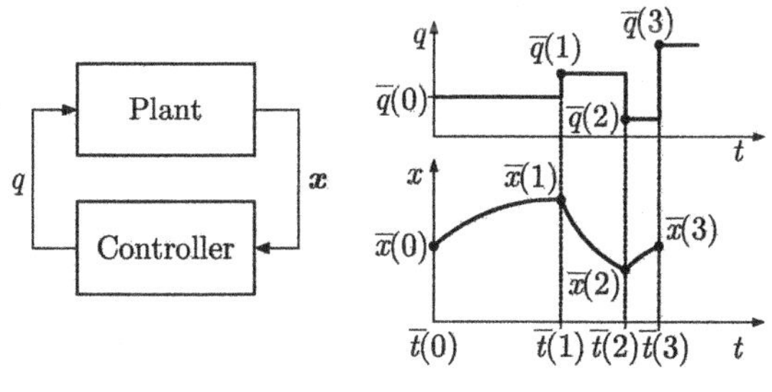

Fig. 1: Discretely controlled continuous system

A lot of practical examples fall well into this class of dynamical systems like DC-DC converters, combustion engines or simulating moving bed chromatographic processes. These examples have the additional characteristics that their function can only be maintained if the mode switching is carried out for ever.

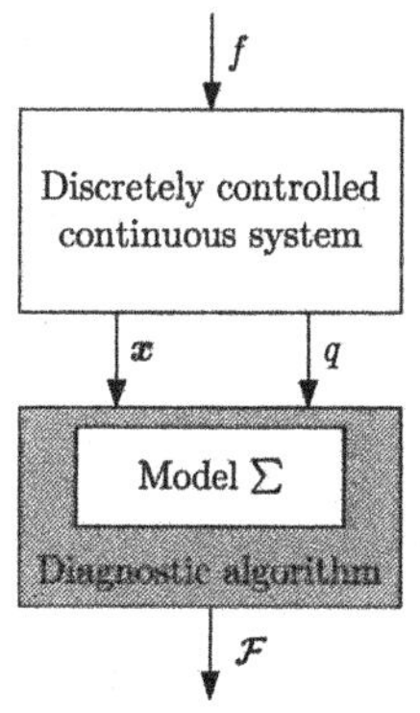

Fig. 2: Diagnosis of discretely controlled systems

Fault Diagnosis. Fault diagnosis has to use the measured switching sequence and the measured continuous movement to detect and identify faults (Fig. 2). The main idea is to test the consistency of these measurements with the behaviour of models of the faultless or the faulty system. Inconsistencies indicate that faults have occurred.

In order to get the diagnostic algorithm with the lowest possible complexity, the kind of measurement information and the granularity of the models used have to be chosen in accordance with the faults to be detected. The "best" algorithm for an application is the one with the lowest complexity that detects or identifies the fault.

This paper is concerned with the adaptation of the measurement and modelling information to the diagnostic task. Five models are described which differ in their granularity. The model abstraction process starts with the hybrid model consisting of continuous state-space models for all operation modes and leads to a nondeterministic automaton as the most abstract description. A hierarchy of diagnostic algorithms is obtained. The validity of the diagnostic results is ensured by the claim that all models should be complete in the sense of Definition 10.

Literature. There are no specific references to the diagnosis of discretely controlled systems, but several starting points to solve this diagnostic problem. Within a given operation mode, the discretely controlled system can be diagnosed by methods elaborated for continuous systems (Gertler 1998), (Blanke et al. 2006). On the other hand, if a purely discrete-event model is used, diagnostic methods for automata or Petri nets can be applied (Sampath et al. 1995).

The aim of this paper is to develop a connection between these two extremes. Starting from the hybrid model, different more abstract representations of discretely controlled systems are described together with the corresponding diagnostic algorithms. The paper uses the idea of ordered abstractions that has been described in (Raisch and O'Young 1997) for controller design and in (Förstner and Lunze 2001) for the diagnosis of quantised systems and represents it for discretely controlled continuous systems.

2. DISCRETELY CONTROLLED CONTINUOUS SYSTEMS

2.1 *Hybrid model*

The paper considers hybrid systems shown in Fig. 1. The plant is described by the state-space model

$$\dot{\mathbf{x}} = \mathbf{g}(\mathbf{x}(t), q(t)), \quad \mathbf{x}(0) = \mathbf{x}_0, \qquad (1)$$

where $\mathbf{x} \in \mathcal{R}^n$ denotes the state vector and $q \in \mathcal{Q} = \{1, 2, ..., n_q\}$ the operation mode. For constant operation mode ($q(t) = \bar{q}$) the model (1) is assumed to have a unique solution $\mathbf{x}(\bar{q}, \mathbf{x}_0, t)$.

The controller evaluates the state trajectory and generates an event $e \in \mathcal{E}$ whenever

$$\Phi_e(\mathbf{x}(t), q(t), t) = 0 \qquad (2)$$

is satisfied. The time $\bar{t}$ denotes the time t for which Eqn. (2) holds. At this time, the state $\mathbf{x}(t)$ has the value $\bar{\mathbf{x}}'$. Then the next operation mode $\bar{q}'$ is determined according to a discrete state transition function C of the controller:

$$\bar{q}' = C(\bar{q}, e). \qquad (3)$$

The control law (2), (3) can be lumped together

$$\bar{q}' = H_q(\bar{q}, \bar{\mathbf{x}}', \bar{t}) \qquad (4)$$
$$\bar{t}' = H_t(\bar{q}, \bar{\mathbf{x}}', \bar{t}), \qquad (5)$$

where $\bar{q}$ and $\bar{t}$ denote the "old" operation mode and the time instant that this operation mode has been assumed, and $\bar{q}'$ and $\bar{t}'$ are the "next" operation mode and event time. The overall model (1), (4) is referred to as the model Σ_H.

2.2 *Behaviour*

The model Σ_H has, for every initial state $\mathbf{x}_0$ and initial operation mode q_0, a unique solution. The time instances of the operation mode changes are enumerated by the counter k and denoted by $\bar{t}(k)$. The continuous state at these time points is $\bar{\mathbf{x}}(k) = \mathbf{x}(\bar{t}(k))$ and the k-th operation mode is $\bar{q}(k)$, where $q(t) = \bar{q}(k)$ for $\bar{t}(k) \le t < \bar{t}(k+1)$ holds. The continuous state movement between two consecutive switching time points is denoted by $\mathbf{x}_{[\bar{t}(k), \bar{t}(k+1)]}$. In summary, the behaviour of the discretely controlled system is described by the *hybrid state sequence*:

$$\mathrm{Traj}_H(q_0, \mathbf{x}_0) = \qquad (6)$$
$$\left(\begin{pmatrix} \bar{t}(0) \\ \bar{\mathbf{x}}(0) \\ \bar{q}(0) \end{pmatrix}, \mathbf{x}_{[\bar{t}(0), \bar{t}(1)]}, \begin{pmatrix} \bar{t}(1) \\ \bar{\mathbf{x}}(1) \\ \bar{q}(1) \end{pmatrix}, \mathbf{x}_{[\bar{t}(1), \bar{t}(2)]}, \ldots \right)$$

2.3 *Abstract descriptions of the behaviour*

A more abstract representations focuses on the time instants at which the operation mode is changed. It is referred to as the discrete-event behaviour, which is described by the sequences of switching time points, operation modes and the values of the continuous state at the switching time instances:

$$\mathrm{Traj}_E(q_0, \mathbf{x}_0) = \left(\begin{pmatrix} \bar{t}(0) \\ \bar{\mathbf{x}}(0) \\ \bar{q}(0) \end{pmatrix}, \begin{pmatrix} \bar{t}(1) \\ \bar{\mathbf{x}}(1) \\ \bar{q}(1) \end{pmatrix}, \ldots \right) \quad (7)$$

Traj_E is called the *timed event/state sequence.* A more abstract behaviour results if the continuous state is ignored:

$$\mathrm{Traj}_S(q_0, \mathbf{x}_0) = \left(\begin{pmatrix} \bar{t}(0) \\ \bar{q}(0) \end{pmatrix}, \begin{pmatrix} \bar{t}(1) \\ \bar{q}(1) \end{pmatrix}, \ldots \right) \quad (8)$$

Traj_S describes the *timed event sequence.* If the temporal information is ignored, a *logic event sequence*

$$\mathrm{Traj}_N(q_0, \mathbf{x}_0) = (\bar{q}(0),\ \bar{q}(1),\ \bar{q}(2), \ldots) \quad (9)$$

is obtained. The relation among the behaviours (6) – (9) can be represented by projection operators P_E, P_S and P_N:

$$\mathrm{Traj}_E = P_E(\mathrm{Traj}_H)$$
$$\mathrm{Traj}_S = P_S(\mathrm{Traj}_H)$$
$$\mathrm{Traj}_N = P_N(\mathrm{Traj}_H).$$

3. PROBLEM STATEMENT

3.1 *Diagnostic problem*

A diagnostic algorithm to be developed has to decide whether a fault has occurred (fault detection) and which fault has occurred (fault identification). It uses a model Σ and measurement information, which describe the current system trajectory Traj. Four approaches will be presented that use the measured trajectories Traj_H, Traj_E, Traj_S and Traj_N together with appropriate models to solve the following problem, where the index $_*$ is replaced by E, S, T or N:

Given: Model Σ_*
Trajectory Traj_*
Find: Set $\mathcal{F}_*$ of fault candidates

For the simplicity of presentation, it is assumed that the fault f does not change if the diagnostic algorithm is applied and that the initial states $\mathbf{x}_0$, q_0 are known.

3.2 *Model hierarchy*

Four abstract representations of discretely controlled systems will be developed, which comprise a model hierarchy (Fig. 3). It will become clear that abstract representations, which will be introduced in the next sections, may be nondeterministic. To get a unified notation, the solutions of all models are represented by some sets that are called the model behaviour and denoted by $\mathcal{B}_H$, $\mathcal{B}_E$, $\mathcal{B}_S$, $\mathcal{B}_T$ or $\mathcal{B}_N$, respectively.

Model	Measurement	Diagnostic result
Nondeterministic automaton Σ_N	Logic event sequence	$\mathcal{F}_N$
Timed automaton Σ_T	Timed event sequence	$\mathcal{F}_T$
Semi-Markov process Σ_S	Timed event sequence	$\mathcal{F}_S$
Embedded map Σ_E	Timed event/state sequence	$\mathcal{F}_E$
Hybrid model Σ_H	Hybrid state sequence	$\mathcal{F}_H$

Abstraction ↑

Fig. 3: Model hierarchy

The modelling aim is to set up these models such that the relation $\mathcal{B}_E = P_E(\mathcal{B}_H)$, $\mathcal{B}_S = P_S(\mathcal{B}_H)$, $\mathcal{B}_T = P_T(\mathcal{B}_H)$ or $\mathcal{B}_N = P_N(\mathcal{B}_H)$, respectively, holds. The application of the operator P_* to the set $\mathcal{B}_*$ means to apply the operator to all elements of the set. However, as the models to be set up are abstract representations, these equalities cannot be satisfied. In order to ensure the validity of the diagnostic results obtained by means of these models, all models have to be complete:

Definition 1. A model Σ_A with the behaviour $\mathcal{B}_A$ is called **complete** with respect to the discretely controlled system and the abstraction operator P_B if it satisfies the relation

$$\mathcal{B}_A \supseteq P_B(\mathcal{B}_H). \quad (10)$$

3.3 *Consistency-based diagnosis*

The idea of consistency-based diagnosis can be briefly explained as follows (Blanke et al. 2003). A fault is known to occur if the model of the faultless system cannot generate the trajectory measured (fault detection). Then it is said that the model and the trajectory are inconsistent. Similarly, fault identification can be accomplished by testing the consistency of the trajectory with the model of the system subject to some faults f. The result is the set $\mathcal{F}$ of *fault candidates,* which include all faults f for which the trajectory is consistent with the model subject to fault f.

Definition 2. Consider a system whose trajectory Traj_* has been measured over some time horizon t_h. The model is represented by its behaviour $\mathcal{B}_*$ over the same time horizon. The trajectory is called **consistent** with the model, if

$$\mathrm{Traj}_* \in \mathcal{B}_*. \quad (11)$$

In order to determine the set of fault candidates, the consistency of the measured trajectory is tested with respect to the behaviour of models that describe the discretely controlled continuous system subject to the faults $f \in \bar{\mathcal{F}}$ where $\bar{\mathcal{F}}$ is the set of all faults under consideration. The behaviour of these models are denoted by $\mathcal{B}_*(f)$. Then the set of fault candidates is defined by

$$\mathcal{F}_* = \{f \mid \text{Traj}_* \in \mathcal{B}_*(f)\}. \tag{12}$$

It is important to see that although the models used represent the discretely controlled continuous system with different granularity, the diagnostic results are clearly related to each other:

Corollary 1. If the model Σ_* is complete, the following relation holds:

$$f \in \mathcal{F}_H \Longrightarrow f \in \mathcal{F}_* \,. \tag{13}$$

That is, if f is a fault candidate with respect to the model Σ_H it is also a fault candidate obtained by any other model Σ_* discussed below.

4. DIAGNOSIS BY MEANS OF EMBEDDED MAPS

Model. If merely the discrete-event trajectory Traj_E should be used for diagnosis, the system state has only to be known at the switching instances. Then the diagnosis refers to the time instances and the points in the state space where the discretely controlled system generates events.

It has been investigated, for example in (Krupar et al. 2004), that it is possible to find a mapping G_E that allows to to determine the $(k+1)$-st triple $(\bar{t}(k+1), \bar{\mathbf{x}}(k+1), \bar{q}(k+1))$ from the knowledge of the k-th triple $(\bar{t}(k), \bar{\mathbf{x}}(k), \bar{q}(k))$. This mapping is called an *embedded mapping* of the hybrid system. It has the general form

$$\begin{pmatrix} \bar{\tau}(k+1) \\ \bar{\mathbf{x}}(k+1) \\ \bar{q}(k+1) \end{pmatrix} = \tilde{G}\left(\begin{pmatrix} \bar{\tau}(k) \\ \bar{\mathbf{x}}(k) \\ \bar{q}(k) \end{pmatrix}\right). \tag{14}$$

where $\bar{\tau}$ is the time span between consecutive switching times $\bar{\tau}(k) = \bar{t}(k) - \bar{t}(k-1)$ for $k \geq 1$ and $\bar{\tau}(0) = 0$. The embedded map is written here in the implicit form

$$G_E(\bar{\mathbf{x}}', \bar{q}', \bar{t}' - \bar{t}, \bar{\mathbf{x}}, \bar{q}) \in \{0, 1\}, \tag{15}$$

where G_E has the value 1 if two consecutive switchings of the discretely controlled system occur at the states $\bar{\mathbf{x}}(k) = \bar{\mathbf{x}}$ and $\bar{\mathbf{x}}(k+1) = \bar{\mathbf{x}}'$. The operation mode $\bar{q}(k+1) = \bar{q}$ is entered at time $\bar{t}(k) = \bar{t}$ and left towards the mode $\bar{q}(k+1) = \bar{q}'$ at the time instance $\bar{t}(k+1) = \bar{t}'$.

Determination of G_E. The embedded map G_E can be determined from the hybrid model Σ_H by integrating the differential equation (1) from the state $\mathbf{x}(\bar{t}) = \bar{\mathbf{x}}$ between two switching times $\bar{t}$ and $\bar{t}'$ to determine the state $\mathbf{x}(\bar{t}') = \bar{\mathbf{x}}'$ and by applying the control law (4) to find the next operation mode $\bar{q}'$. Hence, for the next switching the following relations hold:

$$\begin{aligned} \bar{\mathbf{x}}' &= \mathbf{x}(\bar{q}, \bar{\mathbf{x}}, \bar{t}' - \bar{t}) \\ \bar{q}' &= H_q(\bar{q}, \bar{\mathbf{x}}', \bar{t}) \\ \bar{t}' &= H_t(\bar{q}, \bar{\mathbf{x}}', \bar{t}) \end{aligned}$$

The embedded map obviously yields the same timed event/state trajectory Traj_E as the hybrid model Σ_H. For a given initial state $(q_0, \mathbf{x}_0)$ this trajectory is unique and, hence, the behaviour

$$\mathcal{B}_E = \{\text{Traj}_E(q_0, \mathbf{x}_0)\} \tag{16}$$

is a singleton. Consequently, the model Σ_H is complete: $P_E(\mathcal{B}_H) = \mathcal{B}_E$.

Consistency test. To check the consistency of the timed event/state sequence Traj_E with the model Σ_E, every single state transition from $(\bar{q}, \bar{\mathbf{x}})$ towards $(\bar{q}', \bar{\mathbf{x}}')$ after the time span $\bar{t}' - \bar{t}$ is consistent with the model if

$$G_E(\bar{\mathbf{x}}', \bar{q}', \bar{t}' - \bar{t}, \bar{\mathbf{x}}, \bar{q}) = 1$$

holds. A sequence (7) is consistent with Σ_E if

$$\prod_{k=0}^{k_e-1} G_E(\bar{\mathbf{x}}(k+1), \bar{q}(k+1), \ldots \\ \ldots \bar{t}(k+1) - \bar{t}(k), \bar{\mathbf{x}}(k), \bar{q}(k)) = 1. \tag{17}$$

k_e denotes the number of state changes that are recorded in the trajectory Traj_E.

5. DIAGNOSIS BY MEANS OF SEMI-MARKOV PROCESSES

Model. If the diagnosis should be carried out by only using the sequence of the discrete state $\bar{q}(k)$, the best representation is the timed event sequence Traj_T. Then, it suffices to represent the system by a semi-Markov process Σ_S. Then

$$G_S(\bar{q}', \bar{t}' - \bar{t}, \bar{q}) \in [0, 1] \tag{18}$$

describes the probability that the system, which has reached the operation mode $\bar{q}$ at time $\bar{t}$, switches at time $\bar{t}'$ into the mode $\bar{q}'$

$$G_S(\bar{q}', \bar{t}' - \bar{t}, \bar{q}) = \\ \text{Prob}(q_p(k+1) = \bar{q}', \bar{\tau}_p(k) = \bar{t}' - \bar{t} \mid q_p(k) = \bar{q}),$$

where the index "p" identifies stochastic variables. In particular, $\bar{\tau}_p(k)$ is the stochastic variable that describes the sojourn time in the operation mode $q_p(k)$. For details of this model cf. (Lunze 1999).

For a given initial state $\bar{q}(0) = q_0$ and initial time $\bar{t}(0) = 0$ the model (18) generates several timed event sequences Traj_S all of which are lumped into the set $\mathcal{B}_S$.

Determination of G_S. As shown in (Lunze 1999), the state transition relation G_S of the semi-Markov process can be determined from the given hybrid model Σ_H such that the model is complete. The relation to the embedded map is stated as follows:

$$G_E(\bar{\mathbf{x}}', \bar{q}', \bar{t}' - \bar{t}, \bar{\mathbf{x}}, \bar{q}) = 1 \Rightarrow G_S(\bar{q}', \bar{t}' - \bar{t}, \bar{q}) > 0.$$

As the behaviour $\mathcal{B}_S$ of the semi-Markov process includes all timed event sequences that occur with a positive probability, this relation implies

$$P_S(\mathrm{Traj}_H) \in \mathcal{B}_S, \qquad (19)$$

i. e. the model is complete.

Consistency test. Any trajectory that may occur with positive probability is called consistent with the model Σ_S. Hence the timed event sequence Traj_S is consistent with this model if and only if

$$\prod_{k=0}^{k_e-1} G_S(\bar{q}(k+1), \bar{t}(k+1) - \bar{t}(k), \bar{q}(k)). > 0 \quad (20)$$

6. DIAGNOSIS BY MEANS OF TIMED AUTOMATA

Model. A more abstract representation Σ_T of the discrete-event behaviour of the system uses a timed automaton, which needs only to have a single clock, which is reset for every state transition. The state transition function is

$$G_T(\bar{q}', \bar{t}' - \bar{t}, \bar{q}) \in \{0, 1\}, \qquad (21)$$

where G_T has the value 1 if the discretely controlled system can reside in the operation mode $\bar{q}$ for the time duration $\bar{t}' - \bar{t}$ before the operation mode is changed to become $\bar{q}'$.

Determination of G_T. The function G_T can be determined from G_S as follows:

$$G_S(\bar{q}', \bar{t}' - \bar{t}, \bar{q}) > 0 \;\Rightarrow\; G_T(\bar{q}', \bar{t}' - \bar{t}, \bar{q}) = 1.$$

Hence, all trajectories $\mathrm{Traj}_S(q_0)$ that the semi-Markov model generates with a positive probability are also generated by the timed automaton and $\mathcal{B}_S = \mathcal{B}_T$ holds. The completeness of the model Σ_T follows directly from the completeness of the model Σ_S.

Consistency test. The consistency test is the same as for the semi-Markov model Σ_S but it does not yield the additional information about the probability that the timed event sequence is consistent with the model Σ_T. The sequence Traj_S is consistent with the timed automaton if

$$\prod_{k=0}^{k_e-1} G_T(\bar{q}(k+1), \bar{t}(k+1) - \bar{t}(k), \bar{q}(k)) > 0. (22)$$

7. DIAGNOSIS BY MEANS OF NONDETERMINISTIC AUTOMATA

Model. If the temporal information is ignored and the diagnostic task should be solved by the logical event sequence Traj_N, a nondeterminstic automaton Σ_N is the suitable model of the discretely controlled system. Its state transition function

$$G_N(\bar{q}', \bar{q}) \in \{0, 1\} \qquad (23)$$

has the value 1 if the discretely controlled system can change its operation mode from $\bar{q}$ towards $\bar{q}'$.

Determination of G_N. The function G_N results from the state transition relation of the timed automaton by ignoring the temporal information:

$$\exists \bar{t}, \bar{t}' : G_T(\bar{q}', \bar{t}' - \bar{t}, \bar{q}) = 1 \;\Rightarrow G_N(\bar{q}', \bar{q}) = 1. \quad (24)$$

Due to the completeness of the timed model Σ_T and the relation (24), the model Σ_N is complete.

Consistency test. For the nondeterministic automaton the temporal information is ignored. Hence, from the consistency test (22) the following conditions are obtained. The logic event sequence Traj_N is consistent with the model Σ_N if

$$\prod_{k=0}^{k_e-1} G_N(\bar{q}(k+1), \bar{q}(k)) > 0. \qquad (25)$$

8. HIERARCHIES OF MODELS AND DIAGNOSTIC RESULTS

This section summarises the relation among the different models and the diagnostic results obtained by these models.

Model hierarchy. The more abstract the representation Σ_* is the more trajectories are included in the model behaviour $\mathcal{B}_*$. As these trajectories describe the system on different abstraction levels, a comparison can only be done by projecting them to the same abstraction level.

Theorem 1. The models of the discretely controlled continuous system form the hierarchies

$$\begin{aligned} &\mathcal{B}_N \supseteq P_{NT}(\mathcal{B}_T) \supseteq P_{NS}(\mathcal{B}_S) \supseteq P_{NE}(\mathcal{B}_E) \\ &\quad \supseteq P_{NH}(\mathcal{B}_H). \\ &\mathcal{B}_T \supseteq \mathcal{B}_S \supseteq P_{SE}(\mathcal{B}_E) \supseteq P_{SH}(\mathcal{B}_H) \\ &\mathcal{B}_E \supseteq P_{EH}(\mathcal{B}_H), \end{aligned}$$

where P_{N*}, P_{S*} and P_{EH} denote the projection operators that maps the behaviour of the model Σ_* towards the set of logic event sequences, timed event sequences or timed state/event sequences, respectively, if the state transition functions G_* are chosen according to eqns. (16), (19), and (24).

Diagnostic results. Due to the completeness of all models used and the model hierarchy, the set of fault candidates obtained by the diagnostic algorithm with the different models form a hierarchy.

Theorem 2. If the models are complete, the diagnostic results obtained by means of these models satisfy the following relations:

$$\mathcal{F}_N \supseteq \mathcal{F}_T \supseteq \mathcal{F}_S \supseteq \mathcal{F}_E \supseteq \mathcal{F}_H. \qquad (26)$$

In the theorem, $\mathcal{F}_N$,..., $\mathcal{F}_H$ are the sets of fault candidates obtained by means of the models Σ_N,..., Σ_H. The proof of this theorem uses the relations among the models given in the last sections. For example, the relation $\mathcal{F}_N \supseteq \mathcal{F}_T$ follows directly from the consistency conditions (25) and (22) for both models and the relation (24) between both models.

The theorem shows that the more information about the system is used for diagnosis, the fewer elements are in the set of fault candidates. Although an improvement of the diagnostic result by using more information about the system is an intuitively clear result, the strict inclusion described by eqn. (26) can only be obtained only due to the additional requirement that all models should be complete.

9. CONCLUSION

The paper has shown that discretely controlled continuous systems can be diagnosed by using different levels of information. If the model used is complete, the diagnostic result is valid in the sense that it includes all possible fault candidates.

The hierarchy of models and diagnostic results shown make it possible to adapt the measurement and modelling information used to the practical circumstances of the system under consideration.

Due to space limitations, the diagnostic methods have been explained under two assumptions, which can be released. First, if the initial state is unknown, the diagnostic algorithms have to be applied by considering all possible initial states $q_0 \in \mathcal{Q}$ and $\mathbf{x}_0 \in \mathcal{R}^n$. Then the diagnostic algorithm includes a state observer. Second, if the fault changes during the run of the diagnostic algorithm, fault models have to be used to restrict the temporal behaviour of the fault.

REFERENCES

M. Blanke, M. Kinnaert, J. Lunze, M. Staroswiecki, *Diagnosis and Fault-Tolerant Control* (2nd edition), Springer-Verlag, Heidelberg 2006.

D. Förstern, J. Lunze, Discrete-event models of quantized systems for diagnosis, *Intern. J. Control* ßbf 74 (2001), pp. 690-700.

J. Gertler, *Fault Detection and Diagnosis in Enginnering Systems*, Marcel Dekker, New York 1998.

J. Krupar, A. Mögel, W. Schwarz, Continuous-discrete systems - modelling and statistical analysis, *Intern. Symposium on Nonlinear Theory and its Applications (NOLTA)*, Fukuoka 2004, pp 55-58.

J. Lunze, A timed discrete-event abstraction of continuous-variable systems, *Intern. Journal of Control* **72** (1999), 1147-1164.

J. Lunze, Diagnosis of quantized systems based on a timed discrete-event model, *IEEE Trans. on Systems, Man, and Cybernetics - Part A* **SMC-30** (2000), 322-335.

J. Raisch, S. O'Young, A totally ordered set of discrete abstractions for a given hybrid or continuous system, In *Hybrid Systems IV*, Springer-Verlag, Berlin 1997, pp. 342-360.

M. Sampath, R. Sengupta, S. Lafortune, K. Sinnamohedeen, D. Teneketzis, Diagnosability of discrete event systems, *IEEE Trans.* **AC-40** (1995), 1555-1575.

P. Supavatanakul, *Modelling and Diagnosis of Timed Discrete-Event Systems*, Shaker-Verlag, Aachen 2004.

C. K. Tse, *Complex Behaviour of Switching Power Converters*, CRC Press, Boca Raton, 2003.

Copyright © Analysis and Design of Hybrid Systems
Alghero, Italy 2006

ELSEVIER
IFAC
PUBLICATIONS

USE OF AN OBJECT ORIENTED DYNAMIC HYBRID SIMULATOR FOR THE MONITORING OF INDUSTRIAL PROCESSES

Nelly Olivier, Gilles Hétreux, Jean-Marc Le Lann, Marie-Véronique Le Lann

Laboratoire de Génie Chimique (LGC),Département Procédés et Systèmes Industriels (PSI)
117 route de Narbonne, 31077 Toulouse, France

Abstract: *PrODHyS* is a dynamic hybrid simulation environment, which offers extensible and reusable object oriented components dedicated to the modelling of processes. The purpose of this communication aims at presenting the main concepts of *PrODHyS* through the modelling and the simulation of a hydraulic system. Then, the feasible use of hybrid simulation in a supervision system is underlined. *Copyright © 2006 IFAC*

Keywords: Object oriented software components, modelling and dynamic simulation of hybrid systems, monitoring of semi-continuous process.

1. INTRODUCTION

In a very competitive economic context, the flexibility of the system of production can be a decisive advantage. Generally, this flexibility lies on the search for a greater reactivity to a fluctuating demand, but also to many risks occurring during the manufacture. In this context, a simple failure is considered as prejudicial. This is why the fault diagnosis is the purpose of a particular attention in the scientific and industrial community. The major idea is that the defect must not be undergone but must be controlled. Nowadays, the fault diagnosis remains a large research field. The literature quotes as many diagnosis methods as many domains of application (Venkatasubramanian *and al.*, 2003).
In our case, we deal with batch and semi-continuous processes which are the prevalent mode of production for low volume of high added value products. Such systems are composed of interconnected and shared resources, in which a continuous treatment is carried out. For this reason, they are generally considered as *hybrid systems* in which discrete aspects mix with continuous ones. Moreover, the recipe is more often described with state events (temperature or composition threshold, etc.) than with fixed processing times. As a consequence, the simulation of unit operations and physico-chemical evolution of products often necessitates the implementation of *phenomenological* models. In this context, the traditional tools such as continuous dynamic simulation or discrete event simulation are not well adapted to these problems and the use of hybrid dynamic simulators seems to be a better solution (Zaytoon, 2001).
In this framework, the first part of this communication focuses on the main fundamental concepts of the simulation library *PrODHyS*. These are illustrated through the simulation of a hydraulic system used as benchmark. Then, the potentialities of *PrODHyS* for process control are underlined. Finally the design of the supervision and diagnosis module currently developped is described.

2. *PrODHyS* ENVIRONMENT

The simulation of hybrid system has led to the development of several software such as *gPROMS* (Barton *and al.*, 1994), *BASIP* (Wöllhaf *and al.* 1996), in which the hybrid aspect is described via an imperative language. In parallel, various hybrid formalisms have been defined or obtained by extension of existing discrete or continuous formalisms (Zaytoon, 2001).
In this context, the research works performed for several years within the *PSE* research department *(LGC)* on process modelling and simulation have led to the development of *PrODHyS*. This environment provides a library of classes dedicated to the dynamic hybrid simulation of processes. Based on *object concepts*, *PrODHyS* offers extensible and reusable software components allowing a rigorous and systematic modelling of processes. The primal

contribution of these works consisted in determining and designing the foundation buildings classes. The last important evolution of *PrODHyS* is the integration of the dynamic hybrid simulation kernel (Hétreux *and al*, 2003, Perret *and al*, 2004, Olivier *and al*., 2005). Indeed, the nature of the studied phenomena involves a rigorous description of the continuous and discrete dynamic. The use of differential and algebraic equations (*DAE*) systems seems obvious for the description of continuous aspects. Moreover the high sequential aspect of the considered systems justifies the use of Petri nets model. This is why the *Object Differential Petri Nets (ODPN)* formalism is used to describe the simulation model associated with each component. It combines in the same structure a set of *DAE* systems and high level Petri nets (defining the legal sequences of commutation between states) and has the ability to detect *state* and *time events*.

A detailed description of this formalism can be found in (Hétreux *and al*., 2003). Figure 1 shows an example of evolution in the *ODPN*.

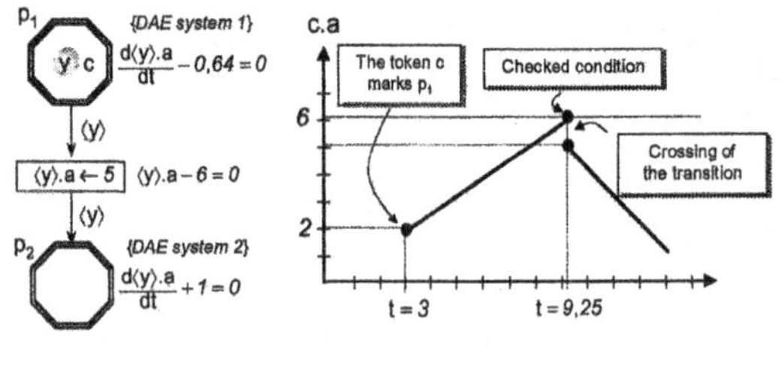

Fig. 1. *ODPN* evolution

Let us only notice that the object concepts and the Petri nets have been exploited through an *extended combined approach*. It consists in making interact these features either by "introducing objects into Petri nets" (use of individualized object tokens carrying properties and methods) or by "introducing Petri nets into objects" (description of the internal behaviour of an object).

3. PROCESS MODELING WITH *PrODHyS*

3.1 General structure of the simulation model

The simulation of a discontinuous process necessitates to model separately the *command* part (the *supervisor*) and the *operative* part (the *process*). Concerning the operative part, the specification of any device of *PrODHyS* is always defined according to two axes: a *topological* axis and a *phenomenological* axis. The topological axis defines the structure of the process (*system* vision): physical connections (material, energy, information) between the different parts of the process and hierarchical decomposition of the devices. The phenomenological axis rests on a mathematical model based on mass and energy balances and thermodynamic and physicochemical laws. Thus, the models of devices are reusable whatever the context. In addition, the *combined approach* is used to dissociate the model of material from the model of devices which contains the material. Thus, object tokens are reusable and reduce the complexity of the devices Petri nets. More details on the modelling of devices and material can be found in previous communications (Perret *and al*., 2004). On the other hand, the model of the command part is specific to the recipe and the process topology. It consists in describing the procedure of manufacture of each product. So, it specifies the assignments of resources and the sequence of tasks ordered in time necessary to the realization of each batch. All these models are merged only when a simulation model is instantiated. Thus, the size and the structure of the resulting *DAE* systems change all along the simulation, according to the actual state of the process.

3.2 Connections between « devices » PN and « recipe » PN

The exchanged signals, between the *command* part and the *operative* part, are modelled by a discrete place. The state of a signal state is associated to the marking of the corresponding place. In this framework, an entity is either an *active* device if it has one or more *signal* places (such as valves, pumps, feeds, column, captors) or a *passive* device if there is no direct relation with the recipe (such as simple tanks or reactors). These notions are illustrated on figure 2. It represents an operative sequence which permits the feed of a tank until a fixed volume is reached. The marking of the signal place of an active entity induces the evolution of its Petri net. This Petri net can itself induce the evolution of active or passive entities in cascade through the net composed with the connection of different material or energy ports.

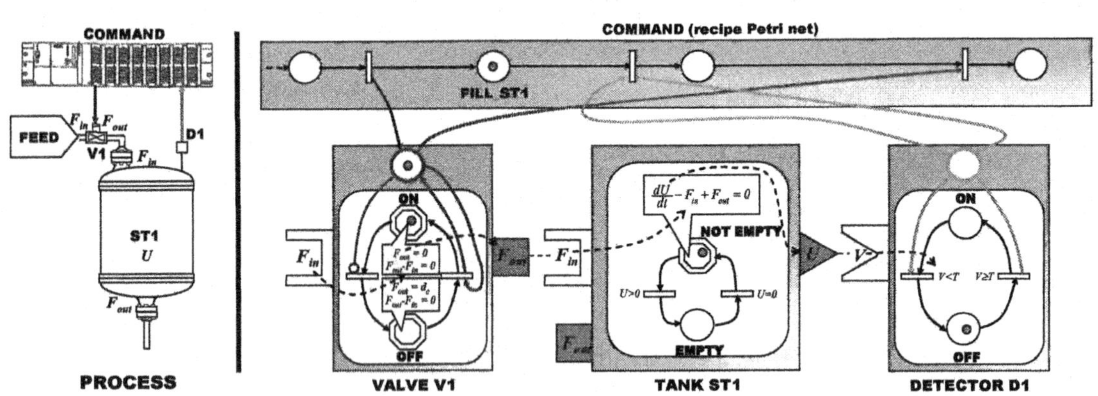

Fig. 2. Interactions between the command level and the process level

4. SIMULATION OF A HYDRAULIC SYSTEM

The considered hydraulic system (cf. figure 3) is inspired by a benchmark defined by the *AS193* "Diagnosis of the hybrid systems" (cf. www.univ-lille1.fr/lail/AS193/).

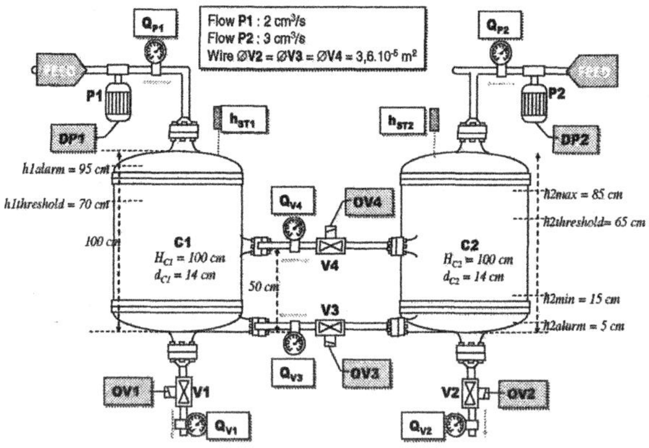

Fig. 3. *Flowsheet of the benchmark*

This system consists of two cylindrical tanks **C1** and **C2**, connected by two pipes with "on/off" valves **V3** and **V4**. The feed of the tanks is maintained by the "on/off" pumps **P1** and **P2**. The tank **C2** can be drained through the "on/off" valve **V2**. The valve **V1** is not used here. The instrumentation of the process is composed (in a maximal configuration) of 6 flow sensors and 2 level sensors.

The goal of the control device consists in maintaining the liquid level *h2* in **C2** between the heights *h2min* and *h2max* by controlling the valve **V4**. The valve **V3** is opened only when the level in **C2** is such *h2≤h2alarm.*

The implemented command is voluntarily simple. Because the command law does not take into account the level *h1* in the tank **C1**, the objective can not always be ensured. The Petri net associated with the command level is presented on figure 4:

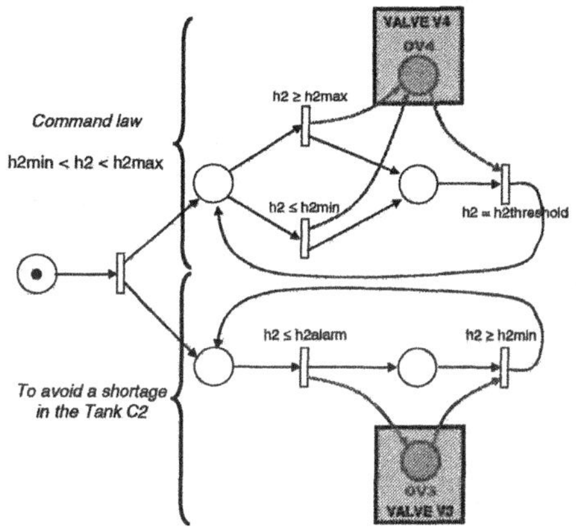

Fig. 4. *Command Petri net*

In this context, various scenarii can be simulated by action on the pumps **P1** and **P2** and the valve **V2**. For the set of parameters indicated on figure 3 and the scenario shown on the followed figures, the simulation results are presented on figure 5.

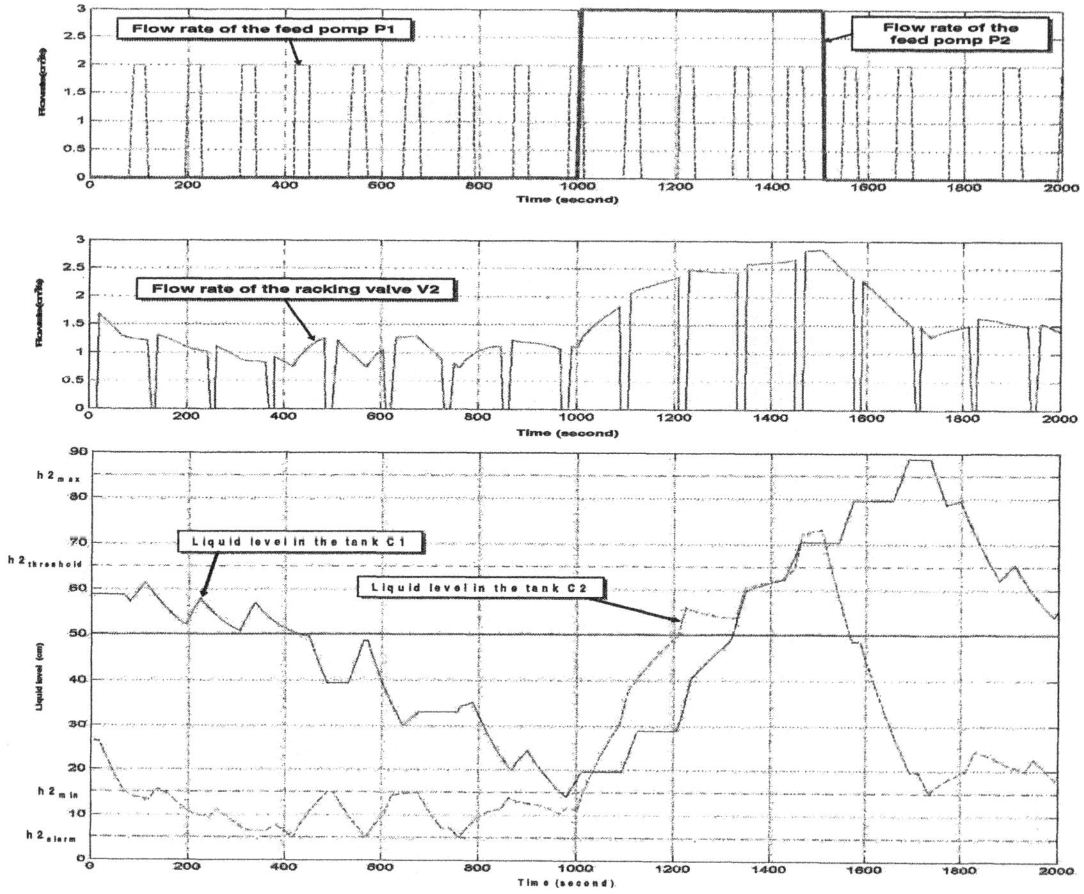

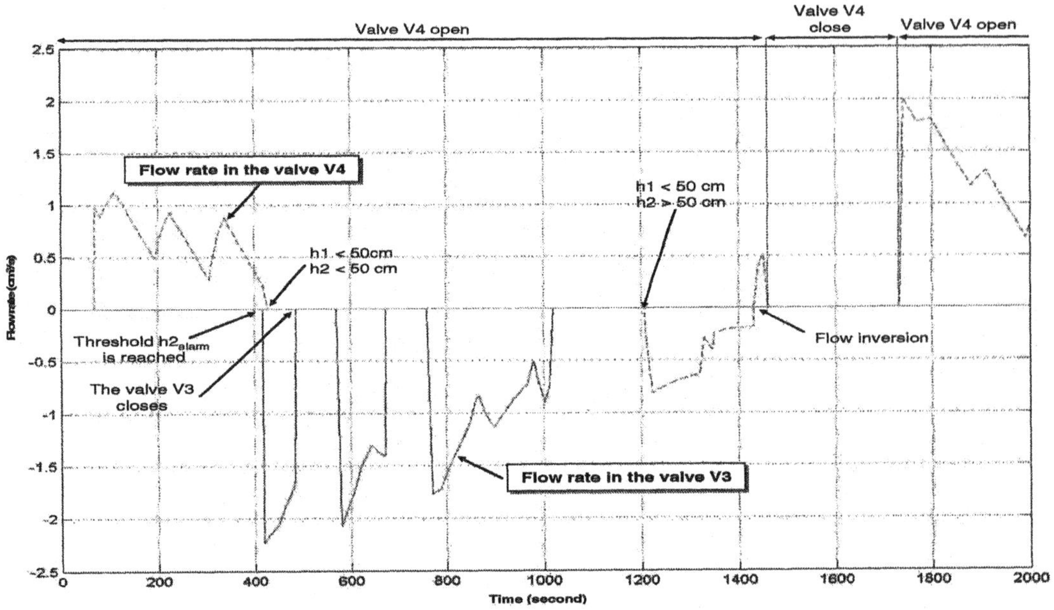

Fig. 5. *Simulation results*

5. USE OF DYNAMIC HYBRID SIMULATION IN A MODEL BASED DIAGNOSIS SYSTEM

Nowadays, for reasons of safety and performance, monitoring and supervision have an important role in process control. The complexity and the size of industrial systems induce an increasing number of process variables and make difficult the work of operators. In this context, a computer aided decision-making tool seems to be wise. For this purpose, the model of simulation of *PrODHyS* is used as a *reference* model to implement the functions of detection and diagnosis. The global principle of this system is shown on the Figure 6. In order to obtain an *observer* of the physical system, a real-time simulation is done in parallel. So, a complete state of the system will be available at any time.

5.1 Fault detection

The supervision module must be able to treat the faults of the physical systems (leak, energy loss, etc.) and the faults of the control/command devices (actuators, captors, etc.). As defined in (De Kleer *and al.* 1987), our approach is based on the hypothesis that the reference model is presumed to be correct. Thus, it is based on the comparison between the predicted behaviour obtained thanks to the simulation of the reference model (values of state variables) and the real observed behaviour (measurements from the

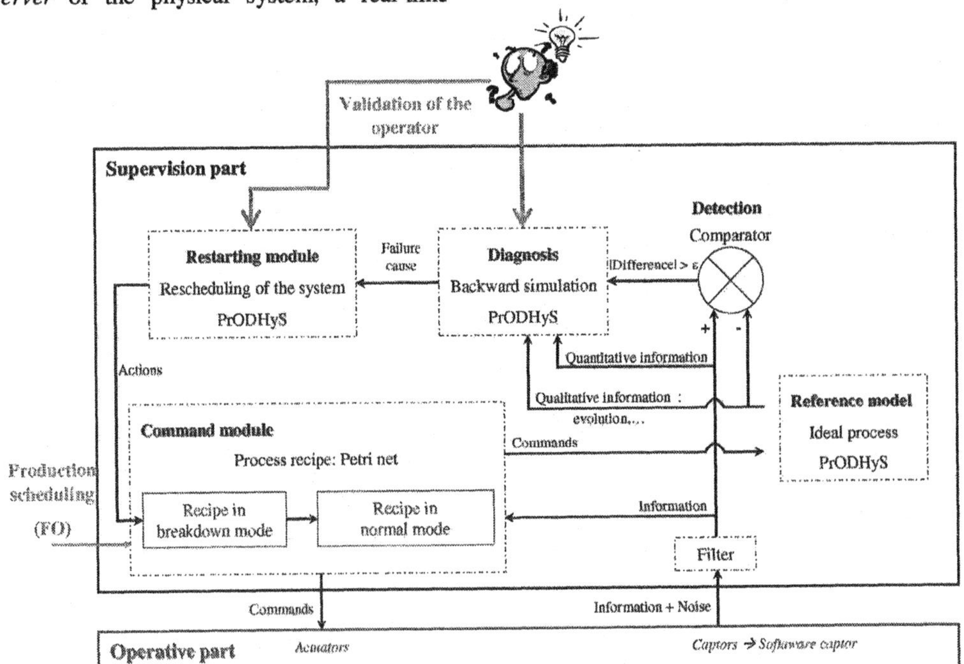

Fig. 6. *Principle of the supervision module*

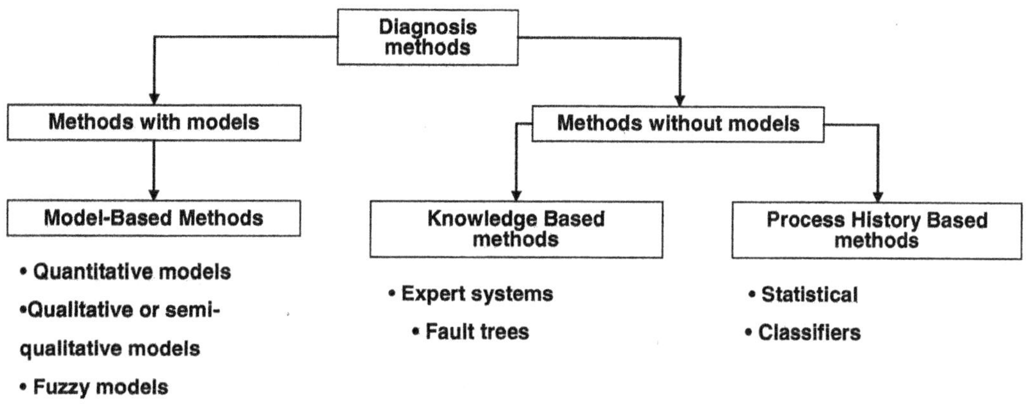

Fig. 7. *Classification of diagnosis methods*

process). Detection is realized by comparison with fixed thresholds. This method lies in the detection of the instant for which the monitored variables go beyond the bounds defined by the thresholds of the objective value (values of the variable within the reference model). Although some methods, such as detection with a variable threshold, can minimize the number of erroneous detections, (Duviella, 2005) indicates that the detection with a fixed threshold is more robust with the disturbances. In addition, a result can be obtained rather quickly and it is easy to implement and to parameterize this method.

For a consistent execution of this task, the measurements must be filtered in order to eliminate the noise. Nevertheless, the value ε, beyond which the difference between measurements and model variables is considered as a defect, remains a delicate point to evaluate. This value is often obtained from a compromise of a series of simulations, in which his value is customized.

Moreover, the evolution of the simulation must be synchronous with the evolution of the real process. However, this feature can not be always ensured. When the reference model is either ahead or late, a retiming of the reference model on the real process is then necessary in order to be able to validate a detection test.

Despite these mecanisms, let us underline the difficulty of the decision stage in the detection. Indeed, some particular states of a system can be found where a defect behaviour seems to be similar to a normal behaviour; in this condition, it is impossible to affirm the absence of defect. This is why the uncertainty on the veracity of the detection tests may lead to unsuccessful diagnosis.

5.2 Fault diagnosis

When the occurence of a defect is proven, the diagnosis stage is launched. This task must be able to determinate a possible cause, compatible with information coming from the process and the reference model. It consists generally in determining the defect components or organs of the physical system. A logical reasoning based on the knowledge available of the system is used. The fault diagnosis for the industrial processes is a large research field. Many methods and many classifications exist in the literature. Figure 7 presents a part of the classification proposed in (Venkatasubramanian *and al.* 2003). Our objective is to couple a model based method (Mosterman *and al.* 1999) with a method based on historical data. In this framework, two research ways are emphasized.

The first approach is based on a tool which integrates qualitative knowledge of the process (knowledge base). It obeys to the rules of the type « IF condition THEN cause » (classification methods). However, to refine the discrimination and the interpretation of the noticed defects, it seems to be attractive to exploit information from the simulation of the reference model, in particular the values of the state variables and the values of their derivatives. As a matter of fact, even if the measurements of the process are pertinent for the command, they are often insufficient for the diagnosis. Moreover, the simulation of the reference model allows the analysis of different scenarios and the validation (or not) of the deductions from the knowledge base.

The second approach consists to exploit the backward hybrid simulation of the process. Let us note that the Petri net is not bijective. Indeed the same final marking does not lead to the same initial marking. This induces the possibility of a conflict of transitions. The historical data of the process (from the real process and from the reference model) and the classification can solve this indeterminism by introducing probabilities in the transitions.

5.3 Modelling of defect

The implementation of this system is rather complex and it is still in development. So, in order to test our prototype, the supervised process is currently simulated with *PrODHyS*.

In this condition, the reference process (without defects) and the monitored process (with possible defects) have the same recipe: the command part doesn't change. Thus the defect appears in the modelling of the devices. The associated simulation model is build from specific *Device* objects in which the faults are defined in an intrinsic way.

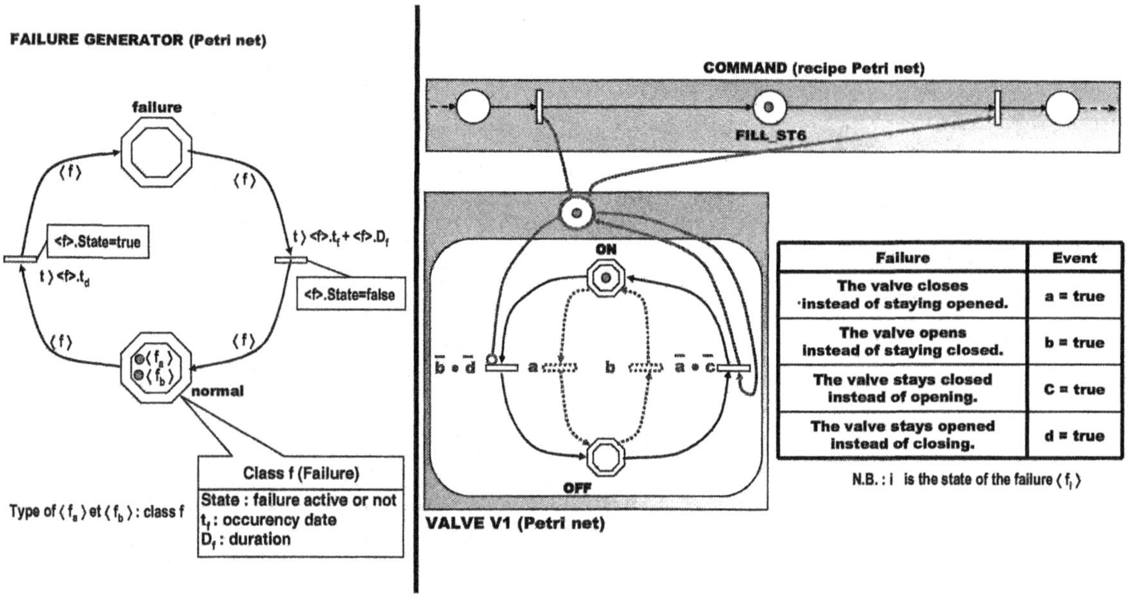

Fig. 8. *Modelling of a faulty valve*

In order to simplify the analysis of the system, the failures are generated by the simulation manager thanks to a calendar which lists the defect, its occurency time and its duration.

The figure 8 represents the modelling of a faulty valve. In the studied case, as we consider only the failures "a" and "b", the place *Normal* of the failure generator is only marked by two object tokens $<f_a>$ and $<f_b>$ (type of the objects: the class failure).

In a second time, during the validation stage, the failures are generated on autonomous and random commutations.

6. CONCLUSION

The object oriented approach brings many advantages in terms of software quality (extensibility, reutilisability, flexibility), but especially in terms of modelling thanks to a hierarchical and modular description which is both abstracted and close to reality. Based on these concepts, *PrODHyS* provides software components intended to model and simulate more specifically the industrial processes. The implementation of a formalism on high level of abstraction associated with powerful numerical methods of integration led to the construction of a robust hybrid dynamic simulator.

In this communication, the potentialities of *PrODHyS* are illustrated through the modelling and the simulation of a hydraulic process. The works in progress aim at integrating this simulation model within a model based diagnosis system. The coupling between a method of classification and a method of backward simulation is able to reduce the number of possibilities for the cause of the observed defect and to validate the possible fault.

REFERENCES

Barton P.I., Pantelides C.C. The Modelling of Combined Discrete/Continuous Processes. AIChE Journal, No. 40, pp.966-979 , 1994

De Kleer J., Williams B. C., Diagnosing multiple faults, Artificial Intelligence, Vol. 32, pp. 97-130, 1987

Duviella E. Conduite réactive des systèmes dynamiques étendus à retards variables. Cas des réseaux hydrographiques, Thèse de Doctorat, ENI Tarbes, France, 2005

Hétreux G., Perret J., LeLann J.M., Object hybrid formalism for modelling and simulation of chemical processes, *ADHS'03*, Saint-Malo, France, 2003

Mosterman P.J., Biswas G., Diagnosis of continuous valued systems in transient operating regions, IEEE Transactions on Systems, Man, and Cybernetics Part A, Vol. 29, No. 6, pp. 554-565, 1999

Olivier N., Hétreux G., Lelann J.M., Formal modelling and simulation for control of batch processes, Conference on Conceptual Modelling and Simulation CMS'05; Marseille, France, 2005

Perret J., Hétreux G., LeLann J.M., Integration of an object formalism within a hybrid dynamic simulation environment, Control Engineering Practice, Vol. 12/10, pp. 1211-1223, 2004

Perret J., Thery R., Hétreux G., Le Lann J.M., 2003. Object-oriented components for dynamic hybrid simulation of a reactive column process, ESCAPE 13

Venkatasubramanian V., Rengaswamy R., Yin K., Kavuri S. N. A review of process fault detection and diagnosis, Computers & Chemical Engineering (Elsevier), Vol. 27, pp. 293-346, 2003.

Wöllhaf K., Fritz M., Schulz C. and Engell S., BaSiP — Batch process simulation with dynamically reconfigured process dynamics, Computers & Chemical Engineering, Vol. 20, Supplement 2, pp. S1281-S1286, 1996

Zaytoon J. Systèmes dynamiques hybrides. HERMES Sciences publications, 2001

ELSEVIER
IFAC
PUBLICATIONS

MODEL PREDICTIVE CONTROL OF NONLINEAR MECHATRONIC SYSTEMS: AN APPLICATION TO A MAGNETICALLY ACTUATED MASS SPRING DAMPER [1]

Stefano Di Cairano * Alberto Bemporad *
Ilya Kolmanovsky ** Davor Hrovat **

* *Dip. Ingegneria dell'Informazione, Università di Siena, Italy. Email:* dicairano,bemporad@dii.unisi.it.
** *Ford Motor Company, Dearborn, Michigan, USA. Email:* ikolmano,dhrovat@ford.com.

Abstract: Mechatronic systems in the automotive applications are characterized by significant nonlinearities and tight performance specifications further exacerbated by state and input constraints. Model Predictive Control (MPC) in conjunction with hybrid modeling can be an attractive and systematic methodology to handle these challenging control problems. In this paper, we focus on a mass spring damper system actuated by an electromagnet, which is one of the most common elements in the automotive actuators, with fuel injectors representing a concrete example. We present two designs which are based, respectively, on a linear MPC approach in cascade with a nonlinear state-dependent saturation, and on a hybrid MPC approach. The performance and the complexity of the two MPC controllers are compared. *Copyright © 2006 IFAC*

Keywords: Model Predictive Control, Automotive Applications, Hybrid systems.

1. INTRODUCTION

Automotive actuators, such as fuel injectors, are examples of mechatronic systems (Hrovat *et al.*, 2000; Barron and Powers, 1996) that are characterized by tight operating requirements (such as high precision, low power consumption, fast transition time), significant nonlinearities, as well as input and state constraints which need to be enforced during the system operation. On the other hand, their dynamics may often be characterized by relatively low-dimensional models.

Model Predictive Control (MPC) (Qin and Badgwell, 2003) is a systematic feedback control design technique which determines the control input via receding horizon optimal control. Its main appeal is in being able to enforce pointwise-in-time constraints while providing the control designer with direct capability to shape the transient response by adjusting the weights in the objective function being minimized. MPC controllers can handle continuous-valued and discrete-valued control inputs, accommodate system parameter changes or subsystem faults, as long as they are reflected in the model used for on-line optimization.

Automotive actuators can often be adequately characterized by low dimensional models, and in this case an explicit implementation of the MPC controller becomes possible (see e.g. (Giorgetti *et al.*, 2005)), whereby the solution is pre-computed off-line and its representation is stored for on-

[1] Work (partially) done in the framework of the HYCON Network of Excellence, contract number FP6-IST-511368.

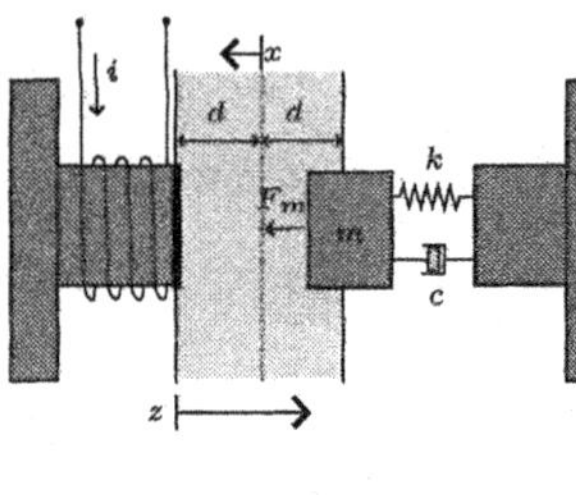

Fig. 1. The schematics of a magnetically actuated mass spring damper system.

line application. The on-line optimization is not required and the computational effort can be reduced to the point where the implementation of these control algorithms becomes feasible within stringent memory and chronometric constraints of automotive micro-controllers.

In this paper, we discuss and illustrate this approach in more detail in application to an electromagnetically actuated mass-spring damper system. Such a system arises very frequently in automotive actuation mechanisms (see (Hrovat *et al.*, 2000; Guzzella and Sciarretta, 2005) and references therein), including fuel injectors. In an actuation system of this kind, there is typically a moving mass which operates against a spring and a damper, while being controlled by a magnetic force from an electromagnetic coil. The force from the coil is unidirectional (i.e., the coil can only attract but not repel the mass), and this force decays inversely proportionally to the square of the distance between the mass and the coil. This force is also proportional to the square of the current, which is controlled to the desired value by an inner loop controller. By neglecting the much faster electrical dynamics, in this paper we consider a second order mechanical system with the magnetic force being the control input. The effect of electrical dynamics is incorporated in an extension of this work, see (Di Cairano *et al.*, 2006).

We present two controller designs, a linear MPC, where a nonlinear constraint on the magnetic force is neglected in the design and subsequently enforced by cascading a nonlinear state-dependent saturation, and a hybrid MPC, which considers also the nonlinear force constraints through a piecewise affine approximation.

2. PHYSICAL MODEL AND CONSTRAINTS

We consider a linear model for the moving mass dynamics in the form

$$\begin{bmatrix} \dot{x}_1(t) \\ \dot{x}_2(t) \end{bmatrix} = \begin{bmatrix} 0 & 1 \\ -\frac{k}{m} & -\frac{c}{m} \end{bmatrix} \begin{bmatrix} x_1(t) \\ x_2(t) \end{bmatrix} + \begin{bmatrix} 0 \\ \frac{1}{m} \end{bmatrix} u(t), \quad (1)$$

where the states are, respectively, the position [m] of the moving mass x_1, and the velocity of the moving mass x_2 [m/sec]. The magnetic force [N] acting on the moving mass is denoted by u. The neutral position of the spring corresponds to $x_1 = 0$ while the coil is located at $x_1 = x_{1c} = 4 \cdot 10^{-3}$ m. The magnetic force is given by

$$u(t) = k_a \frac{i^2(t)}{(z(t) + k_b)^2},$$

where $z = x_{1c} - x_1$ is the distance between the moving mass and the coil, i is the current [A] through the coil and k_a, k_b are constant parameters.

In this paper we consider the case when the inner-loop controller is capable of controlling the current to the desired set-point on a faster time-scale than the mechanical dynamics of the system, so that we may view $u(t)$ as the control input in (1). In order to realize a given $u(t)$ by the current, $i(t)$, $u(t)$ must satisfy the following constraints

$$u(t) \geq 0, \quad (2a)$$

$$u(t) \leq k_a \frac{i_{max}^2}{(z(t) + k_b)^2} \quad (2b)$$

where $i_{max} = 10$ A denotes the maximum current through the coil. The first of these constraints reflects the fact that the magnetic force is unidirectional (i.e., the electromagnet can only attract and not repel the moving mass). The second constraint is due to the limitations of the current which can be delivered by the power electronic circuits in the system. Note that (2) defines a non-convex set in the input+state space, being the intersection of a halfspace and of the hypograph of a convex function.

Besides (2), additional constraints are introduced to bound the moving mass position between the coil and a symmetric stop on the other end, i.e., at $-4 \cdot 10^{-3}$ m,

$$-4 \cdot 10^{-3} \leq x_1(t) \leq 4 \cdot 10^{-3}. \quad (3)$$

Even though the moving mass cannot penetrate into the coil or into the symmetric stop on the other end, these constraints have to be imposed explicitly to preserve the validity of model (1). If this is not done, undesirable moving mass bouncing can create noise and increase wear of the parts.

In a number of practical applications it is actually desirable to control the moving mass so that it is positioned against the coil with $x_1 = x_{1c}$. As the moving mass approaches the coil, its velocity needs to be carefully controlled to avoid high collision velocities (this is called *soft-landing*). In addition, maintaining the velocity of approach relatively low reduces the disturbance to the current control loop. In this paper we use the constraint

$$-\varepsilon - \beta(x_{1c} - x) \leq \dot{x} \leq \varepsilon + \beta(x_{1c} - x) \quad (4)$$

to limit the velocity of approach, where $\beta = 2500\ s^{-1}$ and $\varepsilon = 0.2$ m/s. The maximum allowed absolute value of the moving mass velocity at the neutral position (i.e., at $x_1 = 0$) is $v_{max} = 10.2$ m/s and it is 0.2 m/s at the contact position with the coil (i.e., at $x_1 = x_{1c}$).

To facilitate the application of Model Predictive Control, system model (1) is translated into the discrete-time model

$$\begin{bmatrix} x_1(t+1) \\ x_2(t+1) \end{bmatrix} = \begin{bmatrix} 0.89 & 4.7\cdot10^{-4} \\ -42.77 & 0.85 \end{bmatrix} \begin{bmatrix} x_1(t) \\ x_2(t) \end{bmatrix} + \begin{bmatrix} 7\cdot10^{-7} \\ 2.85\cdot10^{-3} \end{bmatrix} u(t), \quad (5)$$

where we have considered the sampling period $T_s = 0.5$ ms. With the discrete-time approach, constraints (2), (3) and (4) are enforced only at the sampling instants kT_s, $k \in \mathbb{N}$, for system (5).

3. LINEAR MODEL PREDICTIVE CONTROL

Model Predictive Control (Qin and Badgwell, 2003) is an optimization-based closed-loop control strategy in which pointwise-in-time design constraints on system's state, input and output can be explicitly embedded into the controller and, at the same time, it is a closed-loop strategy, since at each time instant the optimization is repeated using the most recent measurements.

The MPC strategy is based on the solution of the optimal control problem

$$\begin{aligned} \min_{\{y_k,u_k\}_{k=0}^{N-1}} \ & \sum_{k=0}^{N_J-1} (y_k - r_y(t))'Q_y(y_k - r_y(t)) + \\ & \Delta u_k' Q_{\Delta u} \Delta u_k \\ \text{subject to}\ & y_{\min} \le y_k \le y_{\max},\ k = 1, ..., N_C \\ & u_{\min} \le u_k \le u_{\max},\ k = 0, ..., N_U \\ & \Delta u_{\min} \le \Delta u_k \le \Delta u_{\max},\ k = 0, ..., N_U \\ & \Delta u_k = 0,\ k \ge N_U \\ & x_{k+1} = Ax_k + Bu_k \\ & y_k = Cx_k + Du_k,\ k = 0, \dots, N_J - 1 \end{aligned} \quad (6)$$

where $\Delta u_k = u_k - u_{k-1}$, $u_{-1} = u(t-1)$ is the previous input, and $r_y(t)$ is the output reference at time step t. N_J is the prediction horizon along which performance is computed, N_C is the horizon along which the output constraints are enforced, and N_U is the number of free control actions, so that $N_U \le N_J$ and $u_k = u_{N_U}$, $\forall k = N_u + 1, \dots, N_J$[2].

The MPC algorithm can be summarized as follows: at each sampling instant t

(1) Set $x_0 = x(t)$.
(2) Solve (6) obtaining $u^*(x(t)) = [u_0^*, \dots, u_{N-1}^*]$.
(3) Apply the input $u(t) = u_0^*$ and discard the remaining elements of $u^*(x(t))$.

The complexity of the MPC algorithm clearly depends on the structure of the optimization problem. In particular, if the system dynamics and the design constraints are linear and Problem (6) involves only continuous variables, the MPC algorithm requires, at each time step t, the solution of a Quadratic Program (QP), for which solution algorithms of polynomial complexity exist. On the other hand if some variables in Problem (6) are integer-valued, which is the case when the system model in (6) is a hybrid model, mixed-integer programming (MIP) techniques are required, which have combinatorial complexity.

When designing model-based control systems, there is a natural trade-off between model complexity and computation required. In particular, the more complex (and presumably accurate) is the model, the more complex Problem (6) may become. In view of this trade-off, in the sequel we first design an MPC controller, disregarding constraint (2b). Since dynamics (5) are linear and constraints (2a), (3), and (4) are also linear, the MPC algorithm requires the solution of QP only. If constraint (2b) is almost never active, the resulting MPC controller, cascaded by a state-dependent input-saturation, may be sufficient for adequately controlling the system, and the controller based on linear MPC solution can be simple and fast. On the other hand, if constraint (2b) is often active, the predicted trajectory will largely differ from the actual one, because of the unmodeled state-dependent input saturation. In the latter case, the system performance will most likely be degraded.

For the electromagnetically actuated mass-spring-damper system, the linear-MPC controller was designed using the Hybrid Toolbox (Bemporad, 2003). To make the moving mass position track a given reference signal and to enforce the constraints in (4) as output constraints, we define the output equation

$$y(t) = \begin{bmatrix} 1 & 0 \\ 2500 & 1 \\ -2500 & 1 \end{bmatrix} x(t) \quad (7)$$

Accordingly, we set

$$Q_y = \begin{bmatrix} 10^4 & 0 & 0 \\ 0 & 0 & 0 \\ 0 & 0 & 0 \end{bmatrix},\ Q_u = 10^{-10},$$
$$y_{\min} = \begin{bmatrix} -4\cdot10^{-3} \\ -\infty \\ -10.2 \end{bmatrix},\ y_{\max} = \begin{bmatrix} 4\cdot10^{-3} \\ 10.2 \\ +\infty \end{bmatrix},$$
$$u_{\min} = 0,\ u_{\max} = 10^4,\ \Delta u_{\min} = -\Delta u_{\max} = \infty,$$
$$N_J = 30,\ N_C = 5,\ N_U = 3,$$

and use the linear dynamic model (5), (7) as prediction model.

Figure 2 shows the behavior of closed-loop formed by the linear model and the MPC controller when tracking a desired reference profile over a

[2] In the MPC literature and in many MPC algorithms usually $N_C = N_J = N_U = N$.

simulation time interval of 0.1 seconds from the initial state $x(0) = \left[\begin{smallmatrix}0\\0\end{smallmatrix}\right]$. Figure 2(a) reports the position, velocity and input profiles with respect to time, and Figure 2(b) reports the phase plane in which satisfaction of velocity constraint (4) is shown. This velocity constraint only becomes active near the contact position x_{1c}. Because the controller cannot provide quick decelerations due to unidirectionality of the magnetic force, it keeps the constraint (4) inactive in parts of the trajectory away from the contact point.

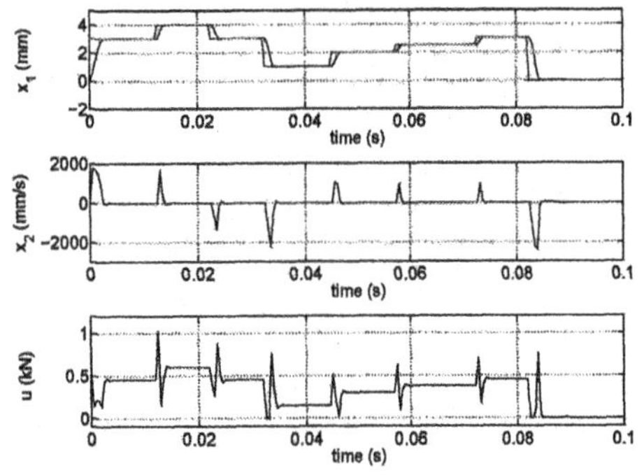

(a) State and input trajectories.

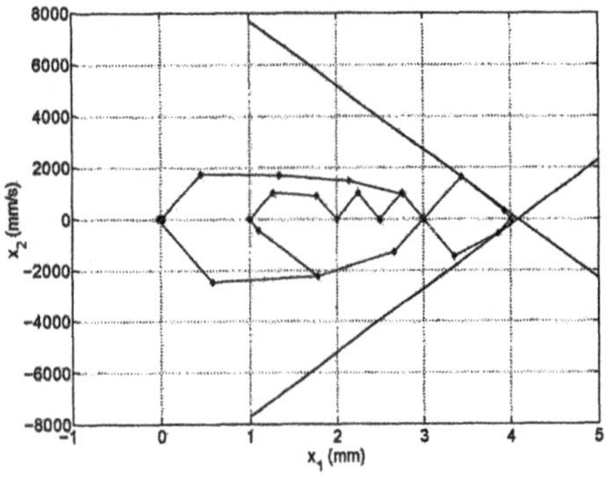

(b) Phase-plane trajectories.

Fig. 2. Ideal linear MPC simulation.

Figure 3 shows the behavior of the closed-loop system when a position-dependent saturation block, enforcing constraint (2b), is cascaded with the MPC controller. The performance clearly degrades, especially when the reference is decreasing. The reason for such degradation is that the linear MPC controller does not recognize that braking the mass at large distances away from the coil is impossible because of the state-dependent input saturation. This is in fact seen from the input plot in Figure 3 where the dashed line corresponds to the output of the MPC controller, while the solid line corresponds to the output of the saturation block.

To avoid wide oscillations and long settling periods, the saturation constraint (2b) should be taken into account in the MPC setup. Unfortunately, (2b) is a nonconvex constraint that cannot be handled by standard linear MPC. Next section shows how such a constraint can be handled by a hybrid MPC approach.

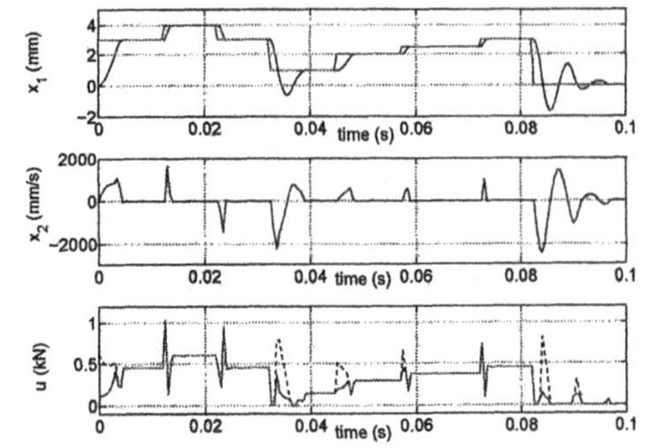

Fig. 3. Effects of saturation (2b).

4. HYBRID MODEL PREDICTIVE CONTROL

We approximate constraint (2b) by a (continuous) piecewise linear function $f(x_1) = r_i x_1 + q_i$, if $x_1 \in [\bar{x}_i, \bar{x}_{i+1})$, $i = 0, \dots \ell$ where $\{\bar{x}_i\}_{i=1}^{\ell-1}$ are the breakpoints of the function profile. In this paper we consider a piecewise linear approximation with three segments ($\ell = 3$), where the breakpoints are $\bar{x}_0 = -x_{1c}$, $\bar{x}_1 = 1.58 \cdot 10^{-4}$ m, $\bar{x}_2 = 1.82 \cdot 10^{-3}$ m and $\bar{x}_3 = x_{1c}$. Next, we introduce two binary variables $\delta_1, \delta_2 \in \{0, 1\}$ defined by the logical conditions

$$\begin{aligned} [\delta_1 = 1] &\leftrightarrow [x_1 \le \bar{x}_1] \\ [\delta_2 = 1] &\leftrightarrow [x_1 \le \bar{x}_2] \end{aligned} \tag{8}$$

and two continuous variables $z_1, z_2 \in \mathbb{R}$ defined by the logical conditions

$$z_1 = \begin{cases} (r_1 - r_2)x_1 + (q_1 - q_2) & \text{if } \delta_1 = 1 \\ 0 & \text{otherwise} \end{cases} \tag{9a}$$

$$z_2 = \begin{cases} r_2 x_1 + q_2 & \text{if } \delta_2 = 1 \\ r_3 x_1 + q_3 & \text{otherwise,} \end{cases} \tag{9b}$$

and impose that

$$u \le z_1 + z_2, \tag{10}$$

where clearly $z_1 + z_2 = f(x_1)$. Constraints (2a), (3), (4), (8), (9), (10), together with (5), are easily modeled in HYSDEL (Torrisi and Bemporad, 2004), and the equivalent Mixed Logical Dynamical (MLD) hybrid model (Bemporad and Morari, 1999)

$$x(k+1) = Ax(k) + B_1 u(k) + B_2 \delta(k) + B_3 z(k), \tag{11a}$$

$$E_2 \delta(k) + E_3 z(k) \le E_1 u(k) + E_4 x(k) + E_5, \tag{11b}$$

corresponding to the saturated magnetic actuator is obtained, where the matrices A, B_i, $i = 1 \dots 3$, E_j, $j = 1, \dots 5$, are generated automatically using the Hybrid Toolbox (Bemporad, 2003).

The hybrid MPC optimization problem is formulated as

$$\min_{\{u_k\}_{k=0}^{N-1}} \Delta x_N^T Q_N \Delta x_N + \sum_{k=0}^{N-1} \Delta x_k^T Q_x \Delta x_k + u_k Q_u u_k \qquad (12a)$$

$$\text{subject to MLD dynamics (11),} \qquad (12b)$$

where $Q_x = Q_N = \left[\begin{smallmatrix} 2\cdot 10^6 & 0 \\ 0 & 0 \end{smallmatrix}\right]$, $Q_u = 10^{-7}$, $N = 3$, and $\Delta x_1(k) = x_1(k) - r_y$. Because of the binary variables δ, the hybrid MPC strategy (12) requires the solution of mixed-integer quadratic programs. Note that only two binary variables are considered for each prediction step, so that the resulting optimization problem is of very small size.

The resulting closed-loop trajectories for the simulation scenario described in Section 3 when the hybrid MPC is applied are reported in Figure 4.

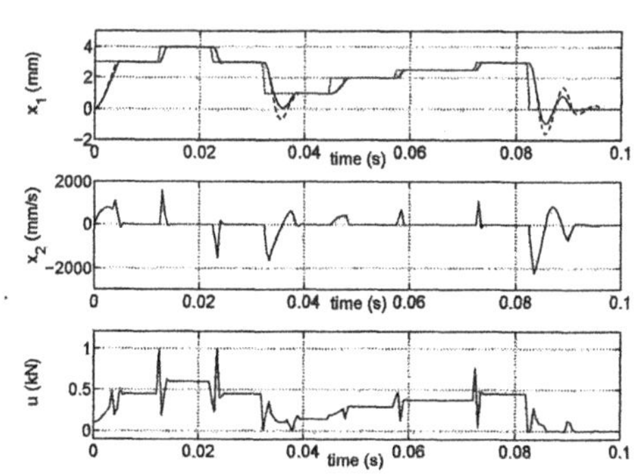

(a) State and input trajectories.

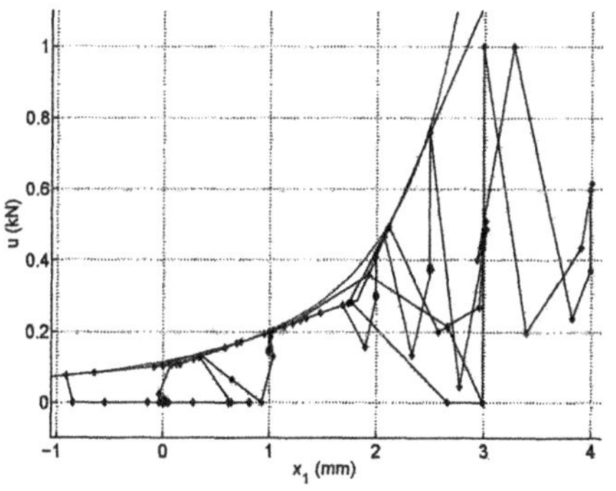

(b) Nonlinear force constraint, its PWL approximation, and input values generated by hybrid MPC.

Fig. 4. Closed-loop system response using the hybrid MPC controller (12).

Note that the PWL approximation (8), (9) is a lower bound to the maximum force profile, so that the force generated by the hybrid MPC algorithm never exceeds the saturation limits. With respect to the simulation of the linear MPC cascade by the saturation block (reported as dashed line in the position trajectory plot in Figure 4) we note that the system reacts a little slower when starting from the neutral position $x_1(0) = 0$, with null velocity $x_2(0) = 0$. This is the effect of the conservative approximation of the force constraint. While such negative effect can be eliminated by introducing a more refined approximation, the positive effects of the hybrid MPC controller are clear when the reference decreases. Both the overshoot and the settling period are reduced, because the controller is now aware of the limited available force and it provides the braking action in the region where a larger magnetic force is available.

MPC controller	Cumulative position error (mm^2)	Input energy (kN^2)
Linear (ideal)	51.4679	29.1918
Linear saturated	97.8608	26.6314
Hybrid	83.1005	26.6588

Table 1. Comparison of the three MPC scenarios

Table 1 compares the cumulative square position error, $\sum_k (x_1(k) - r(k))^2$, and cumulative square inputs (=actuator's energy), $\sum_k u(k)^2$, for the different MPC control scenarios. The tracking performance clearly degrades from the linear MPC controller in the ideal case of no force saturation, to the saturating one, while the hybrid controller has better performance (15%) with respect to the linear-saturated one, despite the slightly conservative approximation of constraint (2b). Moreover one must consider that a certain component of the tracking error is due to the one-step delay in reacting to reference changes, due to the non-anticipative implementation of the MPC algorithms. Such an error, that with respect to data in Table 1 has a value of 25.5, is independent of the controller applied, and thus should not be considered in comparing performances. Following this reasoning, the increase of net performance of the hybrid MPC algorithm is about 20% with respect to the linear-saturated one.

5. EXPLICIT IMPLEMENTATION OF THE CONTROLLER

The implementation of the MPC controllers described in the previous sections in a typical automotive micro-controller with the sampling time $T_s = 0.5$ ms can be very difficult because of the time required for the online solution of the underlying optimization problem. With the motivation to complete off-line a large part of the computations we developed explicit versions of the MPC controllers.

In (Bemporad *et al.*, 2002) it is shown that the solution to Problem (6) can be obtained as a function of the parameters x_0 and r_y (i.e., the actual state and output reference) by using multi-parametric quadratic programming (mp-QP). Using the mp-QP solver in the Hybrid Toolbox, we obtain an explicit feedback law $u(x, r_y)$ in continuous piecewise affine form consisting of 80 regions, which can be evaluated on-line very quickly. The mp-QP algorithm also returns the value function $V(x, r_y) = J^*(x, r_y)$, which is a piecewise quadratic function.

It must be stressed that the implicit MPC controller and the explicit one produce the same results, but there is a difference in the amount of computation required at each sampling step. More specifically, this difference is between the solution of an online optimization problem versus the evaluation of a set of inequalities and the computation of an affine state feedback term.

Figure 5 shows a section of the three-dimensional polyhedral partition of the explicit linear MPC controller, for $r_y = 0$. There is an affine state feedback controller associated to each region in the partition. Figure 5 also shows the state trajectory superimposed over the polyhedral partition.

In the case of hybrid MPC, we use the algorithm of (Bemporad, 2003) to obtain a representation of the MPC controller as a set of (possibly overlapping) piecewise affine controllers. During the on-line operation, at each step for each controller the value function is evaluated, and the input corresponding to the minimum cost is applied. Thus the explicit hybrid MPC solution involves the additional operation of comparing online the value functions. In addition, the number of regions increases to 671 regions, thus the controller requires a larger storage memory in the micro-controller and a larger number of comparison operations to find the active region.

6. CONCLUSIONS

We have compared two MPC solutions to the problem of controlling an electromagnetic actuator: one is based on linear MPC and handles one of the constraints via a-posteriori saturation; the other one is based on a hybrid model of the system and accounts for all the constraints in the design phase.

The hybrid MPC, which takes into account a piecewise linear approximation of the position dependent force constraint, achieves better performance than the linear MPC solution with superimposed saturation. Its main drawback is in higher complexity of the controller. Thus, the ultimate choice between these two MPC solutions can only be made once considering available computing resources and the aggressiveness of performance specifications.

An electromagnetically actuated mass-spring damper laboratory experiment is currently under construction at the University of Siena.

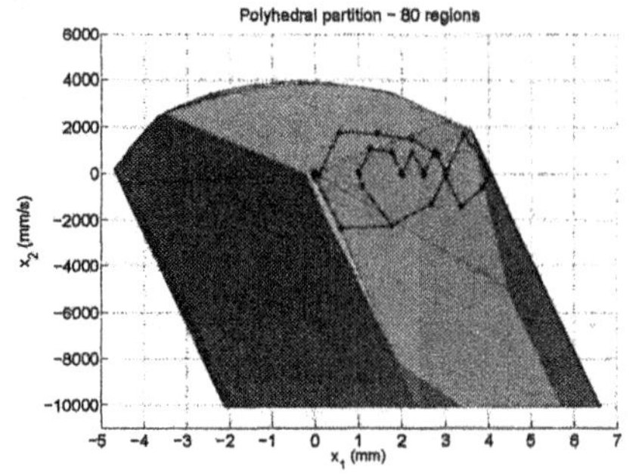

Fig. 5. Section of the linear explicit controller partitions obtained for $r_y = 0$.

REFERENCES

Barron, M.B. and W.F. Powers (1996). The role of electronic controls for future automotive mechatronic systems. *IEEE/ASME Transactions on Mechatronics* **1**(1), 80–88.

Bemporad, A. (2003). *Hybrid Toolbox – User's Guide.* `http://www.dii.unisi.it/hybrid/toolbox.`

Bemporad, A. and M. Morari (1999). Control of systems integrating logic, dynamics, and constraints. *Automatica* **35**(3), 407–427.

Bemporad, A., M. Morari, V. Dua and E.N. Pistikopoulos (2002). The explicit linear quadratic regulator for constrained systems. *Automatica* **38**(1), 3–20.

Di Cairano, S., A. Bemporad, I.V. Kolmanovsky and D. Hrovat (2006). Model predictive control of a magnetically actuated mass-spring-damper for automotive applications. In: *Proc. 45th IEEE Conf. on Decision and Control.* San Diego, CA. Submitted for publication.

Giorgetti, N., A. Bemporad, E. H. Tseng and D. Hrovat (2005). Hybrid model predictive control application towards optimal semi-active suspension. In: *Proc. IEEE Int. Symp. on Industrial Electronics.* Dubrovnik, Croatia.

Guzzella, L. and A. Sciarretta (2005). *Vehicle Propulsion Systems Introduction to Modeling and Optimization.* Springer Verlag.

Hrovat, D., J. Asgari and M. Fodor (2000). Automotive mechatronic systems. In: *Mechatronic Systems, Techniques and Applications, Volume 2.* pp. 1–98. Gordon and Breach Science.

Qin, S.J. and T.A. Badgwell (2003). A survey of industrial model predictive control technology. *Control Engineering Practice* **93, no. 316**, 733–764.

Torrisi, F.D. and A. Bemporad (2004). HYSDEL — A tool for generating computational hybrid models. *IEEE Trans. Contr. Systems Technology* **12**(2), 235–249.

ELSEVIER
IFAC
PUBLICATIONS

SUBTLETIES IN THE AVERAGING OF HYBRID SYSTEMS WITH APPLICATIONS TO POWER ELECTRONICS

Luigi Iannelli [*] **Karl Henrik Johansson** [**] **Ulf T. Jönsson** [***]
Francesco Vasca [*]

[*] *Dept. of Engineering, University of Sannio, Benevento, Italy*
[**] *Electrical Engineering School, Royal Institute of Technology, Stockholm, Sweden*
[***] *Division of Optimization and Systems Theory, Royal Institute of Technology, Stockholm, Sweden*

Abstract: Dither signals are commonly used in electronics for implementing different type of modulations in power converters, which represent a very interesting class of hybrid systems. It was recently shown that a nonsmooth dithered system can be approximated by an averaged system provided that the dither frequency is sufficiently high and that the amplitude distribution function of the dither is absolutely continuous and has bounded derivative. This result is exploited in this paper for power converters. Averaged models corresponding to various shapes of dither signal are analyzed, showing that dither with Lipschitz continuous amplitude distribution function can be used to adapt the equivalent gain of the power converter. *Copyright © 2006 IFAC*

1. INTRODUCTION

In power electronics a large class of systems can be modeled as hybrid dynamical systems due to the presence of switching devices (for instance diodes or transistors) that "instantaneously" change the dynamical behavior of the system. This class of systems can be adequately represented by hybrid systems or nonlinear differential equations with discontinuous nonlinearities. One of the standard approaches for designing control algorithms for power converters, is to use modulation schemes (e.g. pulse width modulation, PWM). In this case the feedback controlled power converter is a hybrid system with one (possible more) external forcing signal(s).

[*] The work by L. Iannelli and F. Vasca was supported by the European Commission within the SICONOS project IST2001-37172. The work by K. H. Johansson and U. Jönsson was supported by the European Commission within the Network of Excellence HYCON and by the Swedish Research Council. K. H. Johansson was also supported by the Swedish Foundation for Strategic Research.

Analysis of forced nonlinear feedback systems can be very difficult. One approach for the analysis is to accept some approximation and base the analysis on the averaged dynamics. Averaging theory provides theoretical justification for this idea in many important cases. Averaging is applied, for example, in studying effects of dither signals that are commonly used to compensate for nonlinearities in feedback control systems. The idea is that by injecting a suitably chosen high-frequency signal in the control loop, the nonlinear sector is effectively narrowed and the system can thereby be stabilized. For the case when the original nonlinearity is Lipschitz continuous, it was shown in the 70s that if the dither frequency is sufficiently high, the behavior of the dithered system will be qualitatively the same as an averaged system, see Zames and Shneydor (1976). However, the Lipschitz continuity assumption on the nonlinearity of the dithered system is often violated in practice. This is the case in many pulse width modulated systems and power converters. These systems can be modeled as nonlinear dithered systems in which the high frequency signal is added at the input of a discontinuous nonlinearity of signum or step function type. It is well known that complex phenomena such as bifurcations and chaos can appear in these systems, e.g., di Bernardo and Tse (2002), Tse et al. (2000), Banerjee and Verghese (2001).

Rigorous analysis of nonsmooth systems with dither excitation is complicated. It is only recently that rigorous results on stability and approximation have been obtained, e.g., Gelig and Churilov (1998), Teel et al. (2004), Iannelli et al. (2003), Iannelli et al. (2006). In this paper we will recap and apply to power electronics systems the averaging result in Iannelli et al. (2006). The result shows that a general class of nonsmooth dithered systems can be approximated by a corresponding class of averaged systems, provided that the dither has sufficiently high frequency and its amplitude distribution function is absolutely continuous with bounded derivative. We will discuss how usual modulation schemes (that use dither signals satisfying the averaging result) work well while other modulation schemes that use dither signals not satisfying the assumptions in Iannelli et al. (2006) may fail. Simulations on a buck converter show the effects of different (dither) signals used in the modulation, while experiments show that the averaging result does not hold when zero-slope signals (like square or trapezoidal waveforms) are used for feedback control of a DC motor. The experiments illustrate how it is possible to control the motor shaft at a desired angular position by applying high frequency dither signals that satisfy the assumption of the averaging theorem, while other dither signals that do not fulfill the conditions do not give a stable closed-loop system, despite their high frequency content.

2. PRELIMINARIES

Consider the dithered feedback system

$$\dot{x}(t) = f_0(x(t),t) + \sum_{i=1}^{m} f_i(x(t),t)n_i(g_i(x(t),t) - \delta_i(t)), \tag{1}$$

where $x(0) = x_0$, $f_i : \mathbb{R}^q \times \mathbb{R} \to \mathbb{R}^q$, $i = 1,\dots,m$, are assumed to be globally Lipschitz with respect to both x and t, and f_0 is piecewise continuous with respect to t, Lipschitz in x, and $f_0(0,t) = 0$ for all $t \geq 0$. Similarly, the functions $g_i : \mathbb{R}^q \times \mathbb{R} \to \mathbb{R}$, $i = 1,\dots,m$, are globally Lipschitz with respect to both x and t. The nonlinearities $n_i : \mathbb{R} \to \mathbb{R}$, $i = 1,\dots,m$, are supposed to be of bounded variation. Each dither signal $\delta_i : [0,\infty) \to \mathbb{R}$ is a p-periodic measurable and bounded function.

The averaged system is defined as

$$\dot{w}(t) = f_0(w(t),t) + \sum_{i=1}^{m} f_i(w(t),t)N_i(g_i(w(t),t)), \tag{2}$$

where $w(0) = w_0$, N_i is the averaged nonlinearity

$$N_i(z) \triangleq \int_{\mathbb{R}} n_i(z-\xi)\mathrm{d}F_{\delta_i}(\xi).$$

Here the integral is a Lebesgue–Stieltjes integral and

$$F_\delta(\xi) \triangleq \frac{1}{p}\mu(\{t \in [0,p) : \delta(t) \leq \xi\})$$

is the amplitude distribution function with Lebesgue measure μ and dither period p.

The following averaging result was proved in Iannelli et al. (2006).

Theorem 2.1. (Iannelli et al. (2006)). Consider the dithered system (1) and the averaged system (2) under the following assumptions:

- (*i*) the dithered system has an absolutely continuous solution,
- (*ii*) f_i and g_i are globally Lipschitz,
- (*iii*) f_0 is globally Lipschitz with respect to x and $f_0(0,t) = 0$,
- (*iv*) n_i is a function of bounded variation,
- (v) each dither δ_i is p-periodic, $|\delta_i| \leq M_\delta$, and has absolutely continuous amplitude distribution function F_{δ_i} with bounded derivative.

Then, the averaged nonlinearities N_i are globally Lipschitz continuous and the averaged system (2) has a unique absolutely continuous solution on $[0,\infty)$. Moreover, for any compact set $\mathcal{K} \subset \mathbb{R}^n$ and any $T > 0$, there exists a positive constant $c(\mathcal{K},T)$ such that

$$|x(t,x_0) - w(t,x_0)| \leq c(\mathcal{K},T)p, \quad \forall x_0 \in \mathcal{K},\ t \in [0,T]. \tag{3}$$

The main result of this paper is to show that if the conditions of the theorem are violated, then the averaged system does not necessarily approximate the dithered system. It suffices to consider a subclass of the general nonsmooth systems introduced above, that will be discussed in the following and that represent a wide class of power electronics systems.

3. AVERAGED MODELS OF POWER ELECTRONICS SYSTEMS

3.1 PWM power converters

A typical approach for modeling converters in the power electronics framework is based on the assumption that switches (such as diodes, thyristors, transistors, mosfets) are ideal in the sense that their current-voltage relationship can be modeled as a piecewise linear characteristic. Under such hypothesis a wide class of PWM power converters can be modeled in the following way:

$$\dot{x}(t) = A_0x(t) + b_0 + \sum_{i=1}^{m}(A_ix(t) + b_i)n(r_i(t) - c_ix(t) - \delta_i(t)) \tag{4}$$

where $x(0) = x_0$, A_i, b_i and c_i are constant matrices of appropriate dimensions and m is the number of modes of the converter. The step nonlinearity $n : \mathbb{R} \to \mathbb{R}$ is given by

$$n(z) = \begin{cases} 1, & z > 0 \\ 0, & z < 0. \end{cases} \tag{5}$$

The external reference signal $r_i(t)$ is assumed to be Lipschitz continuous and the $\delta(t)$ signal is a high frequency signal of period p.

It is now possible to derive the averaged system for the feedback PWM system (4) (see Iannelli et al. (2004) for the details).

Proposition 3.1. The averaged system of the pulse width modulated feedback system (4) is given by

$$\dot{w}(t) = A_0 w(t) + b_0 + \sum_{i=1}^{m} (A_i w(t) + b_i) N_i(r_i(t) - c_i w(t)), \tag{6}$$

where $w(0) = w_0$ and $N_i(z) = F_{\delta_i}(z)$.

PROOF. Follows simply from (2) and

$$N(z) = \int_{\mathbb{R}} n(z-a) \mathrm{d}F_\delta(a) = \int_{-\infty}^{+z} \mathrm{d}F_\delta(a) = F_\delta(z).$$

Note that, since $n(z)$ is discontinuous in 0, the averaged nonlinearity $N(z)$ is well defined except at possible discontinuity points of F_δ. It means that the right-hand side of (6) is not necessarily well-defined everywhere. The averaged system can have a well-defined generalized solution even if the right-hand side is almost everywhere well-defined. This is the case in the sequel of the paper.

3.2 *Dither effects interpreted through averaged models*

An interesting class of PWM power converters which can be modeled through (4) is represented by DC/DC converters. For instance the voltage-mode controlled buck converter reported in Fig. 1 can be modeled by considering $m = 1, A_1 = 0, b_0 = 0, r_1 = k_p V_{ref}$ and

$$A_0 = \begin{bmatrix} -\frac{R_1}{L} & -\frac{1}{L} \\ \frac{1}{C} & -\frac{1}{R_2 C} \end{bmatrix}, \; b_1 = \begin{bmatrix} \frac{E}{L} & 0 \end{bmatrix}^T, \; c_1 = \begin{bmatrix} 0 & k_p \end{bmatrix}$$

where V_{ref} is the output reference voltage, k_p the gain of the proportional controller and $\delta(t)$ may be the sawtooth signal reported in Fig. 2. It is simple to show that the DC/DC buck converter can be represented by using the block scheme reported in Fig. 3 with $\alpha = 0$ and $G(s) = c_1(sI - A_0)^{-1} b_1$. The corresponding block scheme of the averaged model is reported in Fig. 4. Let us now consider a DC/DC buck converter with the following parameters: $R_1 = 0.1\Omega$, $L = 1mH$, $C = 220\mu F$, $R_2 = 8.9\Omega$ $E = 6V$, $p = 100\mu s$, $k_p = 1.1$ and the reference voltage equal to $6V$ and after $0.03s$ to $8V$. By using as dither signals a sawtooth the results reported in Fig. 5 are obtained. By increasing

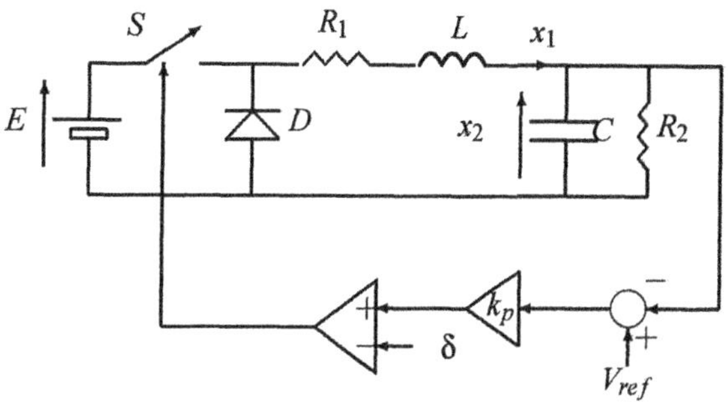

Fig. 1. DC/DC buck converter.

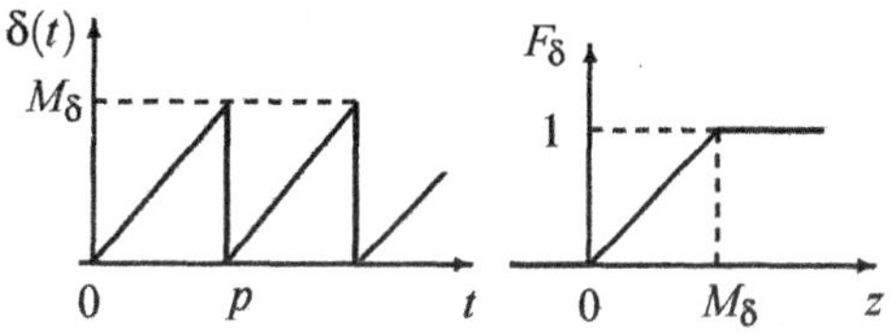

Fig. 2. Sawtooth dither with its amplitude distribution function.

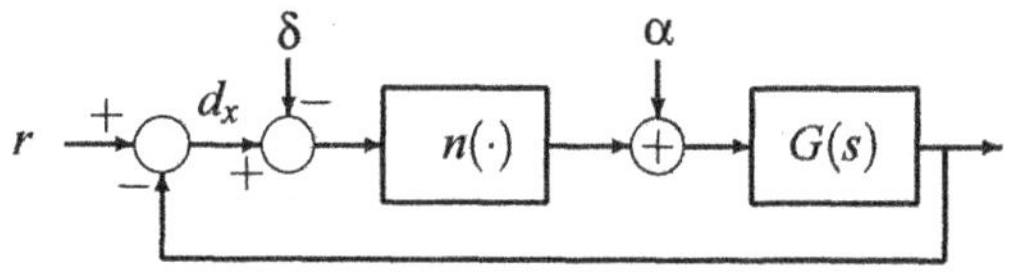

Fig. 3. Equivalent block scheme of a class of power electronic systems; for power converters d_x represents the so-called duty cycle.

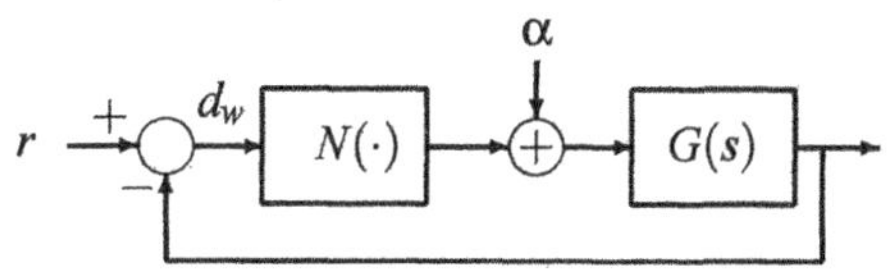

Fig. 4. Block scheme of the averaged model corresponding to Fig. 3.

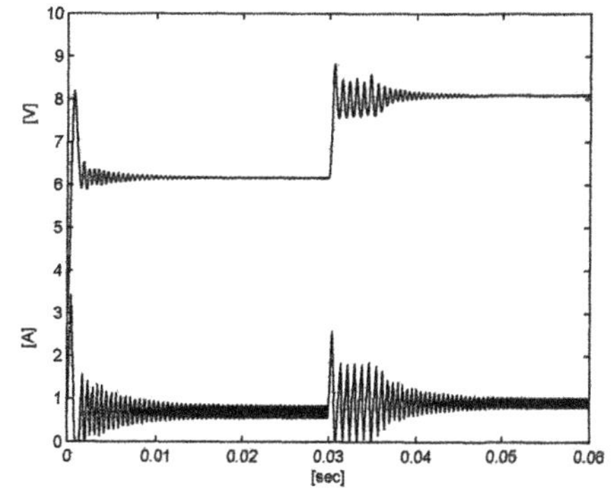

Fig. 5. Capacitor voltage and inductor current of the voltage mode controlled DC/DC buck converter by using a sawtooth dither signal with $M_\delta = 0.4$.

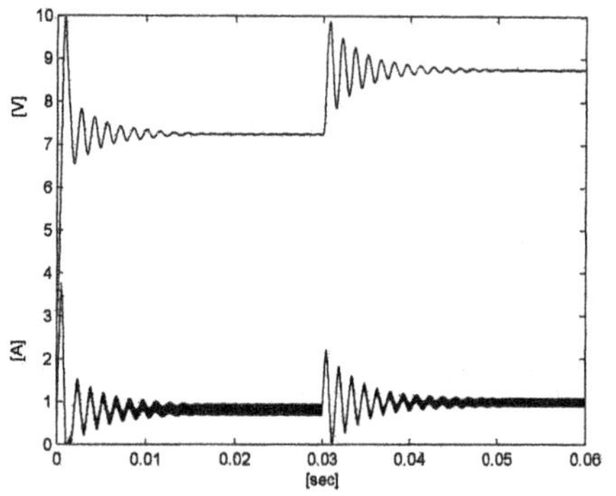

Fig. 6. State variables for the buck converter by using a sawtooth dither signal with $M_\delta = 4$.

the amplitude of the sawtooth dither signal, the average analysis predicts that the average gain will decrease. Therefore one might expect a degradation of the closed loop performance. This is confirmed by the simulation results reported in Fig. 6. By simulating the power converter and the corresponding averaged model in the presence of a sinusoidal dither, similar results are obtained. Thus, the average analysis seems to work properly also if the averaged nonlinearity is non Lipschitz continuous, so as for the case of the sinusoidal dither to which corresponds an amplitude distribution function with unbounded derivative at $\pm M_\delta$. This motivates a further investigation in the need for the assumptions made on the dither signal for the proof of Theorem 2.1.

4. SUBTLETIES IN AVERAGING RESULTS DUE TO INAPPROPRIATE CHOICE OF DITHER SHAPE

In this section we will show that when conditions on the amplitude distribution function are not satisfied, it is possible to find systems for which conclusions of Theorem 2.1 do not hold. Our first result shows that we cannot ensure the existence of a unique solution to the averaged system unless we impose the boundedness of the derivative of F_δ.

Proposition 4.1. Suppose the amplitude distribution function F_δ is absolutely continuous but its derivative is not bounded. Then there exists a dithered system (1) for which the corresponding averaged system (2) does not have a unique solution.

PROOF. Consider the dithered nonsmooth feedback system (4) with $m = 1$, $A_0 = A_1 = 0$, $b_0 = 0$, $b_1 = 4$, $c_1 = -1$, $x(0) = -1$, $r \equiv 0$ and δ is the following $p-$periodic quadratic dither signal

$$\delta(t) = \begin{cases} -4M_\delta\left(\dfrac{t}{p}\right)^2 + M_\delta, & t \bmod p \in (0, \frac{p}{2}] \\ +4M_\delta\left(\dfrac{t}{p}\right)^2 - 8M_\delta\dfrac{t}{p} + 3M_\delta, & t \bmod p \in [\frac{p}{2}, p). \end{cases}$$

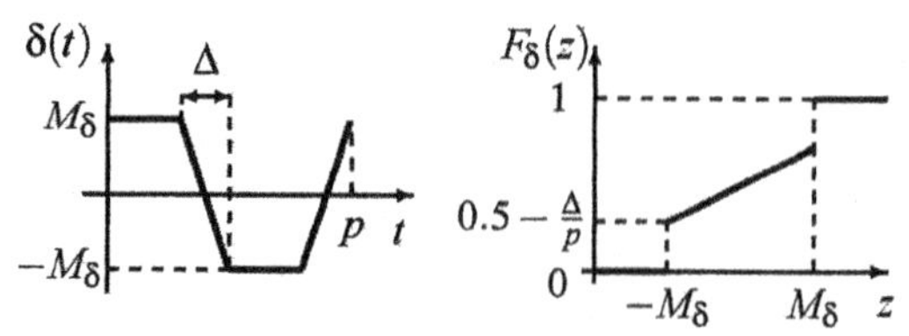

Fig. 7. Trapezoidal dither and the corresponding amplitude distribution function.

The corresponding amplitude distribution function is

$$F_\delta(z) = \begin{cases} 0 & z \le -M_\delta \\ \dfrac{1}{2}\sqrt{1 + \dfrac{z}{M_\delta}} & -M_\delta \le z \le 0 \\ 1 - \dfrac{1}{2}\sqrt{1 - \dfrac{z}{M_\delta}} & 0 \le z \le M_\delta \\ 1 & z \ge M_\delta. \end{cases} \tag{7}$$

Note that the amplitude distribution function is absolutely continuous but its derivative is not bounded. The averaged system is then (6) where $N(z)$ can be obtained from (7) and Proposition 3.1. Note that this averaged system is not globally Lipschitz because F_δ is not globally Lipschitz. It is easy to show that when $M_\delta = 1$ the averaged system has a nonunique solution. In fact, there are infinitely many solutions parameterized by $\tau \in [0, \infty)$ and given by

$$w(t) = \begin{cases} -1 & 0 \le t \le \tau \\ (t-\tau)^2 - 1 & t \ge \tau. \end{cases}$$

The next result states that if the amplitude distribution function is not absolutely continuous then the conclusions of Theorem 2.1 cannot be guaranteed.

Proposition 4.2. Suppose the amplitude distribution function F_δ is discontinuous. Then there exists a dithered system (1) for which the uniform bound in (3) of Theorem 2.1 does not hold.

PROOF. Let us consider the nonsmooth feedback system (4) with $m = 1, A_1$ is a null matrix of adequate dimensions,

$$A_0 = \begin{bmatrix} -1 & -1 \\ 0 & -2 \end{bmatrix}, \; b_0 = \begin{bmatrix} 0 \\ -1 \end{bmatrix}, \; b_1 = \begin{bmatrix} 0 \\ 2 \end{bmatrix}, \; c_1 = \begin{bmatrix} 1 & 0 \end{bmatrix}.$$

The external constant reference is $r_1(t) = R = 0.5$ and δ is a square wave dither of amplitude $M_\delta = 0.5$. It is easy to show that the averaged system is (6) where N is the averaged nonlinearity corresponding to the amplitude distribution function in Fig. 7 with $\Delta = 0$ and $M_\delta = 0.5$. Let us partition the state space of the dithered and the averaged systems into the following three regions, see Fig. 8:

- Region $\Omega_1 = \{x : x_1 < 0\}$. In this region $n(R - c_1 x - \delta) = 1$. The dithered system coincides with the averaged system and they have dynamics $\dot{x} = A_0 x - b_0$. The equilibrium point is $P_1 = +A_0^{-1} b_0 = (-0.5, 0.5)^T$.

- Region $\Omega_2 = \{x : x_1 > 1\}$. In this region $n(R - c_1x - \delta) = 0$. The dithered system coincides with the averaged system and they have dynamics $\dot{x} = A_0x + b_0$. The equilibrium point is $P_2 = -A_0^{-1}b_0 = (0.5, -0.5)^T$.
- Region $\Omega_0 = \{x : 0 < x_1 < 1\}$ with subsets $\Omega_0^+ = \{x : 0 < x_1 < 1,\ x_2 > 0.5\}$ and $\Omega_0^- = \{x : 0 < x_1 < 1,\ x_2 < -0.5\}$. In Ω_0 the state does not affect the output of the step nonlinearity. The dithered system can be represented by the linear system

$$\dot{\zeta}(t) = A_0\zeta(t) - b_0u(t) \quad (8)$$

with u a periodic signal that switches between -1 (when $R - \delta(t) = 0$) and 1 (when $R - \delta(t) = 1$). The averaged system has an input equal to zero in this region, i.e., $\dot{w}(t) = A_0w(t)$.

Consider $x(0) = w(0)$, with $0 < x_1(0) < 1, 0 < x_2(0) < 0.5$ and $x_1(0) > x_2(0)$. It is easy to show that the average trajectory will tend to the origin without leaving the set indicated for the possible initial conditions. The dithered trajectory will oscillate about the averaged solution. By considering the vector fields indicated in Fig. 8, it follows that the dithered trajectory cannot leave the set $\Omega_0 - \{\Omega_0^+ \cup \Omega_0^-\}$ but by crossing the segment $\{x : x_1 = 0,\ 0 \le x_2 \le 0.5\}$. This can, for example, be seen by inspecting the phase-plane in Fig. 8. Moreover in Ω_0 the solution of the dithered system can be represented as

$$x(t) = e^{A_0t}(x(0) - \zeta_0) + \zeta_{ss}(t),$$

where ζ_{ss} is the steady-state p-periodic solution of (8) and

$$\zeta_0 = (I - e^{A_0p})^{-1}\int_0^p e^{A_0(p-s)}bu(s)\mathrm{d}s.$$

Since A_0 is Hurwitz, $x(t)$ will thus converge to $\zeta_{ss}(t)$, which is a counter clockwise oscillation around the origin. It is always possible to choose a small enough dither period p such that $\zeta_{ss}(t)$ never intersects Ω_2, since $\zeta_{ss} \to 0$ when $p \to 0$. It is then clear that $x(t)$ eventually will cross the x_2 axis for some $0 \le x_2 \le 0.5$. ¿From Fig. 8, it is easy to see that the second orthant is an invariant set under the dynamics of the dithered and averaged systems. Moreover, since the system matrix A_0 is Hurwitz, the dithered solution $x(t)$ will tend toward the equilibrium point P_1.

We have thus shown that the dithered and the averaged systems behave qualitatively very different since they converge to two different points, P_1 and the origin, respectively. This is a contradiction to the condition (3) of Theorem 2.1. Indeed, if the compact set $\mathcal{K}$ includes the origin, we would need to make p smaller and smaller the closer x_0 is to the origin (on the trajectory indicated in Fig. 8) in order to get the inequality satisfied, because it always exists a p such that (3) does not hold. Hence, there is no uniform bound on p that holds for all $x_0 \in \mathcal{K}$.

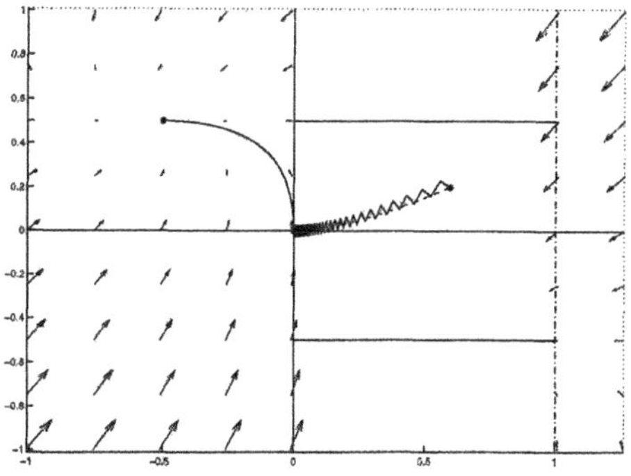

Fig. 8. Phase plane trajectories and vector fields for the dithered system (4) and the averaged system (6) with $p = 0.1$ and initial conditions $x_0 = w_0 = [0.6\ 0.2]^T$. The trajectory of the dithered system converges to P_1 while the trajectory of the averaged system converges to the origin.

5. EXPERIMENTAL RESULTS

In this section we illustrate how the complex behaviors discussed in previous section may also appear in practical power electronics applications. The experimental setup consists of a power electronic drive with a position controlled DC motor and a full bridge DC/DC power converter. The control objective is to put the motor shaft at a desired angular position θ_{ref}. The angular position of the shaft θ is measured by using a rotational potentiometer whose gain is k_{pot}. The motor supply voltage $\pm V_a$ is obtained through a full bridge DC/DC converter. By introducing the state vector $x = (\theta, \omega, i_a)^T$, the dynamic model of the whole power electronic system can be represented as the dithered nonsmooth feedback system (4) with $m = 1$, A_1 is a null matrix of adequate dimensions,

$$A_0 = \begin{bmatrix} 0 & 1 & 0 \\ 0 & -\frac{\beta}{J} & \frac{k_t}{J} \\ 0 & -\frac{k_e}{L_a} & -\frac{R_a}{L_a} \end{bmatrix},\ b_0 = \begin{bmatrix} 0 \\ 0 \\ -\frac{V_a}{L_a} \end{bmatrix},$$
$$b_1 = -2b_0,\ c_1 = \begin{bmatrix} k_{\text{pot}} & 0 & 0 \end{bmatrix}.$$

It is simple to show that such system can be represented through the block diagram reported in Fig. 3 with $\alpha = -0.5$ and $G(s) = c_1(sI - A_0)^{-1}b_1$. The external constant reference is $r(t) = R = V_{\text{ref}}$. The DC motor has the following parameters: $R_a = 2.510\,\Omega$, $L_a = 0.530\,\text{mH}$, $k_t = k_e = 5.700\,\text{mV}/(\text{rad}\cdot\text{s}^{-1})$, $\beta = 0.411\,\text{mN}\cdot\text{cm}/(\text{rad}\cdot\text{s}^{-1})$, $J = 31.400\,\text{g}\cdot\text{cm}^2$, $k_{\text{pot}} = 3/(2\pi)\,\text{V/rad}$, $V_a = 2.500\,\text{V}$. Two dither shapes are considered: a sawtooth signal and a trapezoidal signal. The dither amplitude is in all cases equal to $M_\delta = 0.070$. It can be shown (e.g., using the Popov criterion) that the averaged systems corresponding to the sawtooth and trapezoidal dither cases are both asymptotically stable. For sawtooth dither, the approximation error between the dithered system and the averaged system tends to zero as the dither frequency goes to infinity, in accordance with Theorem 2.1. Hence, since the averaged system is asymptotically stable, the system output goes to zero

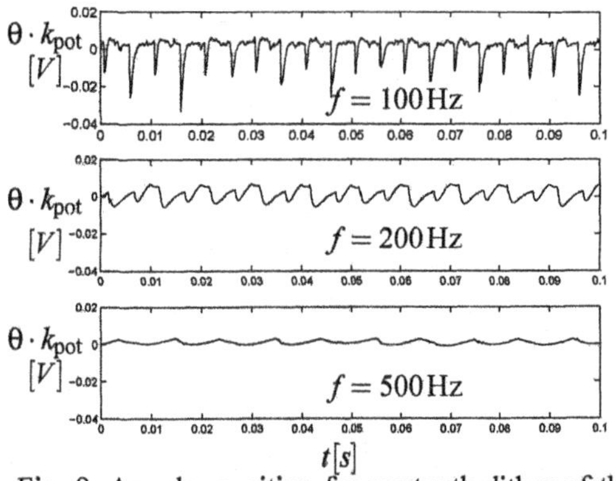

Fig. 9. Angular position for sawtooth dither of three frequencies.

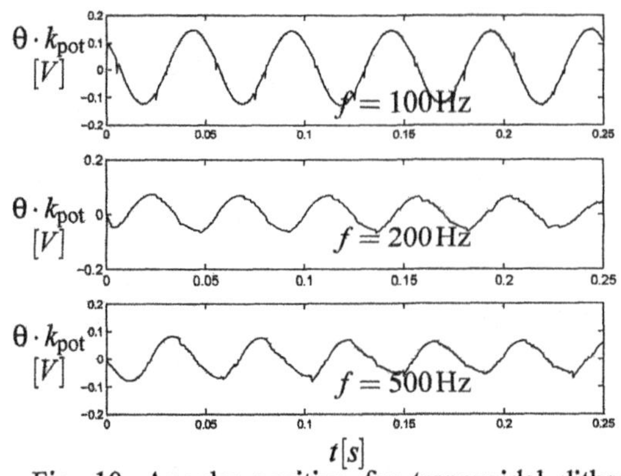

Fig. 10. Angular position for trapezoidal dither of three frequencies. Note the different scale compared to Fig. 9.

as we increase the dither frequency. For trapezoidal dither, the assumptions of the theorem are not fulfilled, since trapezoidal dither has a discontinuous amplitude distribution function.

The following experiments support these theoretical conclusions. The system is stabilized with sawtooth dither, but not with trapezoidal dither. Experiments were carried out using sawtooth dither of frequencies 100, 200, and 500 Hz. Fig. 9 reports the angular position of the motor shaft under steady-state conditions. Note that by increasing the dither frequency the behavior of the dithered system converges to the behavior of the (stable) averaged system (i.e., the system output goes to zero). The ratio between consecutive averages of the peak-to-peak values of the output signal is equal to 3.33 and 2.84 for 100–200 Hz and 200–500 Hz, respectively. These ratios somehow indicates the convergence rate. The averaging effect of the dither thus works properly in this case. Fig. 10 shows experiments with trapezoidal dither. In this case, the system output shows a slow oscillation with a substantial amplitude for all three dither frequencies. (Note that the axes are not the same as in Fig. 9.) The frequency of the oscillation is low compared to the dither frequency, and it seems to be relatively independent of the dither frequency. In particular, note that by increasing the dither frequency, the system output does not converge to zero, as was the case with sawtooth dither. Instead the ratio between consecutive averages of the peak-to-peak values of the output signal is equal to 1.97 and 0.86 for 100–200 Hz and 200–500 Hz, respectively, so going from 200 to 500 Hz, the amplitude of the oscillation is even increasing.

6. CONCLUSIONS

In this paper we have shown that an important class of power electronics systems (namely pulse width modulated systems), usually modeled as hybrid dynamical systems with external periodic forcing, can be recasted in the framework of dithered nonsmooth systems. By exploiting recent theory presented by authors on the averaging of dithered nonsmooth systems, we have analyzed how the shape of the external forcing might affect the averaging result and the behavior of the overall system. Simulations and experiments supported the theoretical discussion and proved the importance of averaging theory and its subtleties in the case of its application to hybrid dynamics.

REFERENCES

S. Banerjee and G.C. Verghese, editors. *Nonlinear Phenomena in Power Electronics: Bifurcations, Chaos, Control, and Applications.* Wiley-IEEE Press, New York, 2001.

M. di Bernardo and C.K. Tse. Complex behavior in switching power converters. *Proceedings of the IEEE*, 90(5):768–781, 2002.

A.Kh. Gelig and A. Churilov. *Stability and Oscillations of Nonlinear Pulse Modulated Systems.* Birkhäuser, Berlin, 1998.

L. Iannelli, K. H. Johansson, U. Jönsson, and F. Vasca. Averaging of nonsmooth systems through dithered switchings. *Automatica*, 42(4):669–676, 2006.

L. Iannelli, K.H. Johansson, U. Jönsson, and F. Vasca. Dither for smoothing relay feedback systems: an averaging approach. *IEEE Transactions on Circuits and Systems, Part I*, 50(8):1025–1035, 2003.

L. Iannelli, K.H. Johansson, U. Jönsson, and F. Vasca. On the averaging of a class of hybrid systems. In *Proc. of IEEE Conference on Decision and Control*, volume 2, pages 1400–1405, Bahamas, 2004.

A. Teel, L. Moreau, and D. Nesic. Input-to-state set stability of pulse width modulated systems with disturbances. *Systems and Control Letters*, 51, 2004.

C.K. Tse, Y.M. Lai, and H.H.C. Iu. Hopf bifurcation and chaos in a free-running current-controlled Ćuk switching regulator. *IEEE Transactions on Circuits and Systems, Part I*, 47(4):448–457, 2000.

G. Zames and N. A. Shneydor. Dither in non-linear systems. *IEEE Transactions on Automatic Control*, 21(5):660–667, 1976.

ELSEVIER
IFAC
PUBLICATIONS

ADAPTIVE CRUISE CONTROLLER DESIGN: A COMPARATIVE ASSESSMENT FOR PWA SYSTEMS

Daniele Corona * **Bart De Schutter** *

* *Delft Center for System and Control, Delft University of Technology, Mekelweg 2, 2628 CD Delft, The Netherlands*

Abstract: We propose the design of an *adaptive cruise controller* (ACC) of a Smart vehicle as a benchmark set up for methods developed for piecewise affine (PWA) systems based on model predictive control (MPC) arguments. The control law of the system aims at achieving a trade-off between tracking and fuel consumption while guaranteeing specific *constraints*, related to physical limitations, safety/comfort issues, environment protection and energy saving. In this paper we consider some PWA MPC control design methods, an on-line, an off-line and a robust on-line and compare them with an on-line linear approximation and an off-line gain scheduling approach. The results will be briefly described and the algorithms will be tested in terms of key issues for real implementation and accuracy of the solution. *Copyright © 2006 IFAC.*

Keywords: piecewise affine systems, model predictive control, engine control, multi-parametric mixed integer linear programming.

1. INTRODUCTION

Recently, significant efforts of the control system and computer science communities have been devoted to the study and design of hybrid systems. Within this class particular attention was paid to piecewise affine (PWA) systems, namely, a finite set of affine system and a switching signal that switches, internally or externally forced, from one affine mode to another.

PWA systems arise from modeling processes that integrate integer/logical behavior with continuous variables or from quantized inputs (Elia and Mitter, 1999), or from the linear spline approximation of nonlinearities (Sontag, 1981). The discontinuities implicitly hidden in the discrete behavior of these systems make the control design a nontrivial task, the complexity of which is additionally increased if constraints are considered.

Nevertheless several methods that aim to design the control law for this class were proposed in the literature and some of them are MPC based. An equivalent model, the mixed logical dynamical (MLD) model (Bemporad and Morari, 1999), is used to compute the control law implicitly or

[1] Supported by (1) the European 6th Framework Network of Excellence "HYbrid CONtrol: Taming Heterogeneity and Complexity of Networked Embedded Systems (HYCON)", contract number FP6-IST-511368, (2) the BSIK project "Transition Sustainable Mobility (TRANSUMO)", and (3) the Transport Research Centre Delft program "Towards Reliable Mobility", and (4) the Dutch Science Foundation (STW), Grant "Model Predictive Control for Hybrid Systems" (DMR. 5675).

Table 1. *Entries of equation* (1).

m: Mass of vehicle	800 kg
c: Viscous coefficient	0.5 kg/m
μ: Friction coefficient (dry asphalt)	0.01
b: Traction force	3700 N
g: Gravity acceleration	9.8 m/s^2
α: Switching velocity	18.75 m/s

explicitly (Borrelli, 2003). Several variants that consider robustness (Kerrigan and Mayne, 2002) or stability properties (Lazar *et al.*, 2005) were also considered. Methods based on the construction of a piecewise Lyapunov function have been developed in (Hedlund and Rantzer, 1999).

Despite the presence of several methods, an applicative comparison test bed that highlights their main features is, to our best knowledge, missing. Therefore in this paper a benchmark set up for the design of MPC for a PWA system is proposed, arising from an application that is focused on the design of an *adaptive cruise controller* (ACC) for a Smart vehicle within physical/safety constraints.

The paper is organized as follows: we first define the PWA model of the system and the problem. This is transformed into a minimization problem with a mixed integer objective function. We then describe the control methods that we use and we present, in the last section, a comparison table.

2. MODEL AND PROBLEM DESCRIPTION

Model. In a basic ACC application 2 cars are driving one after the other (see Figure 1.a). The aim of an ACC is to ensure a minimal separation between the vehicles and a speed adaptation. We assume that the front vehicle communicates its speed and position to the rear vehicle, which has to track them. For the control design purpose we only consider the dynamics of the rear vehicle, the model of which is:

$$m\ddot{s}(t) + (c\dot{s}^2(t) + \mu mg)\mathrm{sgn}(\dot{s}(t)) = bu(t), \quad (1)$$

where $s(t)$ is the position, b is the traction force, $u(t)$ is the normalized throttle/brake. Viscous and static frictions are considered, braking will be simulated by applying a negative u. Numerical values are given in Table 1.

Model (1) is valid as long as the speed is significantly different from zero, hence we impose that the velocity is always above a minimum value.

A least square approximation (Figure 1.b) of the nonlinear friction curve $V = cv^2$ leads to a PWA system:

$$m\ddot{s}(t) + c_i\dot{s}(t) + f_i = bu(t), \quad i = 1, 2.$$

Table 2. *Values of the constraints.*

$x_{2,\min}$: Min. velocity	5.0 m/s
$x_{2,\max}$: Max. velocity	37.5 m/s
d_{safe}: Max. position overshoot	5.0 m
a_{acc}: Max. acceleration	2.5 m/s^2
a_{dec}: Min. acceleration	-1 m/s^2
ξ: Comfort jerk	2.0 m/s^3
$u_{\max}$: Max. throttle/brake	1
Δ_u: Max. throttle/brake variation	0.2

Mode $i = 1(2)$ is active when $\dot{s}(t) < (\geq)\alpha$. The coefficients c_i, f_i are derived using the data shown in Figure 1.b [2].

The discrete-time state-space representation (sampling time $T = 1s$, *zero order hold*) is

$$x(k+1) = A_i x(k) + B_i u(k) + F_i, \quad x_2(k) < \alpha \quad (2)$$

with

$$A_1 = \begin{bmatrix} 1 & 0.97 \\ 0 & 0.99 \end{bmatrix}, B_1 = \begin{bmatrix} 2.31 \\ 4.61 \end{bmatrix}, F_1 = -\begin{bmatrix} 0.05 \\ 0.10 \end{bmatrix},$$
$$A_2 = \begin{bmatrix} 1 & 0.98 \\ 0 & 0.96 \end{bmatrix}, B_2 = \begin{bmatrix} 2.28 \\ 4.54 \end{bmatrix}, F_2 = \begin{bmatrix} 0.22 \\ 0.44 \end{bmatrix},$$

where $x_1(k)$ is the position and $x_2(k)$ is the velocity of the rear vehicle.

Constraints. Safety, comfort and economy or environmental issues, as well as limitations on the model, constrain the performance of the system. Since some methods require bounded variables we assume a minimum ($x_{1,\min} = 0\,m$) and a maximum ($x_{1,\max} = 2000\,m$) position. In particular we consider limitations on the state $x(k)$ and on the input $u(k)$. More precisely, $\forall\, k$,

$$\begin{aligned} x_{\min} \leq x(k) &\leq x_{\max} \\ x_1(k) &\leq \eta_1(k) + d_{\text{safe}} \\ a_{\text{dec}}T \leq x_2(k+1) - x_2(k) &\leq a_{\text{acc}}T \\ |x_2(k+1) - 2x_2(k) + x_2(k-1)| &\leq \xi T^2. \end{aligned} \quad (3)$$

The above equations express, respectively, the operative range of the state, the tracking of the leading vehicle trajectory $\eta(k) = [\eta_1(k), \eta_2(k)]^{\mathrm{T}}$ within a given overshoot d_{safe}, bounds on acceleration and jerk. We have, $\forall\, k$, for the control input:

$$\begin{aligned} |u(k)| &\leq u_{\max} \\ |u(k+1) - u(k)| &\leq \Delta_u. \end{aligned} \quad (4)$$

Numerical values are listed in Table 2. Note that although some of these constraints may be violated without causing major damages, i.e., collision or engine breakdown, we decided to consider all of them as *hard*.

[2] For the sake of simplicity we only consider one breakpoint, leading to a PWA composed of two operating modes. A finer approximation is also possible, by setting more than one breakpoints on the nonlinear curve.

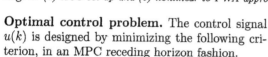

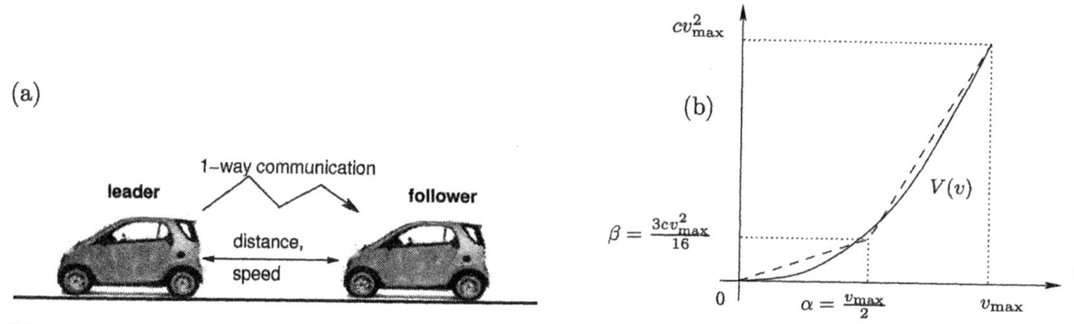

Fig. 1. *(a) ACC set up and (b) nonlinear to PWA approximation.*

Optimal control problem. The control signal $u(k)$ is designed by minimizing the following criterion, in an MPC receding horizon fashion.

$$\min_{\tilde{u}(N_p)} J(\theta(k), \tilde{u}(N_p)) \triangleq \sum_{j=1}^{N_p} ||Q\varepsilon(k+j)||_1 + ||Ru(k+j-1)||_1, \tag{5}$$

s.t., (2), (3) and (4). Here $\varepsilon(k+j) = x(k+j) - \eta(k+j)$ is the *tracking error*, $\tilde{u}(N_p) = [u(k), \ldots, u(k+N_p-1)]$ the sequence of control inputs, Q, R are weight matrices, $\theta(k) \triangleq [u(k-1), x(k-1)^T, x(k)^T, \eta(k+1)^T, \ldots, \eta(k+N_p)^T]^T$ is a set of parameters. In this framework the front vehicle communicates the prediction of N_p samples ahead of its trajectory.

Note that an appropriately tuned shorter control horizon $N_c < N_p$ may also be studied, i.e., $u(k+j) = u(k+N_c-1)$, $j = N_c, \ldots, N_p-1$. This choice has the general advantage of reducing the number of variables and of providing a smoother solution. Nevertheless here we only consider $N_p = N_c$ because a tuning issue was beyond the scope of this paper. The ℓ_1-norm allows the use of (mixed integer) linear programming and usually results in better performance than the ℓ_∞-norm (Borrelli, 2003).

3. DESIGN METHODS

In this section we describe the design methods that we have used on the problem above. For all methods we employed the weight matrices $Q = \mathbf{diag}(0.8, 0.1)$ and $R = 0.01$.

Since some of these methods require the transformation of the PWA system into a *mixed logical dynamical* (MLD) form, we show for completeness how this was performed in our setup.

PWA to MLD transformation. The PWA system (2) is transformed into an MLD system by the introduction of a binary variable $\delta(k)$ (Bemporad and Morari, 1999). The value of $\delta(k)$ equals 0(1) when the active moden is system 1(2). Hence the new model of the system is:

$$x(k+1) = A_1 x(k) + Lv(k) + F_1, \tag{6}$$

where $L = [A_2 - A_1 | B_2 - B_1 | F_2 - F_1 | B_1]$ and $v(k) = [z(k)^T, y(k), \delta(k), u(k)]^T$ (with $z(k) = x(k)\delta(k)$, $y(k) = u(k)\delta(k)$, $\delta(k) \in \{0, 1\}$) is the auxiliary *mixed logical* control input. The variables $z(k), y(k)$ are nonlinear, but they can be converted into equivalent mixed-integer linear inequalities (Bemporad and Morari, 1999):

$$\begin{gathered} x_{\min}\delta(k) \le z(k) \le x_{\max}\delta(k) \\ -x_{\max}(1-\delta(k)) \le z(k) - x(k) \le -x_{\min}(1-\delta(k)) \\ |y(k)| \le u_{\max}\delta(k) \\ |y(k) - u(k)| \le u_{\max}(1-\delta(k)). \end{gathered} \tag{7}$$

The switching condition leads to

$$\begin{gathered} -\delta(k)(v_{\min} - \alpha) \le x_2(k) - v_{\min} \\ \delta(k)(\alpha - v_{\max}) \le -x_2(k) + \alpha. \end{gathered} \tag{8}$$

Method 1: on-line direct approach. This method (Bemporad and Morari, 1999) solves problem (5) s.t. the MLD model (6), the additional constraints (7) and (8). This is appropriately converted into a mixed integer linear program (MILP) as it follows:

$$\begin{aligned} J(\theta(k))^* = {} & \min_{\tilde{y}} c'\tilde{y} \\ & \text{s.t.} \quad E\tilde{y} \le G + E_\theta \theta(k), \end{aligned} \tag{9}$$

where the variable $\tilde{y}$ includes the prediction of the control variable $\tilde{v}$ and the dummy variables introduced to convert the ℓ_1 problem into a linear problem. In a receding horizon fashion, the problem is solved on-line and the best control action is computed at step k. At the next step $k+1$ the set of parameters is updated and a new optimization problem is formulated.

Method 2: off-line explicit approach. Problem (9) can be solved parametrically, as proposed in (Bemporad *et al.*, 2000), based on the theoretical results described in (Dua and Pistikopoulos, 2000). This leads to an mp-MILP (multi parametric mixed integer linear programming) as described in (Borrelli, 2003), Chapter 8.

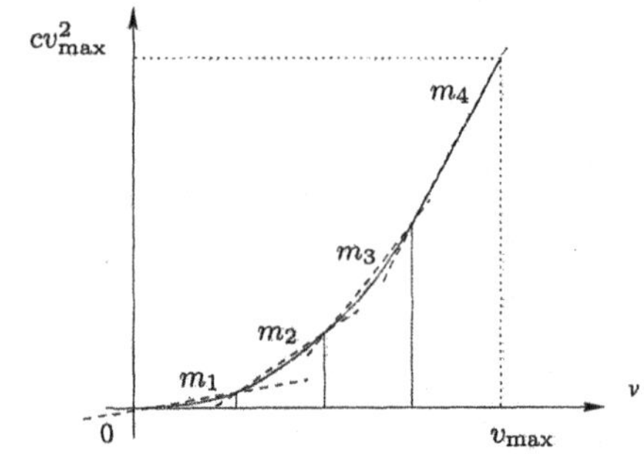

Fig. 2. *Approximation of the nonlinear curve via a gain scheduling approach;* m_1, m_2, m_3, m_4 *indicate the different affine models.*

The mp-MILP solver provides a partition of convex polyhedra in the parameter space θ, and a PWA control law of the form $\tilde{y}(\theta) = F_i\theta + G_i$, that serves as a *look-up table* during the real time evolution. The software we used for this case is part of the `Multi-Parametric-Toolbox release 2.5` (Kvasnica *et al.*, 2004).

Method 3: on-line linear approximation. In alternative to the hybrid optimization we also considered a locally linear approximation approach. The method consists in using a *locally* linearized prediction model. As detailed in (Beccuti *et al.*, 2005), this technique is appealing for relatively slow processes. The manipulated variables are computed on the base of the linear model, that approximates the nonlinear friction force $V = cv^2$ with its tangent line, i.e., $V = cv^2 \approx 2cv_t v - cv_t^2$.

Method 4: off-line linear approximation. The previous method also suggests an off-line version, in a gain scheduling fashion. The nonlinear curve depicted in Figure 1.b, is approximated into $M = 4$ affine models m_1, m_2, m_3, m_4 in point to point secant approximation, as illustrated in Figure 2.

For each affine model m_i we solve an off-line mp-LP problem of the form (9). More precisely we construct $M = 4$ look-up tables, each valid for a given range of velocity. In the simulation the controller selects the table according to the current value of the speed.

Method 5: Positively invariant set. This on-line approach (Lazar *et al.*, 2005) is aimed to determine a positively invariant set $\mathbb{X}_{N_p}$ for the PWA system, within which a strictly linear control action may be used. The cost function of the form

$$J(\theta(k), \tilde{u}(N_p)) \triangleq \sum_{j=1}^{N_p} ||Q_j \varepsilon(k+j)||_1 + ||Ru(k+j-1)||_1, \quad (10)$$

where $Q_j = Q \; \forall j = 1, \ldots, N_p - 1$, is employed as a candidate Lyapunov function to construct (a) an additional end point constraint and (b) a terminal cost matrix Q_{N_p}. The method (Lazar *et al.*, 2005) is conceived for the ℓ_∞-norm, but it can be easily adapted to the ℓ_1-norm.

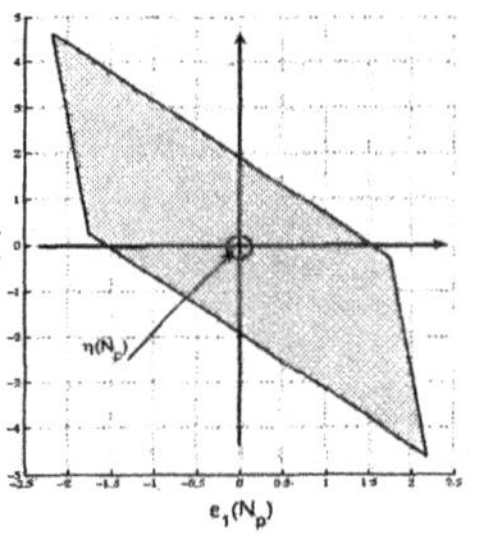

Fig. 3. *Positively invariant set, used as end point constraint in method 5, centered on the last point of the predicted reference trajectory.*

The reformulation of problem (5) with the additional conditions (a) and (b) has the important advantage of providing a *stabilizing* control law for the hybrid system. Hence we compute an on-line MPC controller over the MLD model (6), enriched with an end point constraint.

In this specific tracking problem, the end point constraint (Figure 3) computed off-line, is *centered* in the N_p-th point of the reference trajectory. The terminal cost matrix is $Q_{N_p} = \begin{bmatrix} 4.58 & 0.45 \\ 5.14 & 4.15 \end{bmatrix}$.

4. COMPARISON ISSUES AND RESULTS

The five methods are implemented in `Matlab 7`, `Linux 2.4.22` *OS* on an `INTEL pentium 4, 3GHz`. All optimizations, LP and MILP, are performed via the solvers embedded in `TOMLAB v5.1`, except for method 2 (`CDD Criss Cross` solver).

The prediction model for methods 3 and 4 is a linear approximation and an MLD model for methods 1, 2 and 5. Simulation is carried out over the nonlinear continuous time system (1). Methods 3 and 4 are really competitive as long as the system and prediction model are close[3]. On the other hand, methods 1, 2 and 5, based on PWA approximation of the nonlinearity, are not significantly affected by the model mismatch. The integration of equation (1) is done with `Matlab ode45` with `relative` and `absolute tolerances` 1.0×10^{-8}, `max step size` 1.0×10^{-3}.

We have considered two references: (a) a smooth trajectory, with constant velocity and (b) irregular trajectory, with variable velocity, depicted in Figure 4. Some other features of the common setup

[3] The sampling time plays a fundamental role in this specific application.

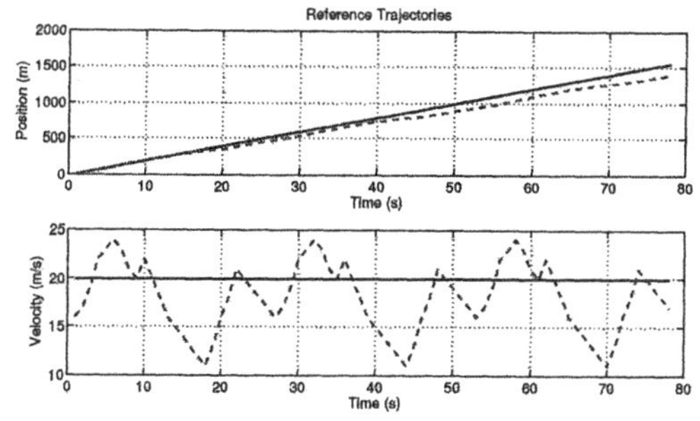

Fig. 4. *Reference trajectories used in simulation, constant (solid) and disturbed (dashed) velocity.*

Table 3. *General data and initial conditions.*

Prediction horizon N_p	3
Control horizon N_c	3
Sample time T	$1\,s$
Simulation time	$75\,s$
Number of variables	MLD: 24, Linear: 12
of which dummy	9
of which integer	MLD: 3, Linear: –
Number of parameters	11
Number of constraints (MLD)	85 (89 for method 5)
Number of constraints (Linear)	78
Input initial condition	$u(-1) = 0$
State initial condition	$x(0) = [0, 5]^T$
State past condition	$x(-1) = [-5, 5.3]^T$

and initial condition of the simulations are listed in Table 3.

Tables 4 and 5 collect the data obtained by running the methods for both references, constant and time varying velocity. In particular Table 4 shows the performance of the methods with the former and Table 5 with the latter.

We believe that the most interesting aspect (from an application point of view) in Tables 4 and 5 is the computational time, both on-line (average and maximum along the simulation period) and off-line, as well as the memory requirements. These entities are related to the size of the optimization problem, in our particular case to the length of the prediction horizon.

Let us observe that the on-line methods based on the MLD transformation are reasonably fast for low values of N_p, and their velocity is comparable with the purely linear method. For higher values of the horizon, the computational time of method 3 is not significantly affected as it is for methods 1 and 5, due to their inner mixed integer structure.

In terms of on-line memory usage we point out that the off-line methods perform quite poorly. In fact the number of the faceted polyhedra grows exponentially with N_p (as the number of parameters depends on N_p) and each number should be in double precision. However it should be remarked that off-line methods do not require the optimizer on board, which can be relevant in an *integrated design* that also takes into account budget issues.

We also provide the number of times when infeasibility occurred during the simulation period. This index is related to the length of the prediction horizon, and it should decrease as the value of N_p increases. The next row of the tables report the cost of the evolution of the obtained control law.

For the on-line methods the item *Maximum tractable* N_p is obtained by choosing the smallest N_p for which the computational time required by the optimizer is lower than the sampling time and the solution is 100% feasible for the whole simulation period. We observe that this is possible for PWA approximations, but not for methods 3 and 4, for which increasing the prediction horizon does not provide significant improvement on the feasibility of the solution.

Method 5 contains an additional terminal constraint, and it is completely infeasible for low values of N_p. The minimum N_p that guarantees 100% feasibility is $N_p = 17(19)$ (reference b), but unfortunately for this size the computation time becomes longer than the sample time ($T = 1\,s$).

For the off-line methods, namely 2 and 4, the value of *Maximum tractable* N_p is based on reasonable on-line computational time and data storage requirements. As it is shown in Tables 4 and 5 the look-up process to the partition table requires more time than the sampling time T. Nevertheless, the look-up process may be improved by use of appropriate binary search algorithms (Tøndel *et al.*, 2003), the complexity of which is logarithmic.

One major limitation of off-line methods is that the look-up table is only valid for a specific configuration of the system. Thus these methods are not able to handle possible variation of the system parameters values, and a recomputation of the table may be required.

5. CONCLUSIONS

Some methods conceived for computing control laws of PWA systems were used to design an ACC for a Smart vehicle. We have made a comparison among several methods. In particular, methods based on MLD transformation (1 and 5) appear to be very reliable and accurate although they require to perform an on-line optimization. On the other side, off-line methods seem to be more efficient in terms of feasibility, but they suffer of high complexity of the data structure, and their exponential complexity is practically prohibitive

Table 4. *Methods applied on the ACC design, with a constant velocity reference.*

Method	1	2	3	4	5
Time on-line (avg) (s)	0.09	0.90	0.19	1.16	0.09
Time on-line (max) (s)	0.57	1.64	0.16	8.31	0.18
Time off-line (s)	0.16	6.84×10^3	0.002	1.13×10^5	0.21
Memory usage on-line (Mb)	4.43+optimizer	28.5	3.5+optimizer	62	4.53+optimizer
Memory usage off-line (Mb)	0.05	28.5	0.01	62	0.06
Infeasibility (%)	62	1.3	60	27	100
Cost of evolution	337.71	5.38×10^3	474.15	365.32	–
Maximum tractable N_p	18	2	> 30	2	17
Number of regions	–	4541	–	19311	–

Table 5. *Methods applied on the ACC design, with time-varying velocity reference.*

Method	1	2	3	4	5
Time on-line (avg) (s)	0.09	1.13	0.19	1.79	0.07
Time on-line (max) (s)	0.17	1.91	0.39	8.56	0.44
Time off-line (s)	0.09	6.84×10^3	0.003	1.13×10^5	0.21
Memory usage on-line (Mb)	4.45+optimizer	28.5	3.5+optimizer	62	4.45+optimizer
Memory usage off-line (Mb)	0.05	28.5	0.01	62	0.05
Infeasibility (%)	84	1.3	77	56	100
Cost of evolution	334.68	1.59×10^3	456.83	391.41	–
Maximum tractable N_p	15	2	> 30	2	19
Number of regions	–	4541	–	19311	–

for a high number of variables. In a further development we intend to consider also a model with discrete input, the gear shift. We will also consider the performance of some other interesting techniques, (Hedlund and Rantzer, 1999; Raković *et al.*, 2004), that appear to be suitable for the described problem.

Acknowledgment. Authors are thankful to M. Kvasnica (ETH Zürich) for his important advices with the minimization tools and to M. Lazar (TU Eindhoven) for providing the end point constraint for Method 5.

REFERENCES

Beccuti, A.G., T. Geyer and M. Morari (2005). A hybrid system approach to power systems voltage control. In: *Proc. 44th IEEE Conf. on Dec. and Contr.*, Seville, Spain. pp. 6774–6779.

Bemporad, A. and M. Morari (1999). Control of systems integrating logic, dynamics, and constraints. *Automatica* **35**(3), 407–427.

Bemporad, A., F. Borrelli and M. Morari (2000). Piecewise linear optimal controllers for hybrid systems. In: *Proc. American Contr. Conf.*, Chicago, USA. pp. 1190–1194.

Borrelli, F. (2003). *Constrained Optimal Control of Linear and Hybrid Systems*. LNCIS 290. Springer–Verlag. Berlin.

Dua, V. and E.N. Pistikopoulos (2000). An algorithm for the solution of multiparametric mixed integer linear programming problems. *Annals of Op. Research* **99**(1–4), 123–139.

Elia, N. and S. Mitter (1999). Quantization of linear systems. In: *Proc. 38th IEEE Conf. on Dec. and Contr.*, Phoenix, Arizona USA. pp. 3428–3435.

Hedlund, S. and A. Rantzer (1999). Optimal control of hybrid systems. In: *Proc. 38th IEEE Conf. on Dec. and Contr.*, Phoenix, USA. pp. 3972–3976.

Kerrigan, E.C. and D.Q. Mayne (2002). Optimal control of constrained, piecewise affine systems with bounded disturbances. In: *Proc. 41th IEEE Conf. on Dec. and Contr.*, Las Vegas, USA. pp. 1552–1557.

Kvasnica, M., P. Grieder, M. Baotić and F.J. Christophersen (2004). *Multi-Parametric Toolbox MPT: User's Manual.* (ETH) Zurich. See: `http://control.ee.ethz.ch/~mpt`.

Lazar, M., W.P.M.H. Heemels, S. Weiland, A. Bemporad and O. Pastravanu (2005). Infinity norms as Lyapunov functions for model predictive control of constrained PWA systems. In: *LNCS: Hybrid Systems: Computation and Control.* number 3414. Springer Verlag. Zürich, Switzerland. pp. 417–432.

Raković, S., E.C. Kerrigan and D.Q. Mayne (2004). Optimal control of constrained piecewise affine systems with state and input–dependent disturbances. In: *Proc. Math. Theory of Networks and Sys.*, Leuven, Belgium.

Sontag, E.D. (1981). Nonlinear regulation: the piecewise affine approach. *IEEE Trans. Automatic Contr.* **26**(2), 346–357.

Tøndel, P., T.A. Johansen and A. Bemporad (2003). Evaluation of piecewise affine control via binary search tree. *Automatica* **39**(3), 945–950.

ELSEVIER
IFAC
PUBLICATIONS

IDLE SPEED CONTROL – A BENCHMARK FOR HYBRID SYSTEM RESEARCH [1]

Andrea Balluchi [*] **Luca Benvenuti** [**,*]
Maria D. Di Benedetto [***] **Tiziano Villa** [****]
Alberto L. Sangiovanni–Vincentelli [*,†]

[*] *PARADES, Via di S. Pantaleo, 66, 00186 Roma, Italy.*
{balluchi, alberto}@parades.rm.cnr.it
[**] *DIS, University of Rome La Sapienza, Via Eudossiana 18, 00184 Rome, Italy.* luca.benvenuti@uniroma1.it
[***] *DEWS, University of L'Aquila, Poggio di Roio, 67040 L'Aquila, Italy.* dibenede@ing.univaq.it
[****] *DIEGM, University of Udine, Via delle Scienze, 208, 33100 Udine, Italy.* villa@uniud.it
[†] *EECS, Univ. of California at Berkeley, CA 94720, USA.*
alberto@eecs.berkeley.edu

Abstract: The design of engine control systems has been traditionally carried out using a mix of heuristic techniques validated by simulation and prototyping with approximate mean–value models. However, the ever increasing demands on passengers' comfort, safety, emissions and fuel consumption imposed by car manufacturers and regulations call for more robust techniques and the use of cycle–accurate models. The use of hybrid methodologies is then natural because of the rich combination of time and event-based behaviors exhibited by a controlled engine. While there is no doubt that hybrid modeling is relevant for this application, its efficiency in providing industrial strength solutions is still debated. For this reason, it is important to corral the hybrid system research community to provide evidence of the quality of the proposed control solutions. In this perspective, we present a hybrid benchmark problem on "Idle Speed Control" proposed by the Network of Excellence HYCON. We hope this benchmark problem will also serve as the basis for comparison of different approaches, thus helping industry to identify the best solutions available. *Copyright © 2006 IFAC*

Keywords: Automotive, engine modeling, hybrid systems.

1. INTRODUCTION

In the automotive industry, increased performance, safety and time-to-market pressure require the use of complex control algorithms with guaranteed properties. Best practices in this industry are based on extensive experimentation and tuning. This procedure needs a substantial overhaul to eliminate long re-design cycles and potential safety problems after the car is introduced in the market. Using more accurate models and control algorithms with guaranteed properties reduces greatly the need for extensive experimen-

[1] This work is supported by the Network of Excellence HYCON, E.C. IST-511368.

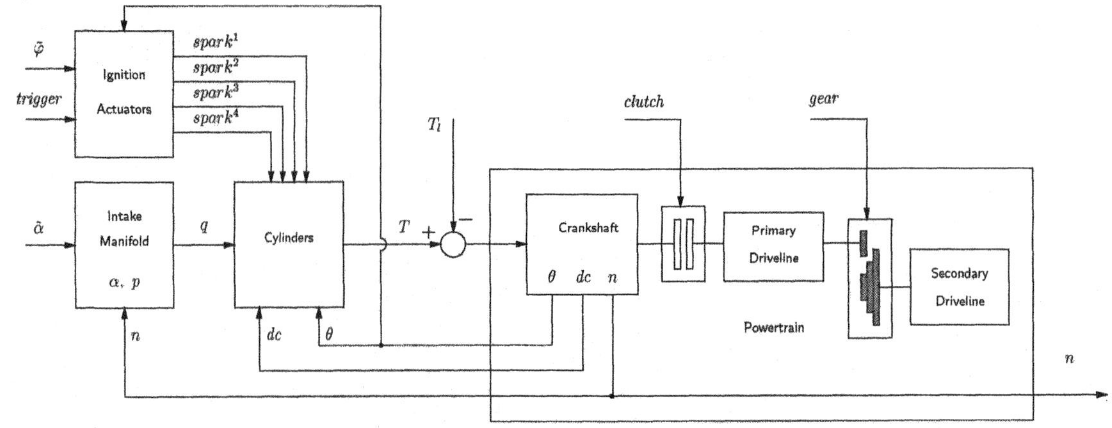

Fig. 1. Engine hybrid model.

tation and points to potential problems early in the design cycle.

In this general scenario, the synthesis of a control strategy for spark–ignition engines at idle speed is one of the most challenging problems. The goal is to maintain the engine speed as close as possible to a reference constant engine speed despite load torque disturbances (due to e.g. the air conditioning system, the steering wheel servo-mechanism) and engagements and disengagements of the transmission occurring when the driver operates on the clutch. In order to achieve the best fuel economy, the reference engine speed is chosen at the minimum value that yields acceptable combustion and emission quality, and noise, vibration and harshness (NVH) characteristics.

A survey on different engine models and control design methodologies for idle control is given in (Hrovat and Sun, 1997). Both time–domain (e.g. (Butts *et al.*, 1999)) and crank–angle domain (e.g. (Yurkovich and Simpson, 1997)) mean–value models have been proposed in the literature. More recently, hybrid system techniques have been applied to the idle speed control problem. The hybrid nature of the problem of engine control comes not only from the digital controllers used to manage an analog plant, but also from the behavior of the plant to be controlled. In fact, an accurate model of a four–stroke gasoline engine has a "natural" hybrid representation because: pistons have four modes of operation corresponding to the stroke they are in, while power–train and air dynamics are continuous–time processes. In addition, these processes interact tightly. In fact, the timing of the transitions between two phases of the pistons is determined by the continuous motion of the power–train, which, in turn, depends on the torque produced by each piston. This problem has been attacked with hybrid approaches in (Balluchi *et al.*, 2000*b*; Balluchi *et al.*, 2002; Balluchi *et al.*, 2004), using hybrid games, formal verification and *control-to-facet* techniques. Other approaches are being developed in several institutions. Given the difficulty and industrial relevance of the engine control problem, together with the availability of models and control algorithms, we believe that offering a complete description and the collateral material will allow researchers to apply their methods to the problem and as a consequence this will push the state of the art considerably.

In this paper, we present a particular benchmark problem in this domain: "Idle Speed Control". This benchmark has been developed under the sponsorship of the Network of Excellence HYCON. The documentation and the simulation files related to this benchmark problem are available at the HYCON web–page www.ist-hycon.org, under "WP2: Performance Evaluation Platform".

2. THE ENGINE HYBRID MODEL

In this section, a hybrid model of a 4–cylinder 4–stroke spark ignition engine equipped with an electronic–throttle is presented. The proposed hybrid model represents accurately the behavior of the engine during idle speed control. The overall system is composed of four blocks, namely the *ignition actuators*, the *intake manifold*, the *cylinders* and the *powertrain* (Figure 1) (to satisfy emission requirements, fuel injection is regulated so that the air and fuel mixture is stoichiometric). The *ignition actuators* deliver the sparks $spark^i$ to the cylinders with a timing defined by the desired spark advance angle $\tilde{\varphi}$. The latter represents the spark ignition control input. When the controller issues a new value $\tilde{\varphi}$, it emits the synchronization event *trigger*. The mass of air q loaded in the cylinders depends on the dynamics of the *intake manifold*. The manifold pressure p is controlled by a throttle valve powered by an electrical motor; α and $\tilde{\alpha}$ denote, respectively, the throttle valve position and the reference to the throttle valve controller. The air charge q is a function of the pressure p and of the crankshaft revolution speed n. The *cylinders* model describes the engine torque generation process. The engine torque T depends on the air charge q and on the timing of spark ignition. The timing sequence of the four strokes of each cylinder is determined by the motion of the piston between the top and the bottom *dead centers* (dc), i.e. the piston uppermost and lowermost positions. The position of the piston is determined by the crankshaft angle θ. Finally, the crankshaft revolution speed n depends on the powertrain dynamics. In idle speed control it is assumed that the gear is idle, while the clutch can be either open or closed. The powertrain is powered by the balance between the engine torque T and the load torque T_l, due to the auxiliary systems driven by the crankshaft (such as e.g. the generator).

2.1 Intake manifold

The intake manifold model gives the cylinder air charge q. A sufficiently accurate model of the air charge can be obtained abstracting away the intake manifold pumping fluctuations due to the periodic motion of pistons and valves. In fact, they are usually filtered out in air charge estimation algorithms for engine torque control. Denoting by α and p the throttle-valve position and the intake manifold pressure p, respectively, we have (see (Hendricks and Sorenson, 1990)):

$$\dot{\alpha} = -\frac{1}{\tau_{thr}}(\alpha - \tilde{\alpha}) \tag{1}$$

$$\dot{p} = \frac{RT_{air}}{V_{pln}}[\, f_{thr}(\alpha) - f_{cyl}(p,n)\,] \tag{2}$$

$$q = \frac{30}{n} f_{cyl}(p,n)\,. \tag{3}$$

Equation (1) represents the actuation dynamics of the throttle valve, with $\tilde{\alpha}$ being the reference command. The intake manifold pressure dynamics (2) depends on the balance between the air-mass flow through the throttle valve $f_{thr}(\alpha)$ and through the cylinder valves $f_{cyl}(p,n)$, modeled as follows:

$$f_{thr}(\alpha) = s_0 + s_1\,\alpha + s_2\,\alpha^2 \tag{4}$$

$$f_{cyl}(p,n) = c_0 + c_1\,p + c_2\,n + c_3\,p\,n\,. \tag{5}$$

2.2 Cylinders.

The engine torque T is given by the sum of the contributions T^i by the four cylinders:

$$T = \sum_{i=1}^{4} T^i\,. \tag{6}$$

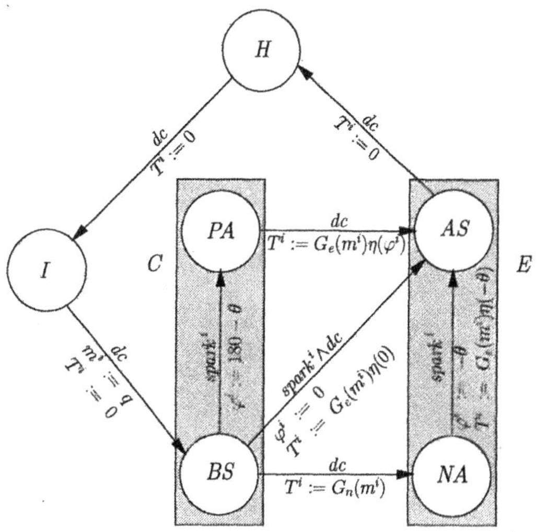

Fig. 2. Hybrid system describing the behavior of the i–th cylinder.

The profile of T^i depends on the current stroke of the cylinder (either *intake*, *compression*, *expansion*, or *exhaust*), the piston position, the mass of air loaded in the cylinder during the intake stroke, and the spark ignition timing. The contributions to the engine torque in the intake, compression and exhaust strokes are quite small and, for small throttle valve angles, only weakly dependent on the air charge. Then, the cylinder torque T^i is modeled as a piecewise constant signal, nonzero in the expansion stroke only, whose value in expansion takes into account the effects of the other three cylinders that evolves in the intake, compression and exhaust strokes.

At every cycle, the mixture is ignited by the spark. Ideally, heat release should occur instantaneously when the piston reaches the compression stroke top–dead–center. However, due to the nonzero combustion time, the maximum engine torque is achieved when spark ignition is given before the piston completes the compression stroke (*positive spark advance*). Delaying spark ignition to the expansion stroke (*negative spark advance*) reduces drastically the engine torque. The spark control input has a very short delay and can be used to reduce the torque much faster than using only the throttle valve. The spark ignition time is commonly defined in terms of the spark advance angle φ^i, which denotes the difference between the angle of the crankshaft at the compression top–dead–center and the one at the ignition time. Hence, the air-fuel mixture is loaded in the cylinder during the intake stroke, while the spark is ignited when the piston is around the compression stroke top dead–center (Hrovat *et al.*, 1998). The delay between mixture intake / ignition and torque generation is represented by the hybrid system depicted in Figure 2, where I, C, E and H stand respectively for the intake, compression,

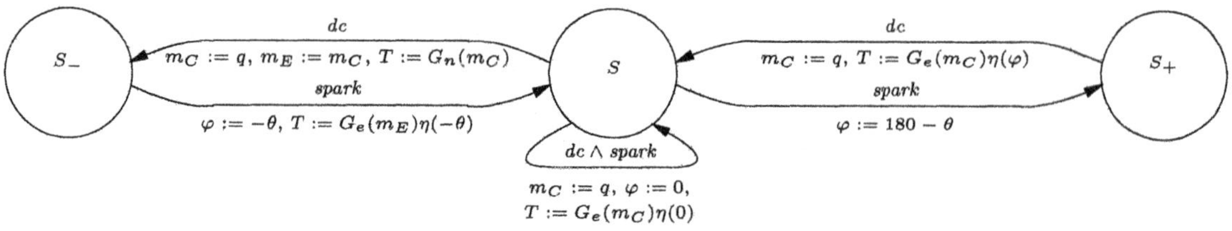

Fig. 3. Hybrid system describing the behavior of a 4-cylinder engine.

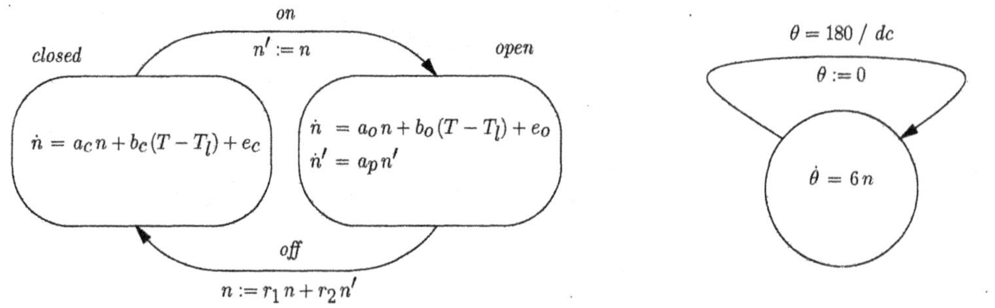

Fig. 4. Powertrain (left) and crankshaft angle (right) hybrid models.

expansion and exhaust strokes. Since spark ignition may occur either during compression or expansion, the macro–states C and E are split as follows:

- *BS* (*Before Spark*). The cylinder is in compression and no spark has been ignited yet.
- *PA* (*Positive Advance*). The cylinder is in compression and the spark has been ignited.
- *NA* (*Negative Advance*). The cylinder is in expansion and the spark has not been ignited yet.
- *AS* (*After Spark*). The cylinder is in expansion and the spark has been ignited.

The hybrid system makes a transition either when the piston reaches a dead–center (*dc*) or when the spark is ignited (*spark*^{i}). The spark advance angle φ^i is evaluated when the spark is ignited (*BS* $\to$ *PA*, *BS* $\to$ *AS*, *NA* $\to$ *AS*) and is expressed in terms of the crankshaft angle θ.

At the end of the intake stroke (transition $I \to$ *BS*), the air charge for the current engine cycle, m^i, is set. In case of negative spark advance, the hybrid system enters the state *NA*, where T^i is positive due to gas expansion and depends on the loaded air mass m^i:

$$T^i = G_n(m^i) = g_0 + g_1\, m^i + g_2\,(m^i)^2 > 0\,. \quad (7)$$

The hybrid system is in state *AS* either during the entire expansion stroke, in case of positive spark advance, or just after spark ignition, in case of negative spark advance. In state *AS*,

$$T^i = G_e(m^i)\,\eta(\varphi^i)\,, \qquad \text{with} \quad (8)$$

$$G_e(m^i) = h_0 + h_1\, m^i + h_2\,(m^i)^2\,, \quad (9)$$

$$\eta(\varphi) = v_0 + v_1\,\varphi + v_2\,\varphi^2 + v_3\,\varphi^3\,, \quad (10)$$

where $\eta(\varphi) \leq 1$ is the ignition efficiency function and $G_e(m^i)$ is the engine torque produced with loaded air mass m^i and optimal spark advance (i.e. $\eta = 1$). The behavior of a four–cylinder in-line engine can be obtained by composing four cylinder hybrid models as given in Figure 2. However, since at any time each cylinder is in a different stroke of the engine cycle, the model can be significantly simplified and reduced to a three–state hybrid model, with discrete states S, S_+ and S_- as depicted in Figure 3. States S, S_+ and S_- correspond to the following cylinder configurations: $S = (I, BS, AS, H)$, $S_+ = (I, PA, AS, H)$, $S_- = (I, BS, NA, H)$. For details on how the reduced model is obtained see (Balluchi *et al.*, 2000*a*).

2.3 Powertrain

In idle speed control, the gear is fixed in neutral position (idle). Consequently, the secondary driveline is disconnected and does not affect the crankshaft dynamics. Due to the actions of the driver on the clutch pedal, the first part of the driveline is either connected or disconnected from the engine (see Figure 1). The dynamics of the crankshaft speed n is given by the hybrid model depicted in Figure 4, where: the discrete states *open* and *closed* encode the two possible positions of the clutch, the input events *on* and *off* represent the driver action on the clutch pedal, and the continuous dynamics are affine.

When the clutch is *open* the primary driveline speed n' evolves independently from the crankshaft speed n. Instead, when the clutch is *closed*, they evolve at the same speed n. When the clutch pedal is released (*open* $\to$ *closed*), the order of the model is reduced and the common speed state is reset. When the clutch is opened (*closed* $\to$ *open*), the primary driveline speed is appropriately initialized. The continuous dynamics and

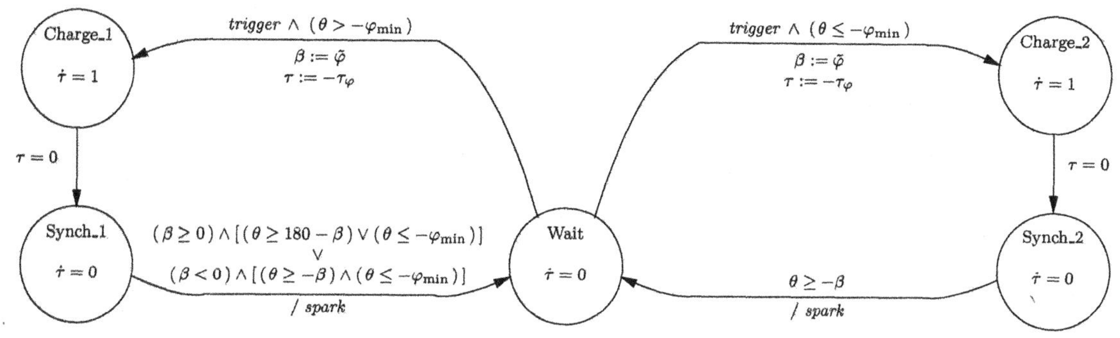

Fig. 5. Hybrid model of the ignition actuator models.

reset parameters depends on inertial momenta and viscous friction coefficients.

The evolution of the crankshaft angle in the interval $[0, 180]$ gives the position of the pistons within each stroke. It is described by the simple hybrid model reported in Figure 4. The dynamics is given by the crankshaft speed n. When the crankshaft angle θ reaches the value 180, it is reset and the dead–center event *dc* is emitted.

2.4 Ignition actuators

The ignition coil charging time introduces a non–negligible delay in spark ignition control. Due to this delay, the desired spark advance $\tilde{\varphi}$ has to be set with a sufficient advance to allow proper spark actuation. The ignition actuator delay is described by the hybrid model depicted in Figure 5.
The system waits in state *Wait* for the desired value of spark advance angle $\tilde{\varphi}$. It is supposed that the desired spark advance angle $\tilde{\varphi}$ is always issued for each cylinder and that it belongs to the feasible range $[\varphi_{\min}, \varphi_{\max}]$, with $\varphi_{\min} < 0$ and $\varphi_{\max} > 0$. The controller provides the desired $\tilde{\varphi}$ at some time between $180 + \varphi_{\min}$ degree in advance and $-\varphi_{\min}$ degree in delay from the compression dead–center. When the controller issues the new value $\tilde{\varphi}$, it emits the event *trigger* to synchronize with the ignition actuator. If $\tilde{\varphi}$ is issued when $\theta \leq -\varphi_{\min}$, then the command is given at the beginning of the expansion stroke (only negative spark advance will be feasible) and the system takes the right cycle. Otherwise, if $\theta > -\varphi_{\min}$, then either positive or negative spark advances could be applied and the system takes the left cycle. After the transition, the system starts charging the ignition coil for a time τ_φ, spent either in state *Charge_1* or state *Charge_2*. When the charging time is elapsed, two cases are in order. (1) the crankshaft angle has not reached the desired spark advance yet ($\theta < 180 - \beta$, for positive spark advance, and $\theta < -\beta$, for negative spark advance): the system remains in state *Synch_1* / *Synch_2* until the desired spark advance is reached; then, it makes the transition to state *Wait* emitting the spark signal. (2) the crankshaft angle has already passed the desired spark advance (the command was issued too late): the system takes the transition to state *Wait* emitting the spark signal.

3. PROBLEM FORMULATION

The purpose of idle speed control is to keep the engine speed n as close as possible to a target value n_0 when the gear is neutral, preventing the engine to stall. For fuel consumption minimization, the target value n_0 is, ideally, the lowest engine speed for which the engine can be robustly controlled avoiding engine stall. More precisely, the goal is to maintain the crankshaft speed n in a specified range, $n_0 \pm \Delta_n$, robustly with respect to two sources of disturbances (a continuous and a discrete one):

- The load torque T_l acting on the crankshaft due to the auxiliary sub-systems;
- The changes on the crankshaft dynamics due to the motion of the clutch.

The control inputs are: the desired spark advance angle $\tilde{\varphi}$ and the throttle valve command $\tilde{\alpha}$. Actuator constraints and dynamics have to be taken into account in the design. Available sensors provide the following feedbacks: throttle valve position α, manifold pressure p, engine speed n.
The design specification can be formalized as a constrained optimal control problem for the hybrid model of the engine described in Section 2. The adoption of a hybrid formalism allows us to represent the cyclic behavior of the engine, thus capturing the effect of each spark command on the generated torque, the interaction between the discrete torque generation and the continuous powertrain and air dynamics, and the discrete changes of the powertrain behavior.
For any action of the torque disturbance $T_l \in [0, T_{l\max}]$ and any switching of the clutch state, the controller has to guarantee that the following constraints are satisfied:

C1 : engine speed

$$|n - n_0| \leq \Delta_n \qquad (11)$$

Problem specification					
n_0	800 [RPM]	$T_{l\,\max}$	22.5 [Nm]	$\varphi_{\min}$	-12 [^{o}]
Δ_n	50 [RPM]	$\alpha_{\max}$	15 [^{o}]	$\varphi_{\max}$	15 [^{o}]
Ignition actuator / Intake manifold		**Powertrain**		***i*–th and 4–th cylinder**	
τ_φ	$5\,10^{-3}$ [sec]			g_0	-2.93 [Nm]
τ_{thr}	$8.35\,10^{-2}$ [sec]	r_1	0.9 [ ]	g_1	$6.03\,10^4$ $[\frac{\text{Nm}}{\text{Kg}}]$
RT_{air}/V_{pln}	$2.152\,10^5$ $[\frac{\text{mbar}}{\text{Kg}}]$	r_2	0.1 [ ]	g_2	$1.36\,10^9$ $[\frac{\text{Nm}}{\text{Kg}^2}]$
s_0	$7\,10^{-4}$ $[\frac{\text{Kg}}{\text{sec}}]$	a_c	-0.5852 $[\frac{1}{\text{sec}}]$	h_0	-4.168 [Nm]
s_1	$3.9\,10^{-4}$ $[\frac{\text{Kg}}{\text{sec}\,^o}]$	b_c	54.26 $[\frac{\text{RPM}}{\text{sec Nm}}]$	h_1	$1.265\,10^5$ $[\frac{\text{Nm}}{\text{Kg}}]$
s_2	$5.78\,10^{-5}$ $[\frac{\text{Kg}}{\text{sec}\,(^o)^2}]$	e_c	-976 $[\frac{\text{RPM}}{\text{sec}}]$	h_2	$2.145\,10^9$ $[\frac{\text{Nm}}{\text{Kg}^2}]$
c_0	$8.279\,10^{-4}$ $[\frac{\text{Kg}}{\text{sec}}]$	a_o	-0.625 $[\frac{1}{\text{sec}}]$	v_0	$7.74\,10^{-1}$ [ ]
c_1	$3.041\,10^{-6}$ $[\frac{\text{Kg}}{\text{sec mbar}}]$	b_o	59.68 $[\frac{\text{RPM}}{\text{sec Nm}}]$	v_1	$1.729\,10^{-2}$ $[(^o)^{-1}]$
c_2	$8.5\,10^{-8}$ $[\frac{\text{Kg}}{\text{sec RPM}}]$	e_o	-1074 $[\frac{\text{RPM}}{\text{sec}}]$	v_2	$-3.65\,10^{-5}$ $[(^o)^{-2}]$
c_3	$2.245\,10^{-9}$ $[\frac{\text{Kg}}{\text{sec mbar RPM}}]$	a_p	-0.1875 $[\frac{1}{\text{sec}}]$	v_3	$-7.24\,10^{-6}$ $[(^o)^{-3}]$

Table 1. Model parameters

C2 : throttle angle

$$0 \le \alpha \le \alpha_{\max} \tag{12}$$

C3 : spark advance control

$$\varphi_{\min} \le \tilde{\varphi} \le \varphi_{\max} \tag{13}$$

$$\varphi_{\min} \le \varphi \le \varphi_{\max} \tag{14}$$

The cost function to minimize is

$$\min \|n - n_0\|^2_{\mathcal{L}_2} = \int_0^\infty (n - n_0)^2 dt \, , \tag{15}$$

in a transient due to a torque load $T_l = T_{l\,\max}$, starting from the steady state point with $T_l = 0$ and assuming that the clutch is open.

4. CONCLUSION

We presented the benchmark problem on "Idle Speed Control" proposed by the Network of Excellence HYCON. The purpose of the benchmark is to promote the application of hybrid system techniques to automotive control problems and demonstrate the effectiveness of hybrid system methodologies for the automotive industry. The description of the benchmark includes: a hybrid model of the engine, formalized system specification and a model for control algorithm validation.

REFERENCES

Balluchi, A., F. Di Natale, A. L. Sangiovanni-Vincentelli and J. H. van Schuppen (2004). Synthesis for idle speed control of an automotive engine. In: *Hybrid Systems: Computation and Control* (R. Alur and G. J. Pappas, Eds.). Vol. 2993 of *Lecture Notes in Computer Science*. pp. 80–94. Springer-Verlag. New York.

Balluchi, A., L. Benvenuti, M. D. Di Benedetto and A. L. Sangiovanni-Vincentelli (2002). Idle speed control synthesis using an assume–guarantee approach. In: *Nonlinear and Hybrid Systems in Automotive Control*. pp. 229–243. Springer-Verlag. London, UK.

Balluchi, A., L. Benvenuti, M. D. Di Benedetto, C. Pinello and A. L. Sangiovanni-Vincentelli (2000*a*). Automotive engine control and hybrid systems: Challenges and opportunities. *Proceedings of the IEEE* **88**(7), 888–912.

Balluchi, A., L. Benvenuti, M. D. Di Benedetto, G. M. Miconi, U. Pozzi, T. Villa, H. Wong-Toi and A. L. Sangiovanni-Vincentelli (2000*b*). Maximal safe set computation for idle speed control of an automotive engine. In: *Hybrid Systems: Computation and Control* (N. Lynch and B.H. Krogh, Eds.). Vol. 1790 of *Lecture Notes in Computer Science*. pp. 32–44. Springer-Verlag. New York, U.S.A.

Butts, K. R., N. Sivashankar and J. Sun (1999). Application of ℓ_1 optimal control to the engine idle speed control problem. *IEEE Trans. on Control Systems Tech.* **7**(2), 258–270.

Hendricks, E. and S. C. Sorenson (1990). Mean value modelling of spark ignition engines. Tech. Rep. 900616. SAE.

Hrovat, D. and J. Sun (1997). Control engineering practice. *Models and control methodologies for IC engine idle speed control design.*

Hrovat, D., D. Colvin and B. K. Powell (1998). Comments on "applications of some new tools to robust stability analysis of spark ignition engine: A case study". *IEEE Trans. on Control Systems Technology* **6**(3), 435–436.

Yurkovich, S. and M. Simpson (1997). Crank-angle domain modeling and control for idle speed. *SAE Journal of Engines* **106**(970027), 34–41.

ELSEVIER
IFAC
PUBLICATIONS

SIMULATION AND VERIFICATION OF HYBRID SYSTEMS USING CHI [1]

D.A. van Beek, J.E. Rooda, R.R.H. Schiffelers

Department of Mechanical Engineering
Eindhoven University of Technology, P.O.Box 513
5600 MB Eindhoven, The Netherlands

INTRODUCTION

The hybrid χ (Chi) formalism (van Beek *et al.*, 2006*a*; Man and Schiffelers, 2006) integrates concepts from dynamics and control theory with concepts from computer science. It integrates ease of modeling with a straightforward, structured operational semantics. Ease of modeling is ensured by means of, among others, the following concepts: 1) different classes of variables: discrete, continuous and algebraic; 2) strong time determinism of alternative composition in combination with delayable guards; 3) integration of urgent and non-urgent actions; 4) differential algebraic equations as in mathematics; 5) concepts for complex system specification: 5a) process terms for scoping that integrate abstraction, local variables, local channels and local recursion definitions; 5b) process definition and instantiation that enable process re-use, encapsulation, hierarchical and/or modular composition of processes; and 5c) different interaction mechanisms: handshake synchronization and synchronous communication that allow interaction between processes without sharing variables, and shared variables that enable modular composition of continuous-time or hybrid processes.

The formal semantics of χ allows the definition of provably correct implementations for simulation and verification. *Copyright © 2006 IFAC*

SIMULATION OF χ MODELS

For simulation of hybrid χ models, two simulators are available: a symbolic simulator and a simulator based on S-functions (The MathWorks, Inc, 2005*b*). Both simulators are defined in terms of a so-called stepper, which computes the set of possible transitions for given a χ process. The stepper consists of three main functions: a function $\mathcal{S}_{\mathrm{a}}$ which returns the set of action steps for given a χ process, a function $\mathcal{S}_{\mathrm{d}}$ which returns the set of time steps for given a χ process, and a function Tr which returns the reduced set of transitions. Action steps and time steps can be seen as symbolic transitions. They contain all information that is needed to determine the transitions that the process from which they are derived can perform without solving predicates. An action step represents zero or more action transitions and a time step represents zero or more time transitions.

In general, the set of transitions of a χ specification is infinite. In particular, the number of time transitions of a χ process is usually infinite: if a process can delay for t time units, then, for every $0 \leq t' \leq t$, it can also delay for t' time units. To get rid of these additional time transitions, instead of returning all time transitions of a time step, for each trajectory only the time transition with longest duration is returned. Although this reduced set of transitions can still be infinite, in practice, this is rarely the case.

Note that an implementation of the stepper functions may impose additional restrictions on the χ syntax. For instance, for an implementation of function Tr, a (symbolic) solver is needed to compute the solutions of action predicates, the solution of delay predicates (differential algebraic equations), and the maximum duration of a time transition. Depending on the solver that is used, additional restrictions may be required.

The stepper functions are defined in such a way that it is easy to define different implementations. In case

[1] Work partially done in the framework of the HYCON Network of Excellence, contract number FP6-IST-511368

of the symbolic simulator, the symbolic solving capabilities of Maple (MapleSoft, n.d.) are used for implementing the Tr function. The simulator based on S-functions interacts with Matlab Simulink (The MathWorks, Inc, 2005*a*) via a so-called DE^+ simulator. The DE^+ simulator performs action transitions until the stepper returns a time step. This time step is then executed by Simulink by solving the delay predicates (ODEs) and monitoring the root functions (possibly resulting in state-events) that are specified in the time step. At the end of the time transition, the DE^+ simulator is called again.

VERIFICATION OF χ MODELS

One of the most successful formalisms for hybrid system verification is the theory of hybrid automata. In (Man and Schiffelers, 2006), formal translations between χ and hybrid automata (in both directions) have been defined. The translation from hybrid automata to χ aims to show that the χ formalism is at least as expressive as the theory of hybrid automata. The translation from (a subset of) χ to hybrid automata enables verification of χ specifications using existing hybrid automata based verification tools.

To the best of our knowledge, none of the hybrid automata definitions from literature is expressive enough to be used as the target for the translation of hybrid χ. Therefore, The translation uses a target hybrid automata definition, called HA_u automata, where the u stands for urgency, that uses features from different hybrid automata definitions. In particular, the definition of the jump predicate in combination with a set of changeable variables is based on (Alur *et al.*, 1996), the solution concept that allows piecewise differentiable functions is based on (van der Schaft and Schumacher, 2000), and the definition of urgent transitions was inspired by (Nicollin *et al.*, 1992).

In (Man and Schiffelers, 2006; van Beek *et al.*, 2006*b*), we use the verification tool PHAVer (Polyhedral Hybrid Automaton Verifyer) (Frehse, 2005) to show that it is indeed possible to verify properties of χ specifications using an existing hybrid automata based verification tool. Since a manual translation is very time consuming and error prone, the translation has been automated by implementing it using the programming language Python (Python website, 2005).

NORMALIZATION OF χ MODELS

The χ process algebra is a rich language that has strong support for modular composition by allowing unrestricted combination of operators such as sequential composition, parallel composition, and scoping of local variables and channels. The fact that the χ process algebra is such a rich language potentially complicates the development of tools for χ, since the implementations have to deal with all possible combinations of the χ atomic statements and the operators that are defined on them. This is where the process algebraic approach of equational reasoning, that allows rewriting models to a simpler form, is essential.

Instead of defining simulation and verification implementations on the full χ language, the process algebraic approach of equational reasoning makes it possible to transform χ models in a series of steps to a (much simpler) normal form (process algebraic linearization), and to define the implementations on the normal form. The original χ model and its normal form are bisimilar, which ensures that relevant model properties are preserved. The normal form has strong syntactical restrictions, no parallel composition operator, and is quite similar to a hybrid automaton. Partial normalization, keeping the top level parallelism intact, is also possible. Currently, generation of the normal form is automated, and correctness of the transformation to the normal form is proved.

REFERENCES

Alur, R., T. A. Henzinger and P. H. Ho (1996). Automatic symbolic verification of embedded systems. *IEEE Transactions on Software Engineering* **22**(3), 181–201.

Frehse, Goran (2005). PHAVer: Algorithmic verification of hybrid systems past HyTech. In: *Hybrid Systems: Computation and Control, 8th International Workshop* (Manfred Morari and Lothar Thiele, Eds.). Vol. 3414 of *Lecture Notes in Computer Science*. pp. 258–273. Springer-Verlag.

Man, K. L. and R. R. H. Schiffelers (2006). Formal Specification and Analysis of Hybrid Systems. PhD thesis. Eindhoven University of Technology.

MapleSoft (n.d.). Maple. http://www.maplesoft.com.

Nicollin, X., A. Olivero, J. Sifakis and S. Yovine (1992). An approach to the description and analysis of hybrid systems. In: *Workshop on Theory of Hybrid Systems*. pp. 149–178.

Python website (2005). http://www.python.org.

The MathWorks, Inc (2005*a*). *Using Simulink, version 6*. http://www.mathworks.com.

The MathWorks, Inc (2005*b*). *Writing S-functions, version 6*. http://www.mathworks.com.

van Beek, D. A., K. L. Man, M. A. Reniers, J. E. Rooda and R. R. H. Schiffelers (2006*a*). Syntax and consistent equation semantics of hybrid Chi. *Journal of Logic and Algebraic Programming* **68**(1-2), 129–210.

van Beek, D. A., K.L. Man, M. A. Reniers, J. E. Rooda and R. R. H. Schiffelers (2006*b*). Formal verification of hybrid Chi models using PHAVer. In: *Proceedings of the conference on Mathematical Modelling 2006*. Vienna, Austria.

van der Schaft, A. J. and J. M. Schumacher (2000). *An Introduction to Hybrid Dynamical Systems*. Vol. 251 of *Springer Lecture Notes in Control and Information Sciences*. Springer.

ELSEVIER
IFAC
PUBLICATIONS

HYBRID SYSTEM SIMULATION WITH SIMEVENTS

Christos G. Cassandras *

* *Center for Information and Systems Engineering*
Boston University, Brookline, MA 02446 USA

Michael I. Clune and Pieter J. Mosterman **

** *The MathWorks*
3 Apple Hill Drive, Natick, Mass. 01760 USA

Abstract: A new simulation product for hybrid systems is described, which combines a time-driven component and an event-driven component (SimEvents). Some of the key issues arising in designing such simulation environments are also discussed. *Copyright © 2006 IFAC*

1. INTRODUCTION

A *Hybrid System* (HS) is often defined as a system that combines continuous with discrete state variables (Levine and (Eds.), 2005). More important, however, is the fact that a HS combines time-driven dynamics (associated with processes modeled through differential or difference equations) with event-driven dynamics (modeled though state automata, Petri nets or other modeling frameworks for *Discrete Event Systems* (DES) (Cassandras and Lafortune, 1999)). Traditionally, simulators (such as Simulink® (MathWorks, 2001)) employ a time-driven execution mechanism, while drastically different ways are employed for DES. SimEvents (MathWorks, 2005) is designed to simulate DES, but is embedded in Simulink and equipped with functionality that enables an effective co-existence of time-driven and event-driven components making up a HS.

2. ARCHITECTURE

Figure 1 highlights the main functional components of the overall architecture. As a DES simulation engine, SimEvents is driven by an Event Calendar where all future events to occur are listed in ascending order of their scheduled time. SimEvents always processes the first event in this list and updates the DES state accordingly. When such an event takes place, the Cooperative Event Driver is responsible for translating it into a Simulink signal. The Data Exchange module passes this signal on to Simulink so that it may trigger a time-driven process or update various model parameters. Conversely, as a time-driven process evolves under the control of Simulink, it may generate events in the form of level-crossing points (from above, from below, or either) that the Data Exchange module appropriately translates so they may be processed by SimEvents blocks. The most challenging aspect of coordinating time-driven and event-driven dynamics is that of proper timing. In the architecture of Fig. 1, the system "clock" is maintained by Simulink and the Cooperative Event Driver is responsible for ensuring consistency between Simulink blocks and SimEvents blocks, which interact with the Event Calendar. Note that when a pure DES is simulated, the only interaction between SimEvents and Simulink is a simple link to the system clock through the Cooperative Event Driver which ensures that the sample times applied are consistent with times in the Event Calendar.

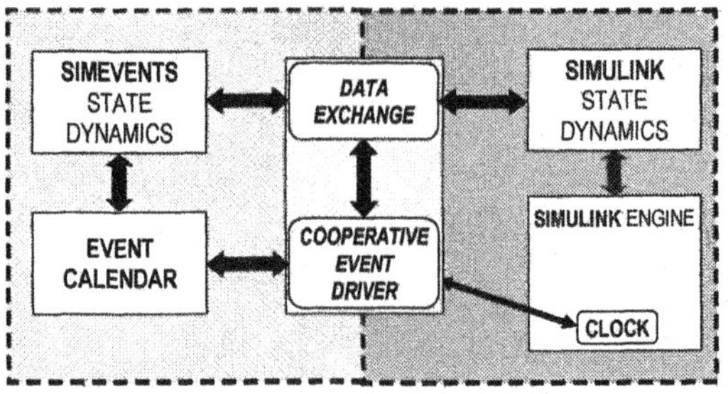

Fig. 1. SimEvents and Simulink collaborative functionality

3. SIMEVENTS FUNCTIONALITY

In Simulink, communication across blocks is based on signals. In SimEvents, it is based on both signals and entities. The "entity" concept is motivated from the view of a DES as an environment consisting of "users" and "resources": users request resources in order to perform various tasks, occupy these resources for a certain amount of time, and then relinquish them so that other users may access them. Examples of users are messages in a communication network and parts in a manufacturing system. Examples of resources are switches in a network and machines in a factory. A typical hybrid system scenario arises when an entity accessing a resource initiates a physical process (thus, defining an event in SimEvents), which is carried out until some termination condition is satisfied (defining another event in Simulink). Based on this approach, SimEvents consists of a number of libraries containing blocks with different system functionalities. The main libraries are the following:

1. **Generators**: Blocks that generate entities, or function calls (i.e., events that call Simulink blocks), or random variates.

2. **Queues**: Blocks where entities can be temporarily stored, while waiting to access a resource.

3. **Servers**: Blocks that model various types of resources.

4. **Routing**: Blocks that control the movement of entities as they access queues and servers.

5. **Gates**: Blocks that control the flow of entities by enabling/disabling access of entities to certain blocks.

6. **Event Translation**: Blocks that enable communication between SimEvents and Simulink by translating events into function calls.

7. **Attributes**: Blocks that assign and modify data to entities. Various control actions are then made based on the values of these data, allowing blocks to differentiate between entities they process.

8. **Subsystems**: These allow a combination of blocks to be executed upon occurrence of specific events (not upon Simulink sample times).

9. **Timers and Counters**: Blocks that measure event occurrence times or time elapsing between events, and blocks that count occurrences of particular event types. These data are supplied to standard display or scope blocks in Simulink or specialized scopes designed specifically for SimEvents.

4. SOME DESIGN PROBLEMS IN HYBRID SYSTEM SIMULATION

We limit ourselves to three key problems that are ubiquitous in combining time-driven and event-driven systems.

1. **Event-driven vs Time-driven Computation.** In a DES or HS setting, the computation of many quantities is not required until a particular event occurs. However, all Simulink computation blocks are executed in a time-driven fashion. This causes an *efficiency* problem as well as a potential *integrity* problem.

To illustrate the efficiency issue, consider two signals x and y and an addition block that generates $z = x + y$. If x and y are constant and only change values at times t_x and t_y respectively, then a time-driven adder needlessly evaluates z at all sample times (as defined by the system clock). The solution provided in SimEvents is to place the adder block in a discrete event subsystem that is triggered only by events occurring at times t_x and t_y.

The integrity issue manifests itself when the period Δ of the sampling mechanism is such that an event, say the one changing x at time t_x, occurs in the interior of an interval $[t, t+\Delta]$. In this case, the value of z over all times in $[t_x, t+\Delta)$ is incorrect. Moreover, if the event at t_x is intended to trigger some other system action, this action can only be taken at time $t+\Delta$. Although most sophisticated simulators are capable of detecting such variable-changing events in their sampling mechanisms, this still imposes a requirement on the timing engine of the simulator instead of making it a process which is naturally triggered by an element of the Event Calendar in Fig. 1.

2. **Declarative vs Imperative Semantics.** A time-driven system naturally employs declarative semantics, i.e., an equation implies a constraint

that the variables involved must satisfy. This fact is crucial when a system includes feedback loops: in Simulink a feedback loop is interpreted as an "algebraic loop" that must be resolved in order for the system simulation to be executed.

In an event-driven system, imperative semantics are used, i.e., an equation implies a strict assignment (which is why the symbol ':=' is sometimes used instead of '='). A feedback loop in this case typically means that an event has been detected in some process at time t, which may change an input variable for that process immediately after t (e.g., disable a process at t as soon as one of its state variables first crosses a given threshold). Combining components from both settings can, therefore, create ambiguities regarding the meaning of loops. In order to resolve this issue, a special SimEvents block termed "Signal Latch" is introduced in such loops to effectively translate a signal from a Simulink block into an event processed by SimEvents blocks, before generating a new signal input to the same Simulink block.

3. **Event concurrency**. In an event-driven system, it is possible for multiple events to occur concurrently. The order in which these events are executed is controlled by means of a "priority" scheme which is part of the underlying DES design. It is also possible for an event to trigger the occurrence of one or more other events, which in turn triggers a number of events, with all this activity taking place in zero time. Such event-ordering capability does not normally exist in a time-driven environment, where events are only defined as level crossings arising in the evolution of a continuous process (although this issue is explicitly addressed in some more advanced simulators, as in (Mosterman, 2002)). This presents a major challenge in a hybrid system setting.

To illustrate this issue, consider a Simulink "enabled subsystem" where some time-driven process $x(t)$ is initiated by an enabling signal, labeled e_1, generated by a SimEvents block event at time t_0 with $x(t_0) = 0$. The subsystem is defined so that the rising edge of an enabling signal $s(t)$ at any time t resets the process to $x(t) = 0$. Further, the process must stop as soon as $x(t) = c$ at some $t_1 > t_0$. This defines a level-crossing event, labeled e_2, which disables this enabled subsystem. The process can subsequently be re-enabled when the next e_1 event takes place. Suppose, however, that, under certain conditions, event e_2 immediately triggers event e_1, i.e., the subsystem is disabled at time t_1 and immediately enabled again. The correct behavior in this case is to re-initialize the process so that $x(t_1^+) = 0$. However, the Simulink timing engine is not designed to process event e_1 followed by event e_2 in zero time; instead, it observes that $s(t_1^-) = s(t_1^+) = 1$ and allows the process to remain enabled with $x(t_1^+) = c$. The only way to resolve this problem is by allowing $s(t)$ to jump from $s(t_1^-) = 1$ to $s(t_1) = 0$ and from $s(t_1) = 0$ to $s(t_1^+) = 1$ again, an operation that basic Simulink blocks are not equipped to perform. In other words, a hybrid system simulator must allow a signal to take multiple values at each point in time. The alternative (without possibly resorting to some advanced functionalities) is to introduce an artificial miniscule delay between the concurrent events e_1 and e_2, allowing the subsystem to process two events at different times.

The broader issue that this problem points to is that of designing a timing mechanism capable of driving differential or difference equations and processing discrete events from an event calendar where concurrent events are possible, hence an ordering scheme is also necessary. This issue boils down to the question "who should control the system clock – the time-driven component or the event-driven component of a hybrid system?"

REFERENCES

Cassandras, C. G. and S. Lafortune (1999). *Introduction to Discrete Event Systems.* Kluwer Academic Publishers.

Levine, W. and D. Hristu (Eds.) (2005). *Handbook of Networked and Embedded Control Systems.* Birkhauser.

MathWorks (2001). *Simulink: A Program for Simulating Dynamic Systems, User Guide.* The MathWorks, Inc.

MathWorks (2005). *SimEvents User's Guide.* The MathWorks, Inc.

Mosterman, P. J. (2002). HyBrSim - a modeling and simulation environment for hybrid bond graphs. *Journal of Systems and Control Engineering* **216**, 35–46. Part I.

ELSEVIER
IFAC
PUBLICATIONS

HYVISUAL: A HYBRID SYSTEM MODELING FRAMEWORK BASED ON PTOLEMY II *

Edward A. Lee Haiyang Zheng

EECS Department, University of California at Berkeley

Keywords: hybrid systems, simulation, modal models, heterogeneous models

Extended Abstract

HyVisual is a hybrid systems modeling framework providing a block diagram visual syntax for specifying continuous dynamics and a bubble-and-arc syntax for specifying modal behavior. It is based on Ptolemy II, is written in Java, and is distributed open-source at http://ptolemy.eecs.berkeley.edu/hyvisual/.

HyVisual has a rigorous operational semantics described in Lee and Zheng (2005). A key property is that it internally uses *superdense time*, where signals are modeled as partial functions of the form $f: \mathbb{R}_+ \times \mathbb{N} \rightharpoonup V$, where $\mathbb{R}_+$ is the non-negative real numbers and represents time, V is the value set (a data type, such as $\mathbb{R}^n$), and $\mathbb{N}$ is the set of natural numbers. Continuous-time functions are total, whereas discrete-event functions are defined only on a discrete subset of $\mathbb{R}_+$. The $\mathbb{N}$ in the domain permits signals to have multiple values in a well-defined order at a particular time. Using this framework, HyVisual gives a rigorous semantics to discontinuous signals (which have multiple values at the point of discontinuity), to discrete-event signals with multiple events at the same time, and to transient states, where the time spent in the state is zero.

An example of a HyVisual model that leverages this is shown in fig. 1, which shows many features of HyVisual. This models Newton's cradle, an apparatus with three (or more) balls hanging from strings (inspired by a one dimensional version in Mosterman (1999)). If the model is initialized with one of the balls displaced as shown in the HyVisual graphical animation at the lower left, then when the ball collides with the middle ball, a transient state results. At that time, the right ball transfers its momentum to the middle ball, and then, without any time elapsing, the middle ball transfers its momentum to the left ball. The two events (state transitions in the state machine at the middle left) are simultaneous but ordered. Other initial conditions can be chosen where two simultaneous events are unordered (e.g., starting with two balls appropriately displaced). HyVisual allows a model to permit nondeterministic choice of enabled transitions.

* This work was supported in part by the Center for Hybrid and Embedded Software Systems (CHESS) at UC Berkeley, which receives support from the National Science Foundation (NSF award No. CCR-0225610), the State of California Micro Program, and the following companies: Agilent, DGIST, General Motors, Hewlett Packard, Infineon, Microsoft, and Toyota.

REFERENCES

Edward A. Lee and Haiyang Zheng. Operational semantics of hybrid systems. In Manfred Morari and Lothar Thiele, editors, *HSCC*, volume LNCS 3414, pages pp. 25–53, Zurich, 2005. Springer.

P. Mosterman. An overview of hybrid simulation phenomena and their support by simulation packages. In F. Varager and J. H. van Schuppen, editors, *HSCC*, volume LNCS 1569, page 165177. Springer-Verlag, 1999.

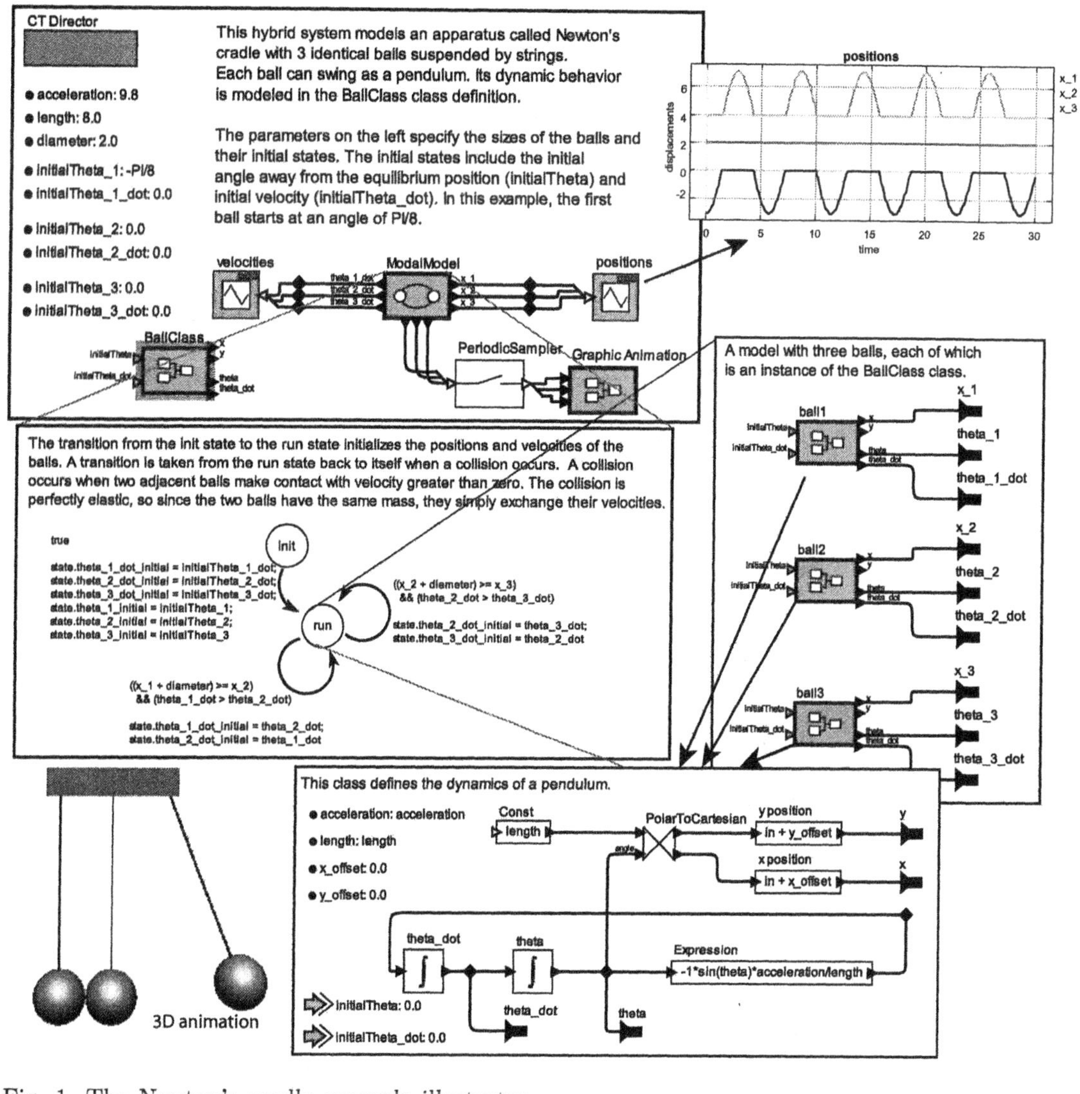

Fig. 1. The Newton's cradle example illustrates many features of HyVisual.

ELSEVIER
IFAC
PUBLICATIONS

TRUETIME: SIMULATION OF NETWORKED COMPUTER CONTROL SYSTEMS

Dan Henriksson, Anton Cervin, Martin Andersson, Karl-Erik Årzén

Department of Automatic Control
Lund University
Box 118, SE-22100 Lund, Sweden
Email: {dan,anton,mandersson,karlerik}@control.lth.se

1. INTRODUCTION

Computer-based control systems and networked control systems are hybrid systems where continuous time-driven dynamics and discrete event-driven dynamics interact. The temporal non-determinism introduced by computing and communication in the form of delays and jitter can lead to significant performance degradation. Software tools are needed to analyze and simulate how the timing affects the control performance.

Timed automata and piecewise linear systems are common modeling formalisms for hybrid systems. TrueTime is a MATLAB/Simulink-based simulation tool that takes a completely different approach. Using TrueTime it is possible to simulate the temporal aspects of multi-tasking real-time kernels and wired or wireless networks within Simulink together with the continuous-time dynamics of the controlled plant. The approach allows simulation at the same level of detail as in the true system. For complete TrueTime descriptions, see (Andersson *et al.*, 2005a; Cervin *et al.*, 2003; Andersson *et al.*, 2005b). TrueTime is available for free download at `http://www.control.lth.se/user/dan/truetime`

2. SIMULATION ENVIRONMENT

TrueTime consists of a block library with a computer kernel block and wired and wireless network blocks, as shown in Figure 1. The blocks are variable-step, discrete, MATLAB S-functions written in C++. The kernel block executes user-defined tasks and interrupt handlers representing, e.g., I/O tasks, control algorithms, and network interfaces. The scheduling policy

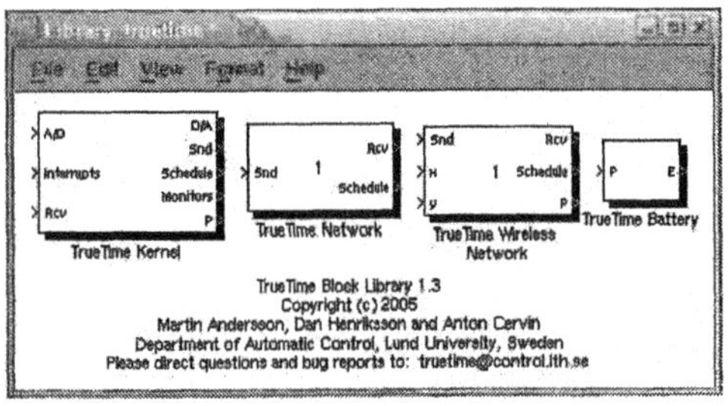

Fig. 1 The TrueTime block library.

of the kernel block is arbitrary and decided by the user. The network blocks distribute messages between computer nodes according to a chosen network model. The blocks are connected with ordinary continuous-time Simulink blocks to form a real-time control system.

All blocks are event-driven, with the execution determined both by internal and external events. Internal events correspond to events such as "a timer has expired," "a task has finished its execution," or "a message has completed its transmission." External events correspond to external interrupts, such as "a message arrived on the network" or "the crank angle passed zero degrees." All outputs are discrete-time signals. The Schedule and Monitors outputs display the allocation of common resources (CPU, monitors, network) during the simulation.

2.1 *The Kernel Block*

The kernel block S-function simulates a computer with a simple but flexible real-time kernel, A/D and D/A converters, a network interface, and external interrupt channels. Internally, the kernel maintains several data structures that are commonly found in a real-time kernel: a ready queue, a time queue, and records for tasks, interrupt handlers, monitors and timers that have been created for the simulation. The execution of tasks and interrupt handlers is defined by code functions, written either in C++ or MATLAB code. Control algorithms may also be defined graphically using ordinary discrete Simulink block diagrams.

Tasks and Interrupt Handlers Tasks are used to simulate both periodic activities, such as controller and I/O tasks, and aperiodic activities, such as communication tasks and event-driven controllers. Each task is defined by a set of attributes (priority, deadline, period, etc.) and a code function. Interrupts may be generated in two ways: externally or internally. An external interrupt is associated with one of the external interrupt channels of the kernel block. The interrupt is triggered when the signal of the corresponding channel changes value. This type of interrupt may be used to simulate engine controllers that are sampled against the rotation of the motor or distributed controllers that execute when measurements arrive on the network. Internal interrupts are associated with timers. Both periodic timers and one-shot timers can be created

Code The code associated with tasks and interrupt handlers is divided in segments, where the code of each segment is executed instantaneously during simulation. The code can interact with other tasks and with the environment at the beginning of each code segment. This execution model makes it possible to model input-output delays, blocking when accessing shared resources, etc. The simulated execution time of each segment is returned by the code function. Besides A/D and D/A conversion, many other kernel primitives exist that can be called from the code functions, e.g., functions to send and receive messages over the network, create and remove timers, perform monitor operations, and change task attributes.

2.2 *The Network Blocks*

The TrueTime network blocks distribute messages between computer nodes according to chosen network models. For wired networks, six of the most common medium access control protocols are supported (CSMA/CD (Ethernet), switched Ethernet, CSMA/CA (CAN), token-ring, FDMA, and TDMA). The wireless network block supports simulation of the IEEE 802.11 WLAN and IEEE 802.15.4 ZigBee standards. For a description of the simple radio model used for simulation of wireless communication, see (Andersson *et al.*, 2005b).

In the same way that code execution is modelled by segments as opposed to execution of individual statements, the network transmissions are not modelled on bit level. Rather, only the interactions between nodes relevant for the timing behavior of the transmissions are modelled. That includes pre- and post-processing delays, collision detection and collision avoidance mechanisms, and probabilities of lost packets. A message contains information about the sending and the receiving computer node, arbitrary user data (typically measurement signals or control signals), the length of the message, and optional real-time attributes such as a priority. When the simulated transmission of a message has completed, it is put in a buffer at the receiving computer node, which is notified by a hardware interrupt.

3. SIMULINK TIMING DETAILS

The TrueTime blocks are event-driven and support external interrupt handling. Therefore, the blocks have a continuous sample time, and the timing of the block is implemented using the Simulink zero-crossing functionality. The next time the kernel (or network block) should wake up (e.g., because a task is to be released from the time queue, a task has finished its execution, or a message transmssion has been completed) is denoted `nextHit`. If there is no known wake-up time, this variable is set to infinity. The basic structure of the zero-crossing function is

```
static void mdlZeroCrossings(SimStruct *S) {
   Store all inputs;
   if (any external interrupt input has
                              changed value) {
      nextHit = ssGetT(S);
   }
   ssGetNonsampledZCs(S)[0] = nextHit - ssGetT(S);
}
```

This will ensure that the Simulink call-back function `mdlOutputs` executes every time an internal or external event has occurred. The kernel and network functions are only called from `mdlOutputs` since this is where the outputs (D/A, schedule, network) can be changed. Further, since several kernel and network blocks may be connected in a circular fashion, *direct feedthrough* is not allowed. We exploit the fact that, when an input changes as a step, `mdlOutputs` is called, followed by `mdlZeroCrossings`. Since direct feedthrough is not allowed, the inputs may only be checked for changes in `mdlZeroCrossings`. There, the zero-crossing function is changed so that the next major step occurs at the current time.

REFERENCES

Andersson, M., D. Henriksson, and A. Cervin (2005a): *TrueTime 1.3—Reference Manual.* Department of Automatic Control, Lund University, Sweden.

Andersson, M., D. Henriksson, A. Cervin, and K.-E. Årzén (2005b): "Simulation of wireless networked control systems." In *Proceedings of the 44th IEEE Conference on Decision and Control and European Control Conference ECC 2005.* Seville, Spain.

Cervin, A., D. Henriksson, B. Lincoln, J. Eker, and K.-E. Årzén (2003): "How does control timing affect performance?" *IEEE Control Systems Magazine,* **23:3**, pp. 16–30.

ELSEVIER
IFAC
PUBLICATIONS

CODIS – A FRAMEWORK FOR CONTINUOUS/DISCRETE SYSTEMS CO-SIMULATION

Gabriela Nicolescu, Faouzi Bouchhima, Luiza Gheorghe

Ecole Polytechnique de Montreal

Abstract: This paper presents CODIS, a co-simulation framework for continuous/discrete systems. Based on a well defined synchronization model and a generic architecture for continuous/discrete simulation models, this framework enables easy specification and automatic generation of simulation models. The supported simulators are Simulink for continuous components and SystemC for discrete components. *Copyright © 2006 IFAC*

Keywords: models, simulation, heterogeneity, accuracy, performance.

1. INTRODUCTION

Modern systems that may be found in various domains like automotive, defense, medical and communications, integrate continuous and discrete models. In a recent ITRS study covering the domain of mixed continuous discrete systems, the conclusion is a "*shortage of design skills and productivity* arising from lack of training and poor automation with needs for basic design tools" as one of the most daunting challenges in this domain (ITRS, 2003).
One of the main difficulties in the definition of CAD tools for continuous/discrete (C/D) systems is due to the heterogeneity of concepts manipulated by the discrete and the continuous components. Therefore, in the case of validation tools, several execution semantics have to be taken in consideration in order to perform global simulation:

- In discrete models (DM), the time represents a global notion for the overall system and advances discretely when passing by time stamps of events, while in continuous models (CM), the time is a global variable involved in data computation and it advances by integration steps that may be variable.
- In discrete models, processes are sensitive to events while in continuous models processes are executed at each integration step.

Currently, co-simulation is a popular validation technique for heterogeneous systems. This technique was successfully applied for discrete systems, but very few applied it for C/D systems. The co-simulation allows joint simulation of heterogeneous components. This requires the elaboration of a global execution model, where the different components communicate through a co-simulation bus via simulation interfaces performing adaptation.
For C/D systems co-simulation, the simulation interfaces have to provide efficient synchronization models in order to cope with the heterogeneous aspects cited above. This implies a complex behavior for these interfaces; their design is time consuming and an important source of error. Simulation interfaces play also an important role in accuracy and performance of global simulation. Consequently, the definition of new co-simulation tools able to provide simulation interfaces is mandatory.
This paper presents CODIS (COntinuous/DIscrete Systems simulation), a co-simulation framework for C/D systems validation. This framework assists designers in building global simulation models. The supported simulators are Simulink for continuous models and OSCI SystemC simulator for discrete models.

2. CODIS FRAMEWORK

1.1 *Synchronisation and generic architecture for C/D simulation in CODIS*

For an accurate synchronisation, each simulator involved in a C/D co-simulation must consider the events coming from the external world and it must reach accurately the time stamps of these events. We refer to this as *events detection.* These time stamps are the synchronization and communication points between the simulators involved in the co-simulation. Therefore, the continuous simulator, Simulink, must detect the next discrete event (timed event) scheduled by the discrete simulator, once the latter has completed the processing corresponding to the current time. In case of SystemC, these events are: clock events, timed notified events, events due to the *wait* function. This detection requires the adjustment of integration steps in Simulink (see Fig. 1).
The discrete simulator, SystemC, must detect the *state events.* A state event is an unpredictable event, generated by the continuous simulator, whose time stamp depends on the values of the state variables (ex: a zero-crossing event, a threshold overtaking

event, etc.). This implies the control of the discrete simulator advancement in time: in stead of advancing with a normal simulation step, the simulator has to advance precisely until the time stamp of the state event (see Fig. 1).

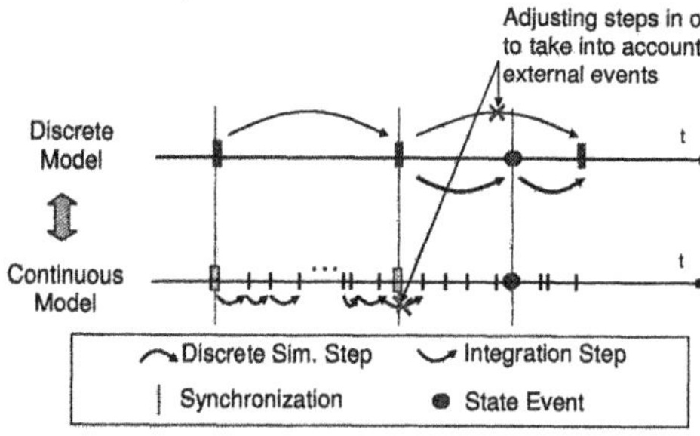

Fig. 1. C/D Synchronisation in CODIS

Fig.2 illustrates the generic architecture used in CODIS for the C/D simulation. CM and DM communicate through a co-simulation bus via simulation interfaces. Each simulation interface presents two main *layers*:

- The *synchronization layer* provides the synchronisation requirements discussed above. For both CM and DM, this layer is composed of three sub-layers, each of them achieving an elementary functionality for synchronisation.
- The *communication layer* is in charge of sending/receiving data between CM and DM.

More details on synchronization and CODIS simulation architecture may be found in (Bouchimma et al., 2005)

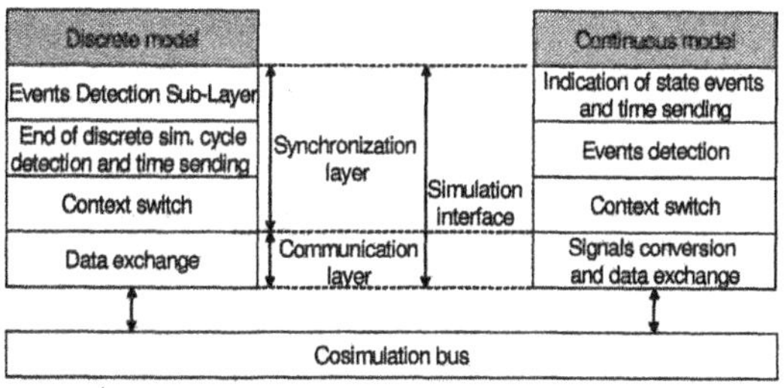

Fig 2. Generic architecture for accurate C/D simulation

1.2 Simulation model generation in CODIS

Based on the presented synchronization model and the generic architecture, a flow for automatic generation of simulation models was implemented in CODIS (see Fig. 3). The inputs of the flow are the CM in Simulink and the DM in SystemC. The output of the flow is the global simulation model, including the co-simulation bus and the simulation interfaces. The interfaces are generated by composing elements from the CODIS library. These elements implement the layers in Fig.2. They may be customized in terms of number of ports and their data type.

The interfaces for DM are automatically generated by a script generator that has as input user defined parameters. The model is compiled and the link editor calls the library from SystemC and a static library called "simulation library" (see Fig. 3).

The interfaces for Simulink are functional blocks programmed in C++ using S-Functions. These blocks are manipulated like all other components of the Simulink library. The user starts by dragging the interfaces from the library into the model's window, then parameterizes them and finally connects them to the model inputs and outputs. Before the simulation, the functionalities of these blocks are loaded by Simulink from the ".dll" dynamically linked libraries (see Fig. 3).

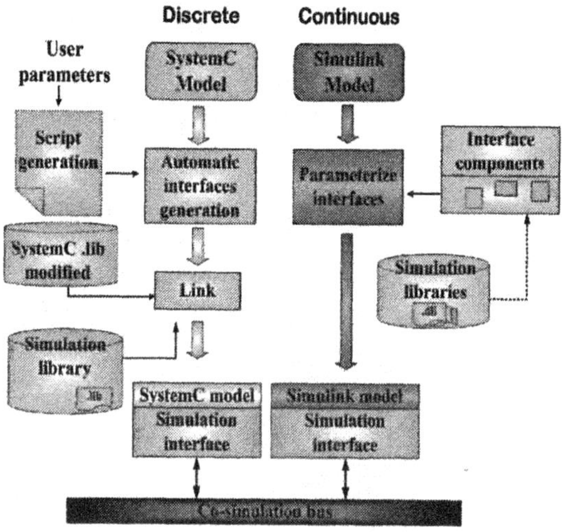

Fig. 3. Flow for simulation models generation

3. EXPERIMENTAL RESULTS

To analyze the capabilities of the proposed framework, we used two illustrative applications: a manipulator arm controller and a Δ/Σ converter.

To evaluate the performances of simulation models generated in CODIS, we measured the overhead given by the simulation interfaces. The overhead caused by the Simulink integration step adjustment when detecting a SystemC event has been measured in a maximum of 10% of total simulation time. The overhead caused by IPC (Inter Process Communication) used for the context switch and the communication layers has been measured in order of maximum 20% of the total simulation time. The cost of the added synchronization functionality in the case of SystemC is negligible and does not exceed 0.02% of the total simulation time.

4. CONCLUSION

This paper presented a simulation framework enabling continuous and discrete models integration. These models may be described using powerful tools for the two domains: Simulink and SystemC.

The experiments have shown a synchronization overhead of less than 30 % in simulation time.

REFERENCES

International Technology Roadmap for Semiconductor Design (2003), available at http://public.itrs.net/.

F. Bouchhima, *et al.* (2005) In: Discrete–Continuous Simulation Model for Accurate Validation in Component-Based Heterogeneous SoC Design, *Proceeding of RSP Conference*, (IEEE)

ELSEVIER
IFAC
PUBLICATIONS

ON THE FINITE-TIME STABILIZATION OF A NONLINEAR UNCERTAIN DYNAMICS VIA SWITCHED CONTROL

Giorgio Bartolini * Alessandro Pisano * Elio Usai *

* *Department of Electrical and Electronic Engineering, University of Cagliari (ITALY).*
{giob,pisano,eusai}@diee.unica.it

Abstract: This paper concerns the finite-time control problem for a class of nonlinear uncertain SISO dynamics with relative degree three. We assume that the measurements are corrupted by uncertain nonlinearities so that the *sign* of the phase variables is the only reliable information available for feedback. The stabilizing properties of a switching control scheme commuting between two unstable structures are demonstrated. Constructive proof and computer simulations are provided. *Copyright © 2006 IFAC*

Keywords: Finite-time control, Nonlinear uncertain systems, Switching controllers

1. INTRODUCTION

A usual approach to the output-feedback control of nonlinear systems consists in combining an observer and a controller such that the latter stabilizes a suitable estimated output having relative degree one (Atassi and Khalil, 1999; Teel and Praly, 1995; Oh and Khalil, 1997; Krishnamurthy *et al.*, 2001).

This approach may fail, especially when the observers are used as differentiators, if sensors are not linear and sufficiently accurate. In fact, sensor saturation and, more seriously, uncertain nonlinearities of sensor devices, limit and often prevent the plane use of the previously mentioned methods even in a linear context (Tao and Kokotovic, 1996; Cao *et al.*, 2003; Zuo, 2005). It is well known (Anosov, 1959) that even the simple triple integrator cannot be stabilized if the only information about the output is its sign.

In this paper we consider the finite-time stabilization problem for an output variable, says s, whose third time derivative is affected by the control variable u. Taking into account the well-known Anosov's result about the stabilization problem for a triple integrator (Anosov, 1959), we assume that the measurement of the phase variables $(s, \dot{s}, \ddot{s})$ is corrupted by an uncertain memory-less nonlinearity preserving the sign of the measured quantity.

We shall demonstrate in this work that a proper switching logic between two Variable Structure Controllers (VSC) can guarantee the semi-global convergence of a class of nonlinear uncertain dynamics towards the set $(s, \dot{s}, \ddot{s}) = (0, 0, 0)$.

The first VSC, $u = -k_0 sign(s)$, is referred to as the "Anosov Unstable" (AU),and the second, $u = -k_1 sign(\dot{s}) - k_2 sign(\ddot{s})$, is a peculiar realization of the Twisting algorithm (Levant, 1993), that uses $\dot{s}$ and $\ddot{s}$, instead of s and $\dot{s}$, as feedback signals. In this note it is named the "Modified Twisting" (M-TW) algorithm.

These two algorithms are strongly inadequate if they act alone. Actually, the Anosov unstable algorithm causes oscillations of the phase variables

(Anosov, 1959) while the Modified Twisting steers the system at some point $P_e \equiv (s_e, 0, 0)$ of the $\dot{s} = \ddot{s} = 0$ axis (Levant, 1993), with, in general, $s_e \neq 0$.

The paper is organized as follows: Section II contains the problem formulation. Section III contains the constructive derivation of the main result, summarized in Theorem 1. Section IV reports some simulation examples, and the concluding remarks are given in Section V. Some details related to the convergence proof are discussed in the Appendix.

2. PROBLEM FORMULATION

Consider the following class of nonlinear uncertain dynamics in regular form (Utkin, 1992)

$$\dot{\mathbf{x}} = \mathbf{h}(\mathbf{x}, \xi) \tag{1}$$

$$\dot{\xi} = \begin{bmatrix} \xi_2 \\ \xi_3 \\ f(\mathbf{x}, \xi) \end{bmatrix} + \begin{bmatrix} 0 \\ 0 \\ g(\mathbf{x}, \xi) \end{bmatrix} u(t) \tag{2}$$

where $\mathbf{x} \in R^m$ is the internal dynamics state-vector, $\xi = [\xi_1, \xi_2, \xi_3] \equiv [s, \dot{s}, \ddot{s}] \in R^3$ is the vector collecting the output $s \in R$ and its first and second derivative, $u \in R$ is the scalar plant control input.

The control aim is to define a control input u guaranteeing that vector ξ is steered to the origin in finite time. Assume what follows:

A1 The internal dynamics (1) are input-to-state stable (ISS) with linear gain (Sontag, 1998).

A2 There are known positive constants F_0, F_1, F_2, G_m, G_M such that

$$|f(\mathbf{x}, \xi)| \leq F_0 + F_1\|\mathbf{x}\| + F_2\|\xi\| \tag{3}$$

$$0 < G_m \leq g(\mathbf{x}, \xi) \leq G_M \tag{4}$$

Assumption $A1$ implies that there exist constants K_1, K_2, $K_3 \in R^+$ such that, for every $\mathbf{x}_0 \in R^m$ and every bounded and continuous $\xi(t)$, the unique maximal solution of the initial value problem (1), with $\mathbf{x}(t_0) = \mathbf{x}_0$, has interval of existence R^+ and, for any $t \in [t_1, t_2] \subseteq [t_0, \infty)$, satisfies the following condition

$$\begin{aligned} \|\mathbf{x}(t)\| \leq \|\mathbf{x}(t_1)\| + K_1 \sup_{t_1 \leq t \leq t_2} |s| + \\ + K_2 \sup_{t_1 \leq t \leq t_2} |\dot{s}| + K_3 \sup_{t_1 \leq t \leq t_2} |\ddot{s}| \end{aligned} \tag{5}$$

A3 Constants K_1, K_2, K_3 in (5) are known

A4 The initial conditions $\mathbf{x}(t_0)$, $\xi(t_0)$ belong to known compact domains X_0 and Ω_0, respectively:

$$\mathbf{x}(t_0) \in X_0, \qquad \xi(t_0) \in \Omega_0 \tag{6}$$

Let signals s, $\dot{s}$, $\ddot{s}$ be measured by sensors featuring a nonlinear uncertain characteristics (see Fig. 1) such that

$$s_m = h_1(s) \quad \dot{s}_m = h_2(\dot{s}) \quad \ddot{s}_m = h_3(\ddot{s}) \tag{7}$$

with the subscript m standing for "measured".

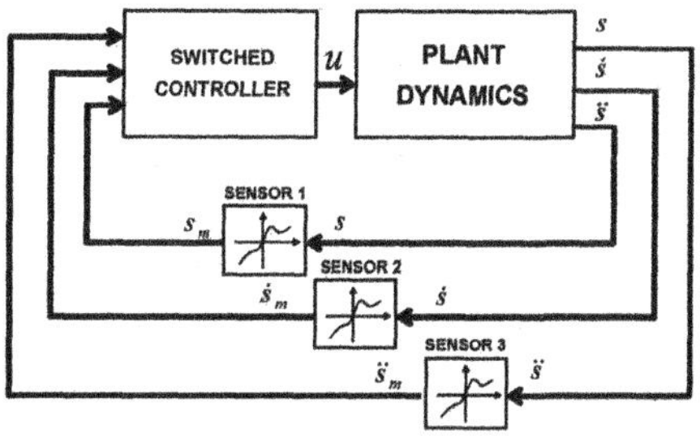

Fig. 1. Block scheme of the control system with nonlinear measurement characteristics

A5 The uncertain sensor characteristics satisfy the following relationship

$$xh_i(x) > 0 \quad \forall x \neq 0, \quad h_i(0) = 0, \quad i = 1, 2, 3 \tag{8}$$

Assumption A5 means that the sign of the measured signals matches the sign of the "true" ones. Nothing is required regarding the slope of the measurement nonlinearity.

3. MAIN RESULT

The finite-time stabilization problem for the uncertain system (1)-(2) satisfying Assumptions A1-A5 can be solved by means of a switched controller implementing a switching policy between the following two "Structures":

$$\begin{aligned} &\textbf{Anosov Unstable (AU)}: \\ &u = -U_0 sign(s) \end{aligned} \tag{9}$$

$$\begin{aligned} &\textbf{Modified Twisting (M-TW)}: \\ &u = -U_2 sign(\dot{s}) - (U_1 - U_2) sign(\ddot{s}) \end{aligned} \tag{10}$$

where U_0, U_1 and U_2 are constant tuning parameters.

Because of both the AU and M-TW structures use only the sign of the phase variables s, $\dot{s}$, $\ddot{s}$,then by Assumption 5 we can disregard the measurement nonlinearities in the stability analysis.

We shall demonstrate that it is possible to evaluate a-priori a constant F and set, correspondingly, the controller parameters U_0, U_1 and U_2 such that the following inequality holds

$$|f(\mathbf{x}, \xi)| \leq F \qquad t \geq t_0 \tag{11}$$

and variables s, $\dot{s}$ and $\ddot{s}$ tend to zero in finite time.

Let parameters U_0, U_1, U_2 satisfy the following inequalities.

$$\begin{array}{c} G_m U_i - F > 0, \quad i = 0, 1, 2 \\ G_m U_1 - F > G_M U_2 + F \end{array} \tag{12}$$

Under the tuning condition (12) for its parameter U_0, the AU control (9), (12) enforces a sequence of time instants $t_{z,i}$ $(i = 1, 2, \ldots)$ such that $s(t_{z,i}) = 0$. We denote as "*zero crossing of s*" the occurrence of such a condition. Unfortunately, the sequences $(t_{z,i} - t_{z,i-1})$, $|\dot{s}(t_{zi})|$ and $|\ddot{s}(t_{zi})|$ are all diverging (Anosov, 1959).

The M-TW control (10), (12) gives the closed-loop trajectories a finite-time converging "Twisting" behaviour in the $\dot{s} - \ddot{s}$ plane (see Fig. 2). Then vector $[s, \dot{s}, \ddot{s}]$ converges in finite time to a stable "*equilibrium point*" $P_e \equiv (s_e, 0, 0)$, with, in general, $s_e \neq 0$. (Levant, 1993).

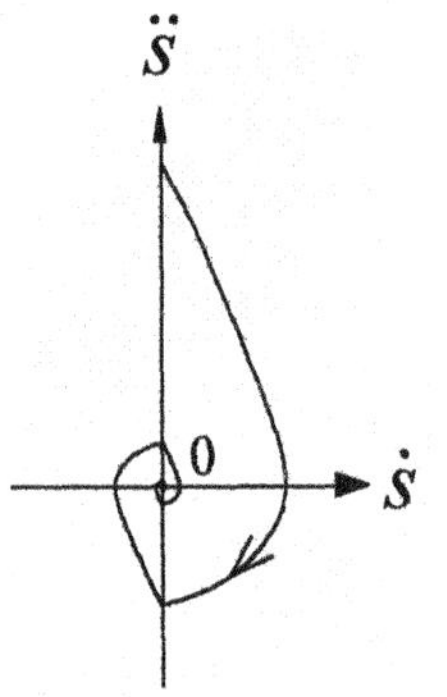

Fig. 2. Trajectories of the Modified Twisting (M-TW) in the $\dot{s} - \ddot{s}$ plane

Thus, when used alone, the two control laws (9) and (10) are not effective. On the contrary, the hereafter described switching policy between the AU and M-TW structures causes the occurrence of a sequence of equilibrium points

$$P_{e,i} \equiv (s_{e,i}, 0, 0) \qquad i = 1, 2, \ldots \tag{13}$$

such that the strict contraction property

$$|s_{e,i+1}| \leq \varepsilon |s_{e,i}|, \quad 0 < \varepsilon < 1, \quad i = 1, 2, \ldots \tag{14}$$

is fulfilled and the convergence to zero of s, $\dot{s}$ and $\ddot{s}$ is guaranteed. Fig. 4 shows a typical closed-loop time evolution of the s variable which fulfills the contraction condition (14).

The proposed controller can be schematized as in Fig. 3.

Remark 1: On-line detection of the transition-enabling conditions, zero-crossings and equilibrium points, is now discussed. Zero-crossing detection implies detection of both positive-to-negative

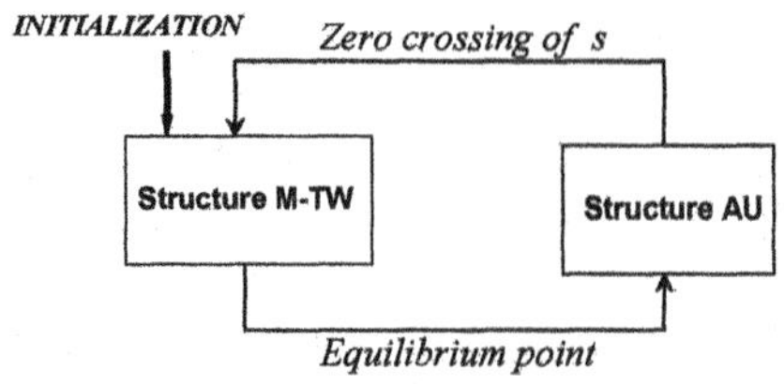

Fig. 3. The proposed switching mechanism

and negative-to-positive sign changings of s. Detection of the equilibrium points is less immediate. In fact, checking condition $\dot{s} = \ddot{s} = 0$ is unpractical in real systems. Following the idea proposed in (Bartolini *et al.*, 2002), the attainment of the equilibrium points $P_{e,i}$ can be easily and efficiently detected by monitoring on-line the switching frequency of the discontinuous control variable in a receding-horizon time interval. The switching frequency of the control signal tends to infinity while approaching the sliding condition which establishes at the equilibrium points. This criterion is also suitable for digital implementation: the sequence $\{u[k], u[k-1], \ldots, u[k-N]\}$ $(u[k] = u(kT_s)$, T_s is the sampling period, N is an integer number) can be stored and processed, and updated, at each sampling time instant. A minimum number of sign variations between the adjacent elements of the sequence can reveal the approaching of an equilibrium point.

The above treatment is summarized in the following Theorem:

Theorem 1 *Consider system (1)-(2) satisfying assumptions A1-A5. Perform the following sequence of steps*

A. Set the desired contraction rate $\varepsilon \in (0,1)$ *appearing in (14) and the arbitrary coefficients* $\eta > 1$, $\gamma \in (0,1)$, $\theta_1 \in (0,1)$, $\theta_2 \in (0,1)$.

B. Compute the unique positive root F^* *of equation (46)-(49) and set* $F \geq F^*$.

C. Compute the unique positive root x^* *of equation* $x^*(3 + 2x^*) + \gamma\varepsilon(1 + x^*)^{3/2} = \varepsilon$.

D. Set the controller parameters U_0, U_1, U_2 *as*

$$U_0 = \eta \frac{F}{G_m}, \quad U_1 = \frac{1}{G_m}\left(\frac{\eta\frac{G_M}{G_m} + 1}{\rho} + 1\right) F$$
$$U_2 = \frac{1}{G_m}\left(\frac{\eta\frac{G_M}{G_m} + 1}{\frac{1}{4}\theta_1\gamma^2\varepsilon^2} - 1\right) F \tag{15}$$

$$\rho = min\left\{\frac{1}{4}\theta_1\gamma^2\varepsilon^2, \theta_2 x^*\right\} \tag{16}$$

and apply the switched controller described in Fig. 3, where the Structures AU and M-TW are defined *in (9), (10). Then, s, $\dot{s}$ and $\ddot{s}$ are steered to zero in finite time.*

Proof. See the Appendix.

Remark 2: By (15), tuning of constant parameters U_0, U_1, U_2 depends on the unknown F. Theorem 1 guarantees that a constant F^* exists such that for any $F \geq F^*$ the desired performance is achieved, and gives a procedure to compute a conservative overestimation of F^*. For practical purposes, the calibration of the control parameters can be performed by simply increasing the single parameter F until satisfactory behaviour is observed in the closed-loop system. This method of experimental tuning of a single gain parameter is not unusual in the VSC context (Utkin, 1992).

4. SIMULATION RESULTS

To validate the present analysis consider the following fifth-order nonlinear system

$$\begin{aligned} \dot{\mathbf{x}} &= \mathbf{A}\mathbf{x} + \mathbf{B}(\xi_1 + \xi_2) \qquad \mathbf{x} = [x_1, x_2]^T \\ \dot{\xi}_i &= \xi_{i+1}, \quad i = 1,2 \\ \dot{\xi}_3 &= \frac{\xi_2}{1+\xi_2^2} + \xi_1 + \xi_3 + \|\mathbf{x}\| + (2 + cos(x_1 + \xi_2)u \end{aligned} \tag{17}$$

$$\mathbf{A} = \begin{bmatrix} 0 & 1 \\ -1 & -2 \end{bmatrix} \qquad \mathbf{B} = \begin{bmatrix} 1 \\ 1 \end{bmatrix} \tag{18}$$

The considered dynamics meets Assumptions A1-A5. Constants appearing in (3)-(5) can be evaluated as: $F_0 = 0.5$, $F_1 = F_2 = 1$, $G_m = 1$, $G_M = 3$, $K_1 = K_2 = 1$, $K_3 = 0$. Initial conditions are taken as follows: $[\xi_1(0), \xi_2(0), \xi_3(0)] = [1, 1, 1]$, $[x_1(0), x_2(0)] = [1, 1]$. The free design parameters are set as $\eta = 1.1$, $\theta_1 = \theta_2 = \gamma = 0.9$, and constant F is taken as 15.

Let us choose a "contraction factor" $\varepsilon = 0.5$. The resulting control parameters are $U_0 = 16.5$, $U_1 = 3598$, $U_2 = 451$. Euler integration algorithm with step $T_s = 0.0005ms$ has been used in the Matlab-Simulink environment.

Occurrence of at least 5 sign changings among the most recent 40 samples of u (see Remark 1) was chosen as the criterion to enable the transition from the M-TW to the AU structure. Fig. 4 reports the time profile of s. It is apparent that the local maxima of s feature the imposed contraction rate of 0.5 in perfect accordance with (14).

Choosing $\varepsilon = 0.7$, and leaving unchanged the free parameters, we obtained the following control magnitudes from the tuning procedure: $U_0 = 16.5$, $U_1 = 2576$, $U_2 = 184$. The plot of the s variable vs. time is shown in Fig. 5. As expected, increasing the value of ε, i.e. reducing the prescribed contraction rate, the required control magnitudes decrease. Again the observed contraction rate of 0.7 is according to (14), and, at the same time, the transient length obviously increases.

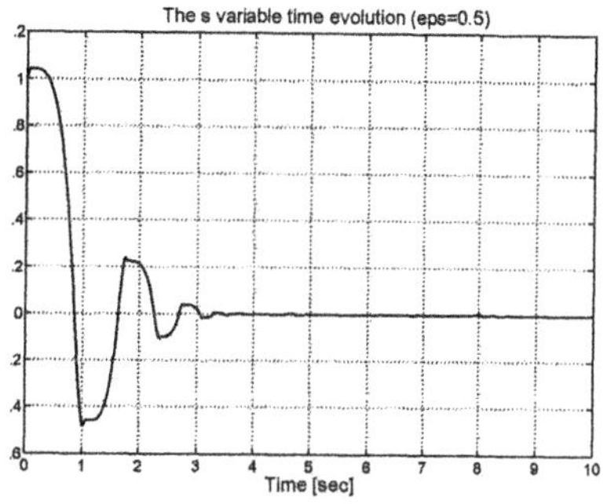

Fig. 4. The output quantity $s(t)$ when $\epsilon = 0.5$

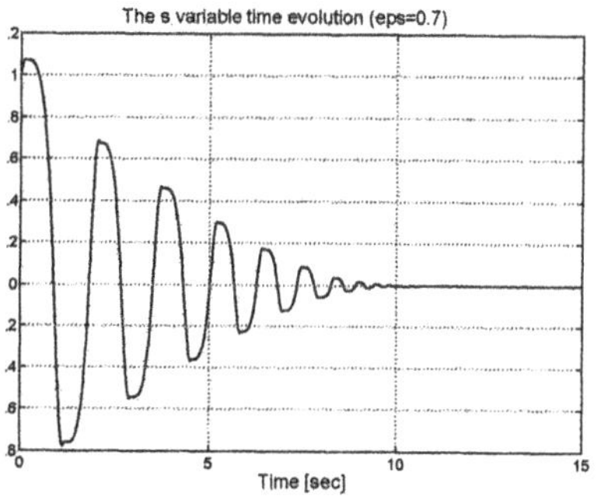

Fig. 5. The output quantity $s(t)$ when $\epsilon = 0.7$.

The dependence of the accuracy on the discretisation step has been analyzed. The sampling step was changed in order to investigate the accuracy order of the proposed method with respect to the sampling period. By comparing the two plots in Fig. 6, it follows that by using a discretization step which is ten times smaller (reducing it to $T_s = 0.05ms$) the accuracy is improved by a factor of almost 1000, which means that the accuracy is $O(T_s^3)$.

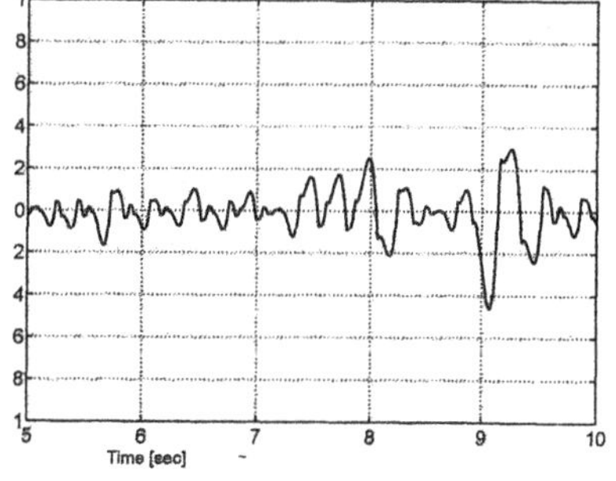

Fig. 6. The sliding accuracy with $\tau = 0.5ms$

The robustness of the proposed technique against the additive measurement noise corrupting the phase variables has been checked as well. We considered a uniformly-distributed random noise with a maximal magnitude of 0.1 superimposed

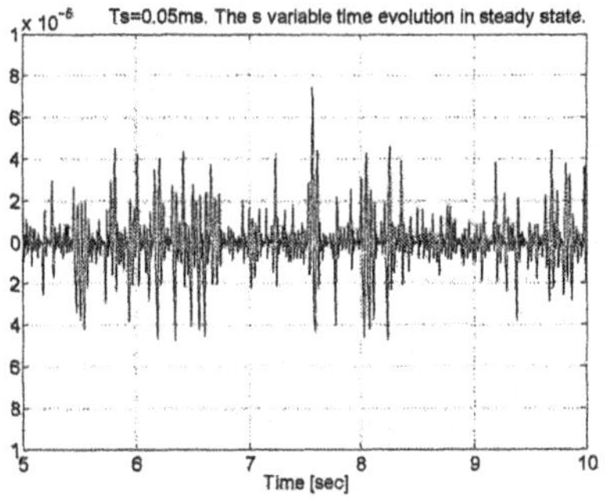

Fig. 7. The sliding accuracy with $\tau = 0.25ms$

to the three feedback variables. The noise causes a detriment of sliding accuracy in the steady state. Fig. 8 shows the time evolution of s is the steady state (the control parameters and sampling period are the same as those used in Fig. 6).

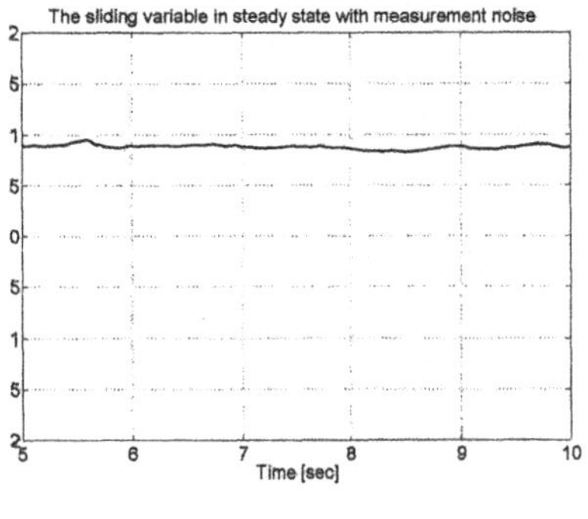

Fig. 8. The sliding accuracy in steady-state in the presence of noise

5. CONCLUSIONS

In this paper a novel control algorithm is presented which stabilizes a class of nonlinear uncertain systems in finite time. We assume that uncertain nonlinearities corrupt the measurement of the phase variables such that their sign of the unique reliable information. The control law consists of two VSC and a supervisor that implements a proper switching logic. The accuracy-order under discretization and sampling, and the robustness against measurement noise, are investigated by simulation.

REFERENCES

Anosov, D.V. (1959). On stability of equilibrium points of relay systems. *Automatica i telemechanica (Automation and Remote Control)* **2**, 135–149.

Atassi, N.A. and H.K. Khalil (1999). A separation principle for the stabilization of a class of nonlinear systems. *IEEE Trans. on Aut. Control* **44**, 1672–1687.

Bartolini, G., A. Levant, A. Pisano and E. Usai (2002). A real-sliding criterion for control adaptation. *Proc. 2002 Workshop on Variable Structure Systems (VSS 2002).*

Bartolini, G., Pisano, A. Punta E. and Usai E. (2003). A survey of applications of second order sliding mode control to mechanical systems. *Int. J. Control* **76**, 875–892.

Cao, Y.-Y., Lin, Z. and Chen, B.M. (2003). An output feedback H_∞ controller design for linear systems subjected to sensor nonlinearities. *IEEE Trans. Circ. Syst.* **7**, 914–921.

Krishnamurthy, P., Khorrami, F. and Chandra, R.S. (2001). Global high-gain-based observer and backstepping controller for generalized output-feedback canonical form. *IEEE Trans. Aut. Contr.* **48**, 2277–2284.

Levant, A. (1993). Sliding order and sliding accuracy in sliding mode control. *Int. J. Control* **58**, 1247–1263.

Oh, S. and H.K. Khalil (1997). Nonlinear output feedback tracking using high-gain observers and variable structure control. *Automatica* **33**, 1845–1856.

Orlov Y. (2004). Finite Time Stability and Robust Control Synthesis of Uncertain Switched Systems. *SIAM Journal on Control and Optimization* **43**,4, 1253–127.

Sontag, E. (1998). *Mathematical Control Theory.* Springer-Verlag. New York.

Tao, G. and P. Kokotovic (1996). *Adaptive Control of Systems with Actuator and Sensor Nonlinearities.* Wiley. New York.

Teel, A. and L. Praly (1995). Tools for semiglobal stabilization via partial state and output feedback. *SIAM J. Contr. Opt.* **33**, 1443–1488.

Utkin, V.I. (1992). *Sliding modes in control and optimization.* Springer Verlag. Berlin.

Zuo, Z. (2005). Output feedback H_∞ controller design for linear discrete-time systems with sensor nonlinearities. *Proc. IEE* **152**, 19–26.

Proof of Theorem 1

Closed-loop stability

We assume, temporarily, that condition (11) holds for some known value of F and we demonstrate the semi-global finite-time convergence of $s, \dot{s}, \ddot{s}$ to zero.

The initial structure is the M-TW, so that the system trajectory is similar to that reported in Fig. 2. An equilibrium point $P_{e,1} \equiv (s_{e,1}, 0, 0)$ is reached in finite time $t_{e,1}$. Assume with no loss of generality that $s_{e,1} < 0$. Fig. 9, shows a typical time evolution of the s variable starting from point P_{e1} subject to the switched controller in Fig. 3 stopped at the first iteration.

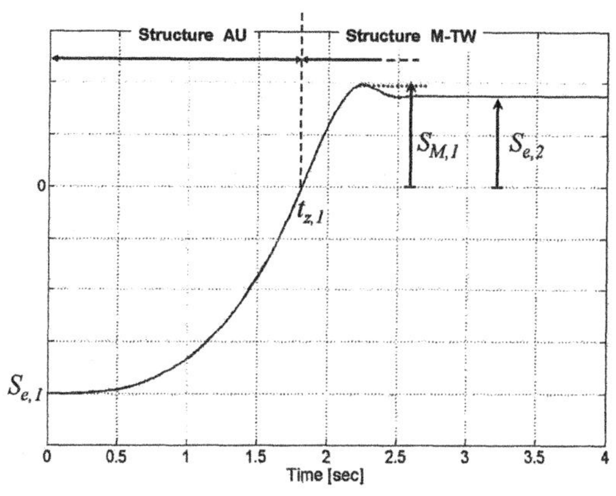

Fig. 9. Time evolution of s leaving from an equilibrium point $P_{e,1}$.

The AU control law is active until a zero crossing of s is attained at the time instant $t_{z1} \geq t_{e1}$. During this time interval the system is governed by the following differential inclusion

$$s^{(3)} \in [-F, F] - [G_m, G_M]U_0 sign(s_{e1}), \quad (19)$$

Since $\dot{s}(t_{e1}) = \ddot{s}(t_{e1}) = 0$, by simple worst-case analysis it follows that the first zero crossing of s is reached after a finite time $t_{z,1}$ such that

$$t_{z,1} - t_{e,1} \in \left[\left(\frac{6|s_{e,1}|}{G_M U_0 + F} \right)^{1/3}, \left(\frac{6|s_{e,1}|}{G_m U_0 - F} \right)^{1/3} \right] \quad (20)$$

By computing the limit solutions of (19), and taking into account (11) and (20), it results that $\ddot{s}(t_{z,1})$ and $\dot{s}(t_{z,1})$ are both positive and such that

$$\begin{aligned} \ddot{s}(t_{z,1}) &\leq \bar{\ddot{s}}_{z,1} \equiv \sqrt[3]{6}(G_M U_0 + F)^{2/3}|s_{e,1}|^{1/3} \\ \dot{s}(t_{z,1}) &\leq \bar{\dot{s}}_{z,1} \equiv \frac{\sqrt[3]{6}}{2}(G_M U_0 + F)^{1/3}|s_{e,1}|^{2/3} \end{aligned} \quad (21)$$

At $t = t_{z,1}$ the controller switches to the M-TW structure, and the origin of the $\dot{s} - \ddot{s}$ phase plane is reached at $t = t_{e,2}$. Fig. 10 depicts the corresponding system trajectory in the $\dot{s} - \ddot{s}$ plane.

Define

$$s_{M,1} = \sup_{t \in [t_{e,1}, t_{e,2}]} |s| \quad (22)$$

Let $t = t_{f,1}$ the time instant at which point $P_{f,1}$ in Fig. 10 is reached. It is easy to show that

$$t_{f,1} \leq \bar{t}_{f,1} \equiv t_{z,1} + \frac{\bar{\ddot{s}}_{z,1}}{G_m U_1 - F} \quad (23)$$

Hence, $s(t_{f,1})$ and $\dot{s}(t_{f,1})$ are bounded as follows

$$\begin{aligned} s(t_{f,1}) &\leq \bar{s}(t_{f,1}) \equiv \bar{\dot{s}}_{z,1}(\bar{t}_{f,1} - t_{z,1}) + \frac{1}{3}\frac{\bar{\ddot{s}}_{z,1}^3}{(G_m U_1 - F)^2} = \\ &= \frac{G_M U_0 + F}{G_m U_1 - F}\left[3 + 2\frac{G_M U_0 + F}{G_m U_1 - F}\right]|s_{e,1}| \end{aligned} \quad (24)$$

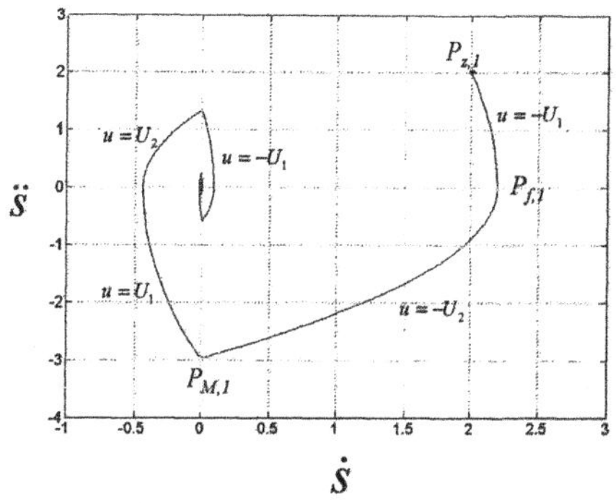

Fig. 10. Phase trajectory of $\dot{s}$ and $\ddot{s}$ leaving from the zero crossing $P_{z,1}$.

$$\begin{aligned} \dot{s}(t_{f,1}) &\leq \bar{\dot{s}}_{f,1} = \bar{\dot{s}}_{z,1} + \frac{1}{2}\frac{\bar{\ddot{s}}_{z,1}^2}{G_m U_1 - F} = \\ &= \frac{1}{2}(G_M U_0 + F)^{1/3}\left[1 + \frac{G_M U + F}{G_m U_1 - F}\right]|6s_{e,1}|^{2/3} \end{aligned} \quad (25)$$

$|s|$ features the maximum overshoot at the point $P_{M,1} \equiv (s_{M_1}, 0, \ddot{s}_{M_1})$ (see Figs. 9 and 10). Then we get

$$|\ddot{s}_{M_1}|^2 \leq 2(G_M U_2 + F)\bar{\dot{s}}_{f,1} \quad (26)$$

The increase of s along the trajectory between points $P_{f,1}$ and $P_{M,1}$ can be expressed as a function of $\bar{\ddot{s}}_{M_1}$ (or, equivalently, of $\bar{\dot{s}}_{f,1}$) according to

$$s_{M_1} \leq s(t_{f,1}) + \frac{1}{3}\frac{\bar{\ddot{s}}_{M_1}^3}{(G_M U_2 + F)^2} \quad (27)$$

Therefore, by combining (24), (26) and (27) we obtain

$$\begin{aligned} s_{M_1} \leq \bar{s}_{M_1} &= \frac{G_M U_0 + F}{G_m U_1 - F}\left[3 + 2\frac{G_M U_0 + F}{G_m U_1 - F}\right]|s_{e,1}| + \\ &+ 2\left[\frac{G_M U_0 + F}{G_M U_2 + F}\right]^{1/2}\left[1 + \frac{G_M U_0 + F}{G_m U - F}\right]|s_{e,1}| \end{aligned} \quad (28)$$

which can be rewritten as

$$s_{M,1} \leq \Sigma(\Delta_1, \Delta_2)|s_{e,1}| \quad (29)$$

$$\begin{aligned} \Sigma(\Delta_1, \Delta_2) &= \Delta_1(3 + 2\Delta_1) + 2\sqrt{\Delta_2}(1 + \Delta_1)^{3/2} \\ \Delta_1 &= \frac{G_M U_0 + F}{G_m U_1 - F}, \quad \Delta_2 = \frac{G_M U_0 + F}{G_M U_2 + F} \end{aligned} \quad (30)$$

Since, by definition, $|s_{M,1}| \geq |s_{e,2}|$, a sufficient condition guaranteeing the contraction condition $|s_{e,2}| \leq \varepsilon|s_{e,1}|$ is to choose U_0, U_1 and U_2 according to the following inequalities

$$G_m U_i - F > 0, \qquad i = 0, 1, 2 \quad (31)$$

$$G_m U_1 - F > G_M U_2 + F \quad (32)$$

$$\Sigma(\Delta_1, \Delta_2) \leq \varepsilon < 1 \quad (33)$$

By iteration, sequence $|s_{e,i}|$ will meet the strict contraction condition (14).

To conclude the design process, a triple U_0, U_1, U_2 which satisfies inequalities (31)-(33) have to be computed . By (31), U_0 can be simply set as in (15) with $\eta > 1$.

Set

$$\Delta_2 = \Delta_2^* = \frac{1}{4}\gamma^2 \varepsilon^2 \qquad 0 < \gamma < 1 \tag{34}$$

Condition (34) derives from imposing $\Sigma(0, \Delta_2^*) = \gamma\varepsilon$, with $\gamma \in (0, 1)$. This makes the inequality (33) satisfied when $\Delta_1 = 0$. Since function $\Sigma(\Delta_1, \cdot)$ is continuous and not decreasing, condition (34) guarantees the existence of a solution interval to (33) in the form $\Delta_1 \in (0, \Delta_1^*)$ where Δ_1^* is the smallest root of equation

$$\Sigma(\Delta_1^*, \Delta_2^*) \equiv \Delta_1^*(3 + 2\Delta_1^*) + \gamma\varepsilon(1 + \Delta_1^*)^{3/2} = \varepsilon \tag{35}$$

Now to guarantee that Δ_1 belongs to the solution set $(0, \Delta_1^*)$ and, at the same time, that $0 \leq \Delta_1 < \Delta_2$ (as required by (32)) we can set

$$\Delta_1 = min(\theta_1 \Delta_2^*, \theta_2\, \Delta_1^*),\ 0 < \theta_1 < 1,\ 0 < \theta_2 < 1, \tag{36}$$

and compute U_1 and U_2 on the basis of the given values of U_0, Δ_1 and Δ_2. Manipulation of the given formulas (34)-(36) taking into account (30) yields the tuning rules (15)-(16).

Finite transient length

To assess the finite time convergence let T_{ei} ($i = 1, 2, \ldots$) be such that $t_{e,i+1} - t_{ei} \leq T_{ei}$. It was shown in (Levant, 1993; Bartolini *et al.*, 2003) that the Twisting algorithm features a finite convergence time that can be overestimated by a bounded function of the initial conditions. By applying such a formula to the actual case it yields

$$T_{e,i} = \lambda |s_{e,i}|^{1/3} \tag{37}$$

where λ is a positive constant. Then, the following condition holds

$$\frac{T_{e,i}}{T_{e,i-1}} = \left(\frac{s_{e,i}}{s_{e,i-1}}\right)^{1/3} \leq \varepsilon^{1/3} < 1 \tag{38}$$

which implies that the total convergence time can be upper bounded by the finite sum of a convergent geometric series.

$$\overline{T} = t_{z,1} + \sum_{i=1}^{\infty} T_{ei} \leq \infty \tag{39}$$

Existence and computation of F

Let

$$X_0 = \sup_{\mathbf{x} \in X_0} \|\mathbf{x}\|, \qquad S_0 = \sup_{\xi \in \Omega_0} |s| \tag{40}$$

$$\dot{S}_0 = \sup_{\xi \in \Omega_0} |\dot{s}| \qquad \ddot{S}_0 = \sup_{\xi \in \Omega_0} |\ddot{s}| \tag{41}$$

and define

$$\overline{s} = \sup_{t \geq t_0} |s|, \qquad \overline{\dot{s}} = \sup_{t \geq t_0} |\dot{s}|, \qquad \overline{\ddot{s}} = \sup_{t \geq t_0} |\ddot{s}|, \tag{42}$$

By combining (3) and (5) it follows that

$$|f(\mathbf{x}, \xi)| \leq H_0 + H_1 \overline{s} + H_2 \overline{\dot{s}} + H_3 \overline{\ddot{s}} \tag{43}$$

with the constants H_0, H_1 and H_2 given as follows:

$$\begin{array}{ll} H_0 = F_0 + F_1 X_0 & H_1 = F_1 K_1 + F_2 \\ H_2 = F_1 K_2 + F_2 & H_3 = F_1 K_3 + F_2 \end{array} \tag{44}$$

Let condition (11) be satisfied for some unknown constant F. Then we can compute the quantities $\overline{s}$, $\overline{\dot{s}}$ and $\overline{\ddot{s}}$, which depend on the unknown F.

If the inequality

$$H_0 + H_1 \overline{s}(F) + H_2 \overline{\dot{s}}(F) + H_3 \overline{\ddot{s}}(F) \leq F \tag{45}$$

admits a semi-infinite solution interval of the type $F \geq F^*$ then the existence of F is proven. Let F^* be the unique positive root of equation

$$H_0 + H_1 \overline{s}(F^*) + H_2 \overline{\dot{s}}(F^*) + H_3 \overline{\ddot{s}}(F^*) = F^* \tag{46}$$

By solving (46) not only the existence of F such that (11) holds is proven, but also one achieves an overestimate of it. The initial conditions of s, $\dot{s}$ and $\ddot{s}$ are taken on the neighbor of the compact domain Ω_0: $s(t_0) = S_0$, $\dot{s}(t_0) = \dot{S}_0$, $\ddot{s}(t_0) = \ddot{S}_0$.

By worst case analysis it can be computed the following overestimates:

$$\begin{aligned} \overline{s}(F) = S_0 &+ \frac{\dot{S}_0 \ddot{S}_0}{G_m U_1 - F} + \frac{1}{3} \frac{\ddot{S}_0^3}{(G_m U_1 - F)^2} + \\ &+ \frac{1}{3} \frac{2\sqrt{2}[G_M U_2 + F]^{3/2} [\dot{S}_0 + \frac{1}{2} \frac{\ddot{S}_0^2}{G_m U_1 - F}]^{3/2}}{(G_m U_2 + F)^2} + \\ &+ \frac{\dot{S}_{z1} \ddot{S}_{z1}}{G_m U_1 - F} + \frac{1}{3} \frac{\ddot{S}_{z1}^3}{(G_m U_1 - F)^2} + \\ &+ \frac{1}{3} \frac{2\sqrt{2}[G_M U_2 + F]^{3/2} [\dot{S}_{z1} + \frac{1}{2} \frac{\ddot{S}_{z1}^2}{G_m U_1 - F}]^{3/2}}{(G_m U_2 + F)^2} \end{aligned} \tag{47}$$

$$\overline{\dot{s}}(F) = \dot{S}_0 + \frac{1}{2} \frac{\ddot{S}_0^2}{G_m U_1 - F} + \dot{S}_{z1} + \frac{1}{2} \frac{\ddot{S}_{z1}^2}{G_m U_1 - F} \tag{48}$$

$$\begin{aligned} \overline{\ddot{s}}(F) = \ddot{S}_0 &+ \sqrt{2(G_M U_2 + F)} \sqrt{\dot{S}_0 + \frac{1}{2} \frac{\ddot{S}_0^2}{G_m U_1 - F}} + \\ &+ \ddot{S}_{z1} + \sqrt{2(G_M U_2 + F)} \sqrt{\dot{S}_{z1} + \frac{1}{2} \frac{\ddot{S}_{z1}^2}{G_m U_1 - F}} \end{aligned} \tag{49}$$

By considering (47)-(49) into (45) it is easy to observe that the left-hand side of the latter is growing with order $O(F^{2/3})$. This guarantees the existence of F^* such that any $F \geq F^*$ fulfills the inequality (11). This concludes the proof. □

ELSEVIER
IFAC
PUBLICATIONS

SEARCH FOR PERIOD-2 CYCLES IN A CLASS OF HYBRID DYNAMICAL SYSTEMS WITH AUTONOMOUS SWITCHINGS. APPLICATION TO A THERMAL DEVICE

Céline Quémard * **Jean-Claude Jolly** *
Jean-Louis Ferrier *

* *Laboratoire d'Ingénierie des Systèmes Automatisés*
Université d'Angers
62, avenue Notre-Dame du Lac - F49000 Angers

Abstract: Our purpose is to complete the analysis of a class of hybrid dynamical systems with autonomous switchings generated by a hysteresis phenomenon. Because we yet have found limit period-1 cycles in paper (Quémard *et al.*, 2005*b*) and because we deal with nonlinear equations systems, the question of the existence of more than one solution for them and so of the existence of cycles with more than one period is rather natural. Equations system for period-2 cycles is determined and a notion of stability is studied. A realistic application to a thermostat with anticipative resistance comes to illustrate theoretical results.

Keywords: Hybrid Dynamical System, Hysteresis, Stability, Period-2 Cycles, Thermostat with Anticipative Resistance, Chaotic Behaviors.

1. INTRODUCTION

In (Quémard *et al.*, 2005*b*), we presented the process for the determination of limit cycles for a hybrid dynamical systems (h.d.s) class with autonomous switchings (see also (Bensoussan and Menaldi, 1997), (Van Der Schaft and Schumacher, 1999), (Zaytoon, 2001)) and we gave the used original method based on formal calculus and on interval analysis for the numerical solution that we applied to a thermal device (Quémard *et al.*, 2005*a*).

Thanks to the relative simplicity of our model, the problem of finding those limit cycles was reduced to the problem of determining the solution (which was unique in (Quémard *et al.*, 2005*b*)) of a non trivial implicit equations system.

But, sometimes, such nonlinear systems can also display two or more simultaneous solutions which generally involve the existence of cycles with more than one period. The varying initial conditions can cause a system to choose one stationary solution rather than other (Zhusubaliyev and Mosekilde, 2003).

Here, we propose to highlight and study this phenomenon determining by the same reasoning than in (Quémard *et al.*, 2005*b*), equations system to solve in order to study period-2 cycles and we highlight those results with the application to a thermostat with anticipative resistance. A main contribution of this paper is to deal with a realistic thermal application of industrial interest whose model, in dimension three, is relatively simple and a little less complex than, for example, the one proposed in (Zhusubaliyev and Mosekilde, 2003), in dimension four.

So, firstly, we present the studied h.d.s class and the equations system to solve in order to obtain period-2 cycles. A study of stability is made using the not original but not usual too *point transformation method of Andronov*. Then, we confirm obtained results with an application to a thermostat with anticipative resistance. We present its mathematical model, temperatures variations simulations and a bifurcation diagram made with *Matlab* which confirm the existence of a period-2 cycle. The study of the sensitivity to initial conditions and of the two parameters variations is also illustrated. We conclude this paper underlining links between properties of such systems illustrated in the numerical study and necessary properties to satisfy in the domain of chaotic behaviors.

2. STUDIED H.D.S CLASS

In $\mathbb{R}^N$, we consider a basis which, in practice, will be either the canonical basis or an eigenvectors basis, generalized or not, which can be useful with calculuses for example. In relation to this basis, we consider the following h.d.s of order N:

$$\dot{X}(t) = AX(t) + q(\xi(t))B + C, \quad \xi(t) = LX(t), \tag{1}$$

where A is a square matrix of order N and X, B, C are columns matrices of order N and L is a row matrix of order N, all these matrices having real entries.

Moreover, we suppose that matrix A has negative and real but not null eigenvalues and that X and so ξ are continuous. Variable q is discrete, taking value 0 or 1 according to ξ. It responds to the hysteresis phenomenon described in Figure 1.

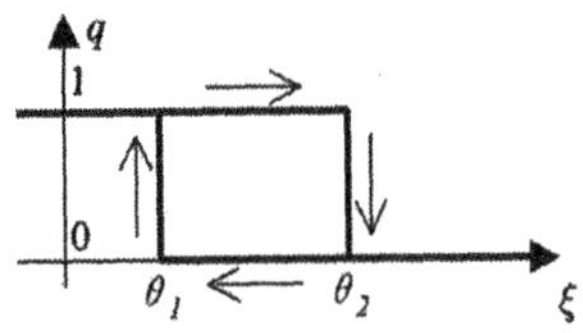

Fig. 1. Hysteresis variable

Values θ_1 and θ_2 are respectively the lower and the upper switching thresholds. Function $q(\xi)$, which is supposed right continuous, is given explicitly by:

$$\begin{cases} q(\xi(t)) = 1 - q(\xi(t_-)) \\ if \begin{cases} \xi(t_-) = \theta_1 \text{ and } q(\xi(t_-)) = 0 \\ \text{or } \xi(t_-) = \theta_2 \text{ and } q(\xi(t_-)) = 1 \end{cases} \\ q(\xi(t)) = q(\xi(t_-) \text{ otherwise} \end{cases}$$

In the first case, t is called switching time.

Finally, we introduce some notations which will be used later. Let $(U_n)_{n\in\mathbb{N}}$ be a suite. We set:

$$U_n^1 \triangleq U_{4n+1}\ , U_n^2 \triangleq U_{4n-2}\ , U_n^3 \triangleq U_{4n-1}\ , U_n^4 \triangleq U_{4n}. \tag{2}$$

3. PERIOD-2 CYCLE EQUATIONS

In (Quémard *et al.*, 2005*b*), we have proved, as it seemed rather natural for such non linear systems (see (Girard, 2003), (Zhusubaliyev and Mosekilde, 2003)), the existence of limit period-1 cycles, that is to say the existence of periodic orbits with two different durations respectively between odd and even switching times.

Because we deal with non linear systems, it is also rather natural to think that there can exist more than one solution for such systems. That is why here, we present, following the same reasoning than in (Quémard *et al.*, 2005*b*), equations for period-2 cycles, that is to say for cycles with four different durations σ^1, σ^2, σ^3, σ^4 between switching times.

Let t_0 be a given initial time. Suite $t_1 < t_2 < ... < t_n < ...$ corresponds to the successive switching times in $]t_0, \infty[$ necessarily distinct since $X(t_{n+1}) \neq X(t_n)$. We suppose that following assumption $Hp(n)$ is implicitly satisfied in calculuses in which t_n appears.

Hypothesis *Hp(n) Instant t_n of the nth switching (later to t_0) exists and is finite.*

Let us set in order to reduce notations $q_n \triangleq q(\xi(t_n))$, $\Delta q_n \triangleq q_n - q_{n-1}$. We have $q(\xi(t)) = q_n$ on $[t_n, t_{n+1}[$, $q_n \in \{0; 1\}$, $q_n = 1 - q_{n-1}$, $\Delta q_n = (-1)^{n-1}\Delta q_1$, $\Delta q_1 = \pm 1$ and with notations (2), we so have $q_n^1 = 1 - q_0$, $q_n^2 = q_0$, $q_n^3 = 1 - q_0$, $q_n^4 = q_0$.

Classical solution of (1) on $[t_n, t_{n+1}[$ gives:

$$X(t) = e^{(t-t_n)A}\Gamma_n - A^{-1}(q_n B + C), \tag{3}$$

with $\Gamma_n \in \mathbb{R}^N$ corresponding to the integration constants, functions of n. Applying the assumption of the state continuity at t_n, $X(t_n) = X(t_n-)$, equation (3) gives:

$$\forall n \geq 1 \quad \Gamma_n = e^{\sigma_n A}\Gamma_{n-1} + \Delta q_n A^{-1}B, \tag{4}$$

where we define $\forall n \geq 1$, $\sigma_n \triangleq t_n - t_{n-1} > 0$ the duration between two successive switching times t_{n-1} and t_n. For $n = 0$ and $t = t_0$, we obtain:

$$\Gamma_0 = X(t_0) + A^{-1}(q_0 B + C). \tag{5}$$

Let $\forall n \geq 1$, $\xi_n = \xi(t_n)$. We remark that $\forall n \geq 1$, $\xi_n = q_n\theta_1 + q_{n-1}\theta_2$ and $\forall n \geq 2$, $\Delta\xi_n = -\Delta q_n \Delta\theta = (-1)^n \Delta q_1 \Delta\theta$ with $\Delta\theta = \theta_2 - \theta_1$. Moreover, we have $\xi_n = LX(t_n-) = L(e^{\sigma_n A}\Gamma_{n-1} - A^{-1}(q_{n-1}B + C))$, and also $\xi_{n-1} = LX(t_{n-1}) = L(\Gamma_{n-1} - A^{-1}(q_{n-1}B + C))$. So, we obtain:

$$\begin{cases} \forall n \geq 2,\ L(e^{\sigma_n A} - I_N)\Gamma_{n-1} - \Delta\xi_n = 0 \\ \text{for } n = 1,\ L(e^{\sigma_1 A}\Gamma_0 - A^{-1}(q_0 B + C)) - \xi_1 = 0. \end{cases} \tag{6}$$

As explained in (Quémard *et al.*, 2005*b*), it is from equations (4) and (6) that we can determine equations of cycles. Let us characterize a possible searched period-2 cycle with four different durations between switching times. With notations (2), let us set:

$$R_n \triangleq (\sigma_n^1, \Gamma_n^1, \sigma_n^2, \Gamma_n^2, \sigma_n^3, \Gamma_n^3, \sigma_n^4, \Gamma_n^4). \tag{7}$$

System of equations (4) and (6) $\forall n \geq 1$, is equivalent to system $H(R_n, R_{n+1}) = 0$, $\forall n, n \geq 1$ where $H = (H_1, H_2, H_3, H_4, H_5, H_6, H_7, H_8)^T$ is a function that we define by:

$$\begin{cases} H_1(R_n, R_{n+1}) = L(e^{\sigma^1_{n+1}A} - I_N)\Gamma^4_{n+1} + \Delta q_1 \Delta\theta \\ H_2(R_n, R_{n+1}) = \Gamma^1_{n+1} - e^{\sigma^1_{n+1}}\Gamma^4_{n+1} - \Delta q_1 A^{-1}B \\ H_3(R_n, R_{n+1}) = L(e^{\sigma^2_{n+1}A} - I_N)\Gamma^1_n - \Delta q_1 \Delta\theta \\ H_4(R_n, R_{n+1}) = \Gamma^2_{n+1} - e^{\sigma^2_{n+1}}\Gamma^1_n + \Delta q_1 A^{-1}B \\ H_5(R_n, R_{n+1}) = L(e^{\sigma^3_{n+1}A} - I_N)\Gamma^2_{n+1} + \Delta q_1 \Delta\theta \\ H_6(R_n, R_{n+1}) = \Gamma^3_{n+1} - e^{\sigma^3_{n+1}}\Gamma^2_{n+1} - \Delta q_1 A^{-1}B \\ H_7(R_n, R_{n+1}) = L(e^{\sigma^4_{n+1}A} - I_N)\Gamma^3_{n+1} - \Delta q_1 \Delta\theta \\ H_8(R_n, R_{n+1}) = \Gamma^4_{n+1} - e^{\sigma^4_{n+1}}\Gamma^3_{n+1} + \Delta q_1 A^{-1}B. \end{cases} \quad (8)$$

Let us suppose that R_n has a limit $R = (\sigma^1, \Gamma^1, \sigma^2, \Gamma^2, \sigma^3, \Gamma^3, \sigma^4, \Gamma^4)$. In those conditions, R is solution to system $H(R, R) = 0$, namely:

$$\begin{cases} L(e^{\sigma^1 A} - I_N)\Gamma^4 + \Delta q_1 \Delta\theta = 0 \\ \Gamma^1 - e^{\sigma^1}\Gamma^4 - \Delta q_1 A^{-1}B = 0 \\ L(e^{\sigma^2 A} - I_N)\Gamma^1 - \Delta q_1 \Delta\theta = 0 \\ \Gamma^2 - e^{\sigma^2}\Gamma^1 + \Delta q_1 A^{-1}B = 0 \\ L(e^{\sigma^3 A} - I_N)\Gamma^2 + \Delta q_1 \Delta\theta = 0 \\ \Gamma^3 - e^{\sigma^3}\Gamma^2 - \Delta q_1 A^{-1}B = 0 \\ L(e^{\sigma^4 A} - I_N)\Gamma^3 - \Delta q_1 \Delta\theta = 0 \\ \Gamma^4 - e^{\sigma^4}\Gamma^3 + \Delta q_1 A^{-1}B = 0. \end{cases} \quad (9)$$

From the second equation of (9) and using the 4th,the 6th and the 8th equations of the same system, we obtain:

$$\Gamma^1 = \Delta q_1 (I_N - e^{\sigma^1 A} + e^{(\sigma^1+\sigma^4)A} - e^{(\sigma^1+\sigma^3+\sigma^4)A})$$

$$(I_N - e^{(\sigma^1+\sigma^2+\sigma^3+\sigma^4)A})^{-1}A^{-1}B.$$

Identically, the 4th,the 6th and the 8th equations of system (9) become:

$$\Gamma^2 = -\Delta q_1 (I_N - e^{\sigma^2 A} + e^{(\sigma^1+\sigma^2)A} - e^{(\sigma^1+\sigma^2+\sigma^4)A})$$

$$(I_N - e^{(\sigma^1+\sigma^2+\sigma^3+\sigma^4)A})^{-1}A^{-1}B.$$

$$\Gamma^3 = \Delta q_1 (I_N - e^{\sigma^3 A} + e^{(\sigma^2+\sigma^3)A} - e^{(\sigma^1+\sigma^2+\sigma^3)A})$$

$$(I_N - e^{(\sigma^1+\sigma^2+\sigma^3+\sigma^4)A})^{-1}A^{-1}B.$$

$$\Gamma^4 = -\Delta q_1 (I_N - e^{\sigma^4 A} + e^{(\sigma^3+\sigma^4)A} - e^{(\sigma^2+\sigma^3+\sigma^4)A})$$

$$(I_N - e^{(\sigma^1+\sigma^2+\sigma^3+\sigma^4)A})^{-1}A^{-1}B.$$

Reinjecting those expressions for Γ^i, $i = 1, .., 4$, in the 1st, 3rd, 5th and 7th equations of (9), we can deduce:

$$\begin{cases} L(I_N - e^{\sigma^1 A}) \\ (I_N - e^{\sigma^4 A} + e^{(\sigma^3+\sigma^4)A} - e^{(\sigma^2+\sigma^3+\sigma^4)A}) \\ (I_N - e^{(\sigma^1+\sigma^2+\sigma^3+\sigma^4)A})^{-1}A^{-1}B + \Delta\theta = 0 \\ \\ L(I_N - e^{\sigma^2 A}) \\ (I_N - e^{\sigma^1 A} + e^{(\sigma^1+\sigma^4)A} - e^{(\sigma^1+\sigma^3+\sigma^4)A}) \\ (I_N - e^{(\sigma^1+\sigma^2+\sigma^3+\sigma^4)A})^{-1}A^{-1}B + \Delta\theta = 0 \\ \\ L(I_N - e^{\sigma^3 A}) \\ (I_N - e^{\sigma^2 A} + e^{(\sigma^1+\sigma^2)A} - e^{(\sigma^1+\sigma^2+\sigma^4)A}) \\ (I_N - e^{(\sigma^1+\sigma^2+\sigma^3+\sigma^4)A})^{-1}A^{-1}B + \Delta\theta = 0 \\ \\ L(I_N - e^{\sigma^4 A}) \\ (I_N - e^{\sigma^3 A} + e^{(\sigma^2+\sigma^3)A} - e^{(\sigma^1+\sigma^2+\sigma^3)A}) \\ (I_N - e^{(\sigma^1+\sigma^2+\sigma^3+\sigma^4)A})^{-1}A^{-1}B + \Delta\theta = 0. \end{cases} \quad (10)$$

But system (10) is not independant because there exists a linear combination using the four equations (the 1st equation minus the 2nd is equal to the 4th equation minus the 3rd). So, here, it just remains three independant equations for four unknowns. The 4th equation will come from the condition of initial switching defined by equation (6) and using equation (4) in the case $n = 1$ and given by:

$$L(\Gamma^1 - A^{-1}((1-q_0)B+C)) - (1-q_0)\theta_1 - q_0\theta_2 = 0.$$

And so, we obtain the final system:

$$\begin{cases} F_1 = L(I_N - e^{\sigma^1 A}) \\ (I_N - e^{\sigma^4 A} + e^{(\sigma^3+\sigma^4)A} - e^{(\sigma^2+\sigma^3+\sigma^4)A}) \\ (I_N - e^{(\sigma^1+\sigma^2+\sigma^3+\sigma^4)A})^{-1}A^{-1}B + \Delta\theta = 0 \\ \\ F_2 = L(I_N - e^{\sigma^3 A}) \\ (I_N - e^{\sigma^2 A} + e^{(\sigma^1+\sigma^2)A} - e^{(\sigma^1+\sigma^2+\sigma^4)A}) \\ (I_N - e^{(\sigma^1+\sigma^2+\sigma^3+\sigma^4)A})^{-1}A^{-1}B + \Delta\theta = 0 \\ \\ F_3 = L(e^{\sigma^2 A}(I_N - e^{\sigma^1 A} + e^{(\sigma^1+\sigma^4)A}) \\ -e^{\sigma^4 A}(I_N - e^{\sigma^3 A} + e^{(\sigma^2+\sigma^3)A})) \\ (I_N - e^{(\sigma^1+\sigma^2+\sigma^3+\sigma^4)A})^{-1}A^{-1}B = 0 \\ \\ F_4 = \Delta q_1 L(I_N - e^{\sigma^1 A} + e^{(\sigma^1+\sigma^4)A} - e^{(\sigma^1+\sigma^3+\sigma^4)A}) \\ (I_N - e^{(\sigma^1+\sigma^2+\sigma^3+\sigma^4)A})^{-1}A^{-1}B \\ -LA^{-1}((1-q_0)B + C) - (1-q_0)\theta_1 - q_0\theta_2 = 0 \end{cases} \quad (11)$$

As proved in theorem 1 in (Quémard *et al.*, 2005*b*) for a period-1 cycle, suite $(\Gamma_n)_{n \geq 0}$ such that $\Gamma_{4p+1} \triangleq \Gamma^1$ for $p \geq 0$, $\Gamma_{4p-2} \triangleq \Gamma^2$ for $p \geq 1$, $\Gamma_{4p-1} \triangleq \Gamma^3$ for $p \geq 1$, $\Gamma_{4p} \triangleq \Gamma^4$ for $p \geq 1$ and suite $(\sigma_n)_{n \geq 0}$ such that $\sigma_{4p+1} \triangleq \sigma^1$ for $p \geq 0$, $\sigma_{4p-2} \triangleq \sigma^2$ for $p \geq 1$, $\sigma_{4p-11} \triangleq \sigma^3$ for $p \geq 1$, $\sigma_{4p} \triangleq \sigma^4$ for $p \geq 1$ define here a trajectory which is a period-2 cycle for $t \geq t_1$.

4. STABILITY STUDY

To study the period-2 cycles stability, we use, like for the period-1 cycles (see (Quémard *et al.*, 2005*b*)), an adapted *point transformation method of Andronov* (Meerov *et al.*, 1979) which is equivalent to the classical idea of the Poincaré map extended to h.d.s (Girard, 2003) what is highlighted in an appendix in (Quémard *et al.*, 2005*b*).

This method consists in a linearization of the problem in the neighbourhood of a fixed point and in an application of the $\mathcal{Z}$-transformation.

From (9), we have $H(R,R) = 0$ with $R = (\sigma^1, \Gamma^1, \sigma^2, \Gamma^2, \sigma^3, \Gamma^3, \sigma^4, \Gamma^4)$. We linearize the problem in the neighbourhood of R replacing $H(R_n, R_{n+1})$ made explicit in (8) by relation:

$$\frac{\partial H}{\partial R_n}(R,R)T_n + \frac{\partial H}{\partial R_{n+1}}(R,R)T_{n+1} = 0, \qquad (12)$$

where $T_n = R_n - R$ and $T_{n+1} = R_{n+1} - R$.

Let us set $U \triangleq \frac{\partial H}{\partial R_n}(R,R)$ and $V \triangleq \frac{\partial H}{\partial R_{n+1}}(R,R)$. Let T^* be the z-transform of suite $(T_n)_{n\in\mathbb{N}}$ (We give an arbitrary value to $\sigma_0 = t_1 - t_0$ in order to have $T_0 = R_0 - R$ well defined). We have $\mathcal{Z}((T_{n+1})_{n\in\mathbb{N}}) = z(T^* - T_0)$. We obtain $UT^* + zV(T^* - T_0) = 0$ and so $T^* = (z^{-1}U + V)^{-1}VT_0$. Let us set $\Delta \triangleq |z^{-1}U + V| = \frac{1}{z^{|K|}}|U + zV|$ where $|K| = 4(N+1)$. Some calculuses lead to:

$$|U + zV| = z^{3N+5}$$

$$.det \begin{pmatrix} 0 & L & 0 & 0 & 0 & 0 & 0 & 0 \\ M_1 & zI_n & 0 & 0 & 0 & 0 & 0 & M_2 \\ 0 & 0 & 0 & L & 0 & 0 & 0 & 0 \\ 0 & M_3 & M_4 & I_N & 0 & 0 & 0 & 0 \\ 0 & 0 & 0 & 0 & 0 & L & 0 & 0 \\ 0 & 0 & 0 & M_5 & M_6 & I_N & 0 & 0 \\ 0 & 0 & 0 & 0 & 0 & 0 & 0 & L \\ 0 & 0 & 0 & 0 & 0 & M_7 & M_8 & I_N \end{pmatrix}$$

$$= z^{3N+5}\Delta',$$

where $M_1 = -Ae^{\sigma^1 A}\Gamma^4$, $M_2 = -e^{\sigma^1 A}$, $M_3 = -e^{\sigma^2 A}$, $M_4 = -Ae^{\sigma^2 A}\Gamma^1$, $M_5 = -e^{\sigma^3 A}$, $M_6 = -e^{\sigma^3 A}\Gamma^2$, $M_7 = -e^{\sigma^4 A}$ and $M_8 = -e^{\sigma^4 A}\Gamma^3$. We can deduce $\Delta = \frac{1}{z^{N-1}}\Delta'$. We remark that the coefficient of z^N in Δ' is null and so, Δ' is an at most $N-1$ degree polynomial. The stability notion corresponds here to the convergence of suite $(R_n)_{n\in\mathbb{N}}$ defined by (7) towards R, that is to say of suite $(T_n)_{n\in\mathbb{N}}$ towards zero. We can say that a cycle is asymptotically stable (see (Meerov *et al.*, 1979)) if all roots of determinant Δ' have moduli less than one, is unstable if there exists at least one root whose modulus is superior to one and we cannot conclude about the cycle stability if there exists one root whose modulus is equal to one and if the others have moduli less than one.

5. PRESENTATION OF A THERMOSTAT WITH ANTICIPATIVE RESISTANCE MODEL

Here, we apply all those previous results to a thermal application treated in (Cébron, 2000) for a problem of optimal control. It deals with a thermostat with anticipative resistance, common in the industrial market (Cyssau, 1990) which controls a convector located in the same room. Figure 2 gives a representation of the physical system and notations used further.

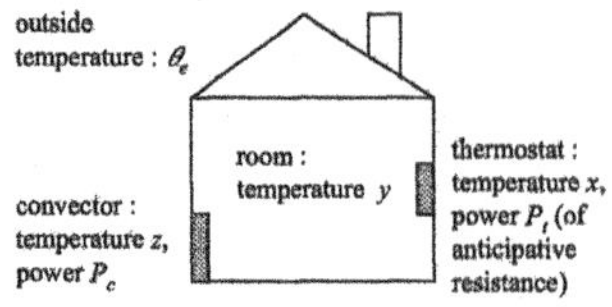

Fig. 2. Thermal process

The functioning principle of such thermostats is the following: powers P_c and P_t are active when $q(x) = 1$ and inactive when $q(x) = 0$. If initially $q = 1$, as P_t is active, the desired temperature is reached by the thermostat temperature x before the room temperature y which makes q changes its value from 1 to 0. Nevertheless, because of the thermal inertia of the room fluid, the room temperature y can eventually continue to increase before decreasing but surely less than without anticipative resistance since active time for P_c is reduced . With this principle, cycle times for the room temperature are shorter and so, as we can wait and verify by simulation or experience, the room temperature variations are reduced. This fact is of industrial interest for energy saving.

Here, we are in dimension $N = 3$ with $X = (x,y,z)^T$ and we consider the $\mathbb{R}^3$ canonical basis. A power assessment and Newton law (Saccadura, 1998) give the following set of equations:

$$\begin{cases} m_t C_t \dot{x} = -\dfrac{x-y}{R_t} + q(x)P_t \\ m_p C_p \dot{y} = -\dfrac{y-z}{R_c} - \dfrac{y-\theta_e}{R_m} \\ m_c C_c \dot{z} = -\dfrac{z-y}{R_c} + q(x)P_c \end{cases} \qquad (13)$$

which is in the form (1). We choose the following realistic numerical values: $q_0 = 1$, $R_t = 1.5$ $K.W^{-1}$, $R_c = 1.35$ $K.W^{-1}$, $R_m = 0.9$ $K.W^{-1}$, $Q_t = m_t C_t = 50$ $J.K^{-1}$, $Q_c = m_c C_c = 732.5$ $J.K^{-1}$, $P_c = 50$ W, $Q_p = m_p C_p = 5000$ $J.K^{-1}$, $\theta_e = 281$ K, $\theta_1 = 293$ K, $\theta_2 = 294$ K, $P_t = 0.8$ W.

Simulations with *Matlab* enable us to obtain time (Figures 3 and 4) and phase (Figure 5) plots.

In Figures 3 and 4, large variations, medium variations and small variations represent respectively temperature variations for the convector, for the thermostat and for the room. Stars and crosses highlight the different switching times. We can

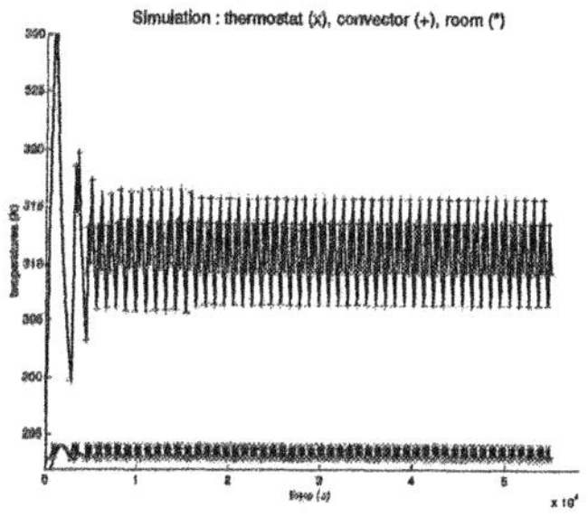

Fig. 3. Time plots

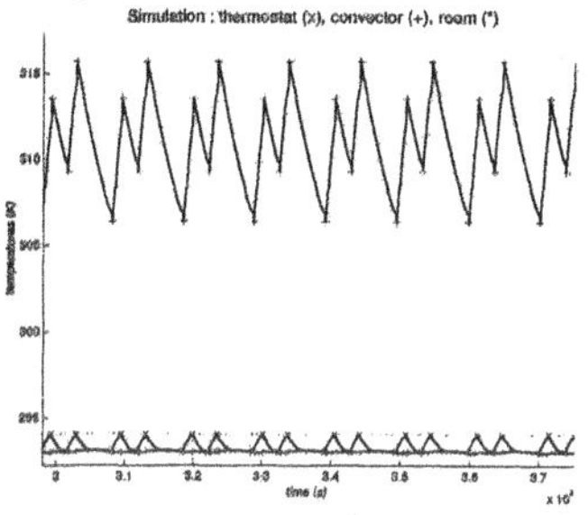

Fig. 4. Time plots (zoom)

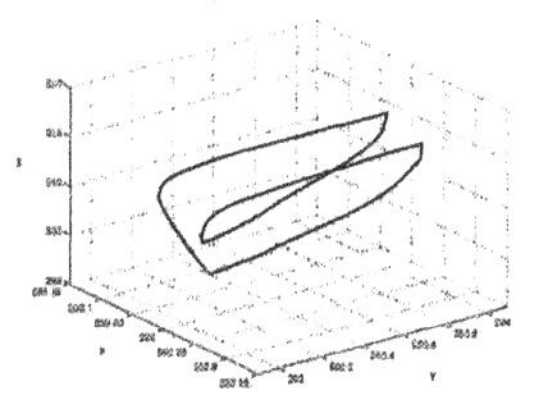

Fig. 5. Phase plots

observe that those variations do not look like those shown in Figure 6 and obtained for other chosen values in (Quémard *et al.*, 2005*b*) and in (Quémard *et al.*, 2005*a*) to illustrate a period-1 cycle.

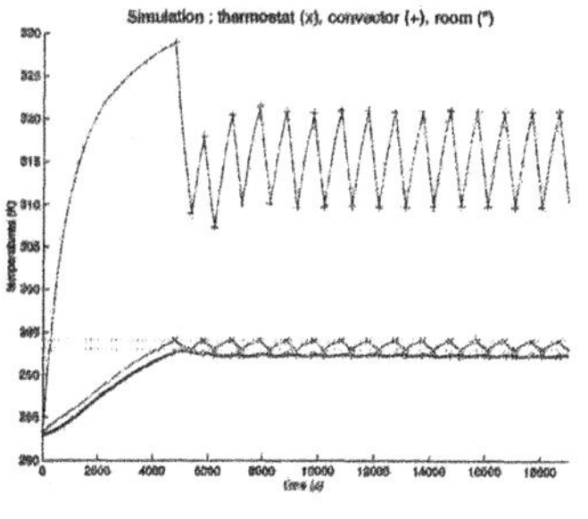

Fig. 6. Illustration of a period-1 cycle

Indeed, it seems that there exists two maxima and two minima for the room and the convector temperatures. The phase plot in space (x, y, z) (Figure 5) confirms this, showing that the system completes two full rotations before repeats itself. This is an illustration of a period-2 cycle (there are four different durations between the different switchings) that one can find in the thermostat with anticipative resistance system.

The following numerical study enables to confirm the existence of period-2 cycles for this studied class of h.d.s computing the possible different durations between two successive switching times.

6. NUMERICAL STUDY

6.1 Determination of the period-2 cycle

We solve system $F = (F_1, F_2, F_3, F_4)^T$ (11) with four unknowns σ^1, σ^2, σ^3, σ^4 programming in *Matlab* a classical algorithm of Newton (Dennis and Schnabel, 1983) taking as initial conditions, values of σ^1, σ^2, σ^3, σ^4 given by simulation: $\sigma^1 = 132.8371\ s$, $\sigma^2 = 529.7769\ s$, $\sigma^3 = 142.3932\ s$, $\sigma^4 = 226.8447\ s$. We also choose those following stop conditions $F^T F < 10^{-15}$ and $i \leq 1000$ (i being the number of iterations) in order that the algorithm can stop if there is no convergence. After 754 iterations, we obtain those following results: $\sigma^1 = 142.1941\ s$, $\sigma^2 = 523.5236\ s$, $\sigma^3 = 157.7412\ s$, $\sigma^4 = 301.2583\ s$ which confirm the existence of four different durations and so the existence of period-2 cycles.

Then, with *Maple* and using the found numerical values for σ^i, $i = 1, .., 4$ given above, we determine the period-2 cycle values:

$$X_{inf1} \simeq (293.00\ 292.9596\ 306.3574)^T,$$

$$X_{inf2} \simeq (293.00\ 292.9696\ 308.7245)^T,$$

$$X_{sup1} \simeq (294.00\ 292.9979\ 315.6553)^T,$$

$$X_{sup2} \simeq (294.00\ 292.9503\ 314.3314)^T,$$

which are well in accordance with Figures 3 and 4.

Moreover, *Maple* is also used for stability study. We compute roots z_i, $i = 1, 2$ of determinant Δ' made explicit in Section 4 and we obtain $z_1 = 0.73569$, $z_2 = 0.00345$ whose moduli are all less than one. So, we can conclude positively with the *point transformation method of Andronov* about the stability of this period-2 cycle.

6.2 Study of the period-2 cycle

From system (11), we can observe that, if we exchange values of σ^1 and σ^3 on the one hand and values of σ^2 and σ^4 on the other hand, we obtain exactly the same system than previously since it is just a question of considered initial discrete state in the description of the orbit. So, in this case, we can conclude that $\sigma^1 = 157.7412\ s$, $\sigma^2 = 301.2583\ s$, $\sigma^3 = 142.1941\ s$, $\sigma^4 = 523.5236\ s$ is also a solution of system (11) which describes the same stable period-2 cycle than before.

Moreover, if now, we set $\sigma^1 = \sigma^3$ and $\sigma^2 = \sigma^4$, equations F_1 and F_2 are identical and F_3 is reduced to zero. So, in this case, it amounts to the system of two equations for two unknowns solved in (Quémard *et al.*, 2005*b*) to find a period-1 cycle. Solving $F_1 = 0$ and $F_4 = 0$, we obtain

$\sigma^1 = \sigma^3 = 147.8814$ s and $\sigma^2 = \sigma^4 = 406.8390$ s which also is a solution for system (11). The application of the *point transformation method of Andronov* with *Maple* enables us to conclude negatively about the stability of this period-1 cycle (determinant Δ' has two roots -1.07178 and -0.05866).

So we can conclude that there is a period-doubling which arises with the lost of stability of the period-1 cycle. As we show computing σ^i, $i = 1,..,4$, the period-1 cycle coninues to exist but now as a saddle cycle (see (Zhusubaliyev and Mosekilde, 2003), (Guckenheimer and Holmes, 1991), (Peitgen *et al.*, 1992)).

With initial conditions chosen above, the Newton algorithm, which is very efficient to estimate zeros of a non linear function, converges towards the period-2 cycle presented above. Nevertheless, with other initial conditions, it can converge towards the period-1 cycle. So, to highlight the sensitivity to initial conditions of this Newton algorithm (see (Zhusubaliyev and Mosekilde, 2003), (Peitgen *et al.*, 1992), (Baker and Gollub, 1990), it is interesting to locate all initial values for which the method converges towards a same zero. Using *Matlab*, we can find a solution to this problem colouring with the same colour all initial points leading to a same zero.

For a question of computing swiftness for the computer and of visibility on figures, we restrict us to plottings in dimension two, setting fixed values for example for σ^1 and σ^3. We plot coulour points in the plane (σ^2, σ^4). For $\sigma^1 = \sigma^3 = 148$ s, we obtain Figure 7, for $\sigma^1 = 135$ s, $\sigma^3 = 148$ s, we obtain Figure 8.

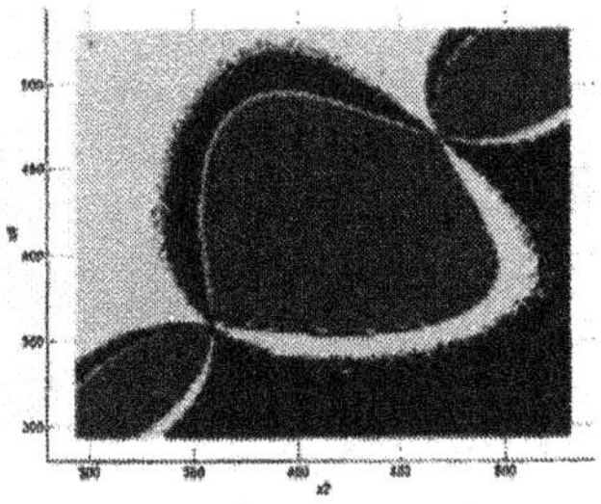

Fig. 7. Influence of initial conditions for fixed values for $\sigma^1 = \sigma^3 = 148$ s

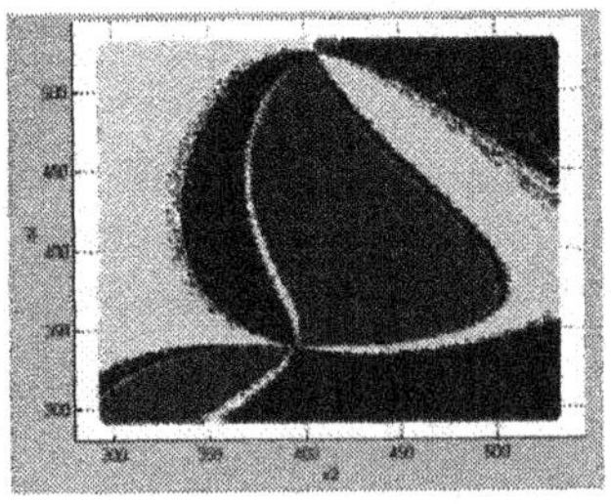

Fig. 8. Influence of initial conditions for fixed values for $\sigma^1 = 135$ s, $\sigma^3 = 148$ s

Light grey set and black set represent all initial points which lead to the period-2 cycle differentiating zero $\sigma^1 = 157.7412$ s, $\sigma^2 = 301.2583$ s, $\sigma^3 = 142.1941$ s, $\sigma^4 = 523.5236$ s (light grey set), and zero $\sigma^1 = 142.1941$ s, $\sigma^2 = 523.5236$ s, $\sigma^3 = 157.7412$ s, $\sigma^4 = 301.2583$ s (dark set). Finally, dark grey set represents all initial points for which the Newton algorithm converges towards the period-1 cycle. Those figures bring out the different fractal attraction basins and the sensitivity to initial conditions principally near the boundaries of the different attraction basins.

6.3 Influence of parameters variations

Vary some parameters can be very useful in order to study the hybrid dynamical system (1). Yet, the only one-parameter variation enables to highlight the period-doubling observed previously. Here, we choose to vary the resistance of the convector R_c. Thus, keeping the same initial conditions and plotting at each iteration of the Newton algorithm σ^2 and σ^4 (we prefer those durations instead of σ^1 and σ^3 because their difference is more important and so we can have a better visibility) while parameter R_c is varying, we obtain in Figure 9 a diagram of bifurcation for our system. It underlines that the curve divides into two branches at approximatively $R_c = 1.21$. Calculuses of roots of Δ' enable us to improve this value approximation to $R_c = 1.20844$ since it is from this value for R_c that one of the real root crosses the boundary of stability at -1 and from this, the period-2 orbit appears.

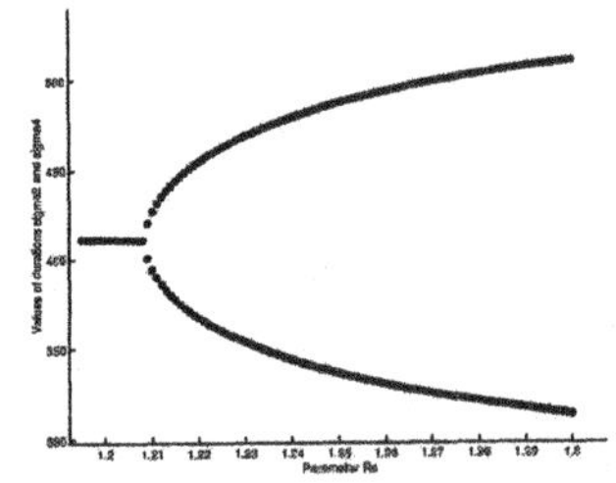

Fig. 9. Bifurcation diagram for thermostat system with the one parameter R_c variation

Now, we can vary simultaneously two parameters. Here, we choose to change values of thermal parameters R_c and P_c. The remaining thermal parameters are supposed to be constant and have the same numerical values than in section 5. At each value for R_c and P_c, from system (11) and using a Newton algorithm, we can determine if numerical values lead to a limit cycle and if it is the case, we can define the nature of the found cycles. Here, we limit our study only detecting period-1 cycles and period-2 cycles but in the future, it would be very interesting to detect cycles with more than two periods if they exist. The division of parameters space for our system (1) into domains of different modes of oscillations is shown in Figure 10. Sets of light grey points

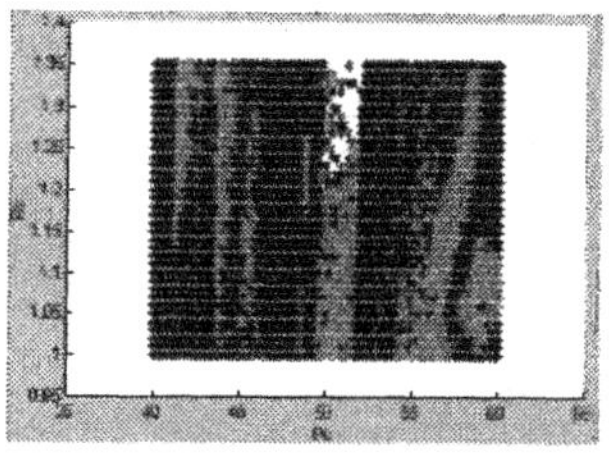

Fig. 10. Influence of parameters variations to the system mode of oscillations

represent sets of values for R_c and P_c which lead to period-2 cycles, those of grey points are values for which the system gives period-1 cycles and finally, sets of black points define R_c and P_c values which lead neither to period-1 cycles nor to period-2 cycles.

7. CONCLUSION

This work completes paper (Quémard *et al.*, 2005*b*) beginning to answer to questions formulated at the end of the study of limit period-1 cycles. Indeed, it establishes that the studied class of h.d.s (1) with autonomous switchings generated by a hysteresis phenomenon can also admit period-2 cycles with four different durations between successive switchings. Some formal and numerical characterizations of this type of cycles are given in this paper. Theoretical results are confirmed by the good adequation with simulations using *Matlab*.

The considered application for that is here interesting because it deals with a thermal device of industrial interest and because its model is a tridimensional model a little less complicated that the one studied in (Zhusubaliyev and Mosekilde, 2003) and from which, some chaotic behaviors have been observed. Nevertheless, a limit of the numerical study is to only detect period-1 and period-2 cycles. Then, some properties like the sensitivity of the model to initial conditions and like the period-doubling with the diagram of bifurcation are illustrated. They concern necessary conditions in the domain of chaotic behaviors. Moreover, Figure 11, which presents a simulation of temperatures variations with *Matlab* for another set of values for the studied thermostat with anticipative resistance model, highlights irregular oscillations and strange behaviors particularly for the room temperature.

So, even if numerical errors which can come from the computer have not to be neglected, the possibility to obtain chaotic behaviors for our h.d.s class is reinforced with yet the existence of multiple cycles and of some necessary properties for that. So, it is an interesting open question which would also show how complexity can arise from simple models.

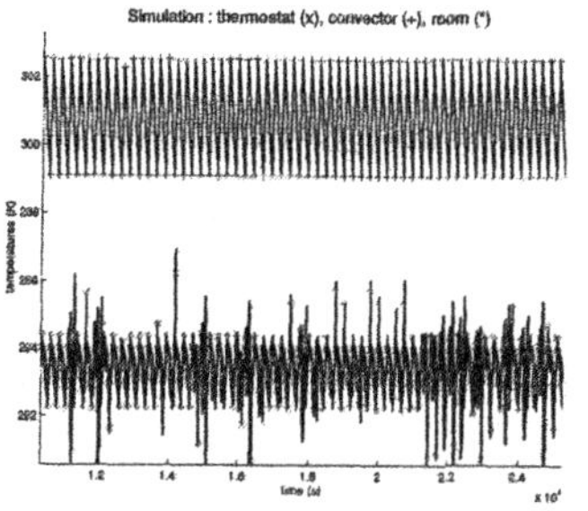

Fig. 11. Illustration of irregular oscillations for some particular values

REFERENCES

Baker, G-L. and J-P. Gollub (1990). *Chaotic dynamics: an introduction.* 2nd ed.. Cambridge University Press.

Bensoussan, A. and J-L. Menaldi (1997). Hybrid control and dynamic programming. *Discrete and Impulsive Systems* **3**, p395–442.

Cébron, B. (2000). Commande de systèmes dynamiques hybrides. PhD thesis. LISA, Angers.

Cyssau, R. (1990). *Manuel de la régulation et de la gestion de l'énergie.* pyc ed.. Association Confort Regulation.

Dennis, J-E. and R-V. Schnabel (1983). *Numerical Methods for Unconstrained Optimization and Nonlinear Equations.* Prentice-Hall, Englewood Cliffs.

Girard, A. (2003). Computation and stability analysis of limit cycles in piecewise linear hybrid systems. In: *ADHS'2003 [Proc. of IFAC Conf.].*

Guckenheimer, J. and P-J. Holmes (1991). *Nonlinear Oscillations, Dynamical Systems, and Bifurcations of Vector Fields.* 3rd ed.. Springer.

Meerov, M., Y. Mikhailov and V. Fiedman (1979). *Principes de la commande automatique.* Mir.

Peitgen, H-O., H. Jurgens and D. Saupe (1992). *Chaos and Fractals.* Springer-Verlag.

Quémard, C., J-C. Jolly and J-L. Ferrier (2005*a*). Mathematical study of a thermal device as a hybrid system. In: *IMACS'2005 World Congress.*

Quémard, C., J-C. Jolly and J-L. Ferrier (2005*b*). Search for cycles in piecewise linear hybrid dynamical systems with autonomous switchings. application to a thermal device. In: *IMACS'2005 World Congress.*

Saccadura, J-F. (1998). *Initiation aux transferts thermiques.* Lavoisier.

Van Der Schaft, A. and H. Schumacher (1999). *An Introduction to Hybrid Dynamical Systems.* Springer.

Zaytoon, J. (2001). *Systèmes dynamiques hybrides.* Hermes.

Zhusubaliyev, Z. and E. Mosekilde (2003). *Bifurcations and Chaos in Piecewise-Smooth Dynamical Systems.* Vol. 44 of *Nonlinear Science.* World World Scientific.

ELSEVIER
IFAC
PUBLICATIONS

STABILIZABILITY OF BIMODAL PIECEWISE LINEAR SYSTEMS WITH CONTINUOUS VECTOR FIELD

Kanat Camlibel [*,**] **Maurice Heemels** [***]
Hans Schumacher [****]

[*] *Dept. of Mechanical Eng., Eindhoven University of Technology, P.O. Box 513, 5600 MB Eindhoven, The Netherlands, e-mail:* k.camlibel@tue.nl
[**] *Dept. of Electronics and Communication Eng., Dogus University, Acibadem 34722, Istanbul, Turkey,*
[***] *Embedded Systems Institute, Den Dolech 2, P.O. Box 513, 5600 MB Eindhoven, The Netherlands, e-mail:* maurice.heemels@esi.nl
[****] *Dept. of Econometrics and Operations Research, Tilburg University, P.O. Box 90153, 5000 LE Tilburg, The Netherlands, e-mail:* j.m.schumacher@uvt.nl

Abstract: This paper studies the open-loop stabilization problem for bimodal systems with continuous vector field. It is based on the earlier work of the authors on the controllability problem for the same class of systems. A full characterization of stabilizability is established by presenting algebraic necessary and sufficient conditions. It turns out that this system class inherits the relationship between controllability and stabilizability of linear systems. *Copyright © 2006 IFAC*

Keywords: Controllability, stabilizability, piecewise linear systems, bimodal systems

1. INTRODUCTION

Controllability and stabilizability of a linear system are two basic concepts which were born in the early sixties. They have played a central role in various problems throughout the history of modern control theory. As such, these concepts have been studied extensively. For instance, in the context of finite-dimensional linear systems given by

$$\dot{x}(t) = Ax(t) + Bu(t) \qquad (1)$$

where $x(t) \in \mathbb{R}^n$ is the state and $u(t) \in \mathbb{R}^m$ is the input at time $t \in \mathbb{R}$, complete algebraic characterizations of stabilizability and controllability are well known. We say that the system (1) is *controllable* if any initial state can be steered to any final state by choosing the input u appropriately. It is said to be *stabilizable* if any initial state can be asymptotically steered to the origin by choosing the input u appropriately. The following theorem summarizes some of the classical results on these concepts (see e.g.(Hautus 1969) for the original results or (Sontag 1998) for an overview).

Theorem 1.1. The linear system (1)

[1] This research is supported by the European Community through the Information Society Technologies thematic programme under the project SICONOS (IST-2001-37172).

- *is controllable if, and only if,* rank $[A - \lambda I \; b] = n$ *for all complex numbers* λ.
- *is stabilizable if, and only if,* rank $[A - \lambda I \; b] = n$ *for all complex numbers* λ *with nonnegative real parts.*

Also in the case of linear systems with constraints on the control set (e.g. $u(t) \in K$, where $K \subset \mathbb{R}^m$ is a closed convex cone) similar connections between controllability and stabilizability are known (Brammer 1972, Smirnov 2000).

This paper focuses on the stabilizability problem for bimodal piecewise linear systems of the form

$$\dot{x}(t) = \begin{cases} A_1 x(t) + b_1 u(t) & \text{if } y(t) \leqslant 0, \\ A_2 x(t) + b_2 u(t) & \text{if } y(t) \geqslant 0 \end{cases} \tag{2a}$$

$$y(t) = c^T x(t) + du(t) \tag{2b}$$

where $A_1, A_2 \in \mathbb{R}^{n \times n}$, $b_1, b_2, c \in \mathbb{R}^n$, and d is a scalar. The characterization of stability and controllability of such a simple class of hybrid systems is already very complex; in (Blondel and Tsitsiklis 1999) it was shown that these problems for a related class of discrete-time systems are NP-hard and undecidable - meaning that there is no algorithm to decide the controllability status of a given system - respectively. In (Blondel and Tsitsiklis 1999) it was advocated that classes should be identified for which these questions are solvable in an efficient way. In case the vector field is continuous (over the switching plane) for (2), algebraic necessary and sufficient conditions for the controllability of this class of systems (and various extensions) are provided by the authors in (Camlibel *et al.* 2003, Camlibel *et al.* 2004, Camlibel *et al.* 2005, Camlibel 2005). The contribution of the current paper is an algebraically verifiable condition for stabilizability for the same class of systems. Interestingly, this result shows that in this class of systems controllability implies stabilizability, as is also true for linear systems but not in general for nonlinear systems.

In the linear case (Hautus 1969) and also in the constrained linear case (Smirnov 2000), one can even show that a linear and Lipschitz continuous, respectively, *state feedback* can be found that does the job. In the piecewise linear case this is still an open issue, although several constructive results for particular feedback structures (e.g. piecewise linear state feedback) based on (control) Lyapunov functions have been proposed in the literature (see e.g. (Hassibi and Boyd 1998)). However, these results give no conclusion on a general level on the stabilizability issue. Only when the design turns out feasible, a stabilizing controller is found and in this sense those papers only present particular instances of sufficient conditions, but not necessary and sufficient cases as is done in this paper.

Also in the case of switched linear systems several results on controllability and stabilizability have appeared, see e.g. (Xie and Wang 2003, Xie and Wang 2005, Sun and Zheng 2001), which construct in addition to a control signal also the switching sequence to stabilize the system. However, since the switching sequence is constructed as well, as opposed to given by a state space partitioning in the piecewise linear case, the case of switched linear systems is essentially different from the case of piecewise linear systems, where a particular switching mechanism is a priori given. Moreover, a full connection between stabilizability and controllability as indicated in this paper for piecewise linear systems is not (yet) available for switched linear systems. However, some partial results are available as, for instance in (Xie and Wang 2003), one proves that controllability implies stabilizability for discrete-time switched linear systems.

The paper is organized as follows. After providing some of the notation used in this paper, the class of systems that we consider and the main result are presented in Section 2. In section 3 a quick review is given of some ingredients from geometric control theory that we need to give the proof of the main results, which can be found in section 4. In section 5 conclusions are given.

1.1 Notation

The set of real numbers is denoted by $\mathbb{R}$, the n-tuples of real numbers by $\mathbb{R}^n$, complex numbers by $\mathbb{C}$, locally integrable functions by L^1. The transpose of a vector x (or matrix M) is denoted by x^T (M^T) and the conjugate transpose by x^* (M^*). For two matrices $M_1 \in \mathbb{R}^{m \times p}$ and $M_2 \in \mathbb{R}^{n \times p}$ with the same number columns, the operator col stacks the matrices in an $(m+n) \times p$ matrix, i.e. $\text{col}(M_1, M_2) = (M_1^T, M_2^T)^T$. All inequalities involving a vector are understood componentwise. A square matrix is said to be *Hurwitz* if the real parts of all its eigenvalues are negative.

2. BIMODAL PIECEWISE LINEAR SYSTEMS

Consider the bimodal piecewise linear system (2) that has a continuous vector field. To be precise, we assume that the dynamics is continuous along the hyperplane $\{(x, u) \mid c^T x + du = 0\}$, i.e.

$$c^T x + du = 0 \Rightarrow A_1 x + b_1 u = A_2 x + b_2 u. \tag{3}$$

This means that

$$A_1 - A_2 = ec^T \tag{4a}$$

$$b_1 - b_2 = ed \tag{4b}$$

for some vector $e \in \mathbb{R}^n$.

As the right hand side of (2) is Lipschitz continuous in the x variable, one can show that for each initial state $x_0 \in \mathbb{R}^n$ and locally-integrable input $u \in L^1$ there exists a unique absolutely continuous function $x^{x_0,u}$ satisfying (2) almost everywhere.

From a control theory point of view, one of the very immediate issues is the controllability of the system at hand. Following the classical literature, we say that the system (2) is *completely controllable* if for any pair of states (x_0, x_f) there exists a locally-integrable input u such that the solution $x^{x_0,u}$ of (2) passes through x_f, i.e. $x^{x_0,u}(\tau) = x_f$ for some $\tau > 0$.

The following theorem on controllability of bimodal systems was proven in (Camlibel 2005).

Theorem 2.1. Suppose that the transfer function $d + c^T(sI - A_1)^{-1}b_1$ is not identically zero. The bimodal system (2) is controllable if, and only if,

(1) the pair $(A_1, \begin{bmatrix} b_1 & e \end{bmatrix})$ is controllable,
(2) the inequality system

$$\mu \geqslant 0 \tag{5}$$

$$\begin{bmatrix} z^T & \mu \end{bmatrix} \begin{bmatrix} A_1 - \lambda I & b_1 \\ c^T & d \end{bmatrix} = 0 \tag{6}$$

$$\begin{bmatrix} z^T & \mu \end{bmatrix} \begin{bmatrix} e \\ 1 \end{bmatrix} \leqslant 0 \tag{7}$$

admits no solution $0 \neq \mathrm{col}(z, \mu) \in \mathbb{R}^{n+1}$ and $\lambda \in \mathbb{R}$.

An equally important concept of system theory is stabilizability. We call the system (2) *(open-loop) stabilizable* if for each initial state x_0 there exists a locally-integrable input u such that the state trajectory satisfies $\lim_{t\to\infty} x^{x_0,u}(t) = 0$.

The following theorem is the main result of this paper. It presents necessary and sufficient conditions for a bimodal system to be stabilizable.

Theorem 2.2. Suppose that the transfer function $d + c^T(sI - A_1)^{-1}b_1$ is not identically zero. The bimodal system (2) is stabilizable if, and only if,

(1) the pair $(A_1, \begin{bmatrix} b_1 & e \end{bmatrix})$ is stabilizable,
(2) the inequality system

$$\mu \geqslant 0 \tag{8a}$$

$$\begin{bmatrix} z^T & \mu \end{bmatrix} \begin{bmatrix} A_1 - \lambda I & b_1 \\ c^T & d \end{bmatrix} = 0 \tag{8b}$$

$$\begin{bmatrix} z^T & \mu \end{bmatrix} \begin{bmatrix} e \\ 1 \end{bmatrix} \leqslant 0 \tag{8c}$$

admits no solution $0 \neq \mathrm{col}(z, \mu) \in \mathbb{R}^{n+1}$ and $0 \leqslant \lambda \in \mathbb{R}$.

Before proceeding to the proof, we need to introduce some terminology.

3. A QUICK REVIEW OF BASIC GEOMETRIC CONTROL THEORY

Consider the linear system $\Sigma(A, B, C, D)$

$$\dot{x}(t) = Ax(t) + Bu(t) \tag{9a}$$

$$y(t) = Cx(t) + Du(t) \tag{9b}$$

where $x(t) \in \mathbb{R}^n$ is the state, $u(t) \in \mathbb{R}^m$ is the input, $y(t) \in \mathbb{R}^p$ is the output at time $t \in \mathbb{R}$, and the matrices A, B, C, D are of appropriate sizes.

We define the *controllable subspace* and *unobservable subspace* as

$$\langle A \mid \mathrm{im}\, B \rangle := \mathrm{im}\, B + A\, \mathrm{im}\, B + \cdots + A^{n-1}\, \mathrm{im}\, B$$

and

$$\langle \ker C \mid A \rangle := \ker C \cap A^{-1} \ker C \cap \cdots \cap A^{1-n} \ker C,$$

respectively. It follows from these definitions that

$$\langle A \mid \mathrm{im}\, B \rangle = \langle \ker B^T \mid A^T \rangle^{\perp} \tag{10}$$

where $\mathcal{W}^{\perp}$ denotes the orthogonal space of $\mathcal{W}$.

We say that a subspace $\mathcal{V}$ is *output-nulling controlled invariant* if for some matrix K the inclusions $(A - BK)\mathcal{V} \subseteq \mathcal{V}$ and $\mathcal{V} \subseteq \ker(C - DK)$ hold. As the set of such subspaces is non-empty and closed under subspace addition, it has a maximal element $\mathcal{V}^*(\Sigma)$ (also written as $\mathcal{V}^*(A, B, C, D)$). Whenever the system Σ is clear from the context, we simply write $\mathcal{V}^*$. The notation $\mathcal{K}(\mathcal{V})$ stands for the set $\{K \mid (A - BK)\mathcal{V} \subseteq \mathcal{V} \text{ and } \mathcal{V} \subseteq \ker(C - DK)\}$. Moreover, we write $\mathcal{K}(A, B, C, D)$ for $\mathcal{K}(\mathcal{V}^*(A, B, C, D))$.

One can compute $\mathcal{V}^*$ as a limit of the subspaces

$$\mathcal{V}^0 = \mathbb{R}^n \tag{11a}$$

$$\mathcal{V}^i = \{x \mid Ax + Bu \in \mathcal{V}^{i-1} \text{ and } Cx + Du = 0 \text{ for some } u\}. \tag{11b}$$

In fact, there exists an index $i \leqslant n - 1$ such that $\mathcal{V}^j = \mathcal{V}^*$ for all $j \geqslant i$.

Dually, we say that a subspace $\mathcal{T}$ is *input-containing conditioned invariant* if for some matrix L the inclusions $(A - LC)\mathcal{T} \subseteq \mathcal{T}$ and $\mathrm{im}(B - LD) \subseteq \mathcal{T}$ hold. As the set of such subspaces is non-empty and closed under subspace intersection, it has a minimal element $\mathcal{T}^*(\Sigma)$ (also written as $\mathcal{T}^*(A, B, C, D)$). Whenever the system Σ is clear from the context, we simply write $\mathcal{T}^*$. The notation $\mathcal{L}(\mathcal{T})$ stands for the set $\{L \mid (A - LC)\mathcal{T} \subseteq \mathcal{T} \text{ and } \mathrm{im}(B - LD) \subseteq \mathcal{T}\}$. Moreover, we write $\mathcal{L}(A, B, C, D)$ for $\mathcal{L}(\mathcal{T}^*(A, B, C, D))$. Note that

$$\langle A \mid \mathrm{im}\, B \rangle \supseteq \mathcal{T}^*(A, B, C, D). \tag{12}$$

We quote some standard facts from geometric control theory in what follows. The first one presents certain invariants under state feedbacks and output injections. Besides the system Σ (9), consider the linear system $\Sigma_{K,L}$ given by

$$\dot{x} = (A - BK - LC + LDK)x + (B - LD)v \quad (13a)$$
$$y = (C - DK)x + Dv. \quad (13b)$$

This system can be obtained from Σ (9) by applying both state feedback $u = -Kx + v$ and output injection $-Ly$.

Proposition 3.1. Let $K \in \mathbb{R}^{m\times n}$ and $L \in \mathbb{R}^{n\times p}$ be given. The following statements hold.

(1) $\langle A \mid \operatorname{im} B\rangle = \langle A - BK \mid \operatorname{im} B\rangle$.
(2) $\langle \ker C \mid A\rangle = \langle \ker C \mid A - LC\rangle$.
(3) $\mathcal{V}^*(\Sigma_{K,L}) = \mathcal{V}^*(\Sigma)$.
(4) $\mathcal{T}^*(\Sigma_{K,L}) = \mathcal{T}^*(\Sigma)$.

The next proposition relates the invertibility of the transfer matrix to the controlled and conditioned invariant subspaces.

Proposition 3.2. (cf. (Aling and Schumacher 1984)). *The transfer matrix $D+C(sI-A)^{-1}B$ is invertible as a rational matrix if, and only if, $\mathcal{V}^* \oplus \mathcal{T}^* = \mathcal{X}$, $\begin{bmatrix} C & D\end{bmatrix}$ is of full row rank, and $\operatorname{col}(B, D)$ is of full column rank. Moreover, the inverse is polynomial if, and only if, $\mathcal{V}^* \cap \langle A \mid \operatorname{im} B\rangle \subseteq \langle \ker C \mid A\rangle$ and $\langle A \mid \operatorname{im} B\rangle \subseteq \mathcal{T}^* + \langle \ker C \mid A\rangle$.*

The following proposition presents sufficient conditions for the absence of invariant zeros. It can be proved by using (11).

Proposition 3.3. Consider the linear system (9) with $p = m$. Suppose that $\mathcal{V}^* = \{0\}$ and the matrix $\operatorname{col}(B, D)$ is of full column rank. Then, the system matrix

$$\begin{bmatrix} A - \lambda I & B \\ C & D \end{bmatrix}$$

is nonsingular for all $\lambda \in \mathbb{C}$.

4. PROOF OF THEOREM 2.2

'only if': Suppose that the bimodal system (2) is stabilizable.

We start by proving the first statement in Theorem 2.2. Let the complex number λ with a nonnegative real part and the complex vector z be such that $z^*A_1 = \lambda z^*$, $z^*b_1 = 0$, $z^*e = 0$. By left multiplying (2) by z^*, one gets $z^*\dot{x} = \lambda z^*x$. Hence, one gets $z^*x(t) = \exp(\lambda t)z^*x(0)$ irrespective of the choice of input signal. Due to stabilizability of (2), for any initial state $x(0)$ one can choose the input u so that $\lim_{t\to\infty} x^{x_0,u}(t) = 0$. This means that z must be zero, i.e. the pair $(A_1, \begin{bmatrix} b_1 & e\end{bmatrix})$ is stabilizable.

We now prove the second statement in Theorem 2.2. Suppose that $\operatorname{col}(z, \mu) \in \mathbb{R}^{n+1}$ is a solution to (8) for $\lambda \geqslant 0$ which means that

$$z^TA_1 = z^T\lambda - \mu c^T \quad (14a)$$
$$z^Tb_1 + \mu d = 0 \quad (14b)$$
$$z^Te + \mu \leqslant 0. \quad (14c)$$

By left multiplying (2) by z^T and using the above relations and (4) we obtain

$$z^T\dot{x} = \begin{cases} \lambda(z^Tx) - \mu y & \text{if } y \leqslant 0 \\ \lambda(z^Tx) - (z^Te + \mu)y & \text{if } y \geqslant 0 \end{cases} \quad (15)$$
$$y = c^Tx + du \quad (16)$$

which implies that

$$z^T\dot{x} \geqslant \lambda z^Tx \quad (17)$$

The Bellman-Gronwall lemma (Desoer and Vidyasagar 1975, p. 252) implies that

$$z^Tx(t) \geqslant \exp(\lambda t)z^Tx(0) \quad (18)$$

Since the bimodal system (2) is stabilizable, $z^Tx(0)$ must be zero. As $x(0)$ is arbitrary, one concludes that $z = 0$. Note that this implies via (14) in turn that $\mu c^T = 0$ and $\mu d = 0$. This yields that $\mu = 0$ due to invertibility of $d + c^T(sI - A_1)^{-1}b_1$. This proves the second statement.

'if': We begin with the following observations

$$\mathcal{V}^*(A_1, b_1, c^T, d) = \mathcal{V}^*(A_2, b_2, c^T, d) \quad (19a)$$
$$\mathcal{T}^*(A_1, b_1, c^T, d) = \mathcal{T}^*(A_2, b_2, c^T, d) \quad (19b)$$
$$\mathcal{K}(A_1, b_1, c^T, d) = \mathcal{K}(A_2, b_2, c^T, d) \quad (19c)$$
$$\mathcal{L}(A_1, b_1, c^T, d) - \{e\} = \mathcal{L}(A_2, b_2, c^T, d) \quad (19d)$$

where $X - \{e\} = \{y \mid y = x - e \text{ for some } x \in X\}$. To see the first one, note that $\mathcal{V}^* := \mathcal{V}^*(A_1, b_1, c^T, d)$ is an output-nulling controlled invariant subspace for the system $\Sigma(A_2, b_2, c^T, d)$ as

$$\mathcal{V}^* \subseteq \ker(c^T - dk^T) \quad (20)$$
$$(A_2 - b_2k^T)\mathcal{V}^* \overset{(4)}{=} (A_1 - ec^T - b_1k^T + edk^T)\mathcal{V}^*$$
$$\overset{(20)}{=} (A_1 - b_1k^T)\mathcal{V}^*$$
$$\subseteq \mathcal{V}^*$$

for any $k^T \in \mathcal{K}(A_1, b_1, c^T, d)$. Since $\mathcal{V}^*(A_2, b_2, c^T, d)$ is the largest of such subspaces, one gets

$$\mathcal{V}^* = \mathcal{V}^*(A_1, b_1, c^T, d) \subseteq \mathcal{V}^*(A_2, b_2, c^T, d).$$

By symmetry, one arrives at (19a). The other relations follow in a similar fashion.

Let $\mathcal{V}^*$ and $\mathcal{T}^*$ denote $\mathcal{V}^*(A_1, b_1, c^T, d)$ and $\mathcal{T}^*(A_1, b_1, c^T, d)$, respectively. Let

$$k^T \in \mathcal{K}(A_1, b_1, c^T, d) = \mathcal{K}(A_2, b_2, c^T, d).$$

Apply the feedback $u = -k^Tx + v$ to the system (2). Then, one gets

$$\dot{x} = \begin{cases} (A_1 - b_1 k^T)x + b_1 v & \text{if } y \leqslant 0, \\ (A_2 - b_2 k^T)x + b_2 v & \text{if } y \geqslant 0. \end{cases} \tag{21a}$$

$$y = (c^T - dk^T)x + dv \tag{21b}$$

Due to Proposition 3.1, the two subspaces $\mathcal{V}^*$ and $\mathcal{T}^*$ remain unchanged. Since $d + c^T(sI - A_1)^{-1}b_1$ is not identically zero and hence invertible as a rational function, it follows from Proposition 3.2 that

(1) $\mathcal{V}^* \oplus \mathcal{T}^* = \mathbb{R}^n$,
(2) $\mathrm{col}(b_1, d)$ is of full column rank, and
(3) $[c^T \; d]$ is of full row rank.

Let $\ell^i \in \mathcal{L}(A_i, b_i, c^T, d)$, $i = 1, 2$, be such that $\ell^1 - \ell^2 = e$. Note that $A_i - b_i k^T - \ell^i[c^T - dk^T]$, $i = 1, 2$ leave both $\mathcal{V}^*$ and $\mathcal{T}^*$ invariant. Moreover, the restrictions of the mappings $A_i - b_i k^T - \ell^i[c^T - dk^T]$ to the subspace $\mathcal{V}^*$ coincide.

Therefore, $A_1 - b_1 k^T - \ell^i[c^T - dk^T]$ must be block diagonal in a basis that is adapted to the decomposition $\mathcal{V}^* \oplus \mathcal{T}^*$. If we further decompose the space $\mathcal{V}^*$ by using the real Jordan decomposition (Lütkepohl 1996, p. 71)) of $\bar{A} := A_i - b_i k^T \mid_{\mathcal{V}^*}$ to separate the eigenspaces of the eigenvalues with nonnegative and negative real parts one gets in these new coordinates for $i = 1, 2$

$$\left[\begin{array}{c|c|c|c} A_i - b_i k^T & b_i & e & l \\ \hline c^T - dk^T & d & 0 & 0 \end{array}\right]$$
$$\parallel$$
$$\left[\begin{array}{ccc|c|c|c} A_- & 0 & \ell_1^i c_3^T & \ell_1^i d & e_1 & \ell_1^i \\ 0 & A_+ & \ell_2^i c_3^T & \ell_2^i d & e_2 & \ell_2^i \\ 0 & 0 & A_3^i & b_3^i & e_3 & \ell_3^i \\ \hline 0 & 0 & c_3^T & d & 0 & 0 \end{array}\right] \tag{22}$$

where $\ell_j^1 - \ell_j^2 = e_j$ for $j \in \{1, 2, 3\}$, $A_3^1 - A_3^2 = e_3 c_3^T$ due to (4a), $b_3^1 - b_3^2 = e_3 d$ due to (4b), and the numbers of the rows of the blocks at the right hand side are, respectively, n_1, n_2, n_3, and 1. Note that

$$\mathcal{T}^*(A_3^i, b_3^i, c_3^T, d) = \mathbb{R}^{n_3} \tag{23a}$$

$$\mathcal{V}^*(A_3^i, b_3^i, c_3^T, d) = \{0\}. \tag{23b}$$

Note also that all eigenvalues of A_- (A_+) have negative (nonnegative) real parts.

Suppose that the two conditions of Theorem 2.2 hold. Let

$$\left[\begin{array}{c|c|c} \bar{A}_i & \bar{b}_i & \bar{e} \\ \hline \bar{c}^T & d & 0 \end{array}\right] = \left[\begin{array}{cc|c|c} A_+ & \ell_2^i c_3^T & \ell_2^i d & e_2 \\ 0 & A_3^i & b_3^i & e_3 \\ \hline 0 & c_3^T & d & 0 \end{array}\right]. \tag{24}$$

Note that $\bar{A}_1 - \bar{A}_2 = \bar{e}\bar{c}^T$ and $\bar{b}_1 - \bar{b}_2 = \bar{e}d$. We claim that the bimodal system

$$\dot{\bar{x}} = \begin{cases} \bar{A}_1 \bar{x} + \bar{b}_1 u & \text{if } \bar{c}^T \bar{x} + du \leqslant 0, \\ \bar{A}_2 \bar{x} + \bar{b}_2 u & \text{if } \bar{c}^T \bar{x} + du \geqslant 0 \end{cases} \tag{25}$$

is controllable. To prove this, we want to invoke Theorem 2.1.

Since A_- is Hurwitz, the first condition of Theorem 2.2 is equivalent to saying that the pair

$$\left(\begin{bmatrix} A_+ & \ell_2^1 c_3^T \\ 0 & A_3^1 \end{bmatrix}, \begin{bmatrix} \ell_2^1 d & e_2 \\ b_3^1 & e_3 \end{bmatrix} \right) \tag{26}$$

is stabilizable. Note that (A_3^1, b_3^1) is controllable as

$$\langle A_3^1 \mid \mathrm{im}\, b_3^1 \rangle \overset{(12)}{\supseteq} \mathcal{T}^*(A_3^1, b_3^1, c_3^T, d) \overset{(23a)}{=} \mathbb{R}^{n_3}.$$

Together with the fact that A_+ has only eigenvalues with nonnegative real parts, this means that the pair (26) is actually controllable. Consequently, the bimodal system (25) satisfies the first condition in Theorem 2.1.

Since A_- is Hurwitz, the second condition of Theorem 2.2 is equivalent to saying that the inequality system

$$\mu \geqslant 0 \tag{27a}$$

$$[z_2^T \; z_3^T \; \mu] \begin{bmatrix} A_+ - \lambda I & \ell_2^1 c_3^T & \ell_2^1 d \\ 0 & A_3^1 - \lambda I & b_3^1 \\ 0 & c_3^T & d \end{bmatrix} = 0 \tag{27b}$$

$$[z_2^T \; z_3^T \; \mu] \begin{bmatrix} e_2 \\ e_3 \\ 1 \end{bmatrix} \leqslant 0 \tag{27c}$$

admits no solution $0 \leqslant \lambda \in \mathbb{R}$ and $0 \neq \mathrm{col}(z_2, z_3, \mu) \in \mathbb{R}^{n_2+n_3+1}$. Since $\mathcal{V}^*(A_3^1, b_3^1, c_3^T, d) = 0$ and $\mathrm{col}(b_3^1, d)$ is of full column rank, it follows from Proposition 3.3 that the system matrix

$$\begin{bmatrix} A_3^1 - \lambda I & b_3^1 \\ c_3^T & d \end{bmatrix}$$

is nonsingular for all complex numbers λ. This implies, with the fact that A_+ has no nonnegative (real) eigenvalues, the inequality system (27) admits no solution for any $\lambda \in \mathbb{R}$ and $0 \neq \mathrm{col}(z_2, z_3, \mu) \in \mathbb{R}^{n_2+n_3+1}$. As a result, the second condition in Theorem 2.1 is satisfied by the bimodal system (25). Therefore, Theorem 2.1 implies that the system (25) is controllable. Let $x_0 := \mathrm{col}(x_{10}, x_{20}, x_{30}) \in \mathbb{R}^{n_1+n_2+n_3}$ be an arbitrary initial state for the system (21) in the coordinates given by (22). Since the system (25) is controllable, x_0 can be steered to a state $\bar{x}_0 = \mathrm{col}(\bar{x}_{10}, 0, 0)$ in finite time t_*. Apply the zero input after reaching this state. Since $c^T \bar{x}_0 = 0$ and A_- is Hurwitz, we can conclude that the state trajectory converges to the origin as t tends to infinity (note that after time t_* the state trajectory remains in $\mathcal{V}^*$ and thus the state-input trajectory is on the switching plane given by $c^T x + du = 0$). ■

5. CONCLUDING REMARKS

The paper has presented necessary and sufficient conditions for the stabilizability of bimodal piecewise linear systems with a continuous vector field.

To the best of the authors' knowledge it is the first time that a full *algebraic* characterization of stabilizability for a class of piecewise linear systems appears in the literature. Interestingly, the relationship between the well-known controllability and stabilizability conditions for linear and for input-constrained linear systems is recovered for this class of hybrid systems as well.

The proofs for these results rely on geometric control theory and controllability results for piecewise linear systems and input-constrained linear systems. The structure present in the model class enables the use of this well-known theory in the context of piecewise linear systems. We believe that this forms a basis for solving various system- and control-theoretic problems like observability, detectability, observer and controller design for this class of systems. The investigation of these problems is one of the major issues of our future work.

REFERENCES

Aling, H. and J.M. Schumacher (1984). A nine-fold canonical decomposition for linear systems. *Int. J. Contr.* **39**, 779–805.

Blondel, V.D. and J.N. Tsitsiklis (1999). Complexity of stability and controllability of elementary hybrid systems. *Automatica* **35**(3), 479–490.

Brammer, R.F. (1972). Controllability in linear autonomous systems with positive controllers. *SIAM J. Control* **10**(2), 329–353.

Camlibel, M.K. (2005). Popov-Belevitch-Hautus type controllability tests for linear complementarity systems. submitted for publication.

Camlibel, M.K., W.P.M.H. Heemels and J.M. Schumacher (2003). Stability and controllability of planar bimodal complementarity systems. In: *Proc. of the 42th IEEE Conference on Decision and Control.* Hawaii (USA).

Camlibel, M.K., W.P.M.H. Heemels and J.M. Schumacher (2004). On the controllability of bimodal piecewise linear systems. In: *Hybrid Systems: Computation and Control* (R. Alur and G.J. Pappas, Eds.). pp. 250–264. Springer. Berlin.

Camlibel, M.K., W.P.M.H. Heemels and J.M. Schumacher (2005). Algebraic necessary and sufficient conditions for the controllability of conewise linear systems. submitted for publication.

Desoer, C.A. and M. Vidyasagar (1975). *Feedback Systems: Input-Output Properties.* Academic Press. London.

Hassibi, A. and S. Boyd (1998). Quadratic stabilization and control of piecewise-linear systems. In: *Proc. of the American Control Conference.* pp. 3659 – 3664.

Hautus, M.L.J. (1969). Controllability and observability conditions of linear autonomous systems. *Ned. Akad. Wetenschappen, Proc. Ser. A* **72**, 443–448.

Lütkepohl, H. (1996). *Handbook of Matrices.* Wiley. New York.

Smirnov, G.V. (2000). *Introduction to the Theory of Differential Inclusions.* American Mathematical Society. Rhode Island.

Sontag, E.D. (1998). *Mathematical Control Theory: Deterministic Finite Dimensional Systems.* 2nd ed.. Springer. New York.

Sun, Z. and D. Zheng (2001). On reachability and stabilization of switched linear systems. *IEEE Trans. Automatic Control* **46**(2), 291–295.

Xie, G. and L. Wang (2003). Controllability and stabilizability of switched linear-systems. *Systems & Control Letters* **48**, 135–155.

Xie, G. and L. Wang (2005). Controllability implies stabilizability for discrete-time switched linear systems. In: *Proc. Hybrid Systems: Computation and Control. Lecture Notes in Computer Science 3414.* pp. 667–682.

ELSEVIER
IFAC
PUBLICATIONS

GLOBAL INPUT-TO-STATE STABILITY AND STABILIZATION OF DISCRETE-TIME PIECE-WISE AFFINE SYSTEMS

M. Lazar *[,1] **W.P.M.H. Heemels** **

* *Department of Electrical Engineering, Eindhoven University of Technology, E-mail:* m.lazar@tue.nl
** *Embedded Systems Institute, Eindhoven*

Abstract: This paper presents sufficient conditions for global Input-to-State (practical) Stability (ISpS) and stabilization of discrete-time, possibly discontinuous, Piece-Wise Affine (PWA) systems. Piece-wise quadratic candidate ISpS (ISS) Lyapunov functions are employed for both analysis and synthesis purposes. This enables us to obtain sufficient conditions based on linear matrix inequalities, which can be solved efficiently. One of the advantages of using the ISpS framework is that the additive disturbance inputs are explicitly taken into account in the analysis and synthesis procedures, and the results apply to PWA systems in their full generality, i.e. non-zero affine terms are allowed in the regions in the partition whose closure contains the origin. *Copyright ©2006 IFAC*

Keywords: Piece-wise affine systems, Hybrid systems, Robust stability, Stabilization methods.

1. INTRODUCTION

Several results on nominal stability analysis of Piece-Wise Affine (PWA) systems are available in the literature, see for example (Mignone *et al.*, 2000), (Ferrari-Trecate *et al.*, 2002), (Feng, 2002) for results in discrete-time. These works employ the Lyapunov stability framework and consider Piece-Wise Quadratic (PWQ) candidate Lyapunov functions. Recently, in (Lazar and Heemels, 2006) the authors showed that nominally exponentially stable discrete-time PWA systems can have zero robustness to arbitrarily small additive disturbances, mainly due to the absence of a continuous Lyapunov function. Therefore, in discrete-time, it is important that disturbances are taken into account when analyzing stability of PWA systems, since robustness is relevant for practical applications.

Robust stability results for discrete-time PWA systems were presented in (Ferrari-Trecate *et al.*, 2002, Section 3), which deals with Linear Matrix Inequalities (LMI) based l_2-gain analysis for PWA systems; and in (Grieder, 2004, Chp. 8.5), where it was observed that, if a robust positively invariant set can be calculated for a nominally asymptotically stable PWA system, then *local* robust convergence is ensured. For *continuous-time* input-to-state stability (Sontag, 1989) results for switched systems and hybrid systems we refer the reader to the recent works (Vu *et al.*, 2005) and (Cai and Teel, 2005). However, to the best of the authors' knowledge, a global robust stability analysis methodology for *discrete-time* PWA systems that can be used for both analysis and synthesis purposes is missing from the literature.

As such, we consider discrete-time PWA systems subject to *unbounded* additive disturbance inputs and we employ the Input-to-State (practical) Stability (ISpS) framework (Sontag, 1989), (Jiang, 1993) in order to obtain *global* robust stability results. For simplicity

[1] Research supported by the Dutch Science Foundation (STW), Grant "Model Predictive Control for Hybrid Systems" (DMR. 5675) and by the European Union through the Network of Excellence HYCON (contract FP6-IST-511368).

and clarity of exposition, only PWQ candidate ISpS (ISS) Lyapunov functions are considered, but the results can also be extended *mutatis mutandis* to piece-wise polynomial or piece-wise affine candidate functions. The paper consists of two parts: the first part deals with ISpS (ISS) analysis, while the second part provides techniques for input-to-state stabilizing controllers synthesis. In both sections the sufficient conditions for ISpS (ISS) are expressed in terms of LMIs, which can be solved efficiently (Boyd *et al.*, 1994).

One of the advantages of using the ISpS (ISS) framework for studying robust stability of discrete-time PWA systems is that the additive disturbance inputs are explicitly taken into account in the analysis and synthesis procedures. Also, the ISpS framework enables us to obtain robust stability results for PWA systems in their full generality, i.e. non-zero affine terms are allowed in the regions in the state-space partition whose closure contains the origin. Note that this situation is often excluded in other works. In this paper we develop a new LMI technique for dealing with non-zero affine terms, which does not rely on a system transformation and the S-procedure, e.g. as done in (Ferrari-Trecate *et al.*, 2002, Remark 3). This technique makes it possible to obtain LMI based sufficient conditions for input-to-state stabilizing controllers synthesis as well, and not just for analysis.

1.1 Notation and basic definitions

Let $\mathbb{R}$, $\mathbb{R}_+$, $\mathbb{Z}$ and $\mathbb{Z}_+$ denote the field of real numbers, the set of non-negative reals, the set of integer numbers and the set of non-negative integers, respectively. We use the notation $\mathbb{Z}_{\geq c}$ to denote the set $\{k \in \mathbb{Z}_+ \mid k \geq c\}$ for some $c \in \mathbb{Z}_+$. Let $\|\cdot\|$ denote the Euclidean norm. For a matrix $Z \in \mathbb{R}^{m\times n}$ let $\|Z\| \triangleq \sup_{x\neq 0} \frac{\|Zx\|}{\|x\|}$ denote its induced Euclidean norm. For a positive definite matrix $Z \in \mathbb{R}^{n\times n}$, $\lambda_{\min}(Z)$ and $\lambda_{\max}(Z)$ denote the smallest and the largest eigenvalue of Z, respectively. For a sequence $\{z_p\}_{p\in\mathbb{Z}_+}$ with $z_p \in \mathbb{R}^l$ let $\|\{z_p\}_{p\in\mathbb{Z}_+}\| \triangleq \sup\{\|z_p\| \mid p \in \mathbb{Z}_+\}$. Let $z_{[k]}$ denote the truncation of $\{z_p\}_{p\in\mathbb{Z}_+}$ at time $k \in \mathbb{Z}_+$, i.e. $z_{[k],p} = z_p$, $p \leq k$. For a set $\mathscr{P} \subseteq \mathbb{R}^n$, we denote by $\partial\mathscr{P}$ the boundary of $\mathscr{P}$, by $\mathrm{int}(\mathscr{P})$ its interior and by $\mathrm{cl}(\mathscr{P})$ its closure. A polyhedron (or a polyhedral set) is a set obtained as the intersection of a finite number of open and/or closed half-spaces. A function $\varphi : \mathbb{R}_+ \to \mathbb{R}_+$ *belongs to class* $\mathscr{K}$ if it is continuous, strictly increasing and $\varphi(0) = 0$. A function $\beta : \mathbb{R}_+ \times \mathbb{R}_+ \to \mathbb{R}_+$ *belongs to class* $\mathscr{KL}$ if for each fixed $k \in \mathbb{R}_+$, $\beta(\cdot, k) \in \mathscr{K}$ and for each fixed $s \in \mathbb{R}_+$, $\beta(s, \cdot)$ is non-increasing and $\lim_{k\to\infty} \beta(s, k) = 0$.

2. INPUT-TO-STATE STABILITY AND PROBLEM STATEMENT

Consider the discrete-time autonomous perturbed nonlinear system described by

$$x_{k+1} = G(x_k, v_k), \quad k \in \mathbb{Z}_+, \tag{1}$$

where $x_k \in \mathbb{R}^n$ is the state, $v_k \in \mathbb{R}^{d_v}$ is an unknown *disturbance input* and $G : \mathbb{R}^n \times \mathbb{R}^{d_v} \to \mathbb{R}^n$ is an arbitrary nonlinear function. For simplicity of notation, we assume that the origin is an equilibrium in (1) for zero disturbance input, meaning that $G(0,0) = 0$.

Next, we define the notions of Input-to-State practical Stability (ISpS) (Jiang, 1993), (Jiang *et al.*, 1996) and Input-to-State Stability (ISS) (Sontag, 1989), (Jiang and Wang, 2001) for the discrete-time perturbed nonlinear system (1).

Definition 1. The system (1) is said to be *globally ISpS* if there exist a $\mathscr{KL}$-function β, a $\mathscr{K}$-function γ and a non-negative constant d such that, for each $x_0 \in \mathbb{R}^n$ and all $\{v_p\}_{p\in\mathbb{Z}_+}$ with $v_p \in \mathbb{R}^{d_v}$ for all $p \in \mathbb{Z}_+$, it holds that the corresponding state trajectory satisfies

$$\|x_k\| \leq \beta(\|x_0\|, k) + \gamma(\|v_{[k-1]}\|) + d, \quad \forall k \in \mathbb{Z}_{\geq 1}. \tag{2}$$

If the above condition holds for $d = 0$, the system (1) is said to be *globally ISS.*

In what follows we state a *discrete-time* version of the *continuous-time* ISpS sufficient conditions of Proposition 2.1 of (Jiang *et al.*, 1996). This result will be used throughout the paper to establish ISpS and ISS for the particular case of PWA systems. For the proof we refer the reader to (Lazar *et al.*, 2006).

Theorem 2. Let d_1, d_2 be non-negative constants, let a, b, c, λ be positive constants with $c \leq b$ and let $\alpha_1(s) \triangleq as^\lambda$, $\alpha_2(s) \triangleq bs^\lambda$, $\alpha_3(s) \triangleq cs^\lambda$ and $\sigma \in \mathscr{K}$. Furthermore, let $V : \mathbb{R}^n \to \mathbb{R}_+$ be a function such that

$$\alpha_1(\|x\|) \leq V(x) \leq \alpha_2(\|x\|) + d_1 \tag{3a}$$

$$V(G(x,v)) - V(x) \leq -\alpha_3(\|x\|) + \sigma(\|v\|) + d_2 \tag{3b}$$

for all $x \in \mathbb{R}^n$ and all $v \in \mathbb{R}^{d_v}$. Then it holds that:

(i) The system (1) is globally ISpS;

(ii) If inequalities (3) hold for $d_1 = d_2 = 0$, the system (1) is globally ISS.

Definition 3. A function V that satisfies the hypothesis of Theorem 2 is called an *ISpS (ISS) Lyapunov function.*

Remark 4. The hypothesis of Theorem 2 allows that both G and V are discontinuous. If inequality (3a) holds for $d_1 = 0$, then the hypothesis of Theorem 2 *only* implies continuity at the point $x = 0$, and *not* necessarily on a neighborhood of $x = 0$.

In this paper we focus on perturbed discrete-time, possibly discontinuous, PWA systems of the form

$$x_{k+1} = G(x_k, v_k) \triangleq A_j x_k + f_j + D_j v_k \text{ if } x_k \in \Omega_j, \tag{4}$$

where $A_j \in \mathbb{R}^{n\times n}$, $f_j \in \mathbb{R}^n$, $D_j \in \mathbb{R}^{n\times d_v}$ for all $j \in \mathscr{S}$ and $\mathscr{S} \triangleq \{1, 2, \ldots, s\}$ is a *finite set* of indexes. The collection $\{\Omega_j \mid j \in \mathcal{S}\}$ defines a partition of $\mathbb{R}^n$, meaning that $\cup_{j\in\mathcal{S}}\Omega_j = \mathbb{R}^n$ and $\text{int}(\Omega_i)\cap\text{int}(\Omega_j) = \emptyset$ for $i \neq j$. Each Ω_j is assumed to be a polyhedron. Let $\mathcal{S}_0 \triangleq \{j \in \mathcal{S} \mid 0 \in \text{cl}(\Omega_j)\}$, $\mathcal{S}_1 \triangleq \{j \in \mathcal{S} \mid 0 \notin \text{cl}(\Omega_j)\}$ and let $\mathcal{S}_{\text{aff}} \triangleq \{j \in \mathcal{S} \mid f_j \neq 0\}$, $\mathcal{S}_{\text{lin}} \triangleq \{j \in \mathcal{S} \mid f_j = 0\}$, so that $\mathcal{S}_0 \cup \mathcal{S}_1 = \mathcal{S}_{\text{aff}} \cup \mathcal{S}_{\text{lin}} = \mathcal{S}$.

The aim of this paper is to derive sufficient conditions for global ISpS and global ISS, respectively, of system (4). In order to do so, we consider PWQ candidate ISpS (ISS) functions of the form

$$V : \mathbb{R}^n \to \mathbb{R}_+,\ V(x) = x^\top P_j x \text{ if } x \in \Omega_j, \quad (5)$$

where P_j, $j \in \mathcal{S}$, are positive definite and symmetric matrices. It is easy to observe that V satisfies condition (3a) with $\alpha_1(\|x\|) \triangleq \min_{j\in\mathcal{S}} \lambda_{\min}(P_j)\|x\|^2$, $\alpha_2(\|x\|) \triangleq \max_{j\in\mathcal{S}} \lambda_{\max}(P_j)\|x\|^2$ and $d_1 = 0$.

3. ANALYSIS

In this section we present LMI based sufficient conditions for global ISpS (ISS) of system (4). Let Q be a known positive definite and symmetric matrix and let γ_1, γ_2 be known positive numbers with $\gamma_1\gamma_2 > 1$. For any $(j,i) \in \mathcal{S} \times \mathcal{S}$ consider now the following LMI:

$$\Delta_{ji} \triangleq \begin{pmatrix} \Xi_{ji} & -A_j^\top P_i & -A_j^\top P_i \\ -P_i A_j & \gamma_1 P_i & -P_i \\ -P_i A_j & -P_i & \gamma_2 P_i \end{pmatrix} > 0, \quad (6)$$

where

$$\Xi_{ji} \triangleq P_j - A_j^\top P_i A_j - E_j^\top U_{ji} E_j - Q - M_{ji}.$$

The matrix E_j, $j \in \mathcal{S}$, defines the cone $\mathscr{C}_j \triangleq \{x \in \mathbb{R}^n \mid E_j x \geq 0\}$ that satisfies $\Omega_j \subseteq \mathscr{C}_j$. The role of these matrices is to introduce an *S*-procedure relaxation (Johansson and Rantzer, 1998). The unknown variables in (6) are the matrices P_j, $j \in \mathcal{S}$, which are required to be positive definite and symmetric, the matrices U_{ji}, $(j,i) \in \mathcal{S} \times \mathcal{S}$, which are required to have non-negative elements, and the matrices M_{ji}, $(j,i) \in \mathcal{S}_{\text{aff}} \times \mathcal{S}$, which are required to be positive definite and symmetric. For all $(j,i) \in \mathcal{S}_{\text{lin}} \times \mathcal{S}$ we take $M_{ji} = 0$. For any $(j,i) \in \mathcal{S}_{\text{aff}} \times \mathcal{S}$, define

$$\mathscr{E}_{ji} \triangleq \{x \in \mathbb{R}^n \mid x^\top M_{ji} x < (1+\gamma_1) f_j^\top P_i f_j\}.$$

Theorem 5. Let system (4), the matrix $Q > 0$ and the numbers $\gamma_1, \gamma_2 > 0$ with $\gamma_1\gamma_2 > 1$ be given. Suppose that the LMIs

$$\Delta_{ji} > 0, \quad (j,i) \in \mathcal{S} \times \mathcal{S} \quad (7)$$

are feasible. Then, it holds that:

(i) The system (4) is globally ISpS;

(ii) If[2] $(\cup_{i\in\mathcal{S}}\mathscr{E}_{ji}) \cap \Omega_j = \emptyset$ for all $j \in \mathcal{S}_{\text{aff}}$, then system (4) is globally ISS;

(iii) If system (4) is Piece-Wise Linear (PWL), i.e. $\mathcal{S}_{\text{lin}} = \mathcal{S}$, then system (4) is globally ISS.

[2] Note that this implies $\mathcal{S}_0 \subseteq \mathcal{S}_{\text{lin}}$.

PROOF. The proof consists in showing that V, as defined in (5), is an ISpS (ISS) Lyapunov function.

(i) As by the hypothesis $\Delta_{ji} > 0$ for all $(j,i) \in \mathcal{S} \times \mathcal{S}$, it follows that:

$$\begin{pmatrix} x^\top & f_j^\top & (D_j v)^\top \end{pmatrix} \Delta_{ji} \begin{pmatrix} x \\ f_j \\ D_j v \end{pmatrix} \geq 0,$$

for all $x \in \Omega_j$, $(j,i) \in \mathcal{S} \times \mathcal{S}$ and all $v \in \mathbb{R}^{d_v}$. The above inequality yields:

$$\begin{aligned}
&(A_j x + f_j + D_j v)^\top P_i (A_j x + f_j + D_j v) - x^\top P_j x \\
&\leq -x^\top Q x + (1+\gamma_2)(D_j v)^\top P_i (D_j v) - x^\top E_j^\top U_{ji} E_j x + \\
&(1+\gamma_1) f_j^\top P_i f_j - x^\top M_{ji} x \leq -\lambda_{\min}(Q)\|x\|^2 + \\
&(1+\gamma_2) \max_{i\in\mathcal{S}} \lambda_{\max}(P_i) \max_{j\in\mathcal{S}} \|D_j\|^2 \|v\|^2 + \\
&(1+\gamma_1) \max_{i\in\mathcal{S}} \lambda_{\max}(P_i) \max_{j\in\mathcal{S}} \|f_j\|^2. \quad (8)
\end{aligned}$$

Hence,

$$V(A_j x + f_j + D_j v) - V(x) \leq -\alpha_3(\|x\|) + \sigma(\|v\|) + d_2$$

for all $x \in \Omega_j$, $(j,i) \in \mathcal{S} \times \mathcal{S}$ and all $v \in \mathbb{R}^{d_v}$, where

$$\begin{aligned}
\alpha_3(\|x\|) &\triangleq \lambda_{\min}(Q)\|x\|^2, \\
\sigma(\|v\|) &\triangleq (1+\gamma_2) \max_{i\in\mathcal{S}} \lambda_{\max}(P_i) \max_{j\in\mathcal{S}} \|D_j\|^2 \|v\|^2, \\
d_2 &\triangleq (1+\gamma_1) \max_{i\in\mathcal{S}} \lambda_{\max}(P_i) \max_{j\in\mathcal{S}} \|f_j\|^2.
\end{aligned}$$

From (6) we also have that for all $(j,i) \in \mathcal{S} \times \mathcal{S}$, $\Delta_{ji} > 0 \Rightarrow \Xi_{ji} > 0 \Rightarrow x^\top (P_j - Q) x \geq 0$ for all $x \in \Omega_j$. Then, it follows that for all $j \in \mathcal{S}$ and all $x \in \Omega_j$:

$$\lambda_{\min}(Q)\|x\|^2 \leq x^\top Q x \leq x^\top P_j x \leq \max_{j\in\mathcal{S}} \lambda_{\max}(P_j)\|x\|^2,$$

which yields $\lambda_{\min}(Q) \triangleq c \leq b \triangleq \max_{j\in\mathcal{S}} \lambda_{\max}(P_j)$. Hence, the function V defined in (5) satisfies the hypothesis of Theorem 2 with $d_1 = 0$ and $d_2 = (1+\gamma_1)\max_{i\in\mathcal{S}} \lambda_{\max}(P_i) \max_{j\in\mathcal{S}} \|f_j\|^2$. Then, the statement follows from Theorem 2.

(ii) To establish global ISS, we need to prove that in the above setting, we obtain $d_2 = 0$ under the additional hypothesis. For $j \in \mathcal{S}_{\text{lin}}$, if $x \in \Omega_j$ we obtain $d_2 = 0$ due to $f_j = 0$. For any $j \in \mathcal{S}_{\text{aff}}$, if $x \in \Omega_j$ it holds that $x \notin \cup_{i\in\mathcal{S}}\mathscr{E}_{ji}$. This yields:

$$(1+\gamma_1) f_j^\top P_i f_j - x^\top M_{ji} x \leq 0,$$

and thus, from the first inequality in (8) it follows that the function V defined in (5) satisfies the hypothesis of Theorem 2 with $d_1 = d_2 = 0$. Then, the statement follows from Theorem 2.

(iii) This is a special case of part (ii). □

The matrix Q gives the gain of the $\mathcal{K}$-function α_3 and is related to the decrease of the state norm, and hence, to the transient behavior. If ISpS (ISS) is the only goal, Q can be chosen less positive definite to reduce conservativeness of the LMI (7). The numbers γ_1, γ_2 and the matrices $\{P_j \mid j \in \mathcal{S}\}$ yield the constant

$d_2 = (1+\gamma_1)\max_{i\in\mathscr{S}}\lambda_{\max}(P_i)\max_{j\in\mathscr{S}}\|f_j\|^2$ and the gain of the $\mathscr{K}$-function

$$\sigma(s) = (1+\gamma_2)\max_{i\in\mathscr{S}}\lambda_{\max}(P_i)\max_{j\in\mathscr{S}}\|D_j\|^2 s^2.$$

Note that a necessary condition for feasibility of the LMI (7) is $\gamma_1\gamma_2 > 1$. As it would be desirable to obtain a constant d_2 and gain of the function σ as small as possible, one has to make a trade-off in choosing γ_1 and γ_2. One could add a cost criterion to (7) and specify γ_1, γ_2 as unknown variables in the resulting optimization problem, which might solve the trade-off. Although in this case (7) is a bilinear matrix inequality (i.e. due to $\gamma_1 P_i$, $\gamma_2 P_i$), since the unknowns γ_1, γ_2 are scalars, this problem can be solved efficiently via semi-definite programming solvers (software), e.g. (Sturm, 2001), (Löfberg, 2002), by setting lower and upper bounds for γ_1, γ_2 and doing bisections.

Remark 6. If the disturbance inputs are bounded, which is a reasonable assumption in practice, it can be proven that ISpS implies global ultimate boundedness. This means that the ISpS property also implies the usual robust stability (convergence) property, e.g. as the one defined in (Grieder, 2004, Chp. 8.5), while the result of Theorem 5 part (i) applies to a more general class of PWA systems.

4. SYNTHESIS

In this section we address the problem of input-to-state (practically) stabilizing controllers synthesis for perturbed discrete-time non-autonomous PWA systems:

$$x_{k+1} = g(x_k,u_k,v_k) \triangleq A_j x_k + B_j u_k + f_j + D_j v_k \quad \text{if } x_k \in \Omega_j, \qquad (9)$$

where $u_k \in \mathbb{R}^m$ is the control input and $B_j \in \mathbb{R}^{n\times m}$ for all $j \in \mathscr{S}$. The nomenclature in (9) is similar with the one used in Section 2 for system (4).

In this paper we take the control input as a PWL state-feedback control law of the form:

$$u_k \triangleq h(x_k) \triangleq K_j x_k \quad \text{if} \quad x_k \in \Omega_j, \qquad (10)$$

where $K_j \in \mathbb{R}^{m\times n}$ for all $j \in \mathscr{S}$. The aim is to calculate the feedback gains $\{K_j \mid j \in \mathscr{S}\}$ such that the PWA closed-loop system (9)-(10) is globally ISpS and ISS, respectively. For this purpose we make use again of PWQ candidate ISpS (ISS) Lyapunov functions of the form (5).

For any $(j,i) \in \mathscr{S}\times\mathscr{S}$, consider now the following LMI:

$$\Delta_{ji} \triangleq \begin{pmatrix} \Delta_{ji}^{11} & \Delta_{ji}^{12} \\ \Delta_{ji}^{21} & \Delta_{ji}^{22} \end{pmatrix} > 0, \qquad (11)$$

where

$$\Delta_{ji}^{11} \triangleq \begin{pmatrix} Z_j & * & * \\ -(A_jZ_j+B_jY_j) & \gamma_1 Z_i & -Z_i \\ -(A_jZ_j+B_jY_j) & -Z_i & \gamma_2 Z_i \end{pmatrix},$$

the term $*$ denotes $-(A_jZ_j+B_jY_j)^\top$ and, for $j \in \mathscr{S}_{\text{aff}}$

$$\Delta_{ji}^{22} \triangleq \text{diag}(\left[\begin{pmatrix} Z_i & 0 & 0 \\ 0 & Z_i & 0 \\ 0 & 0 & Z_i \end{pmatrix}, \begin{pmatrix} Q^{-1} & 0 & 0 \\ 0 & Q^{-1} & 0 \\ 0 & 0 & Q^{-1} \end{pmatrix}, \begin{pmatrix} N_{ji} & 0 & 0 \\ 0 & N_{ji} & 0 \\ 0 & 0 & N_{ji} \end{pmatrix}\right]),$$

$$\Delta_{ji}^{12} = \Delta_{ji}^{21\top} \triangleq$$

$$\triangleq \begin{pmatrix} (A_jZ_j+B_jY_j)^\top & 0 & 0 & Z_j & 0 & 0 & Z_j & 0 & 0 \\ 0 & 0 & 0 & 0 & 0 & 0 & 0 & 0 & 0 \\ 0 & 0 & 0 & 0 & 0 & 0 & 0 & 0 & 0 \end{pmatrix},$$

while for $j \in \mathscr{S}_{\text{lin}}$,

$$\Delta_{ji}^{22} \triangleq \text{diag}(\left[\begin{pmatrix} Z_i & 0 & 0 \\ 0 & Z_i & 0 \\ 0 & 0 & Z_i \end{pmatrix}, \begin{pmatrix} Q^{-1} & 0 & 0 \\ 0 & Q^{-1} & 0 \\ 0 & 0 & Q^{-1} \end{pmatrix}\right]),$$

$$\Delta_{ji}^{12} = \Delta_{ji}^{21\top} \triangleq$$

$$\triangleq \begin{pmatrix} (A_jZ_j+B_jY_j)^\top & 0 & 0 & Z_j & 0 & 0 \\ 0 & 0 & 0 & 0 & 0 & 0 \\ 0 & 0 & 0 & 0 & 0 & 0 \end{pmatrix}.$$

The operator $\text{diag}([L_1,\ldots,L_n])$ denotes a diagonal matrix of appropriate dimensions with the matrices $L_1,\ldots,L_n$ on the main diagonal, and the element 0 denotes everywhere a zero matrix of appropriate dimensions. The unknown variables in (11) are the matrices $Z_j \in \mathbb{R}^{n\times n}$, $j \in \mathscr{S}$, which are required to be positive definite and symmetric, the matrices $Y_j \in \mathbb{R}^{m\times n}$, $j \in \mathscr{S}$, and the matrices N_{ji}, $(j,i) \in \mathscr{S}_{\text{aff}}\times\mathscr{S}$, which are required to be positive definite and symmetric. The matrix Q is a known positive definite and symmetric matrix and the numbers $\gamma_1,\gamma_2 > 0$ with $\gamma_1\gamma_2 > 1$ play the same role as described in Section 3. For any $(j,i) \in \mathscr{S}_{\text{aff}}\times\mathscr{S}$, define

$$\mathscr{E}_{ji} \triangleq \{x \in \mathbb{R}^n \mid x^\top N_{ji}^{-1} x < (1+\gamma_1) f_j^\top P_i f_j\}.$$

Theorem 7. Let system (9), the matrix $Q > 0$ and the numbers $\gamma_1,\gamma_2 > 0$ with $\gamma_1\gamma_2 > 1$ be given. Suppose that the LMIs

$$\Delta_{ji} > 0, \quad (j,i) \in \mathscr{S}\times\mathscr{S} \qquad (12)$$

are feasible and let $\{Z_j, Y_j \mid j \in \mathscr{S}\}$ and $\{N_{ji} \mid (j,i) \in \mathscr{S}_{\text{aff}}\times\mathscr{S}\}$ be a solution. For all $j \in \mathscr{S}$ let $P_j \triangleq Z_j^{-1}$ and let $K_j \triangleq Y_j Z_j^{-1}$. For all $(j,i) \in \mathscr{S}_{\text{lin}}\times\mathscr{S}$ take $M_{ji} = 0$. For all $(j,i) \in \mathscr{S}_{\text{aff}}\times\mathscr{S}$ take $M_{ji} = N_{ji}^{-1}$. Then, it holds that:

(i) The closed-loop system (9)-(10) is globally ISpS;

(ii) If $(\cup_{i\in\mathscr{S}}\mathscr{E}_{ji}) \cap \Omega_j = \emptyset$ for all $j \in \mathscr{S}_{\text{aff}}$, then the closed-loop system (9)-(10) is globally ISS;

(iii) If system (9) is PWL, i.e. $\mathscr{S}_{\text{lin}} = \mathscr{S}$, then the closed-loop system (9)-(10) is globally ISS.

PROOF. By applying the Schur complement (Boyd *et al.*, 1994) to (12), for any $(j,i) \in \mathscr{S}\times\mathscr{S}$ we obtain

$$\Delta_{ji}^{11} - \Delta_{ji}^{21\top}\Delta_{ji}^{22^{-1}}\Delta_{ji}^{21} > 0,$$

which yields the equivalent matrix inequality:

$$\Phi_{ji} \triangleq \begin{pmatrix} \Gamma_{ji} & * & * \\ -(A_jZ_j+B_jY_j) & \gamma_1 Z_i & -Z_i \\ -(A_jZ_j+B_jY_j) & -Z_i & \gamma_2 Z_i \end{pmatrix} > 0, \quad (13)$$

where the term $*$ denotes $-(A_jZ_j+B_jY_j)^\top$ and

$$\Gamma_{ji} \triangleq Z_j - (A_jZ_j+B_jY_j)^\top Z_i^{-1}(A_jZ_j+B_jY_j) - Z_jQZ_j - Z_jN_{ji}^{-1}Z_j.$$

By pre- and post-multiplying (13) with $\begin{pmatrix} Z_j^{-1} & 0 & 0 \\ 0 & Z_i^{-1} & 0 \\ 0 & 0 & Z_i^{-1} \end{pmatrix}$ and by substituting Z_j^{-1} with P_j, $Y_jZ_j^{-1}$ with K_j and N_{ji}^{-1} with M_{ji} turns inequality (13) into the equivalent matrix inequality:

$$\begin{pmatrix} \Xi_{ji} & * & * \\ -P_i(A_j+B_jK_j) & \gamma_1 P_i & -P_i \\ -P_i(A_j+B_jK_j) & -P_i & \gamma_2 P_i \end{pmatrix} > 0,$$

for all $(j,i) \in \mathscr{S} \times \mathscr{S}$, where the term $*$ denotes $-(A_j+B_jK_j)^\top P_i$ and

$$\Xi_{ji} \triangleq P_j - (A_j+B_jK_j)^\top P_i(A_j+B_jK_j) - Q - M_{ji}.$$

Then, it follows that the LMI (7) is feasible for the closed-loop system (9)-(10) for all $(j,i) \in \mathscr{S} \times \mathscr{S}$. The rest of the proof is analogous to the proof of Theorem 5. □

5. ILLUSTRATIVE EXAMPLE

In this example we illustrate the result of Theorem 7 part (ii). Let

$$A(T_s) \triangleq \begin{pmatrix} 1 & T_s & \frac{T_s^2}{2!} & \frac{T_s^3}{3!} \\ 0 & 1 & T_s & \frac{T_s^2}{2!} \\ 0 & 0 & 1 & T_s \\ 0 & 0 & 0 & 1 \end{pmatrix}, \; B(T_s) \triangleq \begin{pmatrix} \frac{T_s^4}{4!} \\ \frac{T_s^3}{3!} \\ \frac{T_s^2}{2!} \\ T_s \end{pmatrix}$$

denote the dynamics corresponding to a discrete-time quadruple integrator, i.e. $x_{k+1} = A(T_s)x_k + B(T_s)u_k$, obtained from a continuous-time quadruple integrator via a sampled-and-hold device with sampling period $T_s > 0$. Let x_i, $i = 1,2,3,4$, denote the i-th component of the state vector. Let $\mathbb{X} \triangleq \{x \in \mathbb{R}^4 \mid -2 < x_4 < 2\}$, let $\Omega_1 \triangleq \{x \in \mathbb{R}^4 \mid x_4 \geq 2\}$ and let $\Omega_4 \triangleq \{x \in \mathbb{R}^4 \mid x_4 \leq -2\}$. Let $\Omega_2 \triangleq \{x \in \mathbb{X} \mid x_4 \geq 0\}$ and $\Omega_3 \triangleq \{x \in \mathbb{X} \mid x_4 < 0\}$. Consider now the following perturbed piecewise affine system:

$$x_{k+1} = \begin{cases} A_1x_k + B_1u_k + f_1 + D_1v_k & \text{if} \quad x_k \in \Omega_1 \\ A_2x_k + B_2u_k + f_2 + D_2v_k & \text{if} \quad x_k \in \Omega_2 \\ A_3x_k + B_3u_k + f_3 + D_3v_k & \text{if} \quad x_k \in \Omega_3 \\ A_4x_k + B_4u_k + f_4 + D_4v_k & \text{if} \quad x_k \in \Omega_4, \end{cases} \quad (14)$$

where $A_1 = A_4 = A(1.2), B_1 = B_4 = B(1.2)$, $A_2 = A(0.9), B_2 = B(0.9)$, $A_3 = A(0.8), B_3 = B(0.8)$, $f_2 = f_3 = 0$, $f_1 = f_4 = [0.1 0.1 0.1 0.1]^\top$ and $D_1 = D_2 = D_3 = D_4 = [1 1 1 1]^\top$. The LMIs (12) were solved[3] for $Q = 0.01I_4$, $\gamma_1 = 2$ and $\gamma_2 = 4$, yielding the following weights of the PWQ ISS Lyapunov function $V(x) = x^\top P_j x$ if $x \in \Omega_j$, $j = 1,2,3,4$, feedbacks $\{K_j \mid j = 1,2,3,4\}$ and matrix M:

$$P_1 = P_4 = \begin{bmatrix} 0.3866 & 0.7019 & 0.5532 & 0.1903 \\ 0.7019 & 1.5632 & 1.3131 & 0.4688 \\ 0.5532 & 1.3131 & 1.2255 & 0.4552 \\ 0.1903 & 0.4688 & 0.4552 & 0.1955 \end{bmatrix},$$

$$P_2 = \begin{bmatrix} 0.3574 & 0.6052 & 0.4420 & 0.1407 \\ 0.6052 & 1.2725 & 0.9894 & 0.3278 \\ 0.4420 & 0.9894 & 0.8812 & 0.3046 \\ 0.1407 & 0.3278 & 0.3046 & 0.1328 \end{bmatrix},$$

$$P_3 = \begin{bmatrix} 0.3779 & 0.6410 & 0.4597 & 0.1453 \\ 0.6410 & 1.3414 & 1.0298 & 0.3390 \\ 0.4597 & 1.0298 & 0.9007 & 0.3118 \\ 0.1453 & 0.3390 & 0.3118 & 0.1334 \end{bmatrix},$$

$$K_1 = K_4 = \begin{bmatrix} -0.3393 & -1.1789 & -1.8520 & -1.7028 \end{bmatrix},$$
$$K_2 = \begin{bmatrix} -0.5584 & -1.7607 & -2.4729 & -2.0012 \end{bmatrix},$$
$$K_3 = \begin{bmatrix} -0.6814 & -2.0895 & -2.8249 & -2.1705 \end{bmatrix},$$

$$M = \begin{bmatrix} 0.0156 & 0.0075 & 0.0023 & 0.0005 \\ 0.0075 & 0.0212 & 0.0082 & 0.0016 \\ 0.0023 & 0.0082 & 0.0146 & 0.0044 \\ 0.0005 & 0.0016 & 0.0044 & 0.0081 \end{bmatrix}.$$

One can easily establish that the hypothesis of Theorem 7 part (ii) is satisfied, i.e. $\mathscr{E}_{1i} \cap \Omega_1 = \emptyset$ and $\mathscr{E}_{4i} \cap \Omega_4 = \emptyset$ for all $i = 1,2,3,4$, by observing that

$$\begin{aligned} \min_{x\in\Omega_1} x^\top Mx = \min_{x\in\Omega_4} x^\top Mx &= 0.4340 \\ &> 0.3221 = \max_{i=1,2,3,4} (1+\gamma_1) f_1^\top P_i f_1 \\ &= \max_{i=1,2,3,4} (1+\gamma_1) f_4^\top P_i f_4. \end{aligned}$$

Hence, system (14) in closed-loop with (10) is globally ISS. The gain of the σ function corresponding to $\gamma_2 = 4$ is 15.8772. This yields an ISS gain equal to 42.52 for system (14)-(10) via the relation $\gamma(s) = \alpha_1^{-1}\left(\frac{2\sigma(s)}{1-\rho}\right) = 42.52s$ (see (Lazar *et al.*, 2006) for details), where $\rho = \frac{c}{b} \in [0,1)$ and γ is the $\mathscr{K}$-function from (2). The closed-loop states trajectories obtained for initial state $x_0 = [6 6 4 4]^\top$ are plotted in Figure 1 together with the additive disturbance input history. The disturbance input was randomly generated in the interval [0 1] until sampling time 60 and then it was set equal to zero. As guaranteed by Theorem 7, the closed-loop system (14)-(10) is globally ISS, which ensures asymptotic stability in the Lyapunov sense when the disturbance inputs converges to zero, as it can be observed in Figure 1.

---

[3] For simplicity we used a common matrix N for all possible mode transitions that can occur when the state is in mode one or mode four, i.e. $N = N_{11} = N_{12} = N_{13} = N_{44} = N_{42} = N_{43}$, which yields $M = N^{-1}$.

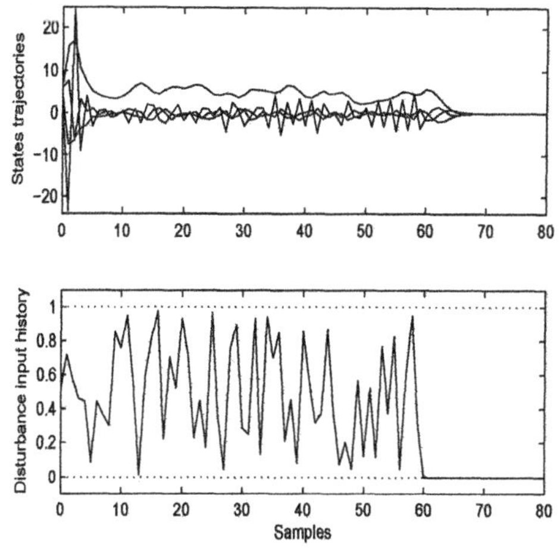

Fig. 1. States trajectories and disturbance input histories for the closed-loop system (14)-(10).

6. CONCLUDING REMARKS

In this paper we presented LMI based sufficient conditions for global input-to-state (practical) stability and stabilization of discrete-time perturbed, possibly discontinuous, PWA systems. The importance of these results cannot be overstated since, recently, in (Lazar and Heemels, 2006) the authors showed that nominally exponentially stable discrete-time PWA systems can have zero robustness to arbitrarily small additive disturbances and hence, special precautions must be taken when implementing stabilizing controllers for PWA systems in practice.

State and input constraints have not been considered in order to obtain global ISpS (ISS) results. However, the usual LMI techniques (Boyd *et al.*, 1994) for specifying state and/or input constraints can be added to the sufficient conditions presented in this paper, resulting in local ISpS (ISS) of constrained PWA systems. Also, a local (i.e. in some subset of $\cup_{j\in\mathcal{S}_0}\Omega_j$) ISS result is obtained under the hypothesis of Theorem 5 (Theorem 7) part (ii), in the case when $\mathcal{E}_{ji}\cap\Omega_j\neq\emptyset$ for some $(j,i)\in\mathcal{S}_{\text{aff}}\times\mathcal{S}$.

The future work deals with extensions to PWA systems affected by parametric uncertainties and the use of norm based candidate ISS Lyapunov functions.

REFERENCES

Boyd, S., L. El Ghaoui, E. Feron and V. Balakrishnan (1994). *Linear Matrix Inequalities in Control Theory*. Vol. 15 of *Studies in Applied Mathematics*. SIAM.

Cai, C. and A.R. Teel (2005). Results on input-to-state stability for hybrid systems. In: *Proceedings of the 44th IEEE Conference on Decision and Control*. Seville, Spain. pp. 5403–5408.

Feng, G. (2002). Stability analysis of piecewise discrete-time linear systems. *IEEE Transactions on Automatic Control* **47**(7), 1108–1112.

Ferrari-Trecate, G., F.A. Cuzzola, D. Mignone and M. Morari (2002). Analysis of discrete-time piecewise affine and hybrid systems. *Automatica* **38**(12), 2139–2146.

Grieder, P. (2004). Efficient computation of feedback controllers for constrained systems. PhD thesis. Swiss Federal Institute of Technology (ETH) Zürich. Switzerland.

Jiang, Z.-P. (1993). Quelques résultats de stabilisation robuste. Application á la commande. PhD thesis. École des Mines de Paris. France.

Jiang, Z.-P. and Y. Wang (2001). Input-to-state stability for discrete-time nonlinear systems. *Automatica* **37**, 857–869.

Jiang, Z.-P., I.M.Y. Mareels and Y. Wang (1996). A Lyapunov formulation of the nonlinear small-gain theorem for interconnected ISS systems. *Automatica* **32**(8), 1211–1215.

Johansson, M. and A. Rantzer (1998). Computation of piecewise quadratic Lyapunov functions for hybrid systems. *IEEE Transactions on Automatic Control* **43**(4), 555–559.

Lazar, M. and W. P. M. H. Heemels (2006). An example of zero robustness in piecewise affine systems. In: *25th BeNeLux Meeting on Systems and Control*. Heeze, The Netherlands. pp. 43–44.

Lazar, M., D. Muñoz de la Peña, W. P. M. H. Heemels and T. Alamo (2006). On the stability of min-max nonlinear model predictive control. In: *17th Symposium on Mathematical Theory of Networks and Systems*. Kyoto, Japan.

Löfberg, J. (2002). YALMIP: Matlab toolbox for rapid prototyping of optimization problems. Web: http://control.ee.ethz.ch/ joloef/yalmip.msql.

Mignone, D., G. Ferrari-Trecate and M. Morari (2000). Stability and stabilization of piecewise affine and hybrid systems: An LMI approach. In: *39th IEEE Conference on Decision and Control*. pp. 504–509.

Sontag, E.D. (1989). Smooth stabilization implies coprime factorization. *IEEE Transactions on Automatic Control* **AC–34**, 435–443.

Sturm, J.F. (2001). SeDuMi: Matlab toolbox for solving optimization problems over symmetric cones. Web: http://fewcal.kub.nl/.

Vu, L., D. Chatterjee and D. Liberzon (2005). ISS of switched systems and applications to switching adaptive control. In: *44th IEEE Conference on Decision and Control*. Seville, Spain. pp. 120–125.

ELSEVIER
IFAC
PUBLICATIONS

FEASIBLE MODE ENUMERATION AND COST COMPARISON FOR EXPLICIT QUADRATIC MODEL PREDICTIVE CONTROL OF HYBRID SYSTEMS

Alessandro Alessio Alberto Bemporad

Dip. Ingegneria dell'Informazione, University of Siena, Italy. Email: alessio,bemporad@dii.unisi.it.

Abstract For hybrid systems in piecewise affine (PWA) form, this paper presents a new methodology for computing the solution, defined over a set of (possibly overlapping) polyhedra, of the finite-time constrained optimal control problem based on quadratic costs. First, feasible mode sequences are determined via backward reachability analysis, and multiparametric quadratic programming is employed to determine candidate polyhedral regions of the solution and the corresponding value functions and optimal control gains. Then, the value functions associated with overlapping regions are compared in order to discard those regions whose associated control law is never optimal. The comparison problem is, in general, nonconvex and is tackled here as a DC (Difference of Convex functions) programming problem. *Copyright © 2006 IFAC*

1. INTRODUCTION

In the recent years, different methods for the design and the analysis of controllers for hybrid systems have been studied (see e.g. (Corona, 2005) and references therein). In particular, multiparametric programming techniques were proposed to synthesize state-feedback controllers defined over a set of polyhedral regions, by solving a finite-time optimal control problems explicitly with respect to the state and reference vectors.

(Bemporad *et al.*, 2000) proposed a procedure for synthesizing piecewise affine optimal controllers for discrete-time linear hybrid systems. A state feedback solution of a finite-time optimal control problem with performance criteria based on linear (1 or ∞) norms is obtained using multiparametric mixed-integer linear programming. A different approach based on dynamic programming was proposed in (Baotic *et al.*, 2003). The use of linear norms has some practical disadvantages, due to the fact that typically good performance can only be achieved with long time horizons. Moreover, the resulting state-space partition is typically very complex, because of the large number of regions.

Quadratic costs allow one to achieve better performances with shorter horizons, although the partition associated with the fully explicit optimal solution to a finite time constrained optimal control (FTCOC) problem for hybrid systems may not be polyhedral (Borrelli *et al.*, 2005).

(Borrelli *et al.*, 2005) proposed an algorithm for computing the solution to the FTCOC problem with quadratic costs. The procedure is based on dynamic programming (DP) iterations. Multiparametric quadratic programs (mpQP) (Bemporad *et al.*, 2002) are solved at each iteration, and quadratic value functions are compared to possibly eliminate regions that are proved to never be optimal. In typical situations the total number of solved mpQPs (as well as of generated polyhedral regions) grows exponentially, and suffers the drawback of an excessive partitioning of the state space.

A different approach was proposed in (Mayne, 2001; Mayne and Rakovic, 2002), where the authors propose to enumerate all possible switching sequences, and for each sequence convert the PWA dynamics into a time-varying system and solve an optimal control problem explicitly via mpQP. As any given initial state may lie in more than one polyhedral region, the associated control gain giving the smallest cost needs to be selected by on-line comparison. This leads to an exponential number of mpQPs that need to be solved and a possibly large *on-line* CPU time spent for comparing the cost functions.

In this paper we propose a different approach that exploits dynamic programming ideas (more precisely, backwards reachability analysis) to obtain all the feasible mode sequences (therefore avoiding an explicit enumeration of all of them), and that, after solving an mpQP for each sequence, post-processes the resulting polyhedral partitions to eliminate all the regions (and their associated control gains) that never provide the lowest cost, using a novel DC (Difference of Convex functions) algorithm. The resulting number of total regions that needs to be stored is minimized, and therefore the CPU time needed by the on-line procedure for searching the region with minimum cost is reduced.

2. HYBRID MPC SETUP

Consider the *Piecewise Affine System* (PWA) described by the relations

$$\begin{aligned} x(k+1) &= A_i x(k) + B_i u(k) + f_i, \\ y(k) &= C_i x(k) + g_i \end{aligned} \quad \text{if } \begin{bmatrix} x(k) \\ u(k) \end{bmatrix} \in \mathcal{X}_i \tag{1}$$

where $\{\mathcal{X}_i\}_{i=1}^{s}$ is a polyhedral partition of the state+input set. Suppose there are no binary states and inputs so that $x(k) \in \mathbb{R}^n$, $u(k) \in \mathbb{R}^m$, and A_i, B_i, f_i, C_i, g_i are matrices of suitable dimension [1]. Hybrid systems of the form (1) can be obtained for instance by system identification tools or by converting HYSDEL models using the method of (Bemporad, 2004).

Model Predictive Control (MPC) ideas can be applied to control hybrid models of the form (1). Here, at each sampling time, an open-loop optimal control problem is solved over a finite horizon N. Only the first sample of the optimal sequence is then applied to the plant at time k. At the next time step, a new optimal control problem based on new measurements of the state is solved over a shifted horizon. The solution relies on the hybrid model (1) of the system dynamics, minimizes a performance figure, and respects all input, output and state constraints.

For simplicity of notation, assume that we want to regulate the system state to the origin. So, the MPC open-loop optimal control problem can be formulated as follows

$$V^*(x(0)) = \min_U J(U, x(0))$$
$$J(U, x(0)) = x'(N)Px(N) + \sum_{k=0}^{N-1} x'(k)Qx(k) + u'(k)Ru(k) \tag{2a}$$

$$\begin{aligned} \text{s.t. } & \text{PWA Model (1)} \\ & x_{\min} \le x(k) \le x_{\max}, \\ & y_{\min} \le y(k) \le y_{\max}, \quad k = 0, \ldots, N-1, \\ & u_{\min} \le u(k) \le u_{\max}, \\ & x(N) \in \mathcal{X}^N, \end{aligned} \tag{2b}$$

where N is the control horizon, and $U \triangleq \{u(k),\ u(k+1),\ u(k+N-1)\}$ is the input sequence to be optimized. The bounds $u_{\min}$, $u_{\max}$, $x_{\min}$, $x_{\max}$, $y_{\min}$, $y_{\max}$ impose limits on inputs, states, and outputs, respectively, P is a weight on the terminal state, and $\mathcal{X}^N$ is a terminal set contained in the box $\{x : x_{\min} \le x \le x_{\max}\}$.

3. EXPLICIT SOLUTION FOR A FIXED MODE SEQUENCE

Problem (2) is usually referred to as the Finite Time Constrained Optimal Control (FTCOC) based on quadratic costs (Borrelli *et al.*, 2005). With an MPC synthesis in mind, our goal is to find the first optimal move $u^*(x(0))$ as a function of the initial state $x(0)$. While for *a given* $x(0)$ the input $u^*(0)$ can be determined *on-line* by solving a mixed-integer quadratic program (Bemporad and Morari, 1999), determining the solution *for all* vectors $x(0)$ within a given polytopic set $\mathcal{X}(0)$ of states of interest and *off-line* is a much harder one (Mayne, 2001; Borrelli *et al.*, 2005; Mayne and Rakovic, 2002). Once the optimal control law is obtained explicitly, on-line computation is reduced to a simple function evaluation.

The problem can be decomposed in a certain number of sub-problems that are easier to solve by exploiting the properties of the hybrid model (1). Starting from a given initial state $x(0)$ and by applying a given input sub-sequence $\{u(0),\ \ldots,\ u(k-1)\}$, the state of the system $x(k)$ belongs to a certain polyhedron $\mathcal{X}_{i(k)}$ of the partition, where $i(k)$ is the mode entered by the hybrid model at time k, $i(k) \in \{1, \ldots, s\}$. We refer to $v = \{i(0),\ \ldots,\ i(N-1)\}$ as the *switching* (or *mode*) *sequence*, and to $v^k = i(k)$ as the $(k+1)$-th element of that sequence, so that $v^k = j$

[1] The formulation and the results of this paper can be immediately extended when some of the input/state components are binary.

means that $\begin{bmatrix} x(k) \\ u(k) \end{bmatrix} \in \mathcal{X}_j$. The maximum number of possible switching sequences is $q \triangleq s^N$. Once a switching sequence v_i is fixed, system (1) is forced to enter the modes defined by v_i and becomes a linear time-varying system.

For a fixed switching sequence v_i, $i \in \{1, \ldots, q\}$, problem (2) becomes

$$J_{v_i}(x(0)) \triangleq \min_U J(U, x(0))$$
$$\text{s.t.} \quad \begin{cases} x(k+1) = A_{v_i^k} x(k) + B_{v_i^k} u(k) + f_{v_i^k}, \\ y(k) = C_{v_i^k} x(k) + D_{v_i^k} u(k) + g_{v_i^k}, \\ \begin{bmatrix} x(k) \\ u(k) \end{bmatrix} \in \mathcal{X}_{v_i^k}, \quad k = 0, \ldots, N-1, \\ x_{\min} \leq x(k) \leq x_{\max} \\ y_{\min} \leq y(k) \leq y_{\max}, \; k = 0, \ldots, N-1, \\ u_{\min} \leq u(k) \leq u_{\max}, \\ x(N) \in \mathcal{X}^N. \end{cases} \tag{3}$$

Problem (3) is an optimal control problem with finite horizon N for a constrained time-varying system, and can be solved via multiparametric quadratic programming, where U are the optimization variables and $x(0) \in \mathcal{X}(0)$ are the parameters.

For all $i = 1, \ldots, q$, the solution of the optimal control problem (3) is a PPWA state feedback control law of the form (Bemporad *et al.*, 2002)

$$u_i^*(x(0)) = F_j^i x(0) + G_j^i, \quad \forall x(0) \in T_j^i, j = 1, \ldots, N_{ri} \tag{4}$$

where $\mathcal{D}_i \triangleq \bigcup_{j=1}^{N_{ri}} T_j^i$ is a convex polyhedron, corresponding to the set of states $x(0)$ for which problem (3) admits a feasible solution. The sub/superscript i in (4) means that this solution is valid for a certain fixed sequence v_i. The optimal solution $u^*(x(0))$ to Problem (2) can be found by solving problem (3) for all feasible sequences v_i, as suggested in (Mayne and Rakovic, 2002), (Mayne, 2001), and then by comparing the costs $J_{v_i}(x(0))$ on-line, given the current state $x(0)$. The optimal set $\mathcal{D}^0$ of the states $x(0)$ for which (2) admits a feasible solution is

$$\mathcal{D}_0 = \bigcup_{i=1}^{q} \mathcal{D}_i, \tag{5}$$

and, in general, is not convex.

All polyhedra T_j^i needs to be analyzed. If $T_j^i \cap T_m^l = \emptyset$ for all $l \neq i$, $l = 1, \ldots, q$, $m = 1, \ldots, N_{rl}$, then the switching sequence v_i is the only feasible one for all the states $x(0) \in T_j^i$, and so the optimal solution $u^*(x(0))$ is given by (4). We will refer to T_j^i as a *polyhedron of single feasibility*. It can happen, however, that some initial states belong to more than one set $\mathcal{D}_i$, so we need to compare the cost functions J_{v_i} in order to choose the optimal control gains (F_j^i, G_j^i). If T_j^i intersects one or more polyhedra, then the states belonging to the intersection are feasible for more than one switching sequence and the corresponding value functions need to be compared in order to compute the optimal control law. In the simple case when only two polyhedra overlap, for all states belonging to $T_j^i \cap T_m^l$ the optimal move $u^*(x(0))$ for problem (2) is

$$u^*(x(0)) = \begin{cases} F_j^i x(0) + G_j^i & \text{if} J_{v_i}^*(x(0)) < J_{v_l}^*(x(0)) \\ F_m^l x(0) + G_m^l & \text{if} J_{v_i}^*(x(0)) > J_{v_l}^*(x(0)) \\ \begin{cases} F_j^i x(0) + G_j^i \\ \text{or} \\ F_m^l x(0) + G_m^l \end{cases} & \text{if} J_{v_i}^*(x(0)) = J_{v_l}^*(x(0)). \end{cases} \tag{6}$$

A polyhedron of *multiple feasibility* [2] . on which n value functions intersect may be split into at most n possibly nonconvex subsets where in each one of them a certain value function is smaller than all the others. Because $J_{v_i}^*(x(0))(i = 1, \ldots, q)$ are quadratic functions on $T_j^i (j = 1, \ldots, N_{ri})$ the closure of the sets corresponding to the optimal state partition, in general, has the form (Borrelli *et al.*, 2005)

$$\bar{\mathcal{R}}_k^i \triangleq \{x : x' L_k^i(j) x + M_k^i(j) x \leq N_k^i(j)\}. \tag{7}$$

In this paper we avoid splitting regions that overlap and storing non-polyhedral sets, but rather keep all polyhedra T_j^i for which the corresponding cost $J_{v_i}^*(x(0))$ is optimal for at least one state $x(0)$, leaving the cost comparison to the on-line procedure. This approach allows one to save memory space (no split implies less regions to store), at the price of a slightly increased on-line CPU time for the evaluation of the control move, because if $x(0)$ belongs to a region of multiple feasibility, the costs corresponding to all overlapping regions where $x(0)$ belong must be computed and compared.

4. ENUMERATION OF FEASIBLE MODE SEQUENCES VIA BACKWARDS REACHABILITY ANALYSIS

Computing the optimal solution via enumeration of all possible switching sequences can be too onerous, as the number of mp-QPs that need to be solved is $q = s^N$. Also, the set $\mathcal{D}_i$ of states $x(0)$ for which problem (3) has a solution may be empty for many switching sequences v_i.

The list of all (and only) sequences that are feasible for problem (3) can be obtained by solving a backwards reachability analysis problem as described below.

Assume that the terminal polyhedral set $\mathcal{X}^N$ is contained in one of the regions $\mathcal{X}_j$ of the polyhedral partition of system (1), i.e., $\mathcal{X}^N \subseteq \mathcal{X}_j$,

---

[2] In general, we say that a polyhedron T_j^i is of *multiplefeasibility* if it has a non-empty intersection with one or more polyhedra $T_m^l, (i, j \neq l, m)$ belonging to a different solution of the form (4).

for some $j \in \{1,\ldots,s\}$ (in case $\mathcal{X}^N$ overlaps with more than one region of the PWA partition, one needs to consider all the nonempty intersections $\mathcal{X}_N \cap \mathcal{X}_i$, $i = 1,\ldots,s$). Next, for each mode $i = 1,\ldots,s$ we determine which polyhedral subsets of $\mathbb{R}^{n+m}$ defined by the linear inequalities

$$\begin{cases} A_i x_{N-1} + B_i u_{N-1} + f_i \in \mathcal{X}^N \\ [x'_{N-1}\ u'_{N-1}]' \in \mathcal{X}_i \\ x_{\min} \le x_{N-1} \le x_{\max} \\ y_{\min} \le C_i x_{N-1} + D_i u_{N-1} + g_i \le y_{\max} \\ u_{\min} \le u_{N-1} \le u_{\max} \end{cases} \quad (8)$$

are nonempty (this just requires a phase-1 of a linear program). Let $\mathcal{X}_j^{N-1}$, $j = 1,\ldots,k_{N-1}$ be such nonempty sets. At the next step of the backwards reachability analysis, for each $j = 1,\ldots,k_{N-1}$ and for each mode $i = 1,\ldots,s$ we determine which polyhedral subsets of $\mathbb{R}^{n+2m}$ defined by the linear inequalities

$$\begin{cases} [(A_i x_{N-2} + B_i u_{N-2} + f_i)'\ u'_{N-1}]' \in \mathcal{X}_j^{N-1} \\ [x'_{N-2}\ u'_{N-2}]' \in \mathcal{X}_i \\ x_{\min} \le x_{N-2} \le x_{\max} \\ y_{\min} \le C_i x_{N-2} + D_i u_{N-2} + g_i \le y_{\max} \\ u_{\min} \le u_{N-2} \le u_{\max} \end{cases} \quad (9)$$

are nonempty. By letting $\mathcal{X}_j^{N-2}$, $j = 1,\ldots,k_{N-2}$ be such nonempty sets, the procedure is repeated backwards until the time index reaches 0.

The switching sequences $v_1,\ldots,v_{\bar{q}}$, where $\bar{q} = k_0 \le s^N$ are all and only the switching sequences for system (1) that satisfy the constraints in (3) for at least one initial state $x(0)$ and input sequence $u(0),\ldots,u(N-1)$, and that will be referred to as the *feasible* switching sequences.

The above procedure is successfully implemented in the Hybrid Toolbox for Matlab (Bemporad, 2003).

5. COST COMPARISONS AND REGION ELIMINATION

The main problem (FTCOC) has been decomposed in $\bar{q}$ subproblems, depending on the number of feasible switching sequences.

Every subproblem (3), once solved via multiparametric quadratic programming, gives a PPWA control law of the form (4) and an associated optimal cost function $J^*_{v_i}(x(0))$ that is convex, continuous, and piecewise quadratic (PWQ) on the same partition.

By solving the problem for every feasible switching sequence v_i, we obtain $\bar{q}$ state partitions $\mathcal{D}_i$ that need to be compared in order to find the optimal solution of Problem (2). For a given $x(0)$, the optimal input $u^*(0)$ is obtained comparing every cost function $J^*_{v_1}(x(0)),,\ldots,J^*_{v_{\bar{q}}}(x(0))$, and find the associated control input at minimum cost.

The main problem is that, in the worst case, the number of possible comparisons that need to be made on line in order to find the minimum cost is $(\bar{q}-1)$, and so the main advantage of saving on-line CPU time by calculating the control law off line may be lost. In addition, typically there are several regions whose associated control law is never the optimal one. A region of multiple feasibility T^i_j is *dominated* if

$$\forall x \in T^i_j, \exists l \in \{1,\ldots,\bar{q}\}, m \in \{1,\ldots,N_{rl}\} : \\ x \in T^i_j \cap T^l_m, J^*_{v_l}(x) < J^*_{v_i}(x). \quad (10)$$

otherwise it is considered *optimal*, since it exists at least one vector where its corresponding function $J^*_{v_i}(x)$ is optimal. It is desirable to eliminate all dominated regions T^i_j and the related cost functions in order to avoid a useless waste of CPU time for searching the region with minimum cost, and of memory for storing dominated regions. In other words, we want to keep only the regions T^i_j that are certainly optimal in a certain subset of the state set $\mathcal{D}(0)$.

5.1 Determination of Polyhedra of Single Feasibility

We first locate all the regions T^i_j for which

$$\exists \bar{x} \in \mathbb{R}^m : \bar{x} \in T^i_j, \bar{x} \notin T^l_m, \forall l \neq i,\ l = 1,\ldots,\bar{q}, \quad (11)$$

where clearly $\bar{x} \in T^i_j \Rightarrow \bar{q} \notin T^i_m$, $\forall m = 1,\ldots,N_{ri}$, $m \neq j$.

Regions T^i_j satisfying (11) do not need to be tested for domination by other regions, as they are clearly optimal in at least one point $\bar{x} \in \mathbb{R}^m$. Condition (11) can be tested by solving the following MILP for all $i = 1,\ldots,\bar{q}$, $\forall j = 1,\ldots,N_{ri}$:

$$\begin{aligned} \min_{x,\delta}\ & 0 \\ \text{s.t.}\ & \sum_{r=1}^{N_h} \delta_{hr} \le N_h - 1, \quad \forall h = 1,\ldots,\bar{q},\ h \neq i \\ & A^i_j x \le b^i_j, \\ & H^r_h x - K^r_h > m\delta, \\ & H^r_h x - K^r_h \le M(1-\delta), \\ & \delta_{hr} \in \{0,1\},\ \forall h = 1,\ldots,\bar{q},\ \forall r = 1,\ldots,N_h \end{aligned} \quad (12)$$

where $\bar{q}$ is the total number of *envelopes* $\mathcal{D}_i$ (that is, of the switching sequences v_i for which (3) is feasible), (A^i_j, b^i_j) defines the region T^i_j that needs to be tested, N_h represents the total number of facets of the envelope of the h-th partition $\mathcal{D}_h$, and $r = 1,\ldots,N_h$ is its r-th facet and finally m, M are chosen such that

$$m < \min_{x \in \mathcal{X}(0)} H^r_h x - K^r_h, \quad M \ge \max_{x \in \mathcal{X}(0)} H^r_h x - K^r_h.$$

The binary variables δ_{hr} satisfy the condition $[\delta_{hr} = 1] \leftrightarrow [H^r_h x \le K_h]$, i.e., $\delta_{hr} = 1$ iff $x \in T^i_j$

satisfies the constraint that defines that r-th facet of the h-th envelope $\mathcal{D}_h$, and the first constraint in (12) imposed that at least one facet inequality is violated, so that $x \notin \mathcal{D}_h$. Regions T_j^i for which (12) is feasible are regions of single feasibility and therefore optimal, so they must be retained in the final hybrid MPC control law.

5.2 DC Programming Approach

In order to find the optimal regions, we need to compare quadratic functions over certain convex sets of parameters, for this reason it can be recast as a DC (Difference of Convex functions) problem (Horst and Thoai, 1999).

For every region T_m^l for which (12) is infeasible, we need to determine all the partitions $\mathcal{D}_k$, $k \neq l$, such that $T_m^l \cap \mathcal{D}_k \neq \emptyset$. Clearly, $T_m^l \subseteq \bigcup_{k \neq l} \mathcal{D}_k$. In general, T_m^l intersects $s_m^l < \bar{q}$ partitions $\mathcal{D}_k$. By letting $\mathcal{S} = \{1, .., \bar{q}\}$, we will refer to $\mathcal{S}_m^l$ as the subset of $\mathcal{S}$ of indices $k \neq l$, such that $T_m^l \cap \mathcal{D}_k \neq \emptyset$. Clearly $s_m^l = \text{card}\ (\mathcal{S}_m^l)$. We assume here for simplicity that $T_m^l \subseteq \mathcal{D}_k$, $\forall k \in \mathcal{S}_m^l$. This assumption will be removed shortly.

For every fixed switching sequence v_l, the optimal solution $J_{v_l}^*$ obtained by solving the associated mpQP problem (3) is piecewise quadratic (PWQ) over the polyhedral partition $\mathcal{D}_l$, and quadratic in every single region T_m^l. In the sequel we will refer to $V_m^l(x)$ as the quadratic term of the value function $J_{v_l}^*$ in the m-th region of the l-th partition. A region T_m^l is *optimal* if

$$\exists x^* \in T_m^l : V_m^l(x^*) - J_{v_k}^*(x^*) < 0, \forall k \in \mathcal{S}_m^l. \quad (13)$$

Condition (13) can be verified by solving the following DC programming problem for all $k \in \mathcal{S}_m^l$

$$T_{lmk}^* = \min_{x \in T_m^l} V_m^l(x) - J_{v_k}^*(x). \quad (14)$$

If $T_{lmk}^* > 0$ for some $k \in \mathcal{S}_m^l$, region T_j^i is certainly dominated (i.e., not optimal) and can be safely discarded. The DC problem (14) is a nonconvex problem. On the other hand, we do not necessarily need to find its optimal solution, but a positive lower bound on the minimum would suffice for checking condition (13). In the next sections we describe a procedure for computing such a lower bound in an arbitrarily tight manner.

In the more general case where $T_m^l \nsubseteq \mathcal{D}_k$ for some $k \in \mathcal{S}_m^l$, for all such indices k the quadratic and piecewise quadratic costs are compared over the subset $\Omega_m^{l,k} \triangleq T_m^l \cap \mathcal{D}_k$. In this case, one can conclude that the region T_m^l is dominated if and only if all its subsets $\Omega_m^{l,k}$ are dominated.

5.3 DC Algorithm

In order to simplify the notation, given a region T_m^l of multiple feasibility and a partition $\mathcal{D}_k$, $k \in \mathcal{S}_m^l$, we will refer to $V_m^l(x)$ and $J_{v_k}^*(x)$ as $f_1(x)$ and $f_2(x)$, respectively.

Now, suppose to compute two PPWA functions $\bar{f}_1$ and $\bar{f}_2$ such that

$$\exists \epsilon_i > 0 : f_i(x) \leq \bar{f}_i(x),\ \epsilon_i = \max_{x \in \Omega_m^{l,k}} (f_i(x) - \bar{f}_i(x)),\ i = 1, 2. \quad (15)$$

Clearly, the following relations

$$f_1(x) - \bar{f}_2(x) \leq f_1(x) - f_2(x) \leq \bar{f}_1(x) - f_2(x) \quad (16)$$

are verified $\forall x \in \Omega_m^{l,k}$.

Now define LB_k, $UB_k \in \mathbb{R}$ as the solutions of the quadratic programs

$$LB_k = \min_{x \in \Omega_m^{l,k}} J_1(x) = f_1(x) - \bar{f}_2(x) \quad (17a)$$

$$UB_k = \max_{x \in \Omega_m^{l,k}} J_2(x) = \bar{f}_1(x) - f_2(x) \quad (17b)$$

By (16)–(17), it follows that

$$LB_k \leq \min_{x \in \Omega_m^{l,k}} f_1(x) - f_2(x) \leq UB_k, \quad \forall k \in \mathcal{S}_m^l. \quad (18)$$

If $UB_k < 0$, $\forall k \in \mathcal{S}_m^l$, then Condition (13) is satisfied (region T_m^l is optimal), while if $LB_{\bar{k}} > 0$ for some $k \in \mathcal{S}_m^l$, then T_m^l is dominated by partition $\mathcal{D}_k$. In the other cases, one needs to subpartition $\Omega_m^{l,k}$ in order to obtain tighter upper and lower bounds, as described in the following algorithm.

---

function SignTest ($f1$, $f2$, $\Omega_m^{l,k}$)

(1) Obtain an initial triangulation of $\Omega_m^{l,k}$ in simplices S_i, ($i = 1, \ldots, N_s$) via Delaunay triangulation (Yepremyan and Falk, 2005);
(2) Optimal := True;
(3) For ($k \in \mathcal{S}_m^l$) and (Optimal=True) Do,
  (a) Optimal := False;
  (b) For $i = 1$ to N_s Do, *loops over S_i *\
    (i) $\{LB_k, UB_k\}$ = Bounds(S_i, True);
      * Compute UB_k, LB_k as in (17a)-(17b) over S_i; *\
    (ii) If $UB_k < 0$ then
      Optimal=true; * $f_1(x) < f_2(x), \forall x \in S_i$ *\
      Choose another partition $\mathcal{D}_k$, $k \in \mathcal{S}_m^l$;
      break;
(4) If (Optimal = False) then 'T_m^l is dominated by $\mathcal{D}_k$, otherwise T_m^l is optimal

---

function Bounds(Simplex S,boolean turn)

(1) If (Turn = true)
  (a) Solve problem (17a).
  (b) Set LB = LB_k and $\bar{x} = x^* = \arg\min_x J_1(x)$.
  (c) If $LB > 0$, return;
  (d) for (k=0 to n) do
    (i) Substitute the k-th vertex of S with $\bar{x}$, and obtain $n+1$ new simplices $S_0, \ldots, S_n$;

```
     (ii) Turn := false;
    (iii) Bounds(S_k, Turn);

(2) Otherwise
   (a) Solve problem (17b).
   (b) Set UB = UB_k and x̄ = x* = arg max_x J_2(x).
   (c) If UB < 0, stop
       the region is optimal;
   (d) Else for (k=0 to n) do
       (i) Substitute the k-th vertex of S with x̄,
           and obtain n+m+1 new simplices S_0,...,S_n
      (ii) Turn := true;
     (iii) Bounds(S_k, Turn);
```

Remark 1. The algorithm computes an initial simplicial partition $S_0, .., S_n$ of the given set $\Omega_m^{l,k}$ and solves the two QPs defined in (17a)-(17b) over S_i, $i = 0, .., n$. Whatever none of the two conditions is satisfied over the current simplex, it proceeds recursively, by splitting every simplex into $n+1$ simplices, adding a new vertex $\bar{x} = \arg\min_x\{J_1(x), \max_x J_2(x)\}$, until any of the conditions (17a)-(17b) is satisfied. The algorithm stops splitting the initial set of simplices when it finds a point where $UB < 0$. Note that the generated simplices are only needed to compare cost functions, and hence are discarded immediately after the comparison. In particular, they are not at all needed to store the control law.

5.4 Upper-approximation of the Value Function

Under the assumptions made in (3), the optimal k-th mpQP solution $J^*_{v_k}(x)$, $k \in S^l_m$, is a convex (piecewise quadratic) function, defined over the convex full-dimensional set of parameters $\mathcal{D}_k \subseteq \mathbb{R}^n$ (Mangasarian and Rosen, 1964). A complete reference for the algorithm used for computing an upper-approximation in piecewise affine form over a simplicial partition of a convex (piecewise quadratic) function can be found in (Bemporad and Filippi, 2006)

5.5 Reduction of Partially Dominated Regions

When a region is only *partially* dominated, that is, if a polyhedral subset $\bar{\Omega}$ of a *optimal* region T^l_m is dominated by a certain partition $\mathcal{D}_k$, it may be desirable to reduce T^l_m to a smaller region that does not contain $\bar{\Omega}$.

Definition 1. A matrix-vector pair (A, b) is a *minimal representation* of a polyhedron $P = \{x : Ax \le b\}$, if there does not exist a pair (A_1, b_1) defining the same polyhedron and such that $\dim(b_1) < \dim(b)$.

Lemma 1. Given two nonempty polyhedra P, Q and their minimal representations $P = \{x \in \mathbb{R}^n : Ax \le b\}, Q = \{x \in \mathbb{R}^n : Cx \le d\}$, the set $P \backslash Q \triangleq \{x \in \mathbb{R}^n : x \in P, x \notin Q\}$ is nonconvex if and only if the number of hyperplanes $c_j x \le d_j \in \partial G$ which intersect the interior of P is greater or equal than two.

Proof: Suppose that $c'_1 x = d_1$ and $c'_2 x = d_2$ are two hyperplanes of Q which intersect the interior of P in two points x_1, x_2, respectively.

Let $[x_1, x_2]$ denote the line segment $\{x \in \mathbb{R}^n : x = \lambda x_1 + (1-\lambda)x_2, 0 \le \lambda \le 1\}$, which is entirely contained in Q. Since x_1, x_2 belong to the interior of P then there exist two scalars $\theta_1 > 1$, $\theta_2 < 0$ such that $\bar{x}_1$ and $\bar{x}_2$ defined as $\bar{x}_1 = \theta_1 x_1 + (1-\theta_1)x_2, \bar{x}_2 = \theta_2 x_1 + (1-\theta_2)x_2$ belong to the interior of P. Since $c'_1\bar{x}_1 = \theta_1 c'_1 x_1 + (1-\theta_1)c'_1 x_2 > \theta_1 d_1 + (1-\theta_1)d_1 = d_1$, then $\bar{x}_1 \notin Q$, and similarly one can show that $x_2 \notin Q$. Hence, $\bar{x}_1$, $\bar{x}_2 \in P\backslash Q$. Consider the convex combination of $\bar{x}_1$, $\bar{x}_2$ defined as $\gamma\bar{x}_1 + (1-\gamma)\bar{x}_2$, $\gamma \in (0,1)$. We want to show that there exists a $0 < \gamma < 1$ such that $\gamma\bar{x}_1 + (1-\gamma)\bar{x}_2 \in Q$, and hence does not belong to $P\backslash Q$. Since $\gamma\bar{x}_1 + (1-\gamma)\bar{x}_2 = \gamma(\theta_1 x_1 + (1-\theta_1)x_2) + (1-\gamma)(\theta_2 x_1 + (1-\theta_2)x_2) = (\gamma\theta_1 + (1-\gamma)\theta_2)x_1 + (1-(\gamma\theta_1+(1-\gamma)\theta_2))x_2$ is a linear combination of x_1 and x_2, and since the open segment (x_1, x_2) is contained in the interior of Q, there exists $\bar{\alpha}$ such that $x = \bar{\alpha}x_1 + (1-\bar{\alpha})x_2 \in Q$, $0 < \bar{\alpha} < 1\}$. By setting $\bar{\alpha} = \gamma\theta_1 + (1-\gamma)\theta_2$ and by choosing any γ such that

$$0 < \frac{-\theta_2}{\theta_1 - \theta_2} < \gamma < \frac{1-\theta_2}{\theta_1 - \theta_2} < 1 \qquad (19)$$

it follows that $x \notin P\backslash Q$, which proves that $P\backslash Q$ is not convex.

In the same way we can show that $P\backslash Q$ is not convex if the number of hyperplanes is more than two, since it is enough to repeat the above argument for every pair of inequalities defined by $(c_i, d_i), (c_k, d_k)$. On the other hand, if only one hyperplane of Q intersects the interior of P the resulting set $P\backslash Q$ is the intersection of convex sets, and therefore convex, or if no hyperplane of Q intersects P then $P\backslash Q = P$ is also convex. □

Thanks to Lemma 1, one can reduce all regions T^l_m that are partially dominated by partitions $\mathcal{D}_k$ that intersect T^l_m with at most two hyperplanes. In this way, on-line computations are possibly simplified because of the reduced overlaps among the regions of the controller's partition.

6. EXAMPLE

Consider the following system

$$x(k+1) = \begin{cases} A_1 x(k) + B_1 u(k) & \text{if } x_2(k) + x_3(k) < 0, \\ & |x_1(k)| < 2 \\ A_2 x(k) + B_2 u(k) & \text{if } x_2(k) + x_3(k) \geq 0, \\ & |x_1(k)| < 2 \\ A_3 x(k) + B_3 u(k) & \text{otherwise} \end{cases} \tag{20}$$

where $x(k) \in \mathbb{X} = [-10, 10]^3$, $u(k) \in \mathbb{U} = [-2, 2]$, $A_1 = \left[\begin{smallmatrix} 1 & .4 & .08 \\ 0 & 1 & .4 \\ 0 & 0 & 1 \end{smallmatrix}\right]$, $A_2 = \left[\begin{smallmatrix} 1 & .7 & .245 \\ 0 & 1 & .7 \\ 0 & 0 & 1 \end{smallmatrix}\right]$, $A_3 = \left[\begin{smallmatrix} 1 & .8 & .32 \\ 0 & 1 & .8 \\ 0 & 0 & 1 \end{smallmatrix}\right]$, $B_1 = [\, .0107 \; .08 \; .4 \,]$, $B_2 = [\, .0572 \; .245 \; .7 \,]$, and $B_3 = [\, .0853 \; .32 \; .8 \,]$. The PWA system has three dynamic modes, defined over 6 regions.

We want to regulate the state of the system to the origin, and find the explicit control law using the quadratic cost defined by the weights $Q = I$, $P = I$, $R = .1$, and control horizon $N = 3$. We obtain 119 feasible switching sequences, instead of the $6^3 = 216$ possible ones, and 632 polyhedral regions T^l_m. The preliminary inclusion test (12) finds 129 regions of single feasibility. The remaining 503 regions T^l_m of multiple feasibility need to be compared with the corresponding s^l_m partitions $\mathcal{D}_k$ in order to detect their optimality. After running the algorithm described in Section 5.3, 283 regions are found to be totally dominated while 36 can be reduced by using the results of Lemma 1. In this way we have reduced the number of regions in the final control law by 40%, therefore decreasing the number of comparisons that needs to be made on line, without any loss of optimality.

7. CONCLUSIONS

In this paper we have proposed an approach for solving hybrid optimal control problems based on quadratic costs explicitly with respect to the initial state. The method lists all feasible switching sequences using backwards reachability analysis, solves the associated multiparametric quadratic programs, and then reduces the total number of regions via a comparison of the value functions. The latter is computed by using a recursive partition of the parameter space in simplices, by making a linear approximation of the convex value functions in each simplex, and by calculating an upper and a lower bound to their difference. The procedure allows one to discard all those regions whose associated value function is never optimal.

REFERENCES

Baotic, M., F.J. Christophersen and M. Morari (2003). Infinite time optimal control of hybrid systems with a linear performance index. In: *Proc. 42th IEEE Conf. on Decision and Control.* Maui, Hawaii, USA. pp. 3191–3196.

Bemporad, A. (2003). *Hybrid Toolbox – User's Guide.* `http://www.dii.unisi.it/hybrid/toolbox`.

Bemporad, A. (2004). Efficient conversion of mixed logical dynamical systems into an equivalent piecewise affine form. *IEEE Trans. Automatic Control* **49**(5), 832–838.

Bemporad, A. and C. Filippi (2006). Approximate multiparametric convex programming. *Computational Optimization and Applications.* to appear.

Bemporad, A. and M. Morari (1999). Control of systems integrating logic, dynamics, and constraints. *Automatica* **35**(3), 407–427.

Bemporad, A., F. Borrelli and M. Morari (2000). Piecewise linear optimal controllers for hybrid systems. in American Control Conference. Chicago, IL. pp. 1190–1194.

Bemporad, A., M. Morari, V. Dua and E.N. Pistikopoulos (2002). The explicit linear quadratic regulator for constrained systems. *Automatica* **38**, 3–20.

Borrelli, F., M. Baotic, A. Bemporad and M. Morari (2005). Dynamic programming for constrained optimal control of discrete-time linear hybrid systems. *Automatica* **41**(10), 1709–1721.

Corona, D. (2005). Optimal Control of Linear Affine Hybrid Automata. PhD thesis. Dipartimento di Ingegneria Elettrica ed Elettronica. University of Cagliari, Italy.

Horst, R. and N.V. Thoai (1999). DC programming: Overview. *Journal of Optimization Theory and Applications* **103**(1), 1–43.

Mangasarian, O.L. and J.B. Rosen (1964). Inequalities for stochastic nonlinear programming problems. *Operations Research* **12**, 143–154.

Mayne, D.Q. (2001). Constrained optimal control. European Control Conference, Plenary Lecture.

Mayne, D.Q. and S. Rakovic (2002). Optimal control of constrained piecewise affine discrete time systems using reverse transformation. *Conference on Decision and Control, Las Vegas, Nevada* pp. 1564–1551.

Yepremyan, L. and J.E. Falk (2005). Delaunay partitions in $\mathbb{R}^n$ applied to non-convex programs and vertex/facet enumeration problems. *Computers and Operations Research* (32), 793–812.

ELSEVIER
IFAC
PUBLICATIONS

AN EFFICIENT ALGORITHM FOR PREDICTIVE CONTROL OF PIECEWISE AFFINE SYSTEMS WITH MIXED INPUTS

Sylvain Leirens * **Jean Buisson** *

* *Supélec–IETR, Cesson-Sévigné, France*

Abstract: This paper presents a mixed optimization algorithm devoted to predictive control of hybrid systems belonging to the PieceWise Affine (PWA) class with mixed (*i.e.* continuous and discrete) inputs. By using the particular structure of the optimization problem, the number and the dimension of the subproblems to solve in order to find the optimum are significantly reduced.
This approach is applied to a classical case study in the field of hybrid systems: the control of water levels of a three tank system. *Copyright © 2006 IFAC*

Keywords: predictive control, hybrid systems, PWA systems, mixed optimization

1. INTRODUCTION

A great part of the literature concerning *advanced control* involves the concept of optimality and needs suitable optimization methods. To address the control problems of complex and constrained industrial processes, predictive control has become a standard, although it requires an optimization procedure which has to be carried out on line. At each sampling time, an optimal control sequence (with respect to a given cost function) on a finished prediction horizon is found, whose first control only is applied to the system. The whole procedure is then repeated at the next time step (*receding* horizon).

Hybrid systems are dynamic systems involving the presence of continuous and discrete variables in the models. Consequently, the optimization methods used by the predictive control scheme are confronted to combinatorial aspects. Among the various paradigms allowing to build hybrid models, some classes of systems lend themselves particularly to the development of predictive control laws, especially in their discrete time formulation. A commonly used class of hybrid systems is piecewise affine systems (PWA) which are defined by a piecewise affine dynamic over non-overlapping regions of the state-input space.

The effort of research is mainly devoted to techniques leading to a reduction of the combinatorial complexity involved in the optimization problem which is associated with the predictive control law. The purpose of these techniques consists in avoiding the enumeration of all the possibilities for the discrete variables on the prediction horizon, which generates a (very) great number of subproblems to be solved to find the searched optimum at each sampling time.

For continuous PWA systems (*i.e.* no discrete state nor discrete control inputs), reach set computation based algorithms are proposed, coupled to a branch & bound strategy in (Bemporad *et al.*, 2000), and using a state transition graph in (Peña *et al.*, 2003). In (Stursberg, 2004), nonlinear programming and hybrid system simulation are embedded into a graph search algorithm. For switched affine systems (SAS) which define a subclass of PWA systems involving mixed control inputs, but no partitioning of the continuous state-input space, a partial enumeration algorithm is proposed in (Leirens *et al.*, 2005).

The Mixed Logical Dynamical (MLD) formalism, whose equivalence with PWA systems has been proved in (Heemels *et al.*, 2001), leads to a model made up of discrete time linear equations and a set of linear inequalities. They involve real as well as binary variables (Bemporad and Morari, 1999). The implementation of predictive control for hybrid systems modeled in a MLD form requires the solving of mixed linear or quadratic optimization problems, called MILP/MIQP, which need a significant amount of computation. These kinds of optimization problems are solved by using branch & bound techniques with relaxation of the binary variables and linear or quadratic programming, called LP/QP.

In this paper, the approach suggested in (Leirens *et al.*, 2005) is extended to PWA systems with mixed control inputs. It is organized as follows. PWA systems are presented in section 2. Hybrid predictive control principles are introduced in section 3 and section 4 is devoted to the solving of the associated mixed optimization problem. Section 5 presents the implementation of the proposed approach and the results obtained for the control of water levels of a classical hybrid control case study: the three tank system. The conclusions are given in section 6.

2. PWA SYSTEMS

The class of systems, called PWA systems with mixed inputs (but whose state is continuous), that are used in this paper is defined by the following discrete time equations:

$$\mathbf{x}(k+1) = \mathbf{A}_i\mathbf{x}(k) + \mathbf{B}_i\mathbf{u}_c(k) + \mathbf{a}_i \quad (1)$$

$$\mathbf{y}(k) = \mathbf{C}_i\mathbf{x}(k) + \mathbf{D}_i\mathbf{u}_c(k) + \mathbf{c}_i \quad (2)$$

where $\mathbf{x} \in \mathbb{X} \subset \mathbb{R}^{n_x}$ is the state vector, $\mathbf{u}_c \in \mathbb{U} \subset \mathbb{R}^{n_{uc}}$ is the continuous control input vector and $\mathbf{y} \in \mathbb{Y} \subset \mathbb{R}^{n_y}$ system output vector. A partition χ_j ($j \in \mathbb{J} \subset \mathbb{N}$) of the continuous state-input space, independent of the discrete input vector $\mathbf{u}_d$ ($\mathbf{u}_d \in \mathbb{U}_d$ with e.g., $\mathbb{U}_d \subset \mathbb{N}^{n_{ud}}$), is defined as follows :

$$\chi_j = \left\{(\mathbf{x}, \mathbf{u}_c) \,|\, \mathbf{F}_j\mathbf{x} + \mathbf{G}_j\mathbf{u}_c \le \mathbf{f}_j\right\} \quad (3)$$

The subscript i in (1)–(2) denotes the mode $i(k) \in \mathbb{I} \subset \mathbb{N}$ with respect to the kth time step and is defined by a value of the discrete inputs $\mathbf{u}_d(k)$ and the membership of a partition χ_j. A function $\varphi : \mathbb{X} \times \mathbb{U}_c \times \mathbb{U}_d \longrightarrow \mathbb{I}$ is then defined such that :

$$i(k) = \varphi(\mathbf{x}(k), \mathbf{u}_c(k), \mathbf{u}_d(k)) \quad (4)$$

The control inputs $\mathbf{u}$ of the system are composed of the continuous ones ($\mathbf{u}_c$) and discrete ones ($\mathbf{u}_d$) with $n_u = n_{uc} + n_{ud}$. The whole number of modes for the system is given by :

$$p = \text{card}(\mathbb{I}) = \text{card}(\mathbb{J}) \times \text{card}(\mathbb{U}_d) \quad (5)$$

3. PREDICTIVE CONTROL

The principle of predictive control consists in solving an open loop optimal control problem on a receding prediction horizon. The loop is closed by the use of a new measurement of the system state at every sampling time. The prediction model (PWA) is time invariant thus the time origin can be used as the current time step in the equations without loss of generality. Let $\mathbf{U}_N$ be the sequence of controls on the prediction horizon, chosen of length N :

$$\mathbf{U}_N = (\mathbf{u}^T(0)\ \mathbf{u}^T(1)\ \cdots\ \mathbf{u}^T(N-1))^T \quad (6)$$

with $\mathbf{u} = (\mathbf{u}_c^T\ \mathbf{u}_d^T)^T$ and given the following cost function :

$$J_N(\mathbf{x}(0), \mathbf{U}_N) = F(\mathbf{x}(N)) + \sum_{k=0}^{N-1} L(\mathbf{x}(k), \mathbf{u}(k)) \quad (7)$$

In practice, the cost function (or performance index to be optimized) includes a term based on the state $\mathbf{x}$ compared to a reference and a term based on the control inputs $\mathbf{u}$. Given :

$$L(\mathbf{x}(k), \mathbf{u}(k)) = \|\mathbf{x}(k+1) - \mathbf{x}_r\|_{\mathbf{Q}_x} + \|\mathbf{u}(k)\|_{\mathbf{Q}_u} \quad (8)$$

$$F(\mathbf{x}(N)) = \|\mathbf{x}(N) - \mathbf{x}_r\|_{\mathbf{Q}_f} \quad (9)$$

where $\mathbf{x}_r$ is the state reference. Using 2-norm, the cost function is quadratic, *i.e.* $\|\mathbf{w}\|_{\mathbf{Q}} \triangleq \mathbf{w}^T\mathbf{Q}\mathbf{w}$. The weighing matrices are such that $\mathbf{Q}_x \geq 0$, $\mathbf{Q}_u > 0$ (by taking account of constraints on $\mathbf{u}$, typically actuator constraints, it is sufficient for the matrix $\mathbf{Q}_u$ to be semi-positive definite) and $\mathbf{Q}_f \geq 0$.

At every time step, the following optimization problem $\mathcal{P}_N$ has to be solved, where superscript o means optimality :

$$\mathcal{P}_N(\mathbf{x}(0)) \ :\ J_N^o(\mathbf{x}(0)) = \min_{\mathbf{U}_N} J_N(\mathbf{x}(0), \mathbf{U}_N) \quad (10)$$

fulfilling the model constraints (1)–(3). A final state constraint $\mathbf{x}(N) = \mathbf{x}_f$ can be added, especially to guarantee the stability of the closed loop system (Mayne *et al.*, 2000).

A characteristic of the optimization problem which is formulated above is its mixed nature. The presence of continuous inputs ($\mathbf{u}_c$) and discrete ones ($\mathbf{u}_d$) leads to consider the searched control sequence $\mathbf{U}_N$ as a sequence of continuous inputs denoted $\mathbf{U}_{cN}$ and a sequence of discrete inputs denoted $\mathbf{U}_{dN}$ on the prediction horizon N. Then :

$$J_N^o(\mathbf{x}(0)) = \min_{\mathbf{U}_{dN}} \left(\min_{\mathbf{U}_{cN}} J_N(\mathbf{x}(0), (\mathbf{U}_{cN}, \mathbf{U}_{dN})) \right) \quad (11)$$

subject to the model constraints, for $k = 0, 1, \cdots, N-1$:

$$\mathbf{x}(k+1) = \mathbf{A}_i\mathbf{x}(k) + \mathbf{B}_i\mathbf{u}_c(k) + \mathbf{a}_i \quad (12)$$

$$\mathbf{y}(k) = \mathbf{C}_i\mathbf{x}(k) + \mathbf{D}_i\mathbf{u}_c(k) + \mathbf{c}_i \quad (13)$$

$$\mathbf{F}_i\mathbf{x}(k) + \mathbf{G}_i\mathbf{u}_c(k) \leq \mathbf{f}_i \quad (14)$$

with $i(k) = \varphi(\mathbf{x}(k), \mathbf{u}_c(k), \mathbf{u}_d(k))$.

Let $\mathbf{I}_N = \left(i(0)\ i(1)\ \cdots\ i(N-1)\right)^T \in \mathbb{I}^N$ be a sequence of modes on the horizon N. It defines a sequence $\mathbf{U}_{dN}$ and N sets of constraints (12)–(14) for $k = 0, 1, \cdots N-1$. The problem $\mathcal{P}_N$ can be rewritten this way :

$$J_N^o(\mathbf{x}(0)) = \min_{\mathbf{I}_N}\left(\min_{\mathbf{U}_{cN}} J_N(\mathbf{x}(0), (\mathbf{U}_{cN}, \mathbf{I}_N))\right) \quad (15)$$

For a given sequence of modes $\mathbf{I}_N$, the cost :

$$J_N^*(\mathbf{x}(0), \mathbf{I}_N) = \min_{\mathbf{U}_{cN}} J_N(\mathbf{x}(0), (\mathbf{U}_{cN}, \mathbf{I}_N)) \quad (16)$$

is the optimal cost that is found by solving a continuous optimization subproblem (the sequence $\mathbf{I}_N$ – then $\mathbf{U}_{dN}$ – is known) subject to constraints. This subproblem can easily be reformulated in a standard problem of quadratic programming (QP). The superscript $*$ means optimality regarding to a sequence of modes: the continuous control sequence called $\mathbf{U}_{cN}^*$ is said optimal with respect to $\mathbf{I}_N$. However, the constrained optimization problem (16) is not necessarily feasible, *i.e.* for the given sequence $\mathbf{I}_N$, no solution can satisfy the model constraints (12)–(14) on the horizon.

4. MIXED OPTIMIZATION

The optimization problem associated with the predictive control of PWA systems is now formulated, this section is then devoted to its solving.

4.1 Exhaustive enumeration

The simplest way to find the searched optimum consists in enumerating all the possible sequences of modes on the prediction horizon and then to solve the QP subproblems associated with the corresponding sequences. For a given sequence of modes, two situations may occur: either there is no solution satisfying the constraints (infeasibility) or the QP subproblem is feasible. For each feasible sequence $\mathbf{I}_N$ (defining a discrete control sequence $\mathbf{U}_{dN}$) an optimal continuous control sequence $\mathbf{U}_{cN}^*$ and a corresponding cost $J_N^*(\mathbf{x}(0), \mathbf{I}_N)$ can be obtained. The searched optimum is then given by the sequence of modes which minimizes (16):

$$J_N^o(\mathbf{x}(k)) = \min_{\mathbf{I}_N}\left(J_N^*(\mathbf{x}(k), \mathbf{I}_N)\right) \quad (17)$$

This method which is called exhaustive enumeration is quickly useless when the number of modes and/or the length of the prediction horizon increase because the problem is NP-hard.

Using a tree shape representation, the depth of the tree of possibilities grows with the length of the horizon. For a given depth, the width of the tree is fixed by the number of possible modes for the system. Each leaf is a QP subproblem to solve and one of them is the searched optimum. The dimension of all the QP subproblems is identical, *i.e.* the dimension of the optimization vector is $\dim(\mathbf{U}_c) = n_{uc} \times N$.

4.2 Partial enumeration

Exhaustive enumeration consists of completely covering the tree of possibilities. The optimization problem associated with the predictive control has a particular structure: the cost is additive with positive terms. The key idea of the suggested partial enumeration algorithm is knowing a suboptimum of (11) – (14), evaluate partial costs in order to prune the tree, *i.e.* cut branches that cannot lead to the optimum. It is a kind of branch & bound algorithm.

For a partial horizon P ($P < N$), *i.e.* at a depth P in the tree, a partial cost is defined as follows:

$$J_P(\mathbf{x}(0), \mathbf{U}_P) = \sum_{k=0}^{P-1} L(\mathbf{x}(k), \mathbf{u}(k)) \quad (18)$$

The proposed approach is a recursive algorithm which is composed of a descent strategy to explore the tree of possibilities and a criterion of branch cutting.

Descent strategy: suppose to be P ($< N$) time steps in the future, *i.e.* at a depth P in the tree of possibilities. The proposed strategy a kind of *best first* one :

- compute the optimal costs J_{P+1} associated with the feasible subproblems for the possible choices of mode i;
- begin with the branch which gives the minimal cost on the horizon $P+1$ to continue the exploration.

Branch cutting: suppose to have a first suboptimum (all the leaves for which the associated QP subproblem is feasible are suboptima, the searched optimum is the best one). Prune the tree by cutting the branches for which:

- the optimal cost on a partial horizon is greater than the cost of the known suboptimum;
- the subproblem is infeasible.

To cut a branch means to eliminate all the branches following it. The *known* suboptimum is updated when a leaf is evaluated and whose cost is lower than the one obtained for the previous suboptimum.

Given a sequence of N modes $\mathbf{I}_N$ (the associated QP subproblem is assumed to be feasible) and a horizon P with $P < N$. The following notations are used :

- $\mathbf{I}_P^{(N)}$ is the sequence of the P first modes extracted from the sequence $\mathbf{I}_N$;
- $\mathbf{U}_{cP}^{(N)}$ is the continuous control sequence of length P extracted from the sequence $\mathbf{U}_{cN}$.

It is recalled that the superscript $*$ means optimality regarding to a given sequence of modes: *i.e.* the sequence $\mathbf{U}_{cN}^*$ is optimal regarding to a given sequence of modes $\mathbf{I}_N$. However the extracted sequence $\mathbf{U}_{cP}^{*(N)}$ is not necessarily optimal over the horizon P.

Proposition: Given a sequence of modes $\mathbf{I}_N$, for all $P < N$, the optimal cost that is obtained for the sequence $\mathbf{I}_N$ is greater than the optimal cost that is obtained for an extracted sequence $\mathbf{I}_P^{(N)}$:

$$\forall P < N, \quad J_N^*(\mathbf{x}(0), \mathbf{I}_N) \geq J_P^*(\mathbf{x}(0), \mathbf{I}_P^{(N)}) \quad (19)$$

Proof: The cost (16) can be split in two terms, for all $P < N$:

$$J_N^*(\mathbf{x}(0), \mathbf{I}_N) = J_P(\mathbf{x}(0), (\mathbf{U}_{cP}^{*(N)}, \mathbf{I}_P^{(N)})) + \sum_{k=P}^{N-1} L(\mathbf{x}(k), (\mathbf{u}_c^*(k), i(k))) + F(\mathbf{x}(N)) (20)$$

The cost is additive with respectively positive and semi-positive definite functions L and F:

$$J_N^*(\mathbf{x}(0), \mathbf{I}_N) \geq J_P(\mathbf{x}(0), (\mathbf{U}_{cP}^{*(N)}, \mathbf{I}_P^{(N)})) \quad (21)$$

The cost $J_P(\mathbf{x}(0), (\mathbf{U}_{cP}^{*(N)}, \mathbf{I}_P^{(N)}))$ is the cost obtained with an extracted sequence $\mathbf{U}_{cP}^{*(N)}$ and is then suboptimal over the horizon P:

$$\underbrace{J_P(\mathbf{x}(0), (\mathbf{U}_{cP}^{*(N)}, \mathbf{I}_P^{(N)}))}_{\text{suboptimal}} \geq \underbrace{J_P^*(\mathbf{x}(0), \mathbf{I}_P^{(N)})}_{\text{optimal}} \quad (22)$$

□

Suppose being at a depth $P < N$ and that the associated cost J_P is greater than a known suboptimum, with respect to the preceding proposition, the corresponding branch and all the following ones can be cut.

The partial enumeration algorithm is a branch & bound algorithm which leads to the optimal solution by taking advantage of the particular structure of the optimization problem associated with predictive control. Mixed integer programming, which was developed to solve standard problems, purely combinatorial or mixed, cannot exploit this feature. In the framework of predictive control of MLD systems, the associated optimization problem is reformulated in such a standard MIQP problem (Bemporad and Morari, 1999; Mignone, 2002).

The suggested descent strategy is a *heuristic* which allows to obtain a first suboptimum that is hoped of good quality, *i.e.* not far from the optimum. The suboptimal character comes from the choice of the sequence of modes (best first strategy at one prediction step) but the suboptimum is obtained by solving a QP subproblem on the full horizon N.

In this approach by partial enumeration, the dimension of QP subproblems to be solved starts with n_{uc} at the top of the tree (one step time prediction) to grow with the horizon until $n_{uc} \times N$ at the bottom of the tree (horizon N).

This algorithm is illustrated on figure 1. The numbers indicate the way followed out in the tree, *i.e.* how the tree has been explored. Bold lines gives the path to the first suboptimum (best first strategy). The presence of a cross means the result of a branch cutting: either the cost is greater than the one of the known suboptimum or the subproblem is infeasible. The searched optimum is marked out by a triangle.

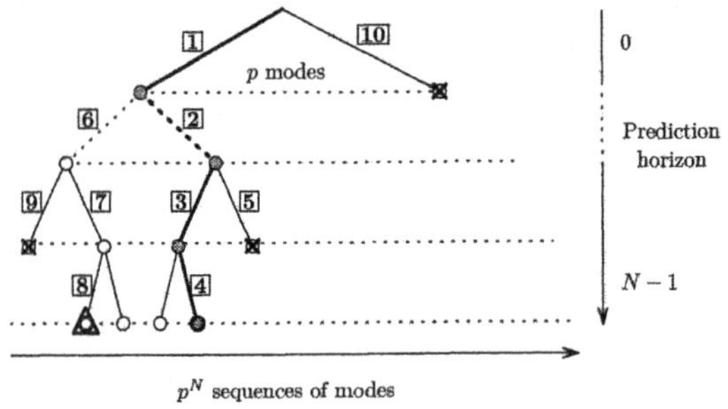

Fig. 1. Partial enumeration

Remarks:

(1) The presence of an equality constraint on the final state $\mathbf{x}(N) = \mathbf{x}_r$ has to be considered only for the evaluation of a cost on the complete horizon (N). This constraint does not exist for the evaluation of a partial cost ($P < N$).
(2) In the case of a purely combinatorial optimization problem, the cost at a node in the tree of possibilities is obtained by the sum of the costs associated with the branches leading to this node (the cost is evaluated downward). The problem considered here has a mixed nature. The cost of a branch is not known *a priori* since it depends on the continuous control $\mathbf{u}_c^*$ found by solving the QP subproblem associated with the path in the tree.

5. APPLICATION EXAMPLE

5.1 Description of the case study

The proposed algorithm is applied to the three tank system (Lunze, 1998) whose diagram is represented figure 2. The system is composed of three

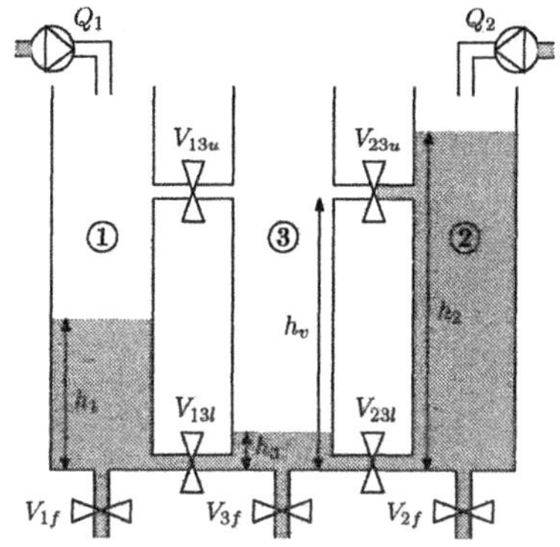

Fig. 2. Three tank system

tanks numbered from 1 to 3 with a maximal height h_{max}. Tanks 1 and 2 are supplied with water by two pumps whose respective flows Q_1 and Q_2 can vary between 0 and Q_{max}. Four valves V_{13l}, V_{23l}, V_{13u} and V_{23u} allow to control flows between the tanks. It is assumed that they can take only two states : opened (1) or closed (0). The upper valves are located at a height h_v. The valves V_{1f}, V_{2f} and V_{3f} define outgoing flows. The levels of water in the three tanks, which are the variables to be controlled, are noted h_1, h_2 and h_3 respectively.

We do not give details about the model in this paper: a complete review of the equations used to build the prediction model can easily be found, e.g. in (Mignone, 2002). We focus on the results obtained for a step change in the water height references. The state space generated by h_1, h_2 and h_3 is divided in eight partitions. Considering both positions, open or closed, of the four valves V_{13l}, V_{23l}, V_{13u} and V_{23u}, the total number of modes for this case study is $8 \times 2^4 = 128$.

Using a sample time $T = 10\,\mathrm{s}$, simulations of $20 \times T = 200\,\mathrm{s}$ have been carried out for various prediction horizons ($2 \leq N \leq 6$). The first column of table 1 gives the numbers of QP subproblems which would have been necessary to solve at each sample step while proceeding by exhaustive enumerations. The three following columns show the minimum (QPmin), average (QPmoy) and maximum (QPmax) numbers of solved subproblems at each sample step obtained using partial enumerations (the number of solved QP may vary at each sample step during the simulation). The minimum number of QP can not be less than the one given by a straight descent in the tree, just enumerating the p possible modes at each time

N	p^N	QP min	QP moy	QP max
2	16384	256	420	1024
3	$2.1\,10^6$	384	1390	6528
4	$2.7\,10^8$	512	4815	44928
5	$3.4\,10^{10}$	1664	9045	86912
6	$4.4\,10^{12}$	2688	15104	139264

Table 1. Performances of the partial enumeration algorithm

	part. enum. (PWA)		miqp (MLD)	
N	QP max	time (s)	QP max	time (s)
2	1024	4.34	589	35.1
3	6528	27.8	6821	591
4	44928	233.3	27609	3583

Table 2. Comparison with miqp–MLD

step while choosing the path with the smallest partial cost: QPmin$\geq p \times N$.

Comparisons have been made between the results obtained by using partial enumeration and miqp algorithms. Although miqp is not recognized as an efficient MIQP solver (such as commercial ones, e.g. CPLEX), it is used as a basis to make the comparison. Both algorithms are Matlab scripts using quadprog from the optimization toolbox. The miqp algorithm is based on branch & bound and relaxation strategies and is associated with the MLD formalism, quoted in introduction. The MLD model of the three tank system has been obtained using the software Hysdel 2.05 (Torrisi *et al.*, 2002). The results are presented in table 2 for $2 \leq N \leq 4$ with the maximum numbers of solved subproblems at each time step and the associated computation times (not taken into account in the control loop). These results have been obtained with Matlab 7.0 running on a 2 GHz clocked PC computer.

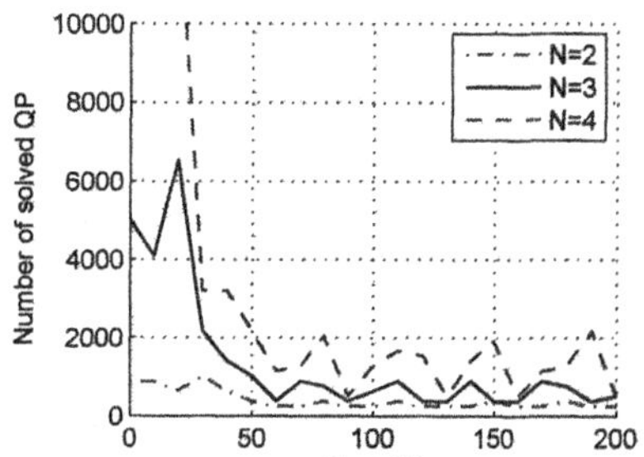

Fig. 3. Number of solved QP versus simulation time ($N = 2, 3, 4$) with partial enumeration

In addition to the number of subproblems solved to obtain the optimum, the computation time is also related to the dimension of the subproblems. The MLD formalism requires the addition of $n_\delta = 3$ auxiliary binary variables and $n_z = 7$ auxiliary real variables. The maximum dimension of a subproblem is then $(n_{uc} + n_{ud} + n_\delta + n_z) \times N$. For the partial enumeration, the maximal dimen-

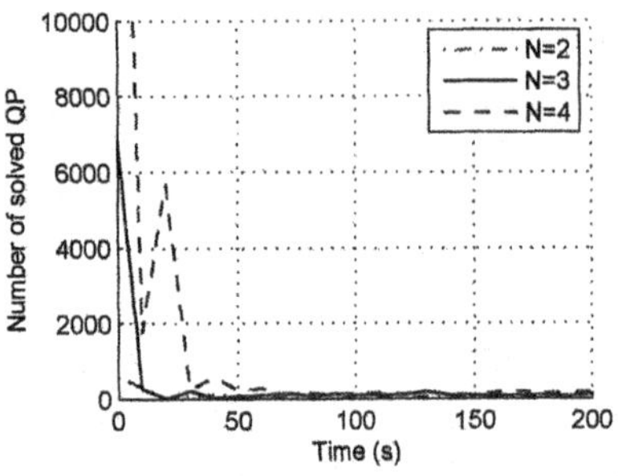

Fig. 4. Number of solved QP versus simulation time ($N = 2, 3, 4$) with miqp

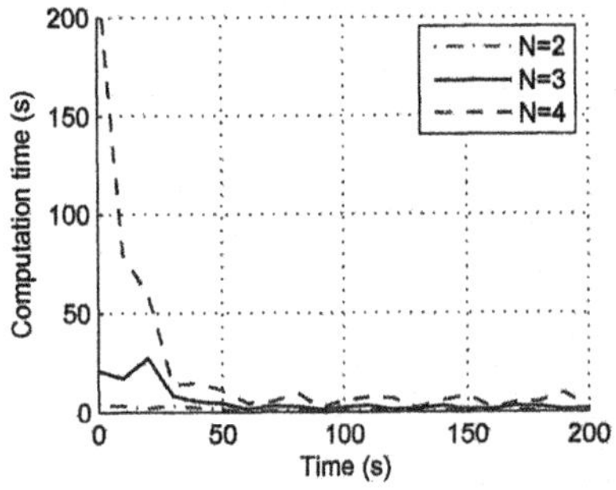

Fig. 5. Computation time versus simulation time ($N = 2, 3, 4$) with partial enumeration

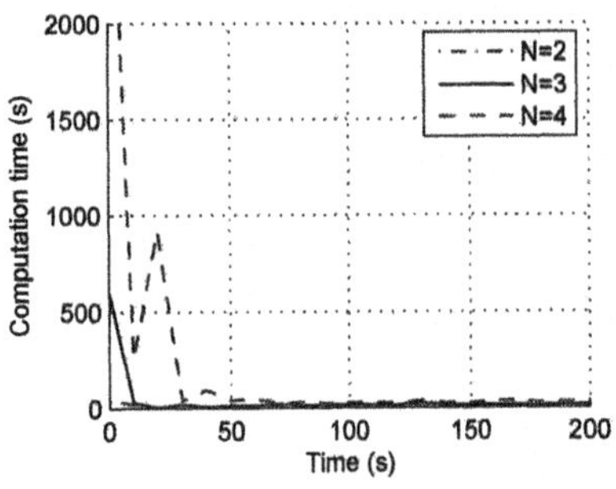

Fig. 6. Computation time versus simulation time ($N = 2, 3, 4$) with miqp – vertical time scale factor = 10 compared to figure 5

sion of a subproblem is $n_{uc} \times N$. The interest of the suggested approach also lies in the small dimension of the subproblems to solve, which induces small computation times. At the beginning of the optimization procedure, it is necessary to solve a significant number of subproblems but of very small dimensions. To cut branches in the tree allows to restrict the number of subproblems with a more significant size to be solved. In this example, it should be noticed that the farther from its reference the state is (actually at the beginning of the simulation), the bigger the computational effort required to solve QP subproblems is.

6. CONCLUSIONS

This article presents a kind of branch & bound algorithm to solve the mixed optimization problem associated with the predictive control of PWA systems with mixed inputs. Knowing a suboptimum, the key idea of this algorithm is to evaluate partial costs (*i.e* at intermediate or partial horizons) in order to prune the tree of the possibilities by cutting the branches which cannot lead to the searched optimum.

REFERENCES

Bemporad, A. and M. Morari (1999). Control of systems integrating logic, dynamics, and constraints. *Automatica* **35**(3), 407–427.

Bemporad, A., L. Giovanardi and F.D. Torrisi (2000). Performance driven reachability analysis for optimal scheduling and control of hybrid systems. In: *Proceedings of the 39th Conference on Decision and Control.* Sydney, Australia.

Heemels, W. P. M. H., B. De Schutter and A. Bemporad (2001). Equivalence of hybrid dynamical models. *Automatica* **37**(7), 1085–1091.

Leirens, S., J. Buisson, P. Bastard and J.-L. Coullon (2005). An efficient algorithm for solving model predictive control of switched affine systems. In: *Proceedings of the Scientific Computation, Applied Mathematics and Simulation World Congress.* Paris, France.

Lunze, J. (1998). Laboratory three tanks system benchmark for the reconfiguration problem. Technical report. Technical University of Hamburg-Harburg, Institute of Control Engineering. Hamburg, Germany.

Mayne, D. Q., J. B. Rawlings, C. V. Rao and P. O. M. Scocaert (2000). Constrained model predictive control: Stability and optimality. *Automatica* **36**(6), 789–814.

Mignone, D. (2002). Control and Estimation of Hybrid Systems with Mathematical Optimization. PhD thesis. Swiss Federal Institute of Technology (ETH). Zürich.

Peña, M., E. F. Camacho and S. Piñón (2003). Hybrid systems for solving model predictive control of piecewise affine system. In: *Proceedings of the IFAC Conference on Analysis and Design of Hybrid Systems.* Saint-Malo, France.

Stursberg, O. (2004). A graph search algorithm for optimal control of hybrid systems. In: *Proceedings of the 43rd Conference on Decision and Control.* Nassau, Bahamas.

Torrisi, F., A. Bemporad and D. Mignone (2002). Hysdel - a tool for generating hybrid models. Technical Report AUT00-03. Automatic Control Laboratory, ETH. Zürich, Switzerland.

ELSEVIER
IFAC
PUBLICATIONS

EXPLICIT MODEL PREDICTIVE CONTROL OF THE BOOST DC-DC CONVERTER

A. Giovanni Beccuti * Georgios Papafotiou * Manfred Morari *

* *Automatic Control Lab., Physikstrasse 3, ETH Zurich, 8092 Zurich, Switzerland*
beccuti,papafotiou,morari@control.ee.ethz.ch

Abstract: This paper extends previous work done on the optimal control of the dc-dc converter boost circuit topology. The same control problem formulation is maintained but a simpler piecewise affine approximation of the system dynamics is derived and employed to obtain an explicit solution whose complexity renders it viable for implementation on a realistic circuit setup. *Copyright© 2006 IFAC.*

Keywords: predictive control, power circuits, piecewise linear controllers.

1. INTRODUCTION

Fixed frequency switch-mode dc-dc converters are a class of electronic power circuits extensively used in regulated dc power supplies and dc motor drive applications due to advantageous features in terms of size, weight and reliable performance. Their role in these applications is to transfer power from a dc input to a load, achieving output voltage regulation in the presence of voltage source and output load variations. The principal control challenge derives from their hybrid nature as their switched circuit topology implies different modes of operation, each with its own related affine continuous-time dynamics. The input variable (duty cycle) also features hard constraints, and safety measures may impose additional constraints such as current limiting.

This paper analyzes the modelling and controller synthesis of the fixed-frequency boost dc-dc converter, in which the semiconductor switch is operated by a pulse sequence with constant switching frequency f_s (resp. period T_s). It is then possible to regulate the dc component of the output voltage through the duty cycle $d = \frac{t_1}{T_s}$, where t_1 denotes the interval within the switching period during which the switch is in the first mode of operation.

By employing this operation principle, the main control objective is to act on the semiconductor switch with a duty cycle such that the dc component of the output voltage reaches the given reference. As the name suggests, for the boost converter this reference value is higher than that of the voltage source, and must be maintained despite variations in the load or the voltage source. Control techniques that are used in practice typically have in common the employment of PI-type controllers tuned on the basis of linearized averaged models (Erickson *et al.*, 1982), (Middlebrook and Cuk, 1976) and that commonly use two control loops for the inductor current and output voltage. Improved controller design strategies involving nonlinear and feed-forward control methods have been formulated (Hiti and Borojevic, 1995), and (Kazimierczuk and Massarini, 1997), but these employ models that do not capture the hybrid dynamics of dc-dc converters. Additionally, none of them allow to directly incorporate constraints in the controller synthesis.

The recent past has witnessed an increased interest in the direction of other alternative control methodologies; passivity-based control design for switched-mode power converters has garnered significant attention (Sira-Ramrez *et al.*, 1997) and has also been employed for the boost circuit in (Jeltsema and Scherpen, 2004), in a way such that the physical structure (energy dissipation and interconnection) is explicitly featured in the model and control scheme, but neglecting variations on the voltage source or on the output load. A hybrid approach is described in (Senesky *et al.*, 2003) and a hybrid automaton synthesized to opportunely switch a boost converter among operating modes, in a way such that however a constant switching frequency, which is crucial from the application point of view, cannot be guaranteed.

Motivated by these issues, this paper extends work recently done on the numerical performance of optimal controller schemes for boost dc-dc converters in (Beccuti *et al.*, 2005); herein the same control problem formulation is maintained but a simpler model is derived in order to allow for the explicit solution of the control problem, so that determining the control law on-line reduces to searching in a look-up table of viable complexity and evaluating an affine function of the state. The approach featured in (Beccuti *et al.*, 2005) is in turn based on the work done in (Geyer *et al.*, 2004*b*) for the dc-dc buck converter topology.

This paper is organized as follows. Section 2 presents the physical setup of the circuit and in Section 3 a model for the boost converter is derived by employing a least square fitting (LSF) approximation of the exact converter dynamics over several regions of the state space to derive a piecewise affine (PWA) system. In Section 4, an optimal control problem incorporating the appropriate control objectives is formulated, and an outline of the employed load estimation setup is given. Section 5 contains simulation results illustrating the performance of the proposed model predictive control (MPC) scheme. Finally, conclusions and further research directions are outlined in Section 6. In the sequel, a normalized time scale will be used, with the time unit being equal to the switching period T_s, and the discrete time instant $t = k$ referring to $t = kT_s$.

2. PHYSICAL MODEL OF THE BOOST CONVERTER

The circuit topology of the boost converter is shown in Fig.1; only the continuous conduction mode will be considered, that is operating points for which the inductor current remains positive.

Using normalized quantities, r_o denotes the output load resistance, r_c the equivalent series resistance (ESR) of the capacitor x_c and r_ℓ is the internal resistance of the inductor x_ℓ. The boost converter features two operation modes with two different affine dynamics. The controller selects the control input, the duty cycle $d(k)$, for each period k, determining when the switch from the first mode to the second takes place. During the time interval $k \le t < k + d(k)$ the switch S is in the s_1 position and the inductor is charged. At the end of this interval S is switched to s_2 and power is transferred to the load. The switch is set back to the s_1 position at the end of the period. As a consequence of this principle of operation, the duty cycle lies in the interval $[0, 1]$ by definition.

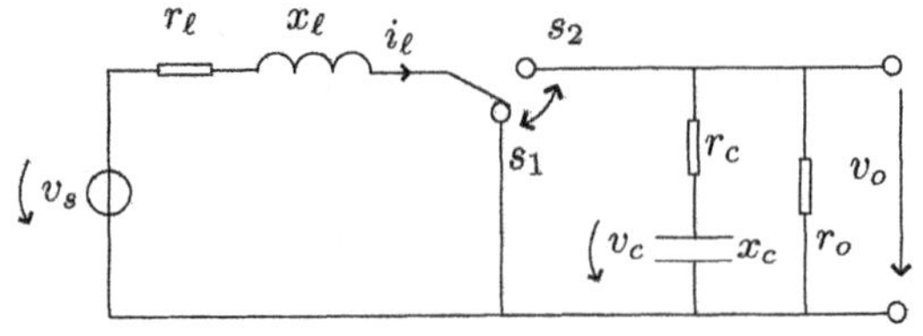

Fig. 1. Topology of the boost converter

By taking $x(t) = [i_\ell(t)\ v_c(t)]^T$ as the state vector, where $i_\ell(t)$ is the inductor current and $v_c(t)$ the capacitor voltage, the system is described by the following pair of affine continuous time state-space equations. The following equations hold

$$\dot{x}(t) = \begin{cases} F_1 x(t) + f_1 v_s, & k \leqslant t < k + d(k) \\ F_2 x(t) + f_2 v_s, & k + d(k) \leqslant t < k + 1 \end{cases} \tag{1a}$$

$$v_o(t) = \begin{cases} g_1^T x(t), & k \leqslant t < k + d(k) \\ g_2^T x(t), & k + d(k) \leqslant t < k + 1 \end{cases} \tag{1b}$$

Matrices F_1 and F_2 and vectors f_1, f_2, g_1 and g_2 are not given here for the sake of brevity but can be easily obtained by elementary circuit theory.

The state vector of the boost converter model is a continuous function of time, since it comprises the inductor current and the capacitor voltage and there are no degenerate loops or nodes in the circuit topology. However, an important feature is the fact that the output voltage is a discontinuous function of time, due to the existence of the two different output vectors g_1 and g_2. The discontinuity occurs at the time instants of the switch commutations and can be exactly calculated as a function of the inductor current as follows. At the beginning of the switching period, when the model switches from mode 2 to mode 1, the jump of the output voltage is equal to

$$v_o(k^+) - v_o(k^-) = -\frac{r_o r_c}{r_o + r_c} i_\ell(k), \tag{2}$$

while at time $t = k + d(k)$ the model switches from mode 1 to mode 2 and the jump of the output voltage amounts to

$$v_o(k + d(k)^+) - v_o(k + d(k)^-) = \frac{r_o r_c i_\ell(k + d(k))}{r_o + r_c}. \tag{3}$$

3. MODELLING FOR CONTROL DESIGN

3.1 Reformulated Continuous-Time Model

From an implementation point of view, it is preferable that all states used in the prediction model be directly measurable. Thus, the capacitor voltage is replaced by the output voltage in the state vector which leads to setting $x(t) = [i_\ell(t)\ v_o(t)]^T$. Additionally, to account for variations in the voltage source v_s directly, the (to be derived) optimal control law would need to be parameterized over v_s. To obviate this requirement and as will further be explained in sections 3.2 and 4.4, the voltage source v_s is removed from the model equations by redefining the scaled state vector $x'(t) = [i'_\ell(t) v'_o(t)] = [\frac{i_\ell(t)}{v_s}\ \frac{v_o(t)}{v_s}]$. This yields the reformulated state-space equations

$$\dot{x}'(t) = \begin{cases} F'_1 x'(t) + f'_1, & k \leqslant t < k + d(k) \\ F'_2 x'(t) + f'_2, & k + d(k) \leqslant t < k + 1 \end{cases} \tag{4a}$$

$$v'_o(t) = g'^T x'(t) \,. \tag{4b}$$

where again matrices and vectors F'_1, F'_2, f'_1, f'_2 and g' can easily be computed from elementary circuit theory.

The voltage source v_s is always considered to be measurable in accordance with common industrial practice, so that the state vector is consistently defined.

An important feature of the reformulated state-space model is the fact that the state vector is now a discontinuous function of time, since it includes the output voltage of the converter. However, the employed hybrid modelling framework can directly incorporate such discontinuities, as shown in the next section.

3.2 Piecewise Affine Discrete-Time Model

The formulation of an adequate model for the boost converter is of fundamental importance for the subsequent derivation and implementation of the optimal control problem. Given the discrete time variation of the input variable, a natural choice is to formulate the model in the discrete time domain by employing a sampling interval equal to the switching period T_s. The employed method considers a direct LSF approximation over several regions of the exact system update equations, yielding a PWA description of the associated non-linear expressions. These can be written as

$$x'(k+1) = \Phi(d(k))x'(k) + \Gamma(d(k)) \tag{5}$$

where $\Phi(d(k))$ and $\Gamma(d(k))$ are matrices that depend nonlinearly on the duty cycle, calculated by integrating (4) from $t = k$ to $t = k+1$, taking into account the discontinuity of the output voltage discussed above.

Expression (5) is approximated by determining the matrices $\bar{A}_i$, $\bar{B}_i$ and $\bar{f}_i$ that describe the system in terms of

$$x'(k+1) = \bar{A}_i x'(k) + \bar{B}_i d(k) + \bar{f}_i \tag{6a}$$

$$\text{if } d(k) \in D_i \quad i = 1, \ldots, \nu \tag{6b}$$

$$0 \leq d(k) \leq 1 \tag{6c}$$

and that minimize the sum of quadratic error terms

$$(\Phi(d(k))x'(k) + \Gamma(d(k)) - (\bar{A}_i x'(k) + \bar{B}_i d(k) + \bar{f}_i))^2 \tag{7}$$

over a gridded series of points $x'(k)$ in the state space $[0, i'_{\ell,max}] \times [0, v'_{o,max}]$, where D_i are the ν intervals $[0, \frac{1}{\nu}]$, ..., $[\frac{\nu-1}{\nu}, 1]$, and $i'_{\ell,max}$, $v'_{o,max}$ are the maximum values of the scaled inductor current and output voltage, respectively.

It should be noted that the choice of normalizing over v_s allows one to obtain matrices $\bar{A}_i$, $\bar{B}_i$, $\bar{f}_i$ that are independent of the voltage source and thus valid for any of its values, since it does not appear in the parameters of the original non-linear update expression (5). However, what needs to be stressed is that the derived PWA model is valid for the given nominal load resistance. In Section 4.5 an estimation scheme connected to the MPC design to account for (unmeasured) changes in r_o is briefly outlined; full details can be found in (Beccuti *et al.*, 2006).

4. THE CONTROL PROBLEM

4.1 Control Issues And Objectives

The main control objective for the boost dc-dc converter is to regulate the dc component of the output voltage v_o to its reference $v_{o,ref}$. This regulation has to be achieved in the presence of the hard constraints on the manipulated variable (the duty cycle) which is bounded between 0 and 1, and needs to be maintained despite the changes in the load r_o and the voltage source v_s whilst rendering steady state operation with constant duty cycle, thus avoiding the occurrence of subharmonic oscillations. As described in (Beccuti *et al.*, 2005) however, and as often done in actual industrial practice (Mohan *et al.*, 1989), it is more convenient to formulate the control problem of the boost dc-dc converter as a current (rather than a voltage) regulation problem, aiming at steering the value of the scaled inductor current i'_ℓ to a reference $i'_{\ell,ref}$. The value $i'_{\ell,ref}$ corresponding to the desired $v_{o,ref}$ can be explicitly calculated on the basis of the known parameters of the circuit during nominal system conditions (Kostakis *et al.*, 2000), including variations on v_s. In the case of a load variation, the Kalman filter (cf. 4.5) appropriately updates $i'_{\ell,ref}$ to restore the system to the desired operating point.

4.2 Model Predictive Control

Model Predictive Control (MPC) has been traditionally and successfully employed in the process industry and recently also for hybrid systems. The control action is obtained by minimizing an objective function at each time step over a finite horizon subject to the equations and constraints of the model. The major advantage of MPC is its straight-forward design procedure. Given a model of the system, including constraints, one only needs to set up an objective function that incorporates the control objectives. Further details about MPC can be found in (Maciejowski, 2002).

4.3 Constrained Finite Time Optimal Control Scheme

The control objectives are to regulate the average output voltage to its reference as fast and with as little overshoot as possible, or equivalently, to minimize the absolute scaled inductor current error $i'_{\ell,err}(k) = |i'_\ell(k) - i'_{\ell,ref}|$. Let $\Delta d(k) = |d(k) - d(k-1)|$ indicate the absolute value of the difference between two consecutive duty cycles. This term is introduced in order to reduce the presence of unwanted chattering in the input when the system has almost reached stationary conditions. Define the penalty matrix $Q = \text{diag}(q_1, q_2)$ with $q_1, q_2 \in \mathbb{R}^+$ and the vector $\varepsilon(k) = [i'_{\ell,err}(k), \Delta d(k)]^T$. Consider the objective function

$$J(D(k), x'(k), d(k-1)) = \sum_{\ell=0}^{L-1} \|Q\,\varepsilon(k+\ell|k)\|_1 \quad (8)$$

penalizing the predicted evolution of $\varepsilon(k+\ell|k)$ from k over the horizon L using the 1-norm.

The control input at time-instant k is then obtained by minimizing the objective function (8) over the sequence of control moves $D(k) = [d(k), \ldots, d(k+L-1)]^T$ subject to the model equations and constraints (6a), (6b), (6c); the resulting optimization program is referred to as the constrained finite time optimal control (CFTOC) problem.

4.4 The State Feedback Law

Multi-parametric programming is employed to solve an optimization problem off-line for a range of parameters. In (Baotic *et al.*, 2003) and (Borrelli, 2003) the authors show how to reformulate a discrete-time CFTOC problem for a PWA system as a multi-parametric program by treating the state vector as a parameter and propose an algorithm for its solution.

Note that the CFTOC problem is not only a parametric function of $x(k)$, but also of the last control move $d(k-1)$, as the changes of the duty cycle are penalized in the objective function; furthermore, as it is necessary to solve the CFTOC problem for all possible values of $i'_{\ell,ref}$, the scaled inductor current reference also enters the augmented state vector, which therefore results in being 4-dimensional. Again, it should be noticed that normalizing the system equations over v_s allows to define a model independently of the voltage source, and therefore an explicit state-feedback law that depends on one parameter less (Papafotiou *et al.*, 2004).

Overall the proposed approach, in accordance with common practice, requires the measurement of the inductor current i_ℓ, output voltage v_o and source voltage v_s [1].

As proven in (Borrelli, 2003) the optimal state-feedback control law $d^*(k)$ is a PWA function of the (augmented) state vector defined on a polyhedral partition of the feasible (augmented) state space.

As a result, such a state-feedback controller can be implemented online, since computing the control input amounts to determining the polyhedron in which the measured state lies and then simply evaluating the corresponding affine control law. In many cases, polyhedra with the same control law form a convex union and can thus be optimally merged (Geyer *et al.*, 2004*a*) and replaced by their union, leading to an equivalent PWA control law of reduced complexity.

4.5 Load Variations

An estimation scheme to account for variations in the load resistance has been derived and coupled with the previously obtained state-feedback controller (for a time-invariant and nominal load) through an external loop. More specifically, this loop adjusts the scaled inductor current reference $i'_{\ell,ref}$ by a corrective term $\hat{i}'_e$ (equal to zero during nominal system operation) and feeds the obtained value $\tilde{i}'_{\ell,ref} = i'_{\ell,ref} - \hat{i}'_e$, into the controller; the adjustment is done in a way such that the error between the inductor current and its *actual* reference is made small.

This can be achieved through the use of a Kalman filter (Jazwinski, 1970) that yields a zero steady-state inductor current error due to its integrating nature. To address the hybrid nature of the model, a filter with two modes is employed, and switching between the two is done according to the switch transitions in the converter. Such an approach is possible as the mode transitions, which are imposed by the duty cycle, are precisely known. To allow for an easier implementation of the filter, two constant Kalman gains are used (one

[1] Due to the reformulation of the model the knowledge of v_c, which is unmeasurable, is not needed.

for each mode). It should be noted that the employment of a switched filter setup requires specific provisions to be made in order to ensure the stability of the estimation scheme (Alessandri and Coletta, 2001); such provisions, together with a complete description of the filter, can be found in (Beccuti *et al.*, 2006).

5. SIMULATION RESULTS

In this section, simulation results demonstrating the performance of the proposed control methodology are presented. The circuit parameters expressed in the per unit system are given by $x_c = 70\,\text{p.u.}$, $x_\ell = 3\,\text{p.u.}$, $r_c = 0.01\,\text{p.u.}$ and $r_\ell = 0.05\,\text{p.u.}$ If not otherwise stated the output resistance is given by $r_o = 1$ p.u. and the voltage source is $v_s = 0.75$ p.u.; the output voltage reference is set to $v_{o,ref} = 1$, to which, for the given circuit parameters, an inductor current reference $i_{\ell,ref} = 1.25$ is associated. The model was derived for a range of values of $[0, 4]$ for the scaled inductor current and $[0, 3]$ for the scaled output voltage; three PWA dynamics were calculated, with the intervals D_i being $[0, \frac{1}{3}]$, $[\frac{1}{3}, \frac{2}{3}]$, and $[\frac{2}{3}, 1]$. For the cost function, the penalty matrix is chosen to be $Q = \text{diag}(10, 1)$ and the prediction horizon is $L = 2$.

As explained in section 4.4 the explicit state-feedback controller is defined in a 4-dimensional space. For the chosen circuit and controller setup its computation yields a polyhedral partition consisting of 239 regions; by utilizing the merging algorithm introduced in (Geyer *et al.*, 2004*a*) the controller can further be simplified to 121 regions.

The first case to be analyzed is that of the transient behaviour during startup. Fig.2(a) and Fig.2(b) depict the step responses of the different schemes during start-up, i.e. $x(0) = [0, 0]^T$. The proposed optimal control schemes yields an output voltage that reaches its stationary conditions with an overshoot of about 4% and within 10 switching periods.

For the second case results stemming from a 25% decrease in the voltage source v_s during steady state operation are shown in Fig.3(a)-3(b); the new value of the voltage source is measured, the current reference updated accordingly and the system restored to its desired output voltage value; as the control problem is formulated in terms of a current tracking scheme the system is steered in such a manner as to quickly reach its required inductor current value.

The third and final case concerns a 100% increase in the load resistance r_o during steady state operation; results are displayed in Fig.4(a)-4(b). The Kalman filter adjusts the current reference and

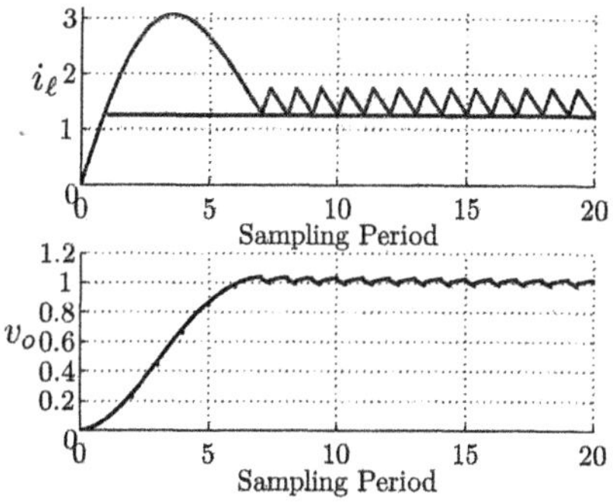

(a) Inductor current (above, with reference) and output voltage (below), in p.u.

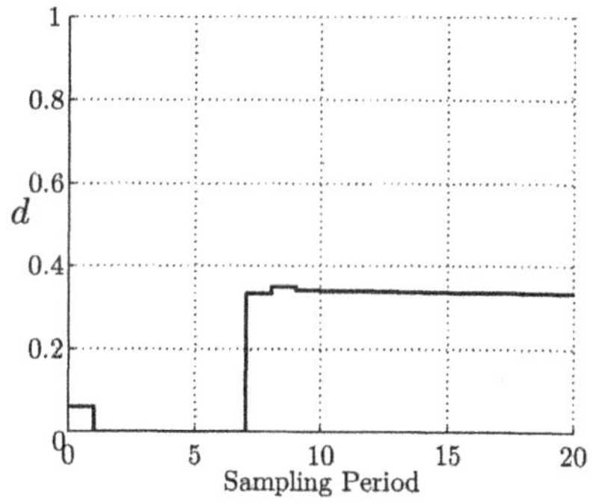

(b) Duty cycle

Fig. 2. Simulation results for the startup scenario

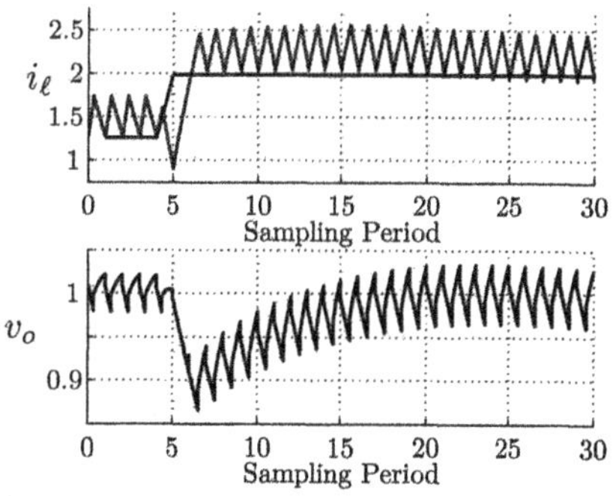

(a) Inductor current (above, with reference) and output voltage(below), in p.u.

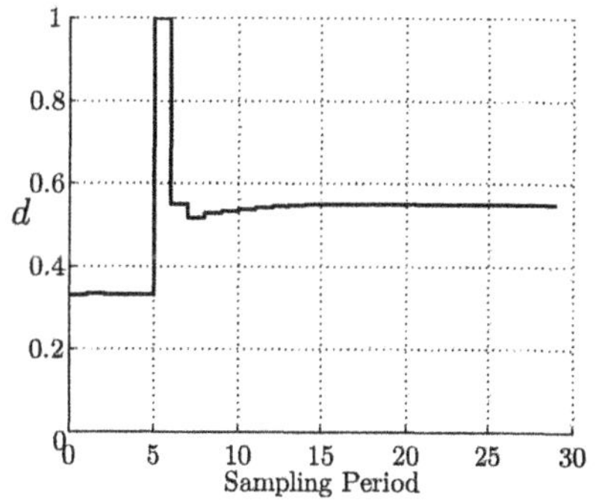

(b) Duty cycle

Fig. 3. Simulation results for the scenario featuring a 25% decrease of v_s

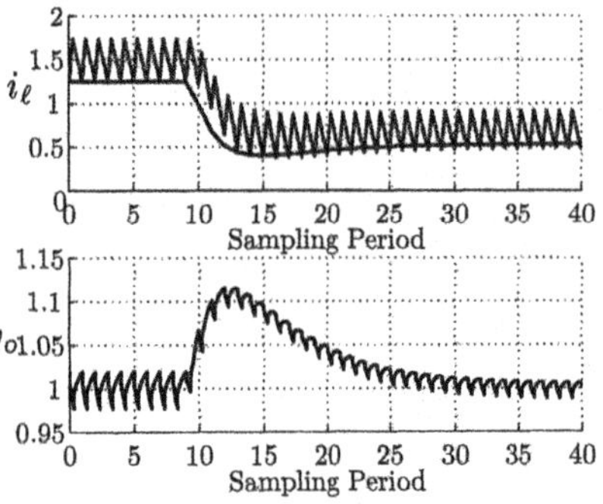

(a) Inductor current (above, with reference) and output voltage(below), in p.u.

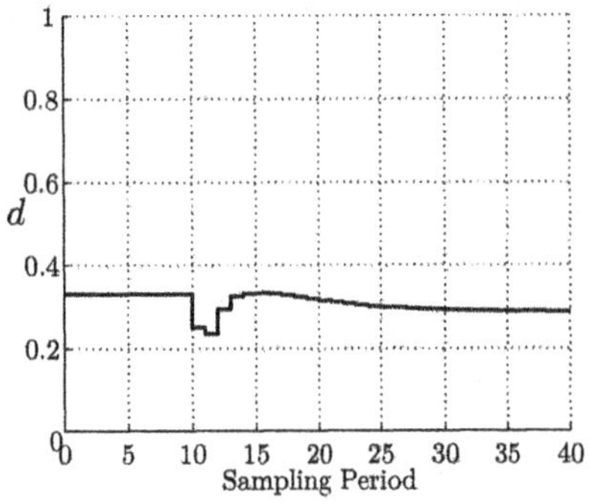

(b) Duty cycle

Fig. 4. Simulation results for the scenario featuring a 100% increase of r_o

the output voltage reaches its desired value after approximately 20 switching periods.

6. CONCLUSIONS AND OUTLOOK

An extension of a previously presented MPC scheme for boost switch mode dc-dc converters has been formulated, whereby the solution is explicitly obtained off-line. This reduces the task of solving the optimal control problem to a search in a look-up table, thus rendering the proposed approach viable for the successive phase of experimental validation on a physical converter. The described setup is additionally integrated with an estimator that allows to consider variations on the load, as typically required by any realistic industrial application. Future research work will be directed at formulating the MPC scheme directly in terms of a voltage reference problem and at considering the case of the discontinuous conduction regime, wherein the inductor current drops to zero and thereafter remains constant, thus introducing a third mode of operation for the circuit.

7. ACKNOWLEDGEMENTS

This work was (partially) done in the framework of the HYCON Network of Excellence, contract number FP6-IST-511368.

REFERENCES

Alessandri, A. and P. Coletta (2001). Switching observers for continuous-time and discrete-time linear systems. *Proc. of the ACC 2001* **3**, 2516–2521.

Baotic, M., F.J. Christophersen and M. Morari (2003). Infinite time optimal control of hybrid systems with a linear performance index. *Proceedings of the CDC 2003* pp. 3191–3196.

Beccuti, A.G., G. Papafotiou and M. Morari (2005). Optimal control of the boost dc-dc converter. *Proceedings of the CDC-ECC 2005.*

Beccuti, A.G., G. Papafotiou and M. Morari (2006). Explicit model predictive control of the boost converter. Technical report. ETH Zurich. Available at: `www.control.ee.ethz.ch`

Borrelli, F. (2003). *Constrained Optimal Control of Linear and Hybrid Systems, Volume 290 of Lecture Notes in Control and Information Sciences.* Springer.

Erickson, R.W., S. Cuk and R.D. Middlebrook (1982). Large signal modeling and analysis of switching regulators. *IEEE Power Electronics Specialists Conference Records* pp. 240–250.

Geyer, T., F.D. Torrisi and M. Morari (2004*a*). Optimal complexity reduction of piecewise affine models based on hyperplane arrangements. *Proceedings of the ACC 2004* pp. 1190–1195.

Geyer, T., G. Papafotiou and M. Morari (2004*b*). On the optimal control of switch-mode dc-dc converters. *Hybrid Systems: Computation and Control* pp. 342–356.

Hiti, S. and D. Borojevic (1995). Robust nonlinear control for the boost converter. *IEEE Transactions on Power Electronics* **10(6)**, 651–658.

Jazwinski, A. H. (1970). *Stochastic Processes and Filtering Theory.* Academic Press.

Jeltsema, D. and J. M. A. Scherpen (2004). Tuning of passivity-preserving controllers for switched-mode power converters. *IEEE Transactions on Automatic Control* **49**, 1333–1344.

Kazimierczuk, M.K. and A. Massarini (1997). Feedforward control dynamic of dc/dc pwm boost converter. *IEEE Transactions on Circuits and Systems-I: Fundamental Theory and Applications* **44(2)**, 143–149.

Kostakis, G. Th., S.N. Manias and N.I. Margaris (2000). A generalized method for calculating the rms values of switching power converters. *IEEE Transactions on Power Electronics* **15**, 616–625.

Maciejowski, J.M. (2002). *Predictive Control.* Prentice Hall.

Middlebrook, R.D. and S. Cuk (1976). A general unified approach to modeling switching power converter stages. *IEEE Power Electronics Specialists Conference Records* pp. 18–34.

Mohan, N., T. M. Undeland, and W. P. Robbins (1989). *Power Electronics: Converters, Applications and Design.* Wiley.

Papafotiou, G., T. Geyer and M. Morari (2004). Hybrid modelling and optimal control of switch-mode dc-dc converters. *IEEE Workshop on Computers in Power Electronics (COMPEL).*

Senesky, M., G. Eirea and T. J. Koo (2003). Hybrid modelling and control of power electronics. *Proceedings of the HSCC 2003 6th International Workshop* pp. 450–465.

Sira-Ramrez, H., R. A. Perez-Moreno, R. Ortega and M. Garcia-Esteban (1997). Passivity-based controllers for the stabilization of dc-to-dc power converters. *Automatica* **33**, 499–513.

ELSEVIER
IFAC
PUBLICATIONS

A NEW DUAL-MODE HYBRID MPC ALGORITHM WITH A ROBUST STABILITY GUARANTEE

M. Lazar *,[1] **W.P.M.H. Heemels** **

* *Department of Electrical Engineering, Eindhoven University of Technology, E-mail:* m.lazar@tue.nl
** *Embedded Systems Institute, Eindhoven*

Abstract: This paper employs the Input-to-State Stability (ISS) framework to investigate the robustness of discrete-time Piece-Wise Affine (PWA) systems in closed-loop with Model Predictive Controllers (MPC), or hybrid MPC for short. We show via an example taken from literature that stabilizing hybrid MPC can generate MPC values functions *that are not ISS Lyapunov functions* for arbitrarily small additive disturbances. As a consequence, it is not easy to prove that nominally stabilizing hybrid MPC schemes are robust. This motivates the need to design MPC schemes for hybrid systems with an a priori robust stability guarantee. A possible solution to this problem was recently developed by the authors for a particular class of PWA systems, i.e. when the origin lies in the interior of one of the regions in the partition. The main contribution of this paper is a novel dual-mode MPC algorithm for hybrid systems with an a priori ISS guarantee. This MPC scheme is applicable to general PWA systems, i.e. when the origin may lie on the boundaries of multiple regions in the partition. *Copyright ©2006 IFAC*

Keywords: Predictive control, Hybrid systems, Robust stability.

1. INTRODUCTION

A certain maturity is reached in the field of Model Predictive Control (MPC) for hybrid systems, regarding computational and nominal stability aspects. This is illustrated by the existing tools for solving hybrid MPC optimization problems, the Hybrid Toolbox (Bemporad, 2003) and the Multi Parametric Toolbox (MPT) (Kvasnica *et al.*, 2004), and by the stability results published in the literature, for example, see (Bemporad and Morari, 1999), (Borrelli, 2003) for attractivity results and (Kerrigan and Mayne, 2002), (Lazar *et al.*, 2005*b*) for asymptotic stability results for Piece-Wise Affine (PWA) systems. In this paper we focus on *inherent robustness* of PWA systems in closed-loop with MPC controllers (or hybrid MPC for short), which is a problem that was not addressed before in the literature. By the inherent robustness property we mean that a nominally stabilizing controller has some robustness in the presence of perturbations. Its importance cannot be overstated, since all controllers designed to be nominally stable are affected by perturbations when applied in practice.

Inherent robustness has been studied in MPC for linear and smooth nonlinear systems. In (Grimm *et al.*, 2004) the authors proved that linear systems in closed-loop with stabilizing MPC are inherently robust due to the presence of a *continuous* MPC value function, which is an Input-to-State Stable (ISS) Lyapunov function (Jiang and Wang, 2001) in this case. However, they also showed via examples that continuous and necessarily nonlinear systems in closed-loop with MPC can actually have zero robustness to arbitrarily small disturbances, in the absence of a continuous MPC

[1] Research supported by the Dutch Science Foundation (STW), Grant "Model Predictive Control for Hybrid Systems" (DMR. 5675) and by the European Union through the Network of Excellence HYCON (contract FP6-IST-511368).

value function[2]. The first contribution of this paper is to issue a warning by presenting an example of a PWA system in closed-loop with a stabilizing MPC controller that generates an MPC value function that is *discontinuous* and, more importantly, it is *not* an ISS Lyapunov function. This indicates that the natural way to ensure ISS (robustness) in MPC fails for PWA systems.

The aim of this paper is to design an MPC scheme for hybrid systems with an a priori robust stability guarantee. Several solutions that rely on *continuous* (or even *Lipschitz continuous*) system dynamics are available in the literature, e.g. see (Limon *et al.*, 2002), (Grimm *et al.*, 2003). However, since hybrid systems are inherently nonlinear and discontinuous, these methods are not applicable to MPC of hybrid systems. Recently, in (Lazar *et al.*, 2005*a*) the authors developed a hybrid MPC scheme with an a priori ISS guarantee, under the assumption that the origin lies in the interior of one of the regions in the state-space partition. However, this method cannot be applied to general PWA systems, i.e. when the origin may lie on the boundaries of multiple regions in the state-space partition. The main contribution of this paper is a new ISS dual-mode MPC algorithm for hybrid systems, which extends the results of (Lazar *et al.*, 2005*a*) to these general PWA systems. The dual-mode MPC scheme uses tightened constraints and it does not require continuity of the MPC value function, nor of the PWA system dynamics. Note that tightened constraints were used before in order to *ensure robust feasibility only*, in smooth nonlinear MPC (Limon *et al.*, 2002). In this paper, however, an extension of this technique is employed for discontinuous PWA systems to achieve *both robust feasibility and ISS* (and thus, robustness to additive disturbance inputs).

A special remark is dedicated to the results presented in (Kerrigan and Mayne, 2002) and (Raković and Mayne, 2004), which deal with dynamic programming and tube based, respectively, approaches for solving feedback *min-max* MPC problems for *continuous* PWA systems, and also provide a robust stability guarantee. These results are opening roads towards feedback *min-max* MPC of hybrid systems. However, in this paper we use a different approach that does not resort to computationally expensive min-max formulations and we specifically include *discontinuous* PWA systems with the origin lying on the boundaries of multiple regions in the partition, which is not the case for the before-mentioned references.

Notation and basic definitions

Let $\mathbb{R}$, $\mathbb{R}_+$, $\mathbb{Z}$ and $\mathbb{Z}_+$ denote the field of real numbers, the set of non-negative reals, the set of integer numbers and the set of non-negative integers, respectively. We use the notation $\mathbb{Z}_{\geq c_1}$ and $\mathbb{Z}_{(c_1,c_2]}$ to denote the sets $\{k \in \mathbb{Z}_+ \mid k \geq c_1\}$ and $\{k \in \mathbb{Z}_+ \mid c_1 < k \leq c_2\}$, respectively, for some $c_1, c_2 \in \mathbb{Z}_+$. Let $\|\cdot\|$ denote an arbitrary Hölder vector p-norm and let $|\cdot|$ denote the absolute value. For a matrix $Z \in \mathbb{R}^{m\times n}$ let $\|Z\| \triangleq \sup_{x\neq 0} \frac{\|Zx\|}{\|x\|}$ denote its corresponding induced matrix norm. For a sequence $\{z_p\}_{p\in\mathbb{Z}_+}$ with $z_p \in \mathbb{R}^l$ let $\|\{z_p\}_{p\in\mathbb{Z}_+}\| \triangleq \sup\{\|z_p\| \mid p \in \mathbb{Z}_+\}$. Let $z_{[k]}$ denote the truncation of $\{z_p\}_{p\in\mathbb{Z}_+}$ at time $k \in \mathbb{Z}_+$, i.e. $z_{[k],p} = z_p$, $p \leq k$. Also, let $z_{[k_1,k_2]}$ denote the truncation of $\{z_p\}_{p\in\mathbb{Z}_+}$ at times $k_1 \in \mathbb{Z}_{\geq 1}$ and $k_2 \in \mathbb{Z}_{\geq k_1}$, i.e. $z_{[k_1,k_2],p} = z_p$, $k_1 \leq p \leq k_2$. For a set $\mathcal{P} \subseteq \mathbb{R}^n$, we denote by $\partial\mathcal{P}$ the boundary of $\mathcal{P}$, by $\text{int}(\mathcal{P})$ its interior and by $\text{cl}(\mathcal{P})$ its closure. For two arbitrary sets $\mathcal{P}_1 \subseteq \mathbb{R}^n$ and $\mathcal{P}_2 \subseteq \mathbb{R}^n$, let $\mathcal{P}_1 \sim \mathcal{P}_2 \triangleq \{x \in \mathbb{R}^n \mid x + \mathcal{P}_2 \subseteq \mathcal{P}_1\}$ and $\mathcal{P}_1 \oplus \mathcal{P}_2 \triangleq \{x+y \mid x \in \mathcal{P}_1, y \in \mathcal{P}_2\}$ denote their Pontryagin difference and Minkowski sum, respectively. For any real $\lambda \geq 0$, the set $\lambda\mathcal{P}$ is defined as $\{x \in \mathbb{R}^n \mid x = \lambda y \text{ for some } y \in \mathcal{P}\}$. A convex and compact set in $\mathbb{R}^n$ that contains the origin in its interior is called a C-set. A polyhedron (or a polyhedral set) is a set obtained as the intersection of a finite number of open and/or closed half-spaces.

2. INPUT-TO-STATE STABILITY PRELIMINARIES

Consider the discrete-time perturbed autonomous nonlinear system described by

$$x_{k+1} = G(x_k, w_k), \quad k \in \mathbb{Z}_+, \tag{1}$$

where $x_k \in \mathbb{R}^n$ is the state, $w_k \in \mathbb{R}^l$ is an unknown disturbance input and $G : \mathbb{R}^n \times \mathbb{R}^l \to \mathbb{R}^n$ is a nonlinear, possibly discontinuous function. For simplicity of notation, we assume that the origin is an equilibrium in (1) for zero disturbance input, meaning that $G(0,0) = 0$.

Definition 1. For a given $0 \leq \lambda \leq 1$, a set $\mathcal{P} \subseteq \mathbb{R}^n$ with $0 \in \text{int}(\mathcal{P})$ is called a *robust λ-contractive set* for system (1) if for all $x \in \mathcal{P}$ it holds that $G(x,w) \in \lambda\mathcal{P}$ for all $w \in \mathbb{W}$. For $\lambda = 1$ a robust λ-contractive set is called a *Robust Positively Invariant (RPI) set.*

Definition 2. A function $\varphi : \mathbb{R}_+ \to \mathbb{R}_+$ *belongs to class $\mathcal{K}$* if it is continuous, strictly increasing and $\varphi(0) = 0$. It *belongs to class $\mathcal{K}_\infty$* if $\varphi \in \mathcal{K}$ and it is radially unbounded (i.e. $\varphi(s) \to \infty$ as $s \to \infty$). A function $\beta : \mathbb{R}_+ \times \mathbb{R}_+ \to \mathbb{R}_+$ *belongs to class $\mathcal{KL}$* if for each fixed k, $\beta(\cdot, k) \in \mathcal{K}$ and for each fixed s, $\beta(s,\cdot)$ is non-increasing and $\lim_{k\to\infty} \beta(s,k) = 0$.

Next, we introduce the notion of input-to-state stability, as defined in (Jiang and Wang, 2001), for the discrete-time nonlinear system (1).

Definition 3. The perturbed system (1) is *globally Input-to-State Stable (ISS)* if there exist a $\mathcal{KL}$-function β and a $\mathcal{K}$-function γ such that, for each initial condition $x_0 \in \mathbb{R}^n$ and all $\{w_p\}_{p\in\mathbb{Z}_+}$ with $w_p \in \mathbb{R}^l$

[2] The value function corresponding to the MPC cost is usually used as the candidate Lyapunov function to prove nominal stability.

for all $p \in \mathbb{Z}_+$, it holds that the corresponding state trajectory satisfies $\|x_k\| \le \beta(\|x_0\|, k) + \gamma(\|w_{[k-1]}\|)$ for all $k \in \mathbb{Z}_{\ge 1}$. Let $\mathbb{X}$ and $\mathbb{W}$ be subsets of $\mathbb{R}^n$ and $\mathbb{R}^l$, respectively, with $0 \in \text{int}(\mathbb{X})$. We call system (1) *ISS for initial conditions in* $\mathbb{X}$ *and disturbances in* $\mathbb{W}$ if there exist a $\mathscr{KL}$-function β and a $\mathscr{K}$-function γ such that, for each $x_0 \in \mathbb{X}$ and all $\{w_p\}_{p \in \mathbb{Z}_+}$ with $w_p \in \mathbb{W}$ for all $p \in \mathbb{Z}_+$, it holds that the corresponding state trajectory satisfies $\|x_k\| \le \beta(\|x_0\|, k) + \gamma(\|w_{[k-1]}\|)$ for all $k \in \mathbb{Z}_{\ge 1}$.

The following sufficient conditions for ISS will be used throughout the paper to establish ISS for the particular case of MPC of hybrid systems.

Theorem 4. Let $\alpha_1(s) \triangleq as^\lambda$, $\alpha_2(s) \triangleq bs^\lambda$, $\alpha_3(s) \triangleq cs^\lambda$ for some $a, b, c, \lambda > 0$ and let $\sigma \in \mathscr{K}$. Let $\mathbb{W}$ be a subset of $\mathbb{R}^l$ that contains the origin. Let $\mathbb{X}$ with $0 \in \text{int}(\mathbb{X})$ be a RPI set for system (1) and let $V : \mathbb{X} \to \mathbb{R}_+$ be a function with $V(0) = 0$. Consider now the following inequalities:

$$\alpha_1(\|x\|) \le V(x) \le \alpha_2(\|x\|), \tag{2a}$$

$$V(G(x,w)) - V(x) \le -\alpha_3(\|x\|) + \sigma(\|w\|). \tag{2b}$$

If inequalities (2) hold for all $x \in \mathbb{X}$ and all $w \in \mathbb{W}$, then system (1) is ISS for initial conditions in $\mathbb{X}$ and disturbances in $\mathbb{W}$. Moreover, the ISS property of Definition 3 holds with

$$\beta(s,k) \triangleq \alpha_1^{-1}(2\rho^k \alpha_2(s)), \quad \gamma(s) \triangleq \alpha_1^{-1}\left(\frac{2\sigma(s)}{1-\rho}\right), \tag{3}$$

where $\rho \triangleq \frac{c}{b} \in [0, 1)$.

PROOF. The proof of this theorem can be based on the proof of Lemma 3.5 in (Jiang and Wang, 2001). Note that although continuity of the candidate ISS Lyapunov function V is assumed in Lemma 3.5 of (Jiang and Wang, 2001), the continuity property is not actually used in the proof. A complete proof, including how the specific form of the β and γ functions given in (3) is obtained, is given in (Lazar *et al.*, 2005*a*). □

Remark 5. The hypothesis of Theorem 4 allows that both G and V are discontinuous. It *only* implies continuity at the point $x = 0$, and *not* necessarily on a neighborhood of $x = 0$.

Definition 6. A function V that satisfies the hypothesis of Theorem 4 is called an *ISS Lyapunov function*.

3. MODEL PREDICTIVE CONTROL OF PWA SYSTEMS PRELIMINARIES

In this paper we consider nominal and perturbed discrete-time PWA systems of the form:

$$x_{k+1} = g(x_k, u_k) \triangleq A_j x_k + B_j u_k + f_j \quad \text{when } x_k \in \Omega_j, \tag{4a}$$

$$\tilde{x}_{k+1} = \tilde{g}(\tilde{x}_k, u_k, w_k) \triangleq A_j \tilde{x}_k + B_j u_k + f_j + w_k \quad \text{when } \tilde{x}_k \in \Omega_j, \tag{4b}$$

where $w_k \in \mathbb{W} \subset \mathbb{R}^n$, $k \in \mathbb{Z}_+$, $A_j \in \mathbb{R}^{n \times n}$, $B_j \in \mathbb{R}^{n \times m}$ and $f_j \in \mathbb{R}^n$, $j \in \mathscr{S}$ with $\mathscr{S} \triangleq \{1, 2, \ldots, s\}$ a *finite set* of indexes. We assume that $\mathbb{W}$ is a bounded polyhedral set that contains the origin, and the state and the input are constrained in some polyhedral C-sets $\mathbb{X}$ and $\mathbb{U}$. The collection $\{\Omega_j \mid j \in \mathscr{S}\}$ defines a partition of $\mathbb{X}$, meaning that $\cup_{j \in \mathscr{S}} \Omega_j = \mathbb{X}$ and $\text{int}(\Omega_i) \cap \text{int}(\Omega_j) = \emptyset$ for $i \ne j$. Each Ω_j is assumed to be a polyhedron (not necessarily closed). Let $\mathscr{S}_0 \triangleq \{j \in \mathscr{S} \mid 0 \in \text{cl}(\Omega_j)\}$ and let $\mathscr{S}_1 \triangleq \{j \in \mathscr{S} \mid 0 \notin \text{cl}(\Omega_j)\}$, so that $\mathscr{S} = \mathscr{S}_0 \cup \mathscr{S}_1$. We assume that the origin is an equilibrium state for (4) with $u = 0$ and therefore we require that $f_j = 0$ for all $j \in \mathscr{S}_0$. Note that this does not exclude PWA systems which are *discontinuous over the boundaries*.

Although we focus on PWA systems of the form (4), the results developed in this paper have a wider applicability since it is known (Heemels *et al.*, 2001) that PWA systems are equivalent under certain mild assumptions with other relevant classes of hybrid systems, such as mixed logical dynamical systems (Bemporad and Morari, 1999) and linear complementarity systems (van der Schaft and Schumacher, 1998).

Next, consider the case when the MPC methodology is used to generate the control input u_k, $k \in \mathbb{Z}_+$, in (4). For a fixed $N \in \mathbb{Z}_{\ge 1}$, let $\mathbf{x}_k(x_k, \mathbf{u}_k) \triangleq (x_{1|k}, \ldots, x_{N|k})$ denote the state sequence generated by the nominal PWA system (4a) from initial state $x_{0|k} \triangleq x_k$ and by applying the input sequence $\mathbf{u}_k \triangleq (u_{0|k}, \ldots, u_{N-1|k}) \in \mathbb{U}^N$, where $\mathbb{U}^N \triangleq \mathbb{U} \times \ldots \times \mathbb{U}$. Furthermore, let $\mathbb{X}_T \subseteq \mathbb{X}$ denote a desired polyhedral target set that contains the origin in its interior. The class of *admissible input sequences* defined with respect to $\mathbb{X}_T$ and state $x_k \in \mathbb{X}$, $k \in \mathbb{Z}_+$, is $\mathscr{U}_N(x_k) \triangleq \{\mathbf{u}_k \in \mathbb{U}^N \mid \mathbf{x}_k(x_k, \mathbf{u}_k) \in \mathbb{X}^N, x_{N|k} \in \mathbb{X}_T\}$. For the rest of the paper let $\|\cdot\|$ denote the ∞-norm for shortness. Consider now the functions $F(x) \triangleq \|P_j x\|$ when $x \in \Omega_j$ and $L(x,u) \triangleq \|Qx\| + \|Ru\|$, where $P_j \in \mathbb{R}^{p_j \times n}$, $j \in \mathscr{S}$, $Q \in \mathbb{R}^{q \times n}$ and $R \in \mathbb{R}^{r \times n}$ are assumed to be known matrices that have full-column rank.

Problem 7. Let $\mathbb{X}_T \subseteq \mathbb{X}$ and $N \in \mathbb{Z}_{\ge 1}$ be given. At time $k \in \mathbb{Z}_+$ let $x_k \in \mathbb{X}$ be given and minimize the cost $J(x_k, \mathbf{u}_k) \triangleq F(x_{N|k}) + \sum_{i=0}^{N-1} L(x_{i|k}, u_{i|k})$, with prediction model (4a), over all sequences $\mathbf{u}_k$ in $\mathscr{U}_N(x_k)$.

In the MPC literature, F, L and N are called the terminal cost, the stage cost and the prediction horizon, respectively. We call an initial state $x_0 \in \mathbb{X}$ *feasible* if $\mathscr{U}_N(x_0) \ne \emptyset$. Similarly, Problem 7 is said to be *feasible* for $x \in \mathbb{X}$ if $\mathscr{U}_N(x) \ne \emptyset$. Let $\mathbb{X}_f(N) \subseteq \mathbb{X}$ denote the set of *feasible states* with respect to Problem 7 and let $\widehat{V} : \mathbb{X}_f(N) \to \mathbb{R}_+$, $\widehat{V}(x_k) \triangleq \inf_{\mathbf{u}_k \in \mathscr{U}_N(x_k)} J(x_k, \mathbf{u}_k)$ denote the MPC value function corresponding to Problem 7. Suppose there exists an optimal sequence of controls $\mathbf{u}_k^* \triangleq (u_{0|k}^*, u_{1|k}^*, \ldots, u_{N-1|k}^*)$ for Problem 7 and any state $x_k \in \mathbb{X}_f(N)$. Then, $\widehat{V}(x_k) = J(x_k, \mathbf{u}_k^*)$ and the *MPC control law* is obtained as

$$\hat{u}(x_k) \triangleq u_{0|k}^*; \quad k \in \mathbb{Z}_+. \tag{5}$$

Consider now an auxiliary state feedback control law $h_{\text{aux}} : \mathbb{R}^n \to \mathbb{R}^m$ with $h_{\text{aux}}(0) = 0$, which is usually employed in proving stability of *terminal cost and constraint set* MPC. In the PWA setting we take this state feedback PWL, i.e. $h_{\text{aux}}(x) \triangleq K_j x$ when $x \in \Omega_j$, $K_j \in \mathbb{R}^{m \times n}$, $j \in \mathscr{S}$. Let $\mathbb{X}_{\mathbb{U}} \triangleq \{x \in \mathbb{X} \mid h_{\text{aux}}(x) \in \mathbb{U}\}$ denote the *safe set* with respect to both state and input constraints for this controller. Let $\mathbb{X}_{\text{PI}}$ with $0 \in \text{int}(\mathbb{X}_{\text{PI}})$ be a Positively Invariant (PI) set for system (4a) in closed-loop with h_{aux} that is contained in $\mathbb{X}_{\mathbb{U}}$. Consider now the following assumption.

Assumption 8. There exist $\{P_j, K_j \mid j \in \mathscr{S}\}$ such that

$$\begin{aligned} \|P_i(A_j + B_j K_j)x + P_i f_j\| - \|P_j x\| \\ + \|Qx\| + \|RK_j x\| \leq 0, \end{aligned} \tag{6}$$

for all $x \in \mathbb{X}_{\text{PI}}$ and all $(j,i) \in \mathscr{S} \times \mathscr{S}$.

Theorem 9. (Lazar *et al.*, 2005*a*) Suppose that Assumption 8 holds and take $\mathbb{X}_T = \mathbb{X}_{\text{PI}}$. Then, the PWA system (4a) in closed-loop with the MPC controller (5) is asymptotically stable in the Lyapunov sense for initial conditions in $\mathbb{X}_f(N)$.

The proof of Theorem 9 relies on the fact that Assumption 8 is equivalent to

$$F(g(x, h_{\text{aux}}(x))) - F(x) + L(x, h_{\text{aux}}(x)) \leq 0, \quad \forall x \in \mathbb{X}_T.$$

This in turn ensures that the hybrid MPC value function $\widehat{V}$ is a *Lyapunov function* for the closed-loop system (4a)-(5), i.e. there exist $\alpha_1(s) \triangleq as^\lambda$, $\alpha_2(s) \triangleq bs^\lambda$, $\alpha_3(s) \triangleq cs^\lambda$, with $a, b, c, \lambda > 0$, such that $\alpha_1(\|x\|) \leq \widehat{V}(x) \leq \alpha_2(\|x\|)$ and $\widehat{V}(g(x, \hat{u}(x))) - \widehat{V}(x) \leq -\alpha_3(\|x\|)$ for all $x \in \mathbb{X}_f(N)$.

In the above setting, Theorem 8.4 of (Borrelli, 2003) states that the MPC control law $\hat{u}$ defined in (5) is a PWA state-feedback. Hence, the resulting *hybrid MPC closed-loop system* is a PWA system, i.e.

$$\begin{aligned} x_{k+1} &= g(x_k, \hat{u}(x_k)) = A_j x_k + B_j \hat{u}(x_k) + f_j, \\ &\quad \hat{u}(x_k) = L_i x_k + l_i \text{ when } x_k \in \Omega_j \cap \overline{\Omega}_i, \quad (7a) \\ \tilde{x}_{k+1} &= \tilde{g}(\tilde{x}_k, \hat{u}(\tilde{x}_k), w_k) = A_j \tilde{x}_k + B_j \hat{u}(\tilde{x}_k) + f_j + w_k, \\ &\quad \hat{u}(\tilde{x}_k) = L_i \tilde{x}_k + l_i \text{ when } \tilde{x}_k \in \Omega_j \cap \overline{\Omega}_i, \quad (7b) \end{aligned}$$

with $(j,i) \in \mathscr{S} \times \overline{\mathscr{S}}$ ($\overline{\mathscr{S}}$ is a finite set of indexes), $k \in \mathbb{Z}_+$, where $L_i \in \mathbb{R}^{m \times n}$, $l_i \in \mathbb{R}^m$, and $\cup_{i \in \overline{\mathscr{S}}} \overline{\Omega}_i = \mathbb{X}_f(N)$ (with $\text{int}(\overline{\Omega}_i) \cap \text{int}(\overline{\Omega}_j) = \emptyset$ for $i \neq j$) is a new partition corresponding to the explicit MPC control law. Moreover, the MPC value function $\widehat{V}$ is a PWA function (recall that $\|\cdot\|$ denotes the ∞-norm), i.e.

$$\widehat{V}(x) \triangleq E_j x + e_j \text{ when } x \in \widehat{\Omega}_j, \tag{8}$$

where $E_j \in \mathbb{R}^{1 \times n}$, $e_j \in \mathbb{R}$, j takes values in some finite set of indexes $\widehat{\mathscr{S}}$, and $\cup_{j \in \widehat{\mathscr{S}}} \widehat{\Omega}_j = \mathbb{X}_f(N)$ (with $\text{int}(\widehat{\Omega}_i) \cap \text{int}(\widehat{\Omega}_j) = \emptyset$ for $i \neq j$) is a new partition corresponding to the MPC value function.

4. NOMINALLY STABILIZING HYBRID MPC VALUE FUNCTIONS ARE NOT NECESSARILY ISS LYAPUNOV FUNCTIONS

In the linear and continuous nonlinear case, nominally stable systems typically have some robustness properties. Note that, as done classically, if V is a *uniformly continuous* (or even stronger, *Lipschitz continuous*) Lyapunov function for the nominal dynamics, i.e. $x_{k+1} = H(x_k)$, and the disturbance acts additively on the state, i.e. $x_{k+1} = H(x_k) + w_k$, then it is easy to prove that the hypothesis of Theorem 4 is satisfied, which ensures ISS. Indeed, uniform continuity implies that for any compact subset $\mathscr{P}$ of $\mathbb{R}^n$ there exists a $\mathscr{K}$-function σ such that for any $x, y \in \mathscr{P}$ it holds that $|V(y) - V(x)| \leq \sigma(\|y - x\|)$. Hence,

$$\begin{aligned} V(H(x) + w) - V(x) &\leq V(H(x)) + \sigma(\|w\|) - V(x) \\ &\leq -\alpha_3(\|x\|) + \sigma(\|w\|) \end{aligned}$$

and thus, V is an ISS Lyapunov function. For more general robust stability results that use *continuous* candidate (ISS) Lyapunov functions see (Grimm *et al.*, 2004). Clearly, the above continuity based robustness (ISS) argument no longer holds if the function V is discontinuous at some points.

Note that discontinuity of the candidate Lyapunov function V does not necessarily obstruct the ISS inequality (2b) to hold. However, we show via an example from literature that stabilizing hybrid MPC can generate discontinuous value functions that are not ISS Lyapunov functions for perturbed systems of the form (7b). As in (7) and (8), the following notation will be used: for $i \in \overline{\mathscr{S}}$ and $j \in \widehat{\mathscr{S}}$, $\hat{u}_i(x) \triangleq L_i x + l_i$ and $\widehat{V}_j(x) \triangleq E_j x + e_j$ for any $x \in \mathbb{X}$.

Example 10. Consider the following discontinuous PWA system, taken from (Mignone *et al.*, 2000):

$$x_{k+1} = \begin{cases} A_1 x_k + B u_k & \text{if} \quad E_1 x_k > 0 \\ A_2 x_k + B u_k & \text{if} \quad E_2 x_k \geq 0 \\ A_3 x_k + B u_k & \text{if} \quad E_3 x_k > 0 \\ A_4 x_k + B u_k & \text{if} \quad E_4 x_k \geq 0 \end{cases} \tag{9}$$

where all inequalities hold componentwise, $A_1 = \begin{bmatrix} -0.04 & -0.461 \\ -0.139 & 0.341 \end{bmatrix}$, $A_2 = \begin{bmatrix} 0.936 & 0.323 \\ 0.788 & -0.049 \end{bmatrix}$, $A_3 = \begin{bmatrix} -0.857 & 0.815 \\ 0.491 & 0.62 \end{bmatrix}$, $A_4 = \begin{bmatrix} -0.022 & 0.644 \\ 0.758 & 0.271 \end{bmatrix}$, $B = \begin{bmatrix} 1 & 0 \end{bmatrix}^\top$, $E_1 = -E_3 = \begin{bmatrix} -1 & 0 \\ 0 & -1 \end{bmatrix}$ and $E_2 = -E_4 = \begin{bmatrix} -1 & 0 \\ 0 & 1 \end{bmatrix}$. The state and the input of system (9) are constrained at all times in the sets $\mathbb{X} = [-10, 10] \times [-10, 10]$ and $\mathbb{U} = [-1, 1]$, respectively. The method presented in (Lazar *et al.*, 2005*b*) was employed to compute a common terminal weight matrix $P = P_1 = P_2 = P_3 = P_4$ and feedbacks $\{K_j \mid j = 1, \ldots, 4\}$ such that inequality (6) of Assumption 8 holds for the stage cost weights $Q = \text{diag}([1\ 1])$ and $R = 0.1$.

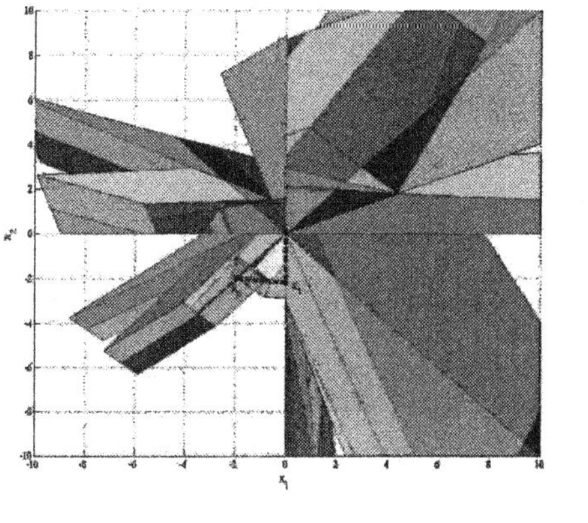

Fig. 1. The feasible set $\mathbb{X}_f(1)$ and state trajectory for the PWA MPC closed-loop system (9)-(5) with $x_0 = [-1.9649\ -1.9649]^\top$.

The following matrices were obtained:

$$P = \begin{bmatrix} 6.7001 & 3.1290 \\ -2.1107 & 4.1998 \end{bmatrix}, K_1 = \begin{bmatrix} 0.2703 & -0.1136 \end{bmatrix},$$
$$K_2 = \begin{bmatrix} -0.8042 & -0.2560 \end{bmatrix}, K_3 = \begin{bmatrix} 1.0122 & -0.7513 \end{bmatrix},$$
$$K_4 = \begin{bmatrix} -0.5548 & -1.1228 \end{bmatrix}. \quad (10)$$

Then, we used the MPT (Kvasnica *et al.*, 2004), which implements the algorithm of (Raković *et al.*, 2004), in order to calculate the terminal constraint set $\mathbb{X}_T$ as the maximal positively invariant set contained in $\mathbb{X}_\mathbb{U}$ for system (9) in closed-loop with h_{aux} with the feedbacks given in (10), and where $\mathbb{X}_\mathbb{U} = \cup_{j=1,\ldots,4}\{x \in \Omega_j \mid K_j x \in \mathbb{U}\}$. By Theorem 9, this is sufficient to guarantee that the MPC closed-loop system (9)-(5) is asymptotically stable in the Lyapunov sense for all $x \in \mathbb{X}_f(N)$, $N \in \mathbb{Z}_{\geq 1}$. Then, the MPT was used to calculate the MPC control law (5) for $N = 1$ as an explicit PWA state-feedback, and to simulate the resulting PWA MPC closed-loop system (7a). The explicit MPC controller is defined over 86 state-space regions $\overline{\Omega}_i$, $i \in \mathcal{S} \triangleq \{1, \ldots, 86\}$ that satisfy $\cup_{i\in\mathcal{S}}\overline{\Omega}_i = \mathbb{X}_f(1)$. The set of feasible states $\mathbb{X}_f(1)$ is plotted in Figure 1 together with the partition corresponding to the explicit MPC control law. □

Lemma 11. For the MPC closed-loop system (7) corresponding to system (9) of Example 10 it holds that:

(i) The value function $\widehat{V}$ and the closed-loop dynamics (7a) *are not continuous*;

(ii) $\widehat{V}$ *is a Lyapunov function* for the closed-loop dynamics (7a) showing asymptotic stability in $\mathbb{X}_f(1)$;

(iii) For any $\varepsilon > 0$, $\widehat{V}$ *is not an ISS Lyapunov function* for the closed-loop dynamics (7b) and disturbances $w \in \mathcal{B}_\varepsilon \triangleq \{w \in \mathbb{W} \mid \|w\| \leq \varepsilon\}$.

PROOF. *(i)* We have chosen the state $x^* = [0\ -2.1830]^\top \in \{\partial\overline{\Omega}_{38} \cap \partial\overline{\Omega}_{58}\} \cap \overline{\Omega}_{38}$ to show that the MPC closed-loop system (7a) and $\widehat{V}$ for the above example are not continuous on $\text{int}(\mathbb{X}_f(1))$. We have obtained the following values:

$$A_2 x^* + B\hat{u}_{38}(x^*) = [0.0130\ 0.1070]^\top;$$
$$A_3 x^* + B\hat{u}_{58}(x^*) = [-0.7791\ -1.3535]^\top;$$
$$\widehat{V}(x^*) = \widehat{V}_{38}(x^*) = 2.6766; \quad \lim_{x\to x^*, x\in\overline{\Omega}_{58}} \widehat{V}(x) = \widehat{V}_{58}(x^*) = 11.7383.$$

(ii) As Assumption 8 is satisfied via the procedure of (Lazar *et al.*, 2005*b*), the statement follows from Theorem 9.

(iii) The MPC closed-loop system (7a) corresponding to system (9) is such that the dynamics active in region $\overline{\Omega}_{38}$ is employed for $x_k = x^*$. The nominal state trajectory obtained for the initial state $x_0 = [-1.9649\ -1.9649]^\top \in \{\partial\overline{\Omega}_{47} \cap \partial\overline{\Omega}_{53}\} \cap \overline{\Omega}_{47}$ reaches the state $x_1 = x^*$ in one step (see Figure 1 for the trajectory plot). Then, for any $\mathcal{K}$-function σ we can take an arbitrarily small disturbance w such that $x^* + w \in \overline{\Omega}_{58}$, for which $\widehat{V}(x^* + w) - \widehat{V}(x_0) = \widehat{V}_{58}(x^* + w) - \widehat{V}(x_0) \approx 0.5970 > \sigma(\|w\|) \geq -\alpha_3(\|x_0\|) + \sigma(\|w\|)$ for any $\alpha_3 \in \mathcal{K}_\infty$. Hence, the ISS inequality (2b) of Theorem 4 does not hold for arbitrarily small w and thus, $\widehat{V}$ is not an ISS Lyapunov function for the closed-loop dynamics (7b). □

The result of Lemma 11-(iii) implies that the most likely and natural candidate (i.e. the MPC value function $\widehat{V}$) for proving ISS for the closed-loop system (7b) fails. Hence, one should be careful in drawing conclusions on robustness from nominal stability (established via $\widehat{V}$) when dealing with hybrid MPC. At least, there is no obvious way to infer ISS from nominal stability in hybrid MPC, or to modify nominally stabilizing MPC schemes for hybrid systems such that ISS is ensured a priori.

5. MAIN RESULT

In this section we present a new technique for setting up ISS MPC schemes for hybrid systems, which uses a dual-mode approach. In the sequel, the nomenclature of Section 3 is employed, i.e. $h_{\text{aux}}(x) = K_j x$ when $x \in \Omega_j$ and, let $\mathbb{X}_{\text{RPI}} \subseteq \mathbb{X}_\mathbb{U}$ with $0 \in \text{int}(\mathbb{X}_{\text{RPI}})$ be a RPI set for system (4b) in closed-loop with h_{aux}. Let $\xi \triangleq \max_{j\in\mathcal{S}} \|P_j\|$, let $\eta \triangleq \max_{j\in\mathcal{S}} \|A_j\|$ and, for any $i \in \mathbb{Z}_{\geq 1}$, let $\mathcal{L}_\mu^i \triangleq \{x \in \mathbb{R}^n \mid \|x\| \leq \mu \sum_{p=0}^{i-1} \eta^p\}$.

Next, choose the terminal set as $\mathbb{X}_T \triangleq \mathbb{X}_{\text{RPI}} \cap \mathbb{X}_N \subset \mathbb{X}_{\text{RPI}}$, where $\mathbb{X}_N \triangleq \cup_{j\in\mathcal{S}}\{\Omega_j \sim \mathcal{L}_\mu^N\} \subseteq \mathbb{X}$, and let $\mathcal{Q}_1(\mathbb{X}_T) \triangleq \{x \in \mathbb{X}_\mathbb{U} \mid g(x, h_{\text{aux}}(x)) \in \mathbb{X}_T\}$. Consider now the following (tightened) set of admissible input sequences:

$$\widetilde{\mathcal{U}}_N(x_k) \triangleq \{\mathbf{u}_k \in \mathbb{U}^N \mid x_{i|k} \in \mathbb{X}_i, i = 1, \ldots, N-1, \quad x_{N|k} \in \mathbb{X}_T\}, k \in \mathbb{Z}_+, \quad (11)$$

where $\mathbb{X}_i \triangleq \cup_{j\in\mathcal{S}}\{\Omega_j \sim \mathcal{L}_\mu^i\} \subseteq \mathbb{X}$ for all $i \in \mathbb{Z}_{[1,N-1]}$ and $(x_{1|k}, \ldots, x_{N|k})$ is a state sequence generated from initial state $x_{0|k} \triangleq \tilde{x}_k$ and by applying the input sequence $\mathbf{u}_k$ to the nominal PWA model (4a). Let $\widetilde{\mathbb{X}}_f(N)$

denote the set of feasible states for Problem 7 with $\widetilde{\mathscr{U}}_N(x_k)$ instead of $\mathscr{U}_N(x_k)$, and let $\widehat{V}$ and $\hat{u}$ denote the corresponding MPC value function and MPC control law, respectively.

We define a dual-mode MPC control law as follows:

$$\hat{u}^{\mathrm{DM}}(x_k) \triangleq \begin{cases} \hat{u}(x_k) \text{ if } x_k \in \widetilde{\mathbb{X}}_f(N) \setminus \mathbb{X}_{\mathrm{RPI}} \\ h_{\mathrm{aux}}(x_k) \text{ if } x_k \in \mathbb{X}_{\mathrm{RPI}} \end{cases} ; k \in \mathbb{Z}_+. \quad (12)$$

Therefore, the set of feasible states corresponding to $\hat{u}^{\mathrm{DM}}$ is $\widetilde{\mathbb{X}}_f(N) \cup \mathbb{X}_{\mathrm{RPI}}$, which contains the origin in its interior due to $0 \in \mathrm{int}(\mathbb{X}_{\mathrm{RPI}})$.

Remark 12. Usually, e.g. see (Kerrigan and Mayne, 2002), in dual-mode robust MPC the terminal set is taken as $\mathbb{X}_{\mathrm{RPI}}$. The terminal state is restricted here to a disconnected subset of $\mathbb{X}_{\mathrm{RPI}}$, i.e. $\mathbb{X}_T = \mathbb{X}_{\mathrm{RPI}} \cap \mathbb{X}_N \subset \mathbb{X}_{\mathrm{RPI}}$, with $0 \notin \mathbb{X}_T$, in order to guarantee robust feasibility of Problem 7 with $\widetilde{\mathscr{U}}_N(x_k)$ instead of $\mathscr{U}_N(x_k)$ and ISS, as it will be shown next. If the state trajectory reaches either $\mathbb{X}_T$ or $\mathbb{X}_{\mathrm{RPI}} \setminus \mathbb{X}_T$, the dual-mode control law switches to the PWL local controller and then the state trajectory remains in $\mathbb{X}_{\mathrm{RPI}}$ (*and not necessarily in* $\mathbb{X}_T$) forever, due to robust positive invariance of $\mathbb{X}_{\mathrm{RPI}}$.

Theorem 13. Take $\mu > 0$ and $N \in \mathbb{Z}_{\geq 1}$ such that $\mathbb{X}_T = \mathbb{X}_{\mathrm{RPI}} \cap \mathbb{X}_N \neq \emptyset$ and let $\mathscr{B}_\mu \triangleq \{w \in \mathbb{W} \mid \|w\| \leq \mu\}$. Suppose that h_{aux} and the terminal cost satisfy (6) for all $x \in \mathbb{X}_{\mathrm{RPI}}$. Then it holds that:

(i) If $\mathbb{X}_T \oplus \mathscr{L}_\mu^N \subseteq \mathscr{Q}_1(\mathbb{X}_T)$ and Problem 7 with $\widetilde{\mathscr{U}}_N(x_k)$ instead of $\mathscr{U}_N(x_k)$ is feasible at time $k \in \mathbb{Z}_+$ for state $\tilde{x}_k \in \mathbb{X}$, then Problem 7 with $\widetilde{\mathscr{U}}_N(x_k)$ instead of $\mathscr{U}_N(x_k)$ is feasible at time $k+1$ for state $\tilde{x}_{k+1} = A_j\tilde{x}_k + B_j\hat{u}^{\mathrm{DM}}(\tilde{x}_k) + f_j + w_k$ for all $w_k \in \mathscr{B}_\mu$ and all $k \in \mathbb{Z}_+$;

(ii) The perturbed PWA system (4b) in closed-loop with $\hat{u}^{\mathrm{DM}}$ is ISS for initial conditions in $\widetilde{\mathbb{X}}_f(N) \cup \mathbb{X}_{\mathrm{RPI}}$ and disturbances in $\mathscr{B}_\mu$.

PROOF. *(i)* There are two situations possible: either $\tilde{x}_k \in \mathbb{X}_{\mathrm{RPI}}$ or $\tilde{x}_k \notin \mathbb{X}_{\mathrm{RPI}}$. If $\tilde{x}_k \in \widetilde{\mathbb{X}}_f(N) \setminus \mathbb{X}_{\mathrm{RPI}}$ for some $k \in \mathbb{Z}_+$, let $(x^*_{1|k}, \ldots, x^*_{N|k})$ denote an optimal predicted state sequence obtained at time k from initial state $x_{0|k} \triangleq \tilde{x}_k \in \widetilde{\mathbb{X}}_f(N) \setminus \mathbb{X}_{\mathrm{RPI}}$ and by applying the input sequence $\mathbf{u}^*_k = (u^*_{0|k}, \ldots, u^*_{N-1|k})$ to the PWA model (4a). Let $(x_{1|k+1}, \ldots, x_{N|k+1})$ denote the state sequence obtained from the perturbed initial state $x_{0|k+1} \triangleq \tilde{x}_{k+1} = x_{k+1} + w_k = x^*_{1|k} + w_k$ and by applying the input sequence $\mathbf{u}_{k+1} \triangleq (u^*_{1|k}, \ldots, u^*_{N-1|k}, h_{\mathrm{aux}}(x_{N-1|k+1}))$ to the nominal PWA model (4a). The state constraints imposed in (11) ensure that: (P1) $(x_{i|k+1}, x^*_{i+1|k}) \in \Omega_{j_{i+1}} \times \Omega_{j_{i+1}}$, $j_{i+1} \in \mathscr{S}$ for all $i = 0, \ldots, N-2$ and, $\|x_{i|k+1} - x^*_{i+1|k}\| \leq \eta^i \mu$ for $i = 0, \ldots, N-1$. Then, as shown in the proof of Theorem 4.3-(i) of (Lazar *et al.*, 2005*a*), we have that $x_{i|k+1} \in \Omega_{j_{i+1}} \sim \mathscr{L}^i_\mu \subset \mathbb{X}_i$ for $i = 1, \ldots, N-2$. Next, $x_{N-1|k+1} = x^*_{N|k} + \prod_{i=1}^{N-1} A_{j_i} w_k$ and $x^*_{N|k} \in \mathbb{X}_T$ imply that $x_{N-1|k+1} \in \mathbb{X}_T \oplus \mathscr{L}^N_\mu$. Since $\mathbb{X}_T \oplus \mathscr{L}^N_\mu \subseteq \mathscr{Q}_1(\mathbb{X}_T)$, it follows that $h_{\mathrm{aux}}(x_{N-1|k+1}) \in \mathbb{U}$ and $x_{N|k+1} \in \mathbb{X}_T$. Hence, $\mathbf{u}_{k+1}$ is feasible at time $k+1$ and the optimization problem as given in Problem 7 with $\widetilde{\mathscr{U}}_N(x_k)$ instead of $\mathscr{U}_N(x_k)$ remains feasible. Consider now the other situation, i.e. $\tilde{x}_k \in \mathbb{X}_{\mathrm{RPI}}$. If the state trajectory enters (or starts in) $\mathbb{X}_{\mathrm{RPI}} \subseteq \mathbb{X}_{\mathbb{U}}$ (note that $\mathbb{X}_T \subset \mathbb{X}_{\mathrm{RPI}}$), feasibility of $\hat{u}^{\mathrm{DM}}(x_k) = h_{\mathrm{aux}}(x_k)$ is ensured due to robust positive invariance of $\mathbb{X}_{\mathrm{RPI}}$ for system (4b) in closed-loop with $u_k = h_{\mathrm{aux}}(x_k)$, $k \in \mathbb{Z}_+$.

(ii) The result of part (i) implies that $\widetilde{\mathbb{X}}_f(N) \cup \mathbb{X}_{\mathrm{RPI}}$ is a RPI set for system (4b) in closed-loop with the dual-mode MPC control $\hat{u}^{\mathrm{DM}}$ and disturbances in $\mathscr{B}_\mu$. To prove ISS, we consider three situations: in Case 1 we assume that $\tilde{x}_k \in \widetilde{\mathbb{X}}_f(N) \setminus \mathbb{X}_{\mathrm{RPI}}$ for all $k \in \mathbb{Z}_+$, in Case 2 we assume that $\tilde{x}_0 \in \mathbb{X}_{\mathrm{RPI}}$, and in Case 3 we assume that $\tilde{x}_0 \in \widetilde{\mathbb{X}}_f(N) \setminus \mathbb{X}_{\mathrm{RPI}}$ and there exists a $p \in \mathbb{Z}_{\geq 1}$ such that $\tilde{x}_k \notin \mathbb{X}_{\mathrm{RPI}}$ for all $k \in \mathbb{Z}_{<p}$ and $\tilde{x}_p \in \mathbb{X}_{\mathrm{RPI}}$.

In Case 1, the hypothesis already ensures that the MPC value function $\widehat{V}$ satisfies the ISS condition (2a) for some $a, b, c > 0$ and $\lambda = 1$ (see Theorem 4.3 of (Lazar *et al.*, 2005*a*) for a proof). Then, it follows that $\alpha_1(\|x\|) \leq \widehat{V}(x) \leq \alpha_2(\|x\|)$ for all $x \in \widetilde{\mathbb{X}}_f(N)$. Let $\tilde{x}_{k+1}$ denote the solution of the perturbed system (4b) in closed-loop with $\hat{u}^{\mathrm{DM}}$ obtained as indicated in part (i) of the proof and let $x^*_{0|k} \triangleq \tilde{x}_k$. Due to full-column rank of Q there exists $\gamma > 0$ such that $\|Qx\| \geq \gamma\|x\|$ for all x. Then, as shown in the proof of Theorem 4.3-(ii) of (Lazar *et al.*, 2005*a*) it holds that

$$\begin{aligned}\widehat{V}(\tilde{x}_{k+1}) - \widehat{V}(\tilde{x}_k) &\leq J(\tilde{x}_{k+1}, \mathbf{u}_{k+1}) - J(\tilde{x}_k, \mathbf{u}^*_k) \\ &\leq -\alpha_3(\|\tilde{x}_k\|) + \sigma(\|w_k\|),\end{aligned}$$

with $\sigma(s) \triangleq (\xi\eta^{N-1} + \|Q\|\sum_{p=0}^{N-2}\eta^p)s$ and $\alpha_3(s) \triangleq \gamma s$. Hence, it follows that $\widehat{V}$ satisfies the hypothesis of Theorem 4, thereby establishing ISS in this particular case for the closed-loop system (4b)-(12), for initial conditions in $\widetilde{\mathbb{X}}_f(N) \setminus \mathbb{X}_{\mathrm{RPI}}$ and disturbances in $\mathscr{B}_\mu$.

In Case 2, we prove that the closed-loop system is ISS by showing that the candidate (discontinuous) ISS Lyapunov function $F(x) = \|P_jx\|$ when $x \in \Omega_j$ satisfies the hypothesis of Theorem 4. Since P_j has full-column rank for all $j \in \mathscr{S}$ there exist positive constants a_j and $b_j \triangleq \|P_j\|$ such that $a_j\|x\| \leq \|P_jx\| \leq b_j\|x\|$ for all $j \in \mathscr{S}$. Hence, the $\mathscr{K}_\infty$-functions $\alpha_1(s) \triangleq \min_{j\in\mathscr{S}} a_js$ and $\alpha_2(s) \triangleq \max_{j\in\mathscr{S}} b_js$ satisfy $\alpha_1(\|x\|) \leq F(x) \leq \alpha_2(\|x\|)$ for all $x \in \mathbb{R}^n$. Next, from the hypothesis we have that inequality (6) holds for all $x \in \mathbb{X}_{\mathrm{RPI}}$ and all $(j, i) \in \mathscr{S} \times \mathscr{S}$, which yields:

$$\begin{aligned}&F((A_j + B_jK_j)x + f_j + w) - F(x) \\ &= \|P_i((A_j + B_jK_j)x + f_j + w)\| - \|P_jx\| \\ &\leq \|P_i(A_j + B_jK_j)x + P_if_j\| + \|P_iw\| - \|P_jx\| \\ &\leq -\|Qx\| + \max_{i\in\mathscr{S}}\|P_i\|\|w\| \leq -\alpha_3(\|x\|) + \sigma(\|w\|),\end{aligned}$$

for all $x \in \mathbb{X}_{\mathrm{RPI}}$, $(j, i) \in \mathscr{S} \times \mathscr{S}$ and disturbances in $\mathscr{B}_\mu$, where $\alpha_3(s) \triangleq \gamma s$ (with $\gamma > 0$ such that $\|Qx\| \geq \gamma\|x\|$) and $\sigma(s) \triangleq \max_{i\in\mathscr{S}}\|P_i\|s$. Then, due to robust positive invariance of $\mathbb{X}_{\mathrm{RPI}}$, ISS for initial conditions in $\mathbb{X}_{\text{RPI}}$ and disturbances in $\mathscr{B}_\mu$ follows from Theorem 4.

In Case 3 there exists a finite $p \in \mathbb{Z}_{\geq 1}$ such that $\tilde{x}_k \notin \mathbb{X}_{\text{RPI}}$ for all $k \in \mathbb{Z}_{<p}$ and $\tilde{x}_p \in \mathbb{X}_{\text{RPI}}$. Then, from Theorem 4, Case 1 and Case 2, it follows that there exist $\mathscr{K}\mathscr{L}$-functions β_1, β_2 and $\mathscr{K}$-functions γ_1, γ_2 such that for all $p \in \mathbb{Z}_{\geq 1}$ it holds:

$$\|\tilde{x}_k\| \leq \beta_1(\|\tilde{x}_0\|, k) + \gamma_1(\|w_{[k-1]}\|), \ \forall k \in \mathbb{Z}_{\leq p},$$
$$\|\tilde{x}_k\| \leq \beta_2(\|\tilde{x}_p\|, k-p) + \gamma_2(\|w_{[k-p,k-1]}\|), \ \forall k \in \mathbb{Z}_{>p},$$

for all $w_{[k-1]} \in \{\mathscr{B}_\mu\}^k$ and all $w_{[k-p,k-1]} \in \{\mathscr{B}_\mu\}^p$, respectively. The functions $\beta_1 \in \mathscr{K}\mathscr{L}$, $\gamma_1 \in \mathscr{K}$ and $\beta_2 \in \mathscr{K}\mathscr{L}$, $\gamma_2 \in \mathscr{K}$ are obtained as in (3) for some constants $\bar{\rho}, \underline{\rho} \in [0,1)$ and some $\mathscr{K}_\infty$-functions $\bar{\alpha}_1(s) \triangleq \bar{a}s$, $\bar{\alpha}_2(s) \triangleq \bar{b}s$ and $\alpha_1(s) \triangleq as$, $\alpha_2(s) \triangleq bs$, with $\bar{a}, \bar{b}, a, b > 0$, respectively. Then, for all $k \in \mathbb{Z}_{>p}$ and all $p \in \mathbb{Z}_{\geq 1}$ it follows that

$$\begin{aligned}
\|\tilde{x}_k\| &\leq \beta_2(\beta_1(\|\tilde{x}_0\|, p) + \gamma_1(\|w_{[p-1]}\|), k-p) \\
&\quad + \gamma_2(\|w_{[k-p,k-1]}\|) \\
&\leq \beta_2(2\beta_1(\|\tilde{x}_0\|, p), k-p) \\
&\quad + \beta_2(2\gamma_1(\|w_{[p-1]}\|), k-p) + \gamma_2(\|w_{[k-p,k-1]}\|) \\
&\overset{(13)}{\leq} \beta_3(\|\tilde{x}_0\|, k) + \beta_2(2\gamma_1(\|w_{[p-1]}\|), 1) \\
&\quad + \gamma_2(\|w_{[k-p,k-1]}\|) \\
&\leq \beta_3(\|\tilde{x}_0\|, k) + \beta_2(2\gamma_1(\|w_{[k-1]}\|), 1) \\
&\quad + \gamma_2(\|w_{[k-1]}\|) \\
&\leq \beta_3(\|\tilde{x}_0\|, k) + \gamma_3(\|w_{[k-1]}\|),
\end{aligned}$$

where $\gamma_3(s) \triangleq \beta_2(2\gamma_1(s), 1) + \gamma_2(s)$ and we used the fact that

$$\begin{aligned}
&\beta_2(2\beta_1(s,p), k-p) \\
&\quad \overset{(3)}{=} \alpha_1^{-1}(2\rho^{k-p}\alpha_2(2\bar{\alpha}_1^{-1}(2\bar{\rho}^p\bar{\alpha}_2(s))) \\
&\quad \leq 8\frac{b\bar{b}}{a\bar{a}}\tilde{\rho}^k s \triangleq \beta_3(s,k), \qquad (13)
\end{aligned}$$

and $\tilde{\rho} \triangleq \max(\rho, \bar{\rho}) \in [0,1)$. Hence, $\beta_3 \in \mathscr{K}\mathscr{L}$ and, since $\beta_2 \in \mathscr{K}\mathscr{L}$ and $\gamma_1, \gamma_2 \in \mathscr{K}$, we obtain that $\gamma_3 \in \mathscr{K}$. Applying Case 1 and Case 2 and combining with the result obtained above for Case 3 it follows that:

$$\|x_k\| \leq \beta(\|\tilde{x}_0\|, k) + \gamma(\|w_{[k-1]}\|),$$

for all $\tilde{x}_0 \in \widetilde{\mathbb{X}}_f(N) \cup \mathbb{X}_{\text{RPI}}$, $w_{[k-1]} \in \{\mathscr{B}_\mu\}^k$ and all $k \in \mathbb{Z}_{\geq 1}$, where

$$\beta(s,k) \triangleq \max(\beta_1(s,k), \beta_2(s,k), \beta_3(s,k))$$

is a $\mathscr{K}\mathscr{L}$-function and $\gamma(s) \triangleq \max(\gamma_1(s), \gamma_2(s), \gamma_3(s))$ is a $\mathscr{K}$-function. Hence, ISS is proven for system (4b) in closed-loop with $\hat{u}^{\text{DM}}$ for all initial conditions in $\widetilde{\mathbb{X}}_f(N) \cup \mathbb{X}_{\text{RPI}}$ and disturbances in $\mathscr{B}_\mu$. □

Illustrative example

Next, we demonstrate the ISS properties of the dual-mode MPC control law (12) on the PWL system (9) of Example 10, introduced in Section 4. The terminal weight matrices $P_j = P$ for $j = 1, \ldots, 4$ and the feedbacks $\{K_j \mid j \in \mathscr{S}\}$ given in (10) are such that inequality (6) holds for all $x \in \mathbb{R}^n$. In order to implement the dual-mode MPC control law one has to compute the terminal set $\mathbb{X}_T$. The MPT (Kvasnica *et al.*, 2004) was employed in order to calculate the maximal RPI set $\mathbb{X}_{\text{RPI}}$ contained in $\mathbb{X}_{\mathbb{U}}$. We choose $\mu = 0.1$ and $N = 1$, for which the terminal constraint set $\mathbb{X}_T = \mathbb{X}_{\text{RPI}} \cap \mathbb{X}_1 \neq \emptyset$ (see Figure 2), where $\mathbb{X}_1 = \cup_{j=1,\ldots,4}\{\Omega_j \sim \mathscr{L}_\mu^1\}$, satisfies the hypothesis of Theorem 13. An explicit solution of Problem 7 with $\widetilde{\mathscr{U}}_N(x_k)$ instead of $\mathscr{U}_N(x_k)$ was calculated with the MPT. The feasible set $\widetilde{\mathbb{X}}_f(1) \cup \mathbb{X}_{\text{RPI}}$ of the dual-mode MPC control law and the state-space partition (138 regions) corresponding to the explicit MPC control law are plotted in Figure 3.

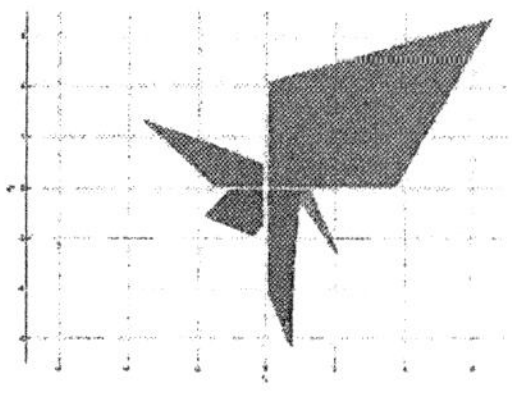

Fig. 2. The terminal constraint set $\mathbb{X}_T = \mathbb{X}_{\text{RPI}} \cap \mathbb{X}_1$.

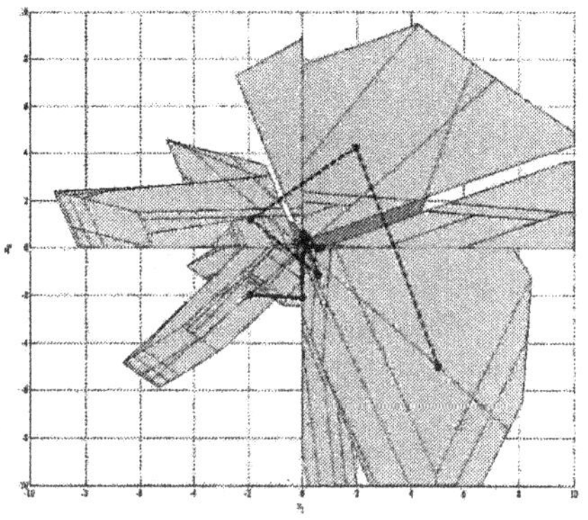

Fig. 3. The feasible set $\widetilde{\mathbb{X}}_f(1) \cup \mathbb{X}_{\text{RPI}}$: $\widetilde{\mathbb{X}}_f(1)$ - light grey; a part of $\mathbb{X}_{\text{RPI}}$ - dark grey.

Note that, by Theorem 13, ISS is ensured for the closed-loop system for initial states in $\widetilde{\mathbb{X}}_f(1) \cup \mathbb{X}_{\text{RPI}}$ and disturbances in $\mathscr{B}_\mu$, without employing a *continuous MPC value function*. Indeed, the dual-mode MPC value function $\widehat{V}$ is discontinuous at any $x \in \partial\Omega_{32} \cap \partial\Omega_{80}$. For example, $\widehat{V}_{32}(x^*) = 2.9038$ and $\widehat{V}_{80}(x^*) = 11.7383$ for $x^* = [0 \ -2.1830]^\top$, i.e. the critical point at which the nominal MPC value function of Example 10 is not an ISS Lyapunov function.

In order to illustrate the ISS property of the dual-mode MPC controller we simulated system (9) in closed-loop with $\hat{u}^{\text{DM}}$ for initial states $x_{01} = [-1.9649 \ -1.9649]^\top$ (solid line) and $x_{02} = [5 \ -5]^\top$ (dashed line) and the disturbance values depicted in Figure 4 - (a), (b) for both x_{01} and x_{02}. The control inputs are also plotted in Figure 4 - (c), (d) for initial states x_{01} and x_{02}, respectively. Once the disturbance converges to zero, the state trajectories also converge to the origin

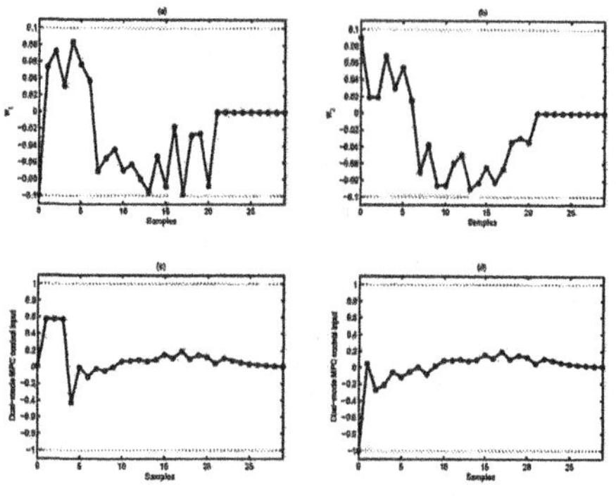

Fig. 4. Disturbance inputs $w = [w_1\ w_2]^\top$ - (a) and (b); $\hat{u}^{\mathrm{DM}}$ for x_{01} - (c); $\hat{u}^{\mathrm{DM}}$ for x_{02} - (d).

for both initial states, due to the ISS property. It is also worth to point out that the initial state x_{01}, which was a problematic initial condition, as shown in the proof of Lemma 11, is contained in the feasible set of the ISS dual-mode MPC controller. This illustrates the effectiveness of the proposed methodology.

6. CONCLUSIONS

This paper considered robust asymptotic stability in terms of ISS for *discontinuous* PWA systems controlled by MPC strategies, as this is an important property from a practical point of view. We presented an example of a PWA system (with the origin lying on the boundaries of multiple regions in the partition) taken from literature for which a nominally stabilizing MPC scheme generates an MPC value function that is not an ISS Lyapunov function. In such cases, there are no systematic ways available for modifying hybrid MPC schemes such that robustness (ISS) is a priori ensured. Therefore, a new method for setting up MPC schemes for general discontinuous PWA systems, with an a priori ISS guarantee, was developed via a dual-mode approach. The dual-mode hybrid MPC algorithm uses tightened constraints and does not require continuity of the system, the MPC control law nor of the MPC value function. An example demonstrated the effectiveness of the developed methodology.

REFERENCES

Bemporad, A. (2003). *Hybrid Toolbox-User's Guide*. http://www.dii.unisi.it/hybrid/toolbox.

Bemporad, A. and M. Morari (1999). Control of systems integrating logic, dynamics, and constraints. *Automatica* **35**, 407–427.

Borrelli, F. (2003). *Constrained optimal control of linear and hybrid systems*. Vol. 290 of *Lecture Notes in Control and Information Sciences*. Springer.

Grimm, G., M.J. Messina, S.E. Tuna and A.R. Teel (2003). Nominally robust model predictive control with state contraints. In: *42nd IEEE Conference on Decision and Control*. Maui, Hawaii. pp. 1413–1418.

Grimm, G., M.J. Messina, S.E. Tuna and A.R. Teel (2004). Examples when nonlinear model predictive control is nonrobust. *Automatica* **40**(10), 1729–1738.

Heemels, W.P.M.H., B. De Schutter and A. Bemporad (2001). Equivalence of hybrid dynamical models. *Automatica* **37**, 1085–1091.

Jiang, Z.-P. and Y. Wang (2001). Input-to-state stability for discrete-time nonlinear systems. *Automatica* **37**, 857–869.

Kerrigan, E.C. and D.Q. Mayne (2002). Optimal control of constrained, piecewise affine systems with bounded disturbances. In: *41st IEEE Conference on Decision and Control*. Las Vegas, Nevada. pp. 1552–1557.

Kvasnica, M., P. Grieder, M. Baotic and M. Morari (2004). Multi Parametric Toolbox (MPT). In: *Hybrid Systems: Computation and Control*. Lecture Notes in Computer Science, Volume 2993. Springer Verlag. Pennsylvania, Philadelphia, USA. pp. 448–462. http://control.ee.ethz.ch/ mpt.

Lazar, M., W.P.M.H. Heemels, A. Bemporad and S. Weiland (2005*a*). On the stability and robustness of non-smooth nonlinear model predictive control. In: *Workshop on Assessment and Future Directions of NMPC*. Freudenstadt-Lauterbad, Germany. pp. 327–334.

Lazar, M., W.P.M.H. Heemels, S. Weiland, A. Bemporad and O. Pastravanu (2005*b*). Infinity norms as Lyapunov functions for model predictive control of constrained PWA systems. In: *Hybrid Systems: Computation and Control*. Vol. 3414 of *Lecture Notes in Computer Science*. Springer Verlag. Zürich, Switzerland. pp. 417–432.

Limon, D., T. Alamo and E. F. Camacho (2002). Input-to-state stable MPC for constrained discrete-time nonlinear systems with bounded additive uncertainties. In: *41st IEEE Conference on Decision and Control*. Las Vegas, Nevada. pp. 4619–4624.

Mignone, D., G. Ferrari-Trecate and M. Morari (2000). Stability and stabilization of piecewise affine systems: An LMI approach. Technical Report AUT00-12. Automatic Control Laboratory, ETH Zürich. Switzerland.

Raković, S.V. and D.Q. Mayne (2004). Robust model predictive control of contrained piecewise affine discrete time systems. In: *6th IFAC NOLCOS*. Stuttgart, Germany.

Raković, S.V., P. Grieder, M. Kvasnica, D.Q. Mayne and M. Morari (2004). Computation of invariant sets for piecewise affine discrete time systems subject to bounded disturbances. In: *43nd IEEE Conference on Decision and Control*. Paradise Island, Bahamas. pp. 1418–1423.

van der Schaft, A. J. and J. M. Schumacher (1998). Complementarity modeling of hybrid systems. *IEEE Transactions on Automatic Control* **43**(3), 483–490.

ELSEVIER
IFAC
PUBLICATIONS

ROBUST MODEL PREDICTIVE CONTROL FOR PIECEWISE AFFINE SYSTEMS SUBJECT TO BOUNDED DISTURBANCES

[1] J. Thomas, [2] S. Olaru, [3] J. Buisson and [2] D. Dumur

[1] *Industrial Education College Beni Swef, Egypt Phone : +20 (0)12 52 63 476*
[2] *Supélec F 91192 Gif sur Yvette cedex , France Phone : +33 (0)1 69 85 13 75*
[3] *Supélec F 35 511 Cesson-Sévigné cedex, France Phone : +33 (0)2 99 84 45 42*
Email : jhh_thomas@yahoo.com, {sorin.olaru, jean.buisson, didier.dumur}@supelec.fr

Abstract: This paper investigates the robust tracking and regulation control problems for discrete-time, piecewise affine systems subject to bounded disturbances. In particular, the main question addressed is related to the existence of a controller such that the closed-loop system exhibits an attainable desired behavior under all possible disturbances. Checking attainability and calculating the state space regions for which a robust control is assured despite the disturbance is performed using a polyhedral approach. A model predictive control law derived from a quadratic cost function minimization is further examined as a fast sub-optimal robust control. An application of the proposed technique to a two-tank benchmark is finally presented.

Keywords: Piecewise affine systems, hybrid systems, bounded disturbances, reachability, model predictive control.

1. INTRODUCTION

Hybrid systems are now of common use in many control applications in industry, e.g. in control of mechanical systems, process control, automotive industry, power systems, aircraft and traffic control. Hybrid systems are heterogeneous dynamical systems, their behavior is determined by interacting continuous variable and discrete event dynamics. Various approaches have been proposed to model hybrid systems (Branicky *et al.*, 1998), such as Automata, Petri nets, Linear Complementary (LC), Piecewise Affine (PWA) (Sontag, 1981), Mixed Logical Dynamical (MLD) models (Bemporad and Morari, 1999a). Different techniques are used to control hybrid systems, for example Model Predictive Control (MPC) (Schutter and van den Boom, 2004; Thomas *et al.*, 2003; Olaru *et al.*, 2003; Olaru *et al.*, 2004) and optimal control (Bemporad, and Morari, 1999a).

An attractive and challenging field of research is currently dealing with hybrid systems subject to uncertainties (Bemporad and Morari, 1999b), either parameters uncertainties or disturbances influences, where problems like safety, reachability, attainability and robust control become interesting questions for researchers.

In this direction, this paper examines a class of discrete-time piecewise affine systems subject to bounded disturbances. For this class of systems, some solutions to the above mentioned problems are already proposed in the literature. For example, in (Lin and Antsaklis, 2003), an attainability checking that employs the predecessor operator, and a controller technique using finite automata and linear programming is presented. In (Necoara *et al.*, 2004, Bemporad *et al.*, 2003), a control technique based on minimizing the worst-case cost function (min-max problem) is proposed to solve the control problem.

The contribution of this paper is based on a polyhedral approach enabling the elaboration of the state space regions for which a robust control exists which drives the plant to a desired behavior despite the disturbances. The safety, reachability and attainability questions are examined through this framework and a robust Model Predictive Control (MPC) with quadratic cost function is presented as a fast suboptimal robust control.

The paper is organized as follows. A brief description of PWA systems and the related class is given in Section 2. Section 3 develops the polyhedral approach which will elaborate the state space regions where reachability, safety and attainability questions can be assured. A fast and suboptimal robust control is then developed in Section 4 for the considered class. An application of the proposed technique to a two-tank benchmark is presented in Section 5. Finally the conclusions and some remarks are given in Section 6.

2. PIECEWISE AFFINE SYSTEMS SUBJECT TO BOUNDED DISTURBANCES

Piecewise affine systems are powerful tools for describing or approximating both nonlinear and hybrid systems, and represent a straightforward extension from linear to hybrid systems. This paper focuses on the particular class of discrete-time piecewise affine systems subject to bounded disturbances, defined as:

$$S^i : \left\{ \mathbf{x}_{k+1} = \mathbf{A}^i \mathbf{x}_k + \mathbf{B}^i \mathbf{u}_k + \mathbf{C}^i \mathbf{d}_k^i + \mathbf{f}^i \right\} \quad \text{for} \begin{bmatrix} \mathbf{x}_k \\ \mathbf{u}_k \end{bmatrix} \in \chi_i \tag{1}$$

$\mathbf{x}_k \in \mathbf{X}, \mathbf{u}_k \in \mathbf{U}, \mathbf{d}_k^i \in \mathbf{D}$ denote the state, input and disturbance vector respectively at instant k (for the ith model) with $\mathbf{X}, \mathbf{U}, \mathbf{D}$ assigned polytopes ($\mathbf{D}$ contains the origin).

$\{\chi_i\}_{i=1}^s$ is the polyhedral coverage of the state and input spaces $\mathbf{X} \times \mathbf{U}$, s being the number of subsystems. Each χ_i is given by:

$$\chi_i = \left\{ \begin{bmatrix} \mathbf{x}_k \\ \mathbf{u}_k \end{bmatrix} \middle| \mathbf{Q}^i \begin{bmatrix} \mathbf{x}_k \\ \mathbf{u}_k \end{bmatrix} \le \mathbf{q}^i \right\} \tag{2}$$

Exact state measurement $\mathbf{x}$ is supposed to be available. Note that the sets χ_i are assumed here to be not disjoint so that the desired model dynamics can be chosen by the bias of switching (logical) decision variables.

Each subsystem S^i defined by the 6-uple $(\mathbf{A}^i, \mathbf{B}^i, \mathbf{C}^i, \mathbf{f}^i, \mathbf{Q}^i, \mathbf{q}^i)$, $i \in I = (1, 2, \cdots, s)$ is a component of the global hybrid system where I is the collection of all subsystems. $\mathbf{A}^i \in \Re^{n \times n}, \mathbf{B}^i \in \Re^{n \times m}, \mathbf{C}^i \in \Re^{n \times r}, \mathbf{Q}^i \in \Re^{p_i(n+m)}$ and $\mathbf{q}^i \in \Re^{p_i}$ is a suitable constant vector, where n, m, r are respectively the dimension of state, input and disturbance vectors, and p_i is the number of hyperplanes defining the χ_i polyhedral.

In this formalism, a logical control input is taken into account by developing an affine model (1) for each input value (1/0), defining linear inequality constraints linking the model with the relevant input value (2).

3. DIRECT REACHABILITY, SAFETY AND ATTAINABILITY: A POLYHEDRAL APPROACH

Let consider the region $\mathbf{R}_k, k > 1$ as a target region in the global state space $\mathbf{X}$. This section considers the robust one-step control region $\mathbf{R}_{k-1}$ as the region in the state space for which there exist a feasible mode (1) and an admissible control signal able to drive the states from $\mathbf{R}_{k-1}$ into $\mathbf{R}_k$ in one-step despite all allowable disturbances, i.e.:

$$\mathbf{R}_{k-1} = \left\{ \begin{array}{l} \mathbf{x}_{k-1} \in \mathbf{X} \left| \exists i \wedge \mathbf{u}_{k-1} \in \mathbf{U}, \begin{bmatrix} \mathbf{x}_{k-1} \\ \mathbf{u}_{k-1} \end{bmatrix} \in \chi_i \right. \\ s.t. \\ \mathbf{A}^i \mathbf{x}_{k-1} + \mathbf{B}^i \mathbf{u}_{k-1} + \mathbf{C}^i \mathbf{d}_{k-1}^i + \mathbf{f}^i \in \mathbf{R}_k, \\ \qquad \forall \mathbf{d}_{k-1}^i \in \mathbf{D} \end{array} \right\} \tag{3}$$

In the following, the computation of this region $\mathbf{R}_{k-1}$ is achieved through a polyhedral approach.
Consider the global state space defined by the following constraints:

$$\mathbf{X} := \left\{ \mathbf{F}_s \mathbf{x} \le \mathbf{g}_s, \ \mathbf{F}_s \in \Re^{p \times n}, \ \mathbf{g}_s \in \Re^p \right\} \tag{4}$$

The control input is supposed to be bounded:

$$\mathbf{U} := \left\{ \mathbf{m}\,\mathbf{u} \le \mathbf{n}, \ \mathbf{m} \in \Re^{p_u \times m}, \mathbf{n} \in \Re^{p_u \times 1} \right\} \tag{5}$$

With disturbance given inside an assigned polytope $\mathbf{d}_{k-1}^i \in \mathbf{D}$, with the target region $\mathbf{R}_k$, defined by:

$$\mathbf{R}_k := \{ \mathbf{F} \mathbf{x}_k \le \mathbf{g} \} \tag{6}$$

and considering each valid model i where $i \in (1, 2, \cdots, s)$ and using the system evaluation equation (1), enables to derive the following region:

$$\begin{aligned} \mathbf{R}_{k-1}^i = \Big\{ & \mathbf{x}_{k-1} \in \Re^n \Big| \forall \mathbf{d}_{k-1}^i \in \mathbf{D}, \\ & \mathbf{F}(\mathbf{A}^i \mathbf{x}_{k-1} + \mathbf{B}^i \mathbf{u}_{k-1} + \mathbf{C}^i \mathbf{d}_{k-1}^i + \mathbf{f}^i) \le \mathbf{g}, \\ & \mathbf{Q}^i [\mathbf{x}_{k-1} \ \ \mathbf{u}_{k-1}]^T \le \mathbf{q}^i, \Big\} \cap \{ \mathbf{F}_s \mathbf{x}_k \le \mathbf{g}_s \} \end{aligned} \tag{7}$$

The presence of disturbances can be in a first step ignored, leading to the computation of the set:

$$\widetilde{\mathbf{R}}_{k-1}^i = \left\{ \begin{bmatrix} \mathbf{F}\mathbf{A}^i & \mathbf{F}\mathbf{B}^i \\ \mathbf{Q}^i & \end{bmatrix} \begin{bmatrix} \mathbf{x}_{k-1} \\ \mathbf{u}_{k-1} \end{bmatrix} \le \begin{bmatrix} \mathbf{g} - \mathbf{F}\mathbf{f}^i \\ \mathbf{q}^i \end{bmatrix} \right\} \cap \{ \mathbf{F}_s \mathbf{x}_{k-1} \le \mathbf{g}_s \} \tag{8}$$

and to the expression of the maximal admissible region for the mode i in the absence of disturbances:

$$\hat{\mathbf{R}}_{k-1}^{i} = \mathrm{Pr}_{\mathbf{X}}\,\tilde{\mathbf{R}}_{k-1}^{i} \tag{9}$$

Remark 1: the projection of polyhedral sets can be efficiently handled in a double representation (generators/ constraints) and related tools can be found as for example - POLYLIB (Wilde, 1994).

Due to the fact that the goal is the construction of a control strategy robust with respect to the entire family of possible disturbances realizations, the previous equation finally becomes:

$$\mathbf{R}_{k-1}^{i} = \hat{\mathbf{R}}_{k-1}^{i} - \mathbf{C}^{i}\mathbf{D} \tag{10}$$

where the subtraction is computed in the Minkowsky sense (exact geometric operation, based on the double representation of polyhedral domains). The set $\mathbf{C}^{i}\mathbf{D}$ is the image of $\mathbf{D}$ by the linear mapping:

$$f:\mathbf{D}\rightarrow \Re^{n}, f(\mathbf{d}) = \mathbf{C}^{i}\mathbf{d}$$

With these sets constructed for each linear sub-model, the global one-step robust controllable region of the state space is thus given by:

$$\mathbf{R}_{k-1} = \bigcup_{i=1}^{s}\mathbf{R}_{k-1}^{i} \tag{11}$$

The procedure presented above can be repeated in a recursive way to find the domain for any limited N steps horizon. Using a dynamic programming approach, after defining the target region $\mathbf{R}_{k+N}$, the state space domain $\mathbf{R}_{k}$ can be recursively calculated, that includes all the states having a feasible control policy that can in N steps derive the states to $\mathbf{R}_{k+N}$ despite the disturbances.

Remark 2: For PWA systems with many sub-models s and for long horizon N, this may imply the exploration of a large number of regions (exponential complexity, Figure 1a). Considering thus "no switch" between sub-models over the N steps horizon (Figure 1b) leads to a lower complexity mechanism. Even if this may imply more conservatism, this suboptimal construction appears to be broad enough for many applications, this will be applied in the following sections.

Safety, a well-known geometric condition for a set to be safe (control invariant) is the following (Lin and Antsaklis, 2002):

the set $\mathbf{R}_{k+1}$ *is safe if and only if* $\mathbf{R}_{k+1} \subseteq \mathbf{R}_{k}$

Attainability, given a finite number of regions $(\mathbf{R}_{k}, \mathbf{R}_{k+1}, \cdots, \mathbf{R}_{k+N}) \in I\times\chi$, the attainability for this sequence of regions is equivalent to the following two different properties:

1. *the direct reachability from region* $\mathbf{R}_{k+j}$ *to* $\mathbf{R}_{k+j+1}$ *for* $0 \le j \le N-1$,
2. *the safety (or control invariance) for region* $\mathbf{R}_{k+N}$.

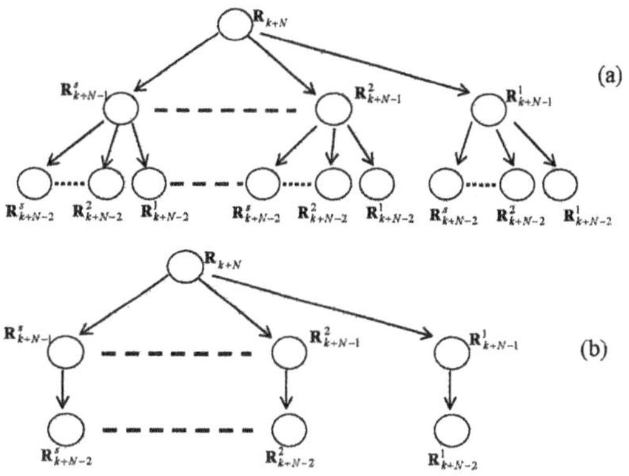

Fig. 1. Regions exploration, (a) complete exploration, (b) exploration with no switch over the N steps.

4. ROBUST MODEL PREDICTIVE CONTROL

The min-max control technique is proposed in the literature as a robust control for such problems, which minimizes the maximum cost, to try to counteract the worst disturbance. This paper focuses on the model predictive control for PWA systems with quadratic cost function as a fast suboptimal robust solution.

The model predictive control proposed here requires solving at each sampling time the following problem:

$$\begin{aligned}&\min_{\mathbf{u}_{k}^{k+N-1}} J(\mathbf{u}_{k}^{k+N-1}, \mathbf{x}_{k}) = \\ &\sum_{j=1}^{N}\left\|\mathbf{x}_{k+j} - \mathbf{x}_{e}\right\|_{\Lambda}^{2} + \sum_{j=0}^{N-1}\left\|\mathbf{u}_{k+j}\right\|_{\Gamma}^{2} \\ &\text{s.t.: } \mathbf{Q}^{i}\begin{bmatrix}\mathbf{x}_{k+j}\\ \mathbf{u}_{k+j}\end{bmatrix} \le \mathbf{q}^{i}, \\ &\mathbf{x}_{k+j} \in \mathbf{R}_{k+j}, \quad \text{for } j = 0,1,\cdots,N\end{aligned} \tag{12}$$

where $\mathbf{x}_{e}$ is the states reference, Λ, Γ are the weighting diagonal matrices in the sense $\|x\|_{\Lambda}^{2} = x^{T}\Lambda x$.

(12) is solved according to the following steps:

1. solve this quadratic problem for each dynamic among the s sub-models, staying on the dedicated branch of Figure 1.b (i.e. as assumed with Remark 2, the open-loop control sequence of the predictive law is elaborated without switching),
2. compare all the resulting costs,
3. retain the model with the lowest cost and the associated control sequence,
4. apply only the first value of this sequence and restart the procedure at the next sampling time.

Remark 3: At each sampling time, the decision process can drive the system to any particular feasible mode due to receding horizon implementation of the optimal open-loop sequence. To sum up, the conservatism is only related to the feasible set coverage and not directly to the chosen performance index.

Remark 4: If the initial state $\mathbf{x}_k$ is included in the union of regions $\mathbf{R}^i_{k-N}$ of different modes (i), the MPC technique can select a suboptimal solution among all feasible modes. The feasibility at instant k implies feasibility at any instant $k+1$ to $k+N$. The longest the prediction, the largest the feasible domain will be.

5. APPLICATION

Let consider as application of the previous theory the following benchmark consisting of two tanks (Figure 2), filled by pump acting on tank 1, continuously manipulated from 0 up to a maximum flow Q_1.

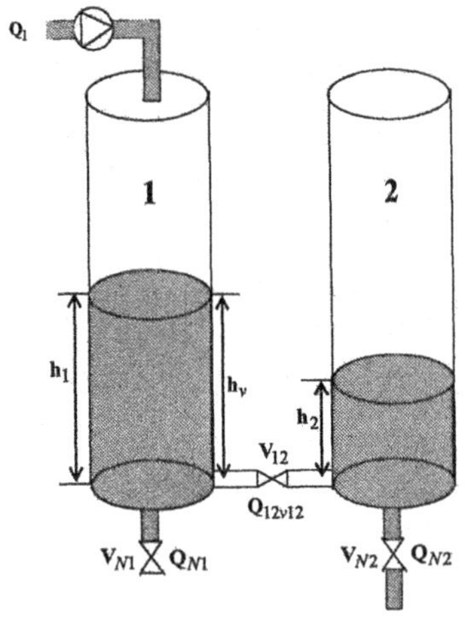

Fig. 2. Two-tank benchmark.

One switching valve V_{12} controls the flow between the tanks, this valve is assumed to be either completely opened or closed ($V_{12}=1$ or 0 respectively). The V_{N2} manual valve controls the nominal outflow of the second tank. It is assumed in further simulations that the manual valves, V_{N1} is always closed and V_{N2} is open. The liquid levels to be controlled are denoted h_1 and h_2 for each tank respectively.

The conservation of mass in the tanks provides the following differential equations:

$$\begin{aligned}\dot{h}_1 &= \frac{1}{A}(Q_1 - Q_{12V12})\\ \dot{h}_2 &= \frac{1}{A}(Q_{12V12} - Q_{N2})\end{aligned} \tag{13}$$

where the Qs denote the flows and A is the cross-sectional area of each of the tanks. The Toricelli law defines the flows in the valves by following expressions:

$$\begin{aligned}Q_{12V12} &= V_{12}aS_{12}sign(h_1-h_2)\sqrt{|2g(h_1-h_2)|}\\ Q_{N2} &= V_{N2}aS_{N2}\sqrt{2gh_2}\end{aligned} \tag{14}$$

where S_i represents the area of valves V_i and a is a constant depending on the liquid. From this, a simplified linear model can be obtained under the form:

$$\begin{aligned}Q_{12V12} &\approx k_{12}V_{12}(h_1-h_2)\\ Q_{N2} &\approx k_{N2}V_{N2}h_2\end{aligned} \tag{15}$$

where: $k_{12} = aS_{12}\sqrt{\frac{2g}{h_{\max}}}$, $k_{N2} = aS_{N2}\sqrt{\frac{2g}{h_{\max}}}$

The Euler discretisation technique is used to further derive the discrete form :

$$\begin{aligned}&h_1(k+1) = h_1(k)+\\ &\quad \frac{T_s}{A}(Q_1(k) - k_{12}V_{12}(h_1(k)-h_2(k)))\\ &h_2(k+1) = h_2(k)+\\ &\quad \frac{T_s}{A}(k_{12}V_{12}(h_1(k)-h_2(k)) - k_{N2}V_{N2}h_2(k))\end{aligned} \tag{16}$$

where T_s is the sampling time, equal to 10 s.

This benchmark can be considered as a piecewise system of form (1), with two subsystems (two modes) described as follows:

For mode one, the valve V_{12} is open:

$$\mathbf{A}^1 = \begin{bmatrix} 0.9269 & 0.07305 \\ 0.07305 & 0.8539 \end{bmatrix},\quad \mathbf{B}^1 = \begin{bmatrix} 649.351 \\ 0 \end{bmatrix}$$

$$\mathbf{Q}^1 = \begin{bmatrix} 1 & 0 & 0 \\ 0 & 1 & 0 \\ -1 & 0 & 0 \\ 0 & -1 & 0 \\ 0 & 0 & 1 \\ 0 & 0 & -1 \end{bmatrix},\quad \mathbf{q}^1 = \begin{bmatrix} 0.62 \\ 0.62 \\ 0 \\ 0 \\ 0.0001 \\ 0 \end{bmatrix}$$

For mode two, the valve V_{12} is closed:

$$\mathbf{A}^2 = \begin{bmatrix} 1 & 0 \\ 0 & 0.9269 \end{bmatrix},\quad \mathbf{B}^2 = \begin{bmatrix} 649.351 \\ 0 \end{bmatrix}$$

$$\mathbf{Q}^2 = \begin{bmatrix} 1 & 0 & 0 \\ 0 & 1 & 0 \\ -1 & 0 & 0 \\ 0 & -1 & 0 \\ 0 & 0 & 1 \\ 0 & 0 & -1 \end{bmatrix},\quad \mathbf{q}^2 = \begin{bmatrix} 0.62 \\ 0.62 \\ 0 \\ 0 \\ 0.0001 \\ 0 \end{bmatrix}$$

The previous constraints have integrated limitations on the global state space:

$$\mathbf{X} := \underbrace{\begin{bmatrix} 1 & 0 \\ 0 & 1 \\ -1 & 0 \\ 0 & -1 \end{bmatrix}}_{\mathbf{F}_s} \mathbf{x} \leq \underbrace{\begin{bmatrix} 0.62 \\ 0.62 \\ 0 \\ 0 \end{bmatrix}}_{\mathbf{g}_s} \tag{17}$$

and limitations on the control signal:

$$\mathbf{U} := \underbrace{\begin{bmatrix} 1 \\ -1 \end{bmatrix}}_{\mathbf{m}} Q_1 \leq \underbrace{\begin{bmatrix} 0.0001 \\ 0 \end{bmatrix}}_{\mathbf{n}} \tag{18}$$

The target region, to which system states will be derived to, is defined by the following constraints:

$$\mathbf{R}_{k+N} := \underbrace{\begin{bmatrix} 1 & 0 \\ 0 & 1 \\ -1 & 0 \\ 0 & -1 \end{bmatrix}}_{\mathbf{F}} \mathbf{x} \leq \underbrace{\begin{bmatrix} 0.55 \\ 0.25 \\ -0.45 \\ -0.15 \end{bmatrix}}_{\mathbf{g}} \tag{19}$$

A polytope for bounded disturbance is finally considered with:

$$\begin{bmatrix} -0.01 \\ -0.01 \end{bmatrix} \leq \begin{bmatrix} d_1 \\ d_2 \end{bmatrix} \leq \begin{bmatrix} 0.01 \\ 0.01 \end{bmatrix} \tag{20}$$

The approach presented above is first applied to elaborate the region $\mathbf{R}_k$ in the state space which includes the states that can be derived in finite N steps to $\mathbf{R}_{k+N}$ despite the disturbance. However, a suboptimal approach is used here (see Remark 2) as a compromise with the computational load.

With this assumption, Figure 3 presents the regions for mode one for $N = 5$, and Figure 4 for mode two with $N = 5$ as well. For both modes, the regions are presented in Figure 5 with $N = 3$.

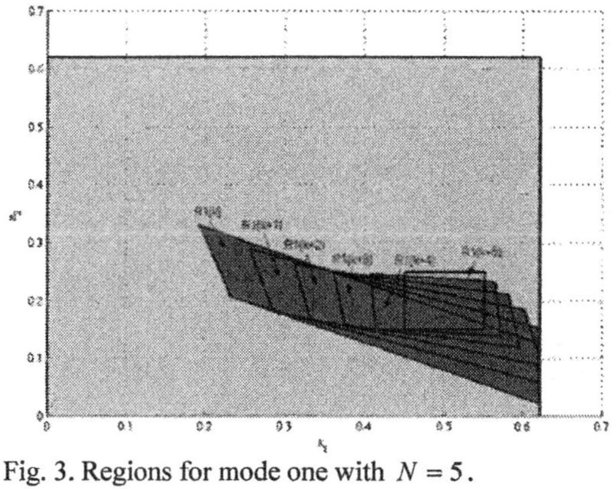

Fig. 3. Regions for mode one with $N = 5$.

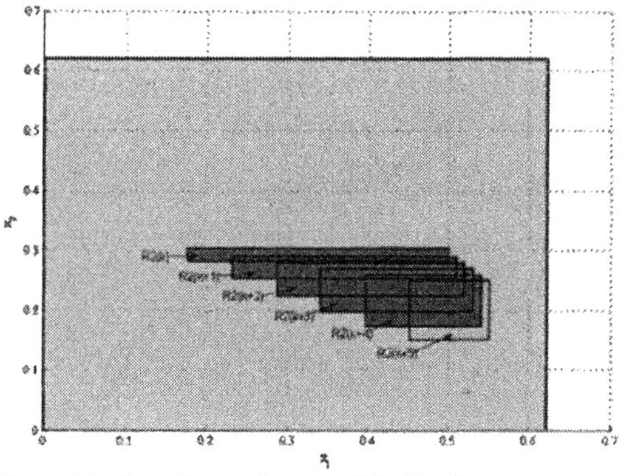

Fig. 4. Regions for mode two with $N = 5$.

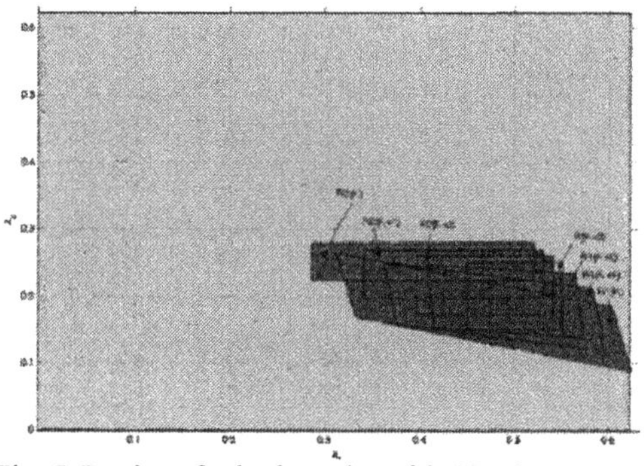

Fig. 5. Regions for both modes with $N = 3$.

The robust model predictive control presented above, is applied considering many different initial states inside the region $\mathbf{R}_k$, and at each time step, a random disturbance is added to the system states. The weighting diagonal terms in the cost function are chosen such that $\Lambda = 1000 * I_2$ and $\Gamma = 1$, and the states reference is $(0.5, 0.2)$.

Figure 6 shows some results of robust MPC with $N = 2$ for extreme initial states inside $\mathbf{R}_k$ with random disturbance, and as Figure 6 shows, all the states in $\mathbf{R}_k$ are derived in two steps ($N = 2$) to the desired region $\mathbf{R}_{k+2}$ despite the disturbance.

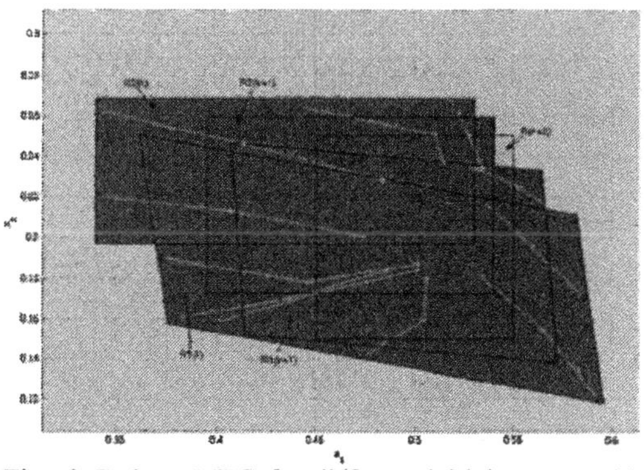

Fig. 6. Robust MPC for different initial states, with $N = 2$.

When talking about complexity, one has to mention that the convex regions computed here, $\mathbf{R}_{k-1}^{i}$ as in (11), are obtained in a dual representation (extreme points/constraints), which does not represent a computational challenge (the MPT toolbox was used here – (Kvasnica *et al.*, 2004)) as long as the number of vertices does not increase (there are polytopic regions with either 4 or 5 vertices). This fact is strongly related to the particular shape of the target region. In this case, neither the projections nor the difference of polyhedral regions should require an important computational effort.

Finally, in Figure 6 one can remark several state trajectories generated based on random disturbance realizations validating any physical extreme combination of states.

6. CONCLUSION

This paper has examined a class of discrete-time piecewise affine systems with bounded disturbance, for which a polyhedral technique has been proposed to find the regions in the state space where a feasible mode and a robust control is assured to derive the system states to the desired region despite the disturbance. Model predictive control technique has been proposed as a fast and suboptimal robust control for the considered problem.

Future work will consider applying the same techniques on discrete-time piecewise affine systems with parameter uncertainties and exogenous disturbances.

REFERENCES

Bemporad A. and M. Morari (1999a). Control of systems integrating logical, dynamics and constraints. Automatica, 35(3):407-427.

Bemporad A. and M. Morari (1999b). Robust model predictive control: A survey. In Robustness in Identification and Control, A. Garulli, A. Tesi, and A. Vicino, Eds., N. 245 in LNCIS, pp. 207-226. Springer Verlag.

Bemporad A., F. Borrelli and M. Morari (2003). Min-max control of constrained uncertain discrete-time linear systems. IEEE Transactions on Automatic Control, 48(9):1600-1606.

Branicky M.S., V.S. Borkar and S.K. Mitter (1998). A unified framework for hybrid control: model and optimal control theory. IEEE Transactions on Automatic Control, 43(1):31-45.

Lin H. and P.J. Antsaklis (2002). Controller synthesis for a class of uncertain piecewise linear hybrid dynamical systems. Proceedings of the 41st IEEE conference on Decision and Control, Las Vegas, Nevada USA, December.

Lin H. and P.J. Antsaklis (2003). Robust tracking and regulation control of uncertain piecewise linear hybrid systems. ISIS Technical report. http://www.nd.edu/~isis/tech.html.

Kvasnica M., P. Grieder and M. Baotic (2004). Multi-Parametric Toolbox (MPT), http://control.ee.ethz.ch/~mpt/

Necoara I., B. De Schutter, T.J.J. van den Boom and J. Hellendoorn (2004). Model predictive control for perturbed continuous piecewise affine systems with bounded disturbances. Proceedings of the 43rd IEEE conference on Decision and Control, Paradise Island, The Bahamas, pp. 1848-1853, December.

Olaru, S., I. Dumitrache and D. Dumur (2003). Modified MLD form for discrete optimization of hybrid systems, IFAC Conference on Analysis and Design of Hybrid Systems ADHS'03, Saint-Malo, France.

Olaru S., J. Thomas, D. Dumur and J. Buisson (2004). Genetic Algorithm based Model Predictive Control for Hybrid Systems under a Modified MLD Form. International Journal of Hybrid Systems. Vol. 4 : 1-2.

De Schutter B. and T.J.J. van den Boom (2004). Model Predictive Control for Discrete-Event and Hybrid Systems – Part II: Hybrid Systems. Proceedings of the 16th International Symposium on Mathematical Theory of Networks and Systems, B. De Moor and B. Motmans, editor Leuven, Belgium, 5-9 July.

Sontag E.D. (1981). Nonlinear regulation: the piecewise linear approach. IEEE Transaction on Automatic Control, 26(2):346-358, April.

Thomas, J., J. Buisson, D. Dumur and H. Guéguen (2003). Predictive Control of Hybrid Systems under a Multi-MLD Formalism. IFAC Conference on Analysis and Design of Hybrid Systems ADHS'03, Saint-Malo, France.

Wilde, D.K. (1994). A Library for Doing Polyhedral Operations. In Technical report 785, IRISA-Rennes, France.

ELSEVIER
IFAC
PUBLICATIONS

STABILIZATION OF SWITCHED LINEAR SYSTEMS WITH UNKNOWN TIME VARYING DELAYS

Laurentiu Hetel, Jamal Daafouz, Claude Iung

*Institut National Polytechnique de Lorraine,
CRAN UMR 7039 CNRS - UHP - INPL,
ENSEM, 2, av. forêt de la Haye,
54516, Vandœvre-Lès-Nancy, Cedex, France.
Email:{Laurentiu.Hetel, Jamal.Daafouz,
Claude.Iung}@ensem.inpl-nancy.fr*

Abstract: We consider continuous time switched systems that are stabilized via a computer. Our goal is to construct a switched digital control for continuous time switched systems that is robust to the varying feedback delay problem. The key idea of this paper is that the control synthesis problem in the case of continuous time systems with uncertain time varying feedback delays can be expressed as a problem of stabilizability for uncertain systems with polytopic uncertainties. For the sake of generality, the problem of switched systems will be considered (the solution for LTI systems can trivially be deduced by eliminating the switching aspect of the problem). *Copyright© 2006 IFAC.*

Keywords: Switched linear systems, time varying feedback delays, polytopic uncertainties, robustness, LMI.

1. INTRODUCTION

Nowadays, many control problems are solved via a computer generated feedback. When dealing with such a digital control, uncertain time varying feedback delays are unavoidable. These delays can affect system performances and lead to instability if they are not taken into account (Wittenmark *et al.*, 1995). The problem engaged scientists from both computer science (Buttazzo, 2002; Martí *et al.*, 2001; Stankovic *et al.*, 2001) and control systems (Åström and Wittenmark, 1997; Wittenmark *et al.*, 1995). It has been studied in the general context of real-time control systems (Nilsson *et al.*, 1998) for embedded and networked control systems (Årzen *et al.*, 2000; Hristu-Varsakelis and Levine, 2005; Montestruque and P.J.Antsaklis, 2004; Tzes *et al.*, 2005; Zhang *et al.*, 2001).

Here, we treat the control synthesis problem for continuous time systems affected by uncertain time varying feedback delays in a digital control context. This paper is organized as follows. In Section 2, we mathematically formalize the problem under study: "the robust stabilizability of continuous time switched systems relative to time varying feedback delays" and we introduce its sampled version. In Section 3, we show that, by using a Taylor series approximation, the system affected by time varying feedback delays can be expressed as a polytopic uncertain system. In Section 4, we present the control synthesis for this problem and we apply it to the polytopic version of the system affected by feedback delays. The approach is illustrated by a numerical example in Section 5.

[1] Work (partially) done in the framework of the HYCON Network of Excellence, contract number FP6-IST-511368

Notations: For a matrix M we denote by $\|M\|$ the induced matrix norm. By $M > 0$ or $M < 0$ we mean that the matrix M is positive or negative definite respectively. By $\mathbf{I}$ (or $\mathbf{0}$) we denote the identity (or the null) matrix with the appropriate dimension.

2. PROBLEM FORMULATION

We consider a continuous time switched system

$$\frac{dx(t)}{dt} = M_\sigma x(t) + N_\sigma u(t), \qquad (1)$$

The switching function $\sigma(t) : \mathbb{R}^+ \to P = \{1, 2, .., N\}$ gives a particular index i indicating the active system regime. It is used to represent sudden changing of system dynamic (for example the switch on/off of a pump in a tank system). $\{M_i \in \mathbf{R}^{n\times n} : i \in P\}$ and $\{N_i \in \mathbf{R}^{n\times m} : i \in P\}$ are two families of matrices. Each pair (M_i, N_i), $i \in P$ describes a continuous time model representing different regimes of system behavior. Here σ will be considered a piecewise constant function that may change its value at $t = kT$, $k \in \mathbb{Z}^+$ with $T > 0$ the sampling period.

In order to design a computer based control, a sampled model of the continuous time system is derived and discrete time control methods are applied. The final digital control is strongly dependent on the sampling period and on the discrete description of the plant.

When sampling and actuation are considered to be periodic and synchronous with the periodicity T, the equivalent discrete representation of the system is given by integrating the solution of the system over one sampling period:

$$x(k+1) = A_\sigma x(k) + B_\sigma u(k), \qquad (2)$$

$$A_\sigma = e^{M_\sigma T}, \qquad B_\sigma = \int_0^T e^{M_\sigma s} ds N_\sigma$$

A more realistic discrete representation should consider that the system is affected by several delays (Wittenmark *et al.*, 1995): delays between the sensor and the digital control $\tau^{sc}(k)$, computing delays in the controller $\tau^c(k)$ and communication delays between the controller and the actuator $\tau^{ca}(k)$. The total delay in the closed-loop is $\Delta T(k) = \tau^{sc}(k) + \tau^c(k) + \tau^{ca}(k)$. These delays have an unknown random varying length. However, the total system delay is bounded, $\Delta T_{min} \leq \Delta T(k) \leq \Delta T_{max}$. Here we will analyze the case where $\Delta T_{max} < T$ and where the switching signal σ and the sampling T are synchronous. When the effect of time delays is taken into account the system input is given by:

$$u(t) = \begin{cases} u(k-1), t \in [kT, kT + \Delta T(k)) \\ u(k), t \in [kT + \Delta T(k), (K+1)T) \end{cases} \qquad (3)$$

Problem: *Assuming that switching signal σ and the sampling T are synchronous, find a switched state feedback that robustly stabilizes the continuous time switched system (1) when the input is affected by time varying delays (3).*

Remark: It is clear that this problem is the same as the classical "timing problem" in the case of LTI systems if we consider N=1.

In order to solve such a problem, we consider the discrete representation of the system (1) over a sampling period:

$$\begin{aligned} x(k+1) = \ & \Phi_\sigma x(k) + \Gamma^0_\sigma(\Delta T(k))u(k) \\ & + \Gamma^1_\sigma(\Delta T(k))u(k-1) \end{aligned} \qquad (4)$$

$$\text{where } \Phi_\sigma = e^{M_\sigma T} = A_\sigma, \qquad (5)$$

$$\Gamma^1_\sigma(\Delta T(k)) = \int_{T-\Delta T(k)}^{T} e^{M_\sigma s} ds N_\sigma$$

$$\Gamma^0_\sigma(\Delta T(k)) = \int_0^{T-\Delta T(k)} e^{M_\sigma s} ds N_\sigma \qquad (6)$$

$$= B_\sigma - \Gamma^1_\sigma(\Delta T(k)). \qquad (7)$$

The proposed system can be written as:

$$z(k+1) = \hat{A}_\sigma z(k) + \hat{B}_\sigma u(k) \qquad (8)$$

$$\text{where } \hat{A}_\sigma(k) = \begin{bmatrix} \Phi_\sigma & \Gamma^1_\sigma(\Delta T(k)) \\ \mathbf{0} & \mathbf{0} \end{bmatrix},$$

$$\hat{B}_\sigma(k) = \begin{bmatrix} \Gamma^0_\sigma(\Delta T(k)) \\ \mathbf{I} \end{bmatrix}$$

$$\text{and } z(k) = \begin{bmatrix} x(k) \\ u(k-1) \end{bmatrix}.$$

This representation has been adapted for switched systems from the real time linear system representation presented by Åström and Wittenmark (Åström and Wittenmark, 1997). In this model $\Gamma^1_\sigma(\Delta T(k))$ and $\Gamma^0_\sigma(\Delta T(k))$ are strongly dependent on the uncertain time varying delay $\Delta T(k)$. Therefore the previous system is an uncertain system with time varying uncertainty.

Now, the problem under study reduces to find a switched state feedback $u(k) = K_\sigma z(k)$ that stabilizes the discrete switched uncertain system (8) when the total system delay $\Delta T_{min} \leq \Delta T(k) \leq \Delta T_{max}$ is time varying and unknown.

In the next section we will show the way the uncertain delay dependent system (8) can be expressed as a switched polytopic uncertain system.

3. EXPRESSING UNCERTAINTIES AS POLYTOPES OF MATRICES

When considering the equivalent discrete representation (8), the two matrices $\Gamma^1_\sigma(\Delta T(k))$ and $\Gamma^0_\sigma(\Delta T(k))$ are delay dependent uncertainties. If they can be expressed as convex polytopes, system (8) can be treated as a switched polytopic uncertain system, for which stability conditions can be expressed in terms of linear matrix inequalities.

Step 1: Taylor series expansion of the uncertainties $\Gamma^1_\sigma(\Delta T(k))$ and $\Gamma^0_\sigma(\Delta T(k))$

Proposition 1. $\Gamma^1_\sigma(\Delta T(k))$ and $\Gamma^0_\sigma(\Delta T(k))$, the elements of system (8) matrices can be expressed as:

$$\Gamma^1_\sigma(\Delta T(k)) = -\sum_{q=1}^{\infty}(-\Delta T(k))^q \frac{M_\sigma^{q-1}}{q!} e^{M_\sigma T} N_\sigma \text{ and} \tag{9}$$

$$\Gamma^0_\sigma(\Delta T(k)) = B_\sigma + \sum_{q=1}^{\infty}(-\Delta T(k))^q \frac{M_\sigma^{q-1}}{q!} e^{M_\sigma T} N_\sigma.$$

Proof. Consider

$$I(x) = \int_{T-x}^{T} e^{M_\sigma s} ds. \tag{10}$$

Using the Taylor series expansion one can write:

$$\begin{aligned} I(x) &= I(0) + \dot{I}(0)x + \ldots + \frac{d^q I}{dt^q}(0)\frac{x^q}{q!} + \ldots \\ &= -\sum_{q=1}^{\infty} \frac{(-x)^q}{q!} M_\sigma^{q-1} e^{M_\sigma T} \end{aligned} \tag{11}$$

From (10), (11) and (5) Proposition 1 is proved for $x = \Delta T(k)$. □

Step 2: h-order approximation of Taylor series for uncertainties

Focusing on the first h terms of the previous formulation, $\Gamma^1_\sigma(\Delta T(k))$ can be expressed as a finite sum and a remainder

$$\Gamma^1_\sigma(\Delta T(k)) = -\sum_{q=1}^{h}(-\Delta T(k))^q \frac{M_\sigma^{q-1}}{q!} e^{M_\sigma T} N_\sigma + \Theta^h_\sigma \tag{12}$$

where the remainder is

$$\begin{aligned} \Theta^h_\sigma &= -\sum_{q=h+1}^{\infty} (-\Delta T(k))^q \frac{M_\sigma^{q-1}}{q!} e^{M_\sigma T} N_\sigma \\ &= -\left(\int_0^{-\Delta T} e^{M_\sigma \tau} d\tau - \sum_{q=1}^{h} (-\Delta T)^q \frac{M_\sigma^{q-1}}{q!}\right) e^{M_\sigma T} N_\sigma. \end{aligned} \tag{13}$$

$\Gamma^{1,h}_\sigma(\Delta T(k))$ and $\Gamma^{0,h}_\sigma(\Delta T(k))$, given by

$$\Gamma^{1,h}_\sigma(\Delta T(k)) = -\sum_{q=1}^{h}(-\Delta T(k))^q \frac{M_\sigma^{q-1}}{q!} e^{M_\sigma T} N_\sigma,$$

$$\Gamma^{0,h}_\sigma(\Delta T(k)) = B_\sigma - \Gamma^{1,h}_\sigma(\Delta T(k)). \tag{14}$$

will be called the h-order approximation.

Step 3: Polytopic form of the h-order approximation

Consider the notations:

$$\begin{aligned} G_{\sigma,q} &= (-1)^{q+1} \frac{M_\sigma^{q-1}}{q!} e^{M_\sigma T} N_\sigma, \ q = 1..h, \text{ and} \\ \phi_1 &= \left[\underline{\rho}^h \mathbf{I}\ \underline{\rho}^{h-1}\mathbf{I} \ldots \underline{\rho}^2 \mathbf{I}\ \underline{\rho}\mathbf{I}\right]', \\ \phi_2 &= \left[\underline{\rho}^h \mathbf{I}\ \underline{\rho}^{h-1}\mathbf{I} \ldots \underline{\rho}^2 \mathbf{I}\ \overline{\rho}\mathbf{I}\right]', \ldots, \\ \phi_{h+1} &= \left[\overline{\rho}^h \mathbf{I}\ \overline{\rho}^{h-1}\mathbf{I} \ldots \overline{\rho}^2 \mathbf{I}\ \overline{\rho}\mathbf{I}\right]' \end{aligned} \tag{15}$$

with $\underline{\rho} = \Delta T_{min}$ and $\overline{\rho} = \Delta T_{max}$.

Proposition 2. The h-order approximation $\Gamma^{1,h}_\sigma(\Delta T(k))$ and $\Gamma^{0,h}_\sigma(\Delta T(k))$ can be expressed as the convex matrix polytopes:

$$\Gamma^{1,h}_\sigma(\Delta T(k)) = \sum_{i=1}^{h+1} \mu_i(k) U^{1,h}_{\sigma,i}, \tag{16}$$

$$\Gamma^{0,h}_\sigma(\Delta T(k)) = \sum_{i=1}^{h+1} \mu_i(k) U^{0,h}_{\sigma,i},$$

$$\sum_{i=1}^{h+1} \mu_i(k) = 1,\ \mu_i(k) > 0\ \forall i = 1,..,h+1,\ \forall k \in \mathbb{Z}^+,$$

where the polytope vertices are

$$U^{1,h}_{\sigma,i} = [G_{\sigma,1} \ldots G_{\sigma,h}]\, \phi_i$$

and

$$U^{0,h}_{\sigma,i} = B_\sigma - U^1_{\sigma,i}$$

respectively.

Proof. From equation (14) and the notations (15) $\Gamma^{1,h}_\sigma(\Delta T(k))$ can be expressed as

$$\Gamma^{1,h}_\sigma(\Delta T(k)) = \sum_{q=1}^{h} \rho^q(k) G_{\sigma,q} \tag{17}$$

with $\rho(k) = \Delta T(k)$. The equation (17) can be written as:

$$\Gamma^{1,h}_\sigma(\Delta T(k)) = [G_{\sigma,1} \ldots G_{\sigma,h}]\, \Psi(k) \tag{18}$$

where

$$\Psi(k) = \left[\rho(k)^h \mathbf{I} \ldots \rho^2(k)\mathbf{I}\ \ \rho(k)\mathbf{I}\right]'$$

In the space of parameters $\Psi(k)$ it can be shown that $\rho(k) \in [\underline{\rho}, \overline{\rho}]$, $\forall k \in \mathbb{Z}^+$

$$\Psi(k) = \sum_{i=1}^{h+1} \mu_i(k) \phi_i, \ \text{with} \tag{19}$$

$$\mu_i(k) > 0 \text{ and } \sum_{i=1}^{h+1} \mu_i(k) = 1\ \forall k \in \mathbf{Z}^+.$$

The uncertain parameters $\mu_i(k)$ are solutions of the linear system:

$$\begin{bmatrix} 1 & \ldots & \ldots & \ldots & 1 \\ \underline{\rho} & \overline{\rho} & \ldots & \ldots & \overline{\rho} \\ \underline{\rho}^2 & \underline{\rho}^2 & \overline{\rho}^2 & \ldots & \overline{\rho}^2 \\ \vdots & & & \ddots & \vdots \\ \underline{\rho}^h & \ldots & \ldots & \underline{\rho}^h & \overline{\rho}^h \end{bmatrix} \begin{bmatrix} \mu_1(k) \\ \mu_2(k) \\ \vdots \\ \mu_{h+1}(k) \end{bmatrix} = \begin{bmatrix} 1 \\ \rho(k) \\ \rho^2(k) \\ \vdots \\ \rho^h(k) \end{bmatrix}$$

Using the classical Gauss method, the solutions can be computed by the recursive formula:

$$\mu_1 = 1 - \frac{\rho - \underline{\rho}}{\overline{\rho} - \underline{\rho}}, \tag{20}$$

$$\mu_k = \frac{\rho^{k-1} - \underline{\rho}^{k-1}}{\overline{\rho}^{k-1} - \underline{\rho}^{k-1}} - \sum_{i=k+1}^{h+1} \mu_i, \; k = 2..h.$$

One can prove that

$$\mu_k = \frac{\rho^{k-1} - \underline{\rho}^{k-1}}{\overline{\rho}^{k-1} - \underline{\rho}^{k-1}} - \frac{\rho^{k} - \underline{\rho}^{k}}{\overline{\rho}^{k} - \underline{\rho}^{k}}, \; k = 2..h.$$

which is strictly positive since the function $f : \Re \to \Re$, $f(x) = \frac{\rho^x - \underline{\rho}^x}{\overline{\rho}^x - \underline{\rho}^x}$ is monotone decreasing for $x \in (0, \infty)$. We can remark that while $\rho \in [\underline{\rho}, \overline{\rho}]$, $0 < \mu_k < 1 \; \forall k = 1..h+1$ therefore all the solutions are barycentric coordinates. From (18,19), with the notations (15) and (14) *Proposition 2* is proved. □

Step 4: System polytopic truncated form

Let

$$\hat{A}^h_{\sigma i} = \begin{bmatrix} \Phi_\sigma & U^{1,h}_{\sigma,i} \\ \mathbf{0} & \mathbf{0} \end{bmatrix}, \hat{B}^h_{\sigma i} = \begin{bmatrix} U^{0,h}_{\sigma,i} \\ \mathbf{I} \end{bmatrix}.$$

Considering the h-order approximation of the uncertainties in equation (14) and Proposition 2, the system (8) can be expressed in the truncated polytopic form:

$$z(k+1) = \hat{A}^h_\sigma z(k) + \hat{B}^h_\sigma u(k) \tag{21}$$

where

$$\hat{A}^h_\sigma(k) = \sum_{i=1}^{h+1} \mu_i(k) \hat{A}^h_{\sigma i}, \quad \hat{B}^h_\sigma(k) = \sum_{i=1}^{h+1} \mu_i(k) \hat{B}^h_{\sigma i}.$$

System (8) truncation is a switched uncertain system with polytopic uncertainty (21).

4. CONTROL SYNTHESIS

In the previous section it was shown that for any switched system with time varying feedback delay there exists a polytopic approximation. Considering the control synthesis in this context, one has to find a switched state feedback that stabilizes the h-order approximation system (21). It is clear that finding such a control law and proving that it is also valid for non truncated form of the system (8) solves the original problem. Robust stability results for switched uncertain systems with polytopic uncertainty are given in this section.

4.1 Switched control design for uncertain switched systems

Until now several control approaches have been presented separately for switched systems (Daafouz *et al.*, 2002) and for polytopic uncertain systems (J.Daafouz and Bernussou, 2001). Here we extend the existing approaches for the following switching uncertain system:

$$z(k+1) = \hat{A}_\sigma(k) z(k) + \hat{B}_\sigma(k) u(k), \tag{22}$$

where

$$\hat{A}_\sigma = \sum_{j=1}^{na_\sigma} \alpha_{\sigma j}(k) A_{\sigma j}, \text{ and } \hat{B}_\sigma = \sum_{l=1}^{nb_\sigma} \beta_{\sigma l}(k) B_{\sigma l},$$

$$\sum_{j=1}^{na_\sigma} \alpha_{\sigma j}(k) = 1, \; \alpha_{\sigma j}(k) \geq 0,$$

$$\sum_{l=1}^{nb_\sigma} \beta_{\sigma l}(k) = 1, \; \beta_{\sigma l}(k) \geq 0, \; \forall \, k \in \mathbb{Z}^+$$

represent the uncertainty on the dynamic and input matrix, respectively. The switching signal is given by σ. Here $\alpha_{\sigma j}$ and $\beta_{\sigma l}$ are the uncertain parameters describing each uncertainty while na_σ and nb_σ represent the number of extreme points in the uncertainty $\hat{A}_\sigma$ and $\hat{B}_\sigma$ respectively. This is a more general form of the switched polyopic system (21); the input and the state matrix are affected by distinct polytopic uncertainty with different uncertain parameters. This system can also be expressed as:

$$z(k+1) = \sum_{i=1}^{N} \xi_i(k) \hat{A}_i(k) z(k) + \sum_{i=1}^{N} \xi_i(k) \hat{B}_i(k) u(k), \tag{23}$$

$$\xi_i : \mathbb{Z}^+ \to \{0,1\}, \; \sum_{i=1}^{N} \xi_i(k) = 1,$$

The closed-loop dynamics with the switched state feedback

$$u(k) = \sum_{i=1}^{N} \xi_i(k) K_i z(k) \tag{24}$$

is described by the equation:

$$z(k+1) = \sum_{i=1}^{N} \xi_i(k) (\hat{A}_i + \hat{B}_i K_i) z(k).$$

With the uncertainty description (22), the equation becomes

$$z(k+1) = \sum_{i=1}^{N} \sum_{j=1}^{na_i} \sum_{l=1}^{nb_i} \xi_i(k) \alpha_{ij}(k) \beta_{il}(k) H_{ijl} z(k)$$

where

$$H_{ijl} = A_{ij} + B_{il} K_i.$$

We use switched parameter dependent Lyapunov functions given by:

$$V(k) = z^T(k) \mathcal{P} z(k) \tag{25}$$

with

$$\mathcal{P} = \sum_{i=1}^{N} \sum_{j=1}^{na_i} \sum_{l=1}^{nb_i} \xi_i(k) \alpha_{ij}(k) \beta_{il}(k) P_{ijl}.$$

The difference along the system trajectories is:

$$V(k+1) - V(k) = z^T(k)(\mathcal{H}^T \mathcal{P}_+ \mathcal{H} - \mathcal{P})z(k),$$

where

$$\mathcal{H} = \sum_{i=1}^{N} \sum_{j=1}^{na_i} \sum_{l=1}^{nb_i} \xi_i(k)\alpha_{ij}(k)\beta_{il}(k)H_{ijl},$$

$$\mathcal{P}_+ = \sum_{i=1}^{N} \sum_{j=1}^{na_i} \sum_{l=1}^{nb_i} \xi_i(k+1)\alpha_{ij}(k+1)\beta_{il}(k+1), \tag{26}$$

$$P_{ijl} = \sum_{m=1}^{N} \sum_{u=1}^{na_m} \sum_{v=1}^{nb_m} \xi_m(k)\alpha_{mu}(k)\beta_{mv}(k)P_{muv}.$$

Theorem 1. System (23) is stabilizable via the control law (24) if there exists symmetric positive definite matrices S_{ijl} and S_{muv}, and matrices G_i and R_i, solutions of the LMI:

$$\begin{bmatrix} G_i + G_i^T - S_{ijl} & G_i^T A_{ij}^T + R_i^T B_{il}^T \\ A_{ij}G_i + B_{il}R_i & S_{muv} \end{bmatrix} > 0, \tag{27}$$

$i, m = 1..N, j = 1..na_i, l = 1..nb_i, u = 1..na_m, v = 1..nb_m$. The switched state feedback control is given by (24) with $K_i = R_i G_i^{-1}$.

Proof. see (Hetel *et al.*, 2005). □

Remark: Theorem 1 gives robust LMI stabilizability conditions based on switched parameter dependent Lyapunov functions and hence less conservative than all the other results in the literature (Ji *et al.*, 2003; Zhai *et al.*, 2003). Moreover, these results obviously apply to the state reconstruction problem for uncertain switched systems (Daafouz *et al.*, 2003).

4.2 Control synthesis for switched systems with uncertain time varying feedback delay

The truncation (21) of the system with uncertain feedback delay (8) can be considered as a switched uncertain system with polytopic uncertainty (22) for which we developed LMI stabilizability conditions (27). The problem of finding a computer control $u(k) = K_\sigma^h z(k)$ for the discrete h-order truncation (21) of the discrete time switched system (8) is the same as the control synthesis problem for a switched uncertain system with polytopic uncertainty (22).

Proposition 3. If there exists positive definite symmetric matrices S_{ij}, S_{uv} and matrices G_i and R_i solutions of the LMI conditions:

$$\begin{bmatrix} G_i + G_i^T - S_{ij} & G_i^T \hat{A}_{ij}^{h^T} + R_i^T \hat{B}_{ij}^{h^T} \\ \hat{A}_{ij}^h G_i + \hat{B}_{ij}^h R_i & S_{uv} \end{bmatrix} > 0 \tag{28}$$

where $i, u = 1 \dots N$, $j, v = 1 \dots h+1$ then the switched state feedback stabilizing gains are given by $K_i^h = R_i G_i^{-1}$.

Proof. see (Hetel *et al.*, 2005) □

A procedure based on considering the neglected remainder Θ_σ^h (see Step 2 and Step 5 in Section 3) as a disturbance and using LMI based conditions for stability analysis can be used to check that the control law obtained using Proposition 3 is valid for the non truncated form of the system (8). Consider the notations:

$$H_{\sigma j}^h = A_{\sigma j}^h + B_{\sigma j}^h K_\sigma^h, \; \hat{H}_\sigma^h = \sum_{j=1}^{h+1} \mu_j H_{\sigma j}^h, \tag{29}$$

$$\mu_i(k) > 0, \; \sum_{i=1}^{h+1} \mu_i(k) = 1 \; \forall k \in \mathbf{Z}^+, \tag{30}$$

$$K_\sigma^h = \begin{bmatrix} K_\sigma^{1,h} & K_\sigma^{2,h} \end{bmatrix}$$

with

$$K_\sigma^{1,h} \in \mathbf{R}^{m \times n}, \quad K_\sigma^{2,h} \in \mathbf{R}^{m \times m}$$

$$D = [\mathbf{I}\ \mathbf{0}]^T, \; E_\sigma^h = [\mathbf{I}\ \mathbf{I}] \begin{bmatrix} -K_\sigma^{1,h} & \mathbf{I} \\ \mathbf{0} & -K_\sigma^{2,h} \end{bmatrix}.$$

Theorem 2. If there exists symmetrical positive definite matrices P_{ij}, and matrices G_{ij} with $i = 1, \dots, N$ and $j = 1, \dots, h+1$ solutions of the *LMI*:

$$\begin{bmatrix} -P_{ij} & H_{ij}^{h^T} G_{pq}^T & H_{ij}^{h^T} G_{pq} D & E_i^{h^T} \\ (\bullet)^T & P_{pq} - G_{pq} - G_{pq}^T & \mathbf{0} & \mathbf{0} \\ (\bullet)^T & (\bullet)^T & D^T(G_{pq} + G_{pq}^T)D - \mathbf{I} & \mathbf{0} \\ (\bullet)^T & (\bullet)^T & (\bullet)^T & -\gamma \mathbf{I} \end{bmatrix} < 0 \tag{31}$$

$\forall\ i, p = 1, \dots, N\ j, q = 1, \dots, h+1$ with $\|\Theta_\sigma^h\| \leq \gamma^{-\frac{1}{2}}$ for all σ, then the feedback gains K_σ^h ensure the stabilizability of the original discrete time system with time varying delays (8), that is system (4) is stable.

Proof. see (Hetel *et al.*, 2005)□

In practical applications, finding a feedback control for the original discrete time system (8) is a compromise between two constraints: high order uncertainty approximation, for a good discrete system representation, and small number of LMIs. These two constraints are opposite because a high order uncertainty approximation implies a large number of LMIs and the numerical LMI solvers are not able to compute a feedback gain when the number of LMIs is too large.

5. NUMERICAL EXAMPLE

Consider the following system:

$$\frac{dx}{dt} = M_\sigma x(t) + N_\sigma u(t), \; \sigma \in \{1, 2\} \text{ with}$$

$$M_1 = \begin{bmatrix} 3 & 2 \\ 1 & -1 \end{bmatrix}, \quad M_2 = \begin{bmatrix} 2 & 1 \\ 3 & -4 \end{bmatrix},$$

$$N_1 = [1\ 0]' \quad \text{and} \quad N_2 = [1.5\ 0]'$$

The system is sampled with $T = 0.3s$ and the feedback delay is bounded by $\Delta T_{min} = 0s$ and $\Delta T_{max} = 0.1s$. When considering a first order approximation of the system, state feedback K_σ^1 gains are found via the LMIs (28) but are not validated by (31). A second order approximation is performed. In this case we obtain the switched feedback gains:

$$K_1^2 = [-2.99\ \ -1.3\ 0.27]$$

and

$$K_2^2 = [-2\ \ -0.38\ 0.21]$$

which can be validated by (31) for the original system.

6. CONCLUSION

This paper was dedicated to the robust control synthesis for continuous time switched systems relative to time varying feedback delays. The sampled version of the system has been considered and the control synthesis in this context has been treated as a problem of stabilizability for uncertain systems with polytopic uncertainties. A method for switched state feedback control synthesis has been presented. A numerical example has been presented to illustrate this approach.

REFERENCES

Årzen, K. E., A. Cervin, J. Eker and L. Sha (2000). An introduction to control. In: *39th IEEE Conference on Decision and Control.* Sydney, Australia.

Åström, K. J. and B. Wittenmark (1997). *Computer-Controlled Systems.* Prentice Hall.

Buttazzo, G. (2002). Real-time operating systems: Problems and novel solutions. In: *Proceedings of the 7th Int. Symposium on Formal Techniques in Real-Time and Fault Tolerant Systems.* Oldenburg, Germany.

Daafouz, J., P. Riedinger and C. Iung (2002). Stability analysis and control synthesis for switched systems: A switched lyapunov function approach. *IEEE Trans. Automat. Contr.* **47**, 1883–1887.

Daafouz, J., P. Riedinger and C. Iung (2003). Observer-based switched control design with pole placement for discrete-time switched systems. *International Journal Of Hybrid System* **3**, 263–282.

Hetel, L., J. Daafouz and C. Iung (2005). Stabilization of arbitrary switched linear systems with unknown time varying delays. *Technical report, December.*

Hristu-Varsakelis, D. and Levine, W. S., Eds. (2005). *Handbook of Networked and Embedded Control Systems Series: Control Engineering.* Birkhäuser.

J.Daafouz and J. Bernussou (2001). Parameter dependent lyapunov functions for discrete time systems with time varying parametric uncertainties. *Systems and Control Letters* **43**, 355–359.

Ji, Z., L.Wang and G.Xie (2003). Robust stability analysis and control synthesis for discrete time uncertain switched systems. In: *Proceedings of the 42th IEEE Conference on Decision and Control.* Maui, Hawaii USA.

Martí, P., G. Fohler, K. Ramamritham and J. M. Fuertes (2001). Jitter compensation for real-time control systems. In: *In 22nd IEEE Real-Time Systems Symposium (RTSS01).* London, UK.

Montestruque, L. and P.J.Antsaklis (2004). Stability of model-based networked control systems with time-varying transmission times. *IEEE Transactions on Automatic Control, Special Issue on Networked Control Systems* **49**, 1562–1572.

Nilsson, J., B. Bernhardsson and B. Wittenmark (1998). Stochastic analysis and control of real-time systems with random time delays. *Automatica* **34**, 57–64.

Stankovic, J. A., T. He, T. F. Abdelzaher, M. Marley, G. Tao, S. H. Son and L. Cenyan (2001). Feedback control scheduling in distributed real-time systems.. In: *IEEE Real-Time Systems Symposium.*

Tzes, A., G. Nikolakopoulos and I. Koutroulis (2005). Development and experimental verification of a mobile client-client networked controlled system. *European Journal of Control.*

Wittenmark, B., J. Nilsson and M. Törngren. (1995). Timing problems in real-time control systems. In: *Proceedings of the 1995 American Control Conference.* Seattle,WA,USA.

Zhai, G., H. Lin and P.J. Antsaklis (2003). Quadratic stabilizability of switched linear systems with polytopic uncertainties. *Int.J.Control* **76**, 747–753.

Zhang, W., M.S. Branicky and S.M. Phillips (2001). Stability of networked control systems. *IEEE Control Systems Magazine.*

ELSEVIER
IFAC
PUBLICATIONS

STABILIZING DYNAMIC CONTROLLER OF SWITCHED LINEAR SYSTEMS

Salim Chaib *,[2], **Abderraouf Benali** *, **Driss Boutat** *, **Jean-Pierre Barbot** **

* *Laboratoire de Vision et Robotique, Ecole Nationale Supérieure d'Ingénieurs de Bourges, 10 Boulevard de Lahitolle, 18020 Bourges, France.*
** *Equipe Commande des Systèmes, Ecole Nationale Supérieure de l'Electronique et de ses Applications, 6 Avenue du Ponceau, 95014 Cergy-Pontoise, France.*

Abstract: This paper is devoted to the problem of design a dynamic controller of switched linear systems. In the first part of this paper, we give some remarks about the influences of switching signal on the asymptotic stability of switched systems. We give a practical example to generate a large class of switching signals. The second part is devoted to the switched dynamic controller design, based on a common Lyapunov function approach. A sufficient condition are formulated as an LMI problem for the switched controller design under arbitrary switching. A stabilizing switched controller with regional pole placements is also formulated as a convex problem, an LMI approach is used to derive the switched dynamic controller with performance limitations.

Keywords: Switched linear systems, Common Lyapunov function, Dynamic controller, LMI, Pole placements.

1. INTRODUCTION

Switched systems are hybrid dynamical systems consisting of a family of continuous-time subsystems and a switching rule that orchestrates the switching among them. The primary motivation for studying switched systems in control theory comes partly from the fact that switched systems have numerous applications in control of mechanical systems, process control, automotive industry, power systems, aircraft, traffic control, biology, network and many other fields [8], [9]. Stability of switched systems is not systematic and we can meet certain strange phenomena, even when all the subsystems are asymptotically stable. For example, the switched system can be unstable under certain switching signals [2], [11]. Thus, the stability of switched systems depends not only on the dynamics of each subsystem but also on the behavior of the switching signals.

The common Lyapunov approach is one of the principal methods to study stability and design a controller of switched systems. This approach is based on the existence of a common quadratic Lyapunov function for all subsystems of the switched system. There have been various attempts to derive conditions for the existence of a common quadratic Lyapunov function. Under the asymptotic stability of each subsystem, a common Lyapunov function exists when the subsystems

[1] This work was supported by "la Région Centre de France" under the grant 2004 0472 A026.

[2] Corresponding author. Tel: +33 2 48 48 40 98. Fax +33 2 48 48 40 50. E-mail address : salim.chaib@ensi-bourges.fr

matrices are pairwise commutative [13]. In [1], [10], the authors proposed a generalization of the commutativity notion, based on the solvability of the Lie algebra generated by the subsystems state matrices, i.e., state matrices are upper-triangularizable in the same reference frame. Note that many other results are based on the multiple Lyapunov (or like Lyapunov) functions approach [2], [5], [11].

In the context of switched systems with linear continuous-time subsystems, the issues of stabilization and control have been studied in many works [5], [11], [12], [14], [15], [16] and [17].

In this paper, we will focus on the study of a dynamic stabilization of switched linear systems. The paper is then organized as follows: In the next Section, we present some remarks on the influence of the switching signal on the stability of switched systems. In section III, a switched dynamic controller is studied based on an LMI approach. A dynamic switched controller with regional pole placements is investigated in section IV. Sufficient conditions for the existence of a dynamic switched controller with performance limitations are then given.

2. STABILITY UNDER SOME SWITCHING SIGNALS

In this section, we are interested in switched linear systems of the form

$$\dot{x}(t) = A_{\sigma(t)}x(t) \tag{1}$$

where $A_\sigma \in R^{n\times n}$, $\sigma \in \mathcal{Q} \triangleq \{1, ..., N\}$, $x \in R^n$ and $\sigma(t) : [0,\ \infty) \to \mathcal{Q}$ is a piecewise constant switching signal.

As mentioned in the introduction, the stability of switched systems are strongly related to the switching signal behavior, except in the case when all subsystems share the same Lyapunov function. Hereafter we will analyze the switched systems by using different class of switching signals. For this, we give four basic classes of switching signals, each class is characterized by some behavior, which can have a significant influence on the asymptotic stability of the switched system. These classes can be summarized by:

Class 0. This class contains all switching signals with a finite number of switchings. We denote this class by $\mathcal{P}_0$. Theoretically for this class, if all the subsystems are asymptotically stable, then at a certain finite time, the switched system evolves as only one subsystem. Therefore the switched system is asymptotically stable.

The remaining classes considered in this paper are those containing an infinite number of switchings.

Class 1. This class defines all switching signals for which any consecutive switching times t_k and t_{k+1}

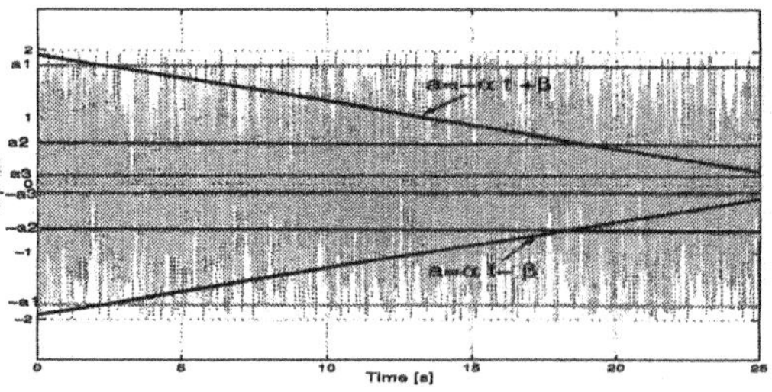

Fig. 1. The random signal used for the generation of $\sigma(t)$

are separated by some *dwell-time* τ_D. Denote by $\mathcal{P}_1$ the set of this class,

$$\mathcal{P}_1 = \{\sigma \in \mathcal{Q} :\ t_{k+1} - t_k \geq \tau_D\} \tag{2}$$

Class 2. This class can be defined as

$$\begin{aligned}\mathcal{P}_2 = \{\sigma \in \mathcal{Q} :\ &\exists \tau > 0 : \forall T > 0,\ \exists\, i > 0\\ &\text{such that}\ \ t_{i+1} - \tau \geq t_i \geq T\}\end{aligned} \tag{3}$$

This class includes the class for which the number of discontinuities of σ in any interval of time is bounded [7]

$$N_\sigma(t,\tau) \leq N_0 + \frac{t-\tau}{\tau_D}, \quad \forall t \geq \tau \geq 0 \tag{4}$$

where $N_\sigma(t,\tau)$ is the number of discontinuities of σ in the open interval (τ, t), τ_D is called the *average dwell time* and N_0 the *chatter bound.*

Class 3. This class is the class of all switching signals who dot not belong to $\mathcal{P}_0$, $\mathcal{P}_1$ or $\mathcal{P}_2$. This class contains the chattering and Zeno switching signals. We can write that

$$\mathcal{P}_1 \subset \mathcal{P}_2 \subset \mathcal{P}_3 \tag{5}$$

Now, we give a procedure to generate each class of switching signals given above. Consider a random signal, with some sample time ΔT, and amplitude $a \in [-a_m, a_m]$. Fig.1 shows an example of such signal. The three classes $\mathcal{P}_1$, $\mathcal{P}_2$, $\mathcal{P}_3$ can be generated for $\mathcal{Q} = \{1, 2\}$ as

$$\begin{aligned}&\mathcal{P}_i = \{\sigma \in \mathcal{Q} :\\ &\sigma = 1 \text{ if } [a \leq -a_i] \vee [(-a_i < a < a_i) \wedge (\sigma^- = 1)]\\ &\sigma = 2 \text{ if } [a \geq a_i] \vee [(-a_i < a < a_i) \wedge (\sigma^- = 2)]\}\end{aligned} \tag{6}$$

where $i = 1, 2, 3$ and $a_3 < a_2 < a_1 < a_m$ are the parameters characterizing each class (see Fig.1), and σ^- is the previous value of σ. The same procedure can be used for $\mathcal{Q} = \{1, ..., N\}$, $N > 2$. We can also generate more complex classes by this procedure. Like a Zeno signal can be generated by setting

$$a_i = \begin{cases} -\alpha t + \beta \text{ if } t \leq \dfrac{\beta}{\alpha} \\ 0 \qquad\quad \text{if } t > \dfrac{\beta}{\alpha} \end{cases}, \quad \alpha, \beta > 0 \tag{7}$$

for an analytic example see [7].

Now, we give some remarks about the stability

under these classes. For this, we limit our remarks to the case of a Lyapunov function equal to $x^T x$. Rewrite each matrix A_σ as $A_\sigma = S_\sigma + M_\sigma$, where S_σ and M_σ are the symmetric and the skew-symmetric part of A_σ respectively, given by $S_\sigma = \frac{1}{2}(A_\sigma + A_\sigma^T)$ and $M_\sigma = \frac{1}{2}(A_\sigma - A_\sigma^T)$. The following theorem is a sufficient condition for stability of the switched system (1).

Theorem 1. *If the following conditions*

i) *S_σ is semi negative for all $\sigma \in \mathcal{Q}$,*
ii) *(S_σ, A_σ) is observable for all $\sigma \in \mathcal{Q}$,*

hold, then the switched system (1) is globally uniformly stable under arbitrary switching. Moreover,

iii) *If $\sigma(t) \in \mathcal{P}_2$, then the switched system is globally uniformly asymptotically stable.*

Proof. The proof can be found in [4]. ∎

Corollary 1.
1) *If the matrices A_σ are Hurwitz, then the pair (S_σ, A_σ) is observable, therefore condition ii) of theorem 1 is fulfilled.*
2) *If the symmetric part S_σ is negative definite for all $\sigma \in \mathcal{Q}$, then the switched system is globally uniformly asymptotically stable under arbitrary switching, and condition iii) of theorem 1 can be relaxed.*
3) *If the linear algebra generated by $\{A_\sigma, A_\sigma^T\}$ is solvable and A_σ are Hurwitz, then S_σ are negative definite, therefore the switched system (1) is globally uniformly asymptotically stable and condition iii) of theorem 1 can be relaxed.*

Example 1. Consider the switched system $\dot{x} = A_{\sigma(t)}x$, with

$$A_1 = \begin{bmatrix} 0 & -1 \\ 1 & -2 \end{bmatrix}, \quad A_2 = \begin{bmatrix} 0 & 1 \\ -1 & -2 \end{bmatrix} \qquad (8)$$

The symmetric part of the matrices A_σ, $\sigma = 1, 2$ are given by

$$S_1 = S_2 = \begin{bmatrix} 0 & 0 \\ 0 & -2 \end{bmatrix} \qquad (9)$$

which are semi-negative definite. Condition *i)* and *ii)* of theorem 1 are satisfied, then the switched system is globally uniformly stable. Now we present some situations in which the switched system can be stable or asymptotically stable depending on condition *iii)* of theorem 1.

a) We use a switching signal of class 3. As stated in theorem 1, in Fig.2, the switched system is stable but does not converge to zero (see [7] for an analytic proof).
b) We use a switching signal $\sigma(t) \in \mathcal{P}_2$. In Fig.3 the switched system is asymptotically stable as mentioned in *iii)* of theorem 1.
c) In the third case, we use a switching signal of class 1. Fig.4 shows that the switched system is asymptotically stable. However the convergence is faster than situation *b)*.

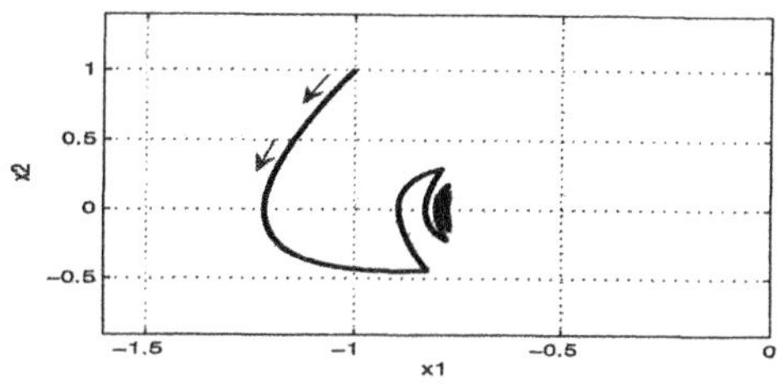

Fig. 2. Trajectory of the switched system under a switching signal of class 3, $\sigma \in \mathcal{P}_3$.

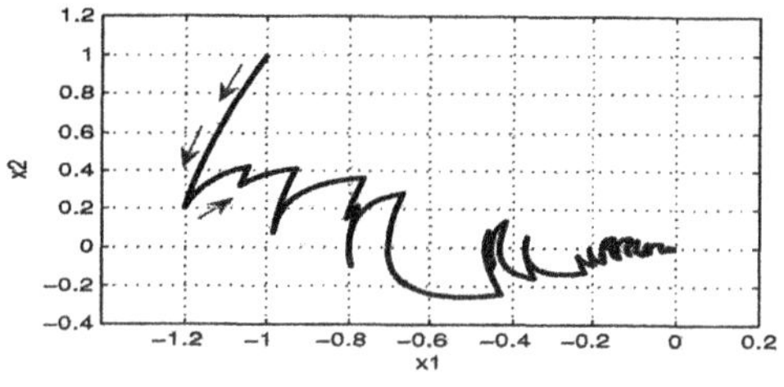

Fig. 3. Trajectory of the switched system under a switching signal of class 2, $\sigma \in \mathcal{P}_2$.

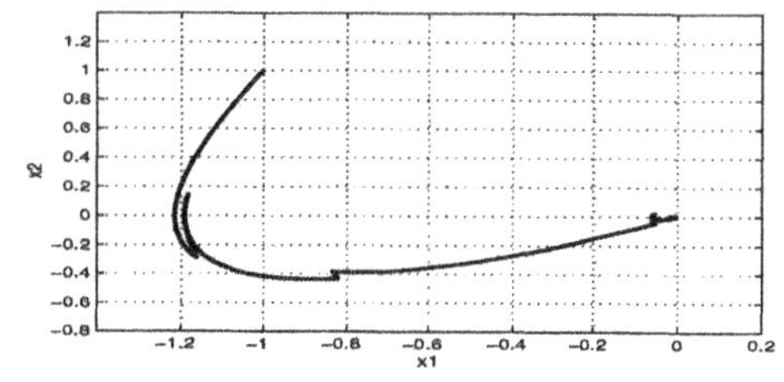

Fig. 4. Trajectory of the switched system under a switching signal of class 1, $\sigma \in \mathcal{P}_1$.

This example confirms that the asymptotic stability depends critically on the class of switching signals considered.
In the sequel, the stability based on the common Lyapunov function will be used to design a stabilizing dynamic controller of switched linear systems.

3. SWITCHED DYNAMIC CONTROLLER

Consider the continuous-time switched linear system described by

$$\begin{aligned} x(t) &= A_{\sigma(t)}x(t) + B_{\sigma(t)}u(t), \\ y(t) &= C_{\sigma(t)}x(t), \\ \sigma(t) &: R^+ \rightarrow \mathcal{Q}, \quad \mathcal{Q} := \{1, ..., N\}. \end{aligned} \qquad (10)$$

where $x(t) \in R^n$ is the continuous state, $u(t) \in R^{n_u}$ is the control input, $y(t) \in R^{n_y}$ is the output, $\sigma(t)$ is the switching signal, and $A_\sigma \in R^{n \times n}$, $B_\sigma \in R^{n \times n_u}$, $C_\sigma \in R^{n_y \times n}$ are the subsystem matrices. We assume that the switching signal $\sigma(t)$ is available in real time. In this section, we search a dynamic stabilizing controller having the following state representation

$$\begin{cases} \dot{x}_c(t) = \mathscr{A}_{\sigma(t)}x_c(t) + \mathscr{B}_{\sigma(t)}y(t) \\ u(t) = \mathscr{C}_{\sigma(t)}x_c(t) + \mathscr{D}_{\sigma(t)}y(t) \\ \sigma(t) \in \mathcal{Q},\ x_c(0) = x_0 \end{cases} \tag{11}$$

where $x_c(t) \in R^{n_c}$ is the state of the controller, $\mathscr{A}_\sigma \in R^{n_c \times n_c}$, $\mathscr{B}_\sigma \in R^{n_c \times n_y}$, $\mathscr{C}_\sigma \in R^{n_u \times n_c}$ and $\mathscr{D}_\sigma \in R^{n_u \times n_y}$, $\sigma \in \mathcal{Q}$ are the state parameters of the switched dynamic controller. A sufficient conditions for the existence of the stabilizing controller of the form (11) are provided by the following theorem.

Theorem 2. *If there exist matrices P_1, Q_1, $(F^1, F^2, F^3, F^4)_\sigma$, $\sigma \in \mathcal{Q}$, and matrices P_2, P_3, Q_2 such that the following LMIs/equation*

$$\begin{bmatrix} A_\sigma P_1 + P_1 A_\sigma^T + B_\sigma F_\sigma^2 + F_\sigma^{2T} B_\sigma^T & A_\sigma + B_\sigma F_\sigma^1 C_\sigma + F_\sigma^{4T} \\ A_\sigma^T + C_\sigma^T F_\sigma^{1T} B_\sigma^T + F_\sigma^4 & Q_1 A_\sigma + A_\sigma^T Q_1 + F_\sigma^3 C_\sigma + C_\sigma^T F_\sigma^{3T} \end{bmatrix} < 0, \tag{12}$$

$$Q_1 P_1 + P_2 Q_2^T = I, \tag{13}$$

and

$$\begin{bmatrix} P_1 & P_2^T \\ P_2 & P_3 \end{bmatrix} > 0 \tag{14}$$

hold, then the dynamic controller (11), with the parameters $(\mathscr{A}, \mathscr{B}, \mathscr{C}, \mathscr{D})_\sigma$ given by

$$\begin{bmatrix} \mathscr{A}_\sigma & \mathscr{B}_\sigma \\ \mathscr{C}_\sigma & \mathscr{D}_\sigma \end{bmatrix} = \begin{bmatrix} Q_2 & Q_1 B_\sigma \\ 0 & I \end{bmatrix}^{-1} \left(\begin{bmatrix} F_\sigma^4 & F_\sigma^3 \\ F_\sigma^2 & F_\sigma^1 \end{bmatrix} + \begin{bmatrix} Q_1 A_\sigma P_1 & 0 \\ 0 & 0 \end{bmatrix} \right) \begin{bmatrix} P_2 & 0 \\ C_\sigma P_1 & I \end{bmatrix}^{-1} \tag{15}$$

ensures the asymptotic stability of the closed loop switched system under arbitrary switching, and all closed loop subsystems share the common Lyapunov function $V(\tilde{x}) = \tilde{x}^T P^{-1} \tilde{x}$, $P := \begin{bmatrix} P_1 & P_2^T \\ P_2 & P_3 \end{bmatrix}$.

Proof. The closed loop switched system is given by

$$\dot{\tilde{x}}(t) = A_{\sigma(t)}^{cl} \tilde{x}(t),\ \tilde{x}(0) = \tilde{x}_0 \tag{16}$$

with

$$A_\sigma^{cl} := \begin{bmatrix} A_\sigma + B_\sigma \mathscr{D}_\sigma C_\sigma & B_\sigma \mathscr{C}_\sigma \\ \mathscr{B}_\sigma C_\sigma & \mathscr{A}_\sigma \end{bmatrix};\ \tilde{x} := \begin{bmatrix} x \\ x_c \end{bmatrix} \tag{17}$$

a sufficient condition for quadratic stability of (16) under arbitrary switching is the existence of a common Lyapunov function $V(\tilde{x}) = \tilde{x}^T W \tilde{x}$, such that

$$W = W^T > 0,\ W A_\sigma^{cl} + A_\sigma^{clT} W < 0,\ \sigma \in \mathcal{Q} \tag{18}$$

Multiplying the second inequality in (18) on the left and right by W^{-1}, and defining a new variable $P = W^{-1}$, we may rewrite (18) as

$$P = P^T > 0,\ \mathcal{V}_\sigma := A_\sigma^{cl} P + P A_\sigma^{clT} < 0,\ \sigma \in \mathcal{Q} \tag{19}$$

This dual inequality is an equivalent condition for quadratic stability. Now, consider that there exists a symmetric matrix $P > 0$, such that (19) holds. Define P and its inverse as

$$P := \begin{bmatrix} P_1 & P_2^T \\ P_2 & P_3 \end{bmatrix},\ P^{-1} := \begin{bmatrix} Q_1 & Q_2 \\ Q_2^T & Q_3 \end{bmatrix} \tag{20}$$

The development of (19) gives

$$\mathcal{V}_\sigma = \begin{bmatrix} \mathcal{V}_\sigma(1,1) & \mathcal{V}_\sigma(1,2) \\ \mathcal{V}_\sigma(2,1) & \mathcal{V}_\sigma(2,2) \end{bmatrix} < 0 \tag{21}$$

where

$$\begin{aligned} \mathcal{V}_\sigma(1,1) &= A_\sigma P_1 + P_1 A_\sigma^T + P_1 C_\sigma^T \mathscr{D}_\sigma^T B_\sigma^T \\ &\quad + P_2^T \mathscr{C}_\sigma^T B_\sigma^T + B_\sigma \mathscr{D}_\sigma C_\sigma P_1 + B_\sigma \mathscr{C}_\sigma P_2 \\ \mathcal{V}_\sigma(1,2) &= A_\sigma P_2^T + B_\sigma \mathscr{D}_\sigma C_\sigma P_2^T + B_\sigma \mathscr{C}_\sigma P_3 \\ &\quad + P_1 \mathscr{C}_\sigma^T B_\sigma^T + P_2^T \mathscr{A}_\sigma^T \\ \mathcal{V}_\sigma(2,1) &= \mathscr{B}_\sigma C_\sigma P_1 + \mathscr{A}_\sigma P_2 + P_2 A_\sigma^T \\ &\quad + P_2 C_\sigma^T \mathscr{D}_\sigma^T B_\sigma^T + P_3 \mathscr{C}_\sigma^T B_\sigma^T \\ \mathcal{V}_\sigma(2,2) &= \mathscr{B}_\sigma C_\sigma P_2^T + \mathscr{A}_\sigma P_3 + P_2 \mathscr{C}_\sigma^T B_\sigma^T \\ &\quad + P_3 \mathscr{A}_\sigma^T \end{aligned}$$

We show well that $\mathcal{V}_\sigma$ with the parameters defined in the above equations, is not affine in the variables P_1, P_2, P_3 and $(\mathscr{A}, \mathscr{B}, \mathscr{C}, \mathscr{D})_\sigma$, then $\mathcal{V}_\sigma < 0$ is not a convex problem. For this we must transform (21) to an affine form in synthesis variables. First, we apply to (21) a transformation, (21) is equivalent to

$$\mathcal{V}_\sigma < 0 \Leftrightarrow \Phi^T \mathcal{V}_\sigma \Phi < 0 \tag{22}$$

where Φ is a matrix transformation to be determined, such that

$$\Phi^T A_\sigma^{cl} P \Phi + \Phi^T P A_\sigma^{clT} \Phi < 0$$

and $\Phi^T A_\sigma^{cl} P \Phi, \Phi^T P A_\sigma^{clT} \Phi =$ indirectly affine $(P_1, P_2, P_3, \mathscr{A}_\sigma, \mathscr{B}_\sigma, \mathscr{C}_\sigma, \mathscr{D}_\sigma)$

Define the matrix transformation Φ as

$$\Phi := \begin{bmatrix} I & Q_1 \\ 0 & Q_2^T \end{bmatrix} \Rightarrow P\Phi = \begin{bmatrix} P_1 & I \\ P_2 & 0 \end{bmatrix} \tag{23}$$

where I is the identity matrix.

Define the following change of variables

$$F_\sigma^1 := \mathscr{D}_\sigma, \tag{24}$$

$$F_\sigma^2 := \mathscr{C}_\sigma P_2 + \mathscr{D}_\sigma C_\sigma P_1, \tag{25}$$

$$F_\sigma^3 := Q_2 \mathscr{B}_\sigma + Q_1 B_\sigma \mathscr{D}_\sigma, \tag{26}$$

$$\begin{aligned} F_\sigma^4 := {} & Q_2 \mathscr{A}_\sigma P_2 + Q_1 A_\sigma P_1 + Q_2 \mathscr{B}_\sigma C_\sigma P_1 \\ & + Q_1 B_\sigma \mathscr{C}_\sigma P_2 + Q_1 B_\sigma \mathscr{D}_\sigma C_\sigma P_1 \end{aligned} \tag{27}$$

by this change of variables, $\Phi^T A_\sigma^{cl} P \Phi, \Phi^T P A_\sigma^{clT} \Phi$ become

$$\Phi^T A_\sigma^{cl} P \Phi = \begin{bmatrix} A_\sigma P_1 + B_\sigma F_\sigma^2 & A_\sigma + B_\sigma F_\sigma^1 C_\sigma \\ F_\sigma^4 & Q_1 A_\sigma + F_\sigma^3 C_\sigma \end{bmatrix}$$

$$\Phi^T P A_\sigma^{clT} \Phi = \begin{bmatrix} P_1 A_\sigma^T + F_\sigma^{2T} B_\sigma^T & F_\sigma^{4T} \\ A_\sigma^T + C_\sigma^T F_\sigma^{1T} B_\sigma^T & A_\sigma^T Q_1 + C_\sigma^T F_\sigma^{3T} \end{bmatrix}$$

which are affine in the variables F_σ^1, F_σ^2, F_σ^3, F_σ^4, P_1, Q_1 and the nonlinear inequality of the synthesis variables P_1,P_2,P_3 and $(\mathscr{A},\mathscr{B},\mathscr{C},\mathscr{D})_\sigma$ can be transformed to an LMI form, with F_σ^1, F_σ^2, F_σ^3, F_σ^4, P_1 and Q_1 as variables of the LMI. Then the problem which consists to find $\mathscr{A}_\sigma, \mathscr{B}_\sigma, \mathscr{C}_\sigma, \mathscr{D}_\sigma$ is transformed to an LMI problem, which consists to compute the variables F_σ^1, F_σ^2, F_σ^3, F_σ^4, P_1, Q_1. These LMIs are given by

$$\begin{bmatrix} A_\sigma P_1 + P_1 A_\sigma^T & A_\sigma + B_\sigma F_\sigma^1 C_\sigma \\ +B_\sigma F_\sigma^2 + F_\sigma^{2T} B_\sigma^T & +F_\sigma^{4T} \\ A_\sigma^T + C_\sigma^T F_\sigma^{1T} B_\sigma^T & Q_1 A_\sigma + A_\sigma^T Q_1 \\ +F_\sigma^4 & +F_\sigma^3 C_\sigma + C_\sigma^T F_\sigma^{3T} \end{bmatrix} < 0, \tag{28}$$

the change of variables (24)-(27) can be putted in the compact form

$$\begin{bmatrix} F_\sigma^4 & F_\sigma^3 \\ F_\sigma^2 & F_\sigma^1 \end{bmatrix} = \begin{bmatrix} Q_2 & Q_1 B_\sigma \\ 0 & I \end{bmatrix} \begin{bmatrix} \mathscr{A}_\sigma & \mathscr{B}_\sigma \\ \mathscr{C}_\sigma & \mathscr{D}_\sigma \end{bmatrix} \begin{bmatrix} P_2 & 0 \\ C_\sigma P_1 & I \end{bmatrix} + \begin{bmatrix} Q_1 \\ 0 \end{bmatrix} A_\sigma \begin{bmatrix} P_1 & 0 \end{bmatrix} \tag{29}$$

LMI (28) do not depend on P_2, Q_2, but these two variables appear in (29), then these variables are necessary to compute the switched controller parameters $(\mathscr{A}_\sigma, \mathscr{B}_\sigma, \mathscr{C}_\sigma, \mathscr{D}_\sigma)$. For this the variables P_2 and Q_2 should be computed a posteriori. Given matrices $Q_1 > 0$ and $P_1 > 0$, the matrices $P_2 > 0$ and $Q_2 > 0$ must be computed from 13, and finally the dynamic switched controller gains are computed using (15). ∎

Remark 1. If $\dim x = \dim x_c$, i.e., the controller has the same order of the plant, then P_2, Q_2 are square and nonsingular matrices. Thus the controller parameters can be calculated by the equation

$$\begin{bmatrix} \mathscr{A}_\sigma & \mathscr{B}_\sigma \\ \mathscr{C}_\sigma & \mathscr{D}_\sigma \end{bmatrix} = \begin{bmatrix} Q_2^{-1} & -Q_2^{-1} Q_1 B_\sigma \\ 0 & I \end{bmatrix} \cdot \begin{bmatrix} F_\sigma^4 - Q_2 A_\sigma P_1 & F_\sigma^3 \\ F_\sigma^2 & F_\sigma^1 \end{bmatrix} \begin{bmatrix} P_2^{-1} & 0 \\ -C_\sigma P_1 P_2^{-1} & I \end{bmatrix}$$

Corollary 2. *In the case of a switched static controller of the form $u(t) = K_\sigma(t)x(t)$, where K_σ, for $\sigma \in \mathcal{Q}$ are the controller gains. Assume that $x(t)$ is available, the conditions in theorem 1 are reduced to the LMI with X_σ as variables of synthesis*

$$PA_\sigma^T + A_\sigma P + B_\sigma X_\sigma + X_\sigma^T B_\sigma^T < 0, \; \sigma \in \mathcal{Q} \tag{30}$$

where P is a symmetric positive definite matrix. The switched static controller gains K_σ, for $\sigma \in \mathcal{Q}$ are given by

$$K_\sigma = X_\sigma P^{-1}, \; \sigma \in \mathcal{Q} \tag{31}$$

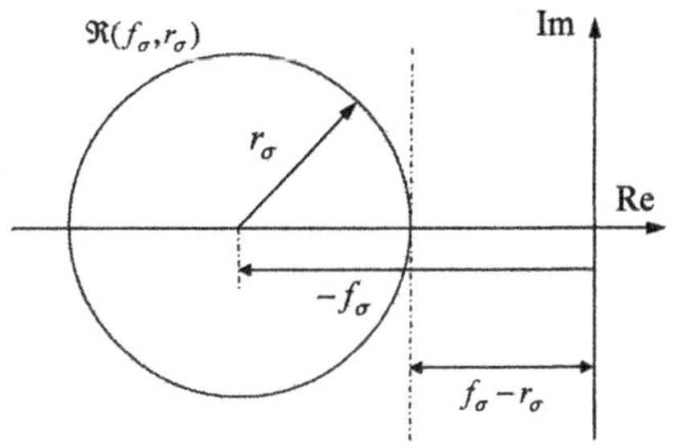

Fig. 5. Circular region $\mathscr{R}(f_\sigma, r_\sigma)$

and the closed loop switched system stability is ensured by the common Lyapunov function $V(x) = x^T P^{-1} x$.

4. DYNAMIC SWITCHED CONTROLLER WITH REGIONAL POLE PLACEMENTS

The goal of this section is to combine the first result of theorem 2 and pole placements, our objective consists then to determine a stabilizing dynamic switched controller of the switched system with some constraint specifications on the poles of the closed loop system, which is defined previously as

$$A_\sigma^{cl} := \begin{bmatrix} A_\sigma + B_\sigma \mathscr{D}_\sigma C_\sigma & B_\sigma \mathscr{C}_\sigma \\ \mathscr{B}_\sigma C_\sigma & \mathscr{A}_\sigma \end{bmatrix} \tag{32}$$

The first objective is then to place the poles of the closed loop switched system inside a circular region $\mathscr{R}(f_\sigma, r_\sigma)$ in the complex plane, with a center at $(-f_\sigma, 0)$, radius $r_\sigma \le f_\sigma$, $\sigma \in \mathcal{Q}$, and distance $(f_\sigma - r_\sigma)$ from the imaginary axis. Fig.5 shows the circular region $\mathscr{R}(f_\sigma, r_\sigma)$ for pole location. The full problem can then be formulated as

$$\text{find } \left[\begin{array}{c|c} \mathscr{A}_\sigma & \mathscr{B}_\sigma \\ \hline \mathscr{C}_\sigma & \mathscr{D}_\sigma \end{array}\right]$$

subject to:

$$\text{eigenvalues}(A_\sigma^{cl}) \in \mathscr{R}(f_\sigma, r_\sigma)$$

$$\dot{\tilde{x}}(t) = A_{\sigma(t)}^{cl} \tilde{x}(t) \text{ is asym-stable } \forall \sigma \in \mathcal{Q}$$

The asymptotic stability of A^{cl} is ensured by the existence of a common Lyapunov function $V(\tilde{x}) = \tilde{x}^T W \tilde{x}$, $W = W^T > 0$. The problem concerning the eigenvalues of A^{cl} is resolved as : In [6] a necessary and sufficient condition assuring that all eigenvalues of a given matrix A_σ^{cl} lie inside a circular region $\mathscr{R}(f_\sigma, r_\sigma)$ is provided by the existence of a symmetric positive definite matrix P such that

$$[A_\sigma^{cl} + (f_\sigma - r_\sigma)I]P + P[A_\sigma^{cl} + (f_\sigma - r_\sigma)I]^T + \frac{1}{r_\sigma}[A_\sigma^{cl} + (f_\sigma - r_\sigma)I]P[A^{cl} + (f_\sigma - r_\sigma)I]^T < 0 \tag{33}$$

Theorem 3. *If there exist matrices P_1, Q_1, $(F^1, F^2, F^3, F^4)_\sigma$, $\sigma \in \mathcal{Q}$, and matrices P_2, P_3, Q_2 such that the following LMIs/equation*

$$\begin{bmatrix} \ell_{11} & \ell_{12} & \ell_{13} & \ell_{14} \\ \ell_{21} & \ell_{22} & \ell_{23} & \ell_{24} \\ \ell_{13}^T & \ell_{23}^T & -r_\sigma P_1 & -r_\sigma I \\ \ell_{14}^T & \ell_{24}^T & -r_\sigma I & -r_\sigma Q_1 \end{bmatrix} < 0, \quad (34)$$

$$Q_1 P_1 + P_2 Q_2^T = I, \quad (35)$$

hold, where

$$\begin{aligned} \ell_{11} &= A_\sigma P_1 + B_\sigma F_\sigma^2 + P_1 A_\sigma^T + F_\sigma^{2T} B_\sigma^T \\ &\quad +2(f_\sigma - r_\sigma)P_1 \\ \ell_{12} &= A_\sigma + B_\sigma F_\sigma^1 C_\sigma + F_\sigma^{4T} + 2(f_\sigma - r_\sigma)I \\ \ell_{13} &= A_\sigma P_1 + B_\sigma F_\sigma^2 + (f_\sigma - r_\sigma)P_1 \\ \ell_{14} &= A_\sigma + B_\sigma F_\sigma^1 C_\sigma + (f_\sigma - r_\sigma)I \\ \ell_{21} &= F_\sigma^4 + A_\sigma^T + C_\sigma^T F_\sigma^{1T} B_\sigma^T + 2(f_\sigma - r_\sigma)I \\ \ell_{22} &= Q_1 A_\sigma + F_\sigma^3 C_\sigma + A_\sigma^T Q_1 + C_\sigma^T F_\sigma^{3T} \\ &\quad +2(f_\sigma - r_\sigma)Q_1 \\ \ell_{23} &= F_\sigma^4 + (f_\sigma - r_\sigma)I \\ \ell_{24} &= Q_1 A_\sigma + F_\sigma^3 C_\sigma + (f_\sigma - r_\sigma)Q_1 \end{aligned}$$

then the dynamic controller (11), with the parameters $(\mathscr{A}, \mathscr{B}, \mathscr{C}, \mathscr{D})_\sigma$ given by (15), is a stabilizing controller for (10), and the stability of the closed loop switched system is ensured by the common Lyapunov function $V(\tilde{x}) = \tilde{x}^T P^{-1} \tilde{x}$, where P is defined as in theorem 2.

Proof. If there exists a matrix transformation $\tilde{\Phi}$ which will be determined, such that $[A_\sigma^{cl} + (f_\sigma - r_\sigma)I]P$, $P[A_\sigma^{cl} + (f_\sigma - r_\sigma)I]^T$ and $r_\sigma P$ can be transformed to an affine form of all the variables synthesis like $(\mathscr{A}_\sigma, \mathscr{B}_\sigma, \mathscr{C}_\sigma, \mathscr{D}_\sigma)$. Then the Schur complement can be applied to nonlinear equation (33), which is equivalent by using the Schur complement to

$$\begin{bmatrix} A_\sigma^{cl} P + P A_\sigma^{clT} + 2(f_\sigma - r_\sigma)P & A_\sigma^{cl} P + (f_\sigma - r_\sigma)P \\ P A_\sigma^{clT} + (f_\sigma - r_\sigma)P & -r_\sigma P \end{bmatrix} < 0 \quad (36)$$

Define P,P^{-1} and the transformation Φ as in (20) and (23), and define the transformation $\tilde{\Phi} := \text{diag}(\Phi, \Phi)$. Apply this transformation to (36), we obtain (34), which is an LMI of the variables synthesis F_σ^1, F_σ^2, F_σ^3, F_σ^4, P_1, Q_1, where $(F^1, F^2, F^3, F^4)_\sigma$ are the change of variables defined as in (24)-(27). ∎

5. CONCLUDING REMARKS

The problem related to the influence of switching signal on the asymptotic stability of switched system is investigated in the first part of this paper, some remarks are driven, based on identity Lyapunov function, an illustrative example is given, to show that stability depends critically on the behavior of the switching signal. The problem of synthesis of a switched dynamic controller has been addressed through LMI approach. A generalization of this result is given, consisting to regional pole placements, an LMI formulation is also given.

REFERENCES

[1] A. A. Agrachev and D. Liberzon, "Lie-algebraic stability criteria for switched systems", SIAM J. Control Optim., vol. 40, pp. 253-270, 2001.

[2] M. S. Branicky, "Multiple Lyapunov functions and other analysis tools for switched and hybrid systems", IEEE Trans. AC, vol. 43, no. 4, pp. 475-482, 1998.

[3] S.Boyd, L.El Ghaoui, E.Feron, V.Balakrishnan, "Linear Matrix Inequalities in System and Control Theory", SIAM, Philadelphia, 1994.

[4] S. Chaib, D. Boutat, A. Benali, J. P. Barbot, "Observer Design of Switched Linear Systems", submited to Journal of Nonlinear Analysis : Hybrid Systems and Applications.

[5] R. A. Decarlo, M. S. Branicky, S. Pettersson, and B. Lennartson, "Perspectives and results on the stability and stabilizability of hybrid systems", Proceedings of the IEEE, Special issue on Hybrid Systems, P. J. Antsaklis Ed., vol. 88, no. 7, pp. 1069-1082, 2000.

[6] W.M. Haddad, D.S. Bernstein, "Controller design with regional pole constraints", IEEE Trans. Automat. Contr, vol. 37, no. 1, pp. 54-69, 1992.

[7] J. Hespanha. "Uniform Stability of Switched Linear Systems: Extensions of LaSalle's Invariance Principle". IEEE Trans. on Automat. Contr., vol. 49, no. 4, pp. 470-482, Apr. 2004.

[8] J. Hespanha, Stochastic hybrid modeling of on-of TCP flows. In proceedings of Hybrid Systems, Computation Control, 2004.

[9] H. de Jong, G. J., C. Hernandez, M. Page, T. Sari, and J. Geiselmann, "Qualitative simulation of genetic regulatory networks using piecewise-linear models", Bulletin of Mathematical Biology, vol.66 pp. 301-340, 2004.

[10] D. Liberzon, J. P. Hespanha and A. S. Morse, "Stability of switched linear systems: A Lie-algebraic condition", Syst. Contr. Lett., vol. 37, no. 3, pp. 117-122, 1999.

[11] D. Liberzon and A. S. Morse, "Basic problems in stability and design of switched systems", IEEE Contr. Syst. Magazine, vol. 19, no. 5, pp. 59-70, 1999.

[12] V.F. Montagner, V. J. S. Leite, R.C.L.F. Olineira, P.L.D. Pers, "State feedback control of switched linear systems : an LMI approach", Journal of Computational and Applied Mathematics, article in press, 2005.

[13] K. S. Narendra and J. Balakrishnan, "A common Lyapunov function for stable LTI systems with commuting A-matrices", IEEE Trans. AC, vol. 39, no. 12, pp. 2469-2471, 1994.

[14] A.V. Savkin, E. Skafidas, R.J. Evans, "Robust output feedback stabilizability via controller switching", Automatica, vol. 35, no. 1, pp. 69-74, 1999.

[15] F. M. Pait and A. S. Morse, "A cyclic switching strategy for parameter adaptive control", IEEE Trans. Automat. Contr., vol. 39, pp. 1172-1183, June 1994.

[16] G. Zhai, B. Hu, K. Yasuda, and A. N. Michel, "Qualitative analysis of discrete-time switched systems," in Proc. 2002 American Contr. Conf., vol. 3, pp. 1880-1885, 2002.

[17] G.S. Zhai, H. Lin, P.J. Antsaklis, "Quadratic stabilizability of switched linear systems with polytopic uncertainties", Internat. J. Control, vol. 76, no. 7, pp. 747-753, 2003.

ELSEVIER
IFAC
PUBLICATIONS

DYNAMIC OUTPUT FEEDBACK STABILIZATION OF CONTINUOUS-TIME SWITCHED SYSTEMS

José C. Geromel[#] and Patrizio Colaneri[*]

[#]DSCE / School of Electrical and Computer Engineering
UNICAMP, CP 6101, 13081 - 970, Campinas, SP, Brazil,
geromel@dsce.fee.unicamp.br
[*] Politecnico di Milano
Dipartimento di Elettronica e Informazione
Piazza Leonardo da Vinci 32, 20133 Milano (Italy)
colaneri@elet.polimi.it

Abstract: This paper considers the closed-loop stabilization problem for continuous time linear switched system. The state variables are assumed to be not accessible so that the feedback strategy hinges on given output variables. The solution of this problem is based on the solution of suitable matrix inequalities for the construction of a full order switched filter and the derivation of the stabilization rule. The main theoretical basis is constituted by the so-called Lyapunov-Metzler inequalities which play a prominent role in the state-feedback stabilization of linear switched systems. Being nonconvex, a more conservative version of the inequalities, expressed in terms of linear matrix inequalities (LMI) plus a line search is given. The theoretical results are illustrated by means of an academic example.

Keywords: Switched systems, Output Feedback, Linear Matrix Inequalities.

1. INTRODUCTION

This paper considers a switched linear system of the following general form

$$\dot{x}(t) = A_{\sigma(t)}x(t) \ , \ x(0) = x_0 \tag{1}$$

$$y(t) = C_{\sigma(t)}x(t) \tag{2}$$

defined for all $t \geq 0$ where $x(t) \in \mathbb{R}^n$ is the state, x_0 is the initial condition, $y(t) \in \mathbb{R}^p$ is the measurement vector and $\sigma(t) \in \{1, 2, \cdots, N\}$ is the switching rule.

Assuming that the two sets of matrices $A_i \in \mathbb{R}^{n\times n}$, $i = 1, \cdots, N$ and $C_i \in \mathbb{R}^{p\times n}$, $i = 1, \cdots, N$ are given, we want to tackle the problem of finding a switching rule $\sigma(t)$ depending on the measurements $y(\cdot)$ in such a way that the closed-loop switched system is globally asymptotically stable. The stability of continuous time linear switched systems have been addressed by many authors, (Branicky, 1998), (Hockerman *et al.*, 1998), (Johansson *et al.*, 1998), (Ye *et al.*, 1998) and (Hespanha, 2004), where the interested reader can find an interesting discussion on a collection of results on uniform stability of switched systems. However, very little attention has been devoted to the design of output feedback control stabilizing laws.

See (Liberzon, 2003) for a rather complete review of stability of continuous time linear switched systems where special attention is given to the case of switching between two linear systems. The same reference also provides a discussion on hybrid feedback based on output measurements which, in our opinion, can not be directly generalized to cope with the problem to be stated afterwards.

In this paper a novel approach is pursued, which is inherited by the recent results obtained for the related state-feedback stabilization problem via the so-called Lyapunov-Metzler inequalities. Precisely, the solution involves a set of symmetric and positive matrices $\{Z_1, \cdots, Z_N\}$, a Metzler matrix Π, a positive matrix X and extra variables $\{L_1, \cdots, L_N\}$. Being the solution of non-convex nature a more conservative but easier to solve asymptotic stability condition is proposed. This condition is expressed in terms of linear matrix inequalities plus a line search, every LMI being solvable in polynomial time, (Boyd *et al.*, 1994).

The notation used throughout is standard. Capital letters denote matrices, small letters denote vectors and small Greek letters denote scalars. For matrices or vectors $(')$ indicates transpose. For symmetric matrices, $X > 0$ (≥ 0) indicates that X is positive definite (nonnegative definite). For square matrices trace(X) denotes the trace function of X being equal to the sum of its eigenvalues. The sets of real and natural numbers are denoted by $\mathbb{R}$ and $\mathbb{N}$ respectively. The $\mathcal{L}_2$ norm of $x(t) \in \mathbb{R}^n$ defined for all $t \geq 0$ equals $\|x(t)\|_2^2 = \int_0^\infty x(t)'x(t)dt$, see (Colaneri *et al.*, 1997) for details.

2. STATE-FEEDBACK

In this section we resume some recent results for the state-feedback stabilization of switched systems, that form the basis for the subsequent achievements on dynamic output feedback. For more information and detailed proofs, the reader is requested to see the reference (Geromel *et al.*, 2005). Precisely, consider the continuous-time linear switched system

$$\dot{x}(t) = A_{\sigma(t)}x(t) \ , \ x(0) = x_0 \tag{3}$$

where all symbols have already been defined earlier. Assume that the state-variable is accessible for feedback. The goal is to determine the function $u(\cdot) : \mathbb{R}^n \to \{1, \cdots, N\}$, such that

$$\sigma(t) = u(x(t)) \tag{4}$$

makes the equilibrium point $x = 0$ of (1) asymptotically stable. Before we proceed, let us recall the class of Metzler matrices denoted by $\mathcal{M}$ and constituted by all matrices $\Pi \in \mathbb{R}^{N \times N}$ with elements π_{ij}, such that

$$\pi_{ij} \geq 0 \ \forall i \neq j \ , \ \sum_{i=1}^{N} \pi_{ij} = 0 \ \forall j \tag{5}$$

It is clear that any $\Pi \in \mathcal{M}$ presents an eigenvalue at the origin of the complex plane since $c'\Pi = 0$ where $c' = [1 \ \cdots \ 1]$. In addition, it is well known that the eigenvector associated to the null eigenvalue of Π is non-negative yielding the conclusion that there exists $\lambda_\infty \geq 0$ with $c'\lambda_\infty = 1$ such that $\Pi\lambda_\infty = 0$. The following result holds.

Lemma 1. Let $Q_i \geq 0$, $i = 1, 2, \cdots, N$, be given. Assume that there exist a set of positive definite matrices $\{P_1, \cdots, P_N\}$ and $\Pi \in \mathcal{M}$ satisfying the Lyapunov-Metzler inequalities

$$A_i'P_i + P_iA_i + \sum_{j=1}^{N} \pi_{ji}P_j + Q_i < 0 \tag{6}$$

for $i = 1, \cdots, N$. The state switching control (4) with

$$u(x(t)) = \arg \min_{i=1,\cdots,N} x(t)P_ix(t) \tag{7}$$

makes the equilibrium solution $x = 0$ of (1) globally asymptotically stable and

$$\int_0^\infty x(t)'Q_{\sigma(t)}x(t)dt \leq \min_{i=1,\cdots,N} x_0'P_ix_0 \tag{8}$$

The result above is important in that it also includes the stability of possible sliding modes. However the numerical determination, if any, of a solution of the Lyapunov-Metzler inequalities with respect to the variables $(\Pi, \{P_1, \cdots, P_N\})$ is not a simple task due to the non-convex nature inherited by the products of variables. Hence, a simpler, although more conservative stability condition can be expressed by means of LMIs being thus solvable by the available machinery plus a line search. The next theorem shows that working with a subclass of Metzler matrices, with the same diagonal elements, this goal is accomplished.

Lemma 2. Let $Q_i \geq 0$, $i = 1, 2, \cdots, N$, be given. Assume that there exist a set of positive definite matrices $\{P_1, \cdots, P_N\}$ and a scalar $\gamma > 0$ satisfying the modified Lyapunov-Metzler inequalities

$$A_i'P_i + P_iA_i + \gamma(P_j - P_i) + Q_i < 0 \tag{9}$$

for all $j \neq i = 1, \cdots, N$. The state switching control (4) with $u(x(t))$ given by (7) makes the equilibrium solution $x = 0$ of (1) globally asymptotically stable and

$$\int_0^\infty x(t)'Q_{\sigma(t)}x(t)dt \leq \sum_{i=1}^{N} x_0'P_ix_0 \tag{10}$$

Remark 1. The Lyapunov - Metzler inequalities have been introduced in order to study the *Mean-Square* (MS) stability of Markov Jump Linear Systems (MJLS), where matrix $\Pi \in \mathcal{M}$ is given and Π' represents the infinitesimal transition matrix of a Markov chain $\sigma(t)$ governing the dynamic system (1). In this respect the vector of probabilities $\lambda_i(t)$ to be in the $i-th$ logical state at any time $t \geq 0$ obeys the differential equation

$\dot{\lambda}(t) = \Pi\lambda(t)$ with initial probability vector λ_0. Hence, λ_∞ represents the stationary probability vector. The stochastic system is MS-stable if and only if there exist a set of positive definite matrices $\{P_1, \cdots, P_N\}$ satisfying the Lyapunov-Metzler inequalities (6), see (Fang *et al.*, 2002).

3. OUTPUT-FEEDBACK

This section contains the new results relative to the output feedback case. Given the measurements

$$y(t) = C_{\sigma(t)}x(t) \tag{11}$$

our main result consists on the determination of a switching rule of the form

$$\sigma(t) = u(y(\cdot)) \tag{12}$$

such that the closed-loop system is asymptotically stable. The function $u(\cdot)$ is indeed a functional of $y(\cdot)$ in the sense that $y(t)$ is viewed as the input of a switched filter that rules out the change of the switching index. To this end, let us introduce the full order switched filter

$$\dot{\hat{x}}(t) = \hat{A}_{\sigma(t)}\hat{x}(t) + \hat{B}_{\sigma(t)}y(t), \quad \hat{x}(0) = \hat{x}_0 \tag{13}$$

where $(\hat{A}_i, \hat{B}_i)$, $i = 1, 2, \cdots, N$ are matrices to be determined. Putting (13) and (1), (2) together we obtain

$$\dot{\tilde{x}}(t) = \tilde{A}_{\sigma(t)}\tilde{x}(t), \quad \tilde{x}(0) = \tilde{x}_0 \tag{14}$$

where $\tilde{x}' = [x' \ \hat{x}']$ and

$$\tilde{A}_i = \begin{bmatrix} A_i & 0 \\ \hat{B}_iC_i & \hat{A}_i \end{bmatrix} \tag{15}$$

Therefore the solution of our problem requires the determination of the switched filter matrices $\hat{A}_i$ and $\hat{B}_i$ and of a switching rule such such that the enlarged switched system is asymptotically stable. However, in doing so, only switching rules that depends exclusively on $\hat{x}(\cdot)$ are permitted. In order to apply the results of the previous section, we limit the structure of the Lyapunov function so as to structurally incorporate switching rules that depends only on $\hat{x}(\cdot)$. Therefore, let

$$\tilde{P}_i = \begin{bmatrix} X & V \\ V' & \hat{X}_i \end{bmatrix}, \quad \det V \neq 0 \tag{16}$$

and notice that

$$\arg\min_i \tilde{x}(t)'\tilde{P}_i\tilde{x}(t) = \arg\min_i \hat{x}(t)\hat{X}_i\hat{x}(t) \tag{17}$$

Hence, to solve the problem under consideration, we need to find a stabilizing rule of the form

$$u(y(\cdot)) = \arg\min_{i=1,\cdots,N} \hat{x}(t)\hat{X}_i\hat{x}(t) \tag{18}$$

Finally consider positive semidefinite matrices Q_i of compatible dimensions and let

$$\tilde{Q}_i = \begin{bmatrix} Q_i & 0 \\ 0 & 0 \end{bmatrix} \tag{19}$$

The following result is a direct consequence of Lemma 1 applied to the composite switched system (14).

Lemma 3. Let $Q_i \geq 0$, $i = 1, 2, \cdots, N$, be given. Assume that there exist a set of positive definite matrices $\{\tilde{P}_1, \cdots, \tilde{P}_N\}$ where each $\tilde{P}_i$ is of the form (16), a set of matrices $\{\hat{A}_1, \cdots, \hat{A}_N\}$, a set of matrices $\{\hat{B}_1, \cdots, \hat{B}_N\}$ and a Metzler matrix $\Pi \in \mathcal{M}$ satisfying the Lyapunov-Metzler inequalities

$$\tilde{A}_i'\tilde{P}_i + \tilde{P}_i\tilde{A}_i + \sum_{j=1}^{N} \pi_{ji}\tilde{P}_j + \tilde{Q}_i < 0 \tag{20}$$

for $i = 1, \cdots, N$. The output switching control (12) with $u(y(\cdot))$ given by (18) makes the equilibrium solution $x = 0$ of (1) globally asymptotically stable and

$$\int_0^\infty x(t)'Q_{\sigma(t)}x(t)dt \leq \min_{i=1,\cdots,N} \tilde{x}_0'\tilde{P}_i\tilde{x}_0 \tag{21}$$

The main problem underlying inequality (20) is that the unknowns of the filter and $\tilde{P}_i$ appear in a nonlinear form. Hence, we try to simplify the inequalities by introducing the square matrices

$$\tilde{T}_i = \begin{bmatrix} I & I \\ \Gamma_i & 0 \end{bmatrix} \tag{22}$$

where Γ_i for $i = 1, \cdots, N$ are suitable extra variables to be determined in order to simplify the problem. To this end, notice that

$$\tilde{T}_i'\tilde{P}_i\tilde{A}_i\tilde{T}_i = \begin{bmatrix} N_{11} & N_{12} \\ N_{21} & N_{22} \end{bmatrix} \tag{23}$$

where

$$\begin{aligned} N_{11} &= (X + \Gamma_i'V')A_i + (V + \Gamma_i'\hat{X}_i)(\hat{B}_iC_i + \hat{A}_i\Gamma_i) \\ N_{12} &= (X + \Gamma_i'V')A_i + (V + \Gamma_i'\hat{X}_i)\hat{B}_iC_i \\ N_{21} &= XA_i + V\hat{B}_iC_i + V\hat{A}_i\Gamma_i \\ N_{22} &= XA_i + V\hat{B}_iC_i \end{aligned}$$

Therefore, a sensible choice of Γ_i is $\Gamma_i = -\hat{X}_i^{-1}V'$ which, by redefining the unknowns as

$$Z_i = X + \Gamma_i'V' \tag{24}$$

$$L_i = V\hat{B}_i \tag{25}$$

$$M_i = -V\hat{A}_i\Gamma_i = V\hat{A}_iV^{-1}(X - Z_i) \tag{26}$$

for all $i = 1, \cdots, N$ leads to the conclusion that the matrix blocks of (23) can alternatively be written as

$$N_{11} = Z_iA_i \tag{27}$$

$$N_{12} = Z_iA_i \tag{28}$$

$$N_{21} = XA_i + L_iC_i - M_i \tag{29}$$

$$N_{22} = XA_i + L_iC_i \tag{30}$$

and, in addition, simple algebraic calculations put in evidence that

$$\tilde{T}_i'\tilde{Q}_i\tilde{T}_i = \begin{bmatrix} Q_i & Q_i \\ Q_i & Q_i \end{bmatrix}, \; \tilde{T}_i'\tilde{P}_i\tilde{T}_i = \begin{bmatrix} Z_i & Z_i \\ Z_i & X \end{bmatrix} \quad (31)$$

The importance of the transformation matrices $\tilde{T}_i$ introduced before is apparent since, as indicated in (31), the product $\tilde{T}_i'\tilde{P}_i\tilde{T}_i$ is linearized and the product $\tilde{T}_i'\tilde{P}_j\tilde{T}_i$ for $j \neq i$ can be expressed in terms of LMIs. Indeed,

$$\begin{aligned} \tilde{T}_i'\tilde{P}_j\tilde{T}_i &= \tilde{T}_i'(\tilde{P}_j - \tilde{P}_i)\tilde{T}_i + \tilde{T}_i'\tilde{P}_i\tilde{T}_i \\ &= \tilde{T}_i' \begin{bmatrix} 0 & 0 \\ 0 & \hat{X}_j - \hat{X}_i \end{bmatrix} \tilde{T}_i + \begin{bmatrix} Z_i & Z_i \\ Z_i & X \end{bmatrix} \\ &= \begin{bmatrix} Z_i + \Gamma_i'(\hat{X}_j - \hat{X}_i)\Gamma_i & Z_i \\ Z_i & X \end{bmatrix} \\ &= \begin{bmatrix} Z_j + (Z_j - Z_i)(X - Z_j)^{-1}(Z_j - Z_i) & Z_i \\ Z_i & X \end{bmatrix} \end{aligned}$$

which calculated for all $i, j = 1, \cdots, N$ imply that

$$\tilde{T}_i' \left(\sum_{j=1}^{N} \pi_{ji}\tilde{P}_j \right) \tilde{T}_i = \sum_{j=1}^{N} \pi_{ji} \begin{bmatrix} Z_j + (Z_j - Z_i)(X - Z_j)^{-1}(Z_j - Z_i) & 0 \\ 0 & 0 \end{bmatrix}$$

Based on these calculations, we are now in position to provide the main result of the paper.

Theorem 1. Let $Q_i \geq 0$, $i = 1, 2, \cdots, N$ be given. There exist matrices $\hat{A}_i$ and $\hat{B}_i$, $i = 1, 2, \cdots, N$ for which inequalities (20) are satisfied for some positive definite matrices $\tilde{P}_i$ of the form (16) if and only if there exist a Metzler matrix Π, a positive definite matrix X, a set of positive matrices Z_i, a set of positive matrices R_{ij} and a set of matrices L_i for all $i, j = 1, 2, \cdots, N$, such that the following matrix inequalities

$$A_i'Z_i + Z_iA_i + \sum_{j=1}^{N} \pi_{ji}R_{ij} + Q_i < 0 \quad (32)$$

$$A_i'X + XA_i + C_i'L_i' + L_iC_i + Q_i < 0 \quad (33)$$

$$R_{ii} < Z_i \quad (34)$$

$$\begin{bmatrix} R_{ij} - Z_j & Z_j - Z_i \\ Z_j - Z_i & X - Z_j \end{bmatrix} > 0, \quad i \neq j \quad (35)$$

hold. Finally, assume that (32)-(35) are satisfied. The output switching control (12) defined by

$$u(y(\cdot)) = \arg\min_i \hat{x}(t)V'(X - Z_i)^{-1}V\hat{x}(t) \quad (36)$$

where V is an arbitrary nonsingular matrix, makes the equilibrium solution $x = 0$ of (1) globally asymptotically stable and

$$\int_0^\infty x(t)'Q_{\sigma(t)}x(t)dt \leq \min_{i=1,\cdots,N} \tilde{x}_0'\tilde{P}_i\tilde{x}_0 \quad (37)$$

Proof: Assume first that inequalities (32)-(35) are satisfied and consider the partitioned matrix (23) whose blocks, from the definitions (24) to (26) together with $\Gamma_i = -\hat{X}_i^{-1}V'$, are written as (27)-(30). Consequently, our main purpose is to investigate the structure of the following symmetric matrix expression

$$S_i := \tilde{T}_i' \left(\tilde{A}_i'\tilde{P}_i + \tilde{P}_i\tilde{A}_i + \sum_{j=1}^{N} \pi_{ji}\tilde{P}_j + \tilde{Q}_i \right) \tilde{T}_i$$

and show that $S_i < 0$ for some choice of the filter matrices. Letting S_{11}, S_{21} and S_{22} the three characterizing blocks of S_i, it follows that

$$\begin{aligned} S_{11} &= A_i'Z_i + Z_iA_i + \sum_{j=1}^{N} \pi_{ji}Y_{ij} + Q_i \\ S_{21} &= A_i'Z_i + XA_i + L_iC_i - M_i + Q_i \\ S_{22} &= A_i'X + XA_i + L_iC_i + C_i'L_i' + Q_i \end{aligned}$$

where

$$Y_{ij} = Z_j + (Z_j - Z_i)(X - Z_j)^{-1}(Z_j - Z_i)$$

The filter matrix $\hat{B}_i$ directly follows from equation (25), where V is a selected nonsingular matrix. Notice now that $S_{22} < 0$ in view of (33). Notice also that assumptions (34) and (35) imply $R_{ij} > Y_{ij}$ and $R_{ii} < Y_{ii}$, so that $S_{11} < 0$. Finally, take the filter matrix $\hat{A}_i$ such that $M_i = A_i'Z_i + XA_i + L_iC_i + Q_i$ which implies $S_{21} = 0$. In conclusion, from (25) and (26), the filter matrices are

$$\hat{B}_i = V^{-1}L_i \quad (38)$$

$$\begin{aligned} \hat{A}_i &= V^{-1}M_i(X - Z_i)^{-1}V \quad (39) \\ &= V^{-1}(A_i'Z_i + XA_i + L_iC_i + Q_i)(X - Z_i)^{-1}V \end{aligned}$$

and $\tilde{P}_i$ as in (16) with $\hat{X}_i = V'(X - Z_i)^{-1}V$. Viceversa, assume that inequalities (20) holds for some $\hat{A}_i$, $\hat{B}_i$ and $\tilde{P}_i$ as in (16). Then, apply the state-space transformation (22) with $\Gamma_i = -\hat{X}_i^{-1}V'$ and consider the expressions (24)-(30). Hence $S_i < 0$ and the filter matrices are

$$\hat{B}_i = V^{-1}L_i \quad (40)$$

$$\hat{A}_i = V^{-1}M_i(X - Z_i)^{-1}V \quad (41)$$

In conclusion, condition (33) coincides with $S_{22} < 0$, and (32) comes from $S_{11} < 0$ by letting $R_{ij} = Y_{ij} + \epsilon I$, $i \neq j$ and $R_{ii} = Z_i - \epsilon I$, for $\epsilon > 0$ small enough. From this (34) and (35) are satisfied.

Finally, assume that (32)-(35) are satisfied, and recall Lemma 3. Therefore, from $\hat{X}_i = V'(X - Z_i)^{-1}V$, it follows that the output switching control (12) with $u(y(\cdot))$ given by (36) makes the equilibrium solution $x = 0$ of (1) globally asymptotically stable and (37) is satisfied. □

It is important to stress the fact that formulas (40) and (41) provide a parametrization of all filters (13) for which (20) is satisfied with $\tilde{P}_i$ as in (16). In order to provide less stringent conditions, the equality $M_i = A_i'Z_i + XA_i + L_iC_i + Q_i$ has been set in the proof of the sufficient part of the theorem so as to put to zero the off-diagonal block entries of S_i.

The structure of the full-order filter is not in the observer form, i.e. $\hat{A}_i \neq A_i - \hat{B}_iC_i$. To recover this condition, an additional constraint, unfortunately non linear, has to be added, namely (the simple check is left to the reader)

$$\begin{aligned} M_i &= (VA_i - L_iC_i)V^{-1}(X - Z_i) \\ &= A_i'Z_i + XA_i + L_iC_i + Q_i \end{aligned} \quad (42)$$

A notable exception can be devised by letting $Q_i = 0$, so overlooking the guaranteed cost property (37). Indeed, in this case, we have the following result.

Theorem 2. Assume that there exist a Metzler matrix Π, a positive definite matrix X, a set of positive matrices L_i and a set of positive matrices Z_i for all $i = 1, 2, \cdots, N$, such that the following matrix inequalities are satisfied:

$$A_i'Z_i + Z_iA_i + \sum_{j=1}^{N} \pi_{ji}Z_j < 0 \quad (43)$$

$$A_i'X + XA_i + C_i'L_i' + L_iC_i < 0 \quad (44)$$

Then, the switching rule

$$u(y(\cdot)) = \arg\min_i \hat{x}(t)Z_i\hat{x}(t) \quad (45)$$

makes the equilibrium solution $x = 0$ of (1) globally asymptotically stable where $\hat{x}$ satisfies the differential equation of the filter (13) in observer form with

$$\hat{B}_i = -X^{-1}L_i \quad (46)$$

$$\hat{A}_i = A_i - \hat{B}_iC_i \quad (47)$$

Proof: The proof relies to Theorem 1, by letting $Z_i \to \epsilon Z_i$ with $\epsilon > 0$ arbitrarily small and $V = -X$. Indeed, notice that the condition (42) for the filter to be in observer form is satisfied for ϵ going to zero and that

$$\begin{aligned} &\arg\min_i \hat{x}(t)V'(X - \epsilon Z_i)^{-1}V\hat{x}(t) = \\ &\arg\min_i \hat{x}(t)(X + (Z_i^{-1}/\epsilon - X^{-1})^{-1})\hat{x}(t) \sim \\ &\arg\min_i \hat{x}(t)\epsilon Z_i\hat{x}(t) \sim \arg\min_i \hat{x}(t)Z_i\hat{x}(t) \end{aligned}$$

holds. □

The conclusion is that if there exist N gains that make the filter quadratically stable, see equation (44), then the usual solution to the Metzler-Lyapunov inequality (state feedback, equation (43)) provides a stabilizing switching rule. It is important to keep in mind that if we want to determine a good switching strategy by minimizing a guaranteed quadratic cost then this solution although stabilizing is not the best that can be done. Moreover, it should be noticed that the output feedback strategies invoked by the theorems presented so far require the existence of a state-observer injection matrices $\hat{L}_i = X^{-1}L_i$, $i = 1, \cdots, N$ that render the set of matrices $A_i + \hat{L}_iC_i$ quadratically stable (see e.g. equation (33)).

Remark 2. There is no difficulty to get the versions of Theorem 1 and Theorem 2 associated to the modified Lyapunov-Metzler inequalities appearing in Lemma 2. The BMIs are replaced by LMIs with an additional parameter that can be determined by line search. The results follow the same pattern of each mentioned theorem, being thus omitted.

The next example illustrates some aspects of the theoretical results obtained so far. Consider the system (1) with $N = 2$ and matrices $\{A_1, A_2\}$ given by

$$A_1 = \begin{bmatrix} 0 & 1 \\ 2 & -9 \end{bmatrix}, \ A_2 = \begin{bmatrix} 0 & 1 \\ -2 & 2 \end{bmatrix} \quad (48)$$

which, as it can be easily verified by inspection, are both unstable. Setting $Q = I$ our goal is to design an output feedback stabilizing control which minimizes the quadratic cost on the left hand side of (37). Considering the filter initial condition $\hat{x}(0) = 0$, it is readily seen that

$$\int_0^\infty x(t)'Q_{\sigma(t)}x(t)dt \leq x_0'Xx_0 \quad (49)$$

whenever the switching rule satisfies the conditions provided by Theorem 1. Setting

$$\Pi = \begin{bmatrix} -20 & 20 \\ 20 & -20 \end{bmatrix} \quad (50)$$

matrices X, $Z_1, \cdots, Z_N$ and $L_1, \cdots, L_N$ have been determined from the solution of the convex programming problem

$$\min\{\mathrm{trace}(X) \ , \ (32) - (35)\} \quad (51)$$

which means that the minimum guaranteed cost (49) is calculated with x_0 being a random variable with zero mean and unitary covariance matrix. The optimal solution provided the minimum guaranteed cost trace$(X) \approx 338.04$. The associated matrix variables and the choice $V = -X$ allow the determination of the stabilizing output feedback switching rule $\sigma(t) = u(y(\cdot))$ given by (36) as well as the switching filter (13).

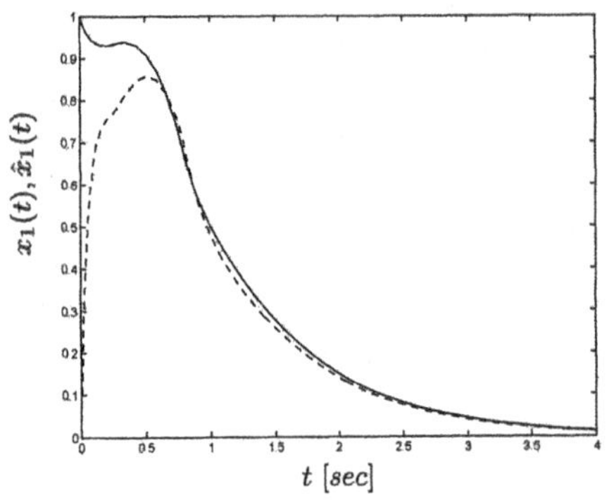

Fig. 1. Time simulation - first state variable

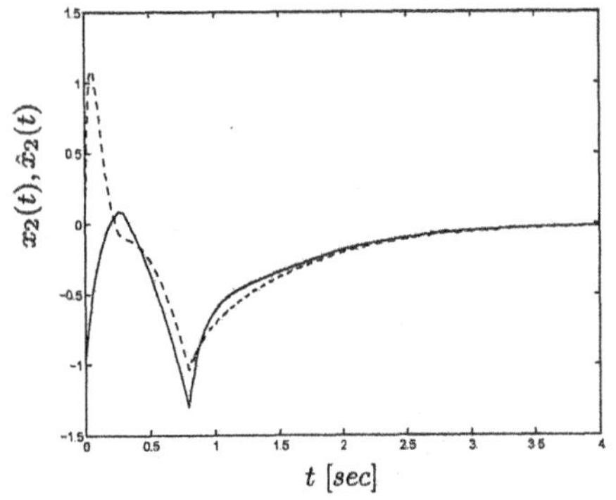

Fig. 2. Time simulation - second state variable

Time simulation has been performed from the initial conditions $x_0 = [1 \ -1]'$ and $\hat{x}_0 = [0 \ 0]'$. Figures 1 and 2 show the trajectories of the state variable $x(t) \in \mathbb{R}^2$ of the system (solid line) and the state variable $\hat{x}(t) \in \mathbb{R}^2$ of the switching filter (dashed line) versus time for the system and filter controlled by the output switching rule $\sigma(t) = u(y(\cdot))$ given by (36). Even though, in this case, the filter does not exhibit an observer structure, it estimates with good precision - under switching - the state variable of the system. Indeed, as it is clearly indicated in the mentioned figures, after a small period of time ($\approx$ 1.0 sec.) $\hat{x}(t)$ provides $x(t)$ from the available measurements $y(\tau), \tau \leq t$. From the same figures, it can be seen, that the proposed control strategy is very effective to stabilize the system under consideration.

4. CONCLUSION

In this paper we have introduced a new stability condition for continuous time switched systems based on output measurements. Furthermore, the determination of a guaranteed cost associated to the proposed control strategy has been addressed. The results reported are necessary and sufficient for the existence of a switching filter and a switching rule based exclusively on the output vector of the original system in such a way the so called Lyapunov-Metzler inequalities have a feasible solution. In a precise case, identifying in the paper, the switching filters are of the form of classical observers. Special attention has been devoted towards the numerical solvability of the design problems by means of methods based on linear matrix inequalities. There are several points that deserve further investigation. Among them, it is important to mention the possible introduction of robustness issues against parameter uncertainty as considered, for instance, in (Geromel *et al.*, 1991) appearing in the system and filters parameters. The example showed that the switching rule is particulary sensitive to parameter uncertainty which may in particular cases lead to instability of the closed loop system.

REFERENCES

S. P. Boyd, L. El Ghaoui, E. Feron and V. Balakrishnan (1994). *Linear Matrix Inequalities in System and Control Theory*, SIAM, Philadelphia, 1994.

M. S. Branicky (1998). Multiple Lyapunov functions and other analysis tools for switched and hybrid systems. *IEEE Trans. Automat. Contr.*, vol. 43, pp. 475–482.

P. Colaneri, J. C. Geromel, and A. Locatelli (1997). *Control Theory and Design - An RH_2 and RH_∞ viewpoint*, Academic Press.

Y. Fang, and K. A. Loparo (2002). Stabilization of continuous-time jump linear systems, *IEEE Trans. Automat. Contr.*, vol. 47, pp. 1590–1603.

J. C. Geromel, P. L. D. Peres, and J. Bernussou (1991). On a convex parameter space method for linear control design of uncertain systems, *SIAM J. Control Optim.*, vol. 29, pp. 381–402.

J. C. Geromel, and P. Colaneri (2005). Stabilization of continuous-time switched systems, *IFAC World Congress*, Prague.

J. P. Hespanha (2004). Uniform stability of switched linear systems : extensions of LaSalle's principle, *IEEE Trans. Automat. Contr.*, vol. 49, pp. 470–482.

J. Hockerman-Frommer, S. R. Kulkarni and P. J. Ramadge (1998). Controller switching based on output predictions errors, *IEEE Trans. Automat. Contr.*, vol. 43, pp. 596–607.

M. Johansson, and A. Rantzer (1998). Computation of piecewise quadratic Lyapunov functions for hybrid systems, *IEEE Trans. Automat. Contr.*, vol. 43, pp. 555–559.

D. Liberzon (2003). *Switching in Systems and Control*, Birkhauser.

H. Ye, A. N. Michel and L. Hou (1998). Stability theory for hybrid dynamical systems, *IEEE Trans. Automat. Contr.*, vol. 43, pp. 461–474.

ELSEVIER
IFAC
PUBLICATIONS

PRACTICAL STABILIZATION OF DISCRETE–TIME LINEAR SISO SYSTEMS UNDER ASSIGNED INPUT AND OUTPUT QUANTIZATION [1]

Bruno Picasso [*,**,2] **Antonio Bicchi** [*]

[*] *Università di Pisa, Centro Interdipartimentale di Ricerca E. Piaggio", Via Diotisalvi 2, 56100 Pisa, Italy*
[**] *Politecnico di Milano, Piazza Leonardo da Vinci 32, 20133 Milano, Italy*

Abstract: This work is concerned with the practical stabilization of discrete–time SISO linear systems under assigned quantization of the input and output spaces. A controller is designed ensuring practical stability properties.
Unlike most of the existing literature, quantization is supposed to be a problem datum rather than a degree of freedom in design. Moreover, in the framework of control under assigned quantization, results are concerned with state quantization only and do not include the quantized output feedback case considered here.
While standard stability analysis techniques are based on Lyapunov theory and invariant ellipsoids, our study involves a particularly suitable family of sets, which are hypercubes in controller form coordinates. *Copyright, © 2006, IFAC*

Keywords: Quantized systems, stabilizability, dynamic output feedback.

1. INTRODUCTION

The need of studying quantized control systems (i.e., dynamical systems with discrete input and/or output variables) arises by many control applications. Commonly encountered examples include the presence of digital sensors and actuators or of finite capacity communication links in the control loop. Quantized control systems have been attracting increasing attention of the control community in the past twenty years (see for instance (Delchamps, 1990; Wong–Brockett, 1999; Brockett–Liberzon, 2000; Elia–Mitter, 2001; Baillieul, 2001; Bicchi et al., 2002; Tatikonda–Mitter, 2004; Fagnani–Zampieri, 2004)).
This paper deals with the control of the linear, time–invariant dynamical system

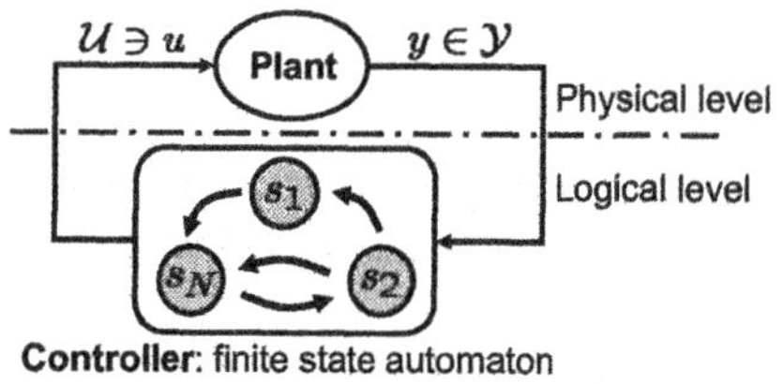

Fig. 1. Graphical illustration of the quantized control system considered in this paper.

$$\begin{cases} x(t+1) = Ax(t) + b\,u(t) \\ y(t) = q\big(Cx(t)\big) \\ x \in \mathbb{R}^n, \quad u \in \mathcal{U} \subset \mathbb{R}, \quad y \in \mathcal{Y}, \quad t \in \mathbb{N} \\ A \in \mathbb{R}^{n \times n}, \quad b \in \mathbb{R}^n, \quad C \in \mathbb{R}^{1 \times n}, \end{cases} \tag{1}$$

where $\mathcal{U}$ is a *given* closed discrete set and $q : \mathbb{R} \to \mathcal{Y}$ is an *assigned* output map taking values in a countable set $\mathcal{Y}$ (finite or infinite). A pictorial representation of the control problem is illustrated in Fig. 1: the system has a hybrid structure and is organized into two levels. At the logical level, the controller manipulates output and input strings from discrete alphabets. At the physical level, the plant is modelled by Eqn. (1).
We focus on the stabilization problem. It has been clarified in (Delchamps, 1990) that "practical"

---

[1] This work was supported by European Commission through the IST-2004-004536 (IP) "RUNES - Reconfigurable Ubiquitous Networked Embedded Systems" and the IST-2004-511368 (NoE) "HYCON - HYbrid CONtrol: Taming Heterogeneity and Complexity of Networked Embedded Systems".
[2] Corresponding author, e–mail: b.picasso@sns.it

stability notions are the suitable stability properties to be considered for quantized systems. Accordingly, our study concerns the construction of symbolic feedback controllers capable of steering the system to within small neighborhoods of the equilibrium, starting from large attraction basins. In most of the existing literature on stabilization, quantization is considered as a degree of freedom in control synthesis: the designer can choose the elements of the control set $\mathcal{U}$, as well as the output map q. Results in this vein have a strong theoretical interest as they allow to identify fundamental limitations in quantized control (Wong–Brockett, 1999; Elia–Mitter, 2001; Ishii–Francis, 2002; Nair–Evans, 2004; Tatikonda–Mitter, 2004; Fagnani–Zampieri, 2004). On the contrary, in this paper we assume that the input and output sets $\mathcal{U}, \mathcal{Y}$, as well as the output map q, are data of the problem: control synthesis for practical stabilization is then subdued to these data. This kind of study provides tools for the analysis of achievable control objectives by using a *given* technology, as e.g., actuators (modelled by $\mathcal{U}$) or sensors (modelled by q).

The main contribution of this paper consists in providing a novel, simple and general technique to solve the practical stabilization problem for quantized SISO systems. The controller is synthesized so as to encapsulate the trajectories into increasingly smaller hypercubes in the controller form coordinates. This approach results in a controller in the form of a finite state automaton. The fundamental observation is that most of the significant information related to the input set $\mathcal{U}$, to the output map q and to the dynamics of the system, are contained in a pair (ρ, H) of scalar functions providing an effective representation of the resolution (or *dispersion*) of the quantizers. The quantization schemes typically encountered in the literature include either innovation (Elia–Mitter, 2001; Tatikonda–Mitter, 2004) or state quantization (Delchamps, 1990; Wong–Brockett, 1999; Liberzon, 2003; Picasso–Bicchi, 2003). In the latter case, the output map $y = q(x)$ is such that $q^{-1}(y)$ is a bounded set. In this work instead, quantized outputs are considered: the output map $q \circ C : \mathbb{R}^n \to \mathcal{Y}$ induces a state space partition $\mathbb{R}^n = \bigcup_{y \in \mathcal{Y}} (q \circ C)^{-1}(y)$ made of *unbounded* sets and this makes the problem considerably different. We also remark that the only contributions addressing quantization as a problem datum do so by fixing either input or state quantization. In our work, we expressly consider the case in which both input and output quantizations are assigned.

The paper is organized as follows: the problem is formulated in Sec. 2, the main result about practical stability is in Sec. 3 (including an example) and its proof is given in Sec. 4. The easy proofs of the most technical results are omitted.

Notation: $Q_n(\Delta) := \left[-\frac{\Delta}{2}; \frac{\Delta}{2}\right]^n = \{x \in \mathbb{R}^n \,|\, \|x\|_\infty \le \frac{\Delta}{2}\}$. Let $E \subseteq \mathbb{R}^k$: E^{ch} and $\#E$ denote respectively its convex hull and its cardinality; $\text{diam}(E) := \sup\limits_{x,y \in E} \|x - y\|_2$ is the diameter of E. Given $v \in \mathbb{R}^k$, $E + v := \{x \in \mathbb{R}^k \,|\, x - v \in E\}$. Let x_i be the ith coordinate of x: given $\Omega \subseteq \mathbb{R}^n$, $\text{Pr}_i(\Omega) := \{\omega_i \,|\, \omega \in \Omega\}$ and $\text{diam}_i(\Omega) := \text{diam}(\text{Pr}_i(\Omega))$. If $A \in \mathbb{R}^{n \times n}$, $A\Omega = \{A\omega \,|\, \omega \in \Omega\}$ while $(Ax)_i$ is the ith coordinate of the vector Ax. x' denotes the transpose of the vector x, x^+ stands for $x(t+1)$.

2. PROBLEM STATEMENT

We deal with the discrete–time, quantized SISO system given in Eqn. (1). We assume that the pair (A, b) is reachable and that the pair (A, C) is observable. Without loss of generality, we assume that the pair (A, b) is in *controller form*:

$$A = \begin{pmatrix} 0 & 1 & \cdots & 0 \\ \vdots & \ddots & \ddots & \vdots \\ 0 & 0 & \cdots & 1 \\ \alpha_1 & \alpha_2 & \cdots & \alpha_n \end{pmatrix}, \; b = \begin{pmatrix} 0 \\ \vdots \\ 0 \\ 1 \end{pmatrix},$$

where $s^n - \alpha_n s^{n-1} - \cdots - \alpha_2 s - \alpha_1$ is the characteristic polynomial of A. Because

$$\|A\|_\infty = \max_{i=1,\ldots,n} \sum_{j=1}^{n} |A_{i,j}| = \max\Big\{1, \sum_{i=1}^{n} |\alpha_i|\Big\},$$

if $\sum_{i=1}^n |\alpha_i| \le 1$ then the system is stable, we hence assume $\sum_{i=1}^n |\alpha_i| > 1$ and we let $\alpha := \sum_{i=1}^n |\alpha_i|$.

The output quantizer $q : \mathbb{R} \to \mathcal{Y}$ is characterized by the induced output space partition: $\mathbb{R} = \bigcup_{y \in \mathcal{Y}} q^{-1}(y)$. The partition is supposed to be locally finite, namely, if $B \subset \mathbb{R}$ is bounded, $\#\{y \in \mathcal{Y} \,|\, B \cap q^{-1}(y) \neq \emptyset\} < +\infty$. We also assume that $\forall y \in \mathcal{Y}$, $q^{-1}(y) \subseteq \mathbb{R}$ is a connected set. In this case, by suitably redefining q without varying the induced output space partition (hence, without loss of generality), we can assume that $\mathcal{Y} \subset \mathbb{R}$ and $q : \mathbb{R} \to \mathcal{Y}$ is such that $\forall y \in \mathcal{Y}$ the closure of $q^{-1}(y)$ is either an interval of length λ_y of the type $[y - \frac{\lambda_y}{2}; y + \frac{\lambda_y}{2}]$ or a half–line. Let $\mathcal{Y}_\star := \{y \in \mathcal{Y} \,|\, q^{-1}(y) \text{ is an interval of finite length}\}$.

With regard to the control set, we assume that $0 \in \mathcal{U}$ so that $(x = 0, u = 0)$ is an equilibrium pair. Given $\mathcal{U}_* \subseteq \mathcal{U}$, let $\rho_{\mathcal{U}_*}$ represent the dispersion (or maximal gap) of $\mathcal{U}_*$, that is:

$$\rho_{\mathcal{U}_*} := \begin{cases} \sup \{\text{diam}(]a;b[) \,|\,]a;b[\subseteq \mathcal{U}_*^{\text{ch}} \text{ and} \\ \qquad]a;b[\cap \mathcal{U}_* = \emptyset\} \quad \text{if } \#\mathcal{U}_* > 1 \\ +\infty \text{ (conventionally)} \quad \text{otherwise.} \end{cases} \tag{2}$$

Definition 1. Given system (1), let

$$\nu : \mathbb{R} \longrightarrow \mathcal{U} \tag{3}$$

be a map which associates to each real number r an element of $\mathcal{U}$ minimizing the Euclidean distance from r. The feedback law $k : \mathbb{R}^n \to \mathcal{U}$ defined by

$$k(x) := \nu\big(-\textstyle\sum_{i=1}^n \alpha_i x_i\big) = \nu\big(-(Ax)_n\big)$$

is called *state feedback quantized deadbeat controller* (state feedback qdb–controller).

The definition of the map ν is well posed because $\mathcal{U}$ is a closed set.

Problem statement: given system (1), we are interested in the synthesis of a controller which,

based on quantized output measurements $y \in \mathcal{Y}$, selects a quantized control $u \in \mathcal{U}$ and ensures practical stability properties. We will consider dynamical controllers and we will study the practical stability notion of (X_0, X_1, Ω)*-stability*.
More precisely, let the controller be described by the following system defined on some set $\mathcal{W}$:

$$\begin{cases} w(t+1) = \gamma(w(t), y(t), t) \\ u(t) = \phi(w(t), y(t), t), \end{cases} \quad (4)$$

where $\gamma : \mathcal{W} \times \mathcal{Y} \times \mathbb{N} \to \mathcal{W}$ and $\phi : \mathcal{W} \times \mathcal{Y} \times \mathbb{N} \to \mathcal{U}$. The closed–loop dynamics induced by the feedback interconnection of such a controller with system (1) is:

$$\begin{cases} x(t+1) = Ax(t) + b\,\phi\Big(w(t), q(Cx(t)), t\Big) \\ w(t+1) = \gamma\Big(w(t), q(Cx(t)), t\Big). \end{cases} \quad (5)$$

Definition 2. (Cf. (Fagnani–Zampieri, 2004)) Let Ω, X_0 and X_1 be subsets of $\mathbb{R}^n$ such that Ω and X_0 are neighborhoods of the origin and $X_1 \supseteq X_0$ is bounded. The controller (4) is said to be (X_0, X_1, Ω)*-stabilizing* iff the corresponding closed–loop dynamics (5) is so that $\forall x(0) \in X_0$ and $\forall w(0) \in \mathcal{W}$, $x(t) \in X_1$ $\forall t \geq 0$ and $\exists \bar{t} \in \mathbb{N}$ such that $\forall t \geq \bar{t}$, $x(t) \in \Omega$.

3. INPUT AND OUTPUT QUANTIZATION

The use of a controller endowed with memory, hence taking the general form in Eqn. (4), allows to treat the quantized output case by taking advantage of some techniques introduced for the quantized state case. In fact, by storing the past inputs and outputs, it is possible to reconstruct a bounded region within which the current state is confined. Nevertheless, the state quantization obtained in this way is *time–varying* so that the results from the quantized state case need to be further elaborated in order to be applied to the quantized output problem.
Following (Picasso et al., 2002; Picasso–Bicchi, 2003), the stabilization problem is studied taking into consideration sets X_0, X_1 and Ω in the form of hypercubes in the controller form coordinates. This choice makes the analysis particularly simple so that explicit results can be provided for arbitrarily assigned input and output quantized sets.

3.1 Preliminaries

Suppose that at time t the current state of the system $x(t)$ is only known to belong to a certain bounded set, more precisely assume that $x(t) \in \mathcal{C}_{x(t)} \subset Q_n(\Delta)$. The existence of a control value $u \in \mathcal{U}$ ensuring that $x(t+1) \in Q_n(\Delta)$ is tantamount to requiring that $\exists u \in \mathcal{U}$ such that $A\mathcal{C}_{x(t)} + bu \subseteq Q_n(\Delta)$. For $u \in \mathcal{U}$, by the controller form of (A, b), it holds that

$$\begin{aligned} x^+ = (x_2, \ldots, x_n, \textstyle\sum_i \alpha_i x_i + u) \in Q_n(\Delta) \\ \iff \big|\textstyle\sum_i \alpha_i x_i + u\big| \leq \tfrac{\Delta}{2}. \end{aligned} \quad (6)$$

Therefore, $x(t+1) \in Q_n(\Delta)$ if and only if

$$\mathrm{Pr}_n(A\mathcal{C}_{x(t)}) + u \subseteq \big[-\tfrac{\Delta}{2}\,;\, \tfrac{\Delta}{2}\big]. \quad (7)$$

It is then important to have an estimate of $\mathrm{diam}_n(A\mathcal{C}_{x(t)})$. Furthermore, the control acts only on the nth component while the others shift upward (see Eqn. (6)) : hence, trajectorics with both the properties of converging to a small neighborhood Ω of the equilibrium and a high speed of convergence towards Ω can be obtained by selecting the control value u so that the middle point of the set $\mathrm{Pr}_n(A\mathcal{C}_{x(t)}) + u$ is as near as possible to 0. This task is accomplished by the state feedback qdb–controller.
Accordingly, the dynamic qdb–controller proposed below is based on the following paradigm:

1- a bounded set within which the current state lies is located;
2- an estimate $\hat{x}$ of the current state is obtained;
3- the control action is selected by the state feedback qdb–controller, namely $u = k(\hat{x})$.

Before explicitly defining the dynamic qdb–controller, we need the following preliminary result.

Lemma 1. If $x \in Q_n(\Delta)$ and u is such that $x^+ \in Q_n(\Delta)$, then $u \in \big[-\frac{\Delta}{2}(\alpha+1)\,;\,\frac{\Delta}{2}(\alpha+1)\big]$.

Hence, if the goal is to find $u \in \mathcal{U}$ such that x^+ remains in $Q_n(\Delta)$, the set of the control values that are relevant to this problem is

$$\mathcal{U}(\Delta) := \mathcal{U} \cap \big[-\tfrac{\Delta}{2}(\alpha+1)\,;\, \tfrac{\Delta}{2}(\alpha+1)\big]. \quad (8)$$

Notice that $\#\mathcal{U}(\Delta) < +\infty$ because $\mathcal{U}$ is a closed discrete set. Let

$$\begin{cases} m(\Delta) := \min \mathcal{U}(\Delta) \\ M(\Delta) := \max \mathcal{U}(\Delta) \end{cases}$$

and, according to (2), let

$$\rho(\Delta) := \rho_{\mathcal{U}(\Delta)} \quad (9)$$

be the dispersion of $\mathcal{U}(\Delta)$.

3.2 The dynamic qdb-controller

In this section the dynamic qdb–controller is defined following the steps listed in the paradigm described in the previous section. The practical stability properties of the corresponding closed loop system are then analyzed in Theorem 1.

Step 1- Derivation of $\mathcal{C}_{x(t)}$

The function $\boldsymbol{q} : \mathbb{R}^n \to \mathcal{Y}^n$ defined by $\boldsymbol{q}(z) := (q(z_1), \ldots, q(z_n))$ induces a partition of $\mathbb{R}^n$ such that $\forall \boldsymbol{y} \in \mathcal{Y}^n_\star$ the closure of $\boldsymbol{q}^{-1}(\boldsymbol{y})$ is $\boldsymbol{y} + \mathcal{P}_{\boldsymbol{y}}$, where $\mathcal{P}_{\boldsymbol{y}} = \prod_{i=1}^n [-\frac{\lambda_{y_i}}{2}\,;\,\frac{\lambda_{y_i}}{2}]$. Let

$$S := \begin{pmatrix} 0 & 0 & \cdots & 0 \\ Cb & 0 & \cdots & 0 \\ CAb & Cb & \ddots & 0 \\ \vdots & \vdots & \ddots & \vdots \\ CA^{n-2}b & CA^{n-3}b & \cdots & Cb \end{pmatrix} \in \mathbb{R}^{n \times (n-1)}.$$

Denote by $\boldsymbol{u}(t)$ and $\boldsymbol{y}(t)$ the vectors collecting respectively the last $n-1$ inputs and the last n outputs at time t ($t \geq n-1$), that is $\boldsymbol{u}(t) := \big(u(t-n+1), \ldots, u(t-1)\big)'$, and $\boldsymbol{y}(t) := \big(y(t-n+1), \ldots, y(t)\big)'$. Let $R := \big[A^{n-2}b \,|\cdots|\, Ab\,|\,b\big] \in \mathbb{R}^{n \times (n-1)}$ and $\mathcal{O} \in \mathbb{R}^{n \times n}$ be the observability

matrix (i.e., the matrix whose i^{th} row is CA^{i-1}): $\mathcal{O}$ is invertible by hypothesis.
By standard theory on observability it holds that

$$\boldsymbol{y}(t) = \boldsymbol{q}\left(\mathcal{O}x(t-n+1) + S\boldsymbol{u}(t)\right),$$

hence

$$x(t-n+1) \in \mathcal{O}^{-1}\Big(\boldsymbol{q}^{-1}(\boldsymbol{y}(t)) - S\boldsymbol{u}(t)\Big)$$

and

$$x(t) \in A^{n-1}\mathcal{O}^{-1}\Big(q^{-1}(\boldsymbol{y}(t))\Big) - -A^{n-1}\mathcal{O}^{-1}S\boldsymbol{u}(t) + R\boldsymbol{u}(t).$$

If moreover $\boldsymbol{y}(t) \in \mathcal{Y}_\star^n$, as the closure of $\boldsymbol{q}^{-1}(\boldsymbol{y})$ is $\boldsymbol{y} + \mathcal{P}_{\boldsymbol{y}}$, then the current state belongs to the following bounded set:

$$\begin{aligned} x(t) \in \mathcal{C}_{x(t)} &:= A^{n-1}\mathcal{O}^{-1}(\mathcal{P}_{\boldsymbol{y}(t)}) + \\ +A^{n-1}\mathcal{O}^{-1}\boldsymbol{y}(t) &+ (R - A^{n-1}\mathcal{O}^{-1}S)\,\boldsymbol{u}(t). \end{aligned} \quad (10)$$

Step 2- State estimation

Let the map $\psi : \mathcal{Y}^n \times \mathcal{U}^{n-1} \to \mathbb{R}^n$ be defined by

$$\psi(\boldsymbol{y}, \boldsymbol{u}) := A^{n-1}\mathcal{O}^{-1}\boldsymbol{y} + (R - A^{n-1}\mathcal{O}^{-1}S)\boldsymbol{u}. \quad (11)$$

If $\boldsymbol{y}(t) \in \mathcal{Y}_\star^n$, then $\hat{x}(t) := \psi(\boldsymbol{y}(t), \boldsymbol{u}(t))$ is the centroid of the parallelogram $\mathcal{C}_{x(t)}$.

Step 3- Control selection

The controller is defined by selecting the control action as if the current state was $\psi(\boldsymbol{y}(t), \boldsymbol{u}(t))$. Naturally, such controller needs to be initialized for $t \leq n-2$, we hence define the *dynamic qdb-controller* as follows: denote by $k(x)$ the state feedback qdb–controller and let [3]

$$u(t) := \begin{cases} 0 & \text{if } t \leq n-2 \\ (k \circ \psi)(\boldsymbol{y}(t), \boldsymbol{u}(t)) & \text{if } t \geq n-1. \end{cases} \quad (12)$$

Practical stability analysis

Let us analyze the resulting closed–loop dynamics. For $t \leq n-1$, by the controller form of A, it holds that $\forall x(0) \in Q_n(\Delta)$ and $\forall t \leq n-1$, $x(t) \in Q_n(\Delta \|A^{n-1}\|_\infty)$. For $t \geq n$, let us determine an upper bound $H(\Delta)$ for $\text{diam}_n(A\mathcal{C}_{x(t)})$: for any $\Delta > 0$, consider $\mathcal{Y}(\Delta) := (q \circ C)(Q_n(\Delta))$. If $\mathcal{Y}(\Delta) \subseteq \mathcal{Y}_\star$, let $\Lambda_\Delta := \max_{y \in \mathcal{Y}(\Delta)} \lambda_y$ and $H(\Delta) := \text{diam}_n\Big(A^n\mathcal{O}^{-1}(Q_n(\Lambda_\Delta))\Big)$, else $H(\Delta) := +\infty$. Let $\Delta > 0$ be such that $\mathcal{Y}(\Delta) \subseteq \mathcal{Y}_\star$ and suppose that $\boldsymbol{y}(t) \in \mathcal{Y}(\Delta)^n$: since $\mathcal{C}_{x(t)}$ is a translation of the set $A^{n-1}\mathcal{O}^{-1}(\mathcal{P}_{\boldsymbol{y}(t)})$ (see Eqn. (10)), and $\mathcal{P}_{\boldsymbol{y}(t)} \subseteq Q_n(\Lambda_\Delta)$, then

$$\begin{aligned} \text{diam}_n(A\mathcal{C}_{x(t)}) &= \text{diam}_n(A^n\mathcal{O}^{-1}(\mathcal{P}_{\boldsymbol{y}(t)})) \leq \\ &\leq \text{diam}_n\Big(A^n\mathcal{O}^{-1}(Q_n(\Lambda_\Delta))\Big) = H(\Delta). \end{aligned} \quad (13)$$

Theorem 1. Let $\Delta_1 > 0$ be such that

$$\begin{cases} m(\Delta_1) < -\dfrac{\Delta_1}{2}(\alpha - 1) & (14) \\ M(\Delta_1) > \dfrac{\Delta_1}{2}(\alpha - 1) & (15) \\ \rho(\Delta_1) + H(\Delta_1) < \Delta_1, & (16) \end{cases}$$

and $\Delta_0 := \frac{\Delta_1}{\|A^{n-1}\|_\infty}$. Consider the following algorithm:

- o Input: $\boldsymbol{\Delta} := \Delta_1$
- • $h := 1$;
- • while $(\rho(\Delta_h) + H(\Delta_h) < \Delta_h)$ do
  $(\Delta_{h+1} := \rho(\Delta_h) + H(\Delta_h)$;
  $\boldsymbol{\Delta} := (\boldsymbol{\Delta}, \Delta_{h+1}) \in \mathbb{R}^{h+1}$; $h := h+1)$ (17)
- o Output: $\boldsymbol{\Delta}$.

The output $\boldsymbol{\Delta} \in \mathbb{R}^f$ for some $f < +\infty$ and $\Delta_h > \Delta_{h+1}\ \forall h = 1, \ldots, f-1$. Moreover, let $k(x)$ be the state feedback qdb–controller with saturated inputs $\mathcal{U} = \mathcal{U}(\Delta_1)$, then the dynamic qdb–controller (12) is $(Q_n(\Delta_0), Q_n(\Delta_1), Q_n(\Delta_f))$–stabilizing.

Proof. The proof is given in next Section 4. ■

Example 1. Consider the unstable system

$$\begin{cases} x^+ = \begin{pmatrix} 0 & 1 \\ 5/4 & 1/4 \end{pmatrix} x + \begin{pmatrix} 0 \\ 1 \end{pmatrix} u \\ y = q(Cx), \end{cases}$$

where $C = (3/2\ 1/3)$, $u \in \mathcal{U} = \{0, \pm1, \pm2, \pm3, \pm4, \pm6, \pm8, \pm12, \pm16, \pm24\}$ and the extremes of the intervals forming the output space partition induced by q are $\{\pm\frac{3}{2}, \pm\frac{9}{2}, \pm\frac{15}{2}, \pm\frac{25}{2}, \pm\frac{39}{2}\}$. According to the developed theory, let $\mathcal{Y} = \mathcal{Y}_\star \cup \{\pm y_s\} = \{0, \pm3, \pm6, \pm10, \pm16, \pm y_s\}$ (where $\mathcal{Y}_\star$ collects the middle points of the output quantization intervals and q takes the values $\pm y_s$ for $|Cx| > \frac{39}{2}$). The values of λ_y for $y \in \mathcal{Y}_\star$ are: $\lambda_0 = \lambda_{\pm3} = \lambda_{\pm6} = 3$, $\lambda_{\pm10} = 5$ and $\lambda_{\pm16} = 7$. The infinity norm of A is $\alpha = \frac{3}{2}$. By direct computations it holds that $\rho(\Delta) + H(\Delta) < \Delta \Leftrightarrow \rho(\Delta) + \frac{6}{7}\Lambda_\Delta < \Delta \Leftrightarrow \Delta \in \left]\frac{25}{7}; \frac{234}{11}\right] := \mathcal{I}$. Also, $M(\frac{234}{11}) = 24 > \frac{234}{11} \cdot \frac{\alpha-1}{2} \simeq 5.32$ and inequalities (14–15) are satisfied $\forall \Delta \in \mathcal{I}$ (see Lemma 4 in Sec. 4), hence Theorem 1 guarantees that $\forall \Delta \in \mathcal{I}$, the dynamic qdb–controller with saturated inputs $\mathcal{U} = \mathcal{U}(\Delta)$ is $(Q_2(\frac{\Delta}{\alpha}), Q_2(\Delta), Q_2(\Delta_f))$–stabilizing with $\Delta_f = \frac{25}{7}$ (see Fig. 2).

4. PROOF OF THEOREM 1

In order to prove Theorem 1 some preliminary results are needed. We will refer to the following notation: $\forall \Delta > 0$ such that $\rho(\Delta) < +\infty$, define the partition $\mathbb{R} = \mathcal{S}_{M(\Delta)} \cup \mathcal{N}_\Delta \cup \mathcal{S}_{m(\Delta)}$, where $\mathcal{S}_{M(\Delta)} :=]-\infty; -M(\Delta) - \frac{\rho(\Delta)}{2}[$, $\mathcal{N}_\Delta := [-M(\Delta) - \frac{\rho(\Delta)}{2}; -m(\Delta) + \frac{\rho(\Delta)}{2}]$ and $\mathcal{S}_{m(\Delta)} :=]-m(\Delta) + \frac{\rho(\Delta)}{2}; +\infty[$. Let $\mathcal{S}_\Delta := \mathcal{S}_{M(\Delta)} \cup \mathcal{S}_{m(\Delta)}$. Let us analyze the main properties of the map ν defining the state feedback qdb–controller.

Lemma 2. (Basic properties of ν). Let $\Delta > 0$:
i) if inequalities (14–15) hold, then $\forall z \in \text{Pr}_n(AQ_n(\Delta))$, $\nu(z) \in \mathcal{U}(\Delta)$;
ii) if $\rho(\Delta) < +\infty$ and $z \in \mathcal{N}_\Delta$, then $|z + \nu(-z)| \leq \frac{\rho(\Delta)}{2}$;
iii) assume $\rho(\Delta) < +\infty$ and let z be such that

[3] This controller can be modelled in the form of Eqn. (4) with $\mathcal{W} := \mathcal{Y}^n \times \mathcal{U}^{n-1}$: see in Appendix.

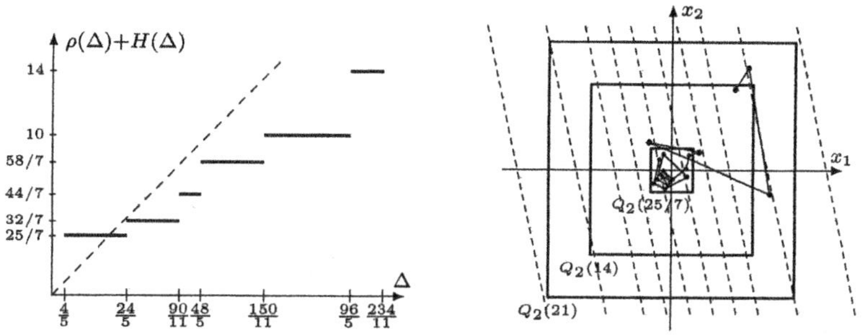

Fig. 2. *Left*: graph of $\rho(\Delta) + H(\Delta)$. *Right*: a trajectory generated by the dynamic qdb–controller for $\Delta_0 = 14$ and $x(0) = (5.42\ 6.60)$. Broken lines identify the state space partition induced by $q \circ C$.

$\nu(-z) \in \mathcal{U}(\Delta)$. If $z \in \mathcal{S}_{M(\Delta)}$, then $\nu(-z) = M(\Delta)$ and $|z + \nu(-z)| = -(z + \nu(-z)) > \frac{\rho(\Delta)}{2}$; if $z \in \mathcal{S}_{m(\Delta)}$, then $\nu(-z) = m(\Delta)$ and $z + \nu(-z) > \frac{\rho(\Delta)}{2}$.

The core of the proof of Theorem 1 is represented by the following result:

Lemma 3. (Main tool). Let $\Delta > 0$ be such that $\rho(\Delta) < +\infty$ and inequalities (14–15) hold. Assume that $x \in Q_n(\Delta)$ and $\hat{x} \in \mathbb{R}^n$ are so that $\left|\big(A(x-\hat{x})\big)_n\right| \leq \frac{\mathscr{H}}{2}$ (for some $\mathscr{H} \geq 0$). Let $k(x)$ be the state feedback qdb–controller and suppose that $k(\hat{x}) \in \mathcal{U}(\Delta)$, then $x^+ = Ax + b\,k(\hat{x})$ is such that $|x_n^+| \leq \max\left\{\frac{\rho(\Delta)+\mathscr{H}}{2},\ \|x\|_\infty - \varphi(\Delta)\right\}$, where

$$\varphi(\Delta) := \min\left\{ M(\Delta) - \tfrac{\Delta}{2}(\alpha - 1),\ -\tfrac{\Delta}{2}(\alpha - 1) - m(\Delta) \right\}. \tag{18}$$

Proof. By definition of k, $x_n^+ = (Ax)_n + \nu\big(-(A\hat{x})_n\big)$. Notice also that, by Lemma 2.*i*, $\nu\big(-(Ax)_n\big) \in \mathcal{U}(\Delta)$. Three cases can occur:

I) Suppose that $(A\hat{x})_n \in \mathcal{N}_\Delta$, then

$$|x_n^+| = \left|\big(A(x-\hat{x})\big)_n + (A\hat{x})_n + \nu\big(-(A\hat{x})_n\big)\right| \leq \leq \left|\big(A(x-\hat{x})\big)_n\right| + \left|(A\hat{x})_n + \nu\big(-(A\hat{x})_n\big)\right| \leq \leq \frac{\mathscr{H}}{2} + \frac{\rho(\Delta)}{2},$$

where the last inequality follows by the hypothesis on $\hat{x}$ and by Lemma 2.*ii*.

II) Suppose that $(A\hat{x})_n \in \mathcal{S}_\Delta$ and x is such that $(Ax)_n \in \mathcal{N}_\Delta$. If $(A\hat{x})_n \in \mathcal{S}_{m(\Delta)}$, then $k(\hat{x}) = m(\Delta)$ thanks to Lemma 2.*iii* which can be applied because, by assumption, $k(\hat{x}) \in \mathcal{U}(\Delta)$. Hence, $x_n^+ = (Ax)_n + m(\Delta) \leq (Ax)_n + \nu\big(-(Ax)_n\big) \leq \frac{\rho(\Delta)}{2}$, where the first inequality holds because $\nu\big(-(Ax)_n\big) \in \mathcal{U}(\Delta)$ and the latter by Lemma 2.*ii*. Moreover, by Lemma 2.*iii*, $(A\hat{x})_n + \nu\big(-(A\hat{x})_n\big) > \frac{\rho(\Delta)}{2}$, and by assumption $\big(A(x-\hat{x})\big)_n \geq -\frac{\mathscr{H}}{2}$, therefore $x_n^+ = \big(A(x-\hat{x})\big)_n + (A\hat{x})_n + \nu\big(-(A\hat{x})_n\big) > -\frac{\mathscr{H}}{2} + \frac{\rho(\Delta)}{2} > -\frac{\mathscr{H}+\rho(\Delta)}{2}$. To sum up, $|x_n^+| \leq \frac{\rho(\Delta)+\mathscr{H}}{2}$. The case $(A\hat{x})_n \in \mathcal{S}_{M(\Delta)}$ is similar.

III) Suppose that $(A\hat{x})_n \in \mathcal{S}_\Delta$ and $(Ax)_n \in \mathcal{S}_\Delta$. If $(A\hat{x})_n \in \mathcal{S}_{m(\Delta)}$, we know by part II that $k(\hat{x}) = m(\Delta)$ and $x_n^+ > -\frac{\mathscr{H}+\rho(\Delta)}{2}$. Assume that $(Ax)_n \in \mathcal{S}_{M(\Delta)}$, since $\nu\big(-(Ax)_n\big) \in \mathcal{U}(\Delta)$, then $x_n^+ = (Ax)_n + m(\Delta) < (Ax)_n + M(\Delta) = (Ax)_n + \nu\big(-(Ax)_n\big) < -\frac{\rho(\Delta)}{2}$, where both the last equality and the last inequality hold by Lemma 2.*iii*. Hence, $|x_n^+| < \frac{\mathscr{H}+\rho(\Delta)}{2}$. If instead $(Ax)_n \in \mathcal{S}_{m(\Delta)}$, then $|x_n^+| \leq \|x\|_\infty - \varphi(\Delta)$. In fact: in this case $k(x) = k(\hat{x}) = m(\Delta)$ and, thanks to inequalities (14–15), we can write $m(\Delta) = -\frac{\Delta}{2}(\alpha - 1) - \varphi(\Delta) - \theta$, with $\theta \geq 0$. Again by Lemma 2.*iii*, $x_n^+ = (Ax)_n + m(\Delta) > \frac{\rho(\Delta)}{2} > 0$, hence $|x_n^+| = (Ax)_n + m(\Delta) \leq \sum_i |\alpha_i|\,|x_i| + m(\Delta) \leq \alpha \cdot \|x\|_\infty + m(\Delta) = \alpha \cdot \|x\|_\infty - \frac{\Delta}{2}(\alpha - 1) - \varphi(\Delta) - \theta \leq \|x\|_\infty - \varphi(\Delta)$ because $\big(\|x\|_\infty - \frac{\Delta}{2}\big)(\alpha - 1) - \theta \leq 0$. The case $(A\hat{x})_n \in \mathcal{S}_{M(\Delta)}$ is similar. ■

Remark 1. The motivation for assuming $k(\hat{x}) \in \mathcal{U}(\Delta)$ (which corresponds to the restriction of the state feedback qdb–controller to the saturated input set $\mathcal{U} = \mathcal{U}(\Delta_1)$ in Theorem 1) is that, by Lemma 1, if $k(\hat{x}) \notin \mathcal{U}(\Delta)$, then $x^+ \notin Q_n(\Delta)$.

Lemma 4. If $\Delta > 0$ satisfies inequalities (14–15), and Δ' is such that $\rho(\Delta) \leq \Delta' < \Delta$, then Δ' satisfies inequalities (14–15).

Proof of Theorem 1. The sequence defined by the algorithm (17) is decreasing by construction, let us show that it is finite: first notice that, by definition, $H(\Delta)$ is a piecewise constant and non–decreasing function. As for $\rho(\Delta)$, $\exists \bar{\Delta} > 0$ such that $\rho(\Delta) = +\infty\ \forall \Delta < \bar{\Delta}$, whilst for $\Delta \geq \bar{\Delta}$, $\rho(\Delta)$ is piecewise constant and non–decreasing with Δ. If $\rho(\Delta_{h+1}) + H(\Delta_{h+1}) < \rho(\Delta_h) + H(\Delta_h)$, then $\rho(\Delta_{h+1}) < \rho(\Delta_h)$ or $H(\Delta_{h+1}) < H(\Delta_h)$: in the first case $\#\mathcal{U}(\Delta_{h+1}) < \#\mathcal{U}(\Delta_h)$, in the latter $\#\mathcal{Y}(\Delta_{h+1}) < \#\mathcal{Y}(\Delta_h)$. Therefore, $f \leq \#\mathcal{U}(\Delta_1) + \#\mathcal{Y}(\Delta_1) < +\infty$ because $\mathcal{U}$ is a closed discrete set and the output space partition induced by q is supposed to be locally finite.

We have already noticed that $\forall x(0) \in Q_n(\Delta_0)$ and $\forall t \leq n-1$, $x(t) \in Q_n(\Delta_1)$, therefore $y(n-1) \in \mathcal{Y}(\Delta_1)^n$. Since $H(\Delta_1) < +\infty$ (see inequality (16)), then $\mathcal{Y}(\Delta_1) \subseteq \mathcal{Y}_\star$ and $\operatorname{diam}_n(A\mathcal{C}_{x(n-1)}) \leq H(\Delta_1)$. Because $\hat{x}(n-1)$ is the centroid of the parallelogram $\mathcal{C}_{x(n-1)}$, then $\left|\Big(A\big(x(n-1) - \hat{x}(n-1)\big)\Big)_n\right| \leq \frac{H(\Delta_1)}{2}$. Moreover, $k\big(\hat{x}(n-1)\big) \in \mathcal{U}(\Delta_1)$ by assumption, therefore

Lemma 3 guarantees that

$$|x_n(n)| \leq \max\left\{\frac{\Delta_2}{2} = \frac{H(\Delta_1)+\rho(\Delta_1)}{2}, \|x(n-1)\|_\infty - \varphi(\Delta_1)\right\} < \frac{\Delta_1}{2} \quad (19)$$

(where $\varphi(\Delta_1) > 0$ is defined in Eqn. (18)). Since $x^+ = (x_2, \ldots, x_n, x_n^+)$, then $x(n) \in Q_n(\Delta_1)$ and $\boldsymbol{y}(n) \in \mathcal{Y}(\Delta_1)^n$: therefore, the arguments which have allowed us to prove that $x(n) \in Q_n(\Delta_1)$ can be repeated so that $\forall t > n$, $x(t) \in Q_n(\Delta_1)$. Furthermore, because $\varphi(\Delta_1) > 0$ does not depend on t, $\exists t_1 > 0$ such that $\forall t \geq t_1$, $x(t) \in Q_n(\Delta_2)$. Therefore $\boldsymbol{y}(t_1+n-1) \in \mathcal{Y}(\Delta_2)^n$ and $\mathrm{diam}_n(AC_{x(t_1+n-1)}) \leq H(\Delta_2)$. If $f > 2$, the arguments above can be iterated until t_{f-1} is found such that $\forall t \geq t_{f-1}$, $x(t) \in Q_n(\Delta_f)$: we have only to check that $\forall h = 2\ldots, f-1$, Δ_h satisfies the hypotheses of Lemma 3.
Indeed, let $h \in \{2, \ldots, f-1\}$. By the algorithm (17), $\rho(\Delta_h) = \Delta_{h+1} - H(\Delta_h) < +\infty$. Inequalities (14–15) are satisfied by Δ_h (this guarantees that $\varphi(\Delta_h) > 0$), in fact: such inequalities hold for Δ_1 by assumption and, since $\rho(\Delta_{h-1}) \leq \Delta_h = \rho(\Delta_{h-1}) + H(\Delta_{h-1}) < \Delta_{h-1}$, then the result follows by recursive application of Lemma 4. As far as the remaining two hypotheses are concerned, that is $k(\hat{x}(t_{h-1}+n-1)) \in \mathcal{U}(\Delta_h)$ and $\left|\left(A(x(t_{h-1}+n-1) - \hat{x}(t_{h-1}+n-1))\right)_n\right| \leq \frac{H(\Delta_h)}{2}$, we have already noticed that $\mathrm{diam}_n(AC_{x(t_1+n-1)}) \leq H(\Delta_2)$ implies that $\left|\left(A(x(t_1+n-1) - \hat{x}(t_1+n-1))\right)_n\right| \leq \frac{H(\Delta_2)}{2}$. Moreover, since $\forall t \geq t_1$, $x(t) \in Q_n(\Delta_2)$, then $\forall t \geq t_1$, $k(\hat{x}(t)) \in \mathcal{U}(\Delta_2)$ thanks to Lemma 1. We then conclude by a recursive argument. ■

5. CONCLUSION

We have introduced a novel technique for the stabilizability analysis of quantized SISO systems. The results hold under very general hypotheses and are of direct applicability. Interesting questions are open for future investigations, especially in the framework of sampled continuous–time systems under communication constraints.

REFERENCES

J. Baillieul, *Feedback Designs in Information–Based Control*, Proc. of the Workshop on Stochastic Theory and Control, Kansas, pages: 35–57. Springer-Verlag. 2001.

A. Bicchi, A. Marigo and B. Piccoli, *On the Reachability of Quantized Control Systems*, IEEE Trans. Autom. Control, 47(4); pages: 546-563. 2002.

R. Brockett and D. Liberzon, *Quantized feedback stabilization of linear systems*, IEEE Trans. Autom. Control, 45(7); pages: 1279–1289. 2000.

D.F. Delchamps, *Stabilizing a Linear System with Quantized State Feedback*, IEEE Trans. Autom. Control, 35(8); pages: 916–924. 1990.

N. Elia and S. Mitter, *Stabilization of Linear Systems With Limited Information*, IEEE Trans. Autom. Control, 46(9); pages: 1384–1400. 2001.

F. Fagnani and S. Zampieri *Quantized stabilization of linear systems: complexity versus performance*, *IEEE Trans. Autom. Control, 49 (9) special issue on "Networked Control Systems"*, pages: 1534–1548. 2004.

H. Ishii and B.A. Francis, *Stabilizing a Linear Systems by Switching Control With Duel Time*, IEEE Trans. Autom. Control, 47(12); pages: 1962–1973. 2002.

D. Liberzon, *Hybrid feedback stabilization of systems with quantized signals*, Automatica, 39; pages: 1543–1554. 2003.

G.N. Nair and R.J. Evans, *Stabilizability of stochastic linear systems with finite feedback data rates*, SIAM Journal on Contr. Optim. Vol. 43 (2), Pages: 413–436. 2004.

B. Picasso, F. Gouaisbaut and A. Bicchi, *Construction of invariant and attractive sets for quantized–input linear systems*, Proc. of the 41st IEEE Conference on Decision and Control "CDC 2002" pages: 824–829. 2002.

B. Picasso and A. Bicchi, *Stabilization of LTI Systems with Quantized State – Quantized Input Static Feedback*, In A. Pnueli and O. Maler, editors, Hybrid Systems: Computation and Control, volume LNCS 2623 of *Lecture Notes in Computer Science*, pages: 405–416. Springer–Verlag, Heidelberg, Germany. 2003.

D.E. Quevedo, G.C. Goodwin, J.A. De Doná, *Finite constraint set receding horizon control*, Int. J. of Robust and Nonlinear Control, 14(4); pages: 355–377. 2004.

S.C. Tatikonda and S. Mitter, *Control Under Communication Constraints*, IEEE Trans. Autom. Control, 49(7); pages: 1056–1068.. 2004.

W. Wong and R. Brockett, *Systems with finite communication bandwidth constraints - part II: Stabilization with limited information feedback*, IEEE Trans. Autom. Control, 44(5); pages: 1049–1053. 1999.

6. APPENDIX

Let us represent the dynamic qdb–controller defined in Eqn. (12) in the form of Eqn. (4). Let

$$\mathcal{W} := \mathcal{Y}^n \times \mathcal{U}^{n-1},$$

the elements $w \in \mathcal{W}$ are denoted either by $w = (\boldsymbol{y}, \boldsymbol{u})$ or $w = (w_1, \cdots, w_{2n-1})$. Let

$$\tilde{\phi}: \mathcal{W} \times (\mathbb{N} \cup \{-1\}) \to \mathcal{U}$$

be defined by

$$\tilde{\phi}((\boldsymbol{y}, \boldsymbol{u}), t) = \begin{cases} 0 & \text{if } t \leq n-2 \\ (k \circ \psi)(\boldsymbol{y}, \boldsymbol{u}) & \text{if } t \geq n-1, \end{cases}$$

and

$$\gamma: \mathcal{W} \times \mathcal{Y} \times \mathbb{N} \to \mathcal{W}$$

be defined by

$$\gamma(w, y, t) = (y, w_1, \cdots, w_{n-1}, \tilde{\phi}(w, t-1), w_{n+1}, \cdots, w_{2n-2}).$$

Finally,

$$\begin{aligned} \phi: \ & \mathcal{W} \times \mathcal{Y} \times \mathbb{N} \to \mathcal{U} \\ & (w, y, t) \mapsto \tilde{\phi}(\gamma(w, y, t), t). \end{aligned}$$

ELSEVIER
IFAC
PUBLICATIONS

BOX INVARIANCE OF HYBRID AND SWITCHED SYSTEMS

Alessandro Abate [1] **Ashish Tiwari** [2]

Abstract: This paper investigates the concept of box invariance for classes of hybrid and switched systems. After motivating and defining the notion, we present a concise summary of results on its characterization for single-domain dynamical systems. The notion is then extended to the case of hybrid and switched systems. We provide sufficient conditions for a hybrid or switched system to be box invariant. Models of many real systems, especially those drawn from biology, have been found to be box invariant. This paper illustrates the concept using a pharmacodynamic model of blood glucose metabolism. *Copyright © 2006 IFAC*

Keywords: Invariant Sets, Hybrid and Switched Dynamical Systems.

1. INTRODUCTION

The concept of box invariance has been recently introduced for classes of dynamical systems (Abate and Tiwari, 2006). The main motivation for this notion comes from biological case studies and models that were investigated with the following question: is the strict concept of Lyapunov stability always descriptive and computationally feasible in the biological realm, which is often characterized by complex dynamics with imprecisely known parameters? As a viable alternative, would it not make sense to look for bounded behavior in the closeness of an equilibrium? Intuitively, it is appealing to consider a notion of stability defined "within certain limits", or "bounds", rather than on ϵ-neighborhoods of equilibria. In other words, we may be interested in the existence of regions around the equilibrium within which a trajectory would indefinitely dwell once it reaches them. We focus here on the simplest possible shape for these regions–that of a box. We have found that models of many natural systems have box shaped regions as invariant sets. Box invariance is also appealing from the automated verification standpoint, since it is computationally tractable for a large class of systems and helps in proving strong safety properties of systems. The notion is related to other concepts in the literature, see Abate and Tiwari (2006) for details.

In this paper, we first formally define the notion of box invariance for continuous dynamical systems (Sec. 2). We then provide characterizations for linear and affine dynamical systems and show how to practically compute an actual box (Sec. 3). A detailed study of affine systems and full proofs of claims are presented elsewhere (Abate and Tiwari, 2006). The main focus of this article is on extending the notion to hybrid and switched systems (Sec. 4). These systems provide a powerful modeling tool, particularly for domains such as biology. As an illustrative example, we shall discuss a switched model for blood glucose concentration in human brain (Sec. 5) and analyze it using the concept of box invariance.

2. THE CONCEPT OF BOX INVARIANCE

We consider general, autonomous and uncontrolled dynamical systems of the form $\dot{x} = f(x), x \in \mathbb{R}^n$. We assume basic continuity and Lipschitz properties for the existence of a unique solution of the vector field, given any possible initial condition. A rectangular box around a point x_0 can be specified using two diagonally opposite points x_{lb} and x_{ub}, where $x_{lb} < x_0 < x_{ub}$ (interpreted component-wise). Such a box has $2n$ surfaces $S^{j,k}(1 \leq j \leq n, k \in \{l,u\})$, where

[1] Department of Electrical Engineering and Computer Sciences, University of California, at Berkeley – {aabate@eecs.berkeley.edu}. Research supported by the grants CCR-0225610 and DAAD19-03-1-0373.
[2] Computer Science Laboratory, SRI International, Menlo Park, CA – {tiwari@csl.sri.com}. Supported in part by NSF CCR-0311348 and DARPA DE-AC03-765F00098.

$S^{j,k} = \{\boldsymbol{y} : (\boldsymbol{x}_{lb})_i \leq \boldsymbol{y}_i \leq (\boldsymbol{x}_{ub})_i \text{ for } i \neq j, \boldsymbol{y}_j = (\boldsymbol{x}_{lb})_j \text{ if } k = l, \boldsymbol{y}_j = (\boldsymbol{x}_{ub})_j \text{ if } k = u\}$.

***Definition** 1.* A dynamical system $\dot{\boldsymbol{x}} = f(\boldsymbol{x})$ is said to be *box invariant* around an equilibrium point $\boldsymbol{x}_0$ if there exists a finite rectangular box around $\boldsymbol{x}_0$, specified by $\boldsymbol{x}_{lb}$ and $\boldsymbol{x}_{ub}$, such that for any point $\boldsymbol{y}$ on any surface $S^{j,k}(1 \leq j \leq n, k \in \{l, u\})$ of this rectangular box, it is the case that $f(\boldsymbol{y})_j \leq 0$ if $k = u$ and $f(\boldsymbol{y})_j \geq 0$ if $k = l$. The system will be said to be *strictly box invariant* if the last inequalities hold strictly.

***Remark** 1.* The concept of box invariance for a dynamical system requires the existence of an *invariant set* with a special (polyhedral) shape. In the case of linear systems, we shall see that this invariant set is also an *ω-limit set* (or a *domain of attraction*).

***Definition** 2.* A system $\dot{\boldsymbol{x}} = f(\boldsymbol{x})$ is said to be *symmetrical box invariant* around the equilibrium $\boldsymbol{x}_0$ if there exists a point $\boldsymbol{u} > \boldsymbol{x}_0$ (interpreted component-wise) such that the system is box invariant with respect to the box defined by $\boldsymbol{u}$ and $(2\boldsymbol{x}_0 - \boldsymbol{u})$.

3. CHARACTERIZATION OF BOX INVARIANCE.

3.1 Linear Systems.

Given a linear system and a box around its equilibrium point, the problem of checking if the system is box invariant with respect to the given box can be solved by verifying the condition only at the 2^n vertices of the box (instead of all points on all the faces of the box).

***Proposition** 1.* A linear system $\dot{\boldsymbol{x}} = A\boldsymbol{x}, \boldsymbol{x} \in \mathbb{R}^n$ is box invariant if there exist two points $\boldsymbol{u} \in (\mathbb{R}^+)^n$ and $\boldsymbol{l} \in (\mathbb{R}^-)^n$ such that for each point $\boldsymbol{c}$, with $c_i \in \{u_i, l_i\}, \forall i$, we have $A\boldsymbol{c} \sim \boldsymbol{0}$, where $\sim_i$ is $\leq$ if $c_i = u_i$ and $\sim_i$ is $\geq$ if $c_i = l_i$.

Proposition 1 shows that box invariance of linear systems can be checked by testing satisfiability of 2^n linear inequality constraints (over $2n$ unknowns given by $\boldsymbol{l}$ and $\boldsymbol{u}$). In two steps, we will show that these 2^n constraints are subsumed by just n linear inequality constraints (over n unknowns). First we prove this fact for symmetric box invariance.

***Theorem** 1.* An n-dimensional linear system $\dot{\boldsymbol{x}} = A\boldsymbol{x}$ is symmetrical box invariant iff there exists a positive vector $\boldsymbol{c} \in \mathbb{R}^{+^n}$ such that $A^m\boldsymbol{c} \leq \boldsymbol{0}$, where $a_{ii}^m = a_{ii} < 0$ and $a_{ij}^m = |a_{ij}|$ for $i \neq j$. This is equivalent to checking if the linear system defined by modified matrix A^m is symmetrical box invariant.

In the second step, we show the surprising result that the property of box invariance and that of symmetrical box invariance are equivalent for linear systems.

***Theorem** 2.* A linear system $\dot{\boldsymbol{x}} = A\boldsymbol{x}$, where $A \in \mathbb{R}^{n \times n}$, is box invariant iff it is symmetrical box invariant.

Putting together Theorem 1 and Theorem 2, we conclude that to check if a linear system $\dot{\boldsymbol{x}} = A\boldsymbol{x}$ is box invariant, we only need to test if the set of n linear inequality constraints, succinctly written as $A^m\boldsymbol{c} \leq 0$ (over the n unknowns $\boldsymbol{c}$) has a nonzero solution. This can be done in *polynomial* time. We can also find a box by generating solutions for the above linear constraint satisfaction problem. In general, it is possible to associate with a dynamical system, defined by system matrix $A \in \mathbb{R}^{n \times n}$, a *cone* in the positive $2^{n^{th}}$-ant described by the set

$$\mathcal{C} = \{\boldsymbol{x} \in \mathbb{R}^{+^n} : A^m\boldsymbol{x} \leq \boldsymbol{0}\}.$$

Any choice of a single point or a pair of distinct points in $\mathcal{C}$ determine, respectively, a symmetric and a non-symmetric box for the system described by A. Box Invariance is a stronger notion than stability for linear systems.

***Theorem** 3.* If a linear dynamical system is box invariant around its equilibrium, then it is stable.

Surprisingly, the well-known concept of P-matrices in linear algebra provides a structural characterization for box invariant linear systems. A matrix A is said to be a *P-matrix* if all of its principal minors are positive.

***Theorem** 4.* Let A be a $n \times n$ matrix such that $a_{ii} \leq 0$ and $a_{ij} \geq 0$ for all $i \neq j$. Then, the following statements are equivalent:

(1) The linear system $\dot{\boldsymbol{x}} = A\boldsymbol{x}$ is strictly symmetrical box invariant.
(2) $-A$ is a *P-matrix*.
(3) For every $i = 1, 2, \ldots, n$, the determinant of the top left $i \times i$ submatrix of $-A$ is positive.

***Remark** 2.* Theorem 4 shows that box invariance of general linear systems can also be tested by checking if the modified matrix $-A^m$ is a P-matrix. It is known that the problem of deciding if a given matrix is a P-matrix is co-NP-hard (Coxson, 1994; Coxson, 1999). Our case is special though, since we know that only the diagonal entries in $-A^m$ are positive. As a result, we can determine if $-A^m$ is a P-matrix using a simple *polynomial* time algorithm; for example, the Fourier-Motzkin elimination method can determine satisfiability and generate the cone $\mathcal{C}$.

Matrices with the shape of those in Theorem 4 (or, equivalently, of A^m in Theorem 1) have actually been studied under the appellative of *Metzler* matrices. There is a wealth of literature on Metzler matrices that can be used to derive results equivalent to those presented above for box invariance; see Abate and Tiwari (2006) for details.

***Remark** 3.* The concept of box invariance, which is closely related to that of classical stability in the linear case, can also be studied via Lyapunov arguments. In our particular instance, to prove box invariance we find a Lyapunov functional which is defined (at least) inside a certain boxed region of the state space. To go from smooth Lyapunov functions to one that is defined on a box, we can intersect proper ellipsoidal functions that have been adequately stretched to the limit (Abate and Tiwari, 2006) or use vector norms (Kiendl *et al.*, 1992). A linear system $\dot{\boldsymbol{x}} = A\boldsymbol{x}$ is box stable iff there exists a diagonal matrix $W \in (\mathbb{R}^+)^{n\times n}$ s.t. $V(\boldsymbol{x}) = ||W\boldsymbol{x}||_\infty$ is a Lyapunov function (cf. Kiendl *et al.* (1992)). The Lyapunov arguments will help in the case of hybrid and switched systems, as we shall see later.

3.2 ***Affine Systems.***

Consider the affine system, $\dot{\boldsymbol{x}} = A\boldsymbol{x}+\boldsymbol{b}$. We relate the box invariance of this system to the positivity of its equilibrium point, $\boldsymbol{x}_0 > \mathbf{0}$. The assumption of $\boldsymbol{x}_0$ being in the positive quadrant is justified both from a technical standpoint and from our applications.

***Theorem** 5.* If the affine system $\dot{\boldsymbol{x}} = A\boldsymbol{x} + \boldsymbol{b}$ is s.t. A is Metzler and $\boldsymbol{b} > \mathbf{0}$, then its equilibrium point $\boldsymbol{x}_0 > \mathbf{0}$ iff the system is box invariant.

***Remark** 4.* The assumptions of the previous theorem can be relaxed to having a non-negative $\boldsymbol{b} \geq \mathbf{0}, \boldsymbol{b} \neq \mathbf{0}$.

In modeling biochemical pathways, the state variables $\boldsymbol{x}$ represent concentration of species such as proteins. When modeling the dynamics of such species, the Metzler form arises naturally since species decay proportionally to their concentration and are created proportionally to their precursor's concentrations. The vector $\boldsymbol{b} > 0$ represents the process of species creation (by transcription and translation, for example). For stable systems, such models typically have positive equilibrium point. By Theorem 5, all such models will be box invariant. Thus, Theorem 5 explains, in part, why many models proposed by biologists tend to be box invariant. For more results on box invariance of affine systems, the reader is referred to Abate and Tiwari (2006).

4. BOX INVARIANCE FOR HYBRID AND SWITCHED SYSTEMS.

In this Section we extend the notion of box invariance to the case of composition of different dynamical systems. We first define hybrid and switched systems and refer the reader to specialized literature for more details (Lygeros *et al.*, 2003; Ames *et al.*, 2005).

***Definition** 3.* A *hybrid system* is a tuple $\mathcal{H} = (Q, E, D, G, R, F)$, where

- $Q = \{1, ..., m\}$ is a finite set of *discrete states.*
- $E \subset Q \times Q$ is a set of *edges* that defines a source-target relation between the domains.
- $D = \{D_i\}_{i\in Q}$ is a set of *domains* where D_i is a compact subset of $\mathbb{R}^n$.
- $G = \{G_e\}_{e\in E}$ is a set of *guards*, where $G_e \subseteq D_{e(1)}$.
- $R = \{R_e\}_{e\in E}$ is a set of *reset maps*; here we assume identity resets.
- $F = \{f_i\}_{i\in Q}$ is a set of *vector fields* such that f_i is Lipschitz on $\mathbb{R}^n$.

A hybrid trajectory (in the state space) is described by adequately specifying a sequence of its initial conditions, switching times and edges, properly related via the guards and reset maps. The behavior of $\mathcal{H}$ allows for possible Zeno trajectories. As the reader may notice, the switching conditions (the "events") are due to spacial restrictions on the various domains. In contrast, *switched systems* specify jumping conditions in time, rather than in the state space.

***Definition** 4.* A *switched system* is a tuple $\mathcal{S} = (Q, E, D, G, R, F)$, where

- Q, E, R, F are characterized as in Def. 3.
- $D = \{D_i\}_{i\in Q}$ is a set of *domains* where $D_i = \mathbb{R}^n$.
- $G = \{0, \tau_1, \tau_2, \ldots\}$ is a set of *guards* in time, where $\tau_i \in \mathbb{R}^+$ are increasing in i. Each τ_i is mapped to a state by a function $g : G \to Q$ such that $(g(\tau_{i-1}), g(\tau_i)) \in E$ for all i.

Consider the hybrid domain $Q\times\mathbb{R}^n$. An *invariant set* of a hybrid or switched system is a subset of $Q \times \mathbb{R}^n$ such that every trajectory originating in this set continues to dwell inside it. The notion of box invariance for hybrid and switched cases is defined so as to emcompass the possibility of multiple equilibria and switchings between the different domains.

***Definition** 5.* A hybrid system $\mathcal{H}$ (a switched system $\mathcal{S}$) is said to be box invariant if there exists a boxed region $B \subset \mathbb{R}^n$ and a subset $Q' \subset Q$ of states such that $Q' \times B$ is an invariant set for $\mathcal{H}$ (or $\mathcal{S}$).

In this paper, we will restrict ourselves to linear vector fields, that is, all f_i's are linear functions of $\boldsymbol{x}$. Hence, in individual domains, boxes will be around the local equilibrium point. For the sake of analysis, we differentiate the following two cases: domains with different equilibria and domains with overlaying equilibria. Furthermore, in the hybrid case, the equilibria may or may not belong to one or more guard sets.

For a hybrid system $\mathcal{H}$, if some discrete state i has an equilibrium $\boldsymbol{x}_0$ that does not belong to any guard, that is $\boldsymbol{x}_0 \notin \bigcup_{e\in E} G_e$, then a sufficient condition for box invariance of $\mathcal{H}$ is obtained from box invariance of the dynamical system of state i.

In such a case, $\mathcal{H}$ is box invariant if there is a small enough box B for the dynamical system of state i that is completely contained in the domain, $B \subset D_i$, and that does not intersect the guard, $B \cap \bigcup_{e \in E} G_e = \emptyset$. This occurs, for example, in the hybrid model of the Delta-Notch lateral inhibition mechanism of Ghosh and Tomlin (2001). The reader should notice that, as in the single-domain nonlinear case, the existence of a box does not imply the existence of boxes of different sizes: expanding a box may cause it to intersect a guard. Next consider the case where $\boldsymbol{x}_0$ is a shared equilibrium that belongs to at least one spacial guard. In this instance, a sufficient condition is the existence of a *single* box for *multiple* domains.

For switching systems, both in the case of shared equilibrium and different equilibria, jumps at possibly any time instant forces us to find a *common* box, which would be invariant in all the domains.

The results in this section resemble those obtained for stability of hybrid and switched systems. In particular, it is known that there are examples of unstable hybrid and switched systems that are the composition of *stable* dynamical systems (Branicky, 1994). We are similarly interested in understanding how the notion of box invariance, which is shown in the linear case to be intrinsically related to that of stability, is translated in the hybrid or switched setting. Furthermore, sufficient conditions for the invariance of the interconnection, according to Def. 5, will be derived. To begin with, let us stress a technique to obtain sufficient conditions for a hybrid dynamical system.

***Proposition** 2.* Let us associate to a hybrid system $\mathcal{H}$ a corresponding switched system $\mathcal{S}$, made up of the same tuple, except for the following two elements: the domains $D_i = \mathbb{R}^n$ and a symbolic non decreasing sequence $G = \{0, \tau_1, \tau_2, \ldots\}$. Given an initial condition, if a universal property $\mathcal{P}$ holds in all trajectories of $\mathcal{S}$ (given by all the possible different sequences with the form of G), then $\mathcal{P}$ holds in $\mathcal{H}$.

The following result deals with the case of switched linear systems. Such systems share the origin as a common equilibrium.

***Theorem** 6.* A switched linear system $\mathcal{S}$, characterized by a set of vector fields of the form:

- $F = \{f_i\}_{i \in Q} = A_i \boldsymbol{x}, \ i \in Q$;

is box invariant around the origin if there exists a single box around which each of the dynamical systems is box invariant. Thus, a sufficient condition for box invariance is that $\bigcap_{i \in Q} C_i \neq \emptyset$, where the cones C_i's are defined as in Section 3.1.

Checking whether two cones, both pivoted on the origin, intersect is equivalent to checking for the intersection of two convex sets. We need to find a $\boldsymbol{c}$ such that $A_i \boldsymbol{c} \leq 0$ for all $i \in \{1, \ldots, m\}$. This is a linear program that can be solved in time polynomial in nm.

***Remark** 5.* There is a lot of work (for instance, see Branicky (1994)) in describing conditions for the Lyapunov stability of switched systems. These conditions either assume the existence of a common Lyapunov function, or require the presence of multiple Lyapunov functions (one for each domain) with certain switching restrictions. Our result can be interpreted as follows: a common box can be thought of as being the equivalent of a "common Lyapunov function" (the observations developed in Remark 3 should make this point clear). The explicit computation of a global Lyapunov function reduces to solving a set of linear matrix inequalities. Many efforts have been made to ease this task (Johansson and Rantzer, 1998). Otherwise, we may intuitively want to come up with conditions on the possible switchings by allowing the trajectory to jump between domains only when it belongs to the intersection of the boxes of these domains: this condition unfortunately seems harder to impose.

Handling box invariance for hybrid systems with general spacial guards is not easy, as we shall later discuss (see for instance Ex. 1). Nevertheless, exploiting the fact that box invariance is a stronger property than stability, we report the following compositional result for proving stability of hybrid systems (illustrated in Fig. 1).

***Theorem** 7.* If a linear hybrid system is composed of box invariant linear systems with a single spacial guard described by a hyperplane crossing the shared equilibrium, then the hybrid system is stable.

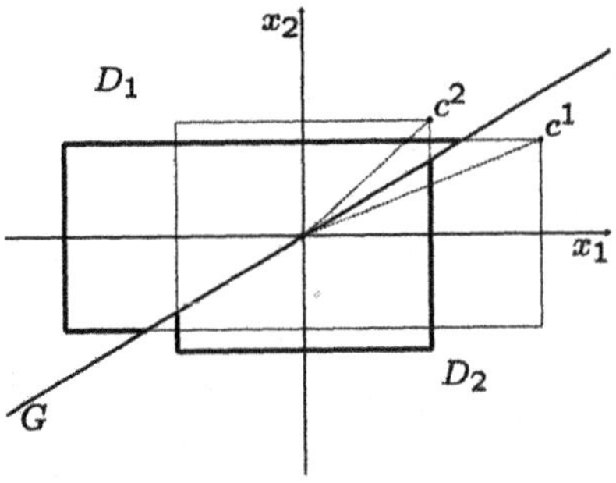

Fig. 1. A planar hybrid system with one single guard (line), as in Th. 7.

The previous theorem holds under rather restrictive conditions. Its generalization to the case of hybrid systems with multiple guards per domain does not hold.

***Example** 1.* Consider the following two-dimensional HS characterized by two modes with domains coinciding with the whole space, $D_1 = D_2 = (\mathbb{R}^+)^2$, and endowed with the following vector fields:

$$\begin{bmatrix} \dot{x_1} \\ \dot{x_2} \end{bmatrix} = A_1 \begin{bmatrix} x_1 \\ x_2 \end{bmatrix} = \begin{pmatrix} -1 & 5 \\ -0.1 & -1 \end{pmatrix} \begin{bmatrix} x_1 \\ x_2 \end{bmatrix};$$
$$\begin{bmatrix} \dot{x_1} \\ \dot{x_2} \end{bmatrix} = A_2 \begin{bmatrix} x_1 \\ x_2 \end{bmatrix} = \begin{pmatrix} -1 & -0.2 \\ 4 & -1 \end{pmatrix} \begin{bmatrix} x_1 \\ x_2 \end{bmatrix}.$$

Assume that there are two edges with the following guards in $\mathbb{R}^2$:

$$G_{1\to 2}(x_1, x_2) = \{\boldsymbol{x} \in \mathbb{R}^2 : x_1 - 5x_2 = 0\};$$
$$G_{2\leftarrow 1}(x_1, x_2) = \{\boldsymbol{x} \in \mathbb{R}^2 : 4x_1 - x_2 = 0\}.$$

Assume again trivial reset maps, and initial condition $(x_1(0), x_2(0)) = (0.1, 0.1) \in D_1$.
In isolation, both linear systems are box invariant and indeed have spiraling convergent trajectories towards the origin. The HS though is evidently unstable (see Fig. 2). Notice that

$$\mathcal{C}_1 = \{(x_1, x_2) : x_1 - 5x_2 \geq 0 \wedge x_2 \geq 0\};$$
$$\mathcal{C}_2 = \{(x_1, x_2) : 4x_1 - x_2 \leq 0 \wedge x_1 \geq 0\};$$

and that $\mathcal{C}_1 \cap \mathcal{C}_2 = \emptyset$. □

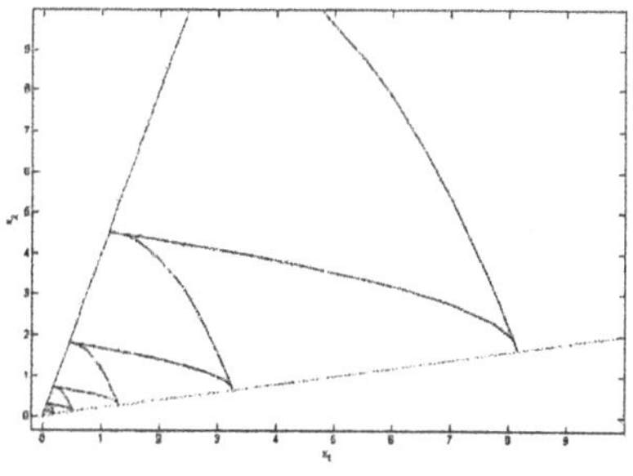

Fig. 2. Simulation for Example 1.

We finally consider the case of affine hybrid (and switched) systems. For a system of the form $\dot{\boldsymbol{x}} = A\boldsymbol{x} + \boldsymbol{b}$ and equilibrium $\boldsymbol{x}^{eq} = -A^{-1}\boldsymbol{b}$, we transform the variables into the new set $\boldsymbol{y} = \boldsymbol{x} - \boldsymbol{x}^{eq}$ and consider the new system $\dot{\boldsymbol{y}} = A\boldsymbol{y}$. Using the modified matrix A^m, we define $\mathcal{C} = \{\boldsymbol{y} \in \mathbb{R}^{+^n} : A^m\boldsymbol{y} \leq \boldsymbol{0}\}$. Let us introduce the "negative cone" $\mathcal{C}^- = \{\boldsymbol{y} \in \mathbb{R}^{-^n} : (-A^m)\boldsymbol{y} \leq \boldsymbol{0}\}$. Note that the actual cones, $\mathcal{C} + \boldsymbol{x}^{eq}$ and $\mathcal{C}^- + \boldsymbol{x}^{eq}$, are obtained by translating these cones to the equilibrium point.

Unlike the linear case, each subsystem could have a different equilibrium point in an affine hybrid system. However, we can still derive sufficient conditions for the existence of a "common box" as in the case of linear hybrid and switched systems (see Fig. 3).

***Theorem** 8.* Consider an affine hybrid system $\mathcal{H}$, where each domain has an equilibrium point $\boldsymbol{x}_i^{eq}, i \in Q$, and all variables are bound to be positive. $\mathcal{H}$ is box invariant if $\bigcap_{i\in Q}(\mathcal{C}_i + \boldsymbol{x}_i^{eq}) \neq \emptyset$.

More generally, if we allow the state space to include negative values, we need to make the condition stronger. Now, the existence of a global box

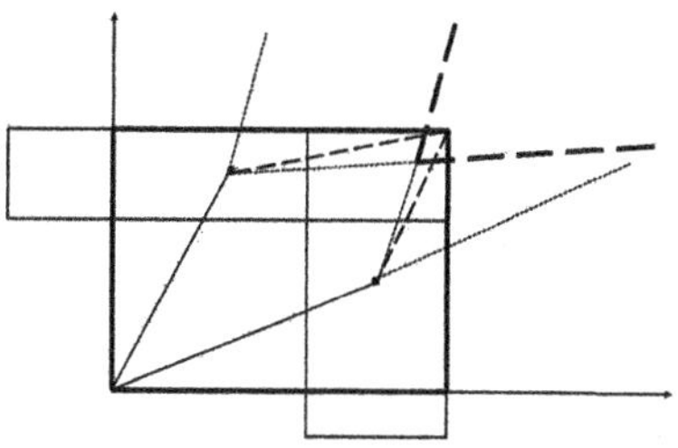

Fig. 3. Common box (shown in dark) for a two mode positive affine switched system.

will depend on having a nonempty intersection of both the positive and negative cones.

***Theorem** 9.* Consider an affine hybrid system $\mathcal{H}$, where each domain has an equilibrium point $\boldsymbol{x}_i^{eq}, i \in Q$. $\mathcal{H}$ will be box invariant if the following holds:

$$\bigcap_{i\in Q}(\mathcal{C}_i + \boldsymbol{x}_i^{eq}) \neq \emptyset \bigwedge \bigcap_{i\in Q}(\mathcal{C}_i^- + \boldsymbol{x}_i^{eq}) \neq \emptyset.$$

In conclusion, the concept of box invariance, even though stronger than that of asymptotic stability, is still not fully compositional in the hybrid case. Nevertheless, the sufficient conditions we have proposed for composability are easy to check and quite general. The comparison with the literature on stability of hybrid and switched systems confirms this fact. Furthermore, our sufficient conditions are enough to establish box invariance in many applications, one of which is described next.

5. A MODEL FOR BLOOD GLUCOSE CONCENTRATION

The following model is taken from (Sorensen, 1985). It is a model of a physiologic compartment, specifically the human brain, and describes the dynamics of the blood glucose concentration. In general, this compartment is part of a network of different parts, which model the concentration in other organs of the body, and which follow some conservation laws that account for the exchange of matter between different compartments. The mass balance equations are the following:

$$V_B\dot{C}_{Bo} = Q_B(C_{Bi} - C_{Bo}) + PA(C_I - C_{Bo}) - r_{RBC}$$
$$V_I\dot{C}_I = PA(C_{Bo} - C_I) - r_T,$$

where V_B describes the capillary volume, V_I the interstitial fluid volume, Q_B the volumetric blood flow rate, PA the permeability-area product, C_{Bi} the arterial blood solute concentration, C_{Bo} the capillary blood solute concentration, C_I the interstitial fluid solute concentration, r_{RBC} the rate of red blood cell uptake of solute, and r_T models the tissue cellular removal of solute through cell membrane. The quantity PA can be expressed as the ratio V_I/T, where T is the transcapillary diffusion time.

		T	10 or 3 [min]
V_B	0.04 [l]	V_I	0.45 [l]
Q_B	0.7 [l/min]	C_{Bi}	0.15 [kg/l]
r_T	2×10^{-6} [kg/min]	r_{RBC}	10^{-5} [kg/min]

We allow T to assume two different values, which correspond to different physiological situations; furthermore, we assume that the switch between these two conditions can happen at any time. This calls for the introduction of a bimodal switched model, composed by the following two dynamics:

$$\dot{x} = A_1 x + b, \text{ if T} = 10[\text{min.}];$$
$$\dot{x} = A_2 x + b, \text{ if T} = 3[\text{min.}].$$

It can be easily checked that both models, considered in isolation, are box invariant. They therefore are stable around two different equilibria. In Fig. 4 we plot trajectories for these two systems, and draw some boxes (that were obtained from the computation of the eigenvalue with the rightmost real part and its corresponding eigenvector). Additionally, pivoted on the equilibria, the cones are shown. In Fig. 5 the two realizations are shown superimposed: the intersection of the two cones is non empty (the cones differ by only one of their boundaries, as expected because of the relative modification of only one row of the state matrix). The smaller one was thus chosen to define the "global" box. Two different simulations, with random switching, starting from opposite initial conditions, are shown in Fig. 6. As can be observed, the box is indeed an invariant for the switched system. This box yields a bound on the values of x, which is the brain blood glucose concentration. These bounds are useful to verify safety of insulin treatments for models of Type I diabetes patients.

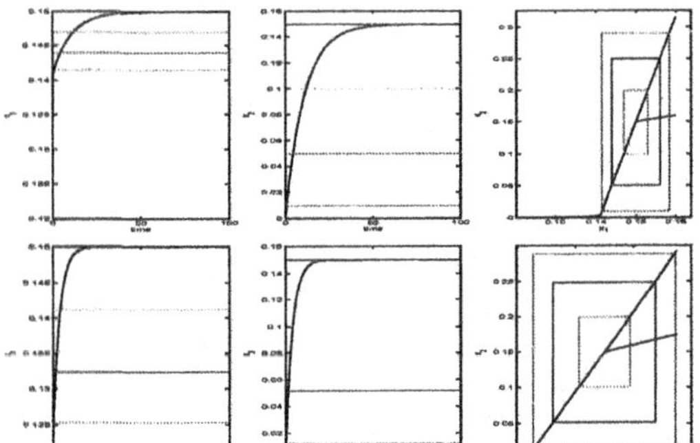

Fig. 4. Simulation of a trajectory for the first(top part) and second(bottom part) system, and computation of some symmetrical boxes.

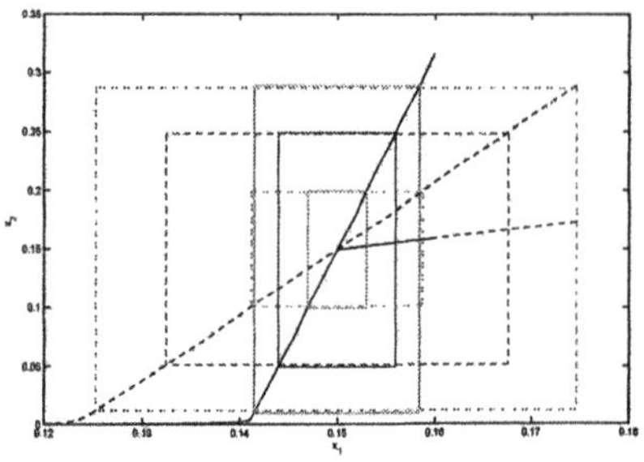

Fig. 5. Comparing the two trajectories, their respective boxes and conical regions. The first in solid, the second in dashed lines.

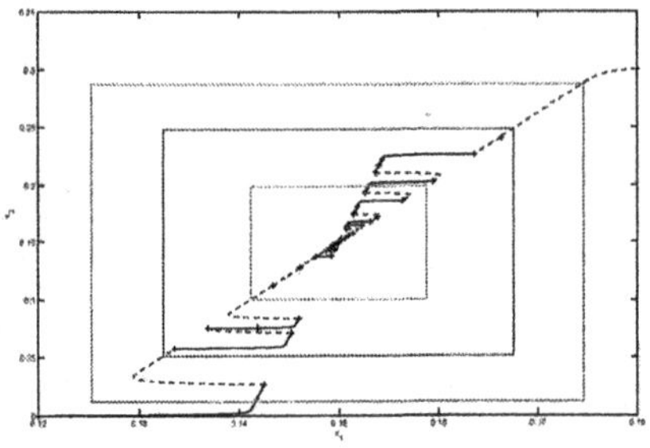

Fig. 6. Randomly Switching System composed by the two dynamical systems, with global boxes.

REFERENCES

Abate, A. and A. Tiwari (2006). The concept of box invariance for biologically-inspired dynamical systems. Technical report. UC Berkeley - EECS.

Ames, A., A. Abate and S. Sastry (2005). Sufficient conditions for the existence of zeno behavior. In: *Proceedings of the 44th Conference on Decision and Control.* IEEE.

Branicky, M. (1994). Stability of switched and hybrid systems. In: *Proceedings of the 33rd Conference on Decision and Control.* IEEE. pp. 3498–3503.

Coxson, G.E. (1994). The p-matrix problem is co-np-complete. *Mathematical Programming* **64**, 173–178.

Coxson, G.E. (1999). Computing exact bounds on elements of an inverse interval matrix is np-hard. *Reliable Computing* **5**, 137–142.

Ghosh, R. and C. J. Tomlin (2001). Lateral inhibition through delta-notch signaling: A piecewise affine hybrid model. In: *Hybrid Systems: Computation and Control, HSCC 2001.* Vol. 2034 of *LNCS.* pp. 232–246.

Johansson, M. and A. Rantzer (1998). Computation of piecewise lyapunov functions for hybrid systems. *IEEE Transactions on Automatic Control* **43**, 555–559.

Kiendl, H., J. Adamy and P. Stelzner (1992). Vector norms as lyapunov functions for linear systems. *IEEE Transactions on Automatic Control* **37**, 839–842.

Lygeros, J., K.H. Johansson, S.N. Simic, J. Zhang and S. Sastry (2003). Dynamical properties of hybrid automata. *IEEE Transactions on Automatic Control* **48**, 2–17.

Sorensen, J.T. (1985). A physiologic model of glucose metabolism in man and its use to design and assess improved insulin therapies for diabetes. PhD thesis. MIT.

ELSEVIER
IFAC
PUBLICATIONS

PERFORMANCE VERIFICATION OF DISCRETE EVENT SYSTEMS USING HYBRID MODEL-CHECKING

Bruno Denis [(1)], **Jean-Jaques Lesage** [(1)], **Zulema Juárez-Orozco** [(1),(2)]

[(1)] *LURPA-ENS de Cachan, France, {denis, lesage, juarez}@lurpa.ens-cachan.fr*
[(2)] *CIATEQ - Querétaro, Mexico, PhD grant financed by CONACYT (Mexico)*

Abstract: The results generated over the past few years on the formal verification of both Discrete Event Systems (DES) and Hybrid Dynamic Systems (HDS) are quite substantial, especially as regards the controller's properties of liveness and safety. In this paper, we will study the range of possibilities offered using the model-checking techniques in order to evaluate DES performances (in terms of quality of service provided by the automated system). This task calls for proceeding with a model-based approach that couples a hybrid model of the plant with a timed discrete model of the controller. We will also show, using a basic example, that by parameterizing the hybrid process model, the model-checker may then be employed to evaluate the robustness of the discrete control to perturbations encountered by the plant. *Copyright © 2006 IFAC*

Keywords: DES controller, hybrid plant model, model-based verification, model-checking, linear hybrid automaton, HyTech.

1. INTRODUCTION

A physical system is not, in most cases, intrinsically either purely discrete or purely continuous; instead, it's the abstraction the control engineer undertakes to ensure automation specification is being met that lends one distinction or the other. In the area of manufacturing systems for example, many processes have mobility axes that enable products to circulate, establish their position, etc.; consequently, some of the major physical variables controlled are either displacements or speeds. Depending not only on the level of quality to be guaranteed for these controlled variables, but also on the aggressiveness or variability in the external environment (e.g. type and importance of perturbations, parameter variation interval for the law of movement) and on the relative weight of economic constraints, the control engineer is required to choose between a servo control or a discrete control for each of these axes. Such automation-related choices will yield the physical variables to be observed and controlled by the controller, with either continuous or discrete control abstraction. Once the requisite displacement axis positioning quality has been achieved and provided that the perturbations encountered remain tolerable, the control engineer will then select a discrete control for obvious cost reduction reasons. This discrete displacement control, despite often being able to accommodate mobile positioning quality requirements, still constitutes an abstraction, and as such necessarily a simplification of the associated physical variables. The logic control of a linear displacement between two extreme positions, observed by means of two limit switch sensors, clearly does not enable ascertaining the precise position of the mobile, nor the time elapsed to complete this displacement.

Satisfying industrial system dependability requirements often necessitates conducting offline analyses, such as formal verification, before placing the automated system into operation (for further information on this topic, see the standard IEC 61508 entitled " Functional safety of electrical / electronic / programmable electronic safety-related systems "). These verifications, which are now frequently performed by means of model-checking (MacMillan, 1993), may be practiced by electing to incorporate or not a plant model.

In this paper, our efforts have focused on checking systems composed of a discrete controller coupled with a continuous (or partially-continuous) physical process whose entire set of observed and controlled variables constitute discrete abstractions of physical variables. To proceed, we will make use of model-checking techniques by coupling the discrete controller model with a hybrid plant model; this set-up will demonstrate that beyond the liveness and safety properties, it is indeed possible to check whether automated system performance is compatible with that stipulated in the specifications. We will also show that by parameterizing the hybrid process model, it becomes possible to use the model-checker to evaluate the robustness of discrete control to perturbations encountered by the plant.

This paper has been organized as follows. After having recalled the possibilities and limitations of both DES and HDS model-checking, we will introduce the expectations derived from a hybrid plant model for verifying a discrete controller. In order to illustrate our approach, the paper's second part will present the example of a positioning axis, which represents a component of a more complex assembly system. We will thus be able to show that use of a model-checker (such as HyTech), by implementation of a hybrid process model, makes it possible to verify the expected performance of this DES. Furthermore, a sensitivity study conducted on model parameters will allow evaluating the robustness of discrete control when confronted with perturbations.

2. ACQUIRED KNOWLEGE AND LIMITATIONS FROM AUTOMATED SYSTEM VERIFICATION

2.1 DES verification.

Formal verification techniques stem from the field of computer science. Only recently have they been adapted and applied to DES verification and, more specifically, to model-checking (Clarke E. M., *et al.*, 1986). The general principle behind model-checking may be expressed as follows (see Figure 1).

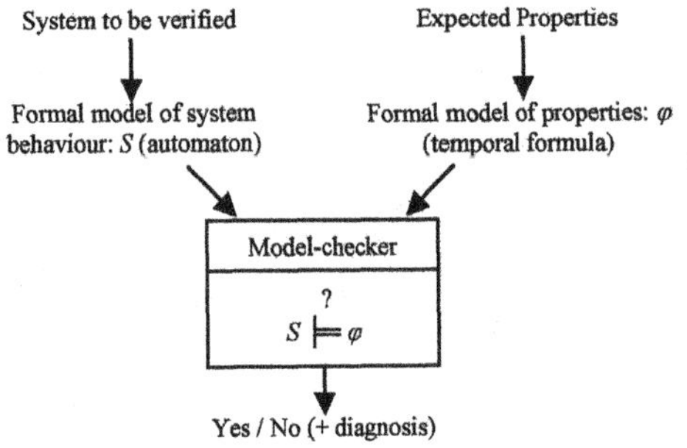

Fig. 1. Model-checking scheme.

Let's start with a system that has been designed to verify an entire array of properties (logical correctness, dependability, liveness, etc.). The first task of model-checking consists of formalizing system behavior in the form of a finite state automaton: S, plus the properties to be verified within a temporal algebra such as CTL (Emerson and Halpern, 1986): φ. The model-checker then conducts a thorough analysis of the state space reachable by S, which serves either to prove that $S \models \varphi$ (this algebraic statement denotes that "the system model satisfies the set of properties φ") or, when such is not the case, to propose a counterexample that revokes those properties not verified by S.

Moreover, a DES may be represented in a generic manner, as shown in Figure 2: a discrete controller acting in a closed loop on a plant. As part of a dependable controller design approach, the system being targeted for verification can thus be (according to Frey. and Litz, 2000) either the controller on its own, presumed to be operating within an open loop on the plant (a non "model-based" verification), or the {controller + plant} assembly set interacting within a closed loop ("model-based" verification).

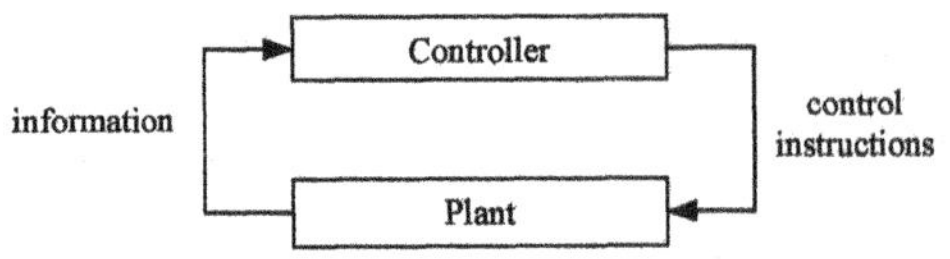

Fig. 2. A generic closed-loop DES.

The research work focusing on DES verification initially favored a non model-based approach (Moon I., 1994). The reachable state space of the controller model is thus to be built in the most permissive manner possible, i.e. such that the evolution of its inputs are in no way constrained by plant behavior. In this case, the safety properties capable of being demonstrated provide the basis for strong proof given that they can be demonstrated regardless of the evolution in controller inputs. On the other hand, a good number of liveness or accessibility properties cannot be demonstrated via a non model-based model-checking approach given the often fast-paced combinatory explosion of the reachable state space.

One means for reducing this combinatory explosion, using realistic constraints that depict the interaction of plant behavior with controller behavior, is to conduct a model-based verification. (Rausch and Krogh, 1998), (Machado, *et al.*, 2003).

Through reliance upon these results, we are now in a position to study the possibilities offered by the model-checking procedure in evaluating DES performance (in terms of quality of service provided by the automated system). To accomplish this task, it is necessary to undertake a model-based approach by selecting a hybrid plant model. The physical variables of the plant, and not their discrete abstraction by the controller, are what give the actual performance measures to be evaluated. The coupling of a hybrid plant model with a discrete controller model thus leads to a Hybrid Dynamic System (HDS) verification problem. We will now proceed by recalling the basic knowledge acquired and current limitations of HDS verification.

2.2 Hybrid systems verification.

While model design using hybrid automata is not exactly straightforward, their semantics is well-adapted to the analysis of HDS behavior. We will then assume that HDS verification is tantamount to exploring the reachable state space of a hybrid automaton. When framed as such, the fundamental

problem of HDS verification becomes one of computing the reachable state space of a hybrid automaton from an initial region. Not only is this computation prone to encountering a rapid combinatory explosion, but in the general case the problem is not-decidable, i.e. impossible to identify a computation algorithm with guaranteed convergence (Henzinger, *et al.*, 1995a). The linear hybrid automata (Alur, et al. 1995) constitute a category of hybrid automata that places verification within the domain of a decidable problem and for this reason, such automata have been selected for the purposes of our work. From an other hand, modular approaches are essential for modeling and verifying of industrial systems. This is why we felt that the HyTech tool (Henzinger, *et al.*, 1995b) was adapted to our needs, by virtue of allowing for the verification of models obtained by composition of linear hybrid automata.
With the context and objectives of our research now established and both the category of hybrid automata and model-checker selected, we will turn our attention next to presenting our approach for verifying discrete control performance via the assessment of a pertinent example.

3. A CASE STUDY

3.1 Presentation of the production system.

The system we are targeting is one of assembling / disassembling bearings within gears; it comprises four workstations (see Figure 3a). The four workstations pertain respectively to the following functions: loading at the beginning of the line (Station 1); identification of the gear material and eventual presence of a bearing (Station 2); insertion or removal of a bearing in the gear (Station 3); and gear sorting, taking into account the material, in anticipation of their unloading (Station 4). This entire assembly line is controlled by a Programmable Logic Controller (PLC).

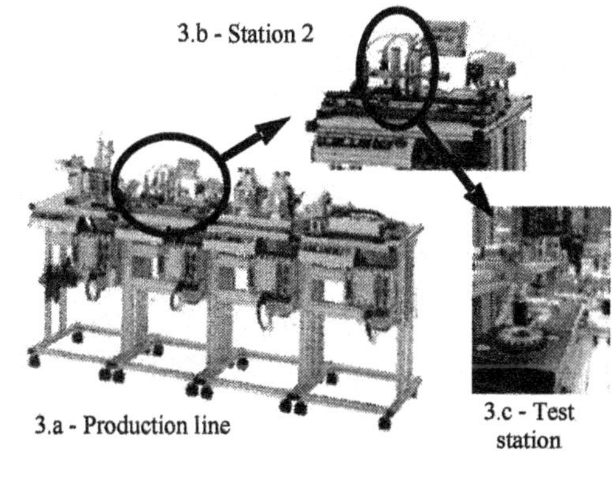

Fig. 3. Assembly/disassembly line.

Control of this assembly system is purely discrete; since this proves to be the most typical case for such systems, seeking more competitive costs leads to avoiding servo controls as much as possible in favor of logic controls. All continuous physical variables of the process being controlled (which for the most part consist of positions) are thus observed and controlled via their discrete abstractions. In all, 82 logical inputs and 50 logical outputs are managed by the PLC. The control model, written using Sequential Function Chart (SFC) syntax according to IEC 61131-3 standard has been verified with a timed model-checking technique (Bel Mokadem, *et al.*, 2005).
The objective of our study herein consists of verifying whether the quality of the positions obtained using such a discrete control is compatible with the precision required by the process. This concept of precision only takes on meaning with respect to the positioning of objects manipulated by either a displacement axis or a group of combined axes. It is thereby unnecessary to include the entire control model and the plant model in order to verify position quality, but rather just the subset that pertains to control of the axis (or group of axes) involved in the studied displacement. Our investigation will focus more heavily on the displacement axis of station 2 (Fig. 3b), with the only relevant information available for displacement control being given in Figure 4.

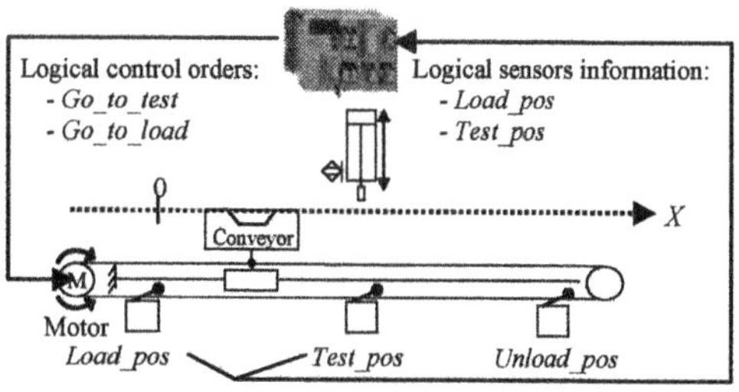

Fig. 4. Displacement axis of station 2.

The gear is initially located on the conveyor at the loading station (*Load_pos*). A motor rotation order in a clockwise direction (*Go_to_test*) implies the conveyor in displacement to the test position (*Test_pos*), a detailed view is provided in Figure 3c. For the test to be successfully conducted, stopping precision must be ≤ 1.5 mm. The conveyor position between these two discrete positions is denoted X. When the presence test of a bearing is completed, a subsequent motor rotation order in the clockwise direction positions the gear, where it gets unloaded (*Unload_pos*). A motor rotation in the counterclockwise direction then brings the conveyor into the loading position, where a new cycle can be launched. We are seeking to quantify the precision of gear positioning at the test station, enabled by this discrete control along the path *Load_pos* $\rightarrow$ *Test_pos*, so as to verify whether this precision is compatible with specification. Execution of this task will entail application of hybrid model-checking using the HyTech tool (Henzinger, *et al.*, 1995b).

3.2 Modelling.

The model of the entire {controller + displacement axis} set, which allows this positioning quality verification, has been laid out in Figure 5.
The hybrid automaton is composed of two synchronized automata.

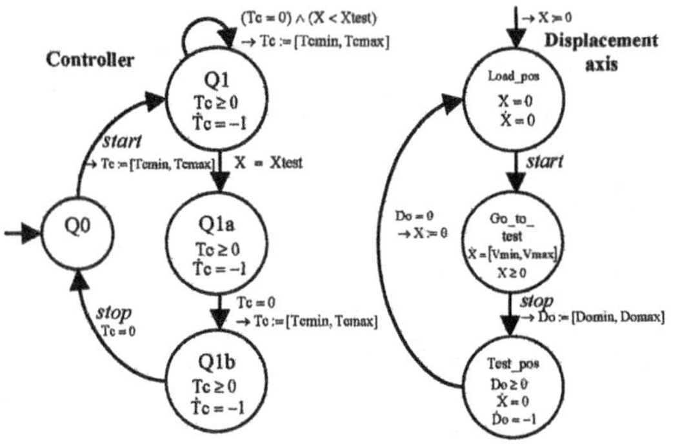

Fig. 5. Hybrid automata of the axis control.

The automaton that model electromechanical axis contains three discrete locations: two correspond with conveyor stopping positions at the extremities (*Load_pos* and *Test_pos*), and the other location corresponds with the movement phase between these extreme positions (*Go_to_test*). The return to loading position, which does not enter into the scope of the present study, is modeled by means of a timer with variable duration D_0. The two continuous state variables associated with this automaton are conveyor position X and timer duration D_0. Upon initialization, the conveyor is placed into the loading position and hence $X = 0$. Both of the stopping situations are associated with the function $dX/dt = 0$ (denoted $\dot{X} = 0$); the movement location is associated with a continuous dynamics of the form $\dot{X} = a$, with a representing a rational constant arbitrarily chosen within an interval: $[V_{min}, V_{max}]$. This arbitrary assignment of constant a enables abstracting the variability in conveyor displacement speed due to perturbations undergone by the displacement axis (mechanical friction, resistant torque on the motor axis, kinematic non-linearities, etc.). Two synchronization events with the automaton used for modeling controller behavior, called *start* and *stop* respectively, are associated with the transitions that correspond to the conveyor movement phase.

The automaton that models controller behavior is composed of four discrete locations and is initialized at location Q_0. The *start* synchronization event with the axis automaton shows that the motor starts up during a PLC cycle change. The monitor of this PLC is of the "periodic" type, indicating that execution of the control program is to take place according to a constant-duration cycle (see Figure 6), which the user is required to setup in PLC. In practice, this cycle duration undergoes slight variations (on the order of 1%), called "jitters". To translate these operations into the hybrid controller automaton, the duration of each PLC processing cycle (Tc) is assigned an arbitrarily-chosen value within an interval: $[Tc_{min}, Tc_{max}]$. Once location Q_1 has been entered, the automaton remains there until the axis has traveled the distance required to position it in alignment with the test station (at which point $X = Xtest$). Locations $Q1_a$ and $Q1_b$ serve to translate that this information is taken into account by the PLC during the subsequent input reading phase, followed by processing and then execution of the motor stop order (Figure 6). The *stop* synchronization event thereby triggers controller automaton re-initialization at the same time it activates the *Test_pos* situation of the displacement axis automaton.

Figure 6 shows the synchronization of the two automata presented in Figure 5 by the *stop* event as well as the evolution in continuous variables Tc and X. The axis positioning error at the level of the test station thus results from the time lag between the moment the axis reaches the *Xtest* position and the moment the PLC is able to react at this event in the input reading phase, with the consequence being that the axis stop order may be issued either within the same PLC cycle or one cycle later. The experimental campaign conducted using HYTECH has revealed this finding and the following discussion is intended to provide greater detail.

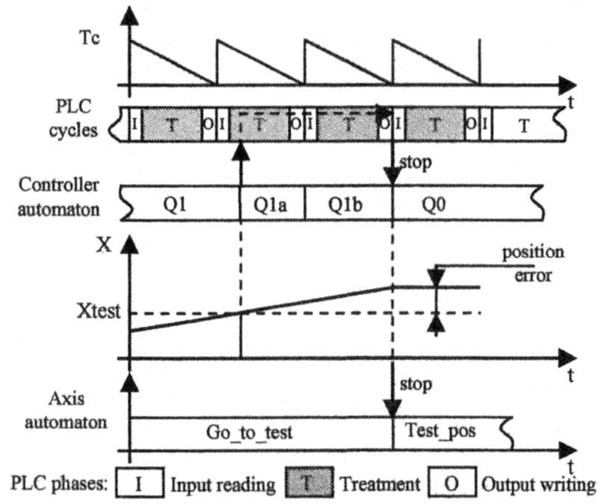

Fig. 6. Synchronous evolutions of automata.

3.3 Implementation and the experimental protocol.

All of our experiments have been carried out by constraining the variation of duration the conveyor's return to the loading position (D_0) over the interval [*3000ms, 3500ms*] (which thus yields *Domin = 3000ms* and *Domax = 3500ms*). Moreover, the abscissa (*Xtest*) of the conveyor at the test station is 80 mm.

Given that these data remain fixed, the HYTECH code obtained from the hybrid automaton shown in Figure 5 is a parametric code, with intervals [*Tcmin, Tcmax*] (for the cycle time to be adjusted on the PLC) and [*Vmin, Vmax*] (for the translation speed of the conveyor subjected to perturbations) constituting the two parameters. For each experiment, numerical values are ascribed to these two intervals. The model-checker then computes the region the automaton is capable of reaching from its initial region. We will initially derive the intersection of this reachable region with the discrete location *Test_pos* so as to build the accessibility domain of the axis position, which is then obtained by projection of this sub-region with respect to the variable X. The set of positions reached by the conveyor thus assumes the form of an interval on X. As an example, for *Vmin* = *Vmax* = *200mm/s*, *Tcmin* = $11ms^{-0.5\%}$ and *Tcmax* = $11ms^{+0.5\%}$, the conveyor stopping position in front of the test station is contained within the interval [*83.182mm,*

84.018mm], which represents a resultant position deviation of 0.836mm.

3.4 Position-stopping quality without perturbations.

For this initial study, the hypothesis has been adopted that the conveyor is not submitted to any perturbation. Its translation speed is thus presumed to be constant and yields: *Vmin = Vmax = 200 mm/s*. The changes in conveyor stopping position may then be observed as a function of the PLC cycle time value when such value varies by more or less than 0.5% around the period adjusted by the user (the so-called jitter phenomenon described in Section 3.2).
Figure 7 reveals that the positioning error evolves in accordance with two phenomena that we will now differentiate. The first corresponds to a systematic error (positive slope line passing through the average error interval value), which can be corrected by an adapted adjustment of sensor position *Test_pos*. The second phenomenon is homogeneous to a random error (the fixed interval deviation on the positions reached for a given *Tc*), whose general amplitude shape increases with PLC cycle time. It can nonetheless be remarked that for certain *Tc* values (e.g. *Tc = 9ms* or *Tc = 13.1ms*), the range of these "random" errors is narrower than that for proximate *Tc* values. The existence of these particular zones of smaller positioning error, while PLC cycle time is increasing, highlights a non-trivial temporal correlation phenomenon between the PLC cycle and occurrence of the test position arrival event depicted in Figure 8.

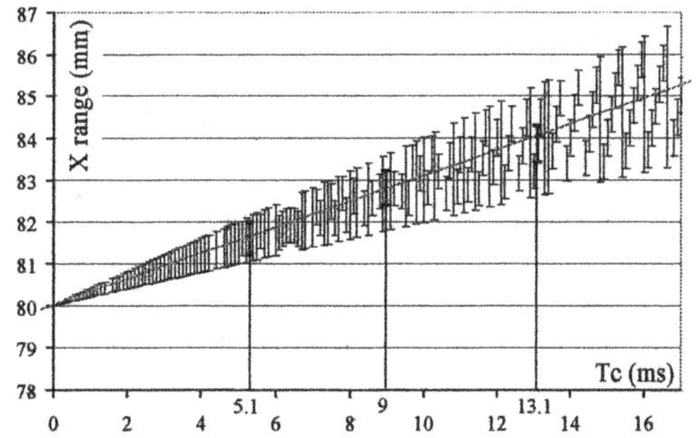

Fig. 7. Conveyor positioning error.

When the temporal detection region of the *X=Xtest* event has been included within a single PLC cycle, conveyor stopping will systematically be controlled during the subsequent cycle output assignment phase (see Figure 8a). In contrast, while the temporal detection region of the *X=Xtest* event straddles two PLC cycles, stopping will intervene during the output assignment phase of one of the two subsequent cycles (Figure 8b). The range of the stopping zone will thus increase.
This temporal dispersion in *stop* order execution is heavily correlated with the PLC cycle time jitter (Figure 8c). Though the ratio between the distance to be traversed and *Tc* is such that, despite the jitter on *Tc*, the number *N* of PLC cycles elapsed between the conveyor start order and detection of the test position (*X=Xtest*) is the same with each new conveyor displacement from *X=0* to *X=Xtest*, the conveyor stop order will always be issued at the end of the same PLC cycle and the error range will always solely depend on the jitter and cycle number *N* (this configuration will be referred to as "Scenario 1" in the following discussion). On the other hand, should the ratio between the distance traveled and *Tc* be such that a variable number of cycles is necessary for conveyor displacement, the *stop* order may then be issued at the end of the various PLC cycles (Figure 8c). The error range is thus merely a function of *Tc* (this configuration will be called "Scenario 2").

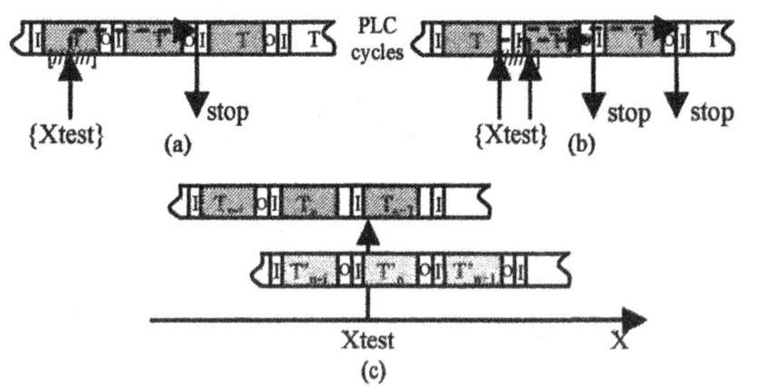

Fig. 8. Correlation between *Tc*, *Xtest* event occurrences and *stop* order.

Figure 9 clearly highlights this phenomenon. For Scenario 2, the stopping range is proportional to the cycle time value. Whereas for Scenario 1 positioning error is independent of cycle time (at least up to the limit of values selected for our experiment).
If specifications call for a maximum positioning error in *Test_pos* of 1.5mm, the PLC cycle period has to be set at a value of below 7.4ms. If the full PLC load (for execution of the overall assembly line program with its 82 inputs and 50 outputs) does not enable parameterizing such a low cycle time, it would then be necessary to target a Scenario 1 parameterization, e.g. 9.0ms or 13.1ms.

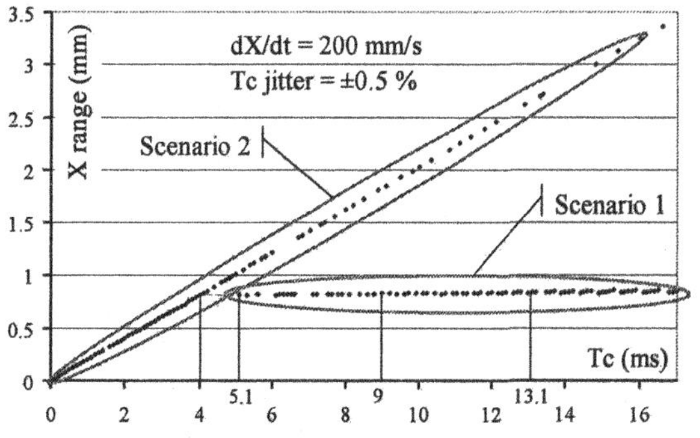

Fig. 9. Conveyor positioning error range.

3.5 Positioning robustness.

It has already been demonstrated that the PLC cycle time may be judiciously chosen in order to guarantee the quality of conveyor positioning under a hypothesis of no translation speed perturbations. Figure 10 presents the results from experiments conducted in order to evaluate the influence of such perturbations on positioning quality for a given cycle time *Tc=13.1ms* (which remains contaminated with a jitter of ± 0.5%). This curve shape is the same for *Tc = 9ms* or *Tc = 5.1ms*.

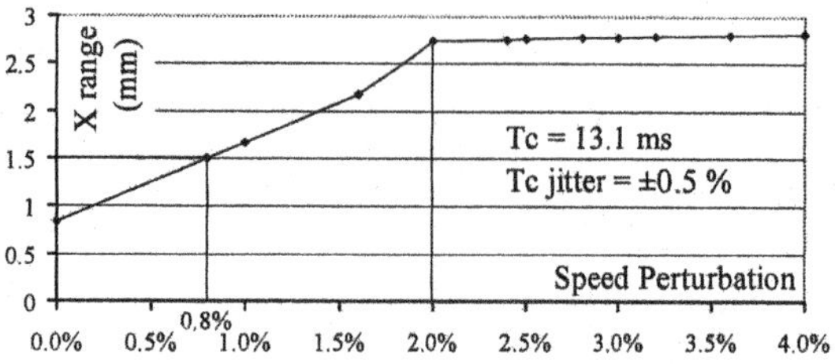

Fig. 10. Effect of conveyor speed perturbation.

Two distinct zones appear on this curve. Above 2% speed perturbation, the positioning error is basically constant; this corresponds to Scenario 2, for which the error range is directly correlated with *Tc* without any major influence on the speed-related perturbation. Below this 2% level, the error increases linearly with perturbations since the system here is behaving according to Scenario 1, which is sensitive to both the jitter on *Tc* and perturbations on the conveyor translation speed.
For a requirement that limits the positioning error to 1.5mm, it thus becomes necessary to ensure that speed perturbations remain less than 0.8% (in the case of a PLC adjustment to *Tc=13.1ms*). Should such not be the case, adopting a servo control in the conveyer position would need to be examined.

3.5 Assessment of the use of HyTech.

The results provided above have necessitated several hundreds of model-checking experiments on the hybrid automaton shown in Figure 5 using the HyTech tool, with different parameter sets being introduced for each experiment. The search for the reachable region of a model never took more than 5 minutes and, in most instances, 30 seconds were sufficient. Computation time did not therefore pose any problem for this study. The primary difficulty encountered had to do with the interruption of reachability computations due to exceeding the integer limit value (2^{64}) during the region computation step. Our case study did necessitate manipulating numerical values with quite varied scales. As an example, the conveyor position progressed over a total path length of 80mm, whereas the calculated stopping error at times amounted to just a few hundredths of a millimeter. For all of the distances manipulated, numerical values out to 4 significant digits thus had to be used. The same difficulty gets magnified when manipulating time since the duration of the conveyor position return lasts approximately *3s*, while a $\pm 0.5\%$ jitter on $Tc = 1ms$ equals $\pm 5\mu s$; seven significant digits therefore proved necessary. This numerical value coding is very detrimental when running HyTech, which interrupts the computation of regions once the integer coding capacity has been exeeded, thereby making it impossible to draw any conclusion on the current verification. In order to avoid this predicament, the units for each model parameter set have been adjusted. For example, depending on the value of *Tc*, the time unit has been fixed at *1μs*, *10μs* or *100μs*, which makes only *5%* of the verifications inconclusive.

4. CONCLUSION AND OUTLOOK

In this paper, we have focused on the verification of systems composed of a discrete controller coupled with a continuous (or partially-continuous) physical process, whose full extent of observed and controlled variables are logical abstractions of physical variables. For that, we employed a model-checking technique that couples the timed discrete model of the controller with a hybrid plant model. By relying upon a case study, we showed the possibility of verifying whether or not system performance is compatible with the specifications. By parameterizing the hybrid system model, it becomes possible for the model-checker to evaluate discrete control robustness to plant perturbations.
According to this approach, HyTech has proven to be an efficient tool. It goes without saying that the size of the model being verified must remain compatible with the difficulties inherent in the well-known combinatory explosion phenomenon typical of model-checking. For this reason, it is necessary to limit performance verification to just the {control + plant} subsystem involved in deriving the evaluated variables. Determining this subsystem being targeted by performance verification along with building the hybrid model used as the abstraction are the current focus of a complementary research project.

REFERENCES

Alur R. et al. (1995). The algorithmic analysis of hybrid systems. Theoritical computer science, **Vol. 138**, p. 3-34.

Bel Mokadem H., Bérard B., Gourcuff V., Roussel J.-M., De Smet O. (2005). Verification of a timed multitask system with UPPAAL. *Proc. of 10th IEEE ETFA*, CDRom paper, 8 pages, September, Catania-Italy.

Clarke, E. M., Emerson, E. A. and A. P. Sistle, (1986). Automatic verification of finite state concurrent system using temporal logic. *ACM Trans. on Programming Languages and Systems* **Vol. 8**, n° 2, p. 244-263.

Emerson E.A. and Halpern J.Y. (1986). Sometimes and Not Never revisited : on branching versus linear time temporal logic. *Journal of the ACM*. **Vol. 33**, n° 1, p. 151-178.

Frey G. and Litz L. (2000). Formal method in PLC programming. *Proc. of IEEE SMC'2000*, CDRom paper, 6 pages, October 8-11, Nashville, USA.

Henzinger T.A., Kopke P.W., Puri A., Varaiya P. (1995a). Waht's decidable about hybrid automata ?. *Proc. of the 27th annual ACM Symposium of Theory of Computing*, p. 373-382.

Henzinger T.A., Ho P.-H., Wong-Toi H. (1995b). A user guide to HyTech. *Proc. of TACAS*, Lecture Notes in Computer Science 1019, Springer-Verlag, p. 41–71.

Mac Millan K.L. (1993). *Symbolic model checking*, Kluwer Academic.

Moon I. (1994). Modeling programmable logic controllers for logic verification. *IEEE Control Systems*, **Vol. 14**, n° 2, 1994, p. 53-59.

Rausch M. and Krogh H. (1998). Formal verification of PLC programs. *Proc. of ACC'98*, Philadelphia, USA.

ELSEVIER
IFAC
PUBLICATIONS

VERIFICATION-INTEGRATED FALSIFICATION OF NON-DETERMINISTIC HYBRID SYSTEMS

Stefan Ratschan * **Jan-Georg Smaus** **

* *Max-Planck-Institut für Informatik, Saarbrücken, Germany*
** *Albert-Ludwigs-Universität Freiburg, Germany*

Abstract: This paper provides a method for coupling safety verification algorithms for non-deterministic (and, in general, non-linear) hybrid systems with the ability of finding concrete counterexamples, i.e., with falsification. Such a tight integration of verification with falsification has the advantage that verification attempts guide the search for concrete counterexamples, and endless attempts to verify unsafe systems or to find counterexamples in safe systems can often be avoided.

Keywords: hybrid systems, verification, falsification, simulation, testing

1. INTRODUCTION

For debugging designs of embedded systems, it is essential to be able to falsify hybrid systems, i.e., to find concrete counterexamples to a given property (e.g., safety). This is usually done by checking the results of simulation runs. However, such an approach has the disadvantage that one might continue looking for counterexamples although there are none, since the system fulfills the sought property. Also, simulation usually requires considerable user intelligence for avoiding redundant runs, determining the necessary time horizon, and canceling obviously useless simulation runs. Moreover, for systems with non-deterministic evolution, even from a given starting point there are uncountably many different trajectories that might be a counterexample. Hence it usually does not suffice to simulate with a fixed instantiation of the non-deterministic values in the specification of the given hybrid system, since these values might change over time (e.g., due to changes of the system input or changes in the environment).

In this paper, we consider non-deterministic systems, and take a step towards avoiding these problems by an approach that is integrated into verification, and that solves for counterexamples instead of blindly searching for them in the space of all possibilities opened by non-determinism.

The verification we have in mind is based on a finite abstraction of the infinite state space of a hybrid system. The safety of the abstract system implies the safety of the original system, and by successively refining the abstraction, it is often (but not always, due to resource and decidability limitations) possible to find an abstraction that *is* safe, provided the original system is safe. Thus, by integrating the method presented here into such an abstraction refinement loop, on the one hand we avoid endless refinement for systems that have an unsafe trajectory, on the other hand we avoid endless search for an unsafe trajectory for safe systems. Moreover, by using information computed by the verification engine, the approach

[1] This work was partly supported by the German Research Council (DFG) as part of the Transregional Collaborative Research Center "Automatic Verification and Analysis of Complex Systems" (SFB/TR 14 AVACS). See www.avacs.org for more information.

can often avoid trying to find counterexamples in parts of the state space that do not lead to an unsafe state. Furthermore, by following the state-space partition of the verification engine instead of the usual time partitioning of simulation, the approach avoids redundant recomputations for the same part of the state space, and is not restricted to a certain time horizon.

By *non-determinism*, we refer to systems where the flow of the real variables within one control mode is not uniquely determined, but rather, this flow has a whole set of evolution possibilities at any time point. In this case, there is hope that a flow described by piecewise polynomial functions fulfills the (in general, non-linear) flow conditions. This observation is the basis of our method. We formalize constraints that imply the existence of a trajectory leading from an initial to an unsafe state, and solve these constraints, arriving at connection points between the pieces.

The limitation of our method to non-deterministic systems can also be interpreted as follows: given a system that is deterministic (the flows are uniquely determined), one could relax the flow constraints to a certain extent to make the system non-deterministic, and then apply our method. If an error path is then found, one can argue that the original system is very close to an unsafe one. From a practical perspective, similar conclusions arise in both cases: one could either say, to be on the safe side, one should assume that systems behave non-deterministically within a certain range, or, to be on the safe side, one should reject systems that are close to an unsafe one.

2. METHOD

We assume that the hybrid system is given using a tuple of families of constraints

$(Init_{m \in M}, Flow_{m \in M}, Jump_{m,m' \in M}, Unsafe_{m \in M})$,

where M denotes the set of discrete modes, $Init_m$ specifies the set of initial states in mode m, the constraint $Flow_m$ the possible continuous flows in mode m, the constraint $Jump_{m,m'}$ the possible discontinuous jumps from mode m to mode m' and $Unsafe_m$ the set of unsafe states in mode m. These constraints may contain elements of a finite set X of continuous state variables. The flow constraints may in addition contain dotted versions of the continuous state variables that denote their differentiation, and the jump constraints may in addition contain primed versions of these variables that denote their value after the jump. As examples, consider the flow constraint $\sin x - 1 \leq \dot{x} \wedge \dot{x} \leq \sin x + 1$, and the jump constraint $x' = x + 1$ which denotes the fact that the jump increases the variable x by one.

The systems we consider are *non-deterministic* in the sense that the flow constraints are typically inequalities, as in the example just given, rather than equations. We do not need to define precisely when a system is (non-)deterministic or give some measure for the degree of non-determinism. It suffices to note that the more non-deterministic the system is, the more likely it is that our method will find an unsafe trajectory.

Concerning the verification, we assume an abstraction refinement algorithm (Clarke *et al.* 2003, Alur *et al.* 2003, Ratschan and She 2005) for verifying safety properties of hybrid systems. Such methods decompose the state space into pieces, and based on that, construct a finite system (the *abstraction*) whose safety implies the safety of the original (*concrete*) system. The abstraction consists of *abstract states*, which are pairs consisting of a mode (i.e., an element of M) and a subset of the continuous state space (i.e., a subset of $\mathbb{R}^{|X|}$). How these sets are represented depends on the verification technique, e.g., they could be polyhedra or hyper-rectangles—however, the material of this section is independent of the chosen representation.

Given an abstract counterexample consisting of a sequence of abstract states $(m_0, B_0), \ldots, (m_k, B_k)$ such that (m_0, B_0) is initial and (m_k, B_k) is unsafe in the abstraction, we try to find points $(m_0, \boldsymbol{x}_0), \ldots, (m_k, \boldsymbol{x}_k)$ such that $\boldsymbol{x}_i \in B_i$ for all $i \in \{0, \ldots, k\}$, and such that there is a piecewise polynomial trajectory of the hybrid system that follows $(m_0, \boldsymbol{x}_0), \ldots, (m_k, \boldsymbol{x}_k)$, with $(m_0, \boldsymbol{x}_0)$ initial, and $(m_k, \boldsymbol{x}_k)$ unsafe. By *piecewise polynomial*, we mean that the trajectory within one mode is a continuous curve composed of (in general) several polynomial pieces.

So we try to find a solution to the constraint

$$\begin{aligned}(\bigwedge_{i \in \{1,\ldots,k\}} \boldsymbol{x}_i \in B_i) \wedge Init_{m_0}(\boldsymbol{x}_0) \wedge \\ \bigwedge_{i \in \{1,\ldots,k\}} (Jump_{m_{i-1},m_i}(\boldsymbol{x}_{i-1}, \boldsymbol{x}_i) \vee \\ (m_{i-1} = m_i \wedge flow_{m_i}(\boldsymbol{x}_{i-1}, \boldsymbol{x}_i))) \wedge \\ Unsafe_{m_k}(\boldsymbol{x}_k) \quad (1)\end{aligned}$$

which we call the *concretization constraint*. This uses the constraints defining the given hybrid system (the ones with uppercase first letter), except the constraint defining the flow, since it contains the differentiation operator, which we would like to avoid. Since we assume that our input system is non-deterministic, the set of trajectories from $\boldsymbol{x}_{i-1}$ to $\boldsymbol{x}_i$ fulfilling $Flow_{m_i}$ will—in general—consist of a central trajectory together with a certain neighborhood around it. Hence, one can find a polynomial approximation of the central trajectory that also fulfills $Flow_{m_i}$. If the neighborhood is sufficiently large or $\boldsymbol{x}_{i-1}$ and $\boldsymbol{x}_i$ sufficiently close, the polynomial trajectory might even be linear.

So we assume a polynomial function $\phi_{\boldsymbol{a}}$ that gives us for a given time point in an interval $[0, \tau]$ the corresponding value of the trajectory. Here, $\boldsymbol{a}$ is a vector of parameters (real numbers) that specifies $\phi_{\boldsymbol{a}}$ in some way that we deliberately leave open at this point. We will later be more specific.

Hence we define $\mathit{flow}_m(\boldsymbol{x}, \boldsymbol{y}) \equiv$

$$\exists \boldsymbol{a} \exists \tau > 0 \, [\mathit{traj}(\boldsymbol{x}, \boldsymbol{y}, \phi_{\boldsymbol{a}}, \tau) \wedge \mathit{constr}_m(\phi_{\boldsymbol{a}}, \tau)] \,. \tag{2}$$

Here $\mathit{traj}(\boldsymbol{x}, \boldsymbol{y}, \phi_{\boldsymbol{a}}, \tau)$ models the fact that function $\phi_{\boldsymbol{a}}$ leads to a trajectory of length τ from $\boldsymbol{x}$ to $\boldsymbol{y}$, and $\mathit{constr}_m(\phi_{\boldsymbol{a}}, \tau)$ models the fact that the trajectory $\phi_{\boldsymbol{a}}$ fulfills the differential constraint Flow_m valid in mode m in the time interval $[0, \tau]$.

So we define $\mathit{traj}(\boldsymbol{x}, \boldsymbol{y}, \phi_{\boldsymbol{a}}, \tau) \equiv$

$$\phi_{\boldsymbol{a}}(0) = \boldsymbol{x} \wedge \phi_{\boldsymbol{a}}(\tau) = \boldsymbol{y},$$

and $\mathit{constr}_m(\phi_{\boldsymbol{a}}, \tau) \equiv$

$$\forall t \in [0, \tau] \left[\mathit{Flow}_m(\phi_{\boldsymbol{a}}(t), \dot{\phi}_{\boldsymbol{a}}(t)) \right], \tag{3}$$

where Flow_m is the constraint defining the flow of the given hybrid system in mode m.

Now observe that we have to solve the constraint $\mathit{flow}_m(\boldsymbol{x}, \boldsymbol{y})$ many times for a given input system, since it occurs k times in (1), and since (1) has to be solved over and over again in the verification/falsification loop sketched in the introduction. Hence we would like to arrive at a formulation of $\mathit{flow}_m(\boldsymbol{x}, \boldsymbol{y})$ for which it is as easy as possible to check whether it has a solution.

Observe that $\mathit{flow}_m(\boldsymbol{x}, \boldsymbol{y})$ contains several quantifiers. Assume that we have a given, fixed input system containing only polynomial constraints with a certain constraint Flow_m defining the possible flow in mode m, and a class of functions $\phi_{\boldsymbol{a}}$ containing only polynomials, in which we search for concrete trajectories. Then, due to the classical result (Tarski 1951) that the first-order theory of the real numbers with addition and multiplication admits quantifier elimination, it is always effectively possible to find an equivalent formula without quantifiers.

In spite of the fact that quantifier-free constraints are easier to check than constraints with quantifiers, we do not opt for automatic quantifier elimination since it can be very costly and lead to a big increase in formula size. Instead, we will try to identify problem classes corresponding to special cases for these constraints for which manual quantifier elimination will yield simple and useful results.

An especially important class is the class where $\phi_{\boldsymbol{a}}$ is linear, i.e., $\phi_{\boldsymbol{a}_1, \boldsymbol{a}_0}(t) = \boldsymbol{a}_1 t + \boldsymbol{a}_0$, where $\boldsymbol{a}_1 \in \mathbb{R}^{|X|}$, $\boldsymbol{a}_0 \in \mathbb{R}^{|X|}$.

So we have: $\mathit{traj}(\boldsymbol{x}, \boldsymbol{y}, \phi_{\boldsymbol{a}_1, \boldsymbol{a}_0}, \tau) \equiv$

$$\begin{aligned} \phi_{\boldsymbol{a}_1, \boldsymbol{a}_0}(0) = \boldsymbol{x} \wedge \phi_{\boldsymbol{a}_1, \boldsymbol{a}_0}(\tau) = \boldsymbol{y} \equiv \\ \boldsymbol{a}_0 = \boldsymbol{x} \wedge \boldsymbol{a}_1 \tau + \boldsymbol{a}_0 = \boldsymbol{y}, \end{aligned} \tag{4}$$

and $\mathit{constr}_m(\phi_{\boldsymbol{a}_1, \boldsymbol{a}_0}, \tau) \equiv$

$$\forall t \in [0, \tau] \, [\mathit{Flow}_m(\boldsymbol{a}_1 t + \boldsymbol{a}_0, \boldsymbol{a}_1)] \tag{5}$$

Now we can solve (4) for $\boldsymbol{a}_0$ and $\boldsymbol{a}_1$, and substitute the resulting terms in (5) which thus becomes

$$\forall t \in [0, \tau] \left[\mathit{Flow}_m(\frac{\boldsymbol{y} - \boldsymbol{x}}{\tau} t + \boldsymbol{x}, \frac{\boldsymbol{y} - \boldsymbol{x}}{\tau}) \right],$$

and (2) becomes

$$\exists \tau > 0 \forall t \in [0, \tau] \left[\mathit{Flow}_m(\frac{\boldsymbol{y} - \boldsymbol{x}}{\tau} t + \boldsymbol{x}, \frac{\boldsymbol{y} - \boldsymbol{x}}{\tau}) \right]. \tag{6}$$

We call Flow *convex* if for every vector $\boldsymbol{c} \in \mathbb{R}^{2|X|}$, and vector $\boldsymbol{d} \in \mathbb{R}^{2|X|}$, if $\mathit{Flow}(\boldsymbol{c})$ and $\mathit{Flow}(\boldsymbol{d})$ hold, then for every $\lambda \in [0, 1]$, also $\mathit{Flow}(\lambda \boldsymbol{c} + (1 - \lambda) \boldsymbol{d})$ holds.

In general, flow constraints of the form $\mathit{Flow}(\boldsymbol{x}, \dot{\boldsymbol{x}}) \equiv \underline{A}\boldsymbol{x} + \underline{b} \leq \dot{\boldsymbol{x}} \leq \overline{A}\boldsymbol{x} + \overline{b}$, where $\underline{A}, \overline{A}$ are a $|X| \times |X|$-matrices, and $\underline{b}, \overline{b}$ are vectors in $\mathbb{R}^{|X|}$, are convex. This follows from the trivial arithmetic facts that $e \leq e'$ implies $\lambda e \leq \lambda e'$, for every $\lambda \geq 0$, and that $e_1 \leq e_1' \wedge e_2 \leq e_2'$ implies $e_1 + e_2 \leq e_1' + e_2'$.

Now if $\phi_{\boldsymbol{a}}$ is a linear function and Flow_m is convex, then (3) is equivalent to

$$\mathit{Flow}_m(\phi_{\boldsymbol{a}}(0), \dot{\phi}_{\boldsymbol{a}}(0)) \wedge \mathit{Flow}_m(\phi_{\boldsymbol{a}}(\tau), \dot{\phi}_{\boldsymbol{a}}(\tau)),$$

i.e., the universal quantification can be replaced by a conjunction referring to the endpoints of the quantified interval. Note that just convexity of Flow_m or linearity of $\phi_{\boldsymbol{a}}$ alone are not sufficient.

Thus (6) can be simplified to

$$\exists \tau > 0 \left[\mathit{Flow}_m(\boldsymbol{x}, \frac{\boldsymbol{y} - \boldsymbol{x}}{\tau}) \wedge \mathit{Flow}_m(\boldsymbol{y}, \frac{\boldsymbol{y} - \boldsymbol{x}}{\tau}) \right] \tag{7}$$

Now we assume, in addition to the linearity of $\phi_{\boldsymbol{a}}$, that Flow_m is of the form $\underline{F}(\boldsymbol{x}) \leq \dot{\boldsymbol{x}} \leq \overline{F}(\boldsymbol{x})$. So (7), i.e. (2), becomes

$$\underline{F}(\boldsymbol{x}) \leq \frac{\boldsymbol{y} - \boldsymbol{x}}{\tau} \leq \overline{F}(\boldsymbol{x}) \wedge \underline{F}(\boldsymbol{y}) \leq \frac{\boldsymbol{y} - \boldsymbol{x}}{\tau} \leq \overline{F}(\boldsymbol{y})$$

that is

$$\tau \underline{F}(\boldsymbol{x}) \leq \boldsymbol{y} - \boldsymbol{x} \leq \tau \overline{F}(\boldsymbol{x}) \wedge \tau \underline{F}(\boldsymbol{y}) \leq \boldsymbol{y} - \boldsymbol{x} \leq \tau \overline{F}(\boldsymbol{y})$$

If we again take the case $\underline{F}(\boldsymbol{x}) = \underline{A}\boldsymbol{x} + \underline{b}$, $\overline{F}(\boldsymbol{x}) = \overline{A}\boldsymbol{x} + \overline{b}$, we get

$$\tau \underline{A}\boldsymbol{x} + \underline{b} \leq \boldsymbol{y} - \boldsymbol{x} \leq \tau \overline{A}\boldsymbol{x} + \overline{b} \wedge \tau \underline{A}\boldsymbol{y} + \underline{b} \leq \boldsymbol{y} - \boldsymbol{x} \leq \tau \overline{A}\boldsymbol{y} + \overline{b}$$

that is

$$\begin{aligned} \boldsymbol{x} + \tau \underline{A}\boldsymbol{x} + \underline{b} \leq \boldsymbol{y} \leq \boldsymbol{x} + \tau \overline{A}\boldsymbol{x} + \overline{b} \wedge \\ \boldsymbol{y} - \tau \overline{A}\boldsymbol{y} + \underline{b} \leq \boldsymbol{x} \leq \boldsymbol{y} - \tau \underline{A}\boldsymbol{y} + \overline{b} \end{aligned}$$

and equivalently

$$(I + \tau\underline{A})\boldsymbol{x} + \underline{\boldsymbol{b}} \leq \boldsymbol{y} \leq (I + \tau\overline{A})\boldsymbol{x} + \overline{\boldsymbol{b}} \wedge$$
$$(I - \tau\overline{A})\boldsymbol{y} + \underline{\boldsymbol{b}} \leq \boldsymbol{x} \leq (I - \tau\underline{A})\boldsymbol{y} + \overline{\boldsymbol{b}}$$

One might want to derive special methods for this specific constraint, however this is beyond the scope of the current paper. Considering the concretization constraint (1) and its special forms shown above, we know: For a given abstract counterexample, if there is a solution to the corresponding concretization constraint, then there is a trajectory from an initial to an unsafe state following this abstract counterexample, i.e., a concrete counterexample. We will show how to solve the concretization constraint in the next section.

3. CONSTRAINT SOLVING

We only consider cases where the universal quantifier of (3) has been eliminated (due to convexity), and where $\phi_{\boldsymbol{a}}$ is linear (although $Flow_m$, while convex, might still be non-linear). Then we are left with (7) whose quantifiers are only existential. For finding solutions, it suffices to drop the existential quantifiers, and finding a solution in all variables, including the ones that were existentially quantified before.

One could solve the concretization constraint using numerical local optimization techniques. However, these need good starting points to be able to find a solution. For this we use interval constraint propagation techniques (Davis 1987, Cleary 1987). These can, given an interval for each variable in a constraint, prune these intervals to smaller ones without losing any solution of the constraint. After constraint propagation we can use the midpoints of the resulting intervals as starting points for local optimization.

The most basic interval constraint propagation method decomposes all atomic constraints (i.e., constraints of the form $t \geq 0$ or $t = 0$, where t is a term) into conjunctions of so-called primitive constraints (i.e., constraints such as $x + y = z$, $xy = z$, $z \in [\underline{a}, \overline{a}]$, or $z \geq 0$, where x, y, z are variables) by introducing additional auxiliary variables (e.g., decomposing $x + \sin y \geq 0$ to $\sin y = v_1 \wedge x + v_1 = v_2 \wedge v_2 \geq 0$). Then it applies interval arithmetic based algorithms (Hickey *et al.* 1998) for pruning the intervals corresponding to the variables occurring in the primitive constraints to smaller ones that still contain the solution set of these primitive constraints. This is done until a fixpoint is reached.

Usually constraint propagation can only compute an overapproximation of the exact bounds on the solution set of constraints (this is also known as the *dependency problem* of interval arithmetic). For example, in the case $x + y = 0 \wedge x - y = 0$ and a starting interval $[-1, 1]$ for both x and y, these intervals cannot be pruned, although there is only one solution—the origin. Still, for certain cases (e.g., for constraints in which no variable occurs more than once), the result of constraint propagation will be tight. Hence, in this case, if constraint propagation yields non-empty intervals, the constraint has a solution, and we do not need numerical local optimization at all.

However, also in our application interval constraint propagation is not necessarily tight: Assume a 2-dimensional system with just one mode m, whose flow constraint $Flow_m$ is given by $-x_1 - 1 \leq \dot{x_1} \leq -x_1 + 1 \wedge \dot{x_2} = 1$, and assume interval vectors $\boldsymbol{x} = ([0, 0], [0, 0])$ and $\boldsymbol{y} = ([-5, 5], [5, 5])$. Then propagation of (7) wrt. the second conjunct of $Flow_m$ will result in the interval $\tau \in [5, 5]$. Further propagation will yield the interval $[-1.25, 1.25]$ for the first component of $\boldsymbol{y}$. However, computation of the explicit solution of the borderline cases of $Flow_m$ shows that this is an overestimation of approximately 25%. For example, for $\boldsymbol{y} = (1.25, 5)$, the trajectory would have the slope $(0.25, 1)$, and these values do not satisfy $Flow_m$. The problem comes from the fact that the slope $(0.25, 1)$ satisfies $Flow_m$ in combination with another point within the computed bounds, namely the point $\boldsymbol{y} = (-1.25, 5)$. However, due to the enclosure of variables in intervals, the precise depency of specific values for $\boldsymbol{y}$ and a specific slope of the trajectory is lost.

4. IMPLEMENTATION AND EXAMPLES

We integrated a prototype implementation of the method into the verification engine HSOLVER (Ratschan and She 2005, Ratschan and She 2004), which does abstraction refinement by incrementally refining a decomposition of the state space into hyper-rectangles, i.e., by bisecting hyper-rectangles. After each refinement step we apply our method to the current abstraction.

However, this abstraction might have a huge number of abstract counterexamples to check. For avoiding this, we observe that many of them share the same sub-sequence (a *fragment*); if there is no trajectory through a given fragment, then there is also no trajectory through any of the abstract counterexamples containing the fragment. Hence we also check fragments, using a version of the concretization constraint that does not contain the first line (specifying an initial state) or last line (specifying an unsafe state), if the corresponding sequence of abstract states does not start with an initial state, or end in an unsafe state, respectively.

In the current version, we produce fragments by recursively transversing the abstraction from initial to unsafe states and checking the corresponding fragments. If the given fragment cannot be concretized we backtrack, refine the abstraction, and avoid checking the abstract counterexamples containing this fragment.

The current prototype does not yet consider jumps, and instead of doing numerical local optimization (which would be difficult to integrate due to the constraint representation used by HSOLVER) to find a solution to the concretization constraint, it simply iterates taking the midpoint of some interval enclosing the solution set and doing interval constraint propagation using the solver RSOLVER (`http://rsolver.sourceforge.net`). In lucky cases this already results in a solution. Since constraint propagation does not enclose the solution set closely, also the unlucky case that this midpoint is not a solution happens quite often. This would make numerical optimization necessary. However, the fact that already such a prototype can solve several examples shows the strength of the method.

We tried the following examples:

Example 1: A very simple example where trajectories follow circles around the origin:

$$x_2 - 1 \leq \dot{x_1} \leq x_2 + 1$$
$$-x_1 - 1 \leq \dot{x_2} \leq -x_1 + 1$$

The set of initial states is given by $x_1 \leq 0, x_2 \geq 2$, the set of unsafe states by $x_1 \geq 2, x_2 \leq -2.5$, and the state space $[-4,4] \times [-4,4]$. The fact that a counterexample has to flow around a half-circle makes this example non-trivial.

After 10 bisections of the state space, in negligible time, the method finds the counterexample $(0,3), (2,2), (2,2), (2.959,0), (2.322,-2), (2,-2.57)$. The repeated point stems from two different, neighboring abstract states.

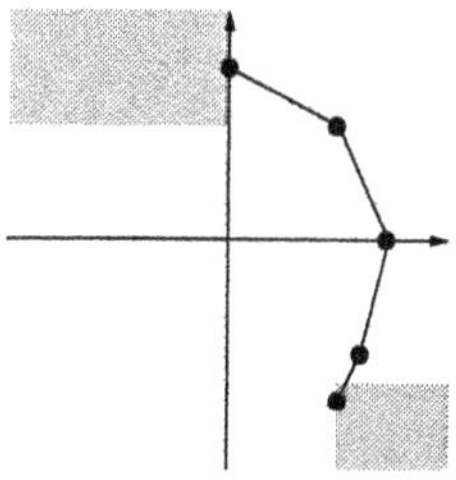

Example 2: A non-deterministic version of an example from (Kapinski *et al.* 2003):

$$x_1 + 3x_2 - 1 \leq \dot{x_1} \leq x_1 + 3x_2 + 1$$
$$x_1 - x_2 - 1 \leq \dot{x_2} \leq x_1 - x_2 + 1$$

The set of initial states is given by $x_1 \leq -3$, the set of unsafe states by $x_1 \geq 1, x_2 \geq 0$, and the state space $[-4,4] \times [-4,4]$. After 75 bisections, in a few seconds, the method finds the counterexample $(-3.5,3.5)$, $(-2.75,2.75)$, $(-2.25,2.288)$, $(-1.75,1.686)$, $(-1.25,1.355)$, $(-0.5,1)$, $(-0.5,1)$, $(-0.5,1)$, $(0.25,0.464)$, $(1,0.440)$.

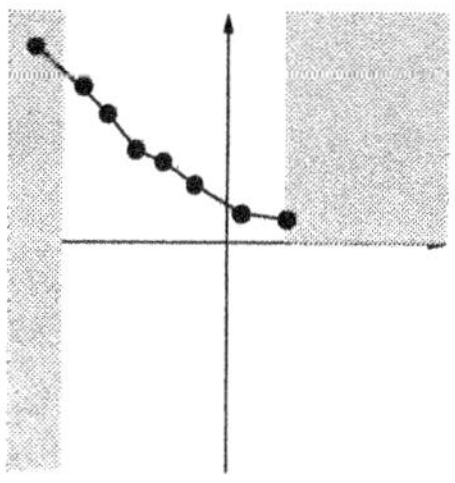

Example 3: A non-deterministic version of another example from (Kapinski *et al.* 2003):

$$x_2 - 1.5 \leq \dot{x_1} \leq x_2 + 1.5$$
$$-8x_1 - 1.5 \leq \dot{x_2} \leq -8x_1 + 1.5$$
$$x_4 - 1.5 \leq \dot{x_3} \leq x_4 + 1.5$$
$$-4x_3 - 1.5 \leq x_4 \leq -4x_3 + 1.5$$

The set of initial states is given by $x_1^2 + x_2^2 + x_3^2 + x_4^2 \leq 4$ the set of unsafe states by $x_1^2 + x_2^2 + x_3^2 + x_4^2 \geq 9$, and the state space by $[-4,4] \times [-4,4] \times [-4,4] \times [-4,4]$. After 16 bisections, but a few minutes of computation, the method finds the counterexample $(2,0,0,0)$, $(2,0,0,0)$, $(1.792,-2.64,-0.155,-0.155)$.

5. RELATED WORK

Apart from manual simulation using according tools (Lee and Zheng 2005, and references therein), the main approach for falsifying safety of dynamical systems is bounded model checking (BMC). This approach is well developed for discrete systems (Biere *et al.* 1999), but only lately receives attention in the hybrid systems cases (Fränzle and Herde 2004, Ábrahám *et al.* 2005). However, the current approaches are limited to dynamics formulated by linear inequalities (i.e., the solution of constant differential inclusions), and there is no systematic way of using bounded model checking for obtaining concrete counterexamples for hybrid systems whose dynamics is given by more interesting (e.g., linear) differential (in)equations. Moreover, BMC might redundantly compute similar information for the same part of the start space, and explores branches that might never lead to an unsafe state. We avoid this by taking advantage of the information generated in the verification algorithm.

For linear discrete-time systems with input, another approach (Kapinski *et al.* 2003) exhaustively simulates the system using a time and state discretization of the input, and tries to avoid redundant computation by merging nearby trajectories. Our approach avoids time discretization by searching for a trajectory between two given points in the state space that may be of unbounded length in time.

Bhatia and Frazzoli (2004) adapt techniques from robotic motion planning to compute an under-approximation of the set of trajectories of a given hybrid system. Prajna and Rantzer (2005) show that—in analogy to Lyanpunov functions—the existence of certain functions implies the reachability of given sets in (non-linear) ODEs. Such functions can be computed using sum-of-squares (SOS) tools (S. Prajna 2002).

Tools for counterexample guided abstraction refinement based on flow pipe computation (Clarke *et al.* 2003) may terminate with a concrete counterexample. They can deal with non-determinism only in the form of parametric systems, where different values for these parameters are tried to search for trajectories that happen to be counterexamples. This fixes the parameter to a certain value between switches. In contrast to that, our approach the trajectory can take *any* solution of the differential constraint, also solutions that might correspond to inputs or disturbances that change between switches.

6. CONCLUSION

We provided a first version of a method for coupling hybrid systems verification algorithms with the ability to find concrete counterexamples for non-deterministic hybrid systems, i.e., with *falsification*. The advantage of the method is that it uses information from system abstractions computed by the verification algorithm to guide the search for a counterexample.

In future work we will provide an efficient implementation of the method based on local optimization, we will try to come up with special, efficient constraint solving algorithms for the special cases of the concretization constraint that we have identified in this paper, we will avoid a full re-check of all the counterexamples after each refinement step by re-using information from earlier checks, and we will provide efficient strategies for the order in which the abstract counterexamples (and their fragments) should be checked. Moreover, we will work on similar methods for hybrid systems without non-determinism.

REFERENCES

Ábrahám, E., B. Becker, F. Klaedtke and M. Steffen (2005). Optimizing bounded model checking for linear hybrid systems. In: *VMCAI* (R. Cousot, Ed.). Vol. 396–412 of *LNCS*. Springer. pp. 396–412.

Alur, R., T. Dang and F. Ivančić (2003). Counter-example guided predicate abstraction of hybrid systems. In: *TACAS* (H. Garavel and J. Hatcliff, Eds.). Vol. 2619 of *LNCS*. Springer. pp. 208–223.

Bhatia, A. and E. Frazzoli (2004). Incremental search methods for reachability analysis of continuous and hybrid systems. In: *HSCC'04* (Rajeev Alur and George J. Pappas, Eds.). number 2993 In: *LNCS*. Springer.

Biere, A., A. Cimatti, E. M. Clarke and Y. Zhu (1999). Symbolic model checking without BDDs. In: *TACAS '99*. Springer. pp. 193–207.

Clarke, E., A. Fehnker, Z. Han, B. Krogh, J. Ouaknine, O. Stursberg and M. Theobald (2003). Abstraction and counterexample-guided refinement in model checking of hybrid systems. *Int. Journal of Foundations of Comp. Science* **14**(4), 583–604.

Cleary, J. G. (1987). Logical arithmetic. *Future Computing Systems* **2**(2), 125–149.

Davis, E. (1987). Constraint propagation with interval labels. *Artif. Intell.* **32**(3), 281–331.

Fränzle, M. and C. Herde (2004). Efficient proof engines for bounded model checking of hybrid systems. In: *FMICS 04, Electr. Notes in Theor. Comp. Sc. (ENTCS)*. Elsevier.

Hickey, T. J., M. H. van Emden and H. Wu (1998). A unified framework for interval constraint and interval arithmetic. In: *CP'98* (M. Maher and J.F. Puget, Eds.). number 1520 In: *LNCS*. Springer. pp. 250–264.

Kapinski, J., B. H. Krogh, O. Maler and O. Stursberg (2003). On systematic simulation of open continuous systems. In: *HSCC'03* (Oded Maler and Amir Pnueli, Eds.). Vol. 2623 of *LNCS*. Springer.

Lee, E. A. and H. Zheng (2005). Operational semantics of hybrid systems. In: Morari and Thiele (2005).

Morari, M. and Thiele, L., Eds. (2005). *Hybrid Systems: Computation and Control*. Vol. 3414 of *LNCS*. Springer.

Prajna, S. and A. Rantzer (2005). Primal-dual tests for safety and reachability. In: Morari and Thiele (2005).

Ratschan, Stefan and Zhikun She (2004). HSolver. `http://hsolver.sourceforge.net`. Software package.

Ratschan, Stefan and Zhikun She (2005). Safety verification of hybrid systems by constraint propagation based abstraction refinement. In: Morari and Thiele (2005).

S. Prajna, A. Papachristodoulou, P. A. Parrilo. (2002). Introducing SOSTOOLS: A general purpose sum of squares programming solver.. In: *CDC'02*.

Tarski, A. (1951). *A Decision Method for Elementary Algebra and Geometry*. Univ. of California Press. Berkeley.

ELSEVIER
IFAC
PUBLICATIONS

AN EVALUATION OF TWO RECENT REACHABILITY ANALYSIS TOOLS FOR HYBRID SYSTEMS

Ibtissem Ben Makhlouf, Stefan Kowalewski *

* *RWTH Aachen University, Chair "Informatik 11" – Embedded Software Laboratory, Ahornstr. 55, 52074 Aachen, Germany*

Abstract: The hybrid systems community is still struggling to provide practically applicable verification tools. Recently, two new tools, PHAVer and Hsolver, were introduced which promise to be a further step in this direction. We evaluate and compare both tools with the help of several benchmark examples. The results show that both have their strengths and weaknesses, and that there still is no all-purpose reachability analysis tool for hybrid systems. *Copyright © 2006 IFAC*

Keywords: Hybrid systems, reachability analysis, Hsolver, PHAVer.

1. INTRODUCTION

Automatic verification of hybrid systems is still a topic of research and far from being established as a standard tool in industry, like, e.g., simulation. To become so, reachability analysis, the core procedure of hybrid systems verification, must be made sufficiently efficient for models of real-world systems, including large, complex and non-linear dynamics.

The first publicly available reachability analysis tool for hybrid systems was HyTech (Henzinger *et al.*, 1997; Henzinger *et al.*, 2000), developed more than a decade ago. It was intended to be a demo implementation of theoretical results and not a prototype of an engineering tool. Consequently, numerical issues were largely neglected which lead to overflow errors even for relatively simple examples. In the following years, however, HyTech became the archetype for numerous attempts to implement reachability tools with better practical applicability, usually extending the basic HyTech algorithm by abstraction techniques.

Recently, the research on reachability analysis of hybrid systems got fresh momentum by the introduction by two different tools, Hsolver (Ratschan and She, 2005) and PHAVer (Frehse, 2005*c*). Both tools build on well tried algorithms while introducing new methodical approaches at the same time. And in both cases, the example computations which were presented in the corresponding publications promised that a new level of practical efficiency has finally been reached.

The aim of this paper is to evaluate the two tools from an independent point of view. For this purpose, we applied both tools to more than ten benchmark examples with several modifications, trying to fathom the boundaries of applicability. An excerpt of the results is presented in the sequel. We concentrate on the examples which revealed qualitative differences rather than quantitative performance differences. Some detailed data including runtime measurements are provided.

The paper is organized as follows. In the next section, we briefly introduce both tools. Section 3 is presenting the experimental results. Numerical results are depicted to make it possible to compare the analysis results to the "real" behavior. Finally, a comparison and some conclusions are given.

2. HSOLVER AND PHAVER

Hsolver is a safety verification tool for non-linear hybrid systems developed by Ratschan and She (Ratschan and She, 2005). The tool is based on interval arithmetics to delimit the trajectories in a piecewise manner to a rectangular grid. The method avoids explicit computation of continuous reachability sets reducing, therefore, rounding errors drastically. To avoid the drawbacks of this approach, namely vast over-approximations, an abstraction refinement framework is developed in which the abstract states are hyper-rectangles in the continuous part of the state space. The abstraction is refined using a splitting strategy, which is optimized by using constraint programming. The splitting method reduces information from possibly incomplete constraint propagation steps in order to avoid new splitting. Jump conditions, initial states, and unsafe states are also described by constraints. Moreover, a pruning function makes it possible to eliminate unreachable sets avoiding inefficient repetitions.

The other tool is PHAVer (Polyhedral Hybrid Automation Verifyer) developed by Frehse (Frehse, 2005*c*) (Frehse, 2005*b*) for verifying safety of linear hybrid automata[1] (LHA). PHAVer uses Hybrid I/O-Automata with affine dynamics (Lynch *et al.*, 2003). Computations use convex polyhedra as the basic data structure as it was already done in HyTech (Henzinger *et al.*, 1997; Henzinger *et al.*, 2000). The implementation is based on the Parma Polyhedra Library (PPL) (Bagnara *et al.*, 2002). The PPL supports closed and non-closed convex polyhedra and infinite precision arithmetic.

Beginning with the initial state, the reachability strategy of PHAVer computes reachable sets by using refinement of locations when affine dynamics are over-approximated by LHA-dynamics. The refinement can be adjusted by parameters. To do so, the constraints are prioritized according to the refinement parameters. The tool offers the possibility to enforce other prioritization criteria. For more details see (Frehse, 2005*a*). Over-approximation using bounding boxes and convex hull abstractions are also possible. The user can combine the different types of over-approximation and control the number of iterations for each type with special parameters given by the tools. PHAVer supports compositional and assume-guarantee reasoning (Frehse *et al.*, 2004).

2.1 Reachability analysis strategy of Hsolver

In Hsolver, a hybrid system is described by:

- a set of modes $s_1, ..., s_n$ each corresponding to a different continuous behavior of the system,
- variables $x_1, ..., x_k$ ranging over closed real intervals $I_1, ..., I_k$ and
- constraints defining the flow in each mode, the jump, the initial and safe conditions.

A hybrid system is then defined as a tuple $(Flow, Jump, Init, UnSafe)$ with:

- the constraint $Flow \subseteq \Phi \times \mathbb{R}^k$ in which Φ is the state space $\{s_1, ..., s_n\} \times I_1 \times ... \times I_k$, and $\mathbb{R}^k$ represents $\{\dot{x}_1, ..., \dot{x}_k\}$,
- the constraint $Jump \subseteq \Phi \times \Phi$, defined over the variables $x_1, ..., x_k$ of the start mode s and the variables $x'_1, ..., x'_k$ of the target mode s', and
- the constraints $Init \subseteq \Phi$ und $UnSafe \subseteq \Phi$ describing the set of the initial and unsafe states.

The original hybrid system described above will be abstracted by a discrete system defined over a new variable state space S.
A discrete system over S is defined as a tuple (Trans, Init, UnSafe) where $Trans \subseteq S \times S$, $Init \subseteq S$ and $UnSafe \subseteq S$ are new constraints deduced from original constraints via the definition of an abstraction relation between the original and the abstracted system as follows:

Definition: An abstraction relation between a hybrid system $(Flow_1, Jump_1, Init_1, UnSafe_1)$ and a discrete system $(Trans_2, Init_2, UnSafe_2)$ over S is a relation $\alpha \subseteq \Phi \times S$ such that:

- for all $q \in \Phi$, if there is a trajectory from an element of $Init_1$ to q according to $Flow_1$ and $Jump_1$, then for all q_α with $\alpha(q, q_\alpha)$ there is a trajectory from an element of $Init_2$ to q_α according to $trans_2$,
- for all $q \in Init_1$, there is a $q_\alpha \in Init_2$, with $\alpha(q, q_\alpha)$ and
- for all $q \in UnSafe_1$, if q is reachable from $Init_1$, then there is a $q_\alpha \in UnSafe_2$, with $\alpha(q, q_\alpha)$.

Hence, Hsolver computes for every hybrid system with a description D an abstraction $Abstract_D(\mathcal{B})$ over sets of abstract states $\mathcal{B}$ containing all elements of the state space reachable from the initial set such that all boxes corresponding to the same mode are non-overlapping. The resulting abstracted system is refined, during the computation, with another reduced abstracted system under the condition that an abstraction relation between the two systems exists, which preserve the reachable states during the transformation. The refinement strategy is repeated until a fixpoint is attained. Using strategies based on constraint propagation programming (Mackworth and Freud, 1985) parts of state space showing not to be reachable are excluded during the ab-

[1] In the sense that the continuous dynamics in each discrete location are linear.

straction process. For this purpose, a constraint $reachable_{B'}(s', z)$ is formulated expressing the conditions under which a point in a box B is reachable. Before that, the flow within a box B and a mode s is described with the constraint $flow_B(s, y, z)$. It expresses the fact that a point $y = (y_1, ..., y_k) \in B$ is reachable from a point $x = (x_1, ..., x_k) \in B$ via a flow in B and s.

The constraint $reachable_{B'}(s', z)$ is defined as the disjunction of three principle constraints:

- Constraints formulating the reachability from the initial set $initflow_B(s, z)$ expressed for a mode s, a box B and a point $z \in B$ reachable from the initial set via a flow in s and B.
- Constraints $jumpflow_{B,B'}(s, s', z)$ describing the reachability from a jump between two modes s and s' corresponding to two boxes B and B' and a point $z \in B'$.
- Constraints $boundaryflow_{B,B'}(s', z)$ expressing the reachability of a point $z \in B'$ from a common point of neighboring boxes B and B' via a flow in B.

After the determination of the reachability constraint $reachable_{B'}(s', z)$, the problem of reachability is transformed in a constraint satisfaction problem. It is solved by constraint propagation programming, applying several techniques for making the analysis more efficient like, e.g., pruning.

2.2 Reachability analysis strategy of PHAVer

PHAVer is a verification framework for linear hybrid systems and compositional reasoning. The automata model differentiates between state, input and output variables. The output can be declared as the inputs for an other automaton. A hybrid I/O-automaton (HIOA) in PHAVer is a tuple $H = (Loc, Var_S, Var_I, Var_O, Lab, \rightarrow, Act, Inv, Init)$ where Loc, Lab, Act, Inv, and $Init$ are defined as usual, and where:

- Var_S, Var_I and $Var_O \subseteq Var_S$ are finite and disjoint sets of state, input and output variables.
- $\rightarrow \subseteq Loc \times Lab \times 2^{V(Var) \times V(Var)} \times Loc$ is a finite set of discrete transitions, where $Var = Var_S \cup Var_I$ and $V(Var)$ denotes the set of evaluations over Var. A transition $(l, a, \mu, l') \in \rightarrow$ means $l \xrightarrow{a,\mu} l'$.

PHAVer deals with I/O-automata with affine dynamics given for each location loc by a conjunction of constraints of the form:

$$a_i^T \dot{x} + \hat{a}_i^T x \bowtie_i b_i, \qquad (1)$$
$$a_i, \hat{a}_i \in \mathbb{Z}^n, \bowtie_i \in \{<, \leq, =\}, i = 1, ..., m.$$

Using linear programming, these constraints are first overapproximated conservatively with linear constraints of the form $\alpha_i^T \dot{x} \bowtie_i \beta_i$, $\alpha_i \in \mathbb{Z}^n$, $\beta_i \in \mathbb{Z}$ which are obtained using the result of an infimum computation of (1) inside $Inv(loc)$ denoted (when it exits) with:

$$p \backslash q = \inf_{x \in Inv(loc)} \alpha_i^T x, \quad p, q \in \mathbb{Z}.$$

The linear constraint on $\dot{x}$ is then given by $\alpha_i = qa_i$ and $\beta_i = qb_i - p$.
This overapproximation introduces a loss of accuracy depending on the size of the location and the angular spread of the derivative vectors in the location. To improve the accuracy, locations are recursively split along hyperplanes which parameters are adequately chosen from the user to minimize the partitions. The splitting of a location is stopped when a minimum size is attained. Otherwise, the splitting process can include a refinement procedure of a location controlled from the user. The user provides a list of candidate normal vectors $a_{h,i}$ of hyperplanes defined by equations having this form $a_h^T x = b_h$ and the minimum and maximum slack $\Delta_{min,h}$, $\Delta_{max,h}$ that the hyperplanes will have in the refined location. The slack is here defined by:

$$\Delta(a_h) = \max_{x \in Inv(loc)} a_h^T x - \min_{x \in Inv(loc)} a_h^T x$$

The candidate hyperplanes are prioritized according to a user controlled list of criteria. For each constraint of type $a^T x \bowtie b$ in a location loc a set of evaluations included in the reachable states N of the location are associated according to a chosen criterium from the following list (Frehse, 2005*c*):

- Prioritize constraints according to their slack.
- Prioritize constraints that have reachable states only on once side.
- Prioritize constraints according to the spread of the derivatives where constraints are discarded if a minimum spread is reached and the slack is smaller than $\Delta_{max,h}$.
- Prioritize constraints according to the derivative spread after the constraint is applied.

Moreover, the number of constraints during the computation is reduced in order to save memory. To select the constraints to be eliminated, PHAVer uses for instance an angle criterion for which the negative cosine of the closest angle between the normal vector of the ith constraint and all others is measured and then compared. Only a predefined number of constraints with the greatest angles are retained. In addition, a procedure for limiting the number of bits necessary to represent a constraints is executed as soon as a given threshold z is exceeded. A new constraint $\alpha_i^T x \bowtie_i \beta_i$ including the original ith constraint $a_i^T x \bowtie_i b_i$ of a polyhedron is computed so that $|\alpha_{i,j}|$, $|\beta_i| \leq 2^{z+1} - 1$. The procedure makes an estimation of the scaling error f. It recomputes β_i using linear programming. If β_i has more than z bits, f will be decreased and the procedure starts again.

3. EXPERIMENTAL RESULTS

3.1 Benchmarks

In order to assess the performance of the tools, we applied them to benchmark examples from the the literature, including examples from (Ratschan and She, 2004). The computations were performed on a PC with a Pentium IV processor at 3,06 GHz and 1 GB of memory. The obtained results summarized in table 1 allow a first comparison between the two tools. A brief description of the studied examples is given here:

Example 1: The two tanks problem. It is a nonlinear problem, formulated in (Ratschan and She, 2005) as follows:

$$\textit{Flow}: \left(s = 1 \rightarrow \begin{pmatrix} \dot{h}_1 \\ \dot{h}_2 \end{pmatrix} = \begin{pmatrix} 1 - \sqrt{h_1} \\ \sqrt{h_1} - \sqrt{h_2} \end{pmatrix}\right) \wedge \left(s = 2 \rightarrow \begin{pmatrix} \dot{h}_1 \\ \dot{h}_2 \end{pmatrix} = \begin{pmatrix} 1 - \sqrt{h_1 - h_2 + 1} \\ \sqrt{h_1 - h_2 + 1} - \sqrt{h_2} \end{pmatrix}\right)$$

Jump: $(s = 1 \wedge 0.99 \leq h_2 \leq 1) \rightarrow (s' = 2 \wedge h_1' = h_1 \wedge h_2' = 1)$

Init: $s = 1 \wedge (h_1 - 5.5)^2 + (h_2 - 0.25)^2 \leq 0.0625$

Unsafe: $(s = 1 \wedge (h_1 - 4.5)^2 + (h_2 - 0.25)^2 < 0.0625)$

State space: $(1, [4,6] \times [0,1]) \cup (2, [4,6] \times [1,2])$

Example 2: The navigation benchmark problem is a linear problem from (Fehnker and Ivancic, 2004). An object at a position $x = (x_1, x_2)^T$ moves within a 3×3 map with a velocity v determined by the differential equation $\dot{v} = A(v - v_d)$, where $A = \begin{pmatrix} -1.2 & 0.1 \\ 0.1 & -1.2 \end{pmatrix}$ and $v_d = (sin(i \cdot \pi/4), cos(i \cdot \pi/4))^T$ is a desired velocity adopting different values depending on the chosen map (i is equal to the value in each cell of the map). We take two different maps:

$$map_1 = \begin{bmatrix} \mathcal{B} & 2 & 4 \\ 2 & 3 & 4 \\ 2 & 2 & \mathcal{A} \end{bmatrix} \text{ and } map_2 = \begin{bmatrix} \mathcal{B} & 2 & 4 \\ 2 & 2 & 4 \\ 1 & 1 & \mathcal{A} \end{bmatrix}.$$

$\mathcal{A}$ and $\mathcal{B}$ correspond to the reachable and forbidden cells. For map_1, we choose $x_0 \in [0,1] \times [0,1]$, $v_0 \in [-0.1, 0.5] \times [-0.05, 0.25]$ as initialization. However, for map_2 the computation begins with $x_0 \in [0,1] \times [0,1]$, $v_0 \in [-0.1, 0.5] \times [-0.05, 0.25]$ as initial sets.

Example 3: The 1-flow problem used in (Preußig *et al.*, 1998) is a simple example without jumps described as follows:

Flow: $\dot{x}_1 = \dot{x}_2 = \dot{x}_3 = 1$

Init: $0 \leq x_1 \leq 2 \wedge 1 < x_2 \leq 2 \wedge 0 < x_3 < 1$

Unsafe: $0 \leq x_1 \leq 2 \wedge 1 < x_2 \leq 2 \wedge 0 < x_3 < 1$

State space: $[0,2] \times [0,2] \times [0,4]$

Example 4: The collision example from a part of the car convoi control problem proposed in (Puri and Varaiya, 1995). It has been transformed in (Ratschan and She, 2004). The problem has no jumps and can be described as follows:

Flow: $\dot{x}_1 = x_3 - x_2$, $\dot{x}_2 = x_4$, $-2 \leq \dot{x}_3 \leq -0.5$, $\dot{x}_4 = -3 \cdot x_4 - 3 \cdot (x_2 - x_3) + (x_1 - x_2 - 10)$

Init: $x_1 = 1$, $x_2 = 2$, $x_3 = 2$, $x_4 = -0.5$

Unsafe: $x_1 \leq 0$

State space: $[0,4] \times [0,2] \times [0,2] \times [-2, -0.5]$

Another version of this problem denoted by *convoi1* is also taken from (Ratschan and She, 2004). The main difference is the dynamics of $\dot{x}_3 = -4x_1 + 3x_2 - 3x_3 + x_4$.

Example 5: A heating example of 3 rooms with 2 heaters from (Fehnker and Ivancic, 2004):

$$\dot{x} = \begin{pmatrix} -0.9 & 0.5 & 0 \\ 0.5 & -1.3 & 0.5 \\ 0 & 0.5 & -0.9 \end{pmatrix} x + \begin{pmatrix} 0.4 \\ 0.3 \\ 0.4 \end{pmatrix} u + diag(6,7,8)h$$

where x is the vector of temperatures, h is indicating whether a heater is on or off and $u = 4$. We performed computations with the minimal temperature threshold equal to 14 and a space state given by $[14,22] \times [14,22] \times [14,22]$.

3.2 Hsolver

Hsolver permits the verifications of nonlinear system from the outset. Difficulties appear, however,when arithmetic operations different than $*, +,$' *sin*, *cos*, *exp* appear, as in the two tank example which involves a *sqrt* operation. In this case, we must transform the equations in order to avoid this operation. The computation results for the different examples are given in table 1. We also present some graphical results of the two tanks and the navigation benchmark examples, in order to understand better the function of the tools. For the two tanks example, in Fig. 1 we see the pruned region in light grey boxes, the 10 boxes corresponding to mode 1 are dark grey and the 12 boxes of mode 2 are medium grey. The trajectory is obtained from a Matlab simulation. The figure shows that the unsafe region (circle left) is in the pruned region. This means, that this region was already excluded from the reachable set at the beginning of the computation which accelerates the termination of the verification process, as we expected. For the navigation benchmark, the computation time for the two instances was very long and the tool could not conclude about the safety in both cases (resulting output "safety unknown"). The problem behind that is illustrated by Fig. 2. It represent the different box decompositions in the three modes of the systems together with one trajectory from Matlab simulations. The issue is that the whole state space needed to be examined during the analysis without ever being able to

	Hsolver						PHAVer		
Example	refinement steps	box recomputation	# calling prune	safety	time(s)	memory(KB)	safety	time	memory
2-tanks	11	13	397	safe	0.33	1548	safe	0.11	1688
map1	362	511	509306	unsafe	2647.75	48048	safe	139.02	126608
map2	342	107	889110	unsafe	4457.7	46892	safe	22.14	54168
flow	1	0	2	safe	0.12	1136	safe	0	868
convoi	367	443	68675	safe	9462	7200	safe	0.10	148
convoi1	0	1	3	safe	0.31	6204	safe	0.11	1652
heating	142	191	469416	unsafe	638.76	37744	safe	0.41	5720

Table 1. Implementation results of the benchmark examples.

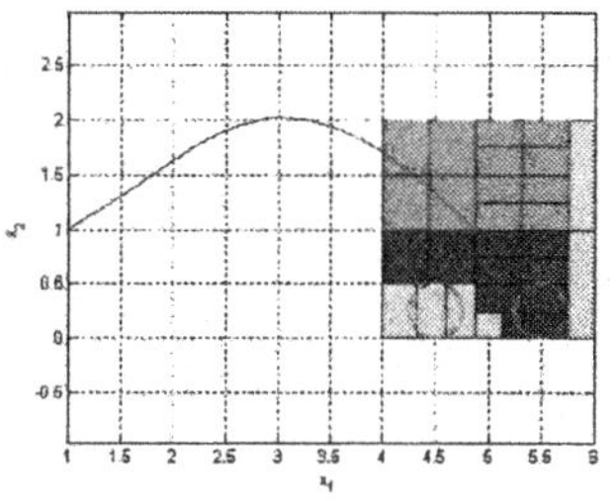

Fig. 1. Hsolver: two tanks problem, initial region: $(x_1 - 5.5)^2 + (x_2 - 0.25)^2 < 0.065$, unsafe region $(x_1 - 4.5)^2 + (x_2 - 0.25)^2 < 0.065$

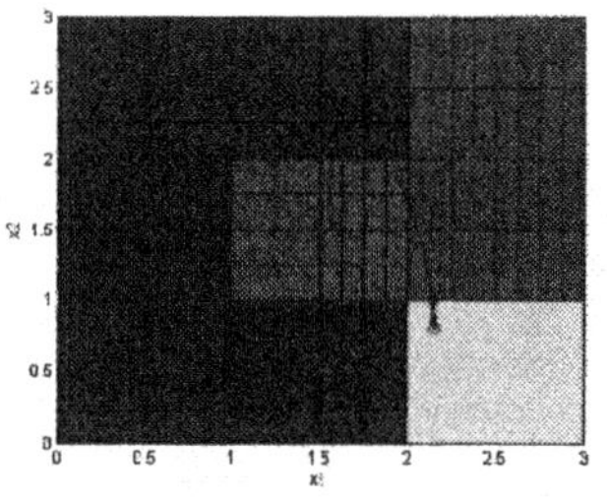

Fig. 2. Hsolver:Navigation benchmark, $map = [B24, 234, 22A]$, $x_0 \in [2,3] \times [1,2]$, $v_0 \in [-0.3, 0.3] \times [-0.3, 0]$

exclude safely the unsafe region. Such a behavior can not be predicted by a standard user.

3.3 PHAVer

Nonlinear problems must be linearized to be analyzable by PHAVer. For the two tanks example, we use the Jacobi method to linearize the system for each box using the center of the boxes as reference point. Of course, this is not an abstraction, and any reachability results for the linearized system will be inconclusive with respect to safety. However, it allows us to gather data for the comparison. We chose two different unsafe sets for the verification. From Matlab simulations we know that the system is safe for the first case and unsafe for the second case. The results of our verification with PHAVer confirm these findings. Fig.3 shows the results for the unsafe case.

The compositional feature of PHAVer is illustrated with the navigation benchmark examples. Here, the automaton is decomposed in two automata, one for the position and another for the

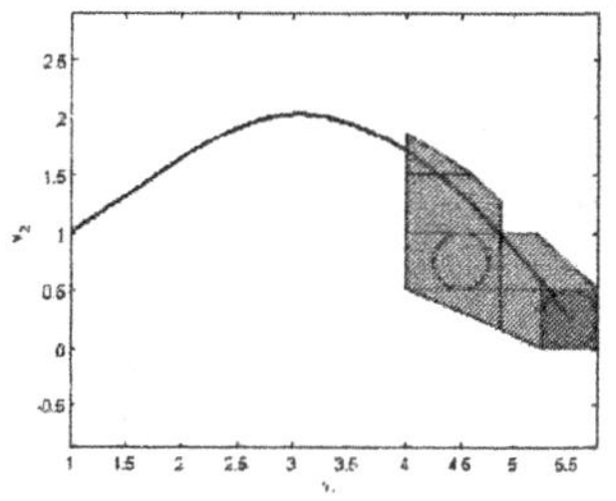

Fig. 3. PHAVer: Two tanks problem, initial region $[5.25, 5.75] \times [0, 0.5]$, unsafe region $[4.25, 4.75] \times [0.5, 1]$

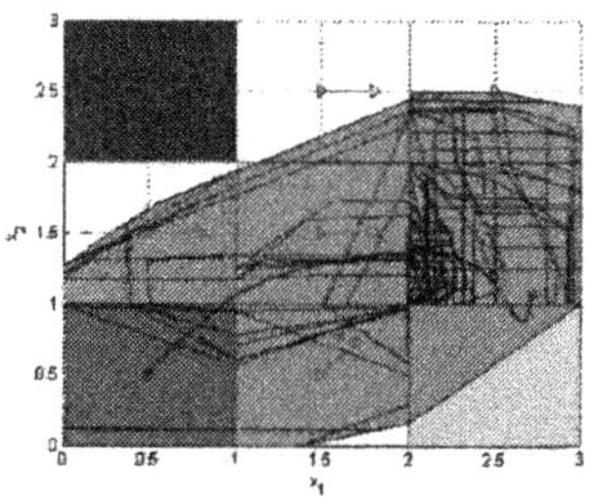

Fig. 4. PHAVer: Navigation benchmark, $map = [B24, 224, 11A]$, $x_0 \in [0,1] \times [0,1]$, $v_0 \in [-0.1, 0.5] \times [-0.05, 0.25]$

velocity. Both automata are synchronized with global labels. The refinement is controlled with the factor $d1min = d2min = 0.25$ and $d1max = d2max = \infty$. For map_1 with the default polyhedra abstraction, the computation time of the verification lies in acceptable range. Using this method for the second map, we get a very long computation time. For such cases PHAVer offers a further feature which permits the combination of different types of abstractions. In our example, we chose a convex hull abstraction during 20 iterations followed by the default abstraction with bounding box iterations until the 39th iteration. We took two different maps: $map1 = [B24, 234, 22A]$ and $map2 = [B24, 224, 11A]$. For the second map the termination is accelerated by the convex hull abstractions for the 20 first iterations, then the processing is switched to normal reachability with bounding box iterations until the 39th iteration. The results for the second map are shown in Fig. 4. The examples are proved to be safe. The figures show the polyhedra decomposition of a restricted region of the position state space. The other examples have been implemented as simple automata because the decomposition is not necessary in this case. The corresponding results are shown in Table 1.

3.4 Comparison of the results of both tools

Comparing the time and the memory requirements in table 1, we can conclude that for all our examples PHAVer is faster than Hsolver. From the memory consumption point of view, we note that, against our expectation, the amount of memory used by Hsolver in examples 3, 2, 4 and 5 is greater than that used by PHAVer. When the safety is unknown, the verification analysis by Hsolver lasts a long time. In the case of non linear systems, we are not able to conclude about the safety of this type of systems because we have no estimation of the linearization error. In summary, Hsolver is an efficient and fast tool, as long as its results are conclusive. In these cases, it is well applicable to non-linear hybrid systems. However, the navigation and heating benchmarks show that the tool still has some limitation. On the other hand, Phaver is an efficient tool for verifying linear hybrid systems. It offers possibilities to control and choose refinement and over-approximation strategies used during computation with a number of functions, commands and parameters. Complicated hybrid systems could be decomposed or simplified and treated by Compositional Reasoning provided by the tool. This could be used to verify non-linear systems, if it would be possible to abstract the original problem by linear hybrid automata. Otherwise, we have to linearize the system and take the linearization error into account. Finally, we could not confirm the conjecture that the use of polyhedra and refinement steps will increase strongly the necessary amount of memory space.

4. CONCLUSIONS

Not quite surprisingly, the application results show that both tools have their strengths and weaknesses. Hsolver is extremely fast for examples where safety can be proven or disproved, but it shows excessive computation times when the results are inconclusive. Since it is able to treat non-linear dynamics directly, it has a clear advantage over PHAVer in this respect since the latter is restricted to affine hybrid systems. PHAVer, on the other side, provided more exact results in the sense that it was able to prove safety correctly for examples for which Hsolver gave the result 'safety unknown'.

Both tools represent a good step in the direction of making reachability analysis applicable for practioneers. However, considering that the benchmarks still were more academic than real world examples and that, nevertheless, there were examples for which computation times exceeded hours or the result was unnecessarily inconclusive, it seems fair to say that the goal has not been reached yet.

REFERENCES

Bagnara, R., E. Ricci, E. Zaffanella and P.M. Hill (2002). Possibly not closed convex polyhedra and the parma polyhedra library. In: *Static Analysis: Proc. Int. Symp., LNCS 2477*. Springer. pp. 213–229.

Fehnker, A. and F. Ivancic (2004). Benchmarks for hybrid systems verification. In: *HSCC'04, LNCS 2993*. Springer. pp. 326–341.

Frehse, G. (2005*a*). Compositional Verification of Hybrid Systems using Simulation Relations. PhD thesis. Cornell University.

Frehse, G. (2005*b*). Phaver. http://www.andrew.cmu.edu/~gfrehse. Software package.

Frehse, G. (2005*c*). PHAVer: Algorithmic verification of hybrid systems past HyTech. In: *HSCC, LNCS 3414*. Springer. pp. 258–273.

Frehse, G., Z. Han and B.H. Krogh (2004). Assume-guarantee reasoning for hybrid I/O-automata by over-approximation of continuous interaction. In: *Proc. CDC'04*. Bahamas.

Henzinger, T.A., B. Horowitz, R. Majumdar and H. Wong-Toi (2000). Beyond HyTech: Hybrid systems analysis using interval numerical methods. In: *HSCC'00, LNCS 1790*. Springer. pp. 130–144.

Henzinger, T.A., P.S. Ho and H. Wong-Toi (1997). Hytech: A model checker for hybrid systems. *Soft. Tools Techn. Transf.* **1**(1,2), 110–122.

Lynch, N., R. Segala and F. Vaandrager (2003). Hybrid i/o automata. *Inf. Comput.* **185**(1), 105–157.

Mackworth, A. K. and E.C. Freud (1985). The complexity of some polynomial network consistency algorithms for constraint satisfaction problems. *Artificial Intelligence* **25**, 65–74.

Preußig, J., S. Kowalewski, H. Wong-Toi and T.A. Henzinger (1998). An algorithm for the approximative analysis of rectangular automata. In: *Proc. FTRTFT'98, LNCS 1486*. Springer. pp. 228–240.

Puri, A. and P. Varaiya (1995). Driving safely in smart cars. In: *Proc. of the 1995 American Control Conference*. pp. 3597–3599.

Ratschan, S. and Z. She (2004). Hsolver. http://www.mpi-sb.mpg.de/ratschan/hsolver. Software package.

Ratschan, S. and Z. She (2005). Safety verification of hybrid systems by constraint propagation based abstraction refinement. In: *HSCC 2005, LNCS 3414*. Springer. pp. 573–589.

ELSEVIER
IFAC
PUBLICATIONS

SAFETY AND RELIABILITY ANALYSIS OF PROTECTION SYSTEMS FOR POWER SYSTEMS

Luca Ferrarini *, Leonardo Ambrosi * and Emanuele Ciapessoni **

** Politecnico di Milano, P.za L. da Vinci 32, 20133 Milan, Italy*
*** CESI RICERCA, Via Rubattino 54, 20134 Milan, Italy*

Abstract: This paper addresses the problem of risk analysis of protection systems and protection scheme of transmission grid. According to IEC 61508, a hybrid model has been developed supporting the analysis of protection systems. The use of analytic models in risk analysis of the elements of system protection scheme allows evaluating the security level associated to different protection strategies and supports the identification of specific ICT criticalities. *Copyright © 2006 IFAC*

Keywords: power systems, simulation, safety, standards, object-modelling technique, hybrid systems

1. INTRODUCTION

Like many other industrial and production systems, the field of production and transmission of energy is facing a trend towards deregulation, in Italy as well as in many European countries. This basically amounts to move from a single-operator, clearly regulated, easily predictable system to a competitive, uncertain, multi-operator system. The introduction of competitive supply and the organizational separation imposed by deregulation has resulted in highly stressed operating conditions and more vulnerable networks. Extremely important in this scenario is the definition of "quality" criteria, to be obeyed by market operators and vendors, in order to be able to positively cooperate to provide a correct service.

Strangely enough, in a market-oriented energy production and management, the "classic" deterministic criteria, adopted and useful in the regulated market, are still adopted to guarantee the security of the system. These are based on the fact that each abnormal operating condition contained in a suitable "contingency set" satisfies predefined performance criteria.

This allows to pragmatically design protection systems without considering all combinations of system configurations and operating conditions, which would be simply unfeasible. The deterministic approach, and the associated simulation methods, provides a simple rule to make decisions: optimize economy within the constraints of the secure operational region. This simplicity has made the deterministic method very attractive and useful in the past. Though effective in practice, the deterministic approach tends to focus on the most severe and credible event, thus producing an oversized and a less agile protection system, which reduces marginal gains and return of investments. Among others, the main deficiencies consist in not taking into account the frequency and the impact of events, and thus the likelihood of unwanted or catastrophic events, and in the negligence of non-limiting events.

This classical approach can be superseded by probabilistic risk based approach: several studies (Dobson, 2004b), (Lucarella et al., 2004), (McCalley and Vittal, 2001) demonstrated the possible gain associated to the use of risk based approach in the analysis and management of electric system security. Accordingly a CIGRE report (Marceau and Endrenyi, 1997) recommended to study probabilistic security assessment methods, and the CIGRE task force 38.02.21 is working on this recommendation. The Electric Power Research Institute (EPRI) was also involved in efforts to develop probabilistic risk assessment methods and tools for risk based security assessment (McCalley and Vittal, 2001). Following the risk-based perspective in the reliability analysis, the international standard IEC 61508 is the main normative reference. However, power systems exhibit a special behavior with respect to the safety and reliability problems, so that the standard needs to be integrated.

In this paper we pursue this approach by advocating the application of risk analysis techniques in the life cycle of protection system and protection scheme of the transmission grid. In particular, the goal is here to obtain a quantitative model in order to evaluate and compare protection systems and protection strategies able to improve the overall reliability of a system. Yu and Singh (Yu and Singh, 2004) followed this approach trying to define a simulation-based method to quantitatively evaluate hybrid stochastic models of power systems. That model was based on stationary continuous models (the so called "load flow model"), and discrete part is a logic model of the operating modes of a suitable combination of more physical components. The work here discussed extends that approach in these facts:

- instead only stationary continuous models, dynamic models are used (DAEs)
- a full modular approach has been adopted, implemented into an object-oriented hybrid dynamic simulator (Modelica/Dymola): any model of the system can be rewritten with more or with less detail without changing anything else of the system , provided that its physical interface is not changed of course;
- the discrete parts have been greatly improved, by separating components
- each discrete model of the physical component has a higher number of states, associated to physical phenomena and not to modes of operation.

The paper first summarize the relevant features of the power systems (Sect. 2) and then discusses the construction of the modular hybrid model in Sect. 3. Sect. 4 hints to the implementation into the simulation environment chosen.

2. CHARACTERISTICS OF POWER SYSTEM

2.1 Interconnections of subsystems

The substantial peculiarity of power systems consists in its complexity, wide area nature and strong interconnection of different subsystems. The system itself is composed of a high number of components belonging to a few types (*lines*, *bars*, *transformers*, *breakers*), but all these elements are electrically interconnected with variable topologies from zone to zone and country to country. This makes that anomalous behaviours of some components have unexpected consequences on others, which are located far away and which operate also in different conditions (Bie And Wang, 2001). Moreover the electrical connection makes the propagation of faults extremely fast.

Under these conditions, the risk assessment of the transmission system becomes a complex matter. While in fact the components are standard, their specific way of working while interconnected in real operating conditions are much more difficult and more uncertain to estimate in design phase.

Therefore, the concept itself of Safety Integrity Level (SIL) defined in the IEC 61508 standard must be revised, being insufficient if applied to single components, and being useless if applied to the entire system.

2.2 Undesired trips

The IEC 61508 basically focuses on the fact that protection systems may fail to intervene when a fault occurs in the system. On the contrary, one of the major causes of outages is constituted by undesired trips, often leading to cascade tripping. These are characterized by the fact that the protection system actually does intervene when it is not requested to do so, which constitutes a possible cause of cascading outages (Bie And Wang, 2001). Notice that the failure does not derive from a specific fault of a device, but from "external" factors to the protection, like the impossibility to measure the real current operating state (Padke, 2002), (Yu and Singh, 2004).

In order to consider these facts, it is necessary to define a common usage of reliability terms. We will use the *dependability* as the main property of a protection system. It concerns the ability to work properly and it is defined as the combination of *reliability*, the ability of a protection system to trip for a fault inside the protection zone, and *security*, the ability to refrain from tripping for fault outside the protection zone.

2.3 Dynamic constraints

This is one of the most important aspects of power systems, aspect that is ignored completely in the IEC standards. It consists in the fact that in many cases those limits that some process variables must satisfy to be in a "safe" condition can vary dynamically in time. This means that a system condition can be estimated as "safe" or not according to other operating conditions of the whole system.

Clearly, it is necessary to take this behaviour into account, both in the design phase of the protection systems and in its evaluation. In any case, it is necessary to develop models somehow adapt to the operating condition of the system.

This allows carrying out a much finer "dynamical" analysis than the classic static analysis of the system, which in turn allows to develop a definitely more precise protection system, on the base of specific risk-based indices of the transmission grid (Makarov, and Hardiman, 2003).

3. A HYBRID MODULAR MODEL FOR TRANSMISSION SYSTEM

One way to evaluate quantitatively the functional safety of the power system is the development of a model capable to capture both continuous dynamics of the system, its probabilistic nature, and its event-driven phenomena.

To this purpose, a suitable modular hybrid model (continuous and discrete event based), sketched in Fig. 1, has been developed. In that figure, the main elements (*lines*, *bars*, *transformers*, *breakers*) of the transmission grid are shown, with their own interfaces and connections.

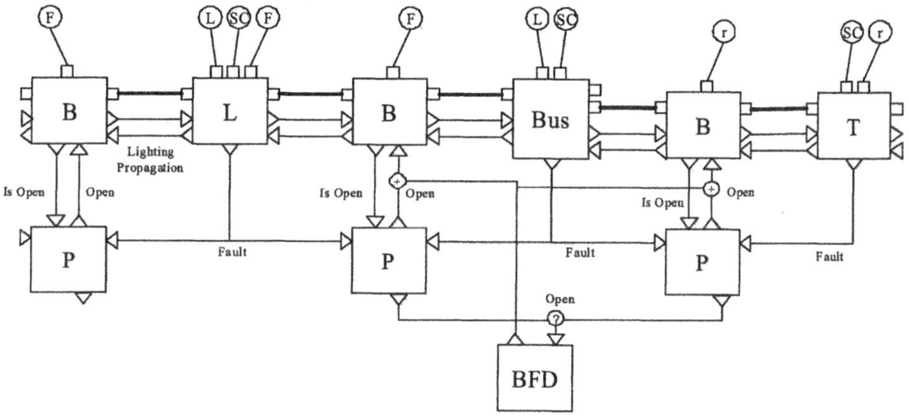

Fig. 1. Sketch of the hybrid model of the transmission systems

As it can be seen in the Fig. 1, each component has some connection with the neighbor. The connections painted with a thick line and a square interface represent the physical connections, i.e. the electrical connections regulated by the classical electrical equations. The connections painted with a thin line and a rectangular interface represent the logical connections, i.e. the way the discrete part of the model "inform" the others about their internal states, dispatching events.
Each module is internally described in a dual way, it has a discrete model and a continuous model that are able to communicate together as we'll see later.

The basic modeling criteria are the following.

- Modularity
It is advisable to have a component-wise modeling and simulation environment, where the user instantiates modules that directly correspond to physical components, instead of writing equations. This means that each module should be thought as the generic behavior of a component under generic constraints; the overall model will come out of the assembly of the basic components, and the user should not intervene in the assembling of equations.

- Hybrid behavior
It's necessary, since the system under investigation shows continuous-time behavior for electrical phenomena, and event-based behavior for discontinuous phenomena like breaks and faults (clearly enough, the modeling of "fast" phenomena as discontinuous is a modeling simplification, if one does not need to go into the atomic details of electricity transmission).
To do so, we implemented the discrete part of a model with Petri net, a well-known formal technique to represent discrete-event systems. Ordinary Petri net models have been extended, with a simple semantic rule, in order to generate outputs and receiving inputs with the continuous part.
The continuous part of the model is quite simply, each element is represented with the classics electrical components (impedance, inductance, switch, and so on), electrically connected with the continuous part of the other models.

- Discrete-Continuous connection.
Some of the transitions of the Petri net model have been endowed with a time delay, stochastically distributed as proposed in the GSPN formalism (Ajmone et al. 1995). This means that a stochastic transition is associated with a value representing the mean time to fire, once enabled, generally represented with the symbol λ_i.

Notice that some of those λ_i are exogenous (i.e. represent physical phenomena that have external causes with respect to the model itself, like the falling of a tree or thunder on the system), but other are not. A typical example relates to the cascade effect. Consider for example the rate of failures of an electrical component. This depends on the instantaneous current circulating in it, as hinted in Fig. 2.

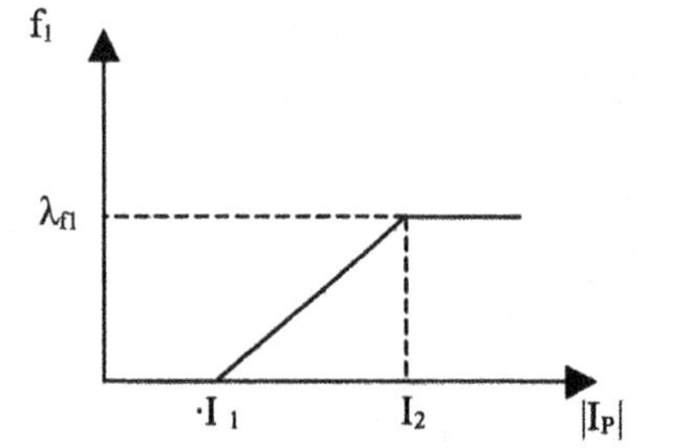

Fig. 2. Fault rate dependence from the circulating current (an example) for an electric device

From the modeling and simulation point of view, this means that the fault rate of a (stochastic) transition of the Petri net model of the discrete-event part of the model of a component depends on the value computed in the continuous time part of the model of the same component. Here clearly arise the connection between the discrete and the continuous part of the model.

This is clearly not a trivial step, since again the continuous part depends on the discrete one, and since the occurrence of future events can not be pre-calculated since other events may occur in the meanwhile.

In fact, if an electrical part should break, the circulating currents and the voltage calculated by the continuous part should change, possibly influencing again in the future the discrete part.

In the sequel, some details are provided on the modeling of the basic components of the power transmission system.

3.1 Line

The line is the basic transmission element for power. The continuous model takes into account the fact that a line can be interrupted (by a breaker or a fault), can be short-circuited (a fault, or an external cause, like a falling tree for example) or can be hit by thunders. The modelling of the lightning effect on lines has been simplified to a logic behaviour (a switch), being the physical modelling out of the scope of this project. Thus, the basic continuous behaviour is sketched in Fig. 3, where Z_{L1}, Z_{L2}, Z_{cc} are impedances.

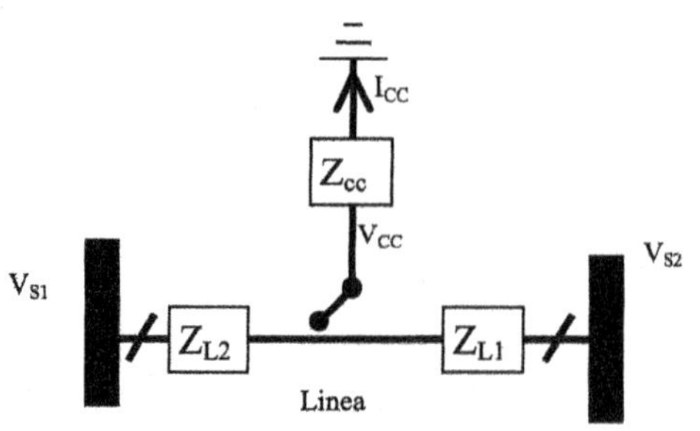

Fig 3.– Basic continuous behaviour of the line

The discrete time part describes all the other behaviour of the net (see Fig. 4).

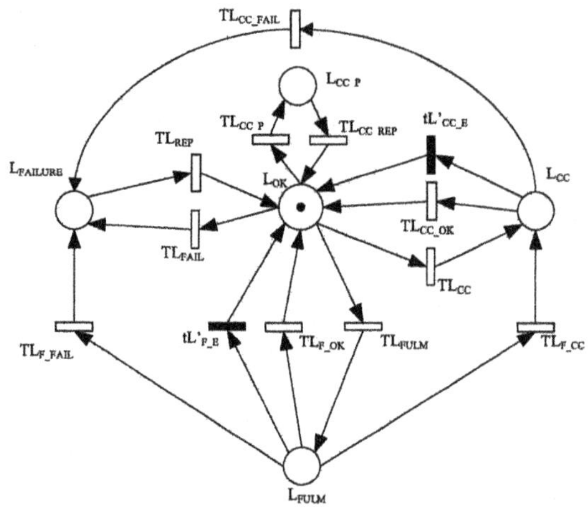

Fig. 4. Petri net model of a line

The meaning of the places and transitions is hinted in the following.

Places:

- L_{OK} = line is healthy.
- L_{CC} = line is short-circuited
- L_{CC_P} = line is permanently short-circuited
- L_{FULM} = line is lightened
- $L_{FAILURE}$ = line is faulty

Transitions:

- TL_{FULM} = the line is struck by a lightning
- TL_{F_OK} = the lightning is extinguished autonomously and the line returns in OK state. (L_{OK})
- tL'_{F_E} = the lightning is extinguished by the intervention of protections,. This immediate transition is conditioned by $|I_{LINE}|<\varepsilon$
- TL_{CC} = the line from the state of OK goes to short-circuit with ground
- TL_{CC_OK} = the short-circuit is autonomously extinguished
- tL'_{CC_E} = the short-circuit is extinguished by the intervention of protections. This transition is conditioned by $|V_{LINE}| <\varepsilon$

- TL_{CC_P} = the line goes from OK state to permanent short-circuit
- TL_{CC_REP} = line is repaired to eliminate the permanent short-circuit
- TL_{FAIL} = failure of a line (normally caused by an object)
- TL_{REP} = line is restored to OK state
- TL_{F_FAIL} = failure of a line caused by a lightning. The effect of the lightning is extinguished
- TL_{F_CC} = the line goes to short-circuit because of a lightning. The energy of the lightning is considered discharged to the ground.
- TL_{CC_FAIL} = the line fails because of a short-circuit (normally caused by the breaking of an insulator)

The overall behaviour of the Petri net model can now be deduced quite straightforwardly.

Finally, the interaction of the two sub-models. It is clear that, if the line change its internal state, the topological view of the net can change, and consequently the electrical calculus. Besides the change of the value of the electrical variables, make change the λ regulating the fire rate of the discrete transition. It is quite obvious, for example, that if an external event occurs, like a short-circuit caused by a falling tree, then the fictitious switch shunting the line to ground is closed and at the same time a transition in the Petri net part is enabled.

2.2 Breaker

The breaker has been modelled with a similar approach. It has four main logical states: open, close, stuck closed and stuck open. The transition from one state to another one are quite simple and be deduced with an easy spot of the name of the transition. The Petri net model is sketched in Fig. 5.

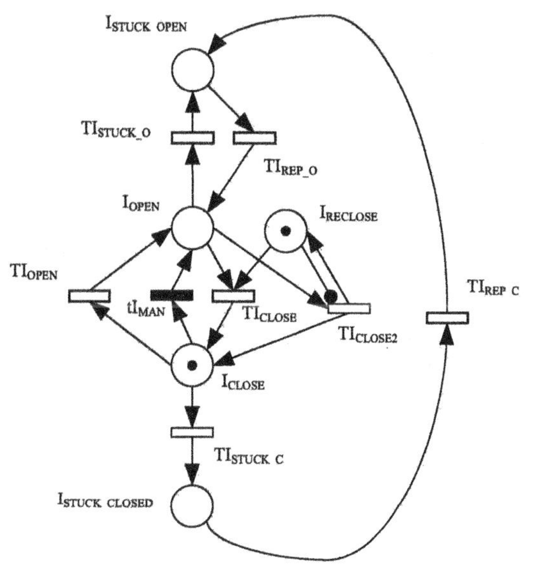

Fig. 5. Petri net model of a breaker.

As it can be argued, also the automatic re-close operation has been modelled. Actually, after an open operation, the breaker can be closed (modelled with a stochastic transition). Should the breaker be opened again, then it can be closed again only with transition Tclose2, only once: such a transition will be conditioned to the fact that the lines attached to it are in "functioning" state.

Transitions:

- tI_{MAN} = opening of the breaker triggered by manual command or by backup devices (Breaker Failure Device)
- TI_{OPEN} = opening of the breaker triggered by the associated protection
- TI_{CLOSE} = fast reclosure of the breaker
- TI'_{CLOSE2} = reclosure of the breaker conditioned by "OK state" of the connected elements
- TI_{STUCK_C} = failure of the breaker from closed to stuck close
- TI_{STUCK_O} = failure of the breaker from open to stuck open
- TI_{REP_O} = reparation of the breaker and restoration in state of stuck open
- TI_{REP_C} = time interval between the failure of the breaker and the moment in which the failure is recognized. The breaker is then opened and restored (TI_{REP_O} transition).

2.3 Bar and transformer

Similarly, also bars and transformers have been modelled. Details are here omitted for the sake of simplicity.

4. SIMULATION

In order to evaluate quantitatively the above model, a suitable simulation environment is to be used. The simulation engine should be accurate enough to deal with continuous-time dynamics, event-based dynamics, and stochastic behaviours. Our choice is the Modelica/Dymola environment. It's an object-oriented modelling and simulation framework, with an easy-to-comprehend component-oriented description, with different physical and communication ports, endowed with a fine symbolic manipulation of equations, which allows both the user to "describe" the equations directly in the continuous time domain and to obtain an optimised simulation code. It's a time-based simulator, which means that the discrete-event part must be suitably treated for efficient and effective implementation, and similarly there is no support for stochastic behaviour, which must then be explicitly introduced.

Currently, the basic models of line, bar, transformer, breaker, generator and load are under final testing. Such specific and reusable components have been modelled as hybrid systems, whose internal behaviour is given by a suitable combination of continuous dynamic sub-module with a discrete-event stochastic sub-module.

In particular, to do so, the stochastic Petri net sub-module has been directly implemented in Modelica, exploiting the capability of using "external functions". Special care has been paid to the modelling of switching behaviours (due to e.g. breakers or faults) and to the mutual interaction between the continuous and the discrete part of each component, and the "transmission" of events between components.

The description of the modelling choices is matter of future works, along with the application to the IEEE Reliability Test System prepared by the Reliability Test System Task Force of the Application of Probability Methods to Power Systems.

5. CONCLUSION AND FUTURE WORK

The application of IEC 61508 to the life cycle of E/E/EP systems for the protection of transmission grid, depends on the identification of appropriate risk analysis techniques, able to manage the complexity and wide area nature of the grid, and supporting the identification of the required security level. To this aim, the paper proposes a hybrid model, based on Generalised Stochastic Petri Net integrated with continuous simulations approach, supporting the risk analysis and evaluation of different protection strategies, on the base of specific risk-based indices.

The paper describes the hybrid model, taking into account both the continuous time dynamics of the power system itself, but also the discrete dynamics involved by disrupt phenomena due to breaks, outages, natural phenomena.

The use of analytic models in SIL analysis of the element of system protection scheme allows to evaluate the security level associated to different protection strategies and to support the identification of specific criticalities of protection system.

Acknowledgement. This work has been supported by MAP (Italian Ministry for Productive Activities) in the frame of Public Interest Energy Research Project named "*Ricerca di Sistema*" (MAP Decree of 28 Feb. 2003).

REFERENCES

Ajmone, M.Marsam, G. Balbo, G. Conte, S. Donatelli and G. Franceschinis (1995). *Modelling with Generalized Stochastic Petri Nets*, Wiley Series in Parallel Computing, John Wiley and Sons.

Bie, Z., and X. Wang (2002). Evaluation of power system cascading outages. *IEEE Power System Technology Intl. Conference..* **Vol. 1**, 13-17 Oct. 2002, p. 415 – 419.

Marceau, R.J., and J. Endrenyi (1997). "*Power System Security Assessment: A Position Paper*". CIGRE Task Force 38.03.12. Electra, **No. 175**, pp. 48-78.

Dobson, I., B. A. Carreras, V. E. Lynch and D. E. Newman (2004a). Complex Systems Analysis of Series of Blackouts: Cascading Failure, Criticality, and Self-organization; *Bulk Power System Dynamics and Control*, Cortina d'Ampezzo, Italy.

Dobson, I., B. A. Carreras and D. E. Newman (2004b). A criticality approach to monitoring cascading failure risk and failure propagation in transmission systems; "*Electricity transmission in deregulated markets" Conference at Carnegie Mellon university*.

Grigg, C. et al. (1999). The IEEE Reliability Test System-1996: a report prepared by the Reliability Test System Task Force of the Application of Probability Methods Subcommittee Power Systems. *IEEE Transactions on Power Systems*, **Vol. 14**, Issue 3, pages 1010 – 1020.

Lucarella, D., M. Pozzi, M. Valisi and G. Vimercati (2004). Un approccio basato sull'analisi di rischio per l'esercizio in sicurezza del sistema elettrico; Italian National Congress on the estimation and management of risk in civil and industrial sites; Pisa, Italy.

Makarov, Y.V., and R.C. Hardiman (2003). On Risk-based Indices for Transmission Systems. *Proc. IEEE PES Annual Meeting*, Toronto, Ontario, Canada, July 13-17, 2003

McCalley, J. and V. Vittal (2001). Risk Based Security Assessment, EPRI Project WO8604-01, 2001, final report.

Padke, A. (2002). Hidden failures in protection systems. *Power systems and communications infrastructures for the future*, International conference, Beijing, 2002.

Yu, X., and C. Singh (2004). A Practical Approach for Integrated Power System Vulnerability Analysis With Protection Failures, *IEEE Transaction on power systems*, **vol. 19**, no. 4, pages 1811 – 1820.

ELSEVIER
IFAC
PUBLICATIONS

A HYBRID APPROACH FOR SAFETY ANALYSIS OF AIRCRAFT SYSTEMS

E. Villani[+], P. E. Miyagi*

[+] *Instituto Tecnológico de Aeronáutica, Department of Mechanical Engineering*
12228-900 São José dos Campos, São Paulo, BRAZIL
** Escola Politécnica, University of São Paulo*
Av. Prof. Mello Moraes, 2231 CEP 05508-900 São Paulo, BRAZIL
e-mail: evillani@ita.br, pemiyagi@usp.br

Abstract: This paper introduces the use of a hybrid modelling and simulation approach for the analysis of safety issues in aircraft systems. Traditionally, safety analysis in aircraft industry is performed without considering the system dynamics. In this paper the dynamics of the aircraft components are modelled using Petri nets and differential equations. Faults are incorporated in the model using probabilistic distributions functions. The reliability of the system under fault is then estimated by simulation. The approach is applied to the landing system of a military aircraft in order to compare two different control strategies for detecting and processing faults. *Copyright © 2006 IFAC*

Keywords: Petri nets, hybrid systems, probabilistic behaviour, safety analysis.

1. INTRODUCTION

Safeness is one of the major concerns in the design of aircraft systems. Most of the aircraft components are provided with redundancy. The degree of redundancy of each component depends on a number of factors such as the kind and probability of fault, per hour of flight. Usually, in order to estimate how safe an aircraft system is, the probabilities of fault in each one of the system components are combined in a static approach using methods such as fault-tree analysis. These methods derive the total probability of fault in the system, which must be within pre-defined limits. The maximum allowed probability of fault depends on how the system deteriorates the level of flight quality and how it affects the aircraft operation and functionality (Stevens; Lewis, 1992).

An important limitation of the methods current under usage in the aeronautic industry is that they do not consider the system dynamics when analysing safeness. When the aircraft behaviour is taken into account, more precise and detailed information can be obtained. Among the issues to be investigated are:

- How a component fault affects the aircraft system behaviour?
- How does it influence the probability of fault in other components?
- How to estimate the probability of critical scenarios that combine a set of component faults?
- How does redundancy affect the system behaviour?
- How to estimate the probability of a wrong diagnostic and its consequences?

In this context, this paper proposes the use of hybrid system modelling and simulation techniques for analysing how the aircraft system dynamics may affect or influence the occurrence, detection and diagnosis of faults. The problem of fault modelling has already been approached by a number of works in many domains. However, most of them consider the problem either from a discrete (Ait-Ameur et al, 2003; Lundqvist, Asplund, 2003) or continuous (Matsuura et al, 2005) point of view. The use of hybrid simulation for the analysis of aircraft safeness is discussed in Pritchett et al (2000).

A hybrid approach is necessary when the system under analysis mixes both continuous and discrete behaviour (Alla, David, 2004). A number of aircraft systems are typically classified as hybrid. They incorporate continuous dynamics such as the continuous positioning of surfaces or the pressure evolution in a hydraulic system, as well as discrete sequence of events, such as switching between components in the case of fault, or executing the command sequences for landing and take-off.

The hybrid modelling formalism considered in this work is the Object-Oriented Differential Predicate Transition nets (OO-DPT net). It combines Petri net for the discrete part and differential equation systems for the continuous one. The object-oriented paradigm is incorporated in order to achieve modularity. The problem of design and analysing aircraft systems using DPT nets has already been presented before (Villani, Miyagi, Valette, 2003). However, previous works have assumed that the system is operating under nominal conditions, i.e., without modelling equipment redundancy and the possibility of fault occurrences. In this paper, faults are incorporated into the models by using probabilistic distribution functions.

The landing system of a military aircraft is presented as an example. Particularly, the approach is used to compare two different control strategies that combine the output of a set of redundant sensors in order to detect fault and avoid dangerous states.

This paper is organized as following. Section 2 describes the proposed approach and the modelling formalism. Section 3 describes the example and Section 4 presents some conclusion.

2. THE APPROACH FOR SAFETY ANALYSIS

2.1 Overview

The proposed approach for safety analysis is divided into the following steps: Step1 - System modelling, Step 2 - Specification of safety requirements, Step 3 – Simulation.

The approach is illustrated in Section 3 using as an example a landing system. Basically, in Step 1, the behaviour of the system under analysis is modelled using the OO-DPT net (presented in Section 2.2). For this purpose the designers must specify the list of all possible faults in each component. A fault may be a discrete event, such as an ON/OFF sensor is blocked in the OFF position. Or faults may also be a continuous activity, such as a leakage in the hydraulic system. In this case the fault is characterized by a set of parameters associated to continuous variables, such as the amount of the leakage per time unit. In the OO-DPT nets faults are modelled by probabilistic distribution functions, which are explained in Section 2.2. The model built in Step 1 must include not only the aircraft components and equipment but also the automatic control system. The strategies used for fault detection, diagnostic and treatment are also modelled as OO-DPT objects.

Step 2 consists of specifying the safety requirements. The system designers must determine what the critical situations are and how is the interaction with the pilot in the case fault. Particular emphasis must be given to the analysis of situations that may result in a wrong diagnostic made either by the automatic control system or the pilot.

Finally, in Step 3 the critical situations are translated into properties that the system must fulfil. An example is to establish a maximum probability to be in a critical state. The requirements are then analysed using Monte Carlo simulation.

2.2 The OO-DPT net

The OO-DPT net is the modelling formalism adopted in this work. It has been introduced in (Villani, Miyagi, Valette, 2005) and it is based on the incorporation of object-oriented (OO) concepts to the Differential Predicate-Transition (DPT) net, proposed in (Champagnat et al, 1998). The OO paradigm assures that an aircraft system can be specified by combining the model of the system components, such as sensors and actuators.

According to the definition of OO-DPT net, the model of a system is composed of the a set of objects organized in 'n' classes (C_1, C_2, ..., C_n). Each class C_i is modelled by a DPT net, which defines an interface between differential equation systems and Petri net elements. Its main features are:

- Each object $O_{w.i}$ of the class C_i is represented by a token or a set of tokens in the DPT net of C_i.
- A set of variables (X_i) is associated with each object of the class C_i: they correspond to the attributes of the class. They are divided in X_{co_i}, X_{int_i}, X_{pb_i} and X_{im_i}. X_{co_i} are constant parameters and do not change their value during the object lifetime. X_{int_i} are internal variables and other objects cannot access their value. X_{pb_i} are public variables, other objects can read their value. X_{im_i} are image variables, they are public variables of other objects that are read by $O_{w.i}$.
- A differential equation system (F_{j_i}) is associated with each place (p_{j_i}): it defines the dynamic of a sub-set of X_i according to the time (θ), when a token of $O_{w.i}$ is in p_{j_i}.
- An enabling function (e_{j_i}) is associated with each transition (t_{j_i}): it triggers the firing of the enabled transitions according to the value of X_i.
- A junction function (j_{j_i}) is associated with each transition (t_{j_i}): it defines the value x_i associated with the tokens of the output places of t_{j_i} after the transition firing.

The communication among objects can be discrete or continuous. The continuous interactions are modelled by sharing the continuous variables among objects (e.g. $O_{w.i}$ and $O_{z.v}$). The value of the shared variables is determined by one object ($O_{w.i}$), where it is defined as a public variable (X_{pb_i}), and can be used in the junction function, the equation systems or the enabling function of another object ($O_{z.v}$), where it is defined as an image variable (X_{im_z}). The discrete interactions are method calls. Each class offers methods that are associated with its transitions and that can be requested by other classes. A method call is modelled as the fusion of two transitions: the transition t_{j_i} of the class C_i that offers the method and the transition t_{w_v} of the class C_v that calls the method. The method call happens when both transitions are enabled in their classes.

In order to model the occurrence of faults, probabilistic behaviour must be introduced into the OO-DPT net. The problem of modelling uncertainty in hybrid system has already been approached in many works of the literature (e.g. Pola et al. (2003)). Among the formalisms that model the discrete dynamics using Petri net is the Fluid Stochastic Petri net (Horton et al, 1996) and Dynamically Coloured Petri (Everdij, Blom, 2005). The introduction of uncertainty into models that merge Petri net and differential equation system is briefly approached in (Khalfaoui, 2003) and is based on Generalized Stochastic Petri net. In the proposed approach, probabilistic behaviour is introduced in junction functions. In this case, a probabilistic distribution (PD) is used to define the value of a variable after the firing of a transition. Each time the transition fires a new value is attributed to the variable according to the probabilistic distribution.

As an example, Figure 2 presents the OO-DPT net of the class C_1 – *Hydraulic Cylinder*, which models the behaviour of the cylinder illustrated in Figure 1.

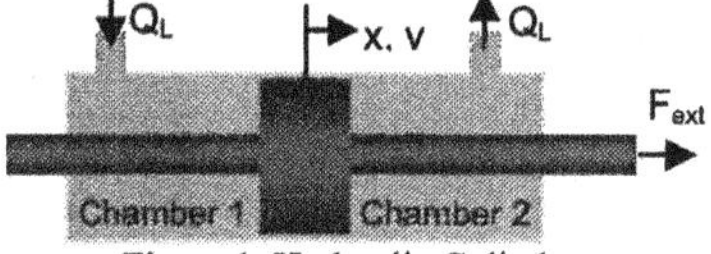

Figure 1. Hydraulic Cylinder.

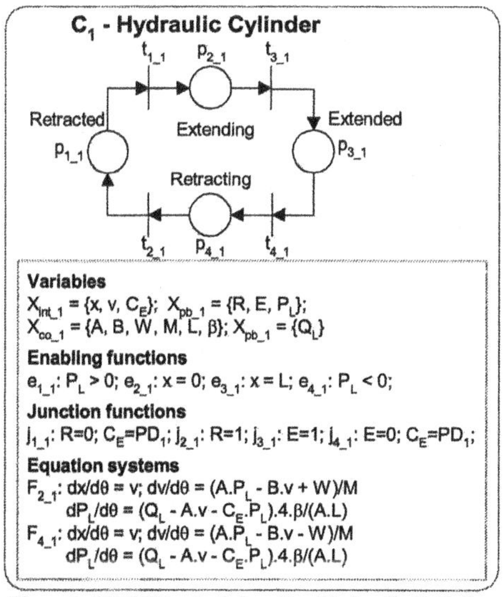

Figure 2. OO-DPT net of class C_1 – Hydraulic Cylinder.

The hydraulic cylinder can be completely extended (p_{3_1}) or retracted (p_{1_1}), or can be extending (p_{2_1}) or retracting (p_{4_1}). When moving, the dynamics of the cylinder is set according to a differential equation system (F_{2_1} and F_{4_1}) composed by 3 equations. The time is represented as 'θ'. The first equation relates the speed of the piston with its position. The second equation is the balance of the forces acting on the piston. The third equation is the balance of mass in Chambers 1 and 2 and it includes the effects of the fluid compressibility on the cylinder dynamics.
The class variables are:

- x, v – position and speed of the piston (x=0 when the piston is completely retracted);
- Q_L – amount of fluid entering Chamber 1 and leaving Chamber 2.
- R, E – auxiliary variables that indicate if the piston is completely extended or retracted;
- C_E – leakage coefficient;
- P_L - difference between the pressure in Chamber 1 and that in Chamber 2;
- V_{PL} – derivative of P_L;
- A, M – area and mass of the piston;
- B – damper coefficient;
- F_{ext} – external force;
- L – maximum value of x;
- β – compressibility coefficient of the fluid;

The communication of the class C_1 - *Hydraulic Cylinder* with other classes is made by sharing variables R, E and Q_L, and by reading variable P_L and V_{PL} from the class C_3 – *Electro-valve* (presented in Section 3).

The OO-DPT net of class C_1 models a fault as a certain amount of hydraulic fluid leakage. The pressure P_L and a leakage coefficient C_E, which is set according to the probabilistic distribution PD_1, determine the amount of leakage. The probabilistic distribution combines the probability of having different kinds of fault with the probability of having a certain amount of leakage when a certain fault occurs. A qualitative example of PD_1 is presented in Figure 3. The low values of C_E are the normal amount of leakage (when no fault occurs), the values on the centre of the graphic are related to problems in the cylinder seal, the high values of C_E are related to structural faults such as cracks in the actuating cylinder.

The next section presents the example considered in this paper: the landing system of a military aircraft.

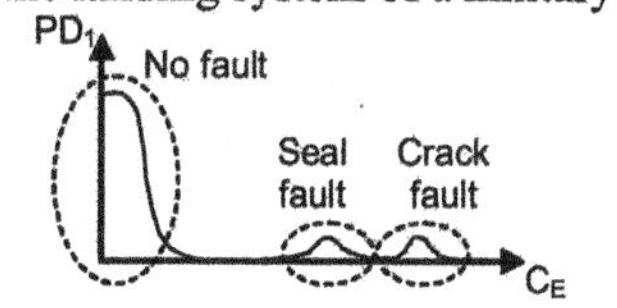

Figure 3. Probabilistic Distribution PD_1 of class C_1 – Hydraulic Cylinder.

3. THE EXAMPLE

The case-study considered in this paper is the landing-system of a military aircraft. It is composed of 3 landing sets (called A, B and C) containing each one a door and a landing-gear. The sequence that must be performed at landing consists of opening the doors of the 3 landing-gear compartments, extending the landing-gears and closing the doors. A similar sequence must be performed at take-off. The landing-gear and door movement is performed by a set of actuating hydraulic cylinders. For each door, a hydraulic cylinder opens and closes the door. For each landing gear, a hydraulic cylinder extends and retracts the landing gear. The hydraulic cylinders are moved by electro-valves. Figure 4 illustrated the hydraulic circuit for the doors.

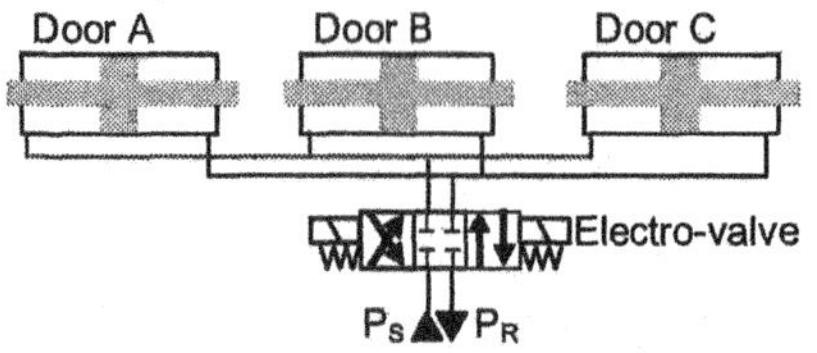

Figure 4. Door hydraulic circuit.

Discrete sensors inform the control system about the positions of the actuating cylinders, the pressure in the hydraulic system, among others. Each discrete sensor is provided with redundancy 3. Sensors signal are used for coordinating the landing system movements during landing and take-off. Furthermore, during cruise and other flight phases, they are constantly read and processed with a certain frequency in order to monitor the landing system and detect problems. In the case of fault the pilot is notified.

The purpose of this example is to analyze how the sensor redundancy augments the system safety and what is the best strategy for combining the signal from the redundant sensors. Because of the limited space, only the door sensors are considered in this paper. Each door has two different kinds of sensor, one indicating if the door is completely open (DO sensor) and one indicating if it is completely closed (DC sensor). Each of them has redundancy 3. In order to unambiguously refer to a sensor, the notation DO_{xy} is used to refer to the DO sensor 'x' (1, 2 or 3) of the door 'y' (A, B or C). Similarly, DC_{xy} is used to identify the DC sensors. The same configuration is adopted for the landing-gears.

Usually, the sensor outputs are processed in 4 levels. In Level 1, the outputs of three redundant sensors (e.g. DO_{1A}, DO_{2A} and DO_{3A}) are compared among them and a combined sensor output is generated. In Level 2, for each door the combined sensor output of door open (DO sensors) is compared with the combined sensor output of door closed (DC sensors) and a door output is provided. It indicates the current state of the door (open, closed, moving or fault). The output of Level 2 is also used in timing monitoring functions. When an order to open or close the door is emitted, the operation must be complete within a time interval. Otherwise, a fault is detected and informed to the pilot. The same approach is performed for each landing-gear. In Level 3, for each landing set, the landing-gear output is compared with the door output and a landing-set output is generated. At this level, faults are detected in situations such as if the door is closed and the landing-gear is moving. Finally, Level 4 compares the three landing-set outputs of Level 3 and a landing-system output indicating the current state of the landing system.

In this example, two different control strategies are considered for processing the sensors at Level 1 and Level 2. In the Level 1 of Strategy 1, if three or two output signals are 'ON', the combined sensor output is 'ON', otherwise the output is 'OFF'. Level 2 of Strategy 1 adopts Table 2 for defining its output.

Table 2 – Rules for Level 2 – Strategy 1.

Level 1 - DO	*Level 1 - DC*	***Level 2***
ON	OFF	*Open*
OFF	ON	*Closed*
OFF	OFF	*Moving*
ON	ON	*Fault*

In the case of Strategy 2, the following rules are considered for Level 1:

- If all the three output signals are 'ON', the combined sensor output is 'ON'. If all the three output signals are 'OFF', the combined sensor output is 'OFF'.
- If two output signals are 'OFF', and one is 'ON', the combined sensor output is 'OFF' and the identity of the sensor with output 'ON' is memorized. If on the next time the sensors are read, the output of this sensor is still different from the other two, this sensor is considered as fault, and from this moment on it is ignored by the control system.
- A similar approach is executed when two output signals are 'ON', and one is 'OFF'.
- If one sensor has been eliminated and the other two are 'OFF', the combined sensor output is 'OFF'. If one sensor has been eliminated and the other two are 'ON', the combined sensor output is 'ON'.
- If one sensor has been eliminated and the other two sensor signals are different from each other, the combined sensor output remains unchanged and an error is memorized. If on the next time the sensors are read, the two outputs are still different, the combined sensor output is fault.

The rules for defining the Level 2 output for Strategy 2 are presented in Table 3.

Table 3 – Rules for Level 2 – Strategy 2.

Level 1 - DO	*Level 1 - DC*	***Level 2***
ON	ON	*Fault*
ON	OFF	*Open*
ON	Fault	*Open*
OFF	ON	*Closed*
OFF	OFF	*Moving*
OFF	Fault	*Fault*
Fault	ON	*Closed*
Fault	OFF	*Fault*
Fault	Fault	*Fault*

This is typically a hybrid problem because an abnormal output of the system can be the result of a sensor fault or of a leakage in a hydraulic cylinder (continuous variable CE of Figure 2). In the second case, the leakage affects the cylinder dynamics, slowing it down or even reverting movement direction if the weight of the landing gear is against it. In this situation the control system must detect a fault and inform the pilot, which can then activate a backup system for the movement of the landing gear. The detection can be affected by a sensor fault and depends on the strategies at Levels 1 and 2.

3.1 System Modelling

Once that the focus of this example is on the processing of door sensor signals, the model of the system is limited to the doors hydraulic cylinders, electro-valve, sensors and controller. Similar objects are used to model the landing-gear hydraulic system.

The system model is composed of a set of 6 classes: *C_1 - Hydraulic Cylinder, C_2 - Discrete Sensor,* C_3 - *Hydraulic Electro-valve, C_4 – Level_1 Processor, C_5 – Level_2 Processor, C_6 - Door Sequence Controller.* The first 3 classes model the behavior of the physical components, while C_4, C_5 and C_6 model the control system. The model of class C_1 has already been presented in Section 3. There are three objects of class C_1: $O_{1.1}$ – Door A, $O_{2.1}$ - Door B, $O_{3.1}$ – Door C.

Model of Class C_2 – Discrete Sensor
Under normal operation, the discrete sensor is either ON (p_{2_2}) or OFF (p_{1_2}). The switching between ON and OFF takes place according to the value of the image variable S_{in}, which corresponds either to the public variable E or R of class *C_1 – Hydraulic Circuit*. However, each time the sensor must switch, a fault may occur and the sensor may go to states 'Blocked ON' (p_{7_2}) or 'Blocked OFF' (p_{6_2}). The fault corresponds to the firing of t_{4_2} or t_{5_2}, and happens with a probability of $(1\text{-}P_{OFF})$ and $(1\text{-}P_{ON})$, respectively. The probabilistic distribution PD_2 sets a random value between 0 and 1 to the variable 'rd', which determines the occurrence of a fault. There are 18 objects of class C_2: $O_{1.2}$ – DO_{1A}, $O_{2.2}$ – DO_{2A}, $O_{3.2}$ – DO_{3A}, $O_{4.2}$ – DO_{1B}, $O_{5.2}$ – DO_{2B}, $O_{6.2}$ – DO_{3B}, $O_{7.2}$ – DO_{1C}, $O_{8.2}$ – DO_{2C}, $O_{9.2}$ – DO_{3C}, $O_{10.2}$ – DC_{1A}, $O_{11.2}$ – DC_{2A}, $O_{12.2}$ – DC_{3A}, $O_{13.2}$ – DC_{1B}, $O_{14.2}$ – DC_{2B}, $O_{15.2}$ – DC_{3B}, $O_{16.2}$ – DC_{1C}, $O_{17.2}$ – DC_{2C}, $O_{18.2}$ – DC_{3C}.

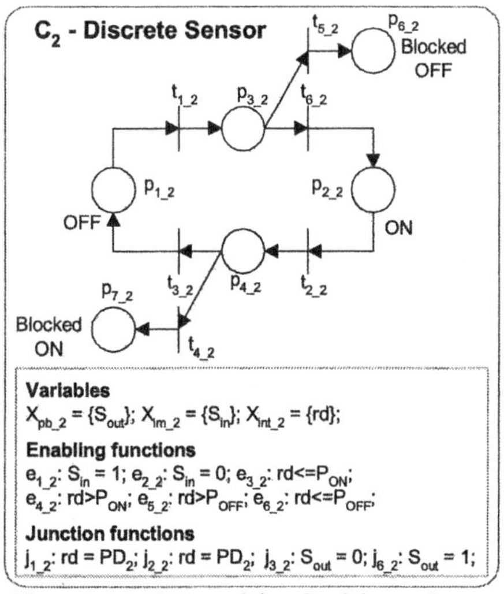

Figure 5. OO-DPT net of class C_2 – Discrete Sensor

Model of Class C_3 – Hydraulic Electro-Valve
This class relates the pressure on the hydraulic circuit (P_L) with the position of the electro-valve (Figure 4). The OO-DPT net is presented in Figure 6. The equation that defines the pressure PL combines the amount of fluid through the valve and the decreasing of pressure when passing through a restriction. No fault is considered for the electro-valve. The class variables are:

- V_{PL} – derivative of the pressure PL;
- Q_T – total amount of fluid to the door cylinders.
- Q_{LA}, Q_{LB}, Q_{LC} – amount of fluid entering Cylinders A, B and C.
- K_1, K_2 – constants of the electro-valve;
- P_S, P_R – Supply and return pressures of the hydraulic system.

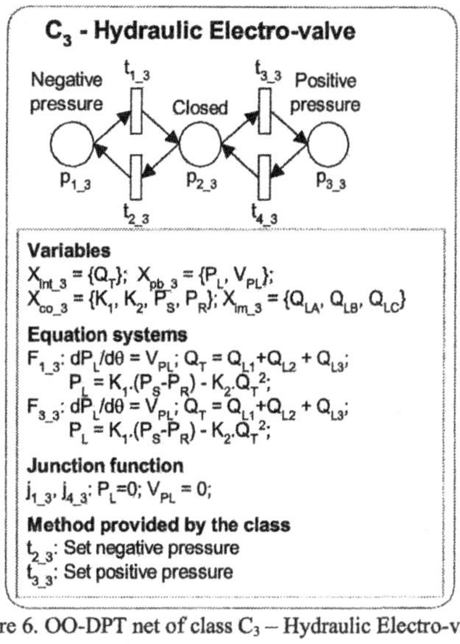

Figure 6. OO-DPT net of class C_3 – Hydraulic Electro-valve.

Model of Class C_4 –Level_1 Processor
This class represents the processing of sensor outputs at Level 1 and has mainly a discrete dynamics. It receives the output of 3 redundant sensors as image variables and provides an output to an object of class *C_5 – Level_2 Processor*. The OO-DPT model of this class varies according to the chosen strategy. Due to limited space, it has not been presented. There are 6 objects of this class: $O_{1.4}$ – DO Output_A, $O_{2.4}$ – DO Output_B, $O_{3.4}$ – DO Output_C, $O_{4.4}$ – DC Output_A, $O_{5.4}$ – DC Output_B, $O_{6.4}$ – DC Output_C.

Model of Class C_5 – Sensor Level_2
The model of this class is similar to the model of class C_4, but instead of processing the sensor signals, this class processes the output of the objects of class C_4 and warns the pilot in the case of fault. There are 3 objects of this class: $O_{1.5}$ – Door Output_A, $O_{2.5}$ – Door Output_B, $O_{3.5}$ – Door Output_C.

Model of Class C_6 – Door Controller
This class controls the operation of the electro-valve and the door opening and closing according to commands emitted by the pilot. It is important to observe that it uses the output of the objects of class C_5 to detect any fault. There is only one object of this class: $O_{1.6}$ – Door Controller.

3.2 Safety Requirements

The number of detected faults and wrong diagnostics are selected as a parameter to evaluate the two control strategies. The best strategy is the one that detects more faults and minimizes the wrong diagnostics. The following situation can be found and are particularly critical to the system operation:

- *Case 1*: Due to sensor faults, the control system indicates the doors are completely open when they are not yet. This is a particular critical situation because the control system will then follow the landing sequence, extending the landing-gear. A crash between the door and landing-gear may occur, damaging the landing system. Similarly, if the control system does not detect that one or more doors have a critical leakage, the pilot is not warned and therefore cannot activate the emergency system.

- *Case 2*: One of the objects $O_{1.5}$, $O_{2.5}$ or $O_{3.5}$ of class C_5 - *Sensor Level_2* of the control system warns the pilot of a fault in the sensors.
- *Case 3*: The object $O_{1.6}$ – Door Controller of the control system detects a fault in the hydraulic actuator (at least one of the doors is not open in maximum time interval) and warns the pilot.

3.3 Simulation

No software is available yet for simulating OO-DPT nets. As a consequence, Monte Carlo simulation has been performed by translating the OO-DPT net into MatLab language. The continuous dynamics is discretized with fixed time steps. Each class corresponds to a different subroutine. Furthermore, due to the confidential nature of aircraft system data, the results published in this section are not based on real values for model parameters and probabilities of faults. They must be considered only as an example of simulation output and not as the behaviour of a real military aircraft.

Referring to Figure 5, the probability of a discrete sensor remaining blocked 'ON' or 'OFF' is $P_{ON} = P_{OFF} = 0.05$. Referring to Figure 2, the PD_1, which assign a value to C_E, is the result of the sum 3 distribution, with the following weight and media:

- Case of no fault: weight=0.8, media=0.01
- Case of seal fault: weight=0.1, media=0.5
- Case of crack fault: weight=0.1, media=0.9

Using this data, simulation is performed by executing sequences of extending and retracting cycles. Table 4 presents the percentage of retracting and extending cycles that corresponds to Case 1, 2 or 3 (described in Section 2). The best strategy must minimize the occurrence of Case 1 (wrong diagnostic) and maximize the occurrence of Cases 2 and 3 (correct fault detections) According to this result the best strategy is Strategy 2.

Table 4 – Simulation results.

	Strategy 1	**Strategy 2**
Case 1	0.01	0.00
Case 2	7.28	7.19
Case 3	0.89	1.42

4. CONCLUSION

This paper presents the application of a hybrid approach for the safety analysis of aircraft systems. The proposed approach is based on OO-DPT net and faults are modelled by means of probabilistic distributions. The approach is applied to the landing system of a military aircraft. As an example the problem of analyzing sensor redundancy and comparing control strategies is detailed.

Results are currently obtained by Monte Carlo simulation. One of the points to be investigated in future work is the development of techniques and strategies to reduce the number of simulations, such as in the Dynamically Petri nets (Everjid, Blom, 2005). For this purpose the modelling flexibility may be restricted in order to satisfy the strong Markov properties. This is a particular critical point in aerospace application because, differently from the data used in Section 3.3, the probabilities of fault of aircraft components and equipment are extremely low, reaching 10^{-5} and even low.

ACKNOWLEDGES

The authors would like to thank the partial financial support of the governmental agencies FAPESP, CNPq and CAPES.

REFERENCES

Ait-Ameur, Y. et al. (2003) "Robustness analysis of avionics embedded systems." *Proceedings of the ACM SIGPLAN Conference on Language, Compiler, and Tool for Embedded Systems*, San Diego.

Alla, H.; David, R. (2004), *Discrete, Continuous, and Hybrid Petri Nets*. Springler Verlag.

Champagnat, R. et al. (1998), "Modelling and Simulation of a Hybrid System through Pr/Tr PN-DAE Model", *3rd International Conference on Automation of Mixed Processes*, Reims.

Everdij, M.H.C.; Blom, H.A.P., (2005) "Piecewise deterministic Markov processes represented by dynamically coloured Petri nets", *Stochastics*, vol.77, n.1, pp.1-29.

Horton, G.; Kulkarni, V.; Nicol, D.; Trivedi, K. "Fluid stochastic Petri nets" *ICASE Report 96-5*, Hampton, 1996.

Lundqvist, K.; Asplund, L. (2003) "A Ravenscar-Compliant Run-time Kernel for Safety-Critical Systems." *Real-Time System*, vol.24, n.1, pp.29-54.

Khalfaoui, S. (2003), "Méthode de Recherche des Scénarios Redoutés pour l'Evaluation de la Sûreté de Fonctionnement des Systèmes Mechatroniques du Monde Automobile", PhD Thesis, *Laboratoire d'Analyse et d'Architecture des Systèmes du CNRS*, Toulouse.

Matsuura, J.P.; Yoneyama, T.; Galvão, R.K.H., (2005) Learning bayesian networks for fault detection: application to the 747 longitudinal motion. *18th Int. Cong. Of Mechanical Engineering*, Ouro Preto.

Pola, G. et al. (2003), "Stochastic hybrid models: an overview". *IFAC Conference on Analysis and Design of Hybrid System (ADHS)*, St Malo.

Pritchett, A.R.; Lee, S.; Huang, D.; Goldsman, D. (2000) Hybrid-system simulation for national airspace system safety analysis, Proc. of the 2000 Winter Simulation Conference, Orlando.

Stevens, B.L.; Lewis, F.L. (1992), *Aircraft control and simulation*. Jonh Wiley & Son.

Villani, E.; Miyagi, P.E., (2003), "Petri Net and OO for the modular analysis of an aircraft landing system", *17th International Congress of Mechanical Engineering*, São Paulo.

Villani, E.; Miyagi, P.E.; Valette, R., (2005), "A Petri-Net based Object-Oriented Approach for the Modelling of Hybrid Productive Systems", *Non-linear Analysis: Theory and Methods*, vol.62, n.8, pp.1394-1418.

ELSEVIER
IFAC
PUBLICATIONS

DETECTING AND ENFORCING MONOTONICITY FOR HYBRID CONTROL SYSTEMS SYNTHESIS [1]

Dmitry Gromov * Jörg Raisch*, **

* *Fachgebiet Regelungssysteme*
Technische Universität Berlin
Email: *{gromov | raisch}@control.tu-berlin.de*
** *Systems and Control Theory Group*
Max-Planck-Institut, Magdeburg, Germany

Abstract: Abstraction based approaches to control of hybrid systems require efficient means of computing outer approximations of reachable continuous state sets. This contribution discusses how the concept of monotonicity can be used for this purpose. It provides an efficient algorithm to check whether a given continuous system is monotone with respect to a (a-priori unknown) partial order and, if not, investigates how to use continuous feedback to enforce monotonicity. In the latter case, the resulting continuous feedback represents a (lower) control level within a hierarchical hybrid control system. *Copyright © 2006 IFAC*

Keywords: Monotone systems, hybrid control, hierarchical control.

1. INTRODUCTION

Abstraction-based approaches to hybrid control systems synthesis have become popular during the last decade (e.g. [Cury et al., 1998, Koutsoukos and Antsaklis, 2003, Chutinan and Krogh, 2000, Lemch and Caines, 1999, Moor and Raisch, 1999]). They essentially "replace" continuous dynamics by discrete abstractions and hence convert the underlying hybrid control problem into a purely discrete one, which can subsequently be addressed using standard methods from discrete event systems (DES) theory. To guarantee that desired closed-loop properties carry over from the approximation level to the underlying hybrid system, one needs to make sure that the behaviour of the abstraction covers the behaviour of the continuous dynamics on a suitable (discrete) external signal space. This, in turn, boils down to computing guaranteed overapproximations for reachability sets in the continuous component's state space. For general nonlinear systems, this represents a highly nontrivial problem. However, if the system under consideration is monotone (e.g. [Smith, 1995, Angeli and Sontag, 2003]) with respect to a partial order in its state space, this becomes a straightforward exercise [Moor and Raisch, 2002]. This paper addresses two problems from this context: (i) it provides a mechanism to efficiently check for the existence of a suitable partial order and hence for monotonicity. (ii) for the case when this test fails, the paper also discusses how continuous feedback can be used to enforce monotonicity.

This contribution is organised as follows: in Section 2, we briefly review the notion of a partial order. Section 3 addresses the concept of monotone, i.e., order preserving, dynamical systems, both for the autonomous and the controlled case. This section is mostly based on [Smith, 1995, Angeli

[1] Work partially done in the framework of the HYCON Network of Excellence, contract number FP6-IST-511368

and Sontag, 2003], but also contains new results on how to efficiently check monotonicity (Propositions 3 and 8). In Section 4, we briefly outline how monotonicity can be used in the context of abstraction based hybrid control synthesis. Finally, in Section 5, we investigate how appropriate continuous feedback on a lower level of a hierarchical hybrid control scheme can enforce monotonicity and hence facilitate the computation of discrete abstractions for higher level control purposes.

2. PARTIAL ORDER RELATIONS

A partial order relation $\preceq$ on a Banach (or, more precisely, ordered metric) space B is defined as an operation satisfying the following three properties:

1. $x \preceq x \quad \forall x \in X$,
2. $(x \preceq y) \wedge (y \preceq z) \Rightarrow x \preceq z \ \forall x, y, z \in X$,
3. $(x \preceq y) \wedge (y \preceq x) \Rightarrow x = y \ \forall x, y \in X$.

We write $x \prec y$ if $x \preceq y$ and $x \neq y$. This relation is no longer reflexive and is referred to as a strict order relation. Usually, to introduce an order relation one uses an auxiliary set $K \subset B$, such that

1. $\alpha k \in K \quad \forall k \in K,\ \alpha \in \mathbb{R}_+$,
2. $k_1 + k_2 \in K \quad \forall k_1, k_2 \in K$,
3. $k \in K \wedge -k \in K \Rightarrow k = 0$.

Thus, K is a convex pointed cone. The relation is defined by $x \preceq y$ if and only if $y - x \in K$.[2] If K has nonempty interior int$K$ then we define $x \prec\prec y$ iff $y - x \in$ int$K$. It is stronger than $\prec$ or $\preceq$ as $x \prec\prec y$ implies $x \prec y$ and therefore $x \preceq y$. In Euclidean space $\mathbb{R}^n$, orthants can play the role of cones. Each orthant $\mathbb{R}^n_\delta \subset \mathbb{R}^n$ is characterised by its signature, i.e. the n-tuple $\delta = \{\delta_1, \ldots, \delta_n\}$ whose elements take values from the two-element set $\{0, 1\}$. $\mathbb{R}^n_\delta$ is defined as $\mathbb{R}^n_\delta = \{x \in \mathbb{R}^n | (-1)^{\delta_i} x_i \geq 0\}$. Hence, the zero signature corresponds to the positive orthant. We use notation $\preceq_\delta$ (resp., $\prec_\delta$ and $\prec\prec_\delta$) to show that the corresponding relation is defined with respect to the orthant $\mathbb{R}^n_\delta$. Relation symbols without index refer to relations w.r.t. the positive orthant.

3. MONOTONE DYNAMICAL SYSTEMS

A monotone dynamical system is a dynamical system on an ordered metric space which has the property that ordered states remain ordered when time progresses. In other words, monotone systems are order preserving dynamical systems. In this section we give some conditions for an arbitrary autonomous dynamical system to be monotone. Furthermore, these results are extended to dynamical systems with inputs.

---

[2] $x \prec y$ iff $y - x \in K \backslash \{0\}$

3.1 Autonomous systems

Consider the dynamical system:

$$\dot{x}(t) = f(x(t)), \tag{1}$$

where $x(t) \in X \subset \mathbb{R}^n$, $f : X \to \mathbb{R}^n$ is a continuously differentiable vector field. The solution of (1) that starts at the point x_0 at $t = 0$ is defined as $\phi_t(x_0)$ and referred to as the flow of (1). To make an assertion about qualitative properties of the above dynamical system we have to introduce some classification.

Definition 1. A vector field $f : X \to \mathbb{R}^n$ is said to be of type K_δ on an open subset $D \subset X$ if for each $i \in \{1, \ldots, n\}$, $(-1)^{\delta_i} f_i(a) \leq (-1)^{\delta_i} f_i(b)$ for any two points a and b in D satisfying $a \preceq_\delta b$ and $a_i = b_i$.

The following proposition ([Smith, 1995], Chapt.3, Prop.5.1) asserts that the type K_δ condition is necessary and sufficient for the order preserving property to hold.

Proposition 2. Let f be of type K_δ on D and $x_0, y_0 \in D$. If $x_0 \preceq_\delta y_0$ (resp., $x_0 \prec_\delta y_0$ or $x_0 \prec\prec_\delta y_0$), $t > 0$, and if $\phi_t(x_0)$ and $\phi_t(y_0)$ are defined and in D, then $\phi_t(x_0) \preceq_\delta \phi_t(y_0)$ (resp., $\phi_t(x_0) \prec_\delta \phi_t(y_0)$ or $\phi_t(x_0) \prec\prec_\delta \phi_t(y_0)$).

The most natural way to decide whether a vector field f is of type K_δ is to analyse the sign structure of the Jacobian matrix of f. More specifically, it can be shown ([Smith, 1995]) that the vector field $f(x)$ is of type K_δ on the convex subset D if and only if

$$(-1)^{\delta_i + \delta_j} \frac{\partial f_i}{\partial x_j}(x) \geq 0, \quad i \neq j,\ x \in D. \tag{2}$$

Condition (2) can be checked in two steps: **Step 1:** Check whether the off-diagonal elements of the Jacobian matrix are *sign-stable*, i.e.

$$\left\{ \frac{\partial f_i(x)}{\partial x_j} \geq 0\ \forall x \in D \right\} \vee \left\{ \frac{\partial f_i(x)}{\partial x_j} \leq 0\ \forall x \in D \right\} \tag{3}$$

and *sign-symmetric*, i.e.

$$\frac{\partial f_i(x)}{\partial x_j} \cdot \frac{\partial f_j(x)}{\partial x_i} \geq 0\ \forall x \in D \tag{4}$$

for all $i, j \in \{1, \ldots, n\}$ such that $i \neq j$.

Step 2: If the tests in Step 1 are satisfied, we need to check whether the (Boolean) equalities

$$\delta_i \oplus \delta_j = s_{ij}, \quad i < j \tag{5}$$

hold, where $\oplus$ represents "exclusive OR" and the $n(n-1)/2$ variables s_{ij}, $i < j$, $j = 2, \ldots, n$ are defined as follows:

$s_{ij} =$

$$\begin{cases} 0 \text{ if } \frac{\partial f_i(x)}{\partial x_j} > 0 \vee \left(\frac{\partial f_i(x)}{\partial x_j} = 0 \wedge \frac{\partial f_j(x)}{\partial x_i} > 0 \right) \\ 1 \text{ if } \frac{\partial f_i(x)}{\partial x_j} < 0 \vee \left(\frac{\partial f_i(x)}{\partial x_j} = 0 \wedge \frac{\partial f_j(x)}{\partial x_i} < 0 \right) \\ \text{arbitrary in}\{0,1\} \text{ if } \frac{\partial f_i(x)}{\partial x_j} = \frac{\partial f_j(x)}{\partial x_i} = 0. \end{cases} \quad (6)$$

Often, one wants to check whether a given vector field is of type K_δ for some (yet unknown) sign structure δ. Step 1 obviously remains the same, but in Step 2 we need to decide whether (5) is solvable for the unknown $\delta = \{\delta_1, \ldots, \delta_n\}$. The following proposition presents an easy way to do this. Moreover, it shows that if the answer is positive, the orthant signature can be easily extracted from the sign structure of the Jacobian matrix.

Proposition 3. The system of Boolean equations (5) is solvable w.r.t. δ_i if and only if the following condition is satisfied:

$$s_{ij} \oplus s_{ik} = s_{jk}, \quad i < j,\ j < k,\ i,j,k \leq n. \quad (7)$$

Proof (necessity) . Let us rewrite expression $s_{ij} \oplus s_{ik}$ using (5):

$$s_{ij} \oplus s_{ik} = \delta_i \oplus \delta_j \oplus \delta_i \oplus \delta_k.$$

By definition, $a \oplus b \equiv b \oplus a$, $a \oplus a \equiv 0$ and $a \oplus 0 \equiv a$. Thus, $s_{ij} \oplus s_{ik} = \delta_j \oplus 0 \oplus \delta_k = \delta_j \oplus \delta_k = s_{jk}$, and we have shown that (7) follows from (5). (**sufficiency**). We now show that (7) implies that

$$\delta = \{0, s_{12}, \ldots, s_{1n}\} \quad (8)$$

is a solution of (5). $\delta_1 \oplus \delta_j = 0 \oplus s_{1j} = s_{1j}$ holds trivially for $j = \{2, \ldots, n\}$, and $\delta_i \oplus \delta_j = s_{1i} \oplus s_{1j} = s_{ij}$, $i, j \in \{2, n\}, j > i$, where the last equality follows from (7). □

Similarly, it can be shown that

$$\tilde{\delta} = \{1, s_{12} \oplus 1, \ldots, s_{1n} \oplus 1\} \quad (9)$$

is also a solution of (5) if (7) holds. Furthermore, (8) and (9) represent the only solutions. This can be shown by considering a vector δ' with $\delta'_i \neq \delta_i$ (i.e. $\delta'_i = \delta_i \oplus 1$) for some $i \in \{1, \ldots, n\}$ and $\delta'_j = \delta_j$ for some $j \neq i$. Hence,

$$\delta'_i \oplus \delta'_j = \delta_i \oplus 1 \oplus \delta_j = s_{ij} \oplus 1 \neq s_{ij},$$

which shows that δ' is not a solution of (5).

Remark 4. (8) and (9) signify orthants that are symmetric w.r.t. the origin.

For linear systems

$$\dot{x}(t) = Ax(t),$$

the Jacobian matrix is the A matrix, i.e. $J(x,u) = A$. Thus, the sign structure of the Jacobian is completely determined by the signs of the elements a_{ij}. Obviously, they are sign-stable, so we need to check only conditions (4) and (5). Condition (4) (*sign-symmetry*) holds if $a_{ij}a_{ji} \geq 0$, $i \neq j$. The second step is to check the corresponding Boolean equation (5) using the method described in Prop. 3.

Example 5. Let us consider the Jacobian matrix with the following sign structure

$$J = \begin{bmatrix} * & + & 0 & - \\ + & * & + & 0 \\ 0 & + & * & 0 \\ - & 0 & 0 & * \end{bmatrix}.$$

Here we use asterisks to stress the fact that diagonal elements do not affect the monotonicity property. According to (6), we have $\{s_{12}, s_{14}, s_{23}\} = \{0, 1, 0\}$ while s_{13} ,s_{24}, and s_{34} are arbitrary. From Proposition 3 we can deduce that (5) is solvable iff $\{s_{13}, s_{24}, s_{34}\} = \{0, 1, 1\}$. The corresponding signature is $\delta = \{0, 0, 0, 1\}$.

3.2 Controlled systems

Some of the previous results can be extended to dynamical systems driven by an exogenous input signal. A system

$$\dot{x}(t) = f(x(t), u(t)), \quad (10)$$

where $x(t) \in X \subset \mathbb{R}^n$, $u(t) \in U \subset \mathbb{R}^m$, $f : X \times U \to \mathbb{R}^n$, generates a flow $\phi_t(x_0, u_\tau)$, $u_\tau = u(\tau)$, $0 \leq \tau \leq t$, which represents a solution of (10) with initial condition $x(0) = x_0$ and external input signal u.

Definition 6. A controlled dynamical system (10) is monotone w.r.t. the orthants $\mathbb{R}^n_\delta$ and $\mathbb{R}^m_\gamma$ if the following implication holds for all $t \geq 0$:

$$x_1 \preceq_\delta x_2,\ u_1(\tau) \preceq_\gamma u_2(\tau),\ 0 \leq \tau \leq t \Rightarrow$$
$$\phi_t(x_1, u_{1\tau}) \preceq_\delta \phi_t(x_2, u_{2\tau}).$$

In [Angeli and Sontag, 2003], a condition for the controlled system (10) to be monotone w.r.t. the orthants $\mathbb{R}^n_\delta$ and $\mathbb{R}^m_\gamma$ has been proposed.

Proposition 7. ([Angeli and Sontag, 2003]). The system (10) is monotone w.r.t. the orthants $\mathbb{R}^n_\delta$ and $\mathbb{R}^m_\gamma$ if and only if the following properties hold for all $x \in D$ and all $u \in U$:

$$(-1)^{\delta_i + \delta_j} \frac{\partial f_i}{\partial x_j}(x, u) \geq 0, \quad i \neq j,\ i,j \leq n$$

$$(-1)^{\delta_i + \gamma_j} \frac{\partial f_i}{\partial u_j}(x, u) \geq 0, \quad i \leq n,\ j \leq m.$$

The above conditions are, in fact, the extended variant of condition (2) from the previous section. Hence, in addition to conditions (3), (4) and (5), which are used to check (2), the following tests need to be performed:

First, the partial derivatives w.r.t. the control variables need to be *sign stable*, i.e.

$$\frac{\partial f_i(x,u)}{\partial u_j} \geq 0 \text{ or } \frac{\partial f_i(x,u)}{\partial u_j} \leq 0, \quad \forall x \in D, \forall u \in U \tag{11}$$

for all $i \leq n,\ j \leq m$. Moreover, the set of Boolean equations

$$\delta_i \oplus \gamma_j = q_{ij}, \quad i \leq n,\ j \leq m, \tag{12}$$

where

$$q_{ij} = \begin{cases} 0 \text{ if } \dfrac{\partial f_i(x)}{\partial u_j} > 0, \\ 1 \text{ if } \dfrac{\partial f_i(x)}{\partial u_j} < 0, \\ \text{arbitrary in } \{0,1\} \text{ if } \dfrac{\partial f_i(x)}{\partial u_j} = 0, \end{cases} \tag{13}$$

needs to be solvable with respect to the vector $\gamma = \{\gamma_1, \ldots, \gamma_m\}$.

The following proposition gives a necessary and sufficient condition for equations (5) and (12) to be solvable.

Proposition 8. The systems of Boolean equations (5) and (12) are solvable if and only if the following conditions are satisfied:

$$s_{ij} \oplus s_{ik} = s_{jk}, \quad i < j,\ j < k,\ i,j,k \leq n, \tag{14}$$

$$q_{ij} \oplus q_{kj} = s_{ik}, \quad i \neq k,\ i,k \leq n,\ j \leq m. \tag{15}$$

Moreover,

$$\delta = \{0, s_{12}, \ldots, s_{1n}\},$$
$$\gamma = \{q_{11}, \ldots, q_{1m}\}$$

is a solution.

Proof. The proof can be carried out according to the same scheme as in Prop. 3. □

Remark 9. It can be shown that the solution is also defined up to inversion, i.e.

$$\tilde{\delta} = \{1, s_{12} \oplus 1, \ldots, s_{1n} \oplus 1\},$$
$$\tilde{\gamma} = \{q_{11} \oplus 1, \ldots, q_{1m} \oplus 1\}$$

is the only other solution of (5), (12).

Remark 10. Condition (15) can be represented as

$$q_{k1} \oplus q_{l1} = s_{kl}, \quad k < l,\ k,l \leq n,$$
$$col_1(Q) \stackrel{\oplus}{=} col_j(Q) \quad \forall i,j \leq m,$$

where $Q = q_{ij}, i \in \{1, \ldots, n\},\ j \in \{1, \ldots, m\}$, and $\stackrel{\oplus}{=}$ denotes an equality up to the inversion w.r.t. $\oplus$.

3.3 Special cases

In the following we point out two special cases of a controlled system (10) where the simpler Proposition 3 suffices to check monotonicity.

a) If $u(t)$ is entirely defined by the present state $x(t)$, i.e. $u(t) = u(x(t))$, the Jacobian of the closed loop system is

$$J_{ij}(x) = \frac{\partial f_i(x, u(x))}{\partial x_j} + \sum_{l=1}^{m} \frac{\partial f_i(x,u(x))}{\partial u_l} \frac{\partial u_l(x)}{\partial x_j}. \tag{16}$$

and the procedure described in Proposition 3 can be applied to (16).

b) In a hybrid control context, the control vector often consists of two components, i.e. $u'(t) = [u_1'(t),\ u_2'(t)]$, where $u_1(t) \in \mathbb{R}^k$, $k < m$ is determined by continuous state feedback, i.e. $u_1(t) = u_1(x(t))$, and u_2 is a piecewise constant signal with finite range $\mathcal{U} \subset \mathbb{R}^{m-k}$, $|\mathcal{U}| = N \in \mathbb{N}$. In this case, the system can be treated separately on intervals, where u_2 is constant, i.e. $u_2(t) = u_\kappa \in \mathcal{U}\ t \in [t_\kappa, t_{\kappa+1})$, and Proposition 3 can be applied again. The value u_κ is interpreted as a parameter, and the Jacobian is given by

$$J_{ij}^{\kappa}(x, u_\kappa) =$$
$$\frac{\partial f_i(x, u_1(x), u_\kappa)}{\partial x_j} + \sum_{l=1}^{k} \frac{\partial f_i(x, u_1(x), u_\kappa)}{\partial u_{1l}} \frac{\partial u_{1l}(x)}{\partial x_j}$$

$\forall \kappa \in \mathbb{N}$.

Note that in our hybrid systems context, monotonicity is only needed to compute safe abstraction. Hence, having monotonicity w.r.t. different orthants for different values of κ will not pose any problems.

4. THE ROLE OF MONOTONICITY IN ABSTRACTION BASED HYBRID CONTROL SYNTHESIS

To put the previous discussion into context, we will now briefly describe the specific hybrid systems scenario we envisage as an application. Consider a continuous system

$$\dot{x}(t) = g_{u_2(t)}(x(t)) \tag{17}$$

where, as indicated before, u_2 is a piecewise constant signal with finite range $\mathcal{U}$, $|\mathcal{U}| = N$.

$$z(t) = h(x(t)) \tag{18}$$

is a discrete-valued output signal with finite range, i.e. $h : \mathbb{R}^n \to \mathcal{Z}$, $|\mathcal{Z}| = M < \infty$. Let us further assume that the system (17), (18) is sampled, either on a regular sampling grid ("time-driven sampling") or on the sampling grid defined by the output signal z switching values ("event-driven sampling"). In the latter case, the input may only be switched at the time instances where the output changes. In both cases, eqns. (17), (18) and the considered sampling device form a continuous system (with state set $X \subset \mathbb{R}^n$) evolving in discrete time $\mathbb{N}$ on a discrete external signal space $\mathcal{U} \times \mathcal{Z}$. Let $\mathcal{B} \subset (\mathcal{U} \times \mathcal{Z})^{\mathbb{N}_0}$ denote its behaviour, i.e. the set of all pairs of discrete input and output signals compatible with the model assumptions. For abstraction based control synthesis, we need a discrete approximation, evolving on the same external signal space and exhibiting behaviour $\mathcal{B}_{\text{ab}} \supseteq \mathcal{B}$. In [Moor and Raisch, 1999], strongest ℓ-complete approximation was advocated as a particularly suitable abstraction. It is characterised by the behaviour

$$\mathcal{B}_\ell := \left\{(u_2, z) : \mathbb{N}_0 \to \mathcal{U}\times\mathcal{Z} \,|\, (u_2, z)|_{[k,k+\ell]} \in \mathcal{B}|_{[0,\ell]} \,\forall k\in\mathbb{N}_0\right\},$$

where the restriction operator $(\cdot)|_{[k,k+\ell]} : (\mathcal{U} \times \mathcal{Z})^{\mathbb{N}_0} \to (\mathcal{U} \times \mathcal{Z})^{(\ell+1)}$ picks the finite string ranging from the k-th to the $(k+\ell)$-th pair of external events and disregards its absolute location on the time axis; it can naturally be extended to sets of signals.

From a computational point of view, determining the strongest ℓ-complete approximation boils down to deciding whether a given string of input and output symbols $(u_{20}, \ldots, u_{2\ell}, z_0, \ldots, z_\ell)$ is an element in $\mathcal{B}|_{[0,\ell]}$. To obtain a precise answer, we would need to compute the evolution of the quantisation cell $h^{-1}(z_0)$ under the flow $\phi_{u_{20}}$ associated with $g_{u_{20}}$, intersect the result with $h^{-1}(z_1)$, track the evolution of the result under the flow $\phi_{u_{21}}$ associated with $g_{u_{21}}$ etc. To obtain safe approximation, or abstraction, it is sufficient to compute outer approximations of these solutions. Clearly, if g is monotone w.r.t. the partial order $\preceq$, and quantisation cells are "boxes" w.r.t. $\preceq$, then $\phi_{u_{2i}}(h^{-1}(z_i))$ is "trapped" within the the evolution of "external points", i.e. $a \preceq h^{-1}(z_i) \preceq b$ implies

$$\phi_{u_{2i}}(a) \preceq \phi_{u_{2i}}(h^{-1}(z_i) \preceq \phi_{u_{2i}}(b).$$

It is then a straightforward exercise to compute the required outer approximations and hence the desired safe abstraction [Moor and Raisch, 2002]. On the basis of such an abstraction, one can compute a discrete non-blocking supervisor enforcing a language-type specification. In [Moor and Raisch, 1999] it has been shown that the resulting supervisor will also be non-blocking and enforce the specification when connected to the underlying continuous model (17), (18).

5. MONOTONISATION THROUGH FEEDBACK

We consider linear control systems

$$\begin{cases} \dot{x}(t) = Ax(t) + Bu(t), \\ y(t) = Cx(t) \end{cases} \tag{19}$$

where $x(t) \in \mathbb{R}^n$, $u(t) \in \mathbb{R}^m$, $y(t) \in \mathbb{R}^l$, B has full column rank and C has full row rank. If the monotonicity test fails, we may still be able to enforce monotonicity by appropriate feedback. For this purpose, we divide the vector of control inputs, $u' = [u_1', u_2']$, where $'$ means "transpose" and $u_1(t) \in \mathbb{R}^k, k < m$, is the part of the control input devoted to enforce monotonicity.

The system (19) then takes the form

$$\dot{x}(t) = Ax(x) + B^1u_1(t) + B^2u_2(t). \tag{20}$$

Defining the control input $u_1(t)$ as a linear function of the current output, $u_1(t) = Ky(t) = KCx(t)$, we change the Jacobian to $J = A + B^1KC$ and, therefore, alter its sign structure accordingly. But we still do not have a clear algorithm to solve this problem in general because of the large number of degrees of freedom (recall that the number of orthants for an n-dimensional system is equal to 2^{n-1}).

The proposed semiformal algorithm uses an approach based on the successive reduction of the number of available degrees of freedom.

(1) If either i-th row of B^1 or the j-th column of C is identical to zero, the elements a_{ij} of the Jacobian remain unchanged.We can now check these elements for consistency by investigating whether Conditions (4) and (5) are satisfied. Clearly, if this is not the case, the monotonicity condition cannot be enforced by feedback from $y(t)$ to $u_1(t)$.
(2) If the result in Step 1 is positive, we can deduce the signs of some other elements of the Jacobian from (7). Note the following "extreme" case: suppose, as above, that the i-th row of B^1 (resp., the j-th column of C) are zero and that all the elements in the corresponding row (resp., column) of A are nonzero (apart possibly from the entry on the diagonal). Then, the corresponding s_{ik}, $k \neq i$ (resp., s_{kj}, $k \neq j$) completely determine the required sign structure of J, as (see (7))

$$s_{k_1k_2} = s_{ik_1} \oplus s_{ik_2}, \tag{21}$$

resp.,

$$s_{k_1k_2} = s_{k_1j} \oplus s_{k_2j}. \tag{22}$$

If, on the other hand, elements in the corresponding row (resp., column) of A are zero, there may be several admittable orthants.
(3) In the next step, we isolate the entries of the Jacobian exhibiting inappropriate signs. We

now need to determine a feedback matrix K to adjust these elements without changing the signs of the other entries. For this, the elements of the real $[k \times l]$-matrix K have to satisfy

$$(-1)^{s_{qp}}(a_{qp} + \sum_{i=1}^{k}\sum_{j=1}^{l} b^1_{qi}k_{ij}c_{jp}) \geq 0, q \neq p.$$

The extension of the proposed algorithm to the class of nonlinear control systems is not straightforward. Usually, an arbitrary nonlinear control system admits monotonisation only in some subset of the state space, if it does at all. Let's denote by $X_\delta \subset X$ a subset of the state space X where the system can be rendered monotone w.r.t. the orthant with signature δ. It is quite common that some subspaces have nonempty intersection, i.e. $X_{\delta_1} \cap X_{\delta_2} \neq \emptyset$. Then, one must choose between several orthants. In this case, a decision can be made on the basis of heuristic considerations and can hardly be formalised. However, in some special cases (e.g. positive systems) the procedure can be successfully applied as is illustrated in the following example.

5.1 Example

To illustrate the applicability of the developed approach we consider a model of the biological processes in an activated sludge process, the so-called IAWQ's[3] Activated Sludge Model No.1 (see [Henze et al., 1987, Lindberg, 1997]). This model describes the three following biological processes: removal of organic matter, nitrification, and denitrification. The considered process is an ideally mixed bioreactor with three components, which can be described by the following differential equations:

$$\begin{aligned}
\frac{dX_b}{dt} &= \frac{Q_{in}}{V}X_{b,in} - \frac{Q_{out}}{V}X_b + \mu(S_s)X_b - bX_b \\
\frac{dS_s}{dt} &= \frac{Q_{in}}{V}S_{s,in} - \frac{Q_{out}}{V}S_s - \frac{1}{Y}\mu(S_s)X_b \\
\frac{dS_o}{dt} &= \frac{Q_{in}}{V}S_{o,in} - \frac{Q_{out}}{V}S_o - \frac{1-Y}{Y}\mu(S_s)X_b - bX_b
\end{aligned} \tag{23}$$

where X_b, S_s and S_o represent the concentrations of biomass, soluble substrate and dissolved oxygen in the reactor. $X_{b,in}$, $S_{s,in}$ and $S_{o,in}$ are the influent concentrations of biomass, soluble substrate and dissolved oxygen. $\mu(S_s)$ is the specific growth rate of the biomass. It is described by Monod's equation,

$$\mu(S_s) = \frac{\bar{\mu}S_s}{K_s + S_s},$$

where $\bar{\mu}$ is the maximum specific growth rate and K_s is the half-velocity constant. The tank volume is denoted V, and the incoming and outgoing flows are Q_{in} and Q_{out}, respectively. The growth yield is Y and b is the decay rate. It is worth noting that all concentrations, input and output variables as well as parameters, are positive. Moreover, the growth yield Y is always less than one.

Using the conventional notation $x := [X_b, S_s, S_o]'$ and $u := [X_{b,in}, S_{s,in}, S_{o,in}, Q_{in}]'$ one can rewrite (23) as

$$\begin{aligned}
\dot{x}_1 &= \frac{u_1 u_4}{V} - \frac{Q_{out}}{V}x_1 + \mu(x_2)x_1 - bx_1 \\
\dot{x}_2 &= \frac{u_2 u_4}{V} - \frac{Q_{out}}{V}x_2 - \frac{1}{Y}\mu(x_2)x_1 \\
\dot{x}_3 &= \frac{u_3 u_4}{V} - \frac{Q_{out}}{V}x_3 - \frac{1-Y}{Y}\mu(x_2)x_1 - bx_1.
\end{aligned} \tag{24}$$

The Jacobian matrix has the following form

$$\frac{Df}{Dx} = \begin{bmatrix} * & \frac{\bar{\mu}K_s x_1}{(K_s+x_2)^2} & 0 \\ -\frac{1}{Y}\frac{\bar{\mu}x_2}{K_s+x_2} & * & 0 \\ -\frac{1-Y}{Y}\frac{\bar{\mu}x_2}{K_s+x_2} - b & -\frac{1-Y}{Y}\frac{\bar{\mu}K_s x_1}{(K_s+x_2)^2} & * \end{bmatrix}. \tag{25}$$

We see that the partial derivatives $\frac{\partial f_1}{\partial x_2}$ and $\frac{\partial f_2}{\partial x_1}$ do not satisfy the *sign-symmetry* condition. Now one has to determine, which one has the "right" sign. The remaining elements of the Jacobian matrix satisfy conditions (3) and (4). The corresponding variables are $s_{13} = 1$, $s_{23} = 1$. Then, from (22) $s_{12} = 0$. This means that both $\frac{\partial f_1}{\partial x_2}$ and $\frac{\partial f_2}{\partial x_1}$ must be nonnegative. The easiest way to change the sign of $\frac{\partial f_2}{\partial x_1}$ is to use the control u_2, because it does not enter the remaining equations. Considering the control u_2 as a function of the state variables, $u_2 = u_2(x)$, we can rewrite the Jacobian (25) as:

$$\frac{Df}{Dx} = \begin{bmatrix} * & \frac{\bar{\mu}K_s x_1}{(K_s+x_2)^2} & 0 \\ \frac{u_4}{V}\frac{\partial u_2(x)}{\partial x_1} - \frac{1}{Y}\frac{\bar{\mu}x_2}{K_s+x_2} & * & \frac{u_4}{V}\frac{\partial u_2(x)}{\partial x_3} \\ -\frac{1-Y}{Y}\frac{\bar{\mu}x_2}{K_s+x_2} - b & -\frac{1-Y}{Y}\frac{\bar{\mu}K_s x_1}{(K_s+x_2)^2} & * \end{bmatrix}$$

Hence, the control $u_2(x)$ has to be chosen to satisfy the following conditions:

---

[3] International Association for Water Quality.

$$\frac{\partial u_2}{\partial x_1}(x) \geq \frac{V}{u_4 Y}\frac{\mu x_2}{K_s + x_2}, \quad (26)$$

$$\frac{\partial u_2}{\partial x_3}(x) \leq 0, \quad (27)$$

$$\forall x \in \mathbb{R}^3_{\geq 0},\ u_4 \neq 0.$$

Conditions (26), (27) define a family of control laws. In particular, a control law can be chosen as

$$u_2(x) = c_1 x_1,$$

where $c_1 = \frac{V\bar{\mu}}{u_4^* Y}$, $u_4^* = \min u_4$. Thus, the system can be rendered monotone by a simple linear feedback. Moreover, it turns out, that the monotonisation procedure does not require the measurement of all state variables.

6. CONCLUSION

We discussed the question how the concept of monotonicity can be used in the context of hybrid systems. We provided a simple and efficient algorithm to check whether an arbitrary continuous system is monotone with respect to some (a priori unknown) partial order relation. This algorithm was extended to the case of control systems. It was also shown how to enforce monotonicity with the help of feedback. The developed semiformal approach was illustrated by an example. We considered a nonlinear model of an ideally mixed bioreactor and showed that this system can be rendered monotone by a very simple linear feedback.

7. ACKNOWLEDGEMENTS

The authors gratefully acknowledge helpful discussions with D. Flockerzi, MPI Magdeburg.

REFERENCES

David Angeli and Eduardo D. Sontag. Monotone control systems. *IEEE Transactions on Automatic Control*, 48(10):1684 – 1698, October 2003.

A. Chutinan and B. H. Krogh. Computing approximating automata for a class of hybrid systems. *Mathematical and Computer Modeling of Dynamical Systems: Special Issue on Discrete Event Models of Continuous Systems*, 6:30 – 50, March 2000.

J.E.R. Cury, B.H. Krogh, and T. Niinomi. Synthesis of supervisory controllers for hybrid systems based on approximating automata. *IEEE Transactions on Automatic Control*, 43(4):564 – 568, April 1998.

M. Henze, C.P.L. Grady, W. Gujer, G.v.R Marais, and T. Matsuo. *Activated sludge model No.1.* IAWPRC Scientific and technical reports, No.1. International Association on Water Pollution Research and Control, IAWPRC, London, 1987.

X.D. Koutsoukos and P.J. Antsaklis. Safety and reachability of piecewise linear hybrid dynamical systems based on discrete abstractions. *Discrete Event Dynamic Systems*, 13(3):203 – 243, July 2003.

E.S. Lemch and P.E. Caines. Hybrid partition machines with disturbances: hierarchical control via partition machines. In *Proceedings of the 38th IEEE Conference on Decision and Control*, volume 5, pages 4909 – 4914, 1999.

C.-F. Lindberg. *Control and estimation strategies applied to the activated sludge process.* PhD thesis, Uppsala University, 1997.

T. Moor and J. Raisch. Abstraction based supervisory controller synthesis for high order monotone continuous systems. In S. Engell, G. Frehse, and E. Schnieder, editors, *Modelling, Analysis, and Design of Hybrid Systems*, LNCIS 279, pages 247 – 265. Springer-Verlag, 2002.

T. Moor and J. Raisch. Supervisory control of hybrid systems within a behavioural framework. *Systems & Control Letters*, 38(3):157 – 166, October 1999.

Hal L. Smith. *Monotone dynamical systems: an introduction to the theory of competitive and cooperative systems*, volume 41 of *Mathematical surveys and monographs.* American Mathematical Society, Providence, RI, 1995.

ELSEVIER
IFAC
PUBLICATIONS

HYBRID SYSTEM CONTROL USING AN ON-LINE DISCRETE EVENT SUPERVISORY STRATEGY

James Millan *, **Siu O'Young** **.

* *Institute for Ocean Technology, National Research Council, St.John's, NL, Canada*
** *Memorial University of Newfoundland, St.John's, NL, Canada*

Abstract: This paper describes a technique for synthesizing controllers for hybrid plants. Our modeling framework allows for the efficient online construction of limited lookahead discrete abstractions of the nonlinear continuous dynamics of the plant model. Discrete event supervisory controller synthesis techniques are used to construct a controller based on a DES specification and the abstracted plant model. The controller is advanced in a moving horizon approach. The modeling framework, synthesis techniques and the on-line computational strategy are discussed. A simple illustrative example is presented in detail and a realistic industrial application is outlined.

Keywords: Discrete-Event Systems, Control, Hybrid Systems, On-line Control, Supervisory Control

1. INTRODUCTION

Theoretical developments in the area of hybrid system control have yet to lead to the widespread solution of any practical industrial problems. While hybrid system modeling is recognized clearly as being central to future control development, the intractability of computations, coupled with a steep learning curve for control system designers, have acted as a barrier to the adoption of hybrid system theory by industry.

A variety of software tools are available for both hybrid system analysis and verification such as HyTech (T.A.Henzinger *et al.* 1997), and *CheckMate* (Chutinan and Krogh 2003). In verification problems, the controller is assumed to be given. Thus, the controller synthesis is a manual process that relies upon the domain knowledge and intuition of the designer. A MATLAB toolbox for simulation and control synthesis for mixed logical dynamical (MLD) hybrid systems in discrete time is available (Torrisi and Bemporad 2004). In the case of simulation tools such as MATLAB Stateflow, the controller design is tested under a variety of conditions to evaluate its safety and correctness. Due to the ad-hoc choice of the test conditions, this technique may miss the particular combination of conditions that leads to design failure. With hybrid verification tools, the computational burden of an exhaustive reachability requires the use of simplified continuous dynamical models. Due to these limitations, the ad-hoc simulation technique is the accepted industry solution, which leaves two problems unsolved: How does the designer take a specification and produce the design? And, how can the resulting design be verified to be correct?

This work attempts to answer both of these questions, by seeking a solution that is familiar to system designers, automates the controller synthesis, and verifies the resulting controller design. The approach is to harness the power of industrially proven system modeling (simulation tools). The nonlinear continuous models are wrapped in a discrete abstraction layer that is based on event detection. The discrete abstraction combined with a DES specification in a limited horizon reachability computation, produces a discrete event (DE) controller that is, within this limited space and time, locally safe. Furthermore, a reduction in computational complexity is achieved by exploiting a lazy (just-in-time) synchronous composition of the specification and plant models at design time. This scheme is implemented as an online computation, in order for the controller operation to be extended into an infinite time horizon.

Limited lookahead (LL) supervisory control has been extensively studied in a DES setting by (Chung *et al.* 1992). In (Raisch and O'Young 1998), discrete abstractions based on the truncated time history of discrete-time LTI continuous models were used to synthesize DES supervisory controllers. Others, (Su *et al.* 2003) and (Abdelwahed *et al.* 2005), have also used discrete abstractions of switched continuous systems in a LL framework to effect control over hybrid systems. Similar to our work is (Stursberg 2004), in which the nonlinear continuous dynamics are retained as embedded simulations, and a graph search algorithm has been described for optimal hybrid control. Our approach differs in that we do not limit switching to discrete time intervals and controller graph pruning is done in a maximally permissive sense with respect to safety as is commonly done in optimal DES supervisory control (Ramadge and Wonham 1987).

Although the discrete event controller size is exponential in lookahead horizon, our computational approach can significantly reduce the complexity of computing a controller, provided that the specification is sufficiently restrictive, since many trajectories are eliminated during the reachability sweep. On the other hand, an overly restrictive specification may lead to the failure of the online synthesis to find a suitable control solution within the lookahead horizon.

This paper is organized as follows: section 2 develops the continuous system modeling and the associated discrete event abstraction technique; section 3 develops a switched continuous model that provides a means of switching between abstractions of continuous model simulations; section 4 develops controller synthesis in this hybrid framework; and finally in section 5 presents an illustrative controller design process.

2. CONTINUOUS SYSTEM ABSTRACTION

It is desirable to utilize a natural and expressive continuous modeling framework, overlaying it with a discrete event, input/output (I/O) interface. This approach has been adopted by (Koutsoukos *et al.* 2000). For now, we will consider the output aspects of the interface, or the conversion of the continuous dynamics to that of discrete event dynamics.

Let the continuous dynamics of a system be described by a nonlinear ordinary differential equation (ODE),

$$\dot{x}(t) = f(x,t) \tag{1}$$

For now, the dynamics described by equation 1 will serve as a placeholder for the complex continuous dynamics that may be produced by industrial/commercial simulation packages. A solution x of the system modeled by equation 1 on a time interval $[t_0, t_1]$ and for some initial condition is a solution to an initial value problem (IVP). We now define the continuous system model with abstraction framework, as follows:

Definition 1. Let a continuous system model (CSM), s, be defined as a triple $s = (f, \Psi, x_0)$, where:

f is a Lipschitz-continuous ordinary differential equation, $\dot{x} = f(x,t)$,
Ψ is a finite set of partitioning functionals, $\Psi = \{F_i : \mathbb{R}^n \to \mathbb{R}, 1 \le i \le N\}$, where each F_i is a continuously differentiable functional,
x_0 is the initial condition, $x(t_0)$.

The set of functionals Ψ, establishes an equivalency, for $x_1, x_2 \in \mathbb{R}^n$:

$$x_1 \sim_p x_2 \iff sign(F_i(x_1)) \times sign(F_i(x_2)) = 1, \quad 1 \le i \le N$$

The state space of the CSM is partitioned into a finite quotient set, $\mathcal{X}$ of equivalence classes Q_j (referred to as regions):

$$\mathcal{X} = \mathbb{R}^n / \sim_p = \{Q_j \subseteq \mathbb{R}^n\}$$

A state transition occurs when a continuous trajectory of the CSM crosses the hypersurface $\mathcal{N}(F) = \{x \in \mathbb{R}^n : F(x) = 0\}$, that lies between adjacent regions. For example, a trajectory x on the time interval $[t_0, t_1]$, such that $x(t_0) \in Q_1$ and $x(t_1) \in Q_2$, the state transition from Q_1 to Q_2 is notationally indicated as $Q_1 \rightsquigarrow Q_2$. The CSM communicates with outside discrete event processes via output events $\sigma \in \Sigma_{out}$ that are uniquely associated with these state transitions. Note that the transition direction is important, that is $Q_1 \rightsquigarrow Q_2 \neq Q_2 \rightsquigarrow Q_1$. For any CSM s_i, it can be shown that a unique trajectory x_i exists for a finite time interval and that this trajectory will generate a finite number of output events.

3. SWITCHED CONTINUOUS MODEL

In (Koutsoukos *et al.* 2000), input symbols to the continuous abstraction are translated into actuator actions, or sampled inputs (similar to a D/A converter). In this work, we have chosen to have each input event $\sigma_i \in \Sigma_{in}$ map to the selection (or choice) of a CSM s_i from a set of available continuous system models $\mathcal{F}$. Thus, control is achieved by switching amongst a set of continuous systems that represent either different operating modes of a system or different systems (hot swapping).

Definition 2. Let a switched continuous model (SCM) be defined as an automaton-like triple $G = (\mathcal{F}, \Gamma, s_0)$, where:

$\mathcal{F}$ is an infinite set of CSMs each with its own discrete abstraction, as in definition 1,

Γ is the enabled system function, which embodies an implementation specific selection mechanism. Let $s' \in \mathcal{F}$ be the currently selected model, and let $\mathcal{A} = \{a \subset \mathcal{F} : 1 \le |a| < \infty\}$ be the set of non-empty finite subsets of $\mathcal{F}$, then $\Gamma : \mathcal{F} \to \mathcal{A}$.

s_0 is the initial continuous system model.

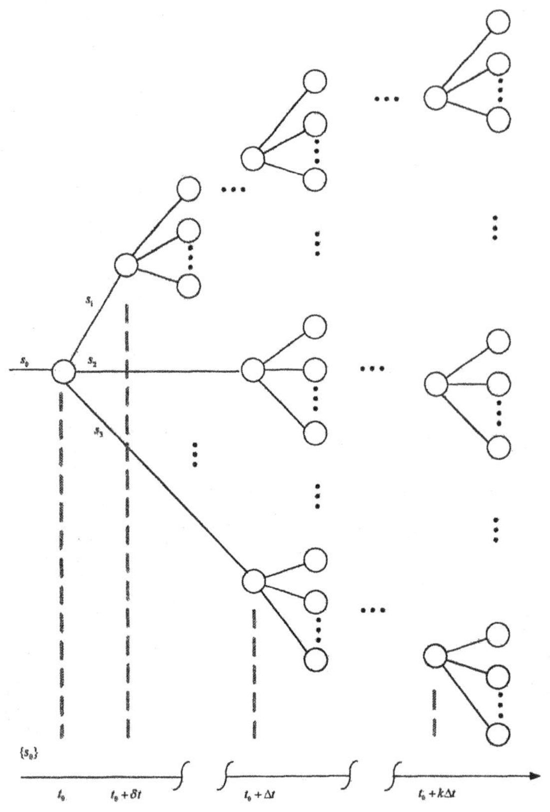

Fig. 1. Reachable continuous system models.

3.1 SCM Execution

An *execution* v of a SCM is a set of sequentially selected CSM, $v = \{s_0, s_1, \ldots s_\alpha \ldots\}$ starting with the initial model s_0. The point in the execution at which the execution changes from one system to another is known as a *choice point.* The term *choice* refers to the ability of the controller at this point to influence the future dynamics of the system, by the selection of the next CSM. In this framework, the choice points occur on some predictable (not necessarily regular) timed schedule, that is governed by a universal timebase. These choice points will be associated with the `tick` (output) event. Additionally, choice points occur whenever an output event is generated by a CSM, so that the controller is able to respond asynchronously to modeled events as they occur. Note that this theoretical modeling framework of the SCM allows both time and state-dependent switching. Because each CSM s_i has its own partitioning set Ψ_i, and dynamics f_i, in the most general case, then our framework admits both partitioning and dynamics changes within a region, due to time dependent switching. This is analogous to the operation of industrial control systems in which a synchronous control cycle is augmented by interrupt-driven control.

3.2 Prediction

Predicting the future execution of an SCM $G = (\mathcal{F}, \Gamma, s_0)$ consists of extending simulations (solutions) for each of the continuous system models, creating new choice points, and again extending the simulations. Thus, the set of future enabled CSMs $\mathcal{S}_R$, can be computed in either a depth or breadth-first reachability. This is the set of CSMs reachable from s_0, or the set of all future executions v of G that originate with s_0 (figure 1). The choice points are identified by the nodes of the graph and the edges are the selected continuous system models. To ensure finiteness of set $\mathcal{S}_R$, it is computed for a limited time lookahead, which can be measured by an integer number p, $p \neq \infty$, of `tick` events. Also, the number of choices (branches at any choice point) is finite and is bounded above by r defined as the maximum $|\Gamma(s_i)|, \forall s_i$. If the number of asynchronous events within one `tick` is bounded above by q, $1 \le q < \infty$, (a non-zeno condition on the execution) then it can be shown that $\mathcal{S}_R$ is finite and has bounds as follows:

$$pq + 1 \le |\mathcal{S}_R| \le \frac{r^{pq+1} - 1}{r - 1}, \; r > 1 \qquad (2)$$

3.3 Discrete Event Equivalent

The set of $s_i \in \mathcal{S}_R$ uniquely and concisely describes both the future continuous and discrete event dynamics of the SCM in dense time. Thus, the SCM is capable of generating a language (with timing) that consists of a finite number of finite length strings. Examining figure 1, at any choice point, there are a finite number of choices of CSM. Each choice (or branch) corresponds to a continuous simulation with event detection and each simulation has an associated discrete event equivalent transition defined as follows.

Definition 3. Let $G = (\mathcal{F}, \Gamma, s_0)$ be a SCM and let $s_a \in \mathcal{F}$, $s_a = (f, \Psi, x_0)$ be a CSM. Let $x_a \in \mathbb{R}^n$ be a solution to the IVP posed by s_a on a time interval $t \in [t_0, t_1)$ then let the DE equivalent transition be $\tau_a = (q_0, \sigma_\alpha, \sigma_\beta, q_1)$. Where $q_0 = (x_a(t_0), t_0), q_1 = (x_a(t_1), t_1) \in \mathbb{R}^n \times \mathbb{R}$ are timed continuous states, the endpoints of the solution x_a, and $\sigma_\alpha \in \Sigma_{in}$ and $\sigma_\beta \in \Sigma_{out}$ are discrete events.

The input event σ_α is the selection mechanism or guard event for the transition. The output event σ_β occurs as a result of the transition of the continuous solution into another state.(crossing a hypersurface), or as a result of reaching the end of the designated simulation time interval, Δt, in which case the output event is `tick`. Thus, the input event can be seen as initiating the occurrence of the output event. The set of transitions $\mathcal{T}_R$, corresponding to the set of reachable CSMs $\mathcal{S}_R$, forms a DE transition structure similar to a Mealy implementation of a finite state automaton. The transition structure $\mathcal{T}_R$ also gives rise to a language $L(G)$ based on the output events of the transitions $\tau_i \in \mathcal{T}_R$. Each string $u_i \in L(G)$ such that $p \leq |u_i| \leq pq < \infty$, represents the discrete event behaviour of a particular future execution v_i of the SCM out to the time horizon $p\Delta t$ in the future. The set $\mathcal{T}_R$ (and likewise $\mathcal{S}_R$) is valid for the current state and time only, and must be recomputed (propagated) from control point to control point as the online control is exercised. Clearly, the language $L(G)$ is also valid only for the current state and time as well.

4. CONTROLLER SYNTHESIS

Having established a framework that allows the discrete abstraction of a continuous model to coexist with discrete event models, we will look at the controller synthesis including the computational model.

4.1 Control

The previous section (3.3) outlined how a limited horizon DE representation of the SCM dynamics can be constructed. Let P be a switched continuous model of a plant. And at any point in time and space, there exists a LL plant language where each string $u_i \in L(P)$ is a discrete abstraction of a possible controlled execution in the future (out to some lookahead horizon). The decision of which execution to use must be made based on our knowledge of the plant dynamics represented by this LL model language. A specification can be used to partially implement this decision. Let S be a DE automaton model of a specification, such that $K = L(S)$, the legal language. From a DES perspective, we wish to remove from $L(P)$ any strings that may carry the system to an illegal state. In its simplest form, this can be achieved by taking the reachable part of synchronous product of the plant and specification, $P \parallel S$. Then, by ensuring that only strings that can carry the system to the LL horizon remain in $L(P \parallel S)$, we are left with a nonblocking controller. This problem was studied extensively by (Chung *et al.* 1992).

It should be noted that there is a possibility that no control solution exists, that is the pruned $L(P \parallel S) = \emptyset$, known as a run-time block. While this is a serious problem for an online controller, it is possible to handle it gracefully through the use of special states that represent safe, but nonuseful states of the system; essentially an emergency shutdown system, which is consistent with industrial practice (Millan 2006*a*).

If $L(P \parallel S) \neq \emptyset$, then the remaining strings represent the legal traces available to the system. The choice of which particular event (or CSM) that will be selected to advance the system is left to another process or a person (in the event of human in the loop control), to make the final decision. This is the nature of an underspecified system for which control must be implemented (Dietrich *et al.* 2002).

4.2 Complexity

The philosophy of the computational and modeling framework is to compute the controller in an efficient manner, that allows for real-time computation. The DE behaviour of a SCM on a finite lookahead horizon is finite, and so it is therefore computable. Unfortunately, the number of states required to represent this language is bounded above by an expression that is exponential in the event lookahead horizon (see equation 2). However, the number of states is bounded below by a linear expression, implying that a range of potentially sub-exponential complexity computations exist.

The controller state complexity is reduced from the unconstrained plant size by the inclusion of the specification S, at design time (which is also runtime). This is because many unsafe traces are invalidated in the joint behaviour of the plant and specification due to the requirement for synchronization on common events. The extent to which the state complexity will be reduced by the inclusion of the specification is difficult to predict, since it is dependent on the plant and specification models as well as the state and time of the execution. This reduction in complexity due to the inclusion of a specification has been noted for validation of hybrid control systems (Stursberg *et al.* 2003).

The controller graph represents the set of Such a controller merely disables unsafe transitions. As was stated earlier some sort of decision mechanism is still required to choose the actual event (actuation) that will be used to send the system forward in time. We will assume that this decision mechanism is implemented by either another controller module, or possibly even a human in the loop (HIL). In any case, the system will find itself at a new state, and the controller will have to be recomputed before the next decision is taken.

4.3 Encapsulation of Simulation Tools

Suppose there exists a simulation tool that, given an initial condition and some parameters, produces a numerical solution for a particular system model. Then without loss of generality, this is comparable to the ODE solver of the IVP of equation 1. Indeed, our only stipulations on the simulation tool are, given a set of parameters: a) it always produces an output (solution existence), b) the output is repeatable for the same parameters (solution uniqueness) and c) the solution is computed in less time than it takes the actual system to execute (real-time implementation). Whether the latter requirement (c) is met, hinges on the extent to which the specification limits the legal trajectories of the plant. If the simulation tool meets each of the above requirements, then with suitable wrapper functions (object methods), an SCM can be built around it, and an online hybrid controller is feasible.

Based on the computation structure outlined here, an experimental software package that computes DE controllers for hybrid systems, called HYSYNTH has been developed. HYSYNTH was developed as a MATLAB class structure to enable developers to leverage the high-level simulation capabilities of the MATLAB environment.

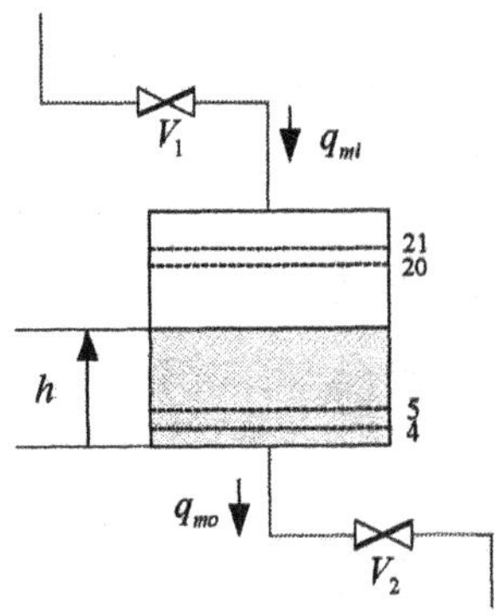

Fig. 2. Tank control schematic, V_1 and V_2 are control valves.

5. EXAMPLES

5.1 Tank Control

In this example, a SCM will represent the model of the tank, which has predominantly continuous dynamics, and a finite state machine will model the specification. Together, the specification and plant models will be used to form a DES controller that can be propagated to enforce the desired behaviour of the tank. In figure 2, the tank has both a fill and drain pipe, which can be controlled independently. If there is turbulent flow from the tank, then the liquid level dynamics are described by the differential equation $\rho A\dot{h} = q_{mi} - q_{mo}$, where h is the liquid level, ρ is the density of the liquid, and A is the cross sectional area of the tank. The liquid mass flow into the tank is q_{mi}. Assuming turbulent flow from the tank, and choosing appropriate values, the resulting nonlinear relation $\dot{h} = q_{mi} - \sqrt{h}$ applies when the tank is draining . If the tank can be switched between filling only, draining only, and simultaneously filling and draining. there are three different CSM dynamics.

$$f_1 : \dot{h} = q_{mi}, \text{ filling only}$$
$$f_2 : \dot{h} = -\sqrt{h} \text{ draining only}$$
$$f_3 : \dot{h} = q_{mi} - \sqrt{h}, \text{ both filling and draining}$$

Each continuous system model $s_i \in \mathcal{F}$ will contain one of these dynamics. The family of continuous system models $\mathcal{F}$ will be infinite if the initial condition, $h_0 \in \mathbb{R}$ is inherited from the preceding CSM. Table 1 defines a common set of partitioning functionals, shared by all CSMs, and associates a set of output events with them. The specification will be designed so that the tank fills to the overfill mark (21) after one `tick`, draining it back through the overfill 2 mark (20), again after one `tick`, repeating this cycle *ad infinitum.* The finite state machine that represents this specification is given in figure 3. This example was coded in HYSYNTH, and the tank was given an initial level of $h = 20$.

Table 1. Functionals with related output events for the tank control example.

Functional	Alarm	Output Events
$F_1(h) = h - 21$	overfill 1	$of1^+, of1^-$
$F_2(h) = h - 20$	overfill 2	$of2^+, of2^-$
$F_3(h) = h - 5$	underfill 1	$uf1^+, uf1^-$
$F_4(h) = h - 4$	underfill 2	$uf2^+, uf2^-$

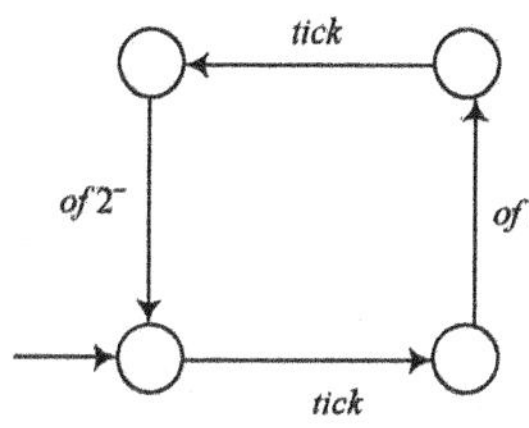

Fig. 3. The tank controller specification.

Using an event horizon lookahead of 15 events, a controller was synthesized repeatedly, propagating the solution by simulation and random selection of controller action. The evolution of the controller structure is illustrated in figure 4(a)-(c); each of the subgraphs in the figure represents the controller at successive time steps. The states are tiny dots connected by edges and the initial state of the graph, i.e. the current state of the controlled system, is indicated by the larger dot. An arrow marks the same state in (a)-(c) and is used to illustrate the growth of the graph ends as the horizon is extended. Any path from the initial state to the end of a branch is 15 events, and represents a safe nonblocking execution of the plant. The past path of the system is indicated by the light line as the initial state moves along. In (b), a choice between the upper and lower left-hand branches must be made. In (c) the lower left-hand branch has been discarded through a choice made by a runtime choice mechanism.

For this example, the number of controller states varied from 51 to 231 over 40 iterations of controller synthesis. Clearly, from equation 2, without the specification, the plant would have a theoretical upper limit of over 21 million states (equation 2, $r = 3$ and $pq = 15$). In this example, the inclusion of the specification at design time has dramatically reduced the computational (state) complexity of the controller. Based on the empirical results of this example, figure 5 clearly illustrates the dramatic improvement in state complexity as a function of lookahead that can be achieved. The upper line is the theoretical state size of the unconstrained plant, while the lower line is a fit through the mean of empirically derived data.

5.2 Oil Offloading

Figure 6 illustrates a control problem that inspired the control techniques that have been developed by the authors. In the offshore oil industry, oil is produced and stored by one vessel, and then transferred to a second vessel that takes it to shore. The goal is to encode the complex and extensive operations manual for this offloading task as a DE specification which will be used to enforce a safe subset of operations for a HIL control. In (Millan and O'Young 2000), the authors developed a simplified version of this problem as a linear hybrid automaton and verified an ad hoc emergency shutdown controller. HySynth has since been used successfully to synthesize a controller for this problem with a complex nonlinear continuous dynamics (Millan 2006*b*).

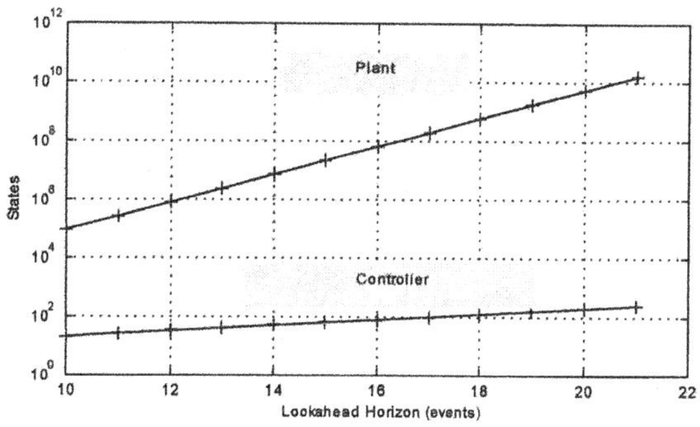

Fig. 5. Comparative complexity of controller and unspecified plant model.

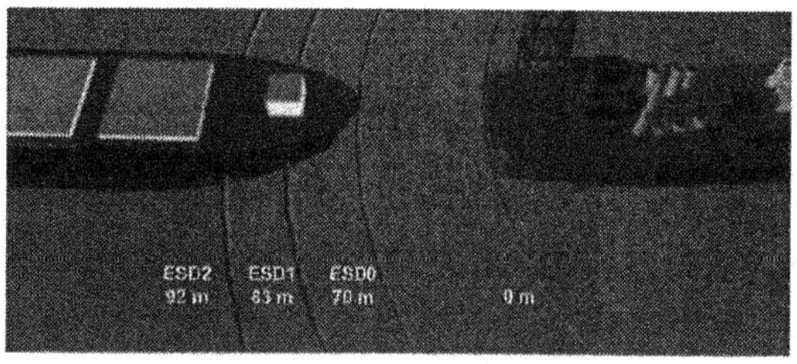

Fig. 6. The offloading of oil from a production vessel (at right) to a shuttle tanker (at left).

6. CONCLUSION

In this paper a switched continuous modeling framework was described that allows generalized nonlinear continuous models to be included seamlessly in a discrete event supervisory control synthesis process. Control is effected by switching between multiple continuous models that may represent either differing operating modes of a system, control inputs or differing systems. Controller nonblocking is identified as the existence of at least one complete string (one that carries the system to the lookahead horizon) in the controller language. A significant reduction in computational complexity is achieved by the inclusion of the specification at design time. An experimental software tool (HySynth) designed for the MATLAB environment, allows for a high level com-

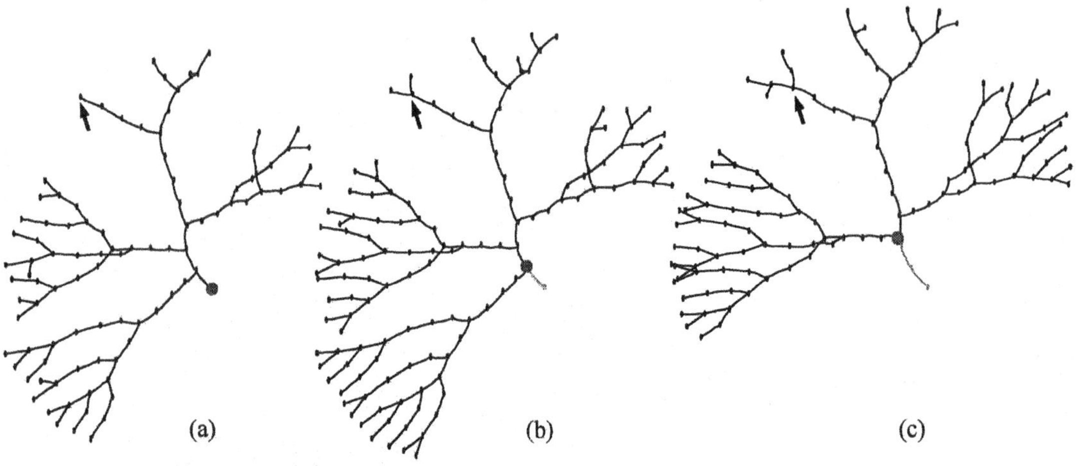

Fig. 4. Propagation of the controller through three updates.

mand interface, coupled with the native MATLAB simulation tools.

Future work will consist of improving the software to allow for enhancing design flexibility and improvements in efficiency. As of now, the control scheme has only been tested in simulation. It would be desirable to put it to a real-world test. Future areas of particular interest are developing techniques to deal with unmodeled effects such as disturbance, and modeling errors.

REFERENCES

Abdelwahed, S., R. Su and S. Neema (2005). A feasible lookahead control for systems with finite control set. In: *Proceedings of the 2005 IEEE Conference on Control Applications*. IEEE. pp. 663–668.

Chung, S., S. Lafortune and F. Lin (1992). Limited lookahead policies in supervisory control of discrete event systems. *IEEE Transactions on Automatic Control* **37**(12), 1921–1935.

Chutinan, A and B. H. Krogh (2003). Computational techniques for hybrid system verification. *IEEE Transactions on Automatic Control* **48**(1), 64–75.

Dietrich, P., R. Malik, W.M. Wonham and B.A. Brandin (2002). Implementation considerations in supervisory control. *Synthesis and control of discrete event systems* pp. 185–201.

Koutsoukos, X.D., P.J. Antsaklis, J.A. Stiver and M.D. Lemmon (2000). Supervisory control of hybrid systems. In: *Proceedings of the IEEE*. pp. 1026–1048. IEEE.

Millan, J. P. (2006*a*). On-line supervisory control of hybrid systems using embedded simulations. In: *Proceedings of the 8th International Workshop on Discrete Event Systems WODES06*.

Millan, J. P. (2006*b*). Online Discrete Event Control of Hybrid Systems. PhD thesis. Memorial University of Newfoundland. Expected July 2006.

Millan, James and Siu O'Young (2000). Hybrid modeling of tandem dynamically positioned vessels. In: *Proceedings of the 39th IEEE Conference on Decision and Control*.

Raisch, J. and S.D. O'Young (1998). Discrete approximation and supervisory control of continuous systems. *IEEE Transactions on Automatic Control* **43**(4), 569–573.

Ramadge, P.J. and W.M. Wonham (1987). Supervisory control of a class of discrete event processes. *SIAM Journal on Control Optimization* **25**(1), 206–230.

Stursberg, O. (2004). A graph search algorithm for optimal control of hybrid systems. In: *Proceedings of the 43rd IEEE Conference on Decision and Control*. pp. 1412–1417.

Stursberg, O., A. Fehnker, Z. Han and B. H. Krogh (2003). Specification-guided analysis of hybrid systems using a hierarchy of validation methods. In: *IFAC Conference on Analysis and Design of Hybrid Systems*. IFAC.

Su, R., S. Abdelwahed, G. Karsai and G. Biswas (2003). Discrete abstraction and supervisory control for switching systems. In: *IEEE International Conference on Systems, Man, and Cybernetics*. Vol. 1. IEEE. pp. 415–421.

T.A.Henzinger, P. Ho and H. Wong-Toi (1997). HyTech: A model checker for hybrid systems. *Software Tools for Technology Transfer* **1**, 110–122.

Torrisi, F.D. and A. Bemporad (2004). HYSDEL - a tool for generating computational hybrid models. *IEEE Transactions on Control Systems Technology* **12**(2), 235–249.

ELSEVIER
IFAC
PUBLICATIONS

NON-DETERMINISTIC REACTIVE SYSTEMS, FROM HYBRID SYSTEMS AND BEHAVIOURAL SYSTEMS PERSPECTIVES

J.M. Davoren * and **Thomas Moor** **

* *Department of Electrical & Electronic Engineering, The University of Melbourne, VIC 3010 AUSTRALIA*
** *Lehrstuhl für Regelungstechnik, Friedrich-Alexander-Universität, Erlangen, 91023 GERMANY*

Abstract: A reactive system is an entity which takes as inputs signals from a certain set, and transforms them to produce as outputs signals in some further set, where a signal is modeled as a function from a time domain to a value space, and time domains are linearly ordered sets. Building on our previous work which generalizes Aubin's Evolutionary Systems model, we develop a general formulation of non-deterministic input-output reactive systems that is uniform for discrete, continuous and hybrid time, and which extends Willems' Behavioural Systems input-output model by allowing hybrid time, and by allowing different time domains for input and output signals. However, our approach differs from these existing frameworks in that instead of taking as primitive signals over infinite-length time domains, we work with finite-length paths and utilize their algebraic and partial-order structure. We illustrate our framework with a generic model of event-driven A/D conversion, transforming a continuous-time input signal into a discrete-time output signal via an intermediate hybrid-time signal. *Copyright © 2006 IFAC*

Keywords: hybrid systems, input-output systems, behavioural systems.

1. INTRODUCTION

In previous work (Davoren *et al.*, 2004), we identify and investigate a class of non-deterministic dynamical systems we call *general flow systems* which include discrete-time transition systems, continuous-time differential inclusions, and hybrid-time systems such as hybrid automata and impulse differential inclusions. Over any value space X and a (non-negative) time line L, a general flow system Φ maps each point $x \in X$ to the set $\Phi(x)$ of all signals or paths $\gamma : T \to X$ of the system with $\gamma(0) = x$ and time domain $T \subseteq L$; in the non-deterministic setting, there may be none, exactly one, or many possible Φ-paths starting from x. In (Davoren *et al.*, 2004), we adapt Aubin's model of *Evolutionary Systems* (Aubin and Dordan, 2002; Aubin *et al.*, 2002) to provide semantics for a temporal logic that is uniform for discrete-time, continuous-time and hybrid-time systems. Over the hybrid time line $L = \mathbb{N} \times \mathbb{R}_0^+$ (ordered lexicographically), hybrid paths $\gamma : T \to X$ have time domains $T \subset L$ of the form $T = \bigcup_{i<N} [(i,0), (i,\Delta_i)]$, with $\Delta_i \in \mathbb{R}_0^+$ the duration of the i-th interval. Within T, time $(i+1,0)$ is the immediate *discrete successor* of time (i,Δ_i), but in the underlying line L, there is a continuum-length "gap" in between. To deal with hybrid signals, two moves were crucial. The

[1] This research is supported by the Australian Research Council, Grant No. DP0208553.

first was to work with finite-length paths having a start-point and an end-point, instead of infinite-length paths over the whole time line, as used for the usual discrete and continuous time lines in the work of Aubin, and also within the *Behavioural Systems* framework (Willems, 1989; Willems, 1991; Moor and Raisch, 1999). The second was to develop a theory of finite-length paths $\gamma : T \rightarrow X$ with time-domains $T \subset L$ that are not simply intervals $T = [0, b]$ but rather finite disjoint unions of intervals, with "gaps" in between. We then build up a theory of *maximal extensions* of finite paths to compare with signal models over unbounded time domains. By working with finite-length paths, we also get to use monoidal and partial-order structure that is not accessible within frameworks based on infinite-length signals. In this respect, our approach is close to that of (Tabuada *et al.*, 2004).

In the present paper, we develop a general formulation of non-deterministic input-output reactive systems that is uniform for discrete, continuous and hybrid time. In broad terms, a *reactive system* Γ takes as input a signal η from a certain set, and transforms it to produce as output one or more signals $\gamma \in \Gamma(\eta)$ in some further set. While such input-output systems have been well-studied within Behavioural Systems theory (Willems, 1989; Willems, 1991) for the usual discrete or continuous time lines, that framework does not extend to hybrid time or to systems where the time-line L_{out} for output paths may differ from the time-line L_{in} for input paths. Basic examples are A/D converters transforming continuous-time signals into discrete-time signals (particularly event-driven rather than fixed periodic sampling), and D/A converters transforming discrete-time signals into continuous-time signals. Our approach is also commensurate with the *Tagged Signal* "Models of Computation" meta-model (Lee and Sangiovanni-Vincentelli, 1998), restricted to the timed case of linearly ordered signal domains, in that it provides a general framework which includes within it diverse model classes, and within which various system properties can be formulated and compared.

In a follow-on paper, we will extend the present work with a study of *compositions* of input-output reactive systems by sequential, parallel and feedback constructions, with the aim of giving an explicit set-theoretic semantics for block-diagram based modeling and description frameworks, such as CHARON (Alur *et al.*, 2000), that allow non-determinism, hybrid signals and time-translations. Additionally, we will give a representation within our framework of Lynch and Vandraager's *hybrid I/O automata* model (Lynch *et al.*, 1996), as well as of Aubin's *impulse differential inclusion* hybrid model (Aubin *et al.*, 2002), extended with inputs and outputs, which are not possible within this short paper.

The body of the paper is as follows. Section 2 covers preliminaries on set-valued maps and linear orders. In Section 3, we develop some basic theory of paths with "gappy" time domains, and tersely review the theory of general flows and their infinitary extensions, enabling a first comparison in Section 4 with Aubin's Evolutionary Systems and with state behaviours in Willems' Behavioural Systems Theory. In Section 5, we develop input-output reactive systems and some basic system properties and their inter-relationships. The main result concerns the relationship between Willems' concept of *non-anticipation*, which is a non-deterministic analogue of *causality*, and the system property of being *extension-preserving* with respect to the natural partial order of extension of finite paths (that is not available in frameworks based on signals with infinite time domains). We also including several illustrative examples, including event-driven A/D conversion. In Section 6, we compare our work with other input-output system models. In particular, we show that for the usual discrete and continuous time lines, our finite-path formulations of system properties such as *non-anticipation*, *input freedom*, and *time-invariance* (suffix-closure), all have an exact correspondence with their infinite signal formulations in Willems' input-output Behavioural Systems theory.

2. PRELIMINARIES: RELATIONS/SET-VALUED MAPS, MONOIDS, AND LINEAR ORDERS

We write $\mathbb{R}_0^+ := [0, \infty)$ and $\mathbb{N}$ for the non-negative reals and natural numbers, respectively.

We write $r : X \rightsquigarrow Y$ to mean $r : X \rightarrow 2^Y$ is a *set-valued map*, with $r(x) \subseteq Y$ for every $x \in X$ (possibly $r(x) = \varnothing$); equivalently, $r \subseteq X \times Y$ is a *relation* – we don't distinguish between a set-valued map and its graph. Let $[X \rightsquigarrow Y] := 2^{X \times Y}$ denote the set of all maps $r : X \rightsquigarrow Y$ from X to Y; it is partially ordered by the subset relation, with least element ϵ (empty map), and closed under *relational converse*: for a map $r : X \rightsquigarrow Y$, the converse $r^{-1} : Y \rightsquigarrow X$ is given by: $x \in r^{-1}(y)$ iff $y \in r(x)$. The *domain* is $\text{dom}(r) := \{x \in X \mid r(x) \neq \varnothing\}$, and the *range* is $\text{ran}(r) := \text{dom}(r^{-1}) \subseteq Y$. A map $r : X \rightsquigarrow Y$ is *total on* X if $\text{dom}(r) = X$. Given $r_1 : X \rightsquigarrow Y$ and $r_2 : Y \rightsquigarrow Z$, their *sequential composition* is $(r_1 \circ r_2) : X \rightsquigarrow Z$, defined by $(r_1 \circ r_2)(x) := \{z \in Z \mid (\exists y \in Y)\, y \in r_1(x) \wedge z \in r_2(y)\}$.

We distinguish several sub-classes of *deterministic maps*. We write $r : X \rightarrow Y$ to mean r is a single-valued *function* that is *total* on domain X, with values in Y, written $r(x) = y$ (rather than

$r(x) = \{y\}$). Let $[X \to Y]$ denote the set of all $r : X \to Y$. We also need *partial functions*, and write $r : X \dashrightarrow Y$ to mean that on $\mathrm{dom}(r) \subseteq X$, r is a single-valued function; let $[X \dashrightarrow Y]$ denote the set of all such maps. For partial functions, we also write $r(x) = y$ when $x \in \mathrm{dom}(r)$, and may write $r(x) = \text{UNDEF}$ when $x \notin \mathrm{dom}(r)$. As sets of maps, $[X \to Y] \subseteq [X \dashrightarrow Y] \subseteq [X \rightsquigarrow Y]$.

On any set X, we define a *monoidal family over* X to be a structure $(X, M, \cdot, \iota)$ where the functions $\cdot : X \to [(M \times M) \dashrightarrow M]$ and $\iota : X \to M$, are such that, if $\cdot_x := \cdot(x) : (M \times M) \dashrightarrow M$ and $\iota_x := \iota(x) \in M$, for each $x \in X$, the following conditions are satisfied, for all $m, m', m'' \in M$:
(a) *family associativity*: if $(m, m') \in \mathrm{dom}(\cdot_x)$ and $(m', m'') \in \mathrm{dom}(\cdot_y)$, then $(m \cdot_x m', m'') \in \mathrm{dom}(\cdot_y)$ and $(m, m' \cdot_y m'') \in \mathrm{dom}(\cdot_x)$ and $(m \cdot_x m') \cdot_y m'' = m \cdot_x (m' \cdot_y m'')$;
(b) *right identity*: if $(m, m') \in \mathrm{dom}(\cdot_x)$ then $(m, \iota_x) \in \mathrm{dom}(\cdot_x)$ and $m \cdot_x \iota_x = m$;
(c) *left identity*: if $(m', m) \in \mathrm{dom}(\cdot_x)$ then $(\iota_x, m) \in \mathrm{dom}(\cdot_x)$ and $\iota_x \cdot_x m = m$.

Let $(L, <, 0)$ be a *linear order* with least element 0 and no largest element, and let $\leqslant$ be the reflexive closure of $<$. We use usual interval notation: for $a, b \in L$, the *bounded intervals* in L include $[a, b] := \{l \in L \mid a \leqslant l \leqslant b\}$ and $(a, b) := \{l \in L \mid a < l < b\}$. The right *unbounded intervals* are: $[a, \infty) := \{l \in L \mid a \leqslant l\}$ and $(a, \infty) := \{l \in L \mid a < l\}$. Given $(L_1, <_1, 0_1)$ and $(L_2, <_2, 0_2)$, a map $g : L_1 \dashrightarrow L_2$ is:

- [*strictly*] *order-preserving* if
  $(\forall l, k \in \mathrm{dom}(g))$, if $l \leqslant_1 k$ [if $l <_1 k$]
  then $g(l) \leqslant_2 g(k)$ [then $g(l) <_2 g(k)$];
- an *order-isomorphism* if g is a total function and a bijection, and both g and g^{-1} are strictly order-preserving.

We take a (future) *time line* to be any linear order $(L, <, 0)$ that is *shift-invariant* in the sense that it is equipped with a family of maps $\{\sigma^{-a}\}_{a \in L}$ such that $\sigma^{-0} \colon L \to L$ is the identity map, and for each $a \in L$, $\sigma^{-a} \colon [a, \infty) \to L$ (the *left a-shift*) is an order-isomorphism, with inverse $\sigma^{+a} := (\sigma^{-a})^{-1} \colon L \to [a, \infty)$ (the *right a-shift*).

The basic examples are the discrete time line $\mathbb{N}$, and the dense continuum $\mathbb{R}_0^+$, the non-negative cones of linearly ordered abelian groups under addition, $\mathbb{Z}$ and $\mathbb{R}$ respectively. The group operation trivially gives shift-invariance: take $\sigma^{-a}(l) = l - a$ for $l \in [a, \infty)$, and inverse $\sigma^{+a}(l) = l + a$.

The hybrid time line $\mathbb{H} := \mathbb{N} \times \mathbb{R}_0^+$ is linearly ordered *lexicographically*: $(i, t) <_{\text{lex}} (j, s)$ iff $i < j$ or $i = j$ and $t < s$. The least element is $\mathbf{0} := (0, 0)$. This ordering does *not* admit any natural addition operation to make it a linearly ordered semigroup, but it is shift-invariant: for $a = (k, r) \in \mathbb{H}$, define $\sigma^{-a} \colon [a, \infty) \to \mathbb{H}$ by $\sigma^{-a}(i, t) := (0, t - r)$ if $i = k$ and $\sigma^{-a}(i, t) := (i - k, t)$ if $i > k$, for all $l = (i, t) \in [a, \infty)$.

We will also require some additional properties of time lines. A time line $(L, <, 0)$ will be called *ω-compact* if there exists an ω-length sequence $\{b_n\}_{n<\omega}$ such that for all $n < \omega$, $b_n \in L$ and $0 < b_n < b_{n+1}$, and $L = \bigcup_{n<\omega}[0, b_n]$. A time line $(L, <, 0)$ is called *Dedekind-complete* if for every subset $A \subseteq L$ such that $A \neq \varnothing$, if A has an upper bound $b \in L$ such that $l \leqslant b$ for all $l \in A$, then $\sup(A)$ exists in L, and likewise for lower bounds and inf's. The time lines $\mathbb{N}$, $\mathbb{R}_0^+$ and $\mathbb{H}$ each have both these properties.

3. PATHS, GENERAL FLOW SYSTEMS, AND THEIR EXTENSIONS

Let $(L, <, 0)$ be a time line. We define a *bounded time domain* in L to be a subset $T \subset L$ such that $T = \bigcup_{n<N}[a_n, b_n]$ with $N \in \mathbb{N}^+$, $a_0 = 0$, $b_{N-1} = b_T$ the maximum element, and $a_n \leqslant b_n < a_{n+1} \leqslant b_{n+1}$ for all $n < N - 1$. Let $\mathsf{BT}(L) \subset 2^L$ denote the set of all bounded time domains in L, and let $\mathsf{BT}_\varnothing(L) = \mathsf{BT}(L) \cup \{\varnothing\}$. Define $\mathsf{BI}(L) := \{T \in \mathsf{BT}(L) \mid (\exists b \in L)\,\mathrm{dom}(\gamma) = [0, b]\}$ and $\mathsf{BI}_\varnothing(L) = \mathsf{BI}(L) \cup \{\varnothing\}$.

Over any set $X \neq \varnothing$ (an arbitrary *value space* or *signal space*), define the set of *L-paths in X*, and the set of *interval L-paths in X*:

$$\mathsf{Path}(L, X) := \{\gamma : L \dashrightarrow X \mid \mathrm{dom}(\gamma) \in \mathsf{BT}(L)\}$$
$$\mathrm{IPath}(L, X) := \{\gamma : L \dashrightarrow X \mid \mathrm{dom}(\gamma) \in \mathsf{BI}(L)\}$$

For $\gamma \in \mathsf{Path}(L, X)$, define $b_\gamma := b_{\mathrm{dom}(\gamma)}$ to be the maximum of $\mathrm{dom}(\gamma)$, so $\gamma(b_\gamma) \in X$ is the end-point and $\gamma(0) \in X$ the start-point of γ. Where ϵ is the empty path, let $\mathsf{Path}_\epsilon(L, X) := \mathsf{Path}(L, X) \cup \{\epsilon\}$ and $\mathcal{P}_\epsilon := \mathcal{P} \cup \{\epsilon\}$ for path sets $\mathcal{P} \subseteq \mathsf{Path}(L, X)$.

We utilise three operations on $\mathsf{Path}_\epsilon(L, X)$; for $\gamma, \gamma' \in \mathsf{Path}_\epsilon(L, X)$, $t \in L$ and $x \in X$, define:

- *restriction* or *prefix* ending at $t \in \mathrm{dom}(\gamma)$:
  $\gamma|_t \in \mathsf{Path}_\epsilon(L, X)$ where $\gamma|_t := \gamma \restriction_{[0,t] \cap \mathrm{dom}(\gamma)}$.
- *translation* or *suffix* starting at $t \in \mathrm{dom}(\gamma)$:
  ${}_t|\gamma \in \mathsf{Path}_\epsilon(L, X)$ where $({}_t|\gamma)(l) := \gamma(\sigma^{+t}(l))$ for all $l \in \mathrm{dom}({}_t|\gamma) := \sigma^{-t}([t, b_\gamma] \cap \mathrm{dom}(\gamma))$.
- *point-concatenation* at $x \in X$:
  $\gamma *_x \gamma' \in \mathsf{Path}_\epsilon(L, X)$ where, for all $l \in L$:
  $(\gamma *_x \gamma')(l) := \gamma(l)$ if $l \in \mathrm{dom}(\gamma)$ and $\gamma'(0) = x = \gamma(b_\gamma)$;
  $(\gamma *_x \gamma')(l) := \gamma'(\sigma^{-b_\gamma}(l))$ if $l \in \sigma^{+b_\gamma}(\mathrm{dom}(\gamma'))$ and $\gamma'(0) = x = \gamma(b_\gamma)$; and
  $(\gamma *_x \gamma')(l) := \text{UNDEF}$ for all other l and x.

For each $x \in X$, the *trivial path* $\theta_x : [0, 0] \to X$ is given by $\theta_x(0) := x$. Then the map $\theta : X \to \mathsf{Path}(L, X)$ given by $\theta(x) := \theta_x$ is an embedding (injective function) of X into $\mathsf{Path}(L, X)$.

For any subset of paths $\mathcal{P} \subseteq \mathrm{Path}_\epsilon(L, X)$, we say $\mathcal{P}$ is *closed under suffixes* (*closed under prefixes*) if for all $\gamma \in \mathcal{P}$ and all $t \in \text{dom}(\gamma)$, the path ${}_t|\gamma \in \mathcal{P}$ (respectively, $\gamma|_t \in \mathcal{P}$). Define $X_\mathcal{P} := \{x \in X \mid (\exists\gamma \in \mathcal{P})\ \gamma(0) = x \vee \gamma(b_\gamma) = x\}$. The structure $(X_\mathcal{P}, \mathcal{P}, *, \theta)$ constitutes a *monoidal family* over the set $X_\mathcal{P}$ if for all $\gamma, \gamma' \in \mathcal{P}$ and all $x \in X_\mathcal{P}$, if $(\gamma, \gamma') \in \text{dom}(*_x)$, then $\gamma *_x \gamma' \in \mathcal{P}$ and $\theta_x \in \mathcal{P}$.

For discrete time $L = \mathbb{N}$, the interval path set $\text{IPath}_\epsilon(\mathbb{N}, X) = X^*$ is the set of all finite *words* or sequences over X. The usual operation of *word-concatenation* from automata and DES theory equips the set X^* as a (total) monoid with identity ϵ. Denoting it by $\cdot$, that operation is definable in terms of point-concatenation as follows: for all $\gamma, \gamma' \in \text{IPath}(\mathbb{N}, X) = X^+$, we have $\gamma \cdot \gamma' = \gamma *_x (\nu_{xy}) *_y \gamma'$, where $\text{dom}(\nu_{xy}) = \{0, 1\}$, $\nu_{xy}(0) = x = \gamma(b_\gamma)$, and $\nu_{xy}(1) = y = \gamma'(0)$.

Note that for all non-empty $\gamma \in \text{Path}(L, X)$ and all $x, y, z \in X$ such that $\gamma(0) = z$ and $\gamma(b_\gamma) = y$, we have: $\epsilon *_x \epsilon = \epsilon$, $\gamma *_y \epsilon = \gamma$, and $\epsilon *_z \gamma = \gamma$. Hence ϵ as well as θ_x functions as a monoid identity for the $*_x$ operation. The following structures are monoidal families over X: $(X, \text{Path}(L, X), *, \theta)$, $(X, \text{Path}_\epsilon(L, X), *, \varepsilon)$, $(X, \text{IPath}(L, X), *, \theta)$ and $(X, \text{IPath}_\epsilon(L, X), *, \varepsilon)$.

The point-concatenation operation is usefully related to the notion of *extensions* of paths which continue from the end value of a given path. Define a partial order on $\mathsf{Path}_\epsilon(L, X)$ from the subset relation and the underlying linear order on L; (re-using notation) we define: $\gamma < \gamma'$ iff $\gamma \subset \gamma'$ and $t < t'$ for all $t \in \text{dom}(\gamma)$ and for all $t' \in \text{dom}(\gamma') - \text{dom}(\gamma)$, in which case we say γ' is a *proper extension* of γ. As usual, $\gamma \leqslant \gamma'$ iff either $\gamma < \gamma'$ or $\gamma = \gamma'$.

Proposition 3.1. Let L be any time line, and X any value space. Then for all $\gamma, \gamma' \in \mathsf{Path}(L, X)$, the following are equivalent:

(i) $\gamma < \gamma'$;

(ii) for all $t \in \text{dom}(\gamma)$, $\gamma|_t = \gamma'|_t$ and for all $t' \in \text{dom}(\gamma') - \text{dom}(\gamma)$, $t < t'$;

(iii) $\gamma' = \gamma *_x \gamma''$ for some $\gamma'' \in \text{Path}(L, X)$ and $x \in X$ with $\gamma''(0) = x$ and $\gamma(b_\gamma) = \gamma'(b_\gamma) = x$ and $\gamma'' \neq \theta_x$.

Now consider the hybrid time line $\mathbb{H} = \mathbb{N} \times \mathbb{R}_0^+$. Define $DS := \text{IPath}_\epsilon(\mathbb{N}, \mathbb{R}_0^+)$ to be the set of all (finite) *duration sequences.* For $\Delta \in DS$, define $HT(\Delta)$ to be the *hybrid time domain* determined by Δ, and over any $X \neq \varnothing$, define $\mathsf{HPath}_\epsilon(X) \subset \mathsf{Path}_\epsilon(\mathbb{H}, X)$ to be the set of all *hybrid paths* over X, as follows:

$$HT(\Delta) := \bigcup_{i<\text{length}(\Delta)} [(i, 0), (i, \Delta_i)]$$

$$\mathsf{HT} := \{ HT(\Delta) \in \mathsf{BT}(\mathbb{H}) \mid \Delta \in DS \}$$

$$\mathsf{HPath}(X) := \{\xi \in \mathsf{Path}(\mathbb{H}, X) \mid \text{dom}(\xi) \in \mathsf{HT}\}$$

General flow systems are dynamical systems over a state space or signal value space. As expected from Behavioural Systems theory (Willems, 1989; Willems, 1991), and as we shall see in examples in Section 5, we can for certain input-output systems associate a general flow dynamical system over an input-state-output product space.

Definition 3.2. Let $(L, <, 0)$ be a time line and let $X \neq \varnothing$ be any space. A *general flow system* over X with time line L is a map $\Phi\colon X \rightsquigarrow \mathsf{Path}(L, X)$ satisfying, for all $x \in \text{dom}(\Phi)$, for all $\gamma \in \Phi(x)$, and for all $t \in \text{dom}(\gamma)$:

(GF0) the *initialization property*: $\gamma(0) = x$;

(GF1) the *time-invariance* or *suffix-closure property*: ${}_t|\gamma \in \Phi(\gamma(t))$;

(GF2) the *point-concatenation property*: $\gamma|_t *_y \gamma' \in \Phi(x)$ for all $\gamma' \in \Phi(y)$ with $y = \gamma(t)$.

- Φ is *non-blocking* if $\Phi(x) \neq \{\theta_x\}$ for all $x \in \text{dom}(\Phi)$;
- Φ is *prefix-closed* if $\gamma|_t \in \Phi(x)$ for all $x \in \text{dom}(\Phi)$, $\gamma \in \Phi(x)$ and $t \in \text{dom}(\gamma)$;
- Φ is *deterministic* if for every $x \in \text{dom}(\Phi)$, the path set $\Phi(x)$ is linearly-ordered by $<$.
- Φ is *$<$-unbounded* if for all $x \in \text{dom}(\Phi)$ and $\gamma \in \Phi(x)$, there exists $\gamma' \in \Phi(x)$ such that $\gamma < \gamma'$.

Among other results (Davoren *et al.*, 2004), it is readily established that Φ is non-blocking iff Φ is $<$-unbounded. Examples of general flow systems include state transition systems over $L = \mathbb{N}$, differential inclusions over $L = \mathbb{R}_0^+$, and hybrid automata and impulse differential inclusions over $L = \mathbb{H}$. Any path set $\mathcal{P} \subseteq \mathsf{Path}(L, X)$ that is monoidal and suffix-closed determines a general flow system.

It is clear that if a flow Φ is non-blocking, then for each $x \in dom(\Phi)$ and $\gamma \in \Phi(x)$, there exists an infinite sequence of paths $\{\gamma_n\}$ with $\gamma_0 = \gamma$ and $\gamma_n \in \Phi(x)$ and $\gamma_n < \gamma_{n+1}$ for all n. Motivated by this fact, we view "maximal extensions" or "completions" of paths as infinitary objects, arising as limits of infinite ordered sequences of finite bounded paths.

Definition 3.3. Let L be an ω-compact time line. For any path set $\mathcal{P} \subseteq \mathsf{Path}_\epsilon(L, X)$, define the *$\omega$-extension* of $\mathcal{P}$, and the *maximized ω-extension* of $\mathcal{P}$, as follows:

$$\mathsf{Ext}^\omega(\mathcal{P}) := \{\beta \in [L \dashrightarrow X] \mid (\exists\overline{\gamma} \in [\omega \to \mathsf{Path}(L, X)])\ (\forall k < \omega)\ \gamma_k := \overline{\gamma}(k) \ \wedge\ \gamma_k \in \mathcal{P} \ \wedge\ \gamma_k < \gamma_{k+1} \ \wedge\ \beta = \bigcup_{k<\omega} \gamma_k \}$$

$$\mathsf{M}^\omega(\mathcal{P}) := \{\beta \in \mathsf{Ext}^\omega(\mathcal{P}) \mid (\forall\gamma \in \mathcal{P})\ \beta \not< \gamma\}$$

Define $\mathsf{EPath}^\omega(L, X) := \mathsf{Ext}^\omega(\,\mathsf{Path}_\epsilon(L, X)\,)$, and $\mathsf{EIPath}^\omega(L, X) := \mathsf{Ext}^\omega(\,\mathsf{IPath}_\epsilon(L, X)\,)$.

The ω-extension $\mathsf{Ext}^\omega(\mathcal{P})$ contains all the partial functions $\beta : L \dashrightarrow X$ that arise as the limit of an

ω-length strictly extending sequence of paths in $\mathcal{P}$. The *maximised* ω-extension $\mathsf{M}^\omega(\mathcal{P})$ throws out from $\mathsf{Ext}^\omega(\mathcal{P})$ all limit paths β that are properly extended by some finite-length path γ in $\mathcal{P}$. (The path extension partial ordering on bounded paths straightforwardly lifts to limit paths.)

Definition 3.4. Let L be an ω-compact time line. Given a general flow system $\Phi\colon X \rightsquigarrow \mathsf{Path}(L,X)$, define the maximised ω-extension $\mathsf{M}^\omega\Phi\colon X \rightsquigarrow \mathsf{EPath}^\omega(L,X)$ by $(\mathsf{M}^\omega\Phi)(x) := \mathsf{M}^\omega(\Phi(x))$ for all $x \in dom(\Phi)$. A flow Φ will be called:

- *ω-extendible* if for all $x \in dom(\Phi)$ and $\gamma \in \Phi(x)$, there exists $\alpha \in (\mathsf{M}^\omega\Phi)(x)$ such that $\gamma < \alpha$.
- *ω-full* if for all $x \in dom(\Phi)$ and $\beta \in \mathsf{Ext}^\omega(\Phi(x))$, there exists $\alpha \in (\mathsf{M}^\omega\Phi)(x)$ such that $\beta \leqslant \alpha$.

In general, $dom(\mathsf{M}^\omega\Phi) \subseteq dom(\Phi)$, and we have $dom(\mathsf{M}^\omega\Phi) = dom(\Phi)$ iff Φ is ω-extendible.

Proposition 3.5. Given a general flow system $\Phi\colon X \rightsquigarrow \mathsf{Path}(L,X)$ over any time line L,

Φ is ω-extendible
iff Φ is $<$-unbounded and ω-full.

For $L = \mathbb{N}$, all limit paths $\beta \in \mathsf{ElPath}^\omega(\mathbb{N},X)$ have infinite time domain, so $(\mathsf{M}^\omega\Phi)(x) = \mathsf{Ext}^\omega(\Phi(x))$ for any non-blocking Φ. For $L = \mathbb{H}$ and ω-extendible Φ, limit paths $\alpha \in \mathrm{ran}(\mathsf{M}^\omega\Phi)$ include those with time domains $\mathrm{dom}(\alpha) = \bigcup_{n<\omega}\{n\} \times [0,\Delta_n]$, as well as those with $\mathrm{dom}(\alpha) = T_0 \cup (\{i\} \times [0,d))$ for some $T_0 \in \mathsf{BT}(\mathbb{H})$ and $(i,0) >_{\mathrm{lex}} b_{T_0}$ and $d \in \mathbb{R}_0^+ \cup \{\infty\}$.

4. COMPARISON WITH EVOLUTIONARY AND BEHAVIOURAL SYSTEM MODELS

Aubin's evolutionary system's (Aubin and Dordan, 2002) are over the usual time lines $L = \mathbb{R}_0^+$ or $L = \mathbb{N}$, and consist of a map $\Psi : X \rightsquigarrow [L \to X]$, with extended whole line paths $\beta : L \to X$, where Ψ is closed under the operations of translation/suffix and point-concatenation, analogous to clauses **(GF1)** and **(GF2)** of Definition 3.2.

In Willems' Behavioural Systems theory (Willems, 1989; Willems, 1991), a *dynamical system* is a structure $\Sigma = (L, X, \mathfrak{B})$ where $L \subseteq \mathbb{R}$ is the time line, X is the signal space, and $\mathfrak{B} \subseteq [L \to X]$ is the *behaviour*. A behaviour $\mathfrak{B}$ is called *time invariant* if for all $\beta : L \to X$ and all $t \in L$, if $\beta \in \mathfrak{B}$ then the t-translation/suffix ${}_t|\beta \in \mathfrak{B}$; and is called a *state behaviour* if for all $t \in L$ and for all $\beta, \beta' \in \mathfrak{B}$, if $x = \beta(t) = \beta'(t)$ then the point-concatenation $(\beta|_t) *_x ({}_t|\beta') \in \mathfrak{B}$. A behaviour $\mathfrak{B}$ is called *complete* if for all $\beta : L \to X$, if $\big[(\forall a,b \in L)\, \beta\restriction_{[a,b]} \in \mathfrak{B}\restriction_{[a,b]}\big]$ then $\beta \in \mathfrak{B}$ (– and note that the reverse implication is trivially true).

Theorem 4.1. Let the time line be either $L = \mathbb{N}$ or $L = \mathbb{R}_0^+$, and $X \neq \varnothing$. Let $\Psi\colon X \rightsquigarrow [L \to X]$ and let $\mathfrak{B} \subseteq [L \to X]$. Then:

Ψ is an Aubin evolutionary system;
iff $\mathfrak{B}$ is a time-invariant and complete state behaviour in the sense of Willems;
iff there exists an ω-extendible interval path general flow system $\Phi\colon X \rightsquigarrow \mathsf{IPath}(L,X)$ such that $\Psi = \mathrm{M}^\omega\Phi$ and $\mathfrak{B} = \mathrm{ran}(\mathrm{M}^\omega\Phi)$.

5. INPUT-OUTPUT REACTIVE SYSTEMS

To allow that the time-line L_{out} for output paths may differ from the time-line L_{in} for input paths, we need to use at least *order-preserving* partial functions $\tau : L_{\mathrm{out}} \dashrightarrow L_{\mathrm{in}}$ which do a *time translation* by *looking in reverse*, starting from a "now" instant $t \in \mathrm{dom}(\gamma) \subset L_{\mathrm{out}}$ in an output path γ, and asking at what time $\tau(t) \in \mathrm{dom}(\eta) \subset L_{\mathrm{in}}$ is the corresponding "now" time point in the input path η which generates γ among its output.

Definition 5.1. Let L_{in} and L_{out} be time-lines, and let U, Y be any non-empty sets. Define

$$\mathsf{TT}(L_{\mathrm{out}}, L_{\mathrm{in}}) := \{\, \tau \in \mathsf{Path}(L_{\mathrm{out}}, L_{\mathrm{in}}) \mid \tau \textit{ is order-preserving} \wedge \tau(0_{\mathrm{out}}) = 0_{\mathrm{in}} \wedge \mathrm{ran}(\tau) \in \mathsf{BT}(L_{\mathrm{in}}) \,\}$$

A *reactive system* is any map Γ of type

$$\Gamma\colon \mathsf{Path}_\epsilon(L_{\mathrm{in}}, U) \rightsquigarrow (\mathsf{Path}_\epsilon(L_{\mathrm{out}}, Y) \times \mathsf{TT}_\epsilon(L_{\mathrm{out}}, L_{\mathrm{in}}))$$

such that for all inputs $\eta \in \mathrm{dom}(\Gamma)$, and for all outputs $(\gamma,\tau) \in \Gamma(\eta)$, we have $\mathrm{ran}(\tau) \subseteq \mathrm{dom}(\eta)$ and $\mathrm{dom}(\tau) = \mathrm{dom}(\gamma)$. A reactive system Γ will be called:

- *time-homogeneous* if: $L_{\mathrm{out}} = L_{\mathrm{in}} = L$, and $(\forall \eta \in \mathrm{dom}(\Gamma))(\forall(\gamma,\tau) \in \Gamma(\eta))$, $\mathrm{dom}(\gamma) = \mathrm{dom}(\eta)$ and τ is the identity function restricted to $\mathrm{dom}(\gamma)$, in which case we treat Γ as a map of type $\Gamma : \mathsf{Path}(L,U) \rightsquigarrow \mathsf{Path}(L,Y)$;
- *[strictly] extension-preserving* if: $(\forall \eta \in \mathrm{dom}(\Gamma))(\forall \eta' \in \mathrm{dom}(\Gamma))$, *if* $\eta \leqslant \eta'$ [*if* $\eta < \eta'$] *then* $(\forall(\gamma,\tau) \in \Gamma(\eta))(\exists(\gamma',\tau') \in \Gamma(\eta'))$ such that $\tau \leqslant \tau'$ and $\gamma \leqslant \gamma'$ [$\tau < \tau'$ and $\gamma < \gamma'$];
- *time-invariant/suffix-closed* if: the path set $\mathrm{dom}(\Gamma)$ is suffix-closed, and $(\forall \eta \in \mathrm{dom}(\Gamma))(\forall(\gamma,\tau) \in \Gamma(\eta))(\forall t \in \mathrm{dom}(\gamma))$, $(\exists \tau' \in \mathsf{TT}(L_{\mathrm{out}}, L_{\mathrm{in}}))$, such that $\tau' = ({}_t|\tau) \circ \sigma^{-t}$ and $({}_t|\gamma, \tau') \in \Gamma({}_{\tau(t)}|\eta)$;
- *prefix-closed* if: the path set $\mathrm{dom}(\Gamma)$ is prefix-closed, and $(\forall \eta \in \mathrm{dom}(\Gamma))(\forall(\gamma,\tau) \in \Gamma(\eta))$ $(\forall t \in \mathrm{dom}(\gamma))$, $(\gamma|_t, \tau|_t) \in \Gamma(\eta|_{\tau(t)})$;
- *non-anticipating* (the output has a non-anticipatory dependence on the input) if: $(\forall \eta, \eta' \in \mathrm{dom}(\Gamma))(\forall s \in \mathrm{dom}(\eta) \cap \mathrm{dom}(\eta'))$ *if* $\eta|_s = \eta'|_s$ *then* $(\forall(\gamma,\tau) \in \Gamma(\eta))(\exists(\gamma',\tau') \in \Gamma(\eta'))$ $(\forall t \in \mathrm{dom}(\gamma) \cap \mathrm{dom}(\gamma'))$ *if* $\tau(t) \leqslant s$ *then* $\tau|_t = \tau'|_t$ and $\gamma|_t = \gamma'|_t$;
- *strictly non-anticipating* if:

$(\forall \eta, \eta' \in \text{dom}(\Gamma))(\forall s \in \text{dom}(\eta) \cap \text{dom}(\eta'))$
if $D_s = [0,s) \cap \text{dom}(\eta) = [0,s) \cap \text{dom}(\eta')$
and $\eta \upharpoonright_{D_s} = \eta' \upharpoonright_{D_s}$
then $(\forall(\gamma,\tau) \in \Gamma(\eta))(\exists(\gamma',\tau') \in \Gamma(\eta'))$
$(\forall t \in \text{dom}(\gamma) \cap \text{dom}(\gamma'))$ *if* $\tau(t) \leqslant s$
then $\tau|_t = \tau'|_t$ and $\gamma|_t = \gamma'|_t$;

- *input-time-unbounded* if: for every input path $\eta \in \text{dom}(\Gamma)$, there exists an $\alpha \in \mathsf{M}^\omega(\text{dom}(\Gamma))$ such that $\eta < \alpha$ and the time domain $\text{dom}(\alpha)$ is an unbounded set in L_{in};
- *totally free in input* if: $\text{dom}(\Gamma) = \mathsf{Path}_\epsilon(L_{\text{in}}, U)$; i.e. Γ is total as a map;
- *monoidal* if: **(a)** $(U_{\mathcal{D}}, \mathcal{D}, *, \theta)$ and $(Y_{\mathcal{R}}, \mathcal{R}, *, \theta)$ are both monoidal families, where $\mathcal{D} := \text{dom}(\Gamma)$ and $\mathcal{R} := \pi_Y(\text{ran}(\Gamma))$; and **(b)** $(\forall \eta, \eta' \in \text{dom}(\Gamma))$ $(\forall u \in U_{\mathcal{D}})$ *if* $\eta(b_\eta) = u = \eta'(0)$ *then* $(\forall(\gamma,\tau) \in \Gamma(\eta))$, $(\forall(\gamma',\tau') \in \Gamma(\eta'))$, if $\gamma(b_\gamma) = \gamma'(0)$, then $(\gamma'',\tau'') \in \Gamma(\eta *_u \eta')$, where $\gamma'' = \gamma *_y \gamma'$, $y = \gamma(b_\gamma)$, $\tau'' = \tau *_t (\tau' \circ \sigma^{+t})$ and $t = b_\gamma$.

Given any non-empty set X, a map Ψ will be called a *parameterized reactive system* if it is of type $\Psi : (X \times \mathsf{Path}_\epsilon(L_{\text{in}}, U)) \rightsquigarrow (\mathsf{Path}_\epsilon(L_{\text{out}}, Y) \times \mathsf{TT}_\epsilon(L_{\text{out}}, L_{\text{in}}))$, and for all $(x,\eta) \in \text{dom}(\Psi)$, and for all $(\gamma,\tau) \in \Psi(x,\eta)$, $\text{ran}(\tau) \subseteq \text{dom}(\eta)$ and $\text{dom}(\tau) = \text{dom}(\gamma)$. Most of these properties of reactive systems can be simply extended to parameterized reactive systems by substituting quantification over $\eta \in \text{dom}(\Gamma)$ with $(x,\eta) \in \text{dom}(\Psi)$, and substituting $(\gamma,\tau) \in \Gamma(\eta)$ with $(\gamma,\tau) \in \Psi(x,\eta)$.

In (Willems, 1989; Willems, 1991), over the discrete and the continuous time lines, the *non-anticipating* property for input-output behaviours is proposed as the non-deterministic analogue of the *causality* property in deterministic systems. Our first major result is that, under the assumption of prefix-closure, the properties of being *non-anticipating* and of being *extension-preserving* are equivalent.

Theorem 5.2. Let $\Gamma : \mathsf{Path}_\epsilon(L_{\text{in}}, U) \rightsquigarrow (\mathsf{Path}_\epsilon(L_{\text{out}}, Y) \times \mathsf{TT}_\epsilon(L_{\text{out}}, L_{\text{in}}))$ be a reactive system, where the time lines L_{in} and L_{out} are ω-compact and Dedekind-complete.
(1.) If Γ is strictly non-anticipating, then Γ is non-anticipating.
(2.) If Γ is strictly extension-preserving, then Γ is extension-preserving
(3.) If Γ is extension-preserving and prefix-closed, then Γ is non-anticipating.
(4.) If Γ is non-anticipating, then Γ is extension-preserving.
From (3.) and (4.), if Γ is prefix-closed, then Γ is extension-preserving iff Γ is non-anticipating.
(5.) If Γ is monoidal and suffix-closed, then Γ is strictly extension-preserving.
(6.) If Γ is totally free in input, then Γ is input-time-unbounded.

Before discussing some examples of reactive systems, we first apply ω-extension constructions to the input-output setting.

Definition 5.3. Given a reactive system Γ with time lines L_{in} and L_{out} both ω-compact, define the *ω-extension* of Γ to be the map $\mathsf{Ext}^\omega\Gamma \colon \mathsf{EPath}^\omega(L_{\text{in}}, U) \rightsquigarrow \mathsf{EPath}^\omega(L_{\text{out}}, Y \times L_{\text{in}})$ such that:

$$
\begin{aligned}
&(\mathsf{Ext}^\omega\Gamma)(\alpha) \\
&:= \{\, (\beta, v) \in \mathsf{Ext}^\omega(\text{ran}(\Gamma)) \mid \\
&\quad (\exists \overline{\eta} \in [\omega \to \mathsf{Path}(L_{\text{in}}, U)]\,) \\
&\quad (\exists (\overline{\gamma}, \overline{\tau}) \in [\omega \to \mathsf{Path}(L_{\text{out}}, Y \times L_{\text{in}})]\,) \\
&\quad (\forall k < \omega)\ \eta_k := \overline{\eta}(k) \ \wedge\ \gamma_k := \overline{\gamma}(k) \ \wedge \\
&\quad \tau_k := \overline{\tau}(k) \ \wedge\ \eta_k \in \text{dom}(\Gamma) \ \wedge\ (\gamma_k, \tau_k) \in \Gamma(\eta_k) \\
&\quad \wedge\ \eta_k < \eta_{k+1} \ \wedge\ (\gamma_k, \tau_k) < (\gamma_{k+1}, \tau_{k+1}) \ \wedge \\
&\quad \alpha = \bigcup_{k<\omega} \eta_k \ \wedge\ \beta = \bigcup_{k<\omega} \gamma_k \ \wedge\ v = \bigcup_{k<\omega} \tau_k \,\}
\end{aligned}
$$

for all $\alpha \in \text{dom}(\mathsf{Ext}^\omega\Gamma) := \mathsf{Ext}^\omega(\text{dom}(\Gamma))$.

In defining the *maximized ω-extension* $\mathsf{M}^\omega\Gamma$ of a reactive system Γ, we want to restrict to limit input paths $\alpha \in \text{dom}(\mathsf{Ext}^\omega\Gamma) \subseteq \mathsf{EPath}^\omega(L_{\text{in}}, U)$ that are not only maximized ω-extensions of finite input paths, so $\alpha \in \mathsf{M}^\omega(\text{dom}(\Gamma))$, but that are also *unbounded* in the length of their time domain. This rules out limit input paths with finite escape time. When a system Γ is input-time-unbounded, then there are no limit input paths with bounded time duration, hence for such systems, there will be no loss in the move from Γ to the maximized ω-extension $\mathsf{M}^\omega\Gamma$.

Note, however, that on the output side, we will *not* be assuming that all resulting limit output paths must be time-unbounded; rather, we will identify that condition into a system property of *output-time-unboundedness.*

Definition 5.4. Let L be an ω-compact time line, and let $X \neq \varnothing$ be any value space. For any path set $\mathcal{P} \subseteq \mathsf{Path}_\epsilon(L, X)$, define the *time-unbounded maximized ω-extension* of $\mathcal{P}$ to be the limit path set $\mathsf{TM}^\omega(\mathcal{P})$, with $\mathsf{TM}^\omega(\mathcal{P}) \subseteq \mathsf{M}^\omega(\mathcal{P}) \subseteq \mathsf{Ext}^\omega(\mathcal{P}) \subseteq \mathsf{EPath}^\omega(L, X)$, defined by:

$$
\begin{aligned}
\mathsf{TM}^\omega(\mathcal{P}) := \{\alpha \in \mathsf{M}^\omega(\mathcal{P}) \mid (\forall t \in L) \\
(\exists s \in \text{dom}(\alpha))\ s \geqslant t\}
\end{aligned}
$$

So the limit path set $\mathsf{TM}^\omega(\mathcal{P})$ is that subset of $\mathsf{M}^\omega(\mathcal{P})$ obtained by throwing away all limit paths α where the time domain $\text{dom}(\alpha)$ is a bounded subset of L. We shall require time-unbounded maximized extensions on the input side.

Definition 5.5. Given a reactive system Γ with ω-compact time lines L_{in} and L_{out}, define the *maximized ω-extension* of Γ to be the map $\mathsf{M}^\omega\Gamma$ of type:

$$\mathsf{M}^\omega\Gamma \colon \mathsf{EPath}^\omega(L_{\text{in}}, U) \rightsquigarrow \mathsf{EPath}^\omega(L_{\text{out}}, Y \times L_{\text{in}})$$

such that $\text{dom}(\mathsf{M}^\omega\Gamma) := \mathsf{TM}^\omega(\text{dom}(\Gamma))$, and $(\mathsf{M}^\omega\Gamma)(\alpha) := (\mathsf{Ext}^\omega\Gamma)(\alpha)$ for all $\alpha \in \text{dom}(\mathsf{M}^\omega\Gamma)$.

A reactive system Γ will be called:
• *ω-responsive* if it is input-time-unbounded and $(\forall \eta \in \mathrm{dom}(\Gamma))(\forall \alpha \in \mathrm{dom}(\mathsf{M}^\omega\Gamma))$, *if* $\eta < \alpha$ *then* $(\forall (\gamma,\tau) \in \Gamma(\eta))(\exists(\beta,v) \in (\mathsf{M}^\omega\Gamma)(\alpha))$ such that $(\gamma,\tau) < (\beta,v)$
• *output-time-unbounded* if every limit output path has an unbounded time domain, which means $\mathrm{ran}(\mathsf{M}^\omega\Gamma) = \mathsf{TM}^\omega(\mathrm{ran}(\Gamma))$.

The following theorem gives a first characterisation of the ω-responsiveness property.

Theorem 5.6. For any reactive system Γ,
(7.) Γ is ω-responsive iff Γ is strictly extension-preserving and input-time-unbounded.
(8.) If Γ is monoidal, suffix-closed and input-time-unbounded, then Γ is ω-responsive.

Example 5.7. A *state machine* is a structure $\mathcal{S} = (X, U, Y, \mathit{UpDt})$ with state set X, input set U, output set Y, and $\mathit{UpDt} : (X \times U) \rightsquigarrow (X \times Y)$ the state/output update map; equivalently, $\mathcal{S}$ is a non-deterministic Mealy machine. Associate with $\mathcal{S}$ three maps on interval paths over $L = \mathbb{N}$.

The first map is the *full input-state-output map* $\Phi_{\mathcal{S}} : (U \times X \times Y) \rightsquigarrow \mathsf{IPath}(\mathbb{N}, U \times X \times Y)$ with:

$$\begin{aligned}
&\Phi_{\mathcal{S}}(u,x,y) \\
&:= \{ (\eta,\xi,\gamma) \in \mathsf{IPath}(\mathbb{N}, U \times X \times Y) \mid \\
&\quad u = \eta(0) \ \wedge \ x = \xi(0) \ \wedge \ y = \gamma(0) \ \wedge \\
&\quad (\forall i < b_\eta)\ (\xi(i+1), \gamma(i)) \in \mathit{UpDt}(\xi(i), \eta(i)) \ \wedge \\
&\quad (\exists x' \in X)\ (x', \gamma(b_\eta)) \in \mathit{UpDt}(\xi(b_\eta), \eta(b_\eta)) \quad \}
\end{aligned}$$

It is readily seen that $\Phi_{\mathcal{S}}$ is a prefix-closed general flow system and that $\Phi_{\mathcal{S}}$ is non-blocking (and ω-extendible) iff the map UpDt is total.

We can then define two reactive systems from the flow $\Phi_{\mathcal{S}}$ via projections. First, the *input-state* system: an parameterized time-homogeneous reactive system

$$\Psi_{\mathcal{S}} : (X \times \mathsf{IPath}(\mathbb{N}, U)) \rightsquigarrow \mathsf{IPath}(\mathbb{N}, X)$$

and second, the (external behaviour) *input-output* system: a time-homogeneous reactive system

$$\Gamma_{\mathcal{S}} : \mathsf{IPath}(\mathbb{N}, U) \rightsquigarrow \mathsf{IPath}(\mathbb{N}, Y)$$

transforming input-paths into output-paths. These two maps are defined as follows:

$$\Psi_{\mathcal{S}}(x,\eta) := \{\xi \in \mathsf{IPath}(\mathbb{N}, X) \mid (\exists \gamma \in \mathsf{IPath}(\mathbb{N}, Y))\ (\eta,\xi,\gamma) \in \Phi_{\mathcal{S}}(\eta(0), x, \gamma(0)) \}$$

$$\Gamma_{\mathcal{S}}(\eta) := \{\gamma \in \mathsf{IPath}(\mathbb{N}, Y) \mid (\exists \xi \in \mathsf{IPath}(\mathbb{N}, X))\ (\eta,\xi,\gamma) \in \Phi_{\mathcal{S}}(\eta(0), \xi(0), \gamma(0)) \}$$

In particular, $(x,\eta) \in \mathrm{dom}(\Psi_{\mathcal{S}})$ iff $(x, \eta(0)) \in \mathrm{dom}(\mathit{UpDt})$, and $\mathrm{dom}(\Gamma_{\mathcal{S}}) = \pi_U(\mathrm{dom}(\Psi_{\mathcal{S}}))$. Observe also that $\Psi_{\mathcal{S}}$ and $\Gamma_{\mathcal{S}}$ are both suffix-closed and prefix-closed, and if the map UpDt is total, then both $\Psi_{\mathcal{S}}$ and $\Gamma_{\mathcal{S}}$ are monoidal, extension-preserving, non-anticipating, totally free in input, ω-responsive and output-time-unbounded.

Example 5.8. Over continuous time, an input-state-output system of *differential inclusions* is a structure $\mathcal{DI} = (X, U, Y, \mathcal{U}, F, G)$ where $X = \mathbb{R}^n$, $U = \mathbb{R}^m$, $Y = \mathbb{R}^p$, $\mathcal{U} \subseteq \mathsf{IPath}(\mathbb{R}_0^+, U)$ is a set of input paths, and the maps $F : (X \times U) \rightsquigarrow \mathbb{R}^n$ and $G : (X \times U) \rightsquigarrow Y$ are subject to regularity assumptions to guarantee existence of solutions without finite escape time (Aubin *et al.*, 2002). The resulting solution map $\mathsf{Sol}_{\mathcal{DI}} : (X \times \mathcal{U}) \rightsquigarrow \mathsf{IPath}(\mathbb{R}_0^+, X)$ is an initialized time-homogeneous reactive system which is suffix-closed, prefix-closed and strictly non-anticipating. The time-homogeneous external I/O map $\Gamma_{\mathcal{DI}} : \mathsf{IPath}(\mathbb{R}_0^+, U) \rightsquigarrow \mathsf{IPath}(\mathbb{R}_0^+, Y)$ also incorporates the G output constraint, and is suffix-closed, prefix-closed, and non-anticipating. These two maps can be combined to define the input-state-output general flow $\Phi_{\mathcal{DI}} : (U \times X \times Y) \rightsquigarrow \mathsf{IPath}(\mathbb{R}_0^+, U \times X \times Y)$.

As our basic example of differing time lines, consider *analog-to-digital conversion* based on *event-driven* sampling rather than fixed periodic sampling.

Example 5.9. Let $Y \subseteq \mathbb{R}^n$ be a continuous *measurement space*, and the input signals will be continuous-time interval paths $\eta \in \mathsf{IPath}(\mathbb{R}_0^+, Y)$. Let $Z \neq \varnothing$ be a finite set of *event symbols*, and let $\mathbf{A} : Z \rightsquigarrow Y$ be any total map from Z to Y which associates with each event symbol $z \in Z$ a non-empty subset $\mathbf{A}(z) \subseteq Y$ of measurement values y which trigger the event symbol z. We require $\mathbf{A}(z) \cap \mathbf{A}(z') \neq \varnothing$ for some $z \neq z'$, as it is in these overlap regions (e.g. common boundaries) that the discrete event output signal $\gamma \in \mathsf{IPath}(\mathbb{N}, Z)$ can switch from value z to value z' (or vice-versa).

We define reactive systems $\mathsf{AnIn}_{\mathbf{A}}$ and $\mathsf{DigOut}_{\mathbf{A}}$:

$$\mathsf{AnIn}_{\mathbf{A}} : \mathsf{IPath}(\mathbb{R}_0^+, Y) \rightsquigarrow (\mathsf{HPath}(Z \times Y) \times \mathsf{TT}(\mathbb{H}, \mathbb{R}_0^+))$$

$$\mathsf{DigOut}_{\mathbf{A}} : \mathsf{HPath}(Z \times Y) \rightsquigarrow (\mathsf{IPath}(\mathbb{N}, Z) \times \mathsf{TT}(\mathbb{N}, \mathbb{H}))$$

with

$$\begin{aligned}
&\mathsf{AnIn}_{\mathbf{A}}(\eta) \\
&:= \{ (\xi, \bar{\tau}) \in \mathsf{HPath}(Z \times Y) \times \mathsf{TT}(\mathbb{H}, \mathbb{R}_0^+) \mid \\
&\quad \mathrm{ran}(\bar{\tau}) \subseteq \mathrm{dom}(\eta) \ \wedge \ \mathrm{dom}(\bar{\tau}) = \mathrm{dom}(\xi) \ \wedge \\
&\quad (\forall (i,t) \in \mathrm{dom}(\xi)) \ (\ \bar{\tau}(i,t) = t + \Sigma_{j<i}\Delta_j \\
&\quad \text{where } \Delta = \mathrm{ds}(\xi) \ \wedge \ \pi_Y\xi(i,t) = \eta(\bar{\tau}(i,t))\ ) \\
&\quad \wedge\ (\forall i < \mathrm{dl}(\xi))(\exists z_i \in Z)(\forall t \in [0, \Delta_i]\,) \\
&\quad \pi_Z\xi(i,t) = z_i \ \wedge \ \eta(\bar{\tau}(i,t)) \in \mathbf{A}(z_i) \qquad \}
\end{aligned}$$

$$\begin{aligned}
&\mathsf{DigOut}_{\mathbf{A}}(\xi) \\
&:= \{ (\gamma, \hat{\tau}) \in \mathsf{IPath}(\mathbb{N}, Z) \times \mathsf{TT}(\mathbb{N}, \mathbb{H}) \mid \\
&\quad \mathrm{ran}(\hat{\tau}) \subseteq \mathrm{dom}(\xi) \ \wedge \ \mathrm{dom}(\hat{\tau}) = \mathrm{dom}(\gamma) \ \wedge \\
&\quad \mathrm{length}(\gamma) = \mathrm{dl}(\xi) \ \wedge \ (\forall k \in \mathrm{dom}(\gamma)) \\
&\quad \hat{\tau}(k) = (k,0) \ \wedge \ \gamma(k) = \pi_Z\xi(\hat{\tau}(k)) \quad \}
\end{aligned}$$

Applied to a real-time input path η with values in Y, the map $\mathsf{AnIn}_{\mathbf{A}}$ will produce as output hybrid paths ξ over the product space $X = Z \times Y$, where the Y-projection $\pi_Y\xi$ reproduces the input η on the hybrid output time line, in the sense that $\pi_Y \xi = \bar{\tau} \circ \eta$, and the time translation $\bar{\tau}$: $\mathrm{dom}(\xi) \to \mathrm{dom}(\eta)$ maps a hybrid time point (i,t) back to $\bar{\tau}(i,t)$, the (real-valued) total duration of the hybrid path ξ to this point. The further constraint on ξ is that the Z-projection $\pi_Z \xi$ is constant with some value z_i for all positions (i,t) between $(i,0)$ and (i,Δ_i), and $(\bar{\tau} \circ \eta)(i,t) \in \mathbf{A}(z_i)$. This means for each $i < \mathrm{dl}(\xi)$, the input path η remains continuously within the region $\mathbf{A}(z_i) \subseteq Y$ for all times $s \in \mathrm{dom}(\eta)$ such that $\tau(i,0) \leqslant s \leqslant \tau(i,\Delta_i)$, and $\eta(s) \in \mathbf{A}(z_i) \cap \mathbf{A}(z_{i+1})$ at a switching time $s = \tau(i,\Delta_i) = \tau(i+1,0)$. The second map $\mathsf{DigOut}_{\mathbf{A}}$ takes as input hybrid paths ξ with values in $X = Z \times Y$ and returns as output a discrete-time path γ with values in Z obtained from ξ by simply projecting on to discrete time and discrete values in Z.

The overall analog-to-digital conversion system is a map $\mathsf{AnDig}_{\mathbf{A}} \colon \mathsf{IPath}(\mathbb{R}_0^+, Y) \rightsquigarrow (\mathsf{IPath}(\mathbb{N}, Z) \times \mathsf{TT}(\mathbb{N}, \mathbb{R}_0^+))$, induced by the map $\mathbf{A}$, and obtained from the two components $\mathsf{AnIn}_{\mathbf{A}}$ and $\mathsf{DigOut}_{\mathbf{A}}$ by a compound sequential composition operation. The systems $\mathsf{AnIn}_{\mathbf{A}}$ and $\mathsf{DigOut}_{\mathbf{A}}$ and $\mathsf{AnDig}_{\mathbf{A}}$ are each extension-preserving, non-anticipating, prefix-closed, suffix-closed and monoidal.

6. COMPARISON WITH OTHER INPUT-OUTPUT SYSTEM MODELS

Consider an input-output behaviour $\mathfrak{B} \subseteq [L \to (U \times Y)]$; we write $(\alpha, \beta) \in \mathfrak{B}$ where $\alpha \in [L \to U]$ and $\beta \in [L \to Y]$. An input-output behaviour $\mathfrak{B}$ has *free input* if $\pi_U(\mathfrak{B}) = [L \to U]$. The output of $\mathfrak{B}$ is *non-anticipating* of the input, if for all $(\alpha', \beta'), (\alpha'', \beta'') \in \mathfrak{B}$, and for all $t \in L$, if $\alpha'|_t = \alpha''|_t$, then there exists $\beta \in [L \to Y]$ such that $\beta|_t = \beta'|_t$ and $(\alpha'', \beta) \in \mathfrak{B}$. The following result establishes a basic correspondence, for interval paths with time lines $L = \mathbb{N}$ and $L = \mathbb{R}_0^+$, between these behavioural properties and the corresponding properties of reactive systems.

Theorem 6.1. Let the time line be either $L = \mathbb{N}$ or $L = \mathbb{R}_0^+$ and let $\Gamma \colon \mathsf{IPath}_\epsilon(L, U) \rightsquigarrow \mathsf{IPath}_\epsilon(L, Y)$ be a time-homogeneous, ω-extendible, output-time-unbounded, interval-path reactive system, in which case, $\mathrm{dom}(\mathsf{M}^\omega \Gamma) \subseteq [L \to U]$ and $\mathrm{ran}(\mathsf{M}^\omega \Gamma) \subseteq [L \to Y]$. Further suppose that $\mathfrak{B} = \mathsf{M}^\omega \Gamma = \{(\alpha, \beta) \mid \beta \in (\mathsf{M}^\omega \Gamma)(\alpha)\}$. Then:

(1.) $\mathfrak{B}$ is a non-anticipating behaviour with free input iff Γ non-anticipating with totally free input;

(2.) $\mathfrak{B}$ is a time-invariant behaviour iff Γ is suffix-closed/time-invariant.

7. CONCLUSION

We have developed a general formulation of non-deterministic input-output reactive systems, based on finite length paths, that is uniform for discrete, continuous and hybrid time, and that allows for the time line of output paths to differ from that of input paths. The work is intended as a first installment of a larger project on the study of compositions of input-output reactive systems by sequential, parallel and feedback constructions, with the aim of providing explicit set-theoretic semantics for non-deterministic dynamics and for time-translations within block-diagram based modeling and description frameworks such as CHARON (Alur *et al.*, 2000), among others.

REFERENCES

Alur, R., R. Grosu, Y. Hur, V. Kumar and I. Lee (2000). Modular specifications of hybrid systems in CHARON. In: *Hybrid Systems: Computation and Control (HSCC'00)* (N.A. Lynch and B.H. Krogh, Eds.). LNCS 1790. Springer-Verlag. pp. 6–19.

Aubin, J.-P. and O. Dordan (2002). Dynamical qualitative analysis of evolutionary systems. In: *Hybrid Systems: Computation and Control (HSCC'02)* (C.J. Tomlin and M.R. Greenstreet, Eds.). LNCS 2289. Springer-Verlag. pp. 62–75.

Aubin, J.-P., J. Lygeros, M. Quincampoix, S. Sastry and N. Seube (2002). Impulse differential inclusions: A viability approach to hybrid systems. *IEEE Trans. Automatic Control* **47**, 2–20.

Davoren, J.M., V. Coulthard, N. Markey and T. Moor (2004). Non-deterministic temporal logics for general flow systems. In: *Hybrid Systems: Computation and Control (HSCC'04)* (R. Alur and G.J. Pappas, Eds.). LNCS 2993. Springer-Verlag. pp. 280–295.

Lee, E.A. and A. Sangiovanni-Vincentelli (1998). A framework for comparing models of computation. *IEEE Trans. Computer-Aided Design of Integrated Circuits & Sys* **17**, 1217–1229.

Lynch, N.A., R. Segala, F. Vaandrager and H.B. Weinberg (1996). Hybrid I/O Automata. In: *Hybrid Systems III* (R. Alur, T.A. Henzinger and E. Sontag, Eds.). LNCS 1066. Springer-Verlag. pp. 496–510.

Moor, T. and J. Raisch (1999). Supervisory control of hybrid systems within a behavioural framework. *System and Control Letters* **38**, 157–166.

Tabuada, P., G. J. Pappas and P. Lima (2004). Compositional abstractions of hybrid control systems. *Discrete Event Dynamic Systems* **14**, 203–238.

Willems, J.C. (1989). Models for dynamics. *Dynamics Reported* **2**, 171–269.

Willems, J.C. (1991). Paradigms and puzzles in the theory of dynamical systems. *IEEE Trans. Automatic Control* **36**, 259–294.

ELSEVIER
IFAC
PUBLICATIONS

CONTROL-INVARIANCE OF SAMPLED-DATA HYBRID SYSTEMS WITH PERIODICALLY CLOCKED EVENTS AND JITTER

Yoshiyuki Tsuchie * **Toshimitsu Ushio** *

* *Graduate School of Engineering Science, Osaka University, Toyonaka, Osaka 560-8531, Japan*

Abstract: Silva and Krogh formulate a sampled-data hybrid automaton to deal with time-driven events and discuss its verification. In this paper, we consider a state feedback control problem of the automaton. First, we introduce two transition systems as semantics of the automaton. Next, using these transition systems, we derive necessary and sufficient conditions for a predicate to be control-invariant. Finally, we show that there always exists the supremal control-invariant subpredicate for any predicate.

Keywords: Hybrid automaton; Jitter; Control-invariance; State feedback;

1. INTRODUCTION

A hybrid automaton is widely used as a model of hybrid systems (Henzinger, 1996). A computer-controlled system is an example of hybrid systems since it has both continuous and discrete variables associated with the physical process (the plant) and the logical dynamics (the control logic and external environment), respectively. In the computer-controlled systems, the measurements and subsequent discrete control actions are usually time-driven events and there exist the jitter variations in their occurrence times. Silva and Krogh proposed an extension of a hybrid automaton called a sampled-data hybrid automaton (SDHA) to model explicitly discrete transitions that are based on time-driven sampling of the continuous state and define a transition system called a sampled-trace transition system (STTS) as semantics to verify its dynamics (Silva and Krogh, 2000; Silva and Krogh, 2001). The SDHA is a pair of a clock structure and a hybrid automaton with clocked and unclocked events. Unclocked events are enabled when its continuous states satisfy their guards while clocked events are time-driven, that is, they are enabled only at specified sampling times in addition to constraints for their guards. A clock structure, which is given by variation interval in the initial phase, a period of clocked times, and a sampling jitter, specifies sequences of sampling times that can be generated in the system. The SDHA can be used as a model of various controlled systems with time-driven events. For example, in networked control systems, the samplings and subsequent control actions through the network can be associated with clocked events while the changes of external environments and internal model changes of plants are associated with unclocked events. Silva and Krogh (2001) propose a verification method for the SDHA using approximated quotient transition systems. But, to the best of our knowledge, a control problem of the SDHA has not been studied.

In discrete event systems, a state feedback controller is often used as a logical control problem, where a control specification is given by a predicate on their states(Ramadge and Wonham, 1987). Its control action is determined by their current states. A discrete event system is called control-invariant if there exists a state feedback controller such that all reachable states in the closed-loop system controlled by the controller satisfy the predicate. A necessary and sufficient condition for the system to be control-invariant is derived(Ramadge and Wonham, 1987). The state feedback control for the discrete event systems is extended to hybrid systems(Chen and

Hanisch, 1999) and hybrid automata with forcible events (Ushio and Takai, 2005). Forcible events are events that can be forced to occur by the control so that temporal performance can be improved (Brandin and Wonham, 1994). Ushio and Takai extend transition semantics of uncontrolled hybrid systems (Henzinger, 1996) to controlled hybrid systems with forcible events and show necessary and sufficient conditions for a predicate to be control-invariant. They also show that there always exists the supremal control-invariant subpredicate for any predicate.

In this paper, we consider state feedback control of the SDHA. Since enablingness of the clocked events depends on the clock structure, a state feedback controller is time-varying in general while it is time-invariant in both hybrid systems without clocked events and discrete event systems. On the other hands, in conventional sampled-data control systems where controllers and sensors activate periodically, the sampling times are periodic but data transmission time may be fluctuated so that a jitter must be taken into consideration. So, we introduce a slight modification of the clock structure to represent sequences of periodic sampling times with jitter so that a periodic state feedback controller is designed.

The rest of this paper is organized as follows: Section 2 reviews transition systems, several predicate transformations, and a concept of control-invariance. Section 3 introduces a controlled SDHA with forcible events, which is given by a pair of a clock structure and a hybrid automaton with clocked and unclocked events. Two labeled transition systems are introduced to define its semantics and necessary and sufficient conditions for existence of state feedback controllers based on the transition systems are shown. Section 3 shows that there always exists the supremal control-invariant subpredicate for any predicate.

2. PRELIMINARIES

We use a labeled transition system $T=(Q, Act, \mathscr{T}, Q_0)$ in order to define semantics of controlled hybrid systems, where Q is a set of states, Act is a set of labels, $\mathscr{T}\subseteq Q\times Act\times Q$ is a state transition relation, $Q_0\subseteq Q$ is a set of initial states. $Act(T;q)\subseteq Act$ is defined by $Act(T;q)=\{a\in Act|^{\exists}q' \in Q \text{ s.t } (q,a,q')\in\mathscr{T}\}$. Let $\mathscr{P}(Q)$ be the set of all predicates on Q. A predicate P is *true* at state $q\in Q$ if $P(q)=1$, and *false* if $P(q)=0$. Denoted by $\vee$, $\wedge$, and $\neg$ are disjunction, conjunction, and negation of predicates, respectively. The term "predicate" and "subset" ($=\{q\in Q|P(q)=1\}$) can be used interchangeably. A partial order "$\leqslant$" for $\mathscr{P}(Q)$ is defined as follows: for $P_1, P_2\in\mathscr{P}(Q)$, $P_1\leqslant P_2\Leftrightarrow P_1(q)\leq P_2(q)$ for $^{\forall}q\in Q$. For each $a\in Act$, a predicate D_a is defined by

$$D_a(q) = \begin{cases} 1 & \text{if } a \in Act(T;q), \\ 0 & \text{otherwise.} \end{cases} \quad (1)$$

We define predicate transformations $wp_a:\mathscr{P}(Q)\rightarrow\mathscr{P}(Q)$ and $wlp_a:\mathscr{P}(Q)\rightarrow\mathscr{P}(Q)$ as follows:

$$wp_a(P)(q) = \begin{cases} 1 \text{ if } Post(q,a)\neq\emptyset \text{ and} \\ \quad (\text{for } ^{\forall}q'\in Post(q,a)) \ P(q')=1, \\ 0 \text{ otherwise,} \end{cases}$$

$$wlp_a(P) = wp_a(P) \ \vee \ \neg D_a,$$

where $Post(q,a)=\{q'\in Q|(q,a,q')\in\mathscr{T}\}$. For a subset $A\subseteq Act$, we define $wp_A(P)=\bigvee_{a\in A} wp_a(P)$. For a subset $A\subseteq Act$, $P\in\mathscr{P}(Q)$ is said to be $(T;A)$-invariant iff, for $^{\forall}a\in A$, $P\leqslant wlp_a(P)$. Let $\Re_{\geq 0}$ and $\Re_{>0}$ be the sets of non-negative and positive reals, respectively. For a piecewise continuous function $h : \Re_{\geq 0} \rightarrow A$, where A is an arbitrary set, $d(h) = d_0(h)d_1(h)d_2(h)\ldots$ is the sequence of points where h is discontinuous. For $t \in \Re_{\geq 0}$, $h(t^-)$ and $h(t^+)$ denote the values for the limits of h at t from the left and right, respectively.

3. SAMPLED-DATA HYBRID AUTOMATON AND STATE FEEDBACK CONTROL

Silva and Krogh proposed a sampled-data hybrid automaton (SDHA) which is modeled by a pair of a hybrid automaton with clocked and unclocked events and a clock structure (Silva and Krogh, 2001). First, we modify the SDHA to introduce a control mechanism with forcible events which are forced to occur by external control action. We define a controlled hybrid automaton with forcible events as follows: $H=(V, E, \Sigma, \Sigma_{con}, \Sigma_{uncon}, \Sigma_{forc}, \Sigma_{cl}, \Sigma_{uncl}, X, init, Flow, jump)$.

- V, Σ are sets of nodes, events, respectively;
- Σ_{con} and Σ_{uncon} are sets of controllable and uncontrollable events, respectively.
- Σ_{forc} is the set of forcible events and, for simplicity, we assume that $\Sigma_{forc}\subseteq\Sigma_{con}$;
- Σ_{cl} and Σ_{uncl} are sets of clocked and unclocked events, respectively;
- $E\subseteq V\times\Sigma\times V$ is the set of edges with associated events, that is, $e(v,\sigma,v')$ is an edge $e \in E$ from v to v' labeled by event σ and corresponds to a discrete transition by the occurrence of σ;
- $X \subseteq \Re^n$ is the set of continuous variables;
- $init : V \rightarrow 2^X$ assigns the initial continuous states, that is, $init(v)$ is the set of all possible initial continuous states in node v;
- $Flow=\{f_v:X\rightarrow\Re^n|^{\forall}v\in V\}$ is the set of flows defining the continuous state equation $\dot{x}= f_v(x)$ for each discrete state $v\in V$. Then let $x=\zeta_{v,x_0}(t)$ be a trajectory which starts from discrete state v at time t=0 and the initial continuous state $x(0)=x_0$, on which no event occurs; and
- $jump : E\rightarrow 2^{X\times X}$ is the jump relation, that is, $(x,x')\in jump(e)$ means that the continuous state x jumps to x' when σ occurs.

Note that $\Sigma_{con} \cap \Sigma_{uncon} = \Sigma_{cl} \cap \Sigma_{uncl} = \emptyset$ and $\Sigma=\Sigma_{con}\cup\Sigma_{uncon}=\Sigma_{cl}\cup\Sigma_{uncl}$. In addition, $\Sigma_{i,j}$ denotes $\Sigma_i \cap \Sigma_j$, where $i\in\{con, uncon, forc\}$ and $j\in\{cl, uncl\}$. The state set Q_H of hybrid automaton H is given by $Q_H=\{(v,x)|^{\forall}v\in V,^{\forall}x\in X\}$. Let $guard(e)$ be an occurrence condition of the discrete transition by edge $e \in E$: $guard(e)=\{x\in X|^{\exists}x'\in X \text{ s.t. } (x,x')\in jump(e)\}$.

Assumption 1 We assume: **(1)** If, for $e(v,\sigma,v')\in E$ and $x,x'\in X$, $(x,x')\in jump(e)$, then $x'\notin guard(\tilde{e})$ for any $\tilde{e}(v',\sigma',v'')\in E$, **(2)** for any $e(v,\sigma,v')\in E$ and $\sigma\in\Sigma_{forc}$, $guard(e)$ is a closed set, and **(3)** if $e(v,\sigma,v')\in E$, then $v\neq v'$.

The enablingness of clocked events does not depend on only the continuous states but also sampling times. To model the sequence of the sampling times, Silva and Krogh defines the clock structure such that durations between each sampling times are specified by the clock period interval and the sampling jitter interval (Silva and Krogh, 2001). In conventional computer controlled systems and networked control systems, controllers and sensors are activated periodically with a specified sampling period and computational delay in processors and/or data transmission delay in networks may cause a jitter in occurrence of clocked events. So, we will modify the clock structure. Let T_θ be a specified sampling period and J the maximum jitter. We assume for simplicity that $J<T_\theta$. Denoted by θ_n is the n-th "nominal" sampling times for the sampling process. Then, we have $\theta_n=\theta_0+nT_\theta$. θ_0 is the initial nominal sampling time. Let $\theta=\{\theta_n|\theta_n=\theta_0+nT_\theta\}$. Thus, a possible sequence of the sampling times $c=c_0c_1\ldots$ are given by the following modified clock structure:
$C(\theta,J)=\{c_0c_1\ldots|\text{for } \forall i\geq 0,\ c_i=\theta_i+J_i,\ J_i\in[0,J]\}$.
Thus, the controlled SDHA H_C is defined by a pair $(H,C(\theta,J))$ of the hybrid automaton with forcible events and the modified clocked structure.

We introduce a state feedback controller $f(q,t)$ with $q\in Q_H$ and $t\in\Re_{\geq 0}$ taking control of forcible and clocked events into consideration, which is an extension of a state feedback controller with forcible events (Ushio and Takai, 2005). Let $\Gamma_{cl}=\{\gamma|\Sigma_{cl,uncon}\subseteq\gamma\subseteq\Sigma_{cl}\}$ be the set of control patterns for clocked events. A state feedback controller f is described by 4-tuple $f=(f_{cl,1},f_{cl,2},f_{uncl,1},f_{uncl,2})$, where

- $f_{cl,1}: Q_H\times\Re_{\geq 0}\rightarrow\Gamma_{cl}$ gives a set control-enabled clocked events;
- $f_{cl,2}: Q_H\times\Re_{\geq 0}\rightarrow 2^{\Sigma_{cl,forc}}$ gives a set of forcible clocked events which are control-enabled and forced to occur; and
- $f_{uncl,1}$ and $f_{uncl,2}$ are given in similar ways to above two definitions, respectively.

Note that a state feedback controller f does not only depend on state, but also time since the sampling times have an effect on the behavior of the controlled SDHA and the control action depends on when clocked events are enabled. In addition, the nominal sampling times θ_i are periodic, the fluctuation of the actual sampling times c_i are in a time interval given by the jitter, the continuous flows are determined by a time-invariant system in each node, and a control specification is given by a predicate on Q_H independently of time. So, it is sufficient to consider a periodic state feedback controller f as follows: for $\forall q\in Q_H$ and $\forall t\geq\theta_0$, $f_{i,j}(q,t)=f_{i,j}(q,t+T_\theta)$, where $i\in\{cl,uncl\}$ and $j\in\{1,2\}$. Denoted by H_C^f is the SDHA controlled by the state feedback controller f.

Henzinger (1996) introduces two transition systems, called timed and timed-abstract transition systems, in order to represent semantics of the hybrid automaton and Ushio and Takai (2005) extended them to a controlled hybrid automaton with a control specification given by a predicate. Next, we extend them to H_C^f. This is also an extension of STTS (Silva and Krogh, 2001). To define semantics of the controlled SDHA by a transition system, the transition system must have a state variable which indicates duration between the current time and sampling times since both the sampling times and the current state in Q_H determine its behavior. In the STTS, the state is composed of the state variable $q\in Q_H$ and two variables ρ and ω which indicate time and elapsed time from the latest sampling time, respectively. Since we modify the clock structure, we introduce a new variable η as a state variable, which indicates if a clocked event may occur at the current time. Thus, we can define two semantics for the controlled SDHA H_C^f given by a controlled timed/time-abstract transition system.

A controlled timed transition system is defined by

$$S_C^t(H_C^f,P)=(Q_t,Act_t,\mathscr{T}_t^f,Q_{t0}). \qquad (2)$$

$Q_t\subseteq Q_H\times\{-1,0,1\}\times[0,\max(\theta_0,T_\theta)]\times\{0,1\}$ is the state set of the transition system. Each element of a state $(q,\rho,\omega,\eta)\in Q_t$ is defined as follows: q indicates a state of H_C. ρ indicates that the current time is before the first nominal sampling time θ_0 ($\rho=-1$), at θ_0 ($\rho=0$), after θ_0 ($\rho=1$). ω indicates absolute time until the current time becomes θ_0, where the system starts at $\omega=0$. When the current time is equal to or greater than θ_0, ω is an elapsed time from the latest nominal sampling time and is reset to zero at each nominal sampling time. Note that ω is zero if ρ is zero. η is reset to one at each nominal sampling time and to zero if $\omega>J$ or after a clocked event occurs. $Q_{t0}=\{(v,x,-1,0,0)\in Q_t|x\in init(v),\ \forall v\in V\}$ is the set of initial states in Q_t. $Act_t=\Sigma\cup\Re_{>0}$. To define the transition relation $\mathscr{T}_t^f$ of $S_C^t(H_C^f,P)$, we introduce a set of clock sequence $STS_C(q_t,\delta)$ as follows: for $q_t=(q,\rho,\omega,\eta)\in Q_t$ and $\delta\in\Re_{\geq 0}$,
$STS_C(q_t,\delta)=$

$$\begin{cases}\{t_0t_1\ldots t_N\mid t_i\in[\theta_i-\omega,\theta_i-\omega+J],\\ \quad i=0,1,\ldots,\ t_N<\delta\leq t_{N+1}\} \\ \hfill \text{if } \rho=-1,\\ \{t_0t_1\ldots t_N\mid t_{i-1}\in[iT_\theta-\omega,iT_\theta-\omega+J],\\ \quad i=1,2,\ldots,\ t_N<\delta\leq t_{N+1}\} \\ \hfill \text{if } \rho\neq-1,\eta=0,\\ \{t_0t_1\ldots t_N\mid t_i\in[iT_\theta,\ iT_\theta+J],\ i=0,1,2,\ldots,\\ \quad t_N<\delta\leq t_{N+1}\} \\ \hfill \text{if } \rho\neq-1,\omega=0,\eta=1,\\ \{t_0t_1\ldots t_N\mid t_{i-1}\in[iT_\theta-\omega,iT_\theta-\omega+J],\\ \quad i=1,2,\ldots,\ t_N<\delta\leq t_{N+1}\}\cup\{t_0t_1\ldots t_N\mid\\ \quad t_i\in\ [iT_\theta-\omega,\ iT_\theta-\omega+J],\ i=1,2,\ldots,\\ \quad t_0\in[0,J-\omega],\ t_N<\delta\leq t_{N+1}\} \\ \hfill \text{if } \rho=1,\omega\neq0,\eta=1.\end{cases}$$

$\mathscr{T}_t^f \subseteq Q_t \times Act_t \times Q_t$ is defined as follows: Consider states $q_t = (q, \rho, \omega, \eta)$ and $q_t' = (q', \rho', \omega', \eta') \in Q_t$.

(A) For each $\sigma \in \Sigma$, $(q_t, \sigma, q_t') \in \mathscr{T}_t^f$ iff the following conditions are satisfied:

(1) $\omega = \omega'$, $\rho = \rho'$, and $e(v, \sigma, v') \in E$.

(2) $(x, x') \in jump(e)$.

(3) if $\rho = \rho' = -1$, then
(i) $\sigma \in f_{uncl,1}(q,\ \omega)$ and (ii) $\eta = \eta' = 0$.

(4) if $\rho = \rho' = 0$, then
- if $\sigma \in \Sigma_{cl}$, then (i) $\sigma \in f_{cl,1}(q, \theta_0)$ and (ii) $\eta = 1$, $\eta' = 0$,
- otherwise (i) $\sigma \in f_{uncl,1}(q, \theta_0)$ and (ii) $\eta = \eta' = 1$.

(5) if $\rho = \rho' = 1$, then
- if $\sigma \in \Sigma_{cl}$, then (i) $\omega \in [0, J]$, (ii) $\sigma \in f_{cl,1}(q, \theta_0 + \omega)$, and (iii) $\eta = 1$, $\eta = 0$,
- otherwise (i) $\sigma \in f_{uncl,1}(q, \theta_0 + \omega)$ and (ii) $\eta = \eta'$.

(B) For $\delta \in \Re_{>0}$, $(q_t,\ \delta,\ q_t') \in \mathscr{T}_t^f$ iff the following conditions are satisfied:

(1) $v = v'$, $x' = \zeta_q(\delta)$.

(2) For $^\forall \alpha, \beta \in (0, \delta)$, $P(v, \zeta_q(\alpha)) = P(v, \zeta_q(\beta))$.

(3) One of the following conditions is satisfied:
- if $\rho = \rho' = -1$, then (i) $\delta = \omega' - \omega$, (ii) $\eta = 0$, $\eta' = 0$, and (iii) for $^\forall e(v, \tilde{\sigma}, \tilde{v}) \in E$, $^\forall t \in [0, \delta)$, $\zeta_q(t) \in guard(e) \Rightarrow \tilde{\sigma} \notin f_{uncl,2}(v, \zeta_q(t), \omega + t)$.
- if $\rho = -1, \rho' = 0$, then (i) $\delta = \theta_0 - \omega$, (ii) $\eta = 0$, $\eta' = 1$, and (iii) for $^\forall e(v, \tilde{\sigma}, \tilde{v}) \in E$, $^\forall t \in [0, \delta)$, $\zeta_q(t) \in guard(e) \Rightarrow \tilde{\sigma} \notin f_{uncl,2}(v, \zeta_q(t), \omega + t)$.
- if $\rho = -1$, $\rho' = 1$, then (i) $^\exists K \geq 0$ s.t. $\delta = \omega' + \theta_K - \omega$, (ii) if $K = 0$, $\omega' \neq 0$, (iii) if $\omega' \in [0, J]$, $\eta' = 1$, (iv) for $^\forall e(v, \tilde{\sigma}, \tilde{v}) \in E$, $^\forall t \in [0, \delta)$, $\zeta_q(t) \in guard(e) \Rightarrow \tilde{\sigma} \notin f_{uncl,2}(v, \zeta_q(t), \omega + t)$, and (v) $^\exists \{t_k\} \in STS_C(q_J, \delta)$ s.t. for $^\forall e(v, \tilde{\sigma}, \tilde{v}) \in E$, $^\forall t \in \{t_k\}$, $\zeta_q(t) \in guard(e) \Rightarrow \tilde{\sigma} \notin f_{cl,2}(v, \zeta_q(t), \omega + t)$.
- if $\rho \neq -1$, $\rho' = 1$, then (i) $^\exists n \in Z_{\geq 0}$ s.t. $\delta = \omega' - \omega + nT_\theta$, (ii) for $^\forall e(v, \tilde{\sigma}, \tilde{v}) \in E$, $^\forall t \in [0, \delta)$, $\zeta_q(t) \in guard(e) \Rightarrow \tilde{\sigma} \notin f_{uncl,2}(v, \zeta_q(t), t + \omega + \theta_0)$, (iii) $^\exists \{t_k\} \in STS_C(q_t, \delta)$ s.t for $^\forall e(v, \tilde{\sigma}, \tilde{v}) \in E$, $^\forall t \in \{t_k\}$, $\zeta_q(t) \in guard(e) \Rightarrow \tilde{\sigma} \notin f_{cl,2}(v, \zeta_q(t), t + \omega + \theta_0)$, (iv) if $\eta = 1$, $\omega' \in [0, J] \Rightarrow \eta' = 1$, and (v) if $\eta = 0$, for any n satisfying (i), the following equations holds. if $n \geq 1$ and $\omega' \in [0, J]$, $\eta' = 1$. if $n = 0$, $\eta' = 0$.

Let $SDS_{C,f}(q_t, \delta, q_t')$ be the set of sampled state sequences on a transition relation $(q_t, \delta, q_t') \in \mathscr{T}_t^f$ defined as follows: $SDS_{C,f}(q_t, \delta, q_t') = \{(v, \zeta_q(t_0)), (v, \zeta_q(t_1)) \ldots (v, \zeta_q(t_N)) \mid \{t_k\}_{k=0}^N \in STS_C(q_t, \delta)$, $^\exists q_{t,k} \in Q_t$, $k = 0, 1, \ldots, N-1$ s.t. $(q_t, t_0, q_{t,0}) \in \mathscr{T}_t^f$, $(q_{t,k}, t_{k+1} - t_k, q_{t,k+1}), (q_{t,N}, \delta - t_N, q_t') \in \mathscr{T}_t^f, q_k = (v, \zeta_q(t_k))\}$, where $q_{t,k} = (q_k, \rho_k, \omega_k, \eta_k)$. In the controlled timed transition system, $\Re_{>0} \subseteq Act_t$ is the set of elapsed times from the latest occurrence of events. So, by aggregating such events into two events indicating that time elapses, a controlled time-abstract transition system is defined by

$$S_C^a(H_C^f, P) = (Q_a, Act_a, \mathscr{T}_a^f, Q_{a0}). \quad (3)$$

The sets Q_a and Q_{a0} are the same as those of S_C^t, that is, $Q_a = Q_t$ and $Q_{a0} = Q_{t0}$. $Act_a = \Sigma \cup \{\tau_{con}, \tau_{uncon}\}$, where τ_{con} and $\tau_{uncon} \notin \Sigma$ are events. $\mathscr{T}_a^f \subseteq Q_a \times Act_a \times Q_a$ is defined as follows: Transitions related to events in Σ are the same as those in $\mathscr{T}_t^f$ in $S_C^t(H_C^f, P)$. Consider two states $q_a = (q, \rho, \omega, \eta)$, $q_a' = (q', \rho', \omega', \eta') \in Q_a$, where $q = (v, x)$, $q' = (v', x') \in Q_H$. Let $\Delta(q_a, q_a') = \{\delta \in \Re_{>0} \mid (q_a, \delta, q_a') \in \mathscr{T}_t^f\}$.

(A) $(q_a, \tau_{con}, q_a') \in \mathscr{T}_a^f$ iff $\Delta(q_a, q_a') \neq \emptyset$ and for any $\delta \in \Delta(q_a, q_a')$,

$$wp_{\Sigma_{cl,forc}}(P)(q_a) \vee \left[\bigwedge_{\{q_k\} \in SDS_{C,f}} \left\{ \bigvee_{\hat{q} \in \{q_k\}} (wp_{\Sigma_{cl,forc}}(P)(\hat{q})) \right\} \right] \vee \left\{ \bigvee_{\epsilon \in [0,\delta)} (wp_{\Sigma_{uncl,forc}}(P)(v, \zeta_q(\epsilon))) \right\} = 1, \quad (4)$$

where $SDS_{C,f} = SDS_{C,f}(q_a, \delta, q_a')$, $P(q_a) = P(q)$.

(B) $(q_a, \tau_{uncon}, q_a') \in \mathscr{T}_a^f$ iff $\Delta(q_a, q_a') \neq \emptyset$ and there exists $\delta \in \Delta(q_a, q_a')$ such that Eq. 4 does not hold.

Let $open = (open_{cl,1}, open_{cl,2}, open_{uncl,1}, open_{uncl,2})$ be a state feedback controller defined as follows: for any $q \in Q_H$ and any $t \in \Re_{\geq 0}$, $open_{cl,1}(q, t) = \Sigma_{cl}$, $open_{cl,2}(q, t) = \emptyset$, $open_{uncl,1}(q, t) = \Sigma_{uncl}$, $open_{uncl,2}(q, t) = \emptyset$. Two transition systems controlled by the controller $open$ are denoted as follows:

$$S_C^t(P) = (Q_t, Act_t, \mathscr{T}_t, Q_{t0}) = S_C^t(H_C^{open}, P),$$
$$S_C^a(P) = (Q_a, Act_a, \mathscr{T}_a, Q_{a0}) = S_C^a(H_C^{open}, P).$$

Note that $S_C^t(P)$ and $S_C^a(P)$ correspond to semantics of the uncontrolled system. From the above definitions, the following lemma is easily shown.

Lemma 1 Let $f = (f_{cl,1}, f_{cl,2}, f_{uncl,1}, f_{uncl,2})$ be a state feedback controller for a controlled SDHA $H_C(H, C(\theta,\ J))$. Then, for two states $q_a = (q,\ \rho,\ \omega,\ \eta)$, $q_a' \in Q_a = Q_t$, and event $\sigma \in \Sigma$,
- $(q_a,\ \sigma,\ q_a') \in \mathscr{T}_t^f \Rightarrow (q_a,\ \sigma,\ q_a') \in \mathscr{T}_t$.
- $(q_a,\ \sigma,\ q_a') \in \mathscr{T}_t$ and
$$\begin{cases} \sigma \in f_{uncl,1}(q,\ \omega) & \text{if } \rho = -1 \\ \sigma \in f_{cl,1}(q,\ \theta_0 + \omega) & \text{if } \sigma \in \Sigma_{cl} \\ \sigma \in f_{uncl,1}(q,\ \theta_0 + \omega) & \text{otherwise} \end{cases}$$
$\Rightarrow (q_a,\ \sigma,\ q_a') \in \mathscr{T}_t^f$, where $q = (v, x) \in Q_H$.

Let r be a run for H_C^f, and r is defined as follows:

$$r = (v, x, c), \quad (5)$$

where $v(t) \in V$ is a trajectory for the discrete variable, $x(t) \in X$ is a trajectory for the continuous variable, and $c \in C$ is the sampling times synchronizing with H_C^f. It is said that r is a run for H_C^f if the following conditions hold:

(1) $v(0) \in V$, $x(0) \in init(v(0))$.

(2) if $t \in d(v)$, then there exists $\sigma \in \Sigma$ such that the following three conditions hold: (i) $e(v(t^-), \sigma, v(t^+)) \in E$, (ii) if $\sigma \in \Sigma_{cl}$, $\sigma \in f_{cl,1}((v(t^-), x(t^-)), t)$ and $t \in c$, (iii) if $\sigma \in \Sigma_{uncl}$, $\sigma \in f_{uncl,1}((v(t^-), x(t^-)), t)$.

(3) if $t \notin d(v)$, then (i) $\dot{x}(t) = f_{v(t)}(x(t))$, (ii) for $^\forall e(v, \tilde{\sigma}, \tilde{v}) \in E$, if $x(t) \in guard(e)$, $\tilde{\sigma} \notin f_{uncl,2}(v(t), x(t), t)$, and (iii) (if $t \in c$) for $^\forall e(v, \tilde{\sigma}, \tilde{v}) \in E$, $x(t) \in guard(e) \Rightarrow \tilde{\sigma} \notin f_{cl,2}(v(t), x(t)), t)$.

The following propositions indicate that the transition systems are the semantics of the controlled SDAH H_C^f.

Proposition 1 Consider a controlled SDHA H_C^f and its associated timed transition system $S_C^t(H_C^f, P)$. If r is a run for H_C^f, then there exists a corresponding sequence of states $q_t^r = q_{t,0}^r q_{t,1}^r \cdots$ that is a state trajectory for $S_C^t(H_C^f, P)$.

Proposition 2 Consider a controlled SDHA H_C^f and its associated timed transition system $S_C^t(H_C^f, P)$. If $q_t = q_{t,0} q_{t,1} q_{t,2} \ldots$ is a state trajectory for $S_C^t(H_C^f, P)$, there exists a corresponding run r^t for H_C^f.

The above propositions are also true for the time-abstract transition system $S_C^a(H_C^f, P)$.

We extend a predicate P on Q_H to $Q_t = Q_a$ as follows: for each state $q_t = (q, \rho, \omega, \eta) \in Q_t$, $P : Q_t \to \{0, 1\}$ is defined by $P(q_t) = P(q)$.

4. CONTROL-INVARIANCE

A concept of control-invariance plays an important role in state feedback control of discrete event systems (Ramadge and Wonham, 1987). A predicate $P \in \mathscr{P}(Q_a)$ is said to be control-invariant if there exists a state feedback controller f such that P is $(S_C^t(H_C^f, P); Act_t)$-invariant. Such a controller f is called a permissive feedback controller. We show necessary and sufficient conditions for P to be control-invariant in the controlled SDHA. We define predicates for trajectories of the continuous variables as follows: for a predicate $P \in \mathscr{P}(Q_a)$, states $q_a = (q, \rho, \omega, \eta) \in Q_a$, $q_a' = (v', \zeta_q(\delta), \rho', \omega', \eta') \in Q_a$, and time $\delta \in \Re_{>0}$,

$$pc_{C,\delta}(P)(q_a) = \begin{cases} 1 & \text{if } P(v, \zeta_q(\epsilon)) = 1, \forall \epsilon \in (0, \delta), \\ 0 & \text{if } P(v, \zeta_q(\epsilon)) = 0, \forall \epsilon \in (0, \delta), \end{cases}$$

$$pwp_{\Sigma_{forc},\delta}(P)(q_a) =$$

$$\left\{ \bigvee_{\epsilon \in [0,\ \delta)} (wp_{\Sigma_{uncl,forc}}(P)(v,\ \zeta_q(\epsilon))) \right\}$$

$$\vee \left[\bigwedge_{\{q_k\} \in SDS_C} \left\{ \bigvee_{\hat{q} \in \{q_k\}} (wp_{\Sigma_{cl,forc}}(P)(\hat{q})) \right\} \right],$$

$$twp_{C,\delta}(P)(q_a) = P(q') \vee pwp_{\Sigma_{forc},\delta}(P)(q_a),$$

where $SDS_C = SDS_{C,open}(q_a, \delta, q_a')$. If a predicate P is a closed set, twp can be rewritten as follows: $twp_{C,\delta}(P)(q_a) = pwp_{\Sigma_{forc},\delta}(P)(q_a)$.

A predicate P is said to be $(S_C^t(P); \Sigma_{uncon}, \Re_{>0}, \Sigma_{forc})$-invariant if the following conditions hold:

(1) P is $(S_C^t(P); \Sigma_{uncon})$-invariant, and

(2) in $S_C^t(P)$, for any $\delta \in \Re_{>0}$,

$$P \leqslant \neg D_\delta \vee wp_{\Sigma_{forc}}(P) \vee (twp_{C,\delta}(P) \wedge pc_{C,\delta}(P)). \quad (6)$$

We show necessary and sufficient conditions for the control-invariance.

Theorem 1 Consider the controlled SDHA H_C and a predicate $P \in \mathscr{P}(Q_a)$. Then, the following three statements are equivalent:

(i) P is control-invariant.

(ii) P is $(S_C^t(P)\ ;\ \Sigma_{uncon}, \Re_{>0}, \Sigma_{forc})$-invariant.

(iii) P is $(S_C^a(P)\ ;\ \Sigma_{uncon} \cup \{\tau_{uncon}\})$-invariant.

Proof: (i)⇒(ii). Suppose that P is control-invariant. Let f be a permissive feedback controller. Suppose that P is not $(S_C^t(P); \Sigma_{uncont}, \Re_{>0}, \Sigma_{forc})$-invariant. Then, we have the following cases:

- Consider the case that there exist $q_t, q_t' \in Q_t$, and $\sigma \in \Sigma_{uncon}$ such that $(q_t, \sigma, q_t') \in \mathscr{T}_t$, $P(q_t) = 1$, and $P(q_t') = 0$. Since $\Sigma_{uncon,cl} \subseteq f_{cl,1}$ and $\Sigma_{uncon,uncl} \subseteq f_{uncl,1}$, we have $(q_t, \sigma, q_t') \in \mathscr{T}_t^f$ by Lemma 1. Since f is a permissive feedback controller, we have $P(q_t) = 0$, which is a contradiction.
- Consider the case that there exist $\delta \in \Re_{>0}$ and $q_t = (q, \rho, \omega, \eta) \in Q_t$ which do not satisfy Eq. (6). Then, $P(q_t) = 1$, there exists $q_t' = (q', \rho', \omega', \eta') \in Q_t$ such that $(q_t, \delta, q_t') \in \mathscr{T}_t$, one of the following conditions is satisfied:
  - **(A)** $P(v, \zeta_q(\epsilon)) = 0$ for any $\epsilon \in (0, \delta)$ and $wp_{\Sigma_{forc}}(P)(q_t) = 0$. Since $wp_{\Sigma_{forc}}(P)(q_t) = 0$, there exist $0 < \tilde{\epsilon} < \delta$ and $\tilde{q}_a = (v, \zeta_q(\tilde{\epsilon}), \tilde{\rho}, \tilde{\omega}, \tilde{\eta}) \in Q_t$ such that $(q_t, \epsilon, \tilde{q}_a) \in \mathscr{T}_t^f$. Then we have $P(\tilde{q}_a) = 0$, which is a contradiction since f is a permissive feedback controller.
  - **(B)** $P(v, \zeta_q(\epsilon)) = 1$ for any $\epsilon \in (0, \delta)$, $P(q_t') = 0$, and $pwp_{\Sigma_{forc},\delta}(P)(q_t) = 0$. Then, we have $(q_t, \delta, q_t') \in \mathscr{T}_t^f$, which is also a contradiction.

From the above contradictions, P is shown to be $(S_C^a(P); \Sigma_{uncon}, \Re_{>0}, \Sigma_{forc})$-invariant.

(ii)⇒(iii) For $q_a, q_a' \in Q_a = Q_t$, and $\sigma \in \Sigma_{uncon}$, the following implication is easily shown: $(q_a, \sigma, q_a') \in \mathscr{T}_t \Rightarrow (q_a, \sigma, q_a') \in \mathscr{T}_a$. Thus, we have P is $(S_C^a(P)\ ;\ \Sigma_{uncon})$-invariant if P is $(S_C^t(P); \Sigma_{uncon})$-invariant. Suppose that P is not $(S_C^a(P); \tau_{uncon})$-invariant. Then, there exist q_a and q_a' such that $(q_a, \tau_{uncon}, q_a') \in \mathscr{T}_a$, $P(q_a) = 1$, and $P(q_a') = 0$. From the definition of τ_{uncon}, there exits $\delta \in \Re_{>0}$ such that $(q_a, \delta, q_a') \in \mathscr{T}_t$, which is a contradiction since P is $(S_C^t(P); \Sigma_{uncon}, \Re_{>0}, \Sigma_{forc})$-invariant. Thus, P is $(S_C^a(P); \Sigma_{uncon} \cup \{\tau_{uncon}\})$-invariant.

(iii)⇒(i) Suppose that P is $(S_C^a(P); \Sigma_{uncon} \cup \{\tau_{uncon}\})$-invariant. Then, we consider the following state feedback controller $f = (f_{cl,1}, f_{cl,2}, f_{uncl,1}, f_{uncl,2})$: for each $q \in Q_H$, $t \in \Re_{\geq 0}$, $f_{cl,1}(q,t) = \Sigma_{cl,uncon} \cup \{\sigma \in \Sigma_{cl,con} \mid wp_\sigma(P)(q, \rho, \omega, 1) = 1$ in $S_C^t(P)\}$, $f_{uncl,1}(q,t) = \Sigma_{uncl,uncon} \cup \{\sigma \in \Sigma_{uncl,con} \mid wp_\sigma(P)(q, \rho, \omega, 0) = 1$ in $S_C^t(P)\}$, where ρ and ω satisfy **(a)** if $t < \theta_0$, then $\rho = -1$, and $\omega = t$. **(b)** if $t = \theta_0$, then $\rho = 0$, and $\omega = 0$. **(c)** if $t > \theta_0$, then $\rho = 1$, and $\omega = t - \theta_0 - \lfloor (t - \theta_0)/T_\theta \rfloor T_\theta$. $f_{cl,2}(q,t) = \Sigma_{cl,forc} \cap f_{cl,1}$. $f_{uncl,2}(q,t) = \Sigma_{uncl,forc} \cap f_{uncl,1}$. Thus, it is easy to prove that P is $(S_C^t(H_C^f, P); Act_t)$-invariant. ∎

5. SUPREMAL CONTROL-INVARIANT SUBPREDICATE

In general, a given predicate $P \in \mathscr{P}(Q_a)$ is not necessarily control-invariant. In this section, we propose a procedure for computation of the supremal control-invariant subpredicate. We introduce some definitions for the predicate P as follows:

- $\mathscr{C}I(P)$ is the set of all control-invariant subpredicates of $P \in \mathscr{P}(Q_a)$.

- $\mathbf{0}\in\mathscr{P}(Q_a)$ is the predicate such that for any $q_a\in Q_a$, $\mathbf{0}(q_a)=0$. Since $\mathbf{0}\in\mathscr{C}I(P)$, $\mathscr{C}I(P)\neq\emptyset$.
- A predicate $P^{\uparrow}\in\mathscr{C}I(P)$ called the supremal control-invariant subpredicate of P is defined as follows: for any $P'\in\mathscr{C}I(Q_a)$, $P'\leqslant P^{\uparrow}$.

Ushio and Takai (2005) showed that there always exists $P^{\uparrow}$ for the hybrid systems with forcible event. In this section, we show the same property also holds for the controlled SDHA.

Since $P'\leqslant P\Rightarrow \text{for}^{\forall}\sigma\in\Sigma, wp_{\sigma}(P')\leqslant wp_{\sigma}(P)$, the following lemma is easily shown.

Lemma 2 Let P and $P'\in\mathscr{P}(Q_a)$ be predicates such that $P'\leqslant P$ and P' is $(S_C^a(P');\Sigma_{uncon}\cup\{\tau_{uncon}\})$-invariant. For any $q_a\in\{q_a=(q,\rho,\omega,\eta)\in Q_a|P'(q_a)=1\}$, if there exists $q'_a\in Q_a$ such that $(q_a,\ \tau_{uncon},\ q'_a)\in\mathscr{T}_a$ in $S_C^a(P)$, then $(q_a,\tau_{uncon},q'_a)\in\mathscr{T}_a$ also holds in $S_C^a(P')$

Using Lemma 2, we prove the following theorem.

Theorem 2 Let I be any index set. if $P_i\in\mathscr{P}(Q_a)$ is $(S_C^a(P_i)\mid\Sigma_{uncon}\cup\{\tau_{uncon}\})$ -invariant for each $i\in I$, then, $P_I=\bigvee_{i\in I}P_i$ is $(S_C^a(P_I)\ ;\ \Sigma_{uncon}\cup\{\tau_{uncon}\})$ -invariant.

By Theorem 2, there exists its supremal control-invariant subpredicate $P^{\uparrow}$ for any predicate $P\in\mathscr{P}(Q_a)$.

The following theorem gives an iterative scheme for computing the supremal control-invariant subpredicate.

Theorem 3 For any $P\in\mathscr{P}(Q_a)$, consider the following iterative computation: $P_{j+1}=P_j\wedge\Psi(P_j)\ (^{\forall}j\geq 0)$, where $P_0:=P$.

Then the following implication holds:

$$^{\exists}k\geq 0 \text{ s.t. } P_{k+1}=P_k\Rightarrow P^{\uparrow}=P_k, \qquad (7)$$

where $\Psi:\mathscr{P}(q_a)\rightarrow\mathscr{P}(q_a)$ is defined as follows:

$$\Psi(P)(q_a)=\begin{cases}1 & \text{if} \displaystyle\bigwedge_{\sigma\in\Sigma_{uncon}\cup\{\tau_{uncon}\}} wlp_{\sigma}(P)(q_a)=1 \text{ in } S_C^a(P)\\ 0 & \text{otherwise.}\end{cases}$$

Proof: Assume that there exists k such that $P_{k+1}=P_k$. For the above iterative scheme, we have $P_k=P_{k+1}=P_k\wedge\Psi(P_k)\leqslant wlp_a(P_k)$ for any $a\in\Sigma_{uncon}\cup\{\tau_{uncon}\}$ and $P_{j+1}=P_j\wedge\Psi(P_j)\leqslant P_0=P$ for any $j\geq 0$. So P_k is a control-invariant subpredicate of P i.e. $P_k\in\mathscr{C}I(P)$, which implies $P_k\leqslant P^{\uparrow}$.

Next, we prove that $P^{\uparrow}\leqslant P_l$ for $l=0,1,\ldots,k$ by induction. **(1)** $l=0$. Since $P^{\uparrow}$ is a subpredicate of P, we have $P^{\uparrow}\leqslant P_0$. **(2)** Suppose that $P^{\uparrow}\leqslant P_l$ holds. Then, we show by a contradiction that $P^{\uparrow}\leqslant P_{l+1}$ holds. If $P^{\uparrow}\leqslant P_{l+1}$ does not holds, then there exists $q_a\in Q_a$ such that $P^{\uparrow}(q_a)=1$ and $\Psi(P_l)(q_a)=0$ since $P^{\uparrow}(q_a)=1\Rightarrow P_l(q_a)=1$ for any $q_a\in Q_a$. If $\Psi(P_l)(q_a)=0$ holds, One of the following cases always holds: **(a)** In $S_C^a(P_l)$, there exists $\sigma\in\Sigma_{uncon}$ such that $wlp_{\sigma}(P_l)(q_a)=0$. Since D_{σ} in $S_C^a(P^{\uparrow})$ is equivalent to that in $S_C^a(P_l)$, there exists $q'_a\in Q_a$ such that $(q_a,\sigma,q'_a)\in\mathscr{T}_a$ in $S_C^a(P^{\uparrow})$ and $P_l(q'_a)=0$, which implies that $P^{\uparrow}(q'_a)=0$. This contradicts the assumption that $P^{\uparrow}$ is control-invariant. **(b)** $wlp_{\tau_{uncon}}(P_l)(q_a)=0$ holds in $S_C^a(P_l)$. Then there exists $q'_a\in Q_a$ such that $(q_a,\tau_{uncon},q'_a)\in\mathscr{T}_a$ in $S_C^a(P_l)$ and $P_l(q'_a)=0$. By Lemma 2, we have $(q_a,\tau_{uncon},q'_a)\in\mathscr{T}_a$ in $S_C^a(P^{\uparrow})$. Since $P^{\uparrow}(q'_a)=0$, this contradicts the assumption that $P^{\uparrow}$ is control-invariant.
For the above cases, we have $P^{\uparrow}\leqslant P_{l+1}$ and $P^{\uparrow}\leqslant P_k$. Therefore, we have $P_k=P^{\uparrow}$. ■

Note that $P^{\uparrow}$ computed by the above scheme depends on time in general while the control specification $P\in\mathscr{P}(Q_H)$ is independent of time.

6. CONCLUSION

This paper considered state feedback control of a sampled-data hybrid automaton as a model of computer-controlled systems where control specifications are given by predicates.

We introduced two transition systems as semantics for the controlled sampled-data hybrid automata and proved necessary and sufficient conditions for the control-invariance, and showed that there always exists the supremal control-invariant subpredicate for any predicate.

In general, the procedure for computation of the supremal control-invariant subpredicate is not decidable. So it is future work to obtain an approximation method for the computation.

REFERENCES

Brandin, B. A. and Wonham (1994). Supervisory control of timed discrete-event systems. *IEEE Transactions on Automatic Control* **39**(2), 329–342.

Chen, H. and H.-M. Hanisch (1999). Control synthesis of hybrid systems based on predicate invariance. In: *Hybrid Systems V*. pp. 1–15.

Henzinger, T. A. (1996). The theory of hybrid automata. *Proceedings of the 11th Symposium on Logic in Computer Science* pp. 278–292.

Ramadge, P. J. and W. M. Wonham (1987). Modular feedback logic for discrete event systems. *SIAM Journal on Control and Optimization* **25**(5), 1202–1218.

Silva, B. I. and B. H. Krogh (2000). Modeling and verification of sampled-data hybrid systems. *ADPM 2000* pp. 237–242.

Silva, B. I. and B. H. Krogh (2001). Modeling and verification of hybrid systems with clocked and unclocked events. *Proc. 40th IEEE CDC* pp. 762–767.

Ushio, T. and S. Takai (2005). Control-invariance of hybrid systems with forcible events. *Automatica* **41**, 669–675.

Author Index

ADHS'06

2nd IFAC Conference on Analysis and Design of Hybrid Systems
Alghero, Italy - June 7-9, 2006

Abate A.	359
Alessio A.	302
Ambrosi L.	383
Andersson M.	272
Årzén K. E.	272
Axelsson H.	95
Azhmyakov V.	89
Bacconi F.	83
Bakule L.	130
Balluchi A.	259
Barbot J.-P.	124
	341
Bartolini G.	276
Bastin G.	191
Bauso D.	77
Beccuti A. G.	315
Beers C. D.	71
Bemporad A.	241
	302
Ben Makhlouf I.	377
Benali A.	341
Benvenuti L.	259
Blanchini F.	77
Bicchi A	353
Birouche A.	12
Biswas G.	71
Blom H. A. P.	160
Boccadoro M.	95
	101
Bordons C.	199
Bouchhima F.	274
Boutat D.	341
Buisson J.	309
	329
Bujorianu M. L.	160
Bukkems B.	142
Caines P.E.	166
Camacho E. F.	199
Camlibel K.	290
Casavola A.	83
Cassandras C. G.	136
	267
Cervin A.	272
Chaib S.	341
Ciapessoni E.	383
Clune M. I.	267
Colaneri P.	347
Corona D.	253
Cuijpers P. J. L.	56
Daafouz J.	12
	335
Davoren J. M.	409
de Best J.	142
De Santis E.	112
De Schutter B.	148
	253
Denis B.	365
Di Benedetto M. D.	24
	112
	259
Di Cairano S.	241
Di Gennaro S.	24
D'Innocenzo A.	24
Djemaï M.	124
Dochain D.	191
Dotoli M.	44
Dumur D.	329
Egerstedt M.	95
	101
El Moudni A.	50
Engell S.	211
Fanti M. P.	44
Ferrarini L.	383
Ferrier J.-L.	283
Floquet T.	124
García-Gabín W.	199
Geist S.	205
Geromel J. C.	347
Gheorghe L.	274
Girard A.	106
Giua A.	37
Gromov D.	205
	395
Guéguen H.	118
Hagander P.	185
Haugwitz S.	185
Hayat S.	50
Heemels W. P. M. H.	290
	296
	321
Hellendoorn J.	148
Henriksson D.	272
Hermanns H.	160
Hetel L.	335
Hétreux G.	235
Hrovat D.	241
Iannelli L.	247
Inagaki S.	64

Iung C. 4
12
335
Johansson K. H. 247
Jolly J-C. 283
Jönsson U. T. 247
Juárez-Orozco Z. 365
Julius A. A. 106
Kaakai F. 50
Kolmanovsky I. 241
Kouretas P. 172
Koutoumpas K. 172
Kowalewski S. 377
Lazar M. 296
321
Le Lann M.-V. 235
Le Lann J.-M. 235
Lecœuche S. 154
Lee E.A. 270
Lee H. 1
Lefebvre M.-A. 118
Leirens S. 309
Lesage J. J. 365
Lichtenberg G. 179
Lunze J. 229
Lygeros J. 172
Ma Z. 166
Mahulea C. 37
Malhame R. 166
Manamanni N. 124
Manders E.-J. 71
Mangini A. M. 44
Messai N. 217
Millan J. 402
Miyagi P. E. 389
Moor T. 409
Morari M. 315
Mosca E. 83
Mosterman P. J. 71
267
Nasri O. 118
Necoara I. 148
Nicolescu G. 274
O'Young S. 402
Olaru S. 329
Olivier N. 235
Papafotiou G. 315
Pappas G. J. 106
Pekpe K. M. 154
Pesenti R. 77
Pettersson S. 18
Picasso B. 353
Pisano A. 276
Pochet Y. 191
Pogromsky A. Y. 6
Pola G. 112
Quémard C. 283
Raisch J. 89
205
395
Ratschan S. 371
Recalde L. 30
37
Reniers M. A. 56
Riedinger P. 4
Riera B. 217
Rodrigues M. 223
Rooda J.E. 6
265
Rostalski P. 179
Saadaoui H. 124
Sangiovanni-Vincentelli A. L. 2
259
Sauter D. 223
Schiffelers R. R. H. 265
Schumacher H. 290
Seatzu C. 37
Silva M. 30
37
Simeonova I. 191
Smaus J. G. 371
Sonntag C. 211
Steinbuch M. 142
Stursberg O. 211
Suzuki T. 64
Theilliol D. 223
Thomas J. 329
Tiwari A. 359
Tsuchie Y. 417
Usai E. 276
Ushio T. 417
Utkin V. 1
Valigi P. 101
van Beek D. A. 265
van de Molengraft R. 142
van den Berg R. A. 6
van den Boom T. J. J. 148
Vasca F. 247
Villa T. 259
Villani E. 389
Wardi Y. 95
101
Warichet F. 191
Wesselowski K. S. 136
Xu J. 30
Yamada N. 64
Zambrano D. 199
Zaytoon J. 217
Zheng H. 270